Lecture Notes in Computer Science 12249

More information about this series at http://www.springer.com/series/7407

Osvaldo Gervasi · Beniamino Murgante ·
Sanjay Misra · Chiara Garau ·
Ivan Blečić · David Taniar ·
Bernady O. Apduhan · Ana Maria A. C. Rocha ·
Eufemia Tarantino · Carmelo Maria Torre ·
Yeliz Karaca (Eds.)

Computational Science and Its Applications – ICCSA 2020

20th International Conference
Cagliari, Italy, July 1–4, 2020
Proceedings, Part I

Springer

Editors
Osvaldo Gervasi Ⓘ
University of Perugia
Perugia, Italy

Sanjay Misra Ⓘ
Chair- Center of ICT/ICE
Covenant University
Ota, Nigeria

Ivan Blečić Ⓘ
University of Cagliari
Cagliari, Italy

Bernady O. Apduhan
Department of Information Science
Kyushu Sangyo University
Fukuoka, Japan

Eufemia Tarantino Ⓘ
Polytechnic University of Bari
Bari, Italy

Yeliz Karaca Ⓘ
Department of Neurology
University of Massachusetts
Medical School
Worcester, MA, USA

Beniamino Murgante Ⓘ
University of Basilicata
Potenza, Potenza, Italy

Chiara Garau Ⓘ
University of Cagliari
Cagliari, Italy

David Taniar Ⓘ
Clayton School of Information Technology
Monash University
Clayton, VIC, Australia

Ana Maria A. C. Rocha Ⓘ
University of Minho
Braga, Portugal

Carmelo Maria Torre Ⓘ
Polytechnic University of Bari
Bari, Italy

ISSN 0302-9743   ISSN 1611-3349 (electronic)
Lecture Notes in Computer Science
ISBN 978-3-030-58798-7   ISBN 978-3-030-58799-4 (eBook)
https://doi.org/10.1007/978-3-030-58799-4

LNCS Sublibrary: SL1 – Theoretical Computer Science and General Issues

This Springer imprint is published by the registered company Springer Nature Switzerland AG
The registered company address is: Gewerbestrasse 11, 6330 Cham, Switzerland

Preface

These seven volumes (LNCS volumes 12249–12255) consist of the peer-reviewed papers from the International Conference on Computational Science and Its Applications (ICCSA 2020) which took place from July 1–4, 2020. Initially the conference was planned to be held in Cagliari, Italy, in collaboration with the University of Cagliari, but due to the COVID-19 pandemic it was organized as an online event.

ICCSA 2020 was a successful event in the conference series, previously held in Saint Petersburg, Russia (2019), Melbourne, Australia (2018), Trieste, Italy (2017), Beijing, China (2016), Banff, Canada (2015), Guimaraes, Portugal (2014), Ho Chi Minh City, Vietnam (2013), Salvador, Brazil (2012), Santander, Spain (2011), Fukuoka, Japan (2010), Suwon, South Korea (2009), Perugia, Italy (2008), Kuala Lumpur, Malaysia (2007), Glasgow, UK (2006), Singapore (2005), Assisi, Italy (2004), Montreal, Canada (2003), and (as ICCS) Amsterdam, The Netherlands (2002) and San Francisco, USA (2001).

Computational science is the main pillar of most of the present research, industrial and commercial applications, and plays a unique role in exploiting ICT innovative technologies. The ICCSA conference series has provided a venue for researchers and industry practitioners to discuss new ideas, to share complex problems and their solutions, and to shape new trends in computational science.

Apart from the general track, ICCSA 2020 also included 52 workshops in various areas of computational science, ranging from computational science technologies to specific areas of computational science, such as software engineering, security, machine learning and artificial intelligence, blockchain technologies, and of applications in many fields. We accepted 498 papers, distributed among 6 conference main tracks, which included 52 in workshops and 32 short papers. We would like to express our appreciation to the workshops chairs and co-chairs for their hard work and dedication.

The success of the ICCSA conference series in general, and of ICCSA 2020 in particular, vitaly depends on the support from many people: authors, presenters, participants, keynote speakers, workshop chairs, session chairs, Organizing Committee members, student volunteers, Program Committee members, Advisory Committee members, international liaison chairs, reviewers, and others in various roles. We take this opportunity to wholeheartedly thank them all.

We also wish to thank our publisher, Springer, for their acceptance to publish the proceedings, for sponsoring part of the Best Papers Awards, and for their kind assistance and cooperation during the editing process.

We cordially invite you to visit the ICCSA website http://www.iccsa.org where you can find all the relevant information about this interesting and exciting event.

July 2020

Osvaldo Gervasi
Beniamino Murgante
Sanjay Misra

Welcome to the Online Conference

The COVID-19 pandemic disrupted our plans for ICCSA 2020, as was the case for the scientific community around the world. Hence, we had to promptly regroup and rush to set in place the organization and the underlying infrastructure of the online event.

We chose to build the technological infrastructure using only open source software. In particular, we used Jitsi (jitsi.org) for the videoconferencing, Riot (riot.im) together with Matrix (matrix.org) for chat and asynchronous communication, and Jibri (github.com/jitsi/jibri) for live streaming sessions on YouTube.

Six Jitsi servers were set up, one for each parallel session. The participants of the sessions were helped and assisted by eight volunteer students (from the Universities of Cagliari, Florence, Perugia, and Bari), who assured technical support and smooth running of the conference proceedings.

The implementation of the software infrastructure and the technical coordination of the volunteers was carried out by Damiano Perri and Marco Simonetti.

Our warmest thanks go to all the volunteering students, to the technical coordinators, and to the development communities of Jitsi, Jibri, Riot, and Matrix, who made their terrific platforms available as open source software.

Our heartfelt thanks go to the keynote speakers: Yaneer Bar-Yam, Cecilia Ceccarelli, and Vincenzo Piuri and to the guests of the closing keynote panel: Mike Batty, Denise Pumain, and Alexis Tsoukiàs.

A big thank you goes to all the 454 speakers, many of whom showed an enormous collaborative spirit, sometimes participating and presenting in almost prohibitive times of the day, given that the participants of this year's conference come from 52 countries scattered over many time zones of the globe.

Finally, we would like to thank Google for letting us livestream all the events via YouTube. In addition to lightening the load of our Jitsi servers, that will allow us to keep memory and to be able to review the most exciting moments of the conference.

We all hope to meet in our beautiful Cagliari next year, safe from COVID-19, and finally free to meet in person and enjoy the beauty of the ICCSA community in the enchanting Sardinia.

July 2020

Ivan Blečić
Chiara Garau

Organization

ICCSA 2020 was organized by the University of Cagliari (Italy), University of Perugia (Italy), University of Basilicata (Italy), Monash University (Australia), Kyushu Sangyo University (Japan), and University of Minho (Portugal).

Honorary General Chairs

Antonio Laganà	Master-UP, Italy
Norio Shiratori	Chuo University, Japan
Kenneth C. J. Tan	Sardina Systems, UK
Corrado Zoppi	University of Cagliari, Italy

General Chairs

Osvaldo Gervasi	University of Perugia, Italy
Ivan Blečić	University of Cagliari, Italy
David Taniar	Monash University, Australia

Program Committee Chairs

Beniamino Murgante	University of Basilicata, Italy
Bernady O. Apduhan	Kyushu Sangyo University, Japan
Chiara Garau	University of Cagliari, Italy
Ana Maria A. C. Rocha	University of Minho, Portugal

International Advisory Committee

Jemal Abawajy	Deakin University, Australia
Dharma P. Agarwal	University of Cincinnati, USA
Rajkumar Buyya	The University of Melbourne, Australia
Claudia Bauzer Medeiros	University of Campinas, Brazil
Manfred M. Fisher	Vienna University of Economics and Business, Austria
Marina L. Gavrilova	University of Calgary, Canada
Yee Leung	Chinese University of Hong Kong, China

International Liaison Chairs

Giuseppe Borruso	University of Trieste, Italy
Elise De Donker	Western Michigan University, USA
Maria Irene Falcão	University of Minho, Portugal
Robert C. H. Hsu	Chung Hua University, Taiwan

Tai-Hoon Kim	Beijing Jaotong University, China
Vladimir Korkhov	Saint Petersburg University, Russia
Sanjay Misra	Covenant University, Nigeria
Takashi Naka	Kyushu Sangyo University, Japan
Rafael D. C. Santos	National Institute for Space Research, Brazil
Maribel Yasmina Santos	University of Minho, Portugal
Elena Stankova	Saint Petersburg University, Russia

Workshop and Session Organizing Chairs

Beniamino Murgante	University of Basilicata, Italy
Sanjay Misra	Covenant University, Nigeria
Jorge Gustavo Rocha	University of Minho, Portugal

Award Chair

| Wenny Rahayu | La Trobe University, Australia |

Publicity Committee Chairs

Elmer Dadios	De La Salle University, Philippines
Nataliia Kulabukhova	Saint Petersburg University, Russia
Daisuke Takahashi	Tsukuba University, Japan
Shangwang Wang	Beijing University of Posts and Telecommunications, China

Technology Chairs

| Damiano Perri | University of Florence, Italy |
| Marco Simonetti | University of Florence, Italy |

Local Arrangement Chairs

Ivan Blečić	University of Cagliari, Italy
Chiara Garau	University of Cagliari, Italy
Ginevra Balletto	University of Cagliari, Italy
Giuseppe Borruso	University of Trieste, Italy
Michele Campagna	University of Cagliari, Italy
Mauro Coni	University of Cagliari, Italy
Anna Maria Colavitti	University of Cagliari, Italy
Giulia Desogus	University of Cagliari, Italy
Sabrina Lai	University of Cagliari, Italy
Francesca Maltinti	University of Cagliari, Italy
Pasquale Mistretta	University of Cagliari, Italy
Augusto Montisci	University of Cagliari, Italy
Francesco Pinna	University of Cagliari, Italy

Davide Spano	University of Cagliari, Italy
Roberto Tonelli	University of Cagliari, Italy
Giuseppe A. Trunfio	University of Sassari, Italy
Corrado Zoppi	University of Cagliari, Italy

Program Committee

Vera Afreixo	University of Aveiro, Portugal
Filipe Alvelos	University of Minho, Portugal
Hartmut Asche	University of Potsdam, Germany
Ginevra Balletto	University of Cagliari, Italy
Michela Bertolotto	University College Dublin, Ireland
Sandro Bimonte	CEMAGREF, TSCF, France
Rod Blais	University of Calgary, Canada
Ivan Blečić	University of Sassari, Italy
Giuseppe Borruso	University of Trieste, Italy
Ana Cristina Braga	University of Minho, Portugal
Massimo Cafaro	University of Salento, Italy
Yves Caniou	Lyon University, France
José A. Cardoso e Cunha	Universidade Nova de Lisboa, Portugal
Rui Cardoso	University of Beira Interior, Portugal
Leocadio G. Casado	University of Almeria, Spain
Carlo Cattani	University of Salerno, Italy
Mete Celik	Erciyes University, Turkey
Hyunseung Choo	Sungkyunkwan University, South Korea
Min Young Chung	Sungkyunkwan University, South Korea
Florbela Maria da Cruz Domingues Correia	Polytechnic Institute of Viana do Castelo, Portugal
Gilberto Corso Pereira	Federal University of Bahia, Brazil
Alessandro Costantini	INFN, Italy
Carla Dal Sasso Freitas	Universidade Federal do Rio Grande do Sul, Brazil
Pradesh Debba	The Council for Scientific and Industrial Research (CSIR), South Africa
Hendrik Decker	Instituto Tecnológico de Informática, Spain
Frank Devai	London South Bank University, UK
Rodolphe Devillers	Memorial University of Newfoundland, Canada
Joana Matos Dias	University of Coimbra, Portugal
Paolino Di Felice	University of L'Aquila, Italy
Prabu Dorairaj	NetApp, India/USA
M. Irene Falcao	University of Minho, Portugal
Cherry Liu Fang	U.S. DOE Ames Laboratory, USA
Florbela P. Fernandes	Polytechnic Institute of Bragança, Portugal
Jose-Jesus Fernandez	National Centre for Biotechnology, CSIS, Spain
Paula Odete Fernandes	Polytechnic Institute of Bragança, Portugal
Adelaide de Fátima Baptista Valente Freitas	University of Aveiro, Portugal

Noelia Faginas Lago	University of Perugia, Italy
Giuseppe Modica	University of Reggio Calabria, Italy
Josè Luis Montaña	University of Cantabria, Spain
Maria Filipa Mourão	IP from Viana do Castelo, Portugal
Louiza de Macedo Mourelle	State University of Rio de Janeiro, Brazil
Nadia Nedjah	State University of Rio de Janeiro, Brazil
Laszlo Neumann	University of Girona, Spain
Kok-Leong Ong	Deakin University, Australia
Belen Palop	Universidad de Valladolid, Spain
Marcin Paprzycki	Polish Academy of Sciences, Poland
Eric Pardede	La Trobe University, Australia
Kwangjin Park	Wonkwang University, South Korea
Ana Isabel Pereira	Polytechnic Institute of Bragança, Portugal
Massimiliano Petri	University of Pisa, Italy
Maurizio Pollino	Italian National Agency for New Technologies, Energy and Sustainable Economic Development, Italy
Alenka Poplin	University of Hamburg, Germany
Vidyasagar Potdar	Curtin University of Technology, Australia
David C. Prosperi	Florida Atlantic University, USA
Wenny Rahayu	La Trobe University, Australia
Jerzy Respondek	Silesian University of Technology, Poland
Humberto Rocha	INESC-Coimbra, Portugal
Jon Rokne	University of Calgary, Canada
Octavio Roncero	CSIC, Spain
Maytham Safar	Kuwait University, Kuwait
Francesco Santini	University of Perugia, Italy
Chiara Saracino	A.O. Ospedale Niguarda Ca' Granda, Italy
Haiduke Sarafian	Penn State University, USA
Marco Paulo Seabra dos Reis	University of Coimbra, Portugal
Jie Shen	University of Michigan, USA
Qi Shi	Liverpool John Moores University, UK
Dale Shires	U.S. Army Research Laboratory, USA
Inês Soares	University of Coimbra, Portugal
Elena Stankova	Saint Petersburg University, Russia
Takuo Suganuma	Tohoku University, Japan
Eufemia Tarantino	Polytechnic University of Bari, Italy
Sergio Tasso	University of Perugia, Italy
Ana Paula Teixeira	University of Trás-os-Montes and Alto Douro, Portugal
Senhorinha Teixeira	University of Minho, Portugal
M. Filomena Teodoro	Portuguese Naval Academy, University of Lisbon, Portugal
Parimala Thulasiraman	University of Manitoba, Canada
Carmelo Torre	Polytechnic University of Bari, Italy
Javier Martinez Torres	Centro Universitario de la Defensa Zaragoza, Spain
Giuseppe A. Trunfio	University of Sassari, Italy

Pablo Vanegas	University of Cuenca, Ecuador
Marco Vizzari	University of Perugia, Italy
Varun Vohra	Merck Inc., USA
Koichi Wada	University of Tsukuba, Japan
Krzysztof Walkowiak	Wroclaw University of Technology, Poland
Zequn Wang	Intelligent Automation Inc., USA
Robert Weibel	University of Zurich, Switzerland
Frank Westad	Norwegian University of Science and Technology, Norway
Roland Wismüller	Universität Siegen, Germany
Mudasser Wyne	SOET National University, USA
Chung-Huang Yang	National Kaohsiung Normal University, Taiwan
Xin-She Yang	National Physical Laboratory, UK
Salim Zabir	France Telecom Japan Co., Japan
Haifeng Zhao	University of California, Davis, USA
Fabiana Zollo	University of Venice, Italy
Albert Y. Zomaya	The University of Sydney, Australia

Workshop Organizers

Advanced Transport Tools and Methods (A2TM 2020)

| Massimiliano Petri | University of Pisa, Italy |
| Antonio Pratelli | University of Pisa, Italy |

Advances in Artificial Intelligence Learning Technologies: Blended Learning, STEM, Computational Thinking and Coding (AAILT 2020)

Valentina Franzoni	University of Perugia, Italy
Alfredo Milani	University of Perugia, Italy
Sergio Tasso	University of Perugia, Italy

Workshop on Advancements in Applied Machine Learning and Data Analytics (AAMDA 2020)

Alessandro Costantini	INFN, Italy
Daniele Cesini	INFN, Italy
Davide Salomoni	INFN, Italy
Doina Cristina Duma	INFN, Italy

Advanced Computational Approaches in Artificial Intelligence and Complex Systems Applications (ACAC 2020)

Yeliz Karaca	University of Massachusetts Medical School, USA
Dumitru Baleanu	Çankaya University, Turkey, and Institute of Space Sciences, Romania
Majaz Moonis	University of Massachusetts Medical School, USA
Yu-Dong Zhang	University of Leicester, UK

Affective Computing and Emotion Recognition (ACER-EMORE 2020)

Valentina Franzoni   University of Perugia, Italy
Alfredo Milani   University of Perugia, Italy
Giulio Biondi   University of Florence, Italy

AI Factory and Smart Manufacturing (AIFACTORY 2020)

Jongpil Jeong   Sungkyunkwan University, South Korea

Air Quality Monitoring and Citizen Science for Smart Urban Management. State of the Art And Perspectives (AirQ&CScience 2020)

Grazie Fattoruso   ENEA CR Portici, Italy
Maurizio Pollino   ENEA CR Casaccia, Italy
Saverio De Vito   ENEA CR Portici, Italy

Automatic Landform Classification: Spatial Methods and Applications (ALCSMA 2020)

Maria Danese   CNR-ISPC, Italy
Dario Gioia   CNR-ISPC, Italy

Advances of Modelling Micromobility in Urban Spaces (AMMUS 2020)

Tiziana Campisi   University of Enna KORE, Italy
Giovanni Tesoriere   University of Enna KORE, Italy
Ioannis Politis   Aristotle University of Thessaloniki, Greece
Socrates Basbas   Aristotle University of Thessaloniki, Greece
Sanja Surdonja   University of Rijeka, Croatia
Marko Rencelj   University of Maribor, Slovenia

Advances in Information Systems and Technologies for Emergency Management, Risk Assessment and Mitigation Based on the Resilience Concepts (ASTER 2020)

Maurizio Pollino   ENEA, Italy
Marco Vona   University of Basilicata, Italy
Amedeo Flora   University of Basilicata, Italy
Chiara Iacovino   University of Basilicata, Italy
Beniamino Murgante   University of Basilicata, Italy

Advances in Web Based Learning (AWBL 2020)

Birol Ciloglugil   Ege University, Turkey
Mustafa Murat Inceoglu   Ege University, Turkey

Blockchain and Distributed Ledgers: Technologies and Applications (BDLTA 2020)

Vladimir Korkhov   Saint Petersburg University, Russia
Elena Stankova   Saint Petersburg University, Russia
Nataliia Kulabukhova   Saint Petersburg University, Russia

Bio and Neuro Inspired Computing and Applications (BIONCA 2020)

Nadia Nedjah   State University of Rio de Janeiro, Brazil
Luiza De Macedo Mourelle State University of Rio de Janeiro, Brazil

Computer Aided Modeling, Simulation and Analysis (CAMSA 2020)

Jie Shen   University of Michigan, USA

Computational and Applied Statistics (CAS 2020)

Ana Cristina Braga   University of Minho, Portugal

Computerized Evidence Based Decision Making (CEBDEM 2020)

Clarice Bleil de Souza   Cardiff University, UK
Valerio Cuttini   University of Pisa, Italy
Federico Cerutti   Cardiff University, UK
Camilla Pezzica   Cardiff University, UK

Computational Geometry and Applications (CGA 2020)

Marina Gavrilova   University of Calgary, Canada

Computational Mathematics, Statistics and Information Management (CMSIM 2020)

Maria Filomena Teodoro   Portuguese Naval Academy, University of Lisbon, Portugal

Computational Optimization and Applications (COA 2020)

Ana Rocha   University of Minho, Portugal
Humberto Rocha   University of Coimbra, Portugal

Computational Astrochemistry (CompAstro 2020)

Marzio Rosi   University of Perugia, Italy
Cecilia Ceccarelli   University of Grenoble, France
Stefano Falcinelli   University of Perugia, Italy
Dimitrios Skouteris   Master-UP, Italy

Cities, Technologies and Planning (CTP 2020)

Beniamino Murgante   University of Basilicata, Italy
Ljiljana Zivkovic   Ministry of Construction, Transport and Infrastructure
                            and Institute of Architecture and Urban & Spatial
                            Planning of Serbia, Serbia
Giuseppe Borruso   University of Trieste, Italy
Malgorzata Hanzl   University of Łódź, Poland

Data Stream Processing and Applications (DASPA 2020)

Raja Chiky   ISEP, France
Rosanna VERDE   University of Campania, Italy
Marcilio De Souto   Orleans University, France

Data Science for Cyber Security (DS4Cyber 2020)

Hongmei Chi   Florida A&M University, USA

Econometric and Multidimensional Evaluation in Urban Environment (EMEUE 2020)

Carmelo Maria Torre   Polytechnic University of Bari, Italy
Pierluigi Morano   Polytechnic University of Bari, Italy
Maria Cerreta   University of Naples, Italy
Paola Perchinunno   University of Bari, Italy
Francesco Tajani   University of Rome, Italy
Simona Panaro   University of Portsmouth, UK
Francesco Scorza   University of Basilicata, Italy

Frontiers in Machine Learning (FIML 2020)

Massimo Bilancia   University of Bari, Italy
Paola Perchinunno   University of Bari, Italy
Pasquale Lops   University of Bari, Italy
Danilo Di Bona   University of Bari, Italy

Future Computing System Technologies and Applications (FiSTA 2020)

Bernady Apduhan   Kyushu Sangyo University, Japan
Rafael Santos   Brazilian National Institute for Space Research, Brazil

Geodesign in Decision Making: Meta Planning and Collaborative Design for Sustainable and Inclusive Development (GDM 2020)

Francesco Scorza   University of Basilicata, Italy
Michele Campagna   University of Cagliari, Italy
Ana Clara Mourao Moura   Federal University of Minas Gerais, Brazil

Geomatics in Forestry and Agriculture: New Advances and Perspectives (GeoForAgr 2020)

Maurizio Pollino	ENEA, Italy
Giuseppe Modica	University of Reggio Calabria, Italy
Marco Vizzari	University of Perugia, Italy

Geographical Analysis, Urban Modeling, Spatial Statistics (GEOG-AND-MOD 2020)

Beniamino Murgante	University of Basilicata, Italy
Giuseppe Borruso	University of Trieste, Italy
Hartmut Asche	University of Potsdam, Germany

Geomatics for Resource Monitoring and Management (GRMM 2020)

Eufemia Tarantino	Polytechnic University of Bari, Italy
Enrico Borgogno Mondino	University of Torino, Italy
Marco Scaioni	Polytechnic University of Milan, Italy
Alessandra Capolupo	Polytechnic University of Bari, Italy

Software Quality (ISSQ 2020)

Sanjay Misra	Covenant University, Nigeria

Collective, Massive and Evolutionary Systems (IWCES 2020)

Alfredo Milani	University of Perugia, Italy
Rajdeep Niyogi	Indian Institute of Technology, Roorkee, India
Alina Elena Baia	University of Florence, Italy

Large Scale Computational Science (LSCS 2020)

Elise De Doncker	Western Michigan University, USA
Fukuko Yuasa	High Energy Accelerator Research Organization (KEK), Japan
Hideo Matsufuru	High Energy Accelerator Research Organization (KEK), Japan

Land Use Monitoring for Sustainability (LUMS 2020)

Carmelo Maria Torre	Polytechnic University of Bari, Italy
Alessandro Bonifazi	Polytechnic University of Bari, Italy
Pasquale Balena	Polytechnic University of Bari, Italy
Massimiliano Bencardino	University of Salerno, Italy
Francesco Tajani	University of Rome, Italy
Pierluigi Morano	Polytechnic University of Bari, Italy
Maria Cerreta	University of Naples, Italy
Giuliano Poli	University of Naples, Italy

Machine Learning for Space and Earth Observation Data (MALSEOD 2020)

Rafael Santos   INPE, Brazil
Karine Ferreira   INPE, Brazil

Building Multi-dimensional Models for Assessing Complex Environmental Systems (MES 2020)

Marta Dell'Ovo   Polytechnic University of Milan, Italy
Vanessa Assumma   Polytechnic University of Torino, Italy
Caterina Caprioli   Polytechnic University of Torino, Italy
Giulia Datola   Polytechnic University of Torino, Italy
Federico dell'Anna   Polytechnic University of Torino, Italy

Ecosystem Services: Nature's Contribution to People in Practice. Assessment Frameworks, Models, Mapping, and Implications (NC2P 2020)

Francesco Scorza   University of Basilicata, Italy
David Cabana   International Marine Center, Italy
Sabrina Lai   University of Cagliari, Italy
Ana Clara Mourao Moura   Federal University of Minas Gerais, Brazil
Corrado Zoppi   University of Cagliari, Italy

Open Knowledge for Socio-economic Development (OKSED 2020)

Luigi Mundula   University of Cagliari, Italy
Flavia Marzano   Link Campus University, Italy
Maria Paradiso   University of Milan, Italy

Scientific Computing Infrastructure (SCI 2020)

Elena Stankova   Saint Petersburg State University, Russia
Vladimir Korkhov   Saint Petersburg State University, Russia
Natalia Kulabukhova   Saint Petersburg State University, Russia

Computational Studies for Energy and Comfort in Buildings (SECoB 2020)

Senhorinha Teixeira   University of Minho, Portugal
Luís Martins   University of Minho, Portugal
Ana Maria Rocha   University of Minho, Portugal

Software Engineering Processes and Applications (SEPA 2020)

Sanjay Misra   Covenant University, Nigeria

Smart Ports - Technologies and Challenges (SmartPorts 2020)

Gianfranco Fancello   University of Cagliari, Italy
Patrizia Serra   University of Cagliari, Italy
Marco Mazzarino   University of Venice, Italy
Luigi Mundula   University of Cagliari, Italy

| Ginevra Balletto | University of Cagliari, Italy |
| Giuseppe Borruso | University of Trieste, Italy |

Sustainability Performance Assessment: Models, Approaches and Applications Toward Interdisciplinary and Integrated Solutions (SPA 2020)

Francesco Scorza	University of Basilicata, Italy
Valentin Grecu	Lucian Blaga University, Romania
Jolanta Dvarioniene	Kaunas University of Technology, Lithuania
Sabrina Lai	University of Cagliari, Italy
Iole Cerminara	University of Basilicata, Italy
Corrado Zoppi	University of Cagliari, Italy

Smart and Sustainable Island Communities (SSIC 2020)

Chiara Garau	University of Cagliari, Italy
Anastasia Stratigea	National Technical University of Athens, Greece
Paola Zamperlin	University of Pisa, Italy
Francesco Scorza	University of Basilicata, Italy

Science, Technologies and Policies to Innovate Spatial Planning (STP4P 2020)

Chiara Garau	University of Cagliari, Italy
Daniele La Rosa	University of Catania, Italy
Francesco Scorza	University of Basilicata, Italy
Anna Maria Colavitti	University of Cagliari, Italy
Beniamino Murgante	University of Basilicata, Italy
Paolo La Greca	University of Catania, Italy

New Frontiers for Strategic Urban Planning (StrategicUP 2020)

Luigi Mundula	University of Cagliari, Italy
Ginevra Balletto	University of Cagliari, Italy
Giuseppe Borruso	University of Trieste, Italy
Michele Campagna	University of Cagliari, Italy
Beniamino Murgante	University of Basilicata, Italy

Theoretical and Computational Chemistry and its Applications (TCCMA 2020)

| Noelia Faginas-Lago | University of Perugia, Italy |
| Andrea Lombardi | University of Perugia, Italy |

Tools and Techniques in Software Development Process (TTSDP 2020)

| Sanjay Misra | Covenant University, Nigeria |

Urban Form Studies (UForm 2020)

| Malgorzata Hanzl | Łódź University of Technology, Poland |

Urban Space Extended Accessibility (USEaccessibility 2020)

Chiara Garau	University of Cagliari, Italy
Francesco Pinna	University of Cagliari, Italy
Beniamino Murgante	University of Basilicata, Italy
Mauro Coni	University of Cagliari, Italy
Francesca Maltinti	University of Cagliari, Italy
Vincenza Torrisi	University of Catania, Italy
Matteo Ignaccolo	University of Catania, Italy

Virtual and Augmented Reality and Applications (VRA 2020)

Osvaldo Gervasi	University of Perugia, Italy
Damiano Perri	University of Perugia, Italy
Marco Simonetti	University of Perugia, Italy
Sergio Tasso	University of Perugia, Italy

Workshop on Advanced and Computational Methods for Earth Science Applications (WACM4ES 2020)

Luca Piroddi	University of Cagliari, Italy
Laura Foddis	University of Cagliari, Italy
Gian Piero Deidda	University of Cagliari, Italy
Augusto Montisci	University of Cagliari, Italy
Gabriele Uras	University of Cagliari, Italy
Giulio Vignoli	University of Cagliari, Italy

Sponsoring Organizations

ICCSA 2020 would not have been possible without tremendous support of many organizations and institutions, for which all organizers and participants of ICCSA 2020 express their sincere gratitude:

Springer International Publishing AG, Germany
(https://www.springer.com)

Computers Open Access Journal
(https://www.mdpi.com/journal/computers)

IEEE Italy Section, Italy
(https://italy.ieeer8.org/)

Centre-North Italy Chapter IEEE GRSS, Italy
(https://cispio.diet.uniroma1.it/marzano/ieee-grs/
index.html)

Italy Section of the Computer Society, Italy
(https://site.ieee.org/italy-cs/)

University of Cagliari, Italy
(https://unica.it/)

University of Perugia, Italy
(https://www.unipg.it)

University of Basilicata, Italy
(http://www.unibas.it)

Monash University, Australia
(https://www.monash.edu/)

Kyushu Sangyo University, Japan
(https://www.kyusan-u.ac.jp/)

University of Minho, Portugal
(https://www.uminho.pt/)

Scientific Association Transport Infrastructures, Italy
(https://www.stradeeautostrade.it/associazioni-e-organizzazioni/asit-associazione-scientifica-infrastrutture-trasporto/)

Regione Sardegna, Italy
(https://regione.sardegna.it/)

Comune di Cagliari, Italy
(https://www.comune.cagliari.it/)

Referees

A. P. Andrade Marina	ISCTE, Instituto Universitário de Lisboa, Portugal
Addesso Paolo	University of Salerno, Italy
Adewumi Adewole	Algonquin College, Canada
Afolabi Adedeji	Covenant University, Nigeria
Afreixo Vera	University of Aveiro, Portugal
Agrawal Smirti	Freelancer, USA
Agrawal Akshat	Amity University Haryana, India
Ahmad Waseem	Federal University of Technology Minna, Nigeria
Akgun Nurten	Bursa Technical University, Turkey
Alam Tauhidul	Louisiana State University Shreveport, USA
Aleixo Sandra M.	CEAUL, Portugal
Alfa Abraham	Federal University of Technology Minna, Nigeria
Alvelos Filipe	University of Minho, Portugal
Alves Alexandra	University of Minho, Portugal
Amato Federico	University of Lausanne, Switzerland
Andrade Marina Alexandra Pedro	ISCTE-IUL, Portugal
Andrianov Sergey	Saint Petersburg State University, Russia
Anelli Angelo	CNR-IGAG, Italy
Anelli Debora	University of Rome, Italy
Annunziata Alfonso	University of Cagliari, Italy
Antognelli Sara	Agricolus, Italy
Aoyama Tatsumi	High Energy Accelerator Research Organization, Japan
Apduhan Bernady	Kyushu Sangyo University, Japan
Ascenzi Daniela	University of Trento, Italy
Asche Harmut	Hasso-Plattner-Institut für Digital Engineering GmbH, Germany
Aslan Burak Galip	Izmir Insitute of Technology, Turkey
Assumma Vanessa	Polytechnic University of Torino, Italy
Astoga Gino	UV, Chile
Atman Uslu Nilüfer	Manisa Celal Bayar University, Turkey
Behera Ranjan Kumar	National Institute of Technology, Rourkela, India
Badsha Shahriar	University of Nevada, USA
Bai Peng	University of Cagliari, Italy
Baia Alina-Elena	University of Perugia, Italy
Balacco Gabriella	Polytechnic University of Bari, Italy
Balci Birim	Celal Bayar University, Turkey
Balena Pasquale	Polytechnic University of Bari, Italy
Balletto Ginevra	University of Cagliari, Italy
Balucani Nadia	University of Perugia, Italy
Bansal Megha	Delhi University, India
Barazzetti Luigi	Polytechnic University of Milan, Italy
Barreto Jeniffer	Istituto Superior Técnico, Portugal
Basbas Socrates	Aristotle University of Thessaloniki, Greece

Berger Katja	Ludwig-Maximilians-Universität München, Germany
Beyene Asrat Mulatu	Addis Ababa Science and Technology University, Ethiopia
Bilancia Massimo	University of Bari Aldo Moro, Italy
Biondi Giulio	University of Firenze, Italy
Blanquer Ignacio	Universitat Politècnica de València, Spain
Bleil de Souza Clarice	Cardiff University, UK
Blečić Ivan	University of Cagliari, Italy
Bogdanov Alexander	Saint Petersburg State University, Russia
Bonifazi Alessandro	Polytechnic University of Bari, Italy
Bontchev Boyan	Sofia University, Bulgaria
Borgogno Mondino Enrico	University of Torino, Italy
Borruso Giuseppe	University of Trieste, Italy
Bouaziz Rahma	Taibah University, Saudi Arabia
Bowles Juliana	University of Saint Andrews, UK
Braga Ana Cristina	University of Minho, Portugal
Brambilla Andrea	Polytechnic University of Milan, Italy
Brito Francisco	University of Minho, Portugal
Buele Jorge	Universidad Tecnológica Indoamérica, Ecuador
Buffoni Andrea	TAGES sc, Italy
Cabana David	International Marine Centre, Italy
Calazan Rogerio	IEAPM, Brazil
Calcina Sergio Vincenzo	University of Cagliari, Italy
Camalan Seda	Atilim University, Turkey
Camarero Alberto	Universidad Politécnica de Madrid, Spain
Campisi Tiziana	University of Enna KORE, Italy
Cannatella Daniele	Delft University of Technology, The Netherlands
Capolupo Alessandra	Polytechnic University of Bari, Italy
Cappucci Sergio	ENEA, Italy
Caprioli Caterina	Polytechnic University of Torino, Italy
Carapau Fermando	Universidade de Evora, Portugal
Carcangiu Sara	University of Cagliari, Italy
Carrasqueira Pedro	INESC Coimbra, Portugal
Caselli Nicolás	PUCV Chile, Chile
Castro de Macedo Jose Nuno	Universidade do Minho, Portugal
Cavallo Carla	University of Naples, Italy
Cerminara Iole	University of Basilicata, Italy
Cerreta Maria	University of Naples, Italy
Cesini Daniele	INFN-CNAF, Italy
Chang Shi-Kuo	University of Pittsburgh, USA
Chetty Girija	University of Canberra, Australia
Chiky Raja	ISEP, France
Chowdhury Dhiman	University of South Carolina, USA
Ciloglugil Birol	Ege University, Turkey
Coletti Cecilia	Università di Chieti-Pescara, Italy

Coni Mauro	University of Cagliari, Italy
Corcoran Padraig	Cardiff University, UK
Cornelio Antonella	Università degli Studi di Brescia, Italy
Correia Aldina	ESTG-PPorto, Portugal
Correia Elisete	University of Trás-os-Montes and Alto Douro, Portugal
Correia Florbela	Polytechnic Institute of Viana do Castelo, Portugal
Costa Lino	Universidade do Minho, Portugal
Costa e Silva Eliana	ESTG-P Porto, Portugal
Costantini Alessandro	INFN, Italy
Crespi Mattia	University of Roma, Italy
Cuca Branka	Polytechnic University of Milano, Italy
De Doncker Elise	Western Michigan University, USA
De Macedo Mourelle Luiza	State University of Rio de Janeiro, Brazil
Daisaka Hiroshi	Hitotsubashi University, Japan
Daldanise Gaia	CNR, Italy
Danese Maria	CNR-ISPC, Italy
Daniele Bartoli	University of Perugia, Italy
Datola Giulia	Polytechnic University of Torino, Italy
De Luca Giandomenico	University of Reggio Calabria, Italy
De Lucia Caterina	University of Foggia, Italy
De Morais Barroca Filho Itamir	Federal University of Rio Grande do Norte, Brazil
De Petris Samuele	University of Torino, Italy
De Sá Alan	Marinha do Brasil, Brazil
De Souto Marcilio	LIFO, University of Orléans, France
De Vito Saverio	ENEA, Italy
De Wilde Pieter	University of Plymouth, UK
Degtyarev Alexander	Saint Petersburg State University, Russia
Dell'Anna Federico	Polytechnic University of Torino, Italy
Dell'Ovo Marta	Polytechnic University of Milano, Italy
Della Mura Fernanda	University of Naples, Italy
Deluka T. Aleksandra	University of Rijeka, Croatia
Demartino Cristoforo	Zhejiang University, China
Dereli Dursun Ahu	Istanbul Commerce University, Turkey
Desogus Giulia	University of Cagliari, Italy
Dettori Marco	University of Sassari, Italy
Devai Frank	London South Bank University, UK
Di Francesco Massimo	University of Cagliari, Italy
Di Liddo Felicia	Polytechnic University of Bari, Italy
Di Paola Gianluigi	University of Molise, Italy
Di Pietro Antonio	ENEA, Italy
Di Pinto Valerio	University of Naples, Italy
Dias Joana	University of Coimbra, Portugal
Dimas Isabel	University of Coimbra, Portugal
Dirvanauskas Darius	Kaunas University of Technology, Lithuania
Djordjevic Aleksandra	University of Belgrade, Serbia

Duma Doina Cristina	INFN-CNAF, Italy
Dumlu Demircioğlu Emine	Yıldız Technical University, Turkey
Dursun Aziz	Virginia Tech University, USA
Dvarioniene Jolanta	Kaunas University of Technology, Lithuania
Errico Maurizio Francesco	University of Enna KORE, Italy
Ezugwu Absalom	University of KwaZulu-Natal, South Africa
Fattoruso Grazia	ENEA, Italy
Faginas-Lago Noelia	University of Perugia, Italy
Falanga Bolognesi Salvatore	ARIESPACE, Italy
Falcinelli Stefano	University of Perugia, Italy
Farias Marcos	National Nuclear Energy Commission, Brazil
Farina Alessandro	University of Pisa, Italy
Feltynowski Marcin	Lodz University of Technology, Poland
Fernandes Florbela	Instituto Politecnico de Bragança, Portugal
Fernandes Paula Odete	Instituto Politécnico de Bragança, Portugal
Fernandez-Sanz Luis	University of Alcala, Spain
Ferreira Ana Cristina	University of Minho, Portugal
Ferreira Fernanda	Porto, Portugal
Fiorini Lorena	University of L'Aquila, Italy
Flora Amedeo	University of Basilicata, Italy
Florez Hector	Universidad Distrital Francisco Jose de Caldas, Colombia
Foddis Maria Laura	University of Cagliari, Italy
Fogli Daniela	University of Brescia, Italy
Fortunelli Martina	Pragma Engineering, Italy
Fragiacomo Massimo	University of L'Aquila, Italy
Franzoni Valentina	Perugia University, Italy
Fusco Giovanni	University of Cote d'Azur, France
Fyrogenis Ioannis	Aristotle University of Thessaloniki, Greece
Gorbachev Yuriy	Coddan Technologies LLC, Russia
Gabrielli Laura	Università Iuav di Venezia, Italy
Gallanos Theodore	Austrian Institute of Technology, Austria
Gamallo Belmonte Pablo	Universitat de Barcelona, Spain
Gankevich Ivan	Saint Petersburg State University, Russia
Garau Chiara	University of Cagliari, Italy
Garcia Para Ernesto	Universidad del Pais Vasco, EHU, Spain
Gargano Riccardo	Universidade de Brasilia, Brazil
Gavrilova Marina	University of Calgary, Canada
Georgiadis Georgios	Aristotle University of Thessaloniki, Greece
Gervasi Osvaldo	University of Perugia, Italy
Giano Salvatore Ivo	University of Basilicata, Italy
Gil Jorge	Chalmers University, Sweden
Gioia Andrea	Polytechnic University of Bari, Italy
Gioia Dario	ISPC-CNT, Italy

Giordano Ludovica	ENEA, Italy
Giorgi Giacomo	University of Perugia, Italy
Giovene di Girasole Eleonora	CNR-IRISS, Italy
Giovinazzi Sonia	ENEA, Italy
Giresini Linda	University of Pisa, Italy
Giuffrida Salvatore	University of Catania, Italy
Golubchikov Oleg	Cardiff University, UK
Gonçalves A. Manuela	University of Minho, Portugal
Gorgoglione Angela	Universidad de la República, Uruguay
Goyal Rinkaj	IPU, Delhi, India
Grishkin Valery	Saint Petersburg State University, Russia
Guerra Eduardo	Free University of Bozen-Bolzano, Italy
Guerrero Abel	University of Guanajuato, Mexico
Gulseven Osman	American University of The Middle East, Kuwait
Gupta Brij	National Institute of Technology, Kurukshetra, India
Guveyi Elcin	Yildiz Teknik University, Turkey
Gülen Kemal Güven	Namk Kemal University, Turkey
Haddad Sandra	Arab Academy for Science, Technology and Maritime Transport, Egypt
Hanzl Malgorzata	Lodz University of Technology, Poland
Hegedus Peter	University of Szeged, Hungary
Hendrix Eligius M. T.	Universidad de Málaga, Spain
Higaki Hiroaki	Tokyo Denki University, Japan
Hossain Syeda Sumbul	Daffodil International University, Bangladesh
Iacovino Chiara	University of Basilicata, Italy
Iakushkin Oleg	Saint Petersburg State University, Russia
Iannuzzo Antonino	ETH Zurich, Switzerland
Idri Ali	University Mohammed V, Morocco
Ignaccolo Matteo	University of Catania, Italy
Ilovan Oana-Ramona	Babeş-Bolyai University, Romania
Isola Federica	University of Cagliari, Italy
Jankovic Marija	CERTH, Greece
Jorge Ana Maria	Instituto Politécnico de Lisboa, Portugal
Kanamori Issaku	RIKEN Center for Computational Science, Japan
Kapenga John	Western Michigan University, USA
Karabulut Korhan	Yasar University, Turkey
Karaca Yeliz	University of Massachusetts Medical School, USA
Karami Ali	University of Guilan, Iran
Kienhofer Frank	WITS, South Africa
Kim Tai-hoon	Beijing Jiaotong University, China
Kimura Shuhei	Tottori University, Japan
Kirillov Denis	Saint Petersburg State University, Russia
Korkhov Vladimir	Saint Petersburg University, Russia
Koszewski Krzysztof	Warsaw University of Technology, Poland
Krzysztofik Sylwia	Lodz University of Technology, Poland

Kulabukhova Nataliia	Saint Petersburg State University, Russia
Kulkarni Shrinivas B.	SDM College of Engineering and Technology, Dharwad, India
Kwiecinski Krystian	Warsaw University of Technology, Poland
Kyvelou Stella	Panteion University of Social and Political Sciences, Greece
Körting Thales	INPE, Brazil
Lal Niranjan	Mody University of Science and Technology, India
Lazzari Maurizio	CNR-ISPC, Italy
Leon Marcelo	Asociacion de Becarios del Ecuador, Ecuador
La Rocca Ludovica	University of Naples, Italy
La Rosa Daniele	University of Catania, Italy
Lai Sabrina	University of Cagliari, Italy
Lalenis Konstantinos	University of Thessaly, Greece
Lannon Simon	Cardiff University, UK
Lasaponara Rosa	CNR, Italy
Lee Chien-Sing	Sunway University, Malaysia
Lemus-Romani José	Pontificia Universidad Católica de Valparaiso, Chile
Leone Federica	University of Cagliari, Italy
Li Yuanxi	Hong Kong Baptist University, China
Locurcio Marco	Polytechnic University of Bari, Italy
Lombardi Andrea	University of Perugia, Italy
Lopez Gayarre Fernando	University of Oviedo, Spain
Lops Pasquale	University of Bari, Italy
Lourenço Vanda	Universidade Nova de Lisboa, Portugal
Luviano José Luís	University of Guanajuato, Mexico
Maltese Antonino	University of Palermo, Italy
Magni Riccardo	Pragma Engineering, Italy
Maheshwari Anil	Carleton University, Canada
Maja Roberto	Polytechnic University of Milano, Italy
Malik Shaveta	Terna Engineering College, India
Maltinti Francesca	University of Cagliari, Italy
Mandado Marcos	University of Vigo, Spain
Manganelli Benedetto	University of Basilicata, Italy
Mangiameli Michele	University of Catania, Italy
Maraschin Clarice	Universidade Federal do Rio Grande do Sul, Brazil
Marigorta Ana Maria	Universidad de Las Palmas de Gran Canaria, Spain
Markov Krassimir	Institute of Electrical Engineering and Informatics, Bulgaria
Martellozzo Federico	University of Firenze, Italy
Marucci Alessandro	University of L'Aquila, Italy
Masini Nicola	IBAM-CNR, Italy
Matsufuru Hideo	High Energy Accelerator Research Organization (KEK), Japan
Matteucci Ilaria	CNR, Italy
Mauro D'Apuzzo	University of Cassino and Southern Lazio, Italy

Mazzarella Chiara	University of Naples, Italy
Mazzarino Marco	University of Venice, Italy
Mazzoni Augusto	University of Roma, Italy
Mele Roberta	University of Naples, Italy
Menezes Raquel	University of Minho, Portugal
Menghini Antonio	Aarhus Geofisica, Italy
Mengoni Paolo	University of Florence, Italy
Merlino Angelo	Università degli Studi Mediterranea, Italy
Milani Alfredo	University of Perugia, Italy
Milic Vladimir	University of Zagreb, Croatia
Millham Richard	Durban University of Technology, South Africa
Mishra B.	University of Szeged, Hungary
Misra Sanjay	Covenant University, Nigeria
Modica Giuseppe	University of Reggio Calabria, Italy
Mohagheghi Mohammadsadegh	Vali-e-Asr University of Rafsanjan, Iran
Molaei Qelichi Mohamad	University of Tehran, Iran
Molinara Mario	University of Cassino and Southern Lazio, Italy
Momo Evelyn Joan	University of Torino, Italy
Monteiro Vitor	University of Minho, Portugal
Montisci Augusto	University of Cagliari, Italy
Morano Pierluigi	Polytechnic University of Bari, Italy
Morganti Alessandro	Polytechnic University of Milano, Italy
Mosca Erica Isa	Polytechnic University of Milan, Italy
Moura Ricardo	CMA-FCT, New University of Lisbon, Portugal
Mourao Maria	Polytechnic Institute of Viana do Castelo, Portugal
Mourão Moura Ana Clara	Federal University of Minas Gerais, Brazil
Mrak Iva	University of Rijeka, Croatia
Murgante Beniamino	University of Basilicata, Italy
Muñoz Mirna	Centro de Investigacion en Matematicas, Mexico
Nedjah Nadia	State University of Rio de Janeiro, Brazil
Nakasato Naohito	University of Aizu, Japan
Natário Isabel Cristina	Universidade Nova de Lisboa, Portugal
Nesticò Antonio	Università degli Studi di Salerno, Italy
Neto Ana Maria	Universidade Federal do ABC, Brazil
Nicolosi Vittorio	University of Rome, Italy
Nikiforiadis Andreas	Aristotle University of Thessaloniki, Greece
Nocera Fabrizio	University of Illinois at Urbana-Champaign, USA
Nocera Silvio	IUAV, Italy
Nogueira Marcelo	Paulista University, Brazil
Nolè Gabriele	CNR, Italy
Nuno Beirao Jose	University of Lisbon, Portugal
Okewu Emma	University of Alcala, Spain
Oluwasefunmi Arogundade	Academy of Mathematics and System Science, China
Oppio Alessandra	Polytechnic University of Milan, Italy
P. Costa M. Fernanda	University of Minho, Portugal

Parisot Olivier	Luxembourg Institute of Science and Technology, Luxembourg
Paddeu Daniela	UWE, UK
Paio Alexandra	ISCTE-Instituto Universitário de Lisboa, Portugal
Palme Massimo	Catholic University of the North, Chile
Panaro Simona	University of Portsmouth, UK
Pancham Jay	Durban University of Technology, South Africa
Pantazis Dimos	University of West Attica, Greece
Papa Enrica	University of Westminster, UK
Pardede Eric	La Trobe University, Australia
Perchinunno Paola	Uniersity of Cagliari, Italy
Perdicoulis Teresa	UTAD, Portugal
Pereira Ana	Polytechnic Institute of Bragança, Portugal
Perri Damiano	University of Perugia, Italy
Petrelli Marco	University of Rome, Italy
Pierri Francesca	University of Perugia, Italy
Piersanti Antonio	ENEA, Italy
Pilogallo Angela	University of Basilicata, Italy
Pinna Francesco	University of Cagliari, Italy
Pinto Telmo	University of Coimbra, Portugal
Piroddi Luca	University of Cagliari, Italy
Poli Giuliano	University of Naples, Italy
Polidoro Maria João	Polytecnic Institute of Porto, Portugal
Polignano Marco	University of Bari, Italy
Politis Ioannis	Aristotle University of Thessaloniki, Greece
Pollino Maurizio	ENEA, Italy
Popoola Segun	Covenant University, Nigeria
Pratelli Antonio	University of Pisa, Italy
Praticò Salvatore	University of Reggio Calabria, Italy
Previtali Mattia	Polytechnic University of Milan, Italy
Puppio Mario Lucio	University of Pisa, Italy
Puttini Ricardo	Universidade de Brasilia, Brazil
Que Zeli	Nanjing Forestry University, China
Queiroz Gilberto	INPE, Brazil
Regalbuto Stefania	University of Naples, Italy
Ravanelli Roberta	University of Roma, Italy
Recanatesi Fabio	University of Tuscia, Italy
Reis Ferreira Gomes Karine	INPE, Brazil
Reis Marco	University of Coimbra, Portugal
Reitano Maria	University of Naples, Italy
Rencelj Marko	University of Maribor, Slovenia
Respondek Jerzy	Silesian University of Technology, Poland
Rimola Albert	Universitat Autònoma de Barcelona, Spain
Rocha Ana	University of Minho, Portugal
Rocha Humberto	University of Coimbra, Portugal
Rocha Maria Celia	UFBA Bahia, Brazil

Šurdonja Sanja	University of Rijeka, Croatia
Sviatov Kirill	Ulyanovsk State Technical University, Russia
Sánchez de Merás Alfredo	Universitat de Valencia, Spain
Takahashi Daisuke	University of Tsukuba, Japan
Tanaka Kazuaki	Kyushu Institute of Technology, Japan
Taniar David	Monash University, Australia
Tapia McClung Rodrigo	Centro de Investigación en Ciencias de Información Geoespacial, Mexico
Tarantino Eufemia	Polytechnic University of Bari, Italy
Tasso Sergio	University of Perugia, Italy
Teixeira Ana Paula	University of Trás-os-Montes and Alto Douro, Portugal
Teixeira Senhorinha	University of Minho, Portugal
Tengku Izhar Tengku Adil	Universiti Teknologi MARA, Malaysia
Teodoro Maria Filomena	University of Lisbon, Portuguese Naval Academy, Portugal
Tesoriere Giovanni	University of Enna KORE, Italy
Thangeda Amarendar Rao	Botho University, Botswana
Tonbul Gokchan	Atilim University, Turkey
Toraldo Emanuele	Polytechnic University of Milan, Italy
Torre Carmelo Maria	Polytechnic University of Bari, Italy
Torrieri Francesca	University of Naples, Italy
Torrisi Vincenza	University of Catania, Italy
Toscano Domenico	University of Naples, Italy
Totaro Vincenzo	Polytechnic University of Bari, Italy
Trigo Antonio	Instituto Politécnico de Coimbra, Portugal
Trunfio Giuseppe A.	University of Sassari, Italy
Trung Pham	HCMUT, Vietnam
Tsoukalas Dimitrios	Centre of Research and Technology Hellas (CERTH), Greece
Tucci Biagio	CNR, Italy
Tucker Simon	Liverpool John Moores University, UK
Tuñon Iñaki	Universidad de Valencia, Spain
Tyagi Amit Kumar	Vellore Institute of Technology, India
Uchibayashi Toshihiro	Kyushu University, Japan
Ueda Takahiro	Seikei University, Japan
Ugliengo Piero	University of Torino, Italy
Valente Ettore	University of Naples, Italy
Vallverdu Jordi	University Autonoma Barcelona, Spain
Vanelslander Thierry	University of Antwerp, Belgium
Vasyunin Dmitry	T-Systems RUS, Russia
Vazart Fanny	University of Grenoble Alpes, France
Vecchiocattivi Franco	University of Perugia, Italy
Vekeman Jelle	Vrije Universiteit Brussel (VUB), Belgium
Verde Rosanna	Università degli Studi della Campania, Italy
Vermaseren Jos	Nikhef, The Netherlands

Vignoli Giulio	University of Cagliari, Italy
Vizzari Marco	University of Perugia, Italy
Vodyaho Alexander	Saint Petersburg State Electrotechnical University, Russia
Vona Marco	University of Basilicata, Italy
Waluyo Agustinus Borgy	Monash University, Australia
Wen Min	Xi'an Jiaotong-Liverpool University, China
Westad Frank	Norwegian University of Science and Technology, Norway
Yuasa Fukuko	KEK, Japan
Yadav Rekha	KL University, India
Yamu Claudia	University of Groningen, The Netherlands
Yao Fenghui	Tennessee State University, USA
Yañez Manuel	Universidad Autónoma de Madrid, Spain
Yoki Karl	Daegu Catholic University, South Korea
Zamperlin Paola	University of Pisa, Italy
Zekeng Ndadji Milliam Maxime	University of Dschang, Cameroon
Žemlička Michal	Charles University, Czech Republic
Zita Sampaio Alcinia	Technical University of Lisbon, Portugal
Živković Ljiljana	Ministry of Construction, Transport and Infrastructure and Institute of Architecture and Urban & Spatial Planning of Serbia, Serbia
Zoppi Corrado	University of Cagliari, Italy
Zucca Marco	Polytechnic University of Milan, Italy
Zullo Francesco	University of L'Aquila, Italy

Contents – Part I

General Track 2: High Performance Computing and Networks

General Track 5: Information Systems and Technologies

General Track 1: Computational Methods, Algorithms and Scientific Applications

A Convergence Analysis of a Multistep Method Applied to an Advection-Diffusion Equation in 1-D

D. T. Robaina[1]($\boxtimes$), S. L. Gonzaga de Oliveira[2], M. Kischinhevsky[3], C. Osthoff[4], and A. Sena[5]

[1] ESPM, Rio de Janeiro, RJ, Brazil
`professor.robaina@gmail.com`
[2] Universidade Federal de Lavras, Lavras, MG, Brazil
`sanderson@ufla.br`
[3] UFF, Niterói, RJ, Brazil
`kisch@ic.uff.br`
[4] Laboratório Nacional de Computação Científica, Petrópolis, RJ, Brazil
`osthoff@lncc.br`
[5] UERJ, Rio de Janeiro, RJ, Brazil
`{asena,mario}@ime.uerj.br`

Abstract. This paper describes a method for convection-dominated fluid or heat flows. This method relies on the Hopmoc method and backward differentiation formulas. The present study discusses the convergence of the method when applied to a convection-diffusion equation in 1-D. The convergence analysis conducted produced sufficient conditions for the consistency analysis and used the von Neumann analysis to demonstrate that the method is stable. Moreover, the numerical results confirmed the conducted convergence analysis.

Keywords: Hopscotch method · Modified method of characteristics · Convergence analysis · Semi-lagrangian approach · Backward differentiation formulas · Numerical method

1 Introduction

The efficient numerical solution of evolutionary differential equations is essential in several areas in engineering and science. The Hopmoc method (see [1] and references therein) is a fast and accurate method to solve convection-dominated fluid or heat flows. The method is in an implicit numerical algorithm. Thus, the nodal update formulas employed are independent and can be used simultaneously at all mesh nodes. Consequently, the Hopmoc method can benefit from parallel computing, which is currently a crucial characteristic in the modern multicore architecture [2].

The Hopmoc method employs finite-difference techniques similar to the Hopscotch method [3,4], which is applied to solve parabolic and elliptic partial differential equations. The Hopmoc method divides the set of unknowns into two

O. Gervasi et al. (Eds.): ICCSA 2020, LNCS 12249, pp. 3–18, 2020.
https://doi.org/10.1007/978-3-030-58799-4_1

subsets alternately approximated. Thus, the method divides each time step into two semi-steps. For example, consider the use of a quadrangular mesh for the solution of a 2D problem. In this scenario, an internal mesh node belongs to one of the subsets, and its four adjacent mesh nodes belong to the other subset. At each time semi-step, the variables contained in a subset are alternately updated using symmetrical explicit and implicit approaches. More specifically, the first time semi-step updates a subset of unknowns using an explicit approach. Since the second implicit time semi-step uses the solution calculated in the previous time semi-step, no linear system is solved. Moreover, the method evaluates semi-steps along with characteristic lines employing concepts of the modified method of characteristics (MMOC) [5]. Specifically, the Hopmoc method uses approximated solutions from previous time steps along the directional derivative following the characteristic line comparably to the MMOC. Therefore, the Hopmoc method employs a semi-Lagrangian approach. Moreover, the method uses a Eulerian structure, but the discrete equations come from a Lagrangian frame of reference. More precisely, the Hopmoc method employs a spacial discretization along a characteristic line from each mesh node. The method is first-order accurate in space and time.

Backward differentiation formulas (BDFs) are implicit methods for the numerical integration of differential equations. BDFs are linear multistep methods that use information from previous time steps to increase the accuracy of an approximation to the derivative of a given function and time.

A recent publication integrated the Hopmoc method and BDFs [6]. We referred to the approach as the BDFHM method.

This paper presents a convergence analysis of the BDFHM method. The consistency and stability analyses rely on the techniques employed by Oliveira et al. [1] and Zhang [7], respectively.

This paper is organized as follows. In Sect. 2, the BDFHM method is presented. In Sect. 3, the convergence analysis of the method is provided. Finally, In Sect. 4, the conclusions and future directions in this investigation are discussed.

2   The BDFHM Method

In this section, we show a solution of the 1-D advection-diffusion equation

$$u_t + v \cdot u_x = d \cdot u_{xx} \tag{1}$$

using the BDFHM method, where d is the diffusion coefficient, v is a constant positive velocity, and $0 \leq x \leq 1$. The advection-diffusion equation is of main importance in many physical systems, especially those involving fluid flow. For instance, in computational hydraulics and fluid dynamics problems, the advection-diffusion equation may be seen to represent quantities such as mass, heat, energy, vorticity, etc. [8]. In Eq. (1), u_t refers to the time derivative and not u evaluated at the discrete-time step t. Nevertheless, we abuse the notation and now use t to denote a discrete-time step so that $0 \leq t \leq T$, for T time steps.

The unit vector $\tau = (v \cdot \delta t, \delta t) = \left(x - \overline{x}, t_{n+\frac{1}{2}} - t_n\right)$ represents the characteristic line associated with the transport $u_t + v \cdot u_x$, and $\overline{x}$ $(\overline{\overline{x}})$ is the "foot" of the characteristic line in the second (first) time semi-step. The derivative in the direction of τ is given by

$$u_\tau = \nabla u \times \frac{\tau}{\|\tau\|} = u_\tau = (u_x, u_t) \times \frac{(v\delta t, \delta t)}{\sqrt{(v\delta t)^2 + (\delta t)^2}}$$

$$= (u_x, u_t) \times \left(\frac{v\delta t}{\delta t\sqrt{(v)^2 + 1}}, \frac{\delta t}{\delta t\sqrt{(v)^2 + 1}}\right)$$

$$= (u_x, u_t) \times \left(\frac{v}{\sqrt{(v)^2 + 1}}, \frac{1}{\sqrt{(v)^2 + 1}}\right) = \left(\frac{v \cdot u_x}{\sqrt{(v)^2 + 1}}, \frac{u_t}{\sqrt{(v)^2 + 1}}\right)$$

$$= \frac{1}{\sqrt{(v)^2 + 1}} \times (v \cdot u_x + u_t) \Rightarrow u_\tau \times \sqrt{(v)^2 + 1} = v \cdot u_x + u_t = du_{xx}.$$

From the diffusion equation along the characteristic line [9], the modified method of characteristics approximates the directional derivative in explicit form through the equation

$$\left(\frac{\overline{u}_i^{n+\frac{1}{2}} - \overline{\overline{u}}_i^n}{\|\tau\|}\right) \times \sqrt{(v)^2 + 1} = d\left(\frac{\overline{\overline{u}}_{i-1}^n - 2\overline{\overline{u}}_i^n + \overline{\overline{u}}_{i+1}^n}{(\Delta x)^2}\right). \tag{2}$$

For clarity, we use a notation with overlines (to indicate for example the foot of the characteristic line that is calculated by an interpolation method) and superscript (to indicate time semi-step), i.e., $\overline{\overline{u}}^n$ $\left(\overline{u}^{n+\frac{1}{2}}\right)$ $\left[\overline{\overline{u}}^{n-\frac{1}{2}}\right]$ represents u evaluated in the foot of the characteristic line calculated in [the first time semi-step from] the previous time (semi-) step, and u^{n+1} represents u evaluated in the second time semi-step of the BDFHM method.

Substituting $\|\tau\| = \delta t \cdot \sqrt{(v)^2 + 1}$ in Eq. (2) yields $\overline{u}_i^{n+\frac{1}{2}} = \overline{\overline{u}}_i^n + \delta t \cdot d \cdot \left(\frac{\overline{\overline{u}}_{i-1}^n - 2\overline{\overline{u}}_i^n + \overline{\overline{u}}_{i+1}^n}{(\Delta x)^2}\right)$. Similarly, one can obtain a discretization of the Laplace operator on an implicit form by means of the equation $u_i^{n+\frac{1}{2}} = \overline{\overline{u}}_i^n + \delta t \cdot d\left(\frac{\overline{u}_{i-1}^n - 2\overline{u}_i^n + \overline{u}_{i+1}^n}{(\Delta x)^2}\right)$. Likewise, the difference operator for the BDFHM can be defined as

$$L_h \overline{\overline{u}}_i^n = d \cdot \left[\frac{\overline{\overline{u}}_{i-1}^n - 2\overline{\overline{u}}_i^n + \overline{\overline{u}}_{i+1}^n}{(\Delta x)^2}\right]. \tag{3}$$

When using operator (3), two time semi-steps can be represented as $\frac{3}{2}\overline{u}_i^{n+\frac{1}{2}} = 2\overline{\overline{u}}_i^n - \frac{1}{2}\overline{\overline{u}}_i^{n-\frac{1}{2}} + \delta t \left(\theta_i^n L_h \overline{\overline{u}}_i^n + \theta_i^{n+\frac{1}{2}} L_h \overline{u}_i^{n+\frac{1}{2}}\right)$ and $\frac{3}{2}u_i^{n+1} = 2\overline{u}_i^{n+\frac{1}{2}} - \frac{1}{2}\overline{\overline{u}}_i^n + \delta t \left(\theta_i^n L_h \overline{u}_i^{n+\frac{1}{2}} + \theta_i^{n+\frac{1}{2}} L_h u_i^{n+1}\right)$, where $\delta t = \frac{\Delta t}{2}$ and $\theta_i^n = 1$ (0) if $n + i$ is odd (even). We describe a complete time step of the BDFHM method as follows.

1. Initialize $\bar{\bar{x}}$ and $\bar{\bar{x}}$ (e.g. using the Hopmoc method) at times steps $t_{-\frac{1}{2}}$ and t_0, respectively.
2. Obtain $\bar{u}$ for all N mesh nodes ($1 \leq i \leq N$) $\bar{x}_i$ (e.g. using an interpolation method). In particular, the BDFHM method uses two interpolations in the first time semi-step, i.e., it uses an interpolation method to calculate $\bar{\bar{u}}^n$ and $\bar{\bar{u}}^{n-\frac{1}{2}}$ to obtain $\bar{u}^{n+\frac{1}{2}}$. In the second time semi-step, no interpolation method is used since $\bar{u}^{n+\frac{1}{2}}$ and $\bar{\bar{u}}^n$ (that are used to obtain u^n) were calculated in the first time semi-step.
3. Calculate (alternately) $\bar{u}_i^{n+\frac{1}{2}}$ using the explicit (implicit) operator for mesh nodes $n + 1 + i$ that belong to the odd (even) subset.
4. Calculate (alternately) u_i^{n+1} using the implicit (explicit) operator for mesh nodes $n + 2 + i$ that belong to the odd (even) subset.

In steps (3) and (4), the explicit approach uses the values from adjacent mesh nodes. The strategy updated these adjacent mesh nodes in the previous time step. The implicit approach uses values from adjacent mesh nodes that were updated in the current time step using the explicit approach. Therefore, no linear system is solved when applying the BDFHM method.

The Hopmoc method takes $\mathcal{O}(T \cdot n)$ time since the 1-D Hopmoc method computes five loops that iterate n 1-D stencil points inside a time loop [10]. The use of BDFs does not modify the complexity of the 1-D Hopmoc method. The reason is that, similar to the MMOC, the loop that computes a BDF takes $\mathcal{O}(n)$ time.

3   Convergence Analysis of the BDFHM Method

In Sect. 3.1, we demonstrate a consistency analysis of the BFD-Hopmoc method using Taylor's series expansion of an expression involving two consecutive time semi-steps. In Sect. 3.2, we show a von Neumann stability analysis of the BFD-Hopmoc method.

3.1   Consistency Analysis

A numerical method is convergent if the numerical solution approximates the analytical solution of the partial differential equation when $\Delta x \rightarrow 0$. In this section, we demonstrate the consistency of the BDFHM method using Taylor's series expansions of the terms from the implicit formula. Let's

$$u_i^{n+1} = \frac{4}{3}\bar{u}_i^{n+\frac{1}{2}} - \frac{1}{3}\bar{\bar{u}}_i^n + \frac{2}{3}\delta t \cdot d \left[\frac{\bar{u}_{i-1}^{n+\frac{1}{2}} - 2\bar{u}_i^{n+\frac{1}{2}} + \bar{u}_{i+1}^{n+\frac{1}{2}}}{(\Delta x)^2} \right] \tag{4}$$

and

$$\bar{u}_i^{n+\frac{1}{2}} = \frac{4}{3}\bar{\bar{u}}_i^n - \frac{1}{3}\bar{\bar{u}}_i^{n-\frac{1}{2}} + \frac{2}{3}\delta t \cdot d \left[\frac{\bar{u}_{i-1}^{n+\frac{1}{2}} - 2\bar{u}_i^{n+\frac{1}{2}} + \bar{u}_{i+1}^{n+\frac{1}{2}}}{(\Delta x)^2} \right] \tag{5}$$

be two consecutive time semi-steps of the BDFHM method. Isolating the unknown variable $\overline{u}_i^{n+\frac{1}{2}}$ in Eq. (5) yields

$$\overline{u}_i^{n+\frac{1}{2}} = \frac{\frac{4}{3}\overline{\overline{u}}_i^n - \frac{1}{3}\overline{\overline{u}}_i^{n-\frac{1}{2}} + \frac{2}{3}\delta t \cdot d \left[\frac{\overline{u}_{i-1}^{n+\frac{1}{2}} + \overline{u}_{i+1}^{n+\frac{1}{2}}}{(\Delta x)^2} \right]}{1 + \frac{4}{3}\frac{\delta t \cdot d}{(\Delta x)^2}}. \tag{6}$$

Substituting Eq. (6) in Eq. (4) yields

$$u_i^{n+1} = \frac{\frac{4}{3}\overline{\overline{u}}_i^n - \frac{1}{3}\overline{\overline{u}}_i^{n-\frac{1}{2}} + \frac{2}{3}\delta t \cdot d \left[\frac{\overline{u}_{i-1}^{n+\frac{1}{2}} + \overline{u}_{i+1}^{n+\frac{1}{2}}}{(\Delta x)^2} \right]}{1 + \frac{4}{3}\frac{\delta t \cdot d}{(\Delta x)^2}} \left(\frac{4}{3} - \frac{4}{3}\frac{\delta t \cdot d}{(\Delta x)^2} \right)$$
$$- \frac{1}{3}\overline{\overline{u}}_i^n + \frac{2}{3}\delta t \cdot d \left[\frac{\overline{u}_{i-1}^{n+\frac{1}{2}} + \overline{u}_{i+1}^{n+\frac{1}{2}}}{(\Delta x)^2} \right]$$

and

$$u_i^{n+1} - \frac{13}{9}\overline{\overline{u}}_i^n + \frac{4}{9}\overline{\overline{u}}_i^{n-\frac{1}{2}}$$
$$= \frac{\delta t \cdot d}{(\Delta x)^2}\left(-\frac{4}{3}u_i^{n+1} - \frac{20}{9}\overline{\overline{u}}_i^n + \frac{4}{9}\overline{\overline{u}}_i^{n-\frac{1}{2}} + \frac{14}{9}\overline{u}_{i-1}^{n+\frac{1}{2}} + \frac{14}{9}\overline{u}_{i+1}^{n+\frac{1}{2}} \right). \tag{7}$$

When using Taylor's series expansions in each term regarding to the node mesh (x_i, t_n), one obtains

$$\overline{\overline{u}}_i^n = u_i^{n+1} - 2v \cdot \delta t \cdot u_x|_i^{n+1} - 2\delta t \cdot u_t|_i^{n+1} + 2v^2(\delta t)^2 u_{xx}|_i^{n+1} + 4v(\delta t)^2 u_{xt}|_i^{n+1}$$
$$+ 2(\delta t)^2 u_{tt}|_i^{n+1} - \frac{2}{3}v^3(\delta t)^3 u_{xxx}|_i^{n+1} - 4v^2(\delta t)^3 u_{xxt}|_i^{n+1} - 4v(\delta t)^3 u_{xtt}|_i^{n+1}$$
$$- \frac{2}{3}(\delta t)^3 u_{ttt}|_i^{n+1} + \mathcal{O}\left((\gamma_1)^4\right); \tag{8}$$

$$\overline{\overline{u}}_i^{n-\frac{1}{2}} = u_i^{n+1} - 3v \cdot \delta t u_x|_i^{n+1} - 3\delta t \cdot u_t|_i^{n+1} + \frac{9}{2}v^2(\delta t)^2 u_{xx}|_i^{n+1} + 9v(\delta t)^2 u_{xt}|_i^{n+1}$$
$$+ \frac{9}{2}(\delta t)^2 u_{tt}|_i^{n+1} - \frac{9}{2}v^3(\delta t)^3 u_{xxx}|_i^{n+1} - \frac{27}{2}v^2(\delta t)^3 u_{xxt}|_i^{n+1}$$
$$- \frac{27}{2}v(\delta t)^3 u_{xtt}|_i^{n+1} - \frac{9}{2}(\delta t)^3 u_{ttt}|_i^{n+1} + \mathcal{O}\left((\gamma_2)^4\right); \tag{9}$$

$$\overline{u}_{i-1}^{n+\frac{1}{2}} = u_i^{n+1} - \delta t \cdot u_t|_i^{n+1} - (h + v \cdot \delta t)\delta t \cdot u_x|_i^{n+1} + \frac{(h + v \cdot \delta t)^2}{2}u_{xx}|_i^{n+1}$$
$$+ (h + v \cdot \delta t)\delta t \cdot u_{xt}|_i^{n+1} + \frac{1}{2}(\delta t)^2 u_{tt}|_i^{n+1} - \frac{(h + v \cdot \delta t)^3}{6}u_{xxx}|_i^{n+1}$$
$$- \frac{(h + v \cdot \delta t)^2 \delta t}{2}u_{xxt}|_i^{n+1} - \frac{(h + v \cdot \delta t)(\delta t)^2}{2}u_{xtt}|_i^{n+1} - \frac{1}{6}(\delta t)^3 u_{ttt}|_i^{n+1}$$
$$+ \mathcal{O}\left((\gamma_3)^4\right); \tag{10}$$

$$\bar{u}_{i+1}^{n+\frac{1}{2}} = u_i^{n+1} - \delta t \cdot u_t|_i^{n+1} + (h - v \cdot \delta t) u_x|_i^{n+1} + \frac{(h - v \cdot \delta t)^2}{2} u_{xx}|_i^{n+1}$$

$$- (h - v \cdot \delta t) \delta t \cdot u_{xt}|_i^{n+1} + \frac{1}{2}(\delta t)^2 u_{tt}|_i^{n+1} - \frac{(h + v \cdot \delta t)^3}{6} u_{xxx}|_i^{n+1}$$

$$- \frac{(h - v \cdot \delta t)^2 \delta t}{2} u_{xxt}|_i^{n+1} + \frac{(h - v \cdot \delta t) \delta t^2}{2} u_{xtt}|_i^{n+1} - \frac{1}{6}(\delta t)^3 u_{ttt}|_i^{n+1}$$

$$+ \mathcal{O}\left((\gamma_4)^4\right); \tag{11}$$

where:

- $h = \Delta x$;
- $\mathcal{O}\left((\gamma_1)^4\right) = -\frac{1}{4!}\left[-2v \cdot \delta t \cdot u_x - 2\delta t \cdot u_t|_i^{n+1}\right]^4 (a_1, b_1)$;
- $\mathcal{O}\left((\gamma_2)^4\right) = -\frac{1}{4!}\left[-3v \cdot \delta t \cdot u_x - 3\delta t \cdot u_t|_i^{n+1}\right]^4 (a_2, b_2)$;
- $\mathcal{O}\left((\gamma_3)^4\right) = -\frac{1}{4!}\left[-(h + v \cdot \delta t)u_x - \delta t \cdot u_t|_i^{n+1}\right]^4 (a_3, b_3)$;
- $\mathcal{O}\left((\gamma_4)^4\right) = -\frac{1}{4!}\left[(h - v \cdot \delta t)u_x - \delta t \cdot u_t|_i^{n+1}\right]^4 (a_4, b_4)$;
- (a_1, b_1) belongs to the segment (x_i, t_{n+1}) and $\left(\bar{\bar{u}}_i^n, t_n\right)$;
- (a_2, b_2) belongs to the segment (x_i, t_{n+1}) and $\left(\bar{\bar{u}}_i^{n-\frac{1}{2}}, t_{n-\frac{1}{2}}\right)$;
- (a_3, b_3) belongs to the segment (x_i, t_{n+1}) and $\left(\bar{u}_{i-1}^{n+\frac{1}{2}}, t_{n+\frac{1}{2}}\right)$;
- (a_4, b_4) belongs to the segment (x_i, t_{n+1}) and $\left(\bar{u}_{i+1}^{n+\frac{1}{2}}, t_{n+\frac{1}{2}}\right)$.

The consistency analysis based on Taylor's series expansion of terms (8)–(11) required a number of simplifications. Substituting Eqs. (8)–(11) in Eq. (7) yields

$$\overbrace{\frac{13}{9}vu_x|_i^{n+1}} + \frac{13}{9}u_t|_i^{n+1} - \frac{13}{9}v^2\delta t \cdot u_{tt}|_i^{n+1} - \frac{26}{9}v \cdot \delta t \cdot u_{xt}|_i^{n+1}$$

$$- \frac{13}{9}\delta t \cdot u_{xx}|_i^{n+1}$$

$$+ \frac{13}{27}v^3(\delta t)^2 u_{xxx}|_i^{n+1} + \frac{26}{9}v^2(\delta t)^2 u_{xxt}|_i^{n+1} + \frac{26}{9}v(\delta t)^2 u_{xtt}|_i^{n+1}$$

$$+ \frac{13}{9}(\delta t)^3 u_{ttt}|_i^{n+1}$$

$$- \frac{13}{18\delta t}\overbrace{\mathcal{O}\left((\gamma_1)^4\right)} - \frac{2}{3}\overbrace{vu_x|_i^{n+1} - \frac{2}{3}u_t|_i^{n+1}} + v^2\delta t \cdot u_{xx}|_i^{n+1} + 2v \cdot \delta t \cdot u_{xt}|_i^{n+1}$$

$$+ \delta t \cdot u_{tt}|_i^{n+1} - v^3(\delta t)^2 u_{xxx}|_i^{n+1} - 3v^2(\delta t)^2 u_{xxt}|_i^{n+1} - 3v(\delta t)^2 u_{xtt}|_i^{n+1} - \delta t \cdot u_{ttt}|_i^{n+1}$$

$$+ \frac{2}{9\delta t}\mathcal{O}\left((\gamma_2)^4\right) = \frac{20}{9}\frac{d}{(\Delta x)^2}v \cdot \delta t \cdot u_x|_i^{n+1} + \frac{20}{9}\frac{d}{(\Delta x)^2}\delta t \cdot u_t|_i^{n+1}$$

$$- \frac{20}{9}\frac{d}{(\Delta x)^2}v^2(\delta t)^2 u_{xx}|_i^{n+1} - \frac{40}{9}\frac{d}{(\Delta x)^2}v(\delta t)^2 u_{xt}|_i^{n+1} - \frac{20}{9}\frac{d}{(\Delta x)^2}(\delta t)^2 u_{tt}|_i^{n+1}$$

$$+ \frac{20}{27}\frac{d}{(\Delta x)^2}v^3(\delta t)^3 u_{xxx}|_i^{n+1} + \frac{40}{9}\frac{d}{(\Delta x)^2}v^2(\delta t)^3 u_{xxt}|_i^{n+1} + \frac{40}{9}\frac{d}{(\Delta x)^2}v(\delta t)^3 u_{xtt}|_i^{n+1}$$

$$+ \frac{20}{27}\frac{d}{(\Delta x)^2}(\delta t)^3 u_{ttt}|_i^{n+1} - \frac{10}{9\delta t}\frac{d}{(\Delta x)^2}\mathcal{O}\left((\gamma_1)^4\right) - \frac{2}{3}\frac{d}{(\Delta x)^2}v \cdot \delta t \cdot u_x|_i^{n+1}$$

$$-\frac{d}{(\Delta x)^2}\frac{2}{3}\delta t \cdot u_t|_i^{n+1} + \frac{d}{(\Delta x)^2}v^2(\delta t)^2 u_{xx}|_i^{n+1} + 2\frac{d}{(\Delta x)^2}v(\delta t)^2 u_{xt}|_i^{n+1}$$

$$+\frac{d}{(\Delta x)^2}(\delta t)^2 u_{tt}|_i^{n+1} - \frac{d}{(\Delta x)^2}v^3(\delta t)^3 u_{xxx}|_i^{n+1} - 3v^2(\delta t)^2 u_{xxt}|_i^{n+1}$$

$$-3\frac{d}{(\Delta x)^2}v(\delta t)^3 u_{xtt}|_i^{n+1} - \frac{d}{(\Delta x)^2}(\delta t)^3 u_{ttt}|_i^{n+1} + \frac{2}{9\delta t}\frac{d}{(\Delta x)^2}\mathcal{O}\left((\gamma_2)^4\right)$$

$$-\frac{7}{9}\frac{d}{(\Delta x)^2}\delta t \cdot u_t|_i^{n+1} - \frac{7}{9}\frac{d}{(\Delta x)^2}(h+v\cdot\delta t)u_x|_i^{n+1} + \overbrace{\frac{7}{9}\frac{d}{(\Delta x)^2}\frac{h^2}{2}u_{xx}|_i^{n+1}}$$

$$+\frac{7}{9}\frac{d}{(\Delta x)^2}\frac{(2h\cdot v\cdot\delta t)}{2}u_{xx}|_i^{n+1} + \frac{7}{9}\frac{d}{(\Delta x)^2}\frac{(v\cdot\delta t)^2}{2}u_{xx}|_i^{n+1}$$

$$+\frac{7}{9}\frac{d}{(\Delta x)^2}(h+v\cdot\delta t)\delta t\cdot u_{xt}|_i^{n+1} + \frac{7}{18}\frac{d}{(\Delta x)^2}(\delta t)^2 u_{tt}|_i^{n+1}$$

$$-\frac{7}{9}\frac{d}{(\Delta x)^2}\frac{(h+v\cdot\delta t)^3}{6}u_{xxx}|_i^{n+1} - \frac{7}{9}\frac{d}{(\Delta x)^2}\frac{(h+v\cdot\delta t)^2\delta t}{2}u_{xxt}|_i^{n+1}$$

$$-\frac{7}{9}\frac{d}{(\Delta x)^2}\frac{(h+v\cdot\delta t)(\delta t)^2}{2}u_{xtt}|_i^{n+1} - \frac{7}{9}\frac{d}{(\Delta x)^2}\frac{1}{6}(\delta t)^3 u_{ttt}|_i^{n+1}$$

$$+\frac{7}{9\delta t}\frac{d}{(\Delta x)^2}\mathcal{O}\left((\gamma_3)^4\right) - \frac{7}{9}\frac{d}{(\Delta x)^1}\delta t\cdot u_t|_i^{n+1} + \frac{7}{9}\frac{d}{(\Delta x)^2}(h-v\cdot\delta t)u_x|_i^{n+1}$$

$$+\overbrace{\frac{7}{9}\frac{d}{(\Delta x)^2}\frac{h^2}{2}u_{xx}|_i^{n+1}} + \frac{7}{9}\frac{d}{(\Delta x)^2}\frac{(2h\cdot v\cdot\delta t)^2}{2}u_{xx}|_i^{n+1} + \frac{7}{9}\frac{d}{(\Delta x)^2}\frac{(v\cdot\delta t)^2}{2}u_{xx}|_i^{n+1}$$

$$-\frac{7}{9}\frac{d}{(\Delta x)^2}(h-v\cdot\delta t)u_{xt}|_i^{n+1} + \frac{7}{9}\frac{d}{(\Delta x)^2}\frac{1}{2}(\delta t)^2 u_{tt}|_i^{n+1}$$

$$+\frac{7}{9}\frac{d}{(\Delta x)^2}\frac{(h-v\cdot\delta t)^3}{6}u_{xxx}|_i^{n+1} - \frac{7}{9}\frac{d}{(\Delta x)^2}\frac{(h-v\cdot\delta t)^2\delta t}{2}u_{xxt}|_i^{n+1}$$

$$-\frac{7}{9}\frac{d}{(\Delta x)^2}\frac{(h-v\cdot\delta t)(\delta t)^2}{2}u_{xtt}|_i^{n+1} - \frac{7}{9}\frac{d}{(\Delta x)^2}\frac{1}{6}(\delta t)^3 u_{ttt}|_i^{n+1} + \frac{7}{9}\frac{d}{(\Delta x)^2}\mathcal{O}\left((\gamma_4)^4\right).$$

Rewriting the selected terms (highlighted with overbraces) in this equation yields

$$\frac{13}{9}vu_x|_i^{n+1} + \frac{13}{9}u_t|_i^{n+1} - \frac{2}{3}vu_x|_i^{n+1} - \frac{2}{3}u_t|_i^{n+1}$$
$$= \frac{7}{9}\frac{d}{(\Delta x)^2}\frac{h^2}{2}u_{xx}|_i^{n+1} + \frac{7}{9}\frac{d}{(\Delta x)^2}\frac{h^2}{2}u_{xx}|_i^{n+1}. \tag{12}$$

Simplifying Eq. (12) and, since $h^2 = (\Delta x)^2$, one obtains

$$u_t|_i^{n+1} = -vu_x|_i^{n+1} + du_{xx}|_i^{n+1}. \tag{13}$$

When observing the resulting terms, the error ϵ associated with the BDFHM method can be written as

$$\epsilon = \overbrace{+\frac{13}{9}v^2\delta t\cdot u_{xx}|_i^{n+1} + \frac{26}{9}v\cdot\delta t\cdot u_{xt}|_i^{n+1} + \frac{13}{9}\delta t\cdot u_{tt}|_i^{n+1}} - \frac{13}{27}v^3(\delta t)^2 u_{xxx}|_i^{n+1}$$

$$-\frac{26}{9}v^2(\delta t)^2 u_{xxt}|_i^{n+1} - \frac{26}{9}v(\delta t)^2 u_{xtt}|_i^{n+1} - \frac{13}{9}(\delta t)^3 u_{ttt}|_i^{n+1} + \frac{13}{18}\mathcal{O}\left((\gamma_1)^4\right)$$

$$\overbrace{\phantom{-\frac{26}{9}v^2(\delta t)^2 u_{xxt}|_i^{n+1} - \frac{26}{9}v(\delta t)^2 u_{xtt}|_i^{n+1}}}\quad\overbrace{\phantom{-\frac{13}{9}(\delta t)^3 u_{ttt}|_i^{n+1}}}$$

$$-v^2\delta t \cdot u_{xx}|_i^{n+1} - 2v\cdot\delta t\cdot u_{xt}|_i^{n+1} - \delta t\cdot u_{tt}|_i^{n+1} + v^3(\delta t)^2 u_{xxx}|_i^{n+1}$$

$$+3v^2(\delta t)^2 u_{xxt}|_i^{n+1} + 3v(\delta t)^2 u_{xtt}|_i^{n+1} + \delta t^2 u_{ttt}|_i^{n+1} - \frac{2}{9}\mathcal{O}\left((\gamma_2)^4\right)$$

$$+\frac{20}{9}\frac{d}{(\Delta x)^2}v\cdot\delta t\cdot u_x|_i^{n+1} + \frac{20}{9}\frac{d}{(\Delta x)^2}\delta t\cdot u_t|_i^{n+1} - \frac{20}{9}\frac{d}{(\Delta x)^2}v^2(\delta t)^2 u_{xx}|_i^{n+1}$$

$$-\frac{40}{9}\frac{d}{(\Delta x)^2}v(\delta t)^2 u_{xt}|_i^{n+1} - \frac{20}{9}\frac{d}{(\Delta x)^2}(\delta t)^2 u_{tt}|_i^{n+1}$$

$$+\frac{20}{27}\frac{d}{(\Delta x)^2}v^3(\delta t)^3 u_{xxx}|_i^{n+1}$$

$$+\frac{40}{9}\frac{d}{(\Delta x)^2}v^2(\delta t)^3 u_{xxt}|_i^{n+1} + \frac{40}{9}\frac{d}{(\Delta x)^2}v(\delta t)^3 u_{xtt}|_i^{n+1}$$

$$+\frac{20}{27}\frac{d}{(\Delta x)^2}(\delta t)^3 u_{ttt}|_i^{n+1} - \frac{10}{9}\frac{d}{(\Delta x)^2}\mathcal{O}\left((\gamma_1)^4\right) - \frac{2}{3}\frac{d}{(\Delta x)^2}v\cdot\delta t\cdot u_x|_i^{n+1}$$

$$-\frac{2}{3}\frac{d}{(\Delta x)^2}\delta t\cdot u_t|_i^{n+1} + \frac{d}{(\Delta x)^2}v^2(\delta t)^2 u_{xx}|_i^{n+1} + 2\frac{d}{(\Delta x)^2}v(\delta t)^2 u_{xt}|_i^{n+1}$$

$$+\frac{d}{(\Delta x)^2}(\delta t)^2 u_{tt}|_i^{n+1} - \frac{d}{(\Delta x)^2}v^3(\delta t)^3 u_{xxx}|_i^{n+1} - 3v^2(\delta t)^2 u_{xxt}|_i^{n+1}$$

$$-3\frac{d}{(\Delta x)^2}v(\delta t)^3 u_{xtt}|_i^{n+1} - \frac{d}{(\Delta x)^2}(\delta t)^3 u_{ttt}|_i^{n+1} + \frac{2}{9}\frac{d}{(\Delta x)^2}\mathcal{O}\left((\gamma_2)^4\right)$$

$$-\frac{7}{9}\frac{d}{(\Delta x)^2}\delta t\cdot u_t|_i^{n+1} - \frac{7}{9}\frac{d}{(\Delta x)^2}\left(h + v\cdot\delta t\right)u_x|_i^{n+1}$$

$$+\frac{7}{9}\frac{d}{(\Delta x)^2}\frac{(2h\cdot v\cdot\delta t)}{2}u_{xx}|_i^{n+1} + \frac{7}{9}\frac{d}{(\Delta x)^2}\frac{(v\cdot\delta t)^2}{2}u_{xx}|_i^{n+1}$$

$$+\frac{7}{9}\frac{d}{(\Delta x)^2}\left(h + v\cdot\delta t\right)\delta t\cdot u_{xt}|_i^{n+1} + \frac{7}{18}\frac{d}{(\Delta x)^2}(\delta t)^2 u_{tt}|_i^{n+1}$$

$$-\frac{7}{9}\frac{d}{(\Delta x)^2}\frac{(h + v\cdot\delta t)^3}{6}u_{xxx}|_i^{n+1} - \frac{7}{9}\frac{d}{(\Delta x)^2}\frac{(h + v\cdot\delta t)^2\,\delta t}{2}u_{xxt}|_i^{n+1}$$

$$-\frac{7}{9}\frac{d}{(\Delta x)^2}\frac{(h + v\cdot\delta t)(\delta t)^2}{2}u_{xtt}|_i^{n+1} - \frac{7}{9}\frac{d}{(\Delta x)^2}\frac{1}{6}(\delta t)^3 u_{ttt}|_i^{n+1}$$

$$+\frac{7}{9}\frac{d}{(\Delta x)^2}\mathcal{O}\left((\gamma_3)^4\right)$$

$$-\frac{7}{9}\frac{d}{(\Delta x)^2}\delta t\cdot u_t|_i^{n+1} + \frac{7}{9}\frac{d}{(\Delta x)^2}\left(h - v\cdot\delta t\right)u_x|_i^{n+1}$$

$$+\frac{7}{9}\frac{d}{(\Delta x)^2}\frac{(2h\cdot v\cdot \delta t)^2}{2}u_{xx}|_i^{n+1}$$

$$+\frac{7}{9}\frac{d}{(\Delta x)^2}\frac{(v\cdot \delta t)^2}{2}u_{xx}|_i^{n+1}$$

$$-\frac{7}{9}\frac{d}{(\Delta x)^2}(h-v\cdot \delta t)\,\delta t\cdot u_{xt}|_i^{n+1}+\frac{7}{9}\frac{d}{(\Delta x)^2}\frac{(\delta t)^2 u_{tt}|_i^{n+1}}{2}$$

$$+\frac{7}{9}\frac{d}{(\Delta x)^2}\frac{(h-v\cdot \delta t)^3}{6}u_{xxx}|_i^{n+1}-\frac{7}{9}\frac{d}{(\Delta x)^2}\frac{(h-v\cdot \delta t)^2\,\delta t}{2}u_{xxt}|_i^{n+1}$$

$$+\frac{7}{9}\frac{d}{(\Delta x)^2}\frac{(h-v\cdot \delta t)\,(\delta t)^2}{2}u_{xtt}|_i^{n+1}-\frac{7}{9}\frac{d}{(\Delta x)^2}\frac{1}{6}(\delta t)^3 u_{ttt}|_i^{n+1}$$

$$+\frac{7}{9}\frac{d}{(\Delta x)^2}\mathcal{O}\left((\gamma_4)^4\right). \tag{14}$$

In Eq. (14), we highlighted terms with Δx in the denominator with an overbrace. In the same equation, we highlighted with a double overbrace the terms that can be simplified. Similarly, we highlighted with overlines the terms that we will develop. Simplifying Eq. (14) produces

$$\epsilon=+\frac{4}{9}v^2\delta t\cdot u_{xx}|_i^{n+1}+\frac{8}{9}v\cdot \delta t\cdot u_{xt}|_i^{n+1}+\frac{4}{9}\delta t\cdot u_{tt}|_i^{n+1}-\frac{13}{27}v^3(\delta t)^2 u_{xxx}|_i^{n+1}$$

$$-\frac{26}{9}v^2(\delta t)^2 u_{xxt}|_i^{n+1}-\frac{26}{9}v(\delta t)^2 u_{xtt}|_i^{n+1}-\frac{13}{9}(\delta t)^3 u_{ttt}|_i^{n+1}+\frac{13}{18}\mathcal{O}\left((\gamma_1)^4\right)$$

$$+v^3(\delta t)^2 u_{xxx}|_i^{n+1}+3v^2(\delta t)^2 u_{xxt}|_i^{n+1}+3v(\delta t)^2 u_{xtt}|_i^{n+1}+(\delta t)^2 u_{ttt}|_i^{n+1}$$

$$-\frac{2}{9}\mathcal{O}\left((\gamma_2)^4\right)-\frac{20}{9}\frac{d}{(\Delta x)^2}v^2(\delta t)^2 u_{xx}|_i^{n+1}-\frac{40}{9}\frac{d}{(\Delta x)^2}v(\delta t)^2 u_{xt}|_i^{n+1}$$

$$-\frac{20}{9}\frac{d}{(\Delta x)^2}(\delta t)^2 u_{tt}|_i^{n+1}+\frac{20}{27}\frac{d}{(\Delta x)^2}v^3(\delta t)^3 u_{xxx}|_i^{n+1}$$

$$+\frac{40}{9}\frac{d}{(\Delta x)^2}v^2(\delta t)^3 u_{xxt}|_i^{n+1}$$

$$+\frac{40}{9}\frac{d}{(\Delta x)^2}v(\delta t)^3 u_{xtt}|_i^{n+1}+\frac{20}{27}\frac{d}{(\Delta x)^2}(\delta t)^3 u_{ttt}|_i^{n+1}-\frac{10}{9}\frac{d}{(\Delta x)^2}\mathcal{O}\left((\gamma_1)^4\right)$$

$$+\frac{d}{(\Delta x)^2}v^2(\delta t)^2 u_{xx}|_i^{n+1}+2\frac{d}{(\Delta x)^2}v(\delta t)^2 u_{xt}|_i^{n+1}+\frac{d}{(\Delta x)^2}(\delta t)^2 u_{tt}|_i^{n+1}$$

$$-\frac{d}{(\Delta x)^2}v^3(\delta t)^3 u_{xxx}|_i^{n+1}-3v^2(\delta t)^2 u_{xxt}|_i^{n+1}-3\frac{d}{(\Delta x)^2}v(\delta t)^3 u_{xtt}|_i^{n+1}$$

$$-\frac{d}{(\Delta x)^2}(\delta t)^3 u_{ttt}|_i^{n+1}+\frac{2}{9}\frac{d}{(\Delta x)^2}\mathcal{O}\left((\gamma_2)^4\right)+\frac{7}{9}\frac{d}{(\Delta x)^2}\frac{(2h\cdot v\cdot \delta t)}{2}u_{xx}|_i^{n+1}$$

$$+\frac{7}{9}\frac{d}{(\Delta x)^2}\frac{(v\cdot \delta t)^2}{2}u_{xx}|_i^{n+1}+\frac{7}{9}\frac{d}{(\Delta x)^2}(h+v\cdot \delta t)\,\delta t\cdot u_{xt}|_i^{n+1}$$

$$+\frac{7}{18}\frac{d}{(\Delta x)^2}(\delta t)^2 u_{tt}|_i^{n+1}-\frac{7}{9}\frac{d}{(\Delta x)^2}\frac{h^3}{6}u_{xxx}|_i^{n+1}-\frac{7}{9}\frac{d}{(\Delta x)^2}\frac{3h^2 v\cdot \delta t}{6}u_{xxx}|_i^{n+1}$$

$$-\frac{7}{9}\frac{d}{(\Delta x)^2}\frac{3h\cdot v^2\delta t^2}{6}u_{xxx}|_i^{n+1}-\frac{7}{9}\frac{d}{(\Delta x)^2}\frac{v^3\delta t^3}{6}u_{xxx}|_i^{n+1}-\frac{7}{9}\frac{d}{(\Delta x)^2}\frac{h^2}{2}\delta t\cdot u_{xxt}|_i^{n+1}$$

$$-\frac{7}{9}\frac{d}{(\Delta x)^2}\frac{2h\cdot v\cdot\delta t}{2}\delta t\cdot u_{xxt}|_i^{n+1}-\frac{7}{9}\frac{d}{(\Delta x)^2}\frac{v^2\delta t^2}{2}\delta t\cdot u_{xxt}|_i^{n+1}$$

$$+\frac{7}{9}\frac{d}{(\Delta x)^2}\frac{(h-v\cdot\delta t)(\delta t)^2}{2}u_{xtt}|_i^{n+1}-\frac{7}{9}\frac{d}{(\Delta x)^2}\frac{1}{6}(\delta t)^3u_{ttt}|_i^{n+1}+\frac{7}{9}\frac{d}{(\Delta x)^2}\mathcal{O}\left((\gamma_3)^4\right)$$

$$+\frac{7}{9}\frac{d}{(\Delta x)^2}\frac{(2h\cdot v\cdot\delta t)^2}{2}u_{xx}|_i^{n+1}+\frac{7}{9}\frac{d}{(\Delta x)^2}\frac{(v\cdot\delta t)^2}{2}u_{xx}|_i^{n+1}$$

$$-\frac{7}{9}\frac{d}{(\Delta x)^2}(h-v\cdot\delta t)\delta t\cdot u_{xt}|_i^{n+1}+\frac{7}{9}\frac{d}{(\Delta x)^2}\frac{1}{2}(\delta t)^2u_{tt}|_i^{n+1}+\frac{7}{9}\frac{d}{(\Delta x)^2}\frac{h^3}{6}u_{xxx}|_i^{n+1}$$

$$\overbrace{-\frac{7}{9}\frac{d}{(\Delta x)^2}\frac{3h^2v\cdot\delta t}{6}u_{xxx}|_i^{n+1}}+\frac{7}{9}\frac{d}{(\Delta x)^2}\frac{3h\cdot v^2\delta t^2}{6}u_{xxx}|_i^{n+1}$$

$$-\frac{7}{9}\frac{d}{(\Delta x)^2}\frac{v^3\delta t^3}{6}u_{xxx}|_i^{n+1}\overbrace{-\frac{7}{9}\frac{d}{(\Delta x)^2}\frac{h^2}{2}\delta t\cdot u_{xxt}|_i^{n+1}}$$

$$+\frac{7}{9}\frac{d}{(\Delta x)^2}\frac{2h\cdot v\cdot\delta t}{2}\delta t\cdot u_{xxt}|_i^{n+1}-\frac{7}{9}\frac{d}{(\Delta x)^2}\frac{v^2\delta t^2}{2}\delta t\cdot u_{xxt}|_i^{n+1}|_i^{n+1}$$

$$+\frac{7}{9}\frac{d}{(\Delta x)^2}\frac{(h-v\cdot\delta t)(\delta t)^2}{2}u_{xtt}|_i^{n+1}-\frac{7}{9}\frac{d}{(\Delta x)^2}\frac{1}{6}(\delta t)^3u_{ttt}|_i^{n+1}+\frac{7}{9}\frac{d}{(\Delta x)^2}\mathcal{O}\left((\gamma_4)^4\right).\ (15)$$

Rewriting the selected terms (highlighted with overbraces) in (15) produces

$$+\frac{4}{9}v^2\delta t\cdot u_{xx}|_i^{n+1}+\frac{8}{9}v\cdot\delta t\cdot u_{xt}|_i^{n+1}+\frac{4}{9}\delta t\cdot u_{tt}|_i^{n+1}$$

$$-\frac{7}{9}\frac{d}{(\Delta x)^2}\frac{3\cdot h^2v\cdot\delta t}{6}u_{xxx}|_i^{n+1}-\frac{7}{9}\frac{d}{(\Delta x)^2}\frac{3h^2\cdot v\cdot\delta t}{6}u_{xxx}|_i^{n+1}$$

$$-\frac{7}{9}\frac{d}{(\Delta x)^2}\frac{h^2}{2}\delta t\cdot u_{xxt}|_i^{n+1}-\frac{7}{9}\frac{d}{(\Delta x)^2}\frac{h^2}{2}\delta t\cdot u_{xxt}|_i^{n+1}=$$

$$+\frac{4}{9}v^2\delta t\cdot u_{xx}|_i^{n+1}+\frac{4}{9}v\cdot\delta t\cdot u_{xt}|_i^{n+1}+\frac{4}{9}v\cdot\delta t\cdot u_{xt}|_i^{n+1}+\frac{4}{9}\delta t\cdot u_{tt}|_i^{n+1}$$

$$-\frac{7}{9}\frac{d}{(\Delta x)^2}h^2v\cdot\delta t\cdot u_{xxx}|_i^{n+1}-\frac{7}{9}\frac{d}{(\Delta x)^2}h^2\delta t\cdot u_{xxt}|_i^{n+1}=$$

$$+\frac{4}{9}v^2\delta t\cdot u_{xx}|_i^{n+1}+\frac{4}{9}v\cdot\delta t\cdot u_{xt}|_i^{n+1}+\frac{4}{9}v\cdot\delta t\cdot u_{xt}|_i^{n+1}+\frac{4}{9}\delta t\cdot u_{tt}|_i^{n+1}$$

$$-\frac{4}{9}dv\cdot\delta t\cdot u_{xxx}|_i^{n+1}-\frac{3}{9}dv\cdot\delta t\cdot u_{xxx}|_i^{n+1}-\frac{4}{9}d\delta t\cdot u_{xxt}|_i^{n+1}-\frac{3}{9}d\delta t\cdot u_{xxt}|_i^{n+1}.$$

Taking into account Eq. (13), one obtains $+\frac{4}{9}v^2\delta t\cdot u_{xx}|_i^{n+1}+\frac{4}{9}v\cdot\delta t\cdot u_{xt}|_i^{n+1}-\frac{4}{9}dv\cdot\delta t\cdot u_{xxx}|_i^{n+1}=+\frac{4}{9}v\cdot\delta t\cdot u_{xt}|_i^{n+1}+\frac{4}{9}\delta t\cdot u_{tt}|_i^{n+1}-\frac{4}{9}d\delta t\cdot u_{xxt}|_i^{n+1}=0$. When adding $\frac{1}{3}d^2\delta t\cdot u_{xxxx}|_i^{n+1}$ to the resulting terms $-\frac{1}{3}dv\cdot\delta t\cdot u_{xxx}|_i^{n+1}-\frac{1}{3}d\delta t\cdot u_{xxt}|_i^{n+1}$, one obtains $-\frac{1}{3}dv\cdot\delta t\cdot u_{xxx}|_i^{n+1}-\frac{1}{3}d\delta t\cdot u_{xxt}|_i^{n+1}+\frac{1}{3}d^2\delta t\cdot u_{xxxx}|_i^{n+1}=0$. When replacing this expression by Eq. (13) in ϵ, the error can be written as

$$\epsilon=-\frac{13}{27}v^3(\delta t)^2u_{xxx}|_i^{n+1}-\frac{26}{9}v^2(\delta t)^2u_{xxt}|_i^{n+1}-\frac{26}{9}v(\delta t)^2u_{xtt}|_i^{n+1}$$

$$-\frac{13}{9}(\delta t)^3 u_{ttt}|_i^{n+1} + \frac{13}{18}\mathcal{O}\left((\gamma_1)^4\right) + v^3(\delta t)^2 u_{xxx}|_i^{n+1}$$

$$+3v^2(\delta t)^2 u_{xxt}|_i^{n+1} + 3v(\delta t)^2 u_{xtt}|_i^{n+1} + (\delta t)^2 u_{ttt}|_i^{n+1} - \frac{2}{9}\mathcal{O}\left((\gamma_2)^4\right)$$

$$-\frac{20}{9}\frac{d}{(\Delta x)^2}v^2(\delta t)^2 u_{xx}|_i^{n+1} - \frac{40}{9}\frac{d}{(\Delta x)^2}v(\delta t)^2 u_{xt}|_i^{n+1}$$

$$-\frac{20}{9}\frac{d}{(\Delta x)^2}(\delta t)^2 u_{tt}|_i^{n+1} + \frac{20}{27}\frac{d}{(\Delta x)^2}v^3(\delta t)^3 u_{xxx}|_i^{n+1}$$

$$+\frac{40}{9}\frac{d}{(\Delta x)^2}v^2(\delta t)^3 u_{xxt}|_i^{n+1} + \frac{40}{9}\frac{d}{(\Delta x)^2}v(\delta t)^3 u_{xtt}|_i^{n+1}$$

$$+\frac{20}{27}\frac{d}{(\Delta x)^2}(\delta t)^3 u_{ttt}|_i^{n+1} - \frac{10}{9}\frac{d}{(\Delta x)^2}\mathcal{O}\left((\gamma_1)^4\right)$$

$$+\frac{d}{(\Delta x)^2}v^2(\delta t)^2 u_{xx}|_i^{n+1} + 2\frac{d}{(\Delta x)^2}v(\delta t)^2 u_{xt}|_i^{n+1}$$

$$+\frac{d}{(\Delta x)^2}(\delta t)^2 u_{tt}|_i^{n+1} - \frac{d}{(\Delta x)^2}v^3(\delta t)^3 u_{xxx}|_i^{n+1} - 3v^2(\delta t)^2 u_{xxt}|_i^{n+1}$$

$$-3\frac{d}{(\Delta x)^2}v(\delta t)^3 u_{xtt}|_i^{n+1} - \frac{d}{(\Delta x)^2}(\delta t)^3 u_{ttt}|_i^{n+1} + \frac{2}{9}\frac{d}{(\Delta x)^2}\mathcal{O}\left((\gamma_2)^4\right)$$

$$+\frac{7}{9}\frac{d}{(\Delta x)^2}\frac{(2h \cdot v \cdot \delta t)}{2}u_{xx}|_i^{n+1} + \frac{7}{9}\frac{d}{(\Delta x)^2}\frac{(v \cdot \delta t)^2}{2}u_{xx}|_i^{n+1}$$

$$+\frac{7}{9}\frac{d}{(\Delta x)^2}(h + v \cdot \delta t)\delta t \cdot u_{xt}|_i^{n+1} + \frac{7}{18}\frac{d}{(\Delta x)^2}(\delta t)^2 u_{tt}|_i^{n+1}$$

$$-\frac{7}{9}\frac{d}{(\Delta x)^2}\frac{h^3}{6}u_{xxx}|_i^{n+1} - \frac{7}{9}\frac{d}{(\Delta x)^2}\frac{3h \cdot v^2(\delta t)^2}{6}u_{xxx}|_i^{n+1}$$

$$-\frac{7}{9}\frac{d}{(\Delta x)^2}\frac{v^3(\delta t)^3}{6}u_{xxx}|_i^{n+1} - \frac{7}{9}\frac{d}{(\Delta x)^2}\frac{2h \cdot v \cdot \delta t}{2}\delta t \cdot u_{xxt}|_i^{n+1}$$

$$-\frac{7}{9}\frac{d}{(\Delta x)^2}\frac{v^2(\delta t)^2}{2}\delta t \cdot u_{xxt}|_i^{n+1} + \frac{7}{9}\frac{d}{(\Delta x)^2}\frac{(h - v \cdot \delta t)(\delta t)^2}{2}u_{xtt}|_i^{n+1}$$

$$-\frac{7}{9}\frac{d}{(\Delta x)^2}\frac{1}{6}(\delta t)^3 u_{ttt}|_i^{n+1} + \frac{7}{9}\frac{d}{(\Delta x)^2}\mathcal{O}\left((\gamma_3)^4\right) + \frac{1}{3}d^2\delta t \cdot u_{xxxx}|_i^{n+1}$$

$$+\frac{7}{9}\frac{d}{(\Delta x)^2}\frac{(2h \cdot v \cdot \delta t)^2}{2}u_{xx}|_i^{n+1} + \frac{7}{9}\frac{d}{(\Delta x)^2}\frac{(v \cdot \delta t)^2}{2}u_{xx}|_i^{n+1}$$

$$-\frac{7}{9}\frac{d}{(\Delta x)^2}(h - v \cdot \delta t)\delta t \cdot u_{xt}|_i^{n+1} + \frac{7}{9}\frac{d}{(\Delta x)^2}\frac{1}{2}(\delta t)^2 u_{tt}|_i^{n+1}$$

$$+\frac{7}{9}\frac{d}{(\Delta x)^2}\frac{h^3}{6}u_{xxx}|_i^{n+1} + \frac{7}{9}\frac{d}{(\Delta x)^2}\frac{3h \cdot v^2\delta t^2}{6}u_{xxx}|_i^{n+1}$$

$$-\frac{7}{9}\frac{d}{(\Delta x)^2}\frac{v^3\delta t^3}{6}u_{xxx}|_i^{n+1} + \frac{7}{9}\frac{d}{(\Delta x)^2}\frac{2h\cdot v\cdot \delta t}{2}\delta t\cdot u_{xxt}|_i^{n+1}$$

$$-\frac{7}{9}\frac{d}{(\Delta x)^2}\frac{v^2\delta t^2}{2}\delta t\cdot u_{xxt}|_i^{n+1}|_i^{n+1} + \frac{7}{9}\frac{d}{(\Delta x)^2}\frac{(h-v\cdot\delta t)(\delta t)^2}{2}u_{xtt}|_i^{n+1}$$

$$-\frac{7}{9}\frac{d}{(\Delta x)^2}\frac{1}{6}(\delta t)^3 u_{ttt}|_i^{n+1} + \frac{7}{9}\frac{d}{(\Delta x)^2}\mathcal{O}\left((\gamma_4)^4\right) \qquad (16)$$

Therefore, the truncation error is $\epsilon = \mathcal{O}(\delta t) + \mathcal{O}\left((\delta t)^2\right) + \mathcal{O}\left((\delta t)^3\right) + \mathcal{O}(\delta t \cdot \Delta x) + \mathcal{O}\left((\Delta x)^2\right) + \mathcal{O}\left(\frac{\delta t}{\Delta x}\right) + \mathcal{O}\left(\frac{(\delta t)^2}{(\Delta x)^2}\right) + \mathcal{O}\left(\frac{(\delta t)^3}{(\Delta x)^2}\right) + \mathcal{O}\left(\frac{(\delta t)^4}{(\Delta x)^2}\right)$. The analysis allowed us to conclude that the BDFHM method has the following truncation error

$$\epsilon = \mathcal{O}(\delta t) + \mathcal{O}\left((\delta t)^2\right) + \mathcal{O}\left((\delta t)^3\right) + \mathcal{O}(\delta t \cdot \Delta x) + \mathcal{O}\left((\Delta x)^2\right)$$

$$+ \mathcal{O}\left(\frac{\delta t}{\Delta x}\right) + \mathcal{O}\left(\frac{(\delta t)^2}{(\Delta x)^2}\right) + \mathcal{O}\left(\frac{(\delta t)^3}{(\Delta x)^2}\right) + \mathcal{O}\left(\frac{(\delta t)^4}{(\Delta x)^2}\right). \qquad (17)$$

The BDFHM method is consistent if $\lim_{\Delta x \to 0} \epsilon = 0$. More specifically, the local truncation error ϵ [i.e., each term in Eq. (16)] must tend to zero when $\Delta x \to 0$ for the discretization to be consistent with the partial differential equation. Using the technique employed by Oliveira et al. [1], which is based on the parametrization of the time-space discretization, a sufficient consistency condition is that δt tends to zero faster than Δx, i.e., $\lim_{\delta t, \Delta x \to 0} \frac{\delta t}{\Delta x} = 0$.

Similar to the analysis conducted by Oliveira et al. [1], consider $m = \epsilon^p$ and $r = \epsilon^q$, where p and q are positive integers, $0 < \epsilon < 1$, and $\epsilon \to 0$. Assume that m tends to zero faster than r. Thus, $\epsilon^p < \epsilon^q$, that is, $p > q$, which is required for the consistency analysis [1]. Consider, for example, the term $\frac{\delta t}{\Delta x}$. Note that

$$\lim_{r,m\to 0}\frac{\delta t}{\Delta x} = \lim_{r,m\to 0}\frac{m}{r} = \begin{cases} c & if\ p = q; \\ \infty & if\ p < q; \\ 0 & if\ p > q; \end{cases}$$

for a constant c, i.e.

- if $r \to 0$ similar to $m \to 0$, i.e $p = q$, then the limit is a constant non-null value;
- if $r \to 0$ faster than $m \to 0$, i.e., $p < q$, then the limit diverges;
- if $r \to 0$ slower than $m \to 0$, i.e., $p > q$, then the limit is zero.

Thus, consistency is not verified if $p \leq q$. Therefore, supposing that $p > q$, one can analyze under which conditions occurs $\epsilon \to 0$. Substituting $m = \epsilon^p$ and $r = \epsilon^q$ in each term of ϵ in Eq. (17) (following the strategy proposed by Oliveira et al. [1]) yields

1. for the term $\mathcal{O}\left(\frac{(\delta t)^2}{(\Delta x)^2}\right)$: $\frac{(\delta t)^2}{(\Delta x)^2} \approx \frac{\epsilon^{2p}}{\epsilon^{2q}} < 1 \Rightarrow \epsilon^{2p} < \epsilon^{2q}$; this inequality is true if $p > q \Leftrightarrow \epsilon^p < \epsilon^q \Rightarrow m < r$;

2. for the term $\mathcal{O}\left(\frac{(\delta t)^3}{(\Delta x)^2}\right)$: $\frac{(\delta t)^3}{(\Delta x)^2} \approx \frac{\epsilon^{3p}}{\epsilon^{2q}} < 1 \Rightarrow \epsilon^{3p} < \epsilon^{2q}$; this inequality is true if $p > \frac{2}{3}q \Leftrightarrow \epsilon^p < \epsilon^{\frac{2q}{3}} \Rightarrow m < r^{\frac{2}{3}}$;

3. for the term $\mathcal{O}\left(\frac{(\delta t)^3}{\Delta x}\right)$: $\frac{(\delta t)^3}{\Delta x} \approx \frac{\epsilon^{3p}}{\epsilon^q} < 1 \Rightarrow \epsilon^{3p} < \epsilon^q$; this inequality is true if $p > \frac{1}{3}q \Leftrightarrow \epsilon^p < \epsilon^{\frac{q}{3}} \Rightarrow m < r^{\frac{1}{3}}$;

4. for the term $\mathcal{O}\left(\frac{(\delta t)^4}{(\Delta x)^2}\right)$: $\frac{(\delta t)^4}{(\Delta x)^2} \approx \frac{\epsilon^{4p}}{\epsilon^{2q}} < 1 \Rightarrow \epsilon^{4p} < \epsilon^{2q}$; this inequality is true if $p > \frac{1}{2}q \Leftrightarrow \epsilon^p < \epsilon^{\frac{q}{2}} \Rightarrow m < r^{\frac{1}{2}}$.

This study safely allows us to conclude that $r < 1$, $m < r < r^{\frac{2}{3}} < r^{\frac{1}{2}} < r^{\frac{1}{3}} \Rightarrow m < r$. Therefore, the sufficient consistency condition for the BDFHM method is that $\delta t \to 0$ faster than $\Delta x \to 0$, i.e., $\lim\limits_{\delta t, \Delta x \to 0} \frac{\delta t}{\Delta x} = 0$.

3.2 Von Neumann Stability Analysis Applied to the BDFHM Method

Different techniques can be used to evaluate the stability of a method. For example, Oliveira *et al.* [1] demonstrated the stability of the Hopmoc method utilizing a spectral condition involving the values of Δt and Δx. Another example is a technique proposed by Zhang [7]. This technique is based on the decomposition of operators to prove the stability of a second order time multistep method, which was applied to the parallel solution of the wave equation. Moreover, this strategy uses the von Neumann stability analysis [11]. The von Neumann stability analysis determines the stability (or instability) of a finite-difference approximation. In a finite-difference approximation, a solution of the finite-difference equation is expanded in a Fourier series. Thereby, this present stability analysis is based on the techniques presented by Zhang [7]. For clarity, we show Eqs. (4) and (5) again below. Thus, taking into account the strategy proposed by Zhang [7], consider two time semi-steps of the BDFHM method,

$$\overline{u}_i^{n+\frac{1}{2}} = \frac{4}{3}\overline{\overline{u}}_i^n - \frac{1}{3}\overline{\overline{u}}_i^{n-\frac{1}{2}} + \frac{2}{3}\delta t \cdot d \left[\frac{\overline{u}_{i-1}^{n+\frac{1}{2}} - 2\overline{u}_i^{n+\frac{1}{2}} + \overline{u}_{i+1}^{n+\frac{1}{2}}}{(\Delta x)^2}\right], \tag{18}$$

$$u_i^{n+1} = \frac{4}{3}\overline{u}_i^{n+\frac{1}{2}} - \frac{1}{3}\overline{\overline{u}}_i^n + \frac{2}{3}\delta t d \left[\frac{\overline{u}_{i-1}^{n+\frac{1}{2}} - 2\overline{u}_i^{n+\frac{1}{2}} + \overline{u}_{i+1}^{n+\frac{1}{2}}}{(\Delta x)^2}\right], \tag{19}$$

and the difference operator $\Delta_h \overline{\overline{u}}_i^n = d \left[\frac{\overline{\overline{u}}_{i-1}^n - 2\overline{\overline{u}}_i^n + \overline{\overline{u}}_{i+1}^n}{(\Delta x)^2}\right]$, recalling that $h = \Delta x$. Both time semi-steps (18) and (19) can be rewritten in function of Δ_h as

$$\overline{u}_i^{n+\frac{1}{2}} = \frac{4}{3}\overline{\overline{u}}_i^n - \frac{1}{3}\overline{\overline{u}}_i^{n-\frac{1}{2}} + \frac{2}{3}\delta t \cdot \Delta_h \overline{u}_i^{n+\frac{1}{2}} \tag{20}$$

and

$$u_i^{n+1} = \frac{4}{3}\overline{u}_i^{n+\frac{1}{2}} - \frac{1}{3}\overline{\overline{u}}_i^n + \frac{2}{3}\delta t \Delta_h \overline{u}_i^{n+\frac{1}{2}}. \tag{21}$$

Let's ν be a auxiliary component, such that $\overline{\overline{\nu}}_i^n = \overline{\overline{u}}_i^{n-\frac{1}{2}}$ and $\overline{\nu}_i^{n+\frac{1}{2}} = \overline{\overline{u}}_i^n$. The system of equations

$$\begin{cases} \overline{u}_i^{n+\frac{1}{2}} = \left(\frac{4}{3} + \frac{2}{3}\delta t \cdot \Delta_h\right) \overline{\overline{u}}_i^n - \frac{1}{3}\overline{\overline{\nu}}_i^n \\ \overline{\nu}_i^{n+\frac{1}{2}} = \overline{\overline{u}}_i^n \end{cases} \tag{22}$$

presents an alternative compatible with Eq. (20), where in each equation is presented only values from two time steps. If $V^{n+\frac{1}{2}} = \left(\overline{u}_i^{n+\frac{1}{2}}, \overline{\nu}_i^{n+\frac{1}{2}}\right)^T$ and $\overline{V}^n = \left(\overline{\overline{u}}_i^n, \overline{\overline{\nu}}_i^n\right)^T$, then system (22) can be rewritten as $V^{n+\frac{1}{2}} = \begin{bmatrix} \frac{4}{3} + \frac{2}{3}\delta t \cdot \Delta_h & -\frac{1}{3} \\ 1 & 0 \end{bmatrix} \overline{V}^n$.

Similarly, considering $\overline{\nu}_i^{n+\frac{1}{2}} = \overline{\overline{u}}_i^n$ and $\nu_i^{n+1} = \overline{u}_i^{n+\frac{1}{2}}$, Eq. (21) can be rewritten as

$$\begin{cases} \left(1 - \frac{2}{3}\delta t \cdot \Delta_h\right) u_i^{n+1} = \frac{4}{3}\overline{u}_i^{n+\frac{1}{2}} - \frac{1}{3}\overline{\nu}_i^{n+\frac{1}{2}} \\ \nu_i^{n+1} = \overline{u}_i^{n+\frac{1}{2}} \end{cases}. \tag{23}$$

Considering $V^{n+1} = (u_i^{n+1}, \nu_i^{n+1})^T$, then system (23) can be rewritten as $\begin{bmatrix} 1 - \frac{2}{3}\delta t \cdot \Delta_h & 0 \\ 0 & 1 \end{bmatrix} V^{n+1} = \begin{bmatrix} \frac{4}{3} & -\frac{1}{3} \\ 1 & 0 \end{bmatrix} V^{n+\frac{1}{2}}$ or

$$V^{n+1} = \begin{bmatrix} \frac{3}{3-2\delta t \cdot \Delta_h} & 0 \\ 0 & 1 \end{bmatrix} \begin{bmatrix} \frac{4}{3} & -\frac{1}{3} \\ 1 & 0 \end{bmatrix} V^{n+\frac{1}{2}}. \tag{24}$$

Substituting system (22) in (24) yields

$$V^{n+1} = \begin{bmatrix} \frac{3}{3-2\delta t \cdot \Delta_h} & 0 \\ 0 & 1 \end{bmatrix} \begin{bmatrix} \frac{4}{3} & -\frac{1}{3} \\ 1 & 0 \end{bmatrix} \begin{bmatrix} \frac{4}{3} + \frac{2}{3}\delta t \cdot \Delta_h & -\frac{1}{3} \\ 1 & 0 \end{bmatrix} \overline{V}^n$$

and

$$V^{n+1} = \begin{bmatrix} \frac{13+8\delta t \cdot \Delta_h}{9-6\delta t \cdot \Delta_h} & \frac{-4}{9-6\delta t \cdot \Delta_h} \\ \frac{4+2\delta t \cdot \Delta_h}{3} & -\frac{1}{3} \end{bmatrix} \overline{\overline{V}}^n. \tag{25}$$

When considering $\left(\overline{u}_i^{n+\frac{1}{2}}, \overline{v}_i^{n+\frac{1}{2}}\right)^T = (w_1, w_2)^T e^{ik_x h}$, where k_x is the number of wavelengths per unit distance, expression (25) can be rewritten as

$$V^{n+1} = \begin{bmatrix} \frac{13-32\Lambda}{9+24\Lambda} & \frac{-4}{9+24\Lambda} \\ \frac{4-8\Lambda}{3} & -\frac{1}{3} \end{bmatrix} \overline{\overline{V}}^n. \tag{26}$$

Solving (26) with $\Lambda = \frac{\delta t}{(\Delta x)^2} d \sin^2\left(\frac{k_x h \sqrt{-1}}{2}\right)$, the eigenvalues of V^{n+1} are $\lambda_1 = \frac{-4\sqrt{-7(\Lambda)^2-26\Lambda+1}-20\Lambda+5}{3(8\Lambda+3)}$ and $\lambda_2 = \frac{4\sqrt{-7(\Lambda)^2-26\Lambda+1}-20\Lambda+5}{3(8\Lambda+3)}$. It is concluded that $\Lambda \neq 0$ because $\delta t, \Delta x > 0$ and $k_x > 0 \Rightarrow \sin^2\left(\frac{ik_x h}{2}\right) > 0$. Then, the analysis of the radicals of each eigenvalue allows us to conclude that: $i)$ if $\Lambda > 0$, then $\lambda_{1,2} \leq \frac{\pm 4\sqrt{+1}-20\Lambda+5}{3(8\Lambda+3)} \leq 1$; $ii)$ if $\Lambda < 0$, then $\lambda_{1,2} \leq \frac{\pm 4\sqrt{-1}\cdot\sqrt{+32}-15}{3(8+3)}$. Thus, $|\lambda_{1,2}| \leq \sqrt{\frac{737}{1089}} \leq 1$. Therefore, based on the von Neumann stability analysis [11], we conclude that the BDFHM method is stable.

4 Conclusions

The BDFHM method integrates the Hopmoc method and backward differentiation formulas. The consistency and stability analyses of the BDFHM method confirmed that the BDFHM method is convergent. Similar to the consistency analysis presented by Oliveira et al. [1], we used the parametrization of the error as a function of time and space discretization. Additionally, the stability analysis follows a technique employed by Zhang [7]. In particular, the use of backward differentiation formulas did not improve the accuracy of the standard Hopmoc method, i.e., the BDFHM method is first-order accurate in space and time.

We intend to study the BDFHM method in conjunction with total variation diminishing techniques [12] and flux-limiting procedures to improve the accuracy results. We also intend to analyze the 2-D BDFHM method applied to stiff problems.

References

1. Oliveira, S., Kischinhevsky, M., Gonzaga de Oliveira, S.L.: Convergence analysis of the Hopmoc method. Int. J. Comput. Math. **86**(8), 1375–1393 (2009)
2. Cabral, F., Osthoff, C., Costa, G., Brandão, D.N., Kischinhevsky, M., Gonzaga de Oliveira, S.L.: Tuning up TVD Hopmoc method on Intel MIC Xeon Phi architectures with Intel Parallel Studio Tools. In: Proceedings of the International Symposium on Computer Architecture and High Performance Computing Workshops (SBACPADW), Campinas, São Paulo, 17–20 October 2017, pp. 19–24 (2017)
3. Gordon, P.: Nonsymmetric difference equations. SIAM J. Appl. Math. **13**, 667–673 (1965)
4. Gourlay, P.: Hopscotch: a fast second order partial differential equation solver. IMA J. Appl. Math. **6**(4), 375–390 (1970)

5. Douglas, J.J., Russell, T.F.: Numerical methods for convection-dominated diffusion problems based on combining the method of characteristics with finite element or finite difference procedures. SIAM J. Numer. Anal. **19**(5), 871–885 (1955)

6. Robaina, D., Gonzaga de Oliveira, S., Kischinhevsky, M., Osthoff, C., Sena, A.: Numerical simulations of the 1-D modified Burgers equation. In: Proceedings of the 2019 Winter Simulation Conference (WSC), National Harbor, EUA, pp. 3231–3242 (2019)

7. Zhang, Z.: The multistep finite difference fractional steps method for a class of viscous wave equations. Math. Methods Appl. Sci. **34**, 442–454 (2011)

8. Ding, H.F., Zhang, Y.X.: A new difference scheme with high accuracy and absolute stability for solving convection diffusion equations. J. Comput. Appl. Math. **230**(2), 600–606 (2009)

9. Abbott, M.B.: An Introduction to the Method of Characteristics. Thames and Hudson Ltda, London (1996)

10. Cabral, F., Osthoff, C., Kischinhevsky, M., Brandão, D.: Hybrid MPI/OpenMP/OpenACC implementations for the solution of convection diffusion equations with Hopmoc method. In: Apduhan, B., Rocha, A.M., Misra, S., Taniar, D., Gervasi, O., Murgante, B., (eds.) 14th International Conference on Computational Science and its Applications (ICCSA), CPS, pp. 196–199. IEEE (2014)

11. Charney, J.G., Fjortoft, R., von Neumann, J.: Numerical integration of barotropic vorticity equation. Tellus **2**, 237–254 (1950)

12. Harten, A.: High resolution schemes for hyperbolic conservation laws. J. Comput. Phys. **49**(3), 357–393 (1983)

The Influence of Reordering Algorithms on the Convergence of a Preconditioned Restarted GMRES Method

S. L. Gonzaga de Oliveira[1(✉)], C. Carvalho[1], and C. Osthoff[2]

[1] Universidade Federal de Lavras, Lavras, MG, Brazil
sanderson@ufla.br
[2] Laboratório Nacional de Computação Científica, Petropólis, RJ, Brazil
osthoff@lncc.br

Abstract. This paper concentrates on applying reordering algorithms as a preprocessing step of a restarted Generalized Minimal Residual (GMRES for short) solver preconditioned by three ILU-type preconditioners. This paper investigates the effect of 13 orderings on the convergence of the preconditioned GMRES solver restarted every 50 steps when applied to nine real large-scale nonsymmetric and not positive definite matrices. Specifically, this paper shows the most promising combination of preconditioners and reordering for each linear system used.

Keywords: Bandwidth reduction · Profile reduction · Reordering algorithms · Graph labeling · GMRES method

1 Introduction

This paper focus on the solution of linear systems in the form $Ax = b$, where $A = [a_{ij}]$ is an $n \times n$ large-scale sparse matrix, x is the unknown n-vector solution, and b is a known n-vector. The solution of linear systems is an important step in various scientific and engineering simulations. The solution of linear systems comprised of large-scale matrices requires long computing times. Specifically, this paper show experiments using real nonsymmetric (structurally as well as numerically) matrices that are not positive definite. As examples, these matrices emerge from the circuit simulation, electromagnetics, and optimization problems. These problems can cause serious complications for preconditioned iterative methods [1,2].

Iterative methods are applied to solve large-scale problems in scientific computing. Since the sizes of the problems in this study are larger than 400,000 unknowns, we applied an iterative method to solve the linear systems. The GMRES method of Saad and Schultz [3] is a leading Krylov subspace method for solving non-Hermitian linear systems $Ax = b$, where $A \in \mathbb{C}^{n \times n}$ is a nonsingular matrix and $b \in \mathbb{C}^n$ (see [1] and references therein). The method uses the Arnoldi algorithm. Hence, its execution costs grow with each iteration. Thus,

© Springer Nature Switzerland AG 2020
O. Gervasi et al. (Eds.): ICCSA 2020, LNCS 12249, pp. 19–32, 2020.
https://doi.org/10.1007/978-3-030-58799-4_2

researchers typically restart the GMRES method every m iterations [3], using the last computed residual as the next initial one. This restarted approach may stagnate [3]. Even if stagnation does not occur, convergence can be slow (see [1]). Hence, preconditioning and data preprocessing are fundamental steps for the use of the restarted GMRES solver.

It is common sense that if the matrix has nonzero coefficients close to the main diagonal, the number of floating-point operations in the ILU factorization can be smaller. Consequently, reordering the unknowns and equations influence the rate of convergence of preconditioned Krylov subspace methods [4,5]. We studied the potential of the preconditioned restarted GMRES method with symmetric permutations of the matrix A for solving linear systems. We used the structure of $A + A^T$ in the experiments. Specifically, we investigated the effects of 13 symmetric permutations to further improve incomplete (approximate) factorization preconditioning for sequential calculations with the restarted GMRES(50) solver. Incomplete factorization preconditioners also depend on the ordering of unknowns and equations.

There are important reasons for finding an appropriate order for the row and columns of the coefficient matrix A. Modern hierarchical memory architecture and paging policies favor programs that consider the locality of reference into account. Heuristics for bandwidth and profile reductions are usually applied to yield an arrangement of graph (corresponding to the matrix A) vertices with the spatial locality. Therefore, these heuristics are typically employed to achieve low running times for solving large-scale sparse linear systems by iterative methods [6].

The present study employs the restarted GMRES(50) solver preconditioned by three preconditioners: ILUT [7], ILUC [8], and Algebraic Recursive Multilevel Solver (ARMS) [9]. Specifically, we employed in the experiments these preconditioners and the Flexible GMRES solver available in ITSOL v.2.0, a package of iterative solvers [10]. Particularly, this is a state-of-the-art implementation of the restarted GMRES method and its preconditioners.

This paper shows a set of numerical experiments on the influence of 13 low-cost reordering strategies on the convergence of the GMRES solver restarted every 50 steps when using three ILU-type preconditioners for a set of nine large-scale sparse matrices. The matrices are not positive definite, but real nonsymmetric matrices.

We selected low-cost heuristics for bandwidth and profile reductions because an adequate vertex labeling of a graph associated with a matrix contained in a linear system may reduce the computational times of an iterative linear solver, such as the GMRES method [4,5]. However, the bandwidth and profile reductions yielded by the reordering algorithms are not directly proportional to the computational time reduction of solving linear systems by an iterative solver. Furthermore, at least when the simulation solves only a single linear system, the total running time, including the time of the reordering procedure, is to be minimized. Thus, an application should employ a low-cost reordering algorithm [6,11]. The reordering strategies evaluated in this study are graph-theory algorithms that perform only symmetric permutations, preserve the set of diagonal entries, and neither affect the coefficients nor the eigenvalues of the nonsymmetric matrix A.

The remainder of this manuscript is structured as follows. In Sect. 2, important works in this field are reviewed. In Sect. 3, we describe how the experiments were conducted in this study. In Sect 4, we comment on the parameters evaluated in the preconditioned restarted GMRES solver. In Sect. 5, the results and analyses are presented. Finally, in Sect. 6, the conclusions are addressed.

2 Related Work

A previous publication reviewed the most important works in this context [1]. We provide only a brief review of the most important contributions in the field.

Benzi *et al.* [4] performed simulations with reordering nonsymmetric matrices for solving linear systems by the GMRES method. Camata *et al.* [5] investigated the use of ordering strategies in tandem with the ILU algorithm used as a preconditioner to the restarted GMRES solver. The authors compared the results yielded by the natural, quotient minimum degree, and RCM-GL [12] orderings. In general, Benzi *et al.* [4] and Camata *et al.* [5] recommended employing the RCM-GL [12] as a preprocessing step for the preconditioned restarted GMRES solver. In particular, the RCM-GL method yielded better results than other methods when applied to reduce the execution times of a linear system solver [13,14].

A recent publication [1] evaluated the effect of nine heuristics for bandwidth reduction on the convergence of the GMRES(50) solver preconditioned by the ILU, ILUT, VBILU, and VBILUT algorithms when applied to 11 linear systems. In the present study, we employed the same solver preconditioned by two other algorithms (ILUC and ARMS, as well as the ILUT preconditioner) applied to nine different linear systems. Additionally, the present study employs four different reordering strategies.

Previous publications (see [6] and references therein) recognized seven promising low-cost heuristics for bandwidth and profile reductions: KP-band [15], reverse breadth-first search (BFS) with starting vertex given by the George-Liu (GL) algorithm (see [6]), RCM-GL [12], MPG [16], Sloan [17], NSloan [18], and Sloan-MGPS [19] heuristics. Similar to the previous work [1], this paper also incorporate in the experiments the BFS, reverse BFS [4], BFS-GL, Cuthill-McKee (CM) [20], Reverse Cuthill-McKee (RCM) [21], and CM-GL procedures.

3 Description of the Tests

We wrote the codes of the heuristics for bandwidth and profile reductions in the C++ programming language using the g++ version 5.4.0 compiler, with the optimization flag -O3. To evaluate the performance of the preconditioned restarted GMRES solver [3] computed after executing reordering algorithms, we selected real nonsymmetric linear systems ranging from 400,000 to 1,000,000 unknowns contained in the SuiteSparse matrix collection [22].

In the experiments, the GMRES(50) solver stopped either when the norm of the residual vector was less than 1e-8 or when a maximum number of iterations (n) was reached. Since the preconditioned GMRES method [10] with 50 as the restart parameter is a fast iterative solver, we aborted an execution that computed more than 20 min supposing that either the solver would not converge or the solver would converge after a much longer time. When using the more difficult linear systems, however, a timeout of 3,600 s was established. The solver executed if the condest associated with the matrix was smaller than 1e+15.

The workstation used in the execution of the simulations featured an Intel® Core™ i7-4770 (16 GB of main memory DDR3 1.333 GHz) (Intel; Santa Clara, CA, United States). The machine used the Ubuntu 16.04.3 LTS 64-bit operating system with Linux kernel-version 4.13.0-39-generic.

4   Parameters Evaluated in the Preconditioned Restarted GMRES Solver

In this section, we show how we chose the initial parameters in the restarted GMRES solver preconditioned by the ILUT, ILUC, and ARMS preconditioners. A previous publication [1] showed that the best results were achieved when using the preconditioned restarted GMRES solver with 50 vectors in the Krylov basis, denoted GMRES(50) solver, among the three other parameters evaluated (30, 100, and 250). With this value for the GMRES solver, we evaluated the most suitable parameters on average for the three preconditioners when applied to the same 35 linear systems (with sizes ranging from 12,111 to 150,102).

Table 1 shows the parameters used for the three preconditioners evaluated in this study. We considered that the best combinations of parameters are those that led to the convergence of the GMRES solver at the lowest cost.

Table 1. Preliminary parameters studied in the preconditioned restarted GMRES solver [3, 10].

Method	Parameter	Values
GMRES	m	30, 50, 100, 250
ILUT,	τ	1e-1, 1e-3, 1e-6, 1e-9
ILUC	p	15, 50, 100, max. degree
ARMS	Number of levels	3, 5, 10
	Tol_{dd}	0.1, 0.4, 0.7
	Block size	30, 100, m

The choice of parameters may not be ideal for every matrix used due to the variety of characteristics present in the linear systems used in this study (i.e., the large number of unknowns, large number of non-null coefficients, etc.). Furthermore, the linear systems arise from different application areas. This generic choice of parameters means that a generic application of the preconditioned restarted GMRES solver will be unsuccessful for some instances contained in the dataset used. Nevertheless, these values made effective the set of algorithms for eight out of nine linear systems used in this study, recalling that the objective was to investigate the performance of reordering algorithms employed as a preprocessing step for the preconditioned restarted GMRES solver. Thus, we decided to set a "generic" preconditioned restarted GMRES solver to study the performance of the reordering algorithms.

For each parameter, we calculated the average times of the preconditioned restarted GMRES solver. Then, we calculated the metric $\rho = \sum_{i=1}^{N} \frac{t_H(i) - t_{min}(i)}{t_{min}(i)}$ [11] with these average times, where $t_H(i)$ is the execution time yielded when using an algorithm H applied to the instance i, $t_{min}(i)$ is the lowest computing time achieved in the instance i, and N is the number of instances.

Usually, searching for satisfactory values, researchers empirically chose the parameters τ and p for a few sample matrices from a specific application. The optimal values are heavily problem-dependent. Similar to the previous publication [1], exploratory investigations showed that the preconditioners yielded better results when using the parameter τ established as 1e-3 both when using the results yielded by applying only the GMRES(50) solver and the solver with the parameters used in the previous publication [1]. The experiments also showed that the preconditioners yielded better results when using the parameter p established as 50. Thus, we employed the ILUT and ILUC algorithms with the parameters (1e-3, 50), (1e-6, 75), (1e-9, 100), and (1e-15,250) when preconditioning the GMRES(50) solver. We used the parameters (1e-15,250) when the preconditioned iterative linear solver did not converge with the other parameters.

The ARMS preconditioner available in the ITSOL package [10] has three parameters: number of levels, a threshold value used for the relative diagonal dominance (tol_{dd}) criterion (ranging from 0 to 1), and block size [10]. The ρ values in Table 2 shows that the best results were achieved when defining the parameters in the ARMS preconditioner related to number of levels, tol_{dd}, and block size as ARMS(5,0.4,30) and ARMS($3, 0.1, \mathfrak{m}$), where $\mathfrak{m} = \left\lfloor \max_{v \in V}[degree(v)]/2 \right\rfloor$, in two cases: when $\mathfrak{m}$ is equal to i) 30, 50, 100, and 250, and ii) 50, respectively. Considering the size of the linear systems used in the experiments, we defined the number of levels as 5 to apply ARMS(5,tol_{dd},30) with tol_{dd} set as 0.1, 0.4, and 0.7, and ARMS($5, 0.7, \mathfrak{m}$) preconditioners.

Table 3 shows the definitive parameters that we used for the preconditioners. We used the parameters described in the following round if the preconditioned GMRES(50) solver did not converge with the parameters set in the previous experiment.

5   Results and Analysis

Tables 4 and 5 show the instance's name, preconditioner used, the average running time of the preconditioned GMRES(50) solver without the use of a reordering algorithm (see column Original), and the running times of this solver in

Table 2. Metric ρ calculated for several combinations of parameters in the ARMS algorithm when preconditioning the GMRES solver [3, 10].

GMRES(m)	No. levels	3	5	10
30,50,100,250	ρ	4	**2**	31
50	ρ	4	8	30
GMRES	Tol$_{dd}$	0.1	0.4	0.7
30,50,100,250	ρ	10.8	**10.6**	48.9
50	ρ	**4.2**	21.5	47.9
GMRES	Block size	30	100	m
30,50,100,250	ρ	**5.1**	5.2	8.4
50	ρ	26.0	4.2	**3.4**

Table 3. Parameters used in the experiments with the preconditioned GMRES(50). The program applied values of the following trial only to linear systems that the solver did not achieve convergence in the previous experiment.

Preconditioner	Parameter	Round			
		1st	2nd	3rd	4th
ILUT,	τ	1e-3	1e-6	1e-9	1e-15
ILUC	p	50	75	100	250
ARMS	Tol$_{dd}$	0.1	0.4	0.7	0.7
	Nr. of levels	5			
	Block size	30			m

tandem with several reordering algorithms. The symbol * in these tables indicates that the solver did not converge because of some inconsistency during preconditioning, such as a diagonal or pivot null or exceedingly small pivots, which is a possibility for matrices that do not have diagonal dominance and for highly unstructured problems. This kind of failure occurs when the preconditioner drops many large fill-ins from the incomplete factors. Ill-conditioning of the triangular factors also results in the instability of the recurrences involved in the forward and backward solves when executing the preconditioning [4]. Additionally, "cond" indicates that we aborted the execution because of the condest is greater than 1e+15. The symbol † in the same tables means that we stopped the run because it computed for 20 min. A superscript indicates the trial that the solver converged. It also means that the solver did not converge in the previous experiment(s). The symbol ‡ denotes that we performed the fourth trial with the solver applied to the linear system, and the simulation terminated for exceeding 3,600 s. Numbers in light gray color indicate the run that required the least amount of work when using the preconditioner. This case is not always the same as the run that demanded the lowest number of iterations. The reason is that, in general, different orderings result in a preconditioner with a different

Table 4. Execution times (s) of the preconditioned GMRES(50) solver applied to five linear systems containing nonsymmetric matrices.

Preconditioner		Original	RBFS-GL	RCM-GL	RBFS	RCM	BFS-GL	CM-GL	BFS	CM	KP	NSloan	Sloan	Sloan-MGPS	MPG
rajat21	β	411571	410770	411032	410023	406325	410770	411032	410023	406325	411571	286495	286495	286495	-
	Profile(·10⁶)	20769	1285	2351	3259	4275	1285	2351	3259	4275	3703	3869	3869	3869	
	ILUT	9.4[4]	14.6[4]	13.9[4]	25.2[4]	12.9[4]	12.0[4]	20.9[4]	11.0[4]	27.1[4]	*	cond	*	*	*
	ILUC t(s)/iter	cond	4.0[12]	†	3.9[24]	†	†	8.5[24]	†	7.4[24]	8.8[22]	†	cond	cond	†
	ARMS t(s)	136.9[3]	371.7[3]	1115.1[3]	386.0[3]	65.0[3]	62.7[3]	746.6	61.5[3]	389.7[3]	699.7	72.8[3]	76.1[3]	12.1[3]	107.9[3]
	iter.	13	14	7	7	12	12	71	12	7	49	5	11	5	12
largebasis	β	400019													
	Profile(·10⁶)	72006													
	ILUT t(s)/iter	*	17	17	19	19	*	17	19	19	21	39	36	39	41
	ILUC	*	10	10	10	10	10	10	10	10	11	13	12	13	12
	ARMS t(s)	5.4	13.3	17.3	9.3	14.2	12.4	17.0	9.4	14.0	18.0	20.2	19.2	19.3	219.9
	iter.	15	15	15	15	15	15	15	15	15	15	15	15	15	15
cage13	β	206333	43926	67522	66483	68262	43267	43699	66483	68262	45337	244025	244255	158310	261828
	Profile(·10⁶)	17473	1943	2273	2115	2277	1936	2146	2115	2277	2999	1680	655	2986	582
	ILUT t(s)	8.3	16.0	21.6	13.5	18.3	17.0	20.7	13.6	18.4	23.2	24.5	24.7	24.6	225.6
	(1e-3,50) iter.	4	4	4	4	4	4	4	4	4	4	4	4	4	4
	ILUC t(s)	5.4	13.3	17.3	9.3	14.2	12.4	17.0	9.4	14.0	18.0	20.2	19.2	19.3	219.9
	(1e-3,50) iter.	15	15	15	15	15	15	15	15	15	15	15	15	15	15
	ARMS	729.7	†	†	†	†	†	†	†	†	†	†	†	†	†
rajat29	β	643985	638910	638900	628692	621471	638910	638900	628692	621471	638901	-			
	Profile(·10⁶)	154707	7350	4251	4234	1061	7350	4251	4234	1061	816				
	ILUT	*	†	cond	†	cond	†	†	†	†	*				‡
	ILUC	†	*	*	*	*	*	*	*	*	*				‡
	ARMS t(s)	133.5[4]	56.3[4]	30.8[4]	54.3[4]	29.1[4]	246.8[4]	66.9[4]	157.0[4]	65.2[4]	cond	-	‡	‡	‡
	iter.	56	57	57	59	57	59	58	59	57	*		‡	‡	‡
rajat30	β	643985	641890	641882	618146	608973	641890	641882	618146	608973	641883				
	Profile(·10⁶)	154740	3029	2740	4076	2281	3029	2740	4076	2281	3437				
	ILUT	†	†	cond	†	cond	†	cond	†	†	cond				‡
	ILUC	†	*	*	*	*	*	*	*	*	*				‡
	ARMS t(s)	228.1	85.3	93.4	81.2	91.3	350.8	90.1	319.7	87.9	359.9[3]	†	‡	‡	‡
	iter	56	57	57	59	57	59	58	59	57	56	†	‡	‡	‡

Table 5. Execution times (s) of the preconditioned GMRES(50) solver applied to four linear systems containing nonsymmetric matrices.

	Preconditioner	Original	RBFS-GL	RCM-GL	RBFS	RCM	BFS-GL	CM-GL	BFS	CM	KP	NSloan	Sloan	Sloan-MGPS	MPG
pre2	β	117686	122060	106715	111213	106715	194193	136572	180310	202917	167760	122676	122676	122676	102002
	Profile($\cdot 10^6$)	7034	3606	1282	1602	1282	1843	1313	1602	2849	2253	3990	3990	3990	4228
	ILUT	*	*	*	*	*	*	*	*	*	*	*	*	*	*
	ILUC	cond	cond	cond	cond	cond	cond	cond	cond	cond	cond	cond	cond	cond	cond
	ARMS	+	+	+	+	+	+	+	+	+	+	+	+	+	+
ASIC_680ks	β	681277	124668	152927	108366	200311	124668	152927	108366	112761	149106	640088	640088	640088	629324
	Profile($\cdot 10^6$)	3728	3425	1895	1895	2415	3728	3425	1895	2415	1400	143	143	143	842
	ILUT t(s)	0.3	2.4	3.4	1.4	2.8	3.0	3.4	2.4	2.1	3.9	7.9	8.0	8.0	53.1
	(1e-3,50) iter.	2	4	2	4	12	13	2	20	4	7	23	18	18	12
	ILUC	+	+	+	+	+	+	+	+	+	+	+	+	+	+
	ARMS t(s)	14.7	3.6	11.7	2.5	3.3^{2}	3.7	4.3	2.8	3.3	4.8	11.5	12.0	12.0	54.8^{2}
	iter.	121	6	11	6	7	5	6	7	6	9	12	12	12	9
ASIC_680k	β	682779	682384	682389	682383	682385	682384	682389	682383	682385	682389				
	Profile($\cdot 10^6$)	147368	3425	4154	550	4139	552	4154	550	4139	808				
	ILUT t(s)/iter.	*	*	*	*	$6.9/4$	*	*	$4.2/25$	*	*	+	+	+	+
	ILUC	*	*	*	*	*	*	*	*	*	*				
	ARMS t(s)	141.6	182.2	187.1	180.8	185.5	132.7	158.3	131.2	157.0	187.1	+	+	+	+
	iter.	19	5	6	5	6	6	5	5	6	5				
tmt_unsym	β	2161	948	946	1081	1081	948	946	1081	1081	948	26926	1659	1654	1528
	Profile($\cdot 10^6$)	842	1278	1277	1535	1535	1278	1277	1535	1535	1278	1184	1277	1277	1192
	ILUT t(s)	90.8^{2}	80.0^{2}	87.2^{2}	77.5^{2}	85.2^{2}	82.7^{2}	84.9^{2}	80.9^{2}	82.4^{2}	90.3^{2}	90.3^{2}	84.4^{2}	84.3^{2}	82.8^{2}
	(1e-6,75) iter.	50	39	42	39	42	42	40	42	39	48	48	39	39	39
	ILUC, ARMS	+	+	+	+	+	+	+	+	+	+	+	+	+	+

number of non-null coefficients [4]. Results in dark gray color are the best results achieved in the linear system. This case is also not necessarily the same as the run that required the lowest number of iterations.

This GMRES(50) solver only converged when applied to the *rajat21*, *rajat29*, and *rajat30* instances using the ARMS preconditioner. The GMRES(50) solver preconditioned with the same preconditioner yielded the best results when applied to the original ordering of the instance *rajat21*. The GMRES(50) solver preconditioned with the ARMS(5,0.7,m) algorithm in tandem with the RCM ordering delivered the best results when applied to the instance *rajat29*. The GMRES(50) solver preconditioned with the ARMS(5,0.1,30) algorithm in conjunction with the BFS ordering yielded the best results when applied to the instance *rajat30*.

Table 4 shows that the RBFS ordering in conjunction with the ILUT(1e-3,50) preconditioner yielded the best results (3.9 s) when using the GMRES(50) solver applied to the instance *largebasis*. The simulations with the GMRES(50) solver preconditioned with the ARMS(5,0.7,30) algorithm converged with the 13 orderings evaluated in this study.

Table 4 also shows that the three preconditioners evaluated in tandem with the GMRES(50) solver yielded better results when applied to the original ordering of the instance *cage13* than employing the 13 reordering algorithms analyzed in the present study. The ILUC(1e-3,50) preconditioner delivered the best results when applied to this linear system.

Despite the number of preconditioners used with different sets of parameters (listed in Table 3) and the use of 13 orderings, the solver did not converge when applied to the instance *pre2*. One could examine better parameters for the preconditioned restarted GMRES solver to achieve convergence with this linear system. However, the principal objective was to study symmetrical permutations used as a preprocessing step of this preconditioned iterative solver.

The GMRES(50) solver preconditioned with the ILUT(1e-3,50) algorithm yielded the best results to the instance *ASIC_680ks* applied to the original ordering of the instance. The GMRES(50) solver preconditioned with the ILUT(1e-3,50) algorithm in tandem with the BFS ordering yielded the best results when applied to the instance *ASIC_680k*.

Table 5 shows the results of the algorithms applied to the instance *tmt_unsym*. The GMRES(50) solver only converged to this instance when preconditioned with the ILUT(1e-6,75) preconditioner. The simulation with the RBFS ordering yielded the shortest running time when applied alongside the ILUT(1e-6,75) preconditioner.

Table 6 provides characteristics (name, size, number of non-null coefficients ($|E|$), application area, the best preconditioner employed, the best ordering applied to the linear system, and the best time computed) of the linear systems composed of real nonsymmetric matrices used in this computational experiment. The RBFS ordering [4] benefited the convergence of the GMRES(50) solver when applied in conjunction with the ILUT [(1e-3,50) and (1e-6,75)] preconditioner in two cases (instances *largebasis* and *tmt_unsym*). The GMRES(50) solver only converged to the instance *tmt_unsym* when applied in tandem with the ILUT preconditioner. Additionally, the RBFS [RCM] ordering favored the convergence of the GMRES(50) solver in conjunction with the ARMS(5,0.1,30) [ARMS(5,0.7,m)] preconditioner when applied to the instance rajat30 [rajat29]. In particular, the use of a reordering algorithm favored the preconditioned GMRES(50) solver to converge much faster than using the original ordering when applied to the instances *largebasis*, *rajat29*, *rajat30*, and *ASIC_680k*. Table 6 highlights these instances in gray color. The BFS procedure had a positive effect when applied to the instance *ASIC_680k* regarding the convergence of the GMRES(50) solver with the ILUT(1e-3,50) preconditioner.

Table 6. Best orderings and preconditioners used in conjunction with the GMRES(50) solver.

Instance	Size	$|E|$	Preconditioner	Ordering	t(s)
rajat21	411676	1876011	ARMS(5,0.7,m)	Original	9.0
largebasis	440020	5240084	ILUT(1e-3,50)	RBFS	3.9
cage13	445315	7479343	ILUC(1e-3,50)	Original	5.3
rajat29	643994	3760246	ARMS(5,0.7,m)	RCM	29.1
rajat30		6175244	ARMS(5,0.1,30)	RBFS	81.2
pre2	659033	5834044	—	—	—
ASIC_680ks	682712	1693767	ILUT(1e-3,50)	Original	0.3
ASIC_680k	682862	2638997		BFS	4.2
tmt_unsym	917825	4584801	ILUT(1e-6,75)	RBFS	77.5

The GMRES(50) solver, when preconditioned by at least one of the three preconditioners employed in this study and applied to the original ordering of the linear system, solved all the eight test problems computed. There was no need to circumvent convergence through reordering in any of these eight problems. Table 6 shows that the GMRES(50) solver yielded the best results when employing a specific preconditioner with the original ordering itself in three out of these nine linear systems. Additionally, the experiments showed that the use of the preconditioned GMRES(50) solver applied to the original sequence of unknowns and equations of the linear system yielded similar times to the best results when a reordering algorithm was employed in simulations using the instance *tmt_sym*.

The experiments showed that the number of iterations is not the best metric to analyze the results. For example, when applied to the instance *largebasis*, the solver in tandem with the ILUT preconditioner took 4.0 and 3.9 s with the RBFS-GL and RBFS orderings in 12 and 24 iterations, respectively. The solver in tandem with the ARMS preconditioner, when applied to the instance *rajat30*, took respectively 81 and 360 s with the RBFS and KP orderings in 59 and 56 iterations. When applied to the instance *ASIC_680k*, the solver in conjunction with the ILUT preconditioner took 4 and 7 s with the BFS and RCM orderings in 25 and 4 iterations, respectively.

It is common sense that an important metric to be evaluated is the bandwidth of the matrices before and after the reordering. In this sense, reordering cannot be useful for preprocessing matrices that have small bandwidth. For example, the reordering methods did not reduce enough the bandwidth of the instance *pre2*. The solver did not converge when applied to this linear system. On the other hand, although the reordering algorithms did not reduce enough the bandwidth of the instances *rajat29*, *rajat30*, *ASIC_680k*, the reordering algorithms favored the computing times of the restarted GMRES solver. Nevertheless, in three cases (*cage13*, *rajat21*, *ASIC_680ks*), although the heuristics reduced the bandwidth of the matrices, the best results were achieved with the original ordering.

The GMRES(50) solver preconditioned by the ARMS preconditioner converged to seven out of nine linear systems used in this study. Including the linear system that the GMRES(50) solver did not converge, this solver preconditioned by the ARMS preconditioner did not converge when applied to the instance *tmt_unsym*, whose simulations terminated for exceeding the timeout stipulated of 20 min.

The GMRES(50) solver preconditioned by the ILUT preconditioner converged to five out of nine linear systems used in this study. The GMRES(50) solver preconditioned by the ILUC preconditioner converged to only one linear system.

6   Conclusions

This paper applied 13 low-cost symmetric permutations to reduce the computing time of the preconditioned GMRES(50) solver. Consistent with the current findings, the experiments conducted in this computational experiment show that the choice of the best combination of reordering algorithm and preconditioned GMRES solver is strongly problem-dependent. This paper indicated promising combinations of preconditioners and reordering heuristics (or the use of the original ordering of the instance) for fast convergence of a GMRES(50) method in nine large-scale linear systems.

A suitable parameter m for the restarted GMRES solver could be very different for each matrix analyzed. Similar to several publications in this field, however, and since the objective was to evaluate the influence of reordering algorithms on the convergence of the restarted GMRES solver, after exploratory investigations (see Sect. 4), we fixed the number of restart vectors for the GMRES solver for all matrices used.

In experiments using nine linear systems (with sizes ranging from 411,676 to 917,825 unknowns and from 1,876,011 to 4,584,801 non-null coefficients) comprised of real nonsymmetric matrices contained in the SuiteSparse matrix collection [22], the RBFS [4] (in three instances), BFS (in one instance), and Reverse Cuthill-McKee [21] (in one instance) orderings yielded the best results to reduce the processing times of the preconditioned GMRES(50) solver. In particular, the George-Liu algorithm [23] did not favor the reordering algorithms in the experiments performed in this study. The experiments included four problems (*largebasis*, *rajat29*, *rajat30*, and *ASIC 680k*) in which the GMRES(50) solver preconditioned by a specific preconditioner converged much faster after a permutation of the coefficient matrix (see Table 6) than using the original ordering of the instance.

Previous works in this field (e.g., [2,4,5]) found that the natural ordering rarely provides the best results, and was usually the worst. The experiments conducted in this study also evaluated the reordering algorithms when applied to the original sequence of unknowns and equations given in the linear systems. The results showed that reordering the matrices did not favor the computing time of the GMRES(50) solver (at least with the parameters used) when applied to three out of nine large-scale test matrices used. Additionally, the computing time of the preconditioned GMRES(50) solver applied to the original ordering yielded similar times to the best results achieved when reordering the rows and columns of another instance (*tmt_unsym*). Some characteristics can explain these different results from previous publications in this subject [2,4,5]. First, the experiments performed in the present study employed different preconditioners. We used the ILUC and ARMS [9] preconditioners, including the ILUT algorithm. Second, the size of the dropped entries strongly influences the effectiveness of incomplete factorizations. The choice of parameters of the GMRES method in tandem with each preconditioner was also conducted quite differently in this computational experiment from those previous publications. Subtle differences in the parameter values of incomplete factorization preconditioners are determinants for the convergence of the GMRES solver. Further analysis of the parameters of the preconditioners could show better results. However, the experiments conducted in the present study showed that the choice of the best parameters is not trivial. Third, the instances included in the experiments are much larger than those used by those researchers. Fourth, the proper set of instances can explain these results, since the best combination of reordering algorithm and preconditioned GMRES solver is highly problem-dependent, as previously mentioned. Fifth, similar to those previous publications, we limited the experiments to symmetric permutations, and we used nonsymmetric linear systems.

The results of the experiments showed that the heuristics for profile reduction used are not recommended (i.e., Sloan's algorithm [17] and its variants, NSloan [18], Sloan-MGPS [19], and MPG [16] heuristics). Moreover, heuristics that yield further profile reductions than Sloan's algorithm [17] and its variants at higher execution costs may not favor the computing time of the preconditioned restarted GMRES method when applied to large-scale linear systems.

One can employ the BFS or RCM procedure or a variant (e.g., a reverse BFS [4]) for preprocessing matrices for the GMRES(50) solver. The RBFS ordering yielded, in general, better performance than the classical RCM ordering [21]. Furthermore, the experiments conducted in the present study showed that the use of the George-Liu algorithm [23] did not favor the convergence of the pre-conditioned GMRES(50) solver.

We plan to investigate the effects of parallel orderings on the convergence of parallel implementations of the restarted preconditioned GMRES solver using OpenMP, Galois, and Message Passing Interface systems. The literature on heuristics for bandwidth reduction presents many works with sequential algorithms and just a small fraction of parallel strategies. A systematic review of parallel heuristics for bandwidth reduction is another future step in our study. Concerning massively parallel computing, we plan to evaluate heuristics for bandwidth reduction implemented within the Intel® Math Kernel Library running on Intel® Scalable processors.

References

1. Gonzaga de Oliveira, S.L., Carvalho, C., Osthoff, C.: The effect of symmetric permutations on the convergence of are started GMRES solver with ILU-type preconditioners. In: 2019 Winter Simulation Conference (WSC), National Harbor, MA,USA, pp. 3219–3230. IEEE (2019)
2. Benzi, M., Haws, J.C., Tuma, M.: Preconditioning highly indefinite and nonsymmetric matrices. SIAM J. Sci. Comput. **22**(4), 1333–1353 (2000)
3. Saad, Y., Schultz, M.H.: GMRES: a generalized minimal residual algorithm for solving nonsymmetric linear systems. SIAM J. Sci. Stat. Comput. **7**, 856–869 (1986)
4. Benzi, M., Szyld, D.B., Van Duin, A.: Orderings for incomplete factorization preconditioning of nonsymmetric problems. SIAM J. Sci. Comput. **20**(5), 1652–1670 (1999)
5. Camata, J.J., Rossa, A.L., Valli, A.M.P., Catabriga, L., Carey, G.F., Coutinho, A.L.G.A.: Reordering and incomplete preconditioning in serial and parallel adaptive mesh refinement and coarsening flow solutions. Int. J. Numer. Meth. Fl. **69**(4), 802–823 (2012)
6. Gonzaga de Oliveira, S.L., Bernardes, J.A.B., Chagas, G.O.: An evaluation of reordering algorithms to reduce the computational cost of the incomplete Cholesky-conjugate gradient method. Comput. Appl. Math. **37**(3), 2965–3004 (2017). https://doi.org/10.1007/s40314-017-0490-5
7. Saad, Y.: ILUT: a dual threshold incomplete LU factorization. Numer. Linear Algebra Appl. **1**(4), 387–402 (1994)
8. Li, N., Saad, Y., Chow, E.: Crout versions of ILU for general sparse matrices. SIAM J. Sci. Comput. **25**(2), 716–728 (2003)
9. Saad, Y., Suchomel, B.: ARMS: an algebraic recursive multilevel solver for general sparse linear systems. Numer. Linear Algebra Appl. **9**(5), 359–378 (2002)
10. Saad, Y.: ITSOL vol 2.0: Iterative solvers package (2016)
11. Gonzaga de Oliveira, S.L., Bernardes, J.A.B., Chagas, G.O.: An evaluation of low-cost heuristics for matrix bandwidth and profile reductions. Comput. Appl. Math. **37**(2), 1412–1471 (2016). https://doi.org/10.1007/s40314-016-0394-9
12. George, A., Liu, J.W.: Computer Solution of Large Sparse Positive Definite Systems. Prentice-Hall, Englewood Cliffs (1981)
13. Gonzaga de Oliveira, S.L., Abreu, A.A.A.M., Robaina, D.T., Kischnhevsky, M.: An evaluation of four reordering algorithms to reduce the computational cost of the Jacobi-preconditioned conjugate gradient method using high-precision arithmetic. Int. J. Bus. Intell. Data Min. **12**(2), 190–209 (2017)

14. Gonzaga de Oliveira, S.L., Abreu, A.A.A.M.: An evaluation of pseudoperipheral vertex finders for the reverse Cuthill-Mckee method for bandwidth and profile reductions of symmetric matrices. In: Proceedings of the 37th International Conference of the Chilean Computer Science Society (SCCC), Santiago, Chile, pp. 1–9. IEEE (2018). https://doi.org/10.1109/SCCC.2018.8705263

15. Koohestani, B., Poli, R.: A hyper-heuristic approach to evolving algorithms for bandwidth reduction based on genetic programming. In: Bramer, M., Petridis, M., Nolle, L. (eds.) Research and Development in Intelligent Systems XXVIII, SGAI 2011, pp. 93–106. Springer, London (2011). https://doi.org/10.1007/978-1-4471-2318-7_7

16. Medeiros, S.R.P., Pimenta, P.M., Goldenberg, P.: Algorithm for profile and wavefront reduction of sparse matrices with a symmetric structure. Eng. Comput. **10**(3), 257–266 (1993)

17. Sloan, S.W.: A Fortran program for profile and wavefront reduction. Int. J. Numer. Meth. Eng. **28**(11), 2651–2679 (1989)

18. Kumfert, G., Pothen, A.: Two improved algorithms for envelope and wavefront reduction. BIT Numer. Math. **37**(3), 559–590 (1997)

19. Reid, J.K., Scott, J.A.: Ordering symmetric sparse matrices for small profile and wavefront. Int. J. Numer. Meth. Eng. **45**(12), 1737–1755 (1999)

20. Cuthill, E., McKee, J.: Reducing the bandwidth of sparse symmetric matrices. In: Proceedings the 1969 24th International Conference, pp. 157–172. ACM (1969)

21. George, A.: Computer implementation of the finite element method. PhD thesis, Stanford University, Stanford (1971)

22. Davis, T.A., Hu, Y.: The University of Florida sparse matrix collection. ACM Trans. Math. Softw. **38**(1), 1–25 (2011)

23. George, A., Liu, J.W.H.: An implementation of a pseudoperipheral node finder. ACM Trans. Math. Softw. **5**(3), 284–295 (1979)

Generalized Invariants of Multiple Scattered Polarized Radiation at Arbitrary Non-symmetrical Optical Levels of a Uniform Slab

Oleg I. Smokty[✉]

St. Petersburg Institute for Informatics and Automation, Russian Academy of Sciences, 39, 14th Line, V.O., St. Petersburg 199178, Russia
soi@iias.spb.su

Abstract. Polarimetric invariants of a uniform slab of a finite optical thickness, which were initially introduced at optical levels symmetrical to its middle in mirror directions, for downgoing and upgoing polarized radiation have been generalized for the case of virtual non-symmetrical optical levels τ_1 and τ_2. In the case of the vertical uniformity of an extended uniform slab, there deterministically exist invariant constructions different from and adequate to standard polarimetric invariants at symmetrical optical levels). These constructions express the properties of the local spatial angular symmetry of intensities of multiple scattered radiation internal fields in a more general analytical form of representation. For such new constructions of polarized radiation, which are called generalized polarimetric invariants, a basic boundary problem of the polarized radiation transfer theory has been formulated and linear Fredholm integral equations of the second kind and linear singular integral equations have been obtained. These generalized polarimetric invariants have proved to be effective for numerical modeling of intensities of multiple scattered polarized radiation in a uniform slab of an arbitrary optical thickness $\tau_0 \leq \infty$, with and without a reflecting bottom.

Keywords: Generalized polarimetric invariants · Non-symmetrical optical levels · Fredholm linear integral equations · Linear singular equations · Basic boundary value problems · Semi-group transformations · Singular eigenfunction

1 Introduction

The concept of polarized radiation field invariants was initially developed for downgoing and upgoing vision directions at optical levels that are mirror symmetrical relative to the middle of a uniform slab of finite optical thickness. Below it will be shown that symmetry restrictions can be eliminated, and generalized photometrical invariants can be introduced at virtual optical levels within any local layers in a given uniform slab. For that, we make required linear semigroup transformations of spatial shift and rotation relative to the virtual boundaries of local layers inside the initial uniform slab.

© Springer Nature Switzerland AG 2020
O. Gervasi et al. (Eds.): ICCSA 2020, LNCS 12249, pp. 33–46, 2020.
https://doi.org/10.1007/978-3-030-58799-4_3

2   The Basic Boundary Value Problem and Linear Fredholm Integral Equations of the Second Kind for Generalized Polarimetric Invariants

If we select an arbitrary local layer $[\tau_1 < \tau_2]$ in some initial uniform slab of a finite optical thickness $\tau_0 < \infty$, then we can eliminate the restriction specified in [1] regarding the symmetry of base optical levels τ and $\tau_0 - \tau$. In this case, at its virtual boundaries τ_1 and τ_2 in the mirror directions of vision η and $-\eta$ at fixed values of the solar zenith angle ξ we can introduce new objects of the radiative transfer theory, namely generalized polarimetric invariants $\hat{\tilde{I}}_{\pm}^m(\tau_1, \tau_2, \eta, \xi, \tau_0)$ of Fourier harmonics for matrix intensities of polarized radiation which are determined by the following relations [2]:

$$\hat{\tilde{I}}_{\pm}^m(\tau_1, \tau_2, \eta, \xi, \tau_0) = \hat{D}\hat{\tilde{I}}^m(\tau_2, -\eta, \xi, \tau_0) \pm \hat{\tilde{I}}^m(\tau_1, \eta, \xi, \tau_0),$$
$$\eta \in [-1, 1] \cap \xi \in [0, 1], \quad 0 \le \tau_1 < \tau_2 \le \tau_0 < \infty, \quad m = \overline{0, M}. \tag{1.1}$$

Azimuthal harmonics of matrix intensities $\hat{\tilde{I}}^m(\tau_2, -\eta, \xi, \tau_0)$ and $\hat{\tilde{I}}^m(\tau_1, \eta, \xi, \tau_0)$ in the downgoing ($\eta > 0$) and upgoing ($\eta < 0$) vision line directions are determined by special Fourier transformations [3], while $\hat{D}(1, 1, -1, -1)$ is a numerical diagonal matrix of mirror transformations of polarized radiation fields.

The matrix values $\hat{\tilde{I}}_{\pm}^m(\tau_1, \tau_2, \eta, \xi, \tau_0)$ introduced according to (1.1) possess the basic property of spatial angular invariance for mutual inversions of the virtual boundaries τ_1 and τ_2 of the selected local layers $[\tau_1, \tau_2]$ and for simultaneous substitution of vision line directions $\eta \leftrightarrow -\eta$:

$$\hat{\tilde{I}}_{\pm}^m(\tau_1, \tau_2, \eta, \xi, \tau_0) = \pm \hat{D}\hat{\tilde{I}}^m(\tau_2, \tau_1 - \eta, \xi, \tau_0),$$
$$\eta \in [-1, 1] \cap \xi \in [0, 1], \quad 0 \le \tau_1 < \tau_2 \le \tau_0 \le \infty, \quad m = \overline{0, M}. \tag{1.2}$$

From relations (1.1)–(1.2) it follows that the new polarimetric invariants $\hat{\tilde{I}}_{\pm}^m(\tau_1, \tau_2, \eta, \xi, \tau_0)$ are a generalization of the earlier introduced analogous values $\hat{\tilde{I}}_{\pm}^m(\tau, \eta, \xi, \tau_0)$ at the mirror symmetrical optical levels τ_1 and $\tau_2 - \tau$ of a uniform slab in "mirror" vision line directions η and $-\eta$ [1]. In that case, the azimuthal harmonics of generalized polarimetric invariants $\hat{\tilde{I}}_{\pm}^m(\tau_1, \tau_2, \eta, \xi, \tau_0)$, unlike standard matrix values $\hat{\tilde{I}}_{\pm}^m(\tau, \eta, \xi, \tau_0)$, exist outside the restrictions associated with the principle of mirror reflection (symmetry) and, as it will be demonstrated below, are determined by linear semi-group transformations of the sought for matrix intensities $\hat{\tilde{I}}^m(\tau_1, \eta, \xi, \tau_0)$ and $\hat{\tilde{I}}^m(\tau_2, -\eta, \xi, \tau_0)$ in relation to vertical spatial shifts of the virtual optical depths τ_1 and τ_2 with the simultaneous rotation of the vision lines $\eta \Leftrightarrow -\eta$.

Note also that exact functional relations and integral equations (including the basic boundary value problem), which are required for finding azimuthal harmonics of generalized matrix polarimetric invariants $\hat{I}_{\pm}^m(\tau_1, \tau_2, \eta, \xi, \tau_0)$, are a result of applying two different approaches. The first one is connected with linear transformations of regular solutions for the initial boundary value problem in the polarize radiation transfer theory [2] and subsequent application of respective representations of generalized polarimetric invariants $\hat{I}_{\pm}^m(\tau_1, \tau_2, \eta, \xi, \tau_0)$ within arbitrary selected local layers $\tau \in [\tau_1, \tau_2]$. The second approach is based on the results obtained earlier [4] while solving the said boundary value problem according to the Case method [5] and subsequently used for considering problems of multiple polarized radiation scattering in a uniform semi-infinite slab ($\tau_0 = \infty$) [6]. The following strict linear Fredholm integral equations of the second kind and linear singular integral equations for the generalized polarimetric invariants $\hat{I}_{\pm}^m(\tau_1, \tau_2, \eta, \xi, \tau_0)$ can be utilized in the case of uniform media of either a finite ($\tau_0 < \infty$) or infinite ($\tau_0 = \infty$) optical thickness.

To determine the azimuthal harmonics of polarized radiation matrix intensities $\hat{I}^m(\tau_1, \eta, \xi, \tau_0)$ and $\hat{I}^m(\tau_2, -\eta, \xi, \tau_0)$ at non-symmetrical optical levels τ_1 and τ_2, we initially apply the first approach. Following this approach, let's consider regular solutions $\hat{I}^m(\tau, \eta, \xi, \tau_0), \eta \in [-1, 1] \cap \xi \in [0, 1] \cap \tau \in [0, \tau_0], \ m = \overline{0, M}$ [1, 2], from the perspective of their linear semi-group transformations associated with the shift of

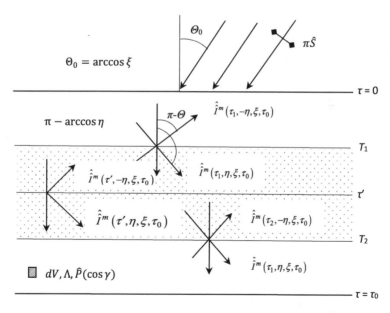

Fig. 1. Azimuthal harmonics of polarized radiation matrix intensities $\hat{I}^m(\tau_1, \eta, \xi, \tau_0)$ and $\hat{I}^m(\tau_2, -\eta, \xi, \tau_0)$ of an arbitrary local layer $[\tau_2, \tau_1]$ embedded in the initial uniform slab $[0, \tau_0]$, for downgoing ($\eta > 0$) and upgoing ($\eta < 0$) vision line directions, with the shift parameter $\alpha = 0$.

optical depths $\tau \in [\tau_1, \tau_2]$ within the arbitrary selected local layers with virtual boundaries τ_1 and τ_2 and the simultaneous rotation of the vision lines $\eta \Leftrightarrow -\eta$ (Fig. 1).

Now let's represent these solutions for downgoing ($\eta > 0$) and upgoing ($\eta < 0$) vision line directions as follows:

$$\hat{I}^m(\tau', \eta, \xi, \tau_0) = \hat{I}^m(\tau_1, \eta, \xi, \tau_0)e^{-\frac{\tau' - \tau_1}{\eta}} + \int_{\tau_1}^{\tau'} \hat{B}^m(\tau', \eta, \xi, \tau_0)e^{-\frac{\tau' - \tau''}{\eta}}\frac{d\tau''}{\eta},$$

$$\eta \in [0, 1] \cap \xi \in [0, 1], \quad [\tau_1, \tau_2] \in [0, \tau_0], \quad \tau_1 < \tau_2, \quad m = \overline{0, M}, \tag{1.3}$$

$$\hat{I}^m(\tau', -\eta, \xi, \tau_0) = \hat{I}^m(\tau_2, -\eta, \xi, \tau_0)e^{-\frac{\tau_2 - \tau'}{\eta}} + \int_{\tau'}^{\tau_2} \hat{B}^m(\tau', -\eta, \xi, \tau_0)e^{-\frac{\tau'' - \tau'}{\eta}}\frac{d\tau''}{\eta},$$

$$\eta \in [-1, 1] \cap \xi \in [0, 1], \quad [\tau_1, \tau_2] \in [0, \tau_0], \quad \tau_1 < \tau_2, \quad m = \overline{0, M}, \tag{1.4}$$

where the values $\hat{B}^m(\tau', \pm\eta, \xi, \tau_0), (\eta, \xi) > 0$ are Fourier harmonics of the initial matrix source functions [2].

From (1.3) and (1.4), at $\tau = \tau_2$ and $\tau = \tau_1$, we obtain the following relations:

$$\hat{I}^m(\tau_2, \eta, \xi, \tau_0) = \hat{I}^m(\tau_1, \eta, \xi, \tau_0)e^{-\frac{\tau_2 - \tau_1}{\eta}} + \int_{\tau_1}^{\tau_2} \hat{B}^m(\tau', \eta, \xi, \tau_0)e^{-\frac{\tau_2 - \tau'}{\eta}}\frac{d\tau'}{\eta}, \tag{1.5}$$

$$\hat{I}^m(\tau_1, -\eta, \xi, \tau_0) = \hat{I}^m(\tau_2, -\eta, \xi, \tau_0)e^{-\frac{\tau_2 - \tau_1}{\eta}} + \int_{\tau_1}^{\tau_2} \hat{B}^m(\tau', -\eta, \xi, \tau_0)e^{-\frac{\tau' - \tau_1}{\eta}}\frac{d\tau'}{\eta}. \tag{1.6}$$

Next we make replacements of the current spatial variables $\tau_2 - \tau' = \tau''$ in (1.5) and $\tau' - \tau_1 = \tau''$ in (1.6), which allow to pass in the above relations to integrating over the virtual parameter $\alpha = \tau'' \geq 0$ that determines the vertical shift of the current optical thickness τ' downward or upward in respect of the boundaries of the arbitrary selected local layer $[\tau_1, \tau_2]$:

$$\hat{I}^m(\tau_2, \eta, \xi, \tau_0) = \hat{I}^m(\tau_1, \eta, \xi, \tau_0)e^{-\frac{\tau_2 - \tau_1}{\eta}} + \int_0^{\tau_2 - \tau_1} \hat{B}^m(\tau_2 - \alpha, \eta, \xi, \tau_0)e^{-\frac{\alpha}{\eta}}\frac{d\alpha}{\eta}, \tag{1.7}$$

$$\hat{I}^m(\tau_1, -\eta, \xi, \tau_0) = \hat{I}^m(\tau_2, -\eta, \xi, \tau_0)e^{-\frac{\tau_2 - \tau_1}{\eta}} + \int_0^{\tau_2 - \tau_1} \hat{B}^m(\tau_1 + \alpha, -\eta, \xi, \tau_0)e^{-\frac{\alpha}{\eta}}\frac{d\alpha}{\eta}. \tag{1.8}$$

The azimuthal harmonics of matrix source functions $\hat{B}^m(\tau_1 + \alpha, \eta, \xi, \tau_0)$ and $\hat{B}^m(\tau_2 - \alpha, -\eta, \xi, \tau_0)$ at arbitrary shifts $\alpha \geq 0$ of the virtual boundaries τ_1 and τ_2 of the current local layers $[\tau_1, \tau_2]$ are determined as follows:

$$\hat{B}^m(\tau_2 - \alpha, \eta, \xi, \tau_0) = \frac{\Lambda}{2}\int_{-1}^{1}\hat{p}^m(\eta, \eta')\hat{I}^m(\tau_2 - \alpha, \eta', \xi, \tau_0)d\eta' + \frac{\Lambda}{4}\hat{p}^m(\eta, \xi)e^{-\frac{\tau_2 - \alpha}{\xi}},$$

$$\eta \in [0, 1] \cap \xi \in [0, 1], \quad 0 \leq \tau_1 < \tau_2 \leq \tau_0, \quad \alpha \geq 0, \quad m = \overline{0, M}, \tag{1.9}$$

$$\hat{B}^m(\tau_1+\alpha,-\eta,\xi,\tau_0) = \frac{\Lambda}{2}\int_{-1}^{1}\hat{D}\hat{p}^m(\eta,\eta')\hat{D}\hat{I}^m(\tau_1+\alpha,-\eta',\xi,\tau_0)d\eta' + \frac{\Lambda}{4}\hat{p}^m(-\eta,\xi)e^{-\frac{\tau_1+\alpha}{\xi}},$$

$$\eta \in [0,1]\cap\xi\in[0,1], \quad 0\le\tau_1<\tau_2\le\tau_0, \quad \alpha>0, \quad m=\overline{0,M}.$$

$$(1.10)$$

The values $\hat{p}^m(\eta,\eta')$ correspond to the azimuthal harmonics of the initial phase matrix functions of scattering $\hat{P}(\cos\gamma)$ and possess the following properties: $\hat{p}^m(-\eta,-\eta') = \hat{D}\hat{p}^m(\eta,\eta')\hat{D}$ and $\hat{p}^m(-\eta,\eta') = \hat{D}\hat{p}^m(\eta,-\eta')\hat{D}$.

Using then (1.5)–(1.6) and (1.9)–(1.10) for the values $\hat{I}^m(\tau_2,\alpha,\eta,\xi,\tau_0)$ and $\hat{I}^m(\tau_1,\alpha,-\eta,\xi,\tau_0)$ we get the following exact linear Fredholm integral equations of the second kind:

$$\hat{I}^m(\tau_2,0,\eta,\xi,\tau_0) = \hat{I}^m(\tau_1,0,\eta,\xi,\tau_0)e^{-\frac{\tau_2-\tau_1}{\eta}}$$

$$+ \frac{\Lambda}{2}\int_0^{\tau_2-\tau_1}\left\{\int_0^1\left[\begin{array}{c}\hat{p}^m(\eta,\eta')\hat{I}^m(\tau_2-\alpha,\eta',\xi,\tau_0)\\ +\hat{p}^m(\eta,-\eta')\hat{I}^m(\tau_2-\alpha,-\eta',\xi,\tau_0)\end{array}\right]d\eta' + \frac{1}{2}\hat{p}^m(\eta,\xi)e^{-\frac{\tau_2-\alpha}{\xi}}\right\}e^{-\frac{\alpha}{\eta}}\frac{d\alpha}{\eta},$$

$$\eta \in [0,1]\cap\xi\in[0,1], \quad 0\le\tau_1<\tau_2\le\tau_0, \quad \alpha\ge0, \quad m=\overline{0,M},$$

$$(1.11)$$

$$\hat{I}^m(\tau_1,0,-\eta,\xi,\tau_0) = \hat{I}^m(\tau_2,0,-\eta,\xi,\tau_0)e^{-\frac{\tau_2-\tau_1}{\eta}}$$

$$+ \frac{\Lambda}{2}\int_0^{\tau_2-\tau_1}\left\{\int_0^1\left[\begin{array}{c}\hat{D}\hat{p}^m(\eta,\eta')\hat{D}\hat{I}^m(\tau_1+\alpha,-\eta',\xi,\tau_0)\\ +\hat{D}\hat{p}^m(\eta,-\eta')\hat{D}\hat{I}^m(\tau_1+\alpha,\eta',\xi,\tau_0)\end{array}\right]d\eta' + \frac{1}{2}\hat{p}^m(-\eta,\xi)e^{-\frac{\tau_1+\alpha}{\xi}}\right\}e^{-\frac{\alpha}{\eta}}\frac{d\alpha}{\eta},$$

$$\eta \in [0,1]\cap\xi\in[0,1], \quad 0\le\tau_1<\tau_2\le\tau_0, \quad \alpha\ge0, \quad m=\overline{0,M}.$$

$$(1.12)$$

Now, introducing instead of (1.1) a more general form of representation of polarimetric invariants $\hat{I}_\pm^m(\tau_1,\tau_2,\alpha,\eta,\xi,\tau_0)$, taking into account the vertical spatial shift $\alpha\ge0$ at virtual optical levels τ_1 and τ_2 in mirror vision line directions η and $-\eta$, we have:

$$\hat{I}_\pm^m(\tau_1,\tau_2,\alpha,\eta,\xi,\tau_0) = \hat{D}\hat{I}^m(\tau_2-\alpha,-\eta,\xi,\tau_0)\pm\hat{I}^m(\tau_1+\alpha,\eta,\xi,\tau_0),\qquad(1.13)$$

$$\hat{I}_\pm^m(\tau_1,\tau_2,\alpha,\eta,\xi,\tau_0) = \pm\hat{D}\hat{I}_\pm^m(\tau_2,\tau_1,-\alpha,-\eta,\xi,\tau_0),$$

$$\eta \in [-1,1]\cap\xi\in[0,1], \quad [\tau_1,\tau_2]\in[0,\tau_0], \quad \alpha\ge0, \quad m=\overline{0,M}.$$

$$(1.14)$$

In the particular case of virtual levels $\tau_1=0$, $\tau_2=\tau_0$ and shift parameter $\alpha=\tau$ from (1.13)–(1.14) we can define polarimetric invariants $\hat{I}_\pm^m(\tau,\eta,\xi,\tau_0)$ and respective invariance relations $\hat{I}_\pm^m(\tau,\eta,\xi,\tau_0) = \pm\hat{D}\hat{I}_\pm^m(\tau_0-\tau,-\eta,\xi,\tau_0)$ obtained in [1] at the symmetrical optical levels τ and $\tau_0-\tau$ for the entire extended layer $\tau\in[0,\tau_0]$, including its external boundaries $\tau=0$ и $\tau=\tau_0$.

Using then (1.13) and (1.14), from (1.11) and (1.12) through some simple transformations we can find exact linear Fredholm integral equations of the second kind for
generalized polarimetric invariants $\hat{I}_{\pm}^{m}(\tau_1, \tau_2, \alpha, \eta, \xi, \tau_0)$ within the arbitrary selected
local layers $[\tau_1, \tau_2]$:

$$
\hat{I}_{\pm}^{m}(\tau_1, \tau_2, 0, -\eta, \xi, \tau_0) = \hat{I}_{\pm}^{m}(\tau_2, \tau_1, 0, -\eta, \xi, \tau_0) e^{-\frac{\tau_2 - \tau_1}{\eta}}
$$

$$
+ \frac{\Lambda}{2} \int_0^{\tau_2 - \tau_1} \left\{ \begin{array}{l} \int_0^1 \left[\begin{array}{l} \hat{p}^{m}(\eta, \eta') \hat{I}_{\pm}^{m}(\tau_1, \tau_2, \alpha, -\eta', \xi, \tau_0) \\ + \hat{p}^{m}(\eta, -\eta') \hat{I}_{\pm}^{m}(\tau_1, \tau_2, \alpha, \eta', \xi, \tau_0) \end{array} \right] d\eta' \\ + \frac{1}{2} \left[\hat{D} \hat{p}^{m}(\eta, \xi) e^{-\frac{\tau_2 - \alpha}{\xi}} \pm \hat{p}^{m}(-\eta, \xi) e^{-\frac{\tau_1 + \alpha}{\xi}} \right] \end{array} \right\} e^{-\frac{\alpha}{\eta}} \frac{d\alpha}{\eta},
$$

$$
\eta \in [-1, 1] \cap \xi \in [0, 1], \quad 0 \le \tau_1 < \tau_2 \le \tau_0, \quad \alpha \in [0, \tau_0], \quad m = \overline{0, M}.
$$

$$(1.15)$$

Note that in the particular case of $\tau_1 = 0$, $\tau_2 = \tau_0$, and $\alpha = \tau_0$ with $\eta < 0$, we have
$\hat{I}_{+}^{m}(\tau_1, \tau_2, \alpha, \eta, \xi, \tau_0) \equiv 0$, and at $\tau_1 = 0$, $\tau_2 = \tau_0$, $\alpha = 0$ and $\eta > 0$ invariance relations
are satisfied for matrix values $\hat{R}_{+}^{m}(\eta, \xi, \tau_0) = \hat{\rho}^{m}(\eta, \xi, \tau_0) + \hat{D}\hat{\sigma}^{m}(\eta, \xi, \tau_0)$ of polarized
radiation external fields, where $\hat{\rho}^{m}(\eta, \xi, \tau_0)$ and $\hat{D}\hat{\sigma}^{m}(\eta, \xi, \tau_0)$ are the values of matrix
brightness coefficient azimuthal harmonics for the initial uniform slab $[0, \tau_0]$.

Therefore, the existence of generalized polarimetric invariants $\hat{I}_{\pm}^{m}(\tau_1, \tau_2, \alpha, \eta, \xi, \tau_0)$
in a uniform slab of an arbitrary optical thickness $\tau_0 \le \infty$ is determined by the general
fundamental property of spatial angular symmetry and invariance of multiple scattered
polarized radiation fields at vertical spatial shifts $\alpha \ge 0$ of the virtual boundaries τ_1 and
τ_2 of local layers $[\tau_1, \tau_2]$ within the entire given extended layer $[0, \tau_0]$ with simultaneous
"mirror" reflections of vision line directions η and $-\eta$.

The virtual parameter $0 \le \alpha \le \tau_0$ of the spatial shift of current optical depths τ
describes the vertical motion of the external boundaries $\tau = 0$ and $\tau = \tau_0$ of the given
uniform slab into the medium $\tau \in [\tau_1, \tau_2]$ to the boundaries of arbitrary selected local
layers $[\tau_1, \tau_2]$ (Fig. 2). Of course, in the above-mentioned particular cases of $\tau_1 = 0$,
$\tau_2 = \tau_0$ and $\alpha = \tau_0$ or $\alpha = 0$, the properties of local spatial angular symmetry for the
entire extended uniform slab $[0, \tau_0]$ are manifested at the mirror symmetrical $(\tau_0/2)$
optical levels τ and $\tau_0-\tau$ in the "mirror" vision line directions η and $-\eta$ in accordance
with the earlier formulated principle of "mirror" reflection (symmetry) for multiple
scattered polarized radiation fields [1, 2].

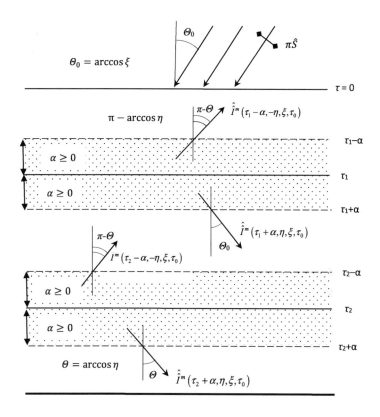

Fig. 2. Azimuthal harmonics of upgoing ($\eta < 0$) and downgoing ($\eta > 0$) polarized radiation in an arbitrary local layer $[\tau_1, \tau_2]$ with a vertical shift up or down of its upper (τ_1) and lower (τ_2) virtual boundaries by the value of parameter $\alpha \geq 0$.

Taking into account the above analysis, let's now formulate the basic boundary value problem for finding generalized polarimetric invariants $\hat{I}^m_{\pm}(\tau_1, \tau_2, \alpha, \eta, \xi, \tau_0)$. For that, in the initial matrix equation for polarized radiation transfer, we make changes of variables $\tau' = \tau_1 + \alpha$, $\eta' \rightarrow -\eta'$, $\eta \rightarrow -\eta$ and $\tau' = \tau_2 - \alpha$, $\eta' \rightarrow -\eta'$, $\eta \rightarrow -\eta$, which is equipotential to introducing new reference levels of current optical depths τ' relative to the virtual boundaries τ_1 and τ_2 which are respectively measured upward to the upper boundary τ_1 and τ_2 of the given layer $\tau = 0$ for upgoing vision line directions ($\eta < 0$) and downward to its lower boundary $\tau = \tau_0$ for downgoing vision line directions ($\eta > 0$):

$$\eta \frac{\partial \hat{I}^m(\tau_1 + \alpha, \eta, \xi, \tau_0)}{\partial \alpha} = -\hat{I}^m(\tau_1 + \alpha, \eta, \xi, \tau_0)$$
$$+ \frac{\Lambda}{2} \int_{-1}^{1} \hat{p}^m(\eta, \eta')\hat{I}^m(\tau_1 + \alpha, \eta', \xi, \tau_0)d\eta' + \frac{\Lambda}{4} \hat{p}^m(\eta, \xi)e^{-\frac{\tau_1 + \alpha}{\xi}},$$
$$\eta \in [-1, 1] \cap \xi \in [0, 1], \quad 0 \le \tau_1 < \tau_2 \le \tau_0, \quad \alpha \in [0, \tau_0], \quad m = \overline{0, M},$$

(1.16)

$$\eta \frac{\partial \hat{I}^m(\tau_2 - \alpha, -\eta, \xi, \tau_0)}{\partial \alpha} = -\hat{I}^m(\tau_2 - \alpha, -\eta, \xi, \tau_0)$$
$$+ \frac{\Lambda}{2} \int_{-1}^{1} \hat{D}\hat{p}^m(\eta, \eta')\hat{D}\hat{I}^m(\tau_2 - \alpha, -\eta', \xi, \tau_0)d\eta' + \frac{\Lambda}{4} \hat{p}^m(-\eta, \xi)e^{-\frac{\tau_2 - \alpha}{\xi}},$$
$$\eta \in [-1, 1] \cap \xi \in [0, 1], \quad 0 \le \tau_1 < \tau_2 \le \tau_0, \quad \alpha \in [0, \tau_0], \quad m = \overline{0, M}.$$

(1.17)

Using then the definition (1.1) of generalized polarimetric invariants $\hat{I}^m_{\pm}(\tau_1, \tau_2, \alpha, \eta, \xi, \tau_0)$ and their basic invariance property (1.2), after obvious linear transformations in the integral parts of Eqs. (1.16) and (1.17), based on the changes of angular variables ($\eta \to -\eta'$, $\eta' \to -\eta'$) and the use of angular symmetry properties of Fourier harmonics of phase scattering matrices $\hat{p}^m(\eta, \eta') = \hat{D}\hat{p}^m(-\eta, -\eta')\hat{D}$, we obtain the sought for matrix equation for polarized radiation transfer in an arbitrary local layer $[\tau_1, \tau_2]$ nested into the initial layer $[0, \tau_0]$:

$$\eta \frac{\partial \hat{I}^m_{\pm}(\tau_1, \tau_2, \alpha, \eta, \xi, \tau_0)}{\partial \alpha} = -\hat{I}^m_{\pm}(\tau_1, \tau_2, \alpha, \eta, \xi, \tau_0)$$
$$+ \frac{\Lambda}{2} \int_{-1}^{1} \hat{p}^m(\eta, \eta')\hat{I}^m_{\pm}(\tau_1, \tau_2, \alpha, \eta, \xi, \tau_0)d\eta'$$
$$+ \frac{\Lambda}{4}\left[\hat{D}\hat{p}^m(-\eta, \xi)e^{-\frac{\tau_2 - \alpha}{\xi}} \pm \hat{p}^m(\eta, \xi)e^{-\frac{\tau_1 + \alpha}{\xi}}\right],$$
$$\eta \in [-1, 1] \cap \xi \in [0, 1], \quad 0 \le \tau_1 < \tau_2 \le \tau_0, \quad \alpha \in [0, \tau_0], \quad m = \overline{0, M}.$$

(1.18)

Matrix integral differential Eqs. (1.18) for each current Fourier harmonic "m" generalize the respective polarized radiation transfer equations (obtained in [2] for the entire initial extended layer $[0, \tau_0]$) to the case of virtual local layers $[\tau_1, \tau_2]$. Formally, the locally symmetrized transfer Eqs. (1.18) show how the multiple scattered polarized radiation field and its invariance and spatial angular symmetry properties are transformed upon the shift from a given extended uniform slab ($0 \le \tau \le \tau_0$) to arbitrary selected local layers $0 \le \tau_1 < \tau_2 \le \tau_0 \le \infty$ inside it.

3   Linear Singular Integral Equations for Generalized Polarimetric Invariants

Now let's consider the second approach for finding strict integral relations and linear integral equations that determine azimuthal harmonics of generalized matrix polarimetric invariants $\hat{I}^m_\pm(\tau_1, \tau_2, \alpha, \eta, \xi, \tau_0)$. As noted above, to solve the initial matrix boundary value problem [2], we can apply the classical Case method [5] based on the series representation of sought for values $\hat{\hat{I}}^m(\tau, \eta, \xi, \tau_0)$ by their singular eigenfunctions of continuous and discrete spectra of the initial polarized radiation transfer equation [4].

By applying this method according to consideration in [4] and [6], we obtained linear singular integral equations for vector matrix intensities of polarized radiation at virtual optical levels τ_1 and τ_2 of a uniform semi-infinite slab $\tau_0 = \infty$. Note that in this case, while preserving the general mathematical structure of equations obtained in [4] and [6], vision line angles $\theta = \arccos\eta$ were used as independent variables in [4] (unlike [6]), and the optical depth τ and solar zenith angle $\theta_0 = \arccos\xi$ played the role of numerical modeling parameters. Furthermore, for the full solution of equations [4] it is necessary first of all to solve the auxiliary boundary value problem for polarized radiation diffusely reflected and transmitted by a given uniform slab. Note also that, unfortunately, in [4] quite obvious linear transformations of the received linear singular integral equations were not made, the omission of which did not allow to introduce the notion of generalized polarimetric invariants $\hat{\hat{I}}^m_\pm(\tau_1, \tau_2, \alpha, \eta, \xi, \tau_0)$ into the polarized radiation transfer theory and as a result to investigate, on their basis, local spatial angular properties of the invariance of multiple scattered radiation fields in a uniform slab of an arbitrary optical thickness $\tau_0 \le \infty$. Later on, this omission was corrected in [6] making use of linear semi-group transformations of Fourier harmonics of the sought for matrix intensities $\hat{\hat{I}}^m(\tau_1 - \alpha, \tau_2 + \alpha, \pm\eta, \xi, \tau_0), (\eta, \xi) > 0$, associated with the spatial shift $\alpha > 0$ of virtual optical depths τ_1 and τ_2 in vertical directions with the simultaneous rotation of vision lines $\eta \Leftrightarrow -\eta$.

By generalizing the analytical results obtained in [4] and [6] to the case of non-symmetrical virtual optical levels τ_1 and τ_2 in a uniform slab of an arbitrary optical thickness $\tau_0 \le \infty$, we can find the following strict relations of spatial and angular equivalence for multiple scattered downgoing $(\eta > 0)$ and upgoing $(\eta < 0)$ polarized radiation with the inversion of optical levels $\tau_1 \leftrightarrow \tau_2$ and simultaneous translation of vision line mirror directions $\eta \leftrightarrow -\eta$:

$$e^{\frac{\tau_1}{\eta}}\left[\hat{T}^m(\eta)\hat{\hat{I}}^m(\tau_1, \eta, \xi, \tau_0) + \frac{1}{2}\int_{-1}^{1}\left[\hat{\Psi}^m(\eta')\right]^T\hat{\hat{I}}^m(\tau_1, \eta', \xi, \tau_0)\frac{\eta'd\eta'}{\eta - \eta'}\right]$$

$$= e^{\frac{\tau_2}{\eta}}\left[\hat{T}^m(\eta)\hat{\hat{I}}^m(\tau_2, \eta, \xi, \tau_0) + \frac{1}{2}\int_{-1}^{1}\left[\hat{\Psi}^m(\eta')\right]^T\hat{\hat{I}}^m(\tau_2, \eta', \xi, \tau_0)\frac{\eta'd\eta'}{\eta - \eta'}\right] \qquad (2.1)$$

$$+ e^{\frac{\tau_1}{\eta}}\hat{f}^m(\tau_1, \eta, \xi, \tau_0) - e^{\frac{\tau_2}{\eta}}\hat{f}^m(\tau_2, \eta, \xi),$$

$$(\eta, \xi) > 0, \ (\tau_1, \tau_2) \in [0, \tau_0], \ m = \overline{0, M},$$

$$e^{\frac{\tau_1}{\eta}}\left[\hat{T}^m(\eta)\hat{I}^m(\tau_2, -\eta, \xi, \tau_0) + \frac{1}{2}\int_{-1}^{1}\left[\hat{\Psi}^m(\eta')\right]^T\hat{I}^m(\tau_2, -\eta', \xi, \tau_0)\frac{\eta' d\eta'}{\eta - \eta'}\right]$$

$$= e^{\frac{\tau_2}{\eta}}\left[\hat{T}^m(\eta)\hat{I}^m(\tau_1, -\eta, \xi, \tau_0) + \frac{1}{2}\int_{-1}^{1}\left[\hat{\Psi}^m(\eta')\right]^T\hat{I}^m(\tau_1, -\eta', \xi, \tau_0)\frac{\eta' d\eta'}{\eta - \eta'}\right] + \quad (2.2)$$

$$+ e^{\frac{\tau_2}{\eta}}\hat{f}^m(\tau_1, -\eta, \xi, \tau_0) - e^{\frac{\tau_1}{\eta}}\hat{f}^m(\tau_2, -\eta, \xi, \tau_0),$$

$$(\eta, \xi) > 0, \ (\tau_1, \tau_2) \in [0, \tau_0], \ m = \overline{0, M}.$$

The matrix transposition operation in (2.1)–(2.2) is denoted by the symbol "T", the characteristic matrix functions $\hat{\Psi}^m(\eta)$ and $\hat{T}^m(\eta)$ are determined from the following relations:

$$\hat{T}^m(\eta) = \hat{E}(1, 1, 1, 1) + \frac{\Lambda}{2}\eta\int_{-1}^{1}\hat{A}(\eta', \eta')\frac{d\eta'}{\eta' - \eta}, \ \hat{\Psi}^m(\eta) = \frac{\Lambda}{2}A^m(\eta, \eta), \quad (2.3)$$

where the parameter Λ is a single scattering albedo, and $\hat{E}(1, 1, 1, 1)$ is a unity matrix. The auxiliary matrix functions $A^m(\eta, \xi)$ are found from the following integral equation:

$$\hat{A}^m(\eta, \xi) = \hat{p}^m(\eta, \xi) + \frac{\Lambda}{2}\eta\int_{-1}^{1}[\hat{p}^m(\eta, \xi) - \hat{p}^m(\eta', \xi)]\frac{\hat{A}^m(\eta, \eta')}{\eta - \eta'}d\eta, \quad (2.4)$$

$$(\eta, \xi) > 0, \ m = \overline{0, M}.$$

Matrix functions $\hat{f}^m(\tau, \eta, \xi, \tau_0)$ are determined by single scattering polarized radiation and can be calculated by the following exact formulas:

$$\hat{f}^m(\tau, \eta, \xi, \tau_0) = \begin{cases} \frac{\Lambda}{4}\hat{A}^m(\eta, \xi)\xi\frac{e^{\frac{\tau}{\xi}} - e^{\frac{\tau}{\eta}}}{\xi - \eta}, & \eta > 0 \\ \frac{\Lambda}{4}\hat{A}^m(\eta, \xi)\xi\frac{e^{\frac{\tau}{\xi}} - e^{-\left(\frac{1}{\xi}\frac{1}{\eta}\right)(\tau_0 - \tau)}e^{\frac{\tau}{\xi}}}{\xi - \eta}, & \eta < 0 \end{cases} \quad (2.5)$$

$$\tau = \tau_1 \geq 0, \ \tau = \tau_2 \leq \tau_0, \ \eta \in [0, \pm 1] \cap \xi \in [0, 1], \ \overline{0, M}.$$

Note that for obtaining the relation (2.2) we have used the angular symmetry properties of matrix functions $\hat{T}^m(\eta)$ and $\left[\hat{\Psi}^m(\eta)\right]^T$:

$$\hat{T}^m(\eta) = \hat{T}^m(-\eta), \ \left[\hat{\Psi}^m(-\eta)\right]^T = \left[\hat{\Psi}^m(\eta)\right]^T. \quad (2.6)$$

The above-obtained integral relations (2.1)–(2.2) of the polarized radiation intensities equivalence at the virtual levels τ_1 and τ_2 in the mirror directions η and $-\eta$ can serve as a formal basis for independent – with regard to (1.1) – introduction of generalized polarimetric invariants $\hat{I}_{\pm}^m(\tau_1, \tau_2, \eta, \xi, \tau_0)$ in a uniform slab of an arbitrary optical thickness $\tau_0 \leq \infty$.

It should, however, be emphasized that in the original work [4] the basic invariance relation (2.1) was considered only in one particular case of $\tau_1 = 0$ and $\tau_2 = \tau$ without considering a possibility of its further use for constructing generalized polarimetric invariants $\hat{I}_{\pm}^m(\tau_1, \tau_2, \eta, \xi, \tau_0)$.

This omission makes it impossible to regard the definition of polarimetric invariants (1.13) and invariance relation (1.14) from a general perspective of semi-group linear transformations of the sought for matrix values $\hat{I}_{\pm}^m(\tau_1, \tau_2, \alpha, \eta, \xi, \tau_0)$ using the actual spatial and angular variables (τ_1, τ_2, η) at the virtual boundaries $\tau_1 < \tau_2$ of arbitrary selected local layers $[\tau_1, \tau_2]$ inside an extended medium $[0, \tau_0]$ with introducing the parameters of their vertical spatial shift $\alpha \geq 0$. By correcting this omission and introducing in (2.1) and (2.2) the parameter of vertical spatial shift $\alpha \geq 0$, we obtain new invariance relations for $\hat{I}^m(\tau_1 + \alpha, \eta, \xi, \tau_0)$ and $\hat{I}^m(\tau_2 - \alpha, -\eta, \xi, \tau_0)$ at virtual boundaries $(\tau_1 + \alpha)$ and $(\tau_2-\alpha)$ with the simultaneous rotation of the vision line $\eta \to \eta$:

$$e^{\frac{\tau_1+\alpha}{\eta}}\left[\hat{T}^m(\eta)\hat{I}^m(\tau_1+\alpha,\eta,\xi,\tau_0) + \frac{1}{2}\int_{-1}^{1}\left[\hat{\Psi}^m(\eta')\right]^T\hat{I}^m(\tau_1+\alpha,\eta',\xi,\tau_0)\frac{\eta'd\eta'}{\eta-\eta'}\right]$$

$$= e^{\frac{\tau_2-\alpha}{\eta}}\left[\hat{T}^m(\eta)\hat{I}^m(\tau_2-\alpha,\eta,\xi,\tau_0) + \frac{1}{2}\int_{-1}^{1}\left[\hat{\Psi}^m(\eta')\right]^T\hat{I}^m(\tau_2-\alpha,\eta',\xi,\tau_0)\frac{\eta'd\eta'}{\eta-\eta'}\right]$$

$$+ e^{\frac{\tau_1+\alpha}{\eta}}\hat{f}^m(\tau_1+\alpha,\eta,\xi,\tau_0) - e^{\frac{\tau_2-\alpha}{\eta}}\hat{f}^m(\tau_2-\alpha,\eta,\xi,\tau_0),$$

$$\eta \in [-1,1] \cap \xi \in [0,1], \quad 0 \leq \tau_1 < \tau_2 \leq \tau_0 \leq \infty, \quad \alpha \geq 0, \quad m = \overline{0, M}.$$

$$(2.7)$$

Similarly, instead of (2.2) we have the following relation:

$$e^{\frac{\tau_1+\alpha}{\eta}}\left[\hat{T}^m(\eta)\hat{I}^m(\tau_2-\alpha,-\eta,\xi,\tau_0) + \frac{1}{2}\int_{-1}^{1}\left[\hat{\Psi}^m(\eta')\right]^T\hat{I}^m(\tau_2-\alpha,-\eta',\xi,\tau_0)\frac{\eta'd\eta'}{\eta-\eta'}\right]$$

$$= e^{\frac{\tau_2-\alpha}{\eta}}\left[\hat{T}^m(\eta)\hat{I}^m(\tau_1+\alpha,-\eta,\xi,\tau_0) + \frac{1}{2}\int_{-1}^{1}\left[\hat{\Psi}^m(\eta')\right]^T\hat{I}^m(\tau_1+\alpha,-\eta',\xi,\tau_0)\frac{\eta'd\eta'}{\eta-\eta'}\right]$$

$$+ e^{\frac{\tau_2-\alpha}{\eta}}\hat{f}^m(\tau_1+\alpha,-\eta,\xi,\tau_0) - e^{\frac{\tau_1+\alpha}{\eta}}\hat{f}^m(\tau_2-\alpha,-\eta,\xi,\tau_0),$$

$$\eta \in [-1,1] \cap \xi \in [0,1], \quad 0 \leq \tau_1 < \tau_2 \leq \tau_0 \leq \infty, \quad \alpha \geq 0, \quad m = \overline{0, M}.$$

$$(2.8)$$

Thus, Eqs. (2.7) and (2.8) form a system of two matrix linear integral singular equations with regard to the sought for values $\hat{I}^m(\tau_1+\alpha, \eta, \xi, \tau_0)$ and $\hat{I}^m(\tau_2 - \alpha, -\eta, \xi, \tau_0)$ at two virtual non-symmetrical optical levels $(\tau_1 + \alpha)$ and $(\tau_2 - \alpha)$ in the mirror vision line directions η и −η.

Let's now introduce into consideration the generalized polarimetric invariants $\hat{I}_{\pm}^m(\tau_1, \tau_2, \alpha, \eta, \xi, \tau_0)$ according the relations (1.13). By making respective transformations in (2.7) and (2.8), instead of the above-mentioned system of equations we obtain separate linear singular integral equations for finding Fourier harmonics of the generalized polarimetric invariants $\hat{I}_{+}^m(\tau_1, \tau_2, \alpha, \eta, \xi, \tau_0)$ и $\hat{I}_{-}^m(\tau_1, \tau_2, \alpha, \eta, \xi, \tau_0)$:

$$
\begin{aligned}
\hat{T}^m(\eta) &\left[e^{\frac{\tau_1+\alpha}{\eta}} \hat{I}_{\pm}^m(\tau_1, \tau_2, \alpha, \eta, \xi, \tau_0) \mp e^{\frac{\tau_2-\alpha}{\eta}} \hat{D}\hat{I}_{\pm}^m(\tau_1, \tau_2, \alpha, -\eta, \xi, \tau_0) \right] \\
&= \frac{1}{2} e^{\frac{\tau_1+\alpha}{\eta}} \int_{-1}^{1} \left[\hat{\Psi}^m(\eta') \right]^T \hat{I}_{\pm}^m(\tau_1, \tau_2, \alpha, \eta, \xi, \tau_0) \frac{\eta' d\eta'}{\eta' - \eta} \\
&\quad \mp \frac{1}{2} e^{\frac{\tau_2-\alpha}{\eta}} \int_{-1}^{1} \left[\hat{\Psi}^m(\eta') \right]^T \hat{D}\hat{I}_{\pm}^m(\tau_1, \tau_2, \alpha, -\eta, \xi, \tau_0) \frac{\eta' d\eta'}{\eta' - \eta} \\
&\quad \pm e^{\frac{\tau_1+\alpha}{\eta}} \hat{f}_{\pm}^m(\tau_1, \tau_2, \alpha, \eta, \xi, \tau_0) \mp e^{\frac{\tau_2-\alpha}{\eta}} \hat{D}\hat{f}_{\pm}^m(\tau_1, \tau_2, \alpha, -\eta, \xi, \tau_0), \\
&\eta \in [-1, 1] \cap \xi \in [0, 1], \quad 0 \le \tau_1 < \tau_2 \le \tau_0 \le \infty, \quad \alpha \ge 0, \quad m = \overline{0, M},
\end{aligned}
\tag{2.9}
$$

where the values $\hat{f}_{\pm}^m(\tau_1, \tau_2, \alpha, \eta, \xi, \tau_0)$ are defined as follows:

$$
\hat{f}_{\pm}^m(\tau_1, \tau_2, \alpha, \eta, \xi, \tau_0) = \hat{D}\hat{f}^m(\tau_2 - \alpha, -\eta, \xi, \tau_0) \mp \hat{f}^m(\tau_1 + \alpha, \eta, \xi, \tau_0).
\tag{2.10}
$$

Note that the invariance property (1.14) allows to consider regular solutions $\hat{I}_{+}^m(\tau_1, \tau_2, \alpha, \eta, \xi, \tau_0)$ and $\hat{I}_{-}^m(\tau_1, \tau_2, \alpha, \eta, \xi, \tau_0)$ in more narrow domains of actual variables (τ_1, τ_2, η) as compared to the system of Eqs. (2.7)–(2.8), namely: $\tau \in (\tau_2 - \tau_1)$, $\eta \in [0, 1]$, $\alpha \ge 0$ or alternatively $\tau \in \frac{1}{2}(\tau_2 - \tau_1)$, $\eta \in [-1, 1]$, $\alpha \ge 0$. Furthermore, in case of numerical solutions of certain matrix linear singular integral Eqs. (2.9), e.g. by the method of spatial angular discretization or by using the Gauss-Seidel relaxation iterative method, the rank of respective linear algebraic systems is 2 times lower than the rank of similar systems of algebraic Eqs. (2.7)–(2.8). However, as in the scalar case [1], matrix linear singular integral Eqs. (2.9) can have non-unique solutions, and to eliminate them it is necessary to apply precise additional integral relations while using characteristic roots $\xi = 1/\hat{k}_m$ of matrix equations $\hat{T}^m(\xi) = 0$, $\xi \in [0, 1]$, $m = \overline{0, M}$.

Now let's consider some formal mathematical peculiarities of Eq. (2.9). Naturally, any regular solution of the initial boundary value problem in the polarized radiation transfer theory, with respective boundary conditions, must satisfy the said matrix linear singular integral equation at arbitrary optical depths τ_1 and τ_2 in the mirror vision line directions η and $-\eta$. Yet this equation cannot directly provide a solution to the initial boundary value problem, since it was derived without using required boundary conditions. Therefore, Eq. (2.9), as well as Eqs. (2.7) and (2.8) can be viewed only as some new calibration integral relations to which any regular solutions of the initial equation for polarized radiation transfer having physical content must satisfy. On the other hand, taking into account relevant boundary conditions, they can serve as an alternative to the initial boundary value problem in the theory of polarized radiation transfer in a uniform slab of an arbitrary optical thickness $\tau_0 < \infty$. The realization of such a capability, however, is a complex mathematical task as it requires additional and cumbersome investigation of the problem of existence and uniqueness of sought for regular solutions.

If we consider one of values $\hat{I}^m(\tau_1 + \alpha, \eta, \xi, \tau_0)$ or $\hat{I}^m(\tau_2 - \alpha, -\eta, \xi, \tau_0)$ to be known, then relations (2.7) and (2.8) will turn into exact matrix linear singular equations for direct finding of the other unknown value. Finally, it should be emphasized that the obvious generalization of the scalar approach [8] to the case of polarization of multiple scattered radiation allows to use the known values $\hat{I}^m(\tau_1, \eta, \xi, \tau_0)$ and $\hat{I}^m(\tau_2, -\eta, \xi, \tau_0)$ at fixed optical depths τ_1 and τ_2 for finding respective azimuthal harmonics of matrix intensities $\hat{I}^m(\tau, \pm\eta, \xi, \tau_0)$ at summarized optical depths $\tau = \tau_1 + \tau_2$. The scalar variant [7], however, did not point to an algorithm of finding the values $\hat{I}^m(\tau_1, \eta, \xi, \tau_0)$ and $\hat{I}^m(\tau_2, -\eta, \xi, \tau_0)$ by the known Fourier harmonics $\hat{I}^m(\tau_1 + \tau_2, \pm\eta, \xi, \tau_0)$, $\eta > 0$. In this respect, linear singular integral Eqs. (2.7) and (2.8) eliminate that omission and enable us to close the scheme for calculating values $\hat{I}^m(\tau_1, \pm\eta, \xi, \tau_0)$, $\hat{I}^m(\tau_2, \pm\eta, \xi, \tau_0)$ and $\hat{I}^m(\tau_1 + \tau_2, \pm\eta, \xi, \tau_0)$, $(\eta, \xi) > 0$, $m = \overline{0, M}$ into a single self-consistent computational cycle.

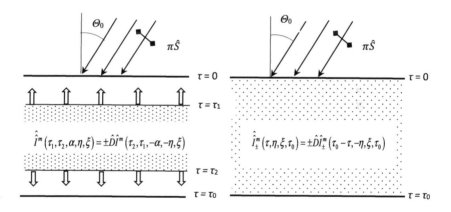

Fig. 3. Preserving the property of spatial angular symmetry and invariance of polarized radiation fields in an arbitrary local layer $[\tau_1, \tau_2]$ when it is extended to the boundaries $\tau_1 = 0$ и $\tau_2 = \tau_0$ of the initial uniform slab of a finite optical thickness $\tau_0 < \infty$.

4 Conclusion

In conclusion, note that the obtained new linear singular integral Eq. (2.9) expresses the general fundamental property of spatial angular invariance of internal polarized radiation fields in a uniform slab of an arbitrary optical thickness ($\tau_0 \le \infty$) relative to a semigroup of linear spatial angular transformations of regular solutions to the initial boundary value problem in the polarized radiation transfer theory, which are associated with a spatial vertical shift of the current optical levels $\tau_i \leftrightarrow \tau_{i+1}$ and angular rotation of respective vision lines $\eta \leftrightarrow -\eta$. In that case, if the basic invariance relation (1.14) is true for a current local layer $[\tau_i, \tau_{i+1}]$ inside the given uniform slab of an arbitrary optical thickness ($\tau_0 \leq \infty$), then it must preserve for the entire extended uniform layer within the limits $\tau \in [0, \infty]$, and vice versa (Fig. 3).

Of course, inside a semi-infinite uniform medium ($\tau_0 = \infty$) this fundamental property of spatial angular symmetry and invariance formally manifests itself only in the case of above-mentioned linear transformations of polarized radiation intensity [6]. In media of a finite optical thickness ($\tau_0 < \infty$), having fixed top and bottom boundaries, this property manifests itself as an obvious consequence of the mirror reflection of polarized radiation fields at symmetrical (with respect to the layer middle $\tau_0/2$) levels (τ) and ($\tau_0 - \tau$) in the mirror vision line directions (η) and ($-\eta$). It should emphasized that the above conception can also be applied for generalized photometrical invariants of the uniform atmospheric slab – arbitrary underlying surface system by using the broad interpretation of the Ambarzumian – Chandrasekhar invariance principle [9].

References

1. Smokty, O.I.: Development of radiative transfer theory methods on the basis of mirror symmetry principle. In: Current Problems in Atmospheric Radiation (IRS 2000), pp. 341–345. A. Deepak Publishing Co., Hampton (2001)
2. Smokty, O.I.: Multiple polarized light scattering in a uniform slab: new invariant constructions and symmetry relations. In: Current Problems in Atmospheric Radiation (IRS 2008), pp. 97–100. Proceedings of American Institute of Physics, New York (2009)
3. Hovenier, T.W., van der Mee, C.V.M.: Fundamental relationships relevant to the transfer of polarized light in a scattering atmosphere. Astron. Astrophys. **128**, 1–12 (1993)
4. Domke, H.: Linear singular integral equations for polarized radiation in isotropic media. Astr. Nachr. **298**(1), 57–69 (1977)
5. Case, K.M., Zweifel, P.F.: Linear Transport Theory. Addison-Wesley, London (1967)
6. Smokty, O.I., Anikonov, A.S.: Light Scattering in Large Optical Thickness Media. Nauka, St. Petersburg (2008)
7. Sobolev, V.V.: Light Scattering in Planetary Atmospheres. Pergamon Press, Oxford (1975)
8. Haan, F.D., Bosma, P.B., Hovenier, T.W.: The adding method for multiple scattering calculations of polarized light. Astron. Astrophys. **183**, 311–389 (1987)
9. Yanovitskij, E.G.: Light Scattering in Inhomogeneous Atmospheres. Springer, Berlin (1997)

User-Friendly Expressions
of the Coefficients of Some
Exponentially Fitted Methods

Dajana Conte[1], Raffaele D'Ambrosio[2], Giuseppe Giordano[1(✉)],
Liviu Gr. Ixaru[3,4], and Beatrice Paternoster[1]

[1] Department of Mathematics, University of Salerno, Fisciano, Salerno, Italy
{dajconte,gigiordano,beapat}@unisa.it
[2] Department of Information Engineering and Computer Science and Mathematics,
University of L'Aquila, L'Aquila, Italy
raffaele.dambrosio@univaq.it
[3] Horia Hulubei National Institute of Physics and Nuclear Engineering,
Bucharest, Romania
[4] Academy of Romanian Scientists,
54 Splaiul Independenţei, 050094 Bucharest, Romania
ixaru@theory.nipne.ro

Abstract. The purpose of this work consists in reformulating the coefficients of some exponentially-fitted (EF) methods with the aim of avoiding numerical cancellations and loss of precision. Usually the coefficients of an EF method are expressed in terms of $\nu = \omega h$, where ω is the frequency and h is the step size. Often, these coefficients exhibit a $0/0$ indeterminate form when $\nu \to 0$. To avoid this feature we will use two sets of functions, called C and S, which have been introduced by Ixaru in [61]. We show that the reformulation of the coefficients in terms of these functions leads to a complete removal of the indeterminacy and thus the convergence of the corresponding EF method is restored. Numerical results will be shown to highlight these properties.

Keywords: Exponential fitting · C and S sets of functions · η_m set of functions

1 Introduction

Exponential fitting is a mathematical procedure to generate numerical methods for different problems with a pronounced oscillatory or hyperbolic behaviour, usually occurring in interpolation, numerical differentiation and quadrature $[30,31,33,35,60,64,65,70,78]$, numerical solution of first order ordinary differential equations $[3,4,27,40,43,49,60,68,72,73,75,76,79]$, second order differential

The authors Conte, D'Ambrosio, Giordano and Paternoster are members of the GNCS group. This work is supported by GNCS-INDAM project and by PRIN2017-MIUR project.

O. Gervasi et al. (Eds.): ICCSA 2020, LNCS 12249, pp. 47–62, 2020.
https://doi.org/10.1007/978-3-030-58799-4_4

equations [54,63], integral equations [15,16], fractional differential equations [1], partial differential equations [14,45,46,50,52]. This procedure has been introduced in [62]. Its central idea consists in determining the coefficients of a numerical method by asking that the method is exact for the following set of functions, which is called a *fitting space*:

$$\mathcal{F} = \{1, x, \ldots, x^K, e^{\pm \mu x}, x e^{\pm \mu x}, \ldots, x^P e^{\pm \mu x}\} \tag{1}$$

where μ may be real or imaginary. The coefficients are functions of the parameter $\nu = \omega h$, where ω is the frequency of the oscillatory or hyperbolic functions and h is the step size. The values of μ to be used in (1) are the imaginary $\mu = i\omega$ and real $\mu = \omega$, respectively.

Often, these coefficients exhibit the indeterminate form $0/0$ when $\nu \to 0$ such that, in order to restore the convergence of the corresponding numerical methods when ν is small (in practice this depends on how small is the step size h), it is necessary to make use of the Taylor series of the coefficients. Expressed in different words, an accurate computation of the EF coefficients requires the knowledge of four different formulas (an analytic formula valid for big ν and a power series for small ν, for each of the trigonometrical or hyperbolic fitting).

In the paper [61] a method was described to replace the four formulas by a single one. The coefficients have been expressed in terms of two sets of particular functions, called $C(Z)$ and $S(Z)$, where $Z = \pm \nu^2$, for real and imaginary μ. A similar method has been introduced in the paper [30], in which the coefficients are expressed in terms of $\eta_m(Z)$ functions.

The work is organized as follows. In Sect. 2 we recall the two sets of C and S functions, and in Sect. 3 the general procedure for the conversion of the coefficient in terms of C and S functions is briefly presented. In Sect. 4 we reformulate the coefficients for the methods in [67,69]. In Sect. 5 numerical experiments are presented to show how the converted coefficients restore the convergence of the method.

2   C and S Functions

The original $\nu = \mu h$ is replaced by the new $Z = \nu^2$ which is negative if ν is imaginary and positive when ν is real. Thus, $Z < 0$ and $Z > 0$ cover the trigonometric and hyperbolic case, respectively.

To define the sets of functions C and S we rely on the family of functions $\eta_m(Z)$ functions, $m = -1, 0, 1, \ldots, \ldots$:

$$\eta_{-1}(Z) = \begin{cases} \cos(|Z|^{1/2}) & \text{if } Z \leq 0 \\ \cosh(Z^{1/2}) & \text{if } Z > 0 \end{cases}, \quad \eta_0(Z) = \begin{cases} \sin(|Z|^{1/2})/|Z|^{1/2} & \text{if } Z < 0 \\ 1 & \text{if } Z = 0 \\ \sinh(Z^{1/2})/Z^{1/2} & \text{if } Z > 0 \end{cases} \tag{2}$$

and, for $Z \neq 0$,

$$\eta_m(Z) = [\eta_{m-2}(Z) - (2m - 1)\eta_{m-1}(Z)]/Z, \quad m = 1, 2, 3, \ldots \tag{3}$$

while for $Z = 0$,

$$\eta_m(0) = 1/(2m+1)!!, \; m = 1, \, 2, \, 3, ... \tag{4}$$

The two sets of C and S functions are defined as follows: $C_{-1}(Z)$ and $S_{-1}(Z)$ are given by the first two η functions,

$$C_{-1}(Z) = \eta_{-1}(Z), S_{-1}(Z) = \eta_0(Z) \tag{5}$$

while the next ones are derived by recurrence for $Z \neq 0$,

$$C_n(Z) = \frac{C_{n-1}(Z) - C_{n-1}(0)}{Z}, S_n(Z) = \frac{S_{n-1}(Z) - S_{n-1}(0)}{Z}, \text{for } n = 0, 1, 2, \ldots, \tag{6}$$

and by the following values at $Z = 0$,

$$C_n(0) = \frac{1}{(2n+2)!}, S_n(0) = \frac{1}{(2n+3)!}, \text{for any } n =, 0, 1, 2, \ldots. \tag{7}$$

An important property is the reverse relations:

$$C_n(Z) = ZC_{n+1}(Z) + C_n(0), S_n(Z) = ZS_{n+1} + S_n(0). \tag{8}$$

for $n = -1, 0, 1, \ldots$. For an accurate computation of these functions, it is necessary to introduce their series expansions:

$$C_n(Z) = \sum_{k=0} \frac{Z^k}{[2(k+n+1)]!}, S_n(Z) = \sum_{k=0} \frac{Z^k}{[2(k+n)+3]!}. \tag{9}$$

Note: We acknowledge with thanks a recent private communication by Prof. Ander Murua that sets C and S are directly related to the Stumpff functions $c_n(Z)(n = 0, 1, 2, 3, ..)$:

$$C_n(Z) = c_{2n+2}(-Z), \; S_n(Z) = c_{2n+3}(-Z).$$

For the Stumpff functions see [81] and references therein.

3   Procedure for the Conversion of Coefficients

Now we describe the procedure introduced in [61] for the conversation of the coefficients. Let $\sigma(z)$, a generic coefficient derived by EF technique:

$$\sigma(z) = \frac{\overline{N}(z)}{\overline{D}(z)}$$

where $\overline{N}(z)$ and $\overline{D}(z)$ contains trigonometrical or hyperbolic functions and tend to 0 when $z \to 0$.

Let us denote generically by $\overline{F}$ any of numerator $\overline{N}$ or denominator $\overline{D}$, and treated separately these two functions. The procedure has two steps.

In the first step, $\overline{F}(z)$ is expressed in terms of Z in the following way:

$$\overline{F}(z) = z^k F(Z) \tag{10}$$

where $k = 0, 1$ and

$$F(Z) = \sum_{m=0}^{M} Z^m F_m(Z). \tag{11}$$

The second step consists in factorizing Z in $F(Z)$ as many times as possible until the form

$$F(Z) = Z^m F^*(Z) \tag{12}$$

where $m \geq 0$ and $F^*(Z) \neq 0$.

To be able to factorize Z in $F(Z)$, we have to first evaluate $F_0(Z)$ at $Z = 0$. If $F_0(0) \neq 0$, then no Z factorization is possible and the procedure is stopped. Instead, if $F_0(0) = 0$, then

$$\mathcal{F}(Z) = F_0(Z) = Z \Delta F_0(Z),$$

take $m = 1$ and the following algorithmic applies:

1. Form $\mathcal{F}_m(Z) = \Delta F_{m-1}(Z) + F_m(Z)$ and evaluate in 0;
2. If $\mathcal{F}_m(0) = 0$, one factorization of Z is possible and we put $\mathcal{F}_m(Z) = Z \Delta F_m(Z)$, then determine $\Delta F_m(Z)$, increase m by 1, and go to (1);
3. If $\mathcal{F}_m(0) \neq 0$, the procedure is stopped and

$$F(Z) = Z^m F^*(Zz) \tag{13}$$

where,

$$F^*(Z) = \begin{cases} \Delta F_{m-1}(Z) + \sum_{i=0}^{M-m} Z^i F_{m+i} & \text{if } m \leq M, \\ \Delta F_{m-1}(Z) & \text{if } m > M \end{cases} \tag{14}$$

In the most cases, $\mathcal{F}_m$ is a linear combination of terms containing functions from sets C and S:

$$\mathcal{F}_m(Z) = K + \sum_{j \geq -1} [a_j C_j(\alpha_j Z) + b_j S_j(\beta_j Z)] \tag{15}$$

where a_j, b_j, K are constant, and $\alpha_j, \beta_j \geq 0$.

By replacing (8) in (15), the following expression is obtained:

$$\mathcal{F}_m(Z) = \mathcal{F}_m(0) + Z \sum_{j \geq -1} [a_j \alpha_j C_{j+1}(\alpha_j Z) + b_j \beta_j S_{j+1}(\beta_j Z)]. \tag{16}$$

If $\mathcal{F}_m(0) = 0$, then the expression of $\Delta_m(Z)$ is thus obtained:

$$\Delta F_m(Z) = \sum_{j \geq -1} [a_j \alpha_j C_{j+1}(\alpha_j Z) + b_j \beta_j S_{j+1}(\beta_j Z)]. \tag{17}$$

After applying the procedure described for the numerator and the denominator of the coefficient, the **user-friendly** reformulation of the coefficient is obtained:

$$\sigma(Z) = \frac{N^*(Z)}{D^*(Z)}, \tag{18}$$

where $N^*(0) \neq 0$ and $D^*(0) \neq 0$.

4   Reformulation of the Coefficients

In this Section we reformulate the coefficients of two relevant classes of EF numerical methods [67,69] in terms of the C and S functions.

Example 1: The first class of methods, developed by Simos et al. in [67] regards the numerical solution of the second-order Initial Value Problems (IVPs) of the form:

$$\begin{cases} y'' = f(x, y(x)) \\ y(x_0) = y_0 \\ y'(x_0) = y_0' \end{cases}. \tag{19}$$

The scheme examined in [67] is of the form:

$$y_{n+1} + d_0 y_n + d_1 y_{n-1} + d_2 y_{n-2} + d_1 y_{n-3} + d_0 y_{n-4} + y_{n-5} =$$
$$h^2 \left(\tilde{d}_0 \tilde{y}''_{n+1} + \tilde{d}_1 y''_n + \tilde{d}_2 y''_{n-1} + \tilde{d}_3 y''_{n-2} + \tilde{d}_2 y_{n-3} + \tilde{d}_1 y_{n-4} + \tilde{d}_0 y''_{n-5} \right) \tag{20}$$

where $\tilde{y}_{n+1}$ is determined by solving

$$\tilde{y}_{n+1} + c_0 y_n + c_1 y_{n-1} + c_2 y_{n-2} + c_1 y_{n-3} + c_0 y_{n-4} + y_{n-5} =$$
$$h^2 \left(\tilde{c}_0 y''_n + \tilde{c}_1 y''_{n-1} + \tilde{c}_2 y''_{n-2} + \tilde{c}_1 y''_{n-3} + \tilde{c}_0 y''_{n-4} \right). \tag{21}$$

and using $\tilde{y}''_{n+1} = f(x_{n+1}, \tilde{y}_{n+1})$ in (20).

The classical version has the constant coefficients:

$$\tilde{c}_0^{\text{class}} = \frac{51484823}{17645880}, \tilde{c}_1^{\text{class}} = \frac{23362512}{735245}, \tilde{c}_2^{\text{class}} = -\frac{723342859}{8822940} \tag{22}$$

$$c_0 = \frac{12519323}{504168}, c_1 = \frac{2712635}{63021}, c_2 = -\frac{551}{4}$$

$$d_0 = -\frac{23362512}{735245}, d_1 = \frac{84437}{105035}, d_2 = -\frac{9}{5} \tag{23}$$

$$\tilde{d}_0 = \frac{1}{15}, \tilde{d}_1 = \frac{209837}{210070}, \tilde{d}_2 = \frac{320221}{315105}, \tilde{d}_3 = \frac{638003}{315105}$$

see [56].

In [67] the exponential fitting procedure is applied to produce the following ν dependent expressions for the $\tilde{c}$ coefficients :

$$
\begin{aligned}
\tilde{c}_0^{ef} = -\frac{1}{2016672\nu^6 \sin^3(\nu)} \Big(& 3(4965191\nu^4 - 82890689\nu^2 + 22589400)\sin(\nu) \\
& - 48(69329\nu^2 - 308970)\sin(2\nu) - (4965191\nu^4 - 8059857\nu^2 - 68357160)\sin(3\nu) \\
& + 7562520(3\nu^2 - 10)\sin(4\nu) - (59935259\nu^2 - 95582232)\nu\cos(\nu) \\
& - 32(437993\nu^2 + 2928636)\nu\cos(2\nu) - 3(4965191\nu^2 + 13671432)\nu\cos(3\nu) \\
& + 75625200\nu\cos(4\nu) + 48(875986\nu^2 - 759933)\nu \Big),
\end{aligned}
\tag{24}
$$

$$
\begin{aligned}
\tilde{c}_1^{ef} = \frac{1}{1008336\nu^6 \sin^3(\nu)} \Big(& -24(875986\nu^4 + 231349\nu^2 - 617940)\sin(\nu) \\
& - 18(14126869\nu^2 - 7562520)\sin(2\nu) + 4(1751972\nu^4 + 9751899\nu^2 - 15198660)\sin(3\nu) \\
& + 3(4965191\nu^2 + 22785720)\sin(4\nu) + 3781260(3\nu^2 - 20)\sin(5\nu) \\
& + 64(875986\nu^2 - 2650563)\nu\cos(\nu) - 12(14126869\nu^2 - 4537512)\nu\cos(2\nu) \\
& - 2(4965191\nu^2 + 13671432)\nu\cos(4\nu) + 60500160\nu\cos(5\nu) \\
& + 6(4965191\nu^2 + 13671432)\nu \Big),
\end{aligned}
\tag{25}
$$

$$
\begin{aligned}
\tilde{c}_2^{ef} = \frac{1}{2016672\nu^6 \sin^3(\nu)} \Big(& 6(42380607\nu^4 + 125300744\nu^2 - 45276960)\sin(\nu) \\
& - 24(2281313\nu^2 - 679290)\sin(2\nu) - 3(28253738\nu^4 - 74403153\nu^2 + 113732280)\sin(3\nu) \\
& - 48(1821301\nu^2 - 5993130)\sin(4\nu) - 3(4965191\nu^2 + 22785720)\sin(5\nu) \\
& - 7562520(\nu^2 - 10)\sin(6\nu) + 160(906559\nu^2 - 2390292)\nu\cos(\nu) \\
& - 16(1751972\nu^2 - 21371481)\nu\cos(2\nu) + 9(33218929\nu^2 + 4596408)\nu\cos(3\nu) \\
& - 32(437993\nu^2 + 6709896)\nu\cos(4\nu) + (4965191\nu^2 + 13671432)\nu\cos(5\nu) \\
& - 45375120\nu\cos(6\nu) - 288(437993\nu^2 - 852624)\nu \Big).
\end{aligned}
\tag{26}
$$

The other coefficients are untouched. They remain the same as in (23). Theoretically we must have

$$
\lim_{\nu \to 0} \tilde{c}_i^{ef} = \tilde{c}_i^{class}
\tag{27}
$$

for $i = 1, 2, 3$, but a direct examination of the EF expressions given in (24–26), shows that these have an indeterminate form $0/0$ for $\nu \to 0$ such that, in a numerical approach, a blow up of each coefficient will be obtained when h is decreased (we remind that $\nu = \omega h$).

This is removed by applying the procedure described in the previous Section. Indeed, now we have:

$$
\begin{aligned}
\tilde{c}_0^{CS} = \frac{1}{2016672 S_{-1}^3(Z)} \Big(& -59935259 C_2(Z) - 897009664 C_2(4Z) - 10858872717 C_2(9Z) \\
& - 95582232 C_3(Z) + 23991386112 C_3(4Z) + 269094796056 C_3(9Z) - 4956173107200 C_3(16Z) \\
& - 14895573 S_1(Z) + 1206541413 S_1(9Z) - 248672067 S_2(Z) - 425957376 S_2(4Z) \\
& 17626907259 S_2(9Z) + 371712983040 S_2(16Z) - 67768200 S_3(Z) - 7593246720 S_3(4Z) \\
& - 1345473980280 S_3(9Z) + 19824692428800 S_3(16Z) \Big);
\end{aligned}
\tag{28}
$$

$$
\begin{aligned}
\tilde{c}_1^{CS} = \frac{1}{252084 S_{-1}^3(Z)} \Big(& -14015776 C_2(Z) + 2712358848 C_2(4Z) + 10168711168 C_2(16Z) \\
& - 42409008 C_3(Z) + 3484809216 C_3(4Z) - 447984483776 C_3(16Z) + 5908218750000 C_3(25Z) \\
& - 5255916 S_1(Z) + 425729196 S_1(9Z) + 1388094 S_2(Z) + 8137076544 S_2(4Z) \\
& - 21327403113 S_2(9Z) - 61012267008 S_2(16Z) - 221558203125 S_2(25Z) + 3707640 S_3(Z) \\
& + 17424046080 S_3(4Z) + 299155224780 S_3(9Z) + 4479854837760 S_3(16Z) \\
& - 36926367187500 S_3(25Z) \Big);
\end{aligned}
\tag{29}
$$

$$\tilde{c}_2^{CS} = \frac{1}{2016672 S_{-1}^3(Z)}\Big(-145049440 C_2(Z) + 1794019328 C_2(4Z) - 217949393169 C_2(9Z)$$

$$+ 57408618496 C_2(16Z) - 77581109375 C_2(25Z) - 382446720 C_3(Z) + 87537586176 C_3(4Z)$$
$$+ 271413295992 C_3(9Z) - 14071671816192 C_3(16Z) + 5340403125000 C_3(25Z)$$
$$- 76212777553920 C_3(36Z) + 254283642 S_1(Z) - 20596975002 S_1(9Z) - 751804464 S_2(Z)$$
$$+ 7008193536 S_2(4Z) - 488159086833 S_2(9Z) + 1432329388030 S_2(16Z) \quad (30)$$
$$+ 1163716640625 S_2(25Z) + 2117021598720 S_2(36Z) - 271661760 S_3(Z) + 8347115520 S_3(4Z)$$
$$- 6715777401720 S_3(9Z) + 75411027394560 S_3(16Z) - 133510078125000 S_3(25Z)$$
$$+ 762127775539200 S_3(36Z)\Big).$$

The new expressions are also quotients of two ν dependent functions but, as expected, they do no longer exhibit indeterminacy when $\nu \to 0$. Also worth mentioning is that the power of Z in (13) for all these three coefficients is $m = 4$. The use of η functions has the same effect. In fact, by applying the procedure described in [30], the coefficients expressed by these functions are (see also [23]):

$$\tilde{c}_0^\eta = \frac{-1071987210\eta_0^4(Z/64) - 535993605\eta_0^2(Z/256)(1 + \eta_0(Z/64)) + \cdots}{13552035840(2 + Z\eta_0^2(Z/4) - 2Z\eta_1(Z))^3} \quad (31)$$

$$\tilde{c}_1^\eta = \frac{-2031821820\eta_0^4(Z/64) - 1015910910\eta_0^2(Z/256)(1 + \eta_0(Z/64)) + \cdots}{13552035840(2 + Z\eta_0^2(Z/4) - 2Z\eta_1(Z))^3} \quad (32)$$

$$\tilde{c}_2^\eta = \frac{-2146293450\eta_0^4(Z/64) - 1073146725\eta_0^2(Z/256)(1 + \eta_0(Z/64)) + \cdots}{6776017920(2 + Z\eta_0^2(Z/4) - 2Z\eta_1(Z))^3} \quad (33)$$

The full expressions can be obtained using the Mathematica modules in [30]. Both ways of deriving single formulae, instead of four, are then acceptable, and the expected theoretical behavior

$$\lim_{Z\to 0} \tilde{c}_i^{CS} = \lim_{Z\to 0} \tilde{c}_i^\eta = \tilde{c}_i^{class}, \quad (34)$$

is preserved.

Example 2: We consider the numerical method developed by Ndukum et al. [69] to solve the first-order IVP:

$$\begin{cases} y' = f(x, y(x)) \\ y(a) = y_0 \end{cases} \quad (35)$$

with $x \in [a, b]$.

The scheme used is a k-step numerical method of the form:

$$y_{n+k} = \sum_{r=0}^{k-1} \alpha_r(\nu) y_{n+r} + h(\beta_k(\nu) f_{n+k} + \beta_{k+1}(\nu) f_{m+k+1}). \quad (36)$$

In the paper [69] the authors presented the coefficients of the method corresponding to $k = 1, 2, 3, 4, 5$. In the paper [61] the case $k = 2$ has been considered.

In the following we consider the case $k = 3$.

By applying the exponential fitting procedure, the coefficients are [69]:

$$\alpha_0^{\text{ef}} = \frac{5\nu - 11\nu\cos\nu + 7\nu\cos 2\nu - \nu\cos 3\nu + 4\sin\nu + 2\nu^2\sin\nu - 2\sin 2\nu}{7\nu\cos\nu - 17\nu\cos 2\nu + 13\nu\cos 3\nu - 3\nu\cos 4\nu + 4\sin\nu + 2\nu^2\sin 2\nu - 2\sin 2\nu} \tag{37}$$

$$\alpha_1^{\text{ef}} = \frac{-12\nu + 23\nu\cos\nu - 9\nu\cos 2\nu - 3\nu\cos 3\nu + \nu\cos 4\nu - 2\sin\nu - 6\nu^2\sin\nu - 2\sin 2\nu + 2\sin 3\nu}{7\nu\cos\nu - 17\nu\cos 2\nu + 13\nu\cos 3\nu - 3\nu\cos 4\nu + 4\sin\nu + 2\nu^2\sin 2\nu - 2\sin 2\nu} \tag{38}$$

$$\alpha_2^{\text{ef}} = \frac{7n\nu - 5\nu\cos\nu - 15\nu\cos 2\nu + 17\nu\cos 3\nu - 4\nu\cos 4\nu + 2\sin\nu + 6\nu^2\sin\nu + 2\sin 2\nu - 2\sin 3\nu}{7\nu\cos\nu - 17\nu\cos 2\nu + 13\nu\cos 3\nu - 3\nu\cos 4\nu + 4\sin\nu + 2\nu^2\sin 2\nu - 2\sin 2\nu} \tag{39}$$

$$\beta_3^{\text{ef}} = \frac{2\nu\cos\nu - 6\nu\cos 2\nu + 6\nu\cos 3\nu - 2\nu\cos 4\nu + 25\sin\nu - 20\sin 2\nu + 5\sin 3\nu}{7\nu\cos\nu - 17\nu\cos 2\nu + 13\nu\cos 3\nu - 3\nu\cos 4\nu + 4\sin\nu + 2\nu^2\sin 2\nu - 2\sin 2\nu} \tag{40}$$

$$\beta_4^{\text{ef}} = \frac{-2\nu + 6\nu\cos\nu - 6\nu\cos 2\nu + 2\nu\cos 3\nu - 13\sin\nu + 12\sin 2\nu - 3\sin 3\nu}{7\nu\cos\nu - 17\nu\cos 2\nu + 13\nu\cos 3\nu - 3\nu\cos 4\nu + 4\sin\nu + 2\nu^2\sin 2\nu - 2\sin 2\nu} \tag{41}$$

which all exhibit the 0/0 indeterminacy.

The coefficients modified by means of the the procedure described in the previous Section are:

$$\tilde{\alpha}_0^{\text{CS}} = \frac{1 + (6(3C_2(Z) - 256C_2(4Z) + 1701C_2(9Z) - 2048C_2(16Z))}{-7C_2(Z) + 1088C_2(4Z) - 9477C_2(9Z) + 12288C_2(16Z) + 2S_1(Z) - 4S_2(Z) + 256S_2(4Z)} \tag{42}$$

$$\tilde{\alpha}_1^{\text{CS}} = \frac{23C_2(Z) - 576C_2(4Z) - 2187C_2(9Z) + 4096C_2(16Z) + 6S_1(Z) - 2S_2(Z) - 256S_2(4Z) + 4371S_2(9Z)}{7C_2(Z) - 1088C_2(4Z) + 9477C_2(9Z) - 2(6144C_2(16Z) + S_1(Z) - 2S_2(Z) + 128S_2(4Z))} \tag{43}$$

$$\tilde{\alpha}_2^{\text{CS}} = \frac{-5C_2(Z) - 960C_2(4Z) + 12393C_2(9Z) - 16384C_2(16Z) - 6S_1(Z) + 2S_2(Z) + 256S_2(4Z) - 4374S_2(9Z)}{7C_2(Z) - 1088C_2(4Z) + 9477C_2(9Z) - 2(6144C_2(16Z) + S_1(Z) - 2S_2(Z) + 128S_2(4Z))} \tag{44}$$

$$\tilde{\beta}_3^{\text{CS}} = \frac{2C_2(Z) - 384C_2(4Z) + 4374C_2(9Z) - 8192C_2(16Z) + 25S_2(Z) - 2560S_2(4Z) + 10935S_2(9Z)}{7C_2(Z) - 1088C_2(4Z) + 9477C_2(9Z) - 2(6144C_2(16Z) + S_1(Z) - 2S_2(Z) + 128S_2(4Z))} \tag{45}$$

$$\tilde{\beta}_4^{\text{CS}} = \frac{3(2C_2(Z) - 128C_2(4Z) + 486C_2(9Z) - 5S_2(Z) + 512S_2(4Z) - 2187S_2(9Z))}{7C_2(Z) - 1088C_2(4Z) + 9477C_2(9Z) - 2(6144C_2(16Z) + S_1(Z) - 2S_2(Z) + 128S_2(4Z))} \tag{46}$$

We observe that in all five coefficient the power of Z in (13) is $m = 3$.

As for the coefficients expressed in terms of $\eta_m(Z)$ functions, these are:

$$\tilde{\alpha}_0^{\eta} = \frac{-210\eta_0^4(Z/16) - 10\eta_0^2(Z/64)(1 + \eta_0(Z/16)) + \ldots}{165\eta_0^2(Z/64)(1 + \eta_0(Z/16)) + \ldots} \tag{47}$$

$$\tilde{\alpha}_1^{\eta} = \frac{630\eta_0^4(Z/16) + 315\eta_0^2(Z/64)(1 + \eta_0(Z/16)) - \ldots}{165\eta_0^2(Z/64)(1 + \eta_0(Z/16)) + \ldots} \tag{48}$$

$$\tilde{\alpha}_2^{\eta} = \frac{-90\eta_0^4(Z/16) - 45\eta_0^2(Z/64)(1 + eta_0(Z/16)) + \ldots}{165\eta_0^2(Z/64)(1 + \eta_0(Z/16)) + \ldots} \tag{49}$$

$$\tilde{\beta}_3^{\eta} = \frac{5(162\eta_0^4(Z/16) + 81\eta_0^2(Z/64)(1 + \eta_0(Z/16)) - \ldots}{165\eta_0^2(Z/64)(1 + \eta_0(Z/16)) + \ldots} \tag{50}$$

$$\tilde{\beta}_4^{\eta} = \frac{-3(90\eta_0^4(Z/16) + 45\eta_0^2(Z/64)(1 + \eta_0(Z/16)) - \ldots}{165\eta_0^2(Z/64)(1 + \eta_0(Z/16)) + \ldots} \tag{51}$$

Similar to the previous case, the full expression of the coefficients can be obtained using the Mathematica modules in [30].

5   A Check on the Effectiveness of the Approach

In this Section we show the graphs of the behavior of the coefficients and how our reformulation restores the convergence of the corresponding method.

On the left column of Fig. 1 we show the h dependence of the coefficients $\tilde{c}_i^{\text{ef}}(\nu)$, $\nu = \omega h$, for $\omega = 10$ and $i = 0, 1, 2$ of Example 1 compared with the reformulated $\tilde{c}_i^{\text{CS}}(Z)$, $Z = -\nu^2 = -(\omega h)^2$, by means of $C(Z)$ and $S(Z)$ functions, and $\tilde{c}_i^{\eta}(Z)$ by means of $\eta_m(Z)$ functions. In particular, we observe that the coefficients expressed in terms of $\eta_m(Z)$ functions and the coefficients expressed in terms of C and S functions converge to the classical value, while the coefficients $\tilde{c}_0^{\text{eg}}, \tilde{c}_1^{\text{eg}}$ and $\tilde{c}_2^{\text{eg}}$ blow up when h is decreased. From the numerical point of view

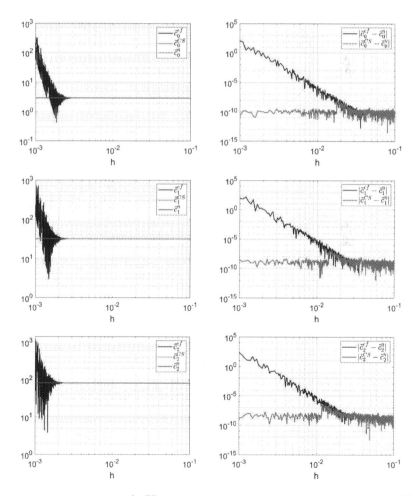

Fig. 1. Left: Coefficients $\tilde{c}_i^{\text{ef}}, \tilde{c}_i^{\text{CS}}, \tilde{c}_i^{\eta}$ of Example 1 for $\omega = 10$; **Right**: Error of $\tilde{c}_i^{\text{bf}}$ and $\tilde{c}_i^{\text{CS}}$ coefficients with respect to $\tilde{c}_i^{\eta}$

the limit tendency (27) is not verified but (34) holds true.

On the right column of the same figure we give additional details. On it we present the deviations of the coefficients computed by the first two approaches with respect to those expressed in terms of the $\eta_m(Z)$ functions. It is seen that the two reformulations (28)–(30) and (31)–(33) differ by a factor of only 10^{-10} irrespective of h while the EF coefficients (24)–(26) exhibit an error which increases as $h \to 0$ to reach a value of about 10^2 for $h = 10^{-3}$.

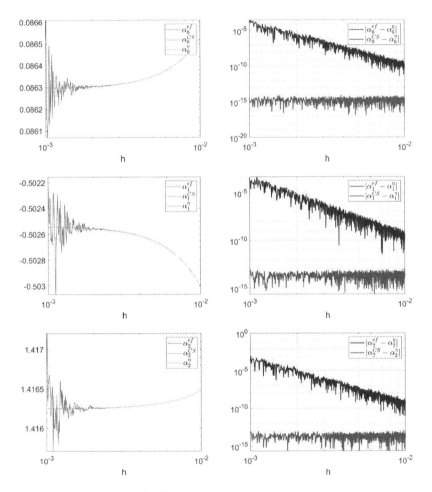

Fig. 2. Left: Coefficients α_i^{ef}, α_i^{CS}, α_i^{η} of Example 2 for $\omega = 10$; **Right**: Error of α_i^{ef}, α_i^{CS} coefficients with respect to α_i^{η}

The same data are presented on Figs. 2 and 3 for the coefficients of Example 2. Again, the coefficients obtained in the frame of the original EF approach of [69] are oscillating and inaccurate when $h \to 0$, in contrast with those in the

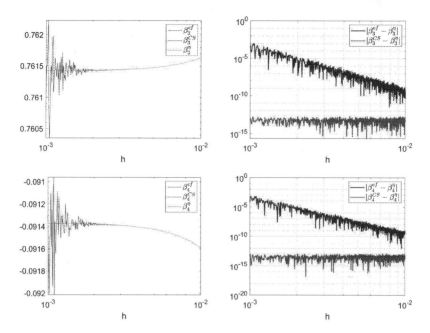

Fig. 3. Left: Coefficients β_i^{ef}, β_i^{CS}, β_i^{η} of Example 2 for $\omega = 10$; **Right**: Error of β_i^{ef} and β_i^{CS} coefficients with respect to β_i^{η}

other two approaches. The results from the latter two approaches are actually in agreement within 10^{-15}.

Also instructive is that, in contrast with Example 1, the blow up of the original EF estimates occurs at values of h much smaller than before, and this is due to the power m in (13) which for the previous case was $m = 4$ while it is $m = 3$ by now. This allows concluding that the need of reformulations in the spirit of the approach presented in this paper is more and more stringent when m is increased.

An important issue is that of checking in what extent the accuracy in the evaluation of the coefficients affects the accuracy of the results when solving numerically a differential equation. To illustrate this we consider the following problem:

$$\begin{cases} y'' = -100y(t) + 99\sin(t) \\ y(0) = 1 \\ y'(0) = 11 \end{cases} \tag{52}$$

with $t \in [0, 20\pi]$, whose analytic solution is $y(t) = \cos(10t) + \sin(10t) + \sin(t)$. In Fig. 4, we compare the method (20)–(21) for three versions of the coefficients; in particular we denote:

– EF: the method with coefficients (23)–(24)–(25)–(26);

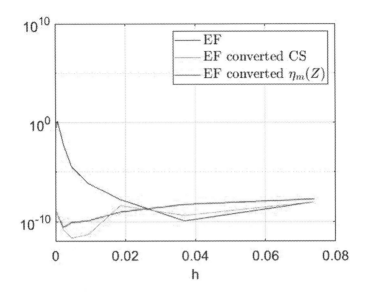

Fig. 4. Absolute error in $t = 20\pi$ of method (20)–(21) on problem (52)

- EF converted CS: the method with coefficients (23)–(28)–(29)–(30) converted by means of C and S functions;
- EF converted $\eta_m(Z)$: the method with coefficients (23)–(31)–(32)–(33) converted by $\eta_m(Z)$ functions.

Figure 4 shows that the reformulation of the coefficients in terms of either C and S or of η functions fully restores the convergence of the method.

6   Conclusions

We have shown that the reformulation of the expressions of the coefficients of EF-based numerical methods for differential equations in terms of the sets of functions C and S has two main advantages:

1. it allows reducing the original set of four expressions for each coefficient to a single one with universal use:
2. it removes completely the potential inaccuracy of the numerical solution when the step size h is small.

The procedure is then recommended as a reliable alternative to that based on η functions.

Further developments of this research will be oriented to the reformulation, through C and S functions, of existing methods for ordinary differential equations [2,17,20,25,26,28,37–39,41,42,44,48,51,53,56,77,80], integral equations [5–8,10,11,24,29,32,34,55,71], stochastic problems [9,12,13,18,19,29,47], fractional equations [12,13,21,22,36], partial differential equations [57–59,66,74].

References

1. Burrage, K., Cardone, A., D'Ambrosio, R., Paternoster, B.: Numerical solution of time fractional diffusion systems. Appl. Numer. Math. **116**, 82–94 (2017)
2. Butcher, J., D'Ambrosio, R.: Partitioned general linear methods for separable Hamiltonian problems. Appl. Numer. Math. **117**, 69–86 (2017)
3. Calvo, M., Franco, J.M., Montijano, J.I., Rández, L.: Explicit Runge-Kutta methods for initial value problems with oscillating solutions. J. Comput. Appl. Math. **76**(1–2), 195–212 (1996)
4. Calvo, M., Montijano, J.I., Rández, L., Van Daele, M.: Exponentially fitted fifth-order two step peer explicit methods. AIP Conf. Proc. **1648**, 150015-1–150015-4 (2015)
5. Capobianco, G., Conte, D.: An efficient and fast parallel method for Volterra integral equations of Abel type. J. Comput. Appl. Math. **189**(1-2), 481–493 (2006)
6. Capobianco, G., Conte, D., Del Prete, I.: High performance numerical methods for Volterra equations with weakly singular kernels. J. Comput. Appl. Math. **228**, 571–579 (2009)
7. Capobianco, G., Conte, D., Paternoster, B.: Construction and implementation of two-step continuous methods for Volterra integral equations. Appl. Numer. Math. **119**, 239–247 (2017)
8. Cardone, A., Conte, D.: Multistep collocation methods for Volterra integro-differential equations. Appl. Math. Comput. **221**, 770–785 (2013)
9. Cardone, A., Conte, D., D'Ambrosio, R., Paternoster, B.: Stability issues for selected stochastic evolutionary problems: a review. Axioms (2018). https://doi.org/10.3390/axioms7040091
10. Cardone, A., Conte, D., D'Ambrosio, R., Paternoster, B.: Collocation methods for Volterra integral and integro-differential equations: a review. Axioms **7**(3), 45 (2018)
11. Cardone, A., Conte, D., Paternoster, B.: A family of multistep collocation methods for Volterra integro-differential equations. AIP Conf. Proc. **1168**(1), 358 (2009)
12. Cardone, A., Conte, D., Patenoster, B.: Two-step collocation methods for fractional differential equations. Discr. Cont. Dyn. Sys. B.**23**(7), 2709–2725 (2018)
13. Cardone, A., D'Ambrosio, R., Paternoster, B.: A spectral method for stochastic fractional differential equations. Appl. Numer. Math. **139**, 115–119 (2019)
14. Cardone, A., D'Ambrosio, R., Paternoster, B.: Exponentially fitted IMEX methods for advection-diffusion problems. J. Comput. Appl. Math. **316**, 100–108 (2017)
15. Cardone, A., D'Ambrosio, R., Paternoster, B.: High order exponentially fitted methods for Volterra integral equations with periodic solution. Appl. Numer. Math. **114C**, 18–29 (2017)
16. Cardone, A., Ixaru, L.G., Paternoster, B.: Exponential fitting direct quadrature methods for Volterra integral equations. Numer. Algor. **55**, 467–480 (2010). https://doi.org/10.1007/s11075-010-9365-1
17. Citro, V., D'Ambrosio, R.: Nearly conservative multivalue methods with extended bounded parasitism. Appl. Numer Math. (2019). https://doi.org/10.1016/j.apnum.2019.12.007
18. Citro, V., D'Ambrosio, R., Di Giovacchino, S.: A-stability preserving perturbation of Runge-Kutta methods for stochastic differential equations. Appl. Math. Lett. **102**, 106098 (2020)
19. Citro, V., D'Ambrosio, R.: Long-term analysis of stochastic theta-methods for damped stochastic oscillators. Appl. Numer Math. **150**, 18–26 (2020)

20. D'Ambrosio, R., Paternoster, B.: P-stable general Nystrom methods for y" = f(x, y). J. Comput. Appl. Math. **262**, 271–280 (2014)

21. Conte, D., Califano, G.: Domain decomposition methods for a class of integro-partial differential equations. AIP Conf. Proc. **1776**, 090050 (2016)

22. Conte, D., Califano, G.: Optimal Schwarz waveform relaxation for fractional diffusion-wave equations. Appl. Numer. Math. **127**, 125–141 (2018)

23. Conte, D., D'Ambrosio, R., Giordano, G., Paternoster, B.: Regularized exponentially fitted methods for oscillatory problems. J. Phy. Conf. Ser. **1564**, 012013 (2020) (in press)

24. Conte, D., D'Ambrosio, R., Izzo, G., Jackiewicz, Z.: Natural volterra Runge-Kutta methods. Numer. Algor. **65**(3), 421–445 (2014)

25. Conte, D., D'Ambrosio, R., Jackiewicz, Z., Paternoster, B.: A pratical approach for the derivation of algebraically stable two-step Runge-Kutta methods. Math. Model. Anal **17**(1), 65–77 (2012)

26. Conte, D., D'Ambrosio, R., Jackiewicz, Z., Paternoster, B.: Numerical search for algebraically stable two-step almost collocation methods. J. Comput. Appl. Math. **239**, 304–321 (2013)

27. Conte, D., D'Ambrosio, R., Moccaldi, M., Paternoster, B.: Adapted explicit two-step peer methods. J. Numer. Math. **27**(2), 69–83 (2019)

28. Conte, D., D'Ambrosio, R., Paternoster, B.: GPU acceleration of waveform relaxation methods for large differential systems. Numer. Algor. **71**(2), 293–310 (2016)

29. Conte, D., D'Ambrosio, R., Paternoster, B.: On the stability of ϑ-methods for stochastic Volterra integral equations. Discr. Cont. Dyn. Sys. - Series B **23**(7), 2695–2708 (2018)

30. Conte, D., Esposito, E., Ixaru, L.G., Paternoster, B.: Some new uses of the $\eta_m(Z)$ functions. Comput. Phys. Commun. **181**, 128–137 (2010)

31. Conte, D., Ixaru, L.G,. Paternoster, B., Santomauro, G.: Exponentially-fitted Gauss-Laguerre quadrature rule for integrals over an unbounded interval. J. Comput. Appl. Math. **255**, 725–736 (2014)

32. Conte, D., Paternoster, B.: A family of multistep collocation methods for Volterra integral equations. AIP Conf. Proc. **936**, 128–131 (2007). https://doi.org/10.1063/1.2790090

33. Conte, D., Paternoster, B.: Modified Gauss–Laguerre exponential fitting based formulae. J. Sci. Comput. **69**(1), 227–243 (2016). https://doi.org/10.1007/s10915-016-0190-0

34. Conte, D., Paternoster, B.: Parallel methods for weakly singular Volterra integral equations on GPUs. Appl. Numer. Math. **114**, 30–37 (2017)

35. Conte, D., Paternoster, B., Santomauro, G.: An exponentially fitted quadrature rule over unbounded intervals. AIP Conf. Proc. **1479**, 1173–1176 (2012). https://doi.org/10.1063/1.4756359

36. Conte, D., Shahmorad, S., Talaei, Y.: New fractional Lanczos vector polynomials and their application to system of Abel-Volterra integral equations and fractional differential equations. J. Comput. Appl. Math. **366**, 112409 (2020)

37. D'Ambrosio, R., De Martino, G., Paternoster, B.: Numerical integration of Hamiltonian problems by G-symplectic methods. Adv. Comput. Math. **40**(2), 553–575 (2014)

38. D'Ambrosio, R., De Martino, G., Paternoster, B.: Order conditions of general Nyström methods. Numer. Algor. **65**(3), 579–595 (2014)

39. D'Ambrosio, R., De Martino, G., Paternoster, B.: General Nyström methods in Nordsieck form: error analysis. J. Comput. Appl. Math. **292**, 694–702 (2016)

40. D'Ambrosio, R., Esposito, E., Paternoster, B.: Exponentially fitted two-step Runge-Kutta methods: construction and parameter selection. Appl. Math. Comp. **218**(14), 7468–7480 (2012)
41. D'Ambrosio, R., Hairer, E.: Long-term stability of multi-value methods for ordinary differential equations. J. Sci. Comput. **60**(3), 627–640 (2014)
42. D'Ambrosio, R., Hairer, E., Zbinden, C.: G-symplecticity implies conjugate-symplecticity of the underlying one-step method. BIT Numer. Math. **53**, 867–872 (2013)
43. D'Ambrosio, R., Ixaru, L.G., Paternoster, B.: Construction of the EF-based Runge-Kutta methods revisited. Comput. Phys. Commun. **182**, 322–329 (2011)
44. D'Ambrosio, R., Izzo, G., Jackiewicz, Z.: Search for highly stable two-step Runge-Kutta methods for ODEs. Appl. Numer. Math. **62**(10), 1361–1379 (2012)
45. D'Ambrosio, R., Moccaldi, M., Paternoster, B.: Adapted numerical methods for advection-reaction-diffusion problems generating periodic wavefronts. Comput. Math. Appl. **74**(5), 1029–1042 (2017)
46. D'Ambrosio, R., Moccaldi, M., Paternoster, B.: Parameter estimation in IMEX-trigonometrically fitted methods for the numerical solution of reaction-diffusion problems. Comp. Phys. Commun. **226**, 55–66 (2018)
47. D'Ambrosio, R., Moccaldi, M., Paternoster, B.: Numerical preservation of long-term dynamics by stochastic two-step methods. Discr. Cont. Dyn. Sys. Ser. B **23**(7), 2763–2773 (2018)
48. D'Ambrosio, R., Moccaldi, M., Paternoster, B., Rossi, F.: Adapted numerical modelling of the Belousov-Zhabotinsky reaction. J. Math. Chem. (2018). https://doi.org/10.1007/s10910-018-0922-5
49. D'Ambrosio, R., Paternoster, B.: Exponentially fitted singly diagonally implicit Runge-Kutta methods. J. Comput. Appl. Math. **263**, 277–287 (2014)
50. D'Ambrosio, R., Paternoster, B.: Numerical solution of a diffusion problem by exponentially fitted finite difference methods. SpringerPlus **3**, 425 (2014)
51. D'Ambrosio, R., Paternoster, B.: A general framework for numerical methods solving second order differential problems. Math. Comput. Simul. **110**(1), 113–124 (2015)
52. D'Ambrosio, R., Paternoster, B.: Numerical solution of reaction-diffusion systems of λ-ω type by trigonometrically fitted methods. J. Comput. Appl. Math. **294**, 436–445 (2016)
53. D'Ambrosio, R., Paternoster, B.: Multivalue collocation methods free from order reduction. J. Comput. Appl. Math. **112515** (2019)
54. D'Ambrosio, R., Paternoster, B., Santomauro, G.: Revised exponentially fitted Runge-Kutta-Nyström methods. Appl. Math. Lett. **30**, 56–60 (2014)
55. De Bonis, M.C., Occorsio, D.: On the simultaneous approximation of a Hilbert transform and its derivatives on the real semiaxis. Appl. Numer. Math. **114**, 132–153 (2017)
56. Fang, J., Liu, C., Hsu, C.-W., Simos, T.E., Tsitouras, C.: Explicit hybrid six-step, sixth order, fully symmetric methods for solving $y'' = f(x, y)$. Math. Meth. Appl. Sci. **42**, 3305–3314 (2019)
57. Francomano, E., Paliaga, M.: The smoothed particle hydrodynamics method via residual iteration. Comput. Methods Appl. Mech. Eng. **352**, 237–245 (2019)
58. Garvie, M.R., Blowey, J.F.: A reaction-diffusion system of λ-ω type Part II: numerical analysis. Euro. J. Appl. Math. **16**, 621–646 (2005)
59. Greenberg, J.M.: Spiral waves for λ-ω systems. Adv. Appl. Math. **2**, 450–455 (1981)
60. Ixaru, L.G.: Runge-Kutta method with equation dependent coefficients. Comput. Phys. Commun. **183**, 63–69 (2012)

61. Ixaru, L.G.: Exponential and trigonometrical fittings: user-friendly expressions for the coefficients. Numer. Algor. **82**, 1085–1096 (2019)
62. Ixaru, L.G., Berghe, G.V.: Exponential Fitting. Kluwer Academic Publishers, Dordrecht (2004)
63. Ixaru, L.G., Paternoster, B.: A conditionally P-stable fourth-order exponential-fitting method for $y'' = f(x, y)$. J. Comput. Appl. Math. **106**(1), 87–98 (1999)
64. Ixaru, L.G., Paternoster, B.: A Gauss quadrature rule for oscillatory integrands. Comput. Phys. Commun. **133**, 177–188 (2001)
65. Kim, J.K., Cools, R., Ixaru, L.G:. Extended quadrature rules for oscillatory integrands. Appl. Numer. Math. **46**, 59–73 (2003)
66. Kopell, N., Howard, L.N.: Plane wave solutions to reaction-diffusion equations. Stud. Appl. Math. **52**, 291–328 (1973)
67. Liu, C., Hsu, C.W., Simos, T.E., Tsitouras, C.: Phase-fitted, six-step methods for solving $x'' = f(t, x)$. Math. Meth. Appl. Sci. **42**, 3942–3949 (2019)
68. Montijano, J.I., Rández, L., Van Daele, M., Calvo, M.: Functionally fitted explicit two step peer methods. J. Sci. Comput. **64**(3), 938–958 (2014)
69. Ndukum, P.L., Biala, T.A., Jator, S.N., Adeniyi, R.B.: On a family of trigonometrically fitted extended backward differentiation formulas for stiff and oscillatory initial value problems. Numer. Algor. **74**, 267–287 (2017)
70. Occorsio, D., Russo, M.G.: Mean convergence of an extended Lagrange interpolation process on $[0, +\infty)$. Acta Math. Hung. **142**(2), 317–338 (2014)
71. Occorsio, D., Russo, M.G.: Nyström methods for Fredholm integral equations using equispaced points. Filomat **28**(1), 49–63 (2014)
72. Ozawa, K.: A functional fitting Runge-Kutta method with variable coefficients. Jpn. J. Ind. Appl. Math. **18**, 107–130 (2001)
73. Paternoster, B.: Present state-of-the-art in exponential fitting. a contribution dedicated to Liviu Ixaru on his 70-th anniversary. Comput. Phys. Commun. **183**, 2499–2512 (2012)
74. Smith, M.J., Rademacher, J.D.M., Sherratt, J.A.: Absolute stability of wavetrains can explain spatiotemporal dynamics in reaction-diffusion systems of lambda-omega type. SIAM J. Appl. Dyn. Syst. **8**, 1136–1159 (2009)
75. Simos, T.E.: A fourth algebraic order exponentially-fitted Runge-Kutta method for the numerical solution of the Schrödinger equation. IMA J. Numer. Anal. **21**, 919–931 (2001)
76. Simos, T.E.: An exponentially-fitted Runge-Kutta method for the numerical integration of initial-value problems with periodic or oscillating solutions. Comput. Phys. Comm. **115**, 1–8 (1998)
77. Vanden Berghe, G., De Meyer, H., Van Daele, M., Van Hecke, T.: Exponentially fitted explicit Runge-Kutta methods. Comput. Phys. Commun. **123**, 7–15 (1999)
78. Van Daele, M., Vanden Berghe, G., Vande Vyver, H.: Exponentially fitted quadrature rules of Gauss type for oscillatory integrands. Appl. Numer. Math. **53**, 509–526 (2005)
79. Van Daele, M., Van Hecke, T., Vanden Berghe, G., De Meyer, H.: Deferred correction with mono-implicit Runge-Kutta methods for first-order IVPs, numerical methods for differential equations. J. Comput. Appl. Math. **111**(1–2), 37–47 (1999)
80. Weiner, R., Biermann, K., Schmitt, B.A., Podhaisky, H.: Explicit two-step peer methods. Comput. Math. Appl. **55**, 609–619 (2008)
81. https://en.wikipedia.org/wiki/Stumpff_function

CFD Prediction of Retractable Landing Gear Aerodynamics

Giuliano De Stefano[1]($\boxtimes$) (ID), Nunzio Natale[1] (ID), Antonio Piccolo[2] (ID), and Giovanni Paolo Reina[3] (ID)

[1] Engineering Department, University of Campania Luigi Vanvitelli, 81031 Aversa, Italy
{giuliano.destefano,nunzio.natale}@unicampania.it
[2] Leonardo Aircraft Company, 80038 Pomigliano d'Arco, Italy
antonio.piccolo@leonardocompany.com
[3] Altran Italy, 10135 Torino, Italy
gianpaolo.reina@libero.it

Abstract. CFD analysis is carried out to evaluate the mean aerodynamic loads on the retractable main landing-gear of a regional transport commercial aircraft. The mean flow around the landing-gear system including doors is simulated by using the Reynolds-averaged Navier-Stokes modelling approach, the governing equations being solved with a finite volume-based numerical technique. The computational grid is automatically adapted to the time-changing geometry by means of a dynamic meshing technique, while following the deployment of the landing-gear system. The present computational modelling approach is verified to have good practical potential by making a comparison with reference experimental data provided by the Leonardo Aircraft Company aerodynamicists.

Keywords: Aircraft landing-gears · Computational Fluid Dynamics · Industrial aerodynamics

1 Introduction

The retractable main landing-gear (MLG) stands for a highly critical subsystem of commercial aircrafts, which must be accurately designed so to have minimum weight and volume, while meeting all the prescribed regulatory and safety requirements. Along with experimental studies, the MLG design can be drastically improved by means of preliminary numerical simulations, which reduce the cost of further analyses as well as the risk of late design fixes. Among the different methodologies that are employed for the purpose, Computational Fluid Dynamics (CFD) methods have been becoming more and more important. In fact, CFD has been strongly emerging as an effective tool for industrial aerodynamics research [1,2]. In this framework, the CFD studies that are typically conducted make use of the Finite Volume (FV) method, while using the Reynolds-averaged

© Springer Nature Switzerland AG 2020
O. Gervasi et al. (Eds.): ICCSA 2020, LNCS 12249, pp. 63–74, 2020.
https://doi.org/10.1007/978-3-030-58799-4_5

Navier Stokes (RANS) turbulence modelling approach, e.g. [3]. Typically, an eddy viscosity-based diffusion term is introduced into the momentum equations to model the effect of turbulence, while solving additional evolution equations for the turbulence variables. In this study, dealing with complex geometries and unsteady flow configurations, the more sophisticated unsteady RANS (URANS) method is utilized [4–7].

The numerical simulation of the turbulent flow around a retractable landing-gear, due to the presence of a number of moving bluff bodies with different sizes and shapes, is very challenging. The turbulent wakes behind the different parts need to be simulated, along with the interaction of the wakes generated by upstream components impinging on downstream ones. The complex airflow can lead to large fluctuations in the aerodynamic forces acting, for instance, on the landing-gear doors and the resulting unsteady loads can cause serious issues when deploying/retracting the landing-gear system. CFD methods can be effectively used to simulate the air flow field around these complex systems in order to determine the temporal evolution of the aerodynamic forces [8], while the predicted mean loads can be used as input for the aircraft structures design. For instance, CFD calculations can provide the initial evaluation of the hinge moments of the landing-gear doors, which helps to size the hydraulic actuators.

The main goal of the present work is the computational evaluation of the mean aerodynamic loads on a retractable MLG system. As the landing-gear moves from stowed to fully-deployed position, the doors are opened and the landing-gear bay is occupied by fresh air. Also considering the inherently unsteadiness, the airflow around a deploying landing gear system is extremely complex, and can be only partially reproduced by simplified studies using generic models. Here, differently from similar previous works solving the same problem, where CFD simulations were conducted to understand the flow physics around rudimentary landing-gears, e.g. [9,10], the prototypical MLG geometry of a new regional transport aircraft is exploited. Moreover, the simulation of the turbulent incompressible flow field around the aircraft MLG model is not performed for the fully deployed configuration, as is usually done, but for the entire extension cycle, while using a dynamically adaptive unstructured grid that is automatically modified in time, while following the deployment of the system. The numerical experiments are conducted using the solver ANSYS Fluent, which is commonly and successfully employed for building virtual wind tunnels in the industrial aerodynamics research, e.g. [11–14]. The present results are validated by comparison with reference experimental data that are provided by the industrial researchers.

2   CFD Model

The design and development of the MLG system requires the knowledge of the aerodynamic loads associated with the time-dependent positions of its various components. However, given the motion laws of the different parts, the overall geometric configuration can be related to the angle of rotation of the landing-gear strut, say θ. Also, depending on the simulated flight conditions, the aerodynamic loading must be determined for different pitch and yaw angles of the aircraft, say α and β, respectively. By assuming these angles as independent variables, whereas $\theta = \theta(t)$, the following linear approximation, for a generic force coefficient C_F, is considered

$$C_F(\alpha, \beta, \theta(t)) \cong C_{F0}(t) + C_{F\alpha}(t)\alpha + C_{F\beta}(t)\beta, \tag{1}$$

where $C_{F0}(t) = C_F(0, 0, \theta(t))$ corresponds to the basic configuration at zero pitch and yaw angles. The partial derivatives $C_{F\alpha}(t)$ and $C_{F\beta}(t)$ have to be evaluated for suitable values of pitch and yaw angles. This way, the unsteady aerodynamic loading is completely characterized by the three time-dependent parameters C_{F0}, $C_{F\alpha}$ and $C_{F\beta}$. To numerically determine these coefficients by using a virtual wind tunnel, one possibility would consist in performing a number of different calculations with different values of (α, β), for some prescribed instantaneous configurations of the landing-gear system. Very simply, once a computation with zero pitch and yaw angles has been conducted to estimate C_{F0}, two additional calculations would suffice to approximate $C_{F\alpha}$ and $C_{F\beta}$, as fractional incremental ratios, for each desired MLG position.

As an alternative, in this study, numerical CFD simulations with moving boundaries are conducted for given couples of pitch and yaw angles, where the body-fitted numerical grids are continuously modified during the calculation, following the synchronous motion of the different components of the landing-gear system under examination. The present dynamic approach allows the numerical solution to more closely follow the unsteady flow evolution and, thus, to better represent the temporal evolution of the stresses acting on the various bodies surfaces and, in particular, on the landing-gear doors. In the following, the overall computational modelling and simulation approach is introduced. The geometry of the simplified system under investigation, namely, the MLG of a regional transport commercial aircraft, is given, along with the main numerical settings of the simulations.

2.1   Geometry

Starting from the complex geometry provided by the industrial partner, the present geometric model for the MLG system has been simplified to make the computational cost of the numerical experiments reasonable. However, in order to achieve a meaningful comparison with reference experimental data, the main features of the original model are maintained, differently form similar studies where rudimentary generic systems were investigated, e.g. [15, 16]. The three-dimensional geometric model is based on the generic structure of a short range narrow fuselage aircraft with MLG compartment. The flow is described in a Cartesian coordinate system (x, y, z), where the three directions correspond to the roll, pitch and yaw axis, respectively. The MLG system consists of two symmetric components that fully retract inside the main body of the aircraft, while rotating around axes that are parallel to the x–axis. The reduced model for each of the two specular sub-systems is comprised of a bay with an opening/closing door, a landing-gear strut rotatable between stowed and deployed positions, and two wheels, as illustrated in Fig. 1. The computational domain is represented by a square prism whose height (130 m long) is aligned with the fuselage symmetry axis, while the side length of the cross section is 60 m long. It is worth noting that, for the calculations with no yaw angle, half the above computational domain is actually employed, while imposing a symmetry condition at the midspan plane. For the simulations with non-zero yaw angle, the mesh mirroring procedure is exploited to build the whole FV grid. The simulations are performed for the complete deployment of the MLG system that lasts 10 s.

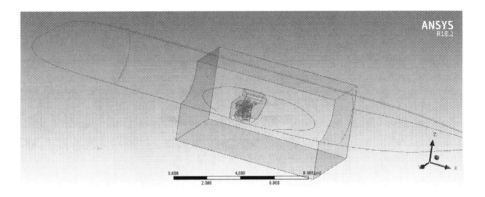

Fig. 1. Geometric model for one of the two symmetric MLG sub-systems. The sub-domain around the landing-gear, with active dynamic meshing, is evidenced.

2.2 Dynamic Meshing

The use of the dynamic meshing methodology allows to obtain accurate time-dependent results because the computational grid varies in time consistently with the changing positions of the different moving MLG parts, e.g. [17]. In order to avoid the deterioration of the mesh quality and/or the degeneration of existing FV cells, during the grid adaptation process, two different methods are used, which are referred to as smoothing and remeshing. The former technique consists in moving the interior nodes of the mesh, without changing their number and connectivity, where appropriate. The latter technique allows for the local update of the mesh by either adding or deleting cells, where the boundary displacement would be otherwise too large with respect to the local mesh size. Following previous studies for bluff body flows [12], the remeshing and the diffusion-based smoothing techniques are simultaneously used in the present CFD analysis.

The computational cost of such a dynamic approach is high, due to the requirement of fine grids and small time steps, in order to ensure both the desired numerical accuracy and the stability of the calculation. However, the application of the dynamic meshing procedure is not necessary in the whole computational domain, but only in flow regions that are expected to be actually influenced by the deployment of the landing-gear system. Therefore, a computational sub-domain with effective dynamic meshing is properly defined, while the FV grid in the rest of the computational domain remains unaltered over time. Practically, this space region contains the bay, the doors and the landing-gear structure, regardless of the instantaneous system configuration, as evidenced in Fig. 1. Initially, two unstructured meshes are generated in the two different sub-domains, while imposing the grid conformity at the interface between them.

2.3   Unsteady RANS

The aerodynamics of the aircraft MLG system is numerically predicted by solving the URANS equations, which describe the unsteady mean turbulent flow field around the aircraft, while using the dynamic meshing technique discussed above. The governing equations are supplied with the Spalart-Allmaras turbulence model [3], where the closure is achieved by solving an additional evolution equation for a modified eddy viscosity variable, e.g. [7]. The present method involves modelling the turbulent boundary layer using a wall-function approach, which is suitable for bluff bodies, where the flow undergoes geometry-induced separation [18]. This way, the boundary conditions are implemented so that relatively coarser meshes can be used. The equations governing the present URANS approach, which are not reported here for brevity, can be found, for instance, in [19].

3   Results

Following the computational modelling approach described above, the numerical simulations are performed by imposing velocity inlet and pressure outlet boundary conditions. The integration time step is prescribed as $\Delta t = 2.5 \times 10^{-3}$ s. A number of different calculations at Mach number 0.27 are conducted with freestream velocities representing realistic flight conditions. The flow Reynolds-number based upon the wheels diameter is $Re = 5.7 \times 10^6$. In the following, the results corresponding to three baseline configurations of particular interest, namely, $(\alpha, \beta) = (0°, 0°)$ (I), $(4°, 0°)$ (II) and $(0°, 5.7°)$ (III), are presented. For each configuration, a steady solution for the initial geometry, corresponding to the stowed position of the landing-gear with doors closed, is preliminarily obtained. The dynamic mesh based calculations are conducted starting from this initial solution.

The initial grid, associated with the stowed configuration, is illustrated in Fig. 2, where a global view at the symmetry plane is reported, along with the close-up view of the landing-gear bay. The mesh resolution that is used represents a fair compromise between numerical accuracy and computational cost, in terms of both memory and time, as empirically found through a grid independency study. As the present computational model employs a dynamic meshing

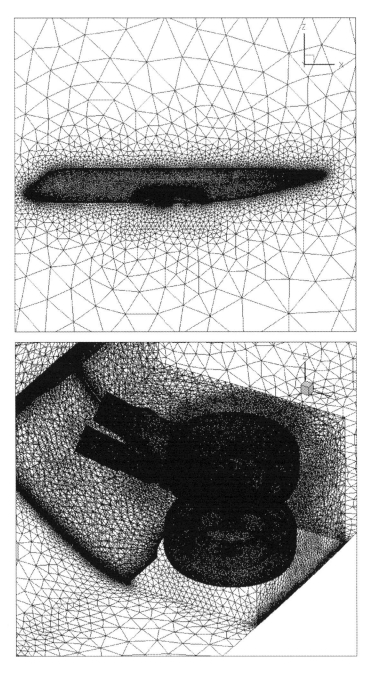

Fig. 2. Initial mesh associated with the stowed configuration: whole domain (top), and close-up view of the landing-gear bay (bottom).

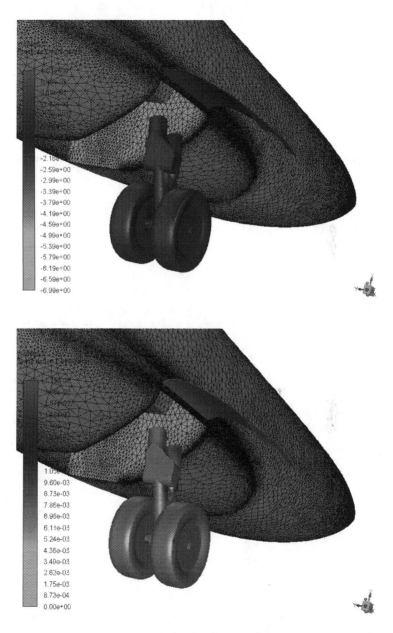

Fig. 3. Surface distributions of pressure (top) and skin friction (bottom) coefficients on the MLG sub-system, for fully deployed configuration.

procedure, a preliminary analysis was conducted for a limited number of static meshes, associated with different positions of the MLG system, by using three different FV grids with increasing resolution. The grids are naturally refined in the landing-gear zone, with the overall number of computational cells resulting in being the same order as for similar studies [20,21]. The two calculations with $\beta = 0°$ are carried out by employing half the computational domain, and about 2.5 million FV cells, whereas the third simulation uses about 5 million cells for the whole domain. Due to the dynamic meshing approach, the quality of the time-dependent mesh is ensured by controlling the associated quality parameters during the calculations.

To illustrate the baseline solution with $(\alpha, \beta) = (0°, 0°)$, the distributions of pressure and skin friction coefficients on the different parts of the MLG system are drawn in Fig. 3, for example, at the final instant that is for the fully deployed configuration. The time histories of the mean aerodynamic loads on the MLG structure, including strut and wheels, are reported in Fig. 4, in terms of force coefficients per unit reference area, for the three different simulations I, II and III. In the latter case, due to the flow asymmetry, the two different sub-systems are separately considered (while being labelled as seen by the pilot). During the deployment phase, the drag increases with the rotation of the gear leg, while yielding its maximum value at $t \approx 6$ s, when the system is completely immersed in the oncoming air flow. The angle of yaw is demonstrated to have a great effect in this case. In fact, the increased drag force for the case III can be attributed to the enlarged wake, and related pressure drag, associated with the positive yaw angle. The lift force shows a negative peak value between $t \approx 3$ and 4 s. As expected, once the deployment phase has terminated at $t = 10$ s, both the force components maintain statistically steady values.

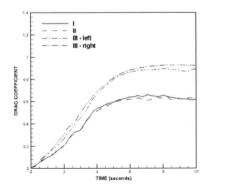 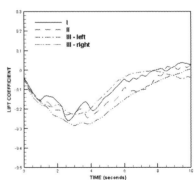

Fig. 4. Temporal evolution of drag (left) and lift (right) force coefficients for the gear structure (strut and wheels), for the three different simulations.

The hinge moment on the doors, which is the component of the moment of the aerodynamic force along the rotation axis, is presented in Fig. 5, for the three

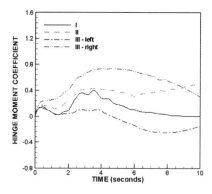

Fig. 5. Temporal evolution of hinge moment coefficient for the doors, for the three different simulations.

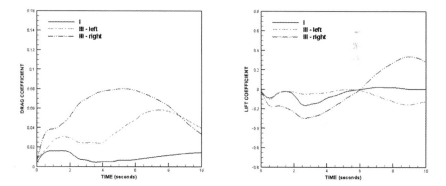

Fig. 6. Temporal evolution of drag (left) and lift (right) force coefficients for the doors, for the simulations I and III.

baseline calculations. Note that the moment is assumed positive in the direction of opening the doors. It can be seen that, initially, the hinge moment is minimal. This fact is due to the presence of the gaps along the doors that allow for the balance of the static pressure force acting on the outer surface and that one exerted by the fluid occupying the landing-gear bay. In fact, a similar result was found in [20] for the large front doors of a nose landing-gear. When looking at the effect of the pitch angle α, the hinge moment increases with this parameter, as it is apparent by making a comparison between the results of simulations I and II. As to simulation III, since it corresponds to a crosswind from left to right (as seen by the pilot), the hinge moment is different for the two different MLG sub-systems. Specifically, the aerodynamic load on the downwind right door is more relevant and the hinge moment takes relatively higher values. The effect of increasing the yaw angle results the most noticeable, as also emphasized in similar studies [21]. This is further shown by inspection of the drag and the lift force coefficients for the MLG doors, which are depicted in Fig. 6. In fact, due to the significant value of the lift force acting on the right door, the large MLG doors can be interpreted as a sort of small additional wings for the aircraft.

In order to validate the proposed computational evaluation procedure, the present CFD results are compared to corresponding data predicted by the structural loads group at Leonardo Aircraft Company, during the preliminary design phase of a small, commercial, regional transport aircraft. The reference data were obtained through empirical extrapolations, by using geometrical shape scaling parameters, along with proper combinations of pitch and yaw angles, which were selected among various flight conditions of industrial interest. The different configurations were expressed in terms of different aircraft weights and centre of gravity excursions, load factors, flap positions and speeds, which are associated with the prototypical model under examination. Here, to make a meaningful comparison with reference empirical data, the predicted normal force acting on the MLG doors is derived from the knowledge of the corresponding numerical hinge moments, known the geometry of the system. The normal force coefficient is obtained by employing the simplified linear model discussed in Sect. 2. Namely, a number of CFD calculations corresponding to a subset of the flight parameters combinations considered by the empirical industrial method are conducted for determining the coefficients C_{F0}, $C_{F\alpha}$ and $C_{F\beta}$ in the expression (1). In Fig. 7, the maximum and minimum values for the scaled normal forces acting on the landing-gear doors are reported along with the envelopes of the empirical data. The comparison with experimental findings appears quite successful for both the extreme values, even if some small discrepancies in the range of low opening angles exist. It is worth stressing that these results are shown in non-dimensional form, without expressly indicating the reference values, because the empirical data are not classified for public diffusion.

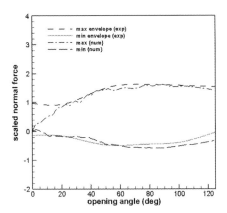

Fig. 7. Maximum and minimum values of the scaled normal force on the doors, compared to reference experimental data.

4 Conclusions

The present study has to be intended as the proof-of-concept, namely, the preliminary development of a CFD based prediction tool in the aerodynamic design of the retractable main landing-gear of a regional transport commercial aircraft. Differently from similar studies, where the CFD results were obtained for generic and rudimental landing-gear geometries, this work deals with simplified models directly derived from real geometries provided by industrial aerodynamicists. Dynamic mesh calculations have been performed with the aim of demonstrating the practical potential of the proposed computational methodology. The CFD analysis of the aerodynamic loads on the retractable landing-gear and doors has been performed in different operating conditions for the deployment cycle. The acceptable agreement with the empirical data made available by the Leonardo Aircraft Company in terms of normal forces has been achieved. There remains the possibility of developing more sophisticated computational models for particular industrial applications, depending on the level of accuracy that is required and the available computational resources.

Acknowledgements. This work was supported by the Italian Regione Campania under the research project SCAVIR (POR Campania FESR 2014/2020), presented by the Campania Aerospace Technological District (DAC). This support is gratefully acknowledged. The authors would like to thank the Leonardo Aircraft Company for providing the landing gear data.

References

1. Johnson, F.T., Tinoco, E.N., Yu, N.J.: Thirty years of development and application of CFD at Boeing Commercial Airplanes. Comput. Fluids **34**, 1115–1151 (2005)
2. Spalart, P.R., Venkatakrishnan, V.: On the role and challenges of CFD in the aerospace industry. Aeronaut. J. **120**(1223), 209–232 (2016)
3. Spalart, P.R., Allmaras, S.R.: A one-equation turbulence model for aerodynamic flows. AIAA Paper 92–0439 (1992)
4. Fröhlich, J., von Terzi, D.: Hybrid LES/RANS methods for the simulation of turbulent flows. Prog. Aerosp. Sci. **44**, 349–377 (2008)
5. Langtry, R.B., Spalart, P.R.: Detached eddy simulation of a nose landing-gear cavity. Solid Mech. Appl. **14**, 357–366 (2009)
6. De Stefano, G., Vasilyev, O.V., Brown-Dymkoski, E.: Wavelet-based adaptive unsteady Reynolds-averaged turbulence modelling of external flows. J. Fluid Mech. **837**, 765–787 (2018)
7. Ge, X., Vasilyev, O.V., De Stefano, G., Hussaini, M.Y.: Wavelet-based adaptive unsteady Reynolds-averaged Navier-Stokes simulations of wall-bounded compressible turbulent flows. AIAA J. **58**(4), 1529–1549 (2020)
8. De Stefano, G., Vasilyev, O.V.: Wavelet-based adaptive simulations of three-dimensional flow past a square cylinder. J. Fluid Mech. **748**, 433–456 (2014)
9. Imamura, T., Hirai, T., Amemiya, K., Yokokawa, Y., Enomoto, S., Yamamoto, K.: Aerodynamic and aeroacoustic simulations of a two-wheel landing-gear. Procedia Eng. **6**, 293–302 (2010)

10. Spalart, P.R., Mejia, K.M.: Analysis of experimental and numerical studies of the rudimentary landing gear. AIAA Paper 2011-355 (2011)
11. Escobar, J.A., Suarez, C.A., Silva, C., López, O.D., Velandia, J.S., Lara, C.A.: Detached-eddy simulation of a wide-body commercial aircraft in high-lift configuration. J. Aircr. **52**(4), 1112–1121 (2015)
12. Reina, G.P., De Stefano, G.: Computational evaluation of wind loads on sun-tracking ground-mounted photovoltaic panel arrays. J. Wind Eng. Ind. Aerodyn. **170**, 283–293 (2017)
13. Rapagnani, D., Buompane, R., Di Leva, A., et al.: A supersonic jet target for the cross section measurement of the $12C(\alpha, \gamma)16O$ reaction with the recoil mass separator ERNA. Nucl. Instrum. Methods Phys. Res. B **407**, 217–221 (2017)
14. Benaouali, A., Kachel, S.: Multidisciplinary design optimization of aircraft wing using commercial software integration. Aerosp. Sci. Technol. **92**, 766–776 (2019)
15. Hedges, L.S., Travin, A.K., Spalart, P.R.: Detached-eddy simulations over a simplified landing gear. J. Fluids Eng. **124**, 413–420 (2002)
16. Xiao, Z., Liu, J., Luo, K., Huang, J., Fu, S.: Investigation of flows around a rudimentary landing gear with advanced detached-eddy-simulation approaches. AIAA J. **51**(1), 107–125 (2013)
17. Rhee, S.H., Koutsavdis, E.K.: Unsteady marine propulsor blade flow - A CFD validation with unstructured dynamic meshing. AIAA Paper 2003–3887 (2003)
18. De Stefano, G., Nejadmalayeri, A., Vasilyev, O.V.: Wall-resolved wavelet-based adaptive large-eddy simulation of bluff-body flows with variable thresholding. J. Fluid Mech. **788**, 303–336 (2016)
19. Wilcox, D.C.: Turbulence Modeling for CFD, 3rd edn. DCW Industries Inc., La Canada (2006)
20. Pavlenko, O.V., Chuban, A.V.: Numerical investigation of the hinge moments of the nose landing gear doors in a passenger aircraft in the process of opening. TsAGI Sci. J. **47**(5), 513–523 (2016)
21. Pavlenko, O.V., Chuban, A.V.: Determining hinge moments of the main landing gear fuselage door by means of numerical flow simulation. TsAGI Sci. J. **49**(7), 781–792 (2018)

CFD Prediction of Aircraft Control Surfaces Aerodynamics

Nunzio Natale[1]([⊠]) [iD], Teresa Salomone[1] [iD], Giuliano De Stefano[1] [iD], and Antonio Piccolo[2] [iD]

[1] Engineering Department, University of Campania Luigi Vanvitelli, 81031 Aversa, Italy
{nunzio.natale,teresa.salomone,giuliano.destefano}@unicampania.it
[2] Leonardo Aircraft Company, 80038 Pomigliano d'Arco, Italy
antonio.piccolo@leonardocompany.com

Abstract. The effectiveness of prototypical control surfaces for a modern regional transport commercial aircraft is examined by means of numerical simulations. The virtual experiments are performed in operational conditions by resolving the mean turbulent flow field around a suitable model of the whole aircraft. The Reynolds-averaged Navier-Stokes modelling approach is used, where the governing equations are solved with a finite volume-based numerical technique. The aerodynamic performance of the flight control surfaces, during an hypothetical conceptual design phase, is evaluated by conducting simulations at different deflections. The present computational modelling approach is verified to have good practical potential by making a comparison with reference industrial data.

Keywords: Computational Fluid Dynamics · Flight control surfaces · Industrial aerodynamics

1 Introduction

Recent progresses in meshing technology, convergence acceleration and calculation performance have allowed Computational Fluid Dynamics (CFD) methods to be intensively used to generate relevant aerodynamic databases, which are particularly useful in the pre-design process of new aircrafts [1,2]. Present trends in industrial aerodynamic design towards cost reduction for product development call for an accurate prediction of the control surfaces aerodynamics, in order to properly arrange the flight control system. The estimation of handling qualities and hinge moments induced by the deflection of the different control surfaces plays a key role in the aircraft sizing process, with a strong impact on the aircraft weight prediction. However, for flows around deployed control surfaces, the complex aerodynamics, as well as the importance of the flight envelope to be covered, makes the numerical simulations particularly challenging.

Differently from typical CFD studies that are performed by employing simplified generic wing-body configurations [3,4], in this study, the computational

© Springer Nature Switzerland AG 2020
O. Gervasi et al. (Eds.): ICCSA 2020, LNCS 12249, pp. 75–89, 2020.
https://doi.org/10.1007/978-3-030-58799-4_6

evaluation of the prototypical configuration of a real regional transport aircraft is performed, with a detailed analysis of the local aerodynamics of the flight control surfaces. The simulations allow to determine the variation of the aerodynamic loads, depending on the deflections of the control surfaces, and thus examine the effectiveness of the control process. Specifically, the effect of ailerons and elevators is studied at a typical flight condition, by means of a number of different calculations. The detailed air flow field is simulated by employing a fully three-dimensional CFD method, while following the Reynolds-averaged Navier-Stokes (RANS) approach. The numerical simulations are conducted using the solver ANSYS Fluent, which is commonly and successfully employed for building virtual wind tunnels in industrial aerodynamics research [5–8]. Indeed, this approach is followed in similar works investigating the aerodynamics of innovative regional aircraft models [9,10]. The feasibility of the proposed computational model for the simulation of complex realistic configurations is assessed through dedicated comparisons with analytical/experimental data provided by the Leonardo Aircraft industrial researchers.

2   Computational Modelling

The preliminary prediction of the global aerodynamic coefficients of interest is obtained using lower fidelity methods, by using either the Vortex Lattice Method (VLM) or a semi-empirical approach [11]. VLM solvers are ideally suited for the preliminary aircraft design environment, where they can be used to predict loads, stability and control data. Some aerodynamic characteristics as lift curve slope, induced drag and lift distribution are obtained through modelling the wing by means of horseshoe vortices distributed along both span and chord, while neglecting the effects of thickness and air viscosity [12].

Furthermore, the aerodynamics of the control surfaces is more accurately predicted by solving the RANS governing equations, which describe the mean turbulent flow field around the aircraft. In this work, the Spalart-Allmaras one-equation model is used, where the turbulence closure is achieved by solving an additional evolution equation for a modified eddy viscosity variable [13]. The turbulent boundary layer is modeled using a wall-function approach. This way, the wall boundary conditions are implemented so that relatively coarser meshes can be used in the calculations. The equations governing the RANS models, which are not reported here for brevity, can be found, for instance, in [14]. The RANS method is commonly used for the simulation of both compressible and incompressible aerodynamic flows, e.g. [15,16].

3   Geometric Models

3.1   Clean Configuration

Isolated Wing. Starting from the computer-aided design (CAD) model of the complete aircraft provided by the Leonardo Aircraft Company, a suitable

Table 1. Clean wings: geometrical data.

Geometrical parameter	Value
Wing area	A
Wing span (without winglets)	b
Wing span (with winglets)	$1.03\ b$
Mean aerodynamic chord, $\bar{c}$	$0.086\ b$
Taper ratio, λ	0.5
Horizontal tailplane area	$0.20\ A$
Horizontal tailplane span	$0.28\ b$
Vertical tailplane area	$0.25\ A$
Vertical tailplane span	$0.16\ b$

geometric sub-model for analysing the isolated wing aerodynamics is initially created. This clean wing model is employed for inviscid calculations conducted to test the in-house methodology that is developed to evaluate the spanwise loading. By removing some features, like nacelles and winglets, the proper comparison with VLM results can be made.

Complete Aircraft. The geometric model for the regional transport aircraft under investigation is comprised of fuselage, wing, empennage, and nacelle groups. Some important geometrical data of the aircraft model are given in Table 1. Here, the different geometrical parameters are expressed in non-dimensional form, by using the wing area A and the wing span b for normalization. The taper ratio λ is defined as the ratio between the tip chord and the root chord of the wing. Note that the actual reference values are not reported in the table, because they are not classified for publication.

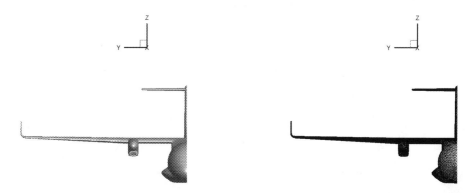

Fig. 1. Clean configuration: frontal views of CAD model (left) and surface mesh (right) for half the aircraft geometry.

A baseline simulation for the clean configuration is first performed. Practically, the gaps between the main aircraft components and the stowed control surfaces are not considered, while the projections of the surfaces edges are maintained, so that the different contributions of the different elements can be evaluated. A hybrid mesh with inflation layers is employed for all the CFD calculations. The numerical simulations for the clean configuration are carried out only for symmetric flow conditions and, thus, half the geometric model is used, while imposing proper symmetry boundary conditions. The FV mesh that is actually used corresponds to the finest mesh produced through a proper sensitivity analysis, being made up of about 7 million elements. Moreover, due to the wall function approach, a suitable first layer thickness is selected, which allows to properly model the boundary layer flow region. The mesh quality, as it holds for all the FV meshes used in this work, can be considered pretty good, especially considering the complexity of the geometrical model. In fact, the maximum skewness of the mesh is less than 0.9 for all the cases, which is within the prescribed limits. The same occurs for other mesh quality parameters, such as the aspect ratio and the orthogonal quality. The frontal view of the aircraft geometry for the clean configuration is presented in Fig. 1, along with the associated surface mesh. Note that the orientation of the reference body axes is such that the x-axis points backward, the y-axis points to the right wing, and the z-axis points upward.

Table 2. Control surfaces: geometrical data.

Geometrical parameter	Value
Aileron starting section, $y_{start}^{aileron}$	$\pm 0.725 \, b/2$
Aileron ending section, $y_{end}^{aileron}$	$\pm 0.992 \, b/2$
Aileron chord ratio, c_a/c	0.3
Elevator starting section, $y_{start}^{elevator}$	$\pm 0.010 \, b/2$
Elevator ending section, $y_{end}^{elevator}$	$\pm 0.277 \, b/2$
Elevator chord ratio, c_e/c	0.33
Horizontal tailplane incidence	i_h
Vertical distance	$z_h - z_w$
Horizontal distance	l_h
Moment calculation reference point	$0.25 \, \bar{c}$

3.2   Control Surfaces

The CAD models for the control surfaces are developed by exploiting the geometrical data provided by industrial aerodynamicists. The analysis is conducted by deflecting only one of the control surfaces, while maintaining the others with no

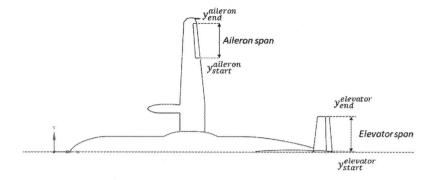

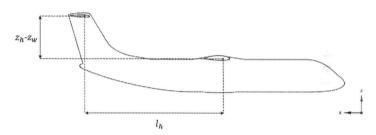

Fig. 2. Control surfaces: geometrical parameters definition.

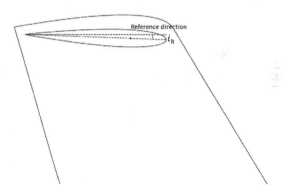

Fig. 3. Aircraft geometry: distances between horizontal tailplane and wing.

Fig. 4. Horizontal tailplane negative incidence.

deflection. An important aspect to be considered is regarding the gaps between the moving surfaces and the correspondent main components. For each control surface, the upper and lower gaps are closed, while the lateral gaps are kept. Some geometrical data associated to the aircraft control surfaces system are summarized in Table 2, and illustrated in Fig. 2. The characteristic high wing T-tail empennage configuration of the regional transport aircraft under study is illustrated in Fig. 3, where $z_h - z_w$ and l_h represent, respectively, the vertical and horizontal distances between the horizontal tailplane and the aircraft wing. The incidence of the horizontal tailplane is represented by i_h, which is the negative setting angle of the tailplane chord c_h with respect to the x–axis direction, as depicted in Fig. 4.

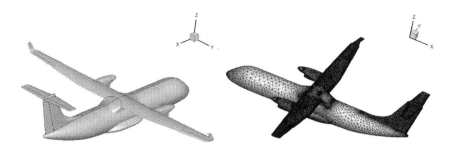

Fig. 5. Ailerons: perspective views of CAD model (left) and surface mesh (right) for the whole aircraft geometry.

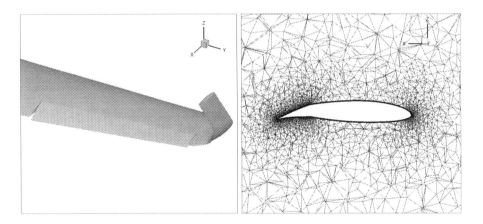

Fig. 6. Right aileron: CAD model (left) and plan view of the volume mesh at the aileron half span section (right).

Ailerons. The CAD model that is developed to study the aerodynamic effect of the ailerons is presented in Figure 5, along with the associated surface mesh. The mesh features are corresponding to those ones considered for the clean configuration, while imposing two opposite deflection angles for the two ailerons, which are $\delta_a = \pm 10°$. Since a symmetrical model can not be employed in this case, the mesh generation is simplified by separately considering the two specular computational sub-domains, while putting a suitable interface between them in

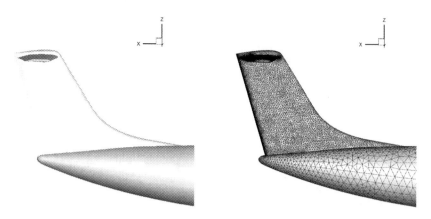

Fig. 7. Right elevator: side views of CAD model (left) and associated surface mesh (right).

the FV mesh used for the calculations. A close-up view of the right aileron is presented on the left side of Fig. 6, while a plan view of the volume mesh corresponding to the aileron half span section is shown on the right side of the same figure. It is worth noting the presence of the inflation layers in the wall region.

Elevators. The model for studying the aerodynamic effect of the elevators is fully consistent with the approach used above. The geometrical characteristics of the elevators are presented in Table 2. In this study, a symmetrical mesh can be used due to the fact that no simulation is performed with non zero sideslip angle. The CAD model for the right elevator is illustrated in Fig. 7, along with the associated surface mesh, for a given positive deflection angle that is $\delta_e = 10°$.

4   Results

The RANS simulations of the mean turbulent flow field around the aircraft model are conducted at Mach number $M_\infty = 0.2$, by employing the Spalart-Allmaras turbulence model, supplied with a wall function approach. The different computational meshes for the different cases are designed to show wall region resolutions that are consistent with the model requirements, as is controlled during the calculations.

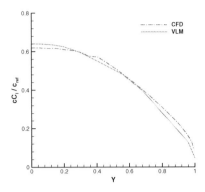

Fig. 8. Isolated wing: spanwise loading distribution for CFD and VLM methods.

4.1 Clean Configuration

Isolated Wing. The loading distribution along the spanwise direction is evaluated through the strip method, by developing a suitable custom field function. Namely, the wing is divided into a series of strips on which pressure and viscous based loads are computed in terms of the product cC_l, where c and C_l stand for the local chord length and the section lift coefficient. As initial verification, the spanwise loading distribution obtained by an inviscid CFD calculation is compared to the VLM-based solution, for a flight condition corresponding to a lift coefficient $C_L = 0.48$. As demonstrated in Fig. 8, where the chord length is normalized by the mean aerodynamic chord ($c_{ref} = \bar{c}$), the agreement between the two different solutions is fully acceptable. In this figure, as well as in the following, the abscissa $Y = y/(b/2)$ represents the normalized coordinate along the spanwise direction.

Complete Aircraft. The model of the complete aircraft in clean configuration is used for a number of numerical tests, whose results are exploited as reference for the following experiments with aerodynamic control surfaces deflected. In the following discussion, the main aerodynamic coefficients and the spanwise loading distributions are examined. In particular, the three different moment coefficients are defined as C_r (rolling), C_m (pitching), and C_n (yawing), where it is assumed $C_r > 0$ when the rolling tends to put the right wing down, $C_m > 0$ when the pitching tends to put the aircraft nose up, and $C_n > 0$ when the yawing tends to put the aircraft nose right. Note that, given the body axes reference introduced above, it practically holds $C_r = -C_{M_x}$, $C_m = C_{M_y}$, and $C_n = -C_{M_z}$, where M_x, M_y and M_z are the moment components along the thee different body-axis directions. The reference point for the moment evaluation is set coincident with the point at one quarter of the mean aerodynamic chord.

A number of steady simulations are conducted at a given flight condition, which is expressed in terms of Mach number and Reynolds number that is $Re_\infty = 11.8 \times 10^6$. The aerodynamic loadings on the aircraft are obtained by

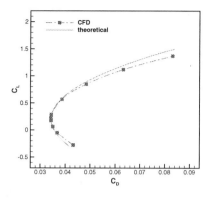

Fig. 9. Clean configuration: drag polar.

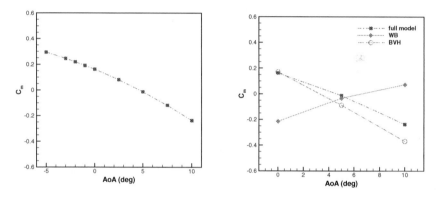

Fig. 10. Clean configuration: pitching moment coefficient (left) and corresponding breakdown (right).

integrating the predicted surfaces forces. The drag polar for the clean configuration is illustrated in Fig. 9, compared with the theoretical parabolic polar, that is

$$C_D = C_{D0} + (C_L - C'_L)^2/(e\pi AR), \tag{1}$$

where C_{D0} stands for the minimum drag coefficient, AR is the aspect ratio and e is the Oswald factor. The present CFD calculations provide C_{D0} equal to 340 drag counts, corresponding to a lift coefficient C'_L of about 0.2. The parabolic curve fitting the numerical data leads to estimate $e = 0.9$ for the Oswald factor. Apparently, the lift coefficient is well predicted for a reasonable operative range of the AoA parameter. The pitching moment coefficient is reported on the left side of Fig. 10, where the negative slope obeys the longitudinal stability criteria. The neutral point has been predicted at 53% of the mean aerodynamic chord, which is consistent with experimental data [17]. The increased C_m slope at high AoA is due to the pendular stability, as is typical for high wing aircrafts [18]. The breakdown of the various contributions to the pitching moment is illustrated on the right side of Fig. 10. The addition of the vertical tail to the wing-body (WB) combination does not affect C_L and C_m, whereas longitudinal stability is introduced by the horizontal tailplane. The body-tailplanes configuration (BVH) is stable but does not produce useful lift. The complete aircraft configuration has the highest C_L slope due to the additional lift provided by the horizontal tailplane, yet it represents an untrimmed configuration.

4.2 Control Surfaces

In the following, in order to show the capability of the proposed computational modelling, the aerodynamic effect of ailerons and elevators deflection is examined. The results are presented and discussed for some suitable deflection angles.

Ailerons. The aerodynamic effect of the ailerons is presented for deflection angles equal to $\delta_a = \pm 10°$, with the right aileron deflected downward. The predicted spanwise loading distribution is presented in Fig. 11, where the loading is calculated up to the sections immediately before the winglet ($Y = \pm 1$). When making a comparison with the result of the baseline simulation with no deflection, the influence of the ailerons is apparent. In Fig. 12, the rolling moment coefficient variation with the lift coefficient is reported, compared to VLM results. A loss in ailerons effectiveness for $C_L > 0.9$ is observed in the CFD solution, while the VLM method is not able to predict it. The corresponding drag breakdown is illustrated in Fig. 13, where each contribution is expressed in percentage to the total value. Finally, the adverse yaw effect is quantified by the derivative $C_{n\delta_a}$, which is equal to $2.6 \times 10^{-3}\ rad^{-1}$ at zero AoA and rises up to $1.6 \times 10^{-2}\ rad^{-1}$ for an AoA of 7.5°.

Elevators. The aerodynamic effect of the elevators is presented for two different deflection angles that are $\delta_e = \pm 10°$. Figure 14 shows the spanwise loading distribution for the wing and the horizontal tailplane, for both deflections. The presence of the vertical tailplane moves the maximum loading slightly away from the symmetry plane. In particular, the interaction seems to be stronger on the lower surface of the horizontal tailplane. In Fig. 15, the spanwise loading distribution on the horizontal tailplane with elevator deflected, compared with the baseline solution with no deflection, is reported. The corresponding pitching moment coefficient plots are presented on the left side of Fig. 16. Apparently, the horizontal tailplane provides a negative lift when the elevator is not deflected, due to its initial negative incidence with respect to the aircraft centerline ($i_h < 0$) and the downwash effect from the wing. Deflecting the elevator by $-10°$ increases the absolute value of the negative lift. On the other hand, deflecting the elevator by $+10°$ provides a positive lift, that is in absolute value lower than the previous case. The drag polar obtained for $\delta_e = -10°$ is shown on the right side of Fig. 16. It is worth noting that the negative deflection of the elevator leads to a relevant increment of 78 drag counts for C_{D0}. By looking at the drag breakdown, which is reported in Fig. 17, it is interesting to note

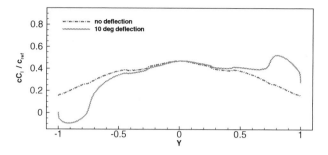

Fig. 11. Ailerons: spanwise wing loading at zero AoA with, and without ailerons deflection.

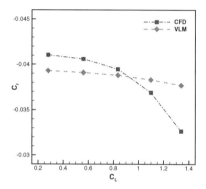

Fig. 12. Ailerons: rolling moment versus lift coefficient.

that deflecting the elevator by $-10°$ (up) leads to higher drag and influences the vertical tailplane as well. This is due to the fact that the elevator leads to higher (negative) load in this case. The pitching moment coefficient variation for the deflection angle $\delta_e = +10°$ is equal to $\Delta C_m = -0.536$, which is comparable to the value obtained using a semi-empirical method that is -0.61 [11]. When making the same analysis for the lift coefficient increment, at zero AoA, it holds $\Delta C_L = 0.109$, while the semi-empirical method providing a value of 0.1. For all the aerodynamic derivatives calculated in the present study, a remarkable matching was found with respect to the values generated by Leonardo Aircraft aerodynamicists. These data were provided as restricted classified data and are not available for public diffusion. Finally, as far as the comparison with the VLM solution is concerned, the discrepancy of the results is large, likely due to differences in the geometries and/or meshes employed. This was also found in similar studies [19].

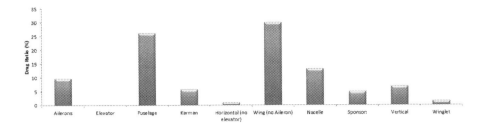

Fig. 13. Ailerons: aircraft drag breakdown.

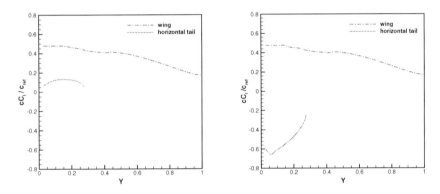

Fig. 14. Elevators: spanwise loading on wing and horizontal tailplane, for $\delta_e = +10°$ (left) and $-10°$ (right).

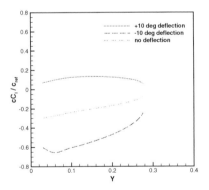

Fig. 15. Elevators: spanwise loading on horizontal tailplane for $\delta_e = +10°$ and $-10°$, compared to clean configuration.

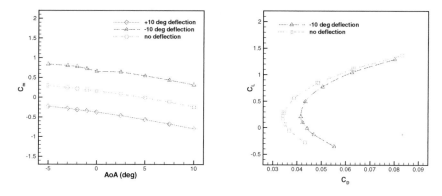

Fig. 16. Elevators: pitching moment coefficient (left) and drag polar (right) for $\delta_e = \pm 10°$, compared to clean configuration.

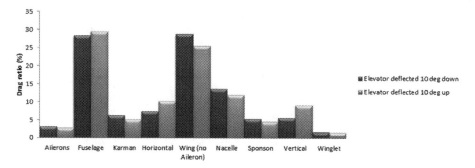

Fig. 17. Elevators: drag breakdown at zero AoA.

5   Conclusions

The present study has to be intended as the proof-of-concept, namely, the preliminary development of a CFD based prediction tool in the aerodynamic design of the control surfaces system of a regional transport aircraft. Differently from similar studies, where the CFD results were obtained for generic geometries, this work deals with simplified models directly derived from real geometries provided by an aeronautical manufacturer that is Leonardo Aircraft Company. Several calculations have been performed with the aim of demonstrating the practical potential of the proposed methodology for the prediction of the control surfaces aerodynamics. The analysis is conducted by performing calculations at a given flight condition and for different deflection angles of ailerons and elevators, while examining the modified aerodynamic force and moment coefficients. It is demonstrated that a complete aerodynamic CFD database can be generated, given the prototypical geometry of the control surfaces, in the pre-design process of new aircrafts. The acceptable agreement with the empirical data made available by the industrial partner has been achieved. There remains the possibility of developing more sophisticated computational models for particular industrial applications, depending on the level of accuracy that is required and the available computational resources.

Acknowledgements. This work was supported by the Italian Regione Campania under the research project SCAVIR (POR Campania FESR 2014/2020), presented by the Campania Aerospace Technological District (DAC). This support is gratefully acknowledged. The authors would like to thank the Leonardo Aircraft Company for providing the CAD models and the design reference data.

References

1. Johnson, F.T., Tinoco, E.N., Yu, N.J.: Thirty years of development and application of CFD at Boeing Commercial Airplanes. Comput. Fluids **34**, 1115–1151 (2005)
2. Spalart, P.R., Venkatakrishnan, V.: On the role and challenges of CFD in the aerospace industry. Aeronaut. J. **120**(1223), 209–232 (2016)
3. Fillola, G., Le Pape, M.-C., Montagnac, M.: Numerical simulations around wing control surfaces. In: 24th International Congress of the Aeronautical Sciences, Yokohama, Japan, 29 August–3 September 2004 (2004)
4. Mertins, R., Elsholz, E., Barakat, S., Colak, B.: 3D viscous flow analysis on wing-body-aileron-spoiler configurations. Aerosp. Sci. Technol. **9**, 476–484 (2005)
5. Escobar, J.A., Suarez, C.A., Silva, C., López, O.D., Velandia, J.S., Lara, C.A.: Detached-eddy simulation of a wide-body commercial aircraft in high-lift configuration. J. Aircr. **52**(4), 1112–1121 (2015)
6. Reina, G.P., De Stefano, G.: Computational evaluation of wind loads on sun-tracking ground-mounted photovoltaic panel arrays. J. Wind Eng. Ind. Aerodyn. **170**, 283–293 (2017)
7. Rapagnani, D., Buompane, R., Di Leva, A., et al.: A supersonic jet target for the cross section measurement of the $12C(\alpha, \gamma)16O$ reaction with the recoil mass separator ERNA. Nucl. Instrum. Methods Phys. Res. B **407**, 217–221 (2017)
8. Benaouali, A., Kachel, S.: Multidisciplinary design optimization of aircraft wing using commercial software integration. Aerosp. Sci. Technol. **92**, 766–776 (2019)
9. Di Marco, A., et al.: Aerodynamic and aeroacoustic investigation of an innovative regional turboprop scaled model: numerical simulations and experiments. CEAS Aeronaut. J. 1–16 (2020)
10. De Stefano, G., Natale, N., Reina, G.P., Piccolo, A.: Computational evaluation of aerodynamic loading on retractable landing-gears. Aerospace **7**(6), 68 (2020)
11. Roskam, J.: Airplane design part VI: preliminary calculation of aerodynamic, thrust and power characteristics. KS, DAR Corporation, Lawrence (2012)
12. Deyoung, J.: Historical evolution of Vortex-Lattice Methods. NASA Langley Res. Center Vortex-Lattice Utilization, 19760021076, pp. 1–9 (1976)
13. Spalart, P.R., Allmaras, S.R.: A one-equation turbulence model for aerodynamic flows. Recherche Aerospatiale **1**, 5–21 (1994)
14. Wilcox, D.C.: Turbulence Modeling for CFD, 3rd edn. DCW Industries Inc., La Canada (2006)
15. Ge, X., Vasilyev, O.V., De Stefano, G., Hussaini, M.Y.: Wavelet-based adaptive unsteady Reynolds-averaged Navier-Stokes simulations of wall-bounded compressible turbulent flows. AIAA J. **58**(4), 1529–1549 (2020)
16. De Stefano, G., Vasilyev, O.V., Brown-Dymkoski, E.: Wavelet-based adaptive unsteady Reynolds-averaged turbulence modelling of external flows. J. Fluid Mech. **837**, 765–787 (2018)

17. Nicolosi, F., Ciliberti, D., Della Vecchia, P., Corcione, S.: Wind tunnel testing of a generic regional turboprop aircraft modular model and development of improved design guidelines. In: AIAA Aviation Forum, Applied Aerodynamics Conference, Atlanta, GA, 25–29 June 2018 (2018)
18. Perkins, C.D., Hage, R.E.: Airplane Performance Stability and Control. Wiley, Hoboken (1949)
19. Mialon, B., Khrabrov, A., Khelil, S.B., et al.: Validation of numerical prediction of dynamic derivatives: the DLR-F12 and the Transcruiser test cases. Prog. Aerosp. Sci. **47**, 674–694 (2011)

Optimal Online Electric Vehicle Charging Scheduling in Unbalanced Three-Phase Power System

Imene Zaidi[1]([⊠])[iD], Ammar Oulamara[2][iD], Lhassane Idoumghar[1][iD], and Michel Basset[1][iD]

[1] Université de Haute-Alsace, IRIMAS UR 7499, 68100 Mulhouse, France
{imene.zaidi,lhassane.idoumghar,michel.basset}@uha.fr
[2] LORIA Laboratory, Université de Lorraine, UMR7503, 54506 Vandoeuvre-lès-Nancy, France
ammar.oulamara@loria.fr

Abstract. This paper studies the electric vehicle (EV) charging scheduling problem where EV arrive at random unknown instants during the day with different charging demands and departure times. We consider single-phase charging EV in a three-phase charging station designed such that each EV has its own parking space. The objective is to build a real-time schedule that minimizes the total tardiness subject to the technical constraints of the charging station. We consider preemptive as well as non-preemptive EV charging. A mixed-integer linear programming (MILP) model is formulated for the offline problem. To solve the online problem, we propose heuristics based on the priority rule. Further, a local search is implemented to improve the objective value of the preemptive EV charging. Simulation results show that the proposed solving approaches outperform the existing heuristics developed in the literature. Moreover, we show that total tardiness is significantly reduced when preemption is exploited.

Keywords: Electric vehicle · Smart charging · Online scheduling · Heuristics · Local search

1 Introduction

The increasing environmental awareness led many governments to establish policies encouraging the adaptation of sustainable and clean energy [14]. Thus, electric vehicles (EV) are seen as promising green technology. In 2018, [6] reported that the worldwide number of EV has surpassed 5.1 million which is double the number of EV in the previous year. Nevertheless, the EV is still less adopted than the internal combustion engine vehicles mostly since battery charging is time-consuming. On the other hand, a high amount of electrical energy drawn from the power grid is needed to meet EV batteries charging demands, especially for fast charging. An excessive electricity demand caused by EV charging will create

© Springer Nature Switzerland AG 2020
O. Gervasi et al. (Eds.): ICCSA 2020, LNCS 12249, pp. 90–106, 2020.
https://doi.org/10.1007/978-3-030-58799-4_7

an extra significant load affecting the power grid stability. Many researchers concluded that uncoordinated EV charging will result in undesirable peaks in the electrical consumption, increases in power losses and voltage deviation [2,13]. Moreover, uncontrolled single-phase EV charging in a three-phase distribution network will cause a higher load imbalance between the phases and a significant increase in neutral current [7,10,12]. Therefore, optimally scheduling the EV charging demands is needed to mitigate these negative impacts and increase EV drivers' satisfaction. Several research studies have been conducted to tackle the EV charging scheduling problem. As an optimization problem, several objective functions were proposed from economical perspective, such as minimizing the electricity costs [19] as well as from technical perspectives such as minimizing the power losses [2,8] and voltage deviation [8] in power grid. Ideally, an optimal charging schedule can achieve these objectives more efficiently if the information about future charging demands were available when the schedule is computed. However, the charging station only knows the demand for already arrived EV. Consequently, an online algorithm is needed to solve the real-time charging scheduling problem.

In this paper, we address the online EV charging scheduling presented in [4]. The proposed optimization methods aim to build an online optimal schedule that minimizes the total tardiness of the charging demands while maintaining the physical constraints of the three-phased charging station. Also, our work investigates the impact of relaxing the non-preemption constraint of charging operations.

The remainder of the paper is organized as follows: in Sect. 2, we briefly review the main works on electric vehicle charging problems; Sect. 3 describes the problem considered by detailing the charging station model and formulating the problem as MILP. In Sect. 4, our proposed optimization methods are presented. Section 5 summarizes the simulation results. Conclusion and future works are given in Sect. 6.

2 Related Work

One of the major challenges in EV charging scheduling is the uncertainly in their charging patterns. In real-life settings, the number of future demands as well as the information about it is not available in advance. Therefore, multiple online and real-time optimization algorithms were established in the literature to address these issues. [17] proposed a dynamic resource allocation system to determine the charging schedule for EV in parking lots where the arrivals of EV can be random or with an appointment. The objective is to minimize the electricity cost bought by the parking service operator from the power grid. The schedule is calculated at the beginning of each 30 min after updating the EV arrival information and charging statues of arrived EV. [19] proposed a charging scheme on a real-time basis to manage the EV charging and incorporated demand response schemes in the parking station. The proposed charging scheduling scheme is designed for the operator of the parking station. It aims to minimize the electricity cost and maximize the number of EV charged simultaneously at each scheduling period through a maximization scheme. The problem is formulated as a 0–1 integer linear programming problem which was approximated via a modified convex relaxation method. A hierarchical control scheme for EV charging across multiple aggregators was proposed in [18] to minimize the electricity cost and system peak load. Each aggregator's charging curve is first solved at the distribution system operator (DSO) level, followed by the power allocation of each EV using a heuristic algorithm. [11] proposes an online adaptive EV charging scheduling framework for utility companies. The objective of the scheduling is to minimize the total electricity cost of EV charging demands subject to battery demand constraint and distribution grid constraints. A MILP is formulated to tackle this problem. The scheduling optimization is solved at each time slot when it detects new arrivals of EV. Furthermore, a three-phase DC MILP optimization strategy is developed since the first optimization can only be applied in single-line systems.

3   Problem Description

3.1   Charging Station Model

Our study is concerned with the EV charging scheduling problem in a charging station designed to be installed as a large public parking or a collective garage as it is described in [15]. This station is fed with a three-phase current power source. Thus, there are three conductors, each carries an alternating current of the same frequency and voltage amplitude from the source to the electrical outlets. These conductors are called lines and each line regroups a number of power outlets where EV can be plugged into for charging. This means that each connected EV is considered as a single-phase load. However, two constraints limit the number of outlets that can deliver power simultaneously. The first constraint is related to the maximum power that can be drawn from each line so that system overload can be avoided. The second constraint maintains the load balance between any two lines. In fact, in three phase power system, the load should be distributed equally between the three lines to minimize power losses and improve the system efficiency.

The design adopted for the charging station is based on master-slave architecture. Figure 1 illustrates the design of such architecture. Each slave commend the switching on or off of two power outlets. It also records the arrival time of EV and communicate it to its master. The masters has a user interface where the EV drivers enter their departure time and the desired energy demand. All these data are communicated to the central server where a scheduler of the EV charging demands is implemented.

To simplify the operating model of the charging station, we make the same assumption as in [4,5] where each EV driver has his own parking place so he can plug his vehicle into the outlet at any time. He also provides the parking duration as well as the charging demand through the user interface. Therefore, there is no queue and the driver does not have to wait before plugging his EV.

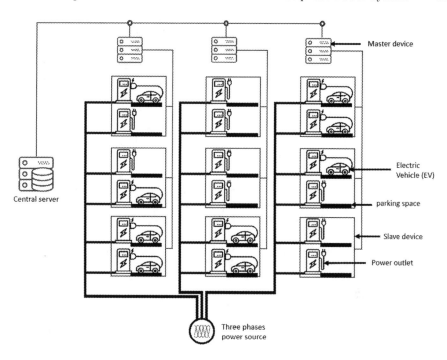

Fig. 1. Scheme of the considered charging station.

However, the EV does not necessarily start recharging immediately and once it does, he cannot unplug it from the power outlet before completing its charging according to the charging demand. It is also assumed that the charging rate is constant and it is the same for all the charging outlets.

3.2   The Charging Scheduling Problem

By controlling the switching on and off of the power outlets, we aim to schedule the charging operations among all the EV plugged in such that the entire charging load in each line does not exceed its capacity. Besides, the maximum imbalanced load between each two lines must be maintained. Since the completion time of each charging operation can exceed the provided departure time, the objective function will be to minimize the total tardiness.

The charging scheduling horizon H starts at 00:00 h for 24 h and is divided into regular intervals of τ which is equal to 6 min in our case. Each EV has its own parking space and arrives at the station at random instants. This means that the arrival time, charging time and departure time are prior unknown before the arrival of the EV. As a result, the schedule must be built iteratively. In literature, there are two main strategies to handle this uncertainty: solving the scheduling problem whenever a new EV is connected or at each time slot. We adopt the second strategy since the time slot we defined is relatively small and it prevents the system from collapsing when a lot of EV arrive in a very short period.

As an initial case study, we consider the non-preemptive scheduling as in [4,5], where the charging of an EV cannot be interrupted until it completes charging i.e the power outlet that was switched on to deliver power to the EV cannot be switched off before the completion of the charging. Thus, the scheduling problem will be to assign a starting time of charging for each arrived EV. Then, we investigate the case study where preemption is allowed. In this case, the charging of EV can be interrupted and another EV will be charged instead. The amount of energy for a preempted EV charging is not lost. When a preempted charging is afterward resumed, it only needs power for its remaining charging time. Such a recharging strategy is highly recommended in smart charging [9]. Moreover, the open charge point protocol (OCPP) currently integrates these operations without human intervention [16].

3.3   A MILP Formulation for the Offline Problem

We first formulate the offline problem as a mixed-integer linear problem (MILP). The indices, parameters, and decision variables are described as follow:

Indexes and Sets. The number of lines $L = 3$. There are N power outlets in the three lines indexed as follow:

- n_1 outlets in line L_1 are indexed $1, \ldots n_1$.
- n_2 outlets in line L_2 are indexed $n_1 + 1, \ldots, n_1 + n_2$.
- n_3 outlets in line L_3 are indexed $n_1 + n_2 + 1, \ldots, n_1 + n_2 + n_3 = N$.
- t time slot index of the horizon H.

Parameters

- r_i the arrival time of EV i
- p_i the charging demand of EV i. As the charging rate is constant, knowing the energy demand p_i of EV i, we can deduce the charging time needed by each EV i. Thus, in the rest of the paper, we consider p_i as the charging time instead of energy demand.
- d_i the desired departure time of EV i
- $\tilde{N}$ the maximum of power outlets that can deliver power simultaneously in any line at any time.
- Δ a parameter that establishes the balance between any two lines, $\Delta \in [0, 1]$

Decision Variables

- $x_{i,t}$ equal to 1 if EV i is charging at the time slot t
- N_k^t the number of EV being charged in line k at time slot t
- T_i the tardiness of EV i, $T_i = \max(0, C_i - d_i)$ where C_i represents the completion time of charging operation i.

Model

– Objective: minimize the total tardiness

$$\sum_{i=1}^{n_1+n_2+n_3} T_i$$

– Constraints:

$$\sum_{t=r_i}^{H} \tau x_{i,t} \geq p_i \qquad \forall i \tag{1}$$

Constraints 1 ensure that EV i charge to its desired charging time p_i after its arrival time r_i.

$$(t+1) \times \tau \times x_{i,t} - d_i \leq T_i \qquad \forall i, t \tag{2}$$

Constraints (2) calculate the tardiness of charging EV i.

$$\tau p_i (x_{i,t} - x_{i,t-1}) \leq \sum_{l=t}^{t+p_i/\tau} \tau x_{i,l} \qquad \forall i, t \tag{3}$$

Constraints (3) ensure the non-preemption of charging in case of non-preemptive scheduling.

$$N_1^t = \sum_{i=1}^{n_1} x_{i,t} \qquad \forall t \tag{4}$$

$$N_2^t = \sum_{i=n_1+1}^{n_1+n_2} x_{i,t} \qquad \forall t \tag{5}$$

$$N_3^t = \sum_{i=n_1+n_2+1}^{n_1+n_2+n_3} x_{i,t} \qquad \forall t \tag{6}$$

Constraints (4), (5) and (6) calculate the number of EV that are charging at the same time at time slot t in lines 1, 2 and 3 respectively.

$$N_j^t \leq \tilde{N} \qquad \forall t, j = 1, 2, 3 \tag{7}$$

Constraints (7) are related to the maximum power that can be drawn from each line. $\tilde{N}$ restricts the number of EV that can be charging simultaneously in each line. Since each outlet delivers power at the same constant rate, the power delivered by each line can be expressed by the number of active outlets.

$$N_j^t - N_k^t \leq \Delta\tilde{N} \qquad \forall t, j \neq k, k, j = 1, 2, 3 \tag{8}$$

$$-N_j^t + N_k^t \leq \Delta\tilde{N} \qquad \forall t, j \neq k, k, j = 1, 2, 3 \tag{9}$$

Constraints (8) and (9) establish the maximum imbalance between any two lines such as the difference between the numbers of EV in any two lines does not exceed $\Delta\tilde{N}$ with $\Delta \in [0, 1]$.

4   Online Optimization Algorithms

Although the MILP model has been developed, finding the optimal schedule with an exact method cannot be done in polynomial time. In the simple case where we have only one line, the problem is equivalent to scheduling jobs on parallel machine $P|r_i| \sum_i T_i$ following the $\alpha|\beta|\gamma$ notation [3] which is NP-Hard. Furthermore, the problem of EV charging scheduling is a dynamic and real-time optimization problem that requires an online high-speed optimization method. Hence, we propose a heuristic to solve the preemptive charging scheduling problem. The heuristic is based on the prtt (Priority Rule for Total Tardiness criterion) dispatching rule used in [1]. To solve the preemptive problem, the same dispatching rule is used. Additionally, we try to improve results by applying a local search procedure.

Algorithm 1: prtt based pseudo-algorithm for non preemptive charging scheduling

---

input : r_i, p_i, d_i for each vehicle, Δ, $\tilde{N}$
output: Total tardiness $\sum T_i$
for *each time slot t* **do**
    for *each line $j = 1, .., L$* **do**
        get the set I' of ready EV in the line j i.e EV that have arrived during the interval $[0, t]$;
    end
    if I' *is not empty* **then**
        For each ready EV we calculate its prtt then we added to the set I_{prtt} ;
        Sort the set I_{prtt} in increasing order of prtt values ;
        Calculate N' the number of EV that can be added to the schedule at time t in line j without breaking the constraints (7), (8) and (9);
        for *each EV i in the first N' EV in I_{prtt}* **do**
            Set the starting time of the EV to t if it doesn't beak the constraints (7), (8 and (9) for the next time slots ie.
            $|N_j^{t'} + 1 - N_k^{t'}| \leq \Delta N$ for t' from t to $t + p_i$, $k = 1, 2, 3k \neq j$
            if the EV i is assigned, update $N_j^{t'}$ for the next time slots t' starting from t to $t + p_i$;
            Calculate the tardiness of the assigned job $T_i = \max(0, t + p_i - d_i)$;
            if the assignment breaks a previous balance constraint redo the time slot t;
        end
    end
end
return $\sum T_i$

4.1   Non-preemptive Scheduling Algorithm

Heuristics based on priority dispatching rules have been widely used in the literature to find near-optimal solutions for NP-Hard scheduling problems since they are simple methods and requires less computation time than sophisticated meta-heuristics. This makes them adequate for real-time and dynamic problems.

We adapt the prtt rule proposed in [1] to the EV scheduling problem presented in previous section. The prtt of a charging operation i at time t is defined by:

$$prtt_i^t = \max(t, r_i) + \max(d_i, max(t, r_i) + p_i) \tag{10}$$

The charging operations are afterward scheduled at each step in ascending order of their prtt values. Since the schedule is built at each time slot considering only the ready EV (*i.e* $r_i \leq t$) without information on the future arrivals, the term $max(r_i, t)$ will always be equal to t. As a result a prtt of EV i is calculated as following:

$$prtt_i^t = \max(d_i, t + p_i) \tag{11}$$

Once an EV starts charging, no interruption is allowed until completion time. Thus, the objective is to assign a starting time st_i for each EV without breaking the constraints (7), (8) and (9). The pseudo-code of the non preemptive charging scheduling is shown in Algorithm 1. At each time slot, for each line, the prtt of ready and not assigned EV are calculated using the equation (11) and they are ordered in increasing order. Then, we calculate N' the number of EV that can be added to the schedule at time t in line j without breaking the constraints (7), (8) and (9). The value of N' can be obtained by:

$$N' = \min(\min_{k=1,..,L, k\neq j}(\Delta \tilde{N} - N_j^t + N_k^t), \tilde{N} - N_j^t) \tag{12}$$

The starting time of the first N' EV that won't break the constraints (7), (8) and (9) in the next time slots is assigned to t. We improve the assignments of EV at the end of each time slot in case that an assignment of job breaks a previous imbalance constraint. This happens when the assignment of an EV i in line j to a starting time t verify that the assignment of another EV i' in line k won't break the imbalance constraints (8) and (9) at time t.

$$\min\{|N_k^t + 1 - N_j^t| - \Delta \tilde{N}; k \in 1, 2, 3, k \neq j\} \leq 0 \tag{13}$$

In this case, we redo the scheduling for the current time slot t to assign this EV.

4.2   Preemptive Scheduling Algorithms

Allowing preemption in scheduling problems is a commonly used relaxation to make the problem easier to solve. Especially in our case where the schedule is built iteratively which makes the decisions of charging some EV at a given time unchangeable even if another EV with higher priority to charge arrives. A high priority EV charging has a smaller prtt value at a given time. Furthermore, preemptive charging is allowed with the new generation of power outlets that use OCPP protocol in which charging service operators can orchestrate the charging of EV.

Heuristic. Unlike the non-preemptive scheduling heuristic proposed, we schedule the charging of the arrived EV at the current time slot only. No charging schedule is built for the next time slots. We calculate the prtt of already and unfinished charging operations at the beginning of each time slot. Then, we schedule the charging operations with smaller prtt values that verify the constraints (7), (8) and (9) for this time slot only.

We modify the prtt function to take into account only the remaining charging time of an EV instead of the whole charging time. The pseudo-code of the preemptive charging scheduling is shown in Algorithm 2.

Algorithm 2: prtt based pseudo-algorithm for preemptive online charging scheduling

---

input : r_i, p_i, d_i for each vehicle, Δ, N
output: Total tardiness $\sum T_i$
// p_i' is the remaining charging time for each EV;
$p_i' \leftarrow p_i$ for each vehicle ;
for *each time slot t* **do**

    for *each line $j = 1, .., L$* **do**

        get the set I' of ready EV in the line j i.e EV that have arrived during the interval $[0, t]$ and didn't complete charging ;

    end

    if *I' is not empty* **then**

        For each ready EV calculate its prtt with p_i' then added it to the set I_{prtt} ;

        Sort the set I_{prtt} in increasing order ;

        Calculate N' the number of EV that can be added to the schedule in line j at time t without breaking the constraints (7), (8) and (9);

        for *the first N' EV in I_{prtt}* **do**

            $N_j^t \leftarrow N_j^t + 1$;

            $p_i' \leftarrow p_i' - 1$;

            if $p_i' = 0$ **then**

                calculate the tardiness of the assigned job

                $T_i = \max(0, t - d_i)$;

            end

            if the assignment breaks a previous balance constraint redo the time slot t;

        end

    end

end
return $\sum T_i$

Local Search. In the previous section, a heuristic based on the prtt dispatching rule is used to obtain the preemptive charging scheduling. Here, further, we propose a local search algorithm to explore solutions with better tardiness. The pseudo algorithm is shown in Algorithm 3.

Algorithm 3: Hill-Climbing Pseudo-Algorithm for preemptive online charging scheduling

---

input : r_i, p_i, d_i for each vehicle, Δ, N
output: Total tardiness $\sum T_i$
initialize the solution S;
for *each time slot t* **do**

    for *each line $j = 1, .., L$* **do**

        get the set I' of ready EV in the line j i.e EV that have arrived during the interval $[0, t]$;

    end

    if *I' is not empty* **then**

        For each ready EV, calculate its prtt with eq (10) then added it to the set I_{prtt} ;

        Sort the set I_{prtt} in increasing order ;

        for *each EV i in I_{prtt}* **do**

            $t' = t$;

            $p'_i = p_i$;

            while $t' \leq H$ *and* $p'_i > 0$ **do**

                if *a job can be added to the schedule at time t' in line j without breaking the constraints (7), (8) and (9)* **then**

                    Add the EV i to the list S^t_j;

                    Decrements p'_i ;

                else

                    shift right a job i' with $prtt(i') > prtt(i)$;

                end

                $t' \leftarrow t' + 1$

            end

        end

    end

    while *not stagnation or time limit not reached* **do**

        $S' \leftarrow$ neighbor of S ;

        if $S' = S$ **then**

            stagnation $\leftarrow$ True

        else

            $S \leftarrow S'$;

        end

    end

end
return $\sum T_i$

At each time slot t, a partial schedule is built using the prtt based heuristic. At this time, we schedule the arrived EV in the time horizon H allowing preemption. Then, a hill-climbing strategy is used to improve the partial schedule. Although this procedure can improve the current partial schedule at the considered time t, it doesn't necessarily improve the whole final schedule. Given that we are in a real-time setting, we can only change the scheduling of EV at the next time slots. Changes cannot be made for previous time slots t' such $t' < t$.

The fundamental part of any local search method is the neighborhood structure. A neighbor is generated by moving a charging operation from a time slot to another time slot. We consider moves that only can improve the current partial schedule. Since the problem is constrained, we have to maintain the feasibility of the schedule each time we need to move a charging operation.

For the description of the neighbor generation moves, we call a time slot t a "source" if we can remove a charging operation from it without breaking the balance constraint (8) and (9). While a time slot t is called a "target" if adding a charging operation at t doesn't exceed its capacity N and doesn't break constraints (8) and (9).

Shift right: a neighbor in shift right is generated by moving a random a charging operation of an EV i in a selected line j from a random source time slot t_1 to a random target time t_2, $t_2 > t_1$. Since we consider only moves that can improve a partial schedule, t_2 should not be greater than its departure time $t_2 \leq d_i$. We guide the search by shifting the charging operations that have a high prtt first so it can be replaced by other charging operations with lower prtt.

Shift left: a neighbor in shift left is generated by moving a charging operation of an EV i in a selected line j from a random source time slot t_1 to a random target time $t_2 < t_1$ such that $t_2 \geq r_i$. A variant of shift left is implemented considering the tardy EV ($T_j > 0$). Thus, we shift left the charging operation at its completion time slot.

Shift right on three lines: a neighbor in shift right on 3 lines is generated by moving three random charging operations i_1, i_2, i_3 from each line $j = 1, 2, 3$ from a random time slot t_1 to a random time t_2 with $t_2 > t_1$ and $N_j^{t_2} + 1 \leq N$. By moving a charging operation per line, we make sure that the balance constraints (8) and (9) are maintained.

Shift left on three lines: a neighbor in shift left on 3 lines is generated by moving three random charging operations i_1, i_2, i_3 of each line $j = 1, 2, 3$ at a random time slot t_1 to a random time t_2 with $t_2 < t_1$, $N_j^{t_2} + 1 \leq N$ and $t_2 \leq \min(r_{i_1}, r_{i_2}, r_{i_3})$.

A local search algorithm starts by initializing an empty solution. A solution S is a feasible partial schedule that describes the set of charging operation scheduled in any line at each time slot $t \in H$. Then, at each time slot t, for each line j, a partial solution S is built by scheduling preemptively the charging operation of arrived EV according to their prtt values in the time slots t' such $t' \geq t$. A shift right move is considered when we have a charging operation of an EV i that has a lower prtt value than an already scheduled charging operation at the time slot t'. After scheduling the charging operations in each line, we start exploring the neighborhood of the solution S. A neighbor solution S' is generated using one of the moves described before. S' will replace S if the total tardiness of S' is less or equal of the total tardiness of S. Since we are in real-time setting, the stopping criterion will be either a given limited time that doesn't exceed τ the length of a time slot or that we cannot generate new solutions using one of the moves.

5 Simulation Results

Table 1. Comparison of results of the first scenario.

$\tilde{N}$	Δ	prtt	prtt pmtn	HC pmtn
Type 1 instances				
20	0,2	10,53%	−35,07%	−39,69%
20	0,4	−3,53%	−26,46%	−31,53%
20	0,6	−3,59%	−27,09%	−31,65%
20	0,8	−3,79%	−27,01%	−31,42%
30	0,2	6,57%	−56,56%	−60,92%
30	0,4	−2,52%	−68,02%	−82,03%
30	0,6	8,30%	−67,63%	−84,94%
30	0,8	8,24%	−67,61%	−84,95%
40	0,2	−3,01%	−68,03%	−69,77%
40	0,4	0,71%	−85,71%	−85,71%
40	0,6	15,00%	−100,00%	−100,00%
40	0,8	15,00%	−100,00%	−100,00%
Type 2 instances				
20	0.2	0,84%	−6,17%	−5,82%
20	0.4	−0,46%	−5,64%	−5,08%
20	0.6	−0,22%	−6,06%	−5,91%
20	0.8	0,79%	−6,30%	−7,28%
30	0.2	−0,31%	−6,07%	−5,53%
30	0.4	−0,34%	−6,66%	−6,46%
30	0.6	0,94%	−8,57%	−9,52%
30	0.8	0,72%	−14,10%	−16,93%
40	0.2	−0,48%	−5,85%	−5,27%
40	0.4	1,20%	−7,11%	−7,99%
40	0.6	2,15%	−14,45%	−18,11%
40	0.8	3,86%	−26,54%	−42,75%

In this section, we evaluate the performance of the proposed methods. We consider the benchmarks proposed in [4]. There are a total number of 180 power outlets in the charging station in which 60 outlets are connected to each line. The arrival, charging demands and departure times are generated following normal distributions with different means and deviations that model the EV charging patterns. Three different scenarios are considered where 60 instances are generated for each scenario. In each instance, 180 EV arrive at the station on a time horizon of 24 h. There are two types of instances according to how the EV are distributed between the three lines. In the first type instances, the EV are distributed uniformly: 60 EV arrive at each line along the day. In the second type instances, 60%, 30% and 10% of EV arrive at the fist, second and third line respectively. This makes the imbalance constraints harder to solve. The instances of the first scenario are generated to represent a normal weekday. The instances of the second scenario are obtained by increasing the arrival rate of EV in a short period of time with different charging and departure times. This makes the charging demands exceed the charging station capacity for these periods. In the third scenario, the instances are generated as in the second scenario but with tighter departure times which makes it the hardest scenario to solve.

Note that the algorithms are implemented in Python 3.7, and run on a CPU Intel Core i5-7440HQ (4 CPUs) operating at 2.8 GHz and 8 GB RAM.

Table 1, Table 2 and Table 3 shows the comparison of results obtained for scenarios 1, 2 and 3 respectively with those obtained in [5] by the heuristic "EV". We refer to the proposed heuristic for non-preemptive charging scheduling by "prtt", by "prtt pmtn" for preemptive one and by "HC pmtn" for the hill-climbing preemptive scheduling. We have 30 instances for each group (Scenario, type). For charging station parameters, $\tilde{N}$ varies between 20, 30 and 40 and Δ varies between 0.2, 0.4, 0.6 and 0.8. The Tardiness is calculated for each instance and then summed up for each group. Then, the decrease (or increase) in the total tardiness of the proposed methods compared to the total tardiness in [5] is calculated as a percentage. The percentage will be a negative number if there was a decrease in total tardiness and positive otherwise.

For the first scenario, the prtt outperforms the heuristic EV in 10 groups out of 24 groups whereas it outperforms the heuristic EV in 12 groups out of 24 for the second scenario and in 18 groups out of 24 in the third scenario. This shows that our heuristic has better objective values scheduling EV that arrive almost at the same time with tighter departure times.

For all scenarios, the prtt pmtn and the HC pmtn always outperform the heuristic EV since scheduling the charging preemptively will allow EV with high priorities that have tighter departure times and smaller charging times to charge instead. Thus, Relaxing the preemption constraints is not redundant and it will have great advantage in minimizing the total tardiness. Comparing the results between the two types of instances, we notice that the decreases in total tardiness in type 1 instances is more important than in type 2 instances. For prtt pmtn, The total tardiness was averagely lower by 36.61% in type 1 instances while it was averagely lower by 7.76% in type 2 instances. For HC pmtn, The total tardiness was averagely lower by 44.70% in type 1 instances while it was averagely lower by 7.83% in type 2 instances.

Table 2. Comparison of results of the second scenario.

$\tilde{N}$	Δ	prtt	prtt pmtn	HC pmtn
Type 1 instances				
20	0,2	5,30%	−19,21%	−20,27%
20	0,4	−0,68%	−17,41%	−18,29%
20	0,6	−0,52%	−17,35%	−18,06%
20	0,8	−0,48%	−17,34%	−18,07%
30	0,2	−0,71%	−31,71%	−42,00%
30	0,4	0,75%	−33,92%	−49,33%
30	0,6	0,41%	−34,51%	−51,19%
30	0,8	0,45%	−34,57%	−51,28%
40	0,2	2,48%	−50,95%	−61,59%
40	0,4	7,72%	−59,98%	−86,77%
40	0,6	7,17%	−60,52%	−89,48%
40	0,8	7,18%	−60,52%	−89,51%
Type 2 instances				
20	0,2	1,57%	−4,60%	−6,10%
20	0,4	−0,38%	−4,42%	−3,62%
20	0,6	−0,45%	−5,31%	−4,12%
20	0,8	−0,57%	−6,90%	−5,23%
30	0,2	0,45%	−4,79%	−4,25%
30	0,4	−0,50%	−5,87%	−4,97%
30	0,6	−0,35%	−8,92%	−8,05%
30	0,8	−0,18%	−12,22%	−11,89%
40	0,2	0,15%	−4,90%	−4,15%
40	0,4	−0,25%	−7,80%	−6,96%
40	0,6	−0,30%	−12,17%	−12,22%
40	0,8	0,28%	−18,22%	−19,13%

About the comparison between the prtt pmtn and the HC pmtn, the tardiness of HC pmtn is better in 39 groups and worse in 33. We can notice that the local search is better in type one instances. Especially in the first and second scenarios and slightly worse in the third scenario. This is because charging operations in the third scenario have tighter departure times so it is harder to find moves that improve the whole schedule total tardiness. However, the overall total tardiness was averagely reduced by 26.26% using HC pmtn whereas it was reduced by 22.2% using prtt pmtn.

Table 3. Comparison of results of the third scenario.

$\tilde{N}$	Δ	prtt	prtt pmtn	HC pmtn
Type 1 instances				
20	0,2	7,20%	−10,36%	−8,54%
20	0,4	−0,68%	−8,57%	−4,87%
20	0,6	−0,61%	−8,58%	−4,08%
20	0,8	−0,57%	−8,75%	−3,97%
30	0,2	−0,30%	−15,33%	−15,81%
30	0,4	−0,76%	−12,13%	−11,82%
30	0,6	−0,17%	−11,90%	−11,01%
30	0,8	−0,17%	−11,87%	−11,01%
40	0,2	0,61%	−23,76%	−31,48%
40	0,4	−0,63%	−14,74%	−29,56%
40	0,6	0,39%	−12,36%	−28,51%
40	0,8	0,39%	−12,36%	−28,51%
Type 2 instances				
20	0,2	1,82%	−5,97%	−7,80%
20	0,4	−1,27%	−4,56%	−3,75%
20	0,6	−0,59%	−4,69%	−2,76%
20	0,8	−0,44%	−5,33%	−2,21%
30	0,2	0,93%	−5,78%	−5,29%
30	0,4	−0,17%	−5,32%	−4,01%
30	0,6	−0,66%	−5,70%	−3,71%
30	0,8	−0,73%	−6,80%	−4,06%
40	0,2	0,09%	−5,62%	−4,95%
40	0,4	−0,60%	−5,93%	−4,28%
40	0,6	−0,47%	−6,24%	−4,91%
40	0,8	−0,59%	−8,05%	−6,29%

6   Conclusion

In this paper, we proposed heuristics and a local search procedure to solve the online charging scheduling problem while maintaining the balance in the three phases system. We first formulate the offline problem as MILP model. To solve the problem in a real-time setting, we proposed a new heuristic based on prtt dispatching rule. Simulation results show that our heuristic outperforms other heuristics for most instances. Moreover, the computation time is relatively negligible which is suitable for real-time scheduling. Also, the effectiveness of preemptive scheduling in reducing the total tardiness is shown through the experimental results.

Future works may involve more realistic objectives as maximizing the delivered charging power between the EV arrival and departure times and minimizing the charging electricity bills paid by the charging station.

References

1. Chu, C., Portmann, M.C.: Some new efficient methods to solve the $n/1/ri/\sum T_i$ scheduling problem. Eur. J. Oper. Res. **58**(3), 404–413 (1992)
2. Clement-Nyns, K., Haesen, E., Driesen, J.: The impact of charging plug-in hybrid electric vehicles on a residential distribution grid. IEEE Trans. Power Syst. **25**(1), 371–380 (2010)
3. Graham, R., Lawler, E., Lenstra, J., Kan, A.: Optimization and approximation in deterministic sequencing and scheduling: a survey. In: Annals of Discrete Mathematics, vol. 5, pp. 287–326. Elsevier (1979)
4. Hernández Arauzo, A., Puente Peinador, J., González, M.A., Varela Arias, J.R., Sedano Franco, J.: Dynamic scheduling of electric vehicle charging under limited power and phase balance constraints. In: Proceedings of SPARK 2013-Scheduling and Planning Applications woRKshop. Association for the Advancement of Artificial Intelligence (2013)
5. Hernández-Arauzo, A., Puente, J., Varela, R., Sedano, J.: Electric vehicle charging under power and balance constraints as dynamic scheduling. Comput. Ind. Eng. **85**, 306–315 (2015)
6. IEA: Global EV Outlook 2019: scaling up the transition to electric mobility. International Energy Agency (IEA) (2019). https://www.iea.org/reports/global-ev-outlook-2019
7. Jiang, C., Torquato, R., Salles, D., Xu, W.: Method to assess the power-qualityimpact of plug-in electric vehicles. IEEE Trans. Power Delivery **29**(2), 958–965 (2014)
8. Kang, Q., Wang, J., Zhou, M., Ammari, A.C.: Centralized charging strategy and scheduling algorithm for electric vehicles under a battery swapping scenario. IEEE Trans. Intell. Transp. Syst. **17**(3), 659–669 (2016)
9. Kara, E.C., Macdonald, J.S., Black, D., Bérges, M., Hug, G., Kiliccote, S.: Estimating the benefits of electric vehicle smart charging at non-residential locations: a data-driven approach. Appl. Energy **155**, 515–525 (2015)
10. Kutt, L., Saarijarvi, E., Lehtonen, M., Molder, H., Niitsoo, J.: A review of the harmonic and unbalance effects in electrical distribution networks due to EV charging. In: 2013 12th International Conference on Environment and Electrical Engineering, Wroclaw, pp. 556–561. IEEE (May 2013)
11. Hua, Lunci., Wang, Jia, Zhou, Chi: Adaptive electric vehicle charging coordination on distribution network. IEEE Trans. Smart Grid **5**(6), 2666–2675 (2014)
12. Moghbel, M., Masoum, M.A., Shahnia, F., Moses, P.: Distribution transformer loading in unbalanced three-phase residential networks with random charging of plug-in electric vehicles. In: 2012 22nd Australasian Universities Power Engineering Conference (AUPEC), pp. 1–6. IEEE (2012)
13. Pieltain Fernandez, L., Gomez San Roman, T., Cossent, R., Mateo Domingo, C., Frias, P.: Assessment of the impact of plug-in electric vehicles on distribution networks. IEEE Trans. Power Syst. **26**(1), 206–213 (2011)
14. Rietmann, N., Lieven, T.: How policy measures succeeded to promote electric mobility - worldwide review and outlook. J. Cleaner Prod. **206**, 66–75 (2019)

15. Sedano Franco, J., Portal Garcia, M., Hernandez Arauzo, A., Villar Flecha, J.R., Puente Peinador, J., Varela Arias, R.: Sistema inteligente de recarga de vehículos eléctricos: diseño y operación. Dyna Ingenieria e Industria **88**(3), 640–647 (2013)
16. Vaidya, B., Mouftah, H.T.: Deployment of secure EV charging system using open charge point protocol. In: 2018 14th International Wireless Communications & Mobile Computing Conference (IWCMC), pp. 922–927. IEEE, Limassol, Cyprus, June 2018
17. Wu, H., Pang, G.K.H., Choy, K.L., Lam, H.Y.: Dynamic resource allocation for parking lot electric vehicle recharging using heuristic fuzzy particle swarm optimization algorithm. Appl. Soft Comput. **71**, 538–552 (2018)
18. Xu, Z., Hu, Z., Song, Y., Zhao, W., Zhang, Y.: Coordination of PEVS charging across multiple aggregators. Appl. Energy **136**, 582–589 (2014)
19. Yao, L., Lim, W.H., Tsai, T.S.: A real-time charging scheme for demand response in electric vehicle parking station. IEEE Trans. Smart Grid **8**(1), 52–62 (2017)

Bounds for Complete Arcs in PG(3, q) and Covering Codes of Radius 3, Codimension 4, Under a Certain Probabilistic Conjecture

Alexander A. Davydov[1]([⊠]) [iD], Stefano Marcugini[2] [iD],
and Fernanda Pambianco[2] [iD]

[1] Institute for Information Transmission Problems (Kharkevich Institute),
Russian Academy of Sciences, Moscow, Russia
`adav@iitp.ru`
[2] Department of Mathematics and Computer Science, Perugia University,
Perugia, Italy
{`stefano.marcugini,fernanda.pambianco`}`@unipg.it`

Abstract. Let $t(N, q)$ be the smallest size of a complete arc in the N-dimensional projective space $PG(N, q)$ over the Galois field of order q. The d-length function $\ell_q(r, R, d)$ is the smallest length of a q-ary linear code of codimension (redundancy) r, covering radius R, and minimum distance d; in particular, $\ell_q(4, 3, 5)$ is the smallest length n of an $[n, n - 4, 5]_q3$ quasi-perfect MDS code. By the definitions, $\ell_q(4, 3, 5) = t(3, q)$. In this paper, a step-by-step construction of complete arcs in $PG(3, q)$ is considered. It is proved that uncovered points are uniformly distributed in the space. A natural conjecture on quantitative estimations of the construction is presented. Under this conjecture, new upper bounds on $t(3, q)$ are obtained, in particular, $t(3, q) < 2.93\sqrt[3]{q \ln q}$.

Keywords: Arcs in projective spaces · Covering codes · The d-length function

1 Introduction

Let $\mathbb{F}_q$ be the Galois field with q elements. Denote by $\mathbb{F}_q^n$ the n-dimensional vector space over $\mathbb{F}_q$. Let $PG(N, q)$ be the N-dimensional projective space over $\mathbb{F}_q$. For

The research of A. A. Davydov was done at IITP RAS and supported by the Russian Government (Contract No. 14.W03.31.0019). The research of S. Marcugini and F. Pambianco was supported in part by the Italian National Group for Algebraic and Geometric Structures and their Applications (GNSAGA - INDAM) and by University of Perugia (Project: Curve, codici e configurazioni di punti, Base Research Fund 2018). This work has been carried out using computing resources of the federal collective usage center Complex for Simulation and Data Processing for Mega-science Facilities at NRC "Kurchatov Institute", http://ckp.nrcki.ru/.

O. Gervasi et al. (Eds.): ICCSA 2020, LNCS 12249, pp. 107–122, 2020.
https://doi.org/10.1007/978-3-030-58799-4_8

an introduction to projective spaces over finite fields and connections between the projective geometry and covering codes, see [10,14,16–22].

An n-arc in $PG(N,q)$ with $n > N+1$ is a set of n points such that no $N+1$ points belong to the same hyperplane of $PG(N,q)$. An n-arc is complete if it is not contained in an $(n+1)$-arc.

Let $t(N,q)$ be *the smallest size of a complete arc in* $PG(N,q)$. The value of $t(N,q)$ is an important open problem. The spectrum of possible sizes of complete arcs is one of the essential tasks in study of projective spaces.

Denote by $[n, n-r, d]_q R$ a q-ary linear code of length n, codimension (redundancy) r, minimum distance d, and covering radius R. The covering radius of a linear code of length n is the least integer R such that the space $\mathbb{F}_q^n$ is covered by spheres of radius R centered at the codewords. For an introduction to coverings of vector Hamming spaces over finite fields, see [7,8,10].

The covering density of an $[n, n-r, d]_q R$-code is the ratio of the total volume of all q^{n-r} spheres of radius R centered at the codewords to the volume of the space $\mathbb{F}_q^n$. The covering quality of a code is better if its covering density is smaller. Codes providing a convenient covering density are called *covering codes* [4,5,7,8,10–14,16,22,23]. Covering codes are applied, e.g. to data compression, decoding errors and erasures, football pools, write-once memories [8]. For fixed q, r, R, the covering density of an $[n, n-r, d]_q R$ code decreases with decreasing n.

Let *the length function* $\ell_q(r, R)$ be the smallest length of a q-ary linear code of codimension (redundancy) r and covering radius R [7,8]. Let *the d-length function* $\ell_q(r, R, d)$ be the smallest length of a q-ary linear code of codimension r, covering radius R, and minimum distance d [5,6,11]. We have $\ell_q(r, R) \leq \ell_q(r, R, d)$.

Columns of a parity check matrix of an $[n, n-N-1, N+2]_q N$ maximum distance separable (MDS) code can be considered as points (in homogeneous coordinates) of a complete n-arc in $PG(N,q)$ [4–6,10,14,16,18,20,22]. If $N = 2, 3$ these codes are quasi-perfect. Arcs and linear MDS codes are equivalent objects. We have $\ell_q(N+1, N) \leq \ell_q(N+1, N, N+2) = t(N,q)$, $\ell_q(4, 3, 5) = t(3, q)$.

The value of $\ell_q(N+1, N, N+2)$ is an open important problem studied insufficiently, see e.g. [6, Introduction], [15, Remark 1].

This paper is devoted to upper bounds on $t(3, q)$ that are also upper bounds on the smallest length $\ell_q(4, 3, 5)$ of an $[n, n-4, 5]_q 3$ quasi-perfect MDS code.

The known results on $t(N,q)$ and $\ell_q(N+1, N, N+2)$, $N \geq 2$, can be found in [1–6,13,26], see also the references therein. In [1–4], the case $N = 2$ is considered. In [6,26], algebraic constructions for $N \geq 3$ are proposed. Computer search results for $N = 3, 4$ are given in [4,5,13].

In this paper, under a natural probabilistic conjecture, we obtain new upper bounds on $t(3, q)$. The new bounds are better than the known ones. In particular, the smallest known complete arcs obtained by algebraic constructions have size of order $\sim q^{3/4}$ [6,26] while the bounds of this paper for $N = 3$ are of order $\simeq (q \ln q)^{1/3}$. Also, in this paper, the computer search is made in a more wide region of q than in [4,5,13]; many results of these papers are improved.

Some results of this paper were briefly presented in [11]. Unfortunately, in [11], some inaccuracies are present in the proof of the main result. In this paper, we *correct the inaccuracies of* [11]; for this we introduce a new tool called "implicit bound", see Sect. 2 (including Remark 1) and Sect. 6. Also, in comparison with [11], in this paper we decrease the constant (2.93 instead of 3) before $\sqrt[3]{q \ln q}$ and increase the range of q values in computer search. This is essential for the initial part of the combined bound on $t(3, q)$ in Theorem 6.

The paper is organized as follows. In Sect. 2, the main results are presented. In Sect. 3, an iterative step-by-step construction of complete arcs is considered. In Sect. 4, it is proved that uncovered points are uniformly distributed in the space. In Sect. 5, Conjecture 2 on quantitative estimations of the construction is presented. Under this conjecture, new upper bounds (implicit and explicit) on $t(3, q)$ are obtained in Sect. 6. In Sect. 7, we give upper bounds based on computer search. In Sect. 8, a combined bound based on the bounds in the previous sections is presented. In Sect. 9, we summarize the results obtained.

2   The Main Results

The main results of the paper are given by Theorem 1 based on the results of Sects. 6–8, see Examples 1, 2, 3, Lemma 3, and Theorems 2, 4, 5, 6.

Theorem 1 (*the main results*). *Let $t(3, q)$ be the smallest size of a complete arc in* PG(3, q). *Let $\ell_q(4, 3)$ be the smallest length of a q-ary linear code of codimension (redundancy) 4 and covering radius 3. Let $\ell_q(4, 3, 5)$ be the smallest length of a q-ary linear code of codimension (redundancy) 4, covering radius 3, and minimum distance 5 (it is a quasi-perfect MDS code).*

(i) *The following upper "computer" bounds are provided by complete arcs obtained with the help of computer search:*

$$\ell_q(4, 3) \leq \ell_q(4, 3, 5) = t(3, q) < \begin{cases} 2.61 \sqrt[3]{q \ln q} \ for \ 13 \leq q \leq 4373 \\ 2.65 \sqrt[3]{q \ln q} \ for \ 4373 < q \leq 7057 \end{cases}.$$

(ii) *The following upper "implicit" and "explicit" bounds hold under Conjecture 2:*

$$\ell_q(4, 3) \leq \ell_q(4, 3, 5) = t(3, q) < \begin{cases} 2.72 \sqrt[3]{q \ln q} \quad for \qquad 7057 < q < 71143 \\ 2 \sqrt[3]{3q \ln q} + 4 < 2.926 \sqrt[3]{q \ln q} \\ \qquad\qquad\qquad\qquad for \ q \geq 71143 \end{cases}.$$

(iii) *The following upper "combined" bound holds under Conjecture 2; it is an union of the bounds* (i) *and* (ii):

$$\ell_q(4, 3) \leq \ell_q(4, 3, 5) = t(3, q) < 2.93 \sqrt[3]{q \ln q} \quad for \ q \geq 11 \ and \ q = 4, 8.$$

Our observations and the results of [4,5] allow us to conjecture the following.

Conjecture 1. The upper bound

$$\ell_q(4, 3) \leq \ell_q(4, 3, 5) = t(3, q) < 2.93 \sqrt[3]{q \ln q}$$

holds for all $q \geq 11$ without any extra conditions and conjectures.

Remark 1. In [11, Abstract, Ths. 6, 20], the final upper bound $3\sqrt[3]{\ln q} \cdot q^{(r-3)/3}$, $r = 3t + 1 \geq 4$, is correct, but its multistep proof contains inaccuracies in the first stage connected with $PG(3, q)$ and codimension (redundancy) $r = 4$.

In particular, it can be checked that in Proof of [11, Th. 12] for $r = 4$, the value $w = \left\lceil 2\sqrt[3]{q + 4}\sqrt[3]{\ln(q^3 + q^2 + q)} + 1 \right\rceil$ satisfies inequality $q - 1 \geq (w-2)(w-3)$ [11, eq. (12)] if and only if $q \geq 71119$. Therefore, see [11, Cor. 13, eq. (4)], the bound $\ell_q(4, 3) \leq \ell_q(4, 3, 5) < 2\sqrt[3]{3(q + 4)\ln(q + 1)} + 4$ also holds only for $q \geq 71119$. However, in [11] this fact is not noted and the mentioned bound is used for $q > 4973$, see [11, Fig. 1, Lem. 14].

In this paper, we prove the bound $\ell_q(4, 3) \leq \ell_q(4, 3, 5) = t(3, q) \leq 2\sqrt[3]{3q\ln q} + 4$ using inequality $q - 13 \geq (w - 2)(w - 3)$ with $w = 2\sqrt[3]{3q\ln q} + 2$, see (11), (19), (20). We show that this inequality holds if and only if $q \geq 71143$. To solve the problem we *introduce an implicit bound*, see Sect. 6.1. The implicit bound is given iteratively by (6)–(9); we calculated it by computer for each $q < 71143$. As the result, we avoided the inaccuracies in the proofs of the upper bounds under Conjecture 2.

3 A Step-By-Step Process to Construct a Complete Arc

In $PG(3, q)$, we construct a complete arc with the help of a greedy iterative step-by-step algorithm (*Algorithm* for short) which, in each step adds one new point to the arc. The added point provides the maximal possible (for the given step) number of new covered points.

If a point of $PG(3, q)$ lies on a plane through three non-collinear points of an arc, we say that *the point is covered by the arc.* Clearly, all the points of the arc and the points of $PG(3, q)$ lying on a bisecant line of the arc are covered.

The space $PG(N, q)$ contains $\theta_{N,q} = \frac{q^{N+1} - 1}{q - 1}$ points.

Assume that after the w-th step of Algorithm, we obtain a w-arc that does not cover exactly U_w points. We denote by $\mathbf{S}(U_w)$ the set of all w-arcs in $PG(3, q)$ each of which does not cover exactly U_w points. The group of collineations $P\Gamma L(4, q)$ preserves $\mathbf{S}(U_w)$.

Remark 2. To increase the volume of $\mathbf{S}(U_w)$ we could include in $\mathbf{S}(U_w)$ all w-arcs in $PG(3, q)$ each of which does not cover $\leq U_w$ points. This does not change estimates below, but some reflections could become more complicated.

The $(w + 1)$-st step of Algorithm starts from a w-arc $\mathcal{K}_w \in \mathbf{S}(U_w)$ chosen from $\mathbf{S}(U_w)$ in a random manner. Every arc of $\mathbf{S}(U_w)$ can be chosen with the probability $\frac{1}{\#\mathbf{S}(U_w)}$. This means that we consider the set $\mathbf{S}(U_w)$ as an ensemble of random objects with the uniform probability distribution.

Let $\mathcal{U}(\mathcal{K})$ be the set of the points of $PG(3, q)$ that are not covered by an arc $\mathcal{K}$. Evidently, $\#\mathcal{U}(\mathcal{K}_w) = U_w$. Let $\mathcal{K}_w$ consist of w points $A_1, A_2, \ldots, A_w$. Let $A_{w+1} \in \mathcal{U}(\mathcal{K}_w)$ be the point that is included into the arc in the $(w + 1)$-st step.

Remark 3. In future, we introduce point subsets with the notation of the form $\mathcal{M}_w(A_{w+1})$. Any uncovered point may be added to $\mathcal{K}_w$. So, there exist U_w distinct subsets $\mathcal{M}_w(A_{w+1})$. When a particular point A_{w+1} is not relevant, we may use the notation $\mathcal{M}_w$. The same concerns the quantities $\Delta_w(A_{w+1})$ and Δ_w.

We call a *bundle* $\mathcal{B}(A_{w+1})$ the set of $\binom{w}{2}$ planes through A_{w+1} and two distinct points of $\mathcal{K}_w$. Let $\pi_{i,j}$, $1 \leq i < j \leq w$, be the plane through A_i, A_j, A_{w+1}. We denote by $\overline{A_i A_j}$ the line through A_i, A_j. We consider the set of the points

$$\mathcal{C}_w(A_{w+1}) = \mathcal{B}(A_{w+1}) \setminus \left(\{A_{w+1}\} \cup \bigcup_{1 \leq i < j \leq w} \overline{A_i A_j} \right) \tag{1}$$

where $\overline{A_i A_j}$ is a bisecant of $\mathcal{K}_w$. The points on the bisecants are covered by $\mathcal{K}_w$; also, A_{w+1} is a new covered point. All the points in $\mathcal{C}_w(A_{w+1})$ are *candidates to be new covered points in the* $(w+1)$-st step. We denote

$$\Omega_w = \binom{w}{2} \left(\theta_{2,q} - 3q - \binom{w-2}{2}(q-2) \right) + \frac{1}{2}\binom{w}{2}\binom{w-2}{2}(q-2) \tag{2}$$

$$+ w(q-1) = \binom{w}{2}(q-1)^2 + w(q-1) - 3\binom{w}{4}(q-2).$$

Lemma 1. *For the set of candidates $\mathcal{C}_w$ we have* $\#\mathcal{C}_w \geq \Omega_w$.

Proof. In $\pi_{i,j}$, a part of points is common with other planes. In (2), we subtract from $\theta_{2,q}$ the volume of these parts. For fixed i, j, there are $\binom{w-2}{2}$ cases when $\{i, j\} \cap \{k, m\} = \emptyset$ and $\pi_{i,j}$ intersects $\pi_{k,m}$. An intersection line contains A_{w+1} and it is skew to $\mathcal{K}_w$. Also, $\pi_{i,j} \cap \pi_{k,m}$ meets $\overline{A_i A_j}$ in a point $P \notin \{A_i, A_j\}$ and $\overline{A_k A_m}$ in $P' \notin \{A_k, A_m\}$. The points P and P' are covered by $\mathcal{K}_w$. Thus, we should subtract 3 from the cardinality $q+1$ of a line. As a result, we have the term $-\binom{w-2}{2}(q-2)$. In principle, for $\{i, j\} \cap \{k, m\} = \emptyset$ and $\{n, p\} \cap \{i, j, k, m\} = \emptyset$, intersection lines $\pi_{i,j} \cap \pi_{k,m}$ and $\pi_{i,j} \cap \pi_{n,p}$ can be the same. But for simplicity, in the estimate of $\#\mathcal{C}_w$, we put that all the intersection lines are distinct.

The term $-3q$ corresponds to the triangle with the vertices A_i, A_j, A_{w+1}.

On the other hand, at least one time, the subtracted values should be included in $\#\mathcal{C}_w$. Therefore, we add $\frac{1}{2}\binom{w}{2}\binom{w-2}{2}(q-2)$ where $\frac{1}{2}$ takes into account that every line $\pi_{i,j} \cap \pi_{k,m}$ is contained at least in two planes of $\mathcal{B}(A_{w+1})$. The addition $w(q-1)$ corresponds to w sets $\overline{A_i A_{w+1}} \setminus \{A_i, A_{w+1}\}$, $i = 1, \ldots, w$.   □

Let $\Delta_w(A_{w+1})$ be the number of new covered points in the $(w+1)$-st step;

$$\Delta_w(A_{w+1}) = \#\mathcal{U}(\mathcal{K}_w) - \#\mathcal{U}(\mathcal{K}_w \cup \{A_{w+1}\}) = \#\{\mathcal{C}_w(A_{w+1}) \cap \mathcal{U}(\mathcal{K}_w)\}. \tag{3}$$

We consider continuous approximations of the discrete functions $\#\mathcal{U}(\mathcal{K}_w)$, $\Delta_w(A_{w+1})$, $\#\mathcal{U}(\mathcal{K}_w \cup \{A_{w+1}\})$, and other ones keeping the same notations.

4   The Uniform Distribution of Uncovered Points

For a point $H \in \mathrm{PG}(3, q)$, let $n_w(H)$ be the number of arcs of $\mathbf{S}(U_w)$ that do not cover it. Each point H is considered as a random object that is not covered by a randomly chosen w-arc $\mathcal{K}_w$ with the probability $p_w(H) = \frac{n_w(H)}{\#\mathbf{S}(U_w)}$.

Lemma 2. *The value $n_w(H)$ and the probability $p_w(H)$ are the same for all points $H \in \mathrm{PG}(3, q)$. They can be considered, respectively, as n_w and*

$$p_w = \frac{n_w}{\#\mathbf{S}(U_w)}.$$

Proof. We denote $\mathbf{K}_w(H)$ the subset of w-arcs in $\mathbf{S}(U_w)$ that do not cover H. By the definition, $n_w(H) = \#\mathbf{K}_w(H)$. Let H_i and H_j be two points of $\mathrm{PG}(3, q)$. In $P\Gamma L(4, q)$, let $\Psi(H_i, H_j)$ be a collineation transforming H_i to H_j. Evidently, $\Psi(H_i, H_j)$ embeds $\mathbf{K}_w(H_i)$ in $\mathbf{K}_w(H_j)$, while $\Psi(H_j, H_i)$ embeds $\mathbf{K}_w(H_j)$ in $\mathbf{K}_w(H_i)$. Thus, we have $\#\mathbf{K}_w(H_i) \leq \#\mathbf{K}_w(H_j)$ and $\#\mathbf{K}_w(H_j) \leq \#\mathbf{K}_w(H_i)$, whence $\#\mathbf{K}_w(H_i) = \#\mathbf{K}_w(H_j)$, i.e. $n_w(H_i) = n_w(H_j)$. □

We have proved that *uncovered points are uniformly distributed in the space* and the probability to be uncovered is independent of a point. Therefore, for a randomly chosen w-arc $\mathcal{K}_w \in \mathbf{S}(U_w)$, the fraction $\#\mathcal{U}_w(\mathcal{K}_w)/\theta_{3,q}$ of uncovered points of $\mathrm{PG}(3, q)$ is equal to the probability p_w that a point of $\mathrm{PG}(3, q)$ is not covered. Thus,

$$p_w = \frac{\#\mathcal{U}_w(\mathcal{K}_w)}{\theta_{3,q}} = \frac{\mathcal{U}_w}{\theta_{3,q}}. \tag{4}$$

Note that by Lemma 2, the multiset consisting of all points that are not covered by all arcs of $\mathbf{S}(U_w)$ has cardinality $n_w\theta_{3,q}$. This cardinality can also be written as $U_w \cdot \#\mathbf{S}(U_w)$. Thus, $n_w\theta_{3,q} = U_w \cdot \#\mathbf{S}(U_w)$ and we have

$$\frac{n_w}{\#\mathbf{S}(U_w)} = \frac{U_w}{\theta_{3,q}}.$$

We obtained another explanation of the equality (4).

Consider a sequence of h random and independent 0/1 trials each of which yields 1 with the probability p_w. The number $s_w(h)$ of ones in this sequens is the random variable with *binomial probability distribution* and the expected value

$$\mathbf{E}[s_w(h)] = hp_w = h\frac{U_w}{\theta_{3,q}}. \tag{5}$$

Remark 4. We can consider the *hypergeometric probability distribution*, which describes the probability of $s'_w(h)$ successes in h random and independent draws without replacement from a finite population of size $\theta_{3,q}$ containing exactly U_w successes. The expected value of $s'_w(h)$ again is

$$\mathbf{E}[s'_w(h)] = h\frac{U_w}{\theta_{3,q}} = \mathbf{E}[s_w(h)].$$

Note also that the average number of uncovered points among h points of PG(3, q) calculated over all $\binom{\theta_{3,q}}{h}$ combinations of h points is

$$\frac{1}{\binom{\theta_{3,q}}{h}} \sum_{i=1}^{h} i \binom{\theta_{3,q} - U_w}{h - i} \binom{U_w}{i} = \frac{U_w}{\binom{\theta_{3,q}}{h}} \sum_{i=1}^{h} \binom{\theta_{3,q} - U_w}{h - i} \binom{U_w - 1}{i - 1}$$

$$= \frac{U_w \binom{\theta_{3,q} - 1}{h - 1}}{\binom{\theta_{3,q}}{h}} = h \frac{U_w}{\theta_{3,q}} = \mathbf{E}[s_w(h)].$$

We denote by $\mathbf{E}_{w,q}$ the expected value of the number of uncovered points among Ω_w randomly taken points in PG(3, q), under the condition that the events to be uncovered are independent. By Lemma 2 and (2), (4), (5), we have

$$\mathbf{E}_{w,q} = \mathbf{E}[s_w(\Omega_w)] = \Omega_w p_w = \Omega_w \frac{U_w}{\theta_{3,q}}.$$

5   The Main Conjecture

The points-candidates of the set (1) cannot be considered as random and independent as they belong to the same bundle $\mathcal{B}(A_{w+1})$. Therefore, in general, $\mathbf{E}[\Delta_w] \neq \mathbf{E}_{w,q}$ where $\mathbf{E}[\Delta_w]$ is the expected value of Δ_w defined in (3).

However, there are *many random aspects of the process*. We note e.g. the relative positions and intersections of planes in distinct bundles and planes through three points of a current arc. Remind also that the starting arc $\mathcal{K}_w$ is taken from $\mathbf{S}(U_w)$ randomly. If q and, hence, the number U_w of distinct bundles and the volume of the ensemble $\mathbf{S}(U_w)$ are relatively large then the random aspects noted are especially essential. Therefore, with a large probability, the variance of the random variable Δ_w implies the existence of bundles providing $\Delta_w(A_{w+1}) > \mathbf{E}[\Delta_w]$ and, moreover, $\Delta_w(A_{w+1}) \geq \mathbf{E}_{w,q}$. By above, Conjecture 2 seems to be reasonable and founded.

Conjecture 2 (**the main conjecture**). For q large enough, in every $(w + 1)$-st step of the step-by-step process in PG(3, q) considered in Sect. 3, there exists a w-arc $\mathcal{K}_w \in \mathbf{S}(U_w)$ such that one can find an uncovered point A_{w+1} providing the inequality

$$\Delta_w(A_{w+1}) \geq \mathbf{E}_{w,q} = \Omega_w \frac{U_w}{\theta_{3,q}}.$$

Remark 5. We could do a "milder" conjecture of the form

$$\Delta_w(A_{w+1}) \geq \frac{\mathbf{E}_{w,q}}{D}, \quad D \geq 1 \text{ is a constant independent of } q,$$

cf. [9, Conjecture 3.3]. This inequality is more easy to achieve than the inequality of Conjecture 2. On the other hand, the new conjecture would give the extra multiplier $\sqrt[3]{D}$ in upper bounds (20). Also, the proofs become more complicated as D is not limited from above.

6   Upper Bounds on $t(3,q)$ Under Conjecture 2

We always have at least one new covered point in the $(w+1)$-st step; it is A_{w+1}. Therefore, $\Delta_w(A_{w+1}) \geq 1$. Also, $\Delta_w(A_{w+1})$ should be integer. By Conjecture 2,

$$\Delta_w(A_{w+1}) \geq \left\lceil \max\left\{1, \Omega_w \frac{U_w}{\theta_{3,q}}\right\}\right\rceil$$

hence, by (3),

$$\#\mathcal{U}_{w+1} = \#\mathcal{U}(\mathcal{K}_w \cup \{A_{w+1}\}) = \#\mathcal{U}(\mathcal{K}_w) - \Delta_w(A_{w+1}) \tag{6}$$
$$\leq U_w - \left\lceil \max\left\{1, \Omega_w \frac{U_w}{\theta_{3,q}}\right\}\right\rceil .$$

In future, we put in $\mathcal{K}_3$ three non-collinear points. Therefore,

$$\#\mathcal{U}_3(\mathcal{K}_3) = \theta_{3,q} - \theta_{2,q} = q^3. \tag{7}$$

6.1   An Implicit Upper Bound $A(w_0, U_{w_0})$

Assume that there exists a w_0-arc $\mathcal{K}_{w_0} \subset \mathrm{PG}(3,q)$ that does not cover exactly U_{w_0} points of $\mathrm{PG}(3,q)$. Starting from the values w_0 and U_{w_0}, one can itera- tively apply relation (6) and obtain eventually $\#\mathcal{U}_{w+1} \leq 1$ for some w, say w^*. Obviously, w^* depends on w_0 and U_{w_0}, i.e. we have a function $w^*(w_0, U_{w_0})$. The obtained complete arc has size k, where

$$k = w^*(w_0, U_{w_0}) + 2 \text{ under condition } \#\mathcal{U}(\mathcal{K}_{w^*(w_0,U_{w_0})} \cup \{A_{w^*(w_0,U_{w_0})+1}\}) \leq 1. \tag{8}$$

We call the value $w^*(w_0, U_{w_0}) + 2$ an *implicit upper bound* $A(w_0, U_{w_0})$. In other words, the following theorem holds.

Theorem 2 (implicit bound). *Let the values w_0, U_{w_0}, and $w^*(w_0, U_{w_0})$ be defined and calculated as above including (8). Then, under Conjecture 2, the following upper bound holds:*

$$t(3,q) \leq w^*(w_0, U_{w_0}) + 2 := A(w_0, U_{w_0}). \tag{9}$$

In future, we use the implicit bound $A(3, q^3)$ based on (7)–(9).

Example 1. The implicit bound $A(3, q^3)$ divided by $\sqrt[3]{q \ln q}$ (i.e. the value $(w^*(3, q^3) + 2)/\sqrt[3]{q \ln q}$) is shown by the top curve in Figs. 1 and 2, for $q \leq 7057$, and by the middle curve in Fig. 3, for $7057 < q < 71143$.

   The bound $A(3, q^3)$ is obtained by computer by iteratively applying (6). By these computations, in the range $7 \leq q \leq 61$, the bound $A(3, q^3)$ tends to increase with the minimal value $A(3, q^3) \sim 3.34 \sqrt[3]{q \ln q}$ for $q = 7$ and the maximal value $A(3, q^3) \sim 7.46 \sqrt[3]{q \ln q}$ for $q = 61$; in the range $61 \leq q \leq 7057$, $A(3, q^3)$ tends to decrease with the minimal value $A(3, q^3) \sim 2.68 \sqrt[3]{q \ln q}$ for $q = 5471$, see Figs. 1 and 2. In the range $7057 < q < 71143$, the bound $A(3, q^3)$ tends to decrease with the maximal value $A(3, q^3) \sim 2.72 \sqrt[3]{q \ln q}$ for $q = 7129$ and the minimal value $A(3, q^3) \sim 2.62 \sqrt[3]{q \ln q}$ for $q = 59863$, see Fig. 3.

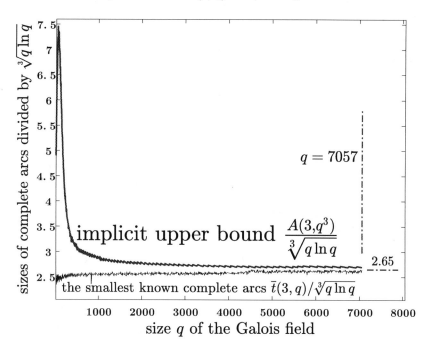

Fig. 1. Implicit upper bound $A(3, q^3)$ on the smallest size of a complete arc in PG(3, q) (*top curve*) vs the smallest known sizes $\bar{t}(3, q)$ of complete arcs in PG(3, q), $q \le 7057$ (*bottom curve*). The bound and the sizes are divided by $\sqrt[3]{q \ln q}$; $\frac{\bar{t}(3,q)}{\sqrt[3]{q \ln q}} < 2.65$

Lemma 3. **(i)** *In the range* $7 \le q \le 7057$, *we have* $A(3, q^3) > 2.68 \sqrt[3]{q \ln q}$.

(ii) *In the range* $7057 < q < 71143$, *we have* $A(3, q^3) < 2.72 \sqrt[3]{q \ln q}$.

Proof. The assertions are obtained by computer search using (6)–(9).   □

6.2   An Explicit Upper Bound

By (2) and (6), we have

$$\#\mathcal{U}_{w+1} = \#\mathcal{U}(\mathcal{K}_w \cup \{A_{w+1}\}) = \#\mathcal{U}(\mathcal{K}_w) - \Delta_w(A_{w+1}) \le U_w \left(1 - \frac{\Omega_w}{\theta_{3,q}}\right) \quad (10)$$

$$= U_w \left(1 - \frac{4(w-1)(q-1)^2 + 8(q-1) - (w-1)(w-2)(w-3)(q-2)}{8\theta_{3,q}} w\right)$$

$$< U_w \left(1 - (4(w-1)(q-1) + 8 - (w-1)(w-2)(w-3))\frac{w(q-1)}{8\theta_{3,q}}\right)$$

$$< U_w \left(1 - (4(q-1) - (w-2)(w-3))\frac{w(w-1)(q-1)}{8\theta_{3,q}}\right).$$

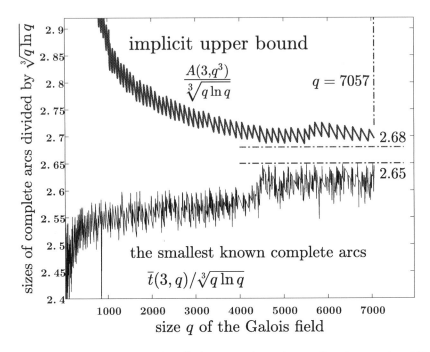

Fig. 2. Implicit upper bound $A(3, q^3)$ (*top curve*) vs the smallest known sizes $\bar{t}(3, q)$ of complete arcs (*bottom curve*) (in another scale than in Fig. 1). The bounds and the sizes are divided by $\sqrt[3]{q \ln q}$; $\frac{\bar{t}(3,q)}{\sqrt[3]{q \ln q}} < 2.65$, $\frac{A(3,q^3)}{\sqrt[3]{q \ln q}} > 2.68$

Assume that

$$q - 13 \geq (w - 2)(w - 3). \tag{11}$$

Under the condition (11) and by (10), we have

$$\#\mathcal{U}(\mathcal{K}_w \cup \{A_{w+1}\}) < U_w \left(1 - \frac{3w(w - 1)(q - 1)(q + 3)}{8(q^3 + q^2 + q + 1)}\right) \tag{12}$$

$$= U_w \left(1 - \frac{3w(w - 1)}{8\left(q - \frac{q^2 - 4q - 1}{q^2 + 2q - 3}\right)}\right) < U_w \left(1 - \frac{3(w^2 - w)}{8q}\right).$$

Starting from (7) and using (12) iteratively, we obtain

$$\#\mathcal{U}_{w+1} \leq q^3 f_q(w) \tag{13}$$

where

$$f_q(w) = \prod_{i=3}^{w} \left(1 - \frac{3(i^2 - i)}{8q}\right). \tag{14}$$

We are finishing the step-by-step process when $\#\mathcal{U}_{w+1} \leq 1$. The size k of an obtained complete arc $\mathcal{K}_k$ is as follows:

$$k = w + 2 \text{ under condition } \#\mathcal{U}_{w+1} \leq 1. \tag{15}$$

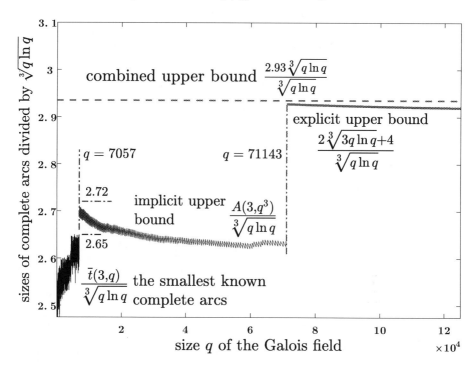

Fig. 3. Upper bounds on the smallest size of a complete arc in PG(3, q) and the smallest known sizes of complete arcs in PG(3, q), $q \leq 125003$. Explicit upper bound $2\sqrt[3]{3q \ln q} + 4$, $71143 \leq q \leq 125003$ (*the right-hand solid curve*); implicit upper bound $A(3, q^3)$, $7057 < q < 71143$ (*the middle curve*); combined upper bound $2.93\sqrt[3]{q \ln q}$ (*the top dashed line* $y = 2.93$); the smallest known sizes $\bar{t}(3, q)$ of complete arcs in PG(3, q), $q \leq 7057$ (*the left-hand curve*). The bounds and the sizes are divided by $\sqrt[3]{q \ln q}$; $\frac{\bar{t}(3,q)}{\sqrt[3]{q \ln q}} < 2.65$, $\frac{A(3,q^3)}{\sqrt[3]{q \ln q}} < 2.72$, $\frac{2\sqrt[3]{3q \ln q}+4}{\sqrt[3]{q \ln q}} < 2.93$, $\frac{2.93\sqrt[3]{q \ln q}}{\sqrt[3]{q \ln q}} = 2.93$

Theorem 3 (the basic inequality). *Let $f_q(w)$ be as in (14). Under Conjecture 2, in $PG(3, q)$ the following holds:*

$$t(3, q) \leq w + 2 \tag{16}$$

where the value w satisfies the "basic" inequality

$$f_q(w) \leq q^{-3}. \tag{17}$$

Proof. By (13), to provide the inequality $\#\mathcal{U}_{w+1} \leq 1$ it is sufficient to find w such that $q^3 f_q(w) \leq 1$. Now (16) follows from (15). $\square$

We find an upper bound on the smallest possible solution of inequality (17). By the Taylor series of $e^{-\alpha}$, we have

$$1 - \alpha < e^{-\alpha} \text{ for } \alpha \neq 0. \tag{18}$$

Lemma 4. *The value*

$$w \geq 2\sqrt[3]{3q\ln q} + 2 \tag{19}$$

satisfies the basic inequality (17).

Proof. By (14),(18), we have

$$f_q(w) = \prod_{i=3}^{w}\left(1 - \frac{3(i^2 - i)}{8q}\right) < e^{-\Upsilon(w)}$$

where

$$\Upsilon(w) = \sum_{i=3}^{w}\frac{3(i^2-i)}{8q} = \frac{3}{8q}\left(-2 + \sum_{i=1}^{w}(i^2-i)\right)$$

$$= \frac{3}{8q}\left(-2 + \frac{w(w+1)(2w+1)}{6} - \frac{w(w+1)}{2}\right) = \frac{w(w^2-1)-6}{8q} > \frac{(w-1)^3}{8q}$$

whence $f_q(w) =< e^{-(w-1)^3/8q}$. Thus, to provide (17) it is sufficient to find w such that $e^{-(w-1)^3/8q} \leq q^{-3}$ whence $(w-1)^3 \geq 8q\ln q^3$ and $w \geq \sqrt[3]{3q\ln q} + 1$. As w should be an integer, in (19) one is added. □

Lemma 5. *The value* $w = 2\sqrt[3]{3q\ln q} + 2$ *of* (19) *satisfies the condition* (11) *if and only if* $q \geq 71143$.

Proof. The assertion can be checked directly by computer. □

Theorem 4 (*explicit bound*). *Under Conjecture 2, in* $PG(3,q)$ *we have*

$$t(3,q) \leq 2\sqrt[3]{3q\ln q} + 4 < 2.926\sqrt[3]{q\ln q}, \quad q \geq 71143. \tag{20}$$

Proof. The assertion follows from (16), (19) and Lemma 5. Note also that $2\sqrt[3]{3q\ln q} + 4 = \sqrt[3]{q\ln q}\left(2\sqrt[3]{3} + \frac{4}{\sqrt[3]{q\ln q}}\right).$ □

Example 2. For $71143 \leq q \leq 125003$, the explicit bound (20) divided by $\sqrt[3]{q\ln q}$ (i.e. the value $\frac{2\sqrt[3]{3q\ln q}+4}{\sqrt[3]{q\ln q}}$) is shown by the right-hand solid curve in Fig. 3.

7   The Smallest Known Complete Arcs in $PG(3,q)$

Let $\bar{t}(3,q)$ be the *smallest **known** size of a complete arc in* $PG(3,q)$.

Obviously, $t(3,q) \leq \bar{t}(3,q)$. The smallest known complete arcs in $PG(3,q)$ are obtained in [4,5] and in this paper by computer search with the help of two types of algorithms. They both build an arc iteratively, step-by-step. Algorithms of these types are considered in [2,3,5], see also the references therein.

Algorithms with Fixed Order of Points (FOP)
We fix an order on points of $PG(3,q)$ as $A_1, A_2, \ldots, A_{\theta_{3,q}}$ where A_1, A_2, A_3 are non-collinear. The set $S^{(1)} = \{A_1, A_2, A_3\}$ is the start of the search. On the j-th

step we obtain the set $S^{(j)}$. To obtain $S^{(j+1)}$, we add to $S^{(j)}$ the first point in the fixed order, which is not covered by $S^{(j)}$.

A point $A_i \in \mathrm{PG}(3, q)$ in homogeneous coordinates is written as $A_i = (x_0^{(i)}, x_1^{(i)}, x_2^{(i)}, x_3^{(i)})$, $x_u^{(i)} \in \mathbb{F}_q$. The number i of A_i is assigned by two ways. In the *lexicographical* order we put $i = \sum_{u=0}^{3} x_u^{(i)} q^{3-u}$. In the *inverse lexicographical* order we have $i = \theta_{3,q} - \sum_{u=0}^{3} x_u^{(i)} q^{3-u}$.

Let q be prime. Then *for an algorithm with fixed order of points, the size of a complete arc and the point list of the arc depend on q only.*

We use "FOP algorithm" with the lexicographical order of points and "2FOP algorithm" where both the lexicographical and the inverse lexicographical orders of points are applied and the best result is taken.

One can consider the algorithm with fixed order of points as a version of the recursive g-parity check matrix algorithm for greedy codes [24, Sect. 7].

Randomized Greedy Algorithms
We use the so-called *d-Rand-Greedy algorithm*. In every step, the d-Rand-Greedy algorithm considers a few randomly chosen points that are not covered in the previous steps. The chosen point providing the most number of new covered points is included to the current arc. This gives smaller complete arcs than algorithms FOP, but takes greater computer time.

The smallest known complete arcs in $PG(3, q)$, $q \leq 7057$, are obtained by computer search in [4,5] and in this paper. These arcs provide the following.

Theorem 5. *(computer bound).* In $\mathrm{PG}(3, q)$, *the following upper bounds hold:*

$$t(3, q) \leq \bar{t}(3, q) \leq \begin{cases} 2.61 \sqrt[3]{q \ln q} & \text{for } 13 \leq q \leq 4373 \\ 2.65 \sqrt[3]{q \ln q} & \text{for } 4373 < q \leq 7057 \end{cases}.$$

Proof. The needed complete arcs are obtained by computer search. To construct arcs, for $q \leq 4451$ the d-Rand-Greedy algorithm is used. For $4451 < q \leq 7057$, the FOP and 2FOP algorithms are applied, in preference, but for $q = 4489, 4679, 4877, 4889, 4913, 5801, 6653$, we used the d-Rand-Greedy algorithm. For $q = 841$, the complete 42-arc of [25] is taken.   □

Example 3. The sizes $\bar{t}(3, q)$ of the smallest known complete arcs in $PG(3, q)$, $q \leq 7057$, divided by $\sqrt[3]{q \ln q}$, are shown by the bottom curves in Figs. 1, 2, 3 and 4.

8   A Combined Upper Bound

We join the upper bounds of the previous sections.

Theorem 6 (combined bound). *Under Conjecture 2, in* $\mathrm{PG}(3, q)$, *the following upper bound holds:*

$$\ell_q(4, 3) \leq \ell_q(4, 3, 5) = t(3, q) < 2.93 \sqrt[3]{q \ln q} \quad \text{for } q \geq 11 \text{ and } q = 4, 8. \quad (21)$$

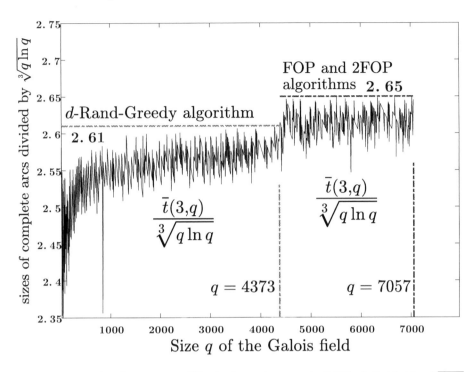

Fig. 4. The smallest **known** sizes $\overline{t}(3,q)$ of complete arcs in $PG(3,q)$ divided by $\sqrt[3]{q \ln q}$

Proof. We use the results of Examples 1, 2, 3, Lemma 3, and Theorems 2, 4, 5; see also Figures 1–4. In particular, Fig. 3 illustrates that the combined bound (21) is provided by the smallest known complete arcs for $13 \leq q \leq 7057$, by the implicit bound for $7057 < q < 71143$, and by the explicit bound for $q \geq 71143$. Finally, for $q = 4, 8, 11$ we take convenient arcs from [5,13].   □

9   Conclusion

We consider the upper bounds on the smallest size $t(3,q)$ of a complete arc in the projective space $PG(3,q)$. This size is equal to the smallest length of an $[n, n-4, 5]_q 3$ quasi-perfect MDS code. We analyse the working mechanism of a step-by-step greedy algorithm constructing a complete arc. It is proved that uncovered points are uniformly distributed in the space. A natural conjecture on an estimate of the number of new covered points in every step of the algorithm is presented. Under this conjecture, new upper bounds on t(3,q) are obtained. The initial part of the bounds is based on computer search, see Theorem 1.

References

1. Bartoli, D., Davydov, A.A., Faina, G., Kreshchuk, A.A., Marcugini, S., Pambianco, F.: Upper bounds on the smallest sizes of a complete arc in PG(2, q) under certain probabilistic conjectures. Probl. Inf. Transm. **50**(4), 320–339 (2014). https://doi.org/10.1134/S0032946014040036

2. Bartoli, D., Davydov, A.A., Faina, G., Kreshchuk, A.A., Marcugini, S., Pambianco, F.: Upper bounds on the smallest size of a complete arc in a finite Desarguesian projective plane based on computer search. J. Geom. **107**(1), 89–117 (2016). https://doi.org/10.1007/s00022-015-0277-z

3. Bartoli, D., Davydov, A.A., Faina, G., Kreshchuk, A.A., Marcugini, S., Pambianco, F.: Tables, bounds and graphics of sizes of complete arcs in the plane PG(2, q) for all $q \leq 321007$ and sporadic q in [323761 ... 430007] obtained by an algorithm with fixed order of points (FOP). arXiv:1404.0469 [math.CO] (2018)

4. Bartoli, D., Davydov, A.A., Giulietti, M., Marcugini, S., Pambianco, F.: New bounds for linear codes of covering radii 2 and 3. Crypt. Commun. **11**(5), 903–920 (2018). https://doi.org/10.1007/s12095-018-0335-0

5. Bartoli, D., Davydov, A.A., Marcugini, S., Pambianco, F.: Tables, bounds and graphics of short linear codes with covering radius 3 and codimension 4 and 5. arXiv:1712.07078 [cs IT] (2020)

6. Bartoli, D., Giulietti, M., Platoni, I.: On the covering radius of MDS codes. IEEE Trans. Inf. Theory **61**(2), 801–811 (2015). https://doi.org/10.1109/TIT.2014.2385084

7. Brualdi, R.A., Litsyn, S., Pless, V.: Covering radius. In: Pless, V.S., Huffman, W.C. (eds.) Handbook of Coding Theory, vol. 1, pp. 755–826. Elsevier, Amsterdam (1998)

8. Cohen, G., Honkala, I., Litsyn, S., Lobstein, A.: Covering Codes, North-Holland Mathematical Library, vol. 54. Elsevier, Amsterdam (1997)

9. Davydov, A.A., Faina, G., Marcugini, S., Pambianco, F.: Upper bounds on the smallest size of a complete cap in PG(N, q), $N \geq 3$, under a certain probabilistic conjecture. Australas. J. Combin. **72**(3), 516–535 (2018)

10. Davydov, A.A., Giulietti, M., Marcugini, S., Pambianco, F.: Linear nonbinary covering codes and saturating sets in projective spaces. Adv. Math. Commun. **5**(1), 119–147 (2011). https://doi.org/10.3934/amc.2011.5.119

11. Davydov, A.A., Marcugini, S., Pambianco, F.: New bounds for linear codes of covering radius 3 and 2-saturating sets in projective spaces. In: Proceedings of the 2019 XVI International Symposium Problems of Redundancy in Information and Control Systems (REDUNDANCY), pp. 47–52. IEEE Xplore, Moscow, Russia, October 2019 (2020). https://doi.org/10.1109/REDUNDANCY48165.2019.9003348

12. Davydov, A.A., Östergård, P.R.J.: Linear codes with covering radius $R = 2, 3$ and codimension tR. IEEE Trans. Inform. Theory **47**(1), 416–421 (2001). https://doi.org/10.1109/18.904551

13. Davydov, A.A., Östergård, P.R.J.: Linear codes with covering radius 3. Des. Codes Crypt. **54**(3), 253–271 (2010). https://doi.org/10.1007/s10623-009-9322-y

14. Etzion, T., Storme, L.: Galois geometries and coding theory. Des. Codes Crypt. **78**(1), 311–350 (2015). https://doi.org/10.1007/s10623-015-0156-5

15. Gabidulin, E.M., Kløve, T.: The Newton radius of MDS codes. In: Proceedings of the IEEE Information Theory Workshop (ITW 1998), pp. 50–51. IEEE Xplore, Killarney, Ireland, June 1998 (2002). https://doi.org/10.1109/ITW.1998.706367

16. Giulietti, M.: The geometry of covering codes: small complete caps and saturating sets in Galois spaces. In: Blackburn, S.R., Holloway, R., Wildon, M. (eds.) Surveys in Combinatorics 2013, London Math. Society Lecture Note Series, vol. 409, pp. 51–90. Cambridge University Press, Cambridge (2013). https://doi.org/10.1017/CBO9781139506748.003

17. Hirschfeld, J.W.P.: Finite Projective Spaces of Three Dimensions. Oxford University Press, Oxford (1985)

18. Hirschfeld, J.W.P.: Projective Geometries Over Finite Fields. Oxford Mathematical Monographs, 2nd edn. Clarendon Press, Oxford (1998)

19. Hirschfeld, J.W.P., Storme, L.: The packing problem in statistics, coding theory and finite projective spaces: update 2001. In: Blokhuis, A., Hirschfeld, J.W.P., Jungnickel, D., Thas, J.A. (eds.) Finite Geometries, Developments of Mathematics, Proceedings of the Fourth Isle of Thorns Conference, Chelwood Gate, vol. 3, pp. 201–246. Kluwer Academic, Boston (2001). https://doi.org/10.1007/978-1-4613-0283-4_13

20. Hirschfeld, J.W.P., Thas, J.A.: Open problems in finite projective spaces. Finite Fields Appl. **32**, 44–81 (2015). https://doi.org/10.1016/j.ffa.2014.10.006

21. Klein, A., Storme, L.: Applications of finite geometry in coding theory and cryptography. In: Crnković, D., Tonchev, V. (eds.) Security, Coding Theory and Related Combinatorics, NATO Science for Peace and Security, Series - D: Information and Communication Security, vol. 29, pp. 38–58. IOS Press (2011). https://doi.org/10.3233/978-1-60750-663-8-38

22. Landjev, I., Storme, L.: Galois geometry and coding theory. In: De. Beule, J., Storme, L. (eds.) Current Research Topics in Galois geometry, chap. 8, pp. 187–214. NOVA Academic, New York (2011)

23. Lobstein, A.: Covering radius, an online bibliography (2019). https://www.lri.fr/~lobstein/bib-a-jour.pdf. Accessed 18 Jun 2020

24. Pless, V.S.: Coding constructions. In: Pless, V.S., Huffman, W.C. (eds.) Handbook of Coding Theory, vol. 1, pp. 141–176. Elsevier, Amsterdam (1998)

25. Sonnino, A.: Transitive $PSL(2,7)$-invariant 42-arcs in 3-dimensional projective spaces. Des. Codes Crypt. **72**(2), 455–463 (2014). https://doi.org/10.1007/s10623-012-9778-z

26. Storme, L.: Small arcs in projective spaces. J. Geom. **58**(1–2), 179–191 (1997). https://doi.org/10.1007/BF01222939

Numerical Simulation of the Reactive Transport at the Pore Scale

Vadim Lisitsa[1(✉)] ![ORCID] and Tatyana Khachkova[2]

[1] Sobolev Institute of Mathematics SB RAS,
4 Koptug Avenue, Novosibirsk 630090, Russia
lisitsavv@ipgg.sbras.ru
[2] Institute of Petroleum Geology and Geophysics SB RAS,
3 Koptug Avenue, Novosibirsk 630090, Russia

Abstract. We present a numerical algorithm for simulation of the chemical fluid-solid interaction at the pore scale, focusing on the core matrix dissolution and secondary mineral precipitation. The algorithm is based on the explicit use of chemical kinetics of the heterogeneous reactions to determine the rate of the fluid-solid interface changes. Fluid flow and chemical transport are simulated by finite differences. In contrast, level-set methods with immersed boundary conditions are applied to account for arbitrary interface position (not aligned to the grid lines).

Keywords: Pore scale · Reactive transport · Level-set · Immersed boundaries · Persistence diagrams

1 Introduction

Reactive transport in porous media has a number of applications, especially related to the Earth sciences. Among others, it is worth mentioning enhanced oil recovery (EOR), especially for unconventional reservoirs [6], acid hydro-fracturing [38]. This reservoir treatment increases the permeability of the reservoir due to the partial dissolution of the rock matrix and the dissolution channels forming. Also, CO_2 sequestration is, probably, the most well-known example of reactive transport. Typically CO_2 is injected in carbonate formations, where it dissolves in a formation fluid, increasing its acidity; thus, initiating chemical reaction with the rock matrix [16,35].

The efficient organization of EOR procedures and CO_2 sequestration requires predictive hydrodynamic modeling accounting for the reactive transport at the reservoir scale. There are a number of models of these processes at the reservoir scale [1,8,13,25,29]. However, these models use empirical relations between the permeability and tortuosity of the pore space, which determine the convective and diffusion components of the process, respectively, with porosity. Moreover, these relations may vary during the dissolution process. We suggest using

The research was supported by the Russian Science Foundation grant no. 19-77-20004.

O. Gervasi et al. (Eds.): ICCSA 2020, LNCS 12249, pp. 123–134, 2020.
https://doi.org/10.1007/978-3-030-58799-4_9

numerical simulation directly at the pore scale to recover these relations for a wide range of dissolution scenarios. Mathematical model, describing the reactive transport at the pore scale includes the equations describing the fluid flow, the transfer of chemical components (convection-diffusion) and the reaction at the interface between the solid and liquid phases, taking into account the chemical-kinetic relations [5,8,10,15,21]. The main difference between the approaches is the approximation of the boundary conditions on a moving boundary. Three different methods can be mentioned. The first is based on the use of rectangular grids and the corresponding finite-difference approximations of the Stokes or Navier-Stokes equations and convection-diffusion. The reaction is taken into account by changing the "concentration" or "relative mass" of the solid phase at the boundary points [14,37]. The advantage of this approach is the ease of implementation. The disadvantages of the method include the semi-empirical specification of the rate of dissolution of the solid phase (change in "concentration") based on some averaging of the chemical properties of the medium in the vicinity of the boundary. The second group of models is characterized by an explicit treatment of the interface. In this case, the position of the boundary is set explicitly with the application of the finite volume method with truncated cells [23,24,30,33]. However, approaches of this type are difficult to implement when the topology of the computational domain changes. The third group of methods includes those with the implicit treatment of the interface position. Such methods include the level-set [17,18,26] and the phase-field method [36]. These approaches allow treating the chemical kinetics of the process explicitly. Moreover, level-set and phase-field methods easily handle topology changes. Such models for solving the Stokes equations and convection-diffusion allow using the finite difference method on a regular rectangular grid in combination with immersed boundary conditions [20,22,27,28,34].

This paper presents an algorithm for modeling chemical interaction with the rock, based on the level-set method to take into account changes in the position of the boundary using the immersed boundary approach. Section 2 provides a statement of the problem and a mathematical model. The numerical algorithms used to solve the problem are discussed in Sect. 3. The main attention is paid to the approximation of boundary conditions for a complex and changing topology of pore space. Numerical experiments are presented in Sect. 4.

2   Statement of the Problem

To simulate reactive transport at the pore scale, we use the splitting concerning physical scales. We assume that the interface changes with the slowest rate among all considered processes. Thus, this rate defines the time scale of the problem in general. Also, the fluid flow rate is considered low, which is always a case for underground flows. At the same time, the flow comes to a steady state almost instantly in case of small geometry changes. As a result, three different time scales are considered in the model separately.

We consider a bounded domain $D \subseteq R^2$, composed of the nonintersecting time-dependent subdomains $D_p(t)$ and $D_m(t)$ corresponding the pore space and the matrix, respectively. The boundary of the domain is $\partial D = \Gamma_{outlet} \cup \Gamma_{inlet} \cup \Gamma_{nf}$. We denote the interface between the pore space and the matrix as $\bar{D}_p(t) \cap \bar{D}_m(t) = \Gamma(t)$ which is a union of the sufficiently smooth curves. A sketch of hte model is presented in Fig. 1.

The fluid flow is defined in the pore space $D_p(t)$ satisfying the steady state Stokes equation:

$$\mu\nabla^2 \boldsymbol{u} - \nabla p = 0,$$
$$\nabla \cdot \boldsymbol{u} = 0. \tag{1}$$

with boundary conditions

$$\boldsymbol{u}(\boldsymbol{x}) = 0, \qquad \boldsymbol{x} \in \Gamma(t) \cup \Gamma_{nf},$$
$$p(\boldsymbol{x}) = p_{bc}(\boldsymbol{x}), \ \boldsymbol{x} \in \Gamma_{inlet}^p \cup \Gamma_{outlet}, \tag{2}$$

where $\boldsymbol{u} = (u_1, u_2)^T \in R^2$ is the velocity vector, p is the pressure, μ is the dynamic viscosity, $p_{bc}(\boldsymbol{x})$ is the pressure at the inlet boundary, $\boldsymbol{x} = (x_1, x_2)^T$ is the vector of spatial coordinates.

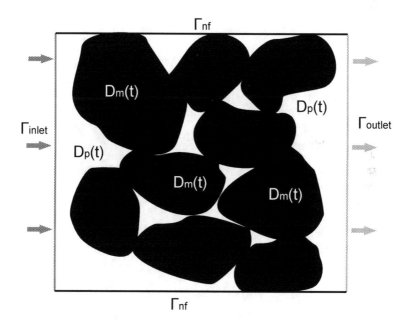

Fig. 1. A sketch of the computational domain.

The active component propagates in the pore space $D_p(t)$ according to the convection-diffusion equation:

$$\frac{\partial C}{\partial t} + \nabla \cdot (\boldsymbol{u}C - D\nabla C) = 0, \tag{3}$$

with boundary conditions:

$$
\begin{aligned}
D\frac{\partial C}{\partial n} &= k_r(C - C_s), \; \boldsymbol{x} \in \Gamma(t), \\
C &= C_{in}, \qquad\quad \boldsymbol{x} \in \Gamma_{inlet}, \\
\frac{\partial C}{\partial n} &= 0, \qquad\qquad\; \boldsymbol{x} \in \Gamma_{outlet} \cup \Gamma_{nf}.
\end{aligned}
\tag{4}
$$

In these notations D is the diffusion coefficient, C is the reactant concentration, $\boldsymbol{n}$ is the inner (with respect to $D_p(t)$) normal vector, k_r is the reaction rate coefficient, Γ_{inlet} is the inlet boundary. We consider only the first order reactions, one active reactant and no reactant sources inside the computational domain; thus the right-hand side in (3) equals to zero.

The changes in the interface $\Gamma(t)$ caused by the chemical reactions governed by the equation:

$$
v_n(\boldsymbol{x}, t) = \frac{K_c k}{\rho}(C - C_s), \; \boldsymbol{x} \in \Gamma(t),
\tag{5}
$$

where v_n is the normal component of the boundary velocity, K_c is the stoichiometric coefficient, ρ is the mass density of the matrix mineral, C_s reactant concentration at equilibrium.

2.1   Level-Set Method

We use the finite differences to solve Eqs. (1) and (3). However, the interface $\Gamma(t)$ vary and may not align to the grid lines. To deal with such an irregular geometry of the interface, we use the level-set method where the interface $\Gamma(t)$ is defined implicitly as the constant level line of a function $\varphi(\boldsymbol{x}, t)$; i.e.

$$
\Gamma(t) = \{\boldsymbol{x}|\varphi(\boldsymbol{x}) = 0\}.
$$

It is natural to require $\Gamma(t)$ as the zero-level. In this case, the subdomains D_p and D_m are defined as

$$
D_p(\boldsymbol{x}, t) = \{\boldsymbol{x}|\varphi(x, t) > 0\}, \; D_m(\boldsymbol{x}, t) = \{\boldsymbol{x}|\varphi(x, t) < 0\}.
$$

Moreover, the level-set function $\varphi(\boldsymbol{x})$ is constructed as the signed distance to the interface; i.e. $\|\nabla_x \varphi(\boldsymbol{x}, t)\| = 1$. This leads to a natural definition of the normal vector

$$
\boldsymbol{n} = \nabla_x \varphi(\boldsymbol{x}, t).
$$

Using the level-set function one may rewrite the equation for the interface changes as follows [9, 26]:

$$
\begin{aligned}
\frac{\partial \varphi(\boldsymbol{x}, t)}{\partial t} + v_n(\boldsymbol{x}, t) &= 0, \\
\varphi(\boldsymbol{x}, 0) &= \varphi_0,
\end{aligned}
\tag{6}
$$

where v_n is the normal velocity, defined by the Eq. (5).

3   Numerical Solution

To solve Eqs. (1), (3), and (6) we introduce the staggered grid, where the grid-functions are defied according to the rule $p_{i,j} = p(ih_1, jh_2)$, $(u_1)_{i+1/2,j} = u_1((i+1/2)h_1, jh_2)$, $(u_2)_{i,j+1/2} = u_2(ih_1, (j+1/2)h_2)$, $C_{i,j}^n = C(ih_1, jh_2, n\tau)$, and $varphi_{i,j}^n = \varphi(ih_1, jh_2, n\tau)$, where h_1 and h_2 are the spatial steps, τ is the grid step with respect to time (either physical or artificial). Note, that the interface $\Gamma(t)$ defined as a constant level of function φ is approximated by a piecewise linear line. Note also that the model is a micro-CT scan of the real rock sample [2–4]; thus, it is extremely detailed for either description of the pore space structure or the physical processes. As a result the low-order finite difference schemes can be used in application to pore scale simulation.

3.1   Solving Stokes Equation

We use the second order finite-difference scheme to approximate the Stokes equation, where pressure is defined in the centers of the grid cells; i.e. in points with integer indices; whereas the components of the velocity vector are defined at the faces of the cells; i.e., in points with one half-integer index:

$$\begin{aligned} \mu D_1^2[u_1]_{i+1/2,j} + D_2^2[u_1]_{i+1/2,j} - D_1^c[p]_{i+1/2,j} &= 0, \text{ if } \varphi_{i+1/2,j} > 0, \\ \mu D_1^2[u_2]_{i,j+1/2} + D_2^2[u_2]_{i,j+1/2} - D_2^c[p]_{i,j+1/2} &= 0, \text{ if } \varphi_{i,j+1/2} > 0, \\ D_1^c[u_1]_{i,j} + D_2^c[u_2]_{i,j} &= 0, \qquad \text{ if } \varphi_{i,j} > 0, \end{aligned} \tag{7}$$

In these notations D_1^c and D_2^c are the central differences approximating first derivatives with respect to spatial variables; whereas D_1^2 and D_2^2 are the second order approximations of the second derivatives:

$$D_1^2[f]_{I,J} = \frac{f_{I+1,J} - 2f_{I,J} + f_{I,J}}{h_1^2} = \left. \frac{\partial^2 f}{\partial x_1^2} \right|_{I,J} + O(h_1^2), \tag{8}$$

$$D_1^c[f]_{I,J} = \frac{f_{I+1/1,J} - f_{I,J-1/2}}{h_1} = \left. \frac{\partial f}{\partial x_1} \right|_{I,J} + O(h_1^2). \tag{9}$$

Here indices I and J can be either integer or half-integer, but i and j are integers only. The operators approximating derivatives with respect to the other spatial direction can be obtained by the permutations of the role of spatial indices. Note that level-set function $\varphi_{i,j}$ is defined at integer points, and its values in half-integer points are interpolated. We are not specifying the iterative method to solve system (1), but focus on the computation of the action of the system matrix on a vector.

Equations (1) are valid only if all the points from a stencil, centered in particular point belong to pore space $D_p(t)$. If a stencil contains points from $D_m(t)$, where the solution is not defined we suggest using the immersed boundary method [12,17,19,20,22,28,34]. The idea of this approach is to extrapolate the solution from $D_p(t)$ to the points from $D_m(t)$ which are needed for simulation. The extrapolation is based on the boundary conditions stated at the interface;

i.e., $\boldsymbol{u} = 0$ on $\Gamma(t)$; thus, velocity vector can be considered as odd function with respect to the interface, whereas p is even. Assume one needs to extrapolate the solution to the point $((x_1)_I, (x_2)_J)$. The distance from this point to the interface is defined by the level-set function $\varphi_{I,J}$ and the normal direction to the interface is $\boldsymbol{n} = \nabla_x \varphi(\boldsymbol{x})|_{I,J}$. Thus, the projection of the point to the interface is

$$((x_1)_c, (x_2)_c) = ((x_1)_I, (x_2)_J) + \varphi_{I,J} \boldsymbol{n},$$

and the orthogonal reflection has the coordinates

$$((x_1)_n, (x_2)_n) = ((x_1)_I, (x_2)_J) + 2\varphi_{I,J} \boldsymbol{n},$$

and moreover

$$\boldsymbol{u}_{I,J} = -\boldsymbol{u}((x_1)_n, (x_2)_n) + O(\varphi_{I,J}^2).$$

A standard approach to compute $\boldsymbol{u}((x_1)_n, (x_2)_n)$ is the use of bilinear interpolation using the four nearest grid points from the regular grid. However, if a pore space is considered, it is not guaranteed that the four nearest points belong to the pore space D_p. Thus, we use all available points to construct the interpolation. We pick the four regular points (those of them which belong to the pore space) and also the point at the interface $((x_1)_c, (x_2)_c)$. It is clear that to compute linear interpolation; three points are enough; thus, if more than three points are available, the problem is overdetermined, and we compute the minimal-norm solution. If less than three points are available, we can only achieve the first order of accuracy. We can use the interpolation weights proportional to the distances from the considered points to point $((x_1)_n, (x_2)_n)$.

3.2 Solution of the Convection-Diffusion Equation

To approximate the convection-diffusion equation we use the first-order scheme

$$\frac{C_{i,j}^{n+1} - C_{i,j}^n}{\tau} + D_1^1[u_1 C]_{i,j}^n + D_2^1[u_2 C]_{i,j}^n - DD_1^2[C]_{i,j}^n - DD_2^2[C]_{i,j}^n = 0, \quad (10)$$
$$\varphi_{i,j} > 0,$$

where

$$D_1^1[u_1 C]_{i,j}^n = \frac{F_{i+1/2,j} - F_{i_1-1/2,j}}{h},$$
$$F_{i+1/2} = \begin{cases} (u_1)_{i+1/2,j} C_{i+1}, & (u_1)_{i_1+1/2,j} < 0 \\ (u_1)_{i+1/2,j} C_i, & (u_1)_{i_1+1/2,j} > 0 \end{cases} \quad (11)$$

Operators, approximating the derivatives with respect to the other spatial direction, can be obtained by the permutation of the spatial indices. Operators D_1^2 and D_2^2 are introduced above in Eq. (8). Note that due to very fine discretization, the first-order scheme provides suitable accuracy.

Same as before, the hardest part of the approximation is the dealing with the boundary conditions at $\Gamma(t)$. The concentration satisfies Robin boundary conditions on $\Gamma(t)$, which can be considered a linear combination of the Dirichlet and Neumann ones to apply the immersed boundary conditions

$$D\nabla C \cdot \boldsymbol{n} + k_r C = k_r C_s.$$

We assume that the point $((x_1)_I, (x_2)_J) \in D_m$ belongs to a stencil centered in D_p. Thus an extrapolation is required. Same as before, $((x_1)_n, (x_2)_n)$ is the orthogonal reflection of the immersed point $((x_1)_I, (x_2)_J)$ over the interface; thus, the boundary condition can be approximated as follows

$$D\frac{C((x_1)_n, (x_2)_n) + C_{I,J}}{2} + k_r\frac{C((x_1)_n, (x_2)_n) - C_{I,J}}{2\varphi_{I,J}} = k_r C_s.$$

where $C((x_1)_n, (x_2)_n)$ is interpolated using available points in the domain $D_p(t)$.

3.3   Change of the Interface

To simulate the interface movement one needs to solve the Eq. (6), which can be done by the finite difference method:

$$\frac{\varphi_{i,j}^{n+1} - \varphi_{i,j}^n}{\tau} = -(v_n)_{i,j}^n, \tag{12}$$

where $(v_n)_{i,j}^n$ is the rate of the interface changes.

Equation (6) is defined everywhere in D, thus its right-hand side should be defined accordingly. However, function $v_n(x)$ is defined only at Γ and should be continued inside subdomains D_p and D_m. Following [7,26] the rate v_n can be continued inside subdomains as a constant along the normal direction to the interface; i.e., the steady-state solution of the following equation should be computed:

$$\frac{\partial q}{\partial t} + sign(\varphi)\left(\frac{\nabla\varphi}{|\nabla\varphi|} \cdot \nabla q\right) = 0,$$
$$q(x, 0) = \tilde{v}_n(x, t_0), \tag{13}$$

where $\tilde{v}_n$ are the initial conditions which coincide with the velocity at the interface and trivial elsewhere.

We use the cental differences to approximate the gradient of φ whereas the upwind scheme is used to approximate the derivatives of q [7]. Note, that accurate solution of Eq. (13) is only needed in a vicinity of the interface; i.e., inside a strip $D_s = \{x : |\varphi(x)| \leq 2\sqrt{h_1^2 + h_2^2}\}$. Thus, only a few iterations are sufficient to get accurate solution there. After that additional redistancing is applied to ensure that φ is the signed distance [9,26,31,32].

4   Numerical Experiments

We use the developed algorithm to simulate chemical fluid-solid interaction. We apply statistical simulation to generate the models. In particular, we used the truncated Gaussian field [11], where the model is represented as a realization of the Gaussian random field $G(v, l)$ with prescribed correlation length l, and point x_0 belongs to D_p, if values of the field in this point exceeds a threshold $G(x_0, l) > R$. Otherwise, the point belongs to the matrix. Porosity or the ratio of the pore space to the whole volume is defined uniquely by the threshold value R. We consider models with a fixed correlation length of $5 \cdot 10^{-5}$ m, which is the typical size of the heterogeneities of carbonate rocks. The porosity is fixed as 65%, which is too high for realistic rocks but ensures the percolation of the pore space in 2D. The size of the model is $[0, 50l_1] \times [0, 50l_2]$, where l_1 and l_2 are the correlation lengths along x_1 and x_2 respectively. We consider the models where $l_1 = l_2$. The grid step is $h_1 = h_2 = 10^{-5}$ m.

The core matrix is calcite with mass density $\rho = 2710$ kg/m^3, and stoichiometric coefficient of the reaction equal to one; i.e., $K = 1$. The fluid is the reservoir water under assumption that changes in the reactant concentration do not affect the physical properties of the fluid, thus we fix the dynamic viscosity $\mu = 0.00028$ Pa·s. To vary dissolution scenarios we change the pressure drop at the outer boundary; i.e., p_{bc} in Eq. (2); diffusion coefficient D and the reaction rate k_r in Eqs. (3) and (4). The values of the parameters are provided in the Table 1. The active component is the cations H^+ with equilibrium concentration corresponding to $pH = 7$, whereas acidity at the inlet was $pH = 2.3$ following laboratory experiments presented in [16].

It is convenient to use dimensionless variables to describe the results of the simulation. These variables are Reynolds number $Re = \frac{LU}{\nu}$; Peclet number $Pe = \frac{UL}{D}$, indicating if convection of diffusion dominates in the reactant transport; Damkohler number $Da = \frac{k_r L}{D}$, indicating if the reaction is diffusion-limited or kinetic limited. In these notations, U is characteristic flow velocity, L is a characteristic length scale, we assume it to be equal to the correlation length of the Gaussian field. During chemical fluid-solid interaction, the matrix of the core dissolved. Thus both L and U may change. In our further description, we will refer to the values corresponding to the initial time.

To illustrate the applicability of the algorithm for simulation of the chemical fluid-solid interaction, we present the results of six numerical experiments corresponding different combinations of the Reynolds, Peclet, and Damkohler numbers, as presented in Table 1. Figures 2, 3 and 4 represent the distribution of the pH at different time instants for six experiments. Localized dark spots correspond to the core matrix. Models 1,3,5 correspond to low Damkohler numbers (diffusion-controlled reactions) if compared with models 2,4,6. The diffusion-controlled reactions are slower, and the reactant is delivered in all channels of the pore space, but its access to the interface is limited. Note that the increase of the Reynolds number causes deeper penetration of the reactant to the sample (experiment 5). If the kinetic-controlled reactions are considered (experiments 2,4,6), the reactant can not penetrate the sample and interacts with the matrix close to the inlet regardless of the values of the Re and Pe numbers.

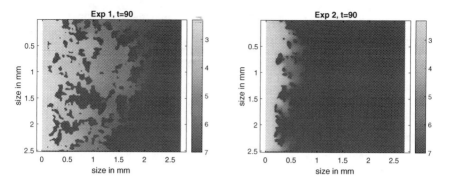

Fig. 2. *pH* in the pore space for experiments 1 (left) and 2 (right) at final time instant.

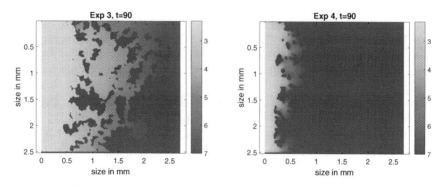

Fig. 3. *pH* in the pore space for experiments 3 (left) and 4 (right) at final time instant.

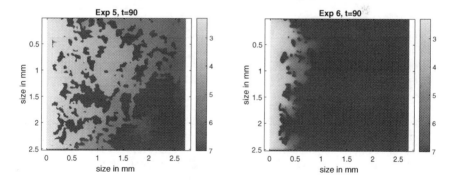

Fig. 4. *pH* in the pore space for experiments 5 (left) and 6 (right) at final time instant.

Table 1. Values of the parameters in the numerical experiments

	Re	Pe	Da	p_{bc}	D	k_r
1	$1.2 \cdot 10^{-6}$	$5.3 \cdot 10^{-3}$	$8 \cdot 10^{-2}$	$5 \cdot 10^{-2}$	$9.3 \cdot 10^{-8}$	$1.5 \cdot 10^{-4}$
2	$1.2 \cdot 10^{-6}$	$5.3 \cdot 10^{-3}$	$8 \cdot 10^{0}$	$5 \cdot 10^{-2}$	$9.3 \cdot 10^{-8}$	$1.5 \cdot 10^{-2}$
3	$1.2 \cdot 10^{-5}$	$5.3 \cdot 10^{-3}$	$8 \cdot 10^{-2}$	$5 \cdot 10^{-1}$	$9.3 \cdot 10^{-7}$	$1.5 \cdot 10^{-3}$
4	$1.2 \cdot 10^{-5}$	$5.3 \cdot 10^{-3}$	$8 \cdot 10^{0}$	$5 \cdot 10^{-1}$	$9.3 \cdot 10^{-7}$	$1.5 \cdot 10^{-1}$
5	$1.2 \cdot 10^{-5}$	$5.3 \cdot 10^{-2}$	$8 \cdot 10^{-2}$	$5 \cdot 10^{-1}$	$9.3 \cdot 10^{-8}$	$1.5 \cdot 10^{-4}$
6	$1.2 \cdot 10^{-5}$	$5.3 \cdot 10^{-2}$	$8 \cdot 10^{0}$	$5 \cdot 10^{-1}$	$9.3 \cdot 10^{-8}$	$1.5 \cdot 10^{-2}$

5   Conclusions

We presented a numerical algorithm for the simulation of chemical fluid-solid interaction at the pore scale. The algorithm is based on splitting with respect to the physical processes. It is assumed that the fluid flow gets steady instantly after small changes of the pore space geometry. Moreover, the flow velocity is low; thus, flow is simulated as a solution to the Stokes equation. The convection-diffusion equation governs chemical transport with Robin boundary conditions. The pore space - matrix interface is defined implicitly by the level-set method where the boundary conditions are approximated using the immersed boundary method. Presented numerical experiments illustrate the applicability of the suggested approach for simulation chemical fluid-solid interaction for a wide range of models, including different types of rocks and artificial porous materials.

References

1. Amikiya, A.E., Banda, M.K.: Modelling and simulation of reactive transport phenomena. J. Comput. Sci. **28**, 155–167 (2018)
2. Andra, H., et al.: Digital rock physics benchmarks - part I: imaging and segmentation. Comput. Geosci. **50**, 25–32 (2013)
3. Andra, H., et al.: Digital rock physics benchmarks - part II: computing effective properties. Comput. Geosci. **50**, 33–43 (2013)
4. Bazaikin, Y., et al.: Effect of CT image size and resolution on the accuracy of rock property estimates. J. Geophys. Res. Solid Earth **122**(5), 3635–3647 (2017)
5. Bouchelaghem, F.: A numerical and analytical study on calcite dissolution and gypsum precipitation. Appl. Math. Model. **34**(2), 467–480 (2010)
6. Emberley, S., Hutcheon, I., Shevalier, M., Durocher, K., Gunter, W.D., Perkins, E.H.: Geochemical monitoring of fluid-rock interaction and CO2 storage at the weyburn CO2-injection enhanced oil recovery site, Saskatchewan, Canada. Energy **29**(9), 1393–1401 (2004)
7. Fedkiw, R.P., Aslam, T., Merriman, B., Osher, S.: A non-oscillatory eulerian approach to interfaces in multimaterial flows (the ghost fluid method). J. Comput. Phys. **152**(2), 457–492 (1999)
8. Ghommem, M., Zhao, W., Dyer, S., Qiu, X., Brady, D.: Carbonate acidizing: modeling, analysis, and characterization of wormhole formation and propagation. J. Pet. Sci. Eng. **131**, 18–33 (2015)

9. Gibou, F., Fedkiw, R., Osher, S.: A review of level-set methods and some recent applications. J. Comput. Phys. **353**, 82–109 (2018)
10. Hao, Y., Smith, M., Sholokhova, Y., Carroll, S.: CO2-induced dissolution of low permeability carbonates. Part II: numerical modeling of experiments. Adv. Water Resour. **62**(Part C), 388–408 (2013)
11. Hyman, J.D., Winter, C.L.: Stochastic generation of explicit pore structures by thresholding Gaussian random fields. J. Comput. Phys. **277**, 16–31 (2014)
12. Johansen, H., Colella, P.: A cartesian grid embedded boundary method for poisson's equation on irregular domains. J. Comput. Phys. **147**(1), 60–85 (1998)
13. Kalia, N., Balakotaiah, V.: Effect of medium heterogeneities on reactive dissolution of carbonates. Chem. Eng. Sci. **64**(2), 376–390 (2009)
14. Kang, Q., Chen, L., Valocchi, A.J., Viswanathan, H.S.: Pore-scale study of dissolution-induced changes in permeability and porosity of porous media. J. Hydrol. **517**, 1049–1055 (2014)
15. Leal, A.M.M., Blunt, M.J., LaForce, T.C.: A robust and efficient numerical method for multiphase equilibrium calculations: application to CO2-brine-rock systems at high temperatures, pressures and salinities. Adv. Water Res. **62**(Part C), 409–430 (2013)
16. Lebedev, M., Zhang, Y., Sarmadivaleh, M., Barifcani, A., Al-Khdheeawi, E., Iglauer, S.: Carbon geosequestration in limestone: pore-scale dissolution and geomechanical weakening. Int. J. Greenhouse Gas Control **66**, 106–119 (2017)
17. Li, X., Huang, H., Meakin, P.: Level set simulation of coupled advection-diffusion and pore structure evolution due to mineral precipitation in porous media. Water Resour. Res. **44**(12), W12407 (2008)
18. Li, X., Huang, H., Meakin, P.: A three-dimensional level set simulation of coupled reactive transport and precipitation/dissolution. Int. J. Heat Mass Transf. **53**(13), 2908–2923 (2010)
19. Luo, K., Zhuang, Z., Fan, J., Haugen, N.E.L.: A ghost-cell immersed boundary method for simulations of heat transfer in compressible flows under different boundary conditions. Int. J. Heat Mass Transf. **92**, 708–717 (2016)
20. Marella, S., Krishnan, S., Liu, H., Udaykumar, H.S.: Sharp interface cartesian grid method i: an easily implemented technique for 3D moving boundary computations. J. Comput. Phys. **210**(1), 1–31 (2005)
21. Meirmanov, A., Omarov, N., Tcheverda, V., Zhumaly, A.: Mesoscopic dynamics of solid-liquid interfaces: a general mathematical model. Siberian Electron. Math. Rep. **12**, 144–160 (2015)
22. Mittal, R., Iaccarino, G.: Immersed boundary methods. Ann. Rev. Fluid Mech. **37**(1), 239–261 (2005)
23. Molins, S., Trebotich, D., Steefel, C.I., Shen, C.: An investigation of the effect of pore scale flow on average geochemical reaction rates using direct numerical simulation. Water Resour. Res. **48**(3), W03527 (2012)
24. Molins, S., et al.: Pore-scale controls on calcite dissolution rates from flow-through laboratory and numerical experiments. Environ. Sci. Technol. **48**(13), 7453–7460 (2014)
25. Mou, J., Zhang, S.: Modeling acid leakoff during multistage alternate injection of pad and acid in acid fracturing. J. Nat. Gas Sci. Eng. **26**, 1161–1173 (2015)
26. Osher, S., Fedkiw, R.P.: Level set methods: an overview and some recent results. J. Comput. Phys. **169**(2), 463–502 (2001)
27. Peskin, C.S.: Flow patterns around heart valves: a numerical method. J. Comput. Phys. **10**, 252–271 (1972)

28. Sotiropoulos, F., Yang, X.: Immersed boundary methods for simulating fluid-structure interaction. Prog. Aerosp. Sci. **65**, 1–21 (2014)

29. Steefel, C.I., et al.: Reactive transport codes for subsurface environmental simulation. Comput. Geosci. **19**(3), 445–478 (2014). https://doi.org/10.1007/s10596-014-9443-x

30. Steefel, C.I., Lasaga, A.C.: A coupled model for transport of multiple chemical species and kinetic precipitation/dissolution reactions with application to reactive flow in single phase hydrothermal systems. Am. J. Sci. **294**(5), 529–592 (1994)

31. Sussman, M., Fatemi, E.: An efficient, interface-preserving level set redistancing algorithm and its application to interfacial incompressible fluid flow. SIAM J. Sci. Comput. **20**(4), 1165–1191 (1999)

32. Sussman, M., Fatemi, E., Smereka, P., Osher, S.: An improved level set method for incompressible two-phase flows. Comput. Fluids **27**(5), 663–680 (1998)

33. Trebotich, D., Adams, M.F., Molins, S., Steefel, C.I., Shen, C.: High-resolution simulation of pore-scale reactive transport processes associated with carbon sequestration. Comput. Sci. Eng. **16**(6), 22–31 (2014)

34. Tseng, Y.H., Ferziger, J.H.: A ghost-cell immersed boundary method for flow in complex geometry. J. Comput. Phys. **192**(2), 593–623 (2003)

35. Vanorio, T., Nur, A., Ebert, Y.: Rock physics analysis and time-lapse rock imaging of geochemical effects due to the injection of CO2 into reservoir rocks. Geophysics **76**(5), O23–O33 (2011)

36. Xu, Z., Meakin, P.: Phase-field modeling of solute precipitation and dissolution. J. Chem. Phys. **129**(1), 014705 (2008)

37. Yoon, H., Valocchi, A.J., Werth, C.J., Dewers, T.: Pore-scale simulation of mixing-induced calcium carbonate precipitation and dissolution in a microfluidic pore network. Water Resour. Res. **48**(2), W02524 (2012)

38. Zimmermann, G., Blöcher, G., Reinicke, A., Brandt, W.: Rock specific hydraulic fracturing and matrix acidizing to enhance a geothermal system - concepts and field results. Tectonophysics **503**(1), 146–154 (2011)

Multivalue Almost Collocation Methods with Diagonal Coefficient Matrix

Dajana Conte[1], Raffaele D'Ambrosio[2], Maria Pia D'Arienzo[1(✉)], and Beatrice Paternoster[1]

[1] Department of Mathematics, University of Salerno, Via Giovanni Paolo II 132, 84084 Fisciano, Italy
{dajconte,mdarienzo,beapat}@unisa.it
[2] Department of Engineering and Computer Science and Mathematics, University of L'Aquila, Via Vetoio, Loc. Coppito, 67100 L'Aquila, Italy
raffaele.dambrosio@univaq.it

Abstract. We introduce a family of multivalue almost collocation methods with diagonal coefficient matrix for the numerical solution of ordinary differential equations. The choice of this type of coefficient matrix permits a reduction of the computational cost and a parallel implementation. Collocation gives a continuous extension of the solution which is useful for a variable step size implementation. We provide examples of A-stable methods with two and three stages and order 3.

Keywords: Multivalue methods · Almost collocation · Schur analysis

1 Introduction

Consider the initial value problem:

$$\begin{cases} y'(t) = f(y(t)), \, t \in [t_0, T], \\ y(t_0) = y_0, \end{cases} \tag{1}$$

$f : \mathbb{R}^k \to \mathbb{R}^k$. Multivalue methods are a large class of numerical methods used to solve (1). Classical methods for the solution of ordinary differential equations, such as Runge Kutta and linear multistep methods, are special cases of these methods [3,5,6,28,45]. Multivalue methods have also been treated as geometric numerical integrators in [4,25,28].

Multivalue methods are characterized by the abscissa vector $\mathbf{c} = [c_1, c_2, ..., c_m]^T$ and four coefficient matrices $\mathbf{A} = [a_{ij}]$, $\mathbf{U} = [u_{ij}]$, $\mathbf{B} = [b_{ij}]$ and $\mathbf{V} = [v_{ij}]$, where:

$$\mathbf{A} \in \mathbb{R}^{m \times m}, \qquad \mathbf{U} \in \mathbb{R}^{m \times r}, \qquad \mathbf{B} \in \mathbb{R}^{r \times m}, \qquad \mathbf{V} \in \mathbb{R}^{r \times r}.$$

O. Gervasi et al. (Eds.): ICCSA 2020, LNCS 12249, pp. 135–148, 2020.
https://doi.org/10.1007/978-3-030-58799-4_10

On the uniform grid $t_n = t_0 + nh, n = 0, 1, ..., N, Nh = T - t_0$, the method takes the form:

$$
\begin{aligned}
Y_i^{[n]} &= h \sum_{j=1}^{m} a_{ij} f\left(Y_j^{[n]}\right) + \sum_{j=1}^{r} u_{ij} y_j^{[n-1]}, \ i = 1, 2, ..., m, \\
y_i^{[n]} &= h \sum_{j=1}^{m} b_{ij} f\left(Y_j^{[n]}\right) + \sum_{j=1}^{r} v_{ij} y_j^{[n-1]}, \ \ i = 1, 2, ..., r,
\end{aligned}
\tag{2}
$$

$n = 0, ..., N$, where m is the number of internal stages and r is the number of external stages.

Multivalue methods are defined by a starting procedure $\mathcal{S}_h$ for the computation of the starting vector, a forward procedure $\mathcal{G}_h$, which updates the vector of the approximations at each step point and a finishing procedure $\mathcal{F}_h$, which permits to compute the corresponding numerical solution. These methods can be extended using collocation in order to obtain a smooth solution. Collocation is a technique which approximates the solution with continuous approximants belonging to a finite dimensional space (usually algebraic polynomials). The approximation satisfies interpolation conditions at the grid points and satisfies the differential equations on the collocation points [46,48,49,51,52,56]. Those methods are very effective because they permit to avoid the order reduction typical of Runge Kutta methods, also in presence of stiffness. Stiff problems arise in many relevant mathematical models [44,50,55], therefore they are object of wide attention in the literature, see [7,43,47,54] and references therein.

Because of the implicitness of such methods, the computational cost of the integration process is strictly connected to the numerical solution of non linear systems of external stages at each time step of dimension mk, where k is the dimension of system (1) and m is the number of stages. We focus on the development of methods with diagonal coefficient matrix $\mathbf{A}$ in (2), for which the nonlinear system of mk equations reduces to m independent systems of dimension k, thus it is possible to reduce the computational effort and to parallelize the method.

In order to build A-stable methods, it is not possible to impose all the collocation conditions and thus we consider almost collocation [27,34]. Collocation and almost collocation methods are widely used also for the solution of integral and integro-differential equations [9,17,23].

The organization of this paper is as follows. In Sect. 2 we summarize multivalue methods, describing their formulation and some results about order conditions. In Sect. 3 we discuss the construction of almost collocation methods with diagonal coefficient matrix. In Sect. 4 we present some examples of methods with two and three stages. In Sect. 5 numerical results are provided. Finally, in Sect. 6 some concluding remarks are given and plans for future research are outlined.

2   Multivalue Collocation Methods

Multivalue collocation methods described in [39] are of the form (2) where the external stages have the Nordsieck form:

$$
y^{[n]} = \begin{bmatrix} y_1^{[n]} \\ y_2^{[n]} \\ \vdots \\ y_r^{[n]} \end{bmatrix} \approx \begin{bmatrix} y(x_n) \\ hy'(x_n) \\ \vdots \\ h^{r-1}y^{r-1}(x_n) \end{bmatrix}
\tag{3}
$$

and the piecewise collocation polynomial:

$$
P_n(t_n + \theta h) = \sum_{i=1}^{r} \alpha_i(\theta) y_i^{[n]} + h \sum_{i=1}^{m} \beta_i(\theta) f(P_n(t_n + c_i h)), \quad \theta \in [0,1],
\tag{4}
$$

provides a dense approximation to the solution of (1). We impose the following interpolation conditions:

$$
P_n(t_n) = y_1^{[n]}, \quad hP_n'(t_n) = y_2^{[n]}, \quad \ldots \quad h^{r-1}P_n^{(r-1)}(t_n) = y_r^{[n]},
$$

and collocation conditions

$$
P_n'(t_n + c_i h) = f(P_n(t_n + c_i h)), \quad i = 1, 2, .., m.
$$

We can observe that the polynomial (4) has globally class C^{r-1} while most interpolants based on Runge-Kutta methods only have global C^1 continuity [41,42].

So, the matrices of multivalue methods assume the following form:

$$
\mathbf{A} = [\beta_j(c_i)]_{i,j=1,\ldots,m}, \qquad \mathbf{U} = [\alpha_j(c_i)]_{i=1,\ldots,m,j=1,\ldots,r},
$$

$$
\mathbf{B} = \left[\beta_j^{(i-1)}(1)\right]_{i=1,\ldots,m,j=1,\ldots,r}, \qquad \mathbf{V} = \left[\alpha_j^{(i-1)}(1)\right]_{i,j=1,\ldots,r}.
$$

We, now, summarize some important results regarding the order of the method [39].

Theorem 1. *A multivalue collocation method given by the approximation* $P_n(t_n + \theta h)$ *in (4),* $\theta \in [0,1]$, *is an approximation of uniform order* p *to the solution of the well-posed problem (1) if and only if*

$$
\alpha_1(\theta) = 1
\tag{5}
$$

$$
\frac{\theta^\nu}{\nu!} - \alpha_{\nu+1}(\theta) - \sum_{i=1}^{m} \frac{c_i^{\nu-1}}{(\nu-1)!} \beta_i(\theta) = 0, \quad \nu = 1, \ldots, r-1,
\tag{6}
$$

$$
\frac{\theta^\nu}{\nu!} - \sum_{i=1}^{m} \frac{c_i^{\nu-1}}{(\nu-1)!} \beta_i(\theta) = 0, \quad \nu = r, \ldots, p.
\tag{7}
$$

Corollary 1. *The uniform order of convergence for a multivalue collocation method* (4) *is* $m + r - 1$.

Theorem 2. *An A-stable multivalue collocation method* (4) *fulfills the constraint* $r \leq m + 1$.

Theorem 3. *The order conditions in* (5)–(7) *imply:*

$$\alpha_j(0) = \delta_{j1}, \ \alpha_j^{(\nu)}(0) = \delta_{j,\nu+1}, \quad j = 1, 2, ..., r, \quad \nu = 1, 2, ..., r - 1, \tag{8}$$

$$\beta_j(0) = \beta_j^{(\nu)}(0) = 0, \quad j = 1, 2, ..., m, \quad \nu = 1, 2, ..., r - 1, \tag{9}$$

$$\alpha_j'(c_i) = 0, \quad i = 1, 2, ..., r, \quad j = 1, 2, ..., m, \tag{10}$$

$$\beta_j'(c_i) = \delta_{ij}, \quad i, j = 1, 2, ..., m, \tag{11}$$

being δ_{ij} *the usual Kronecker delta.*

Proof. The conditions (9) follow immediately by substituting $\theta = 0$ in (7) and in its derivatives. The conditions (8) follow from (5)–(6) and (9), substituting $\theta = 0$. To show (11), we differentiate (7) and replace $\theta = c_i, i = 1, ..., m$, while (10) is derived by the differentiation of (6), putting $\theta = c_i, i = 1, ..., m$, and (11). $\quad\square$

3 Construction of Almost Collocation Multivalue Methods with Diagonal Coefficient Matrix

The computational cost of the method (2) is strictly connected to the structure of the matrix $\mathbf{A}$. In order to reduce this cost, we want to construct a multivalue method with a diagonal matrix $\mathbf{A}$, so we have to determine the functional basis $\{\beta_j(\theta), j = 1, ..., m\}$ such that $\beta_j(c_i) = 0$ for $i \neq j$. In this way, we can not impose all the collocation conditions, but we have to relax some of them.

The following theorem holds.

Theorem 4. *A multivalue collocation method* (4) *has a diagonal coefficient matrix* $\mathbf{A}$ *and order* $p = r - 1$ *if*

$$\beta_j(\theta) = \omega_j(\theta) \prod_{k=1, k \neq j}^{m} (\theta - c_k), \quad j = 1, ..., m, \tag{12}$$

where $\omega_j(\theta)$ *is a polynomial of degree* $r - m + 1$:

$$\omega_j(\theta) = \sum_{k=0}^{r-m+1} \mu_k^{(j)} \theta^k, \tag{13}$$

and

$$\alpha_1(\theta) = 1 \tag{14}$$

$$\frac{\theta^\nu}{\nu!} - \alpha_{\nu+1}(\theta) - \sum_{i=1}^{m} \frac{c_i^{\nu-1}}{(\nu - 1)!} \omega_i(\theta) \prod_{k=1, k \neq i}^{m} (\theta - c_k) = 0, \ \nu = 1, ..., r - 1. \tag{15}$$

Proof. We want **A** to be diagonal, so we have to impose that $\beta_j(c_i) = 0$ for $i \neq j$. If we substitute c_i in (12), we obtain:

$$\beta_j(c_i) = \omega_j(c_i) \prod_{k=1, k \neq j}^{m} (c_i - c_k) = 0, \quad j = 1, ..., m, \tag{16}$$

so (12) is proved. Moreover, (14)–(15) are obtained by replacing (12) in (5)–(6). □

We observe that $\beta_j(\theta), j = 1, ..., m$, are polynomial of degree r and conditions (14)–(15) permit to compute the functions $\alpha_i(\theta), i = 1, ..., r$, from $\beta_j(\theta)$. In the following we will fix $r = m + 1$. The parameters $\mu_k^{(j)}$ are free parameters which can be chosen in order to obtain A-stable methods. In searching for A-stable formulae, we have to analyze the properties of the stability matrix:

$$\mathbf{M}(z) = \mathbf{V} + z\mathbf{B}(I - z\mathbf{A})^{-1}\mathbf{U}, \tag{17}$$

where I is the identity matrix in $\mathbb{R}^{m \times m}$. In particular, we are interested in the computation of the roots of the stability function of the method:

$$p(\omega, z) = det(\omega I - \mathbf{M}(z)). \tag{18}$$

This roots have to be in the unit circle for all $z \in \mathbb{C}$ such that $Re(z) \leq 0$. By the maximul principle, that will happen if the denominator of $p(\omega, z)$ does not have poles in the negative half plane $\mathbb{C}_-$ and if the roots of the $p(\omega, iy)|$ are in the unit circle for all $y \in \mathbb{R}$. The last condition can be verified using the following Schur criterion.

Criterion 5. *Consider the polynomial [47]*

$$\phi(\omega) = c_k\omega^k + c_{k-1}\omega^{k-1} + ... + c_1\omega + c_0, \tag{19}$$

where c_i are complex coefficient, $c_k \neq 0$ and $c_0 \neq 0$, $\phi(\omega)$ is said to be a Schur polynomial if all its roots $\omega_i, i = 1, 2, ..., k$ are inside the unit circle. Define

$$\hat{\phi}(\omega) = \bar{c}_0\omega^k + \bar{c}_1\omega^{k-1} + ... + \bar{c}_{k-1}\omega + \bar{c}_k, \tag{20}$$

where $\bar{c}_i$ is the complex conjugate of c_i. Define also the polynomial

$$\phi_1(\omega) = \frac{1}{\omega}\left(\hat{\phi}(0)\phi(\omega) - \phi(0)\hat{\phi}(\omega)\right) \tag{21}$$

of degree at most $k - 1$. The following theorem holds.

Theorem 6. *(Schur). $\phi(\omega)$ is a Schur polynomial if and only if*

$$\left|\hat{\phi}(0)\right| > |\phi(0)| \tag{22}$$

and $\phi_1(\omega)$ is a Schur polynomial [47].

4   Examples of Methods

In this section we present examples of A-stable methods with two and three stages.

4.1   Two-Stage Methods

According to Theorem 4, we fix $\beta_j(\theta)$ as in (12) with $m = 2$ and $r = 3$, so $\omega_j(\theta)$ are polynomial of degree 2 of the form:

$$\omega_j(\theta) = \mu_0^{(j)} + \mu_1^{(j)}\theta + \mu_2^{(j)}\theta^2, \tag{23}$$

and the collocation polynomial is:

$$P_n(t_n + \vartheta h) = y_1^{[n]} + \alpha_2(\vartheta)y_2^{[n]} + \alpha_3(\vartheta)y_3^{[n]} + h\left(\beta_1(\vartheta)f(P(t_n + c_1 h))\right.$$
$$\left. + \beta_2(\vartheta)f(P(t_n + c_2 h))\right).$$

We choose the values for the parameters $\mu_k^{(j)}$ in (23) by imposing the condition (7) for $\nu = r$ and by performing the Schur analysis of the characteristic polynomial of the stability matrix corresponding to the Butcher tableau:

$$\left[\begin{array}{c|c}\mathbf{A} & \mathbf{U} \\ \hline \mathbf{B} & \mathbf{V}\end{array}\right] = \left[\begin{array}{cc|ccc} \beta_1(c_1) & 0 & 1 & \alpha_2(c_1) & \alpha_3(c_1) \\ 0 & \beta_2(c_2) & 1 & \alpha_2(c_2) & \alpha_3(c_2) \\ \hline \beta_1(1) & \beta_2(1) & 1 & \alpha_2(1) & \alpha_3(1) \\ \beta_1'(1) & \beta_2'(1) & 0 & \alpha_2'(1) & \alpha_3'(1) \\ \beta_1''(1) & \beta_2''(1) & 0 & \alpha_2''(1) & \alpha_3''(1) \end{array}\right] \tag{24}$$

We obtain

$$\mu_0^{(1)} = 0, \quad \mu_1^{(1)} = \frac{1}{3(c_1 - c_2)}, \quad \mu_2^{(1)} = 0,$$
$$\mu_0^{(2)} = 0, \quad \mu_1^{(2)} = -\frac{c_1}{3(c_1 - c_2)c_2}, \quad \mu_2^{(2)} = \frac{1}{3c_2^2},$$

so

$$\alpha_2(\vartheta) = \frac{-\vartheta^3 + \vartheta^2(c_1 + c_2) + \vartheta(2c_2^2 - c_1 c_2)}{3c_2^2},$$

$$\alpha_3(\vartheta) = \frac{-2\vartheta^3 + \vartheta^2(2c_1 + 3c_2) - 2c_1 c_2 \vartheta}{6c_2}, \tag{25}$$

$$\beta_1(\vartheta) = \frac{\vartheta(\vartheta - c_2)}{3(c_1 - c_2)}, \quad \beta_2(\vartheta) = \frac{(\vartheta^2(c_2 - c_1) + c_1 c_2 \vartheta)(c_1 - \vartheta)}{3c_2^2(c_1 - c_2)}.$$

These methods have order 3. Figure 1 shows the region of A-stability in the (c_1, c_2) plane obtained from the Schur analysis of the method (24)–(25).

As an example, we chose $c_1 = 3$ and $c_2 = 29/10$, obtaining:

$$\alpha_2(\vartheta) = \vartheta\left(-\frac{25}{216}\vartheta^2 + \frac{\vartheta}{432} + \frac{359}{360}\right), \quad \alpha_3(\vartheta) = \vartheta\left(-\frac{5}{36}\vartheta^2 + \frac{61}{360}\vartheta + \frac{119}{300}\right),$$

$$\beta_1(\vartheta) = \frac{2}{3}\vartheta\left(\vartheta - \frac{6}{5}\right), \quad \beta_2(\vartheta) = \vartheta\left(\frac{25}{216}\vartheta^2 - \frac{289}{432}\vartheta + \frac{289}{360}\right).$$

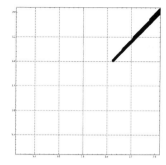

Fig. 1. Region of A-stability in the (c_1, c_2) plane.

which is the continuous C^2 extension of uniform order $p = 3$ of the A-stable multivalue method:

$$
\left[\begin{array}{c|c} \mathbf{A} & \mathbf{U} \\ \hline \mathbf{B} & \mathbf{V} \end{array}\right] =
\left[\begin{array}{cc|ccc}
1 & 0 & 1 & 2 & \dfrac{3}{2} \\[2ex]
0 & \dfrac{29}{30} & 1 & \dfrac{29}{15} & \dfrac{841}{600} \\[2ex]
\hline
\dfrac{209}{15} & -\dfrac{37520}{2523} & 1 & \dfrac{24446}{12615} & \dfrac{1589}{870} \\[2ex]
-\dfrac{62}{5} & \dfrac{11260}{841} & 0 & \dfrac{47}{4205} & -\dfrac{91}{145} \\[2ex]
-\dfrac{91}{15} & \dfrac{5660}{841} & 0 & -\dfrac{8369}{12165} & -\dfrac{46}{145}
\end{array}\right]
$$

4.2   Three-Stage Multivalue Almost Collocation Method

Now we fix $m = 3$ and $r = 4$, so:

$$
P_n(t_n + \vartheta h) = y_1^{[n]} + \alpha_2(\vartheta)y_2^{[n]} + \alpha_3(\vartheta)y_3^{[n]} + \alpha_4(\vartheta)y_4^{[n]}
$$
$$
+h\left(\beta_1(\vartheta)f(P(t_n + c_1 h)) + \beta_2(\vartheta)f(P(t_n + c_2 h)) + \beta_3(\vartheta)f(P(t_n + c_3 h))\right).
$$

According to Theorem 4 we obtain a method of order $p = 3$, by fixing $\beta_j(\theta)$ as in (12) and finding the parameters $\mu_k^{(j)}$ such that the method is A-stable, so we perform Schur analysis of the characteristic polynomial of the stability matrix corresponding to the Butcher tableau:

$$
\left[\begin{array}{c|c} \mathbf{A} & \mathbf{U} \\ \hline \mathbf{B} & \mathbf{V} \end{array}\right] =
\left[\begin{array}{ccc|cccc}
\beta_1(c_1) & 0 & 0 & 1 & \alpha_2(c_1) & \alpha_3(c_1) & \alpha_4(c_1) \\
0 & \beta_2(c_2) & 0 & 1 & \alpha_2(c_2) & \alpha_3(c_2) & \alpha_4(c_2) \\
0 & 0 & \beta_3(c_3) & 1 & \alpha_2(c_3) & \alpha_3(c_3) & \alpha_4(c_3) \\
\hline
\beta_1(1) & \beta_2(1) & \beta_3(1) & 1 & \alpha_2(1) & \alpha_3(1) & \alpha_4(1) \\
\beta_1'(1) & \beta_2'(1) & \beta_3'(1) & 0 & \alpha_2'(1) & \alpha_3'(1) & \alpha_4'(1) \\
\beta_1''(1) & \beta_2''(1) & \beta_3''(1) & 0 & \alpha_2''(1) & \alpha_3''(1) & \alpha_4''(1) \\
\beta_1'''(1) & \beta_2'''(1) & \beta_3'''(1) & 0 & \alpha_2'''(1) & \alpha_3'''(1) & \alpha_4'''(1)
\end{array}\right]
$$

We find:

$$\alpha_2(\vartheta) = \left(-\frac{2\vartheta^4}{c_3^2} + \frac{2\vartheta}{3} + \frac{2\vartheta^3}{c_3^2}(c_1 + c_2 + c_3)\right)(5c_1^2 + 2c_2^2 - 7c_3^2)$$
$$+ \left(-\frac{2\vartheta^2}{c_3^2}(c_1c_2 + c_1c_3 + c_2c_3) + \frac{2\vartheta}{c_3^2}c_1c_2c_3\right)(5c_1^2 + 2c_2^2 - 7c_3^2),$$

$$\alpha_3(\vartheta) = \left(\frac{2\vartheta^4}{c_3} - \frac{2\vartheta^3}{c_3}(c_1 + c_2 + c_3)\right)(5c_1(c_3 - c_1) + 2c_2(c_3 - c_2))$$
$$+ \left(\frac{2\vartheta^2}{c_3}(c_1c_2 + c_1c_3 + c_2c_3) - \frac{2\vartheta}{c_3}c_1c_2\right)(5c_1(c_3 - c_1) + 2c_2(c_3 - c_2))$$
$$+ \frac{\vartheta^2}{6c_3}c_2(c_1c_2 + c_1c_3 + c_2c_3), \quad \alpha_4(\theta) = 0,$$

$$\beta_1(\vartheta) = \omega_1(\vartheta)(\vartheta - c_2)(\vartheta - c_3), \quad \beta_2(\vartheta) = \omega_2(\vartheta)(\vartheta - c_1)(\vartheta - c_3),$$
$$\beta_3(\vartheta) = \omega_3(\vartheta)(\vartheta - c_1)(\vartheta - c_2),$$

where

$$\omega_1(\vartheta) = \frac{\vartheta(30(\vartheta - c_1)(c_1c_2 + c_1c_3 - c_2c_3 - c_1^2) + 1)}{3(c_1 - c_2)(c_1 - c_3)},$$

$$\omega_2(\vartheta) = \frac{\vartheta(12(\vartheta - c_2)(c_1c_2 - c_1c_3 + c_2c_3 - c_2^2) + 1)}{3(c_2 - c_1)(c_2 - c_3)},$$

$$\omega_3(\vartheta) = \frac{\vartheta((\vartheta(30c_1^2 + 12c_2^2) - 30c_1^2c_3 - 12c_2^2c_3)(c_1c_3 - c_1c_2 + c_2c_3 - c_3^2) - c_3^2)}{3c_3^2(c_1 - c_3)(c_3 - c_2)}.$$

For those polynomials, we perform again Schur analysis fixing one value for time of the abscissa coefficients. So Fig. 2 shows the regions of A-stability in the (c_2, c_3) plane for $c_1 = 9/5$, in the (c_1, c_3) plane for $c_2 = 8/5$ and in the (c_1, c_2) plane for $c_3 = 17/10$, respectively.

As an example, we chose $c_1 = 9/5$, $c_2 = 8/5$ and $c_3 = 17/10$, obtaining:

$$\alpha_2(\vartheta) = \vartheta\left(-\frac{218}{289}\vartheta^3 + \frac{327}{85}\vartheta^2 - \frac{47197}{7225}\vartheta + \frac{27794}{6375}\right),$$

$$\alpha_3(\vartheta) = \vartheta\left(-\frac{58}{85}\vartheta^3 + \frac{87}{25}\vartheta^2 - \frac{73217}{12750}\vartheta + \frac{2088}{625}\right), \quad \alpha_4(\vartheta) = 0,$$

$$\beta_1(\vartheta) = \vartheta\left(-10\vartheta^3 + \frac{203}{3}\vartheta^2 - \frac{708}{5}\vartheta + \frac{7072}{75}\right),$$

$$\beta_2(\vartheta) = \vartheta\left(-4\vartheta^3 + \frac{556}{15}\vartheta^2 - \frac{6973}{75}\vartheta + \frac{8823}{125}\right),$$

$$\beta_3(\vartheta) = 8\vartheta\left(\frac{533}{289}\vartheta^3 - \frac{3461}{255}\vartheta^2 + \frac{653246}{21675}\vartheta - \frac{44688}{2125}\right),$$

which is the continuous C^2 extension of uniform order $p = 3$ of the A-stable general linear method:

$$
\left[\begin{array}{c|c} \mathbf{A} & \mathbf{U} \\ \hline \mathbf{B} & \mathbf{V} \end{array}\right] =
\left[\begin{array}{ccc|ccc}
\dfrac{3}{5} & 0 & 0 & 1 & \dfrac{6}{5} & \dfrac{27}{50} & 0 \\[2ex]
0 & \dfrac{8}{15} & 0 & 1 & \dfrac{16}{15} & \dfrac{32}{75} & 0 \\[2ex]
0 & 0 & \dfrac{17}{30} & 1 & \dfrac{17}{15} & \dfrac{289}{600} & 0 \\[2ex]
\hline
\dfrac{259}{25} & -\dfrac{4004}{375} & \dfrac{757088}{36125} & 1 & \dfrac{99718}{108375} & \dfrac{25241}{63750} & 0 \\[2ex]
-\dfrac{1943}{75} & -\dfrac{7561}{375} & \dfrac{5120776}{108375} & 0 & -\dfrac{19637}{108375} & -\dfrac{13822}{31875} & 0 \\[2ex]
\dfrac{14}{5} & -\dfrac{866}{75} & \dfrac{168656}{21675} & 0 & \dfrac{6976}{7225} & \dfrac{7693}{6375} & 0 \\[2ex]
166 & \dfrac{632}{5} & -\dfrac{429712}{1445} & 0 & \dfrac{7194}{1445} & \dfrac{1914}{425} & 0
\end{array}\right]
$$

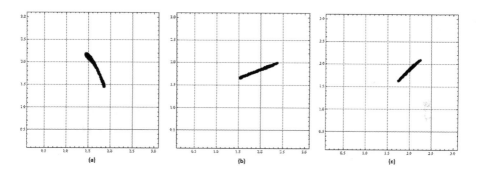

Fig. 2. Region of A-stability: **(a)** in the (c_2, c_3) plane for $c_1 = 9/5$; **(b)** in the (c_1, c_3) plane for $c_2 = 8/5$; **(c)** in the (c_1, c_2) plane for $c_3 = 17/10$.

5   Numerical Results

In this section we present numerical results for the methods introduced previously. In particular we denote with:

- GLM2: the method in Sect. 4.1
- GLM3: the method in Sect. 4.2
- RK2: the two stages Gaussian Runge-Kutta method:

$$
\begin{array}{cc|cc}
\dfrac{1}{2} - \dfrac{\sqrt{3}}{6} & & \dfrac{1}{4} & \dfrac{1}{4} - \dfrac{\sqrt{3}}{6} \\
\dfrac{1}{2} + \dfrac{\sqrt{3}}{6} & & \dfrac{1}{4} + \dfrac{\sqrt{3}}{6} & \dfrac{1}{4} \\
\hline
& & \dfrac{1}{2} & \dfrac{1}{2}
\end{array}
$$

The methods GLM2 and GLM3 have uniform order $p = 3$, while RK2 has order 4 and uniform order 2, therefore it suffers from order reduction when the problem is stiff.

We consider the Prothero-Robinson problem:

$$
\begin{cases}
y'(t) = \lambda(y(t) - sin(t)) + cos(t), \ t \in [0, 10], \\
y(t_0) = y_0,
\end{cases}
\tag{26}
$$

with $Re(\lambda) < 0$ which is stiff when $\lambda \ll 0$.

Table 1 and 2 show the error in the final step point for different values of the step size and the experimental order of methods GLM2, GLM3 and RK2, respectively, for different values of λ in problem (26).

Table 1. Absolute errors (in the final step point) and effective orders of convergence for problem (26) with $\lambda = -10^3$.

h	GLM2		GLM3		RK2	
	Error	p	Error	p	Error	p
1/10	$4.9008 \ 10^{-5}$		$4.1930 \ 10^{-6}$		$1.77 \ 10^{-4}$	
1/20	$3.0606 \ 10^{-6}$	4.0011	$2.6733 \ 10^{-7}$	3.9713	$1.32 \ 10^{-5}$	3.75
1/40	$1.9182 \ 10^{-7}$	3.9960	$1.7166 \ 10^{-8}$	3.9610	$7.82 \ 10^{-7}$	4.08
1/80	$1.2089 \ 10^{-8}$	3.9880	$1.1240 \ 10^{-9}$	3.9328	$4.78 \ 10^{-8}$	4.03

Table 2. Absolute errors (in the final step point) and effective orders of convergence for problem (26) with $\lambda = -10^6$.

h	GLM2		GLM3		RK2	
	Error	p	Error	p	Error	p
1/10	$4.8836 \ 10^{-5}$		$4.1468 \ 10^{-6}$		$1.52 \ 10^{-4}$	
1/20	$3.0403 \ 10^{-6}$	4.0057	$2.6123 \ 10^{-7}$	3.9886	$3.84 \ 10^{-5}$	1.98
1/40	$1.8934 \ 10^{-7}$	4.0052	$1.6450 \ 10^{-8}$	3.9892	$9.99 \ 10^{-6}$	1.94
1/80	$1.1849 \ 10^{-8}$	3.9981	$1.0133 \ 10^{-9}$	4.0210	$2.78 \ 10^{-6}$	1.85

We can notice that the experimental order is consistent with the theoretical one and, even in the case of stiffness, for GLM2 and GLM3.

6   Conclusions

In this paper multivalue collocation methods with diagonal coefficient matrix have been presented. We prove that these methods have at least order $p = r - 1$ and we have constructed A-stable methods with two and three stages with order $p = 3$. Thanks to the structure of the coefficient matrix, those methods can be easily parallelized, so the computational effort can be reduced. In the future we aim to construct such types of methods for different operators such as stochastic differential equations [15,21,31], fractional differential equations [2,10,13,16], partial differential equations [1,11,14,20,30,32,35,38,40], Volterra integral equations [8,9,12,17,23], second order problems [26,37], oscillatory problems [19,22,24,33,36,53], as well as to the development of algebraically stable high order collocation based multivalue methods [18,29].

Acknowledgements. The authors are members of the GNCS group. This work is supported by GNCS-INDAM project and by PRIN2017-MIUR project.

References

1. Adragna, R., Cascaval, R.C., D'Arienzo, M.P., Manzo, R.: Flow simulations in the human cardiovascular system under variable conditions. In: Proceedings of EMSS 2015 (the 27th European Modelling & Simulation Symposium), Bergeggi (SV), Italia, 21–23 September 2015, pp. 228–233 (2015)
2. Burrage, K., Cardone, A., D'Ambrosio, R., Paternoster, B.: Numerical solution of time fractional diffusion systems. Appl. Numer. Math. **116**, 82–94 (2017)
3. Butcher, J.C.: General linear methods. Comput. Math. Appl. **31**(4–5), 105–112 (1996)
4. Butcher, J., D'Ambrosio, R.: Partitioned general linear methods for separable Hamiltonian problems. Appl. Numer. Math. **117**, 69–86 (2017)
5. Butcher, J.C., Jackiewicz, Z.: Diagonally implicit general linear methods for ordinary differential equations. BIT Numer. Math. **33**(3), 452–472 (1993). https://doi.org/10.1007/BF01990528
6. Butcher, J.C., Wright, W.M.: The construction of practical general linear methods. BIT Numer. Math. **43**(4), 695–721 (2003). https://doi.org/10.1023/B:BITN.0000009952.71388.23
7. Butcher, J.C.: Numerical Methods for Ordinary Differential Equations, 2nd edn. Wiley, Chichester (2008)
8. Capobianco, G., Conte, D., Del Prete, I.: High performance parallel numerical methods for Volterra equations with weakly singular kernels. J. Comput. Appl. Math. **228**, 571–579 (2009)
9. Cardone, A., Conte, D., Paternoster, B.: A family of multistep collocation methods for Volterra integro-differential equations. AIP Conf. Proc. **1168**(1), 358–361 (2009)
10. Cardone, A., Conte, D., Patenoster, B.: Two-step collocation methods for fractional differential equations. Discrete Continuous Dyn. Syst. - B **23**(7), 2709–2725 (2018)
11. Cardone, A., D'Ambrosio, R., Paternoster, B.: Exponentially fitted IMEX methods for advection-diffusion problems. J. Comput. Appl. Math. **316**, 100–108 (2017)

12. Cardone, A., D'Ambrosio, R., Paternoster, B.: High order exponentially fitted methods for Volterra integral equations with periodic solution. Appl. Numer. Math. **114C**, 18–29 (2017)
13. Cardone, A., D'Ambrosio, R., Paternoster, B.: A spectral method for stochastic fractional differential equations. Appl. Numer. Math. **139**, 115–119 (2019)
14. Cascaval, R.C., D'Apice, C., D'Arienzo, M.P.: Simulation of heart rate variability model in a network. In: Proceedings of International Conference of Numerical Analysis and Applied Mathematics 2016 (ICNAAM 2016), Rodi, Grecia, 19–25 September 2016, pp. 1–4 (2016). ISBN 978-0-7354-1538-6. vol. 1863, 560054 (2017)
15. Citro, V., D'Ambrosio, R., Di Giovacchino, S.: A-stability preserving perturbation of Runge-Kutta methods for stochastic differential equations. Appl. Math. Lett. **102**, 106098 (2020)
16. Conte, D., Califano, G.: Optimal Schwarz waveform relaxation for fractional diffusion-wave equations. Appl. Numer. Math. **127**, 125–141 (2018)
17. Conte, D., Capobianco, G., Paternoster, B.: Construction and implementation of two-step continuous methods for Volterra Integral Equations. Appl. Numer. Math. **119**, 239–247 (2017)
18. Conte, D., D'Ambrosio, R., Jackiewicz, Z., Paternoster, B.: Numerical search for algebraically stable two-step continuous Runge-Kutta methods. J. Comput. Appl. Math. **239**, 304–321 (2013)
19. Conte, D., D'Ambrosio, R., Moccaldi, M., Paternoster, B.: Adapted explicit two-step peer methods. J. Numer. Math. **27**(2), 69–83 (2019)
20. Conte, D., D'Ambrosio, R., Paternoster, B.: GPU acceleration of waveform relaxation methods for large differential systems. Numer. Algorithms **71**(2), 293–310 (2016)
21. Conte, D., D'Ambrosio, R., Paternoster, B.: On the stability of theta-methods for stochastic Volterra integral equations. Discrete Continuous Dyn. Syst. - B **23**(7), 2695–2708 (2018)
22. Conte, D., Esposito, E., Paternoster, B., Ixaru, L.G.: Some new uses of the $\eta_m(Z)$ functions. Comput. Phys. Commun. **181**, 128–137 (2010)
23. Conte, D., Paternoster, B.: A family of multistep collocation methods for Volterra integral equations. In: Simos, T.E., Psihoyios, G., Tsitouras, Ch. (eds.) Numerical Analysis and Applied Mathematics. AIP Conference Proceedings, vol. 936, pp. 128–131. Springer, New York (2007)
24. Conte, D., Paternoster, B.: Modified Gauss-Laguerre exponential fitting based formulae. J. Sci. Comput. **69**(1), 227–243 (2016)
25. D'Ambrosio, R., De Martino, G., Paternoster, B.: Numerical integration of Hamiltonian problems by G-symplectic methods. Adv. Comput. Math. **40**(2), 553–575 (2014)
26. D'Ambrosio, R., De Martino, G., Paternoster, B.: General Nystrom methods in Nordsieck form: error analysis. J. Comput. Appl. Math. **292**, 694–702 (2016)
27. D'Ambrosio, R., Ferro, M., Jackiewicz, Z., Paternoster, B.: Two-step almost collocation methods for ordinary differential equations. Numer. Algorithms **53**(2–3), 195–217 (2010)
28. D'Ambrosio, R., Hairer, E.: Long-term stability of multi-value methods for ordinary differential equations. J. Sci. Comput. **60**(3), 627–640 (2014)
29. D'Ambrosio, R., Izzo, G., Jackiewicz, Z.: Search for highly stable two-step Runge-Kutta methods for ODEs. Appl. Numer. Math. **62**(10), 1361–1379 (2012)
30. D'Ambrosio, R., Moccaldi, M., Paternoster, B.: Adapted numerical methods for advection-reaction-diffusion problems generating periodic wavefronts. Comput. Math. Appl. **74**(5), 1029–1042 (2017)

31. D'Ambrosio, R., Moccaldi, M., Paternoster, B.: Numerical preservation of long-term dynamics by stochastic two-step methods. Discrete Continuous Dyn. Syst. - B **23**(7), 2763–2773 (2018)

32. D'Ambrosio, R., Moccaldi, M., Paternoster, B.: Parameter estimation in IMEX-trigonometrically fitted methods for the numerical solution of reaction-diffusion problems. Comput. Phys. Commun. **226**, 55–66 (2018)

33. D'Ambrosio, R., Moccaldi, M., Paternoster, B., Rossi, F.: Adapted numerical modelling of the Belousov-Zhabotinsky reaction. J. Math. Chem. **56**(10), 2867–2897 (2018)

34. D'Ambrosio, R., Paternoster, B.: Two-step modified collocation methods with structured coefficients matrix for ordinary differential equations. Appl. Numer. Math. **62**(10), 1325–1334 (2012)

35. D'Ambrosio, R., Paternoster, B.: Numerical solution of a diffusion problem by exponentially fitted finite difference methods. SpringerPlus **3**(1), 1–7 (2014). https://doi.org/10.1186/2193-1801-3-425

36. D'Ambrosio, R., Paternoster, B.: Exponentially fitted singly diagonally implicit Runge-Kutta methods. J. Comput. Appl. Math. **263**, 277–287 (2014)

37. D'Ambrosio, R., Paternoster, B.: A general framework for numerical methods solving second order differential problems. Math. Comput. Simul. **110**(1), 113–124 (2015)

38. D'Ambrosio, R., Paternoster, B.: Numerical solution of reaction-diffusion systems of lambda-omega type by trigonometrically fitted methods. J. Comput. Appl. Math. **294 C**, 436–445 (2016)

39. D'Ambrosio, R., Paternoster, B.: Multivalue collocation methods free from order reduction. J. Comput. Appl. Math. (2019). https://doi.org/10.1016/j.cam.2019.112515

40. D'Apice, C., D'Arienzo, M.P., Kogut, P.I., Manzo, R.: On boundary optimal control problem for an arterial system: existence of feasible solutions. J. Evol. Equ. **18**(4), 1745–1786 (2018). https://doi.org/10.1007/s00028-018-0460-4

41. Enright, W.H., Jackson, K.R., Norsett, S.P., Thomsen, P.G.: Interpolants for Runge-Kutta formulas. ACM Trans. Math. Softw. **12**(3), 193–218 (1986)

42. Enright, W.H., Muir, P.H.: Super-convergent interpolants for the collocation solution of boundary value ordinary differential equations. SIAM J. Sci. Comput. **21**(1), 227–254 (1999)

43. Hairer, E., Wanner, G.: Solving Ordinary Differential Equations II -Stiff and Differential-Algebraic Problems. Springer, Heidelberg (2002)

44. Heldt, F.S., Frensing, T., Pflugmacher, A., Gropler, R., Peschel, B., Reichl, U.: Multiscale modeling of influenza a virus infection supports the development of direct-acting antivirals. PLOS Comput. Biol. **9**(11), e1003372 (2013)

45. Jackiewicz, Z.: General Linear Methods for Ordinary Differential Equations. Wiley, Hoboken (2009)

46. Jackiewicz, Z., Tracogna, S.: A general class of two-step Runge-Kutta methods for ordinary differential equations. SIAM J. Numer. Anal. **32**, 1390–1427 (1995)

47. Lambert, J.D.: Numerical Methods for Ordinary Differential Systems: The Initial Value Problem. Wiley, Chichester (1991)

48. Lie, I., Norsett, S.P.: Superconvergence for multistep collocation. Math. Comput. **52**(185), 65–79 (1989)

49. Lie, I.: The stability function for multistep collocation methods. Numer. Math. **57**(8), 779–787 (1990). https://doi.org/10.1007/BF01386443

50. Noble, D., Varghese, A., Kohl, P., Noble, P.: Improved guinea-pig ventricular cell model incorporating a diadic space, IKr and IKs, and length- and tension-dependent processes. Can. J. Cardiol. **14**, 123–134 (1998)
51. Norsett, S.P.: Collocation and perturbed collocation methods. In: Watson, G.A. (ed.) Numerical Analysis. LNM, vol. 773, pp. 119–132. Springer, Heidelberg (1980). https://doi.org/10.1007/BFb0094168
52. Norsett, S.P., Wanner, G.: Perturbed collocation and Runge Kutta methods. Numer. Math. **38**(2), 193–208 (1981). https://doi.org/10.1007/BF01397089
53. Paternoster, B.: Two step Runge-Kutta-Nyström methods for $y" = f(x,y)$ and P-stability. In: Sloot, P.M.A., Hoekstra, A.G., Tan, C.J.K., Dongarra, J.J. (eds.) ICCS 2002. LNCS, vol. 2331, pp. 459–466. Springer, Heidelberg (2002). https://doi.org/10.1007/3-540-47789-6_48
54. Söderlind, G., Jay, L., Calvo, M.: Stiffness 1952–2012: sixty years in search of a definition. BIT **55**(2), 531–558 (2015)
55. Southern, J., et al.: Multi-scale computational modelling in biology and physiology. Progress Biophys. Mol. Biol. **96**, 60–89 (2008)
56. Wright, K.: Some relationships between implicit Runge-Kutta, collocation and Lanczos τ-methods, and their stability properties. BIT **10**, 217–227 (1970). https://doi.org/10.1007/BF01936868

Characterizing and Analyzing the Relation Between Bin-Packing Problem and Tabu Search Algorithm

V. Landero[1]([⊠]), David Ríos[1], Joaquín Pérez[2], L. Cruz[3], and Carlos Collazos-Morales[4]

[1] Universidad Politécnica de Apodaca, Nuevo León, Mexico
{vlandero, drios}@upapnl.edu.mx
[2] Departamento de Ciencias Computacionales, Centro Nacional de Investigación y Desarrollo Tecnológico (CENIDET), AP 5-164, Cuernavaca 62490, Mexico
jperez@cenidet.edu.mx
[3] División de Estudios de Posgrado e Investigación, Instituto Tecnológico de Ciudad Madero (ITCM), Cd. Madero, Mexico
[4] Vicerrectoría de Investigaciones, Universidad Manuela Beltrán, Bogotá, Colombia
carlos.collazos@docentes.umb.edu.co

Abstract. The relation between problem and solution algorithm presents a similar phenomenon in different research problems (optimization, decision, classification, ordering); the algorithm performance is very good in some cases of the problem, and very bad in other. Majority of related works have worked for predicting the most adequate algorithm to solve a new problem instance. However, the relation between problem and algorithm is not understood at all. In this paper a formal characterization of this relation is proposed to facilitate the analysis and understanding of the phenomenon. Case studies for Tabu Search algorithm and One Dimension Bin Packing problem were performed, considering three important sections of algorithm logical structure. Significant variables of problem structure and algorithm searching behavior from past experiments, metrics known by scientific community were considered (Autocorrelation Coefficient and Length) and significant variables of algorithm operative behavior were proposed. The models discovered in the case studies gave guidelines that permits to redesign algorithm logical structure, which outperforms to the original algorithm in an average of 69%. The proposed characterization for the relation problem-algorithm could be a formal procedure for obtaining guidelines that improves the algorithm performance.

1 Introduction

There exits a great variety of problems as constraint satisfaction, decision, optimization, forecasting, classification, clustering, sorting; where it has found that in certain problem instances the performance of a solution algorithm is very good and in other is very bad [1–4]. This phenomenon has been observed by broad range of disciplines: computational complexity theory, operations research, data mining, machine learning, artificial

© Springer Nature Switzerland AG 2020
O. Gervasi et al. (Eds.): ICCSA 2020, LNCS 12249, pp. 149–164, 2020.
https://doi.org/10.1007/978-3-030-58799-4_11

intelligence and bioinformatics [5]. In the real life situations, there is not an algorithm that outperforms others algorithms in all circumstances [6, 7] on some problem domain. The related works in the majority of cases have focused in building predictive models [8–22] for giving the best solution to new problem instances. Supervised or/and unsupervised learning algorithms have been used by the majority of related works for building that models. However, the built models are difficult to interpret the principal structure and relations between significant variables. These are used for prediction and there is no understanding of why an algorithm is better adequate to solve a set of problem instances than another (phenomenon). A few related works have tried explaining it [23–28]. Table 1 shows some of the latest related works; it emphasizes the information (variables) utilized (problem, behavior, and logical structure of algorithm) and the analysis main purpose. As it can be seen, not all information has been included. Also, it is necessary one formal formulation of phenomenon as a problem statement and one formal procedure for characterize the relation between problem-algorithm, that permits analyze deeply and solve the formulated problem, understanding the phenomenon.

Table 1. Related works

Work	Problem	Algorithm		Analysis purpose
		Behavior	Logical structure	
[13]	✓		✓	Prediction
[14]	✓		✓	Prediction
[19]	✓	✓		Prediction
[22]	✓			Prediction
[26]	✓			Explanation
[27]	✓	✓		Explanation
[28]	✓	✓		Explanation
This paper	✓	✓	✓	Explanation

In this paper, firstly, a nomenclature and the problem statement are formally formulated to previous questioning, where this formulation facilitates the proposal of a formal characterization of relation problem-algorithm (Sect. 2). It permits analyze and understand the relation of One Dimension Bin-Packing problem and Tabu Search algorithm and solve the problem statement performing cases of study. Three parts of algorithm logical structure, significant variables for characterizing structure and space of problem, algorithm searching behavior are considered, and significant variables for characterizing the algorithm operative behavior were proposed (Sect. 3). The models obtained by proposal characterization allowed to redesign Tabu Search algorithm for each case of study (Sect. 3). The results of proposed redesigns performance are analyzed and validated by means a statistical test in Sect. 4. The conclusions and future work are described in Sect. 5.

2   Problem Statement and Proposed Solution

Let the next nomenclature,

$I = \{i_1, i_2, ..., i_m\}$ a set of instances of problem or problem space.

$F =$ the features space, it represents the mapping of each problem instance to a set characterization variables.

$A = \{a_1, a_2, ..., a_n\}$ a set of algorithms.

$a_s = a_s \in A$, where a_s is the algorithm to analyze in each study case for relation problem-algorithm.

$B =$ the features space, it represents the mapping of behavior of algorithm a_s to a set characterization variables.

$D = \{D_1, D_2, ..., D_n\}$ a partition of I, where $|A| = |D|$.

$X = \{(a_q \in A, D_q \in D) \mid d(a_q(i)) > d(\alpha(i)) \; \forall \; \alpha \in (A - \{a_q\}), \; \forall \; i \in D_q\}$,

is a set of ordered pairs (a_q, D_q), where each algorithm $a_q \in A$ solved instances associated to D_q better than others algorithms (see function d as Expression 1). In the following sections each ordered pair (a_q, D_q) will be used to mean the domain region D_q (implicitly inferiority region $(D_q)^c$) of algorithm a_q.

$$d(a_q(i)) = \begin{cases} 1 & \text{if algorithm } a_q \text{ has the rm } quality \text{ among all the algorithms for instance } i \\ 1 & \text{if algorithm } a_q \text{ has the smallest } time \text{ among all the algorithms (when they have the same } quality) \\ 0 & \text{otherwise} \end{cases} \quad (1)$$

$P =$ the performance space, it represents the mapping of performance of algorithm a_s, to a set of performance characterization variables; so too the value of function d $(a_s(i)) \; \forall \; i \in I$.

2.1   Problem Statement

According mentioned in Introduction and before nomenclature, the next research question arises:

Why an algorithm a_s dominates in an instance's region D_s?

The problem of explaining formally why an algorithm outperforms others in solving an instances region and why not in others could be described formally as follows:

For a domination region $(a_s, D_s) \in X$ and inferior region $(D_s)^c$, of an algorithm $a_s \in A$, applied to problem instances I, with problem features F, algorithm features B, find explanation model R, which represents the relations between significant variables that characterize the problem instances, algorithm features, and provides solid foundations to explain why algorithm a_s is superior for solving instances in the domination region D_s and inferior for solving instances in subset $(D_s)^c$.

Figure 1 shows a possible graphical solution to above. The model R, obtained by some mechanism M and sets F, B, P (described previously) as input; where values in F were obtained by variables a, b, c, d and values in B were obtained by variables v, x, y, z. The model R shows important and significant relations between variables that were significant and influence on the values from function d. In this example, b, c, x, y. The variables b and c characterize relevant and significant information about description and space of problem. The variables x and y characterize relevant and significant information from operative and searching behavior of analyzed algorithm.

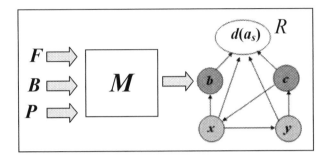

Fig. 1. Example of model R obtained by some mechanism M

The model R could be interpreted as one causal model, where it is Directed Acyclic Graph (DAG) over a set of vertices (variables) and a set of directed edges that connect vertices. These relations can be formally interpreted as causal (cause-effect) if the graph accomplish with the conditions: Causal Markov, Minimality and Faithfulness [29]. The causal relations (cause-effect) are estimated and the model is validated. The interest relations are analyzed and interpreted for obtaining explanations. If the model R accomplish with the above, it could be say that model R could formally answer the above principal question and problem statement. Nevertheless, there can be other models R_2 and R_3 with other variables that characterize: the structure and space of problem (d, e, h, j); the operative and searching of algorithm (w, z, n, v) where there are not influence variables for values of function d (see Fig. 2).

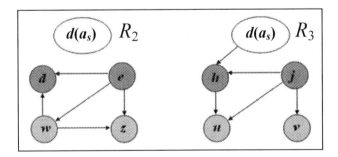

Fig. 2. Examples of models R_2 and R_3

The causal models R_2 and R_3 could not answer to principal question. Therefore, it is not easy to find one model R like Fig. 1. It will depend a lot on a mechanism M that can find from sets F, B, significant variables, whose values and its relation to values from function d allows build such model. The next section a characterization of relation of problem-algorithm is proposed, where its objective is to solve the before problem statement through case studies about the relation Bin-Packing problem and Tabu Search algorithm.

2.2   Proposed Characterization for Relation Problem - Algorithm

The relation between problem – algorithm can be formally characterized by the function M described by Expression 2. The domain is the set of parameters: the set F, which is obtained by the function f; the set B, which is obtained by the function b; the set P, which is obtained by the function p.

$$R = (M \cdot v \cdot e \cdot s \cdot t)(f(I), b(A), p(a_s, A)) \qquad (2)$$

The function f develops the process of mapping each problem instance from set I to a set variable that characterize the problem structure and space. The function b develops the process of mapping the behavior of each algorithm from set A to a set of variables that characterize the searching and operative behavior. The function p develops the process of mapping the performance of algorithm a_s to a set of variables that characterize the time, quality and its domain over other algorithms in A (described in nomenclature). The function t, described by Expression 3, firstly builds the dataset C, where the tuples represent the problem instances and the columns are the values that characterize the problem (set F), the behavior, performance and domination of algorithm a_s. A discretization process is performed on the continuous dataset C. The codomain of the function t is the discretized dataset T. The function s (Expression 4) performs a process of selecting significant variables from discretized dataset T. The codomain of this function is the discrete dataset S.

$$T = t(f(I), b(A), p(A)) \qquad (3)$$

$$s(T) = S \qquad (4)$$

The function e, described by Expression 5, consists in a learning process for obtaining causal relations from dataset S. The codomain of this function is the set of relations E. The function v, described by Expression 6, consists in estimating the causal relations in set E. The codomain of this function is the estimations set V. The codomain of function M is the causal model R of a set significant variables S, which represents a set of causal relations E and a set of estimations of these relations.

$$e(S) = E \qquad (5)$$

$$v(E) = V \qquad (6)$$

3 Characterizing and Analyzing the Relation Bin-Packing Problem and Tabu Search Algorithm: Case Studies

3.1 Description of Framework

The framework consists of characterizing and analyzing the relation between One Dimension Bin-Packing (BPP) problem and Tabu Search algorithm through proposed function M; deepening in the problem structure, solutions space, the algorithm internal logical structure, its behavior operational, behavior during search and performance. Two types of instances for problem BPP were considered. The first, *instances* 1 (set I of nomenclature), for characterization and analysis process; and the second, *instances* 2 (different to *instances* 1) for prediction process. Each type of instances has 324 instances (this sample size has given good results in past experimentations). These were collected randomly from repositories [30, 31].

The internal logical structure of Tabu Search algorithm consists in four important parts. In this paper, the focus will be in three parts (control parameter, initial solution, search methodology), where the last part (stop criterion) is considered as a future work. Each case of study consists of comparing two variants, which only are different by one change in some fundamental part of its internal logical structure (see Table 2). In this paper, the size of Tabu list (control parameter) can be defined in a static (S) or dynamic way (D). It is to say, the size of Tabu list is defined as: 7 [32] or $\sqrt{n}$, where n is the number of the objects or items of the problem instance. The initial solution can be generated randomly (S) or by means deterministic procedure (H). For generating the neighborhood of a solution, one method can be considered (O) or several methods (M), which were proposed in [33]. The stop criterion was considered the same for all variants, it happens after 4000 iterations (divergence). Table 3 shows the cases of study.

Table 2. Variants of tabu search algorithm

Variants	Tabu list		Initial solution		Neighborhood	
	S	D	R	H	O	M
TB1		✓		✓		✓
TB2	✓			✓		✓
TB3		✓	✓			✓
TB4		✓		✓	✓	

3.2 Variables for Problem and Algorithm

Problem Variables. There are two considered variables for characterizing the problem structure, which are proposed in [16]. The first b characterizes the proportion of the total size of the objects that can be assigned to one container. The second f characterizes the proportion of objects where its weight w_i is factor of the container capacity.

Table 3. Cases of study

Case of study	Variants	Internal logical structure
1	TB1 and TB2	Tabu List Size
2	TB1 and TB3	Initial Solution
3	TB1 and TB4	Neighborhood

The problem solutions space is characterized by variable os, it has been one significant variable in past experimentations [28]. A sample of ms (the value 100 has given good results) randomly generated solutions before algorithm execution is built. This variable characterizes the variability of fitness function $f(x)$ [34] of these ms solutions.

Algorithm Variables. In this paper, the characterization of the algorithm operative behavior is proposed. It is characterized by two significant variables, the first is the number of feasible solutions found by algorithm per instance (variable $efac$), the second is the variance of these solutions (variable $evfac$). Also, two ways are considered by characterizing the algorithm behavior on trajectory during the search process (solutions generated during execution). The first way is using the concept fitness landscape with two known metrics, the autocorrelation coefficient ($coef$) and autocorrelation length ($long$), which were described in [35]. The second way is using significant variables proposed in past experiments [25, 28]. These are number of inflexion points nc, number of valleys nv, the average size tm of the valleys and its dispersion vd from algorithm search trajectory.

Performance Variables. The performance variables considered are *time* and *quality*, which are described in [28]. The first is the number of evaluations of the fitness function for feasible and infeasible solutions. The *quality* variable is the ratio of the best solution found by the algorithm (final number of containers) to the theoretical solution; this is the objects sizes divided by the containers capacity.

3.3 Characterizing the Relation Bin Packing Problem – Tabu Search Algorithm

The set of problem instances I (*instances I*) is characterized by function f. This performs the calculation of problem variables described in before section on each problem instance in set I. A set F is built as Expression 7. The algorithms in set A are executed on each problem instance, where the function p performs the calculation of performance variables described in before section and a set P is built. The function b performs the calculation of algorithm variables described in before section for this algorithm, a set B is built as Expression 8.

$$F = \{\{b_1, f_1, os_1\}, \{b_2, f_2, os_2\}, \ldots, \{b_m, f_m, os_m\}\} \tag{7}$$

The set C is obtained by function t, where its information is obtained from sets F, B and P. The Expression 9 shows an example of set C.

$$B = \left\{ \begin{array}{l} \{efac_1, evfac_1, nc_1, nv_1, tm_1, vd_1\}, \\ \{efac_2, evfac_2, nc_2, nv_2, tm_2, vd_2\}, \ldots, \\ \{efac_m, evfac_m, nc_m, nv_m, tm_m, vd_m\} \end{array} \right\} \tag{8}$$

$$C = \left\{ \begin{array}{l} \{b_1, f_1, os_1, efac_1, evfac_1, nc_1, nv_1, tm_1, vd_1, time_1, quality_1, d(a_s(1))\}, \\ \{b_2, f_2, os_2, efac_2, evfac_2, nc_2, nv_2, tm_2, vd_2, time_2, quality_2, d(a_s(2))\}, \\ \qquad\qquad\qquad\qquad \ldots, \\ \{b_m, f_m, os_m, efac_m, evfac_m, nc_m, nv_m, tm_m, vd_m, time_m, quality_m, d(a_m(1))\} \end{array} \right\} \tag{9}$$

The set C is discretized by the method MDL [36] and the set T is obtained. Then, the function s performs the method Correlation-based Feature Selection (CFS) [37]. It is important to emphasize that this selection depends of the values of set T, each study case is different due to the included variants and the sets S are different. The next sections describe which are these sets S, so too, the functions e and v.

Study Case 1

The set S, obtained by function s, is described by Expression 10. In Sect. 3.2 were mentioned two more common ways to characterize the trajectory traced by the algorithm during its execution: fitness landscape variables (*coef, long*) and our variables (*nv, tm, vd*).

$$S = \left\{ \begin{array}{l} \{f_1, os_1, nv_1, tm_1, vd_1, d(a_s(1))\}, \{f_2, os_2, nv_2, tm_2, vd_2, d(a_s(2))\}, \\ \ldots, \{f_m, os_m, nv_m, tm_m, vd_m, d(a_s(m))\} \end{array} \right\} \tag{10}$$

Therefore, we built another different set S_2 with information f, os, *coef* and *long*. The value of function d is considered in two before cases. The function e performs the structure learning algorithm PC [29]. The causal inference software HUGIN (Hugin Expert, www.hugin.com) was used with a confidence level of 95%. The domain of function e, in a first time, was the set S, obtaining the causal structure E (Expression 11); in a second time was the set S_2, obtaining the causal structure E_2 (Expression 12).

$$E = \left\{ \begin{array}{l} \{\{d,f\}, \{d, os\}, \{d, nv\}, \{d, tm\}\}, \{\{vd, os\}, \{vd, nv\}, \{vd, tm\}\}, \\ \{\{tm, f\}, \{tm, os\}\}, \{\{nv, os\}\} \end{array} \right\} \tag{11}$$

$$E_2 = \{\{\{d,f\}, \{d, os\}\}, \{\{os, coef\}, \{os, long\}\}, \{\{coef, long\}\}\} \tag{12}$$

The causal relations in E are estimated by function v, which performs the parameter learning algorithm Counting [29]. Figure 3(a) shows the causal model R, containing the causal structure E and causal relations estimations V (due short space, the complete set V is not shown, only one part in Table 3). So too, Fig. 3(b) shows the causal model R_2, containing the causal structure E_2 and causal relations estimations V_2. This did not yield relevant information about direct causes of the algorithm behavior and performance, in terms of regions of domination (D_s) and inferiority ($(D_s)^c$), using the correlation coefficient (*coef*) and autocorrelation long (*long*). Due before, the set V_2 is not shown. Conversely, the causal model **R** (a) shows that the variables f, os, nv and tm are direct causes. The most important probabilities of direct causes of algorithm behavior and performance, in terms of regions domination and inferiority, are shown in Table 4 (one part of set **V**). The internal logical structure of variant TB1 permits storage more solutions in Tabu List (big) being more restrictive for accepting a new neighbor solution. Variant TB1 is successful with instances where the sizes of objects or items that are multiples of the container, f, is in the range 2 [0.041, 1] and there is a variability of solutions space, os, in the range 3 [0.1695, 1].

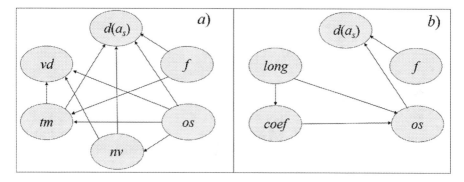

Fig. 3. Causal models **R** and **R**$_2$

This indicates that there is more facility for generating different neighbor solutions and can be difficult that these can be in Tabu List due these are vastly different. Valleys were identified in the search trajectory of variant, the number of valleys, nv, is in the range 2 [0.061, 1] with sizes, tm, in the range 2 [0.3138, 1] (causal relation 1). This variant wins in quality. This variant has disadvantage with instances where the number of the sizes of the objects or items that are not multiple of container, f, is in the range 1 [0, 0.040] and the variability between solutions of problem space, vo, is in the ranges 1 [0, 0.054], 2 [0.055, 0.1694]. So too, a number of valleys were identified, nv, in the range 1 [0, 0.060] with sizes in the range 1 [0, 0.3137] (causal relations 2 and 3). The neighbors that can be generated from these instances will not be very different, generating a flat search trajectory; which can be Tabu, due the size of this list is big. Therefore, variant TB2 has advantage of this situation, due it uses a size of Tabu List more small (7) and is more flexible for accepting generated neighbor solutions. Variant TB1 lost in time.

Case of Study 2
In the development of this experimental test, function s could not select variables. Therefore, this case of study could not be performed.

Case of Study 3
The set S, obtained by function s is described by Expression 13.

Table 4. Relations Estimation, part of set **V**

	%
$P(d = 1 \mid f = 2, os = 3, nv = 2, tm = 2)$	97.82
$P(d = 0 \mid f = 1, os = 1, nv = 1, tm = 1)$	70.78
$P(d = 0 \mid f = 2, os = 2, nv = 1, tm = 1)$	85.71

$$S = \left\{ \begin{array}{l} \{b_1, os_1, efac_1, evfac_1, nc_1, nv_1, tm_1, d(a_s(1))\}, \ldots, \\ \{b_m, os_m, efac_m, evfac_m, nc_m, nv_m, tm_m, d(a_s(m))\} \end{array} \right\} \tag{13}$$

Two different causal structures were built by function e (as case of study 1): the first **E** (Expression 14) with values of these variables and the second E_2 (Expression 15) with b, os, $efac$, $evfac$, $coef$ and $long$. So too, the function d is considered in two before cases (see Fig. 4). Part (a) of Fig. 4 shows the causal model **R**, containing the causal structure **E** and causal relations estimations **V**. The model **R** yields to more complete and relevant information about direct causes of the algorithm behavior and performance, in terms of regions of domination (D_s) and inferiority $((D_s)^c)$, considering information about problem structure. The causal relations found in **E** are estimated by function v and the most important probabilities of direct causes of algorithm behavior and performance, in terms of regions domination and inferiority, are shown in Table 5 (one part of set **V**). The use of several methods for generating neighbor solutions (variant TB1) permits has advantage in problems where the total sum of sizes of objects or items is much greater than the capacity of the container and the arrangement of the objects can be very varied; it is to say, a great diversity of solutions can be generated.

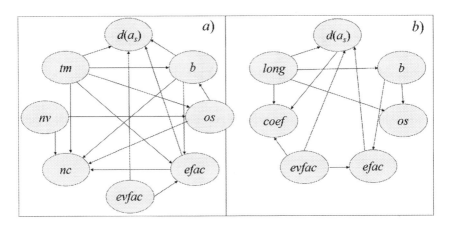

Fig. 4. Causal models **R** and R_2

Table 5. Relations Estimation, part of set **V**

	%
$P(d = 1 \mid b = 1, \textit{efac} = 1, \textit{evfac} = 2, \textit{nv} = 2)$	98.78
$P(d = 0 \mid b = 2, \textit{efac} = 2, \textit{evfac} = 1, \textit{nv} = 1)$	71.42
$P(d = 0 \mid b = 3, \textit{efac} = 3, \textit{evfac} = 1, \textit{nv} = 1)$	100

$$E = \left\{ \begin{array}{c} \{\{d,b\}, \{d,efac\}, \{d,evfac\}, \{d,nv\}\}, \{\{b,os\}, \{b,tm\}\}, \\ \{\{nc,b\}, \{nc,os\}, \{nc,efac\}, \{nc,nv\}, \{nc,tm\}\}, \\ \{\{efac,b\}, \{efac,evfac\}, \{efac,tm\}\}, \{\{os,nv\}, \{os,tm\}\} \end{array} \right\} \quad (14)$$

$$E_2 = \left\{ \begin{array}{c} \{\{d,efac\}, \{d,evfac\}, \{d,long\}\}, \{\{efac,b\}, \{efac,evfac\}\}, \\ \{\{coef,d\}, \{coef,long\}, \{coef,evfac\}\}, \{\{os,b\}, \{os,long\}\} \end{array} \right\} \quad (15)$$

Its internal logical structure enables intensify the search, generating a number of feasible solutions, *efac*, in the range 1 [0, 0.67], with a variability between feasible solutions, *evfac*, in the range 2 [0.1377, 1]; the search trajectory corresponds better to this problem structure and problem space, because it may enter and exit from a number of valleys, *nv*, in the range 2 [0.0297, 1] (relation 1). This variant wins in quality. The variant TB1 loses in time in problems where the total sum of sizes of objects or items is not much greater than the capacity of the container; it indicates that there is no variety to arrangement the objects; it is to say, there is a number of feasible solutions, *efac*, in the range 1 [0.68, 1] with a variability between feasible solutions, *evfac*, in the range 2 [0, 0.1376]. Therefore, there is no necessity for intensifying the search, the number of valleys, *nv*, is in the range 1 [0, 0.0296]. The variant may take longer to find solutions, where there is little to find (relations 2 and 3); its exhaustive attempt to generate neighbor solutions produces a cost in time and lost by this factor. The variant TB4 has advantage in this situation, because it only uses one method for generating neighbor solutions; it is very limited for searching in the problem space by its own structure and it is better adjusted to this kind of problems.

Redesign Proposals

The analysis of relation Bin-Packing problem and Tabu Search algorithm of each study case permit to find guidelines to redesign the algorithm. Figure 5 shows the proposal of redesign from case of study 1, marked by R1TB1.

It consists to automatically adjust its logical design, in terms of the Tabu list size. It is to say, if value *f* falls in the second interval and *os* falls in the third interval, the Tabu list size will be $\sqrt{n}$; otherwise it will be 7. So too, Fig. 5 shows the proposal of redesign from case of study 3, marked by R3TB1. It consists to adjust automatically its logical design, in terms of the way to build the neighborhood. It is to say, if value *b* falls in the first interval the neighborhood will be built by means several methods; otherwise, one method will be performed.

Tabu Search Algorithm
1 **Begin**
2      $x^* =$ a feasible initial solution; $x = x^*$
3   **if** $(f==2$ and $os==3)$ or $(f==1$ and $os==3)$ **then** $nLTabu = \sqrt{n}$; ⎫ R1TB1
4   **else** $nLTabu = 7$; ⎭
5      $nLTabu = \sqrt{n}$; ⎫ **R3TB1**
6      $tn = nLTabu$, Tenency of one solution in Tabu list
7      $LCANDI = \varnothing$; $LTabu = \varnothing$
8 **Repeat**
9   Building neighborhood of x using swap $(0, 1)$ or swap $(1, 1)$ or ⎫ R1TB1
10   swap $(1, 2)$ or swap $(2, 2)$ or swap $(1, 0)$ or swap $(2, 1)$ ⎭
11   **if** $b==1$ **then** ⎫
12   Building neighborhood $N(x)$ of x using swap $(0, 1)$ or swap $(1, 1)$ ⎬ R3TB1
13   or swap $(1, 2)$ or swap $(2, 2)$ or swap $(1, 0)$ or swap $(2,1)$ ⎪
14   **else** Building neighborhood $N(x)$ of x using swap $(0, 1)$ ⎭
15     $LCANDI =$ list of candidate solutions, taken from neighborhood x
16        $y =$ the best solution of $LCANDI$ and $y \notin LTabu$
17     $LTabu = LTabu \cup \{(y, tn)\}$
18   **For each** $e \in LTabu$ the tenency of e is decremented
19   **If** the tenency e has expired **Then** $LTabu = LTabu - \{(e, tn)\}$
20   **If** $f(y) < f(x)$ **Then** $x = y$
21 **Until** 4000 iterations.
22 **End**

Fig. 5. Redesign proposals

4 Results Analysis

Table 6 shows the experimentation. Column 1 indicates the case of study; Column 2 indicates the prediction percentage of generated model on *instances* 2; for this, the causal inference software NETICA (Norsys Corporation, www.norsys.com) was utilized. Column 3 indicates an outperform percentage of proposed redesign with respect to original algorithm. The function M builds a causal model R that permits give answer to question and problem statement in explanation level for the cases of study 1 and 3. For the case of study 2, function s could not select variables that could be considered for building a causal model; therefore, this case of study could not be performed. One possible interpretation of this result, it may be that the method for generating the initial solution does not impact the algorithm performance. This had already been observed by [38] with the same Tabu Search algorithm for solving another problem, the Job-Shop Scheduling problem. The models generated in the cases 1 and 3 permit to make predictions about the algorithm performance in terms of domination or inferior region $(D_s, (D_s)^c)$ with a percentage higher than 70% over a problem instances test set (*instances* 2) (see Table 6).

On the other hand, the causal relations found by the models generated permit to make redesign proposals; which improve the algorithm performance (see Table 6). To

Table 6. Experimentation results

Redesign proposal	% model prediction	Outperform percentage
R1TB1	79.01%	65%
R3TB1	72.14%	73%

validate the above, the two sample two-side Wilcoxon signed rank test was applied for significance levels 95% and 99% to verify the means of values of performance variable *quality* (it is not assuming a normal distribution). The Table 7 shows the null hypothesis conclusion (means are equal), where is rejected for two significance levels. It means, there is a significant improvement of redesign proposals.

Table 7. Null hypothesis conclusion

Redesign proposal	Significance level	Test statistic	Critical value	Conclusion
R1TB1	95%	8.2966	1.9600	Rejected
	99%	8.2966	2.5758	Rejected
R3TB1	95%	12.2664	1.9600	Rejected
	99%	12.2664	2.5758	Rejected

5   Conclusions

This paper formally formulates the research question "why an algorithm is the best for solving an instances subset and why not in other instances" as a problem statement to solve. This questioning is implicitly observed by scientific community in the solution process of problems such as decision, optimization, forecasting, classification, clustering, sorting. Nevertheless, its common objective, in the majority cases, is to predict the algorithm most adequate to solve a new problem instance. A few related works have tried to obtain explanations, without a formal formulation of this questioning as one problem to solve. As well as, it is necessary to explore more parts of algorithm logical design, analyze its relation to significant variables from problem and algorithm, including also significant variables of algorithm operative behavior.

Therefore, this paper also proposes one characterization of the relation problem-algorithm to solve the proposal problem statement through case studies of One Dimension Bin-Packing problem and Tabu Search algorithm. Significant variables were considered for characterizing the problem structure and space; the behavior of algorithm during its search trajectory and variables proposed for characterizing its operational behavior. As well as, three parts of the algorithm internal logical structure are considered in the analysis. The proposal characterization allowed to found causal models. Such models contribute to the justification for the use of an algorithm for solving a test instances subset, obtaining an average prediction percentage of 76%. Important relations between the algorithm performance, in terms of domination or inferior region and significant variables were identified from these causal models. Such obtained formal explanations permit to propose guidelines to redesign the algorithm internal logical structure for improving its performance. The redesign proposals outperform to original algorithm in an average of 69% out of 324 problem instances. The proposed characterization of relation problem-algorithm could be a formal procedure for obtaining guidelines that improves the algorithms performance. As future work is considered to extend this proposed characterization to other variants of the Tabu Search algorithm to further explore the internal logical structure; as well as to other algorithms (Genetic, Ant Colony Optimization, etc.), firstly to same problem and then to other optimization problems.

References

1. Garey, M.R., Jhonson, D.S.: Computers and Intractability, a Guide to the Theory of NP-completeness. W. H. Freeman and Company, New York (1979)
2. Papadimitriou, C., Steiglitz, K.: Combinatorial Optimization, Algorithms and Complexity. Prentice Hall, Upper Saddle River (1982)
3. Rendell, L., Cho, H.: Empirical learning as a function of concept character. Mach. Learn. **5**, 267–298 (1990)
4. Lagoudakis, M., Littman, M.: Learning to select branching rules in the DPLL procedure for satisfiability. Electron. Notes Discrete Math. **9**, 344–359 (2001)
5. Smith-Miles, K.: Cross-disciplinary perspectives on meta-learning for algorithm selection. ACM Comput. Surv. **41**(1), 1–25 (2009)
6. Wolpert, D., Macready, W.: No free lunch theorems for optimizations. IEEE Trans. Evol. Comput. **1**(1), 67–82 (1996)
7. Vanchipura, R., Sridharan, R.: Development and analysis of constructive heuristic algorithms for flow shop scheduling problems with sequence-dependent setup times. Int. J. Adv. Manuf. Technol. **67**, 1337–1353 (2013)
8. Hutter, F., Xu, L., Hoos, H., Leyton-Brown, K.: Algorithm runtime prediction: methods & evaluation. Artif. Intell. **206**, 79–111 (2014)
9. Xu, L., Hoos, H., Leyton-Brown, K.: Hydra: automatically configuring algorithms for portfolio-based selection. In: Proceedings of the 25th National Conference on Artificial Intelligence (AAAI 2010), pp. 210–216 (2010)
10. Cayci, A., Menasalvas, E., Saygin, Y., Eibe, S.: Self-configuring data mining for ubiquitous computing. Inf. Sci. **246**, 83–99 (2013)
11. Pavón, R., Díaz, F., Laza, R., Luzón, M.: Experimental evaluation of an automatic parameter setting system. Expert Syst. Appl. **37**, 5224–5238 (2010)
12. Hutter, F., Hoos, H.H., Leyton-Brown, K.: Sequential model-based optimization for general algorithm configuration. In: Coello, C.A.C. (ed.) LION 2011. LNCS, vol. 6683, pp. 507–523. Springer, Heidelberg (2011). https://doi.org/10.1007/978-3-642-25566-3_40
13. Yeguas, E., Luzón, M., Pavón, R., Laza, R., Arroyo, G., Díaz, F.: Automatic parameter tuning for evolutionary algorithms using a Bayesian case-based reasoning system. Appl. Soft Comput. **18**, 185–195 (2014)
14. Ries, J., Beullens, P.: A semi-automated design of instance-based fuzzy parameter tuning for metaheuristics based on decision tree induction. J. Oper. Res. Soc. **66**(5), 782–793 (2015)
15. Yong, X., Feng, D., Rongchun, Z.: Optimal selection of image segmentation algorithms based on performance prediction. In: Proceedings of the Pan-Sydney Area Workshop on Visual Information Processing, pp. 105–108. Australian Computer Society, Inc. (2003)

16. Pérez, J., Pazos, R.A., Frausto, J., Rodríguez, G., Romero, D., Cruz, L.: A statistical approach for algorithm selection. In: Ribeiro, C.C., Martins, S.L. (eds.) WEA 2004. LNCS, vol. 3059, pp. 417–431. Springer, Heidelberg (2004). https://doi.org/10.1007/978-3-540-24838-5_31

17. Nikolić, M., Marić, F., Janičić, P.: Instance-based selection of policies for SAT solvers. In: Kullmann, O. (ed.) SAT 2009. LNCS, vol. 5584, pp. 326–340. Springer, Heidelberg (2009). https://doi.org/10.1007/978-3-642-02777-2_31

18. Yuen, S., Zhang, X.: Multiobjective evolutionary algorithm portfolio: choosing suitable algorithm for multiobjective optimization problem. In: 2014 IEEE Congress on Evolutionary Computation (CEC), Beijing, China, pp. 1967–1973 (2014)

19. Munoz, M., Kirley, M., Halgamuge, S.: Exploratory landscape analysis of continuous space optimization problems using information content. IEEE Trans. Evol. Comput. **19**(1), 74–87 (2015)

20. Leyton-Brown, K., Hoos, H., Hutter, F., Xu, L.: Understanding the empirical hardness of np-complete problems. Mag. Commun. ACM **57**(5), 98–107 (2014)

21. Cruz, L., Gómez, C., Pérez, J., Landero, V., Quiroz, M., Ochoa, A.: Algorithm Selection: From Meta-Learning to Hyper-Heuristics. INTECH Open Access Publisher (2012)

22. Wagner, M., Lindauer, M., Misir, M., et al.: A case of study of algorithm selection for the travelling thief problem. J. Heuristics, 1–26 (2017)

23. Pérez, J., Cruz, L., Landero, V.: Explaining performance of the threshold accepting algorithm for the bin packing problem: a causal approach. Pol. J. Environ. Stud. **16**(5B), 72–76 (2007)

24. Tavares, J.: Multidimensional knapsack problem: a fitness landscape analysis. IEEE Trans. Syst. Man Cybern. Part B Cybern. **38**(3), 604–616 (2008)

25. Pérez, J., et al.: An application of causality for representing and providing formal explanations about the behavior of the threshold accepting algorithm. In: Rutkowski, L., Tadeusiewicz, R., Zadeh, L.A., Zurada, J.M. (eds.) ICAISC 2008. LNCS (LNAI), vol. 5097, pp. 1087–1098. Springer, Heidelberg (2008). https://doi.org/10.1007/978-3-540-69731-2_102

26. Smith-Miles, K., van Hemert, J., Lim, X.Y.: Understanding TSP difficulty by learning from evolved instances. In: Blum, C., Battiti, R. (eds.) LION 2010. LNCS, vol. 6073, pp. 266–280. Springer, Heidelberg (2010). https://doi.org/10.1007/978-3-642-13800-3_29

27. Quiroz, M., Cruz, L., Torrez, J., Gómez, C.: Improving the performance of heuristic algorithms based on exploratory data analysis. In: Castillo, O., Melin, P., Kacprzyk, J. (eds.) Recent Advances on Hybrid Intelligent Systems, Studies in Computational Intelligence, vol. 452, pp. 361–375. Springer, Heidelberg (2013)

28. Landero, V., Pérez, J., Cruz, L., Turrubiates, T., Rios, D.: Effects in the algorithm performance from problem structure, searching behavior and temperature: a causal study case for threshold accepting and bin-packing problem. In: Misra, S., Gervasi, O., Murgante, B. (eds.) ICCSA 2019. LNCS, vol. 11619, pp. 152–166. Springer, Heidelberg (2019). https://doi.org/10.1007/978-3-030-24289-3_13

29. Spirtes, P., Glymour, C., Scheines, R.: Causation, Prediction, and Search, 2nd edn. The MIT Press, Cambridge (2001)

30. Beasley, J., E.: OR-Library. Brunel University (2006). http://people.brunel.ac.uk/~mastjjb/jeb/orlib/binpackinfo.html

31. Scholl, A., Klein, R.: (2003). http://www.wiwi.uni-jena.de/Entscheidung/binpp/

32. Glover, F.: Tabu search - Part I, first comprehensive description of tabu search. ORSA-J. Comput. **1**(3), 190–206 (1989)

33. Fleszar, K., Hindi, K.S.: New heuristics for one-dimensional bin packing. Comput. Oper. Res. **29**, 821–839 (2002)

34. Khuri, S., Schütz, M., Heitkötter, J.: Evolutionary heuristics for the bin packing problem. In: Artificial Neural Nets and Genetic Algorithms. Springer, Vienna (1995). https://doi.org/10.1007/978-3-7091-7535-4_75

35. Merz, P., Freisleben, B.: Fitness landscapes and memetic algorithm design. In: New Ideas in Optimization, pp. 245–260. McGraw-Hill Ltd., UK (1999)

36. Fayyad, U.M., Irani, K.B.: Multi-interval discretization of continuous-valued attributes for classification learning. In: IJCAI, pp. 1022–1029 (1993)

37. Hall, M.A.: Feature selection for discrete and numeric class machine learning (1999)

38. Watson, J., Darrell, W., Adele, E.: Linking search space structure, run-time dynamics, and problem difficulty: a step toward demystifying tabu search. J. Artif. Intell. Res. **24**, 221–261 (2005)

Efficient Speed-Up of Radial Basis Functions Approximation and Interpolation Formula Evaluation

Michal Smolik$^{(\boxtimes)}$ (iD) and Vaclav Skala (iD)

Faculty of Applied Sciences, University of West Bohemia, Pilsen, Czech Republic
{smolik,skala}@kiv.zcu.cz

Abstract. This paper presents a method for efficient Radial basis function (RBF) evaluation if compactly supported radial basis functions (CSRBF) are used. Application of CSRBF leads to sparse matrices, due to limited influence of radial basis functions in the data domain and thus non-zero weights (coefficients) are valid only for some areas in the data domain. The presented algorithm uses space subdivision which enables us to use only relevant weights for efficient RBF function evaluation. This approach is applicable for $2D$ and $3D$ case and leads to a significant speed-up. This approach is applicable in cases when the RBF function is evaluated repeatably.

Keywords: Radial basis functions · Space subdivision · Function evaluation · Interpolation · Approximation

1 Introduction

Interpolation and approximation are well-known techniques in many scientific disciplines, i.e. mostly in physical sciences [7]. There are two main approaches dealing with an interpolation, as well as an approximation. The first one is the mesh-based approach and the second one is the mesh-less approach.

The mesh-based approaches need to know the connectivity of the interpolated or approximated dataset. However, the triangulation of the input dataset can be highly time consuming in higher dimensions, as the computational time complexity of Delaunay triangulation [20] is $O(N^{\lceil d/2 \rceil + 1})$, i.e. for $d = 2$ is $O(N^2)$ and for $d = 3$ is $O(N^3)$ [23]. Of course, there are some algorithms [5,6,15] that are dealing with decreasing of the time complexity of Delaunay triangulation. However, the computational time to construct the triangulation is still high.

On the other hand, the mesh-less techniques do not require any tessellation, which is a significant advantage over the mesh-based techniques, i.e. one can directly start to compute the mesh-less approximation or interpolation from the

The research was supported by projects Czech Science Foundation (GACR) No. GA17-05534S, by Ministry of Education, Youth, and Sport of Czech Republic - University spec. research - 1311, and partially by SGS 2019-016.

O. Gervasi et al. (Eds.): ICCSA 2020, LNCS 12249, pp. 165–176, 2020.
https://doi.org/10.1007/978-3-030-58799-4_12

scattered or unsorted input dataset. There are many algorithms for mesh-less approximation of scattered data. Many of them are described in the book [8] or [10]. Other examples of mesh-less approximation are in [1–3].

The mostly used mesh-less technique for interpolation and approximation is the Radial basis function interpolation and approximation. It uses only the distance between the pairs of centers of radial basis functions and input points. An advantage of the Radial basis function interpolation and approximation is time complexity which is independent of the dimension, which is useful in higher dimensions.

There are many research papers focused on speeding-up the interpolation or approximation process and solving the stability of computation. The paper [12] uses the preconditioned Krylov iteration to compute fast interpolation. The paper [28] uses global radial basis functions for interpolation but use them in a local sense to speed-up the interpolation. The paper [4] provides a summary of fast RBF interpolation and approximation. These methods speed-up the approximation or interpolation, i.e. finding weights of RBFs, whereas we would like to speed-up the evaluation of the final Radial basis function formula. In this paper, we focus on the efficient evaluation of the RBFs, i.e. computation of interpolation or approximation, when coefficients (weights) of the RBF have already been computed. The approach is based on the space subdivision application.

2   Radial Basis Functions

The Radial Basis Function (RBF) interpolation [18] and approximation [8] is a meshless technique which was introduced by Hardy [13]. It is commonly used in many scientific disciplines such as solution of partial differential equations [14,32], image reconstruction [29], neural networks [31], vector field [24,26,27], GIS systems [16], optics [19] etc.

The formula for computing the function value of RBF is the weighted sum of radial basis functions and has the following form

$$f(\boldsymbol{x}) = \sum_{i=1}^{N} \lambda_i \varphi \left(\|\boldsymbol{x} - \boldsymbol{\xi}_i\| \right), \tag{1}$$

where $\boldsymbol{x}$ is the position for which (1) provides the interpolating value, $\varphi(\dots)$ is the radial basis function, λ_i is the weight of radial basis function, N is the number of radial basis functions and $\boldsymbol{\xi}_i$ is the center of radial basis function.

The radial basis functions used in (1) can be selected from two main groups. The first ones are "global" radial basis functions. These functions influence the whole interval of the data interpolation/approximation domain. The most known examples of "global" radial basis functions are shown below.

$$\text{Thin Plate Spline} \qquad \varphi(r) = r^2 \log r = \frac{1}{2} r^2 \log r^2$$

$$\text{Gauss function} \qquad \varphi(r) = e^{-(\epsilon r)^2}$$

$$\text{Inverse Quadric} \qquad \varphi(r) = \frac{1}{1 + (\epsilon r)^2} \tag{2}$$

$$\text{Inverse Multiquadric} \qquad \varphi(r) = \frac{1}{\sqrt{1 + (\epsilon r)^2}}$$

$$\text{Multiquadric} \qquad \varphi(r) = \sqrt{1 + (\epsilon r)^2}$$

where ϵ is the shape parameter of the radial basis function.

Local RBFs (CSRBF) were introduced by [30]. These radial basis functions have only local influence, i.e. the function value is non-zero only on some limited interval. This is a great advantage when solving the linear system of equations, as the interpolation or approximation matrix is sparse. The CSRBF has the following form

$$\varphi(r) = (1 - r)_+^q P(r), \tag{3}$$

where $P(r)$ is some polynomial, q is some non-zero positive number and

$$(1 - r)_+ = \begin{cases} (1 - r) & (1 - r) \geq 0 \\ 0 & (1 - r) < 0 \end{cases} \tag{4}$$

The well known examples of CSRBF are given by.

$$\varphi_1(r) = (1 - \hat{r})_+$$
$$\varphi_2(r) = (1 - \hat{r})_+^3 (3\hat{r} + 1)$$
$$\varphi_3(r) = (1 - \hat{r})_+^5 (8\hat{r}^2 + 5\hat{r} + 1)$$
$$\varphi_4(r) = (1 - \hat{r})_+^2$$
$$\varphi_5(r) = (1 - \hat{r})_+^4 (4\hat{r} + 1)$$

$$\varphi_6(r) = (1 - \hat{r})_+^6 (35\hat{r}^2 + 18\hat{r} + 3)$$
$$\varphi_7(r) = (1 - \hat{r})_+^8 (32\hat{r}^3 + 25\hat{r}^2 + 8\hat{r} + 1)$$
$$\varphi_8(r) = (1 - \hat{r})_+^3$$
$$\varphi_9(r) = (1 - \hat{r})_+^3 (5\hat{r} + 1)$$
$$\varphi_{10}(r) = (1 - \hat{r})_+^7 (16\hat{r}^2 + 7\hat{r} + 1)$$

$$\tag{5}$$

where $\hat{r} = \epsilon r$ and ϵ is the shape parameter of the radial basis function, see Fig. 1 for a visualization of (5).

2.1   RBF Interpolation

The RBF interpolation was introduced by [13]. It uses the formula (1) and the radial basis functions are placed into the location of input data. The interpolation formula has the following form

$$h_i = f(\boldsymbol{x}_i) = \sum_{j=1}^{N} \lambda_j \varphi \left(\| \boldsymbol{x}_i - \boldsymbol{x}_j \| \right) \tag{6}$$

$$\text{for } \forall i \in \{1, \ldots, N\},$$

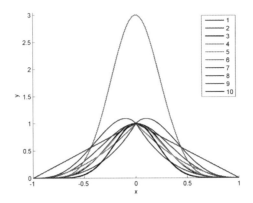

Fig. 1. Examples of CSRBF from (5).

where N is the number of input points and h_i is the function value at point $\boldsymbol{x}_i$. The Eq. (6) can be rewritten in a matrix form as

$$\boldsymbol{A\lambda} = \boldsymbol{h}. \tag{7}$$

As $\varphi\left(\|\boldsymbol{x}_i - \boldsymbol{x}_j\|\right) = \varphi\left(\|\boldsymbol{x}_j - \boldsymbol{x}_i\|\right)$, the matrix $\boldsymbol{A}$ is symmetric.

2.2 RBF Approximation

The RBF approximation [8,17,21] is based on the RBF interpolation. The number of radial basis functions is smaller than the number of input points. It leads to an over-determined system of linear equations

$$h_i = f(\boldsymbol{x}_i) = \sum_{j=1}^{M} \lambda_j \varphi\left(\|\boldsymbol{x}_i - \boldsymbol{\xi}_j\|\right) \tag{8}$$

$$\text{for } \forall i \in \{1, \dots, N\},$$

where $\boldsymbol{\xi}_j$ is the center of radial basis function, M is the number of radial basis functions and $M < N$ (mostly $M \ll N$). It is not possible to fulfill all equations at once in (8), so we need to use the least squares errors method (LSE) using the following formula

$$\boldsymbol{A}^T \boldsymbol{A\lambda} = \boldsymbol{A}^T \boldsymbol{h}, \tag{9}$$

where $\boldsymbol{A}^T \boldsymbol{A}$ is a symmetric matrix (Fig. 2).

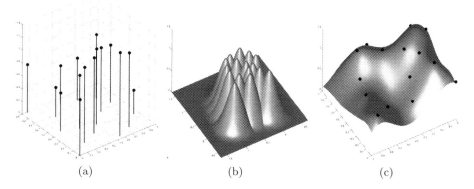

(a)   (b)   (c)

Fig. 2. The visualization of data values (a), the RBF collocation functions (b) and the resulting interpolant (c).

3   Proposed Approach Using CSRBF

The compactly supported radial basis functions (CSRBFs) have limited influence of each radial basis function given by the radius, which is defined as

$$r_{max} = \frac{1}{\epsilon}, \tag{10}$$

where ϵ is the shape parameter of the local radial function.

Due to using CSRBFs, the interpolation or approximation matrix $\boldsymbol{A}$ (in (7) or (9)) is sparse (we assume that the maximum distance of input points is larger than r_{max}). Solving the system of linear equations (7) or (9), the vector of weights $\boldsymbol{\lambda}$ is obtained.

Using the lambda ($\boldsymbol{\lambda}$) coefficients, we can compute the function value of RBF at a location $\boldsymbol{x}$ using the formula

$$f(\boldsymbol{x}) = \sum_{j=1}^{M} \lambda_j \varphi \left(\|\boldsymbol{x} - \boldsymbol{\xi}_j\| \right), \tag{11}$$

where M is the number of radial basis functions and $\boldsymbol{\xi}_j$ is the location of radial basis function, i.e. its center.

The number of radial basis functions can be very high for real-world approximation and interpolation problems. However, the evaluated radial basis functions $\varphi \left(\|\boldsymbol{x} - \boldsymbol{\xi}_j\| \right)$ are mostly equal to zero as the influence of each radial basis function is limited to the maximal radius $r_{max} = 1/\epsilon$. However, evaluation of such zero radial basis functions is wasting computational time and the evaluation of such zero radial basis functions should be omitted.

3.1   Space Subdivision

Considering the processing of very large data sets, it is necessary to deal with the efficiency of the evaluation of the final interpolant, especially if the CSRBFs are used.

The proposed approach presents an algorithm for speeding-up the evaluation of the function value of RBF when using the local radial basis functions, i.e. CSRBF (5), see [22] for interpolation of large datasets with CSRBF.

The proposed approach is advisable to combine with the method for fast interpolation using the space subdivision [25,26]. This method also divides all the input points into a grid, computes the RBF interpolation or approximation of each cell in the grid, and finally blends all the interpolations or approximations together using simple but efficient formula. This approach is suitable for large datasets that cannot be interpolated or approximated using the standard RBF method.

To speed-up the RBF function evaluation, we need to know some relation between the position of a point (where the RBF is to be evaluated) and the location of all radial basis functions. Only the radial basis functions that are closer than $r_{max} = 1/\epsilon$ should be evaluated and used for the RBF evaluation.

Finding all of those functions every time without any special data structure could increase the computational time on the contrary. A much better solution is to use space subdivision to divide all centers of radial basis functions into a rectangular grid.

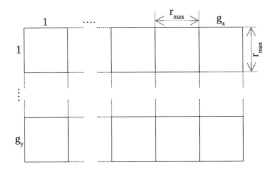

Fig. 3. Visualization of grid used for space subdivision.

Let us split the data domain into a rectangular grid, see Fig. 3. The size of each cell in the grid should be $r_{max} \times r_{max}$, i.e. the size depends on the shape parameter which can be selected as described in [9,11,13]. The reason for this size will be clarified later in the following chapter.

The total number of cells is

$$G = g_x g_y \tag{12}$$

in the $2D$ case or likewise

$$G = g_x g_y g_z \tag{13}$$

in the $3D$ case of RBF interpolation or approximation. The size g_x is computed as

$$g_x = \left\lceil \frac{x_{max} - x_{min}}{r_{max}} \right\rceil \tag{14}$$

and equations for g_y or g_z are straightforward.

Dividing all radial basis functions according to their centers into the grid is of $O(N)$ time complexity and has no influence on the final computational time of the RBF interpolation or approximation.

The next step is a standard computation of the RBF interpolation or approximation, i.e. solving a system of linear equations. As a result, we obtain the weighting coefficients of radial basis functions, i.e. the lambda coefficients ($\boldsymbol{\lambda}$). Each lambda coefficient is associated with exactly one radial basis function. These coefficients are also divided into the grid and associated with corresponding radial basis functions.

The additional time complexity required by the proposed algorithm to the standard RBF computation can be considered as the preprocessing and is

$$O(N + N) \approx O(N), \tag{15}$$

where the first $O(N)$ is for the division of points into the grid. The second $O(N)$ is for the division of the calculated lambda values (λ_i) into the grid, i.e. associating them with the corresponding radial basis functions.

3.2   RBF Function Value Computation

When the RBF is traditionally evaluated, all the radial basis functions are evaluated for the input point $\boldsymbol{x}$ and multiplied with the weighting lambda coefficients. However, many radial basis functions for the input point $\boldsymbol{x}$ are equal zero because the input point $\boldsymbol{x}$ is more distant than r_{max} ($r_{max} = 1/\epsilon$) from the centers of radial basis functions.

The point $\boldsymbol{x}$ that belongs to one cell of the grid can only affect the value of some radial basis functions, i.e. others will be zero. The visualization of the influence is in Fig. 4. It can be seen, that the point that belongs to one cell can only affect the evaluation of radial basis functions in that cell and all one-neighborhood cells around.

The RBF function value is computed at a point $\boldsymbol{x}$. The first step of the proposed algorithm is the location of the cell in the grid, where this point belongs to. For the RBF function value evaluation, we will use only radial basis functions (with associated computed weights) of the cell where $\boldsymbol{x}$ lies and all the one-neighborhood cells around. All other radial basis functions can be skipped as their value is always zero and their evaluation and multiplication with their weights only increase the computational time while not giving any additional increment for the sum of weighted radial basis functions.

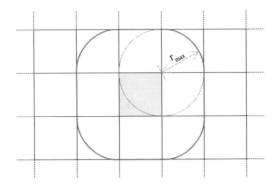

Fig. 4. The area of influence (red color) for evaluation of radial basis functions when the point for function evaluation lies in the marked cell. (Color figure online)

The computation of RBF function value using the proposed approach decreases the computational time compared to the standard RBF function evaluation.

3.3   Theoretical Speed-Up of RBF Function Value Computation

In this chapter, the expected speed-up of our algorithm for RBF function value computation is analyzed. Let us assume a uniform distribution of points. Then the time complexity of the standard evaluation of a function using RBF is

$$O(M),\tag{16}$$

where M is the number of radial basis functions.

The time complexity of the proposed algorithm for evaluation of Radial basis function is

$$O\left(3^d \frac{M}{G}\right),\tag{17}$$

where d is the dimension, i.e. $d = 2$ for 2D and $d = 3$ for 3D, the G represents the total number of cells in grid.

The expected speed-up of our proposed algorithm to the standard one is computed as

$$\upsilon \approx \frac{O(M)}{O\left(3^d \frac{M}{G}\right)},\tag{18}$$

which is equal

$$\upsilon \approx \frac{G}{3^d}.\tag{19}$$

For G higher than 3^d the speed-up is higher than one, i.e. our proposed algorithm is faster.

The value of G, i.e. the number of cells in the grid, depends on r_{max} and thus on the shape parameter ϵ of radial basis function. Usually r_{max} is much

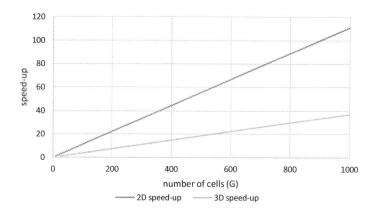

Fig. 5. The theoretical speed-up of evaluation of RBF in $2D$ and $3D$.

smaller than the range of the input data in each axis, i.e. the grid size will be bigger than 3×3 (in $2D$) or $3 \times 3 \times 3$ (in $3D$). Thus the speed-up will be higher than one, see Fig. 5.

4   Experimental Results

The theoretical speed-up needs to be confirmed experimentally for both $2D$ and $3D$ RBF cases. In our experiments, we created data sets with a different number of randomly distributed input points with a uniform distribution and associated function values. We also tested the proposed approach with real large datasets of LiDAR data[1] (see Fig. 6 for visualization of RBF interpolation). Then the RBF interpolation with the following CSRBF radial basis function was used.

$$\varphi_5(r) = (1 - \epsilon r)_+^4 (4\epsilon r + 1), \tag{20}$$

where this radial basis function is C^2 smooth at the origin, i.e. it is continuous and has the first and the second continuous derivative. The experimental results do not depend on the dataset values, it only depends on the number of cells. The number of cells is determined according to the shape parameter. Due to this, we do not discuss the type of interpolated data as the results are valid for any dataset with the same number of input points and the same number of cells.

In the experiments the radial basis functions were divided into grids of different sizes (see Table 1 for $2D$ and Table 2 for $3D$).

We evaluated each RBF in 10^9 random points with uniform distribution to obtain a sufficiently large number of measurements. For the evaluation, we used both our proposed approach with space subdivision and also the standard approach without space subdivision. The speed-up of our proposed algorithm compared to the standard one is summarized in Table 1 for $2D$ and Table 2 for $3D$.

---

[1] https://liblas.org/.

Table 1. The measured speed-up of 2D RBF evaluation using the proposed approach.

	Number of radial basis functions	Number of cells	Speed-up
Synthetic data sets	1000	16	1.80
	5 000	64	6.81
	10 000	100	10.88
	10 000	225	24.53
	100 000	1 089	119.19
	1 000 000	10 000	1097.78
Real	131 044	1 296	141.98
Data sets	756 150	7 350	806.46

(a)   (b)

Fig. 6. RBF interpolation of real datasets. The part of Alps mountain consist of 131 044 points (a) and the mountain of Saint Helens consists of 756 150 points (b).

It can be seen, that even for a small number of radial basis functions and a relatively small number of cells the proposed algorithm is significantly faster. With an increasing number of cells, the speed-up increases as well. For large datasets, the speed-up can be even larger than 100, and thus our proposed algorithm can save a lot of computational time during the evaluation of RBF function. The proposed approach was tested with real datasets as well and proved its ability for high speed-up of RBF function evaluation, see Table 1.

Table 2. The measured speed-up of 3D RBF evaluation using the proposed approach.

	Number of radial basis functions	Number of cells	Speed-up
Synthetic data sets	1000	27	1.01
	5 000	80	2.85
	10 000	125	4.54
	10 000	216	7.86
	100 000	1 000	36.56
	1 000 000	10 648	390.03

5   Conclusion

In this contribution, we have presented a new approach for speeding up the evaluation of Radial basis functions (RBF). The proposed approach uses the space-subdivision to select only appropriate locations of radial basis functions that are used to evaluate the RBF. This proposed approach is easy to implement and can be used for any dimension.

The proposed approach has significant speed-up compared to the standard one evaluation of RBF. This speed-up was confirmed by experiments that were made for $2D$ as well as for $3D$ RBF using Matlab/Octave.

Acknowledgments. The authors would like to thank their colleagues at the University of West Bohemia, Plzen, for their discussions and suggestions. Also, great thanks belong to our colleague Martin Cervenka for his hints and ideas. The research was supported by projects Czech Science Foundation (GACR) No. GA17-05534S, by the Ministry of Education, Youth, and Sport of Czech Republic - University spec. research - 1311, and partially by SGS 2019-016.

References

1. Adams, B., Wicke, M.: Meshless approximation methods and applications in physics based modeling and animation. In: Eurographics (Tutorials), pp. 213–239 (2009)
2. Atluri, S., Liu, H., Han, Z.: Meshless local Petrov-Galerkin (MLPG) mixed finite difference method for solid mechanics. Comput. Model. Eng. Sci. **15**(1), 1 (2006)
3. Belytschko, T., Krongauz, Y., Organ, D., Fleming, M., Krysl, P.: Meshless methods: an overview and recent developments. Comput. Methods Appl. Mech. Eng. **139**(1), 3–47 (1996)
4. Biancolini, M.E.: Fast Radial Basis Functions for Engineering Applications. Springer, Cham (2017). https://doi.org/10.1007/978-3-319-75011-8
5. Chen, M.-B.: A parallel 3D Delaunay triangulation method. In: 2011 IEEE Ninth International Symposium on Parallel and Distributed Processing with Applications, pp. 52–56. IEEE (2011)
6. Cignoni, P., Montani, C., Scopigno, R.: DeWall: a fast divide and conquer Delaunay triangulation algorithm in Ed. Comput. Aided Des. **30**(5), 333–341 (1998)
7. Davis, P.J.: Interpolation and Approximation. Courier Corporation, North Chelmsford (1975)
8. Fasshauer, G.E.: Meshfree Approximation Methods with MATLAB, vol. 6. World Scientific, Singapore (2007)
9. Fasshauer, G.E., Zhang, J.G.: On choosing "optimal" shape parameters for RBF approximation. Numer. Algor. **45**(1–4), 345–368 (2007). https://doi.org/10.1007/s11075-007-9072-8
10. Ferreira, A.J.M., Kansa, E.J., Fasshauer, G.E., Leitão, V.M.A.: Progress on Meshless Methods. Springer, Dordrecht (2009). https://doi.org/10.1007/978-1-4020-8821-6
11. Franke, R.: Scattered data interpolation: tests of some methods. Math. Comput. **38**(157), 181–200 (1982)
12. Gumerov, N.A., Duraiswami, R.: Fast radial basis function interpolation via preconditioned Krylov iteration. SIAM J. Sci. Comput. **29**(5), 1876–1899 (2007)

13. Hardy, R.L.: Multiquadric equations of topography and other irregular surfaces. J. Geophys. Res. **76**(8), 1905–1915 (1971)
14. Larsson, E., Fornberg, B.: A numerical study of some radial basis function based solution methods for elliptic PDEs. Comput. Math. Appl. **46**(5), 891–902 (2003)
15. Liu, Y., Snoeyink, J.: A comparison of five implementations of 3D Delaunay tessellation. In: Combinatorial and Computational Geometry, vol. 52, no. 439–458, p. 56 (2005)
16. Majdisova, Z., Skala, V.: Big geo data surface approximation using radial basis functions: a comparative study. Comput. Geosci. **109**, 51–58 (2017)
17. Majdisova, Z., Skala, V.: Radial basis function approximations: comparison and applications. Appl. Math. Model. **51**, 728–743 (2017)
18. Pan, R., Skala, V.: A two-level approach to implicit surface modeling with compactly supported radial basis functions. Eng. Comput. **27**(3), 299–307 (2011). https://doi.org/10.1007/s00366-010-0199-1
19. Prakash, G., Kulkarni, M., Sripati, U.: Using RBF neural networks and Kullback-Leibler distance to classify channel models in free space optics. In: 2012 International Conference on Optical Engineering (ICOE), pp. 1–6. IEEE (2012)
20. Rajan, V.T.: Optimality of the Delaunay triangulation in $\mathbb{R}^d$. Discrete Comput. Geom. **12**(2), 189–202 (1994). https://doi.org/10.1007/BF02574375
21. Skala, V.: Fast interpolation and approximation of scattered multidimensional and dynamic data using radial basis functions. WSEAS Trans. Math. **12**(5), 501–511 (2013). E-ISSN 2224-2880
22. Skala, V.: RBF interpolation with CSRBF of large data sets. Procedia Comput. Sci. **108**, 2433–2437 (2017)
23. Smolik, M., Skala, V.: Highly parallel algorithm for large data in-core and out-core triangulation in E2 and E3. Procedia Comput. Sci. **51**, 2613–2622 (2015)
24. Smolik, M., Skala, V.: Spherical RBF vector field interpolation: experimental study. In: 2017 IEEE 15th International Symposium on Applied Machine Intelligence and Informatics (SAMI), pp. 431–434. IEEE (2017)
25. Smolik, M., Skala, V.: Large scattered data interpolation with radial basis functions and space subdivision. Integr. Comput. Aided Eng. **25**(1), 49–62 (2018)
26. Smolik, M., Skala, V.: Efficient simple large scattered 3D vector fields radial basis functions approximation using space subdivision. In: Misra, S., et al. (eds.) ICCSA 2019. LNCS, vol. 11619, pp. 337–350. Springer, Cham (2019). https://doi.org/10.1007/978-3-030-24289-3_25
27. Smolik, M., Skala, V., Majdisova, Z.: Vector field radial basis function approximation. Adv. Eng. Softw. **123**, 117–129 (2018)
28. Torres, C.E., Barba, L.A.: Fast radial basis function interpolation with Gaussians by localization and iteration. J. Comput. Phys. **228**(14), 4976–4999 (2009)
29. Uhlir, K., Skala, V.: Reconstruction of damaged images using radial basis functions. In: 2005 13th European Signal Processing Conference, pp. 1–4. IEEE (2005)
30. Wendland, H.: Computational aspects of radial basis function approximation. Stud. Comput. Math. **12**, 231–256 (2006)
31. Yingwei, L., Sundararajan, N., Saratchandran, P.: Performance evaluation of a sequential minimal radial basis function (RBF) neural network learning algorithm. IEEE Trans. Neural Networks **9**(2), 308–318 (1998)
32. Zhang, X., Song, K.Z., Lu, M.W., Liu, X.: Meshless methods based on collocation with radial basis functions. Comput. Mech. **26**(4), 333–343 (2000). https://doi.org/10.1007/s004660000181

Numerical Risk Analyses of the Impact of Meteorological Conditions on Probability of Airport Runway Excursion Accidents

Misagh Ketabdari[(✉)] [ID], Emanuele Toraldo [ID],
and Maurizio Crispino [ID]

Department of Civil and Environmental Engineering, Transportation
Infrastructures and Geosciences Section, Politecnico di Milano,
20133 Milan, Italy
{misagh.ketabdari, emanuele.toraldo,
Maurizio.crispino}@polimi.it

Abstract. A continuous growth in air transport industry over the recent decades increases the probability of occurrence of accidents and consequently the need of aviation risk and safety assessments. In order to assess the risks related to aircraft ground operations, all the influencing variables that can affect the safety of maneuvers should be determined. Generally, catastrophic accidents that occur inside an aerodrome are assigned to the runway incursions and excursions. Incursions are dedicated to all events happening inside the runway (e.g. existence of an obstacle or accident of two aircraft) and excursions are considered for an unsuccessful aircraft operation which leads to surpassing the designated thresholds/borders of the runway. Landing and take-off overruns, as the main excursion events, are responsible for the major recorded incidents/accidents over the past 50 years. Many variables can affect the probability of runway-related accidents such as weather conditions, aircraft braking potential, pilot's level of experience, etc. The scope of this study is to determine weather-wise parameters that amplify the probability of excursion events.

In order to quantify the probability of each type of accidents, the effect of each meteorological variables on the aircraft operation is simulated by RSARA/ LRSARA© simulators, released by Aircraft Cooperative Research Program (ACRP). In this regard, specific airports with diverse characteristics (i.e.; landlocked airport, extreme annual operations, extreme runway geometries and weather conditions) are selected as case studies in the analyses. Gusts and wind forces, specifically cross-wind, turned out to be the most dominant weather-wise influencing parameters on the occurrence of runway excursions.

Keywords: Numerical risk analysis · RSA · Meteorological conditions

1 Introduction

Air transportation has the major influence on facilitating economic growth. Around 2 billion passengers annually are transported by this industry in 10 years period. One of the main reasons is the high demand for these services, as a result of that, enhancing

© Springer Nature Switzerland AG 2020
O. Gervasi et al. (Eds.): ICCSA 2020, LNCS 12249, pp. 177–190, 2020.
https://doi.org/10.1007/978-3-030-58799-4_13

rapid movement of passengers, goods and services to the domestic and global market. The demand for air transport has increased steadily over the past years. Passenger numbers have grown by 50% over the last decade and have more than doubled since the 1990s [1]. Therefore, extra attention should be paid to the safety of aviation because of its tremendous demanding potential.

Aviation risk can be dedicated to the measurements of the events frequency and the severity level of the possible consequences. Therefore, risk assessment is the interaction between aerodrome related hazards, airport's components vulnerability and the total exposure of the system. Risk can be defined as a measurement of the threat to safety of a complex posed by the accident scenarios and their consequences [2].

Identification of aerodrome related scenarios, their probability of occurrence and severity classification of the possible consequences (e.g. fatalities, aircraft damages, etc.) are the main sub-models that risk assessment methodologies in the field of air transport should cover.

Runway related events as results of aircraft operations can be classified into incursion and excursion accidents. Runway incursions are collisions of one aircraft with an accidental obstacle or another aircraft inside the runway.

On the other hand, runway excursions is dedicated to the any exit from the runway borders such as aircraft veers-off or overruns during takeoff or landing phases of flights. According to the recorded accidents/incidents statistics from all over the world, runway excursions are the most common type of events reported annually. These types of events can result in loss of life and/or damage to aircraft, buildings, etc. These facts prove the criticality of excursion events and the necessity to investigate them more. Runway excursions can be divided into following categories:

- Landing Overrun (LDOR), Under-Shoot (LDUS), Veer-Off (LDVO),
- Take-Off Overrun (TOOR), Veer-Off (TOVO) [3].

Runway Safety Areas (RSAs) are essential to be designed in order to mitigate the consequences of aircraft ground-operation events. These specific areas are categorizing depending on the accident scenario that may occur in the proximity of the runway; paved areas located after the runway ends, which are called Runway End Safety Areas (RESAs) and the longitudinal safety strips which are surrounding laterally the runway. RESAs mitigate the possible consequences in events of runway overrun or undershoot, while, lateral safety strips reduce the severity of aircraft veer-off incidents. The main scope of this study is to analyze numerically the impact of various meteorological conditions on the probability of the occurrence of runway-related events and to investigate the mitigating effects of RSAs on the severity of the consequences by simulating three case studies.

2   Background Review and Scope

The safety strips around the runway are designed with the primary motive of mitigating the possible consequences of incidents/accidents when aircraft surpasses the designated runway borders. In particular, RESA should be constructed in order to reduce the severity of the events consequences. Furthermore, aircraft veer-off, as a result of numerous variables such as crosswinds, is probable to occur therefore, longitudinal strips along the runway lateral edges can minimize the consequences. According to the recent updates on the recommendations by International Civil Aviation Organization (ICAO), the 90 m RESA longitudinal dimension is now converted to 240 m. Plus, the total longitudinal protection after the runway threshold, including the 60 m strip, has been doubled from 150 to 300 m [4] (see Fig. 1).

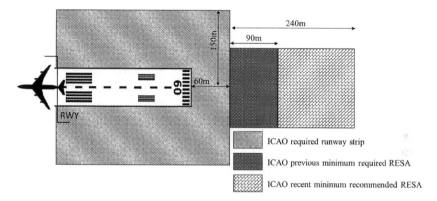

Fig. 1. Reference scheme for ICAO; geometry recommendations for RSAs [5]

Due to spatial circumstances (e.g. landlocked fields), applying the recommended practices prescribed by ICAO is restricted for some airports. In other words, changing the RSA geometry in some cases is not feasible because the runway can be surrounded by different forms of obstacles such as water surfaces (e.g. sea, rivers), residential areas, roads, unfavorable terrain, railroads and etc. Flat and even terrains without interfering obstacles are the ideal field scenarios for application of current RSAs standards. Where it is not possible to respect the RSA geometry by neither standard nor recommended practices, as the matter of safety, other mitigation strategies such as Engineered Material Arresting System (EMAS) can be considered.

According to ICAO Annex 14, any type of objects existing in RESA, which may endanger the safety of aircraft maneuvers, is considered as an obstacle and should be removed as far as practicable. Safety strips that are located before and after the runway should provide a cleared and graded area for aircraft. These areas are intended to serve in the event of undershoot or overrun excursions in various phases of flight. In these scenarios, these safety strips should be designed as to reduce the severity of the possible consequences to the aircraft and passengers by enhancing the aircraft deceleration and facilitate the movement of rescue and fire fighting vehicles [5].

Lateral safety strips that are surrounding the runway should be designed to reduce the severity of the consequences associated with potential lateral aircraft veering-off by providing an even and levelled area, respecting the certain cross slope of the runway surface. These lateral strips should be constructed clear from any obstacles in order to protect aircraft overflying them during failed take-off and landing operations. ICAO prohibits objects in these areas except frangible mountings visual aids [6]. In other words, no fixed objects, other than frangible visual aids required for air navigational purposes, shall be permitted on the lateral runway strips [5].

Different national and international accidents/incidents databases from all over the world are collected to get used in the process of numerical risk analyses in order to determine the contributory meteorological parameters that affect the probability of occurrences of aviation ground operations accidents. In this regard, three airports with different characteristics (e.g. runway geometry, weather conditions, etc.) are simulated, by means of Runway Safety Area Risk analysis (RSARA/LRSARA©) software.

3   Data Collection and Numerical Analyses

In order to analyze the impact of each meteorological condition on the aviation risks, statistical databases consist of accident/incident records in the national and international scales are required that cover the main reasoning for each type of events. The output of this attempt can be also useful for the authorities in charge to prioritize their mitigation actions and setting monitoring operations on the design and execution of airport infrastructures. Accidents/incidents that are recorded annually can be misleading since the chances of irregularity in records are high. In this regard, the trend of the statistics over a period of years is fundamental, not annual variation. This point of view demonstrates the evolution rate of the air transport safety which is necessary in setting right priorities for the required actions and developing the right predictions for the years to come [7].

Figure 2 is extracted from the accidents/incidents reports published by Airbus, Boeing, National Transportation Safety Board (NTSB), and Aviation Safety Network (ASN) as the four major aviation databases for a period of 15 years. In the process of collecting the data, it turned out that as the aircraft operations increase, the number of events are also increase. Event statistical records by these four references aviation databases apparently demonstrate a steady decrease annual rate of incidents/accidents for recent years respect to the past. It should be also noted that hull loss incidents are always more probable to occur respect to the fatal accidents for the corresponding time span. Furthermore, it can be interpreted that the majority of the incidents/accidents occur during the landing/approach and take-off phases of flights and in form of runway excursions [8]. Beside the human errors (e.g. pilot incapability, misleading operational data, etc.), meteorological conditions may noticeably affect the safety of the operations by compromising the ground operations (e.g. low visibility, inadequate skid resistance and pavement friction, etc.). In this regard, further investigation is carried out by this study to assess numerically and separately the impact of each weather parameter and meteorological condition on the safety of the maneuvers.

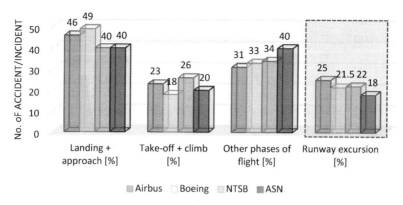

Fig. 2. Accident/incident records from 4 major databases (5-year timespan)

Each of the meteorological variables has a contributory index that needs to be numerically analyzed because of the large number of accidents it has contributed to. For this matter, the statistical records of runway-related events for a five-year time span are collected and the trend is analyzed. NTSB databases are selected for the following numerical analysis according to higher recorded events respect to other databases. Totally, 8,657 aviation incidents/accidents from all over the world are collected that in 1,740 (20%) of these accidents meteorological conditions are contributing factors [9]. According to NTSB [10], Wind, visibility/ceiling, high density altitude, turbulence, carburetor icing, updrafts/downdrafts, precipitation, icing, thunderstorms, temperature extremes, and lightning are the triggering factors in most weather-caused accidents. Weather and non-weather-related accidents for a 5-year time span extracted from NTSB records are depicted in Table 1.

Table 1. Meteorological related and non-related worldwide accidents according to NTSB

Recording year	No. of weather-related accidents	No. of non-weather-related accidents	% of weather-related accidents
2003	409	1463	21.85
2004	378	1344	21.95
2005	336	1444	18.88
2006	303	1289	19.03
2007	314	1377	18.57

According to Table 1, aviation accidents in general, have shown an overall descending trend over 5-year time span. In the final NTSB report for each accident in which meteorological conditions were found to be a triggering cause or contributing factor, a subject modifier is cited to designate the specific weather condition(s) encountered during the accident. In fact, multiple meteorological parameters can overlap and have a complex effect on the occurrence of one event. For total 1,740 accidents records, there are 2,230 citations of meteorological subject modifiers. The breakdown of these statistic records based on various meteorological variables is presented in Fig. 3.

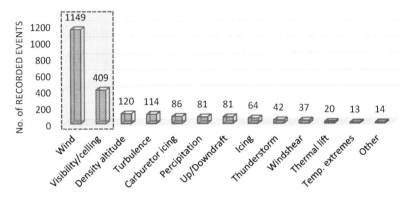

Fig. 3. Recorded meteorological conditions accidents (5-year timespan)

Wind (by 1149 citations) and visibility/ceiling (by 409 citations) impact factors can be considered as dominant contributing parameters in aviation risks compare to other meteorological modifiers and are going to be analyzed separately in continue.

3.1   Meteorological Break Down: Wind Impact Factor

As mentioned in Fig. 3, wind impact factor with 1,149 citations (51.52%) is responsible for more than half of the cause or contributing factor citations in the accidents database for 5-year time span that is broken down in Fig. 4.

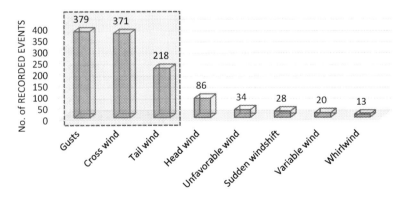

Fig. 4. Break down of recorded wind parameters affecting the accidents (5-year timespan)

According to Fig. 4, gusts, crosswind and tailwind are accounting for approximately 85% of the recorded events for 5-year timespan.

3.2   Meteorological Break Down: Visibility/Ceiling Impact Factors

Visibility radius and ceiling clearance are other principal causes or contributing factors in meteorological-related accidents according to the collected databases, with 409 citations over a 5-year timespan. Visibility refers to the greatest horizontal distance at which prominent objects can be viewed with the naked eye. Visibility can be compromised by precipitation, fog, haze and etc. In the field of aviation, a ceiling is considered as the lowest layer of clouds reported as being broken or overcast, or the vertical visibility into an obscuration like fog or haze. The breakdown of visibility into its related parameters shows low ceiling, clouds and fog to be the principal contributing factors, accounting for approximately 86% of the recorded events for 5-year timespan (See Fig. 5).

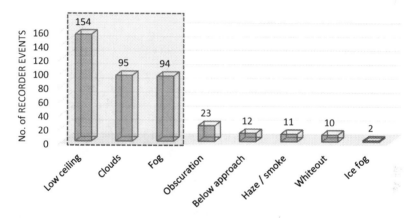

Fig. 5.   Break down of recorded visibility/ceiling parameters (5-year timespan)

Carried out analyses prove that among all meteorological related elements, wind and visibility/ceiling are responsible for the majority of events (51.52% and 18.34% respectively). From those, more than half of wind-related accidents occurred during the landing phase (57.7%) and around half of the visibility/ceiling related accidents occurred during approach, maneauvering and cruise phases (combined 51.43%). Among all wind components, the most influencing ones are gusts, crosswind and tailwind (combined 85%), and the principal components of visibility/ceiling are low ceiling condition and existence of clouds and fog (combined 86%).

4 Numerical Risk Analyses of Aircraft Ground Operations

In order to numerically compute the probability of occurrence of runway-related accidents/incidents RSARA© aircraft ground movement simulator is utilized. This software, which was developed as part of ACRP Project 4-08 [11], evaluates the overruns, veers-off and undershoots risks associated with landing and take-off phases of flight operations.

4.1 Case Studies Selection

Three specific airports are selected according to their unique meteorological charac-teristics in order to carry out sensitivity analyses. These unique modifiers consist of Historical weather data (HWD) covering historical operations period (wind, tempera-ture, visibility, etc.), RSAs (geometry, type of surface, and existence of EMAS and obstacles) and annual traffic volumes. General specifications of the airports are com-piled and displayed in Table 2.

Table 2. Technical specifications summary of selected Case Studies CS1, CS2, and CS3

Case study	CS1	CS2	CS3
RWY dimension	2440 × 30 m	1508 × 30 m	2450 × 46 m
DTHR[1]	RWY 33 (305 m)	RWY 09 (97 m) RWY 27 (72 m)	RWY 06 (120 m)
Elevation	2389 m	6 m	103 m
ASDA[2]	RWY 15 (2135.4 m) RWY 33 (2440.3 m)	RWY 09 (1319 m) RWY 27 (1319 m)	RWY 06 (2450 m) RWY 24 (2450 m)
LDA[3]	RWY 15 (2135.4 m) RWY 33 (2135.4 m)	RWY 09 (1319 m) RWY 27 (1319 m)	RWY 06 (2330 m) RWY 24 (2450 m)
Pavement	Asphalt	Grooved concrete	Concrete
RESA	RWY15 (61 × 43 m) RWY33 (61 × 43 m)	RWY09 (56 × 61 m) RWY27 (90 × 61 m)	RWY06 (90 × 90 m) RWY24 (180 × 90 m)
NAVAIDs[4]	CAT II[5] on side	CAT II both sides	CAT II one side
Selection	Strong wind, gusts, small dimension RESA	Short runway, steep approach, water-locked	Heavy rain/fog, low visibility, valley-locked

1 DTHR: Displaced Threshold
2 ASDA: Accelerate-Stop Distance Available
3 LDA: Landing Distance Available
4 NAVAIDs: Navigational Aid Systems
5 CAT II: Instrumental Landing System category 2

4.2   Input Variables

Following variables are required in order to simulate the aircraft take-off and landing operations and assess numerically the probabilities of occurrence of any possible excursion events.

- Airport characteristics: above sea level elevation, annual traffic and expected annual growth, type of flights (commercial, cargo, taxi/commuter or general aviation), and being international hub or not;
- Risk criteria: the value of 1.0E−6 is selected as the Target Level of Safety (TLS) in this study (one failure in one million operations);
- Runway configuration: runway designation, ASDA, LDA, and approach category (types of available approach instruments);
- RSA geometry, obstacles and OFA distance: designing the RESA geometry and declaring the existing lateral Obstacle Free Area (OFA). This distance is the clearance from the runway edge to the nearest obstacle, fixed or movable. The drawings are in the form of cells matrixes. Each cell corresponds to a coordinate that is referenced to the runway centerline. According to these drawings (see Fig. 6 and 7), the green (#G) is related to the grass, the yellow (#Y) is related to instrumental landing system (frangible obstacles), the pink (#P) is related to unpaved area, the gray (#Gr) is paved area, and the red (#R) area is related to water/mountain (catastrophic obstacles);

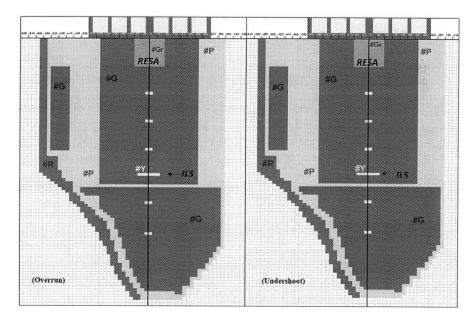

Fig. 6. Example of input scheme of RSA geometry and existing obstacles (CS1 RWY head)

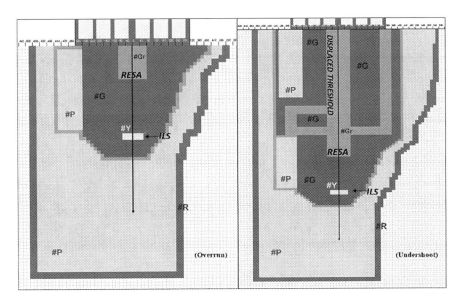

Fig. 7. Example of input scheme of RSA geometry and existing obstacles (CS1 RWY end)

- HOD: for each case studies operation data for a period of one year are collected, which cover the date and time of operation, runway designation, arrival/departure, aircraft FAA code, flight category (commercial, general aviation, commuter/taxi or cargo) and flight type (international or domestic).
- HWD: it should be noted that the periods of the collected weather data and operations data must be synced. Normally weather data are collected on an hourly basis for more accurate results, which cover visibility, wind direction, wind speed, air temperature, ceiling, and existence of rain, snow, snow showers, rain showers, thunderstorms, ice crystals, ice pellets, snow pellets, pellet showers, freezing rain, freezing drizzle, wind gusts and fog. After collecting the required data, each HWD should be coupled with one HOD and the probability of associated accidents should be simulated for each historical operation [12].

4.3   Adopted Frequency Model for Numerical Computation

The probability of occurrence of each runway-related event is adopted from the frequency model in a logistic format that offered by ACRP [11], which is described by the following equation:

$$P = \frac{1}{1 + e^{-(b_0 + b_1 X_1 + b_2 X_2 + b_3 X_3 + ...)}} \tag{1}$$

Where, P is the probability of occurrence of one specific event, X_i is independent binary variable such as ceiling, visibility, crosswind, precipitation, etc.; b_i is the regression coefficient [13]. According to Eq. (1), average probability of the occurrence of the event outside the RSA, average number of years expected between incident/accidents, and the percentage of movements with risk higher than adopted TLS can be achieved as the outputs of the numerical simulations.

5   RSARA/LRSARA© Numerical Sensitivity Analyses

In order to assess the impact level of each meteorological condition various aircraft operation scenarios are defined to get simulated by RSARA© and Lateral Runway Safety Area Risk analysis (LRSARA©) simulators. Meteorological components are altered for each of the scenarios in order to simulate critical weather conditions for sake of sensitivity analyses. The outputs cover the average probability of excursions under selected boundary conditions. Three meteorological scenarios are designed for these sensitivity analyses, which consist of extreme weather conditions:

- Favorable weather condition, which is expected when there is perfect visibility, mild air temperature, no extreme cloud coverage, no precipitation, no thunder-storms, etc.;
- Unfavorable weather condition, which involves extreme forms of precipitation, presence of dense fog, heavy wind gusts, thunderstorms, maximum cloud density, low visibility, wet runways with thick water-film, etc.

Respect to the defined meteorological scenarios, the sensitivity analyses are carried out for each case study separately and the average probability of runway excursion events are presented in the following. According to Fig. 8, the average probabilities of LDOR and LDVO are higher than other types of accidents. Landing is a critical phase, in some cases more critical than take-off. There is a difference of one order of magnitude between the extreme weather scenarios and normal (state of art) condition.

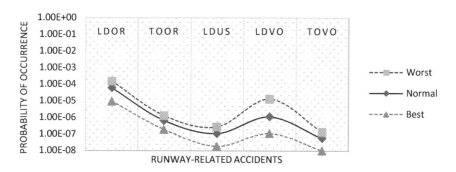

Fig. 8. Meteorological conditions sensitivity analyses for case study A1

According to Fig. 9, various meteorological conditions can be more crucial and have more critical impacts on the overruns and undershoots types of events rather than veer-off accidents, since the divergence of the veer-off probability values are less than the others.

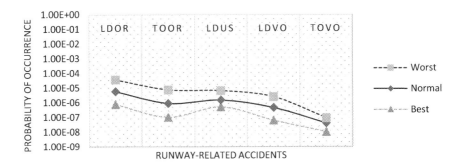

Fig. 9. Meteorological conditions sensitivity analyses for case study A2

According to Fig. 10, the average probability of all types of the runway excursion events undergo significant alteration by moving from unfavorable to favorable meteorological conditions, except for TOVO. Furthermore, LDVO experiences maximum rate of probability divergence due to available distance on either side of the runways.

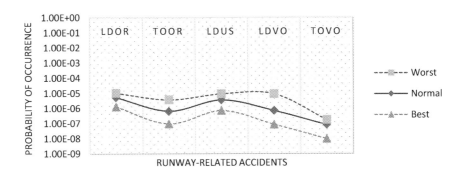

Fig. 10. Meteorological conditions sensitivity analyses for case study A3

6 Conclusion

This study evaluates the impact level of each meteorological parameter on the probability of occurrence or runway excursion events. In this regard, three weather condition scenarios are designed to be evaluated in numerical risk analyses. The principal scope of this investigation is based on general concept of risk and safety and quantifying the probability of each possible runway-related accident by simulating three selected case-studies with unique topographical and meteorological characteristics. RSARA and LRSARA© software, which are developed according to the ACRP methodology of aircraft ground operation risk assessment, are adopted for further sensitivity analyses.

As it can interpreted by analyzing the collected database of worldwide accident/incident records, the total recorded incidents are noticeably lower than the accidents, which can be mainly due to the tendency of airport authorities to avoid releasing minor incidents records to the public for the sake of their publicity. Moreover, it should be noted that meteorological conditions affect noticeably the number of accidents during visual meteorological condition (VMC). Runway excursion records during VMC are higher than those recorded during instrumental meteorological condition (IMC). Airports equipped with Instrumental Landing Systems (ILSs) facilities greatly the aircraft landing efforts therefore ILSs can reduce the probability of landing excursion events during IMC and intense weather conditions [12].

There is a difference of one order of magnitude between the extreme weather scenarios and normal (state of art) condition for landing-related events. Average probability of all types of the runway excursion events undergo significant alteration by moving from unfavorable to favorable meteorological conditions, except for TOVO. Meteorological conditions can have more critical impacts on the overruns and undershoots types of events rather than veer-off accidents, since the divergence of the veer-off probability values are less than the others. Meteorological parameters such as wind gusts (especially crosswinds), rain, ice and snow are the principal factors that cause unfavorable weather conditions. In other words, those factors can increase the chance of specific accident up to three orders of magnitude. Furthermore, wind element and specifically cross-wind turned out to be the most dominant influencing factor on runway excursion risks.

Acknowledgements. There is no conflict of interest to disclose. This research did not receive any specific grant from funding agencies in the public, commercial, or not-for-profit sectors

References

1. Boeing: 777X Airplane characteristics for airport planning (2018). http://www.boeing.com/commercial/airports/plan_manuals.html
2. Preyssl, C.: Safety risk assessment and management—the ESA approach. Reliab. Eng. Syst. Saf. **49**(3), 303–309 (1995)
3. Airport Cooperative Research Program: ACRP report 3, Analysis of Aircraft Overruns and Undershoots for RSA, National Academies of Sciences, Engineering, and Medicine; Transportation Research Board (2008). https://doi.org/10.17226/14137
4. International Civil Aviation Administration: ICAO Annex 6 - Part 1, Operation of Aircraft - International Commercial Air Transport – Aeroplanes (2010)
5. International Civil Aviation Administration: ICAO, Annex 14: aerodromes – aerodrome design and operations, Convention on international civil aviation, 6th edn., vol. 1 (2013)
6. International Civil Aviation Administration: ICAO, Doc 9859: SMM – Safety Management Manual (2013)

7. Ketabdari, M., Giustozzi, F., Crispino, M.: Probability contour map of landing overrun based on aircraft braking distance computation. In: WCPAM2017. CRC Press – Taylor & Francis Group (2019)

8. Guerra, L., Murino, T., Romano, E.: Airport system analysis: a probabilistic risk assessment model. Int. J. Syst. Appl. Eng. Dev. 2(2), 52–65 (2008)

9. National Transportation Safety Board (NTSB): Review of US Civil Aviation Accidents, Calendar Year 2011, Annual Review NTSB/ARA-14/01, Washington, DC (2014)

10. National Transportation Safety Board (NTSB): Risk factors associated with weather-related General Aviation Accidents, Safety study, NTSB Number: SS0501, Washington DC, USA (2005)

11. Airport Cooperative Research Program: ACRP report 50, Improved Models for Risk Assessment of RSA, National Academies of Sciences, Engineering, and Medicine; Transportation Research Board (2011). https://doi.org/10.17226/13635

12. Ketabdari, M., Giustozzi, F., Crispino, M.: Sensitivity analysis of influencing factors in probabilistic risk assessment for airports. Saf. Sci. 107, 173–187 (2018)

13. Ayres, M., et al.: Modelling the location and consequences of aircraft accidents. Saf. Sci. 51(1), 178–186 (2013)

Simplified Models for Electromechanics of Cardiac Myocyte

João Gabriel Rocha Silva[1,3(✉)], Ricardo Silva Campos[1(✉)],
Carolina Ribeiro Xavier[2(✉)], and Rodrigo Weber dos Santos[1(✉)]

[1] Postgraduate Program in Computational Modeling,
Federal University of Juiz de Fora, Juiz de Fora, Brazil
joaogabriel.comp@gmail.com, rodrico.weber@ufjj.edu.br
[2] Postgraduate Program in Computer Science,
Federal University of São João del Rei, São João del-Rei, Brazil
carolinaxavier@ufsj.edu.br
[3] Federal Institute of Education, Science and Technology of Mato Grosso,
Pontes e Lacerda, Brazil
joao.silva@plc.ifmt.edu.br

Abstract. The study of the electromechanical activity of the heart through computational models is important for the interpretation of several cardiac phenomena. However, computational models for this purpose can be computationally expensive. In this work, we present the simplified models at the cellular level which were able to qualitatively reproduce the cardiac electromechanical activity based on the contraction of myocytes. To create these models a parameter adjustment was performed via genetic algorithms. The proposed models with adjusted parameters presented satisfactory results for the reproduction of the active force of the heart with the advantage of being based on only two ordinary differential equations.

Keywords: Simplified models · Electromechanical · Genetic algorithms

1 Introduction

The heart is a vital organ. It is responsible for pumping oxygenated blood throughout the body. Its division is defined by four chambers: two ventricles and two atria. These chambers aid in the procedure of receiving and pumping blood through the organ. The study of the behavior and development of treatments, diagnoses and drugs related to the heart are of high importance by their clinical interest [13].

In Brazil, 20% of all deaths in the adult population over 30 years are caused by cardiovascular diseases [6]. In a global sphere, it is estimated that one-third of all causes of death in the world are from cardiovascular diseases [3]. In addition, in 2010, it was estimated that around U$ 315 billions were applied in research

© Springer Nature Switzerland AG 2020
O. Gervasi et al. (Eds.): ICCSA 2020, LNCS 12249, pp. 191–204, 2020.
https://doi.org/10.1007/978-3-030-58799-4_14

and development of cardiovascular diseases cases [3]. In this way, debates and studies on this subject are increasingly recurrent in the scientific community.

The treatment, diagnosis and interpretation of these diseases are of fundamental importance contributing to the study, for example, of the electrical and mechanical properties of the heart.

In this aspect, mathematical modeling has become a widely used tool in the study of different phenomena, among them, models that describe the behavior of the heart stand out. They can be used for the development and testing of new drugs [2] and for the identification of diseases [8].

The earliest models of electrophysiology were simplified as: [5] with only four differential equations and [10]. Over time the models have become more complex, and to introduce, for example, the mechanical part [9] and [12] have 11 differential equations and 45 algebraic equations, due to the high complexity of these representations the models have become larger, which requires a longer computational time for the simulation.

For larger simulations, less computational expensive models are needed. In this sense [7] and [4] presented simplified models for the connection between the electrical and mechanical part of the heart.

In this work, we evaluated the ability of simple models to reproduce more realistic electromechanical simulations, coming from more complex models. In addition, we propose modifications that significantly improve the results of the models and further reduce it's computational.

2   Cardiac Electromechanics

The cardiac electromechanical activity can be defined, in a simplified way, as: given an electrical stimulus (action potential) the increase of calcium concentration in the cell's intracellular medium begins, this causes the actin filaments to bind to myosin. After this connection, the size of the sarcomere is reduced due to the sliding of these filaments. This slip generates a force, this force results in the process of contraction.

2.1   The Electrical System

The Action Potential (AP) can be defined as a variation in the electrical potential of the cell membrane. This variation is due to the differences in concentrations between the ions that surround the intra and extracellular medium. In the heart, AP has the function of synchronizing the rhythm of contraction and relaxation of the heart. The AP is generated in a region called the sinoatrial node also known as the heart's natural pacemaker.

The AP cycle can be described by the following stages: the initial stage is associated with rest, and consists of the state of equilibrium between the potentials of the intra- and extracellular medium, which guarantees an integrity of the cellular structure. When the cell is stimulated, a sodium influx (Na^+) leads to the accentuated growth of action potential, leaving the inside of the cell less negative, causing depolarization. Quickly the sodium channels close and a potassium efflux (K^+) starts and the potential initiates a repolarization process. Then, a phase called a plateau is initiated, and is characterized by a depolarizing current of calcium, in which the calcium influx equilibrates with the potassium efflux, maintaining for a period the almost constant potential value. It is during this period that the contraction of the cardiomyocyte occurs due to the fact that calcium entry stimulates the release of calcium into the sarcoplasmic reticulum. Over time the calcium channels close and the output of potassium to the extracellular medium intensifies, characterizing the repolarization. In this stage the potential tends to its initial equilibrium (rest).

2.2 Cellular Contraction

The increase of the concentration of an ion in the intracellular medium is allowed during the period of activation of the AP, due to ion channel openings that allow ion exchange between the intra and extracellular media. The contraction is basically due to the increase in internal calcium concentration during an electrical stimulus (AP).

The increase in internal calcium concentration alone is not able to promote the onset of sarcomere contraction. However, this step serves as a flag for the sarcoplasmic reticulum, due to calcium ion channels on the cell membrane. Thus, the sarcoplasmic reticulum (SR), the calcium reservoir of the cell, causes an efflux of the calcium that added to the calcium coming from the extracellular environment characterizes the process of release of calcium induced by calcium. In this way, the contraction process begins. Relaxation is related to the reuptake of calcium into the reticulum. Thus, the concentration of calcium within the cell decreases, characterizing the diastole process.

Cardiac myocytes are composed of structures called myofibrils. In the composition of these structures are contractile substances, the sarcomeres. These are responsible for myocyte contraction and relaxation. Sarcomers are composed of filaments called actin and myosin.

These filaments overlap in stretches of their length. The excess of calcium in the intracellular medium causes the sliding of the myosin filaments to that of actin to generate a reduction in the size of the sarcomere, producing a tension (force), characterizing the process of contraction.

3   Computational Models for Cardiac Physiology

Cellular models for physiology are basically divided into two groups according to their characteristics: simplified models and detailed biophysical models. The simplified models are characterized by the use of few equations to describe a physiological phenomenon based on a phenomenological approach. The detailed biophysical models use numerous considerations on biological factors that influence the activity to be reproduced, such as ion channels, ion exchangers, among others. Thus, these models can count on many equations and consequently have a high computational cost for simulation.

3.1   Cellular Model of TenTuscher (2004)

A detailed biophysical mathematical model for reproduction of human action potential is presented in [14]. For this simulation the model is supported by 19 differential equations.

In [14], the behavior of the cell membrane is associated with the behavior of a capacitor in parallel with variable non linear resistors. The ionic currents that surround the membrane and ion exchange pumps are modeled as non linear resistors. Thus, the electrophysiological behavior for one cell is described by Eq. 1.

$$\frac{dV}{dt} = -\frac{I_{ion} + I_{stim}}{C_m} \tag{1}$$

where I_{stim} refers to the external stimulus current, C_m means the capacitance per surface area and I_{ion} is the sum of all ionic currents (potassium, calcium, sodium, for example).

3.2   Mechanical Models

Mechanical Model of Rice (2008). A model for active force in rat cardiac myocytes is shown in [12]. This model belongs to the class of detailed biophysical models, since it considers numerous biological components.

This model is described by 11 ordinary differential equations and 45 algebraic equations. Among other topics, Rice considers the detailing of sarcomere geometry, filament slip, ionic concentrations of calcium and cross-bridges.

The Rice model is based on states of a Markov chain. Where state of N_{XB} is a non-permissive state, with the task of preventing the formation of cross-bridges, the state P_{XB} refers to a permissive conformation of the proteins. XB_{PreR} is the strongly bonded state, prior to rotation of the myosin head, and the state XB_{PorstR} occurs at the time the myosin is twisted causing force generation. In addition, states transition rates depend on the calcium binding to troponin. A detailed description of the equations and considerations on the parameters used can be found in [12].

Mechanical Model of Nash-Panfilov (2004). In order to present a simplified model for the mechanical activity and consequently for the electromechanical coupling, [7] approximated the active tension of canine cardiac myocytes using only one differential and one algebraic equation.

The work presented in action potential used for coupling the active force equation was presented in [1], another simplified model for dogs AP.

The reproduction of the active force of the model proposed by Nash & Panfilov is given by:

$$\frac{\mathrm{d}Ta}{\mathrm{d}t} = \epsilon(V)(k_{Ta}V - Ta),$$

$$\epsilon(V) = \begin{cases} e_0 & for\ V < 0.05 \\ 10e_0 & for\ V \geq 0.05 \end{cases} \tag{2}$$

where k_{Ta} controls the active force amplitude and the $\epsilon(V)$ moderates the delay in development ($V < 0.05$) and recuperation ($V \geq 0.05$) of the force in relation to the AP.

Mechanical Model of Goktepe and Kuhl (2010). With the objective of smoothing the simplified active force equation proposed by [7]. In [4] a new equation was presented for this approximation in cardiomyocytes, which is given by:

$$\frac{\mathrm{d}Ta}{\mathrm{d}t} = \epsilon(V)(k(V - V_r) - Ta),$$

$$\epsilon(V) = e_0 + (e_\infty - e_0)e^{-e^{(\xi(V - vs))}} \tag{3}$$

where e_0, e_∞, ξ, vs and k are model parameters, which indicate: e_0 and e_∞ constant concentration rates that delimit the value of $\epsilon(V)$, ξ refers to the transition rate, vs the rate of phase change, and k is the saturation control of the active force. In addition, V_r is the value of the resting potential.

4   Evaluation of Electromechanical Models

4.1   TenTuscher/Rice Excitation-Contraction Coupling Model

In [11] a model is presented for coupling between the electrical and mechanical system for human cardiac myocytes. For this, two models of the literature [14] and [12] were used, where the first model is responsible for the electrical system while the second is responsible for the mechanical system.

Figure 1(a) shows the normalized active potential and active force of the coupled model proposed in [11] from the [14] and [12] models.

The advantages of using this coupled model in a simulation are supported by the fact of the reproductive quality of the mechanical activity for human cardiac myocytes. However, it is a more complex model, and consequently more differential equations can lead to a high computational cost for a scenario of large simulations (for example for cardiac tissue), thus characterizing the disadvantage in the use this model.

4.2   Aliev-Panfilov/Nash-Panfilov Excitation-Contraction Coupling Model

An electromechanical model based on the Eqs. 2 and the excitation model proposed by [1] is presented in [7]. Both models were developed to describe canine myocytes, however a small adjustment of parameters was necessary.

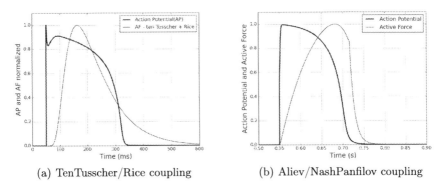

(a) TenTusscher/Rice coupling   (b) Aliev/NashPanfilov coupling

Fig. 1. Action Potential and Active Force for coupled models analyzed.

The Fig. 1(b) shows the action potential and the active force proposed [7]. On the results presented by this coupling it is possible to notice that the model does not present good quality in the reproduction of the active force, based on the physiological behavior described by [11] for the active force, characterizing a disadvantage.

The advantage in choosing this coupled model for a research can then be justified in computational cost issues, since the mechanical model has only one differential equation and another algebraic equations for the generation of the active force.

5   Creation of Models and Results for Each Step

After the study and evaluation process of these coupled models, we propose two new models (Model A and Model B) for electromechanical coupling based on the concepts of these evaluations. After the development of these models, a third model (Model C) is presented, based on empirical considerations during the process of creating the first models.

The purpose of developing new electromechanical models is motivated by the minimization of computational cost (advantage presented by [7]) together with a quality in the simulation of the active force, (benefit presented in [11]). A model that reproduces quickly and with quality the electromechanical coupling can be very useful.

5.1   Step 1 - Direct Coupling

The first phase of the development of the models consists of the direct coupling of the dependent variables between the electric and the mechanical model. Thus, the cellular model of [14] was coupled directly to the mechanical model of [7] (Model A). Analogously, the electric model is coupled to the mechanical model of [4] (Model B).

As a first study, the values of all parameters for both coupled models were maintained according to each of the original proposals [4,7,14].

Considering the fact that the action potential is a dimensionless variable in [7], the action potential used in all the coupled simulations was also normalized.

In Model B, the activation rate controlling function is prepared to receive the parameter V in a non-normalized physiological domain, around $[-86, 31]$, for the cellular model of [14]. Thus, the action potential is readapted in this term from the equation to a normalized domain on a real scale according to Eq. 4:

$$V^* = V * 143 - 94, \tag{4}$$

Where V^* is the action potential adapted to be used in the activation function of Model B and V is the action potential coming from [14].

The Fig. 2 presents the result of the direct coupling between the [14] model and the models [7] in (a) and [4] in (b) the direct coupling.

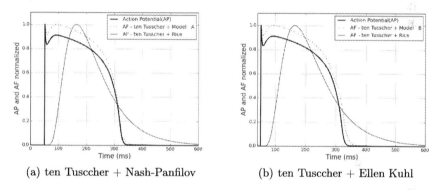

(a) ten Tusccher + Nash-Panfilov   (b) ten Tusccher + Ellen Kuhl

Fig. 2. Action Potential and Active Force (TA) for the coupled models ten Tusscher + Nash-Panfilov and ten Tusscher + Ellen Kuhl.

As it is possible to see, the direct coupling does not present a suitable behavior for the active force in either of the two proposals. This phenomenon can be related to values of unsuitable parameters, making the active force tend to the behavior of the action potential.

5.2   Step 2 - Adjustment of Parameters with Genetic Algorithm

The second step for the creation to models consisted of small considerations in the mathematical model and in the adjustment of model's parameters by a genetic algorithm developed find a set of parameters that show a smaller difference between the simulation of the active force of the presented models and active force coming from to [11].

From this stage the mechanical models underwent changes of parameter values via genetic algorithm. The Eq. 5 presents the equations for the active force in Model A, where it is evidenced, the parameters to be adjusted by the Genetic Algorithm. They are: kTa, e_0 and e_t.

$$\frac{dTa}{dt} = \epsilon(V)(\mathbf{k_{Ta}}V - Ta),$$

$$\epsilon(V) = \begin{cases} \mathbf{e_0} & \text{for } V < \mathbf{e_t} \\ 10\mathbf{e_0} & \text{for } V \geq \mathbf{e_t} \end{cases} \tag{5}$$

In order to use the genetic algorithm it is necessary to define the domain of the values that each parameter can assume (search space). The e_t is the domain of the action potential (since it controls values for phase changes of the action potential, acting as a threshold). The e_0 domain suggests values close to the values shown in [7]. Finally, the domain for kTa follows the domain of action potential due to its normalization. Thus, the parameters e_0, e_t and kTa belong to the domain $[0, 1]$.

The same procedure to define parameters to be adjusted was applied to Model B, where the Eq. 6 presents, highlighted, all the parameters submitted to the Genetic Algorithm. One consideration is the exclusion of the parameter V_r (resting potential) because the domain is normalized, so the resting potential is 0.

$$\frac{dTa}{dt} = \epsilon(V)(\mathbf{k}V - Ta),$$

$$\epsilon(V) = (\mathbf{e_0} + (\mathbf{e_\infty} - \mathbf{e_0})e^{-e^{(-\xi(V^* - \mathbf{vs}))}}) \tag{6}$$

For all parameters, the domains were defined by values close to the original parameter values presented by [4]. Another consideration used is the requirement of a proportionality between e_0 and e_∞, this proportionality follows the example presented in the original proposal $e_0 = 10\ e_\infty$.

The Fig. 3 presents the result for small changes in model considerations and submission of the parameters of each models to the genetic algorithm.

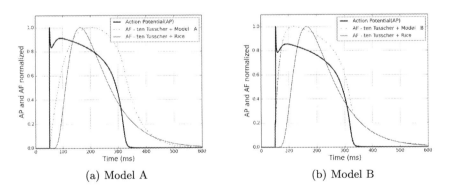

(a) Model A   (b) Model B

Fig. 3. Action Potential and Active Force for coupled models A and B after adjustment via genetic algorithm - Step 2

It is possible to verify that the active force presented better results when compared to the first phase of the model creation process aiming at an approximation to the active force of the example model. However, the quality of the reproduction is still poor, especially for Model B, which still tends to the action potential in parts of its evolution.

The parameter values found for Model A were: $e_0 = 0.001953$, $e_t = 0.000002$ mV and $kTa = 0.760743$ kPa. For the Model B: $e_0 = 0.007813$ mV, $e_\infty = 0.078125$ mV, $\xi = 0.148438$ mV, $vs = 78.749999$, $k = 0.556641$ mV.

5.3 Step 3 - Mathematical Modifications and Addition of a Continuous Delay

In order to correct behavior and add delay between stimulus and contraction, new strategies at the mathematical level were adopted in Phase 3. The first strategy was to change the term $kTa * V$ in Model A and $(k * V)$ for Model B.

Considering the biological existence of a delay between the electrical stimulus and the beginning of the cardiac muscle contraction, we realized the need of a new differential equation. Thus, a new equation is inserted for the active force called Intermediate Active Force (Ta_i) which assumes the previous Ta equation and the Active Force already considered is now expressed as the multiplication of the activation control (even used in the Intermediate Active Force) by the difference between Ta and Ta_i.

The sets of Eqs. 7 and 8 express Models A and B with the addition of the considerations proposed in this phase as well as, highlighted, the parameters to be submitted to the genetic algorithm with the intention of reproducing the force from a more complex model.

$$\frac{dTa_i}{dt} = \epsilon_0(V)(kTa(V) - Ta_i),$$

$$\frac{dTa}{dt} = \epsilon_1(V)(Ta_i - Ta),$$

$$\epsilon_0(V) = \epsilon_1(V) = \begin{cases} \mathbf{e_0} \text{ for } \mathbf{V} < \mathbf{e_t} \\ \mathbf{e_0} \text{ for } \mathbf{V} \geq \mathbf{e_t} \end{cases},$$

$$kTa(V) = \frac{1}{\sigma\sqrt{2\pi}} e^{\frac{-1}{2}(\frac{V-1}{\sigma})^2}$$

(7)

The domains of each parameter for the submission in the genetic algorithm continue as in the previous phase and the new parameter added (σ), follows the domain of the action potential for both models ([0.1]).

$$\frac{dTa_i}{dt} = \epsilon_0(V)(k(V) - Ta_i),$$

$$\frac{dTa}{dt} = \epsilon_1(V)(Ta_i - Ta),$$

$$\epsilon_0(V) = \epsilon_1(V) = (\mathbf{e_0} + (\mathbf{e_\infty} - \mathbf{e_0})e^{-e^{(-\xi(V-\mathbf{vs}))}}),$$

$$k(V) = \frac{1}{\sigma\sqrt{2\pi}} e^{\frac{-1}{2}(\frac{V-1}{\sigma})^2}$$

(8)

Figure 5 presents the results obtained for the third phase of the development of Models A and B. It is possible to verify a significant increase in the quality in the reproduction of the active tension in both models when compared to the curve of a complex realistic model (Fig. 4).

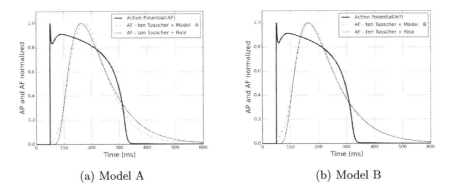

(a) Model A   (b) Model B

Fig. 4. Action Potential and Active Force for coupled models A and B after adjustment via genetic algorithm - Step 3

Therefore, the development of the new models follows a promising path. However, the delay phenomenon (one of the proposals for adding this phase) has not yet obtained the expected result. Based on the fact that even then, when an electrical stimulus happens, even slowly, the contraction begins, not characterizing the biological behavior for this phenomenon.

The parameter values found for Model A were: $e_0 = 0.001953$, $e_t = 0.000977$ mV, $\sigma = 0.066406$. For the Model B: $e_0 = 0.003906$ mV^{-1}, $e_\infty = 0.039063$ mV^{-1}, $\xi = 0.001953$ mV, $vs = -0.078084$ mV, $\sigma = 0.049805$.

5.4   Step 4 - Delay Correction

The fourth and last stage of the development of the models consists in the correction of the multiplier of the differential equation proposed for the delay due to its still inadequate behavior. Thus, the term multiplier is no longer the same as the differential equation for Ta_i. It breaks down into a function defined by Ta_i-dependent parts and the action potential. The fully developed Model A is given by Eq. 9.

$$\frac{dTa_i}{dt} = \epsilon_0(V)(kTa(V) - Ta_i)$$

$$\frac{dTa}{dt} = \epsilon_1(V, Ta_i)(Ta_i - Ta)$$

$$\epsilon_0(V) = \begin{cases} e_0 & \text{for } V < e_t \\ 10e_0 & \text{for } V \geq e_t \end{cases} \tag{9}$$

$$\epsilon_1(V, Ta_i) = \begin{cases} x_1 & \text{for } V > x_2 \text{ and } Ta_i < x_3 \\ e_0(V) & \text{otherwise} \end{cases}$$

$$kTa(V) = \frac{1}{\sigma\sqrt{2\pi}} \ e^{\frac{-1}{2}(\frac{V-1}{\sigma})^2}$$

Equations 10 presents Model B. The changes in the multiplier (speed control) $(e_1(V, Ta_i))$ are the same as those for Model A.

$$\frac{dTa_i}{dt} = \epsilon_0(V)(k(V) - Ta_i)$$

$$\frac{dTa}{dt} = \epsilon_1(V)(Ta_i - Ta)$$

$$\epsilon_0(V) = (e_0 + (e_\infty - e_0)e^{-e^{(-\xi(V-vs))}}) \tag{10}$$

$$\epsilon_1(V, Ta_i) = \begin{cases} x_1 & \text{for } V > x_2 \text{ and } Ta_i < x_3 \\ e_0(V) & \text{otherwise} \end{cases}$$

$$k(V) = \frac{1}{\sigma\sqrt{2\pi}} \ e^{\frac{-1}{2}(\frac{V-1}{\sigma})^2}$$

All the parameters added in this phase belong to the domain $[0, 1]$ in both models. For the last time, after these considerations and modifications presented in this phase the parameters of the models were submitted to the GA developed in order to minimize the difference between the curves of the complex model and the simplified models.

As can be seen in Fig. 5, the proposed models were able to reproduce with excellent quality the active force that generates the contraction based on the comparison of the [11] model, a complex model with numerous considerations.

The parameter values found for Model A were: $e_0 = 0.016602$, $e_t = 0.999024$ mV, $\sigma = 0.050781$, $x_1 = 0.00001$, $x_2 = 0.82594$ mV, $x_3 = 0.42852$. For the Model B: $e_0 = 0.003906$ mV^{-1}, $e_\infty = 0.039063$ mV^{-1}, $\xi = 0.001953$ mV, $vs = -0.312459$ mV, $\sigma = 0.049805$, $x_1 = 0.00001$, $x_2 = 0.8$ mV, $x_3 = 0.22$.

5.5   The Model C

A phenomena observed based on the development of Models A and B is that for both models the smaller the variation of the multiplier $e_0(V)$ the better the result of the reproduction of the active force.

Based on this fact, the creation of a new model for active force generation began: Model C. In this model, all the considerations and phases generated in the process of creating Models A and B are maintained, however, with only the

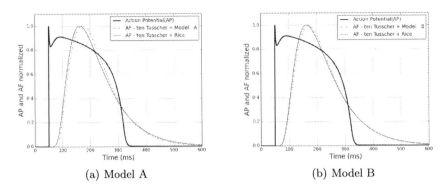

(a) Model A   (b) Model B

Fig. 5. Action Potential and Active Force for coupled models A and B after GA adjustment. - Step 4

modification of the multiplier of the differential equation for Ta_i. This modification consists of replacing the multiplier $e_0(V)$ with an activation constant c_0. The set of Eqs. 11 presents Model C in a complete way with the addition of this change.

$$\frac{dTa_i}{dt} = \mathbf{c_0}(k(V) - Ta_i)$$

$$\frac{dTa}{dt} = \epsilon_1(V, Ta_i)(Ta_i - Ta)$$

$$\epsilon_1(V) = \begin{cases} \mathbf{x_1} & \text{for } V > \mathbf{x_2} \text{ and } Ta_i < \mathbf{x_3} \\ e_0(V) & \text{otherwise} \end{cases} \tag{11}$$

$$k(V) = \frac{1}{\sigma\sqrt{2\pi}} \; e^{\frac{-1}{2}(\frac{V-1}{\sigma})^2}$$

In Fig. 6 the active force is shown for the electromechanical model C from the parameter adjustment performed by the GA.

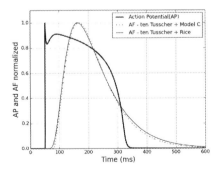

Fig. 6. Potential of Action and Active Force for the electromechanical model C after adjustment via genetic algorithm.

It is possible to observe that just like the previous models, this model was also able to reproduce the active force coming from a more complex model. However, in an even simpler way, eliminating an algebraic equation and two parameters in relation to Model A and three parameters in relation to Model B.

The parameter values found for Model C were: $c_0 = 0.016602$, $\sigma = 0.042969$, $x_1 = 0.0001$, $x_2 = 0.827044$ mV, $x_3 = 0.209961$.

6 Conclusions

In this work three simplified mathematical-computational models are proposed and implemented for the electromechanical coupling of the heart. The objective of the development of these models was to present models with few differential equations and consequently low computational cost for the simulation of the electromechanical activity of the heart without loss of quality in the reproduction of the mechanical activity coming from the active force.

For this, we compared the new models with a complex model presented in [11].

After considering mathematical changes in the equations of the presented models and adjustments using a GA, it was possible to conclude that the simulations achieved were satisfactory for all models developed. The new model was able to reproduce the generation of myocyte force in very clare agreement with a complex and detailed modeled [11] and yet reduced the computational costs by eliminating 11 differential equations and more than 40 algebraic equations.

When dealing with simplified models, we give up allegiance to the biology of the problem. And we are more concerned with the representation of phenomena.

Acknowledgments. This work was partially supported by the Brazilian National Council for Scientific and Technological Development (CNPq - under grant 153465/2018-2), the Foundation for Support of Research of the State of Minas Gerais, Brazil (FAPEMIG), Coordination for the Improvement of Higher Education Personnel, Brazil (CAPES) and Federal University of Juiz de Fora, Brazil (UFJF).

References

1. Aliev, R.R., Panfilov, A.V.: A simple two-variable model of cardiac excitation. Chaos, Solitons Fractals **7**(3), 293–301 (1996)
2. Brennan, T., Fink, M., Rodriguez, B.: Multiscale modelling of drug-induced effects on cardiac electrophysiological activity. Eur. J. Pharm. Sci. **36**(1), 62–77 (2009)
3. Go, A.S., et al.: Heart disease and stroke statistics–2014 update: a report from the American heart association. Circulation **129**(3), e28 (2014)
4. Göktepe, S., Kuhl, E.: Electromechanics of the heart: a unified approach to the strongly coupled excitation-contraction problem. Comput. Mech. **45**(2), 227–243 (2010)
5. Hodgkin, A.L., Huxley, A.F.: A quantitative description of membrane current and its application to conduction and excitation in nerve. J. Physiol. **117**(4), 500–544 (1952)

6. Mansur, A.d.P., Favarato, D.: Mortality due to cardiovascular diseases in brazil and in the metropolitan region of São Paulo: a 2011 update. Arquivos brasileiros de cardiologia **99**(2), 755–761 (2012)
7. Nash, M.P., Panfilov, A.V.: Electromechanical model of excitable tissue to study reentrant cardiac arrhythmias. Prog. Biophys. Mol. Biol. **85**(2), 501–522 (2004)
8. Nataraj, C., Jalali, A., Ghorbanian, P.: Application of computational intelligence techniques for cardiovascular diagnostics. In: The Cardiovascular System-Physiology, Diagnostics and Clinical Implications. InTech (2012)
9. Negroni, J.A., Lascano, E.C.: Simulation of steady state and transient cardiac muscle response experiments with a huxley-based contraction model. J. Mol. Cell. Cardiol. **45**(2), 300–312 (2008)
10. Noble, D.: A modification of the hodgkin–huxley equations applicable to purkinje fibre action and pacemaker potentials. J. Physiol. **160**(2), 317–352 (1962)
11. de Oliveira, B.L., Sundnes, J., dos Santos, R.W.: The development of a new computational model for the electromechanics of the human ventricular myocyte. In: 2010 Annual International Conference of the IEEE Engineering in Medicine and Biology Society (EMBC), pp. 3820–3823. IEEE (2010)
12. Rice, J.J., Wang, F., Bers, D.M., De Tombe, P.P.: Approximate model of cooperative activation and crossbridge cycling in cardiac muscle using ordinary differential equations. Biophys. J. **95**(5), 2368–2390 (2008)
13. Rocha, B.M.: Modelagem da atividade eletromecânica do coração e os efeitos da deformação na repolarização. Ph.D. thesis, Laboratório Nacional de Computação Científica (2014)
14. Ten Tusscher, K., Noble, D., Noble, P., Panfilov, A.V.: A model for human ventricular tissue. Am. J. Physiol. Heart Circulatory Physiol. **286**(4), H1573–H1589 (2004)

Electrotonic Effect on Action Potential Dispersion with Cellular Automata

Ricardo Silva Campos[1,2($\boxtimes$)], João Gabriel Rocha Silva[1,3],
Helio José Corrêa Barbosa[1,2], and Rodrigo Weber dos Santos[1]

[1] Federal University of Juiz de Fora, Juiz de Fora, Brazil
ricardo@ice.ufjf.br, rodrigo.weber@ufjf.edu.br
[2] Laboratório Nacional de Computação Científica, Petrópolis, Brazil
hcbm@lncc.br
[3] Federal Institute of Education, Science and Technology of Mato Grosso,
Pontes e Lacerda, Brazil
joao.silva@plc.ifmt.edu.br

Abstract. In this study we used a simplified electrophysiological simulator of the heart, based on cellular automata, for studying the electrotonic effect. This effect is caused by interaction between cells which changes their behavior. This feature homogenizes the repolarization dispersion, that leads all cells to return to their resting potential at the same time, regardless of their initial action potential duration, in order to avoid undesired excitation.

For achieving this feature on our automata, we proposed a new method for updating the action potential duration, by evaluating the current duration, propagation velocity and tissue conductance. The results suggest that, despite its simplicity, our simulator is suitable to mimic traditional models based on differential equations.

Keywords: Cardiac simulation · Cellular automata · Repolarization dispersion · Electrotonic effect

1 Introduction

Heart diseases are associated with high mortality rates worldwide. The development of medicines, medical devices, and non-invasive techniques for the heart demand multidisciplinary efforts towards the characterization of cardiac function.

In particular, the important phenomenon of electric propagation in the heart involves complex non-linear biophysical processes and structures. Its multi-scale nature spans from genes and nanometre processes, such as ionic movement and protein dynamic conformation, to centimetre phenomena such as whole heart

This work was supported by the National Council for Scientific and Technological Development (CNPq) - Brazil (grants 153465/2018-2 and 312337/2017-5).

O. Gervasi et al. (Eds.): ICCSA 2020, LNCS 12249, pp. 205–215, 2020.
https://doi.org/10.1007/978-3-030-58799-4_15

structure and contraction. The electrical activation triggers the tissue deformation that ejects blood from the ventricles. Abnormal changes in the electrical properties of cardiac cells and in the tissue structure can lead to life-threatening arrhythmia and fibrillation. Cardiac modelling challenges science in many aspects. The multi-scale and multi-physics nature of the heart bring new challenges to the mathematical modelling community, since sophisticated techniques are necessary to cope with them. In addition, a complete whole organ model would have to include all the physics and their interactions. Whereas a lot of research work has been done in the formulation of electrical models, structural-mechanics or fluid-mechanics, mathematical formulations that include all the interactions of the aforementioned physics are still incipient. As a multidisciplinary research field, a background on biology, biophysics, mathematics, and computing is necessary during the development and use of cardiac models. In particular, the implementation of the mathematical models is a time consuming process that requires advanced numerical methods and computer programming skills. In addition, the numerical resolution typically demands high performance computing environments, in the case of large simulations (tissue or organ-level models), and parallel programming expertise may add more complexity to this multidisciplinary area of research.

In general, the cardiac activity is described by non-linear systems of differential equations with tens of variables and close to a hundred parameters. Cardiac cells are connected to each other by gap junctions creating a channel between neighbouring cells and allowing the flux of electrical current in the form of ions. An electrically stimulated cell transmits an electrical signal to the neighbouring cells allowing the propagation of an electrical wave to the whole heart which triggers contraction. An electrical signal is propagated from one cardiac cell to its neighbors, in a process that triggers the mechanical contraction. This phenomenon is usually modelled using partial differential equations (PDEs) [17].

Although the PDEs used for cardiac modelling are able to perform realistic tissue simulations, they involve the simulation of thousands of cells, which make their numerical solution quite challenging. This is an issue for clinical software that may demand accurate results in real-time simulations. Therefore, some effort has been made in speeding up the PDE solvers via parallel computing [14], as well as by different techniques to emulate PDEs simulations with less computational cost [1].

In a previous work [4] we proposed a simple and fast discrete simulator, named FisioPacer. It is a 3D electro-mechanical model of cardiac tissue, described by the coupling of a cellular automata (CA) and mass-spring systems (MSS). It was able to mimic the action potential propagation and tissue contraction obtained by PDE solvers in less computational time. In this work, we propose a modification on the cellular automata, in order to add realism to our model, by changing the action potential features according to mechanical coupling of cells, the so called electrotonic effect [2]. Some features of the tissue are evaluated, such as the different action potential morphology, velocity of propagation and conductance, and then a new action potential duration (APD) is automatically set to each cell. Preliminary results suggest our model is able to mimic experimental results of electrotonic effect from traditional models based on PDEs, easily found on literature [6,10,11].

1.1   Related Studies on Electrotonic Coupling

The heart tissue is composed of three layers: epicardium, myocardium, and endo-cardium. Epicardium is the outer layer and its function is to protect the inner parts. Myocardium is the middle and thickest layer. It is composed of muscle fibers, which are responsible for the heart contraction. Endocardium is the inner layer of the heart. It is thin and covers the heart chambers. The electric pulse starts at the endocardium and flows to the epicardium. The cells in each layer have different physiological properties, in particular, they have different action potential duration (APD). In normal conditions, the mid-myocardium (M-cells) have a longer AP than endo and epicardium cells. This difference is noticed in isolated cells. However, when the three types of cells are arranged together, they are connected by gap junctions. The level of coupling between neighboring cells during deformation can change their physiological properties; that's called the electrotonic effect. It changes the action potential duration [13] and the repolarization dispersion [15,16]. Thus, the change on APD homogenizes the repolarization dispersion. This means cells can be excited in different times, but their AP tend to return to resting potential at the same time, despite the difference in their APD and the distance among cells. This mechanism avoids undesired excitation caused by late repolarized cells.

As a matter of comparison, we used the Luo and Rudy model (LRd) [9], which is built based on experimental data. Its ionic currents reproduce the action potential of mammalians, mostly guinea pigs. Later, the work of Viswanathan et al. [18] proposed a set of experiments with LRd model for observing the impacts of the electrotonic effect. It shows how the flow of electric current changes the action potential on membrane of coupled cells, aiming at evaluating the importance of some ionic currents in the heterogeneity of repolarization.

For instance, in guinea pigs, the APD of isolated epicardial cells, endocardial cells, and M cells are 160, 185, and 250 ms, respectively [18], which means the APD difference between midmyocardium and epicardial cells (ΔAPD) is 90 ms. On the other hand, M, epi and endocardium cells combined resulted in $\Delta APD = 18$ ms. The coupling of cells is determined by conductance of gap junctions and by AP propagation velocity [12]. Therefore, decreasing the cell coupling level increases the ΔAPD.

2   Methods

A cellular automaton is a model of a spatially distributed process that can be used to simulate various real-world processes. A two-dimensional cellular automaton consists of a two-dimensional lattice of cells where all cells are updated synchronously for each discrete time step. The cells are updated according to some updating rules that will depend on the local neighborhood of a particular cell.

The idea of macroscopically simulating the excitation-propagation in the cardiac tissue with a CA was proposed in [8] and extensively used in the literature [3,7]. The CA is built on the idea that a single cell gets excited if the electrical potential exceeds a determined threshold. Once it is excited, it can trigger the excitation of neighboring cells. In this manner, an electrical stimulus will propagate through the CA grid as the time steps are computed. In this work, the CA states are related to the action potential (AP) and force development in a cell. To make CAs more efficient they are usually parameterized using simulated data from accurate models. This means that the states related to the AP in the cell will be related to a specific portion of the cardiac cell potential. Figure 1 Part A presents the AP computed by ODEs, the AP divided into five different states that represent the different physiological stages of the AP.

In state S0 the cell is in its resting potential where it can be stimulated, in S1 the cell was stimulated and can stimulate the neighbors. In S2 the cell is still able to stimulate the neighbors. In S3 the cell is in its absolute-refractory state where it cannot be stimulated and does not stimulate its neighbors. In S4 the cell is in its relative refractory state where it can be stimulated by more than one neighbor but it does not stimulate its neighbors. As described, the states of a cell generate rules for when a cell can stimulate a neighbor and when it can be stimulated. These rules are an important aspect which will allow the stimulus to propagate.

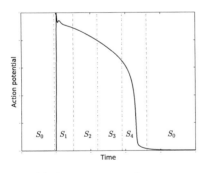
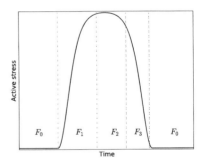

(a) Action potential states.   (b) Active force states.

Fig. 1. Action potential and active force of a cardiac cell computed by ordinary differential equations and their respective representation via CA.

Another important point is how the cells change their states. The AP has a predetermined period so that the states will be spontaneously changed after the AP starts, where the time of each state is a parameter of the system. Our CA is adapted to work with irregular meshes of tetrahedrons. In that case, the cells of the system are the tetrahedrons and a cell is considered a neighbor of other cell if they share at least one vertex. The distance between two neighbors cells is computed as the distance between their barycenters, given by:

$$X_b^i = \frac{1}{4} \sum_{a=1}^{4} x_a^i, \tag{1}$$

where x_a are the coordinates of the vertices from tetrahedron x^i.

Equally important, CA states are updated at every discretized time, dt. Based on the distance, velocity (passed as parameter to the model) and activation time it is possible to calculate the time that a stimulus takes to travel from one CA cell to another, in order to propagate the action potential. An anisotropic tissue was used, so that the propagation velocity is different in the three directions of interest in the heart tissue: fiber, sheet, and normal-sheet. To find the time t for a stimulus to travel from one cell to another, first the direction between the barycenters is computed and then the distance d:

$$D_{ij} = X_b^j - X_b^i \tag{2}$$
$$d = \|D_{ij}\|, \tag{3}$$

where X_b^i and X_b^j are the positions of the barycenter of elements i and its neighbor j. Next, the total velocity of the AP is calculated, based on the velocities in each one of the directions: v_f, v_s and v_n:

$$V = v_f F + v_s S + v_n N, \tag{4}$$
$$v = |V \cdot \hat{D}_{ij}|, \tag{5}$$

where F, S and N are respectively fiber, sheet and normal normalized directions. $\hat{D}_{ij}$ is the normalized direction between elements i and j. Then the traveling time t of the propagation between them is:

$$t = \frac{d}{v}, \tag{6}$$

That verifies if there is enough time to propagate the stimulus by comparing the time since the neighbor has been stimulated and time t.

Finally, electrical potential is coupled with the active force, which is responsible for starting the contraction of the cardiac tissue. When the cell is stimulated, there is an increase in the concentration of calcium ions inside the cell, which triggers the cell contraction. The force development has a delay after the cell is stimulated because of its dependence on the calcium ions. The force development of a cell can be represented in states that change over time like the electrical potential states. Figure 1 Part B presents the force development states and its relation with the electrical states. The force-development states will only pass from state F0 (no contraction force) to state F1 when the electrical state of the cell goes from state S1 to S2. After this change, force development will be time dependent, but will not depend on the electrical state of the cell.

2.1   The AP Repolarization Homogenization

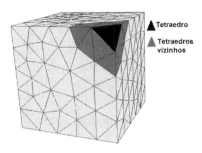

Fig. 2. A cell and its neighbors.

Our implementation changes a cell APD according to its neighbors. Every cell that shares at least one node is considered a neighbor, as is depicted in Fig. 2. So, for each cell, we first compute the average APD of all neighbors:

$$APD_t^{avg} = \sum_{i=0}^{n} APD_t^i, \tag{7}$$

where APD_t^i is the APD at time t of neighbor i. Then we compute the cell coupling level, for quantifying the degree of collaboration among cells varying on time:

$$c_t = p \times f_t \tag{8}$$

where is $p \in [0, 1]$ is parameter to represent the conductance, a user-defined value, and f_t is the normalized active force given by cellular automata in percentage values. When the force reaches its maximum value ($f = 1$), it is on its maximum deformation, so the cells are in the most coupled state, due to their contraction. Therefore, the maximum contribution of neighbors to a cell APD is reached. When that is zero, it means the cell does not accept any contribution and its APD remains unchanged. Finally, the new APD is given by:

$$APD_{t+1} = (1 - c_t) \times APD_t + c_t \times APD_t^{avg}. \tag{9}$$

3   Results

Our experiments have the setup presented in the work of Viswanathan *et al.* [18]. We simulated a tissue segment containing the same amount of the three types of cells. We stimulated cells on endocardium wall $x = 0$, and so the polarization wave flows from endo to epicardium. This configuration is shown in Fig. 3.

Viswanathan *et al.* [18] used two parameters for determining the level of coupling. The velocity of propagation v_p, given in mm/s, and gap junction conductance g_j, given in μS. They both are homogeneous in the tissue. Different

Fig. 3. Distribution of cell types in fiber.

values were tested for each simulation, to represent different levels of coupling. For a healthy tissue, it was used $(v_p, g_j) = (560, 2.5)$, typical values for simulating a normal coupling. Isolated cells represent the absence of gap-junction coupling. Intermediate configurations stand for reduced coupling, where the velocity value is typical for propagation transverse to fiber $(v_p, g_j) = (180, 0.25)$, and poorly coupled, that happens in pathological conditions such as infarction $(v_p, g_j) = (35, 0.025)$. For comparing these experiments, it was used the ΔAPD, that considers the difference of action potential duration of M-cells and epicardium cells, in the middle of respective regions. They have the corresponding maximum and minimum duration. Thus, for normal coupling, ΔAPD is 18 ms. For isolated cells, ΔAPD is 90 ms. As the level of coupling decreases, ΔAPD increases, tending to the value of isolated cells. For reduced and poor coupling, the difference on APD is respectively 44 and 72 ms.

In our FisioPacer simulations we used the same value for propagation velocity, but the gap-junction conductance g_f needed to be different, since in our model it is a percentage value, and not an absolute value of electric conductance as in the LRd model. For tuning the FisioPacer coupling, we manually tested g_f, until we obtain the same ΔAPD for the LRd model. The respective conductance g_f for poor, reduced and normal coupling is 1.64%, 4.30% and 10.5%.

Initially, the cells are set with the isolated cells APDs. During the simulation, the model applies the Eq. 9, in every time step. Therefore the APD of each cell is updated. The initial and final APD for each cell type and experiment can be found in Table 1. These values are measured in the middle of each region.

Table 1. APDs (ms) at the end of simulation, per coupling and cell-type.

Coupling	Cell type		
	Endo	Mid	Epi
Isolated	185	250	160
Poor	191	241	169
Reduced	200	223	178
Normal	206	208	190

For a better visualization, we plotted all APDs over the tissue on Fig. 4 for FisioPacer simulations. When there is no cell coupling, they remain with the same APD. Therefore we can see all cells of region with the same duration, in the solid line in the figure. For a normal tissue, the APD over the tissue is homogeneous, so the curve of normal tissue is smoother, as there is more coupling, so ΔAPD is smaller. For reduced and poor coupling, ΔAPD is greater and therefore the curve is less flat than normal. As the level of coupling decreases, the APD curve tends to be closer to that of isolated cells. These curves are similar to those obtained by the LRd model in Viswanathan *et al.* [18].

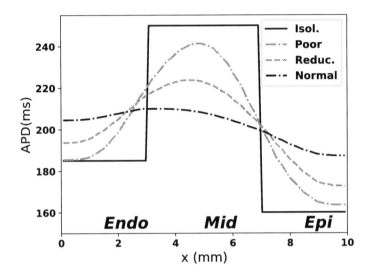

Fig. 4. Action potential duration with different coupling level, over the tissue.

The action potential for each tissue configuration can be found at Fig. 5. For isolated cells, since there is no cell coupling, the AP remains in its original duration (Fig. 5a). For poor coupling, propagation velocity is low, therefore the polarization wave takes more time to reach the epicardium cells. Besides that, the conductance is also low, so there is less interaction on cells and less changes on ΔAPD. For reduced coupling, the velocity is greater and thus the epicardium is excited sooner. The ΔAPD is lesser than poor coupling, since conductance is greater. Finally, the normal coupling results in a very fast depolarization, making cells excited in close moments. The ΔAPD is low, due the high conductance and therefore cells return to resting almost together, i.e., the repolarization dispersion is homogeneous.

The performance was very fast. We used a finite element mesh containing 518 tetrahedrons. We simulated 500 ms of cardiac tissue activity, with a timestep of 0.1 ms. In a Windows 10 desktop computer with an i7 processor and 16 gb RAM, our simulation took in average 420 ms to run.

In our previous works, we performed a comparison with another simulator based on a traditional finite element approach. FisioPacer was up to 90 times faster, depending on simulations configuration [5].

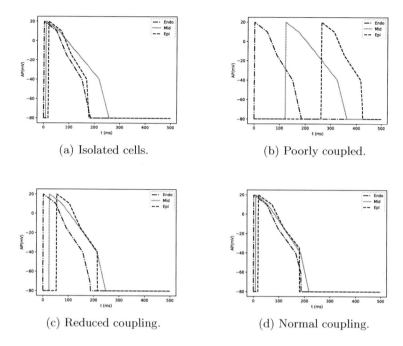

(a) Isolated cells.   (b) Poorly coupled.

(c) Reduced coupling.   (d) Normal coupling.

Fig. 5. Action potential of endo, mid and epicardium cell with different coupling levels.

4   Conclusion

In this work we presented a cardiac simulator based on cellular automata, that
was modified to represent the electrotonic effect. This effect occurs due to the
cell coupling, where cells interact with each other and therefore their behavior
can change.

Specifically, this interaction can change the action potential duration (APD),
in order to prevent undesired propagation waves. For instance, isolated cells
have a very different APD. For guinea pig models the difference between mid-
myocardium and epicardium is around 90ms. However, in a healthy tissue, this
difference decreases 18 ms, 20% of the original value. In this sense, we adapted
our model for reproducing the APD change, according to parameters found in
the literature, such as tissue conductance, propagation velocity and APD of
neighbor cells. Despite the simplicity of our methods, our simulator was able to
reproduce results of another simulator, obtaining the same ΔAPD in different
coupling levels.

References

1. Atienza, F.A., et al.: A probabilistic model of cardiac electrical activity based on a cellular automata system. Revista Española de Cardiología (English Edition) **58**(1), 41–47 (2005)
2. Bazhutina, A., Balakina-Vikulova, N., Solovyova, O., Katsnelson, L.: Mathematical model of electrotonic interaction between mechanically active cardiomyocyte and fibroblasts. In: 2019 Ural Symposium on Biomedical Engineering, Radioelectronics and Information Technology (USBEREIT), pp. 114–117. IEEE (2019)
3. Bora, C., Serinağaoğlu, Y., Tönük, E.: Electromechanical heart tissue model using cellular automaton. In: 2010 15th National Biomedical Engineering Meeting, pp. 1–4. IEEE (2010)
4. Campos, R.S., Rocha, B.M., Lobosco, M., dos Santos, R.W.: Multilevel parallelism scheme in a genetic algorithm applied to cardiac models with mass-spring systems. J. Supercomput. **73**(2), 609–623 (2017)
5. Campos, R.S., Rocha, B.M., da Silva Barra, L.P., Lobosco, M., dos Santos, R.W.: A parallel genetic algorithm to adjust a cardiac model based on cellular automaton and mass-spring systems. In: Malyshkin, V. (ed.) PaCT 2015. LNCS, vol. 9251, pp. 149–163. Springer, Cham (2015). https://doi.org/10.1007/978-3-319-21909-7_15
6. Dos Santos, R.W., Otaviano Campos, F., Neumann Ciuffo, L., Nygren, A., Giles, W., Koch, H.: ATX-II effects on the apparent location of M cells in a computational model of a human left ventricular wedge. J. Cardiovasc. Electrophysiol. **17**(s1), S86–S95 (2006)
7. Gharpure, P.B., Johnson, C.R., Harrison, N.: A cellular automaton model of electrical activation in canine ventricles: a validation study. SCI Institute (1995)
8. Gharpure, P.B.: A cellular automation model of electrical wave propagation in cardiac muscle. Ph.D. thesis, Department of Bioengineering, University of Utah (1996)
9. Luo, C.H., Rudy, Y.: A dynamic model of the cardiac ventricular action potential. II. Afterdepolarizations, triggered activity, and potentiation. Circ. Res. **74**(6), 1097–1113 (1994)
10. MacCannell, K.A., Bazzazi, H., Chilton, L., Shibukawa, Y., Clark, R.B., Giles, W.R.: A mathematical model of electrotonic interactions between ventricular myocytes and fibroblasts. Biophys. J. **92**(11), 4121–4132 (2007)
11. de Oliveira, B.L., Rocha, B.M.R.B.M., Barra, L.P.S., Toledo, E.M., Sundnes, J., dos Santos, R.W.: Effects of deformation on transmural dispersion of repolarization using in silico models of human left ventricular wedge. Int. J. For Numer. Methods Biomed. Eng. 29, 1323–1337 (2013)
12. Rudy, Y.: Electrotonic cell-cell interactions in cardiac tissue: effects on action potential propagation and repolarization. Ann. N. Y. Acad. Sci. **1047**(1), 308–313 (2005)
13. Sampson, K.J., Henriquez, C.S.: Electrotonic influences on action potential duration dispersion in small hearts: a simulation study. Am. J. Physiol. (Heart Circ. Physiol.) **289**, 350–360 (2005)
14. dos Santos, R.W., Plank, G., Bauer, S., Vigmond, E.J.: Parallel multigrid preconditioner for the cardiac bidomain model. IEEE Trans. Biomed. Eng. **51**(11), 1960–1968 (2004)
15. Taggart, P., Sutton, P., Opthof, T., Coronel, R., Kallis, P.: Electrotonic cancellation of transmural electrical gradients in the left ventricle in man. Prog. Biophys. Mol. Biol. **82**(1), 243–254 (2003)

16. Toyoshima, H., Burgess, M.J.: Electrotonic interaction during canine ventricular repolarization. Circ. Res. **43**, 348–356 (1978)
17. Vigmond, E., Dos Santos, R.W., Prassl, A., Deo, M., Plank, G.: Solvers for the cardiac bidomain equations. Prog. Biophys. Mol. Biol. **96**(1–3), 3–18 (2008)
18. Viswanathan, P.C., Shaw, R.M., Rudy, Y.: Effects of i_{Kr} and i_{Ks} heterogeneity on action potential duration and its rate dependence: a simulation study. Circulation **99**(18), 2466–2474 (1999)

Hybrid Model for the Analysis of Human Gait: A Non-linear Approach

Ramón E. R. González[1](✉), Carlos Collazos-Morales[2](✉),
João P. Galdino[1], P. H. Figueiredo[1], Juan Lombana[2],
Yésica Moreno[2], Sara M. Segura[2], Iván Ruiz[2], Juan P. Ospina[2],
César A. Cárdenas[2], Farid Meléndez-Pertuz[3],
and Paola Ariza-Colpas[3]

[1] Departamento de Física, Universidade Federal de Pernambuco, Recife, Brazil
ramayo_g@yahoo.com.br
[2] Vicerrectoria de Investigaciones, Universidad Manuela Beltrán,
Bogotá, Colombia
cacollazos@gmail.com
[3] Universidad de la Costa, Barranquilla, Colombia

Abstract. In this work, a generalization of the study of the human gait was made from already existent models in the literature, like models of Keller and Kockshenev. In this hybrid model, a strategy of metabolic energy minimization is combined in a race process, with a non-linear description of the movement of the mass center's libration, trying to reproduce the behavior of the walk-run transition. The results of the experimental data, for different speed regimes, indicate that the perimeter of the trajectory of the mass center is a relevant quantity in the quantification of this dynamic. An experimental procedure was put into practice in collaboration with the research group in Biomedical Engineering, Basic Sciences and Laboratories of the Manuela Beltrán University in Bogotá, Colombia.

Keywords: Biomechanics · Center of mass · Dynamic · Hybrid model · Perimeters · Reaction force · Walk-run transition

1 Introduction

Modern biomechanics arose with the studies of Archivald V. Hill on the transformation of heat from the mechanical work of muscular contractions [1]. The correlation of thermodynamic parameters with the heat transformed by the muscles led him to win the Nobel Prize in Physiology and Medicine in 1922. With the discovery of Hill, the links between the macroscopic biological systems and the universal characteristics of Physics were tightened.

The complexity of human locomotion comes from the fact that there are diverse and varied interactions between the body and the environment. A simplification of these interactions could be that the chemical potential energy originating from the muscles and the elastic potential energy of the tendons and type of the muscular elasticity end up transforming into work and heat [2]. The cyclic contractions in the active muscles

© Springer Nature Switzerland AG 2020
O. Gervasi et al. (Eds.): ICCSA 2020, LNCS 12249, pp. 216–231, 2020.
https://doi.org/10.1007/978-3-030-58799-4_16

give rise to reaction forces of the soil along the lower extremities. The force resulting from the gravitational action and the strength of resistance accelerates and decelerates the mass center of the body. Studies concerning the mechanical efficiency of loco-motion [3] show that for different constant speeds, the rapidity in which energy is transformed, determined by oxygen consumption and by external work, is different for walking and running, the energy cost is lower in the first. When we deal with bipedal movement in particular, nonlinear effects are often observed, but most of the approx-imate models for this movement ignore such effects. These models can be reliable for primary studies but end up being insufficient in relation to the complexity of the movement, losing the naturalness and clinical accuracy of the locomotion.

1.1   Reaction Force: The Hypothesis of Non-linearity

A reasonable approximation, when describing the process of gait, is to suppose a cyclic pattern of corporal movements that are repeated at each step [4]. The process of normal gait is one in which the erect body in movement is consistent with one leg and then on the other. This process has two phases, the support phase, which begins when the leg is in contact with the ground and lasts until this contact is lost. This phase represents 60% of the entire cycle of the March. The remaining 40% are the balance or balancing phase, since the leg loses contact with the ground until they come into contact again.

In the process of walking there are two very important basic aspects, the continuous reaction forces of the ground that sustain and provide the body with torque and the periodic movements of the feet from one position of support to the next, in the direction of movement. The reaction force of the soil depends intimately on the speed of travel [5]. The entire process is controlled by the neuromuscular system, which is why it is a complex process, which means that locomotion is a system in which changes in small components result in significant changes.

1.2   Metabolism and Mechanical Power for Walking. The Keller Model

People walk naturally in a way that energy consumption is optimized [4]. The meta-bolic rate, the variation of energy per time of physical activity, is generally measured indirectly by the amount of oxygen consumed during bodily activity [6]. In order to minimize energy dissipation, the neuromuscular system selects the patient. Deviations from this normal gait pattern result in increased energy consumption and limit loco-motion [7].

In 1974, J.B. Keller [8] proposed a model based only on variational calculus and elementary dynamics to study the running of extraordinary human performance. In this model, it is assumed that the reaction force of the soil affects the amount $E(t)$ of power per unit mass from muscle that store N_2 of the food and the consumption of O_2 of the individual. The reactions that occur in the body of the individual use these chemical reserves to provide power for locomotion. For prolonged periods of activity, the individual's biology supplies a σ rate of energy supply obtained from respiration and circulation. Keller determined the theoretical relationship between the shortest time T, based on physiological parameters, in which a distance D can be traversed (1).

$$D = \frac{F}{\gamma^2} \left(\gamma T - 1 + e^{-\gamma T} \right) \tag{1}$$

In this calculation, it was considered that the propulsive force of the sprinter cannot exceed a certain maximum value (2).

$$f(t) \leq F \tag{2}$$

The constant γ is a physiological constant and has units of time inverse. So that there is an optimized solution, according to the results of Keller's work, for the sprinter in starting a finite number of velocity curves described according to (3) is obtained.

$$v(t) = \left(\frac{\sigma}{\gamma} + C e^{-2t\gamma} \right)^{\frac{1}{2}} \tag{3}$$

Where the constant C is arbitrary and is determined by the initial velocity.

The important cases in Keller's work are: an acceleration curve during the interval in which the runner exerts the maximum momentum; the constant velocity curve from the moment the runner reaches its maximum speed and the deceleration curve that begins when $E(t) \equiv 0$.

These three curves are combined ensuring continuity throughout the run and that the area under the speed curve must be maximum. A summary of the above can be seen in (4) and in Figs. 1 and 2.

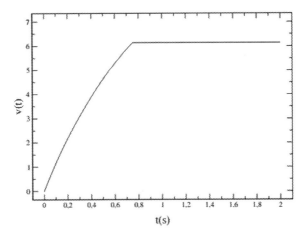

Fig. 1. Speed behavior of $D < D_c$, where D_c is the critical distance. There is an impulse given by the variation of the velocity until the first transition, where, after the transition, the velocity remains constant and equal to V.

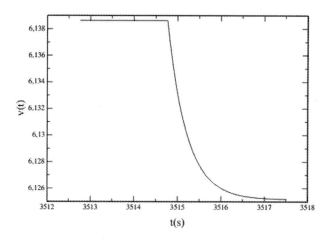

Fig. 2. Behavior of the speed when the energy power is carried to the limit $E(t) = 0$. In the transition, the curve presents a decline in speed referring to the athlete's energy limit.

$$v(t) = \begin{cases} \frac{F}{\gamma}(1 - e^{-\gamma t_1}), & 0 \leq t \leq t_1 \\ V, & t_1 \leq t \leq t_2 \\ \left(\frac{a}{\gamma} + \left(V^2 - \frac{a}{\gamma}\right)e^{-2\gamma(t-t_2)}\right)^{\frac{1}{2}}, & t_2 \leq t \leq T \end{cases} \quad (4)$$

When analyzing competitions between professional athletes and world record data. Keller estimated parameters for a run with optimal or next optimization strategy. These values, corresponding to physiological variables are the following (5).

$$F = 12.2\frac{m}{s^2}, \frac{1}{\gamma} = 0.892\,s, \sigma = 41.56\frac{m^2}{s^3}, T_c = 27.7\,s, E_0 = 2409\frac{m^2}{s^3} \quad (5)$$

The parameters T_C and E_0 are the critical time (the minimum for the distance D) and the initial power per unit mass.

1.3   Kokshenev Model for Constant Speed Walks

For the study of the oscillations that the center of mass experience in a locomotion at constant speed it is important to observe the movement of the center of mass in a sagittal plane. Kokshenev, in his 2004 paper [9], used this plane as a reference to study the movement of the center of mass during human walking at constant speed. In this model, an inertial reference system is considered moving as a virtual mass center defined by the displacement vector $\overrightarrow{R_0}(t)$. The conditions established in the model for displacement are: $x_0(t) = Vt$, where V is a constant and $y_0(t) = H$ is the average height of the center of mass in relation to the ground, where the origin of the inertial coordinate system is. In this way, the displacement vector relative to the movement of the center of mass of the human body with the virtual center of mass is $\overrightarrow{\Delta r}(t) = \vec{R}(t) - \overrightarrow{R_0}(t)$, where $\vec{R}(t)$ is the displacement vector of the center of mass of the human body in relation to the same referential of $\overrightarrow{R_0}(t)$.

In the model, a driving force $\overrightarrow{\Delta F}(t)$ is defined, which is the neuromuscular capacity to exercise work, which is derived from observations of small oscillations close to the weight support of the body. In (6), we can see this force related to the ground reaction force $\vec{F}(t)$ and the force of gravity.

$$\vec{F}(t) = -m\vec{g} + \overrightarrow{\Delta F}(t) \tag{6}$$

The driving force, as well as the velocity and displacement of the center of mass must respect the condition of cyclometry that preserves the amount of movement of the system.

Applying the Lagrangean formalism, and considering harmonic and non-harmonic solutions for the movement of the center of mass, Kokshenev found the force that characterizes the forced oscillatory movement of the center of mass around the point $\overrightarrow{\Delta r} = \overrightarrow{\Delta r_0}$ (7).

$$\vec{F}(t) = \overrightarrow{\Delta F_0}(t) + \overrightarrow{\Delta F_1}(t) \tag{7}$$

The first term of (6) is the force that describes the free movement of the center of masses as a superposition of linear oscillations. With the increase of the speed, the anharmonic effects become important, being necessary the introduction of a potential resulting from the expansion in a Taylor series of the elastic potential of the Hamiltonian. The second term then results from the gradient of this expanded potential up to the order of the anharmonic effects. Result of all this, the components of the force given by (6) are presented in (8) and (9).

$$\overrightarrow{F_x}(t) = -m\omega_0^2(\Delta l_0 \sin(\omega_0 t) - \Delta l_1 \sin(2\omega_0 t)) \tag{8}$$

$$\overrightarrow{F_y}(t) = m\vec{g} + m\omega_0^2(\Delta h_0 \cos(\omega_0 t) - \Delta h_1 \cos(2\omega_0 t)) \tag{9}$$

The coefficients $\Delta l_0(v), \Delta h_0(v), \Delta l_1(v)$ and $\Delta h_1(v)$ are harmonic and anharmonic amplitudes respectively, whose values correspond to experimental data and $\omega_0(v)$ corresponds to the frequency of a cycle on the step cycle.

Finally, the introduction of a locomotive resistive force $\overrightarrow{\Delta F_{res}}(t) = -\gamma\overrightarrow{\Delta r_1}(t)$ where $\gamma(v)$ represents the coefficient of friction, results in the following functions for the respective positions in a steady state with $\omega_0 t \gg 1$ and in a low resistance approach (10) and (11).

$$x(t) = Vt + \Delta x_0(t) + \frac{\Delta l_1}{3} \frac{\sin(2\omega_0 t + \vartheta)}{\sqrt{1 + \tan(\vartheta)^2}} \tag{10}$$

$$y(t) = H + \Delta y_0(t) + \frac{\Delta h_1}{3} \frac{\cos(2\omega_0 t + \vartheta)}{\sqrt{1 + \tan(\vartheta)^2}} \tag{11}$$

These equations are the solution of the following equation of motion (12).

$$m\overrightarrow{\Delta \ddot{r}_1}(t) + \gamma \overrightarrow{\Delta \dot{r}_1}(t) + k_0 \overrightarrow{\Delta r_1}(t) = \overrightarrow{\Delta F_1}(t) \tag{12}$$

The force $\overrightarrow{\Delta F_1}(t)$ is given by (7) and (8) previous.

The results of Kokshenev show a closed orbit given by a hypocycloid $(\Delta r_1 < \Delta r_0 \ll H)$, around a fixed point and clockwise. It is assumed, in the work of Kokshenev, that this orbit is described, for walking, as a characteristic ellipse, with amplitudes $\Delta l = \Delta l_0 + \frac{\Delta l_1}{3} \frac{1}{\sqrt{1 + \tan(\vartheta)^2}}$ and $\Delta h = \Delta h_0 + \frac{\Delta h_1}{3} \frac{1}{\sqrt{1 + \tan(\vartheta)^2}}$, horizontal and vertical, respectively. Given the conditions of the model, the center of mass moves with constant speed V at a certain height H and rotates along a hypocycloid circumscribed by a "flattened" or "shrunken" ellipse of eccentricity e_+ (e_-) given by the following expression (13).

$$e_\pm = \sqrt{1 - \left(\frac{\Delta l_0}{\Delta h_0}\right)^{\pm 2}} \tag{13}$$

2 Experimental Analysis of Walk for Different Speeds

The study was approved by the Ethics Committees of the Manuela Beltrán University. Written informed consent was obtained by the patients.

2.1 Experimental Environment and Appliances

Experimental data were collected in a space of approximately $16 \, \text{m}^2$, where there is a track formed by four platforms of force and six motion detection cameras around the platforms, see in Fig. 3.

The cameras used are part of the data acquisition system for BTS GAITLAB motion analysis. These optoelectronic cameras measure the displacement, with an accuracy of $\pm 10^{-7}$ m, of body segments in a time interval of $\pm 10^{-2}$ s [10]. The experimental data to validate the models are extracted from.

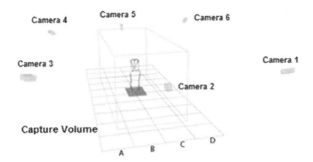

Fig. 3. Scheme that shows the space where the experiments were conducted. The figure portrays the volume occupied by the track and platforms. The motion detection cameras were distributed in the positions of points A, B, C, D, selected to facilitate the orientation of the volunteer.

2.2 Positioning Protocol for Markers and Orientation of Body Segments. Location of the Center of Masses

In this study used only the markers of the hip of the Davis protocol distributed in different segments and corporal regions of the volunteer (Fig. 4). For each range of speed (around 1.03 m/s, 1.81 m/s and 3.2 m/s), In the study was included both female and male volunteers (5 females and 6 males).

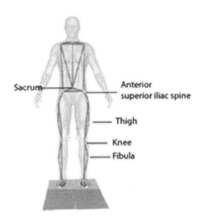

Fig. 4. Location of markers according to the DAVIS protocol (16), Frontal plane (by the authors).

In literature, we can find several references in relation to the position of the center of mass of the human body. For Miralles [11] the center of mass lies behind the lumbar vertebra L5. Yet for Dufour and Pillu [12] it is located before the sacral vertebra S_2. We suggest in this work that the center of mass would be placed just between these two vertebrae without using reaction forces. The Smart TRACKER and Smart Analyzer was used, a program with which we are able see the displacement along the track and to capture the position of each of the markers placed on the volunteer. A simple model was created that virtually simulates the markers and their connection (Fig. 5).

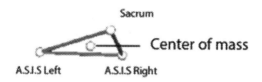

Fig. 5. Position of the center of masses. Midpoint between the referenced locations. *A.S.I.S: Anterior superior iliac spine* (by the authors).

With the Smart TRACKER and Smart ANALYZER, the displacements of the real markers coupled with the volunteer was interpolated, guaranteeing the continuity of the information throughout the capture of the data. With the defined function, a virtual point situated at the midpoint of sacrum and anterior superior iliac spine (left and right) was created. On this point all the clinical analysis corresponding to the March was done.

3   Experimental Analysis of Walk for Different Speeds

3.1   Details of the Model

In order to find results for different walking speed using the Kokshenev model for non-linear running, we used the experimental data conceived in the experiments described above for comparison with important aspects of the model. In our model we use (12), deduced in the Kokshenev model, which, from a simpler mechanical analog, represents oscillations in the two-dimensional plane under the action of a viscous resistant force and the reaction force of the soil acting on favor of the movement, turning the system a two-dimensional pendulum damped and forced.

The diversity of the velocities was issued by (4). The physiological parameters used in optimization by Keller were adopted (5). We noted that, by varying these parameters and implementing them in the equation for speed, there was a dependence between the terms related to the transitions between the different speed regimes. At time t_1, which separates the velocity regime with exponential growth of the constant velocity regime, the following behavior is obtained (14).

$$t_1 = -\frac{1}{\gamma} ln\left(1 - \frac{V\gamma}{F}\right) \tag{14}$$

From this, a maximum value is induced for the ratio F/γ from which we can find t_1. In this way, the relationship between the parameters F and γ is established as the maximum limit for the value of the speed reached (maximum speed). From the previous

Eq. (14) we can see that the maximum speed that the system can reach is equal to the limit in which time tends to infinity in the following relation (15).

$$v(t) = \frac{F}{\gamma}\left(1 - e^{-\gamma t}\right) \tag{15}$$

Making the maximum speed can be expressed according to (16).

$$v_{max} = \frac{F}{\gamma} \tag{16}$$

The minimum speed, on the other hand, refers to the case in which the system works for long periods of time. At the time t_2 when the physiological wear begins, the velocity transition occurs and this time we can obtain it from (4), as follows (17) [woodside].

$$t_2 = \frac{\left\{E_0 + \frac{FV}{\gamma} + \left[\left(\frac{F}{\gamma}\right)^2 - V^2\right]\ln\left[1 - \left(\frac{V\gamma}{F}\right)\right]\right\}}{(V^2\gamma - \sigma)} \tag{17}$$

In order to maintain the coherence of the function from t_2 to T, in case that v is less than the root of the term σ/γ, then we will have a divergence on the function that characterizes this regime transition. We can then interpret the following as the minimum speed (18).

$$v_{min} = \sqrt{\frac{\sigma}{\gamma}} \tag{18}$$

For the values of the physiological parameters cited, the maximum time T that an individual can reach, at the end of the slope curve due to physiological wear is (19).

$$T = t_2 - \frac{1}{2}\ln\left\{\frac{1}{4}\left[1 - \left(\frac{\sigma}{V^2\gamma}\right)\right]\right\} \tag{19}$$

In this equation, we can see that values of $\sigma \ll V^2\gamma$ there will be a value of T while for $\sigma \gg V^2\gamma$ the equation diverges, where T is in that case inaccessible. The connection between the parameters σ and T, as well as the minimum speed allows us to find different values of σ respecting the maximum and minimum values of the speed.

In (12), the terms k_0 and γ are functions of velocity. The form as $\omega_0(v) = \frac{k_0(v)}{m}$ varies was obtained by Kokshenev [9] using the results of experimental data reported in [3]. It was defined that $\omega_0(v)$ is a linear function of velocity, as follows (20).

$$\omega_0(v) = 4.94 + 4.02v \tag{20}$$

With this result, the coefficient of friction per unit of mass $\omega_1(v) = \frac{\gamma(v)}{m}$ was found as follows (21).

$$\omega_1(v) = 6.37 - 6.15v + 2.38v^2 \tag{21}$$

Introducing Keller's optimal speed and substituting the dependence with speed for the dependence with time, for small oscillations we can affirm that (22).

$$\omega_0(t) \approx 4.94 + 4.02\left(1 - e^{-v(t)}\right) \tag{22}$$

In the numerical solution (12), using (20) and (21) e with $v(t)$ being Keller's optimal velocity, we note that it grows rapidly for the constant velocity value and (12) is quickly damped by the growth of $\omega_1(v)$. The model presented here never comes to contemplate the physiological wear due to the quadratic growth of $\omega_1(v)$ and its mathematical complexity.

The physiological parameter chosen to determine the minimum velocity was σ. For each value of σ, the last oscillation of the x and y positions was recorded and parametric curves were constructed. These curves vary between the maximum time t_{max} of the occurrence of the movement and the difference between that time and an R term dependent on the angular velocity ω_0 in t_{max} as follows (23).

$$R = \frac{2\pi}{\omega_0(t_{max})} \tag{23}$$

The registration time of a parametric curve is (24):

$$t_{max} - R < t < t_{max} \tag{24}$$

4   Results

With the equations already defined, the behaviors of the components $x(t)$ e and $y(t)$ for the trajectory of the center of masses were determined. Three velocity values were chosen: v = 1.03 m/s, 1.81 m/s and 3.20 m/s, referring to those obtained in collaboration with the Research Group in Biomedical Engineering, Basic Sciences and Laboratories of the Manuela Beltrán University, in Bogotá, Colombia. The final period of oscillation and comparison with a period taken from the real data was plotted from the model. We chose the data closest to an average of a characteristic behavior of force, for the experimental data. The characteristic behavior of the normalized force of the actual data is related to the average behavior of the data obtained with all the volunteers and platforms. In this way, we selected the data of the individual MF7P1, female, on the P_1 platform, one of the four used to capture the FRS, see Figs. 6 and 7.

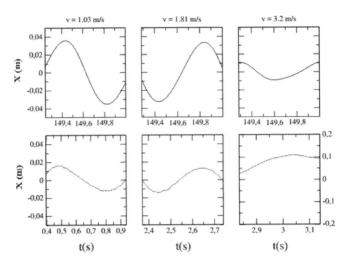

Fig. 6. Behavior of the position $x(t)$ of the center of masses for three different speeds in different periods. The black lines represent the results derived from the model and the red lines, the results from the real data. The first column on the left shows the graphs for the velocity v = 1.03 m/s. These graphs are in phase and the amplitude obtained with the model are twice as large. The middle column shows the results for the velocity v = 1.81 m/s. The curves are also in phase and the ratio between the amplitude is the same as for the lower speed. The third column, for v = 3.20 m/s, shows curves in phase opposition, in this case, the amplitude of the curve referring to the actual data is four times greater. (Color figure online)

We see in Fig. 6, for the first two speed, that the oscillations have the same phase, both in time and in amplitude. Yet for the speed of 3.2 m/s, we see a phase shift between the curves of approximately $\pi/2$ and the values of the amplitude referring to the real data are in a proportion four times greater than the amplitude of the model.

In Fig. 7, the same time interval was plotted. It is observed, to 1.03 m/s a point of maximum similar in both graphs, of the model and of the real data. For the intermediate speed, 1.81 m/s we see extremes of inverted phases and for 3.20 m/s, although the amplitude obtained from the model is three times lower than that obtained from the actual data, the phases of the oscillations are quite similar.

From the model, parametric curves for x(t) and y(t) were obtained by varying the value of the physiological constant σ, reported by Keller (5) from 0.2 m^2/s^3 to 12 m^2/s^3. It was observed that there is a perimeter for the trajectory, associated with each translation speed and that the velocity value that maximizes the perimeter is v = 1.38 m/s.

Figure 8 illustrates the behavior of the perimeter of the trajectories of the center of masses for different translation speed. At speeds up to v = 1.38 m/s, the perimeter of the trajectories increases and higher speed it decreases.

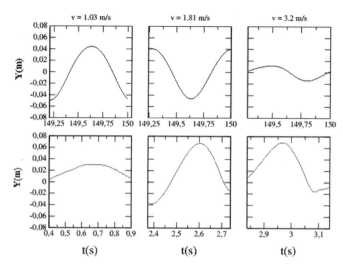

Fig. 7. Behavior of the position $y(t)$ of the center of masses for three different speeds in different periods. The black lines represent the results derived from the model and the red lines, the results from the real data. The first column on the left shows the graphs for the velocity v = 1.03 m/s, where a similar maximum point is seen in both graphs. The central column shows the results for v = 1.81 m/s and is see that the curves have inverted phases. The column on the right represents the velocity v = 3.20 m/s, where it can be seen that the two curves oscillate in phase, although the amplitude of the curve referring to the actual data is three times greater. (Color figure online)

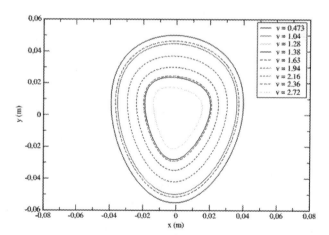

Fig. 8. Parametric curves of the trajectories of the center of masses for different translation speed. The solid lines represent the lowest speeds, for which a gradual increase in the perimeter is seen as a function of the speed up to v = 1.38 m/s. For higher speeds, represented by dashed lines, we see a decrease in the perimeter of the trajectories as the speed of translation increases, showing that the amplitudes in the curves become increasingly smaller with the increase in speed.

For each of the curves in Fig. 8, the perimeter was calculated, corresponding to the trajectory of the center of mass for each velocity. A graph of the perimeter as a function of speed was constructed where the maximum point was easily identified, corresponding to the velocity v = 1.38 m/s. The perimeters corresponding to experimental data found in the literature were also calculated [13] and trajectories of the center of mass obtained from the experimental data, in collaboration with the Research Group on Biomedical Engineering, Basic Sciences and Laboratories of Manuela Beltrán University, in Bogotá, Colombia. These last data are referring to a certain running regime, with a definite velocity of the center of mass of the volunteer. With this data obtained an average of the trajectories of the center of mass of each volunteer and with such means a closed parametric curve was generated, for which its characteristic parameter was calculated. The perimeter of the path of just one individual was likewise calculated, which approximated the result obtained by the average in each speed regime.

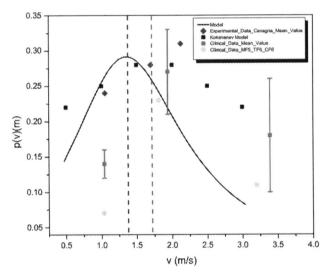

Fig. 9. Perimeter according to the translation speed. The line represents the points obtained from the model. The blue points were available from the literature [ref 22]. The red dots are referring to the average of the actual data, obtained from four volunteers. The green points are the perimeters of a volunteer that was felt better to the curve of the model. Finally, the black points are the perimeters corresponding to the eccentricities of the literature reported by Kokshenev [9]. (Color figure online)

In the graph of Fig. 9, we can see all the information regarding perimeters as a function of speed. From experimental results reported by Kokshenev, it was possible to find the perimeter corresponding to each plotted eccentricity, as a function of the speed and they were represented using black squares. With red squares the average values of the perimeters corresponding to the experimental data were represented in each speed regime. The blue points are referring to real data from the literature [14] and the green points are the perimeters of the trajectories of a single individual, MPF5P1, which is adapted to the values obtained for the average of the parametric trajectories. The dashed red line marks the speed considered by Kokshenev as the transition point between the March and the race, this speed is $v = 1.73$ m/s. The dashed black line represents the maximum perimeter velocity $v = 1.38$ m/s. This speed represents, for our model, what would be the transition between the running and running regimes. It is reasonable to think that close to this point of maximum of the gait is located, said in some way, normal for a healthy adult [15]. This walking speed is the most stable for the mass center of healthy adults.

The model proposed presents peculiar behavior in relation to the perimeter of the trajectory of the center of masses as a function of the translation speed. Trying to solve the model for variable speeds, in the way proposed by Keller, has ineffective implications, since the translation speed has a very short duration in the regime in which the speed grows exponentially, observed in Eq. (4) and lasts a long time in the stationary speed regime, before reaching physiological wear.

In fact, for practical purpose, physiological wear is unattainable for the simulation time that is generated.

Kokshenev work with the hypothesis that the eccentricity varies depending on the speed. In this work we chose to study the perimeter depending on the speed as an approach to obtain the results. This approach is plausible due to the relationship between the eccentricity and the amplitude of the curves. The visibility between the perimeter and the speed was another point for the use of this approach. A result of this relationship is the maximum point found, for $v = 1.38$ m/s. This value coincides with one of the most accepted values in the literature for the "normal" speed of a healthy adult [16–18].

For the topic addressed in this paper there are various methods and models in the literature [10], on the other hand, the model studied here, despite being a simplification of effects of other natures, is acceptable because, using only non-linear mechanics, the results obtained result in a good approximation of reality.

5   Conclusions

The objective of this work was to approach non-linear effects in biomechanics. Using models already known from the literature, such as Kokshenev and Keller, an association of these models was achieved in order to obtain results from a model with more general characteristics. An experimental procedure was adopted in collaboration with the research group in Biomedical Engineering, Basic Sciences and Laboratories of the Manuela Beltrán University in Bogotá, Colombia. These experiments generated data that, finally, were compared with the created model [16–18].

The results of the proposed model for low speeds, works quite well, from $v = 1.0$ m/s to approximately $v = 1.38$ m/s, for which the perimeter of the center of mass calculated from the model coincides or results fairly close to the experimental perimeter. As the speed increases, deviations are observed more and more accentuated. It is also observed that the trajectory is more accentuated for the critical speed $v = 1.38$ m/s and decreases both, with the increase and with the decrease in speed.

It is noticeable that the presented model manages to approximate the experimental results for low speed of the March, in spite of the model using physiological parameters that optimize the gait. Deviations at high speed are hypothetically associated with noise from the central pattern generator (PCG), a biological neural network responsible for locomotion that produces a rhythmic pattern in the absence of sensory responses or descendants that carry specific temporal information [19, 20].

References

1. Basset Jr., D.R.: Scientific contributions of AV. Hill: exercise physiology pioneer. J. Appl. Physiol. **93**, 1567–1582 (2002)
2. Madihally, S.V.: Principles of Biomedical Engineering, 1st edn. Artech House, Norwood (2010)
3. Gatesy, S.M.: Bipedal locomotion: effects of speed, size and limb posture in birds and humans. J. Zool. **224**, 127–147 (1990)
4. Rose, J., Gamble, J.G.: Marcha–Teoria e práctica da marcha humana, 2nd edn., editor Guanabara (2007)
5. Munro, C.F., Miller, D.I., Fuglevard, A.J.: Ground reaction forces in running: a reexamination. J. Biomech. **20**, 147–155 (1987)
6. Weir, J.B.: New methods for calculating metabolic rate with special reference to protein metabolism. J. Physiol. **109**, 1–9 (1949)
7. Blessey, R.: Energy cost of normal walking. Orthop. Clin. North Am. **9**, 356–358 (1978)
8. Keller, J.B.: Optimal velocity in a race. Am. Math. Mon. **81**, 474–480 (1974)
9. Kokshenev, V.B.: Dynamics of human walking at steady speeds. Phys. Rev. Lett. **93**, 20 (2004)
10. Collazos, C.A., Argothy, R.E.: Physical modeling of normal and pathological gait using identification of kinematic parameters. Int. J. Biol. Biomed. Eng. **8** (2014)
11. Marrero, R.C.M.: Biomecanica clinica del aparato locomotor. Masson (1998)
12. Dufour, M., Pillu, M.: Biomecanica functional. Masson (2006)
13. Willems, P.A., Cavanga, G.A., Heglund, N.C.: External, internal and total work in human locomotion. J. Exp. Biol. **198**, 379–393 (1995)
14. Cavanagh, P.R., Lafortune, M.A.: Ground reaction forces in distance running. J. Biomech. **13**, 397–406 (1980)
15. Silveira, M.C: Análise da estabilidade da marcha de adultos em diferentes condições visuais. M.S. thesis, Escola de Educação Física, Universidade Federal do Rio Grande do Sul (2013)
16. Davis, R.B.: A gait analysis data collection and reduction technique. Hum. Mov. Sci. **10**, 575–587 (1991)
17. Ramos, C., Collazos, C.A., Maldonado, A.: Acquisition of lower limb joint variables by an inertial card system. In: Torres, I., Bustamante, J., Sierra, D. (eds.) VII Latin American Congress on Biomedical Engineering CLAIB 2016, Bucaramanga, Santander, Colombia, October 26th–28th, 2016. IP, vol. 60, pp. 369–372. Springer, Singapore (2017). https://doi.org/10.1007/978-981-10-4086-3_93

18. Collazos, C.A., Castellanos, H.E., Cardona, J.A., Lozano, J.C., Gutiérrez, A., Riveros, M.A.: A simple physical model of human gait using principles of kinematics and BTS GAITLAB. In: Torres, I., Bustamante, J., Sierra, D. (eds.) VII Latin American Congress on Biomedical Engineering CLAIB 2016, Bucaramanga, Santander, Colombia, October 26th–28th, 2016. IP, vol. 60, pp. 333–336. Springer, Singapore (2017). https://doi.org/10.1007/978-981-10-4086-3_84

19. Jiménez, G., Collazos Morales, C.A., De-la-Hoz-Franco, E., Ariza-Colpas, P., González, R. E.R., Maldonado-Franco, A.: Wavelet transform selection method for biological signal treatment. In: Tiwary, U.S., Chaudhury, S. (eds.) IHCI 2019. LNCS, vol. 11886, pp. 23–34. Springer, Cham (2020). https://doi.org/10.1007/978-3-030-44689-5_3

Improving Fresh Fruit Bunches Unloading Process at Palm Oil Mills with Discrete-Event Simulation

Jelibeth Racedo-Gutiérrez[1]([⊠]), Carlos Collazos-Morales[1],
Luis Gómez-Rueda[2], Joaquín Sánchez-Cifuentes[1], Fredy A. Sanz[1],
and César A. Cardenas[1]

[1] Universidad Manuela Beltrán, Bogotá, D.C, Colombia
jelibeth.racedo@docentes.umb.edu.co
[2] Universitaria Agustiniana, Bogotá, D.C, Colombia

Abstract. Palm oil industry is an important agricultural activity in Colombia. Agents of this industry have chosen associative strategies and vertical integrations to increase their efficiency. However, reducing logistics costs to compete with countries such as Thailand and Malaysia, is still a challenge, also due to the scarceness of applied research in the palm oil logistics processes. In this paper, we present a continuous-discrete simulation-based approach to evaluate and optimize the fresh fruit bunches unloading process at the mill. Two simulation models representing the current process and alternative process scenarios were developed and compared. The findings show that important improvements in waiting time and supplier satisfaction can be achieved without major investments.

Keywords: Palm oil mills · Unloading process · Simulation

1 Introduction

Palm oil is one of the main sources of oils and fats in the world. It is used in a wide variety of food and cosmetic products, and can be used as source for biofuel or biodiesel. The palm oil market has grown significantly in the last years, reaching an important quantity of 62.8 million metric tons (mt) in 2015 [1]. Indonesia and Malaysia dominate the global palm oil production, accounting for around 85 to 90 percent of the total production. Colombia, as one of the countries included in the remaining producers, has shown a positive growth in recent years, with a significant production volume distributed in crude oils, refined oils, margarines, and hydrogenated mixtures.

In 2017, Colombian crude palm oil production totaled 1.62 millions of tons, with a positive variation (41.9%) compared to 2016. This data shows evidence of a consolidated growth trend, which can be observed in Fig. 1.

This growth is also visible in the area dedicated to cultivate palm oil. According to the 2017 Statistical Yearbook, the Colombian planted area in 2016 was 512,076 ha, with an increase of 2.6% compared to 2015. Out of this total, 113,029 ha were in development and 339,048 were already in the production phase [1].

© Springer Nature Switzerland AG 2020
O. Gervasi et al. (Eds.): ICCSA 2020, LNCS 12249, pp. 232–245, 2020.
https://doi.org/10.1007/978-3-030-58799-4_17

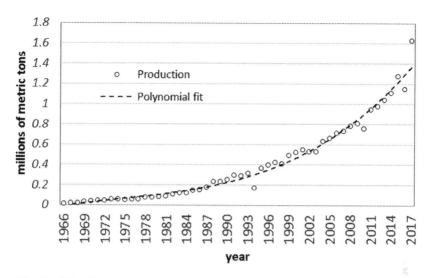

Fig. 1. Colombia palm oil production. Source: Colombian Ministry of Agriculture.

FEDEPALMA (the Colombian National Federation of Palm Oil Cultivators) divides Colombian palm oil planted area in 4 geographical zones, each one with the features described in Table 1.

Table 1. Planted areas with palm oil in Colombia.

Zone	Planted area (hectares)	Municipalities
East	206,559	36
North	124,948	68
Central	161,623	45
South-West	18,946	3

During 2016, all areas presented a remarkable performance in palm oil production, principally the Eastern and Southwestern regions, with a positive annual variation of 3.3% and 5.3%, respectively. Regarding the regional share in domestic production, the Eastern region held the highest contribution with 43.5%, followed by the Central region with 27.8%, the Northern region at 26.6%, and the Southwestern region at 2.2%. In terms of extraction rate, the Eastern and Central regions were more efficient.

Although the gap between Colombian palm oil productivity and the productivity of the leading producers has been narrowing over the years, there are still significant differences. The Colombian palm oil agroindustry faces costs in the processes of cultivation, harvesting, production and marketing that can be higher than those experienced by the international competitors, due to the poor condition of the transportation infrastructure, suboptimal harvesting and transportation processes and underutilization of the installed capacity at the mills [2]. The production costs of crude palm oil

(ACP) is currently experiencing harvesting, transportation of fresh fruit bunches (FBB) and oil extraction with a participation of 16%, 5% and 21%, respectively [3]. Even though 5% may seem insignificant, these transportation costs are relevant to compete in an international market with countries like Indonesia and Malaysia; especially because these countries have proven their superiority in terms of quality and cost efficiency. For this reason, research efforts are required in order to develop strategies for taking advantage of the potential of oil palm crops in Colombia.

This project aims at developing a simulation model approach to evaluate and optimize the unloading process at the mills, by the comparative assessment of the performance of alternative strategies that can be implemented to adjust operations. We describe our methodology by an application case to a mill located in the Northern zone, department of Magdalena. Palm oil industry is one of the economy drivers for that region, and it is considered as a stable source of employment. Currently, there are 17 municipalities with palm oil plantations and eight active palm oil mills located in the four municipalities of Aracataca, Ciénaga, El Retén and Zona Bananera. The presence of many facilities causes a high level of competition between companies since farmers have many options to sell their crops. In term of price, there are certain rules for healthy competition, so the key to retain the suppliers lies on providing them additional benefits. One of the aspects that suppliers value the most is the response time, both the time required for transporting the crop from the farm to the mill, and the unloading time at the mill. In particular, it is evident that weaknesses exist in the unloading process at the mill, where truck can get stuck for a long time inside the facility waiting for been unloaded. This affects the return time to the field, and therefore it will slow the supply chain and increase the transportation cost for suppliers. These undesirable situations lead to supplier dissatisfaction. With our simulation approach, we intend to analyze the waiting time at the mill, both the actual system and for a set of alternative scenarios, seeking to identify cost-effective options for process performance improvement.

The convenience of using the presented simulation approach lies in the fact that it can straightforwardly be replicated to other palm oil mills due to the similarity of the operation mills, with unloading and temporary storage processes taking place to ensure the continuity of extraction oil processes. Additionally, our simulation requires specific data that is available in all these companies, mainly because they are used for billing aspects, and for calculating statistics reported to FEDEPALMA. Thus, collecting the necessary input data to feed simulation models is not a concern. Obviously, some parameters of the simulation will depend on the specific aspects of the supply chain, particularly the unloading rates which must be determined according to the type of vehicles used, rate of palm fruit processing, and the amount of additional resources such as personnel available for manual unloading.

We built our discrete-event simulation models with SIMIO® [4]. A simulation package that facilitates to represent a system from a facility design perspective, and that provides visualization features that makes it easy for decision-makers to understand the model and the proposed improvement alternatives. With SIMIO®, we built a valid model of the processes of interest, and for the considered case study, we evaluate a set of improvement alternative that would allow achieving an estimated reduction of the average time in system by at least 23%. Similarly, the maximum time in system can be reduced at least by 35%.

The paper is organized as follow: first, Sect. 2 briefly describes the literature review on the use of simulation applied to agricultural sectors. In Sect. 3, we describe the palm oil supply process, the performance indicators and the available data. We present details of the simulation model in Sect. 4. In Sect. 5, we detail the result of the evaluation of alternative scenarios and the comparison with the actual configuration. Finally, Sect. 6 offers conclusions and identifies directions for future work.

2 Literature Review

Supply chains performance and logistics processes have been widely studied using different operations research approaches such as mathematical programming models, inventory models, and queuing theory. A comprehensive literature review is presented in [5], and further applications are described in [6–10]. In [11] the authors present a review of the most relevant research works that used mathematical and computational tools for optimizing sugarcane harvesting processes.

Simulation models have been extensively applied for analyzing manufacturing process [12]. For this, a computerized model of a system is created in order to obtain a comprehensive understanding of its behavior for a given set of operational conditions [13]. One clear advantage of simulation over other approaches is that it can predict the performance of a system before actually implementing it, and it allows anticipating the impact of changes before they are applied to the system.

In the context of agro industrial systems, simulation has been used to study mainly manufacturing and logistic processes. A great deal of attention has been paid to sugarcane production. For instance, several simulation studies dealt with the sugarcane production chain, such as [14], which use a detailed simulation model to analyze the receiving system of sugarcane and the unloading processes, without considering field operations. Likewise, [15] study the receiving system and the rate of arrival of the sugarcane from the field to the mill, but do not consider operations in the field and the return of the trucks to the front of the cane harvest. Research conducted by [16] shows how discrete-event simulation allows to identify the impact of different types of transport practices on the overall sugarcane supply system. Later, [17] propose a simulation model to study harvest operations, transportation and unloading at the mill as a whole. The analysis allowed authors to quantify the waiting times of the raw material from harvesting to unloading within the mill and the penalty for poor quality. [18] propose a simulation-based decision support system to analyze the operational performance of Colombian sugarcane supply chain. Recently, [19] use discrete-event simulation to model cutting, loading and transportation processes for a sugarcane plantation located at São Paulo region. Authors use the model to evaluate vehicles allocation and the performance of other logistics operation, considering changes due to climatic variables and vehicle degradation.

With regard to palm oil agroindustry, the literature that proposes using simulation techniques is less abundant, and a review can be found in [20]. A relevant work for the purposes of the study we present in this paper is the one by [12], which uses a discrete-event simulation model to study oil production system in two phases: replicating the actual fresh fruit bunches (FFB) processing system and developing an improved one.

The studies that address the improvement of Colombian palm oil industry have been more commonly using optimization models, sometimes in combination with simulation approaches. For instance, a research presented by [21] evaluated the concept of reverse logistic flows in the palm oil supply chain through the integration of a mathematical model with simulation. The authors demonstrated that integrating practices associated with the implementation of all forward flows and all possible reverse flows can provide significant benefits over the implementation of forward flows only. [22] proposed two mixed-integer programming models aiming to optimize vehicles allocation for transportation of FFB between internal collection centers, external locations and mill. [23] present a description and value of the relationship between echelons of Colombian oil palm supply chain. [24] defined mathematical models to achieve an optimal configuration of a closed-loop palm oil supply chain. Later, [25] developed a MINLP model for planning harvesting and product extraction in the oil palm supply chain at tactical and operational level. Additionally, some authors have used simulation techniques to evaluate palm oil refining process and biodiesel production process from crude palm oil [26–28].

Despite these efforts, little has been done to evaluate logistics operations of palm oil supply chain, such as transportation and unloading process, to the best of our knowledge. Hence, the purpose of this study is to provide a contribution to fill this gap in the existing literature, by proposing a simulation model that captures important aspects of the fresh fruit bunches unloading processes at Colombian palm oil mills, taking as a case study a mill located at the department of Magdalena. Our objective is to study by simulation the current unloading process and to comparatively evaluate possible improvements.

3   Description of Supply Logistics Processes of Palm Oil Mills

Time and quality are two important variables for FFB supply process because they affect the characteristics of crude palm oil extracted. Hence, mills are located near to the plantations, ensuring that once the raw material are harvested, they are processed within a reasonable period of time. Harvesting may occur every day through the year; however, the quantity of FFB harvested tends to be higher during dates close to mid and end of month. This practice makes transportation planning more complex since companies not only have to deal with randomness, but also with the uncertainty in harvesting scheduling.

Supply logistics processes and the type of vehicles used for transportation may vary among companies in the sector. Some companies have their own vehicles, mainly with automatic unloading, for collecting FFB from plantations. This allows a better control of vehicles arrivals to the mill. In other cases, transportation is done by outsourcing, and farmers decide when to send the fruit to the mill. Since different type of vehicles need to be served, companies must hire workers for the manual unloading of non-automatic vehicles.

Despite the above, both basic operations of FFB supply system inside mills and the extraction process are mostly the same for all companies. The manufacturing process of the FFB includes the logistic, crude palm oil, and palm kernel departments. The logistic department is in charge of planning the resources required for receiving the fruit until the FFB are placed into the wagons for sterilization. The crude palm oil processing at the mill is shown in Fig. 2. Upon reception, the FBB are weighted and unloaded to start the oil extraction phases. First, fruits are separated from the bunches and the spikelet (threshing), then squeezed to release the palm oil in the fruits (digestion), then pressed to extract the oil from the digested mash, and the obtained product clarified and dried to remove solid and liquid impurities. The final products are then stored and finally dispatched to customers

In the following, we shall focus on the first three steps of the process described in Fig. 2.

Fig. 2. The crude palm oil process flow.

3.1 Description of the FFB Reception, Weighthing and Unloading Processes

The process begins with the arrival of the vehicle. We identified three different types of vehicles used to transport the fruit: trucks, automatic trucks (trucks with a hydraulic system that eliminates the need to use workers for unloading the cargo) and small wagons towed by tractors. When a vehicle arrives, a ticket number is assigned, and this number is used as an ID to record all information required for the company such as supplier name, place of origin, weight, and type of vehicle, arrival time, and exit time. In order to do that, the drivers must present a document from their respective farm called *reference*. This document includes the code of the place of origin and the number of bunches of fruit. The security guard receives the *reference*, writes down the number of vehicles in the surveillance book and gives the ticket. Once the vehicle enters to the mill, it goes to the weighing area. After the initial weighing, the vehicle goes to the waiting queue to be unloaded. At this point, an employee completes an application form with the arrival time, the number of vehicles, the time of weighing, and the supplier name.

Later, an employee sends a vehicle to the platform for unloading the fruit. When a vehicle goes to the platform, the unloading process begins. First, the vehicle is parked in one of the available places located above the hopper. If the vehicle is an automatic truck, the process takes a little time and no workers are required. Otherwise, the workers will manually remove the fruit using hooks. When the vehicle is empty, it leaves the platform and goes to the weighing area. This activity is necessary in order to calculate the net weight of the cargo.

Each hour, a specific amount of fruit is removed from the hopper by means of a mechanical system, which allows the fruit to fall into the wagons due to the inclination of the hopper. The wagons are then moved to the thresher and sterilizer, starting the oil extraction process. Figure 3 illustrates the events in the palm oil mill, which were described above.

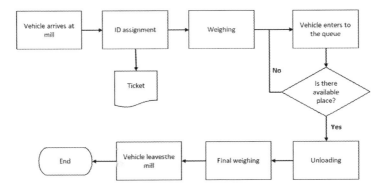

Fig. 3. Flowchart of FFB unloading at mills.

An important aspect of the system lies in the possible truck lines that can be formed in the mill due to the unloading process described above. This situation may occur because of the delay of vehicles during the unloading of FFB, since there are several farmers sending the fruit to the mill. Thus, the process can force a vehicle to be waiting for hours and delay the return of the vehicle to the farm, affecting the functioning of the system as a whole, especially if the vehicle must visit the mill again. This problem may endanger suppliers satisfaction and the relationship between them and companies.

3.2   Data and Performance Indicators

Palm oil companies collect data through the weighing information system using the ID assigned to the vehicles. Since arrival times and net weight transported exhibit a random behavior, this data can be used to estimate the statistical distribution for the interarrival time of vehicles and the load transported for each type of vehicle. It is important to highlight that if a company only uses one type of vehicle, minor changes can be done in the model building to adjust to the actual operations at the mill.

Given that companies want to increase the mill extraction capacity, a stable supply of FFB is required, and it is crucial to maintain a good relationship with suppliers. Because of that, companies are interested in reducing the time that a vehicle remains into the mill. Therefore, we use this variable to analyze the current process and compare it with an alternative strategy based on priority, in order to determinate if it is possible to reduce the time in system of vehicles.

4 Simulation of Palm Oil Supply Logistic Processes at Mills

4.1 Model Building and Assumptions

In this paper, we present a simulation model that combines continuous/discrete modeling in order to increase the results accuracy. Additionally, the methodology proposed by [29] was followed, which includes the next steps: problem formulation; project planning; formulation of the conceptual model; data collection; translation of the model; verification and validation; experimental design; experimentation; setting project interpretation and statistical analysis of the results; comparison and identification of the best solutions; documentation and presentation of results.

The conceptual model of the case study was translated into software SIMIO[®,] using both Standard and Flow Library. The model was built as general as possible, so it considers the main characteristics of Colombian palm oil companies. The assumptions considered into the modelling step are the followings: a) Mills operate 24 h; however, they establish a time window for FFB reception. b) Three types of vehicles are considered into the model (wagons, automatic trucks, and non-automatic trucks). c) Mill is not interrupted and the rate of FFB consumptions is known and constant during the day. d) There are enough wagons for storing FFB removed from hoppers and therefore, they would not be considered as a constraint.

The model built in SIMIO[®] is shown in Fig. 4. As can be seen, the vehicles arrivals are represented by two different sources with interarrival time, one source for trucks and other one for small wagons. The sources create container entities, which content kilograms of FFB. Entities get into a server named *WeighingMach*, and later vehicles are transferred to a station. Entities remain into the station until they can pass to the next stage. We used logic processes to transfer entities from the Weighing area to the station and from to station to the platform (unloading places). In order to properly represent the system, we used an *Emptier object* to model the process that empties the contents of the vehicle (container entity). This object allows us to model the continuous filling of the hopper, while the vehicle is monitored through a Tally statistic that collects the time between the arrival and the unloading ending. After being served, the vehicle goes to the Weighing area for final weighing, and leaves the mill. On the other hand, we used a *Tank object* to model the hopper with specifics weight flow rate for

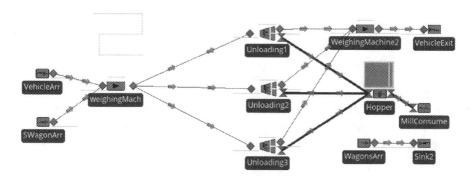

Fig. 4. Simulation model of the current unloading process.

input and output flows. Given that the hopper output flow is not continuous, we implemented a complementary model to enable hourly the regulator flow, so a specific amount of FBB is removed from the hopper.

4.2   Verification and Validation of the Simulation Model

Verification and validation are the most difficult but important step in simulation modeling. While verification is concerned with building the model correctly, validation is concerned with building the correct model [12]. Since the technique of distribution fitting was applied to select the appropriate probability distribution for a series of data of interest such as interarrival time and kilograms of FBB carried by vehicles, we decided to verify the simulation model by comparing the probability distribution of real data with the total number of arrival and the total kilograms of fruit received per day, obtained through 12 replications of the model with a run length of 24 h.

The results indicate that both series of number of arrivals data (real system and simulation model) can be fitted by a normal distribution and the means are not statistically different from each other. A similar outcome was obtained for the variable kilograms of FFB carried by vehicles. These results indicate that the model was appropriately built.

For the validation process, we compared the probability distribution of real time in system with the Tally statistic recorded for all the entities, running the model during 12 days. We established that there are enough similarities between both series of data and hence, we can affirm that the model was correctly built. Once the verification and validation processes were concluded, we analyzed the modification of the current unloading procedure proposed by the company.

The analysis of the results was conducted only after the simulation model was completely verified and validated and assured that the assumptions and simplifications embraced were accurately implemented in the computational model.

4.3   Case Study

We conduct an empirical case study using real data from a Colombian palm oil company. The selected palm oil company is an organization with more than forty years of experience in the market. The company has administrative headquarters in the city of Santa Marta and has a palm oil mill in the municipality of Zona Bananera, which processes the FFB provided by external suppliers and strategic alliances that represent about 63% of the total suppliers. The remaining amount of FFB is supplied by founding partners. The associative structure of inclusive business models known as *Strategic Productive Alliances* integrates small and medium farmers organized in associations, and it has allowed the noteworthy growth of this agroindustrial sector in Colombia, showed in the increase of the number of municipalities with established crops. This type of associative strategy is used by this company and it has enabled it to increase the supply of the fruit, which has a significant effect on the transport and unloading requirements.

The mill has a processing capacity of 34 tons of FFB per hour and has three available places to unload the FFB cargo. It operates 24 h per day; however, FFB load are only received within a time window of 15 h, from 6:00 to 21:00, every day.

To quantify the parameters and the distribution of random variables, the company provided access to the log and the data stored on weighing system. We use information of 12 days of operations in order to estimate. The considered time period corresponds to the arrival of 644 vehicles, including small wagons, trucks and automatic trucks. The vehicles transported a total amount of 3.2 million of kilograms of FFB to be processed at the mill.

Given that different vehicles arrive to the mill, we decided to test if there were statistical differences between trucks and automatic trucks interarrival times. For this, we used a mean comparison test for unpaired data with unequal variances using Welch's. The obtained result indicates that the means are not statistically different from each other (See APPENDIX C). Furthermore, the small wagon arrival was modeled apart from the others type of vehicles due to the number of entities per arrival is a random variable. Table 2 summarizes the parameters and random variables values used as input for the simulation model.

Table 2. Input parameters and distributions of random variables.

Parameter/random variable	Value
Interarrival time for trucks and automatic trucks	Random.Exponential (40)
Entities per arrival	1
Interarrival time for small wagons	Random.Exponential (120)
Entities per arrival	Random.Triangular (2, 4, 7)
Weighing server processing time	3 min
Automatic trucks weight	Random.Weibull (5.43286,12047.72)
Trucks weight	Random.Weibull (3.76144,16158.09)
Small wagons weight	Random.Weibull (9.9244, 7721.85)
Unloading flow rate for automatic truck	4500 kg per minute
Unloading flow rate for other vehicles	500 kg per minute
Hopper output flow rate	34000 kg per minute, hourly

5   Results and Discussion

Our objective is to investigate the potential reduction in the time of the system that could be obtained with a simple modification of the currently unloading procedure. The policy conceived by the company consists of enabling another unloading place for those vehicles that have been parked in the station for over 2 h (Scenario 1). For this, the model was slightly changed to incorporate the new policy, as illustrated in Fig. 5. First, we added an object that empties the contents of the vehicle as we did in the initial model; however, the content is stored in wagons until there is available capacity in hopper. This aspect was not modeled since it is possible to place the FFB in the first hours of each day. The output flow rate for automatic trucks is higher due to the vehicles can unload the box instead of the fruit. Nonetheless, the unloading rate for manual unloading is the same. Second, we modify the logic process used to control the transfer of entities to the emptier objects. The model was prepared to run for 360 h, using an experiment of 10 replications and 95% of confidence level.

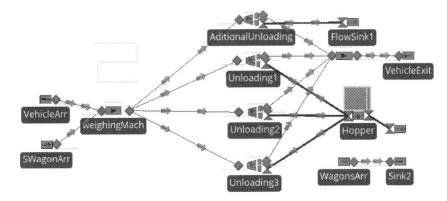

Fig. 5. Simulation model of the scenarios.

As a result of our analysis, we decided to propose a different scenario (Scenario 2) in which the additional emptier could be used as follows: If an automatic truck arrives to the mill and the unloading places are busy, the vehicle could be transferred to the additional unloading. If a small wagon or truck has been parked in the station for over than 2 h, it could be transferred to the additional unloading. An additional scenario is evaluated (Scenario 3), in which the maximum waiting time in station is 1 h. As in the previous scenario, the model was prepared to run for 360 h, using an experiment of 10 replications and 95% of confidence level.

Table 3 provides summary statistics for the variable time in system in each scenario as well as the initial model.

Table 3. Comparision of scenarios.

Model	Measure	Average	Half width
Current procedure	Average time in system	3.0970	0.3496
	Maximum time in system	14.6039	2.3236
Scenario 1	Average time in system	2.3726	0.2154
	Maximum time in system	8.6103	0.9189
Scenario 2	Average time in system	2.2932	0.1873
	Maximum time in system	9.4616	1.0862
Scenario 3	Average time in system	1.7655	0.1333
	Maximum time in system	7.9568	1.1725

It is evident from the previous results that if the company decides to implement any scenario, the average time in system can be reduced at least by 23%. Similarly, the maximum time in the system can be reduced at least by 35%.

This shows that it is possible to decrease the delays in the system and, at the same time, improve the supplier satisfaction. It is expected that readers consider the cost-effectiveness ratio, and for this it is important to clearly explain the following aspects: the unloading process for automatic truck do not require workers, so no additional costs are incurred. The company hires a workers' cooperative for unloading the FFB and it pays it for the total of kilograms moved plus a percentage per number of workers. Since the total of kilograms is the same, the company only needs to assume an additional cost of two workers. There is always a vehicle available for material movements at the mill, and this vehicle can transport the wagons of FFB from the new place to platform for unloading the fruit. In the light of the above, it is far more convenient and less expensive to change the current procedure.

6   Conclusions

This paper describes a simulation model used to analyze the current state and scenarios in the palm oil supply chain. Thus, the verified and validated model had been used to estimate the impact of operative changes in the waiting time for vehicles. In this research, the investigated improvement enhances expanding the number of unloading docks and exploring different priorities in the procedure. Comparison between the simulation of current and alternative systems indicates that the average time in system and the maximum time in system can be reduced at least by 23% and 35%, respectively. The most important finding is that the decrease in waiting time can be reached without a major investment. Due to the similarity between palm oil mills located in the department of Magdalena, this simulation model can be applied to other companies to evaluate improvement alternatives. For future work, a model that includes harvest operations, transportation and unloading at the mill should be carried out. Moreover, the extraction process and its breakdowns should be studied since this causes delays in the system.

References

1. FEDEPALMA: Anuario estadístico 2017: Principales cifras de la agroindustria de la palma de aceite en Colombia 2012–2016, Bogotá D.C. (2017)
2. Superintendencia de Industria y Comercio: Agroindustria de la Palma de Aceite: Diagnóstico de Libre Competencia, pp. 1–9 (2011)
3. Duarte-Guterman Cía. Ltda.: Actualización de Costos de Producción de Aceite de Palma Informe final, Bogotá, Colombia (2009)
4. Joines, J.A., Roberts, S.D.: Simulation Modeling with SIMIO: A Workbook. Simio LLC, Pittsburgh (2015)
5. Beamon, B.M.: Supply chain design and analysis: models and methods. Int. J. Prod. Econ. 55(3), 281–294 (1998). https://doi.org/10.1016/S0925-5273(98)00079-6

6. Van Der Vorst, J.G.A.J., Beulens, A.J.M., Van Beek, P.: Modelling and simulating multi-echelon food systems. Eur. J. Oper. Res. **122**(2), 354–366 (2000). https://doi.org/10.1016/S0377-2217(99)00238-6

7. Hung, W.Y., Samsatli, N.J., Shah, N.: Object-oriented dynamic supply-chain modelling incorporated with production scheduling. Eur. J. Oper. Res. **169**(3), 1064–1076 (2006). https://doi.org/10.1016/j.ejor.2005.02.009

8. Lopez Milan, E., Miquel Fernandez, S., Pla Aragones, L.M.: Sugar cane transportation in Cuba, a case study. Eur. J. Oper. Res. **174**(1), 374–386 (2006). https://doi.org/10.1016/j.ejor.2005.01.028

9. Grunow, M., Günther, H.O., Westinner, R.: Supply optimization for the production of raw sugar. Int. J. Prod. Econ. **110**(1–2), 224–239 (2007). https://doi.org/10.1016/j.ijpe.2007.02.019

10. López-Milán, E., Plà-Aragonés, Lluis M.: Optimization of the supply chain management of sugarcane in Cuba. In: Plà-Aragonés, Lluis M. (ed.) Handbook of Operations Research in Agriculture and the Agri-Food Industry. ISORMS, vol. 224, pp. 107–127. Springer, New York (2015). https://doi.org/10.1007/978-1-4939-2483-7_5

11. Doriguel, F., Crusciol, C.A.C., de O Florentino, H.: Mathematical optimization models in the sugarcane harvesting process. Sugarcane Technol. Res. (2018). https://doi.org/10.5772/intechopen.71530

12. Mohd-Lair, N.-A., Chan, C., Chua, B., Liew, W.Y.H.: Application of simulation in process improvement of palm oil mill FFB production: a case study. In: IEEE 2012 International Conference Statistics Science Business, Engineering, pp. 1–5 (2012). https://doi.org/10.1109/icssbe.2012.6396617

13. Kelton, W.D., Sadowski, R.P., Sturrock, D.T.: Simulation with Arena. McGrawHill Inc., New York (2007)

14. Iannoni, A.P., Morabito, R.: Análise do sistema logístico de recepção de cana-de-açúcar: um estudo de caso utilizando simulação discreta. Gestão Produção **9**(2), 107–128 (2002). https://doi.org/10.1590/S0104-530X2002000200002

15. Prichanont, K., Prichanont, S., Buransri, N.: Improvement guidelines for sugar cane delivery systems. In: 35th International Conference on Computers and Industrial Engineering, pp. 1585–1590 (2005)

16. McDonald, B.C., Dube, E., Bezuidenhout, C.N.: TranSwarm: a sugarcane transport simulation model based on behavioural logistics. In: Proceedings of South African Sugar Technolohy Association, pp. 434–438 (2008)

17. de Assis Rangel, J.J., Cunha, A.P., de Azevedo, L.R., Vianna, D.S.: A simulation model to evaluate sugarcane supply systems. In: Simulation Conference (WSC), Proceedings of the 2010 Winter, pp. 2114–2125 (2010)

18. Bocanegra-Herrera, C.C., Vidal, C.J.: Development of a simulation model as a decision support system for sugarcane supply [Desarrollo de un modelo de simulación como un sistema de soporte de decisiones para el abastecimiento de caña de azúcar]. Dyna **83**(198), 181–187 (2016). https://doi.org/10.15446/dyna.v83n198.52719

19. Das, M., Tomazela, G.J.M., Leandro, C.R., Oliveira, M.C., de Campos, F.C.: Simulação de eventos discretos na avaliação de um processo de corte, carregamento e transporte de cana-de- açúcar. Espacios, vol. 38, no. 21, p. 41 (2017)

20. Lestari, F., Ismail, K., Bakar, A., Hamid, A., Supriyanto, E., Sutupo, W.: Simulation of refinery-supplier relationship. In: International MultiConference Engineers Computer Scientists, vol. II, pp. 1–5 (2016)

21. Alfonso, E., Ferrucho, D., Roldán, A., Vargas, M., González, F.: Scenario analysis for reverse logistics implementation case study of the palm oil supply chain. In: Proceedings of 2009 Winter Simulation Conference, pp. 2310–2319 (2009)

22. Adarme, W., Fontilla, C., Arango, M.D.: Modelos logisticos para la optimizacion del transporte de racimos de fruto de palma de aceite en Colombia. Cienc. e Ing. Neogranadina **21**, 89–114 (2011)
23. García-Cáceres, R., Núñez-Moreno, A., Ramírez-Ortiz, T., Jaimes-Suárez, S.: Caracterización de la Fase Upstream de la Cadena de Valor y Abastecimiento de la Agroindustria de la Palma de Aceite en Colombia. Dyna **179**, 79–89 (2013)
24. Alfonso-Lizarazo, E.H., Montoya-Torres, J.R., Gutiérrez-Franco, E.: Modelling reverse logistics process in the agro-industrial sector: the case of the palm oil supply chain. Appl. Math. Model. **37**, 9652–9654 (2013). https://doi.org/10.1016/j.apm.2013.05.015
25. García-Cáceres, R., Martínez-Avella, M.E., Palacios-Gómez, F.: Tactical optimization of the oil palm agribusiness supply chain. Appl. Math. Model. **39**(20), 6375–6395 (2015). https://doi.org/10.1016/j.apm.2015.01.031
26. Ocampo, M.B., Gutiérrez, L.F., Sánchez, Ó.J.: Simulation of palm oil refining process. In: Vitae, vol. 23 (2016)
27. Velosa, F., Gomez, J.M.: Simulation and optimization of the process used in Colombia for the production of biodiesel from palm oil: a kinetic analysis and an economical approach. Technology **1**, 2 (2008)
28. Zapata, C., Martínez, I., Arenas Castiblanco, E., Henao Uribe, C.: Producción De Biodiesel a Partir De Aceite Crudo De Palma: Diseño Y Simulación De Dos Procesos Continuos. Dyna **151**, 71–82 (2007)
29. Banks, J., Carson, J.S., Nelson, B.L., Nicol, D.M.: Discrete Event System Simulation. New Jersey Prentice Hall Inc., Upper Saddle River (2000)

OpenFOAM Numerical Simulations with Different Lid Driven Cavity Shapes

César A. Cárdenas R.[1(✉)], Carlos Andrés Collazos Morales[1], Juan P. Ospina[1],
Joaquín F. Sánchez[1], Jelibeth Racedo-Gutiérrez[1], Paola Ariza-Colpas[2],
Emiro De-la-Hoz-Franco[2], and Ramón E. R. González[3]

[1] Universidad Manuela Beltrán, Vicerectoría de Investigaciones,
Bogotá, D.C., Colombia
cesar.cardenas@docentes.umb.edu.co
[2] Universidad de la Costa, Barranquilla, Colombia
[3] Departamento de Física, Universidad Federal de Pernambuco, Recife, Brazil

Abstract. The finite volume method have been developed to solve the
Navier-Stokes equations with primitive variables and non dimensional
form. This work examine the classical benchmark problem of the lid-
driven cavity at a different Reynolds range (Re = 10,100,400, 1000, 2000,
3200) and several cavity geometries. The cavity configurations include
square cavity, skewed cavity, trapezoidal cavity and arcshaped cavity.
The flow is assumed laminar and solved in a uniform mesh. A CFD tool
with its solvers (*icoFoam*) will be used for this study.

Keywords: Cavity · OpenFOAM · icoFoam · Vorticity · Lid-driven
cavities

1 Introduction

Cavity flows have been of great importance in many industrial processes applications. These type of flows such as lid-driven cavity has served as a model for
testing and validation. They provide a model to understand more complex flows
with closed recirculation regions. These types of flows contain a wide variety
ranging from rotation near the recirculation region to a strong extent near the
edges of the top cover. Generally, numerical simulations of 2D cavity flows are
made at different Reynolds numbers. The incompressible laminar flow in square,
trapezoidal and skewed cavities whose top wall moves at a uniform speed of 1
m/s are studied in this work. The Reynolds numbers taken range from 10 to
3200. Also, different mesh sizes are determined for all configurations (41×41,
61×61, 81×81, 129×129)[1,5,7].

2 OpenFOAM Solvers

OpenFOAM (Open Field Operation and Manipulation) CFD Toolbox is free tool
produced by a commercial company. It is a CFD package written in C++. It is

© Springer Nature Switzerland AG 2020
O. Gervasi et al. (Eds.): ICCSA 2020, LNCS 12249, pp. 246–260, 2020.
https://doi.org/10.1007/978-3-030-58799-4_18

combined with appropriate implementations of numerical methods and even discretization of partial differential equations and resulting linear systems solutions. The discretization of governing equations in OpenFOAM is based on the finite volume method (FVM). It is formulated with collocated arrangements, pressure and speed results by segregated methods. SIMPLE (Semi-implicit Method for Pressure Linked Equations) or PISO (Pressure Implicit Splitting of Operators), are the most used algorithms for pressure-speed coupling. This software also offers a variety of schemes of interpolation, solvers and preconditioners for the algebraic equation system. To create a OpenFOAM case, three files are required: 0, or initialization, system and constant. In the 0 directory, the initial condition properties of the fluid are established. It also contains two subdirectories P and U, which are the velocity fields and pressure. In the constant directory there are two subdirectories: Polymesh and transport properties. Finally, in the system directory, the methods of discretization and procedure of solution are set. OpenFOAM always operates in 3D and all geometries are generated in 3D. However, for the cavity case, it is possible to instruct it to solve the 2D case specifying a empty boundary condition. Within the system directory there is a subdirectory called fvSchemes which defines the discretization schemes. For this case, Euler is the one used for temporary discretization and Gauss Liner for convection. The fvSolution defines the solution procedure. For the linear equations system, P is defined as $PPCG$, DIC is a preconditioner in which the conjugate gradient PCG is employed as solver. For the system of U linear equations, the following parameters are defined: $DILU$ as a preconditioner with a conjugated gradient $PBiCG$ as solver. Some research works present interesting simulations in which different mesh sizes and different Reynolds numbers are used. Also, the appearance or emergence of vortices is noticed. Several studies have shown that the boundary condition of vorticity has a significant influence on the simulation stability [2,8].

3   Procedure for the Solution of Different Cavities in OpenFOAM

For the rectangular cavity case study, all the border conditions are walls. The upper wall moves in the x direction at a constant speed while the others are stationary. The flow is assumed as laminar which is resolved in a uniform mesh by using the icoFoam approach. This solver is commomly applied for laminar and incompressible flow.

3.1   Governing Equations and Discretization

The moment and continuity equations are incorporated into the mathematical model for the 2D flow cavity problem. It is noted that the Navier-Stokes equation is the moment equation for incompressible flow in 2D [3,4,6,9].

Continuity Equation

$$\frac{\partial u}{\partial x} + \frac{\partial v}{\partial y} = 0 \tag{1}$$

Momemtum Equation

$$\frac{\partial u}{\partial t} + u\frac{\partial u}{\partial x} + v\frac{\partial u}{\partial y} = -\frac{1}{\rho}\frac{\partial p}{\partial x} + \frac{\mu}{\rho}\left(\frac{\partial^2 u}{\partial x^2} + \frac{\partial^2 u}{\partial y^2}\right)$$
$$\frac{\partial u}{\partial t} + u\frac{\partial v}{\partial x} + v\frac{\partial v}{\partial y} = -\frac{1}{\rho}\frac{\partial p}{\partial y} + \frac{\mu}{\rho}\left(\frac{\partial^2 u}{\partial x^2} + \frac{\partial^2 u}{\partial y^2}\right) \tag{2}$$

Where u and v are the velocities in the x and y directions. t is time ρ, is the density and μ is the viscosity. A steady 2D flow is considered for this work. The top border (lid) moves at a certain velocity while the other walls remain fixed. For all the applied geometries, the top wall always moves. Cartesian coordinates are utilized (x, y) and its origin is located in the left lower corner and indicates the unit vectors in the x and y directions. With the non-dimensional variables given as:

$$u = \frac{u'}{U_{lid}}$$

$$v = \frac{v'}{U_{lid}}$$

$$t = \frac{t'U_{lid}}{L_{lid}}$$

$$x = \frac{x'}{L_{lid}}$$

$$y = \frac{y'}{L_{lid}} \tag{3}$$

$$\rho = \frac{\rho'}{\rho_{ref}}$$

$$\mu = \frac{\mu'}{\mu_{ref}}$$

$$p = \frac{p' - p_{ref}}{\rho_{ref}U_{lid}^2}$$

$$Re = \frac{\rho'U_{lid}L_{lid}}{\mu'}$$

$$\frac{\partial u}{\partial t} = \frac{u_{i,j}^{n+1} - u_{i,j}^n}{\Delta t} + O(\Delta t)$$

$$\frac{\partial u^2}{\partial t} = \frac{u_e^2 - u_w^2}{\Delta x_s} + O(\Delta x_s^2)$$

$$u_e = \frac{1}{2}(u_{i,j} + u_{i+1,j})$$

$$u_e = \frac{1}{2}(u_{i-1,j} + u_{i,j})$$

$$\frac{\partial uv}{\partial y} = \frac{(uv)_n - uv_s}{\Delta y_v)} + O(\Delta y^2)(u_n) = \frac{1}{2}(u_{i,j} + u_{i,j+1})$$

$$(u)_s = \frac{1}{2}(u_{i,j-1}) + (u_{i,j})$$

$$(v)_n = \frac{1}{2}(v_{i-1,j+1}) + (v_{i,j+1})$$

$$v_s = \frac{1}{2}(v_{i-1,j} + v_{i,j})$$

$$\frac{\partial p}{\partial x} = \frac{p_{i,j} - p_{i-1,j}}{\Delta x_s} + O(\Delta x_s^2)$$

$$\frac{\partial^2 u}{\partial x^2} = \frac{\frac{\partial u}{\partial x_e} - \frac{\partial u}{\partial x_w}}{\Delta x_s} + O(\Delta x_s)$$

$$\left(\frac{\partial u}{\partial x}\right)_e = \frac{u_{i+1,j} - u_{i,j}}{\Delta x_u} + O(\Delta x_u^2)$$

$$\left(\frac{\partial u}{\partial x}\right)_w = \frac{u_{i,j} - u_{i-1,j}}{\Delta x_u} + O(\Delta x_u^2)$$

$$\frac{\partial^2 u}{\partial y^2} = \frac{\frac{\partial u}{\partial y_n} - \frac{\partial u}{\partial y_s}}{\Delta x_s} + O(\Delta y_v^2)$$

$$\left(\frac{\partial u}{\partial y}\right)_n = \frac{u_{i,j+1} - u_{i,j}}{\Delta y_s} + O(\Delta y_s^2)$$

$$\left(\frac{\partial u}{\partial y}\right)_s = \frac{u_{j,i-1} - u_{i,j-1}}{\Delta y_s} + O(\Delta y_s^2)$$

$$\tag{4}$$

$$\frac{\partial \Omega}{\partial T} + U\frac{\partial \Omega}{\partial X} + V\frac{\partial \Omega}{\partial Y} = \frac{1}{R_e}\left(\frac{\partial^2 \Omega}{\partial X^2} + \frac{\partial^2 \Omega}{\partial Y^2}\right) \tag{5}$$

$$U = \frac{\partial \Psi}{\partial Y}$$

$$V = -\frac{\partial \Psi}{\partial X} \tag{6}$$

The domain is divided into different control volumes (CVs) [10]. A separated Cartesian mesh is used both horizontally and vertically. For each control volume (CV), the momentum and continuity equations are approximated by using some algebraic expressions. These involve the u, v, p values in the center of each control volume and its neighbors. The momentum equations can be represented as discretizations of finite volumes differences in which i and j are the cell indexes within a staggered mesh in x and y directions. Figure 1 shows a control volume P ans its neighbors S, E, N, and W. Likewise, the previous abbreviations will denote the position vectors of CVs centers as well. These three equations can be combined with what is known as stream function and the vorticity equation. Then, the transport vorticity equation (3) ant two more velocity equations (6) are obtained. ω is the voticity and Ψ is the stream function and are non-dimensional variables. The PISO method involves a predictor step and two correcting steps. It means that the velocity fields u and v do not satisfy continuity unless pressure is corrected. The p, u, v are assumed at the beginning. The following is a summary of the PISO algorithm steps:

- Step 1–3. The discretized momentum equations and the correcting equation of pressure are solved. Also, velocities and pressure are adjusted.
- Step 4. The second rectifying pressure equation is solved.
- Step 5. Pressure and speeds correction.
- Step 6. The other discretized transport equations are resolved.
- Converge? Yes → stop.
- No → Return to the starting point (Figs. 3, 4, 5, 7, 8, 9, 11, 12, 14, 15 and 17).

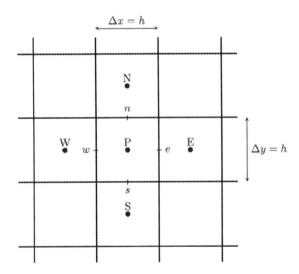

Fig. 1. Control volume - Taken from [10]

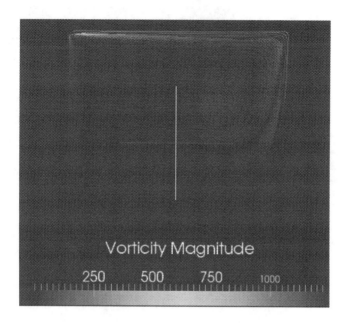

Fig. 2. Vorticity contours

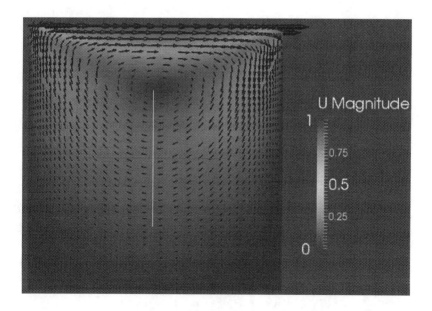

Fig. 3. Velocity ranges m/s

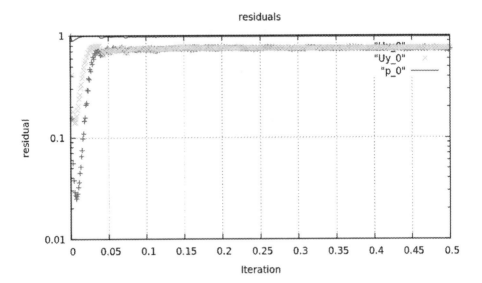

Fig. 4. Residulas vs Iterations - square case

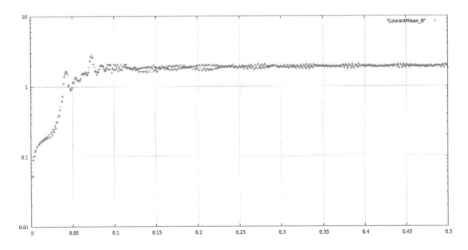

Fig. 5. Courant number mean - square case

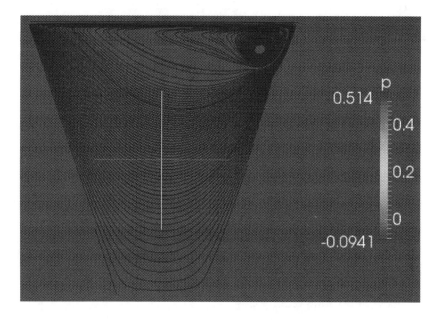

Fig. 6. Pressure contours

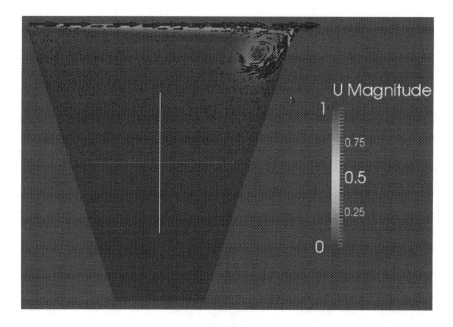

Fig. 7. Velocity ranges (m/s)

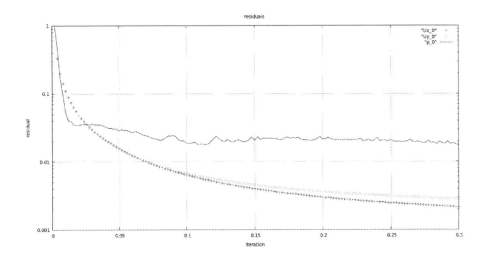

Fig. 8. Residuals vs Iterations - trapezoidal case

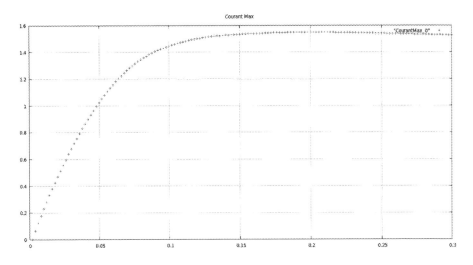

Fig. 9. Courant number mean - trapezoidal case

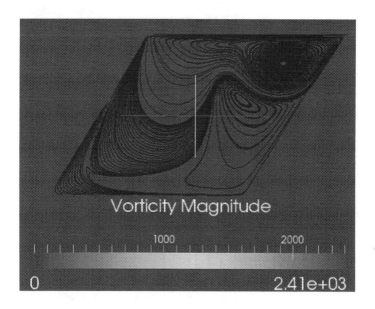

Fig. 10. Vorticity contours

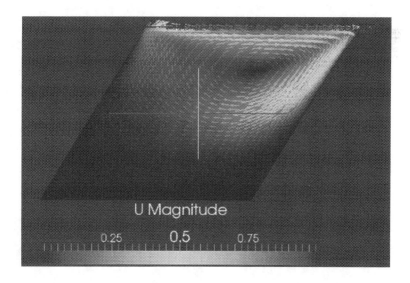

Fig. 11. Velocity ranges m/s

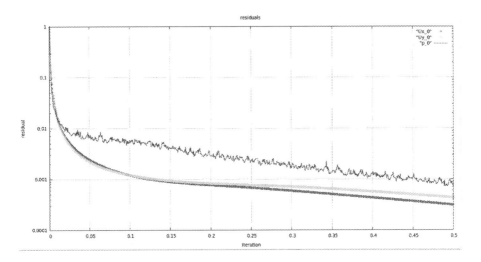

Fig. 12. Residuals vs Iterations - skewed case

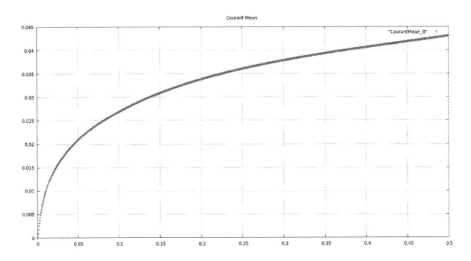

Fig. 13. Courant number mean - skewed case

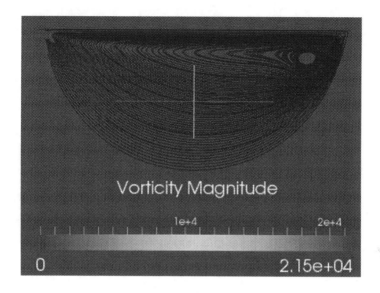

Fig. 14. Vorticity contours

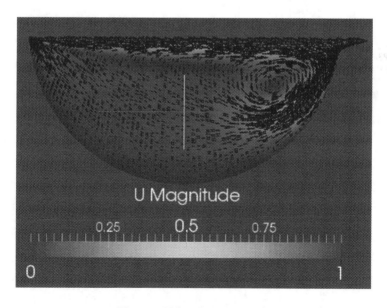

Fig. 15. Velocity ranges m/s

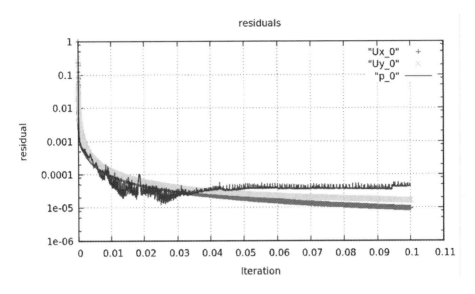

Fig. 16. Residual vs Iterations - arc-shaped case

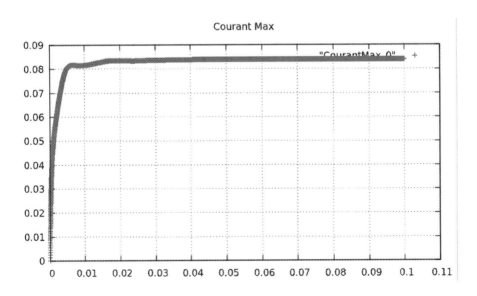

Fig. 17. Courant number mean - arc-shaped case

4 Results

Different geometries (square, skewed and trapezoidal) are applied to study the 2D cavity flow behaviour by using an openFOAM solver. Also, several values for parameters such as reynolds numbers, pressure and velocities are used. Hence, results are obtained for all these conditions to check the influence on the cavity vorticity The other figures present the simulation results. These include Courant numbers variation over time and iterations required for convergence. The results are discussed for the three proposed cases. All the plots are got through gnuplot.

1. Square cavity
   Variables are defined for U_x, U_y, P and Re. Other parameters are taken by default as set up in the CFD OpenFOAM solver (turbulence variables k, epsilon). Hence, in this type of geometry cavity, it is noticed in the residuals plot that P converges at a steady value of 1. Likewise, U_x and U_y converge close to 1. Regarding the Courant number, it is observed that is goes a bit above 1 as the time step also increases. However, it looks like convergence is not affected in any way since the solver can deal with relative large numbers. Also, it can be seen in Fig. 2 that vorticity is generated at the top of the cavity and shifts all over the cavity.
2. Trapezoidal cavity
   Similar to the square case, different Re, speed and pressure values are considered. An important matter noted is that the maximum Courant number is a bit larger than 1 when convergence is reached. With respect to the residuals, it can be seen that the simulation parameters converge at lower ranges. An important issue is that pressure is solved at higher numbers and it does not affect the solution as the solver can handle it. Figure 6 shows pressure p streamlines and some kind of vortex originated at the right top corner and its variations from top to center and all over the cavity.
3. Skewed cavity
   Normally this kind of cavity is inclined at $45°$ concerning the x axis. It is seen in Fig. 13 that the Courant number got is less than 1. Concerning the residuals, velocities U_x, U_y and P achieve small values for convergence. In addition, Fig. 10 depicts vorticity contour at the right top corner and moves to the whole cavity. Some vorticity is also created at the bottom, particularly on the left corner. Re numbers are taken at different values as well.
4. Arc-shaped cavity
   The simulation of this cavity geometry is also carried out by setting up different reynolds numbers R_e as well as velocities. In the residuals (Fig. 16), all the parameters converged at very small values. The Courant number reached is very small even smaller than 1 which facilitates an appropriate solution. Vorticity is obtained at the right top corner which shifts towards the center of the cavity attempting to get the left side and bottom.

5 Conclusions

The OpenFAOM tutorial for the 2D cavity lid driven cases establishes some parameters for simulation purposes of an incompressible and isothermal flow. In this work, a complete mathematical model is developed. It provides an overall insight of the partial differential equations to be resolved. Since different cavity configurations are taken, it can be seen that a solution is given for instance, for Courant ranges over 1. Certainly, the square and trapezoidal cases are the ones that get these values above to 1 to be solved. Likewise, it seems that no matter the velocity value, vorticity is always originated for all cavity shapes using just a uniform mesh. It can be inferred that fine grids or with better resolution may even improve the prediction of vortices. There are also more complex cases such as the arc-shaped type for which a solution is found as well. Hence, the *icofoam* solver is suitable or appropriate to give a solution of complex cavity flows. It would be possible to increase convergence and accuracy employing other solver like *pisofoam* or SIMPLE.

References

1. Akula, B., et al.: Partially-averaged navier-stokes (PANS) simulations of lid-driven cavity flow–part 1: comparison with URANS and LES. In: Progress in Hybrid RANS-LES Modelling, pp. 359–369. Springer (2015)
2. Chen, G., et al.: OpenFOAM for computational fluid dynamics. Not. AMS **61**(4), 354–363 (2014)
3. Farahani, M.S., Gokhale, M.Y., Bagheri, J.: Numerical simulation of creeping flow in Square Lid-Driven Cavity by using Open-Foam
4. Liu, Q., et al.: Instability and sensitivity analysis of flows using OpenFOAM®. Chin. J. Aeronaut. **29**(2), 316–325 (2016). https://doi.org/10.1016/j.cja.2016.02.012, ISSN: 1000–9361. http://www.sciencedirect.com/science/article/pii/S100093-6116300024
5. Marchi, C.H., Suero, R., Araki, L.K.: The lid-driven square cavity flow: numerical solution with a 1024 x 1024 grid. J. Brazilian Soc. Mech. Sci. Eng. **31**(3), 186–198 (2009)
6. Mercan, H., Atalik, K.: Flow structure for power-law fluids in lid-driven arc-shape cavities. Korea-Aust. Rheol. J. **23**(2), 71–80 (2011)
7. Razi, P., et al.: Partially-averaged Navier-Stokes (PANS) simulations of lid-driven cavity flow–part ii: ow structures. In: Progress in Hybrid RANS-LES Modelling, pp. 421–430. Springer (2015)
8. Sousa, R.G., et al.: Lid-driven cavity flow of viscoelastic liquids. J. Nonnewton. Fluid Mech. **234**, 129–138 (2016)
9. Jignesh, A., Thaker, P., Banerjee, B.J.: Numerical simulation of flow in lid-driven cavity using OpenFOAM. In: International Conference on Current Trends in Technology: NUiCONE-2011, Institute of Technology, Nirma University, Ahmedabad (2011)
10. Vaidehi, A.: Simple solver for driven cavity OW problem. Department of Mechanical Engineering, Purdue University, ASME (2010)

Mathematical Modelling and Identification of a Quadrotor

César A. Cárdenas R.[1]([✉]), Carlos Andrés Collazos Morales[1], Juan P. Ospina[1],
Joaquín F. Sánchez[1], Claudia Caro-Ruiz[1], Víctor Hugo Grisales[2],
Paola Ariza-Colpas[3], Emiro De-la-Hoz-Franco[3], and Ramón E. R. González[4]

[1] Universidad Manuela Beltrán, Vicerectoría de Investigaciones,
Bogotá, D.C., Colombia
cesar.cardenas@docentes.umb.edu.co
[2] Departamento de Ingeniería Mecánica y Mecatrónica,
Universidad Nacional de Colombia, Bogotá, D.C., Colombia
[3] Universidad de la Costa, Barranquilla, Colombia
[4] Departamento de Física, Universidad Federal de Pernambuco, Recife, Brazil

Abstract. Motivated by the important growth of VTOL vehicles research such as quadrotors and to a small extent autonomous flight, a quadrotor dynamical model is presented in this work. The purpose of this study is to get a better understanding of its flight dynamics. It is an underactuated system. So, a simplified and clear model is needed to implement controllers on these kind of unmanned aerial systems. In addition, a computational tool is used for validation purposes. For future works embedded or intelligent control systems can be developed to control them. Gyroscopic and some aerodynamics effects are neglected.

Keywords: Quadrotor · VTOL · Flight dynamics · UAV

1 Introduction

Humanity has been having a significant development increase that centuries ago it would not have been possible to believe that the first manned airplane had been built and flown by the Wright brothers in the early 1900s. Aircrafts are not the only ones with flight capacities. There are other flying vehicles like drones or unmanned aerial vehicles (UAVs) which can do so. Hence, there have been many researchers and engineers from different areas interested in developing aerial vehicles without the influence of humans. Several engineering areas such as aerodynamics, control, embedded electronics are associated to this type of systems. These type of vehicles can have small designs that favor their abilities for carrying or payload. The term *drones* has been used because of the autonomy constraints they could have. This is the reason why embedded and guidance control systems are applied to drones which permit autonomous flight tasks. One of the categories of UAVs is the multirotor which has the possibility of vertical takeoff and landing (VTOL). Nowadays, unmanned aerial vehicles

© Springer Nature Switzerland AG 2020
O. Gervasi et al. (Eds.): ICCSA 2020, LNCS 12249, pp. 261–275, 2020.
https://doi.org/10.1007/978-3-030-58799-4_19

play a very meaningful role in the current aerospace industry. They can provide different autonomous flight applications such as environmental research, rescue, traffic monitoring, agricultural inspections, image and video, scientific research, inspections of places with very difficult access and even more recently, home delivery of products. Therefore, it is important to remark that their uses are not just limited to dangerous roles. Due to the great progress of technology, UAVs have been evolving up to autonomous systems. They are capable of running by themselves without any kind of interventions from humans and with a pre-determined flight mission. Therefore, this fact has drawn the attention to research more in-depth about autonomous aircrafts [3, 10, 16, 17].

2   Reference Frames

It is necessary to have at least one reference frame to describe any position or motion. The use of additional reference frames will make the derivation of the equations of motions easier. When using multiple reference frames there is one important issue which is related to the transformation of vector coordinates from one frame to another. The rotation matrices used are based on the Euler angles. A reference fixed frame is applied to determine distance and direction. A coordinate system is used to represent measurements in a frame. In flight dynamics there are two reference frames clearly defined, the Earth Fixed Frame and Body Fixed Frame [5, 12, 14, 17]. The E-frame is chosen as the inertial frame. $(0_E \ X_E \ Y_E \ Z_E)$. The origin is at O_E. In this reference frame the linear position (ξ^E) and angular position (Θ^E) of the quadrotor are defined. The other reference frame required is the body frame $(0_B \ X_B \ Y_B \ Z_B)$ which is attached to the quadrotor body. The origin is at the vehicle's reference point O_B. In this B-frame the linear velocity, the angular velocity and the forces and torques are defined.

3   Quadrotor Assumptions for Modeling

Considering that computation of any model is just an approximation to real conditions found in the real world, some assumptions need to be made for simulation purposes:

- The quadrotor is treated as a rigid structure.
- The propellers are rigid. Aerodynamics effects such as blade flapping are neglected.
- Gyroscopic effects and aerodynamic torques can be ignored.
- The quadrotor structure is treated as symmetrical.
- Wind disturbances are ignored.

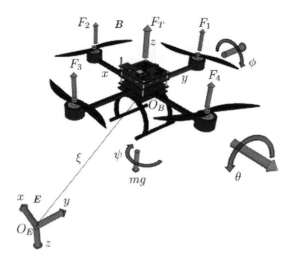

Fig. 1. Reference systems

4   Quadrotor System Variables

The quadrotor motion has six degrees of freedom (6DOF) which are defined as follows: $\xi = $ (x, y, z) that represents the linear position of the quadrotor and $\Theta = (\theta, \phi, \psi)$ is the attitude or orientation. These are also known as the Euler angles *pitch, roll* and *yaw.* Therefore, if $\Theta = (\theta, \phi, \psi)$ and $\xi = $ (x, y, z), so the general position vector Φ is (Fig. 1):

$$\Phi = \begin{bmatrix} \xi \\ \Theta \end{bmatrix} = \begin{bmatrix} x \\ y \\ z \\ \theta \\ \phi \\ \psi \end{bmatrix} \tag{1}$$

Let v and ω be the quadrotor linear and angular velocities $v = (u, v, w)$ $\omega = (p, q, r)$. Therefore, the linear and angular velocities are:

$$\upsilon = \begin{bmatrix} v \\ \omega \end{bmatrix} = \begin{bmatrix} u \\ v \\ w \\ p \\ q \\ r \end{bmatrix} \tag{2}$$

4.1   Quadrotor Kinematics

To be able to transform the vectors from the E-frame to the B-frame a direction cosine matrix is required. If the rotations are done first around x

axis then around y and the final one around z axis, the rotation matrix is: $R_\Upsilon = R(\phi, \theta, \psi) = R_x(\phi) R_y(\theta) R_z(\psi)$ [14, 15, 17, 18].

Where:

Roll rate: $p = \dot{\phi} - \dot{\psi} \sin \theta$
Pitch rate: $q = \dot{\theta} \cos \phi + \dot{\psi} \sin \phi \cos \theta$
Yaw rate: $r = \dot{\psi} \cos \phi \cos \theta - \dot{\theta} \sin \phi$

$$R_x(\phi) = \begin{bmatrix} 1 & 0 & 0 \\ 0 & \cos\phi & -\sin\phi \\ 0 & \sin\phi & \cos\phi \end{bmatrix} \tag{3}$$

$$R_y(\theta) = \begin{bmatrix} \cos\theta & 0 & \sin\theta \\ 0 & 1 & 0 \\ -\sin\theta & 0 & \cos\theta \end{bmatrix} \tag{4}$$

$$R_z(\psi) = \begin{bmatrix} \cos\psi & -\sin\psi & 0 \\ \sin\psi & \cos\psi & 0 \\ 0 & 0 & 1 \end{bmatrix} \tag{5}$$

Then, the complete rotation matrix is the product of the three rotation matrices:

$$R_{\Upsilon_B^E} = \begin{bmatrix} \cos\theta\cos\psi & \cos\psi\sin\theta\sin\phi - \sin\psi\cos\phi & \sin\phi\sin\psi + \cos\phi\cos\psi\sin\theta \\ \sin\psi\cos\theta & \cos\psi\cos\phi + \sin\psi\sin\theta\sin\phi & \sin\psi\sin\theta\cos\phi - \cos\psi\sin\phi \\ -\sin\theta & \cos\theta\sin\phi & \cos\phi\cos\theta \end{bmatrix} \tag{6}$$

The relationship between the angular velocity $\dot{\omega}$ in (E-frame) and the angular velocity in (B-frame) is given by: $\dot{\omega} = T_\chi \omega$. The transformation matrix T_χ is found by computing the attitude rates $(\dot{\phi}, \dot{\theta}, \dot{\psi})$ and body rates (p, q, r) as follows:

$$v^E = \dot{\xi}^E = \begin{bmatrix} \dot{x} \\ \dot{y} \\ \dot{z} \end{bmatrix} = R_\Upsilon v^B \tag{7}$$

$\Rightarrow$

$$v^E = R_\Upsilon = \begin{bmatrix} \cos\theta\cos\psi & \cos\psi\sin\theta\sin\phi - \sin\psi\cos\phi & \sin\phi\sin\psi + \cos\phi\cos\psi\sin\theta \\ \sin\psi\cos\theta & \cos\psi\cos\phi + \sin\psi\sin\theta\sin\phi & \sin\psi\sin\theta\cos\phi - \cos\psi\sin\phi \\ -\sin\theta & \cos\theta\sin\phi & \cos\phi\cos\theta \end{bmatrix} \begin{bmatrix} u \\ v \\ w \end{bmatrix} \tag{8}$$

The angular body rates p, q, r are shown in Fig. 2. First, *roll* rotates around the angle ϕ and angular velocity $\dot{\phi}$. Then, *pitch* which rotates around the θ angle with $\dot{\theta}$ angular velocity. Likewise, *yaw* rotates through angle ψ with angular velocity $\dot{\psi}$. Hence, the relationship between the aerial vehicle body rates and attitude rates is given as:

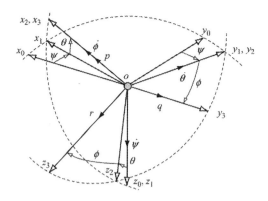

Fig. 2. Angular velocities transformation

Roll rate:
$$p = \dot{\phi} - \dot{\psi}\ sin\ \theta \tag{9}$$

Pitch rate:
$$q = \dot{\theta}\ cos\ \phi + \dot{\psi}sin\phi\ cos\theta \tag{10}$$

Yaw rate:
$$r = \dot{\psi}\ cos\phi + cos\theta - \dot{\theta}sin\ \phi \tag{11}$$

When the quadrotor Euler angles (ϕ, θ, ψ) are considered small $(\phi, \theta, \psi \cong 0)$ the body rates equations can be taken as: [4,17,18].

$$\begin{bmatrix} \dot{\phi} \\ \dot{\theta} \\ \dot{\psi} \end{bmatrix} = \begin{bmatrix} p \\ q \\ r \end{bmatrix} \tag{12}$$

Since $[x,\ y,\ z,\ \phi,\ \theta,\ \psi]^T$ is the vector that contains the linear and angular position of the quadrotor in the E-frame and $[u,\ v,\ w,\ p,\ q,\ r]^T$ is the vector which contains the linear and angular velocities in the B-frame, then these two reference frames are related as follows:

$$v = R_\Upsilon \cdot v_B \tag{13}$$
$$\omega = T_\chi \cdot \omega_B \tag{14}$$

Finally, the kinematic model of a quadrotor is:

$$\begin{cases} \dot{x} &= w[sin(\phi)sin(\psi) + cos(\phi)cos(\psi)sin(\theta)] - v[cos(\phi)sin(\psi) - cos(\phi)sin(\psi)sin(\theta)] + u[cos(\psi)cos(\theta)] \\ \dot{y} &= v[cos(\phi)cos(\psi) + sin(\phi)sin(\psi)sin(\theta)] - w[cos(\psi)sin(\phi) - cos(\psi)sin(\psi)sin(\theta)] + u[cos(\theta)sin(\psi)] \\ \dot{z} &= w[cos(\phi)cos(\theta)] - u[sin(\theta)] + v[cos(\theta)sin(\phi)] \\ \dot{\phi} &= p + r[cos(\phi)tan(\theta)] + q[sin(\phi)tan(\theta)] \\ \dot{\theta} &= qcos(\phi) - rsin(\phi) \\ \dot{\psi} &= q\frac{sin(\phi)}{cos(\theta)} + r\frac{cos(\phi)}{cos(\theta)} \end{cases}$$
$$\tag{15}$$

4.2   Quadrotor Dynamics

The second Newton's law states that a forced applied in the E reference frame is equal to the product of the vehicle's mass and its acceleration. i.e, *Force* = *mass* × *acceleration*. Therefore, it can be inferred that:

$$F = m\left(\frac{dv}{dt}\right) = m\left(\frac{d^2\xi}{dt^2}\right) \tag{16}$$

Force = *mass* × *acceleration* Hence,

$$F = m\left(\frac{dv}{dt}\right) = m\left(\frac{d^2\xi}{dt^2}\right) \tag{17}$$

From Eq. 1:

$$\left(\frac{d^2\xi^E}{dt^2}\right) = \ddot{\xi}^E = \begin{bmatrix} \ddot{x} \\ \ddot{y} \\ \ddot{z} \end{bmatrix} \tag{18}$$

The total force exerted on the quadrotor is given by the following matrix relation:

$$m(\omega^B \times v^B + \dot{v}^B) = F^B \tag{19}$$

The total force exerted on the quadrotor is given by the following matrix relation:

$$m(\omega^B \times v^B + \dot{v}^B) = F^B \tag{20}$$

Where × is the cross product, m is the quadcopter mass and $F^B = [F_x \ F_y \ F_z]^T$ which is the total force. So, a skew symmetric matrix is got and expressed as: [17, 18].

$$\omega^B \times v^B = \begin{bmatrix} qw - rv \\ ru - pw \\ pv - qu \end{bmatrix} \tag{21}$$

And if the differential of the linear velocity is:

$$v = \begin{bmatrix} \dot{u} \\ \dot{v} \\ \dot{w} \end{bmatrix} \tag{22}$$

Therefore, the following equation is obtained:

$$F^B = m \begin{bmatrix} \dot{u} \\ \dot{v} \\ \dot{w} \end{bmatrix} + \begin{bmatrix} qw - rv \\ ru - pw \\ pv - qu \end{bmatrix} \tag{23}$$

If the column vector $F^B = [F_x \; F_y \; F_z]^T$ is substituted and dividing by m, we get:

$$\frac{1}{m}\begin{bmatrix} F_x \\ F_y \\ F_z \end{bmatrix} = \begin{bmatrix} \dot{u} \\ \dot{v} \\ \dot{w} \end{bmatrix} + \begin{bmatrix} qw - rv \\ ru - pw \\ pv - qu \end{bmatrix} \tag{24}$$

Rewriting, the following is acquired:

$$\begin{bmatrix} \dot{u} \\ \dot{v} \\ \dot{w} \end{bmatrix} = \frac{1}{m}\begin{bmatrix} F_x \\ F_y \\ F_z \end{bmatrix} - \begin{bmatrix} qw - rv \\ ru - pw \\ pv - qu \end{bmatrix} \tag{25}$$

Then, the total torque applied to the quadrotor is given by:

$$I\dot{\omega}^B + \omega^B \times (I\omega^B) + M_G = M^B \tag{26}$$

Where $\omega^B \times v^B$ is known as the Coriolis term, $[M_x \; M_y \; M_z] = M^B$ is the total torque, M_G gyroscopic effects due to rotor's inertia and I known as the inertia matrix [7,17,18].
A quadcopter is assumed as symmetric:

$$I_{xy} = I_{xz} = I_{yx} = I_{yz} = I_{zx} = I_{zy} = 0 \tag{27}$$

$$I = \begin{bmatrix} I_{xx} & I_{xy} & I_{xz} \\ I_{yx} & I_{yy} & I_{yz} \\ I_{zx} & I_{zy} & I_{zz} \end{bmatrix} = \begin{bmatrix} I_{xx} & 0 & 0 \\ 0 & I_{yy} & 0 \\ 0 & 0 & I_{zz} \end{bmatrix} \tag{28}$$

The angular acceleration is given by:

$$\dot{\omega}^B = I^{-1}[T_\chi \tau^B - (\omega^B \times I\omega^B)] \tag{29}$$

Where τ^B is given as:

$$\tau^B = [\tau_r \; \tau_p \; \tau_y]^T \tag{30}$$

Therefore $T_\chi \tau^B$ is:

$$T_\chi \tau^B = \begin{bmatrix} \tau_r \\ \tau_p \\ \tau_y \end{bmatrix} = I\dot{\omega}^B + [\omega^B \times (I\omega^B)] \tag{31}$$

The inverse of matrix 28 is:

$$I^{-1} = \begin{bmatrix} \frac{1}{I_{xx}} & 0 & 0 \\ 0 & \frac{1}{I_{yy}} & 0 \\ 0 & 0 & \frac{1}{I_{zz}} \end{bmatrix} \tag{32}$$

If the angular velocities in Eq. 2 are derived we obtain:

$$\dot{\omega}^B = \begin{bmatrix} \dot{p} \\ \dot{q} \\ \dot{r} \end{bmatrix} \tag{33}$$

Now substituting in Eq. 29 results:

$$\dot{\omega}^B = \begin{bmatrix} \dot{p} \\ \dot{q} \\ \dot{r} \end{bmatrix} = I^{-1} \left[T_\chi \tau^B - \left(\begin{bmatrix} p \\ q \\ r \end{bmatrix} \times \begin{bmatrix} I_{xx} & 0 & 0 \\ 0 & I_{yy} & 0 \\ 0 & 0 & I_{zz} \end{bmatrix} \begin{bmatrix} p \\ q \\ r \end{bmatrix} \right) \right] \tag{34}$$

Simplifying:

$$\dot{\omega}^B = \begin{bmatrix} \dot{p} \\ \dot{q} \\ \dot{r} \end{bmatrix} = I^{-1} \left[\begin{bmatrix} \tau_r \\ \tau_p \\ \tau_y \end{bmatrix} - \begin{bmatrix} qrI_{zz} - rqI_{yy} \\ rpI_{xx} - prI_{zz} \\ pqI_{yy} - qpI_{xx} \end{bmatrix} \right] \tag{35}$$

Substituting I^{-1} in Eq. 35 we get:

$$\dot{\omega}^B = \begin{bmatrix} \dot{p} \\ \dot{q} \\ \dot{r} \end{bmatrix} = \begin{bmatrix} \frac{1}{I_{xx}} & 0 & 0 \\ 0 & \frac{1}{I_{yy}} & 0 \\ 0 & 0 & \frac{1}{I_{zz}} \end{bmatrix} \left[\begin{bmatrix} \tau_r \\ \tau_p \\ \tau_y \end{bmatrix} \begin{bmatrix} qrI_{zz} - rqI_{yy} \\ rpI_{xx} - prI_{zz} \\ pqI_{yy} - qpI_{xx} \end{bmatrix} \right] \tag{36}$$

Simplifying:

$$\dot{\omega}^B = \begin{bmatrix} \dot{p} \\ \dot{q} \\ \dot{r} \end{bmatrix} = \begin{bmatrix} \frac{\tau_r}{I_{xx}} \\ \frac{\tau_p}{I_{yy}} \\ \frac{\tau_y}{I_{zz}} \end{bmatrix} - \begin{bmatrix} I_{zz} - I_{yy} \frac{qr}{I_{xx}} \\ I_{xx} - I_{zz} \frac{pr}{I_{yy}} \\ I_{yy} - I_{xx} \frac{pq}{I_{zz}} \end{bmatrix} \tag{37}$$

4.3 Thrust and Moment

It is assumed that the torque and thrust caused by each rotor acts particularly in the z axis of the B-frame. Accordingly, the net propulsive force in the z^B direction is given by:

$$F_t = \begin{bmatrix} x \\ y \\ z \end{bmatrix} = \begin{bmatrix} 0 \\ 0 \\ F_1 + F_2 + F_3 + F_4 \end{bmatrix} \tag{38}$$

Moment results from the thrust action of each rotor around the center of mass that induces a pitch and roll motion. Furthermore, there is a reactive torque of the rotors on the vehicle which produces a yaw reply. The moment vector is therefore: [9,17].

$$M_t = \begin{bmatrix} \tau_r \\ \tau_p \\ \tau_y \end{bmatrix} = \begin{bmatrix} lb(F_2 - F_4) \\ lb(F_3 - F_1) \\ (F_1 - F_2 + F_3 - F_4)d \end{bmatrix} \tag{39}$$

Where l is distance between any rotor and the center of the drone, b is the thrust factor and d is the drag factor [9, 17, 19].

The external forces used are based on the model found in [4, 14]:

$$
\begin{aligned}
F_x &= -W\sin\theta + x = m(\dot{u} + qw - rv)\\
F_y &= W\cos\theta\cos\psi + x = m(\dot{v} + ru - pw)\\
F_z &= W\cos\theta\cos\psi + z = m(\dot{w} + pv - qu)
\end{aligned}
\tag{40}
$$

$$
\begin{aligned}
M_x &= I_{xx}\dot{p} + (I_{zz} - I_{yy})qr\\
M_y &= I_{yy}\dot{q} + (I_{xx} - I_{zz})rp\\
M_z &= I_{zz}\dot{r} + (I_{yy} - I_{xx})pq
\end{aligned}
\tag{41}
$$

The Matrix 42 shows the relationship between the the net torque that is performed on the vehicle and the angular acceleration. The total torque is the summation of the aerodynamic torque τ_a and the torque generated by the rotors [13].

$$
\begin{bmatrix} M_x\\ M_y\\ M_z \end{bmatrix} Aero + \begin{bmatrix} M_x\\ M_y\\ M_z \end{bmatrix} Quad = \begin{bmatrix} I_{xx}\dot{p} - (I_{yz}(q^2 r^2) - I_{zx}(\dot{r} + pq) - I_{xy}(\dot{q} - rp) - (I_{yy} - I_{zz})qr\\ I_{yy}\dot{q} - I_{zx}(r^2 - p^2) - I_{xy}(\dot{p} + qr) - I_{yz}(\dot{r} - pq) - (I_{zz} - I_{xx})rp\\ I_{xx}\dot{p} - I_{yz}(q^2 - r^2) - I_{zx}(\dot{r} + pq) - I_{xy}(\dot{q} - rp) - (I_{yy} - I_{zz})qr \end{bmatrix}
\tag{42}
$$

Therefore, the quadrotor dynamic model in the B-frame is:

$$
\begin{cases}
F_x = m(\dot{u} + qw - rv)\\
F_y = m(\dot{v} - pw + ru)\\
F_z = m(\dot{w} + pv - qu)\\
M_x = I_{xx}\dot{p} + (I_{zz} - I_{yy})qr\\
M_y = I_{yy}\dot{q} + (I_{xx} - I_{zz})rp\\
M_z = I_{zz}\dot{r} + (I_{yy} - I_{xx})pq
\end{cases}
\tag{43}
$$

The rotation of the propellers combined with the rotation of the body results in a gyroscopic torque which is given by:

$$
M_G = J_m(\omega_B \times z^B)(\Omega_1 + \Omega_2 + \Omega_3 + \Omega_4)
\tag{44}
$$

Where Ω $(i = 1, 2, 3, 4)$ is propeller angular velocity, J_m is the inertia of each motor. Taking into an account that $\tau_B = [\tau_p \ \tau_r \ \tau_y]$ are the torques generated by the actuators action and $\tau_a = [\tau_{ax} \ \tau_{ay} \ \tau_{az}]$ are the aerodynamic torques, finally, the complete quadrotor dynamic model is:

$$
\begin{cases}
-mg[\sin(\theta)] + F_{ax} = m(\dot{u} + qw - rv)\\
mg[\cos(\theta)\sin(\phi)] + F_{ay} = m(\dot{v} - pw + ru)\\
mg[\cos(\theta)\cos(\phi)] + F_{az} - ft = m(\dot{w} + pv - qu)\\
\tau_p + \tau_{ax} = \dot{p}I_{xx} - qrI_{yy} + qrI_{zz}\\
\tau_r + \tau_{ay} = \dot{q}I_{yy} + prI_{xx} - prI_{zz}\\
\tau_y + \tau_{az} = \dot{r}I_{zz} - pqI_{xx} + pqI_{yy}
\end{cases}
\tag{45}
$$

$F^B = [F_x \; F_y \; F_z]$ [N], R_Υ is the rotation matrix and m [kg] is the mass of the quadrotor. So, it is possible to combine them as follows: [17,18].

$$F^B = \begin{bmatrix} F_x \\ F_y \\ F_z \end{bmatrix} = mR_\Upsilon \frac{d}{dt} \dot{\xi} = mR_\Upsilon \begin{bmatrix} \ddot{x} \\ \ddot{y} \\ \ddot{z} \end{bmatrix} \tag{46}$$

They can be arranged as:

$$\begin{bmatrix} \ddot{x} \\ \ddot{y} \\ \ddot{z} \end{bmatrix} = \frac{1}{m} R_\Upsilon^{-1} \begin{bmatrix} F_x \\ F_y \\ F_z \end{bmatrix} \tag{47}$$

Making substitutions the equation is simplified as:

$$\begin{bmatrix} \ddot{x} \\ \ddot{y} \\ \ddot{z} \end{bmatrix} = \frac{1}{m} R_\Upsilon^{-1} \begin{bmatrix} -mgsin\theta \\ mgsin\phi cos\theta \\ mgcos\phi cos\theta \end{bmatrix} + \begin{bmatrix} 0 \\ 0 \\ F_1 + F_2 + F_3 + F_4 \end{bmatrix} \tag{48}$$

Therefore:

$$\begin{cases} \ddot{x} = \frac{F_{total}}{m}[sin(\phi)sin(\psi) + cos(\phi)cos(\psi)sin(\theta)] \\ \ddot{y} = \frac{F_{total}}{m}[cos(\phi)sin(\psi)sin(\theta) - cos(\psi)sin(\phi)] \\ \ddot{z} = -g\frac{F_{total}}{m}[cos(\phi)cos(\theta)] \end{cases} \tag{49}$$

As stated earlier, for small angles of movement $[\dot{\phi} \; \dot{\theta} \; \dot{\psi}]^T = [p \; q \; r]^T$, hence, the dynamic model of quadrotor in the E-frame is:

$$\begin{cases} \ddot{x} = \frac{F_{total}}{m}[sin(\phi)sin(\psi) + cos(\phi)cos(\psi)sin(\theta)] \\ \ddot{y} = \frac{F_{total}}{m}[cos(\phi)sin(\psi)sin(\theta) - cos(\psi)sin(\phi)] \\ \ddot{z} = -g\frac{F_{total}}{m}[cos(\phi)cos(\theta)] \\ \ddot{\phi} = \frac{I_{yy}-I_{zz}}{I_{xx}}\dot{\theta}\dot{\psi} + \frac{\tau_x}{I_{xx}} \\ \ddot{\theta} = \frac{I_{zz}-I_{xx}}{I_{yy}}\dot{\phi}\dot{\psi} + \frac{\tau_x}{I_{yy}} \\ \ddot{\psi} = \frac{I_{xx}-I_{yy}}{I_{zz}}\dot{\phi}\dot{\theta} + \frac{\tau_x}{I_{zz}} \end{cases} \tag{50}$$

4.4 Motor and Propeller Dynamics

The following is the differential equation for a DC motor:

$$J_m \dot{\Omega}_m = -\frac{K_E K_q}{R} \Omega_m - \tau_l + \frac{K_q}{R} V \tag{51}$$

Where J_m is the motor inertia $[N \; m \; s^2]$, $\dot{\Omega}_m$ is the motor angular speed $[rad/s^2]$, τ_l is the load torque $[N \; m]$. There is an acceleration or deceleration of the angular speed of the motor Ω_m when the torques τ_m (motor torque) and τ_l are not the same. The motor torque τ_m is proportional to the current i and through the torque constant K_q $[Amps/Nm]$, K_E is known as the voltage motor constant [rad s/V].

The power required to hover is given as:

$$P = TV_i \equiv TV_{rh} = T\sqrt{\frac{T}{2\rho A_b}} = \frac{T^{3/2}}{\sqrt{2\rho A_b}} \qquad (52)$$

Where A_b is the area cleaned out by the rotor blades and ρ is the air density. The above expression can be related to the induced velocity v_{rh} in forward flight as:

$$v_i = \frac{v_{rh}^2}{\sqrt{(v_\infty cos\alpha)^2 + (v_i - v_\infty sin\alpha)^2}} \qquad (53)$$

Where α is rotor angle of attack and v_∞ is velocity stream flow [8,11]. If Eq. 52 is substituted for v_{rh}, it results;

$$fom\ \eta_m \frac{\tau_m}{K_q}V = T\sqrt{\frac{T}{2\rho A_b}} \qquad (54)$$

The rotor power in forward flight is given by:

$$P = T(v_\infty\ sin\ \alpha + v_i) \qquad (55)$$

So, the ideal thrust T when a power input P can be calculated as follows:

$$T = \frac{P}{v_\infty\ sin\ \alpha + v_i} \qquad (56)$$

Where the denominator is the airspeed across the rotors.
The motor torque is proportional to the thrust constant K_t:

$$\tau_m = K_t T \qquad (57)$$

If Eq. 52 is substituted for τ_m, it gives:

$$fom\ \eta_m \frac{K_t T}{K_q}V = T\sqrt{\frac{T}{2\rho A_b}} \qquad (58)$$

Torque τ_m is provided by each motor that is balanced by the drag torque. Accordingly, the torque acting on the propeller is:

$$\tau_m = J_m \dot{\Omega} + \tau_{drag} \qquad (59)$$

Finally, the relationship between voltage and thrust is given by:

$$T = 2\rho A_b \left[\frac{fom\ \eta_m\ K_t}{K_q}\right]^2 V^2 \qquad (60)$$

A quadrotor is controlled by providing four torques $(\tau_1\ \tau_2\ \tau_3\ \tau_4)$ to the rotors which produce four thrust forces $(F_1\ F_2\ F_3\ F_4)$ in the z axis. The net torques acting on the body can be calculated by using the inputs $U = (U_1\ U_2\ U_3\ U_4)$ that can be applied to control the quadrotor. As an aerodynamic consideration forces and torques are proportional to the squared propeller's speed [1].
Therefore the relationship between motions and propellers' squared speed is as follows:

$$F_T = b(\Omega_1^2 + \Omega_2^2 + \Omega_3^2 - \Omega_4^2)$$
$$\tau_{x(\theta)} = bl(\Omega_2^2 - \Omega_4^2)$$
$$\tau_{y(\phi)} = bl(\Omega_1^2 - \Omega_3^2)$$
$$\tau_{z(\psi)} = K_{drag}(\Omega_1^2 + \Omega_2^2 - \Omega_3^2 + \Omega_4^2)$$

$$(61)$$

Where l [m] is the distance between any rotor and the center of the drone, b is the thrust factor $[N\ s^2]$, K_{drag} is a drag constant $[N\ m\ s^2]$ and Ω_i $[rad\ s^{-1}]$ is the propeller angular acceleration [7,17,18]. Then the dynamic model of the quadrotor is:

$$\begin{cases}
-mg[sin(\theta)] + F_{ax} = m(\dot{u} + qw - rv) \\
mg[cos(\theta)sin(\phi)] + F_{ay} = m(\dot{v} - pw + ru) \\
mg[cos(\theta)cos(\phi)] + F_{az} - b(\Omega_1^2 + \Omega_2^2 - \Omega_3^2 - \Omega_4^2) = m(\dot{w} + pv - qu) \\
bl(\Omega_3^2 - \Omega_1^2) + \tau_{ax} = \dot{p}I_x - qrI_y + qrI_z \\
bl(\Omega_4^2 - \Omega_2^2) + \tau_{ay} = \dot{q}I_y - prI_x + prI_z \\
K_{drag}(\Omega_2^2 + \Omega_4^2 - \Omega_1^2 - \Omega_3^2) + \tau_{az} = \dot{r}I_z - pqI_x + pqI_y
\end{cases}$$

$$(62)$$

The overall propeller's velocities Ω $[rad\ s^{-1}]$ and propeller's velocities vector $\boldsymbol{\Omega}$ are defined as: [1].

$$\Omega = -\Omega_1 + \Omega_2 - \Omega_3 + \Omega_4 \tag{63}$$

$$\boldsymbol{\Omega} = \begin{bmatrix} \Omega_1 \\ \Omega_2 \\ \Omega_3 \\ \Omega_4 \end{bmatrix} \tag{64}$$

The following equations system shows the relationship between the control inputs and the propellers' squared speed:

$$\begin{bmatrix} U_1 \\ U_2 \\ U_3 \\ U_4 \end{bmatrix} = \begin{bmatrix} b(\Omega_1^2 + \Omega_2^2 + \Omega_3^2 + \Omega_4^2) \\ bl(\Omega_4^2 - \Omega_2^2) \\ bl(\Omega_3^2 - \Omega_1^2) \\ d(\Omega_2^2 + \Omega_4^2 - \Omega_1^2 - \Omega_3^2) \end{bmatrix} \tag{65}$$

Where l [m] is the distance from the center of the quadrotor and the propeller and $[U_1\ U_2\ U_3\ U_4]$ are the propeller's speed control inputs. U_1 is the combination of thrust forces responsible for the quadrotor altitude z. U_2 is the differential thrust between rotors 2 and 4 which generate the roll moment. U_3 is the thrust differential between rotors 1 and 3 that create the pitch moment. Finally, U_4 is the combination of the individual torques between the clockwise and counter-clockwise rotors in charge of generating yaw rotation. Consequently, U_1 generates the desired quadrotor altitude, U_2 and U_3 generate the respective roll and pitch angles whereas U_4 creates the yaw angle [1,2,6,7,18]. Simplifying and including U variables we get:

$$\mathbf{f(x,u)} = \begin{pmatrix} \dot{\phi} = \dot{\theta}\dot{\psi}\lambda_1 + \dot{\theta}\lambda_2\Omega_r + b_1U_2 \\ \dot{\theta} = \dot{\phi}\dot{\psi}\lambda_3 + \dot{\phi}\lambda_4\Omega_r + b_2U_3 \\ \dot{\psi} = \dot{\theta}\dot{\phi}\lambda_5 + b_3U_4 \\ \ddot{x} = \mu_x\dfrac{1}{m}U_1 \\ \ddot{y} = \mu_y\dfrac{1}{m}U_1 \\ \ddot{z} = g - (cos\phi cos\theta)\dfrac{1}{m}U_1 \end{pmatrix} \tag{66}$$

J_m $[N\ m/\ s^2]$ is the motor inertia.

5   Simulations and Results

Table 1 presents the selected variables for simulation purposes which are based on the AR drone. Since a complete math model is provided, other variables such as thrust and drag coefficients as well as motor inertia are involved in the flight control analysis. Therefore, after performing several control simulations and some gain adjustments, it can be seen that PD control has a good response getting a better and quick stabilization of the vehicle. Also, when the moments of inertia are of very small values, attitude stabilization is more difficult. Some overshoot or peaks are also noticed in attitude angles. For altitude, peaks are not really evident. The PID gains are computed as well. Additional corrections are required as some signal deviations from the desired references are observed. The mean of dynamics values are helpful for more appropriate results (Fig. 3).

Table 1. Quadrotor parameters (AR Drone)

Dynamic variables	Value
m	0.1 kg
I_x	0.45 kg · m^2
I_y	0.51 kg · m^2
I_z	0.95 kg · m^2
l	0.5 m

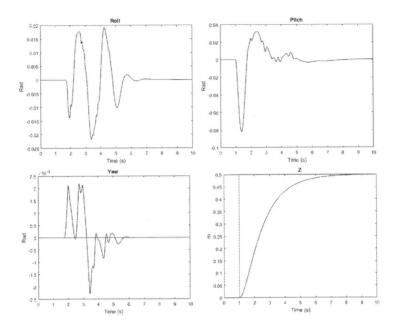

Fig. 3. Altitude/Attitude quadrotor performance

6   Conclusions

Vertical takeoff and landing aircrafts have a complex and highly non-linear dynamics. A complete model is given using the Newton-Euler approach. It could be used to predict changes in position and orientation of a quadcopter at a given point in time. It can be achieved by varying the rotors speed. Equations are inverted to find inputs needed for a certain position. Additionally, there are six degrees of freedom and just four inputs. Therefore, it is not appropriate and a simplified dynamics model would be required to be then implemented for control purposes. Also, there are some sensor measurements that are based on the E-frame and not on sensors frames. This model approach can be adapted to consider voltage inputs and not just U since there is an important association between thrust and voltage. It is necessary to adjust control gains. So, the smaller the moments of inertia and mass the more difficult for the platform to stabilize. State estimation would have a significant role because of an estimation of sensor measurements is necessary for autopilots design.

References

1. Bresciani, T.: Modelling, identification and control of a quadrotor helicopter. MSc theses (2008)
2. Cárdenas Ruiz, C.A.: Performance study of the flight control and path planning for a UAV type Quadrotor. Ingeniería Mecatrónica (2018)

3. Collazos, C., et al.: State estimation of a dehydration process by interval analysis. In: Figueroa-García, J.C., López-Santana, E.R., Rodriguez-Molano, J.I. (eds.) WEA 2018. CCIS, vol. 915, pp. 66–77. Springer, Cham (2018). https://doi.org/10.1007/978-3-030-00350-0_6

4. Cook, M.V.: Flight Dynamics Principles. Elsevier (2007)

5. Cork, L.R.: Aircraft dynamic navigation for unmanned aerial vehicles. Ph.D. thesis. Queensland University of Technology (2014)

6. Cowling, I.: Towards autonomy of a Quadrotor UAV (2008)

7. Habib, M.K., Abdelaal, W.G.A., Saad, M.S., et al.: Dynamic modeling and control of a Quadrotor using linear and nonlinear approaches (2014)

8. Huang, H., et al.: Aerodynamics and control of autonomous Quadrotor helicopters in aggressive maneuvering. In: IEEE International Conference on Robotics and Automation, ICRA 2009, pp. 3277–3282. IEEE (2009)

9. Ireland, M.L.: Investigations in multi-resolution modelling of the Quadrotor micro air vehicle. Ph.D. thesis. University of Glasgow (2014)

10. Jiménez-Cabas, J., et al.: Robust control of an evaporator through algebraic Riccati equations and d-K iteration. In: Misra, S., et al. (eds.) ICCSA 2019. LNCS, vol. 11620, pp. 731–742. Springer, Cham (2019). https://doi.org/10.1007/978-3-030-24296-1_58

11. Leishman, G.J.: Principles of Helicopter Aerodynamics with CD Extra. Cambridge University Press, Cambridge (2006)

12. Lugo-Cárdenas, I., et al.: Dubins path generation for a fixed wing UAV. In: 2014 International Conference on Unmanned Aircraft Systems (ICUAS), pp. 339–346. IEEE (2014)

13. Malang, Y.: Design and Control of a Vertical Takeo and Landing Fixed-wing Unmanned Aerial Vehicle. Ph.D. thesis (2016)

14. Mulder, S.: Flight Dynamics (2007)

15. Swartling, J.O.: Circumnavigation with a group of Quadrotor helicopters (2014)

16. Pharpatara, P.: Trajectory planning for aerial vehicles with constraints. Theses. Université Paris-Saclay; Université d'Evry-Val-d'Essonne, September 2015. https://tel.archives-ouvertes.fr/tel-01206423

17. Poyi, G.T.: A novel approach to the control of quad-rotor helicopters using fuzzy-neural networks (2014)

18. Sabatino, F.: Quadrotor control: modeling, nonlinear control design, and simulation (2015)

19. Voos, H.: Nonlinear control of a Quadrotor micro-UAV using feedback linearization. In: International Conference on Mechatronics, ICM 2009, pp. 1–6. IEEE (2009)

The Effect of a Phase Change During the Deposition Stage in Fused Filament Fabrication

S. F. Costa[1](✉) [iD], F. M. Duarte[2] [iD], and J. A. Covas[2] [iD]

[1] Center for Research and Innovation in Business Sciences and Information Systems (CIICESI), School of Management and Technology, Porto Polytechnic Institute, Felgueiras, Portugal
sfc@estg.ipp.pt
[2] Department of Polymer Engineering, Institute for Polymers and Composites (IPC), University of Minho, Guimarães, Portugal
{fduarte,jcovas}@dep.uminho.pt

Abstract. Additive Manufacturing Techniques such as Fused Filament Fabrication (FFF) produce 3D parts with complex geometries directly from a computer model without the need of using molding tools. Due to the rapid growth of these techniques, researchers have been increasingly interested in the availability of strategies, models or data that may assist process optimization. In fact, 3D parts often exhibit limited mechanical performance, which is usually the result of poor bonding between adjacent filaments. In turn, the latter is influenced by the temperature field history during deposition. This study aims at evaluating the influence of considering a phase change from the melt to the solid state on the heat transfer during the deposition stage, as undergone by semi-crystalline polymers. The energy equation considering solidification is solved analytically and then inserted in the MatLab® code previously developed by the authors to model cooling in FFF. The predictions of temperature evolution during the deposition of a simple 3D part demonstrate the importance of that thermal condition and highlight the influence of the type of material used for FFF.

Keywords: Fused Filament Fabrication (FFF) · Phase change · Modelling · Heat transfer

1 Introduction

Additive Manufacturing (AM) is a group of technologies that produces three-dimensional physical objects by gradually adding material [1]. Since the 1980s these technologies became adopted for a growing number of medical/engineering applications and have gradually entered our lives with the development of low-cost 3D printers. In Fused Filament Fabrication (FFF), parts are made layer by layer, each layer being obtained by the extrusion through a nozzle and continuous deposition of a molten plastic filament. The nozzle movements are controlled by a computer, in accordance with a previously defined deposition sequence. Thus, FFF can produce a prototype or a finished product from a Computer Aided Design (CAD) model without the use of molds.

© Springer Nature Switzerland AG 2020
O. Gervasi et al. (Eds.): ICCSA 2020, LNCS 12249, pp. 276–285, 2020.
https://doi.org/10.1007/978-3-030-58799-4_20

A relative limited range of materials is commercially available for FFF. Acrylonitrile Butadiene Styrene (ABS) and Polylactic Acid (PLA) are the most popular. ABS exhibits good impact resistance and toughness, heat stability, chemical resistance and long service life [2, 3]. PLA is biodegradable (compostable) and has low melting temperature [4, 5]. Several authors compared the performance of the two materials in terms of dimensional accuracy, surface roughness [6] and emission of volatile organic compounds (VOC) [7]. For example, Mudassir [8] proposed a methodology to select among the two materials. Nevertheless, 3D printed parts often show insufficient mechanical performance [9]. This is generally due to insufficient bonding between adjacent filaments, which in turn is determined by the temperature history upon cooling [10]. Consequently, knowledge of the temperature evolution during the deposition/cooling stage is valuable for process set-up and optimization. In a previous work, the authors developed a simulation method to predict the temperature history at any location of a 3D part. This entailed the development of an algorithm to define/up-date automatically contacts and thermal/initial conditions as the deposition proceeds, as well as a criterion to compute the adhesion degree between adjacent filaments [11]. Despite the practical usefulness and the general good agreement between theoretical predictions and experimental data [12], the method can only be used for amorphous materials, such as ABS, which do not exhibit a phase transition from the melt to the solid state. In the case of partially crystalline polymers like PLA, the enthalpy of fusion (also known as latent heat of fusion), must be considered.

The objective of this work is to consider a phase change in the calculation of the temperature evolution during deposition and cooling of FFF parts, and to estimate the magnitude of its effect by comparing the temperature evolution for simple ABS and PLA parts. Once this is accomplished, temperature predictions can be made for the increasing range of materials used in FFF, in order to better understand correlations between processing conditions, material characteristics and printed part properties. The paper is organized as follows. Section 2 presents the algorithm for dealing with cooling materials with a phase change and the resulting computer code. Section 3 considers the deposition and cooling of a single filament and of a simple 3D part to illustrate the differences in cooling of amorphous and semi-crystalline polymers.

2 A Code for the Prediction of Temperatures and Adhesion

2.1 Available Code

It has been shown that during the deposition of a filament, the heat transfers by convection with the environment and by conduction with the support/adjacent filaments control its temperature history [13]. The corresponding energy balance can then be translated into a differential energy equation, which is analytically solved yielding a mathematical expression to compute the temperature evolution with the time [14]. In the simulation method developed by the authors, this expression is used by an algorithm developed using the MatLab, which activates/deactivates automatically contacts and thermal conditions depending on the part geometry, deposition sequence and operating conditions. By coupling an adhesion criteria to the temperature profile history it is also possible to predict the degree of adhesion between adjacent filaments (Fig. 1) [11]. Then, the objective is now to include an algorithm for the phase change, to make possible the study of parts fabricated from partially crystalline polymers.

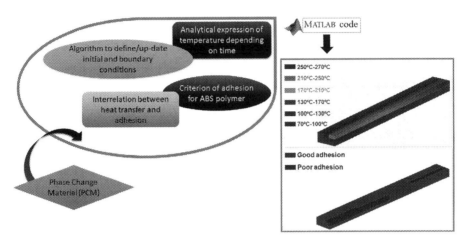

Fig. 1. Diagram of the algorithm to predict temperatures and adhesion of amorphous parts made by FFF.

2.2 Insertion of a Phase Change

When a molten filament made of a partially-crystalline polymer cools down, eventually a phase change from liquid to solid occurs during the time interval $[\tau_l, \tau_s]$, where τ_l is the instant at which the filament reaches the solidification temperature T_{solid} and τ_s is the end of the phase change (Fig. 2).

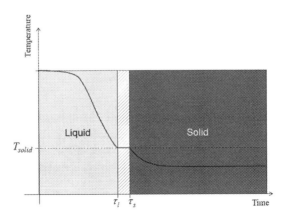

Fig. 2. Typical temperature evolution when a phase change occurs.

If N is the total number of deposited filament segments, the following phase change condition can be defined in the interval $\Delta t = \tau_s - \tau_l$, for each filament:

$$\begin{pmatrix} Heat\ losses \\ by \\ convection \end{pmatrix} + \begin{pmatrix} Heat\ transfer \\ with\ adjacent\ filaments \\ and/or\ with\ support \end{pmatrix} = \begin{pmatrix} Heat\ released \\ during\ solidification \end{pmatrix} \quad (1)$$

An equation can be written for the rth deposited filament segment ($r \in \{1,\dots,N\}$):

$$\left[h_{conv}(A_r)_{conv}(T_{solid} - T_E) + \sum_{i=1}^{n} h_i(A_r)_i(T_{solid} - (T_r)_i) \right] \Delta t = m\lambda \quad (2)$$

where h_{conv} is the convective heat transfer coefficient ($W/m^2\ °C$), T_E is the environment temperature ($°C$), n is the number of physical contacts of the filament with adjacent filament segments or with the support, h_i is the thermal contact conductance for contact i ($W/m^2\ °C$), $(T_r)_i$ is the temperature of the adjacent filament segment or support for contact i ($°C$), m is the mass of the filament segment (kg), λ is the latent heat of fusion (J/kg), $(A_r)_{conv}$ is the area exposed to the environment (m^2) and $(A_r)_i$ is the area of contact i for the rth filament segment (m^2), which are given by:

$$\begin{cases} (A_r)_{conv} = PL\left(1 - \sum_{i=1}^{n}(a_r)_i\alpha_i\right) \\ (A_r)_i = PL(a_r)_i\alpha_i \end{cases}, \ \forall i \in \{1,\dots,n\} \quad (3)$$

P is the cross-section perimeter (m), L is the filament segment length (m), α_i is the fraction of P that is in contact with another segment or with the support and $(a_r)_i$ is defined by, $\forall i \in \{1,\dots,n\}, \forall r \in \{1,\dots,N\}$:

$$(a_r)_i = \begin{cases} 1, & if\ the\ r^{th}\ filament\ segment\ has\ the\ i^{th}\ contact \\ 0, & otherwise \end{cases} \quad (4)$$

By using the expressions of $(A_r)_{conv}$ and $(A_r)_i$:

$$\left[\begin{array}{l} h_{conv}PL\left(1 - \sum_{i=1}^{n}(a_r)_i\alpha_i\right)(T_{solid} - T_E) + \\ + \sum_{i=1}^{n} h_iPL(a_r)_i\alpha_i(T_{solid} - (T_r)_i) \end{array} \right] \Delta t = \rho AL\lambda \quad (5)$$

where ρ is density (kg/m^3) and A is the cross-section area (m^2). Simplifying Eq. 5, an expression for τ_s is obtained:

$$\tau_s = \tau_l + \frac{\rho A\lambda}{h_{conv}P\left(1 - \sum_{i=1}^{n}(a_r)_i\alpha_i\right)(T_{solid} - T_E) + \sum_{i=1}^{n} h_iP(a_r)_i\alpha_i(T_{solid} - (T_r)_i)} \quad (6)$$

When inserting the phase change in the simulation method, some assumptions must be made:

1. When the computed temperature of the rth filament segment reaches T_{solid}, the thermal conditions used to compute the value of τ_s are the thermal conditions activated at the instant τ_l;
2. If $((T_r)_i)_0$ is the temperature of the adjacent filament segment for contact i at instant τ_l, the value of $(T_r)_i$ will be assumed as the average between $((T_r)_i)_0$ and T_E;
3. If during a phase change a filament contacts a new hotter filament, the phase change is interrupted and temperatures are re-computed. When its temperature reaches once more the solidification temperature T_{solid}, τ_s is computed with the new thermal conditions;
4. The crystallization growth does not affect the thermal properties, which are also taken as independent of temperature.

Equation (6) and the assumptions above were inserted in the algorithm. At each time increment, the filaments starting a phase change are identified. The temperature of those filaments suffering a phase change is kept constant unless a new adjacent filament interrupts the process. A simplified flowchart is presented in Fig. 3.

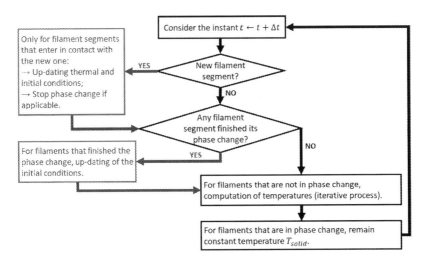

Fig. 3. Simplified flowchart of the section of the code dealing with the phase change.

3   Examples

The first example concerns the deposition of a single filament (Fig. 4.a), while the second deals with the deposition of 10 filaments, where thermal contacts between adjacent filaments develop (Fig. 4.b). Table 1 presents the main properties of the two polymers and Table 2 identifies the process parameters and computational variables.

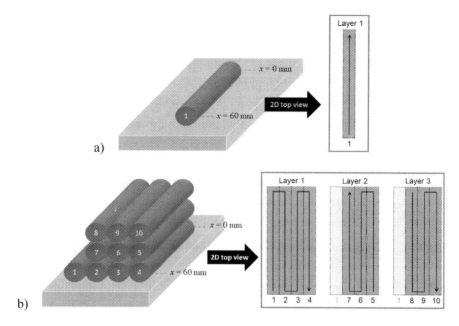

Fig. 4. a) Example 1: deposition of a single filament; b) Example 2: deposition of 10 filaments.

Table 1. Material properties.

Property	ABS	PLA
Density (kg/m³)	$\rho = 1050$	$\rho = 1300$
Thermal conductivity (W/m °C)	$k = 0.2$	$k = 0.1$
Specific heat (J/kg °C)	$C = 2020$	$C = 2100$
Latent heat (J/kg)	–	$\lambda = 30\ 000$
Solidification temperature (°C)	–	$T_{solid} = 150$

Figure 5 shows the temperature evolution of a single PLA filament, with and without phase change (at $x = 30$ mm, i.e., at the middle of filament). The phase change starts at $t = 3.25$ s and lasts 0.75 s. During this period the temperature remains constant, whereas it decreases continuously if the phase change is not considered. This results in a temperature difference of 13.9 °C (at $t = 4$ s) between the two temperatures.

When 10 filaments are deposited, contacts with adjacent filaments and support arise. Figure 6 shows the temperature evolution of filament 2 at $x = 30$ mm, with and without phase change, by considering the PLA material. The phase change occurs at $t = 3.68$ s and lasts 0.4 s, due to the thermal contacts with filaments 1 and 3. The peak initiated at $t = 10$ s is created by the new thermal contact with filament 7. The maximum temperature difference of 18.9 °C between the two curves is observed at approximately $t = 8$ s. This difference is higher than in the previous example due to the contacts between filaments. As the temperature remains constant during the phase

Table 2. Process and computational parameters.

Property	Value
Process parameters Extrusion temperature (°C)	$T_L = 230$
Environment temperature (°C)	$T_E = 25$
Deposition velocity (mm/s)	30
Convective heat transfer coefficient (W/m² °C) – Natural convection	$h_{conv} = 30$
Thermal contact conductance between adjacent filaments (W/m² °C)	$h_i = 200$
Thermal contact conductance between filaments and support (W/m² °C)	$h_i = 10$
Fraction of contact length	$\alpha_i = 0.2$
Filament length (mm)	60
Filament cross-section geometry	*circle*
Filament cross-section diameter (mm)	0.25
Deposition sequence	Unidirectional and aligned
Computational parameters Time increment (s)	0.01
Temperature convergence error (°C)	0.001

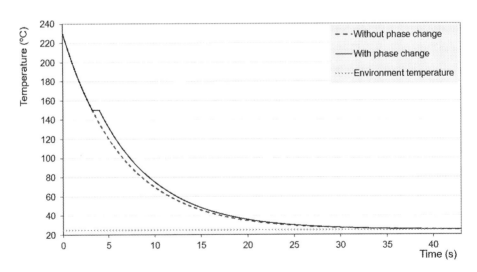

Fig. 5. Evolution of temperature with time for a single PLA filament with and without phase change (at $x = 30$ mm from the edge).

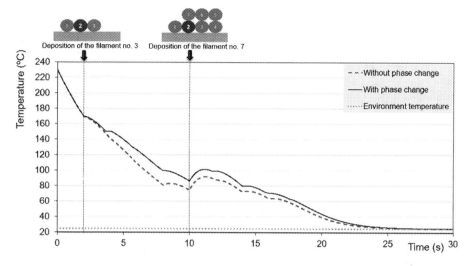

Fig. 6. Evolution of temperature with time for filament no. 2 with and without phase change (at $x = 30$ mm from the edge).

change, those filaments that are in contact with filaments that are under changing phase cool slower, i.e., the temperature of each filament is influenced by its own phase change and by the phase change of the other filaments.

Figure 7 shows the temperature evolution of filament 2, at $x = 30$ mm, for the ABS and PLA materials, with the properties presented in Table 1. The ABS filament cools faster, with a maximum temperature difference of 39.1 °C at approximately $t = 6$ s. This means that this PLA is a better option for FFF, as slower cooling favors adhesion between filaments.

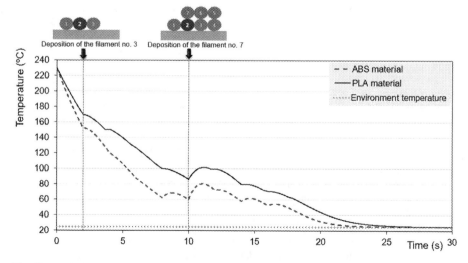

Fig. 7. Evolution of temperature with time for filament no. 2 for ABS and PLA (at $x = 30$ mm from the edge).

4   Conclusions

A simulation method to predict the temperature evolution of filaments during deposition and cooling in FFF, capable of dealing with both amorphous and semi-crystalline polymers, was implemented. The differences between the two types of materials were highlighted with two examples, by using ABS and PLA polymers. The evolution of temperature with time can become significantly different, especially when several filaments become in contact with each other.

The work presented is a first step towards predicting the properties of parts produced by FFF. Knowing the evolution of temperature with time as the deposition stage proceeds, it is possible to predict bonding between contiguous filaments by means of an healing criterion, as well as part shrinkage and warpage arising due to local temperature gradients. As a further step, the mechanical properties of 3D parts could be estimated, for example, with the use of sintering models.

Acknowledgments. This work has been supported by national funds through FCT – Fundação para a Ciência e Tecnologia through project UIDB/04728/2020.

References

1. Gebhardt, A.: Understanding Additive Manufacturing: Rapid Prototyping - Rapid Tooling - Rapid Manufacturing, 1st edn. Edition Hanser, Munich (2012)
2. Izdebska, J.: Printing on Polymers: Theory and Practice. In: Izdebska, J., Thomas, S. (eds.) Printing on Polymers: Fundamentals and Applications, pp. 1–20. William Andrew, Waltham (2016)
3. Rodríguez-Panes, A., Claver, J., Camacho, A.M.: The influence of manufacturing parameters on the mechanical behaviour of PLA and ABS pieces manufactured by FDM: a comparative analysis. Materials 11(8), 1–21 (2018)
4. Milde, J., Hrusecky, R., Zaujec, R., Morovic, L., Gorog, A.: Research of ABS and PLA materials in the process of fused deposition modeling method. In: Katalinic, B. (eds.) Proceedings of the 28th DAAAM International Symposium, Vienna, Austria, pp. 812–820. DAAM International (2017)
5. Kuznetsov, V.E., Solonin, A.N., Urzhumtsev, O.D., Schilling, R., Tavitov, A.G.: Strength of PLA components fabricated with fused deposition technology using a desktop 3D printer as a function of geometrical parameters of the process. Polymers 10(3), 313–323 (2018)
6. Hafsa, M., Ibrahim, M., Wahab, M., Zahid, M.: Evaluation of FDM pattern with ABS and PLA material. Appl. Mech. Mater. 465–466, 55–59 (2013)
7. Wojtyla, S., Klama, P., Baran, T.: Is 3D printing safe? Analysis of the thermal treatment of thermoplastics: ABS, PLA, PET, and nylon. J. Occup. Environ. Hyg. 14(6), 80–85 (2017)
8. Mudassir, A.: Measuring accuracy of two 3D Printing Materials, Department of Engineering Technologies. Bowling Green State University, Bowling Green (2016)
9. Sood, A.K., Ohdar, R.K., Mahapatra, S.S.: Parametric appraisal of mechanical property of fused deposition modelling processed parts. Mater. Des. 31, 287–295 (2010)
10. Sun, Q., Rizvi, G.M., Bellehumeur, C.T., Gu, P.: Effect of processing conditions on the bonding quality of FDM polymer filaments. Rapid Prototyping J. 14, 72–80 (2008)
11. Yang, F., Pitchumani, R.: Healing of thermoplastic polymers at an interface under nonisothermal conditions. Macromolecules 35, 3213–3224 (2002)

12. Costa, S.F., Duarte, F.M., Covas, J.A.: Estimation of filament temperature and adhesion development in fused deposition techniques. J. Mater. Process. Technol. **245**, 167–179 (2017)
13. Costa, S.F., Duarte, F.M., Covas, J.A.: Thermal conditions affecting heat transfer in FDM/FFE: a contribution towards the numerical modelling of the process. Virtual Phys. Prototyping **10**, 1–12 (2014)
14. Costa, S.F., Duarte, F.M., Covas, J.A.: An analytical solution for heat transfer during deposition in extrusion-based 3D printing techniques. In: Proceedings of the 15th International Conference Computational and Mathematical Methods in Science and Engineering, pp. 1161–1172 (2015)

Diameter and Convex Hull of Points Using Space Subdivision in $\mathbf{E}^2$ and $\mathbf{E}^3$

Vaclav Skala$^{(\boxtimes)}$ (ID)

Department of Computer Science and Engineering, Faculty of Applied Sciences,
University of West Bohemia, 301 00 Pilsen, CZ, Czech Republic
skala@kiv.zcu.cz
http://www.VaclavSkala.eu

Abstract. Convex hull of points and its diameter computation is a frequent task in many engineering problems, However, in engineering solutions, the asymptotic computational complexity is less important than the computational complexity for the expected data size to be processed. This contribution describes "an engineering solution" of the convex hulls and their diameter computation using space-subdivision and data-reduction approaches. This approach proved a significant speed-up of computation with simplicity of implementation. Surprisingly, the experiments proved, that in the case of the space subdivision the reduction of points is so efficient, that the "brute force" algorithms for the convex hull and its diameter computation of the remaining points have nearly no influence to the time of computation.

1 Introduction

The Convex Hull (CH) and the Convex Hull Diameter (CH-D) algorithms are applicable in many areas. Those algorithms are deeply analyzed in the *computational geometry* for asymptotic behavior, i.e. $N \mapsto \infty$. However, it is not quite what today's applications need. In engineering problems, there are two main aspects, which have to be respected:

– the number of input elements
– the dimension of the data

Geometrically oriented algorithms usually process two or three dimensional data and the number of elements is usually not higher than 10^8 in the E^2 case ($10^4 \times 10^4$), resp. 10^{12} in the E^3 case ($10^4 \times 10^4 \times 10^4$) in the real applications. It means, that the engineering solutions have to respect several factors:

– the limited size of data sets to be processed, i.e. asymptotically better algorithm is not necessarily the best one for the intended application scope,
– numerical stability and robustness, i.e. the implementation has to respect a limited precision of computation resulting from the IEEE 754-2019 floating-point representation standard, including the fact that the *Quadruple* and *Octuple* precisions are not supported on today's processors,

© Springer Nature Switzerland AG 2020
O. Gervasi et al. (Eds.): ICCSA 2020, LNCS 12249, pp. 286–295, 2020.
https://doi.org/10.1007/978-3-030-58799-4_21

- memory management and data transfer via data-bus from memory to CPU/GPU and vice-versa; also the influence of caching cannot be ignored,
- simplicity and efficiency of algorithms, i.e. too complicated algorithm will not be probably used, especially if its behavior is not "stable" and predicable,

This contribution describes the principle of two basic efficient algorithms, recently designed, implemented and verified, which are based on efficient reduction of points using the space subdivision and significant reduction of points remaining for the final processing, i.e.:

- Convex Hull Diameter (CH-D) of points in E^2 - the algorithms are usually based on the Convex Hull (CH) algorithms, which are more or less based on sophisticated algorithms. However, they are not easy to implement especially in the limited precision of computation for a higher number of points.
- Convex Hull (CH) of points in E^2 - algorithm using a deterministic heuristic approach with space subdivision.

In the following, simple algorithms for finding CH-D and CH of the given points in the E^2 case are described, which are easy to implement and very efficient. It should be noted, that those algorithms can be easily extended to the E^3 case.

2   Finding a Diameter of a Convex Hull

Finding the maximum distance of points in the E^n case is usually done by a simple algorithm that has $O(N^2)$, where N is the number of the given points.

Algorithm 1: Maximum distance of two points in E^2 or E^3

Result: Maximum_Distance
\# N - Number of points $\mathbf{x}_i, i = 1, ..., N$;
d := 0;
i := 0;
while $i <= N - 1$ **do**
    j:= i+1;
    while $j <= N$ **do**
        d0 := $\|\mathbf{x}_i - \mathbf{x}_j\|$; \# Euclidean distance
        if $d0 < d$ **then**
            d := d0 ;
        end
    end
end
Maximum_Distance := d;

However, such an algorithm is quite inefficient due to its $O(N^2)$ computational complexity and also of the $\|.\|$ computation, where $\sqrt{(.)}$ is used. Algorithm 2 presents simple modification using $\|.\|^2$ for the distance comparison.

Algorithm 2: Modified Maximum distance of two points in E^2 or E^3

Result: Maximum_Distance
\# N - Number of points $\mathbf{x}_i, i = 1, ..., N$;
d := 0;
i := 0;
while $i <= N - 1$ **do**
    j:= i+1;
    while $j <= N$ **do**
        d0 := $\|\mathbf{x}_i - \mathbf{x}_j\|^2$; \# Euclidean distance
        if $d0 < d$ **then**
            d := d0 ;
        end
    end
end
Maximum_Distance := sqrt(d);

Such a very simple modification of the algorithm Algorithm1 has a significant influence on computational efficiency. However, for larger values of N, the time of computing is still very high. Finding the maximum distance of two points, generally in the E^n case, is equivalent to the convex hull diameter problem, which has deeply analyzed in computational geometry for a long time. There are

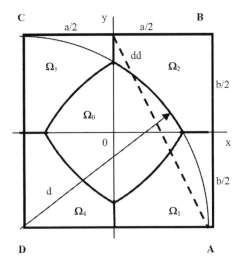

Fig. 1. Splitting data into areas by subdivision

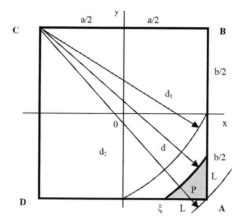

Fig. 2. Timing of the Convex Hull Diameter algorithms with space subdivision

several Convex Hull Diameter (CH-D) algorithms based on Convex Hull (CH) algorithms developed recently, e.g. Toussaint [26]. Some algorithms were deterministic, some of those based on a stochastic approach, e.g. Xue [27]. Efficient CH algorithms get more and more complex as far as implementation aspects are concerned. Recently, a simple algorithm based on the space subdivision was developed Skala [19], Fig. 1. It is primarily based on finding extreme points of the Axis Aligned Bounding Box (AABB) and the closest points to the AABB corners, which forms the first estimation of the convex hull, Fig. 1, where d is its diameter. Then the points are split to five areas $\Omega_i, i = 0, ..., 4$. The points from the central area Ω_0 cannot have any influence to the final convex hull [19,23] and can be removed automatically.

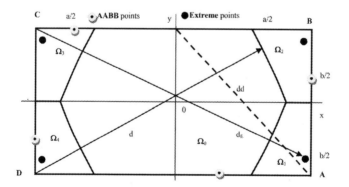

Fig. 3. Splitting data into areas by subdivision

The CH-D algorithm is based on the idea of points reduction first, followed by actual CH algorithm use for finding the diameter of the reduced data-set, i.e. computation of the CH-D value. The influence of the simple reduction step, which is of $O(N)$ complexity, was overwhelming the expectations see Fig. 4 and Fig. 5.

$$\xi = \frac{\Omega}{\Omega_1 + \Omega_2 + \Omega_3 + \Omega_4} \tag{1}$$

where Ω is the area of the AABB box.

The expected reduction ratio Eq. 1 is given by the area bounded by d_1 and d_2, where d_1 is the worst case of the estimated diameter form the AABB box and when the data are in the squared domain, Fig. 2. If the data are in the rectangular area, see Fig. 3, then the Ω_i areas are getting significantly smaller.

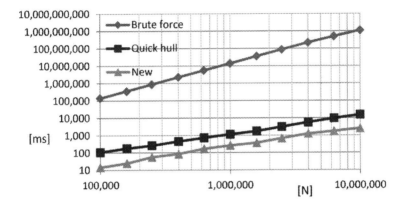

Fig. 4. Timing of the Convex Hull Diameter algorithms with space subdivision

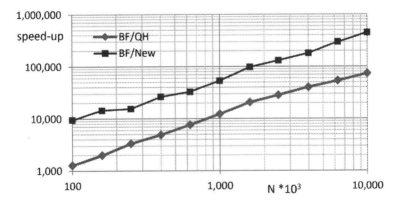

Fig. 5. Speed-up of the Convex Hull Diameter algorithm with space subdivision
BF/QH - BruteForce/ QuickHull, BF/New - BruteForce/proposed

As the final data sets after this reduction step were extremely small, simple CH algorithms were used, practically without any influence to time complexity. The Halton's [11] point generation was used in the experimental evaluation and detailed description and experimental results were described in Skala [19,20,22]. The timing and speed-up are presented in Fig. 4 and Fig. 5.

It can be seen that the space subdivision significantly speed-up the CH-D algorithm. Surprisingly, the presented CH-D algorithm handle close circle generated points efficiently as well. If the AABB is not a square, the presented CH-D algorithm gets even more efficient Skala [19,22]. The extension to the E^3 case is simple, straight forward and easy to implement Skala [21].

3   Convex Hull with Space Subdivision

Several deterministic Convex Hull (CH) algorithms have been described. Table 1 presents some well-known algorithms and their asymptotic computational complexity. The estimation of the lower bound complexity was given by Yao [28]. Some of those were incremental, e.g. Kallay [13], Also a stochastic approach has been described by recently, e.g. Xue [27].

Table 1. Table of Convex Hull algorithms and their complexities

Algorithm	Complexity	Reference
Gift Wrapping	nh	Chand and Kapur, 1970 [5]
Graham Scan	$n\,log(n)$	Graham, 1972 [9]
Jarvis March	nh	Jarvis, 1973 [12]
QuickHull	nh	Barber, 1996 [2], Eddy, 1977 [8]
		Bykat, 1978 [3], Devai, 1979 [7]
Divide & Conquer	$n\,log(n)$	Preparata and Hong, 1977 [18]
Monotone Chain	$n\,log(n)$	Andrew, 1979 [1]
Incremental	$n\,log(n)$	Kallay, 1984 [13]
Marriage before Conquest	$n\,log(h)$	Kirkpatrick & Seidel, 1986 [14]
Chan's algorithm	$n\,log(h)$	Liu and Chen, 2007 [15]

Recently, an interesting QuickhullDisk algorithm, which is based on disks was published by Nguen [17]. Parallel algorithms output-sensitive were described by Chan [4], Gupta [10]. Manual comparison of some algorithms via multimedia exposition can be found at Loffler [16]. In engineering applications, usually a two-dimensional or three-dimensional CH algorithms are required and the number of points to be processed is up to 10^8. Therefore, a simple Smart Convex Hull (S-CH) algorithm with space subdivision was developed for the E^2 case [19]. The modification to the E^3 case was described in Skala [21]. Parallel algorithms

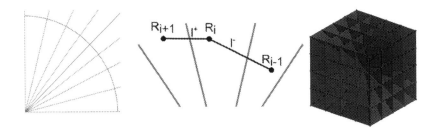

Fig. 6. The S-CH algorithm space subdivision in E^2 and E^3 taken from [21,23]

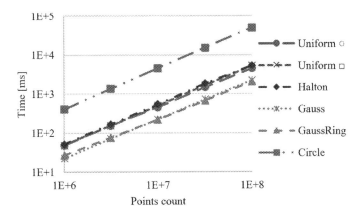

Fig. 7. Timing of the S-CH algorithm for a different distribution of points [20]

were described as well, e.g. by Sugihara[25], Gupta [10], a modification for GPU use by Stein [24].

The S-CH algorithm in the E^2 case is based on the polar space subdivision, Fig. 6, which replaces the orthogonal space subdivision in the CH-D algorithm described above. Detailed algorithm description and experimental results can be found in Skala [20]. The AABB border is split uniformly, so the angles in polar space splitting are different. Figure 7 presents the behavior of the S-CH algorithm for different data distribution including uniform on a circle.

Figure 8 presents the speed-up of the S-CH algorithm using the Graham Scan algorithm after the reduction of points, i.e. in the final step, with the selected convex hull algorithms Fig. 9.

The S-CH algorithm stores maximum distance from the origin in the angular segment and related two segments $R_{i+1}R_i$ and R_iR_{i-1} are updated so they form the approximate convex hull. Each angular segment may contain several points that form the final convex hull. The number of angular segments partially influences the time of computation, see Skala [23].

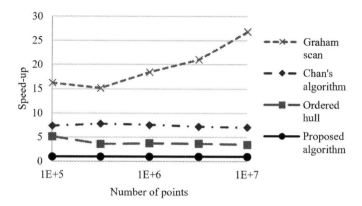

Fig. 8. Speed-up of the S-CH algorithm with the uniform distribution of points [20]

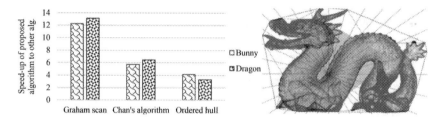

Fig. 9. Speed-up of the S-CH algorithm with the uniform distribution of points [20]

4   Conclusions

This paper presents an "engineering approach" to the Convex Hull and Convex Hull Diameter computations based on the space subdivision techniques designed for processing up to 10^8 points in the E^2 case. The algorithms are easily extensible to the E^3 case. Detailed results of experiments are given in detail in related papers, namely [20,21].

The presented approach is also used in teaching to show the influence of space subdivision to computational complexity. The presented algorithms are expected to be modified for GPU use in the future.

Acknowledgments. The author would like to thank colleagues and students at the University of West Bohemia, Pilsen, for their discussions and suggestions, especially to Zuzana Majdisova and Michal Smolik for recent implementations and experiments made, and to anonymous reviewers for their valuable comments and hints provided.

References

1. Andrew, A.: Another efficient algorithm for convex hulls in two dimensions. Inf. Process. Lett. **9**(5), 216–219 (1979)

2. Bradford, B., Dobkin, D., Dobkin, H., Huhdanpaa, D.P.: The quickhull algorithm for convex hulls. ACM Trans. Mathe. Softw. **22**(4), 469–483 (1996)
3. Bykat, A.: Convex hull of a finite set of points in two dimensions. Inf. Process. Lett. **7**(6), 296–298 (1978)
4. Chan, T.: Optimal output-sensitive convex hull algorithms in two and three dimensions. Discrete Comput. Geom. **16**(4), 361–368 (1996)
5. Chand, D., Kapur, S.: An algorithm for convex polytopes. J. ACM **17**(1), 78–86 (1970)
6. Dobkin, D., Snyder, L.: On a general method for maximizing and minimizing among certain geometric problems. In: FOCS-IEEE Symposium on Foundations of Computer Science Proceedings, pp. 9–17 (1979)
7. Dévai, F., Szendrényi, T.: Comments on convex hull of a finite set of points in two dimensions. Inf. Process. Lett. **9**(3), 141–142 (1979)
8. Eddy, W.: Algorithm 523: convex, a new convex hull algorithm for planar sets. ACM Trans. Math. Softw. (TOMS) **3**(4), 398–403 (1997)
9. Graham, R.: An efficient algorithm for determining the convex hull of a finite planar set. Inf. Process. Lett. **1**(4), 132–133 (1972)
10. Gupta, N., Sen, S.: Faster output-sensitive parallel algorithms for 3D convex hulls and vector maxima. J. Para. Distrib. Comput. **63**(4), 488–500 (2003)
11. Halton, J.: Algorithm 247: radical-inverse quasi-random point sequence. CACM **7**(12), 701–702 (1964)
12. Jarvis, R.: On the identification of the convex hull of a finite set of points in the plane. Inf. Process. Lett. **2**(1), 18–21 (1973)
13. Kallay, M.: The complexity of incremental convex hull algorithms in R^d. Inf. Process. Lett. **19**(4), 197 (1984)
14. Kirkpatrick, D., Seidel, R.: The ultimate planar convex hull algorithm? SIAM J. Comput. **15**(1), 287–299 (1986)
15. Liu, G., Chen, C.: A new algorithm for computing the convex hull of a planar point set. J. Zhejiang Univ. Sci. A **8**(8), 1210–1217 (2007)
16. Löffler, M.: A manual comparison of convex hull algorithms (multimedia exposition). In: 35th International Symposium on Computational Geometry (SoCG 2019), volume 129 of Leibniz International Proceedings in Informatics (LIPIcs), Dagstuhl, Germany, pp. 65:1–65:2. Schloss Dagstuhl-Leibniz-Zentrum fuer Informatik (2019)
17. Nguyen, K., Song, C., Ryu, J., Thanh, P.A., Hoang, N.-D., Kim, D.-S.: Quickhulldisk: a faster convex hull algorithm for disks. Appl. Math. Comput. **363**, 124626 (2019)
18. Preparata, F., Hong, S.: Convex hulls of finite sets of points in two and three dimensions. CACM **20**(2), 87–93 (1997)
19. Skala, V.: Fast $O_{\text{expected}}(N)$ algorithm for finding exact maximum distance in E^2 instead of $O(N^2)$ or $O(NlgN)$. In: ICNAAM 2013, AIP Proceedings, USA, vol. 1558, pp. 2496–2499. AIP Publishing (2013)
20. Skala, V., Majdisova, Z.: Fast algorithm for finding maximum distance with space subdivision in E^2. In: Zhang, Y.-J. (ed.) ICIG 2015. LNCS, vol. 9218, pp. 261–274. Springer, Cham (2015). https://doi.org/10.1007/978-3-319-21963-9_24
21. Skala, V., Majdisova, Z., Smolik, M.: Space subdivision to speed-up convex hull construction in E^3. Adv. Softw. Eng. **91**(C), 12–22 (2016)
22. Skala, V., Smolik, M.: Simple and fast $Oexp(N)$ algorithm for finding an exact maximum distance in E^2 instead of $O(N^2)$ or $O(NlgN)$. In: Misra, S., et al. (eds.) ICCSA 2019. LNCS, vol. 11619, pp. 367–380. Springer, Cham (2019). https://doi.org/10.1007/978-3-030-24289-3_27

23. Skala, V., Smolik, M., Majdisova, Z.: Reducing the number of points on the convex hull calculation using the polar space subdivision in E^2. In: 29th SIBGRAPI Conference on Graphics, Patterns and Images (SIBGRAPI 2016), pp. 40–47. IEEE (2016)
24. Stein, A., Geva, E., El-Sana, J.: Cudahull: fast parallel 3D convex hull on the GPU. Comput. Graph. **36**(4), 265–271 (2012)
25. Sugihara, K.: Robust gift wrapping for the three-dimensional convex hull. J. Comput. Syst. Sci. **49**(2), 391–407 (1994)
26. Toussaint, J., McAlear, G.T.: A simple o(n log n) algorithm for finding the maximum distance between two finite planar sets. Pattern Recogn. Lett. **1**(1), 21–24 (1982)
27. Xue, J., Li, Y., Janardan, R.: On the expected diameter, width, and complexity of a stochastic convex hull. Comput. Geom. **82**, 16–31 (2019)
28. Yao, A., Andrew, C.: A lower bound to finding convex hulls. J. ACM **28**(4), 780–787 (1981)

Automated Selection of the Computational Parameters for the Higher-Order Parabolic Equation Numerical Methods

Mikhail S. Lytaev[✉] [ID]

Saint Petersburg Institute for Informatics and Automation of the Russian Academy of Sciences, 14-th Linia, VI, No. 39, Saint Petersburg 199178, Russia
mikelytaev@gmail.com

Abstract. This study is devoted to the features of the numerical methods for the parabolic wave equation. While seeking a numerical solution, it is necessary to select a set of computational parameters of the numerical method. The choice of the computational parameters affects the speed and accuracy of the calculations. Automation of the choice of computational parameters is useful when applying mentioned numerical methods in complex software systems, where the user cannot select them manually. In this paper, we consider a finite-difference split-step Padé method for the one-way Helmholtz equation. A discrete dispersion relation based algorithm for finding the optimal computational parameters of the numerical method is presented.

Keywords: Parabolic equation · Dispersion analysis · Wave propagation · Computational parameter · Padé approximant · Finite-difference approximation

1 Introduction

Approximation of the elliptic Helmholtz equation by the parabolic equation (PE) has gained a considerable popularity in the wave propagation study. In 1946, Soviet scientists M. A. Leontovich and V. A. Fock [21] proposed a wave-based approach to solving the tropospheric radio wave propagation problem using the PE. The original analysis of the PE method has been based on some physical considerations. Subsequently, a series of papers appeared [3,9,10,15], where the PE method is derived and analyzed more strictly from a mathematical point of view, relying mainly on the theory of operators.

The PE method is well suitable for wave propagation modeling in inhomogeneous waveguides. PE method is used for tropospheric radio wave propagation [22], propagation of the acoustic waves in the underwater sound channel [18,29], acoustic and electromagnetic scattering [17,32], optics [14], quantum mechanics [11,26] and geophysics [35]. The main advantage of this method is the possibility

© Springer Nature Switzerland AG 2020
O. Gervasi et al. (Eds.): ICCSA 2020, LNCS 12249, pp. 296–311, 2020.
https://doi.org/10.1007/978-3-030-58799-4_22

of effective numerical implementation. Several software systems have been implemented based on the PE method: AREPS [5], RAM [13], CARPET, PETOOL [28]. A significant number of PE codes for underwater acoustics are collected in [30].

Any numerical method has many artificial parameters, such as the geometry of the computational grid, the size of the integration area, the approximation order, the convergence threshold and so on. These parameters directly affect the accuracy and speed of calculations. Despite a significant number of works devoted to numerical methods for PE, they practically do not pay attention to the problem of choosing the computational parameters depending on the propagation conditions, as well as describing the limits of applicability of these numerical methods. It is usually assumed that the researcher can pick them up manually based on his experience. However, this approach is completely inapplicable when using these methods in complex software systems. Operators of such systems cannot manually select computational parameters for each propagation case. There is a need for a method that is able to automatically determine the computational parameters depending on the input data and the required accuracy of calculations. The purpose of the present research is to develop a method for automating the finding of the computational parameters of PE numerical methods depending on the input data. The subject of the study is the finite-difference rational approximations of the one-way Helmholtz equation [7,22].

The paper is organized as follows. Section 2 describes the numerical scheme under consideration with emphasis on spectral representation of the propagation operator. In Sect. 3, a discrete dispersion analysis of the numerical scheme is performed. Section 4 is devoted to the synthesis of the automated computational parameter selection algorithm. Numerical examples are given in Sect. 5.

2   Numerical Method

A number of wave propagation problems can be reduced to the Helmholtz equation in the following form [22]

$$\frac{\partial^2 \psi}{\partial x^2} + \frac{\partial^2 \psi}{\partial z^2} + k^2 m^2(x, z)\psi = 0, \tag{1}$$

where

- $\psi(x, z)$ is a complex-valued wave field function;
- $m^2(x, z)$ is the squared refractive index;
- $k = \frac{2\pi}{\lambda}$ is the wavenumber;
- λ is the wavelength.

Equation (1) may be defined in an unbounded domain. We are seeking the solution of Eq. (1) on a finite domain $\Omega = [0, x_{max}] \times [0, z_{max}]$. Nevertheless, the bounds of Ω may be transparent [12,19,25].

The wave field is generated by the initial condition

$$\psi(0, z) = \psi_0(z),$$

where $\psi_0(z)$ is a known function, which usually corresponds to the far-field aperture pattern of the wave field radiator. We proceed from the assumption that the generated waves propagate mostly in a preferred direction along the longitudinal coordinate x.

2.1 Propagation Operator Approximation

Evolution solution of Eq. (1) for the outgoing field can be written as follows [22]

$$u(x + \Delta x, z) = \exp\left(ik\Delta x\left(\sqrt{1 + L} - 1\right)\right) u(x, z) \tag{2}$$

where

$$Lu = \frac{1}{k^2}\frac{\partial^2 u}{\partial z^2} + \left(m^2(x, z) - 1\right) u,$$

$$u(x, z) = e^{-ikx}\psi(x, z). \tag{3}$$

Thus, the step-by-step numerical method is based on the approximation of the pseudo-differential propagation operator (2). The following designations are used in this paper

$$z_j = j\Delta z, \ x_n = n\Delta x, \ u_j^n = u(x_n, z_j).$$

Using the definition of pseudo-differential operator [34] and Fourier transform by variable z, expression (2) can be rewritten as follows

$$\tilde{u}^{n+1}(k_z) = \exp\left(ik\Delta x\left(\sqrt{1 - \frac{k_z^2}{k^2} + (m^2(x, z) - 1)} - 1\right)\right)\tilde{u}^n(k_z), \tag{4}$$

where

$$\tilde{u}^n(k_z) = \frac{1}{\sqrt{2\pi}}\int\limits_{-\infty}^{+\infty} u^n(z)e^{-ik_z z}dk_z.$$

To simplify expression (4), we use the Padé approximant [1,7], written in the following form

$$\exp\left(ik\Delta x\left(\sqrt{1 + \xi} - 1\right)\right) \approx \frac{\prod_{l=1}^{n} 1 + a_l\xi}{\prod_{l=1}^{m} 1 + b_l\xi} = \prod_{l=1}^{p}\frac{1 + a_l\xi}{1 + b_l\xi}. \tag{5}$$

Methods for calculating and analyzing Padé approximation coefficients for various elementary functions are described in [1]. For the numerical calculations of coefficients $a_1, \ldots, a_p, b_1, \ldots, b_p$, it is recommended to use computer algebra and symbolic calculations.

It will be shown later that the approximation (5) affects the maximum propagation angle and the maximum longitudinal step size Δx.

Substituting (5) into (4) and introducing auxiliary functions $v_1^n, v_2^n, \ldots, v_{p-1}^n$, the propagation operator in spectral domain can be approximately written as a system of equations

$$
\begin{cases}
\left(1 + b_1 \left(-\frac{k_z^2}{k^2} + m^2 - 1\right)\right) \tilde{v}_1^n = \left(1 + a_1 \left(-\frac{k_z^2}{k^2} + m^2 - 1\right)\right) \tilde{u}^{n-1} \\
\left(1 + b_l \left(-\frac{k_z^2}{k^2} + m^2 - 1\right)\right) \tilde{v}_l^n = \left(1 + a_l \left(-\frac{k_z^2}{k^2} + m^2 - 1\right)\right) \tilde{v}_{l-1}^n \quad l = 2 \ldots p - 1 \\
\ldots \\
\left(1 + b_p \left(-\frac{k_z^2}{k^2} + m^2 - 1\right)\right) \tilde{u}^n = \left(1 + a_p \left(-\frac{k_z^2}{k^2} + m^2 - 1\right)\right) \tilde{v}_{p-1}^n.
\end{cases}
$$
$$(6)$$

By applying the inverse Fourier transform to each row of system (6), we obtain the following system of differential equations

$$
\begin{cases}
(1 + b_1 L) v_1^n = (1 + a_1 L) u^{n-1} \\
(1 + b_l L) v_l^n = (1 + a_l L) v_{l-1}^n \quad l = 2, \ldots, p - 1 \\
\ldots \\
(1 + b_p L) u^n = (1 + a_p L) v_{p-1}^n.
\end{cases}
$$
$$(7)$$

This system can be formally written as a function of operator L

$$
u^{n+1} = \prod_{l=1}^{p} \frac{1 + a_l L}{1 + b_l L} u^n.
$$

Thus, the use of the propagation operator (2) at each step is reduced to the solution of the system of one-dimensional Eqs. (7).

The system of Eqs. (7) is solved sequentially from top to bottom. Each row is a similar differential equation of the form

$$
(1 + bL) v = (1 + aL) u, \tag{8}
$$

where u is a known function, which is obtained on a previous step, v is unknown function.

Thus, the original differential equation is reduced to a semi-discrete one. Its solution can be represented by a sequence of one-dimensional differential equations of the form (8).

2.2 Transversal Operator Discretization

Further, we construct a numerical scheme for Eq. (8). We take into consideration a uniform partition in the z direction with step size Δz and denote $u(j\Delta z) = u_j$. Consider the second order difference operator

$$\delta^2 u = u(z - \Delta z) - 2u(z) + u(z + \Delta z) = u_{j-1} - 2u_j + u_{j+1}.$$

Operator δ^2 can be represented as follows

$$\delta^2 u = \frac{1}{\sqrt{2\pi}} \int\limits_{-\infty}^{+\infty} e^{ik_z z} \left(e^{-ik_z \Delta z} - 2 + e^{ik_z \Delta z} \right) \tilde{u}(k_z) dk_z =$$

$$\frac{1}{\sqrt{2\pi}} \int\limits_{-\infty}^{+\infty} e^{ik_z z} f(k_z) \tilde{u}(k_z) dk_z,$$

where

$$f(k_z) = 2 \left(\cosh\left(ik_z \Delta z \right) - 1 \right).$$

Thus, owing to the definition of function of differential operator [34], δ^2 can be written as a function

$$\delta^2 u = f\left(\frac{\partial}{\partial z} \right) u.$$

Now we will construct the inverse operator. We will express the differential operator through the second difference operator. Consider the inverse function

$$k_z = f^{-1}(\xi) = \frac{1}{i\Delta z} \cosh^{-1}\left(1 + \frac{\xi}{2} \right) = \frac{1}{i\Delta z} \ln\left(1 + \frac{\xi}{2} + \sqrt{\xi + \frac{\xi^2}{4}} \right).$$

Second derivative operator can be represented in the following form

$$\frac{\partial^2 u}{\partial z^2} = \frac{1}{\sqrt{2\pi}} \int\limits_{-\infty}^{+\infty} e^{ik_z z} (-k_z^2) \tilde{u}(k_z) dk_z =$$

$$\frac{1}{\sqrt{2\pi}} \int\limits_{-\infty}^{+\infty} e^{ik_z z} \left(- \left(f^{-1}\left(f(k_z) \right) \right)^2 \right) \tilde{u}(k_z) dk_z =$$

$$\frac{1}{\Delta z^2} \ln^2 \left(1 + \frac{\delta^2}{2} + \sqrt{\left(1 + \frac{\delta^2}{2} \right)^2 - 1} \right). \quad (9)$$

This expression allows one to construct an arbitrary accuracy finite-difference approximation of the differential operator. Indeed, by decomposing expression (9) according to Taylor's formula, we get

$$\frac{\partial^2 u}{\partial z^2} = \frac{1}{\Delta z^2} \left[\delta^2 - \frac{1}{12} \delta^4 + \frac{1}{90} \delta^6 - \dots \right] u. \quad (10)$$

Leaving the first two terms in the decomposition (10) and using Padé approximation, we get the following expression for the second derivative [20]

$$\frac{\partial^2}{\partial z^2} \approx \frac{1}{\Delta z^2} \frac{\delta^2}{1 + \beta \delta^2},\tag{11}$$

where

$$\beta = \begin{cases} 0, & \text{for 2-d order approximation,} \\ \frac{1}{12}, & \text{for 4-th order approximation.} \end{cases}$$

Substituting this approximation into (8), we obtain the following difference equation

$$\left((k\Delta z)^2 \left(1 + \beta \delta^2 \right) + b \delta^2 + b \left(k\Delta z \right)^2 \left(m_j^2 - 1 \right) \right) v_j +$$
$$b \left(k\Delta z \right)^2 \beta \delta^2 \left(\left(m_j^2 - 1 \right) v_j \right) =$$
$$\left((k\Delta z)^2 \left(1 + \beta \delta^2 \right) + a \delta^2 + a \left(k\Delta z \right)^2 \left(m_j^2 - 1 \right) \right) u_j +$$
$$a \left(k\Delta z \right)^2 \beta \delta^2 \left(\left(m_j^2 - 1 \right) u_j \right), \tag{12}$$

where $m_j^2 = m^2(n\Delta x, j\Delta z)$.

Equation (12) on a finite grid with discrete boundary conditions can be represented as a tridiagonal system of linear algebraic equations that admits an effective numerical solution using the tridiagonal matrix algorithm [31].

Using an approximation of order higher than four in (10) is possible, but results in less sparse matrices.

In the case of a homogeneous medium ($m(x, z) \equiv 1$), longitudinal and transversal discretization can be joined using Padé approximation. Operator L in this case coincides with the second derivative operator. Substituting the dependence of the differential operator on the second difference (9) into the propagation operator (2) and applying Padé approximation to the resulting operator, which now depends on the second difference operator δ^2, we obtain the following expression

$$\exp \left(ik\Delta x \left(\sqrt{1 + L} - 1 \right) \right) =$$
$$\exp \left(ik\Delta x \left(\sqrt{1 + \frac{1}{k^2 \Delta z^2} \ln^2 \left(1 + \frac{\delta^2}{2} + \sqrt{\left(1 + \frac{\delta^2}{2} \right)^2 - 1} \right)} - 1 \right) \right) \approx$$
$$\frac{\prod_{l=1}^{n} 1 + a_l' \delta^2}{\prod_{l=1}^{m} 1 + b_l' \delta^2}. \tag{13}$$

Next, we apply the algorithm described above for a second-order scheme ($\beta = 0$), but using the new coefficients a_l' and b_l'.

It should be noted that another widely used method for solving PE is the split-step Fourier method [28]. The split-step Fourier method does a better job of approximating the diffraction part of the propagation operator and is preferred in cases when the boundary conditions do not have a significant effect on propagation. At the same time, this method meets serious issues with modeling both artificial and natural boundary conditions [24,37]. Also, the split-step Fourier method is not applicable for accounting for strong spatial variations of the refractive index m^2.

Note that Padé-[1/1] approximation of the propagation operator is equivalent to the Crank-Nicolson numerical scheme [22] for PE, thus the considered numerical method is a generalization of it.

The above reasoning can also be applied to the linear Schrodinger equation. It is sufficient to replace the square root operator $\sqrt{1 + L}$ with $1 + L/2$ in the expression (2).

3   Discrete Dispersion Analysis

When constructing the numerical scheme in the previous section, several additional parameters appeared to be defined. Namely: the grid step along the longitudinal and transversal coordinate (Δx and Δz) and Padé approximation coefficients $a_1 \ldots a_p$, $b_1 \ldots b_p$.

It is well known that the Crank-Nicolson numerical scheme and finite-difference Padé approximations considered in this paper are unconditionally stable [22]. Thus, the solution of the finite-difference equation tends to the exact solution as the sampling of the computational grid increases. At the same time, reducing the cells of the computational grid increases the required time and memory. Therefore it is of great interest from a computational point of view to determine the maximum longitudinal and transversal increments, which at the same time provide acceptable accuracy. It is common to use adaptive step size algorithms [31] to solve this problem, which consists in gradually reducing the grid steps until the required accuracy is achieved. However, since the numerical scheme under consideration requires the selection of several parameters simultaneously, an alternative method is required. This study proposes an alternative solution based on the analysis of discrete dispersion relations.

We analyze the dispersion relations [4,16] for the discrete numerical scheme, described in the previous section. Keeping in mind replacement (3), consider the discrete plane wave

$$E_j^n = \exp\left(-ikn\Delta x\right)\exp\left(ik_x n\Delta x + ik_z j\Delta z\right),\tag{14}$$

where

- $k_x = k\cos\theta$ is the horizontal wavenumber;
- $k_z = k\sin\theta$ is the vertical wavenumber;
- θ is the angle between the direction of the plane wave and the positive direction of the x-axis.

Dispersion analysis is performed for the case of a homogeneous medium $(m(x, z) \equiv 1)$. Taking into account that

$$E_j^{n+1} = \exp\left(i\left(k_x - k\right) \Delta x\right) E_j^n,$$

we substitute the expression (14) to the discrete scheme. Then, after some simple calculations, we derive the following dispersion relation

$$\exp\left(ik_x \Delta x\right) = \exp\left(ik\Delta x\right) \prod_{l=1}^{p} \frac{1 - \frac{a_l}{k^2} \frac{4}{\Delta z^2} \left(\sin^2\left(\frac{k_z \Delta z}{2}\right) + 4\beta \sin^4\left(\frac{k_z \Delta z}{2}\right)\right)}{1 - \frac{b_l}{k^2} \frac{4}{\Delta z^2} \left(\sin^2\left(\frac{k_z \Delta z}{2}\right) + 4\beta \sin^4\left(\frac{k_z \Delta z}{2}\right)\right)}.$$

Dispersion relation for the original Helmholtz Eq. (1) is written as follows

$$k_x^2 + k_z^2 = k^2.$$

Discrete horizontal wavenumber as a function of the computational parameters and the propagation angle θ is expressed as follows

$$k_x^d(\Delta x, \Delta z, a_1 \ldots a_p, b_1 \ldots b_p, \theta) =$$
$$k + \frac{1}{i\Delta x} \ln \left[\prod_{l=1}^{p} \frac{1 - \frac{a_l}{k^2} \frac{4}{\Delta z^2} \left(\sin^2\left(\frac{k \sin\theta \Delta z}{2}\right) + 4\beta \sin^4\left(\frac{k \sin\theta \Delta z}{2}\right)\right)}{1 - \frac{b_l}{k^2} \frac{4}{\Delta z^2} \left(\sin^2\left(\frac{k \sin\theta \Delta z}{2}\right) + 4\beta \sin^4\left(\frac{k \sin\theta \Delta z}{2}\right)\right)} \right]. \quad (15)$$

Horizontal wavenumber for the original Helmholtz equation (1) is expressed by the following formula

$$k_x = \begin{cases} \sqrt{k^2 - k_z^2}, & |k_z| \leq k, \\ i\sqrt{k_z^2 - k^2}, & |k_z| > k. \end{cases} \quad (16)$$

Using expressions (15) and (16), one can estimate the phase error of the numerical scheme at each propagation step. The discrete dispersion relation (15) includes all the above mentioned computational parameters of the numerical scheme, which makes it possible to analyze it. Relation (15) gives an opportunity to analyze the accuracy of a particular scheme depending on the propagation angle.

Figure 1 represents the dependence of the value k_x^d on the propagation angle for various approximations of the propagation operator. Corresponding relative error is depicted in Fig. 2. One can easily see that the increased accuracy of the approximation leads to an increased maximum propagation angle.

A numerical scheme based on Padé approximation of the form $[p/q]$ is stable under the condition $0 \leq q - p \leq 2$ [6]. The scheme is A-stable under the condition $p = q$ [6, 22]. Its disadvantage is that the corresponding values of k_x are always real. As one can see from expression (16), in the evanescent part of the spectrum $(k_z > k)$, the horizontal wavenumber of the original Helmholtz equation is purely imaginary. Thus, evanescent waves, which are formed when the waves are

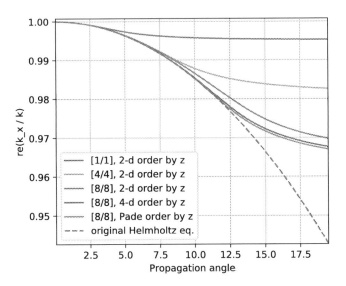

Fig. 1. Dependence of the value k_x^d on the propagation angle for various approximations of the propagation operator. In all the examples $\Delta x = 100\lambda$, $\Delta z = 1\lambda$.

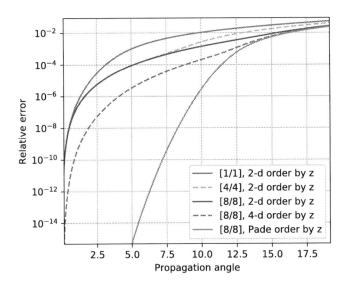

Fig. 2. Dependence of the relative error of value k_x^d on the propagation angle for various approximations of the propagation operator. In all the examples $\Delta x = 100\lambda$, $\Delta z = 1\lambda$.

scattered on vertical inhomogeneities, do not attenuate when using the scheme $[p/p]$. This leads to artificial Gibbs oscillations of the solution [22,25]. Schemes of the form $[p-1/p]$ and $[p-2/p]$ are L-stable, thus they suppress the occurring high-frequency oscillations.

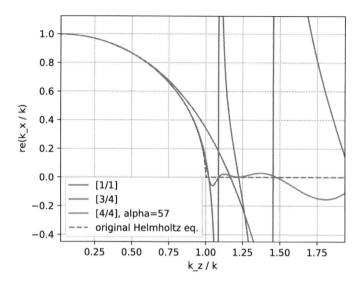

Fig. 3. Behavior of the real part of k_x^d in the evanescent part of the spectrum. In all the examples $\Delta x = 0.1\lambda$, $\Delta z = 0.1\lambda$.

Figure 3 and 4 demonstrate the dependence of the real and imaginary parts of k_x on k_x^d for various approximations. It can be seen that the scheme [3/4] really provides the fulfillment of the condition $\mathrm{Im}(k_x) > 0$ for $k_z > k$, i.e. evanescent waves will be damped. However, the behavior of k_x^d for scheme [3/4] is not the same as for original Helmholtz equation. In many propagation problems, the evanescent spectrum is not significant, and such regularization method is quite suitable. To correctly account for the evanescent spectrum, special Padé approximations of the square root should be applied [27,36]. Their essence is to turn the branch cut of the square root to a certain angle α. Then, based on a real-valued approximation of the form [2]

$$\sqrt{1+\xi} \approx 1 + \sum_{l=1}^{p} \frac{a_l \xi}{1 + b_l \xi},$$

where

$$a_l = \frac{2}{2p+1} \sin^2 \frac{l\pi}{2p+1},$$

$$b_l = \cos^2 \frac{l\pi}{2p+1},$$

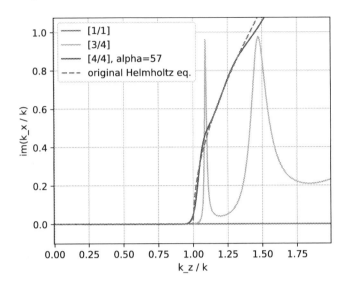

Fig. 4. Behavior of the imaginary part of k_x^d in the evanescent part of the spectrum. In all the examples $\Delta x = 0.1\lambda, \Delta z = 0.1\lambda$.

one can write the following complex valued approximation

$$\sqrt{1+\xi} = e^{i\alpha/2}\sqrt{1+(1+\xi)\,e^{-i\alpha}-1} \approx$$

$$e^{i\alpha/2}\left[1+\sum_{l=1}^{p}\frac{a_l\left[(1+\xi)\,e^{-i\alpha}-1\right]}{1+b_l\left[(1+\xi)\,e^{-i\alpha}-1\right]}\right].$$

Figure 3 and 4 show that the correct choice of parameter α allows to correctly approximate k_x^d in the evanescent part of the spectrum.

4   Parameter Selection Algorithm

The seeking of the optimal computational parameters consists of the following two stages: determining the maximum propagation angle θ_{max} and solving the minimization problem.

The analysis of the discrete dispersion relation demonstrates that increasing the maximum wave propagation angle increases the required density of the calculation grid and, accordingly, the calculation time. It should be noted that the required maximum propagation angle strongly depends on the propagation conditions, especially on boundary conditions. The angle between the propagating wave and the surface does not exceed a few degrees when propagating over a smooth surface, such as the surface of the sea. Inhomogeneities of the landscape require consideration of a wider angular spectrum. However, even in this case, the maximum angle rarely exceeds 10°. Waves with propagation angle up to 90°

and evanescent waves must be considered in the presence of vertical obstacles that usually occur in the urban environment. Thus, based on the geometry of the propagation medium, one can determine the maximum propagation angle θ_{max}.

Next, we need to solve the problem of computational parameter optimization. It is necessary to minimize the number of calculations of the Eq. (8), as well as the number of nodes of its discrete analog (12). It corresponds to the minimization of the following expression

$$g(\Delta x, \Delta z, a_1 \ldots a_p, b_1 \ldots b_p) = \frac{x_{max}}{\Delta x} p \frac{z_{max}}{\Delta z} \rightarrow \min \tag{17}$$

when

$$|k_x(\theta_{max}) - k_x^d(\Delta x, \Delta z, a_1 \ldots a_p, b_1 \ldots b_p, \theta_{max})| < \varepsilon \frac{\Delta x}{x_{max}}, \tag{18}$$

where ε is acceptable phase error at the distance x_{max} from the source of radiation. Obviously, the speed of calculations is directly depend on the value of function g. The minimization algorithm is implemented by simply iterating over the various values of the computational parameters. This approach gives satisfactory results in terms of the time of the minimization procedure, however more advanced optimization methods may be applied in the future.

It should be noted that the dispersion analysis is carried out without taking into account the inhomogeneities of the refractive index. This assumption is not significant when the refractive index does not suffer strong spatial variations. However, the variations of the refractive index are often used to account for the spatial location of the obstacles [8,24]. For path-through integration of the obstacles, it is noteworthy that the wavelength λ' inside the medium with the refractive index n' changes following the ratio

$$\lambda' \sim \frac{1}{n'}.$$

Therefore, it is necessary to proportionally reduce the size of the cells of the computational grid inside the obstacle. A non-uniform computational grid may be used for this purpose [33].

Thus, simplified description of the algorithm is the following:

1. Estimate the maximum propagation angle θ_{max} based on the geometry of the problem;
2. Minimize function (17) under condition (18).

The first part of the algorithm significantly depends on the particular problem and its geometry. The second part does not depend on the specific problem.

5   Numerical Results and Discussion

In this section, we analyze the results of the presented algorithm for various values of the maximum propagation angle θ_{max} and approximation orders. We consider the following Padé approximations: $[1/1]$, $[1/2]$, $[3/4]$, $[7,8]$ and approximation of the form (13). 4-th order transversal approximation is used. The length of the computational domain $x_{max} = 2e4\lambda$, width $z_{max} = 1e2\lambda$. Acceptable phase error $\varepsilon = 10^{-3}$. The results of the algorithm for the maximum propagation angle equal to $3°$, $10°$, $20°$ and $45°$ are given in Tables 1, 2, 3 and 4 respectively. These tables present the optimal cell sizes of the computational grid for various approximations, as well as the value of the function g. "-" means that reasonable computational parameters could not be selected.

Increasing the approximation order of the finite-difference numerical scheme has a significant impact on the performance. It is clearly seen that increasing the approximation order affects both the maximum propagation angle and the calculation speed. Padé-$[7/8]$ approximation yields three orders faster numerical scheme than the standard Crank-Nicolson ($[1/1]$) approximation even in a narrow-angle case. Standard Crank-Nicolson numerical method can only be used for very small propagation angles. Propagation angles greater than $45°$ can only be handled by the scheme (13).

Table 1. Results of the algorithm for the maximum propagation angle $\theta_{max} = 3°$.

Padé order	$[1/1]$	$[1/2]$	$[3/4]$	$[7/8]$	$[7/8]$
order by z	4	4	4	4	Padé
Δx, wavelengths	0.8	20.0	300.0	1000.0	1000.0
Δz, wavelengths	0.2	0.9	1.8	2.6	2.0
g		2.5e7	4.4e5	3.0e4	1.2e4	1.6e4

Table 2. Results of the algorithm for the maximum propagation angle $\theta_{max} = 10°$.

Padé order	$[1/1]$	$[1/2]$	$[3/4]$	$[7/8]$	$[7/8]$
order by z	4	4	4	4	Padé
Δx, wavelengths	–	–	10.0	70.0	80.0
Δz, wavelengths	–	–	0.1	0.2	1.9
g	–	–	1.6e7	2.3e6	2.1e5

The numerical propagation method considered in this paper and the presented method for optimal computational parameter selection are implemented by the author as Python 3 open-source software library [23].

Table 3. Results of the algorithm for the maximum propagation angle $\theta_{max} = 20°$.

Padé order	[1/1]	[1/2]	[3/4]	[7/8]	[7/8]
order by z	4	4	4	4	Padé
Δx, wavelengths	–	–	2.0	10.0	20.0
Δz, wavelengths	–	–	0.01	0.01	1
g	–	–	8.0e8	3.2e8	1.6e6

Table 4. Results of the algorithm for the maximum propagation angle $\theta_{max} = 45°$.

Padé order	[1/1]	[1/2]	[3/4]	[7/8]	[7/8]
order by z	4	4	4	4	Padé
Δx, wavelengths	–	–	–	2.0	4.0
Δz, wavelengths	–	–	–	0.005	0.4
g	–	–	–	3.2e9	2.0e7

6   Conclusion

Discrete dispersion analysis of the finite-difference numerical scheme based on
Padé approximation of the one-way Helmholtz equation was carried out. Based
on this analysis, an algorithm for automatic computational parameter selection
of the specified numerical scheme has been developed. The method is useful for
the numerical solution of a class of wave propagation problems. The method
is based on the theoretical analysis of the numerical scheme and uses a priori
information about the propagation medium. The presented method is easier to
implement and more effective than the adaptive selection of parameters. The
presented method can be used in complex software systems that use wave propagation modeling.

References

1. Baker, G.A., Graves-Morris, P.: Pade Approximants, vol. 59. Cambridge University
   Press, Cambridge (1996)
2. Bamberger, A., Engquist, B., Halpern, L., Joly, P.: Higher order paraxial wave
   equation approximations in heterogeneous media. SIAM J. Appl. Math. **48**(1),
   129–154 (1988)
3. Bamberger, A., Engquist, B., Halpern, L., Joly, P.: Parabolic wave equation approximations in heterogenous media. SIAM J. Appl. Math. **48**(1), 99–128 (1988)
4. Bekker, E.V., Sewell, P., Benson, T.M., Vukovic, A.: Wide-angle alternating-
   direction implicit finite-difference beam propagation method. J. Lightwave. Technol. **27**(14), 2595–2604 (2009)
5. Brookner, E., Cornely, P.R., Lok, Y.F.: Areps and temper-getting familiar with
   these powerful propagation software tools. In: IEEE Radar Conference, pp. 1034–
   1043. IEEE (2007)

6. Butcher, J.C.: Numerical Methods for Ordinary Differential Equations. John Wiley & Sons, Hoboken (2016)

7. Collins, M.D.: A split-step pade solution for the parabolic equation method. J. Acoust. Soc. Am. **93**(4), 1736–1742 (1993)

8. Collins, M.D., Westwood, E.K.: A higher-order energy-conserving parabolic equqation for range-dependent ocean depth, sound speed, and density. J. Acoust. Soc. Am. **89**(3), 1068–1075 (1991)

9. Corones, J.P.: Bremmer series that correct parabolic approximations. J. Math. Anal. Appl. **50**(2), 361–372 (1975)

10. Corones, J.P., Krueger, R.J.: Higher-order parabolic approximations to time-independent wave equations. J. Math. Phys. **24**(9), 2301–2304 (1983)

11. van Dijk, W., Vanderwoerd, T., Prins, S.J.: Numerical solutions of the time-dependent schrödinger equation in two dimensions. Phys. Rev. E **95**(2), 023310 (2017)

12. Ehrhardt, M., Zisowsky, A.: Discrete non-local boundary conditions for split-step padé approximations of the one-way helmholtz equation. J. Comput. Appl. Math. **200**(2), 471–490 (2007)

13. Etter, P.C.: Underwater Acoustic Modeling and Simulation. CRC Press, Boca Raton (2018)

14. Feshchenko, R.M., Popov, A.V.: Exact transparent boundary condition for the parabolic equation in a rectangular computational domain. J. Opt. Soc. Am. A **28**(3), 373–380 (2011)

15. Fishman, L., Gautesen, A.K., Sun, Z.: Uniform high-frequency approximations of the square root helmholtz operator symbol. Wave Motion **26**(2), 127–161 (1997)

16. Hadley, G.R.: Wide-angle beam propagation using padé approximant operators. Opt. Lett. **17**(20), 1426–1428 (1992)

17. He, Z., Chen, R.S.: Frequency-domain and time-domain solvers of parabolic equation for rotationally symmetric geometries. Comput. Phys. Commun. **220**, 181–187 (2017)

18. Jensen, F.B., Kuperman, W.A., Porter, M.B., Schmidt, H.: Computational Ocean Acoustics. Springer, Heidelberg (2014). https://doi.org/10.1007/978-1-4419-8678-8

19. Ji, S., Yang, Y., Pang, G., Antoine, X.: Accurate artificial boundary conditions for the semi-discretized linear schrödinger and heat equations on rectangular domains. Comput. Phys. Commun. **222**, 84–93 (2018)

20. Lee, D., Schultz, M.H.: Numerical ocean acoustic propagation in three dimensions. World scientific (1995)

21. Leontovich, M.A., Fock, V.A.: Solution of the problem of propagation of electromagnetic waves along the earth's surface by the method of parabolic equation. J. Phys. USSR **10**(1), 13–23 (1946)

22. Levy, M.F.: Parabolic Equation Methods for Electromagnetic Wave Propagation. The Institution of Electrical Engineers, UK (2000)

23. Lytaev, M.S.: Python wave proragation library (2020). https://github.com/mikelytaev/wave-propagation

24. Lytaev, M.S., Vladyko, A.G.: Split-step padé approximations of the Helmholtz equation for radio coverage prediction over irregular terrain. In: Advances in Wireless and Optical Communications (RTUWO), pp. 179–184. IEEE (2018)

25. Lytaev, M.S.: Nonlocal boundary conditions for split-step padé approximations of the helmholtz equation with modified refractive index. IEEE Antennas. Wirel. Propag. Lett. **17**(8), 1561–1565 (2018)

26. Majorosi, S., Czirják, A.: Fourth order real space solver for the time-dependent schrödinger equation with singular Coulomb potential. Comput. Phys. Commun. **208**, 9–28 (2016)

27. Milinazzo, F.A., Zala, C.A., Brooke, G.H.: Rational square-root approximations for parabolic equation algorithms. J. Acoust. Soc. Am. **101**(2), 760–766 (1997)

28. Ozgun, O., Apaydin, G., Kuzuoglu, M., Sevgi, L.: PETOOL: MATLAB-based one-way and two-way split-step parabolic equation tool for radiowave propagation over variable terrain. Comput. Phys. Commun. **182**(12), 2638–2654 (2011)

29. Petrov, P.S., Ehrhardt, M.: Transparent boundary conditions for iterative high-order parabolic equations. J. Comput. Phys. **313**, 144–158 (2016)

30. Porter, M.: Ocean acoustics library (2020). https://oalib-acoustics.org/

31. Press, W.H., Teukolsky, S.A., Vetterling, W.T., Flannery, B.P.: Numerical Recipes: The Art of Scientific Computing. Cambridge University Press, Cambridge (2007)

32. Ramamurti, A., Calvo, D.C.: Multisector parabolic-equation approach to compute acoustic scattering by noncanonically shaped impenetrable objects. Phys. Rev. E **100**(6), 063309 (2019)

33. Sanders, W.M., Collins, M.D.: Nonuniform depth grids in parabolic equation solutions. J. Acoust. Soc. Am. **133**(4), 1953–1958 (2013)

34. Taylor, M.: Pseudodifferential Operators and Nonlinear PDE, vol. 100. Springer, Heidelberg (2012). https://doi.org/10.1007/978-1-4612-0431-2

35. Terekhov, A.V.: The laguerre finite difference one-way equation solver. Comput. Phys. Commun. **214**, 71–82 (2017)

36. Yevick, D., Thomson, D.J.: Complex padé approximants for wide-angle acoustic propagators. J. Acoust. Soc. Am. **108**(6), 2784–2790 (2000)

37. Zhang, P., Bai, L., Wu, Z., Guo, L.: Applying the parabolic equation to tropospheric groundwave propagation: a review of recent achievements and significant milestones. IEEE Antennas. Propag. Mag. **58**(3), 31–44 (2016)

A Biased Random-Key Genetic Algorithm for Bandwidth Reduction

P. H. G. Silva[1], D. N. Brandão[1], I. S. Morais[1(✉)],
and S. L. Gonzaga de Oliveira[2]

[1] Centro Federal de Educação Tecnológica Celso Suckow da Fonseca,
Rio de Janeiro, RJ, Brazil
{pegonzalez,diego.brandao,igor.morais}@eic.cefet-rj.br
[2] Universidade Federal de Lavras, Lavras, MG, Brazil
sanderson@ufla.br

Abstract. The bandwidth minimization problem is a well-known $\mathcal{NP}$-hard problem. This paper describes our experience in implementing a biased random-key genetic algorithm for the bandwidth reduction problem. Specifically, this paper compares the results of the new algorithm with the results yielded by four approaches. The results obtained on a set of standard benchmark matrices taken from the SuiteSparse sparse matrix collection indicated that the novel approach did not compare favorably with the state-of-the-art metaheuristic algorithm for bandwidth reduction. The former seems to be faster than the latter. On the other hand, the design of heuristics for bandwidth reduction is a very consolidated research area. Thus, a paradigm shift seems necessary to design a heuristic with better results than the state-of-the-art metaheuristic algorithm at shorter execution times.

Keywords: Bandwidth reduction · Biased random-key genetic algorithms · Ordering · Reordering algorithms · Renumbering · Sparse matrices

1 Introduction

The bandwidth minimization problem consists of labeling the vertices of a graph $G = (V, E)$ (composed of a set of vertices V and a set of edges E) with integer numbers from 1 to n, where $n = |V|$ is the number of vertices such that the maximum absolute difference between labels of adjacent vertices is minimum. The bandwidth minimization problem is a well-known $\mathcal{NP}$-hard problem [1]. Prevalent algorithms for bandwidth reduction are heuristic methods since they attempt to find a labeling that provides a small bandwidth in a reasonable amount of time. Thus, heuristics for bandwidth reduction place nonzero coefficients of a sparse matrix as close to the main diagonal as possible. Figure 1 illustrates an example of applying a heuristic for bandwidth reduction to a matrix.

Let $A = [a_{ij}]$ be an $n \times n$ symmetric adjacency matrix associated with an undirected graph $G = (V, E)$. The bandwidth of matrix A is defined as $\beta(A) =$

© Springer Nature Switzerland AG 2020
O. Gervasi et al. (Eds.): ICCSA 2020, LNCS 12249, pp. 312–321, 2020.
https://doi.org/10.1007/978-3-030-58799-4_23

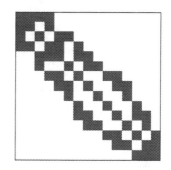

Fig. 1. An example of applying a heuristic for bandwidth reduction to a matrix. The nonzero coefficients (represented in green) of the matrix on the left are placed along the main diagonal in the matrix on the right after executing a heuristic for bandwidth reduction. (color figure online)

$\max_{1 \le i \le n} [\beta_i(A)]$, where $\beta_i(A) = i - \min_{1 \le j \le i} [j : a_{ij} \ne 0]$. Equivalently, the bandwidth of G for a vertex labeling $S = \{s(v_1), s(v_2), \cdots, s(v_{|V|})\}$ (i.e. a bijective mapping from V to the set $\{1, 2, \cdots, |V|\}$) is $\beta(G) = \max_{\{v,u\} \in E} [|s(v) - s(u)|]$.

Since bandwidth reduction is related to a vast range of important scientific and engineering applications, such as computational fluid dynamics and structural problems, researchers have proposed a large number of heuristics to reduce the bandwidth of matrices since the 1960s [2]. The Reverse Cuthill-McKee method [3] is the most classical heuristic for bandwidth reduction [4–7]. For instance, this method is available on MATLAB [8,9] and GNU Octave [10] mathematical softwares as the function $symrcm$[1], and on Boost C++ Library [11][2].

Metaheuristic-based algorithms for bandwidth reduction began to be proposed mainly in the 1990s [12]. As previously mentioned, since bandwidth reduction is a relevant problem in various scientific and engineering fields, practitioners have applied the best-known metaheuristics in the design of heuristics for bandwidth reduction [12], including tabu search [13,14], GRASP [15], genetic algorithm [16,17], ant colony optimization [6,7,18–21], particle swarm optimization [22], simulated annealing [23,24], variable neighbourhood search [25], genetic programming [26–28] and self-organizing maps [29,30].

The VNS-band heuristic [25] yielded better bandwidth results than a heuristic based on the simulated annealing metaheuristic [23], which was considered the previous state-of-the-art heuristic for bandwidth reduction. Torres-Jimenez et al. [24] proposed a Dual Representation Simulated Annealing (DRSA) heuristic for bandwidth reduction. The DRSA heuristic delivered better bandwidth

---

[1] https://www.mathworks.com/help/matlab/ref/symrcm.html?requestedDomain= www.mathworks.com, https://octave.sourceforge.io/octave/function/symrcm.html.

[2] http://www.boost.org/doc/libs/1_38_0/libs/graph/doc/cuthill_mckee_ordering. html.

results than the VNS-band heuristic. Based on the results reported [24], the DRSA heuristic is the state-of-the-art metaheuristic-based algorithm for bandwidth reduction.

Researchers have successfully applied the biased random-key genetic algorithms (BRKGA) to several optimization problems such as routing in OSPF networks [31], cell formation in manufacturing [32], and telecommunications [33]. This paper implements a BRKGA [34] for bandwidth reduction. To the best of our knowledge, this is the first heuristic for bandwidth reduction based on BRKGA. This paper compares the results yielded by the novel algorithm with the results obtained by the DRSA heuristic [24], a GRASP heuristic, the Reverse Cuthill-McKee method [3], and a variant of the breadth-first search procedure.

The remainder of this paper is organized as follows. In Sect. 2, a BRKGA for bandwidth reduction is introduced. In Sect. 3, we describe how the experiments were conducted. In Sect. 4, we discuss the results. Finally, in Sect. 5, the main conclusions are presented.

2   A BRKGA for Bandwidth Reduction

A BRKGA is similar to a genetic algorithm, but the former employs different strategies aiming at surpassing some disadvantages related to the latter. Bean [35] introduced the random-key genetic algorithms (RKGA) for combinatorial optimization problems. In RKGA, a permutation vector represents a solution to the problem in context. Vectors of random real numbers, called keys, represent the solutions. A deterministic algorithm, called a decoder, associates a feasible solution of the combinatorial optimization problem with a solution vector. Thus, the algorithm can compute an objective value or fitness function to the solution of the combinatorial optimization problem.

RKGA randomly selects two parents from the entire population to implement the crossover operation. Parents are allowed to be selected for mating more than once in a given generation. The principal objective of an RKGA is to mitigate the difficulty of genetic algorithm operators in dealing with feasible solutions.

A BRKGA selects parents for crossover differently from RKGA (see [34] for a review). BRKGA generates each element combining one individual selected at random from the elite solutions in the current population, whereas the other is a non-elite solution. The selection is said biased because one parent is always an elite individual. Therefore, elite solutions have a high probability of passing their genes to individuals of the new generation.

We followed the mains steps of the metaheuristic approach to design the BRKGA for bandwidth reduction. The BRKGA for bandwidth reduction evolves a population of chromosomes that consists of vectors of real numbers (called keys). Thus, a vector represents each solution. Each component in this vector is a real number in the range $[0, 1]$. Each solution represented by a chromosome is decoded by a decoding heuristic that receives the vector of keys and builds a solution.

We used a BRKGA framework provided by Toso and Resende [36]. In this framework, the user implements only problem-dependent procedures, such as the decoding method, which is responsible for decoding a vector of random keys to a solution of the problem in context. The decoding method is also responsible for evaluating the solution and treating penalties if the algorithm generates an infeasible solution. Thus, the decoding method interprets a vector of random keys of the BRKGA.

BRKGA obtains each new individual by combining a randomly selected individual from the elite group (p_e) of the current population (p), and another one from the non-elite set $(p\backslash p_e)$ of individuals. Thus, a single individual can be selected multiple times, and therefore can produce more than one offspring. Since BRKGA requires that $|p_e| < |p\backslash p_e|$, the probability of BRKGA to select for reproduction $(1/|p_e|)$ an individual from the elite group is greater than the probability of the algorithm to select a non-elite individual $(1/|p\backslash p_e|)$.

BRKGA does not use the standard mutation operator. This operator changes parts of chromosomes with a small probability. Instead, BRKGA employs the concept of mutants. Specifically, BRKGA introduces a fixed number of mutant solutions in the population in each generation. The algorithm randomly generates mutants solutions, equivalently as performed to construct the initial population. Mutants are equivalent to the mutation operator in traditional genetic algorithms, diversifying the search and helping the procedure to escape from local optimal solutions.

A constructive heuristic generates the initial population in the BRKGA for bandwidth reduction. The method is a random breadth-first search procedure. The procedure randomly labels vertices belonging to an adjacency list.

We used the parametrized uniform crossover scheme proposed by Spears and De Jong [37] to combine two parent solutions and produce an offspring. The offspring inherits each key from the best fit of the two parents with probability $\phi_a > 0.5$ and from the other parent with probability $1 - \phi_a$.

To accelerate convergence, each time the approach calls the decoder, if the solution is feasible, the BRKGA applies a built-in local search procedure. Specifically, the BRKGA applies a decoding method that takes a vector of random keys as input and returns the difference in bandwidth reduction between the new solution and the original bandwidth of the matrix. After executing the decoding method, the BRKGA performs a local search procedure proposed by Limpg et al. [16]. After applying the local search, the BRKGA converts the newly found solution to a chromosome representation and returns its cost.

3   Description of the Tests

We used the C++ programming language to implement the BRKGA for bandwidth reduction. We employed the same programming language to code the Reverse Cuthill-McKee method, GRASP approach, and a variant of the breadth-first search procedure (i.e., the constructive approach used to generate the initial population in the BRKGA). The workstation used in the execution of the simulations featured an Intel® Core™ i7 (CPU 4.2 GHz, 32 GB of main memory) (Intel; Santa Clara, CA, United States).

The BRKGA for bandwidth reduction uses a specific seed for random number generation at each run. After an exploratory investigation, we set the parameters of the serial BRKGA as follows:

– size of chromosomes : n;
– population size : 1,000;
– elite set fraction : 0.5;
– fraction of population to be replaced by mutants : 0.5;
– probability that offspring inherits an allele from elite parent : 0.5;
– number of independent populations : 1;
– current generation : 0 (so that the BRKGA restarts from the beginning of the approach);
– the best two individuals are combined every 100 generations;
– stopping criterion : 1,000 generations.

The results yielded by the DRSA method were taken from the original publication [24]. The DRSA heuristic was executed in a processor AMD Opteron^{TM} 6274 running at 2.2 GHz [24]. The authors of the DRSA algorithm [24] presented the best bandwidth achieved by their method and the running times in seconds required to yield the best bandwidth. However, we were not able to find the timeout set for the heuristic.

4   Results

In this section, we evaluate the results obtained by the BRKGA for bandwidth reductions against the DRSA algorithm, GRASP approach, the Reverse Cuthill-McKee method, and a variant of the breadth-first search (Rand-BFS) procedure when applied to 43 matrices taken from the SuiteSparse matrix collection [38]. Table 1 shows the name, size, and original bandwidth (β_0) of the matrices used in this computational experiment. Additionally, the table shows the bandwidth results and CPU times obtained by the DRSA (taken from the original publication [24]), BRKGA (using the evolutionary algorithm in one execution carried out in each matrix), and GRASP approaches. The Reverse Cuthill-McKee method (RCM) took less than 0.01 seconds to compute each of the matrices used in this study. The variant of the breadth-first search procedure is even faster than the Reverse Cuthill-McKee method.

The last line of Table 1 shows $\sum \beta$, i.e., the sum of bandwidth results yielded by the algorithms. The same line shows the average time of the heuristics evaluated.

As expected, the other algorithms analyzed in this study dominated the Rand-BFS procedure in terms of bandwidth reduction. The metaheuristic algorithms returned better bandwidth results than did the Rand-BFS procedure in 42 matrices. The Rand-BFS procedure yielded the same bandwidth result as the BRKGA and GRASP algorithms when applied to the matrix gr_30_30. In particular, the Rand-BFS procedure delivered a better bandwidth result than the RCM method in this matrix. On the other hand, the RCM method yielded

Table 1. Bandwidth results yielded by several methods when applied to 43 matrices.

Matrix	n	β_0	DRSA(β)	t(s)	BRKGA(β)	t(s)	GRASP(β)	t(s)	RCM	Rnd-BFS
dwt_209	209	184	23	2	33	16	33	1	33	47
gre_216a	216	36	21	5	21	13	21	1	21	44
dwt_221	221	187	13	7	15	14	15	5	15	27
impcol_e	225	92	42	9	63	36	67	2	75	106
dwt_245	245	115	21	9	33	13	34	1	55	56
bcspwr04	274	265	24	3	37	15	37	1	49	70
ash292	292	24	19	4	24	18	24	1	32	37
can_292	292	282	38	6	57	27	60	1	72	81
dwt_310	310	28	12	11	13	20	13	1	15	19
gre_343	343	49	28	4	28	22	28	1	28	81
dwt_361	361	50	14	23	15	24	15	1	15	44
dwt_419	419	356	25	45	32	31	32	1	34	67
bcsstk06	420	47	45	15	48	116	49	3	50	55
bcsstm07	420	47	45	14	49	102	49	3	56	73
impcol_d	425	406	39	844	56	25	55	1	80	65
hor_131	434	421	54	26	66	143	65	5	84	119
bcspwr05	443	435	27	7	49	17	49	1	68	57
can_445	445	403	52	11	74	32	74	1	78	119
494_bus	494	428	28	18	55	17	55	1	82	71
dwt_503	503	452	41	50	51	66	52	2	64	77
gre_512	512	64	36	9	56	25	36	1	36	69
fs_541_1	541	540	270	1	465	181	433	117	533	538
dwt_592	592	259	29	53	40	45	42	1	42	65
662_bus	662	335	39	14	77	26	78	1	118	103
fs_680_1	680	600	17	27	17	50	20	1	20	41
685_bus	685	550	32	145	57	31	57	1	102	77
can_715	715	611	71	182	116	70	116	4	133	155
fs_760_1	760	740	37	39	39	146	41	4	43	81
bcsstk19	817	567	14	97	20	59	20	2	22	31
bp_0	822	820	234	1219	386	92	417	2	422	593
bp_1000	822	820	283	4988	457	140	496	5	479	642
bp_1200	822	820	287	1165	467	142	488	5	506	607
bp_1400	822	820	291	2140	475	141	484	5	536	632
bp_1600	822	820	292	2261	469	144	488	5	539	624
bp_200	822	820	257	1639	471	108	459	3	508	649
bp_400	822	820	267	2341	480	115	479	5	530	578
bp_600	822	820	272	1315	488	121	485	4	541	614
bp_800	822	820	278	1065	454	131	492	5	542	552
can_838	838	837	86	70	111	118	119	4	151	164
dwt_878	878	519	25	85	33	64	35	3	46	77
gr_30_30	900	31	33	57	48	65	48	4	59	48
dwt_918	918	839	32	62	46	65	46	2	57	61
dwt_992	992	513	35	107	52	194	52	5	65	70
$\sum \beta$ and average time			3828	470	6143	71	6258	5	7036	8386

better bandwidth results than did the Rand-BFS procedure in 37 matrices. The latter returned better bandwidth results than did the former in five matrices.

The metaheuristic algorithms dominated the RCM method. The DRSA algorithm yielded better bandwidth results than did the RCM method in 41 matrices. Moreover, the latter delivered the same bandwidth results as the former in two cases. The BRKGA yielded better bandwidth results than did the RCM method in 36 matrices. The latter delivered the same bandwidth results as the former in six test problems. The RCM method delivered a better bandwidth result than did the BRKGA when applied to the matrix gre_512. The GRASP algorithm yielded better bandwidth results than did the RCM method in 35 matrices. The latter provided the same bandwidth results as the former when applied to seven matrices. The RCM method delivered better bandwidth results than did the GRASP algorithm when applied to the matrix bp_1000.

In general, the BRKGA returned better bandwidth results than did the GRASP algorithm at longer running times. The BRKGA yielded better bandwidth results than did the GRASP algorithm in 17 matrices. Furthermore, the latter provided better bandwidth results than did the former in seven test problems. Both algorithms returned the same bandwidth in 19 matrices.

The DRSA algorithm yielded better bandwidth results than did the four other heuristics evaluated in 40 test problems. The DRSA and BRKGA approaches provided the same results in three cases (matrices gre_216a, gre_343, and fs_680_1). We could not compare the running times of the algorithms because the studies used different machines.

We performed statistical tests for pairwise comparison. Specifically, we used the Wilcoxon matched-pairs signed-rank test to analyze the bandwidth reduction yielded by the DRSA and BRKGA approaches. The null hypothesis for the test was that the algorithms yielded identical results, i.e., the median difference was zero. The alternative hypothesis was that the median difference is positive. Thus, we used a two-tailed test because we wanted to determine if there was any difference between the groups we were comparing (i.e., the possibility of positive or negative differences). In this case, $R_+ = 820$ and $R_- = 0$. The test statistic is, therefore, $T = 0$. Thus, $T \leq T_{crit(\alpha=0.01,40)} = 247$. We have statistically significant evidence at $\alpha = 0.01$ (i.e., the level of significance 0.01 is related to the 99% confidence level), to show that the median difference is positive, i.e., that the DRSA algorithm dominated the BRKGA. In this case, p-value equals 3.56056e−8. The use of a corrected significance level of $1 - (1 - \alpha)^{1/2} = 0.005$ (Dunn-Šidák correction) or $\alpha/2 \approx 0.005$ (Bonferroni correction) confirms the previous statement.

5   Conclusions

This paper describes our experience in implementing a BRKGA for bandwidth reduction. The algorithm did not compare favorably with the state-of-the-art metaheuristic-based heuristic for bandwidth reduction. Although the novel approach yielded better bandwidth results than a GRASP approach and classical methods for bandwidth reduction, namely the Reverse Cuthill-McKee and a variant of the breadth-first search procedure, the DRSA algorithm dominated the BRKGA approach. The design of a parallel BRKGA for bandwidth reduction is a future step in this investigation.

References

1. Papadimitriou, C.H.: The NP-completeness of bandwidth minimization problem. Comput. J. **16**, 177–192 (1976)
2. Gonzaga de Oliveira, S.L., Chagas, G.O.: A systematic review of heuristics for symmetric-matrix bandwidth reduction: methods not based on metaheuristics. In: The XLVII Brazilian Symposium of Operational Research (SBPO), Ipojuca-PE, Brazil, Sobrapo, August 2015
3. George, A., Liu, J.W.: Computer Solution of Large Sparse Positive Definite Systems. Prentice-Hall, Englewood Cliffs (1981)
4. Gonzaga de Oliveira, S.L., Bernardes, J.A.B., Chagas, G.O.: An evaluation of low-cost heuristics for matrix bandwidth and profile reductions. Comput. Appl. Math. **37**(2), 1412–1471 (2016). https://doi.org/10.1007/s40314-016-0394-9
5. Gonzaga de Oliveira, S.L., Abreu, A.A.A.M.: An evaluation of pseudoperipheral vertex finders for the reverse Cuthill-Mckee method for bandwidth and profile reductions of symmetric matrices. In: Proceedings of the 37th International Conference of the Chilean Computer Science Society (SCCC), Santiago, Chile, November 2018, pp. 1–9. IEEE (2018). https://doi.org/10.1109/SCCC.2018.8705263
6. Gonzaga de Oliveira, S. L., Silva, L.M.: Evolving reordering algorithms using an ant colony hyperheuristic approach for accelerating the convergence of the ICCG method. Eng. Comput. (2019). https://doi.org/10.1007/s00366-019-00801-5)
7. Gonzaga de Oliveira, S., Silva, L.: An ant colony hyperheuristic approach for matrix bandwidth reduction. Appl. Soft Comput. **94**, 106434 (2020)
8. Gilbert, J.R., Moler, C., Schreiber, R.: Sparse matrices in MATLAB: design and implementation. SIAM J. Matrix Anal. **3**(1), 333–356 (1992)
9. The MathWorks Inc.: MATLAB (1994–2018). http://www.mathworks.com/products/matlab/
10. Eaton, J.W., Bateman, D., Hauberg, S., Wehbring, R.: GNU Octave version 4.0.0 manual: a high-level interactive language for numerical computations (2015)
11. Boost: Boost C++ libraries (2017). http://www.boost.org/. Accessed 28 Jun 2017
12. Chagas, G.O., Gonzaga de Oliveira, S.L.: Metaheuristic-based heuristics for symmetric-matrix bandwidth reduction: a systematic review. Procedia Comput. Sci. 51, 211–220 (2015)
13. Martí, R., Laguna, M., Glover, F., Campos, V.: Reducing the bandwidth of a sparse matrix with tabu search. Eur. J. Oper. Res. **135**(2), 450–459 (2001)
14. Campos, V., Piñana, E., Martí, R.: Adaptive memory programming for matrix bandwidth minimization. Ann. Oper. Res. **183**, 7–23 (2011)
15. Piñana, E., Plana, I., Campos, V., Martí, R.: GRASP and path relinking for the matrix bandwidth minimization. Eur. J. Oper. Res. **153**(1), 200–210 (2004)
16. Lim, A., Rodrigues, B., Xiao, F.: Heuristics for matrix bandwidth reduction. Eur. J. Oper. Res. **174**(1), 69–91 (2006)
17. Czibula, G., Crişan, G.C., Pintea, C.M., Czibula, I.G.: Soft computing approaches on the bandwidth problem. Informatica **24**(2), 169–180 (2013)

18. Lim, A., Lin, J., Rodrigues, B., Xiao, F.: Ant colony optimization with hill climbing for the bandwidth minimization problem. Appl. Soft Comput. **6**(2), 180–188 (2006)
19. Kaveh, A., Sharafi, P.: Nodal ordering for bandwidth reduction using ant system algorithm. Eng. Comput. **26**, 313–323 (2009)
20. Pintea, C.-M., Crişan, G.-C., Chira, C.: A hybrid ACO approach to the matrix bandwidth minimization problem. In: Graña Romay, M., Corchado, E., Garcia Sebastian, M.T. (eds.) HAIS 2010. LNCS (LNAI), vol. 6076, pp. 405–412. Springer, Heidelberg (2010). https://doi.org/10.1007/978-3-642-13769-3_49
21. Pintea, C.M., Crişan, G.C., Shira, C.: Hybrid ant models with a transition policy for solving a complex problem. Logic J. IGPL **20**(3), 560–569 (2012)
22. Lim, A., Lin, J., Xiao, F.: Particle swarm optimization and hill climbing for the bandwidth minimization problem. Appl. Intell. **3**(26), 175–182 (2007)
23. Rodriguez-Tello, E., Jin-Kao, H., Torres-Jimenez, J.: An improved simulated annealing algorithm for bandwidth minimization. Eur. J. Oper. Res. **185**, 1319–1335 (2008)
24. Torres-Jimenez, J., Izquierdo-Marquez, I., Garcia-Robledo, A., Gonzalez-Gomez, A., Bernal, J., Kacker, R.N.: A dual representation simulated annealing algorithm for the bandwidth minimization problem on graphs. Inf. Sci. **303**, 33–49 (2015)
25. Mladenovic, N., Urosevic, D., Pérez-Brito, D., García-González, C.G.: Variable neighbourhood search for bandwidth reduction. Eur. J. Oper. Res. **1**(200), 14–27 (2010)
26. Koohestani, B., Poli, R.: A hyper-heuristic approach to evolving algorithms for bandwidth reduction based on genetic programming. In: Bramer, M., Petridis, M., Nolle, L. (eds.) Research and Development in Intelligent Systems XXVIII, pp. 93–106. Springer, London (2011). https://doi.org/10.1007/978-1-4471-2318-7_7
27. Pop, P.C., Matei, O.: An improved heuristic for the bandwidth minimization based on genetic programming. In: Corchado, E., Kurzyński, M., Woźniak, M. (eds.) HAIS 2011. LNCS (LNAI), vol. 6679, pp. 67–74. Springer, Heidelberg (2011). https://doi.org/10.1007/978-3-642-21222-2_9
28. Pop, P., Matei, O., Comes, C.A.: Reducing the bandwidth of a sparse matrix with a genetic algorithm. Optim. J. Math. Prog. Oper. Res. **63**(12), 1851–1876 (2013)
29. Gonzaga de Oliveira, S.L., de Abreu, A.A.A.M., Robaina, D., Kischinhevsky, M.: A new heuristic for bandwidth and profile reductions of matrices using a self-organizing map. In: Gervasi, O., et al. (eds.) ICCSA 2016. LNCS, vol. 9786, pp. 54–70. Springer, Cham (2016). https://doi.org/10.1007/978-3-319-42085-1_5
30. Gonzaga de Oliveira, S.L., Abreu, A.A.A.M., Robaina, D.T., Kischnhevsky, M.: An evaluation of four reordering algorithms to reduce the computational cost of the Jacobi-preconditioned conjugate gradient method using high-precision arithmetic. Int. J. Bus. Intell. Data Min. **12**(2), 190–209 (2017)
31. Ericsson, M., Resende, M.G.C., Pardalos, P.M.: A genetic algorithm for the weight setting problem in OSPF routing. J. Comb. Optim. **6**, 299–333 (2002)
32. Gonçalves, J.F., Resende, M.G.C.: An evolutionary algorithm for manufacturing cell formation. Comput. Ind. Eng. **47**, 247–273 (2004)
33. Resende, M.G.C.: Biased random-key genetic algorithms with applications in telecommunications. TOP **20**, 130–153 (2012)
34. Gonçalves, J.F., Resende, M.G.C.: Biased random-key genetic algorithms for combinatorial optimization. J. Heuristics **17**, 487–525 (2011)
35. Bean, J.C.: Genetic algorithms and random keys for sequencing and optimization. ORSA J. Comput. **6**, 154–160 (1994)
36. Toso, R.F., Resende, M.G.C.: A C++ application programming interface for biased random-key genetic algorithms. Optim. Methods Softw. **30**, 81–93 (2015)

37. Spears, W.M., De Jong, K.D.: On the virtues of parameterized uniform crossover. Technical report, DTIC Document (1995)
38. Davis, T.A., Hu, Y.: The University of Florida sparse matrix collection. ACM Trans. Math. Softw. **38**(1), 1–25 (2011)

DTN-SMTP: A Novel Mail Transfer Protocol with Minimized Interactions for Space Internet

Donghyuk Lee[1] , Jinyeong Kang[1] , Mwamba Kasongo Dahouda[1] ,
Inwhee Joe[1(✉)] , and Kyungrak Lee[2]

[1] Hanyang University, 222, Wangsimni-ro,
Seongdong-gu, Seoul 04763, Republic of Korea
{shine5601,achieve365,dahouda37,iwjoe}@hanyang.ac.kr
[2] ETRI, 218, Gajeong-ro, Yuseong-gu, Daejeon 34129, Republic of Korea
krlee@etri.re.kr

Abstract. The communication paradigms for Delay/Disruption Tolerant Networks (DTN) have been modeled after email. Supporting email over DTN in a backwards compatible manner in a heterogeneous environment has yet to be fully defined. In the Simple Mail Transfer Protocol (SMTP) based on TCP/IP used in the existing terrestrial Internet, the protocol works through multiple interactions between the sender's mail server and the receiver's mail server. However, in the space Internet environment, since the contact times are limited, reliability cannot be guaranteed for the interaction between the server and the client used in the existing SMTP. Therefore, this paper proposes a novel mail transfer protocol, DTN-SMTP for space Internet over DTN. To minimize the interaction of the existing SMTP, it relies on one-way transmission and optionally performs retransmission mechanisms. Finally, we have built and configured two DTN nodes to implement DTN-SMTP. Also, we have confirmed the mail's reception and the file attachment from the external mail client with minimized interactions between the SMTP messages.

Keywords: Delay/Disruption Tolerant Network · Simple Mail Transfer Protocol · One-way transmission

1 Introduction

Space internet environment presents high propagation delays, low data rates and frequent network disconnections may occur. Therefore, in environments such as the space Internet there are limitations of network protocols such as TCP/IP used in the terrestrial Internet [1]. DTN was designed as a protocol that can be applied in a special network environment as in [2]. DTN is a proposed network

This work was supported by the Electronics and Telecommunication Research Institute [Development of Space Internet Technology for Korea Lunar Exploration] project in 2020.

O. Gervasi et al. (Eds.): ICCSA 2020, LNCS 12249, pp. 322–331, 2020.
https://doi.org/10.1007/978-3-030-58799-4_24

to guarantee the reliability of end-to-end connection in the extreme environments and frequent disconnection of the network. In DTN, each node on the network has a bundle layer that can store messages and uses a bundle protocol to overcome the special environment [3]. The Bundle Protocol transmits data in units of bundles and guarantees reliability between data transmission even when the end-to-end connection is disconnected using Store-and-Forward [4]. In contrast to the DTN environment, SMTP is a protocol used to transmit e-mail in a terrestrial Internet environment and operates based on TCP/IP. Sending and receiving message between the sender's and the receiver's mail servers, the process of transmitting the SMTP header containing the information of the mail the sender wants to send and the SMTP body containing the mail content to the receiver's mail server [5]. It is SMTP that implements the above process as a protocol, and is a core protocol for sending mail in a terrestrial Internet environment where the network infrastructure is well constructed compared to the space Internet environment. In this paper, we propose a DTN-SMTP suitable for the extreme Internet environment, which is an extreme network environment, by securing the reliability of data between interruptions and efficiently reduce interactions using the Bundle Protocol, one of the characteristics of DTN.

The paper is organized as follows: Sect. 2 presents related work where DTN, Bundle Protocol and Postfix will be explained. Section 3 describes our proposed architecture of DTN-SMTP and the flow chart according to the roles of the sender and receiver. The following Sect. 4 explains the implementation of the proposed DTN-SMTP and provides the result by comparing the transmission speed of SMTP and DTN-SMTP. Finally, Sect. 5 concludes this paper.

2   Related Works

2.1   Delay/Disruption Tolerant Network (DTN)

DTN is a network used in lack continuous network connectivity and low data rates for communication between nodes, such as disaster situations or deep space environments where the network infrastructure is not sufficiently equipped. Stable data transmission of end-to-end connection, such as typical network protocol TCP/IP, cannot be guaranteed. DTN is reliability can be guaranteed in situations where high propagation delay, low data rate and frequent network disconnection occur. In addition to the space Internet or interplanetary communication, the DTN technologies can be applied to a variety of challenged networks.

2.2   Bundle Protocol

The Bundle Protocol is implemented in the Bundle Layer between the Application Layer and the Transport Layer, and transmits data in bundle units. DTN cannot guarantee end-to-end connectivity due to frequent network delays and interruptions. However, the Store-and-Forward technique is used in case of network interruption in the Bundle Layer [6]. DTN node uses persistent storage of the Bundle Layer to store data when the link is disconnected and transmit it when the link is restored. As a result, DTN can secure end-to-end data transmission reliability using the Bundle Protocol.

2.3   Simple Mail Transfer Protocol (SMTP)

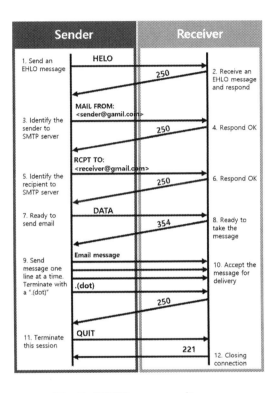

Fig. 1. SMTP sequence diagram

SMTP is a TCP/IP based email transmission protocol and uses port 25. As shown in Fig. 1, the commands MAIL, RCPT and DATA are sent. The MAIL and RCPT commands define the email sender and receiver. Send the contents of the mail to the receiver through the DATA command. The receiver responds by sending a request command complete code of 250. When the data of the sender ends, the message is terminated through .(Dot). Sender can send e-mail to Receiver through the procedure as shown in Fig. 1 [7].

2.4   Postfix

SMTP can transmit e-mail data from the sender's mail server to the receiver's mail server by exchanging simple messages in a TCP/IP-based terrestrial Internet environment. To use SMTP, this paper used Postfix Mail Transfer Agent.

Postfix is an open source Mail Transfer Agent (MTA) written in C language and is used as an alternative to Qmail and Sendmail [8]. It supports SMTP authentication using Simple Authentication and Security Layer and can easily construct a mail server [9].

3   DTN-SMTP

Existing SMTP is a protocol that operates based on TCP/IP. However, if SMTP is used in a space Internet environment where network delays and interruptions occur frequently, the number of interactions between mail servers increases, resulting in network delays and overloads.

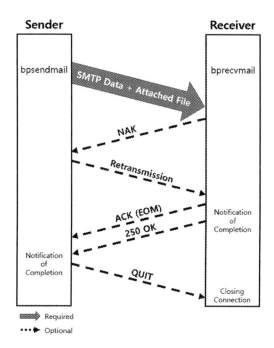

Fig. 2. DTN-SMTP sequence diagram

In this paper, we propose DTN-SMTP that can be applied in the space Internet environment as shown in Fig. 2. Sender and Receiver are DTN nodes, which can send mail between DTN nodes using *bpsendmail* and *bprecvmail* applications respectively. In order to implement the proposed DTN-SMTP, the interplanetary overlay network (ION) that implemented DTN in NASA was used [10]. However, *bpsendmail* and *bprecvmail* applications were added for DTN-SMTP, because none of the applications provided by ION is for sending mail. Each application was created based on *bpsendfile* and *bprecvfile* provided by ION, and was implemented to attach files in email transmission between DTN nodes. In ION, file transmission or reception is possible through *bpsendfile* and *bprecvfile* application. However, there is a limitation in applying SMTP because the file name and extension are not managed.

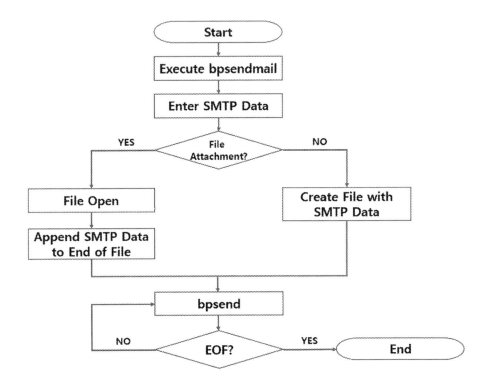

Fig. 3. Sender-side flow chart

To send email between DTN nodes, as shown in Fig. 3, the *bpsendmail* application is executed first on the Sender side. After that, enter the SMTP data. Enter the email address and name of the sender and receiver the information that goes into the SMTP header and enter the contents of the email corresponding to the SMTP body. After that, it will have a branch on whether to attach the file. If a file is attached the SMTP header and body are added to the end of the file through the file descriptor for the attached file. If no file is attached metadata is generated and transmitted through the received SMTP header and body. Depending on whether or not a file is attached the SMTP header and body can be located at the end of the original file or a file consisting of only the SMTP header and body can be created. Send the file or metadata including the SMTP header and body and the role of *bpsendmail* ends.

In the Receiver, SMTP header and body can be received through *bprecvmail*. As shown in Fig. 4, when a file is first received from the sender the operation

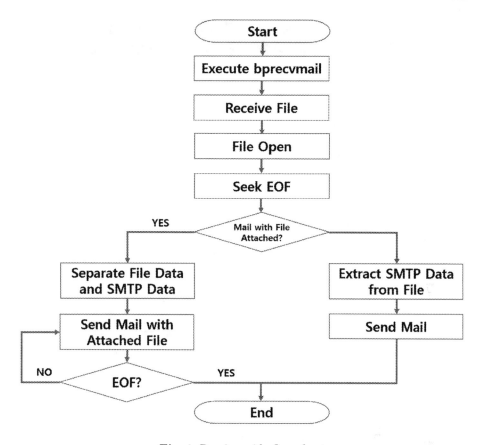

Fig. 4. Receiver-side flow chart

differs depending on whether a file is attached or not. Since the Sender inserts the SMTP header and body at the end of the file, it is necessary to find the EOF and separate the file content and the SMTP data based on the offset of the SMTP data and the file contents. Conversely, if there is no file attachment the received file itself contains only the SMTP header and body, and is transmitted to the external mail client using the corresponding SMTP data. DTN-SMTP using *bpsendmail* and *bprecvmail* is resistant to frequent network delays and interruptions provided by DTN, and operates on the basis of Licklider Transmission Protocol (LTP) [11], which checks whether data is transmitted in hops, thereby guaranteeing long-distance data transmission and data reliability.

When sending an email with DTN nodes including attachments, it can be divided into the roles of Sender and Receiver as shown in Fig. 5 to confirm the sequence diagram. The sender attaches the SMTP data to the attachment using the *bpsendmail* application and sends the file using the Bundle Protocol between DTN nodes. After separating the file received from the receiver into the original file and the SMTP data area, the attachment is transmitted to the external mail

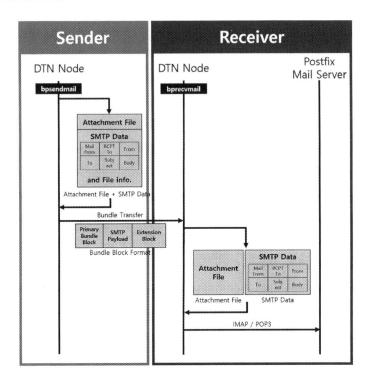

Fig. 5. DTN-SMTP message format & flow

client using the information of the SMTP data. Using the above method, it is possible to effectively communicate SMTP data and attachments in a limited space Internet environment through one-way transmission while guaranteeing reliability using DTN's Bundle Protocol.

4   Implementation

4.1   Test-Bed Configuration

For the implementation of DTN-SMTP, using Interplanetary Overlay Network (ION), which implemented the DTN Bundle Protocol at NASA. The DTN node, which acts as a sender and receiver, was installed and configured by installing ION-3.7.0 version on Ubuntu 16.04 LTS Operating System and Postfix 3.1.0 version was installed and used to send emails from the Receiver node to external mail clients. ION's Convergence Layer Adapter (CLA) uses LTP to ensure reliability between DTN nodes. In addition, the Bundle Protocol operation on of TCP was confirmed to be compatible with the terrestrial Internet.

4.2   Implementation of Suggested DTN-SMTP

In order to implement DTN-SMTP, *bpsendmail* and *bprecvmail* were added to the existing ION. *bpsendmail* expands and positions EOF by the length of inserting the SMTP header and body entered by the user after the point at which the file content ends depending on the presence or absence of an attachment. At this time, if there is no attached file, metadata including only the SMTP header and body is generated, and then the EOF is located. *bprecvmail* uses the received file to separate the original file, the SMTP header and the body if an attachment is included and sends the mail and the attachment to an external mail client. If there is no attachment, only the SMTP data is in the received file, so the mail is sent to the external mail client using the SMTP data.

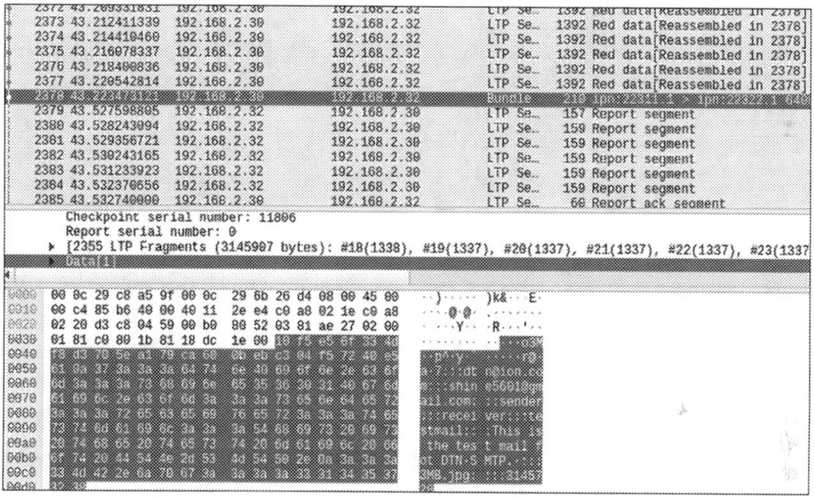

Fig. 6. LTP communication between DTN nodes

Figure 6 shows the LTP packets communication between the DTN nodes through Wireshark. When checking the payload of the Bundle packet, which is a set of LTP Segments, it was confirmed that the SMTP header and body sent by Sender were received by the Receiver. After completing the transmission between DTN nodes, Fig. 7 show packet capture screen that sends mail to an external SMTP server using the attachment and SMTP data received from the Sender. Confirmed that the SMTP data and attachment sent by Sender are sent to an external mail using SMTP.

DTN-SMTP architecture As a result of implementation according to the structure, it was confirmed that the interaction is efficiently reduced when transmitting SMTP data between DTN nodes.

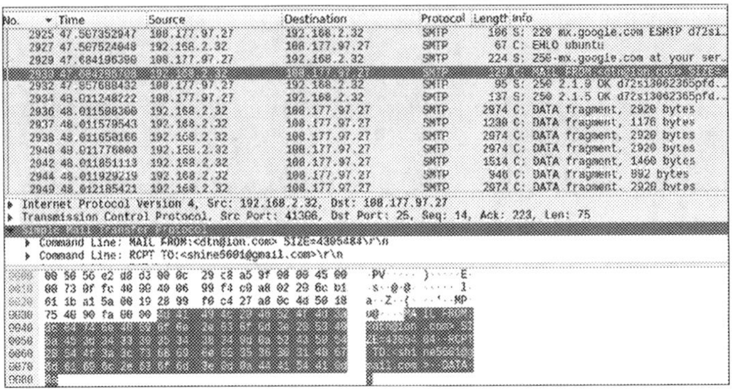

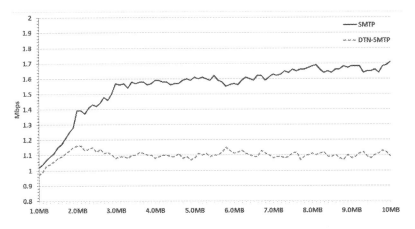

Fig. 7. Communication between DTN node and SMTP server

Fig. 8. SMTP vs. DTN-SMTP data rate

As shown in Fig. 8, This is a graph measuring the transmission speed of SMTP and DTN-SMTP with different attachment file size settings from 1 MB to 10 MB in 100 KB increments. When the size of the attached file was 1.7 MB or less, there was no significant difference in performance between SMTP and DTN-SMTP. However, when sending an email with an attachment exceeding 1.7 MB, the SMTP showed a trend of increasing the data rate as the attachment size increased and DTN-SMTP guaranteed an average data rate of 1.1 Mbps even if the attachment size increased. Due to the nature of DTN, the Store-and-Forward technique is used and the existence of a relay node is essential. As a result, DTN-SMTP has a lower transmission rate than SMTP because it has an additional relay node and LTP retransmission mechanism.

5   Conclusion

Since the existing SMTP operates based on the TCP/IP of the terrestrial Internet environment, it is unsuitable for the space Internet environment where there is lack continuous network connectivity. However, through the application of DTN-SMTP, the interaction of the existing SMTP was effectively reduced and one-way transmission was confirmed. DTN-SMTP is implemented by adding *bpsendmail* and *bprecvmail* application to ION. SMTP data and attachments were transmitted using the Bundle Protocol between DTN nodes. In particular, by determining the presence or absence of an attachment, the original file and SMTP data were transmitted to suit the space Internet environment, and the end-to-end reliability was secured using the Store-and-Forward method and LTP of the Bundle Protocol used by DTN. Through the implementation results, it was confirmed that the proposed DTN-SMTP transmits and receives SMTP data to be suitable for the space Internet environment. As a result of performance evaluation, the transmission speed of SMTP and DTN-SMTP was measured, and the DTN-SMTP guarantees an average transmission speed of 1.1 Mbps. Through future research, the transmission speed of DTN-SMTP can be improved by applying a routing method that can increase the number of forwarding per unit of time in the DTN Store-and-Forward technique. In addition, future DTN-SMTP can conduct efficient data transmission as well as research to enhance security for email transfer in space Internet.

References

1. Fall, K.: A delay-tolerant network architecture for challenged internets. In: Proceedings of the 2003 Conference on Applications, Technologies, Architectures, and Protocols for Computer Communications. ACM (2003)
2. Cerf, V., Burleigh, S.: Delay-tolerant networking architecture. RFC 4838 (2007)
3. Scott, K., Burleigh, S.: RFC 5050: Bundle protocol specification. IRTF DTN Research Group (2007)
4. Venkataraman, V.: Delay tolerant networking-a tutorial, pp. 72–78. Department of Computer Science, The University of Texas (2009)
5. Postel, J.B.: Simple mail transfer protocol. Information Sciences (1982)
6. Farrell, S.: When TCP breaks: delay-and disruption-tolerant networking. IEEE Internet Comput. **10**(4), 72–78 (2006)
7. Klensin, J. (ed.).: RFC 2821: Simple mail transfer protocol (2001)
8. Dent, K.D.: Postfix: The Definitive Guide: A Secure and Easy-to-Use MTA for UNIX. O'Reilly Media Inc., Sebastopol (2003)
9. Myers, J.: RFC 2222: Simple Authentication and Security Layer (SASL) (1997)
10. Burleigh, S.: Interplanetary overlay network: an implementation of the DTN bundle protocol (2007)
11. Ramadas, M., Burleigh, S., Farrell, S.: Licklider transmission protocol-specification. IETF request for comments RFC 5326 (2008)

Spatial Modelling of Black Scabbardfish Fishery Off the Portuguese Coast

Lídia Maria André[1,2(✉)] , Ivone Figueiredo[2,3] , M. Lucília Carvalho[3] ,
Paula Simões[1,4] , and Isabel Natário[1,5]

[1] Centro de Matemática e Aplicações (CMA) da Faculdade de Ciências e Tecnologia,
Universidade NOVA de Lisboa, Caparica, Portugal
lidiamandre@gmail.com

[2] Divisão de Modelação e Gestão de Recursos da Pesca, Instituto Português
do Mar e da Atmosfera, Lisbon, Portugal
ifigueiredo@ipma.pt

[3] Centro de Estatística e Aplicações (CEAUL), Faculdade de Ciências
da Universidade de Lisboa, Lisbon, Portugal
mlucilia.carvalho@gmail.com

[4] Centro de Investigação, Desenvolvimento e Inovação da Academia Militar
(CINAMIL), Lisbon, Portugal
pc.simoes@campus.fct.unl.pt

[5] Departamento de Matemática, Faculdade de Ciências e Tecnologia,
Universidade NOVA de Lisboa, Caparica, Portugal
icn@fct.unl.pt

Abstract. The Black Scabbardfish is a deep-water fish species that lives
at depths greater than 700 m. In Portugal mainland, this is an important commercial resource which is exploited by longliners that operate
at specific fishing grounds located off the coast. The monitoring of the
population status mainly relies on the fishery data as no independent
scientific surveys take place. The present work focus on modelling the
spatial distribution of the BSF species relative biomass. Georeferenced
data given by the location of the fishing hauls and the corresponding
catches are available for a set of different vessels that belong to the longline fishing fleet. A classical geostatistical approach was applied to fit a
variogram and evaluate the isotropy of the data. Then, different regression models with fixed, structured and unstructured random effects were
fitted under a Bayesian framework, considering the Stochastic Partial
Differential Equation (SPDE) methodology under the Integrated Nested
Laplace Approximation (INLA), addressing some practical implementation issues. The models with spatial effects seemed to perform better,
although some practical constraints related to the considered covariates
hindered the choice.

Keywords: Geostatistics · SPDE · INLA · Black Scabbardfish

© Springer Nature Switzerland AG 2020
O. Gervasi et al. (Eds.): ICCSA 2020, LNCS 12249, pp. 332–344, 2020.
https://doi.org/10.1007/978-3-030-58799-4_25

1   Introduction

Expand knowledge about biodiversity and species abundance is an important scientific challenge. One way of obtaining information about fish assemblages, e.g. depth and geographical distributions of species, their density, diversity and effect of fishing, is through research surveys as they provide fishery independent information. For the last 20/30 years, most surveys in Western European waters focused on the study of fishery resources that lived at the continental shelf or upper slope [3]. For monitoring populations of deep-water species, however, the surveys are scarce.

In Portugal, the spatial distribution and abundance of Black Scabbardfish (BSF), which is an important commercial fishery for the country, is mostly unknown. BSF is a deep-water fish species that lives at depths greater than 700 m [6] and, therefore, the monitoring of the population status mainly relies on fishery data as no independent scientific surveys take place [14]. In Portugal mainland, it is exploited by longliners that operate at specific fishing grounds along the coast [2].

The distribution of a given species in a particular region as well as the corresponding patterns occur in space. In order to study the natural structure and to understand the functional processes behind them, identifying the relevant spatial dimension at which these all occur is required. Usually, the biological and environmental phenomena can be monitored and measured only at a limited number of spatial locations, even if it is defined continuously over a region. Due to that, geostatistical models focus on inferring a continuous spatial process based on data observed at a finite number of locations, which means that the process which determines the data-locations and the process modelled are stochastically independent.

Although some aspects of the spatio-temporal modelling of species distribution has been discussed, for example in [12], the inclusion of a spatial tendency has not been considered before for the BSF species. Therefore, the present work focus on modelling the spatial distribution of the BSF species relative biomass along the Portuguese coast. Georeferenced data given by the location of the fishing hauls and the corresponding catches are available for a set of different vessels that belong to the longline fishing fleet.

A preliminary geostatistical analysis was performed using geoR [4] and the study of the isotropy of the data was performed with spTest [18]. RGeostats [13] was also used to reassure that the fitted variogram was the optimal.

Modelling was considered under a Bayesian framework due to its flexibility and because the uncertainty in the model parameters is taken into account. Although extensively used for Bayesian inference, Markov Chain Monte Carlo methods (MCMC) have some computational issues in this context. When dealing with spatial models they may be extremely slow or unfeasible [1]. An alternative method is the Integrated Nested Laplace Approximations approach (INLA) proposed by [16], which is a deterministic algorithm rather than simulation based such as MCMC. Since these methods have proven to provide accuracy and fast results, they were used, mainly through the package R-INLA.

INLA is particularly suited for geostatistical models where it is assumed that a continuous spatial surface underlies the observations, which is very well captured through the Stochastic Partial Differential Equation (SPDE) method proposed by [8]. This methodology represents a continuous Gaussian Field (GF) using a discretely indexed Gaussian Markov Random Field (GMRF), considering a SPDE whose solution is a GF with a Matérn covariance

$$\mathrm{Cov}(\xi_i, \xi_j) = \frac{\sigma^2}{\Gamma(\lambda)2^{\lambda-1}} \left(\frac{\sqrt{8\lambda}}{r} \|s_i - s_j\| \right)^{\lambda} K_{\lambda} \left(\frac{\sqrt{8\lambda}}{r} \|s_i - s_j\| \right), \qquad (1)$$

where $\|s_i - s_j\|$ is the Euclidean distance between two locations $s_i, s_j \in \mathbb{R}^2$, σ^2 is the marginal variance, r is the range and K_{λ} denotes the modified Bessel function of the second kind and order $\lambda > 0$; see [1]. Although the solution is a Markov indexed random field when $-1 \leq \lambda \leq 1$, for R-INLA only $\lambda = 1$ was fully tested; see [7].

In spite of its flexibility, the SPDE approach with INLA has also some issues. For instance, the smoothness parameter λ is fixed at 1, which is somehow restrictive for model fit. There are also some concerns regarding the parameterisation of the range parameter. While in conventional geostatistic applications, the range is the point where the correlation reaches 0.05, in R-INLA, the range is the point where the correlation is 0.01. Moreover, this parameter in R-INLA is related to a scale parameter κ through the following equality $r = \frac{\sqrt{8\lambda}}{\kappa}$. Therefore, special attention has to be given when comparing results from different geostatistical packages, such as geoR or RGeostats. The main fault concerns the choice of the prioris for the hyperparameters of the models and, consequently, good model specifications. In fact, the use of the default priors resulted in a poor fit of the models.

This paper is organised as follows: in Sect. 2 the data are described, in Sect. 3 the models are detailed, results are presented in Sect. 4 and discussed in Sect. 5.

2   Data

The data for this study were provided by the Instituto Português do Mar e da Atmosfera (IPMA) and correspond to the BSF catches (in Kg) by fishing haul of the longline fishing fleet, in the fishing grounds of the South zone of Portugal, from 2009 to 2013. The data were divided into two semesters, with semester 1 including the months from March to August and semester 2 from September to February of the following year. Moreover, since the BSF catches are quite higher during the months corresponding to the second semester, the analysis has considered only these semesters. The data also include the location of each fishing haul, Fig. 1, the corresponding vessel identification and the depth (DEPTH) out of the locations where the fish was captured. Later, the vessels were grouped into three levels according to their tonnage (PRT) that relates to the cargo capacity.

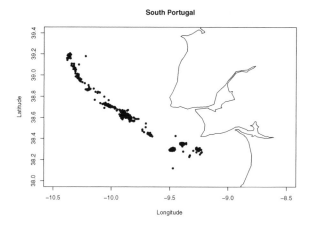

Fig. 1. Locations of the BSF catches.

The left panel of Fig. 2 displays the sample distribution of the BSF catches. In order to meet the Gaussian assumption required by the traditional geostatistical methods, a Box-Cox transformation of the catches with $\lambda = 1/2$ was used (right panel of Fig. 2):

$$\text{BSF}^* = 2\sqrt{\text{BSF}} - 2. \tag{2}$$

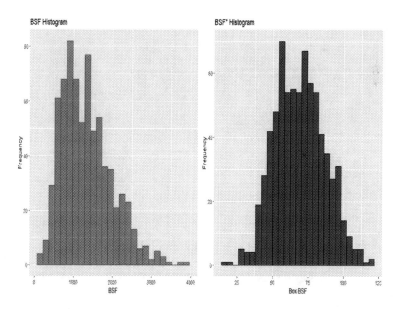

Fig. 2. Sample distribution of BSF catches in the original scale (left) and with the Box-Cox ($\lambda = 1/2$) transformation (right).

The Kolmogorov-Smirnov test for normality of BSF* accepted this hypothesis, resulting in a p-value of 0.486.

If data are isotropic, such that the dependence between sampling locations only relies on the corresponding distance and not on orientation, it greatly simplifies geostatistical models. This has been tested using the non-parametric test proposed by [11] in the R-package spTest [18]. The test is for the null hypothesis which is isotropy, $i.e.$,

$$H_0 : \quad A\boldsymbol{\gamma}(\cdot) = \mathbf{0},$$

where $A = \begin{bmatrix} 1 & -1 & 0 & 0 \\ 0 & 0 & 1 & -1 \end{bmatrix}$ is a full rank matrix [9] and $\boldsymbol{\gamma}(\cdot)$ is the semivariogram evaluated at any two spatial lags. First, one has to choose the lags $\boldsymbol{\Lambda}$ at which anisotropy must be tested. Although this is a subjective choice, it is recommended to choose short lags and to contast points of orthogonal lags. Therefore, $\boldsymbol{\Lambda} = \{\boldsymbol{h}_1 = (2,0), \boldsymbol{h}_2 = (0,2), \boldsymbol{h}_3 = (1,1), \boldsymbol{h}_4 = (-1,1)\}$ was considered since the directions of the variogram estimates start to differ approximately at 4 km, as can be seen in Fig. 3.

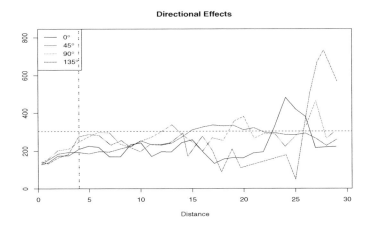

Fig. 3. Variograms at directions 0° (black), 45° (red), 90° (green) and 135° (blue). (Color figure online)

The test statistic is given by

$$TS = b_n^2 (A\hat{G}_n)^T (A\hat{\Sigma}A^T)^{-1}(A\hat{G}_n), \tag{3}$$

where b_n^2 is a normalising constant, G_n is a vector of the true values of the semivariogram at the chosen lags, $\hat{G}_n$ is the vector of the point estimates of the semivariogram at the chosen lags, Σ is the asymptotic variance covariance matrix of $\hat{G}_n$ and $\hat{\Sigma}$ its estimate. The p-value can be obtained from the asymptotic χ^2 distribution with degrees of freedom given by the row rank of A. A grid based block bootstrap approach is applied to estimate Σ. This methodology creates a spatial permutation of sampling locations' blocks [19]. Therefore it is required to choose the window or block size, which can lead to different conclusions as the performance of the test is sensitive to these choices. Moreover, for an irregularly shaped sampling domain, as it is the case, incomplete blocks may complicate the subsampling procedure. Following the recomendations in [18,19], blocks with dimensions 8.342×15.200 were chosen and a p-value of 0.960 was obtained, *i.e.,* the isotropy hypothesis was not rejected.

3 Models

Let observations BSF_i^*, $i = 1, \cdots, n$, be modelled as

$$BSF_i^* \sim N(\eta_i, \sigma_e^2), \tag{4}$$

where n is the total number of fishing hauls, σ_e^2 is the nugget effect and η_i is the response mean. Several linear models were considered to fit the data and variables tonnage (PRT) and depth (DEPTH) were taken as fixed or random effects. The inclusion of a spatial tendency was done through the SPDE approach presented before where an approximated latent GF $\xi(s)$ has a Matérn covariance function as defined in (1). The eleven models considered are presented below and summarised in Table 1.

1. Tonnage is taken as categorical, *i.e.,*

$$\eta_i = \beta_0 + \sum_{m=2}^{M} \beta_m \mathrm{PRT}_{mi}, \tag{5}$$

   where M is the number of groups of PRT.
2. Tonnage is a random effect with a sum to zero constraint, *i.e.,*

$$\eta_{i(m)} = \beta_0 + \mu_m \quad \text{and} \quad \sum_{m=1}^{M} \mu_m = 0. \tag{6}$$

3. Depth is taken as a fixed effect, *i.e.,*

$$\eta_i = \beta_0 + \gamma \mathrm{DEPTH}_i \tag{7}$$

4. Tonnage and depth are fixed effects, *i.e.,*

$$\eta_i = \beta_0 + \sum_{m=2}^{M} \beta_m \mathrm{PRT}_{mi} + \gamma \mathrm{DEPTH}_i \tag{8}$$

5. Depth is taken as a fixed effect whereas tonnage is a random effect with a sum to zero constraint, *i.e.,*

$$\eta_{i(m)} = \beta_0 + \mu_m + \gamma \mathrm{DEPTH}_i \quad \text{and} \quad \sum_{m=1}^{M} \mu_m = 0 \tag{9}$$

6. An approximated latent GF $\xi(s)$ with Matérn covariance function is included, *i.e.*,

$$\eta_i = \beta_0 + \sum_{g=1}^{G} A_{ig}\tilde{\xi}_g, \tag{10}$$

where $\{\tilde{\xi}_g\}$ are zero mean Gaussian distributed weights and A_{ig} is the value $\varphi_g(s_i)$ of the g-th deterministic basis function evaluated in s_i; A_{ig} is also the generic element of the sparse $n \times G$ matrix $\mathbf{A}$ that maps the GMRF $\tilde{\xi}(s)$ from the G triangulation vertices of a mesh to the n observation locations; see [1].

7. Tonnage is categorical and $\xi(s)$ is an approximated latent GF, *i.e.*,

$$eta_i = \beta_0 + \sum_{m=2}^{M} \beta_m \mathrm{PRT}_{mi} + \sum_{g=1}^{G} A_{ig}\tilde{\xi}_i. \tag{11}$$

8. Tonnage is taken as random effect, with a sum to zero constraint, and $\xi(s)$ is an approximated latent GF, *i.e.*,

$$\eta_{i(m)} = \beta_0 + \mu_m + \sum_{g=1}^{G} A_{ig}\tilde{\xi}_g \quad \text{and} \quad \sum_{m=1}^{M} \mu_m = 0. \tag{12}$$

9. Depth is a fixed effect and $\xi(s)$ is an approximated latent GF, *i.e.*,

$$\eta_i = \beta_0 + \gamma \mathrm{DEPTH}_i + \sum_{g=1}^{G} A_{ig}\tilde{\xi}_g. \tag{13}$$

10. Tonnage and depth are fixed effects and $\xi(s)$ is an approximated latent GF, *i.e.*,

$$\eta_i = \beta_0 + \sum_{m=2}^{M} \beta_m \mathrm{PRT}_{mi} + \gamma \mathrm{DEPTH}_i + \sum_{g=1}^{G} A_{ig}\tilde{\xi}_i. \tag{14}$$

11. Tonnage is taken as random effect with a sum to zero constraint, depth is a fixed effect and $\xi(s)$ is an approximated latent GF, *i.e.*,

$$\eta_{i(m)} = \beta_0 + \mu_m + \gamma \mathrm{DEPTH}_i + \sum_{g=1}^{G} A_{ig}\tilde{\xi}_g \quad \text{and} \quad \sum_{m=1}^{M} \mu_m = 0. \tag{15}$$

3.1 Choice of Prior Distributions

The analysis started by considering the default priors of R–INLA, *i.e.*, $\beta_m, \gamma \sim N(0, 0.001^{-1})$ and $\tau_e = 1/\sigma_e^2$, $\tau_{\mathrm{PRT}} \sim Gamma(1, 0.00005)$. However, an extremely high precision for the random effect, τ_{PRT}, was estimated. This indicated that this parameter is very sensitive to the choice of the prior since that

Table 1. Linear models considered to fit the BSF* data

Model	Variables			
	PRT		DEPTH	$\xi(s)$
	Fixed Effect	Random Effect	Fixed Effect	
1	✓			
2		✓		
3			✓	
4	✓		✓	
5		✓	✓	
6				✓
7	✓			✓
8		✓		✓
9			✓	✓
10	✓		✓	✓
11		✓	✓	✓

huge variations on the posterior mean were obtained considering different priors. Following [5], slightly more informative priors on the precision τ_{PRT} were used. However, different priors were considered for this parameter and, in order for the model to converge properly, that is for the INLA algorithm to be able to find the correct posterior mode for the Laplace approximation, the values for the corrected standard deviation for the hyperparameters, which appear in the verbose mode of the model in R-INLA, have to be near the value 1 (between 0.5 and 1.5) [15]. Therefore, for models 2 and 5 (Eqs. (6) and (9), respectively), a truncated normal prior with precision 0.1 was chosen. As for models 8 and 11 (Eqs. (12) and (15), respectively), a Gamma with parameters 0.5 and $0.5 \times \text{Var}(\text{residuals}_{Model6})$ was assumed. For the spatial parameters, the Penalised Complexity Priors (PC Priors) introduced by [17] were used. These priors are a way to overcome the problem of overfitting and intend to penalise the deviation from a base model, where $r = \infty$ and $\sigma^2 = 0$, by shrinking towards it. Moreover, they do not depend on the spatial design neither change if data are made available at new locations. The PC prior for the range was defined according to some aspects, such as the histogram and the median of the distance between the coordinates of the observations, which was approximately 49 km [20], and the practical range obtained in the Matérn variogram, which was approximately 20 km. Therefore, and in order to obtain a proper convergence of the model, $P[r < 30] = 0.2$ was considered the PC prior for the range. In respect of the PC prior for the marginal standard deviation, $P[\sigma > 10] = 0.01$ was applied taking into account the variance obtained in the fitted variogram, which was approximately 127.3 ($\sqrt{127.3} \approx 11$).

3.2   Mesh Construction

The SPDE approach for spatial modelling relies on a finite element method to approximate a GF by a GMRF, which is computationally feasible. This approximation lies on a triangulation of the spatial domain through a mesh in which basis functions are defined for the method. The construction of the mesh is then an important point of the process. It has to be fine enough and computationally feasible, thus some considerations have to be taken into account. For instance, the triangles in the inner domain have to be as regular as possible and extremely small triangles must be avoided. For these motives, the maximum allowed triangle edge length in the inner domain was 1/3 of the range of the study area and five times more for the outer extension. Moreover, the minimum allowed distance between points was 1/5 of the triangle edge length of the inner extension [20]. The resulting mesh is represented in Fig. 4.

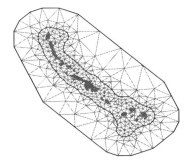

Fig. 4. Mesh considered for the spatial random effects.

4   Results

The preliminary analysis was done in geoR, which does not provide a function to find the optimal variogram. In fact, the choice has to be done by eye. A way to overcome this issue is to resort to RGeostats package, which has a function to determine the optimal variogram. In this case, the optimal variogram for the data obtained was a combination of a Cubic and Spherical covariance families. However, in order to be able to compare the results with the analysis undertaken in R-INLA, the smoothness parameter had to be fixed to 1 and the covariance family had to be Matérn. Yet, when comparing the optimal variogram with the Matérn variogram with the parameter fixed to 1, these were very similar, differing mainly in the nugget effect and in the cubic part of the optimal variogram, as can been seen in Table 2 and in Fig. 5.

The described models in Sect. 3 were fitted in R-INLA and the Deviance Information Criterion (DIC) and the Watanabe-Akaike Information Criterion (WAIC) were used as comparison measures. Results summarised in Table 3.

Table 2. Covariance parameters

Variogram	Total Sill	Nugget Effect	Covariance Family	Range	Sill
Optimal	283.409	132.262	Cubic	4.605	51.406
			Spherical	26.175	99.741
Matérn	284.419	157.116	Matérn	20.433	127.303

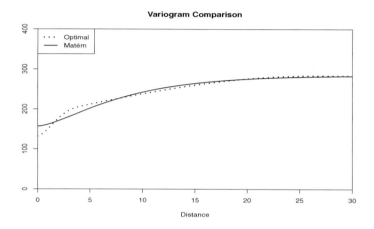

Fig. 5. Variogram comparison.

Table 3. Model estimates

Model	r range	σ st. dev	$\tau_e = 1/\sigma_e^2$	τ_{PRT}	DIC	WAIC
1. FE PRT	–	–	0.005	–	6032.13	6032.32
2. RE PRT	–	–	0.005	0.026	6032.21	6032.40
3. FE Depth	–	–	0.003	–	6273.95	6273.94
4. FE PRT & Depth	–	–	0.005	–	6032.80	6032.98
5. FE Depth + RE PRT	–	–	0.005	0.026	6032.87	6033.04
6. SE	38.656	14.448	0.006	–	5891.80	5896.87
7. SE + FE PRT	32.359	11.530	0.006	–	5867.31	5871.71
8. SE + RE PRT	31.513	11.522	0.006	0.001	5867.21	5871.66
9. SE + FE Depth	38.046	14.621	0.006	–	5892.85	5897.94
10. SE + FE PRT & Depth	30.390	11.515	0.006	–	5867.57	5872.28
11. SE + FE Depth + RE PRT	30.145	11.834	0.006	0.001	5867.82	5872.28

FE - Fixed Effect; RE - Random Effect; SE - Spatial Effect

The values for the criteria measures DIC and WAIC for Models 7 (Eq. (11)), 8 (Eq. (12)), 10 (Eq. (14)) and 11 (Eq. (15)) are very similar. Despite that, the model with spatial effect and the variable tonnage as a random effect (Model 8 Eq. (12)) has the lowest values and, therefore, seems to be the best model to fit the data. Furthermore, the variable depth proved to be not relevant considering a 95% credible interval for its effect.

Finally, predictions for the latent field were computed and the plots with the posterior mean are shown below (Fig. 6):

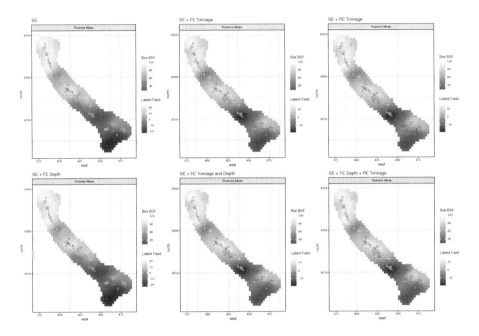

Fig. 6. Estimated posterior latent field for each spatial model and observation data (Box.BSF).

5   Discussion

This work is a first step in modelling the BSF fishery along the Portuguese coast. The spatial nature of the data was perceived and modelled but, being these fishery data, it is most likely that they suffer from preferentiality of sampling locations, reflecting the fishermen preferences. Incorporating this in analysis is the next step of this study.

A few problems arose while adjusting some models and, although they have been addressed in the best way possible, the results presented should be seen with caution. One of the problems concerns the construction of the mesh and how to define the arguments of the function `inla.mesh.2d()`. This is a problem because, if the mesh is not fine enough, the boundary extension is not big enough or the triangles are not as regular as possible in shape and size, results may not be accurate. On the contrary, the more precise is the mesh, the more computational effort is required. Another issue is related to the model with spatial effect and a categorical variable. In order to map the latent field, the function `inla.stack()` is used. However, to the best of our knowledge, `inla.stack()`

does not support categorical variables. Several attempts were performed to over-come this problem. First, the categorical variable was incorporated directly in the model and no intercept was defined, which implied that the first level works as the intercept. Then, a design matrix with dummy variables was created. The latter approach was the selected one as it produced the best results. Also, there was a problem concerning the priors. The variable PRT, either as a categorical variable or taken as a random effect, was found to be highly sensitive to the choice of the priori, both in terms of values obtained for the posterior mean and in terms of obtaining a proper convergence of the model. Furthermore, regarding the PC Priors, little variations on the initial values for the range and/or stan-dard deviation caused large variations on the values of the posterior means for those parameters. In fact, when checking the verbose, it was especially difficult to find correct posteriors for the precision of the random effect, for the standard deviation and for the range hyperparameters. These must be better addressed in the future as this is a work in progress. Finally, only linear models were consid-ered. Inspecting the plots of the residuals against the fitted values, the observed relation given by a non-parametric smoother did not justify fitting non-linear models [10] to these data.

Acknowledgments. This work was supported by national funds through FCT - Fundação para a Ciência e a Tecnologia, I.P., through the project "PREF-ERENTIAL: Improving Spatial Estimation and Survey Design Through Preferen-tial Sampling in Fishery and Biological Applications" with reference PTDC/MAT-STA/28243/2017, project UIDB/00297/2020 (Centro de Matemática e Aplicações) and project UIDB/00006/2020 (Centro de Estatística e Aplicações da Universidade de Lis-boa).

References

1. Blangiardo, M., Cameletti, M.: Spatial and Spatio-Temporal Bayesian Models with R-INLA. Wiley, Chichester (2015). https://doi.org/10.1002/9781118950203
2. Bordalo-Machado, P., Figueiredo, I.: The fishery for black scabbardfish (aphanopus carbo lowe, 1839) in the Portuguese continental slope. Rev. Fish Biol. Fisheries **19**(1), 49–67 (2009). https://doi.org/10.1007/s11160-008-9089-7
3. Committee, R.M.: Report of the international bottom trawl survey working group. Technical report. D:05, International Council for the Exploitation of the Sea (ICES), Lorient, France, March 2003
4. Diggle, P.J., Ribeiro Jr., P.J.: Model-Based Geostatistics. Springer Series in Statis-tics, Springer, New York (2007). https://doi.org/10.1007/978-0-387-48536-2
5. Faraway, J.: Inla for mixed models. https://people.bath.ac.uk/jjf23/inla/
6. Gordo, L.: Black scabbardfish (Aphanopus carbo Lowe, 1839) in the southern northeast atlantic: Considerations on its fishery. Scientia Marina **73**, 11–16 (12 2009). https://doi.org/10.3989/scimar.2009.73s2011
7. Krainski, E., et al.: Advanced Spatial Modeling with Stochastic Partial Differential Equations Using R and INLA. Chapman and Hall/CRC (2018). https://doi.org/10.1201/9780429031892

8. Lindgren, F., Rue, H.v., Lindström, J.: An explicit link between Gaussian fields and Gaussian Markov random fields: the stochastic partial differential equation approach. J. R. Stat. Soc. Ser. B Stat. Methodol. **73**(4), 423–498 (2011). https://doi.org/10.1111/j.1467-9868.2011.00777.x. with discussion and a reply by the authors

9. Lu, N., Zimmerman, D.L.: Testing for directional symmetry in spatial dependence using the periodogram. J. Statist. Plann. Inference **129**(1–2), 369–385 (2005). https://doi.org/10.1016/j.jspi.2004.06.058

10. Ludwig, G., Zhu, J., Reyes, P., Chen, C.-S., Conley, S.P.: On spline-based approaches to spatial linear regression for geostatistical data. Environ. Ecol. Stat. **27**(2), 175–202 (2020). https://doi.org/10.1007/s10651-020-00441-9

11. Maity, A., Sherman, M.: Testing for spatial isotropy under general designs. J. Statist. Plann. Inference **142**(5), 1081–1091 (2012). https://doi.org/10.1016/j.jspi.2011.11.013

12. Martínez-Minaya, J., Cameletti, M., Conesa, D., Pennino, M.G.: Species distribution modeling: a statistical review with focus in spatio-temporal issues. Stoch. Env. Res. Risk Assess. **32**(11), 3227–3244 (2018). https://doi.org/10.1007/s00477-018-1548-7

13. MINES ParisTech / ARMINES: RGeostats: The Geostatistical R Package. Free download from (2020). http://cg.ensmp.fr/rgeostats

14. Natário, I., Figueiredo, I., Lucília Carvalho, M.: A state space model approach for modelling the population dynamics of black scabbardfish in Portuguese mainland waters. Dyn. Games Sci., 499–512 (2015). https://doi.org/10.1007/978-3-319-16118-1_26

15. Rue, H.V.: R-inla discussion group. http://www.r-inla.org

16. Rue, H.V., Martino, S., Chopin, N.: Approximate Bayesian inference for latent Gaussian models by using integrated nested Laplace approximations. J. R. Stat. Soc. Ser. B Stat. Methodol. **71**(2), 319–392 (2009). https://doi.org/10.1111/j.1467-9868.2008.00700.x

17. Simpson, D., Rue, H.V., Riebler, A., Martins, T.G., Sø rbye, S.H.: Penalising model component complexity: a principled, practical approach to constructing priors. Statist. Sci. **32**(1), 1–28 (2017). https://doi.org/10.1214/16-STS576

18. Weller, Z.: spTest: an R package implementing nonparametric tests of isotropy. J. Stat. Softw. **83** (2015). https://doi.org/10.18637/jss.v083.i04

19. Weller, Z.D., Hoeting, J.A.: A review of nonparametric hypothesis tests of isotropy properties in spatial data. Stat. Sci. **31**(3), 305–324 (2016). https://doi.org/10.1214/16-STS547

20. Zuur, A.F., Ieno, E.N., Saveliev, A.A.: Begginer's Guide to Spatial. Temporal and Spatio-Temporal Ecological Data Analysis with R-INLA. Highland Statistics, LTD, Newburgh, UK (2017)

Folding-BSD Algorithm for Binary Sequence Decomposition

Jose Luis Martin-Navarro and Amparo Fúster-Sabater[✉]

Instituto de Tecnologías Físicas y de la Información, C.S.I.C.,
Serrano 144, 28006 Madrid, Spain
jomarna6@inf.upv.es, amparo@iec.csic.es

Abstract. The IoT revolution leads to a range of critical services which rely on IoT devices. Nevertheless, they often lack proper security, becoming the gateway to attack the whole system. IoT security protocols often rely on stream ciphers, where PRNGs are an essential part of them. In this article, a family of ciphers with strong characteristics that make them difficult to be analyzed by standard methods is described. In addition, we will discuss an innovative technique of sequence decomposition and present a novel algorithm to evaluate the strength of binary sequences, key part of the IoT security stack.

Keywords: PRNG · LFSR · Binomial sequences · Stream ciphers · IoT

1 Introduction

Breakthroughs in electronics and telecommunications fields made Internet of Things (IoT) a reality with an every day growing number of sensors and devices around us. Nowadays, diverse critical services as smart-grid, e-health, agricultural or industrial automation depend on an IoT infrastructure. At any rate, as the services around IoT grow dramatically so do the security risks [1,2].

Low-cost IoT devices, currently characterized by their resource constrains in processing power, memory, size and energy consumption, are also recognizable by their minimal or non-existent security. Combining this lack of security with their network dependability, they become the perfect target as a gateway to the whole network [3]. There are already some attacks where a vulnerable IoT sensor was used to gain control over the whole system (in [4], wireless sensors are used to gain access to an automotive system).

Research partially supported by Ministerio de Economía, Industria y Competitividad, Agencia Estatal de Investigación, and Fondo Europeo de Desarrollo Regional (FEDER, UE) under project COPCIS (TIN2017-84844-C2-1-R) and by Comunidad de Madrid (Spain) under project CYNAMON (P2018/TCS-4566), also co-funded by European Union FEDER funds. The first author was supported by JAE "Introduction to research scholarships for students" of the Spanish Ministry of Science and Innovation.

© Springer Nature Switzerland AG 2020
O. Gervasi et al. (Eds.): ICCSA 2020, LNCS 12249, pp. 345–359, 2020.
https://doi.org/10.1007/978-3-030-58799-4_26

This is the reason why general research [5], 5G related research [6] or specific calls such as that of NIST for lightweight cryptography primitives [7], are addressing this concerning topic. Although different protocols of communication and orchestration are being proposed [8], lightweight cryptography in general and stream ciphers in particular are the stepping stones on which such protocols are built to guarantee both device and network security.

In this work, first we will introduce Linear Feedback Shift Registers (LFSR), key components in stream ciphers, often used as Pseudo Random Number Generators (PRNG). Next, we will present the generalized shelf shrinking generator, a particular family of ciphers with strong cryptographic characteristics which remain strong to the standard Berlekamp-Massey Algorithm [9]. Then, we will analyze an innovative sequence decomposition introduced by Cardell et al. in [10] and will show how it can be used to analyze the properties of binary sequences. Finally, we will propose a novel algorithm based on the symmetry of the Binomial Sequences (BS) and discuss the comparison among algorithms.

The study of the generalized shelf shrinking generator is not a random choice. Indeed, it produces not only sequences that are hard to analyzed by the Berlekamp-Massey algorithm, but also it has been implemented in hardware [11] along on RFID devices [12]. Studying the robustness of these sequences could prevent vulnerabilities on the IoT devices and the services built on them.

2 LFSR-Based Sequence Generators: The Generalized Self-shrinking Generator

Linear Feedback Shift Registers (LFSRs) [13] are linear structures currently used in the generation of pseudorandom sequences. The LFSRs are electronic devices included in most of the sequence generators proposed in the literature. Main reasons for a so generalized use are: LFSRs provide high performance when used for sequence generation, they are particularly well-suited to hardware implementations and such registers can be readily analysed by means of algebraic techniques.

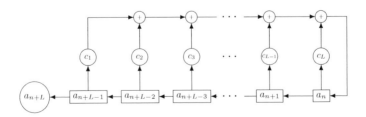

Fig. 1. LFSR of length L

According to Fig. 1, a LFSR consists of L interconnected stages numbered $(0, 1, \ldots, L - 1)$ (from left to right) able to store one bit, the feedback or connection polynomial

$$p(x) = x^L + c_1 x^{L-1} + c_2 x^{L-2} + \ldots + c_{L-1} x + c_L$$

with binary coefficients c_i and the initial state (stage contents at the initial instant). In addition, a clock controls the data shift. LFSRs generate sequences by means of shifts and linear feedbacks.

The output of an LFSR with nonzero initial state is a binary sequence notated $\{a_n\}$ $(n = 0, 1, 2, \ldots)$. If the polynomial $p(x)$ is primitive [13], then the output sequence is called PN-sequence (pseudo-noise sequence). Moreover, a PN-sequence has period $T = 2^L - 1$ bits with 2^{L-1} ones and $2^{L-1} - 1$ zeros. On the other hand, linear complexity (LC) of a sequence is a parameter closely related to the LFSR. In fact, LC is defined as the length of the shortest LFSR able to generate such a sequence.

Although an LFSR in itself is an excellent generator of pseudorandom sequence, nevertheless it has undesirable linearity properties which reduce the security of its use. In practice, the introduction of any type of nonlinearity in the process of generation of the pseudorandom sequence is needed. The irregular decimation of PN-sequences is one of the most popular techniques to destroy the inherent linearity of the LFSRs [14, 15].

Inside the type of irregularly decimated generators, we can enumerate: a) the shrinking generator introduced in [16] that includes two LFSRs, b) the self-shrinking generator [17] that involves just one LFSR and c) the generalized self-shrinking generator proposed in [18] that is considered as a generalization of the self-shrinking generator as well as a simplification of the shrinking generator. In this work, we focus on the last type of generator that is the generalized self-shrinking generator.

2.1   The Generalized Self-shrinking Generator (GSSG)

The generalized self-shrinking generator can be described as follows:

1. It makes use of two PN-sequences: $\{a_n\}$ a PN-sequence produced by an LFSR with L stages and a shifted version of such a sequence denoted by $\{v_n\}$. In fact, $\{v_n\} = \{a_{n+p}\}$ corresponds to the own sequence $\{a_n\}$ but rotated cyclically p positions to the left with $(p = 0, 1, \ldots, 2^L - 2)$.
2. It relates both sequences by means of a simple decimation rule to generate the output sequence.

For $n \geq 0$, we define the decimation rule as follows:

$$\begin{cases} \text{If } a_n = 1 \text{ then } v_n \text{ is output,} \\ \text{If } a_n = 0 \text{ then } v_n \text{ is discarded and there is no output bit.} \end{cases}$$

Thus, for each value of p an output sequence $\{s_n\}_p = \{s_0 \, s_1 \, s_2 \ldots\}_p$ is generated. Such a sequence is called the p generalized self-shrunken sequence (GSS-sequence) or simply generalized sequence associated with the shift p. Recall that

Table 1. GSS-sequences for an LFSR with polynomial $p(x) = x^4 + x + 1$

shift p	$\{v_n\}$ sequences	GSS-sequences	shift p	$\{v_n\}$ sequences	GSS-sequences
0	111100010011010	11111111	8	001101011110001	00111100
1	111000100110101	11100100	9	011010111100010	01101001
2	110001001101011	11000011	10	110101111000100	11011000
3	100010011010111	10001101	11	101011110001001	10101010
4	000100110101111	00011011	12	010111100010011	01010101
5	001001101011110	00100111	13	101111000100110	10110001
6	010011010111100	01001110	14	011110001001101	01110010
7	100110101111000	10010110	$--$	000000000000000	00000000
	111100010011010			111100010011010	

$\{a_n\}$ remains fixed while $\{v_n\}$ is the sliding sequence or left-shifted version of $\{a_n\}$. When p ranges in the interval $[0, 1, \ldots, 2^L - 2]$, then we obtain the $2^L - 1$ members of the family of generalized sequences based on the PN-sequence $\{a_n\}$. Since the PN-sequence has 2^{L-1} ones, the period of any generalized sequence will be 2^{L-1} of divisors, that is a power of 2. Next a simple example is introduced.

Example 1. For an LFSR with primitive polynomial $p(x) = x^4 + x + 1$ and initial state $(1\ 1\ 1\ 1)$, we generate the generalized sequences depicted in Table 1. The bits in bold in the different sequences $\{v_n\}$ are the digits of the corresponding GSS-sequences associated to their corresponding shifts p. The PN-sequence $\{a_n\}$ with period $T = 2^4 - 1$ is written at the bottom of the table.

There are two particular facts that differentiate the GSS-sequences from the PN-sequences generated by LFSRs:

- The period of the GSS-sequences is a power of 2, in contrast to PN-sequences whose period is $2^L - 1$. This difference arises from the fact that PN-sequences cannot have a run of zeros of the length of its internal state.
- The LC of PN-sequences equals L while that of the GSS-sequences is 2^{L-1}. This property is desirable because such sequences exhibit a great LC with low resources. As a result, the Berlekamp-Massey algorithm would need to process 2^L bits with a computational complexity of $O(2^{2*L})$.

3   Binomial Sequences and Sierpinski's Triangle

In this section, we introduce a new representation of the binary sequences whose period is a power of 2 in terms of the binomial sequences. Next, the close relationship among binomial sequences and the Sierpinski's triangle is also analyzed.

3.1   Binomial Sequences

The binomial number $\binom{n}{i}$ is the coefficient of the power x^i in the polynomial expansion of $(1 + x)^n$. For every non-negative integer n, it is a well-known fact that $\binom{n}{0} = 1$ as well as $\binom{n}{i} = 0$ for $i > n$.

(a) Pascal's triangle   (b) Sierpinski's triangle mod 2

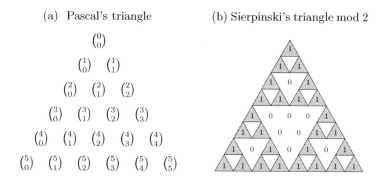

Fig. 2. Pascal and Sierpinski's triangles

The binomial coefficients reduced modulo 2 allow us to define the concept of binomial sequence.

Definition 1. *Given an integer $i \geq 0$, the sequence $\{b_n\}_i$ ($n = 0, 1, 2, \ldots$) whose elements are binomial coefficients reduced modulo 2, that is $b_n = \binom{n}{i} \bmod 2$, is called the i-th binomial sequence.*

Table 2. The eight first binomial sequences, their periods and linear complexities

Binomial coeff.	Binomial sequences $\{\binom{n}{i}\}$	Period	Linear complexity
$\binom{n}{0}$	$\{1,1,1,1,1,1,1,1,\ldots\}$	$T_0 = 1$	$LC_0 = 1$
$\binom{n}{1}$	$\{0,1,0,1,0,1,0,1,\ldots\}$	$T_1 = 2$	$LC_1 = 2$
$\binom{n}{2}$	$\{0,0,1,1,0,0,1,1,\ldots\}$	$T_2 = 4$	$LC_2 = 3$
$\binom{n}{3}$	$\{0,0,0,1,0,0,0,1,\ldots\}$	$T_3 = 4$	$LC_3 = 4$
$\binom{n}{4}$	$\{0,0,0,0,1,1,1,1,\ldots\}$	$T_4 = 8$	$LC_4 = 5$
$\binom{n}{5}$	$\{0,0,0,0,0,1,0,1,\ldots\}$	$T_5 = 8$	$LC_5 = 6$
$\binom{n}{6}$	$\{0,0,0,0,0,0,1,1,\ldots\}$	$T_6 = 8$	$LC_6 = 7$
$\binom{n}{7}$	$\{0,0,0,0,0,0,0,1,\ldots\}$	$T_7 = 8$	$LC_7 = 8$

Table 2 shows the eight first binomial sequences $\{\binom{n}{i}\}$, $i = 0, 1, \ldots, 7$, see [19], with their corresponding periods and linear complexities, denoted by T_i and LC_i, respectively.

Next, different properties of the binomial sequences are introduced.

1. Given the binomial sequence $\left\{\binom{n}{2^L+k}\right\}$, with $0 \leq k < 2^L$, we have that [10, Proposition 1.b]:
   a) Such a binomial sequence has period $T = 2^{L+1}$.

b) The first period of such a binomial sequence has the following structure:

$$
\left\{ \binom{n}{2^L + k} \right\}_{0 \leq n < 2^{L+1}} =
\begin{cases}
0 & \text{if } 0 \leq n < 2^L + k, \\
\binom{n}{k}_{mod\,2} & \text{if } 2^L + k \leq n < 2^{L+1}.
\end{cases}
$$

2. The linear complexity of the sequence $\left\{ \binom{n}{i} \right\}$ with $0 \leq i$ is $LC = i + 1$, see [10, Theorem 3].

3. Every binary sequence whose period is a power of 2 can be written as a linear combination of a finite number of binomial sequences [10, Theorem 1]. Such a combination is called the Binomial Sequence Decomposition.

4. Given a sequence with binomial representation $\sum_{k=0}^{t} \left\{ \binom{n}{i_k} \right\}$, where $i_0 < i_1 < \cdots < i_t$ are integer indexes, then its linear complexity is given by $LC = i_t + 1$, see [19].

5. Given a sequence with binomial representation $\sum_{k=0}^{t} \left\{ \binom{n}{i_k} \right\}$, where $i_0 < i_1 < \cdots < i_t$ are integer indexes, then its period T is that of the binomial sequence $\left\{ \binom{n}{i_t} \right\}$, see [19].

Notice that the generalized sequences are binary sequences whose period is a power of 2. Consequently, they can be written in terms of binomial sequences satisfying all the previous properties.

3.2 The Sierpinski's Triangle

When the binomial coefficients are arranged into rows for the successive values of $n = 0, 1, 2, \ldots$, then the generated structure is the Pascal's triangle (see Fig. 2a). If we color the odd numbers and shade the even ones in such a triangle, then we get the Sierpinski's triangle whose version reduced mod 2 is depicted in Fig. 2b.

Recall that the successive diagonals of the Sierpinski's triangle in Fig. 2a correspond to the successive binomial sequences $\left\{ \binom{n}{i} \right\}$, $(i = 0, 1, 2, \ldots)$ starting at the first 1. That is, the binomial sequences can be found inside the Sierpinski's triangle mod 2, and also are related to certain cellular automata [10].

4 Algorithms to Calculate the Linear Complexity

NOTATION: For the sake of simplicity, in this section the binomial coefficient $\binom{n}{k}$ will denote the corresponding $k-th$ binomial sequence. Then, the term $\binom{n}{k}_{i,j}$ stands for the binary sub-sequence of $\binom{n}{k}$ between the bits i and $j-1$ while $\binom{n}{k}_j$ is just for the case $i = 0$.

The general method of computing the linear complexity of any sequence is the Berlekamp-Massey algorithm [9]. In order to work, this algorithm needs to process $2*n$ bits of the sequence with a computational complexity of $O(n^2)$ [20].

This section first introduces in detail the basic Binomial Sequence Decomposition algorithm (b-BSD) as well as an improvement on the algorithm implementation.

Second, a new approach to the Binomial Sequence Decomposition is developed, giving rise to the folding Binomial Sequence Decomposition algorithm (f-BSD). Such an algorithm improves the throughput of previous methods thanks to the symmetry of the binomial sequences.

Finally, a comparison among Berlekamp-Massey, b-BSD and f-BSD algorithms is presented. Such a comparison allows us to discuss the improvement of the f-BSD algorithm presented in this article.

4.1   Basic Binomial Sequence Decomposition (b-BSD)

Based on the mathematical results provided in the previous section, a basic Binomial Sequence Decomposition algorithm (b-BSD) can be designed in order to calculate the LC of a given sequence. In particular, two facts are used:

– A sequence of length n can be decomposed in $t+1$ binomial sequences (third item in 3.1):

$$seq_n = \binom{n}{k_0} + \cdots + \binom{n}{k_t}.$$

– The lineal complexity of a sequence can be calculated from the maximum binomial sequence of its BSD (second item in Sect. 3.1). Since the binomial sequences are in order, then LC satisfies the following expression:

$$LC = max_{0 \leq i \leq t}\left(\binom{n}{k_i}\right) + 1 = k_t + 1.$$

The resulting algorithm can be seen in Algorithm 1. It takes the sequence to be analyzed as input and checks for every bit equal to 1. When $bit_i == 1$, it sums the sequence with the corresponding binomial sequence $(seq + \binom{n}{i})$ stopping when all the binomial sequences have been found. A step-by-step example of the algorithm decomposing a sequence of length 16 can be seen in Table 3.

Algorithm 1. Basic Binomial Sequence Decomposition (b-BSD)

Require: seq : intercepted bits
   $binom = [\emptyset]$
   for $i = 0$; $i < length(seq)$; $i{+}{+}$ **do**
     if $seq_i \neq 0$ **then**
       $seq{+} = \binom{n}{i}$
       $binom.add(i)$
     end if
   end for
   return $binom$: Binomial decomposition of the intercepted bits

---

Table 3. Step by step b-BSD example on seq_{16}

Step	Op.	Seq.	0 1 2 3	4 5 6 7	8 9 10 11	12 13 14 15
0		seq	0 0 1 0	0 1 0 0	1 0 1 1	1 1 0 1
	$+$	$\binom{n}{2}$	0 0 1 1	0 0 1 1	0 0 1 1	0 0 1 1
1	$=$	seq	0 0 0 1	0 1 1 1	1 0 0 0	1 1 1 0
	$+$	$\binom{n}{3}$	0 0 0 1	0 0 0 1	0 0 0 1	0 0 0 1
2	$=$	seq	0 0 0 0	0 1 1 0	1 0 0 1	1 1 1 1
	$+$	$\binom{n}{5}$	0 0 0 0	0 1 0 1	0 0 0 0	0 1 0 1
3	$=$	seq	0 0 0 0	0 0 1 1	1 0 0 1	1 0 1 0
	$+$	$\binom{n}{6}$	0 0 0 0	0 0 1 1	0 0 0 0	0 0 1 1
4	$=$	seq	0 0 0 0	0 0 0 0	1 0 0 1	1 0 0 1
	$+$	$\binom{n}{8}$	0 0 0 0	0 0 0 0	1 1 1 1	1 1 1 1
5	$=$	seq	0 0 0 0	0 0 0 0	0 1 1 0	0 1 1 0
	$+$	$\binom{n}{9}$	0 0 0 0	0 0 0 0	0 1 0 1	0 1 0 1
6	$=$	seq	0 0 0 0	0 0 0 0	0 0 1 1	0 0 1 1
	$+$	$\binom{n}{10}$	0 0 0 0	0 0 0 0	0 0 1 1	0 0 1 1
end	$=$	seq	0 0 0 0	0 0 0 0	0 0 0 0	0 0 0 0

$$seq = \binom{n}{2} + \binom{n}{3} + \binom{n}{5} + \binom{n}{6} + \binom{n}{8} + \binom{n}{9} + \binom{n}{10}$$

Thus, the b-BSD algorithm is able to calculate the LC, as the Berlekamp-Massey algorithm does, but with only n bits of the sequence instead of $2*n$. The complexity of b-BSD algorithm, which performs the sum of two sequences of n bits (n additions) for every binomial sequence, is $O(t*n)$, t being the number of binomial sequences in which the main sequence is decomposed with $t \ll n$.

Moreover, the logic of the algorithm can be improved by avoiding the sum of the sub-sequence that are zero. On the one hand, thanks to the characteristic 1.b) of Sect. 3.1, we know that $\binom{n}{k} = 0 \ \forall n < k$. On the other hand, at each step of the b-BSD algorithm the sequence begins with zeros. That is, at step i the k_i first terms of the sequence are zeros.

If these two facts are combined, then the number of algorithm operations can be reduced. When the algorithm detects the first 1 in the $i-th$ position of seq, instead of performing the sum of two sequences of n bits ($seq_n + \binom{n}{i}$), it just sums both sequences between the $i-th$ and $(n-1)-th$ bits ($seq_{i,n} + \binom{n}{i}_{i,n}$), as the head of both sequences ($[0, i-1]$) is made up of zeros.

Compared with $t*n$, the number of operations is reduced as follows:

$$\sum_{k_i=k_1}^{k_t} (n - k_i) < t*n.$$

In addition, we do not need to perform the sum of any bit after the max binomial. The reason is that, for every binary sequence produced by the generalized self-shrinking generator, the maximum binomial sequence can be calculated as $k_{max} = n - \log n$. The final number of operations will be:

$$\sum_{k_i=k_1}^{k_t} (k_{max} - k_i) < \sum_{k_i=k_1}^{k_t} (n - k_i) < t * n.$$

To upgrade the code in Algorithm 1 only the sum of both sequences is changed, which will be now $seq = seq_{i,max} + \binom{n}{i}_{i,max}$, with $max = n - \log n$.

Summing it up, for a sequence with a characteristic polynomial of degree σ, with length $n = 2^{\sigma-1}$, the b-BSD algorithm will require $n - \log n$ bits of the sequence to calculate the LC with a complexity of $O(t * n)$.

4.2 Folding Binomial Sequence Decomposition (f-BSD)

Despite the improvement in both complexity and length requirements between b-BSD and Berlekamp-Massey, there is still room for enhancing the decomposition mechanism.

In the next sub-section a new algorithm design is explained, improving the results of the b-BSD algorithm by taking advantage of the symmetry of the binomial sequences.

In order to fully understand the f-BSD algorithm, a particular matrix representation of the decomposition, based on the symmetric properties of binomial sequences, is presented.

Symmetry of $\binom{n}{k}$: There are two properties regarding the symmetric structure of binomial sequences and their relation to the powers of 2 that are explained in the next sub-section.

In all the binomial sequences ($\binom{n}{k}$ with $k < \frac{n}{2}$ and n being the length of the sequence), it is possible to observe the following structure:

$$\binom{n}{k}_n = \left(\binom{n}{k}_{0,\frac{n}{2}}, \binom{n}{k}_{\frac{n}{2},n} \right).$$

In Table 4, where $\frac{n}{2} = 8$, this phenomenon is observable on binomial sequences $\binom{n}{2}, \binom{n}{3}, \binom{n}{5}$ and $\binom{n}{6}$, where the eight first bits repeat themselves.

In fact, this is a simplification of a stronger result, defined in Theorem 1.

Theorem 1. *For every binomial sequence $\binom{n}{k}$ there exists an integer $l \in \mathbb{N}$ such that 2^l is the period of the binomial sequence as well as satisfies the inequality*

$$\forall \binom{n}{k}, \exists\, l \in \mathbb{N} \implies 2^{l-1} < k \leq 2^l.$$

The result follows directly from item 1 in 3.1. On the other hand, recall that the binomial sequences $\binom{n}{k}$ with $\frac{n}{2} \leq k \leq n$ start with k zeros, so they can be divided in the following way: $\binom{n}{k} = zeros + seq = \left(zeros_{0,\frac{n}{2}}, \binom{n}{k}_{\frac{n}{2},n} \right).$

From the item 1.b in Subsect. 3.1 there is an interesting characteristic of the sub-sequence $\binom{n}{k}_{\frac{n}{2},n}$, which can be converted in another binomial sequence with the following expression: $\binom{n}{k}_{\frac{n}{2},n} = \binom{n}{k-\frac{n}{2}}_{\frac{n}{2}}$. Taking into consideration the period of the sequence, the Theorem 2 arises:

Theorem 2. *Every binomial sequence* $\binom{n}{k}_n$, $k \leq n$ *with period* 2^l, $2^{l-1} < k \leq 2^l$, *can be divided into two binomial sequences of length* $\frac{n}{2}$ *as follows:*

$$\binom{n}{k}_n = \left(zeros_{2^{l-1}}, \binom{n}{k-2^{l-1}}_{2^{l-1},n} \right).$$

In particular, as the sequences analyzed have a length of n a power of 2 bits, they can be divided into two sequences of length $\frac{n}{2}$. Again, it can be observed in Table 4 on the binomial sequences $\binom{n}{8}$, $\binom{n}{9}$ and $\binom{n}{10}$.

Table 4. Binomial Sequence Decomposition at a glance

seq	0 0 1 0	0 1 0 0	1 0 1 1	1 1 0 1
$\binom{n}{2} = \binom{n}{2}_{\frac{n}{2}}, \binom{n}{2}_{\frac{n}{2}}$	0 0 1 1	0 0 1 1	0 0 1 1	0 0 1 1
$\binom{n}{3} = \binom{n}{3}_{\frac{n}{2}}, \binom{n}{3}_{\frac{n}{2}}$	0 0 0 1	0 0 0 1	0 0 0 1	0 0 0 1
$\binom{n}{5} = \binom{n}{5}_{\frac{n}{2}}, \binom{n}{5}_{\frac{n}{2}}$	0 0 0 0	0 1 0 1	0 0 0 0	0 1 0 1
$\binom{n}{6} = \binom{n}{6}_{\frac{n}{2}}, \binom{n}{6}_{\frac{n}{2}}$	0 0 0 0	0 0 1 1	0 0 0 0	0 0 1 1
$\binom{n}{8} = zeros_{\frac{n}{2}}, \binom{n}{0}_{\frac{n}{2}}$	0 0 0 0	0 0 0 0	1 1 1 1	1 1 1 1
$\binom{n}{9} = zeros_{\frac{n}{2}}, \binom{n}{1}_{\frac{n}{2}}$	0 0 0 0	0 0 0 0	0 1 0 1	0 1 0 1
$\binom{n}{10} = zeros_{\frac{n}{2}}, \binom{n}{2}_{\frac{n}{2}}$	0 0 0 0	0 0 0 0	0 0 1 1	0 0 1 1
$seq = \binom{n}{2} + \binom{n}{3} + \binom{n}{5} + \binom{n}{6} + \binom{n}{8} + \binom{n}{9} + \binom{n}{10}$				

Putting together the facts regarding the symmetry of the binomial sequences, they can be classified in two groups depending on their division. It is explained in Algorithm 2.

Algorithm 2. Binomial Sequences Classification

For a given sequence $\binom{n}{k}_n$:
if $k \leq \frac{n}{2}$ **then**
$$\binom{n}{k}_n := \left(\binom{n}{k}_{\frac{n}{2}}, \binom{n}{k}_{\frac{n}{2}} \right)$$
else
$$\binom{n}{k}_n := \left(zeros_{\frac{n}{2}}, \left(\binom{n}{k-\frac{n}{2}} \right)_{\frac{n}{2}} \right)$$
end if

If the binomial sequences are divided as seen in Algorithm 2, then it results in the $\left(\begin{array}{c|c} A & B \\ \hline C & D \end{array} \right)$ matrix representation of the BSD, as shown in (1).

$$
\left(\begin{pmatrix} n \\ k_0 \end{pmatrix} \right) = \begin{pmatrix} \begin{pmatrix} n \\ k_0 \end{pmatrix} \\ \vdots \\ \begin{pmatrix} n \\ k_{i-1} \end{pmatrix} \\ \hline \begin{pmatrix} n \\ k_i \end{pmatrix} \\ \vdots \\ \begin{pmatrix} n \\ k_t \end{pmatrix} \end{pmatrix}_{k_{i-1} < \frac{n}{2} \le k_i} = \begin{pmatrix} \begin{pmatrix} n \\ k_0 \end{pmatrix}_{\frac{n}{2}} & \begin{pmatrix} n \\ k_0 \end{pmatrix}_{\frac{n}{2}} \\ \vdots & \vdots \\ \begin{pmatrix} n \\ k_{i-1} \end{pmatrix}_{\frac{n}{2}} & \begin{pmatrix} n \\ k_{i-1} \end{pmatrix}_{\frac{n}{2}} \\ \hline 0 & \begin{pmatrix} n \\ k_i - \frac{n}{2} \end{pmatrix}_{\frac{n}{2}} \\ \vdots & \vdots \\ 0 & \begin{pmatrix} n \\ k_t - \frac{n}{2} \end{pmatrix}_{\frac{n}{2}} \end{pmatrix} = \begin{pmatrix} M_0 & M_1 \\ M_2 & M_3 \end{pmatrix} \quad (1)
$$

Three important characteristics about the matrix representation shown in (1) form the core of the folding BSD algorithm.

- $M_0 = M_1$.
- $M_2 = 0$.
- As the length of the sequences is $n = 2^{\sigma-1}$, the matrix representation can be extended in a recursive way, taking M_3 and repeating the same process until it cannot be divided more (length $= 1$).

$$
\begin{pmatrix} M_0 & M_1 \\ M_2 & M_3 \end{pmatrix} = \begin{pmatrix} M_0 & M_1 \\ M_2 & \begin{matrix} M_{3,0} & M_{3,1} \\ M_{3,2} & M_{3,3} \end{matrix} \end{pmatrix} = \begin{pmatrix} M_0 & M_1 \\ M_2 & \begin{matrix} M_{3,0} & M_{3,1} \\ M_{3,2} & \begin{matrix} M_{3,3,0} & M_{3,3,1} \\ M_{3,3,2} & M_{3,3,3} \end{matrix} \end{matrix} \end{pmatrix} \quad (2)
$$

The following expression (3) is an example of the matrix representation of the sequence decomposition of Table 4.

$$
\left(\begin{pmatrix} n \\ 2 \\ n \\ 3 \\ n \\ 5 \\ n \\ 6 \\ n \\ 8 \\ n \\ 9 \\ n \\ 10 \end{pmatrix} \right) = \begin{pmatrix} 0011\ 0011 & 0011\ 0011 \\ 0001\ 0001 & 0001\ 0001 \\ 0000\ 0101 & 0000\ 0101 \\ 0000\ 0011 & 0000\ 0011 \\ \hline 0000\ 0000 & 1111\ 1111 \\ 0000\ 0000 & 0101\ 0101 \\ 0000\ 0000 & 0011\ 0011 \end{pmatrix} = \begin{pmatrix} A & B \\ C & D \end{pmatrix} \to D = \begin{pmatrix} 1111 & 1111 \\ 0101 & 0101 \\ 0011 & 0011 \\ \emptyset & \emptyset \end{pmatrix} = \dots
$$

$$ (3) $$

In order to calculate the LC of the given sequence, only the highest binomial sequence of its decomposition is needed. Thus, the f-BSD algorithm will benefit from the symmetry of the binomial sequences by reducing recursively the length of the sequence to analyze, as depicted in the matrix expression (2).

4.3 Folding Binomial Sequence Decomposition Algorithm

The previous subsection described all the elements needed by the f-BSD algorithm. In fact, the algorithm locates the maximum binomial sequence to calculate

LC. At every step, it sums the first half of the sequence with the second half. If the result is different from zero, then it continues with the second half of the sequence. Otherwise, it continues with the first half, finishing when only one bit is left.

At every step, the folding mechanism reduces the length of the studied sequence by 2. It performs a sum of half the length of the sequence too, with a total of $\log n$ steps. Given a sequence seq with length n, the number of operations of the algorithm can be calculated as follows:

$$\frac{n}{2} + \frac{\frac{n}{2}}{2} + \cdots = \frac{n}{2} + \frac{n}{4} + \cdots = \sum_{i=0}^{\log n} \frac{n}{2^i} \approx n$$

The final pseudo-code of the algorithm, for a given binary sequence of length n (although we can reduce it to $n - \log n$ as explained in Sect. 4.1) and complexity $O(n)$, can be found in Algorithm 3. The way this algorithm searches for the maximum binomial sequence is similar to that of the binary search algorithm. The difference results in that the binary search only performs one comparison in each step, while our algorithms needs to sum $\frac{length(n)}{2^i}$ operations per step.

Lets see how it works with an example. Taking the sequence of the previous examples $seq = 00100100\ 10111101$

Step 1:
$$\begin{array}{r} 0010\ 0100 \\ +\ 1011\ 1101 \\ \hline 1001\ 1001 \end{array}$$
As $aux = 1001\ 1001 \neq 0_{0,7}$, then $seq = aux = 1001\ 1001$ and $k = 8$.

Step 2:
$$aux = \begin{array}{r} 10\ 01 \\ +\ 10\ 01 \\ \hline 00\ 00 \end{array}$$
As $aux = 0_{0,3}$, then $seq = 1001$.

Step 3:
$$aux = \begin{array}{r} 1\ 0 \\ +\ 0\ 1 \\ \hline 1\ 1 \end{array}$$
As $aux \neq 0_{0,1}$, then $seq = aux = 1\ 1$ and $k = 8 + 2$.

Step 4:
$$aux = \begin{array}{r} 1 \\ +\ 1 \\ \hline 0 \end{array}$$
As $aux = 0$, then $seq = aux = 0$.

- End: the maximum binomial sequence is $\binom{n}{10} \rightarrow LC = k + 1 = 10 + 1 = 11$.

4.4   Algorithm Comparison

When putting together the three algorithms that can calculate the complexity of a sequence (Berlekamp-Massey, b-BSD and f-BSD), it is interesting to compare the computational complexity and length requirements of each of them as shown in Table 5.

Although the Berlekamp-Massey algorithm is able to calculate the linear complexity of any sequence, it is not the best choice for particular sequences as the GSS-sequences $O(n^2)$. It is in that situation where the Binomial Sequence

Algorithm 3. Folding Binomial Sequences Classification

Require: seq : intercepted bits
   $k = 0$
   while $length(seq) > 1$ **do**
     $n = length(seq)$
     aux $= seq_{\frac{n}{2},n} + seq_{0,\frac{n}{2}-1}$
     if $aux \neq 0_{\frac{n}{2}}$ **then**
       $seq = aux$
       $k+ = \frac{n}{2}$
     else
       $seq = seq_{0,\frac{n}{2}}$
     end if
   end while
   return k: maximum binomial sequence

Table 5. Algorithm comparison

Algorithms	Length required	Complexity
Berlekamp-Massey	2 * n	$O(n^2)$
b-BSD	$n - \log n$	$O(t * n)$
f-BSD	$n - \log n$	$O(n)$

Decomposition can be really useful, in particular the folding algorithm presented in this part of the work.

A particular difference between b-BSD and f-BSD is that the later performance does not depend on the number of binomial sequences in the decomposition. That means that in general its performance will be better, but it could depend on the particular sequence under study.

Although it is not the purpose of this work, it is worth saying that the f-BSD algorithm can be parallelized in the calculus of the LC of a given sequence, while b-BSD performs the calculus in a sequential way.

5   Conclusion

In this work, the folding-BSD algorithm has been introduced. It exhibits much better performance on sequences particularly hard to be decomposed by the Berlekamp-Massey algorithm. This is a big step in the study of binary sequences with period a power of two, and makes it easier to find vulnerabilities in this kind of sequences. Detecting such vulnerabilities in a cipher implemented in practical applications could compromise the corresponding IoT devices and the services behind them.

Moreover, the binomial decomposition of sequences as a way to extract information from a given sequence is an innovative but powerful tool, and it is left for future work its application to other kind of binary sequences. Also, there is still room for improvement on how the fractal structure of the binomial sequences can be profited to decompose a binary sequence without handling the whole sequence.

About the f-BSD algorithm presented in this article, it shows a better theoretical characteristics in both complexity and length of the sequence required. Future works may study the algorithm performance in real world scenarios by applying it to different binary sequences, taking advantage of the parallel capabilities of the algorithm.

References

1. Chin, W.-L., Li, W., Chen, H.-H.: Energy big data security threats in IOT-based smart grid communications. IEEE Commun. Mag. **55**(10), 70–75 (2017)
2. Meyer, D., Haase, J., Eckert, M., Klauer, B.: New attack vectors for building automation and IOT. In: IECON 2017–43rd Annual Conference of the IEEE Industrial Electronics Society, pp. 8126–8131
3. Gallegos-Segovia, P.L., et al.: Internet of things as an attack vector to critical infrastructures of cities. In: 2017 International Caribbean Conference on Devices, Circuits and Systems (ICCDCS), pp. 117–120 (2017)
4. Rouf, I., et al.: Security and privacy vulnerabilities of in-car wireless networks: a tire pressure monitoring system case study. In: USENIX Security Symposium, vol. 10 (2010)
5. Cynthia, J., Parveen Sultana, H., Saroja, M.N., Senthil, J.: Security protocols for IoT. In: Jeyanthi, N., Abraham, A., Mcheick, H. (eds.) Ubiquitous Computing and Computing Security of IoT. SBD, vol. 47, pp. 1–28. Springer, Cham (2019). https://doi.org/10.1007/978-3-030-01566-4_1
6. Mavromoustakis, C.X., Mastorakis, G., Batalla, J.M. (eds.): Internet of Things (IoT) in 5G Mobile Technologies. MOST, vol. 8. Springer, Cham (2016). https://doi.org/10.1007/978-3-319-30913-2
7. Nist lightweight cryptography project. https://csrc.nist.gov/Projects/Lightweight-Cryptography. Accessed 30 Mar 2020
8. McGinthy, J.M.: Solutions for internet of things security challenges: trust and authentication. Ph.D. thesis, Virginia Tech (2019)
9. Massey, J.: Shift-register synthesis and BCH decoding. IEEE Trans. Inf. Theory **15**(1), 122–127 (1969)
10. Cardell, S.D., Fúster-Sabater, A.: Binomial representation of cryptographic binary sequences and its relation to cellular automata. Complexity (2019)
11. Chang, K.Y., et al.: Method and apparatus for generating keystream. US Patent 7,587,046, 8 September 2009
12. Kang, Y.-S., Kim, H.-W., Chung, K.-I.: Apparatus and method for protecting RFID data. US Patent 8,386,794, 26 February 2013
13. Golomb, S.W., et al.: Shift Register Sequences. Aegean Park Press, Laguna Hills (1967)
14. Cardell, S.D., Fúster-Sabater, A.: The modified self-shrinking generator? In: Shi, Y., et al. (eds.) ICCS 2018. LNCS, vol. 10860, pp. 653–663. Springer, Cham (2018). https://doi.org/10.1007/978-3-319-93698-7_50
15. Díaz Cardell, S., Fúster-Sabater, A.: Cryptography with Shrinking Generators. SM. Springer, Cham (2019). https://doi.org/10.1007/978-3-030-12850-0

16. Coppersmith, D., Krawczyk, H., Mansour, Y.: The shrinking generator. In: Stinson, D.R. (ed.) CRYPTO 1993. LNCS, vol. 773, pp. 22–39. Springer, Heidelberg (1994). https://doi.org/10.1007/3-540-48329-2_3

17. Meier, W., Staffelbach, O.: The self-shrinking generator. In: Communications and Cryptography, pp. 287–295. Springer (1994). https://doi.org/10.1007/978-1-4615-2694-0_28

18. Hu, Y., Xiao, G.: Generalized self-shrinking generator. IEEE Trans. Inf. Theory **50**(4), 714–719 (2004)

19. Fúster-Sabater, A.: Generation of cryptographic sequences by means of difference equations. Appl. Math. Inf. Sci. **8**(2), 475 (2014)

20. Cusick, T.W., Stanica, P.: Cryptographic Boolean Functions and Applications. Academic Press, Cambridge (2017)

Construction of the Regular Dodecahedron with the MATHEMATICA

Ricardo Velezmoro León[1](✉) ⓘ, Marcela Velásquez Fernández[2](✉) ⓘ,
and Jorge Jimenez Gomez[2](✉) ⓘ

[1] Math Department, National University of Piura, Urb. Miraflores S/N Castilla,
20002 Piura, Peru
rvelezmorol@unp.edu.pe
[2] National University of Piura, Urb. Miraflores S/N Castilla, 20002 Piura, Peru
fmvelasquezf@unp.edu.pe, jorge2020.mate@gmail.com

Abstract. Man is always curious about the patterns and beauty of platonic solids. Throughout history it has shaped them into artistic works, even today it uses them in technological models; likewise nature has adopted its geometric patterns. Through the development of digital technologies, geometric designs are made using the architecture of Platonic solids. The object of study and construction in this work is the Dodecahedron using Wolfram Mathematica 11.2 Software. For this, an efficient logical language will be used, framed within the discipline of geometry, where relationships and hierarchies are established in reasoning through the programming language, allowing its architecture to be associated with creations of man, objects and organisms of nature.

Keywords: Regular dodecahedron · Programming · Wolfram mathematica.

1 Introduction

The exuberant geometry of regular polyhedra attributed to the nature of their geometry, aesthetic, symbolic, mystical and cosmic, has fascinated man [1]. Although the origin of regular polyhedra is not known exactly, there is a conviction that they were known in the time of the Neolithic peoples. However, it was Pythagoras and his school that systematized and rigorously taught solid sayings. They are attributed the construction of the dodecahedron, the hexahedron and associated them with cosmic figures.

A prominent Greek for his interest in geometry was Platón [2], who had a cosmological view of regular polyhedra, saying, "It is necessary to explain what properties the most beautiful bodies should have [...], they must have the property of dividing equally and similar the surface of the sphere in which it is inscribed." It exposes us the existence of the fifth essence interpreted as the dodecahedron, to which no element of matter is associated.

National University of Piura.

The original version of this chapter was revised: the presentation of the authors' name corrected. The correction to this chapter is available at https://doi.org/10.1007/978-3-030-58799-4_75

© Springer Nature Switzerland AG 2020, corrected publication 2021
O. Gervasi et al. (Eds.): ICCSA 2020, LNCS 12249, pp. 360–375, 2020.
https://doi.org/10.1007/978-3-030-58799-4_27

But the one who formalized and consecrated the Platonic solids as mathematical elements inscribing them in the sphere was Euclid [3], who made contributions on his constructions accompanied by demonstrations in book XIII from proposition 13 to 17 in his book Elements.

Interest in regular polyhedra grew, especially for the regular dodecahedron also drawn by Leonardo da Vinci, reflected in a book by Luca Pacioli [4], where he describes the regular dodecahedron: "The hollow flat dodecahedron (see Fig. 1(a)) or solid (see Fig. 1(b)) has thirty lines or equal sides that form at sixty surface angles, has twenty solid angles and twelve bases or surfaces that contain it. These are all pentagonal, with sides and angles equal to each other, as can be deduced from their shape".

The German scholar Durero [5] devised an original proto-topological method, developing it on a flat surface so that their faces form a coherent network, which when cut into paper and properly folded their faces will form a three-dimensional model of the Platonic solids and in 1525 published the plane development of the dodecahedron (see Fig. 1(c)) in his book "The four books of measurement". Later his work is highlighted by Panofsky [6], for his strictly scientific treatment when using construction rules of a perspective drawing based on Euclidean concepts and that perspective is an important branch of mathematics.

Another who was very interested was Kepler, he considered these regular solids capable of being inscribed in a sphere, he believed he saw in them the secret architecture of the universe and affirmed the earth is "the measure for the rest of the orbits, it is circumscribed to a dodecahedron [...]" (see Fig. 1(d)).

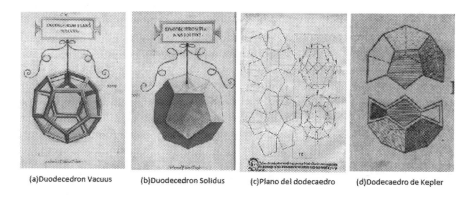

(a)Duodecedron Vacuus   (b)Duodecedron Solidus   (c)Plano del dodecaedro   (d)Dodecaedro de Kepler

Fig. 1. Representations of the dodecahedron throughout history.

Where are the dodecahedron present? This Platonic solid has been inspired by artistic works such as Salvador Dalí: The Last Supper painted in 1955 (see Fig. 2(a)); in architecture such as: the Carbonera in Madrid (see Fig. 2(b)) and the Premium planetarium in Córdoba-Argentina (see Fig. 2(c)); Sculptural works such as: the Cosmographic Model (see Fig. 2(d)) that is seen in the Museum of the History of Science in Oxford. In nature, we can find minerals such as: Pyrite (see Fig. 2(e)): FeS_2, Calcite (see Fig. 2(f)); likewise, in the form of pollen (see Fig. 2(g)). Also, at the molecular level in the crystalline dodecahedral structure that water molecules form when it traps methane forming methane hydrate (see Fig. 2(h)). It is remarkable that the study of geometry develops the capacities of formal abstract thinking: argumentation, demonstration and generalization. In this work the dodecahedron will be studied and for this we will resort to three cognitive processes: visualization, construction and reasoning. These allow us to present concepts, shapes, relationships and properties through the computer using Mathematica 11.2 Software using it as a powerful mathematical and scientific tool.

Fig. 2. The dodecahedron in nature and man's creations.

2   Analyzing and Programming in MATHEMATICA 11.2

The study will start from one of the regular geometric solids: the dodecahedron. We will study this one Geometric entity in a rational way and model it through the computer, after reflecting on its building. The analysis carried out makes it possible to establish links between the knowledge acquired and the new ones, using written communication through symbols and vocabulary typical of Geometry. To begin the study of the dodecahedron, let's visualize Fig. 3. For the construction of dodecahedron in Mathematica software, we start from a question: It is possible to know the coordinates of all the vertices of the dodecahedron starting from a canonical circumference of radius 1 u. which rests on the XY plane? It is observed that the vertices of the dodecahedron are vertices of 4 parallel regular pentagons. We will start from the regular pentagon at the base of the dodecahedron, for this we will build a circumference in its canonical form

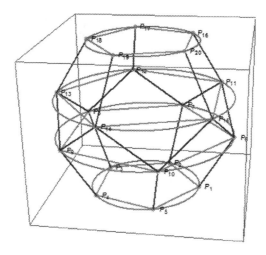

Fig. 3. The dodecahedron.

of radius 1. We can observe the equation of the circumference in its parametric form in (1):

$$cir[t_-] := \{Cos[t], Sin[t], 0\}. \tag{1}$$

To graph (1) we will use the following command that we visualize in (2):

$$cirbase = ParametricPlot3D[\{Cos[t], Sin[t], 0\}, \{t, 0, 2Pi\}]. \tag{2}$$

We will build the pentagon inscribed on the circumference described in (1). Since the pentagon is regular (which we will call base pentagon), its central angle measures 72, to find its vertices we will use (3):

$$PuntosPentagonobase = Delete[Table[cir[t], \{t, 0, 2Pi\}, \frac{2Pi}{5}], -1]. \tag{3}$$

Generating the graphical representations of the points of the base pentagon we will use (4):

$$Ptosbase3d = Graphics3D[\{Hue[0.75], PointSize[0.02], Tooltip/Point/$$
$$@PuntosPentágonobase\}]. \tag{4}$$

To generate labels in the display of points in (4) we will use (5):

$$listPuntosPentágonobase = Graphics3D[Table[Text[P_i, Puntos$$
$$Pentágonobase[[i]] + \{0.2, 0, 0\}], \{i, Length[PuntosPentágonobase]\}]]. \tag{5}$$

To visualize the graphic representation of (4) and (5) we will use (6):

$$Fig. 1 = Show[listPuntosPentágonobase, Ptosbase3d]. \tag{6}$$

Now we will find the vertices of the base pentagon, but in the generated list we repeat the first point at the end in order to close the regular base pentagon. See (7):

$$Puntosbase = Table[cir[t], \{t, 0, 2Pi, \frac{2Pi}{5}\}].\tag{7}$$

With the list (7) we generate the graphic representation of the base pentagon (8):

$$fig2 = Graphics3D[\{Thick, Hue[0.7], Line[Puntosbase]\}].\tag{8}$$

Now we graph the regular pentagon called base, inscribed in the circuit generated by (1), using (9) and we will visualize its graphic representation in Fig. 4:

$$a1 = Show[fig1, cirbase, fig2].\tag{9}$$

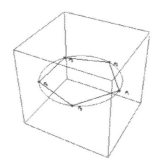

Fig. 4. Regular pentagons base and near base

To find the next regular pentagon parallel to the base of the dodecahedron, we need the measure of its side that coincides with the distance between points P_1 and P_3 with (10):

$$distP1P3 = Norm[PuntosPentágonobase[[1]] - PuntosPentágono$$
$$base[[3]]]//Simplify,\tag{10}$$

getting: $\sqrt{\frac{1}{2}(5 + \sqrt{5})}$.

Let's look at the regular polygon inscribed in a circle, having as data the central angle and the measure of its side given in (10). Note Fig. 5: For the calculation of the radius R of the circumference that inscribes the regular pentagon called pentagon near the base shown in Fig. 4, we will use (11):

$$Solve[\sin[\frac{Pi}{5}] == \frac{\frac{L}{2}}{r}],\tag{11}$$

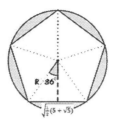

Fig. 5. Second regular pentagon that we will call a pentagon near the base.

getting: $\{\{r \to \sqrt{\{\frac{5+\sqrt{5}}{5-\sqrt{5}}\}}\}\}$.

Simply when the previous expression by (12) we get:

$$R = \sqrt{\frac{5 + \sqrt{5}}{5 - \sqrt{5}}} //FullSimplify, \tag{12}$$

we obtain: $R = \frac{1+\sqrt{5}}{2}$. We will find the parametric equation of the circumference that inscribes the second regular pentagon, previously we will calculate the radius. See Fig. 6:

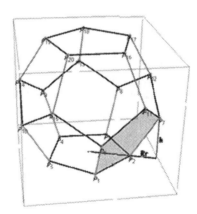

Fig. 6. Visualizing the dodecahedron the distance h and Rr to be calculated

We calculate the measure of the side of the pentagon, for this we will use (13):

$$distP1P2 = Norm[PuntosPentágonobase[[1]] - PuntosPentágono$$
$$base[[2]]]//Simplify, \tag{13}$$

from where we get: $\sqrt{\frac{5-\sqrt{5}}{2}}$. Now let's calculate the difference of the radii of circles that inscribes the pentagon of the base and the pentagon near the base, for them we will use the command (14):

$$Rr = R - 1//FullSimplify,\tag{14}$$

obtaining result $\frac{-1+\sqrt{5}}{2}$.

Using the Pythagorean Theorem we calculate the height of the circle that contains the points of the pentagon near the base, see (15):

$$h = \sqrt{distP1P2^2 - Rr^2}//FullSimplify,\tag{15}$$

getting result 1. Now we will find the parametric equation of the circumference that inscribes the regular pentagon called near the base, whose vertices belong to the dodocahedron in (16):

$$cir1[t] := \{\frac{1}{2}(1 + \sqrt{5})Cos[t], \frac{1}{2}(1 + \sqrt{5})Sin[t], 1\}.\tag{16}$$

From the above equation, we find the vertices of the regular pentagon near the base and generate its graphical representation in the same way as the base regular pentagon. Now we join the two parallel regular pentagons, where we can visualize 10 of the vertices of the dodecahedron, see (17) and its graph in Fig. 7:

$$b1 = Show[a1, a2].\tag{17}$$

Fig. 7. Regular pentagons base and near base

Let's build the third regular pentagon that is close to the ceiling, see Fig. 8:

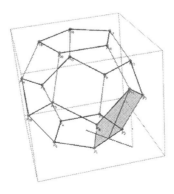

Fig. 8. Regular pentagons base and near base

Let's calculate the midpoint between points P_1 and P_2, using (18):

$$p1 = PuntosPentágonobase[[1]]$$
$$p2 = PuntosPentágonobase[[2]]$$
$$Pm = \frac{p1 + p2}{2},$$

(18)

where it was obtained: $\{\frac{1+\frac{-1+\sqrt{5}}{4}}{2}, \frac{1}{2}\sqrt{\frac{5}{8} + \frac{\sqrt{5}}{8}}, 0\}$.

Since the distance from P_{11} to P_m is the same from P_4 to P_m, we calculate it by (19):

$$P4 = PuntosPentágonobase[[4]]$$
$$distPmP4 = Norm[Pm - P4]//Simplify,$$

(19)

is obtained: $\frac{1}{2}\sqrt{\frac{5(3+\sqrt{5})}{2}}$.

Let us calculate the height of the third regular pentagon with respect to the XY plane, using (20):

$$H = \sqrt{(distPmP4^2 - Rr^2)}//Simplify,$$

(20)

we obtain: $\frac{1}{2}\sqrt{\frac{3(1+\sqrt{5})}{2}}$.

With the value found in (20), the parametric equation of the circle that inscribes the third regular pentagon that we will call the pentagon near the ceiling is defined in (21):

$$cir2[t_1] := \left\{ \frac{(1+\sqrt{5})Cos[t]}{2}, \frac{(1+\sqrt{5})Sin[t]}{2}, \frac{3+9\sqrt{5}}{2} \right\}.$$

(21)

We proceed to make the programming as the first regular pentagon. Now we join the three parallel regular pentagons, whose vertices represents 15 of the 20 vertices of the dodecahedron, using (21) the Fig. 9 is generated:

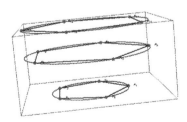

Fig. 9. Regular pentagons: base, base wax, and close to ceiling.

Let's build the last inscribed regular pentagon, representing the roof. It is known that heights of the pentagon near the ceiling is H and the height of the east and the ceiling is 1, see (22)

$$h = 1$$

$$H = \frac{1}{2}\sqrt{\frac{3 + 9\sqrt{5}}{2}}.$$

(22)

Since the measurements of the heights of the pentagon near the ceiling is H and the height from the east to the ceiling pentagon is 1, we can find the parametric equation of the circumference that inscribes the fourth regular pentagon. See (23):

$$cirtecho = ParametricPlot3D[\{Cos[t], Sin[t], 1 + \frac{1}{2}\sqrt{\frac{3 + 9\sqrt{5}}{2}}, \{t, 0, 2Pi\}].$$

(23)

It is programmed as the first regular pentagon called the base. Later we make a graphic representation of the 4 regular pentagons, together visualize the 20 vertices of the dodecahedron. See Fig. 10 which is generated by (24)

$$b4 = Show[b2, a4].$$

(24)

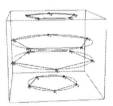

Fig. 10. The 4 parallel regular pentagons whose vertices belong to the dodecahedron.

All the vertices obtained in the constructed regular pentagons are listed, together these are vertices of the dodecahedron. See (25).

$$Pto = \{\{1,0,0\}, \{\frac{1}{4}(-1+\sqrt{5}), \sqrt{\frac{5}{8}+\frac{\sqrt{5}}{8}}, 0\},$$

$$\{\frac{1}{4}(-1+\sqrt{5}), \sqrt{\frac{5}{8}+\frac{\sqrt{5}}{8}}, 0\}, \{\frac{1}{4}(-1-\sqrt{5}), -\sqrt{\frac{5}{8}-\frac{\sqrt{5}}{8}}, 0\},$$

$$\{\frac{1}{4}(-1+\sqrt{5}), -\sqrt{\frac{5}{8}+\frac{\sqrt{5}}{8}}, 0\}, \{\frac{1}{4}(-1+\sqrt{5}), -\sqrt{\frac{5}{8}+\frac{\sqrt{5}}{8}}, 0\},$$

$$\{\frac{1}{8}(-1+\sqrt{5})(1+\sqrt{5}), \frac{1}{2}\sqrt{\frac{5}{8}+\frac{\sqrt{5}}{8}}(1+\sqrt{5}), 1\},$$

$$\{\frac{1}{8}(-1-\sqrt{5})(1+\sqrt{5}), \frac{1}{2}\sqrt{\frac{5}{8}-\frac{\sqrt{5}}{8}}(1+\sqrt{5}), 1\},$$

$$\{\frac{1}{8}(-1-\sqrt{5})(1+\sqrt{5}), -\frac{1}{2}\sqrt{\frac{5}{8}-\frac{\sqrt{5}}{8}}(1+\sqrt{5}), 1\},$$

$$\{\frac{1}{8}(-1+\sqrt{5})(1+\sqrt{5}), -\frac{1}{2}\sqrt{\frac{5}{8}+\frac{\sqrt{5}}{8}}(1+\sqrt{5}), 1\},$$

$$\{\frac{1}{4}(3+\sqrt{5}), \frac{1}{4}\sqrt{\frac{1}{2}(5-\sqrt{5})}(1+\sqrt{5}), \frac{1}{2}\sqrt{\frac{3}{2}(1+3\sqrt{5})}\},$$

$$\{-\frac{1}{2}, \frac{1}{4}(1+\sqrt{5})\sqrt{\frac{1}{2}(5+\sqrt{5})}, \frac{1}{2}\sqrt{\frac{3}{2}(1+3\sqrt{5})}\},$$

$$\{-\frac{1}{2}, \frac{1}{4}(1+\sqrt{5})\sqrt{\frac{1}{2}(5+\sqrt{5})}, \frac{1}{2}\sqrt{\frac{3}{2}(1+3\sqrt{5})}\},$$

$$\{-\frac{1}{2}, \frac{1}{4}(1+\sqrt{5})\sqrt{\frac{1}{2}(5+\sqrt{5})}, \frac{1}{2}\sqrt{\frac{3}{2}(1+3\sqrt{5})}\},$$

$$\{\frac{1}{4}(3+\sqrt{5}), -\frac{1}{4}\sqrt{\frac{1}{2}(5-\sqrt{5})}(1+\sqrt{5}), \frac{1}{2}\sqrt{\frac{3}{2}(1+3\sqrt{5})}\},$$

$$\{\frac{1}{4}(3+\sqrt{5}), -\frac{1}{4}\sqrt{\frac{1}{2}(5-\sqrt{5})}(1+\sqrt{5}), \frac{1}{2}\sqrt{\frac{3}{2}(1+3\sqrt{5})}\},$$

$$\{\frac{1}{4}(1-\sqrt{5}), \sqrt{\frac{5}{8}+\frac{\sqrt{5}}{8}}, \frac{1}{4}(4+\sqrt{6+18\sqrt{5}})\}, \{-1, 0, \frac{1}{4}(4+\sqrt{6+18\sqrt{5}})\},$$

$$\{\frac{1}{4}(1-\sqrt{5}), -\frac{1}{2}\sqrt{\frac{1}{2}(5+\sqrt{5})}, \frac{1}{4}(4+\sqrt{6+18\sqrt{5}})\},$$

$$\{\frac{1}{4}(1+\sqrt{5}), -\sqrt{\frac{5}{8}-\frac{\sqrt{5}}{8}}, \frac{1}{4}(4+\sqrt{6+18\sqrt{5}})\}\}.$$

$$(25)$$

Let's find the distance between the base regular pentagon and the roof regular pentagon. See (26):

$$hz = \frac{1 + \frac{1}{2}\sqrt{\frac{3}{2}(1 + 3\sqrt{5})}}{2}, \tag{26}$$

moving the points of the dodecahedron to the coordinate center. See (27) y (28).

$$Tras[P : \{\_,\_,\_\}] := P - \{0, 0, \sqrt{\frac{3}{2}(1 + 3\sqrt{5})}\}, \tag{27}$$

$$ptotD = Tras/@Pto. \tag{28}$$

Building the dodecahedron with the regular pentagons inscribed in the circles called: base, near the base, near the ceiling and ceiling. See (29) y (30).

$line1 = Graphics3D[Hue[0], Thick, Line[\{ptotD[[1]], ptotD[[6]]\}],$
$Line[\{ptotD[[2]], ptotD[[7]]\}], Line[\{ptotD[[3]], ptotD[[8]]\}],$
$Line[\{ptotD[[4]], ptotD[[9]]\}], Line[\{ptotD[[5]], ptotD[[10]]\}]]$
$line2 = Graphics3D[Hue[0], Thick, Line[\{ptotD[[6]], ptotD[[11]],$
$ptotD[[7]], ptotD[[12]], ptotD[[8]], ptotD[[13]], ptotD[[9]], ptotD[[14]],$
$ptotD[[10]], ptotD[[15]], ptotD[[6]]\}]]$
$line3 = Graphics3D[Hue[0], Thick, Line[\{ptotD[[11]], ptotD[[16]]\}],$
$Line[\{ptotD[[12]], ptotD[[7]]\}], Line[\{ptotD[[13]], ptotD[[18]]\}],$
$Line[\{ptotD[[14]], ptotD[[19]]\}], Line[\{ptotD[[15]], ptotD[[20]]\}]],$

(29)

$pent1 = Graphics3D[\{Hue[0.], Opacity[0.8], Polygon[\{ptotD[[1]],$
$ptotD[[2]], ptotD[[3]], ptotD[[4]], ptotD[[5]]\}]\}]$
$pent2 = Graphics3D[\{Hue[0.4], Opacity[0.4], Polygon[\{ptotD[[6]],$
$ptotD[[7]], ptotD[[8]], ptotD[[9]], ptotD[[10]]\}]\}]$
$pent3 = Graphics3D[\{Hue[0.5], Opacity[0.4], Polygon[\{ptotD[[11]],$
$ptotD[[12]], ptotD[[13]], ptotD[[14]], ptotD[[15]]\}]\}]$
$pent4 = Graphics3D[\{Hue[0.5], Opacity[0.8], Polygon[\{ptotD[[16]],$
$ptotD[[17]], ptotD[[18]], ptotD[[19]], ptotD[[20]]\}]\}]$
$Show[b4, line1, line2, line3, pent1, pent2, pent3, pent4].$

(30)

Let's visualize its graphical representation Fig. 10 by (31).

$$Show[b4, line1, line2, line3, pent1, pent2, pent3, pent4]. \tag{31}$$

We will build the dodecahedron. See (32) (Fig. 11).

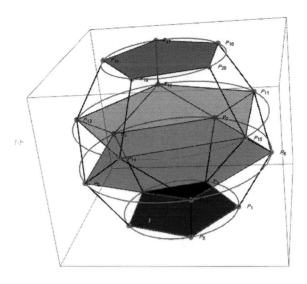

Fig. 11. vertexes of the dodecahedron.

$Dodecaedro =$

$Graphics3D[\{Hue[0], Thick, Line[\{PtotD[[6]], PtotD[[11]],$

$PtotD[[7]], PtotD[[12]], PtotD[[8]], PtotD[[13]], PtotD[[9]], PtotD[[14]],$

$PtotD[[10]], PtotD[[15]], PtotD[[6]]\}], Line[\{PtotDPto[[1]],$

$PtotD[[6]]\}], Line[\{PtotD[[1]], PtotDPto[[2]], PtotD[[3]], PtotD[[4]],$

$PtotD[[5]], PtotD[[1]]\}], Line\{PtotD[[16]], PtotD[[17]], PtotD[[18]],$

$PtotD[[19]], PtotD[[20]], PtotD[[16]]\}], Line[\{PtotD[[2]], PtotD[[7]]\}],$ (32)

$Line[\{PtotD[[15]], PtotD[[20]]\}], Line[\{PtotD[[14]], PtotD[[19]]\}],$

$Line[\{PtotD[[13]], PtotD[[18]]\}], Line[\{PtotD[[12]], PtotD[[17]]\}],$

$Line[\{PtotD[[11]], PtotD[[16]]\}], Line[\{PtotD[[3]], PtotD[[8]]\}],$

$Line[\{PtotD[[4]], PtotD[[9]]\}], Line[\{PtotD[[5]], PtotD[[10]]\}]\}].$

Let's generate the points of the vertices (33) and their (34) labels of the dodecahedron.

$$PtosDodecaedro = Graphics3D[\{Hue[0.75], PointSize[0.02],$$
$$Tooltip/@Point/@PtotD\}], \tag{33}$$

$$listPuntosDodecaedro = Graphics3D[Table[Text[P_i,$$
$$Pto[[i]] + \{0.2, 0, 0\}], \{i, Length[PtotD]\}]]. \tag{34}$$

Let's visualize its graphical representation Fig. 12 by (35).

$$figd = Show[Dodecaedro, PtosDodecaedro, listPuntosDodecaedro]. \quad (35)$$

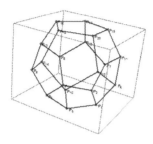

Fig. 12. The dodecahedron

Generating the vertices of the dodecahedron and its labels. See (36).

$PtosDodecaedro12 = Graphics3D[\{Hue[0.75], PointSize[0.02],$
$Tooltip/@Point/ptotD\}]$
$listPtosDodecaedro12 = Graphics3D[Table[Text[Subscript[P, i], \quad (36)$
$PtotD[[i]] + 0.2, 0, 0], i, Length[PtotD]]];.$

Building the Platonic solid called the dodecahedron. See (37).

$Solidoplatonico12 = ConvexHullMesh[ptostrasldode, MeshCellHighlight$
$- > 2, All- > Opacity[0.65, Hue[0.]]].$

$$(37)$$

Calculating the radius of the sphere that inscribes the dodecahedron. See (38):

$$RadioEsferaID = Norm[\{1, 0, -\frac{1}{2}(1 + \frac{1}{2}\sqrt{\frac{3}{2}(1 + 3\sqrt{5})})\}], \quad (38)$$

we obtain: $RadEID = \sqrt{1 + \frac{1}{4}(1 + \sqrt{\frac{3}{2}(1 + 3\sqrt{5})})^2} \sim 1.6800978811170325$.
Using (39) we find the graphical representation of the sphere.

$EsfID = ParametricPlot3D[RadEID\{Cos[u]Cos[v], Sin[u]Cos[v], Sin[v]\},$
$\{u, 0, 2 \quad Pi\}, \{v, -\frac{Pi}{2}, \frac{Pi}{2}\},$
$PlotStyle- > \{Opacity[0.4], Yellow].$

$$(39)$$

Graphing the inscribed dodecahedron inside a sphere Fig. 13 using (40).

$$Show[Solidoplatonico12, EsfID, PtosDodecaedro12, listPuntosDodecaedro12]. \tag{40}$$

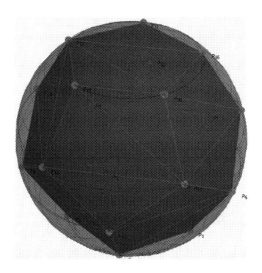

Fig. 13. The dodecahedron

A new question arises: Can we generate the dodecahedron, starting from a sphere of center (0, 0, 0) and of radius R? Using the (41) we can find the list of the vertices of the Dodecahedron with its graphic representation and its circumscribed sphere, given its radius of said sphere

$$PtosDodeER[R_] := Modele[pttos, Pttos = \cfrac{R}{\sqrt{1 + \frac{1}{64}(4 + \sqrt{6 + 18\sqrt{5}})^2}}$$

$ptotD//FullSimplify//PowerExpand;$
$Print[Pttos]; Show[\{ConvexHullMesh[Pttos, MeshCellHighlight$
$-> \{\{2, All\}-> Opacity[0.65, Hue[0.]]\}], Graphics3D[\{PointSize \tag{41}$
$[0.03], Hue[0.75], Point/@Pttos\}], ParametricPlot3D[R\{Cos[u]Cos[v],$
$Sin[u]Cos[v], Sin[v]\}, \{0, 0, 2Pi\}, \{v, -\cfrac{Pi}{2},$
$\cfrac{Pi}{2}\}, PlotStyle-> Opacity[0.5]]\}]].$

In the analysis another question arises, can we find the sphere inscribed in the dodecahedron? Doing the respective analysis, we have to calculate the centroids of the dodecahedron faces using the following (42):

$$C(c_x, c_y, c_z) =$$
$$(\frac{x_1 + x_2 + x_3 + x_4 + x_5}{5}, \frac{y_1 + y_2 + y_3 + y_4 + y_5}{5},$$
$$\frac{z_1 + z_2 + z_3 + z_4 + z_5}{5}).$$

(42)

Using the (43) we can find the list of the vertices of the Dodecahedron with its graphic representation and its inscribed sphere, given its radius of said sphere.

$PtosDodeEsfI[r\_] := Modele[pttosIE, PttosIE =$

$$\frac{r}{\sqrt{\frac{1}{4} + \frac{1}{2\sqrt{5}} + (\frac{1}{2} + \frac{1}{\sqrt{5}})^2 + \frac{(4+3\sqrt{6+18\sqrt{5}})^2}{1600}}}$$

$ptotD//FullSimplify//PowerExpand;$

$Print[PttosIE]; Show[\{ConvexHullMesh[Pttos, MeshCellHighlight$
$-> \{\{2, All\}-> Opacity[0.65, Hue[0.]]\}], Graphics3D[\{PointSize$
$[0.03], Hue[0.75], Point/@PttosIE\}], ParametricPlot3D[r\{Cos[u]Cos[v],$
$Sin[u]Cos[v], Sin[v]\}, \{0, 0, 2Pi\}, \{v, -\frac{Pi}{2},$
$\frac{Pi}{2}\}, PlotStyle-> Opacity[0.5]]\}]].$

(43)

3   Conclusions

In this article algorithms have been developed to calculate characteristics of the geometry of the Regular dodecahedron encoded in the language of Mathematica v. 11.2. The following characteristics have been observed: The dodecahedron can be built from parallel regular pentagons as specified below: The regular pentagon (called the base) inscribed in a circumference of radius 1, is parallel to the regular pentagon (near the base). The latter can be generated by moving the regular pentagon (called the base) one unit up and scaling the radius of the circle that inscribes it to the golden number. The regular pentagon (called near ceiling) of radio number Golden, can be generated by moving the regular pentagon (called near the base) upwards by a distance of approximately 0.7001695429 and rotating 36. The regular pentagon (called the ceiling) of Radius 1 can be generated by moving the regular pentagon (called the ceiling fence) up a distance of 1 and shrinking down to the radius 1 measurement. To construct the inscribed dodecahedron in a sphere with a center at the coordinate origin given its radius, it is necessary to have the information about the center of the original dodecahedron and the distance from the center to any vertex. To construct the circumscribed dodecahedron in a sphere centered at the coordinate origin given its radius, it is necessary to have the information about the centroids of the faces of the original dodecahedron and the distance from the center to any centroid point. The ratio of the radii of the circumscribed and inscribed spheres to the dodecahedron is approximately 1.2725887427162601970982. It is suggested to extend the investigation and create packages of the Platonic solids and others, later applying them to the design of the polyhedral architecture of different artificial objects or existing organisms from nature.

References

1. DivulgaMAT   RSME.   https://www.divulgamat.net/divulgamat15/index.php? option=com_content&view=article&id=3386, Accessed 6 Apr 2020
2. Platón.: Obras completas. In: Aguilar, Madrid-España, Ed. (1986)
3. Pacioli, L.: La divina proporción. In: AKAL (ed.) Translation of Juan Calatrava Escobar, Spain (2008)
4. Euclides: Elementos. In: GREDOS (ed.) Translation of Maria Luisa Puertas Castaños, Spain (1994)
5. Durero, A.: De la medida. In: AKAL (ed.) Translation of Juan Calatrava Escobar and Jesus Espino Nuño, Madrid-Spain (2000)
6. Panofsky, E.: The Life Art of ALBERCHT DÜRER. Princeton University Press. Edición: 1st Princeton Classic (2005)
7. Torrence, F., Torrence, A.: The Student's Introduction to Mathematica and the Wolfram Language. Cambridge University Press, United Kingdom (2019)
8. Couprie, M., Cousty, J., Kenmochi, Y., Mustafa, N. (eds.): DGCI 2019. LNCS, vol. 11414. Springer, Cham (2019). https://doi.org/10.1007/978-3-030-14085-4

GPU Memory Access Optimization for 2D Electrical Wave Propagation Through Cardiac Tissue and Karma Model Using Time and Space Blocking

Christian Willian Siqueira Pires$^{(\boxtimes)}$ (ID), Eduardo Charles Vasconcellos$^{(\boxtimes)}$ (ID), and Esteban Walter Gonzalez Clua$^{(\boxtimes)}$ (ID)

Institute of Computing, Fluminense Federal University, Niterói, RJ, Brazil
{cwspires,esteban}@id.uff.br, charles.edu@gmail.com
http://www.uff.br

Abstract. Abnormal heart function is one of the largest causes of death in the world. Understanding and simulating its contraction and relaxation movements may reduce this mortality and improve specific treatments. Due the high processing time to compute propagation of an electrical wave through cardiac tissue, interactive and real time simulations are a challenge, making unable corrective treatments on the fly. In this paper we propose an approach to reduce the electrophysiology models simulation computational time using the finite difference method to solve the laplacian operator, required by the karma model. Our solution optimizes accesses to global memory by storing more than one instant of time in shared memory within each GPU kernel call. We present tests that validate our proposal and show an increase performance close to 20% when compared with available GPU implementations.

Keywords: Gpu computing · Cardiac electrophysiology models · Parallel cardiac dynamics simulations · Finite difference method

1 Introduction

The contraction and relaxation of the heart are results of the propagation of electrical wave through cardiac tissue. Deficiencies on this wave propagation may cause abnormal behavior of the heart, resulting in problems like ventricular tachycardia and fibrillation, which in a more severe situation can lead to death. Many mathematical models are available for simulating the electrical wave propagation through cardiac tissues. These models help to simulate heart conditions, like tachycardia and fibrillation, and the effects of different treatments.

Electrophysiology models for cardiac tissue have a high computational demand, mainly for three dimensional simulations. Some author address this problems by using high computational resource, such as clusters or supercomputers [5,24,25,29]. More recently, some of these models have been ported to

© Springer Nature Switzerland AG 2020
O. Gervasi et al. (Eds.): ICCSA 2020, LNCS 12249, pp. 376–390, 2020.
https://doi.org/10.1007/978-3-030-58799-4_28

GPUs, which allow their execution in lower cost platforms and opening real time simulation possibilities. Although, the Laplacian operator, frequently used on electrophysiology models, makes difficult to efficiently parallelize them on GPUs [1,26,33].

In this work, we propose a novel approach to reduce computational time when solving the Laplacian with the finite difference method. We overlap time steps computation, reducing the number of global memory accesses. We used a GPU implementation of the 2D Karma model for cardiac tissue [18], and we show that our proposed approach can reduce the complete simulation time by about 30%. Karma is a two variables model that is reliable to reproduce abnormal heart conditions like spiral waves associated with cardiac arrhythmia [10,11,18]. As a base for our evaluations, we have used the classical parallel Laplacian implementation firstly proposed by Micikevicius [23].

This paper is organized as follows. Section 2 gives an overview of the modeling and simulation of cardiac dynamics, including the standard second-order Laplacian discretization. Section 3 describes the classic parallel GPU solution proposed by Micikevicius [23]. Section 4 presents our proposed parallel approach through the GPU architecture. Section 5 presents and discusses the results achieved by our solution. Finally, Sect. 6 presents the conclusions and future work.

2 Computational Simulation of Electrical Dynamics in Cardiac Tissue

The electrical wave propagation in cardiac tissue can be described by a variation in time of the cell membrane's electrical potential U, for each cardiac tissue cell. Under a continuum approximation, this process can be represented using the following reaction-diffusion equation:

$$\frac{\partial U}{\partial t} = \nabla \cdot D\nabla U - \frac{I_{ion}}{C_m} \ . \tag{1}$$

The first term at the right side represents the diffusion component, and the second term on the right side represents the reaction component, where I_{ion} corresponds to the total current across the cell membrane and C_m the constant cell membrane capacitance. The diffusion coefficient D can be a scalar or a tensor and describes how cells are coupled together; it also may contain some information about tissue structure, such as the local fiber orientation [7]. The value of D affects the speed of electrical wave propagation in tissues [1,7].

The reaction term is modeled by a system of nonlinear ordinary differential equations, such as described in Eq. 2:

$$\frac{d\mathbf{y}}{dt} = \mathbf{F}(\mathbf{y}, U(\mathbf{y}, t), t) \ . \tag{2}$$

Its exact form depends on the level of complexity of how the electrophysiology model describes heart tissue cells. For each additional variable $\mathbf{y_j}$, $\mathbf{F_j}(\mathbf{y_j}, U, t)$ is a nonlinear function. Clayton et al. provide a good review of cardiac electrical activity modelling [7].

The system of differential equations in reaction term represents mechanisms responsible for ionic currents across the cell membranes and changes of ion concentrations inside and outside cells [7,11]. There are many mathematical models that describe cellular cardiac electrophysiology [2,3,16,18,28,32]. The main difference among them lies in the reaction term.

2.1   Karma Model

In 1993, Karma proposed one of the simplest models used to describe cardiac electrical dynamic [18]. The model has two variables and consists of the following differential equations:

$$\frac{\partial U}{\partial t} = D\nabla^2 U - U + \left([1 - \tanh(U-3)]\, U/2\right)\left[\gamma - \left(\frac{v}{v^*}\right)^{xm}\right] \tag{3}$$

$$\frac{dv}{dt} = \epsilon\left[\Theta\left(U-1\right) - v\right], \tag{4}$$

where U represents the electrical membrane potential and v is a recovery variable. The main purpose of Karma model is to simulate the behavior of spiral waves, such as those associated with cardiac arrhythmia [10,11,18].

2.2   Numerical Solution

Finite difference methods (FDM) are a common numerical methods used to obtain a numerical solution for many different numerical problems [14]. This method requires a domain discretization for all variables.

In order to solve the differential equations of the Karma model on a GPU using FDM, we adopt a forward difference for time integration at time t_n and a second-order central difference approximation for the spatial derivative at position $p = (x_i, y_j, z_k)$. In addition, we assume $h = \Delta x = \Delta y = \Delta z$ and $t_n = n\Delta T$. Therefore, the finite difference approximation for space (cell tissue) and time is modeled as follows:

$$\begin{aligned}
U_{i,j,k}^{n+1} = {}&(1 - \Delta T)\, U_{i,j,k}^n + \\
&+ \frac{D\Delta T}{h}\left(U_{i+1,j,k}^n + U_{i-1,j,k}^n + U_{i,j+1,k}^n + U_{i,j-1,k}^n\right. \\
&\left. + U_{i,j,k+1}^n + U_{i,j,k-1}^n - 6U_{i,j,k}^n\right) + \\
&+ \Delta T\left(0.5 * \left[1 - \tanh(U_{i,j,k}^n - 3)\right] U_{i,j,k}^n\right)\left[\gamma - \left(\frac{v_{i,j,k}^n}{v_{i,j,k}^{n*}}\right)^{xm}\right],
\end{aligned} \tag{5}$$

where n is an index corresponding to the nth time step. For the numerical experiments presented in this paper, D corresponds to a uniform field that applies the value one to all points in the domain.

3   Classic Laplacian Implementation on a GPU

In this section we discuss some details of the GPU cache memory and present the shared memory approach for dealing with the spatial dependency imposed by the Laplacian discretization on this particular architecture. This approach, which is based on simple row major order [1], can works as base for some three dimensional propagation problems [1, 12, 22, 23, 26].

GPUs are specialized hardware that can be used to process data in a massively parallel way. When the computation may be performed at each memory position independently, GPUs can achieve much better performance than CPUs. The CUDA parallel programming model is a powerful tool to develop general purpose applications for Nvidia GPUs. It is based on threads that can be addressed in one, two, or three dimensional arrays. Despite the arrangement chosen by the developer, GPU executes threads following a linear order in groups of 32 consecutive threads at same time, called warps. Additionally, GPU cache memory considers that consecutive threads access consecutive memory addresses. Thus, CUDA developers take advantage of row major order to store/load data in GPU memory. This approach seeks to minimize non-coalesced memory operations by storing and loading several consecutive data values in cache given a memory address access. The specialized GPU architecture may store or load up to 128 consecutive bytes in one cache line with a single read/write operation.

The usual parallel approach to compute time-explicit FDM on a GPU consists of addressing each point of the mesh using one CUDA thread [1, 12, 22]. For each new time step, the GPU computes thousands of points in parallel, solving Eq. 5 for each point of the mesh, where the only dependency is the temporal domain.

However, 2D or 3D domains present great challenges in minimizing the memory latency, since accessing neighboring data points in these domains may cause many non-coalesced memory operations to obtain nearest neighbors, which are required for the spatial discretization [22]. This compromises GPU performance due to the greater elapsed time to access all neighbors, due the reduction of concurrency features.

The stencil pattern illustrated in Fig. 1 represents the required data to compute the value at the next time step $t + 1$ at a given point p. When computing single precision values, its neighbors in the x-direction are offset by 4 bytes to the left or right, while the offset of neighbors in the y-direction depends on the domain size. Clearly, in this case the goal of accessing only nearby memory locations cannot be achieved.

4   2D Space and Time Blocking on GPU

One possible approach to avoid redundant global memory access is to store all required data of a thread block in its shared memory [23]. Figures 2, 3 and 4 depicts an example on how to load data to shared memory, taking advantage of row major ordination.

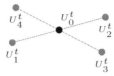

Fig. 1. 2D stencil representing data required for calculating the value at the next numerical time step $t + 1$ at each domain point U_0 using a standard second-order FDM.

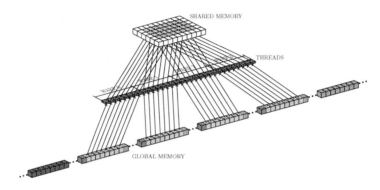

Fig. 2. Access pattern for core data on global memory.

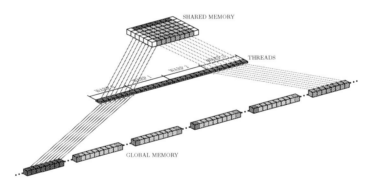

Fig. 3. Access pattern for y neighborhoods.

A possible approach to decrease computing time is balancing computation and accesses of shared memory with global memory accesses [15]. To do this, in each block, all necessary data for t_n time steps is transferred from global memory to shared memory. Doing so, some data need to be recalculated.

The use of shared memory can help avoiding to loose time with redundant global memory access. However, CUDA thread blocks do not have any kind of data synchronization. For this reason, on every time step of the FDM, it is necessary to update global memory with the new values computed by the thread

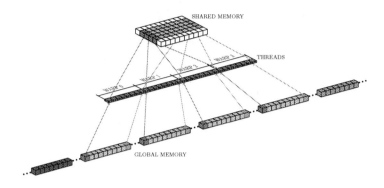

Fig. 4. Access pattern for x neighborhoods.

blocks. This means that for all time step computations it is necessary to read and write data in the global memory. We propose to reduce global memory transactions applying an approach called space-time blocking. The global memory must be updated only every t_n time steps, which is the amount of data necessary to execute n steps calculation. These data are copied from global memory to shared memory in the first time step t, and, in the following $t + t_n$, the result is written back. The time steps between t and $t + t_n$ computations are made accessing only shared memory.

The domain division through blocks, represented by yellow in Fig. 5, shows the shared memory for one time step computation, where the resulting blocks have independent shared memory. In this case, the thread or block communication is not being considered. The threads from each block have to copy the data from global memory to shared memory, execute the required computation and write the results back to the global memory.

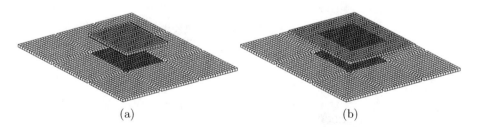

(a)   (b)

Fig. 5. Global memory blocks division in yellow grid, with tile copied to shared memory for one time step computation. The result elements to be computed is represented by dark blue, the border region in red, and unused elements in gray. (a) shows the shared memory size copies for one time step, and, (b) shows for six time steps (Color figure online)

Figure 6 shows a representation of a 2D domain tile, where the green middle element is used for the stencil computation. All the stencil neighbors computation use the green middle element as a neighbor. In this case, for a $k = 2$ stencil calculation, the reuse of an element excluding the border conditions correspond to 5 read and one write operations. When copying the data to shared memory, the global memory accesses is reduced to 2 times, corresponding only to one tile reading an the result writing on the global memory. The remaining memory operations remains only at the shared memory [23]. This memory reuse occurs in all time step computation, and the reuse is multiplied for each additional time step copied to shared memory.

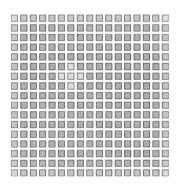

Fig. 6. Shared memory tile for $k = 2$, where green elements represent the neighbors required for the middle stencil computation, the red elements are the halo region and blue elements are the elements to compute in this block (Color figure online)

$$S = 3 * (Bx + k * n) * (By + k * n) \tag{6}$$

Equation 6 represents the memory required for n time step computations, where S is the total shared memory elements, Bx and By are the block dimension, k is the stencil size and n is the number of time steps. The number three corresponds to triplicating the amount of shared memory. This happens due the tile stencil required to store the variables described in Sect. 2.1, which are, the old electrical potential data, the new step potential data and the recovery variable v, data required to the karma model. v data is accessed two times for each time step computation, but its value is not present at the registers, due the disparity between the number of threads and the number of elements to be calculated at the middle time step computations.

The number of threads is equal to the number of elements to be calculated at the final time step computation. This is shown as dark blue elements in Figs. 5 and 7, although the size of elements to be copied is greater than the number of threads. The elements are linearly distributed across the threads, including elements that are not necessary to calculate, avoiding code complexity. The work is distributed to threads in a linearly way using a stride pattern, which means that the work is divided by the number of threads and the result is the number of strides necessary to compute. These elements are show in Fig. 6 and 5 with the gray color. In order to copy the shared memory, copying the size difference between the number of threads and the number of elements require the same process.

The tile copy required for six time steps calculations is show in Fig. 7 as the biggest tile, where red represent border elements required the tile below computation. The light blue elements represents the data to be calculated for the tile below. All data copied at the biggest tile, over the dark blue elements represents the repeated computation that does not occur in the implementation of shared memory, but, the computations .

Fig. 7. Space time blocking memory representation. The yellow color shows the global memory blocks division grid. The dark blue elements is the result of the all computation. The biggest tile represents all memory needed for six time steps computation, where each tile below is the result of calculating the tile above, when the light blue elements shows the required elements for computation. The red elements is the border region considering the next step. The gray elements represents the unused data for the next time step computation. (Color figure online)

The number of time steps added to the blocks size must be multiple of the global memory bandwidth, in order to achieve data coalescence. In the computation stage, all data present at the shared memory is related to thread disparity at the final of the block execution. For solving this all step time calculations have to be multiple of 32, which corresponds to the warp size.

GPU's are limited by the number of threads per block and a size of shared memory. Since the memory access performs in different ways according to the GPU model, it may be relevant to choose the tile size according with the stencil size, and the hardware limitations, such as shared memory size. It is also important to calculate the number of elements to be calculated according to the multiple of the number of threads in a warp. Copies and previous time steps need to be executed at a different number of elements and not necessarily as a multiple of the number of threads. It may be challenging to fit the best tile size and time step size in order to maximize computation time [15].

5   Results

To evaluate the time-space blocking, we ran several experiments in a Nvidia Geforce 1080Ti GPU. The machine host had 32 Gb of RAM and it is powered by an Intel Xeon E5-2620 CPU (8 cores). In our experiments, we variate block size, domain size and the number of time steps computed in a single kernel call, but numerical discretization and initial conditions remain the same. We used a time step of 0.05 ms and our spatial resolution was equal in both axes ($\Delta x = \Delta y$). The initial condition was given by:

$$U_{ij}^0 = 3.0 \quad \text{for } x_k \leq 0.05 * (domain\ size\ x) \tag{7}$$

$$v_{ij}^0 = 0.5 \quad \text{for all } x_k . \tag{8}$$

In our experiments we simulated a total of one second ($1s$); therefore, the time domain discretization generates $20,000$ time steps.

Figure 8a depicts the amount of shared memory demanded by the different configurations we evaluate. The memory was calculated according to Eq. 6 and, as we are using single precision variables, it was multiplied by 4 bytes. The figure shows that, due to our GPU limitation of 48 Kb per thread block, not all configurations could be tested. Figure 8b shows all configurations that fits on our GPU. We discarded all experiments that require more than 48 Kb per block.

We measured execution on the GPU time using *cudaEventRecord*. We are no interested in CPU time; all our computation is performed on GPU. The U and v fields are initialized by a kernel and stored directly on GPU global memory. Figure 9 shows the performance of space-time blocking for the selected experiments. As the number of time steps computed per kernel increases, also the computation time increases. This happens due to the escalation of share memory required to store the variables. By raising the amount of share memory needed by each kernel, we reduce the number of concurrent kernels. The maximum amount of shared memory per multiprocessor, in our GPU, is 64 Kb.

To see how space-time blocking (STB) performs compared to classic shared memory implementation (CSM), we will look more carefully at our faster experiments (Fig. 10). We ran the same simulations with CSM, varying block sizes, and we chose the faster configuration to compare it with STB. The results are presented in Fig. 10. Table 1 depicts time saved by STB in comparison with CSM.

We also evaluate the coherence of our simulation outputs by simulates a spiral wave. We ran a simulation for $20,000$ time steps (1 second of cardiac activity). As in our experiments the initial condition was defined by Eqs. 7 and 8. The spiral wave was generated at time step $8,000$ by reset fields U and v to reset values in the bottom half of the domain. Figure 11 shows some frames of a spiral wave simulation with our GPU space-time blocking implementation.

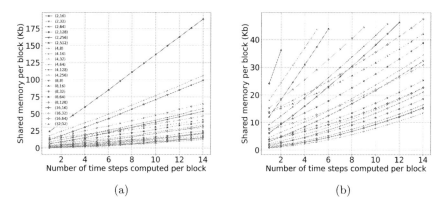

(a)   (b)

Fig. 8. The figure exhibits the amount of shared memory required by different sets of block size and number of time steps computed per kernel call. Figure (a) depicts all considered configurations, while figure (b) shows only the ones that fit the GPU's shared memory limitation.

Table 1. Best simulation times comparison. All simulations did run 20,000 numerical time steps.

Domain size	Block size for STB	CSM elapsed elapsed time (s)	Time saved elapsed time (s)	Space-time blocking (%)
1024^2	16×32	1.074	0.906	18.543%
2048^2	16×32	4.149	3.673	12.959%
4096^2	32×32	17.379	15.655	11.012%
8192^2	32×32	70.754	64.568	9.580%

6   Related Work

Over the last decade, GPU computational power increased in such a way that they may be used in place of supercomputers for specific scientific computational problems [8,27]. Cardiac electrical dynamics simulations is no exception to this trend [6,19,31].

Several authors have tested GPU performance previously for different cardiac tissue models. They evaluated not only the performance of adapted CPU implementations for GPUs [13,20,34], but also different parallel strategies [1,9,21,26,30,36] and numerical methods [4,35].

Bartocci et al. [1] evaluated GPU implementations for cardiac tissue models with different complexities. They tried two different strategies to accelerate the PDE (Partial Differential Equation) solution through memory optimization: one using shared memory and the other using texture memory. Running on a Tesla C2070, their simulation with a 2D mesh composed of 2^{22} cells took almost $10s$ using texture memory and just over $15s$ using shared memory.

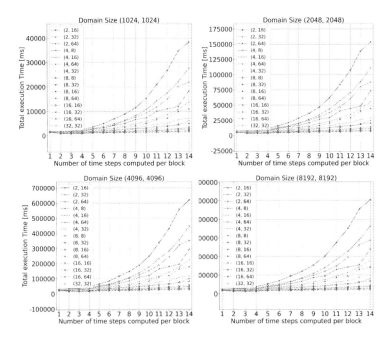

Fig. 9. Elapsed Time for different domain sizes, block sizes and number of time steps solved per kernel call. The distinct color lines means different block sizes.

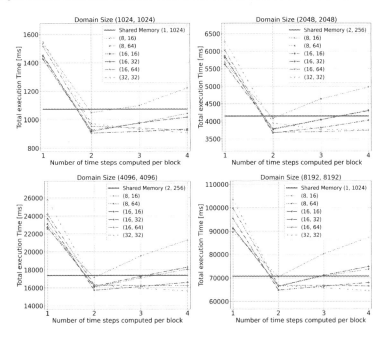

Fig. 10. Elapsed Time for different domain sizes, block sizes and number of time steps solved per kernel call. The distinct color lines means different block sizes.

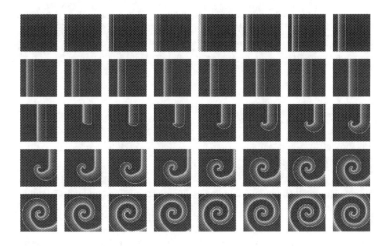

Fig. 11. One second of electrical wave propagation through cardiac tissue.

Nimmagadda [26] explored a linear data structure in GPU simulations using a 2D bidomain model of cardiac electrophysiology. The authors ordered their data to enable WARP coalesced memory access. Using a Tesla C1060, they reported a computation time around $1686s$ to calculate $10,000$ time steps in a $256 \times 256 \times 256$ 3D mesh. The authors also tried a multi-GPU approach and reported an acceleration for the PDE solution of around $14\times$ compared with their single-GPU approach.

Xia et al. [36] used linear structures to accelerate memory access when solving the diffusion term of Eq. 1. The authors' simulations performed $120,000$ time steps in a $317 \times 204 \times 109$ mesh, but only approximately 1 million are valid cell nodes (they omitted empty cells to avoid GPU performance degradation). In this case, they reported a computation time a little over $500s$ to calculate the solution using FDM and double-precision variables (this $500s$ time took into account only the PDE solution).

More recently, Kaboudian et al. (2019) [17] proposed a WebGL based approach to perform cardiac electrophysiology simulations on GPUs, using FDM and single precision. The author compares spiral waves simulated with their WebGL approach with spiral waves from experimental data. For a simulation with a 3-variable model in a 2D domain with 512^2 points, they reported that is possible to run up to $38,000$ time steps per second in a Titan X with Pascal architecture. This means they can process up to 9.96×10^9 domain points (cells) per second. While their work present a good real time visualization result, they are constrained to small domains.

Vasconcellos et al. (2020) [33] proposed a strategy to accelerate the computation of the diffusion term (Eq. 1) through a data-structure and memory access pattern designed to maximize coalescent memory transactions and minimize branch divergence. The authors achieved results approximately 1.4 times faster than the share memory approach proposed by Micikevicius (2009) [23].

They also combine this data structure with a designed communication strategy to take advantage in the case of simulations in multi-GPU platforms. We demonstrate that in the multi-GPU approach performs, simulations in 3D tissue can be just 4× slower than real time.

7   Conclusion

We have shown that with our proposed memory model approach we can reduce close to 20% of the simulation time compared to the classical performance of simulated wave propagation through cardiac tissues using GPU architectures with the karma model. The main feature of our solution is the fact that it copies and calculates multiple time steps through a single copy to the GPU shared memory. Although we have applied directly to cardiac tissue simulation, we believe that many other wave propagation simulations can also benefit.

Kaboudian et al. (2019) [17] reported 38000 time steps per second, corresponding to 9.96×10^9 domain points (cells) per second in a 512^2 2D size domain simulation. Our solution performs 22075 time steps per second, corresponding to 23.14×10^9 domain points in a 1024^2 2D size domain simulation.

We have not achieved yet real-time results, but we believe that in near future it will be possible to increase the performance through machine learning strategies capable of building multi-resolution spaces, according with the wave behavior and tissues characteristics. Improvements on the unified memory of the GPU may also be relevant for future works on the field.

References

1. Bartocci, E., Cherry, E.M., Glimm, J., Grosu, R., Smolka, S.A., Fenton, F.H.: Toward real-time simulation of cardiac dynamics. In: Proceedings of the 9th International Conference on Computational Methods in Systems Biology, CMSB 2011, pp. 103–112. ACM, New York (2011). https://doi.org/10.1145/2037509.2037525
2. Beeler, G.W., Reuter, H.: Reconstruction of the action potential of ventricular myocardial fibres. J. Physiol. **268**(1), 177–210 (1977). https://www.ncbi.nlm.nih.gov/pmc/articles/PMC1283659/
3. Bueno-Orovio, A., Cherry, E.M., Fenton, F.H.: Minimal model for human ventricular action potentials in tissue. J. Theor. Biol. **253**(3), 544–560 (2008). https://www.sciencedirect.com/science/article/pii/S0022519308001690
4. Campos, J.O., Oliveira, R.S., dos Santos, R.W., Rocha, B.M.: Lattice boltzmann method for parallel simulations of cardiac electrophysiology using GPUs. J. Comput. Appl. Math. **295**, 70–82 (2016). https://www.sciencedirect.com/science/article/pii/S0377042715000692
5. Chai, J., et al.: Towards simulation of subcellular calcium dynamics at nanometre resolution. Int. J. High Perform. Comput. Appl. **29**(1), 51–63 (2015). https://doi.org/10.1177/1094342013514465
6. Clayton, R.H., et al.: Models of cardiac tissue electrophysiology: progress, challenges and open questions. Prog. Biophys. Mol. Biol. **104**(1–3), 22–48 (2011). https://www.sciencedirect.com/science/article/pii/S0079610710000362

7. Clayton, R., Panfilov, A.: A guide to modelling cardiac electrical activity in anatomically detailed ventricles. Prog. Biophys. Mol. Biol. **96**(1–3), 19–43 (2008). https://www.sciencedirect.com/science/article/pii/S0079610707000454

8. Dematté, L., Prandi, D.: GPU computing for systems biology. Briefings Bioinf. **11**(3), 323–333 (2010). https://doi.org/10.1093/bib/bbq006

9. Esmaili, E., Akoglu, A., Ditzler, G., Hariri, S., Moukabary, T., Szep, J.: Autonomic management of 3d cardiac simulations. In: 2017 International Conference on Cloud and Autonomic Computing (ICCAC), pp. 1–9, September 2017

10. Fenton, F., Karma, A.: Vortex dynamics in three-dimensional continuous myocardium with fiber rotation: filament instability and fibrillation. Chaos **8**(1), 20–47 (1998). https://scitation.aip.org/content/aip/journal/chaos/8/1/10.1063/1.166311

11. Fenton, F.H., Cherry, E.M.: Models of cardiac cell. Scholarpedia **3**(8), 1868 (2008). https://www.scholarpedia.org/

12. Giles, M., László, E., Reguly, I., Appleyard, J., Demouth, J.: GPU implementation of finite difference solvers. In: Proceedings of the 7th Workshop on High Performance Computational Finance, WHPCF 2014, pp. 1–8. IEEE Press, Piscataway (2014). https://doi.org/10.1109/WHPCF.2014.10

13. Higham, J., Aslanidi, O., Zhang, H.: Large speed increase using novel GPU based algorithms to simulate cardiac excitation waves in 3d rabbit ventricles. In: 2011 Computing in Cardiology, pp. 9–12, September 2011

14. Hoffman, J.D., Frankel, S.: Numerical Methods for Engineers and Scientists. CRC Press, New York (2001)

15. Holewinski, J., Pouchet, L.N., Sadayappan, P.: High-performance code generation for stencil computations on GPU architectures. In: Proceedings of the 26th ACM International Conference on Supercomputing, ICS 2012, pp. 311–320. ACM, New York (2012). https://doi.org/10.1145/2304576.2304619

16. Iyer, V., Mazhari, R., Winslow, R.L.: A computational model of the human left-ventricular epicardial myocyte. Biophys. J. **87**(3), 1507–1525 (2004). https://www.sciencedirect.com/science/article/pii/S0006349504736346

17. Kaboudian, A., Cherry, E.M., Fenton, F.H.: Real-time interactive simulations of large-scale systems on personal computers and cell phones: toward patient-specific heart modeling and other applications. Sci. Adv. **5**(3), eaav6019 (2019). https://advances.sciencemag.org/content/5/3/eaav6019

18. Karma, A.: Spiral breakup in model equations of action potential propagation in cardiac tissue. Phys. Rev. Lett. **71**, 1103–1106 (1993). https://doi.org/10.1103/PhysRevLett.71.1103

19. Lopez-Perez, A., Sebastin, R., Ferrero, J.M.: Three-dimensional cardiac computational modelling: methods, features and applications. Biomed. Eng. OnLine **14**(1), 1–31 (2015). https://doi.org/10.1186/s12938-015-0033-5

20. Mena, A., Rodriguez, J.F.: Using graphic processor units for the study of electric propagation in realistic heart models. In: 2012 Computing in Cardiology, pp. 37–40, September 2012

21. Mena, A., Ferrero, J.M., Matas, J.F.R.: GPU accelerated solver for nonlinear reaction-diffusion systems. Application to the electrophysiology problem. Comput. Phys. Commun. **196**, 280–289 (2015). https://doi.org/10.1016/j.cpc.2015.06.018. https://www.sciencedirect.com/science/article/pii/S0010465515002635

22. Michéa, D., Komatitsch, D.: Accelerating a three-dimensional finite-difference wave propagation code using GPU graphics cards. Geophys. J. Int. **182**(1), 389–402 (2010). https://doi.org/10.1111/j.1365-246X.2010.04616.x

23. Micikevicius, P.: 3D finite difference computation on GPUs using CUDA. In: Proceedings of 2nd Workshop on General Purpose Processing on Graphics Processing Units, pp. 79–84. ACM (2009)

24. Mirin, A.A., et al.: Toward real-time modeling of human heart ventricles at cellular resolution: simulation of drug-induced arrhythmias. In: Proceedings of the International Conference on High Performance Computing, Networking, Storage and Analysis, SC 2012, pp. 2:1–2:11. IEEE Computer Society Press, Los Alamitos (2012). https://dl.acm.org/citation.cfm?id=2388996.2388999

25. Niederer, S., Mitchell, L., Smith, N., Plank, G.: Simulating human cardiac electrophysiology on clinical time-scales. Front. Physiol. **2**, 14 (2011). https://www.frontiersin.org/articles/10.3389/fphys.2011.00014/full

26. Nimmagadda, V.K., Akoglu, A., Hariri, S., Moukabrav, T.: Cardiac simulation on multi-GPU platform. J. Supercomputing **59**(3), 1360–1378 (2012). https://doi.org/10.1007/s11227-010-0540-x

27. Nobile, M.S., Cazzaniga, P., Tangherloni, A., Besozzi, D.: Graphics processing units in bioinformatics, computational biology and systems biology. Briefings Bioinf. **18**(5), 870–885 (2017). https://doi.org/10.1093/bib/bbw058

28. O'Hara, T., Virág, L., Varró, A., Rudy, Y.: Simulation of the undiseased human cardiac ventricular action potential: Model formulation and experimental validation. PLOS Comput. Biol. **7**(5), 1–29 (2011). https://doi.org/10.1371/journal.pcbi.1002061

29. Richards, D.F., et al.: Towards real-time simulation of cardiac electrophysiology in a human heart at high resolution. Comput. Methods Biomech. Biomed. Eng. **16**(7), 802–805 (2013). https://doi.org/10.1080/10255842.2013.795556

30. Rocha, B.M., et al.: Accelerating cardiac excitation spread simulations using graphics processing units. Concurrency Comput. Pract. Experience **23**(7), 708–720 (2011)

31. Szafaryn, L.G., Skadron, K., Saucerman, J.J.: Experiences accelerating matlab systems biology applications. In: Proceedings of the Workshop on Biomedicine in Computing: Systems, Architectures, and Circuits, pp. 1–4 (2009)

32. Ten Tusscher, K.H.W.J., Panfilov, A.V.: Alternans and spiral breakup in a human ventricular tissue model. Am. J. Physiol. Heart Circulatory Physiol. **291**(3), H1088–H1100 (2006). https://ajpheart.physiology.org/content/291/3/H1088

33. Vasconcellos, E.C., Clua, E.W., Fenton, F.H., Zamith, M.: Accelerating simulations of cardiac electrical dynamics through a multi-GPU platform and an optimized data structure. Concurrency Comput. Pract. Experience **32**(5), e5528 (2020). https://onlinelibrary.wiley.com/doi/abs/10.1002/cpe.5528

34. Vigueras, G., Roy, I., Cookson, A., Lee, J., Smith, N., Nordsletten, D.: Toward GPGPU accelerated human electromechanical cardiac simulations. Int. J. Numer. Methods Biomed. Eng. **30**(1), 117–134 (2014). https://doi.org/10.1002/cnm.2593

35. Vincent, K., et al.: High-order finite element methods for cardiac monodomain simulations. Front. Physiol. **6**, 217 (2015). https://www.frontiersin.org/article/10.3389/fphys.2015.00217

36. Xia, Y., Wang, K., Zhang, H.: Parallel optimization of 3d cardiac electrophysiological model using GPU. Comput. Math. Methods Med. **2015**, 1–10 (2015)

Optimization of the Use of Critical Resources in the Development of Offshore Oil Fields

Sérgio Bassi and Debora Pretti Ronconi[(✉)] ⓘ

Production Engineering Department, University of São Paulo (USP),
São Paulo 05508-010, Brazil
dronconi@usp.br

Abstract. This paper focuses on a real case of connection of subsea oil wells to offshore platforms using pipe laying support vessels. The objective of this study is to maximize the oil production curve through an optimized use of the outsourced fleet. Specific features of this scenario are considered, such as technical constraints of each vessel, the availability of the vessels, materials for connection, and the end of the previous phase, called completion. A mixed integer linear programming model is developed considering several constraints that structure this complex situation, among which a relevant characteristic of the problem: the increase of the production curve using injection wells to fight the natural decline of producing wells over time. This mathematical model was tested in small computational instances, showing adequate behavior, which demonstrates that it faithfully represents the situation portrayed and can be used, combined with more advanced computational resources, to achieve better results.

Keywords: Oil wells · Routing problem · Mixed integer linear programming

1 Introduction

In the past decade, Brazil experienced a significant growth in oil production, especially with important discoveries of oil reserves in the pre-salt layer in ultra-deep waters (over 2,000 meters deep). There are 12 billion barrels of proven oil reserves in the country; in 2018, the offshore oil production amounted to approximately 903 million barrels [1]. Figure 1 shows the evolution of oil production in Brazil between 2004 and 2018. The Brazilian industry has faced an increasing challenge to develop technologies involved in the construction of wells to exploit these new oil reserves [2].

In offshore production, highly complex items that operate in severe conditions are considered critical resources. They also have elevated costs and high manufacturing lead times. Therefore, an advanced planning is required to make sure these items are available in a timely fashion. Offshore activities require the use of specific vessels to build offshore wells (the so-called oil drilling rigs that are used for drilling and completion of offshore wells) and to connect these wells to the production platforms (the pipe laying support vessels, PLSVs). The daily costs of a vessel can amount to US$ 500,000 [3]. There is a limited number of vessels and they are outsourced; thus, it is

© Springer Nature Switzerland AG 2020
O. Gervasi et al. (Eds.): ICCSA 2020, LNCS 12249, pp. 391–405, 2020.
https://doi.org/10.1007/978-3-030-58799-4_29

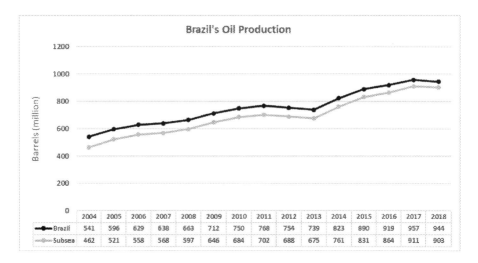

Fig. 1. Volumes of oil production between 2004 and 2018: total country production and subsea oil production (according to ref. [1]).

essential to properly schedule them and to find the best routes to the locations where activities will be carried out to optimize physical and monetary resources.

There are several studies in the literature that deal specifically with situations in the maritime environment that require the use of vessels. Nishi and Izuno [4] tackled a problem of routing and scheduling of crude oil ships with split deliveries, aiming to minimize distances while respecting the capacity of the ships that collect oil in various parts of the world and distribute it to some customers. A mathematical model whose constraints are similar to those used in terrestrial/road routing situations (for example, vessel capacity, subtour elimination restrictions, and route start and end restrictions) was presented; and a column-generation-based heuristic was introduced. Another work that describes the transportation of crude oil, with collection and delivery of products, was written by Henning et al. [5]. The addressed problem is characterized by the heterogeneous fleet and the transportation of multiple products. Other considered constraints were the port loading capacities, split loads, and time windows for both collection and delivery activities. A model considering these and other real-life features in which the objective is to minimize fuel costs and port charges was introduced. Lee and Kim [6] addressed a routing problem of a heterogeneous fleet with split deliveries in the context of a steel manufacturing company. They presented a mixed integer linear programming (MILP) model as well as an Adaptive Large Neighbourhood Search (ALNS) heuristic. Assis and Camponogara [7] dealt with the problem of relief ships that transport products between offshore oil platforms and a land terminal. The operation is depicted as a graph in which vertices represent the terminal and the platforms; while arcs represent the travel times. Possible operations for the ships are moving, offloading, uploading, and waiting. From this characterization, an MILP model was proposed. Another MILP approach in the oil industry was proposed by Lin et al. [8] in a problem of scheduling a fleet of marine vessels for crude oil tanker lightering. This kind of activity is often related to sailing to refineries; it describes the transfer of crude oil from a discharging tanker to smaller vessels to make the tanker lighter. Stanzani et al. [9] addressed a real life multiship routing and scheduling problem with inventory constraints that arises in pickup and delivery operations of crude oil from various offshore platforms to coastal terminals. MILP models were presented to deal with small-to-moderate instances; while a matheuristic was proposed to deal with larger instances.

Some works address problems related to platform supply vessels (PSVs) that support the offshore oil and gas exploration and production activities. The problem consists in delivering goods from an onshore supply base to one or more offshore units and in returning items form these units to the onshore base. Gribkovskaia et al. [10] approached this problem with a single vessel considering the limited loading and unloading capacities at the platforms. A mathematical model, constructive heuristics, and a tabu search algorithm were presented. Kisialiou et al. [11] addressed the determination of the fleet composition and of the vessels' schedules involving flexible departures (multiple voyage departure options every day) from the onshore base. An ALNS algorithm was developed and a comparison with results provided by the resolution of a voyage-based model on small- and medium-sized instances was presented. Cruz et al. [12] proposed a mathematical model and a heuristic solution strategy for the fleet composition and periodic routing problem integrated with the berth allocation problem. In addition to these works, applications of optimization strategies based on MILP models can be found in Mardaneh et al. [13] and Halvorsen-Weare and Fagerholt [14].

Other studies tackle task-scheduling problems in petroleum projects. Bassi et al. [15] considered the problem of planning and scheduling a fleet of rigs to be used for drilling wells or for maintenance activities with uncertain service times. The authors developed a procedure based on simulation and optimization strategies to generate expected solutions. Pereira et al. [16] optimized the usage of PLSVs and drilling rigs in a problem of scheduling oil well development activities. The same problem, taking into account resources' displacement times, was addressed in Moura et al. [17]. In both papers, a Greedy Randomized Adaptive Search Procedure (GRASP) was applied. Also related to the PLSVs programming area, Cunha et al. [18] carried out a study making an analogy between routing activities and scheduling activities in parallel machines. The objective is to reschedule the PLSVs aiming to minimize the impacts caused by operational disruptions. These authors propose an MILP formulation for the PLSVs routing problem that starts from previously created activity blocks (voyages). Additionally, a method based on the Iterated Local Search (ILS) metaheuristic was proposed.

There are three stages involved in the construction of a well, namely, drilling, completion, and connection. This work focuses on the stage of wells' connections by PLSVs, a stage that immediately precedes the beginning of the operation of a well. In general, the connection process consists of three steps: I) loading of the PLSVs with the lines to be launched (this stage occurs in a loading base), II) navigation until the location of a well (where lines will be effectively launched), and III) launching the lines between the well and its associated stationary platform unit (SPU). Figure 2 summarizes this scenario. The problem considered in the present work deals with the usage of

PLSVs for the connection of several wells simultaneously. PLSVs are responsible for connecting flexible ducts and electro-hydraulic umbilical pipelines (EHU) between the SPUs and the producing and injection wells at sea. Thus, the problem consists in allocating activities over time to a heterogeneous fleet of vessels, considering aspects such as the non-concomitant use of different vessels to perform tasks in the same well, the technical viability of a vessel to perform a connection activity, the availability of raw material for a given itinerary, and the relation between vessels' loading constraints and the amount of material that must be loaded into them in order to perform the assigned activities. The goal is to maximize the oil production curve of a given set of wells or project. An innovative feature of the proposed approach is to tackle the decline in the production of producing wells as well as the influence of injection wells in this decline.

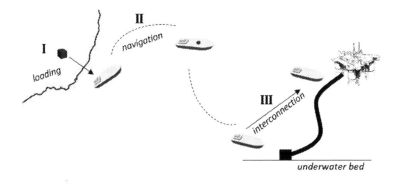

Fig. 2. Schematic representation of the problem: loading, navigation and interconnection of ducts.

This paper is organized as follows. Section 2 describes the problem and presents the problem description and a mixed integer linear programming model. Section 3 presents numerical experiments with a set of small-sized instances, given that optimal solutions to the real instances of the problem are not expected to be obtained within a reasonable execution time. Section 4 presents the final considerations and next steps suggested for this study.

2   Problem Description and a Mathematical Model

In the present section, a detailed description of the considered problem and its MILP formulation are presented. There are two types of wells: producing wells and injection wells. Each injection well is associated with one or more producing well. Producing wells are effectively responsible for oil production; while injection wells are responsible for repressurizing the oil reservoir and, consequently, for increasing the volume generated by the associated producing wells. Before a production well can start producing, there are three ducts or lines to be connected to it: two flexible ducts (one for production and one for service) and one EHU. On the other hand, injection wells must be connected to two or three lines depending on whether they inject one or two types of fluid (one or two for the fluids plus one EHU). All lines must be installed between the wellhead and the associated platform (not necessarily in this order); this means that there is no intermediate equipment working as a line hub. There is only one loading base whose position, as well as the positions of wells and platforms, is fixed. Thus, distances between the loading base, the wells, and the platforms are all known a priori. There is no precedence relationship between the activities and, once started, an activity cannot be interrupted. Concomitance is not allowed, i.e. it is not possible to perform two or more activities at the same time in the same well. Each well has a date (end of its completion stage) at which the connection stage activities at the well can start to be executed. Each duct and EHU has an arrival date to the loading base and, naturally, they can only be loaded into a PLSV after their arrival to the base. Not every PLSV can perform every task (technical viability) since each PLSV has a storage capacity, an upper bound on the water depth of the activities it can perform, and a duct launch capacity.

After the end of the connection stage of a producing well, a testing stage, whose duration (commissioning time) is known, must be executed; and, after the stage of testing, the well starts its production (independently of the other wells being considered). The productivity of each well, in barrels per day (bpd), as a function of time, is also known a priori. One of the key points and differential aspect of this study is considering the decline of the well production and the effect of the inclusion of injection wells over time. A producing well begins to operate at its maximum volume, which is called the initial production potential. Over time, this volume is reduced due to the natural decline of production of the reservoir, which is explained by the decrease in pressure. The injection wells, whether water- or gas-operated, repressurize the reservoir, adding a fraction of oil production to that field in subsequent periods. Figure 3 illustrates the natural decline of a producing well including the effect of an injection well. In the illustrated example, the producing well has an initial production potential of 50,000 bpd and starts operating at time 10. The production decreases at a known linear rate. The injection well starts operating at time 20 and it has the capacity of increasing the productivity of the producing well by 6%. Therefore, if there were no injection well, the production would continue to decline, as depicted by the solid line. More than one injection well might be associated with a producing well. In this case, the production is increased by the sum of the fractions added by each injection well. As a whole, the productivity of a set of wells as a function of time can be computed and the objective is, satisfying all the problem constraints, to assign activities to the PSLVs and to determine the date these activities will be executed in order to maximize the total productivity of the wells within the considered period of time.

Figure 4 shows, with the help of a Gantt chart, a feasible solution for an instance with two vessels and three wells. In the figure, *Well # (act.i)* represents activity i to be performed on well #. The dotted lines represent the instant at which the three lines of a well were launched and, therefore, the well is ready to be commissioned (beginning of the test phase). It is also possible to observe, in this example, that activities are not concomitant, i.e. there is no time intersection between the activities of the same well.

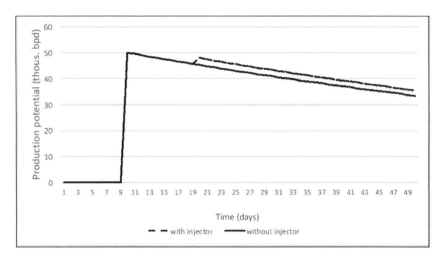

Fig. 3. Natural decline and effect of the use of injection wells on the volume of a producing well.

Fig. 4. Gantt chart of a feasible solution for an instance with two vessels and three wells.

An MILP model for the problem is now presented. In order to avoid a large number of indices in the variables, activities to be performed on the wells are numbered consecutively. This means that, for example, if there are three activities to be performed on wells 1 and 3 and two activities to be performed on well 2 then the activities of the first well will be numbered 1, 2, and 3; the activities of the second well will be numbered 4 and 5; and the activities of the third well will be numbered 6, 7, and 8. We denote by n the total number of activities, $N = \{1, 2, ..., n\}$, and $N^+ = \{0, 1, 2, ..., n, n + 1\}$, where 0 and $n + 1$ represent the activities of leaving from and returning back to the loading base, respectively. Since solutions include the possibility of multiple itineraries for each PLSV, activities 0 and $n + 1$ are "special" activities that might be performed more than once. It is worth noting that an itinerary consists in a closed route leaving from and returning back to the loading base. It is the definition of routes that reflects in the model the fact that the raw material to be used in a set of activities must be loaded in the PLSV before starting the route that contains them. P denotes the number of wells and the sets P^* and P' are a partition of $\{1, ..., P\}$ that represent the indices of the producing wells and the injection wells, respectively. $A = \{(p^*, p') \mid p^* \in P^*$ and $p' \in P'\}$ represents the existent associations between the producing and injection wells. A description of the additional instance data follows. T is the number of days in the analyzed period and $\prod = (1, 2, ..., T)$ is the set of possible operation periods of producing wells; B is the number of PLSVs and $V = (1, 2, \ldots, B)$ is the set of PLSVs; R is an upper bound on the allowed number of routes per PLSV; d_{ij} is the duration (in days) of the transfer between the sites of activities i and j; s_i is the duration (in days) of activity i; m_i^b is a binary constant whose value is one if the PLSV b is able to perform activity i and zero otherwise; if $p*$ is a producing well, W_{p*}^π is the number of barrels per day produced by well $p*$ in its π-th producing day; if p' is an injection well and $(p*, p') \in A$, $Inj_{p'}^\pi$ is the fraction of increment (between zero and one) in the production of well $p*$ in its π-th producing day; DC_i is the time from which activity i can be executed (this constant is identical for all activities to be performed in the same well and it corresponds to the end of the completion stage of the well); DM_i is the time at which the material required to execute activity i becomes available; q_i is the length (in km) of the duct required for activity i; Q_b is the storage capacity of vessel b (in km of ducts); β is the loading time (in days) per kilometer of duct; and c is the duration in days of the testing phase of a duct.

The description of indexes and variables used in the proposed MILP model and the complete model are presented next.

Indexes

i, j, h:	activities;
b:	vessel;
r:	route;
p:	well;
$p*$:	producing well;
p':	injection well;
π :	period of operation (ordinal) of a producing well

Variables

X_{ij}^{rb}:   1, if vessel b, in route r, performs activity j right after activity i; 0, otherwise;

Y_i^{rb}:   1, if activity i is in route r of vessel b; 0, otherwise;

t_i^{rb}:   auxiliary variable to determine the start time of activity i that is performed in route r of the vessel b; especially for activity 0, it is the time when the vessel leaves the loading base in each route r of each vessel b;

g_i^{rb}:   actual start time of activity i if this activity is performed in route r of the vessel b; 0, otherwise;

$a_{i1,i2}$:   1, if activity $i1$ occurs before activity $i2$, being $i1$ and $i2$ activities that will take place at the same well; 0, otherwise;

σ^{rb} :   duration of loading of the ducts used in the activities that make up route r of a vessel b;

k_p: finalization time of the connection stage of well p;

μ_{p*}^π :   1, if producing well $p*$ has its π-th operation period in the analysis interval; 0, otherwise;

$\gamma_{p'}^\pi$:   1, if injection well p' operates in the π-th operation period of its respective producing well in the analysis interval; 0, otherwise

$$\text{maximize} \sum_{\pi=1}^{T} \sum_{\forall p^*} \left(W_{p^*}^{\pi} \left(\mu_{p^*}^{\pi} + \sum_{p'/(p^*,p') \in A} Inj_{p'}^{\pi} \gamma_{p'}^{\pi} \right) \right) \tag{1}$$

subject to:

$$\sum_{j \in N^+ \setminus \{0\}} X_{ij}^{rb} = Y_i^{rb} \qquad i \in N, \forall r, \forall b \tag{2}$$

$$\sum_{b=1}^{B} \sum_{r=1}^{R} Y_i^{rb} = 1 \qquad i \in N \tag{3}$$

$$\sum_{i \in N^+} X_{ih}^{rb} - \sum_{j \in N^+} X_{hj}^{rb} = 0 \qquad h \in N, \forall r, \forall b \tag{4}$$

$$\sum_{i \in N^+} X_{0i}^{rb} = 1 \qquad \forall r, \forall b \tag{5}$$

$$\sum_{i \in N^+} X_{i(I+1)}^{rb} = 1 \qquad \forall r, \forall b \tag{6}$$

$$\sum_{i \in N} q_i Y_i^{rb} \leq Q_b \qquad \forall r, \forall b \tag{7}$$

$$t_i^{rb} + s_i + d_{ij} - M\left(1 - X_{ij}^{rb}\right) \leq t_j^{rb} \qquad i \in N^+, j \in N^+, \forall r, \forall b \tag{8}$$

$$t_0^{rb} \geq DM_i Y_i^{rb} + \sigma^{rb} \qquad i \in N, \forall r, \forall b \tag{9}$$

$$\sigma^{rb} = \beta \sum_{i \in N} q_i Y_i^{rb} \qquad \forall r, \forall b \tag{10}$$

$$t_{n+1}^{rb} + \sigma^{(r+1)b} \leq t_0^{(r+1)b} \qquad r = 1, \ldots, R-1; \forall b \tag{11}$$

$$Y_i^{rb} \leq m_i^b \qquad i \in N, \forall r, \forall b \tag{12}$$

$$t_i^{rb} \geq DC_i \qquad i \in N, \forall r, \forall b \tag{13}$$

$$g_i^{rb} \geq t_i^{rb} - M\left(1 - Y_i^{rb}\right) \qquad i \in N, \forall r, \forall b \tag{14}$$

$$g_i^{rb} \leq t_i^{rb} + M\left(1 - Y_i^{rb}\right) \qquad i \in N, \forall r, \forall b \tag{15}$$

$$g_i^{rb} \leq MY_i^{rb} \qquad i \in N, \forall r, \forall b \tag{16}$$

$$k_p \geq \sum_{r=1}^{R} \sum_{b=1}^{B} g_{(3p-2)}^{rb} + s_{3p-2} + c$$

$$k_p \geq \sum_{r=1}^{R} \sum_{b=1}^{B} g_{3p-1}^{rb} + s_{3p-1} + c \tag{17}$$

$$k_p \geq \sum_{r=1}^{R} \sum_{b=1}^{B} g_{3p}^{rb} + s_{3p} + c \qquad \forall p$$

$$\sum_{b=1}^{B} \sum_{r=1}^{R} g_{3p-2}^{rb} - \sum_{b=1}^{B} \sum_{r=1}^{R} g_{3p-1}^{rb} + s_{3p-2}$$

$$\leq M\left(1 - a_{3p-2,3p-1}\right)$$

$$\sum_{b=1}^{B} \sum_{r=1}^{R} g_{3p-1}^{rb} - \sum_{b=1}^{B} \sum_{r=1}^{R} g_{3p-2}^{rb} + s_{3p-1} \leq M a_{3p-2,3p-1} \tag{18}$$

$$\sum_{b=1}^{B} \sum_{r=1}^{R} g_{3p-1}^{rb} - \sum_{b=1}^{B} \sum_{r=1}^{R} g_{3p}^{rb} + s_{3p-1} \leq M\left(1 - a_{3p-1,3p}\right) \qquad \forall p$$

$$\sum_{b=1}^{B} \sum_{r=1}^{R} g_{3p}^{rb} - \sum_{b=1}^{B} \sum_{r=1}^{R} g_{3p-1}^{rb} + s_{3p} \leq M a_{3p-1,3p}$$

$$\sum_{b=1}^{B} \sum_{r=1}^{R} g_{3p-2}^{rb} - \sum_{b=1}^{B} \sum_{r=1}^{R} g_{3p}^{rb} + s_{3p-2} \leq M\left(1 - a_{3p-2,3p}\right)$$

$$\sum_{b=1}^{B} \sum_{r=1}^{R} g_{3p}^{rb} - \sum_{b=1}^{B} \sum_{r=1}^{R} g_{3p-2}^{rb} + s_{3p} \leq M a_{3p-2,3p}$$

$$\mu_{p^*}^{\pi} \leq \frac{T - k_p}{\pi} \qquad \forall \pi, \forall p^* \in P^* \tag{19}$$

$$\pi \gamma_{p'}^{\pi} + M\left(1 - \gamma_{p'}^{\pi}\right) \geq k_{p'} - k_{p^*} + 1 \qquad \forall \pi, \forall (p^*, p') \in A \tag{20}$$

$$\mu_{p^*}^{\pi} \geq \gamma_{p'}^{\pi} \qquad \forall \pi, \forall (p^*, p') \in A \tag{21}$$

$$t_i^{rb} \geq 0 \qquad i \in N^+, \forall r, \forall b \tag{22}$$

$$g_i^{rb} \geq 0 \qquad i \in N^+, \forall r, \forall b \tag{23}$$

$$k_p \geq 0 \qquad \forall p \tag{24}$$

$$\sigma^{rb} \geq 0 \qquad \forall r, \forall b \tag{25}$$

$$X_{ij}^{rb} \quad binary \qquad i \in N^+, j \in N^+, \forall r, \forall b \tag{26}$$

$$Y_i^{rb} \quad binary \qquad i \in N^+, \forall r, \forall b \tag{27}$$

$$a_{i1,i2} \ \ binary \hspace{3cm} i1, i2 \in N \hspace{2cm} (28)$$

$$\mu^{\pi}_{p^*} \ \ binary \hspace{3cm} \forall \pi, \forall p^* \hspace{2.5cm} (29)$$

$$\gamma^{\pi}_{p'} \ \ binary \hspace{3cm} \forall \pi, \forall p' \hspace{2.5cm} (30)$$

The objective function (1) aims at increasing the production volume in the considered period. At this point, it is crucial to clearly define time π: π is the period of operation of the producing well; and it necessarily starts at period 1. Thus, in each operation period, the objective function multiplies the production potential of the producing well ($W^{\pi}_{p^*}$) by the variables that indicate in which periods the well will be producing ($\mu^{\pi}_{p^*}$) and, moreover, in which of these times the corresponding injection well will be operating, which results in a production increase factor given by $Inj^{\pi}_{p'} \gamma^{\pi}_{p'}$. It is important to point out that π is the operating time of the producing well even when the wells at hand are injectors. In this case, π starts counting when the corresponding producing well begins operating.

Constraints (2) express that, if any activity is performed by a given vessel route ($Y^{rb}_i = 1$), there must be an exit arc for an activity j (which may be an activity in the base, $n + 1$, or for the execution of an activity in any other location), that is, one, and only one $X^{rb}_{ij} = 1$. Constraints (3) imply that all activities must be performed. It is important to note that activities of departure and arrival to the loading base are not considered in this restriction since they must occur once in each itinerary of each vessel, as it will be seen later in Constraints (5) and (6). Constraints (4) are responsible for the existence of flow in the routes of each vessel. Given that an activity h is being performed, equalities (4) ensure that, prior to it, an activity i was performed and, after it, an activity j will be performed. Activities i and j can be activities performed on departure and arrival at the base (0 and $n + 1$, respectively). Constraints (5) and (6) establish the requirement of departure and arrival at the base in each route of each PLSV, respectively. Constraints (7) guarantee that the number of pipelines to be loaded in each route of each vessel does not exceed the capacity Q_b of the PLSV. Constraints (8) ensure the time continuity within the same itinerary of the same vessel. If activity j follows activity i in the route r of the vessel b, then X^{rb}_{ij} is equal to 1. Consequently, $t^{rb}_i + s_i + d_{ij} \leq t^{rb}_j$, i.e., the start time of the next activity, j, is greater than or equal to the start time of the previous activity, i, added to the execution time of activity i (s_i) and the transfer time between the place where activity i is performed and where the activity j will be performed (d_{ij}). Otherwise, the value t^{rb}_j is not limited by the constraint. Constraints (8) also avoid the existence of subtours, given that the following activity must always take place after the previous activity.

Constraints (9) state that a PLSV can only leave the base to perform a route once all necessary lines of the route are available (DM_i) and loaded (σ^{rb}). Constraints (10) calculate the time to be spent in loading the lines of a route (σ^{rb}); the components of this computation are: the length of the line of each activity i (q_i), which is activated in the multiplication if Y^{rb}_i is 1, which represents that activity i is part of route r of vessel b; and a parameter β representing the time, in days, spent to load each km of line.

Constraints (11) refer to the transition between routes in one vessel. The constraints state that the next route can begin at least at the end of the previous route (PLSV returns to the loading base), plus the loading time of the next load. In a scenario where the required resources are already available, the inequality constraint holds as equality. Constraints (12) evaluate if a vessel can perform a given activity. A matrix of parameters m_i^b is provided to check which PLSVs can be used for each one of the tasks to be performed. The set of constraints (13) prevents connection activities from being started before the completion activity of that well has been completed. Constraints (14), (15), and (16) are used to compute the times g_i^{rb} that are, in fact, used in each of the routes. Variables t_i^{rb} are also used for intermediate calculations, so variables g_i^{rb} are established since these variables can only take positive values when the activity is performed. If $Y_i^{rb} = 1$, g_i^{rb} can only take a value equal to t_i^{rb}, on the other hand if $Y_i^{rb} = 0$, Constraints (16) force g_i^{rb} to be zero. These variables are important to establish Constraints (18) that avoid the concomitance of two vessels in the same well. The set of constraints (17) narrows the finalization time of the connection stage of each well (k_p). Knowing that each well is composed of 3 connection activities, these constraints calculate the earliest completion date of each activity as if it was the critical activity of the well. These 3 equations refer to the first, second, and third activity of the well p, respectively. For example, the term $\sum_{r=1}^{R} \sum_{b=1}^{B} g_{(3p-2)}^{rb}$, considering that there is only one positive $g_{(3p-2)}^{rb}$, indicates the time when activity $3p-2$ begins. The duration of the activity s_{3p-2} is added to this sum to determine the end time of the activity. In addition, the commissioning time (c) is computed in each case. The second and third equations of set (17) perform similarly. The earliest date k_p is, therefore, defined as the highest value of these three equations, for each p. The set of constraints (18) prevents more than one vessel from operating at the same time in one well. Since each well has three activities, these inequalities allow one connection activity to either be terminated before the start of another activity or to start after the end of another activity at the same well. Variables $a_{i1,i2}$ indicate the order in which the activities take place; and variables g_i^{rb} are used because they are not influenced by intermediate calculations, as variables t_i^{rb}.

   The set of constraints (19) determines the production periods of the wells, μ_{p*}^{π}, according to the finalization time of the connection stage, k_p. Since the potential of each producing well (considering the decline of production) is function of the period π of production, the establishment of these periods is crucial for the evaluation of the objective function. If a well p produces for 30 periods, variables $\mu_{p*}^{1}, \ldots, \mu_{p*}^{30}$ should be 1, while the others should be 0. Note that the value of the right-hand-side of Constraints (19), $(T - k_p)/\pi$, decreases as π increases. Therefore, the value of the denominator will gradually narrow the value of μ_{p*}^{π}. Since it is a maximization problem, μ_{p*}^{π} tends to always be 1 until the moment of inversion, when $(T - k_p)/\pi$ becomes smaller than 1, which forces μ_{p*}^{π} to become 0. This inversion occurs when π converges to the values greater than $T - k_p$ meaning that the end of the analysis period was reached. Constraints (20) determine the periods in which the influence of the injection wells should be computed. The calculation of $\gamma_{p'}^{\pi}$ is based on the finalization time of the connection stage of their corresponding producing well (k_{p^*}). Observe that π is the operating period of the producing well. If $k_{p'} - k_{p^*} + 1$ (right-hand-side of the constraint) has a value greater than the coefficient π that multiplies $\gamma_{p'}^{\pi}$, the variable $\gamma_{p'}^{\pi}$ will assume value 0 and will not add any value to the objective function. Otherwise, when expression $k_{p'} - k_{p^*} + 1$ is smaller than or equal to the coefficient π, $\gamma_{p'}^{\pi}$ will be 0 or 1 to comply with the considered constraint. In this case, the model forces $\gamma_{p'}^{\pi}$ to be 1 because this decision increases the objective function. For example, if the producing well has its connection finalization at time 5 (i.e. $k_{p^*} = 5$) and its corresponding injection well at time 9 (i.e. $k_{p'} = 9$), the following scenario would occur: the first four production periods of the producing well would not have an injection well in operation ($\gamma_{p'}^{1} = = \gamma_{p'}^{3} = \gamma_{p'}^{4} = 0$). The terms from $\gamma_{p'}^{5}$ on must be positive (up to the limit T of analysis) due to the fact that the injection well is already operating and that it can add value to the objective function.

The set of constraints (21) prevents an injection well from adding value to the objective function before the beginning of the operation of the corresponding producing well or after the end of the analysis period. Constraints (22–30) define the domain of the variables.

3   Numerical Experiments

In this section, we aim to evaluate the performance of the proposed method in a set of small-sized instances in order to obtain its optimal solution. With this purpose, a set of thirty instances, all of them with two wells (one injection and one producing well) and six activities, was generated. Instances are divided into six sets of five instances each. In all the thirty instances, the capacity of the PLSVs is 12 km of ducts; the loading time per kilometer of duct is 0.5 day; all PLSVs can perform all activities; the production potential of the producing well is 180 bpd at its first producing day and it decreases at the rate of one barrel per day; the fraction of increment given by the injection well is 0.03; the duration of the testing phase of a duct (commissioning time) is 3 days; and the considered period is equal to 180 days. Tables 1 and 2 complete the description of the instances in Set I. Table 1 displays the transfer times between the activities; while Table 2 displays, for each activity, the duration of the activity (s_i), the length of the duct (q_i), the completion time (DC_i) of the associated well (i.e. the time at which the activity can start to be executed), and the time at which the duct is made available (DM_i). Sets II to VI correspond to variations of Set I as described in Table 3.

Numerical experiments were run on a Dell Studio XPS 8100 computer with an Intel Core i7 2.93 GHz and 16 Gb of RAM memory. The commercial solver CPLEX (version 12.1), with all its default parameters, was able to find optimal solutions for all the instances. Table 3 shows the results. It is worth noting that, in some instances, CPLEX used more than seven hours of CPU time. The analysis of the optimal solutions shows a stabilization of the objective function values in instances with three to five PLSVs within the same set, especially in Sets V and VI.

Table 1. Transfer times (d_{ij}) between activities (in days).

i \ j	0	1	2	3	4	5	6	7
0	0	2	2	2	3	3	3	0
1	2	0	0	0	1	1	1	2
2	2	0	0	0	1	1	1	2
3	2	0	0	0	1	1	1	2
4	3	1	1	1	0	0	0	3
5	3	1	1	1	0	0	0	3
6	3	1	1	1	0	0	0	3
7	0	2	2	2	3	3	3	0

Table 2. Input parameters.

p	Type	i	s_i	q_i	DC_i	DM_i
1	Producing well	1	3	5	4	3
		2	4	4	4	8
		3	5	6	4	7
2	Injection well	4	5	6	6	2
		5	6	4	6	10
		6	7	5	6	12

4 Final Considerations

This paper addresses the process of interconnecting oil wells to offshore production units in the high seas. Analyzing the features and technical areas involved in this problem, it is possible to understand how complex it is from the geological study phase until the production from the oil wells after each duct is interconnected with the vessels (PLSV). Given the growth in the Brazilian offshore oil production in the past years and, especially, the contribution of wells from the pre-salt layer to this outcome, studies intended to characterize and improve this process are very relevant. Following this trend, our research study developed and evaluated techniques to optimize the use of the physical resources of an oil company, aiming to maximize the oil production curve.

A mixed integer linear programming model was developed to tackle a series of practical constraints of the problem to be solved. It should be highlighted that additional aspects in relation to the literature were considered, mainly with respect to the increase in the production curve due to the operation of injection wells and to the natural decline of producing wells during their operational lifespan. Numerical experiments with small instances were conducted and the proposed model was successfully validated. These numerical experiments showed consistent results, showing an increase in the value of the objective function with a greater number of PLSVs available. This fact indicates that, with a larger number of vessels, more activities can

Table 3. Comparison of the optimal solutions.

Instances description						Optimal results	
Set	#PLSVs	#routes	#binary variables	#real variables	#constraints	main features	obj. function	processing time (s)
I	1	6	786	111	1459	See Tables 1 and 2	16,059.4	1,406.00
	2	3	786	111	1458		16,357.8	776.00
	3	2	786	111	1457		16,401.8	191.00
	4	2	926	145	1750		16,427.3	190.00
	5	2	1066	179	2043		16,432.5	6,162.00
II	1	6	786	111	1459	$d_{0j} = d_{7j} = 3, j = 1, 2, 3$	15,963.0	4,871.00
	2	3	786	111	1458	and $d_{0j} = d_{7j} = 2 \, j = 4, 5, 6$	16,291.9	541.00
	3	2	786	111	1457		16,369.7	155.00
	4	2	926	145	1750		16,396.4	214.00
	5	2	1066	179	2043		16,417.3	1,779.00
III	1	6	786	111	1459	$q_i = 4, i = 1, ..., 6$	16,155.3	6,470.00
	2	3	786	111	1458		16,387.3	5,312.00
	3	2	786	111	1457		16,426.6	407.00
	4	2	926	145	1750		16,452.1	22,031.00
	5	2	1066	179	2043		16,452.1	28,737.00
IV	1	6	786	111	1459	$q_i = 8, i = 1, ..., 6$	15,712.7	3.24
	2	3	786	111	1458		16,268.0	1.06
	3	2	786	111	1457		16,356.2	0.82
	4	2	926	145	1750		16,381.1	1.85
	5	2	1066	179	2043		16,401.6	1.98
V	1	6	786	111	1459	$(s_1, ..., s_6) =$	14,866.9	1,032.00
	2	3	786	111	1458	$(10,12,15,7,9,10)$	15,489.6	218.00
	3	2	786	111	1457		15,567.4	91.17
	4	2	926	145	1750		15,567.4	108.00
	5	2	1066	179	2043		15,567.4	3,925.00
VI	1	6	786	111	1459	$DC_i = 20. \ i = 1, ..., 3$	15,712.1	829.00
	2	3	786	111	1458		16,061.9	798.00
	3	2	786	111	1457		16,129.8	60.02
	4	2	926	145	1750		16,129.8	6.57
	5	2	1066	179	2043		16,129.8	24.37

occur in parallel, allowing the anticipation of the start-up of producing wells and increasing the volume of production in the analyzed period.

As future topic for research, we suggest the elaboration of a constructive heuristic in order to validate the behavior of this model for bigger instances. Also, it is possible the application of neighborhood search strategies to investigate activity movements after the initial allocation indicated by the heuristic procedure. Since our target problem has many constraints, preparing an advanced procedure represents a great challenge because the solution can easily lose its viability characteristics.

Acknowledgements. This research has been partially supported by FAPESP (grants 13/07375-0 and 16/01860-1) and CNPq (grant 306083/2016-7). The authors would like to thank the careful reading of the reviewers, whose comments helped to improve the quality of the manuscript.

References

1. National Agency for Petroleum, Natural Gas and Biofuels (ANP): Oil, Natural Gas and Biofuels Statistical Yearbook (2019). http://www.anp.gov.br/publicacoes/anuario-estatistico/5237-anuario-estatistico-2019. Accessed 25 Apr 2020
2. National Bank for Economic and Social Development (BNDES): BNDES: a bank with a history and a future (2012). https://web.bndes.gov.br/bib/jspui/handle/1408/2101. Accessed 25 Apr 2020
3. Kaiser, M.J., Snyder, B.: The five offshore drilling rig markets. Mar. Policy **39**, 201–214 (2013)
4. Nishi, T., Izuno, T.: Column generation heuristics for ship routing and scheduling problems in crude oil transportation with split deliveries. Comput. Chem. Eng. **60**, 329–338 (2004)
5. Hennig, F., et al.: Maritime crude oil transportation – a split pickup and split delivery problem. Eur. J. Oper. Res. **218**, 764–774 (2012)
6. Lee, J., Kim, B.: Industrial ship routing problem with split delivery and two types of vessel. Expert Syst. Appl. **42**, 9012–9023 (2015)
7. Assis, L.S., Camponogara, E.: A MILP model for planning the trips of dynamic positioned tankers with variable travel time. Transp. Res. Part E **93**, 372–388 (2016)
8. Lin, X., Chajakis, E.D., Floudas, C.A.: Scheduling of tanker lightering via a novel continuous-time optimization framework. Ind. Eng. Chem. Res. **42**, 4441–4451 (2003)
9. Stanzani, A., Pureza, V., Morabito, R., Silva, B.J., Yamashita, D., Ribas, P.C.: Optimizing multiship routing and scheduling with constraints on inventory levels in a Brazilian oil company. Int. Trans. Oper. Res. **25**, 1163–1198 (2018)
10. Gribkovskaia, I., Laporte, G., Shlopak, A.: A tabu search heuristic for a routing problem arising in servicing of offshore oil and gas platforms. J. Oper. Res. Soc. **59**, 1449–1459 (2008)
11. Kisialiou, Y., Gribkovskaia, G., Laporte, G.: The periodic supply vessel planning problem with flexible departure times and coupled vessels. Comput. Oper. Res. **94**, 52–64 (2018)
12. Cruz, R., Mendes, A.B., Bahiense, L., Wu, Y.: Integrating berth allocation decisions in a fleet composition and periodic routing problem of platform supply vessels. Eur. J. Oper. Res. **275**, 334–346 (2019)
13. Mardaneh, E., Lin, Q., Loxton, R., Wilson, N.: Cargo scheduling decision support for offshore oil and gas production: a case study. Optim. Eng. **18**(4), 991–1008 (2017). https://doi.org/10.1007/s11081-017-9348-3
14. Halvorsen-Weare, E.E., Fagerholt, K.: Optimization in offshore supply vessel planning. Optim. Eng. **18**(1), 317–341 (2016). https://doi.org/10.1007/s11081-016-9315-4
15. Bassi, H.V., Ferreira Filho, V.J.M., Bahiense, L.: Planning and scheduling a fleet of rigs using simulation–optimization. Comput. Ind. Eng. **63**, 1074–1088 (2012)
16. Pereira, R.A., Moura, A.V., de Souza, C.C.: Comparative experiments with GRASP and constraint programming for the oil well drilling problem. In: Nikoletseas, S.E. (ed.) WEA 2005. LNCS, vol. 3503, pp. 328–340. Springer, Heidelberg (2005). https://doi.org/10.1007/11427186_29
17. Moura, A.V., Pereira, R.A., Souza, C.C.: Scheduling activities at oil wells with resource displacement. Int. Trans. Oper. Res. **15**, 659–683 (2008)
18. Cunha, V., Santos, I., Pessoa, L., Hamacher, S.: An ILS heuristic for the ship scheduling problem: application in the oil industry. Int. Trans. Oper. Res. **27**, 197–218 (2020)

Extending *Maxima* Capabilities
for Performing Elementary
Matrix Operations

Karina F. M. Castillo-Labán$^{(\boxtimes)}$ and Robert Ipanaqué-Chero

Universidad Nacional de Piura, Urb. Miraflores s/n, Castilla, Piura, Peru
karina17121995@gmail.com, ripanaquec@unp.edu.pe,
http://www.unp.edu.pe/pers/ripanaque

Abstract. The elementary operations on matrices are of great importance and utility since they are applied in obtaining the so-called equivalent matrices, which are related to each other in many important attributes. In this paper we describe a new package for dealing with the elementary matrix operations of a given matrix. The package, developed by the authors in the freeware program *Maxima*, version 5.43.2, incorporates two commands which have a fairly intuitive syntax and allow the user indicating one or more elementary operations together so that they are applied to the rows (or columns) of a given matrix. Such features make it more effective and friendly than the default built-in commands in *Maxima* to perform such operations. In addition, our outputs are consistent with *Maxima*'s notation. Several illustrative examples, aimed to show the good performance of the package, are also given.

Keywords: Elementary matrix operations · Row operations · Column operations · Maxima

1 Introduction

In addition to the usual operations defined in the $\mathcal{M}_{m \times n}$ set: addition, multiplication of a scalar, multiplication, and transposition, there are operations that are performed between the elements of some rows or columns of a matrix, called elementary operations. These operations are of great importance and utility because the application of one or more of them allows obtaining certain matrices called equivalent matrices. These matrices, due to their characteristics, are related to each other in many important attributes such as: range, determinant, solution of a system of linear equations and linear combination. And, of course, when considering an augmented matrix, they allow us to obtain the inverse of said matrix [8]. Regarding the inverse of a matrix, there is a very similar method and, more importantly, a "spirit" of methodology that also uses elementary matrix operations on rows and columns, but specializes in certain types of matrices [6,7].

© Springer Nature Switzerland AG 2020
O. Gervasi et al. (Eds.): ICCSA 2020, LNCS 12249, pp. 406–420, 2020.
https://doi.org/10.1007/978-3-030-58799-4_30

Calculating elementary matrix operations can be tedious when they are of order $m \times n$, such that $m, n > 3$. Furthermore, there is a growing concern to develop commands (in Maple [10], Mathematica [11], MatLab [9], Maxima [4], etc.), applications (in Java [1,5]) and programs (Linear Algebra Decoded [12]) specialized in automating such calculation.

What has been expressed in the previous paragraphs, the fact that *Maxima* includes four commands (with two well-marked flaws: the syntax is not intuitive and they do not allow to indicate several operations together) and the fact that the *Maxima* is distributed under the GPL [14] license has been reason enough to choose it as the programming language in which to develop a new package in *Maxima*. This package, which dealing with the elementary matrix operations of a given matrix, is described in this paper. The package, developed by the authors in the freeware program *Maxima*, version 5.43.2, incorporates two commands which have a fairly intuitive syntax and allow the user indicating one or more elementary operations together so that they are applied to the rows (or columns) of a given matrix. The outputs obtained with the built-in commands in the new package are consistent with the *Maxima* syntax. Several illustrative examples will be discussed later on in order to show the good performance of the package.

The rest of this paper is organized as follows: Sect. 2 outlines concepts related to elementary operations with the rows (columns) of a given matrix. Then, Sect. 3 describes the main standard *Maxima* tools for performing elementary matrix operations. And Sect. 4 describes the new package. Finally, the conclusions are in Sect. 5.

2 Preliminaries

When making calculations with matrices (calculation of inverse, determinant, range, etc.), the operations of the following type are especially important:

1. Swap two rows (or columns).
2. Add a multiple of another row (or column) to one row(or column).
3. Multiply a row (or column) by a non-zero scalar.

They are simple operations, which conserve certain quantities associated with the matrix in which they are operated (example: the range) or modify them in a controlled way (example: the determinant).

It will be useful to use adapted notations to describe the elementary operations that are done on rows.

1. Swap rows i and j:
$$R_i \leftrightarrow R_j$$
(remark: it will be used only for $i \neq j$).
2. Add to row i row j multiplied by scalar λ:
$$R_i + \lambda R_j \rightarrow R_i$$
(remark: it will be used only for $i \neq j$).

3. Multiply row i by $\lambda \neq 0$:

$$\lambda R_i \rightarrow R_i .$$

Changing the R by the C in previous list, the notations to describe the elementary operations that are done on columns are obtained.

3   Standard *Maxima* Tools for Elementary Matrix Operations

Maxima includes the `linearalgebra` package, one of whose commands is `rowop` (similarly includes the `columnop` command) [4]. The syntax for this command is:

$$\text{rowop}(M, i, j, \lambda). \tag{1}$$

And it works like this: if M is a matrix, it returns the resulting matrix after doing the operation:

$$R_i - \lambda \cdot R_j \rightarrow R_i ; \tag{2}$$

if M does not have a row i or j, it returns an error message.

Example 1. Find A^{-1} to

$$A = \begin{pmatrix} 3\,2\,1 \\ 4\,5\,2 \\ 2\,1\,4 \end{pmatrix},$$

using the method of elementary operations on rows [2] with the assistance of *Maxima*.

Solution 1. Let's form the matrix $(A|I)$ and stored it in the variable AI:

(%i1) AI:matrix([3,2,1,1,0,0],[4,5,2,0,1,0],[2,1,4,0,0,1])$

Now, let's do:

Operation. $\frac{1}{3}R_1 \rightarrow R_1$.

(%i2) AI:rowop(AI,1,1,2/3);

(%o2)

$$\begin{pmatrix} 1 & \frac{2}{3} & \frac{1}{3} & \frac{1}{3} & 0 & 0 \\ 4 & 5 & 2 & 0 & 1 & 0 \\ 2 & 1 & 4 & 0 & 0 & 1 \end{pmatrix}$$

Operation. $R_2 - 4F_1 \rightarrow R_2$.

(%i3) AI:rowop(AI,2,1,4);

(%o3)

$$\begin{pmatrix} 1 & \frac{2}{3} & \frac{1}{3} & \frac{1}{3} & 0 & 0 \\ 0 & \frac{7}{3} & \frac{2}{3} & -\frac{4}{3} & 1 & 0 \\ 2 & 1 & 4 & 0 & 0 & 1 \end{pmatrix}$$

Operation. $R_3 - 2F_1 \rightarrow R_3$.

(%i4) AI:rowop(AI,3,1,2);

(%o4)

$$\begin{pmatrix} 1 & \frac{2}{3} & \frac{1}{3} & \frac{1}{3} & 0 & 0 \\ 0 & \frac{7}{3} & \frac{2}{3} & -\frac{4}{3} & 1 & 0 \\ 0 & -\frac{1}{3} & \frac{10}{3} & -\frac{2}{3} & 0 & 1 \end{pmatrix}$$

Operation. $R_1 - \frac{2}{7}R_2 \rightarrow R_1$.

(%i5) AI:rowop(AI,1,2,2/7);

(%o5)

$$\begin{pmatrix} 1 & 0 & \frac{1}{7} & \frac{5}{7} & -\frac{2}{7} & 0 \\ 0 & \frac{7}{3} & \frac{2}{3} & -\frac{4}{3} & 1 & 0 \\ 0 & -\frac{1}{3} & \frac{10}{3} & -\frac{2}{3} & 0 & 1 \end{pmatrix}$$

Operation. $\frac{3}{7}R_2 \rightarrow R_2$.

(%i6) AI:rowop(AI,2,2,4/7);

(%o6)

$$\begin{pmatrix} 1 & 0 & \frac{1}{7} & \frac{5}{7} & -\frac{2}{7} & 0 \\ 0 & 1 & \frac{2}{7} & -\frac{4}{7} & \frac{3}{7} & 0 \\ 0 & -\frac{1}{3} & \frac{10}{3} & -\frac{2}{3} & 0 & 1 \end{pmatrix}$$

Operation. $R_3 + \frac{1}{3}R_2 \rightarrow R_3$.

(%i7) AI:rowop(AI,3,2,-1/3);

(%o7)

$$\begin{pmatrix} 1 & 0 & \frac{1}{7} & \frac{5}{7} & -\frac{2}{7} & 0 \\ 0 & 1 & \frac{2}{7} & -\frac{4}{7} & \frac{3}{7} & 0 \\ 0 & 0 & \frac{24}{7} & -\frac{6}{7} & \frac{1}{7} & 1 \end{pmatrix}$$

Operation. $R_1 - \frac{1}{24}R_3 \rightarrow R_1$.

(%i8) AI:rowop(AI,1,3,1/24);

(%o8)

$$\begin{pmatrix} 1 & 0 & 0 & \frac{3}{4} & -\frac{7}{24} & -\frac{1}{24} \\ 0 & 1 & \frac{2}{7} & -\frac{4}{7} & \frac{3}{7} & 0 \\ 0 & 0 & \frac{24}{7} & -\frac{6}{7} & \frac{1}{7} & 1 \end{pmatrix}$$

Operation. $R_2 - \frac{1}{12}R_3 \rightarrow R_2$.

(%i9) AI:rowop(AI,2,3,1/12);

(%o9)

$$\begin{pmatrix} 1 & 0 & 0 & \frac{3}{4} & -\frac{7}{24} & -\frac{1}{24} \\ 0 & 1 & 0 & -\frac{1}{2} & \frac{5}{12} & -\frac{1}{12} \\ 0 & 0 & \frac{24}{7} & -\frac{6}{7} & \frac{1}{7} & 1 \end{pmatrix}$$

Operation. $\frac{7}{24}R_3 \rightarrow R_3$.

(%i10) `AI:rowop(AI,3,3,17/24);`

(%o10)
$$
\begin{pmatrix}
1 & 0 & 0 & \frac{3}{4} & -\frac{7}{24} & -\frac{1}{24} \\
0 & 1 & 0 & -\frac{1}{2} & \frac{5}{12} & -\frac{1}{12} \\
0 & 0 & 1 & -\frac{1}{4} & \frac{1}{24} & \frac{7}{24}
\end{pmatrix}
$$

Therefore,

$$
A^{-1} =
\begin{pmatrix}
\frac{3}{4} & -\frac{7}{24} & -\frac{1}{24} \\
-\frac{1}{2} & \frac{5}{12} & -\frac{1}{12} \\
-\frac{1}{4} & \frac{1}{24} & \frac{7}{24}
\end{pmatrix} .
$$

□

Example 2. Find A^{-1} to

$$
A =
\begin{pmatrix}
4 & -2 & 5 \\
3 & 0 & 2 \\
1 & 2 & -1
\end{pmatrix} ,
$$

using the method of elementary operations on rows with the assistance of *Maxima*.

Solution 2. Let's form the matrix $(A|I)$ and stored it in the variable AI:

(%i11) `AI:matrix([4,-2,5,1,0,0],[3,0,2,0,1,0],[1,2,-1,0,0,1])$`

Now, let's do:

Operation. $R_1 \leftrightarrow R_3$.

?

The `rowop` command does not allow this operation.   □

Remark 1. The `rowop` syntax is not intuitive, since (1) and (2) are not very similar.

Remark 2. The transformations $\lambda \cdot R_i \rightarrow R_i$ must be expressed in the form:

$$
R_i - (1 - \lambda)R_i \rightarrow R_i ,
$$

as seen in (%i2),(%i6) y (%i10).

Remark 3. The `rowop` command does not allow indicating three operations together, for (%i1), until obtaining an output like the one obtained in (%o4). For this you could think of using an anonymous function and the command `lreduce` in the form:

(%i12) `fa:lambda([x,y],apply(rowop,cons(x,y)))$`
(%i13) `lreduce(fa,[AI,[2,1,4/3],[3,1,2/3],[1,1,2/3]]);`

(%o13)

$$\begin{pmatrix} 1 & \frac{2}{3} & \frac{1}{3} & \frac{1}{3} & 0 & 0 \\ 0 & \frac{7}{3} & \frac{2}{3} & -\frac{4}{3} & 1 & 0 \\ 0 & -\frac{1}{3} & \frac{10}{3} & -\frac{2}{3} & 0 & 1 \end{pmatrix} .$$

However, this requires changing the order of operations and additional knowledge about *Maxima*.

Remark 4. The command `rowop` does not allow to indicate the interchange of rows for (%i11). For this, the `rowswap` command can be used:

(%i14) rowswap(AI,3,1);

(%o14)

$$\begin{pmatrix} 1 & 2 & -1 & 0 & 0 & 1 \\ 3 & 0 & 2 & 0 & 1 & 0 \\ 4 & -2 & 5 & 1 & 0 & 0 \end{pmatrix}$$

Although, this requires the use of another command.

The built-in `columnop` and `columnswap` commands present the same difficulties.

4   The Package `matop`: Some Illustrative Examples

As shown in the previous section, the `rowop` command (just like the `columnop` command) has certain limitations. However, this does not happen with the commands incorporated in the package developed by the authors:

(%i15) load(matop)$

which includes the commands:

$$\texttt{rowoper}(M, R[i] + \lambda * R[j] -> R[i]) .$$

$$\texttt{rowoper}(M, R[i] <-> R[j]) .$$

and

$$\texttt{columnoper}(M, C[i] + \lambda * C[j] -> C[i]) .$$

$$\texttt{columnoper}(M, C[i] <-> C[j]) .$$

to perform elementary operations on the rows and columns of a given matrix, M. For both commands it is possible to indicate several operations at the same time.

Example 3. Find A^{-1} to

$$A = \begin{pmatrix} 3 & 2 & 1 \\ 4 & 5 & 2 \\ 2 & 1 & 4 \end{pmatrix} ,$$

using the method of elementary operations on rows [2] with the assistance of the new *Maxima* package.

Solution 3. Let's form the matrix $(A|I)$ and stored it in the variable AI:

(%i16) AI:matrix([3,2,1,1,0,0],[4,5,2,0,1,0],[2,1,4,0,0,1])$

Now, let's do:

Operations. $\frac{1}{3}R_1 \to R_1$, $R_2 - \frac{4}{3}R_1 \to R_2$, $R_3 - \frac{2}{3}R_1 \to R_3$.

(%i17) rowoper(AI,
             1/3*R[1]->R[1],R[2]-4/3*R[1]->R[2],R[3]-2/3*R[1]->R[3]);

(%o17)
$$\begin{pmatrix} 1 & \frac{2}{3} & \frac{1}{3} & \frac{1}{3} & 0 & 0 \\ 0 & \frac{7}{3} & \frac{2}{3} & -\frac{4}{3} & 1 & 0 \\ 0 & -\frac{1}{3} & \frac{10}{3} & -\frac{2}{3} & 0 & 1 \end{pmatrix}$$

Operations. $R_1 - \frac{2}{7}R_2 \to R_1$, $\frac{3}{7}R_2 \to R_2$, $R_3 + \frac{1}{7}R_2 \to R_3$.

(%i18) rowoper(%,
             R[1]-2/7*R[2]->R[1],3/7*R[2]->R[2],R[3]+1/7*R[2]->R[3]);

(%o18)
$$\begin{pmatrix} 1 & 0 & \frac{1}{7} & \frac{5}{7} & -\frac{2}{7} & 0 \\ 0 & 1 & \frac{2}{7} & -\frac{4}{7} & \frac{3}{7} & 0 \\ 0 & 0 & \frac{24}{7} & -\frac{6}{7} & \frac{1}{7} & 1 \end{pmatrix}$$

Operations. $R_1 - \frac{1}{24}R_3 \to R_1$, $R_2 - \frac{1}{12}R_3 \to R_2$, $\frac{7}{24}R_3 \to R_3$.

(%i19) rowoper(%,
             R[1]-1/24*R[3]->R[1],R[2]-1/12*R[3]->R[2],7/24*R[3]->R[3]);

(%o19)
$$\begin{pmatrix} 1 & 0 & 0 & \frac{3}{4} & -\frac{7}{24} & -\frac{1}{24} \\ 0 & 1 & 0 & -\frac{1}{2} & \frac{5}{12} & -\frac{1}{12} \\ 0 & 0 & 1 & -\frac{1}{4} & \frac{1}{24} & \frac{7}{24} \end{pmatrix}$$

Therefore,

$$A^{-1} = \begin{pmatrix} \frac{3}{4} & -\frac{7}{24} & -\frac{1}{24} \\ -\frac{1}{2} & \frac{5}{12} & -\frac{1}{12} \\ -\frac{1}{4} & \frac{1}{24} & \frac{7}{24} \end{pmatrix}.$$

□

Remark 5. As seen in (%i17), (%i18) and (%i19) the new command `rowoper` allows indicating all the operations that allow one to pivot and zero the rest elements of a column.

Example 4. Find A^{-1} to

$$A = \begin{pmatrix} 4 & -2 & 5 \\ 3 & 0 & 2 \\ 1 & 2 & -1 \end{pmatrix},$$

using the method of elementary operations on rows with the assistance of the new *Maxima* package.

Solution 4. Let's form the matrix $(A|I)$ and stored it in the variable AI:

(%i20) AI:matrix([4,-2,5,1,0,0],[3,0,2,0,1,0],[1,2,-1,0,0,1])$

Now, let's do:

Operations. $\{R_1 \leftrightarrow R_3, R_2 - 3R_1 \to R_2, R_3 - 4R_1 \to R_3\}$,
$\{R_1 + \frac{1}{3}R_2 \to R_1, -\frac{1}{6}R_2 \to R_2, R_3 - \frac{5}{3}R_2 \to R_3\}$,
$\{R_1 - R_3 \to R_1, R_2 + \frac{5}{4}R_3 \to R_2, \frac{3}{2}R_3 \to R_3\}$.

(%i21) rowoper(AI,
        [R[1]<->R[3],R[2]-3*R[1]->R[2],R[3]-4*R[1]->R[3]],
        [R[1]+1/3*R[2]->R[1],-1/6*R[2]->R[2],R[3]-5/3*R[2]->R[3]],
        [R[1]-R[3]->R[1],R[2]+5/4*R[3]->R[2],3/2*R[3]->R[3]]
        );

(%o21)

$$\left[\begin{pmatrix} 1 & 2 & -1 & 0 & 0 & 1 \\ 0 & -6 & 5 & 0 & 1 & -3 \\ 0 & -10 & 9 & 1 & 0 & -4 \end{pmatrix}, \begin{pmatrix} 1 & 0 & \frac{2}{3} & 0 & \frac{1}{3} & 0 \\ 0 & 1 & -\frac{5}{6} & 0 & -\frac{1}{6} & \frac{1}{2} \\ 0 & 0 & \frac{2}{3} & 1 & -\frac{5}{3} & 1 \end{pmatrix}, \begin{pmatrix} 1 & 0 & 0 & -1 & 2 & -1 \\ 0 & 1 & 0 & \frac{5}{4} & -\frac{9}{4} & \frac{7}{4} \\ 0 & 0 & 1 & \frac{3}{2} & -\frac{5}{2} & \frac{3}{2} \end{pmatrix} \right]$$

Therefore,

$$A^{-1} = \begin{pmatrix} -1 & 2 & -1 \\ \frac{5}{4} & -\frac{9}{4} & \frac{7}{4} \\ \frac{3}{2} & -\frac{5}{2} & \frac{3}{2} \end{pmatrix}.$$

Remark 6. As seen in (%i21) the new command rowoper also allows us to indicate all the elementary operations necessary to obtain the inverse of a given matrix.

□

Example 5. Find A^{-1} to

$$A = \begin{pmatrix} 4 & -2 & 5 \\ 3 & 0 & 2 \\ 1 & 2 & -1 \end{pmatrix},$$

using the method of elementary operations on columns with the assistance of the new *Maxima* package.

Solution 5. Let's form the matrix $\begin{pmatrix} I \\ A \end{pmatrix}$ and stored it in the variable IA:

(%i22) IA:matrix(
        [1,0,0],[0,1,0],[0,0,1],
        [3,2,1],[4,5,2],[2,1,4]
)$

Now, let's do:

Operations. $\{\frac{1}{3}C_1 \rightarrow C_1, C_2 - \frac{2}{3}C_1 \rightarrow C_2, C_3 - \frac{1}{3}C_1 \rightarrow C_3\}$,
$\{C_1 - \frac{4}{7}C_2 \rightarrow C_1, \frac{3}{7}C_2 \rightarrow C_2, C_3 - \frac{2}{7}C_2 \rightarrow C_3\}$,
$\{C_1 - \frac{1}{4}C_3 \rightarrow C_1, C_2 + \frac{1}{24}C_3 \rightarrow C_2, \frac{7}{24}C_3 \rightarrow C_3\}$.

```
(%i23)  columnoper(IA,
          [ 1/3*C[1]->C[1],C[2]-2/3*C[1]->C[2],C[3]-1/3*C[1]->C[3] ],
          [ C[1]-4/7*C[2]->C[1],3/7*C[2]->C[2],C[3]-2/7*C[2]->C[3] ],
          [ C[1]-1/4*C[3]->C[1],C[2]+1/24*C[3]->C[2],7/24*C[3]->C[3] ]
          );
```

(%o23)

$$
\left[\begin{pmatrix} \frac{1}{3} & -\frac{2}{3} & -\frac{1}{3} \\ 0 & 1 & 0 \\ 0 & 0 & 1 \\ 1 & 0 & 0 \\ \frac{4}{3} & \frac{7}{3} & \frac{2}{3} \\ \frac{2}{3} & -\frac{1}{3} & \frac{10}{3} \end{pmatrix}, \begin{pmatrix} \frac{5}{7} & -\frac{2}{7} & -\frac{1}{7} \\ -\frac{4}{7} & \frac{3}{7} & -\frac{2}{7} \\ 0 & 0 & 1 \\ 1 & 0 & 0 \\ 0 & 1 & 0 \\ \frac{6}{7} & -\frac{1}{7} & \frac{24}{7} \end{pmatrix}, \begin{pmatrix} \frac{3}{4} & -\frac{7}{24} & -\frac{1}{24} \\ -\frac{1}{2} & \frac{5}{12} & -\frac{1}{12} \\ -\frac{1}{4} & \frac{1}{24} & \frac{7}{24} \\ 1 & 0 & 0 \\ 0 & 1 & 0 \\ 0 & 0 & 1 \end{pmatrix} \right]
$$

Therefore,

$$
A^{-1} = \begin{pmatrix} \frac{3}{4} & -\frac{7}{24} & -\frac{1}{24} \\ -\frac{1}{2} & \frac{5}{12} & -\frac{1}{12} \\ -\frac{1}{4} & \frac{1}{24} & \frac{7}{24} \end{pmatrix}.
$$

Remark 7. As seen in (%i23) the new command `columnoper` has the same functionality as `rowoper`.

□

Example 6. Solve the system of linear equations

$$
\begin{cases} 23.4x_1 - 45.8x_2 + 43.7x_3 = -87.2 \\ 86.4x_1 + 12.3x_2 - 56.9x_3 = -14.5 \\ 93.6x_1 - 50.7x_2 + 12.6x_3 = -44.4 \end{cases},
$$

using the method of elementary operations on rows with the assistance of the new *Maxima* package.

Solution 6. Let's form the matrix $(A|B)$ and stored it in the variable M:

```
(%i24) M:matrix(
          [23.4,-45.8,43.7,87.2],
          [86.4,12.3,-56.9,14.5],
          [93.6,-50.7,12.6,44.4]
       )$
```

Below are indicated and then the respective elementary operations are performed on the rows of M.

Operation. $\{\frac{R_1}{M_{1,1}} \rightarrow R_1, R_2 - \frac{M_{2,1}}{M_{1,1}}R_1 \rightarrow R_2, R_3 - \frac{M_{3,1}}{M_{1,1}} \rightarrow R_3\}$.

(%i25) M:rowoper(M,
        R[1]/M[1,1]->R[1],R[2]-M[2,1]/M[1,1]*R[1]->R[2],
        R[3]-M[3,1]/M[1,1]*R[1]->R[3]);

(%o25)

$$\begin{pmatrix} 1.0 & -1.957264957264954 & 1.867521367521363 & 3.726495726495727 \\ 0.0 & 181.4076923076923 & -218.2538461538462 & -307.4692307692308 \\ 0.0 & 132.5 & -162.2 & -304.4 \end{pmatrix}$$

Operation. $\{R_1 - \frac{M_{1,2}}{M_{2,2}} R_2 \to R_1, \frac{R_2}{M_{2,2}} \to R_2, R_3 - \frac{M_{3,2}}{M_{2,2}} R_2 \to R_3\}$.

(%i26) M:rowoper(M,
        R[1]-M[1,2]/M[2,2]*R[2]->R[1],R[2]/M[2,2]->R[2],
        R[3]-M[3,2]/M[2,2]*R[2]->R[3]);

(%o26)

$$\begin{pmatrix} 1.0 & -2.22044604925031310^{-16} & -0.487288395124549 & 0.4091129674388871 \\ 0.0 & 1.0 & -1.203112411482863 & -1.694907348513766 \\ 0.0 & 0.0 & -2.787605478522664 & -79.82477632192686 \end{pmatrix}$$

Operation. $\{R_1 - \frac{M_{1,3}}{M_{3,3}} R_3 \to R_1, R_2 - \frac{M_{2,3}}{M_{3,3}} R_3 \to R_2, \frac{R_3}{M_{3,3}} \to R_3\}$.

(%i27) M:rowoper(M,
        R[1]-M[1,3]/M[3,3]*R[3]->R[1],R[2]-M[2,3]/M[3,3]*R[3]->R[2],
        R[3]/M[3,3]->R[3]);

(%o27)

$$\begin{pmatrix} 1.0 & -2.22044604925031310^{-16} & 0.0 & 14.36291218500331 \\ 0.0 & 1.0 & 0.0 & 32.75694743391065 \\ 0.0 & 0.0 & 1.0 & 28.63560749070965 \end{pmatrix}$$

Therefore,

$$x_1 \approx 14.36291218500331$$
$$x_2 \approx 32.75694743391065$$
$$x_3 \approx 28.63560749070965$$

Remark 8. In cases like this it is not possible to indicate all the operations because the variable that stores the matrix must be updated immediately after operating a whole row of the matrix.

$\square$

Example 7. Microsoft manufactures electronic circuit boards for PCs. Each fax-modem, data compression, and sound synthesizer board requires a certain amount of machine time to insert, solder, and assemble the chips. These data (in minutes) together with the number of minutes of machine time available for each operation are summarized in Table 1. The profit for each board is $20, $15 and $10 respectively. How many boards of each type must be manufactured to maximize Microsoft's profit, if at least 300 data understanding boards are required to be produced?

Table 1. Data summary for the example 7.

	Fax-modem	Data compression	Sound synthesizer	Available
Chip insertion	4.00	2.50	2.00	5000
Welding	2.00	6.00	4.00	4200
Assemble and tested	1.50	2.00	2.50	2400

Solution 7. First we make the model:

X_1: Fax-Modem boards,
X_2: Data Comprehension dashboards,
X_3: Boards Sound synthesizers.

Maximize:

$$Z = 20X_1 + 15X_2 + 10X_3,$$

Subject to:

$$4X_1 + 2.5X_2 + 2X_3 \leq 5000,$$
$$2X_1 + 6X_2 + 4X_3 \leq 4200,$$
$$1.5X_1 + 2X_2 + 2.5X_3 \leq 2400,$$
$$X_2 \geq 300,$$
$$X_i \geq 0.$$

After standardizing we obtain:

$$4X_1 + 2.5X_2 + 2X_3 + S_1 = 5000,$$
$$2X_1 + 6X_2 + 4X_3 + S_2 = 4200,$$
$$1.5X_1 + 2X_2 + 2.5X_3 + S_3 = 2400,$$
$$X_2 - S_4 + A_1 = 300.$$

Now we proceed to build the first scheme associated with the model as shown in Fig. 1. Let A be the matrix associated with the model, note that $Z = \left(B^\top \cdot A\right)^\top$.

Since $-M - 15$ is the smallest value of $Z - C$ and $300/1.0 = 300.0$ is the smallest quotient of those obtained by dividing, one by one, the elements of D and X_2, then 1.0 is the pivot. And A_1 is replaced by X_2.

Now, we define the matrix associated with example 7 and then perform the elementary operations on the respective rows, taking the pivot as a reference.

$$\begin{array}{c} B \diagdown \overset{C \to}{} \\ \downarrow \end{array} \quad \begin{array}{ccccccccc} & & 20 & 15 & 10 & 0 & 0 & 0 & 0 & -M \\ & D & X_1 & X_2 & X_3 & S_1 & S_2 & S_3 & S_4 & A \end{array}$$

$$\begin{array}{cc} 0 & S_1 \\ 0 & S_2 \\ 0 & S_3 \\ -M & A_1 \end{array} \left(\begin{array}{ccccccccc} 5000 & 4.0 & 2.5 & 2.0 & 1 & 0 & 0 & 0 & 0 \\ 4200 & 2.0 & 6.0 & 4.0 & 0 & 1 & 0 & 0 & 0 \\ 2400 & 1.5 & 2.0 & 2.5 & 0 & 0 & 1 & 0 & 0 \\ 300 & 0.0 & 1.0 & 0.0 & 0 & 0 & 0 & -1 & 1 \end{array} \right)$$

$$\begin{array}{ccccccccc} Z \to -300M & 0 & -M & 0 & 0 & 0 & 0 & M & -M \\ Z - C \to & -20-M & -15-10 & 0 & 0 & 0 & 0 & M & 0 \\ & \uparrow \end{array}$$

Fig. 1. First scheme associated with the model of the example 7.

(%i28) fpprintprec:6$

(%i29) A: matrix(
        [5000, 4.0, 2.5, 2.0, 1, 0, 0, 0, 0],
        [4200, 2.0, 6.0, 4.0, 0, 1, 0, 0, 0],
        [2400, 1.5, 2.0, 2.5, 0, 0, 1, 0, 0],
        [300, 0.0, 1.0, 0.0, 0, 0, 0, -1, 1]
)$

Operation. $\{R_1 - \frac{A_{1,3}}{A_{4,3}} R_4 \to R_1, R_2 - \frac{A_{2,3}}{A_{4,3}} R_4 \to R_2, R_3 - \frac{A_{3,3}}{A_{4,3}} \to R_3\}$.

(%i30) A:rowoper(A,
        R[1]-A[1,3]/A[4,3]*R[4]->R[1],
        R[2]-A[2,3]/A[4,3]*R[4]->R[2],
        R[3]-A[3,3]/A[4,3]*R[4]->R[3]);

(%o30)

$$\begin{pmatrix} 4250.0 & 4.0 & 0.0 & 2.0 & 1 & 0 & 0 & 2.5 & -2.5 \\ 2400.0 & 2.0 & 0.0 & 4.0 & 0 & 1 & 0 & 6.0 & -6.0 \\ 1800.0 & 1.5 & 0.0 & 2.5 & 0 & 0 & 1 & 2.0 & -2.0 \\ 300 & 0.0 & 1.0 & 0.0 & 0 & 0 & 0 & -1 & 1 \end{pmatrix}$$

Figure 2 shows the update of the first scheme associated with the example 7 (second scheme).

Since -20 is the smallest value of $Z - C$ and $4250.0/4.0 = 1062.5$ is the smallest quotient obtained after dividing, one by one, the elements of D and $X1$ (note that divisions by zero are discarded), then 4.0 is the new pivot and S_1 is replaced by X_1.

Next, on the equivalent matrix (previously obtained), we perform the respective elementary operations on the equivalent matrix associated with the model.

Operation. $\{\frac{R_1}{A_{2,2}} \to R_1, R_2 - \frac{A_{2,2}}{A_{1,2}} R_1 \to R_2, R_3 - \frac{A_{3,2}}{A_{1,2}} R_1 \to R_3\}$.

(%i31) A:rowoper(A,
        R[1]/A[1,2]->R[1],

$$
\begin{array}{c}
B\!\diagdown\!\!\begin{array}{c}C\to\end{array}\\ \downarrow\end{array}
\quad
\begin{array}{c|ccccccccc}
 & & 20 & 15 & 10 & 0 & 0 & 0 & 0 & -M\\
 & D & X_1 & X_2 & X_3 & S_1 & S_2 & S_3 & S_4 & A\\
\hline
0\ \ S_1 & 4250.0 & 4.0 & 0.0 & 2.0 & 1 & 0 & 0 & 2.5 & -2.5\\
0\ \ S_2 & 2400.0 & 2.0 & 0.0 & 4.0 & 0 & 1 & 0 & 6.0 & -6.0\\
0\ \ S_3 & 1800.0 & 1.5 & 0.0 & 2.5 & 0 & 0 & 1 & 2.0 & -2.0\\
15\ X_2 & 300 & 0.0 & 1.0 & 0.0 & 0 & 0 & 0 & -1 & 1\\
\end{array}
$$

$Z\to 300\times 15$	0	15	0	0	0	0	-15	15
$Z-C\to$	-20	0	-10	0	0	0	-15	$15+M$
	$\uparrow$							

Fig. 2. Second scheme associated with the model of the example 7.

```
R[2]-A[2,2]/A[1,2]*R[1]->R[2],
R[3]-A[3,2]/A[1,2]*R[1]->R[3]);
```

(%o31)

$$
\begin{pmatrix}
1062.5 & 1.0 & 0.0 & 0.5 & 0.25 & 0 & 0 & 0.625 & -0.625\\
275.0 & 0.0 & 0.0 & 3.0 & -0.5 & 1 & 0 & 4.75 & -4.75\\
206.25 & 0.0 & 0.0 & 1.75 & -0.375 & 0 & 1 & 1.0625 & -1.0625\\
300 & 0.0 & 1.0 & 0.0 & 0 & 0 & 0 & -1 & 1
\end{pmatrix}
$$

Figure 3 shows the update of the second scheme associated with the example 7 (third scheme).

$$
\begin{array}{c}
B\!\diagdown\!\!\begin{array}{c}C\to\end{array}\\ \downarrow\end{array}
\quad
\begin{array}{c|ccccccccc}
 & & 20 & 15 & 10 & 0 & 0 & 0 & 0 & -M\\
 & D & X_1 & X_2 & X_3 & S_1 & S_2 & S_3 & S_4 & A\\
\hline
20\ X_1 & 1062.5 & 1.0 & 0.0 & 0.5 & 0.25 & 0 & 0 & 0.625 & -0.625\\
0\ \ S_2 & 275.0 & 0.0 & 0.0 & 3.0 & -0.5 & 1 & 0 & 4.75 & -4.75\\
0\ \ S_3 & 206.25 & 0.0 & 0.0 & 1.75 & -0.375 & 0 & 1 & 1.0625 & -1.0625\\
15\ X_2 & 300 & 0.0 & 1.0 & 0.0 & 0 & 0 & 0 & -1 & 1\\
\end{array}
$$

$Z\to$ 25750	20	15	5	0	0	0	$-5/2$	$5/2$
$Z-C\to$	0	0	-10	0	0	0	$-5/2$	$5/2+M$
							$\uparrow$	

Fig. 3. Third scheme associated with the model of the example 7.

Since $-\frac{5}{2}$ is the smallest value of $Z-C$ and $275.0/4.75 \approx 57.89$ is the smallest quotient obtained after dividing, one by one, the elements of D and $X1$ (note that negative values are discarded), then 4.75 is the new pivot and S_2 is replaced by S_4.

Next, on the equivalent matrix (previously obtained), we perform the respective elementary operations on the equivalent matrix associated with the model.

Operation. $\{R_1 - \frac{A_{1,9}}{A_{2,9}}R_2 \to R_1, \frac{R_2}{A_{2,9}} \to R_2, R_3 - \frac{A_{3,9}}{A_{2,9}}R_2 \to R_3, R_4 - \frac{A_{4,9}}{A_{2,9}}R_2 \to R_4\}$.

(%i31) A:rowoper(A,
        R[1]-A[1,9]/A[2,9]*R[2]->R[1],
        R[2]/A[2,9]->R[2],
        R[3]-A[3,9]/A[2,9]*R[2]->R[3],
        R[4]-A[4,9]/A[2,9]*R[4]->R[4]);

(%o31)

$$
\begin{pmatrix}
1026.32 & 1.0 & 0.0 & 0.105263 & 0.315789 & -0.131579 & 0 & 0.0 & 0.0 \\
-57.8947 & 0.0 & 0.0 & -0.631579 & 0.105263 & -0.210526 & 0 & -1.0 & 1.0 \\
144.737 & 0.0 & 0.0 & 1.07895 & -0.263158 & -0.223684 & 1 & 0.0 & 0.0 \\
357.895 & 0.0 & 1.0 & 0.631579 & -0.105263 & 0.210526 & 0 & 0.0 & 0.0
\end{pmatrix}
$$

Figure 4 shows the update of the third scheme associated with the example 7 (fourth scheme).

$B \downarrow$	$C \rightarrow$		20	15	10	0	0	0	0	$-M$
		D	X_1	X_2	X_3	S_1	S_2	S_3	S_4	A
20	X_1	1026.32	1.0	0.0	0.105263	0.315789	-0.131579	0	0.0	0.0
0	S_4	-57.8947	0.0	0.0	-0.631579	0.105263	-0.210526	0	-1.0	1.0
0	S_3	144.737	0.0	0.0	1.07895	-0.263158	-0.223684	1	0.0	0.0
15	X_2	357.895	0.0	1.0	0.631579	-0.105263	0.210526	0	0.0	0.0
$Z \rightarrow$		25894.74	20	15	11.579	3.947	0.526	0	0	0
$Z-C \rightarrow$			0	0	1.579	3.947	0.526	0	0	M

Fig. 4. Fourth scheme associated with the model of the example 7.

Since $Z_i - C_i \geqq 0$, the solution is optimal. Therefore, we conclude that:

$$
\begin{aligned}
Z_{max} &\approx 25894.7\,, \\
X_1 &\approx 1026.32\,, \\
X_2 &\approx 357.895\,, \\
X_3 &= 0\,.
\end{aligned}
$$

5 Conclusions and Further Remarks

This paper presents the new matop package that incorporates intuitive rowoper and columnoper commands to perform elementary operations on the rows and columns of a given matrix, respectively. Through examples, the paper shows the ways to indicate the elementary operations. All the examples show the proper functioning of the package; as well as the consistency of the results obtained with the *Maxima* syntax. All the programming has been developed and coded by the authors in version 5.43.2 of the freeware program *Maxima* [13] using programming based on transformation rules, comparison patterns [3] and optimized code. Additionally, considering that the command can be easily implemented in other scientific programs that offer capabilities similar to those of *Maxima*, it would be interesting to do similar research on those programs in the near future.

Acknowledgements. The authors would like to thank to the reviewers for their valuable comments and suggestions.

References

1. Bogacki, P.: http://www.nibcode.com/es/algebra-lineal. Accessed 25 Apr 2020
2. Bolgov, V., et al.: Problemas de las Matemáticas Superiores I. Mir, Moscow (1983)
3. Dodier, R.: http://maxima.sourceforge.net/docs/manual/es/maxima_34.html. Accessed 10 Apr 2020
4. Dodier, R.: http://maxima.sourceforge.net/docs/manual/de/maxima_58.html. Accessed 10 Apr 2020
5. Hudelson, M.: http://www.math.wsu.edu/faculty/hudelson/linalg.html. Accessed 25 Apr 2020
6. Respondek, J.S.: Recursive numerical recipes for the high efficient inversion of the confluent Vandermonde matrices. Appl. Math. Comput. **225**, 718–730 (2013). http://dx.doi.org/10.1016/j.amc.2013.10.018
7. Respondek, J.S.: Incremental numerical recipes for the high efficient inversion of the confluent Vandermonde matrices. Comput. Math. Appl. **71**(2), 489–502 (2015). http://dx.doi.org/10.1016/j.camwa.2015.12.016
8. Larson, R., Falvo D.C.: Elementary Linear Algebra, 6th edn. CENGAGE Learning (2009)
9. Looring, T.: http://www.math.unm.edu/~loring/links/linear_s06/rowOps.html. Accessed 15 Feb 2020
10. Maple Online Help. http://www.maplesoft.com/support/help/Maple/view.aspx?path=LinearAlgebra/RowOperation. Accessed 10 Feb 2020
11. Muneeb, U.: http://mikrotechnica.wordpress.com/2014/01/31/elementary-row-operations-with-mathematica/. Accessed 25 Apr 2020
12. Nibcode Solutions. http://www.maplesoft.com/support/help/Maple/view.aspx?path=LinearAlgebra/RowOperation. Accessed 10 Jan 2020
13. Rodríguez, M.: Manual de Maxima. Maxima en SourceForge (2012)
14. Wikipedia. http://es.wikipedia.org/wiki/GNU_General_Public_License. Accessed 15 Jan 2020

A Graph-Based Clustering Analysis of the QM9 Dataset via SMILES Descriptors

Gabriel A. Pinheiro[1]([⊠]), Juarez L. F. Da Silva[2], Marinalva D. Soares[3],
and Marcos G. Quiles[3]

[1] Associate Laboratory for Computing and Applied Mathematics, National Institute
for Space Research, PO BOX 515, 12227-010 São José dos Campos, SP, Brazil
`gabriel.pinheiro@inpe.br`
[2] São Carlos Institute of Chemistry, University of São Paulo, PO Box 780, 13560-970
São Carlos, SP, Brazil
`juarez_dasilva@iqsc.usp.br`
[3] Institute of Science and Technology, Federal University of São Paulo, 12247-014
São José dos Campos, SP, Brazil
`mdiasoares@gmail.com`, `quiles@unifesp.br`

Abstract. Machine learning has become a new hot-topic in Materials
Sciences. For instance, several approaches from unsupervised and supervised learning have been applied as surrogate models to study the properties of several classes of materials. Here, we investigate, from a graph-
based clustering perspective, the Quantum QM9 dataset. This dataset is
one of the most used datasets in this scenario. Our investigation is two-
fold: 1) understand whether the QM9 samples are organized in clusters,
and 2) if the clustering structure might provide us with some insights
regarding anomalous molecules, or molecules that jeopardize the accuracy of supervised property prediction methods. Our results show that
the QM9 is indeed structured into clusters. These clusters, for instance,
might suggest better approaches for splitting the dataset when using
cross-correlation approaches in supervised learning. However, regarding
our second question, our finds indicate that the clustering structure,
obtained via Simplified Molecular Input Line Entry System (SMILES)
representation, cannot be used to filter anomalous samples in property
prediction. Thus, further investigation regarding this limitation should
be conducted in future research.

Keywords: Clustering · Graph · Quantum-chemistry

1 Introduction

Machine learning methods can be roughly divided into two main paradigms:
supervised and unsupervised [11,17]. Methods belonging to the supervised
paradigm demands a dataset in which all samples are labeled, and the objective
is the map the input features that describe each sample to its respective label. If
the labels are discrete, we have a classification problem; if labels are continuous,

© Springer Nature Switzerland AG 2020
O. Gervasi et al. (Eds.): ICCSA 2020, LNCS 12249, pp. 421–433, 2020.
https://doi.org/10.1007/978-3-030-58799-4_31

we have a regression problem [11]. On the other hand, unsupervised methods do not require labels. In this scenario, the method tries to find hidden patterns using only the available features. For instance, these methods can be used for clustering the dataset into a set of homogeneous groups; select which features are the most important to represent the samples, or even to perform anomaly detection [5].

When dealing with classification or regression problems, unsupervised methods can be adopted as a preprocessing phase [13]. These methods can be used to filter, learn new representation, and also for grouping samples into clusters. Thus, revealing how data samples are distributed, or clustered, into the feature's space. There are several clustering methods in the literature [14]. Among them, graph-based methods have some compelling features. For example, graph-based methods rely on how data samples are drawn. Samples are represented as nodes in the graph and the edges, or links, represent the similarity between samples (nodes). A graph-based method has some advantages in comparison to other clustering methods, such as the K-means [14]. For instance, the graph might represent non-trivial structures in the feature space that are beyond Gaussian structures commonly detected by the K-means [6,8,23]. Moreover, the relation between patterns, such as paths, are also easily obtained by analyzing the graph structure [23].

Here, we apply a graph-based clustering method for analyzing the QM9 dataset [20], which is one of the most used datasets in machine learning applied to materials science. The QM9 dataset provides there basic representation for each molecule: atom positions represented by the X, Y, Z coordinates (XYZ representation); the International Chemical Identifier, or InChI [20]; and the Simplified Molecular Input Line Entry System representations, or SMILES [22], which represent atoms and bonds using a string with a simple grammar. In contrast to the XYZ representation, the SMILES format does not require an optimization of the molecular structure via Density Functional Theory (DFT) or Molecular Dynamics [7,15,16] or a complete knowledge of the experimental structure, and hence, the SMILES representation has a smaller computational cost to be obtained. Here, we take the SMILES format into account to encode the molecules into a feature vector (see Sect. 2.1).

In this work, by conducting this investigation of the clustering structure of the QM9, we expect to find some insights about the QM9 molecules. Specifically, this study aims to answer the following questions:

1. Are the samples in the QM9 organized in clusters?
2. Regarding the property prediction problem, is it possible to identify molecules that jeopardize the accuracy of predictors (anomalous molecules)?

Our results, as depicted in Sect. 3, show that the molecules in the QM9 dataset are indeed organized into clusters. Thus, this structure might be used to set the cross-validation structure when dealing with supervised property predictions. However, regarding our second question, our results have not highlighted any insight on how to isolate anomalous molecules using the cluster structure of the data. Consequently, further investigation must be conducted to evaluate how to filter these molecules that expose the property prediction process.

The remain of this paper is organized as follows. Section 2 introduces the QM9 dataset and the methods taken into account in this research. Results and discussion are depicted in Sect. 3. Finally, some concluding remarks are drawn in Sect. 4.

2 Data and Methods

2.1 The QM9 Dataset

The quantum-chemistry data used in this work came from the QM9 quantum-chemistry data, a public dataset composed of ~134-kilo organic molecules made up of C, H, O, N, and F with up to twenty-nine atoms. This dataset provides 15 properties (geometric, energetic, electronic and thermodynamic) and geometries optimized via the DFT-B3LYP/6-31G($2df, p$) level of theory, as well as the SMILES, XYZ, and InChI representations for every molecule therein [20]. We used the QM9 previously processed by [10], which removed ~3k molecules that failed in the InChI consistent check [20] and reports the atomization energies of the total energies properties from the original QM9. We also eliminated 612 molecules that the RDKit[1]. Python package was not able to read. Thus, after withdrawn all these samples from the QM9, the dataset was reduced to 130 217 molecules.

2.2 Feature Extraction

Several features can be extract from each molecule. In SMILES, each molecule is represented by a string containing its atoms (atomic symbols) and bonds, hydrogen atoms may be omitted in this representation [21]. In this sense, a Propane molecule can be whether of these string CC#C or C#CC, where # is a triple bond. Thus, without an explicit loading order the algorithm can yield various strings for a particular molecule, as shown in the Propane example. Thus, a general approach to overcome this problem is the use of canonical SMILES, which guarantees uniqueness representation. Nevertheless, recent work with SMILES has explored the noncanonical SMILES as data augmentation of neural network models [1,2].

Herein, we utilize the RDKit to read the SMILES derived from the relaxed geometry and the Mordred[2] Python package for feature extraction. Mordred is open-source software with an extensive molecular descriptor calculator for cheminformatics. It computes up to 1826 topological and geometric descriptors, including those available in the RDkit [18]. As the package has integration with RDKit, it was required to convert the molecule in an RDkit molecule object to perform the descriptor extraction. Moreover, since RDKit can understand

---

[1] https://www.rdkit.org/.

[2] https://pypi.org/project/mordred/.

noncanonical and canonical SMILES as the same molecule, the canonization of the SMILES was not required. In this work, we focus only on the topological features, which are SMILES-based descriptors, covering modules as diverse as autocorrelation, atom count, Wiener index, and others.

After extracting all features of the QM9 using the Mordred package, we eliminate three subset of features: 3D descriptors (213 features), features with variance equal to zero (539 features), and features with missing value (194 features). Thus, our final dataset consists of 130217 samples with 880 features.

2.3   Graph Generation

In a graph-based method, the first step consists of generating a network (or graph) $G = (V, E)$ for a given data set [4]. The input is represented as $\mathcal{X} = \{x_1, x_2, ..., x_n\}$, in which n represents the number of samples, and x_i is the feature vector of sample i. The dataset might also provide a set of labels $\mathcal{L} = \{y_1, y_2, ...y_n\}$. When dealing with unsupervised learning, only $\mathcal{X}$ is taken into account.

To generate a graph, we define a set of nodes $V = \{v_1, v_2, ..., v_n\}$, which represent the data samples in $\mathcal{X}$, and a set of edges E representing the similarity (or relation) between pair of samples. There are several manners to define the set of edges, e.g. the $\epsilon-$cut and the $k-$NN [23]. Here we considered the $\epsilon-$cut method. Each edge $e_{ij} \in E$, which represent the relation between samples i and j is defined as follows:

$$e_{ij} = \begin{cases} 1 \text{ if } d(x_i, x_j) \leq \theta \text{ and } i \neq j \\ 0 \text{ if } d(x_i, x_j) > \theta \text{ or } i = j \end{cases} \tag{1}$$

where θ is a connection threshold that defines the minimum similarity to create the connection between samples i and j; and $d(x_i, x_j)$ is a distance function. Here we assume $d(x_i, x_j)$ to be the Euclidean distance:

$$d(x_i, x_j) = \|x_i - x_j\|. \tag{2}$$

Once the graph G is generated, the analysis of the dataset might proceed.

3   Results and Discussion

In this section we present a summary of all computer experiments conducted in this research.

3.1   Feature Space Reduction

As defined in Sect. 2.3, each pair of samples is connected by an edge whether the distance between their vector representations, x_i and x_j is shorter than a threshold θ. To compute this distance between a pair of molecules, we have taken two vector representations into account. First, we assume that each molecule is represented by all available features (880 dimensions). Second, due to the high correlation observed between several pairs of features, we perform the Principal Component Analysis (PCA) and extract the major components to represent the samples.

To evaluate these two representations, we select a subset of 5k random molecules of the QM9 and run several experiments. We found that the two approaches led to quantitatively similar results. Thus, we decided to continue our experiments using only the second approach, which consists in computing the PCA of the data. This selection is three-fold:

1. Accurate representation: with 134 components (out of 880 features), we represent 99 % of the variance of the original data representation;
2. PCA is fast to compute;
3. The generation of the graph demands the computation of $O(n^2)$ pair distances; thus, by reducing the vector representation from 880 to 134 we greatly reduce the computational cost of the process.

3.2   Cluster Structure of the QM9

With each molecule represented by its 134 principal components, we built several graphs varying the threshold. We aim to generate a graph which is sparse albeit connected. Thus, we seek for a threshold that delivers a graph with a lower number of components and with a low average degree ($\langle k \rangle$). If a large number of components are observed, the graph might be too disconnected and an evaluation of the relational structure between molecules is not allowed. Conversely, if the number of components is low, it leads to a very high ($\langle k \rangle$) creating a densely connected graph. In this scenario, the evaluation is also compromised.

To face this trade-off, we run experiments with several thresholds. Table 1 shows some results regarding the generated graph for a given threshold (θ). Specifically, we depict the number of connected components (#-Comp); size of the largest component (L−Comp); number of communities (#-Comm); size of the largest community (L-Comm); number of singletons (#-Singletons); average degree ($\langle k \rangle$). The communities were detected with the Louvain Algorithm [3], albeit similar results were found with other community detection algorithms [19]. It is worth noting that community and graph cluster represents, in the context of this study, the same type of structure: a group of densely connected nodes.

Evaluating the results in Table 1, we find that $\theta = 15$ provides a proper threshold, the number of components is low and $\langle k \rangle \approx 51$, which represents a sparse graph. Moreover, the number of singletons is also small meaning that only a set of 1326 molecules are completely isolated from all the components of the graph. With this graph we can perform a deeper evaluation of the QM9 via the graph structure.

Figure 1 depicts the size of components (a) and communities (b). We can observed that there are a few components and few communities with a large number of nodes and several small components/communities. It represents that the dataset has indeed a cluster structure with a few large homogeneous regions

Table 1. Graph measurements. Number of connected components (#-comp); size of the largest component (L−Comp); number of communities (#-Comm); size of the largest community (L-Comm); Singletons; average degree ($\langle k \rangle$) for each graph generating according to the threshold θ.

Threshold (θ)	#-Comp	L−Comp	#-Comm	L−Comm	#-Singletons	$\langle k \rangle$
1	130009	3	130009	3	129977	≈0
5	114618	16	114903	10	116952	0.27
10	42203	1539	52655	189	58809	3.95
15	2055	87922	4697	12105	1326	51.1
20	264	125957	467	30470	180	734.2
25	59	129939	86	78563	28	6104.0

and many sparse structures composed by a small number of molecules. This cluster structure might be used to address the cross-correlation technique used in supervised machine learning processes, such as in molecular property prediction. In this case, instead of randomly splitting the data set into n folds, the clusters could be taken into account to deliver this division. Thus, we can guarantee that samples from each cluster are represented in the folds, which is not guaranteed when traditional cross-validation algorithm is employed.

Interestingly, in the log-log plot shown in Fig. 1, both curves seem to be fitted by a straight line and might indicate a power-law distribution regarding the size of the components and communities. This type of distribution is widely observed in real graphs (networks) from biology to technological graphs [9].

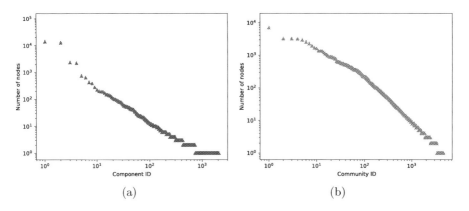

           (a)   (b)

Fig. 1. Connected components and communities of the graph build with $\theta = 15$. (a) Components; (b) Communities.

Out of 4697 clusters, only 182 represent clusters with more than 100 samples. Moreover, these 182 communities sum up 98 096 molecules of a total of 130 218.

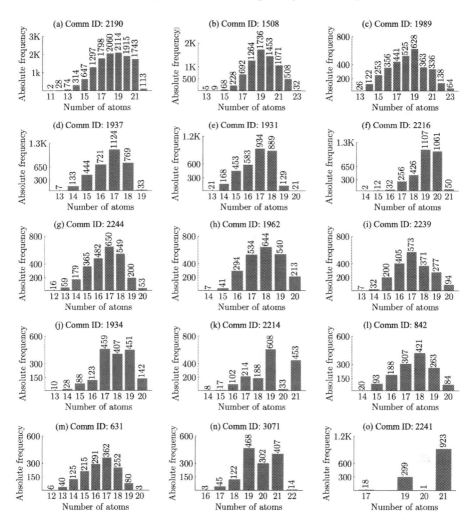

Fig. 2. Histogram of the molecule size grouped into the largest fifteen communities.

The remaining samples are clustered into sets with only a small number of samples. Figures 2 and 3 show histograms of molecule sizes and chemical species of the largest clusters. One could observe that the distribution of molecules regarding their sizes and chemical species is not completely homogeneous, which indicates that each cluster groups molecules with distinct features. For instance, by analyzing the clusters (IDs) 1937 and 2216, depicted in Figs. 2(d) and (f), and 3(d) and (f), we see that the number of atoms and the distribution of chemical species are different between clusters. The same outcome holds for the other clusters. Moreover, the larger the cluster is, the more heterogeneous is its structure. Small clusters contain more similar molecules regarding their sizes and chemical species.

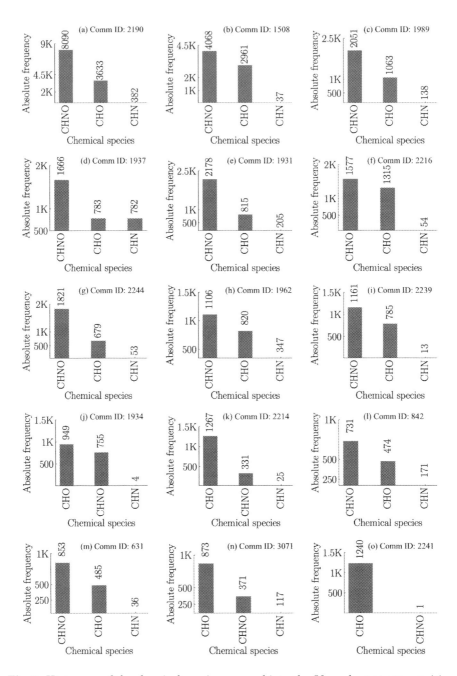

Fig. 3. Histogram of the chemical species grouped into the fifteen largest communities.

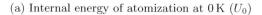

(a) Internal energy of atomization at 0 K (U_0)

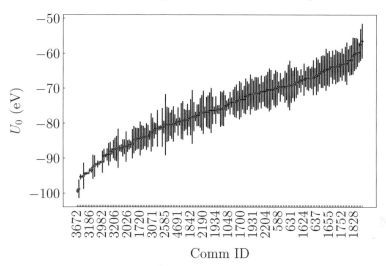

(b) Heat capacity at 298.15 K (C_v)

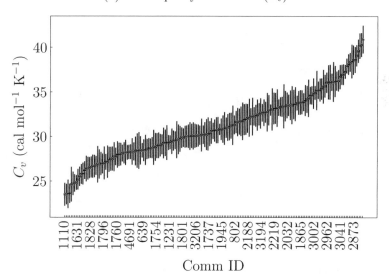

Fig. 4. Internal energy (U_0) and heat capacity C_v in each cluster. Red marker and black line represent the mean and standard deviation of the properties in each cluster, respectively. Only clusters with more than 100 samples are shown. Nevertheless, we labeled only 23 clusters IDs on the x-axis. (Color figure online)

To further investigate the cluster structures of the data, in Fig. 4, we show the mean and the standard deviation of the heat capacity at $298.15\,\mathrm{K}$ (C_v), and the internal energy of atomization at $0\,\mathrm{K}$ (U_0) for each cluster. To facilitate the visual scrutiny of the figures, we present only clusters with more than 100 samples.

For both properties, we can notice that each cluster is assigned to a specific mean value of C_v and U_0 and a lower standard deviation, which means that the clusters, indeed, group similar molecules in the same fashion as observed for molecule sizes and chemical species.

The cluster structure of the network and its relation to the property prediction problem is discussed in the following section.

3.3   Community Evaluation and Anomalous Molecules Identification

To identify the anomalous molecules regarding property prediction, we train Multilayer Perceptron neural networks (MLP) [12] over the QM9 for several available properties. After the learning phase, we evaluate each molecule of the dataset and annotate their errors. Using a cross-correlation approach [17], we select a subset of 1 % of molecules that presented the largest errors. Thus, our task here was to try to identify those molecules, named anomalous, using the clusters structure described in the last section.

Figure 5 illustrates a graph in which nodes represent communities. In this coarse-grained visualization of the whole network we can see how the communities (or clusters) are related to each other. Size of the nodes indicates the size of the communities, color represents the proportion of anomalous molecules clustered in the community. Nodes in gray color indicate that about 1 % of the samples in the community introduces a high error in the predictor. This amount corresponds to the same proportion of anomalous molecules labeled in the QM9. Thus, this outcome opposes our initial hypotheses that anomalous samples could be grouped into specific clusters (groups of high error molecules). Unfortunately, based only on our clustering analysis with SMILES descriptors, we are not able to highlight these anomalous samples and remove them from the QM9 dataset.

In Fig. 5, communities in purple color represents homogeneous communities consisting of only molecules with low error. Although it might suggest a pattern of very accurate regions of the feature space, these communities are quite small and represent just a small fraction of the molecules of the dataset.

By analyzing each community isolated, we observe that most of the small communities (size between 2 and 100), are quite homogeneous regarding the properties of their molecules. Besides, from a total of 3193 communities in this range, only 213 communities contain molecules with high error prediction (anomalous). On the other hand, when considering larger communities (with more than 100 samples), the number of communities containing anomalous molecules is 95 out of 175. This result is sound assuming that the larger the community is, the more heterogeneous is its structure. Regarding the communities holding a single node, which represents a total of 1326 communities, a set of 53 singletons represent anomalous molecules that should be further scrutinized.

Fig. 5. Graph of the community (cluster) structure. Each node represents a community. Size of the circle informs the size of the community (number of clustered molecules). Gray color point communities in which about 1% of molecules are anomalous. Purple color represents communities with absence of anomalous molecules. (Color figure online)

4   Conclusions

In this work, we investigate, from a graph-based clustering perspective, the Quantum-chemistry QM9 dataset. This study was two-fold: 1) investigate if the QM9 is organized in clusters; and 2) whether high-error molecules, regarding the property prediction problem, could be isolated into specific clusters. It would allow us to remove these noise samples from the dataset.

First, we have mapped the QM9 molecules into a graph by encoding each molecule as a feature vector representing descriptors computed from the SMILES representation. Next, we scrutinize the cluster structure of the graph considering several thresholds and by applying the Louvain Community detection algorithm. Once a suitable graph structure was obtained, several analyses were conducted, and the results indicated that: 1) the QM9 is composed of several clusters, in which the molecules are grouped according to their properties; and 2) by using this clustered structure we cannot isolate anomalous molecules that reduce the accuracy supervised models, such as the MLP network used for property prediction.

In the next step, we will evaluate other approaches for graph construction and feature representation for the molecules of the QM9 dataset. In particular, our future research will focus on how to detect and isolate anomalous molecules from a dataset composed of molecules. This task is essential to build accurate property prediction models that can be employed as surrogate models in material analysis.

Acknowledgments. The authors gratefully acknowledge support from FAPESP (São Paulo Research Foundation, Grant Numbers 2016/23642-6, 2017/11631-2, and 2018/21401-7), Shell and the strategic importance of the support given by ANP (Brazil's National Oil, Natural Gas and Biofuels Agency) through the R&D levy regulation. We also thank Johnatan Mucelini for the support with the quantum QM9 dataset.

References

1. Arús-Pous, J., et al.: Randomized smiles strings improve the quality of molecular generative models. J. Cheminformatics **11**(1), 1–13 (2019)
2. Bjerrum, E.J.: Smiles enumeration as data augmentation for neural network modeling of molecules. arXiv preprint arXiv:1703.07076 (2017)
3. Blondel, V.D., Guillaume, J.L., Lambiotte, R., Lefebvre, E.: Fast unfolding of communities in large networks. J. Stat. Mech. Theo. Exp. **2008**(10), P10008 (2008). https://doi.org/10.1088/1742-5468/2008/10/p10008
4. Breve, F.A., Zhao, L., Quiles, M.G.: Particle competition and cooperation for semi-supervised learning with label noise. Neurocomputing **160**, 63–72 (2015). https://doi.org/10.1016/j.neucom.2014.08.082
5. Chandola, V., Banerjee, A., Kumar, V.: Anomaly detection: a survey. ACM Comput. Surv. **41**(3) (2009). https://doi.org/10.1145/1541880.1541882
6. Chen, C., Ye, W., Zuo, Y., Zheng, C., Ong, S.P.: Graph networks as a universal machine learning framework for molecules and crystals. Chem. Mater. **31**(9), 3564–3572 (2019). https://doi.org/10.1021/acs.chemmater.9b01294
7. Chmiela, S., Sauceda, H.E., Müller, K.R., Tkatchenko, A.: Towards exact molecular dynamics simulations with machine-learned force fields. Nat. Commun. **9**(1), 3887 (2018)
8. Cook, D.J., Holder, L.B.: Graph-based data mining. IEEE Intell. Syst. Appl. **15**(2), 32–41 (2000). https://doi.org/10.1109/5254.850825
9. Costa, L.D.F., et al.: Analyzing and modeling real-world phenomena with complex networks: a survey of applications. Adv. Phys. **60**, 329–412 (2011)
10. Faber, F.A., et al.: Prediction errors of molecular machine learning models lower than hybrid DFT error. J. Chem. Theory Comput. **13**(11), 5255–5264 (2017)
11. Hastie, T., Tibshirani, R., Friedman, J.: The Elements of Statistical Learning. SSS. Springer, New York (2009). https://doi.org/10.1007/978-0-387-84858-7

12. Haykin, S.S.: Neural Networks and Learning Machines, 3rd edn. Pearson Education, Upper Saddle River (2009)
13. Hinton, G.E., Sejnowski, T.J., Hughes, H.: Unsupervised Learning: Foundations of Neural Computation. MIT Press, Cambridge (1999)
14. Jain, A.K.: Data clustering: 50 years beyond k-means. Pattern Recogn. Lett. **31**(8), 651–666 (2010). https://doi.org/10.1016/j.patrec.2009.09.011
15. Karplus, M., McCammon, J.A.: Molecular dynamics simulations of biomolecules. Nat. Struct. Biol. **9**(9), 646–652 (2002). https://doi.org/10.1038/nsb0902-646
16. Mardirossian, N., Head-Gordon, M.: Thirty years of density functional theory in computational chemistry: an overview and extensive assessment of 200 density functionals. Mol. Phys. **115**(19), 2315–2372 (2017). https://doi.org/10.1080/00268976.2017.1333644
17. Mitchell, T.M., Thomas, M.: Machine Learning. McGraw-Hill, Maidenhead (1997)
18. Moriwaki, H., Tian, Y.S., Kawashita, N., Takagi, T.: Mordred: a molecular descriptor calculator. J. Cheminformatics **10**(1), 4 (2018)
19. Quiles, M.G., Macau, E.E.N., Rubido, N.: Dynamical detection of network communities. Sci. Rep. **6**(1), 25570 (2016)
20. Ramakrishnan, R., Dral, P.O., Rupp, M., Von Lilienfeld, O.A.: Quantum chemistry structures and properties of 134 kilo molecules. Sci. Data **1**, 140022 (2014)
21. Weininger, D.: Smiles, a chemical language and information system. 1. introduction to methodology and encoding rules. J. Chem. Inf. Comput. Sci. **28**(1), 31–36 (1988)
22. Wu, Z., et al.: MoleculeNet: a benchmark for molecular machine learning. Chem. Sci. **9**(2), 513–530 (2018)
23. Zhu, X.: Semi-supervised learning with graphs. Ph.D. thesis, School of Computer Science, Carnegie Mellon University (2005)

Q-Learning Based Forwarding Strategy
in Named Data Networks

Hend Hnaien[1,2(✉)] and Haifa Touati[1,2]

[1] Hatem Bettaher IResCoMath Research Unit, Gabes, Tunisia
`haifa.touati@cristal.rnu.tn`
[2] Faculty of Science of Gabes, Gabes, Tunisia
`h2II_05@hotmail.com`

Abstract. Named Data Networking (NDN) emerged as a promising new communication architecture aimed to cope with the need for efficient and robust data dissemination. NDN forwarding strategy plays a significant role for efficient data dissemination. Most of the currently deployed forwarding strategies use fixed control rules given by the routing layer. Obviously these simplified rules are inaccurate in dynamically changing networks. In this paper, we propose a novel Interest forwarding scheme called Q-Learning based Forwarding Strategy (QLFS). QLFS embedded a continual and online learning process that ensures quick reaction to sudden disruption during network operation. At each NDN router, forwarding decisions are continually adapted according to delivery times variation and perceived events, i. e. NACK reception, Interest Timeout... Our simulation results show that our proposed approach is more efficient than state of the art forwarding strategy in term of data delivery and number of timeout events.

Keywords: Named data networking · Forwarding strategy · Q-learning

1 Introduction

Named Data Networking [1] is a new data-centric architecture that redesigns the Internet communication paradigm from IP address based model to content name based model. The main idea of the Named Data Networks communication model is to focus on the data itself instead of its location. In Named Data Networks, users insert the name of the requested content into a request packet, called *"Interest"*, that is sent over the network and the nearest node which has previously stored this content, or *"Data"* packet, must deliver it. In NDN, the node that sends the *Interest*, is called *Consumer* and the original *Data* source is called *Producer* [2].

Each NDN node manages three different data structures: The *Pending Interest Table (PIT)*, the *Content Store (CS)* and the *Forwarding Information Base (FIB)*. The *Pending Interest Table (PIT)* tracks the outstanding Interests as

© Springer Nature Switzerland AG 2020
O. Gervasi et al. (Eds.): ICCSA 2020, LNCS 12249, pp. 434–444, 2020.
https://doi.org/10.1007/978-3-030-58799-4_32

well as the interface from which they come, in order to deliver back the Data to the Consumer on the Interest reverse path. The *Content Store (CS)* is used to store the received Data packets in order to serve the upcoming demands of the same Data. Finally, the *Forwarding Information Base (FIB)* stores the Data name prefixes and the interfaces where to forward the *Interest* packets. The FIB is populated and updated by a name-based routing protocol. For each incoming Interest, a longest prefix match lookup is performed on the FIB, and the list of outgoing faces stored on the found FIB entry is an important reference for the strategy module.

To request a content, the consumer sends an Interest and based on the forwarding strategy, the Interest packet is passed from one node to another until it reaches the producer that holds the requested content. The simplest forwarding strategy is the *multicast approach* [3] that forwards incoming Interest to all upstreams indicated by the supplied FIB entry, except the incoming interface. The *best route* [3] strategy forwards interests using the eligible upstream with the lowest routing costs. If there is no available upstream, e.g. the upstream link is down, the Interest is rejected and a *NACK* with the error code *no route* is generated and returned to the downstream/consumer node via the incoming interface of the Interest. If the consumer wants, it can retransmit the Interest, and the *best route* strategy would retry it with another next hop.

However, basic forwarding strategies do not adapt to the dynamic nature of the network. They frequently cause problems when there is a sudden change in the network architecture. Several forwarding strategies have been proposed in the literature to adapt the forwarding strategy to different network environments [4], [5] and [6]. In this context, we have thought to exploit reinforcement learning, which has proven to be efficient for real time decision making in different sectors and even for network traffic management.

The main contribution of this paper is the proposal of a new adaptive interest forwarding strategy for Named Data Networks. This strategy is based on the notion of reinforcement learning [7]. During execution, each node goes through training and exploitation phases. In the training phase, each node tries to have a background on its neighborhood and on the best path, in term of delivery time and reliability, to reach a given content. Once the environment has been explored, the node transits to the exploitation phase, and interest forwarding is done based on the $Q - values$ already calculated in the previous phase.

The remaining of the paper is organized as follows: In Sect. 2, we give a brief presentation of the Q-Learning process. Then, we present our solution called QLFS in Sect. 3. In Sect. 4, we move on to evaluate our solution by comparing its performances to the *best route* strategy. Finally, Sect. 5 concludes the paper and gives some future directions of this work.

2   An Overview of the Q-Learning Technique

Q-Learning [8] is a reinforcement learning algorithm that seeks to find the best action to take given the current state. The Q-Learning process involves 5 key entities: an **Environment**, an **Agent**, a set of **States** S, **Reward** values, and a set of **Actions** per state, denoted A. By performing an *Action* $a_{i,j} \in A$, the *Agent* transits from a *State* i to a *State* j. Executing an *Action* in a specific *State* provides the *Agent* with a *Reward*.

The agent is immersed in an environment, and makes decisions based on his current state. In return, the environment provides the agent with a reward, which can be positive or negative. The agent seeks, through iterated experiences, an optimal decision that maximizes the sum of the rewards over time.

It does this by adding the maximum reward attainable from future states to the reward achieved by its current state. Hence, the current action is influenced by the potential future reward. This potential reward is a weighted sum of the expected rewards values of all future steps starting from the current state.

3   Q-Learning Based Forwarding Strategy (QLFS)

The main idea of our solution is to develop a Q-Learning based forwarding strategy that helps the NDN router to find the suitable next hop and to adapt quickly its forwarding decisions according to network conditions.

As explained in the previous section, to apply Q-Learning we need to define five principals components which are: Environment, Agent, States, Actions and Rewards. In the context of Interest forwarding, we choose to model these components as follows:

- The *Environment* is the Named Data Network.
- Each NDN node is an *Agent*.
- A *State* is the next hop to which a received interest is forwarded.
- An *Action* is forwarding an interest to one of the neighboring nodes.
- *Rewards* are RTT based values.

To decide to which adjacent node should an NDN router forwards the received interest to reach as quickly as possible the content producer, a new $Q - table$ structure is implemented and maintained at each node. This table holds the $Q-value$ corresponding to each decision. For each forwarded interest, the router waits for the corresponding data packet to update the $Q-values$. On the arrival of a Data packet, a reward value is computed for the arrival interface and used to update the $Q - value$ as follows:

$$Q_n(x, c) = (1 - \alpha) * Q_{n-1}(x, c) + \alpha * [R_n(x, c) + \max_{v \in N(x)} Q_n(v, c)] \qquad (1)$$

where:

- $Q_n(x, c)$ is the $Q - value$ to reach a content c from a node x during the nth iteration.

- $R_n(x,c)$ is the reward value for the transition from node x to the next node selected to reach the content c. This value depends on the delivery time and the loss and failure events. The router computes the delay between the transmission of an interest and the reception of the corresponding data packet, denoted RTT. When the data packet is received, the current RTT value is compared to the minimum, average and maximum observed RTT values and the reward value is set according to the interval where the current RTT is. The closer the current RTT is to the minimum RTT, the higher the reward will be and vice versa: the closer the current RTT is to the maximum RTT, the lower the reward will be. Moreover, if a timeout event is detected or a NACK packet is received a penalty is applied by assigning a negative reward value.
- $Q_n(v,c)$ is the $Q - value$ to reach a content c from a neighbor node v during the nth iteration.
- $N(x)$ is the list of neighbors of node x.
- α is the learning rate which controls how quickly the algorithm adapts to the random changes imposed by the environment.

We note, that we modified the Data packet structure, as illustrated in Fig. 1, by adding a new field to hold the maximum $Q - value$ for a given prefix and return it to the downstream neighbor. In fact, before returning a data packet, the router node looks for the next hop that has the highest $Q - value$ and inserts this value in the data packet. For this purpose we have defined a new tag, called $MaxQTag$, so that we can save the maximum $Q - value$ among all the $Q - values$ of neighboring nodes.

When a node receives a data packet, it extracts the tag and uses it to update the $Q - value$ relative to this prefix, according to Eq. 1.

Fig. 1. Modified data packet structure

When an Interest is received, the router forwards it according to the current phase. Three phases are defined in QLFS: the **Initial training phase**, the **Training phase** and the **Exploitation phase**.

- During the **Initial training phase**, each received interest will be:
  - Forwarded to the lowest cost next hop stored in the FIB table. This action assures that the interest reaches the content producer, which reduces the loss probability.
  - Forwarded to one of the unused next hops. There is no random choice of next hop thus we can explore more possible links.

– During the **Training phase**, the router performs two actions:
  - Instead of using the routing cost metric to make decision, the node chooses the face with the maximum $Q - value$ to forward the interest.
  - The node continues to explore other paths by forwarding the interest to one of the unused next hops.

  Before leaving the training phase, the maximum $Q - value$, $Q_{max}(x, c)$, is saved for each node.
– During the **Exploitation phase**, each router forwards the received interest to the next hop that has the maximum $Q - value$. In fact at this step: each node has the necessary knowledge-background about how to reach a given content and which link to choose based on $Q - values$.

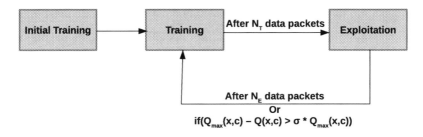

Fig. 2. QLFS learning cycle

An NDN router switches continually between the *training* and the *exploitation* phases. As shown in Fig. 2, after the reception of N_T data packets, the algorithm switches from *training* to *exploitation*. During the exploitation phase, only the maximum $Q - values$ will be updated. To detect eventual change in network conditions (failed link repaired, other new added links to explore...), our system periodically returns to the *training* phase. More precisely, QLFS switches to the *training* step, if N_E data packets are received or the performance of the chosen best link degrades. This latter is detected through the following test:

$$Q_{max}(x, c) - Q(x, c) > \sigma * Q_{max}(x, c) \qquad (2)$$

In summary, during the *exploitation* phase, for each update of the $Q - table$, the QLFS strategy verifies the number of data packets received during the current *exploitation* phase and the gap between the current $Q - value$ and the $Q_{max}(x, c)$ used at the beginning of the current *exploitation* phase. If one of the above tests is true, the QLFS protocol returns to the *training* phase.

4   Performance Evaluation

In this section, we evaluate the proposed QLFS strategy by comparing it with the basic forwarding strategy of the NDN architecture, namely *best route*. To that end, we have developed our proposal on ndnSIM [9], which is an NS-3 based simulator that implements the Named Data Networking (NDN) architecture using C++ language.

4.1 Simulation Model

Our simulation model consists of five nodes: one consumer, one producer and three routers as shown in Fig. 3. The nodes are connected through Point-To-Point links. Each link is characterized by a data rate of 1Mbps, a delay of 10 ms and a queue size of 20 packets.

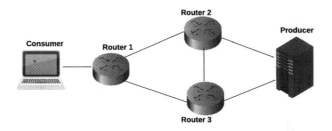

Fig. 3. Simulation model

The consumer sends 100 interests per second and the producer will reply to all requests with data packets of size 1024 KB. Table 1 lists most of the simulation parameters.

Table 1. Simulation Parameters.

Parameter	Value
Network links	P2P links
Data Rate	1 Mbps
Delay	10 ms
Payload Size	1024 Kb
Queue	20 packets
Interest Rate	100 interests per second
Simulation Time	20s
α	0.8
σ	0.5

In order to better set the value of the parameter *learning rate*, denoted α, we have run simulations while varying the value of this parameter from 0.2 to 0.9. Results reported in Fig. 4, show that the best performances are achieved when

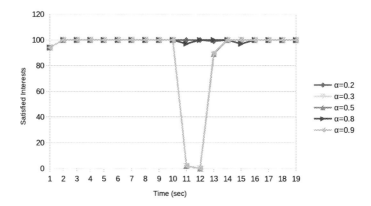

Fig. 4. Setting the learning rate parameter

the *learning rate* is set to 0.8. Thus this value will be used in the subsequent simulations.

To study the performance of the proposed QLFS scheme, we used the following metrics: *Satisfied interests, TimedOut Interests* and *Out Interests per interface*. The evaluation metrics are defined as follows:

- **Satisfied interests**: This metric is computed at the consumer side and is defined as the number of interest packets for which the consumer receives the requested data packets.
- **TimedOut Interests**: This metric is computed at the consumer side and is defined as the number of interests for which the consumer hasn't received the requested data until lifetime expiry.
- **Out Interests per interface**: This metric is computed by router *Router*1 and is defined as the number of interest forwarded through a each interface of the router.

4.2 Simulation Results and Performance Analysis

In a first part, we will compare the two strategies, QLFS and *best route*, following the metrics already mentioned in an ideal scenario, i.e. without link failure. In the second part, we compare the performance of both strategies in a dynamically changing network, i.e. in presence of link failure.

Scenario Without Link Failure. Simulations performed is a scenario without link failure, show that the two strategies have similar performances in terms of *satisfied interests* (Fig. 5) and *timedout interests* (Fig. 6), these results confirm that our solution doesn't degrade the network performance when link quality is stable and no failure or disruptive event occurs.

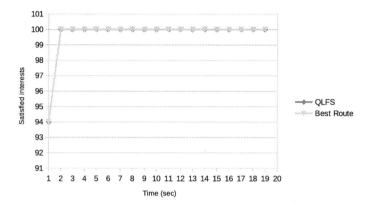

Fig. 5. Satisfied Interest in a scenario without link failure

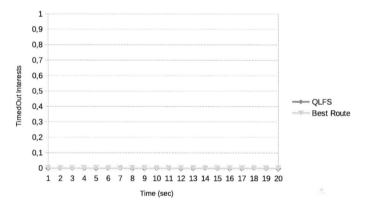

Fig. 6. TimedOut Interest in a scenario without link failure

Scenario with Link Failure. In a real-world scenario, one or more failed links are probably encountered. For this purpose, we provoke a link failure between *Router*1 and *Router*3 and study how each approach will react to this sudden and temporary disruption.

The evaluation of the number of Out Interests per interface at *Router*1, reported in Fig. 7, show that using the *est route* strategy, *Router*1 forwards all interests through interface 258 which is linked to *Router*3. Even when the destination becomes inaccessible from this interface after the link failure, *best route* continues to choose this interface despite the presence of a better alternative through interface 259, which is linked to *Router*2.

However, using the QLFS strategy, and as shown in Fig. 8, when the link failure occurs, *Router*1 immediately switches and forwards all interests through interface 259. The adaptive learning process introduced in QLFS helps the *Router*1 to quickly detect the failure event, through the penalty applied to interface 258. This penalty reduces the $Q - value$ of interface 258 and favours sending the interests through another interface, namely interface 259 which has better $Q - value$.

These results, confirm that the QLFS strategy is more efficient and more reactive to sudden disruptions than the *best route* strategy.

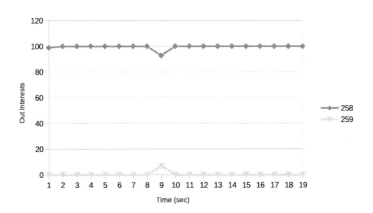

Fig. 7. Best route: out interests from router 1 in a scenario with link failure

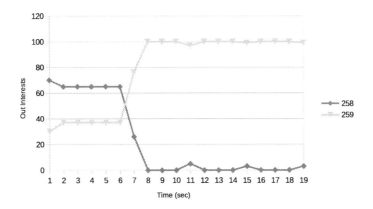

Fig. 8. QLFS: out interests from router 1 in a scenario with link failure

Finally, the evaluation of the number of satisfied interest reported in Fig. 9, reveals that when the link failure is applied, the number of satisfied interest decreases to 0 using *best route*, but using QLFS the number of satisfied interest is not affected by the presence of link failure. Similarly, Fig. 10 shows clearly that using *best route* several interest timeout events occur which induce unnecessary interest retransmissions and waste network bandwidth. However, using QLFS, the number of timed out interests tends towards 0 and thus the network bandwidth is efficiently consumed.

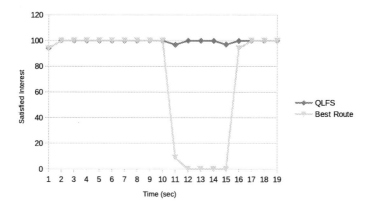

Fig. 9. Satisfied Interests in a scenario with link failure

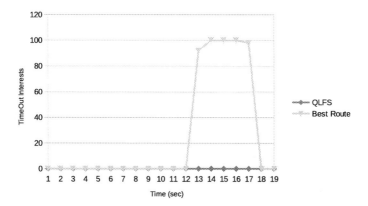

Fig. 10. TimedOut Interests in a scenario with link failure

As a conclusion, all simulations results confirm that the performance of the *best route* strategy degrades as the link already chosen by this strategy is broken, whereas the QLFS strategy is more resistant to this problem and can easily and rapidly find an alternative path to forward interest packets.

5   Conclusion

In this paper, we proposed the QLFS strategy, which brings more efficiency and reactivity to the interest forwarding process in the NDN architecture. Through the use of reinforcement learning and more precisely the Q-Learning algorithm, QLFS performs an online learning to adapt its forwarding decision according to the delivery time and the perceived loss events. Analysis and simulations were conducted to evaluate the performance of the proposed strategy. The results show that QLFS outperforms *best route* in term of speed of finding an alternative path when the initial path is interrupted. Results, also show that QLFS significantly minimizes the number of timed out interests, hence maximizing the number of satisfied interests compared to the *best route* strategy. In a future work, we plan to extend our experimental study to more complex scenarios.

References

1. Jacobson, V., Smetters, D.K., Thornton, J.D., Plass, M.F., Briggs, N.H., Braynard, R.L.: Networking named content. In: Proceedings of the 5th International Conference on Emerging Networking Experiments and Technologies, New York, NY, USA, pp. 1–12 (2009)
2. Wissingh, B., Wood, C., Afanasyev, A., Zhang, L., Oran, D., Tschudin, C.: Information-Centric Networking (ICN): Content-Centric Networking (CCNx) and Named Data Networking (NDN) Terminology. RFC 8793, June 2020
3. Yi, C., Afanasyev, A., Moiseenko, I., Wang, L., Zhang, B., Zhang, L.: A case for stateful forwarding plane. Comput. Commun. **36**, 779–791 (2013)
4. Kardi, A., Touati, H.: NDVN: named data for vehicular networking. Int. J. Eng. Res. Technol. IJERT **4**, 1–6 (2015)
5. Aboud, A., Touati, H.: Geographic interest forwarding in NDN-based wireless sensor networks. In: Proceedings of the 13th ACS/IEEE International Conference on Computer Systems and Applications (AICCSA), Morocco, Agadir, pp. 1–8 (2016)
6. Aboud, A., Touati, H., Hnich, B.: Efficient forwarding strategy in a NDN-based internet of things. Cluster Comput. **22**(3), 805–818 (2018). https://doi.org/10.1007/s10586-018-2859-7
7. Kaelbling, L.P., Littman, M.L., Moore, A.W.: Reinforcement learning: a survey. J. Artif. Intell. Res. **4**, 237–285 (1996)
8. Watkins, C.J., Dayan, P.: Technical note: Q-learning. Mach. Learn. **8**, 279–292 (1992)
9. Mastorakis, S., Afanasyev, A., Moiseenko, I., Zhang, L.: ndnSIM 2: an updated NDN simulator for NS-3. Technical report NDN-0028, Revision 2 (2016)

General Track 2: High Performance Computing and Networks

Comparative Performance Analysis of Job Scheduling Algorithms in a Real-World Scientific Application

Fernando Emilio Puntel[1(✉)], Andrea Schwertner Charão[1(✉)], and Adriano Petry[2(✉)]

[1] Federal University of Santa Maria, Santa Maria, RS, Brazil
{fepuntel,andrea}@inf.ufsm.br
[2] National Institute for Space Research, Santa Maria, RS, Brazil
adriano.petry@inpe.br

Abstract. In High Performance Computing, it is common to deal with substantial computing resources, and the use of a Resource Management System (RMS) becomes fundamental. The job scheduling algorithm is a key part of a RMS, and the selection of the best job scheduling that meets the user needs is of most relevance. In this work, we use a real-world scientific application to evaluate the performance of 4 different job scheduling algorithms: First in, first out (FIFO), Shortest Job First (SJF), EASY-backfilling and Fattened-backfilling. These algorithms worked with RMS SLURM workload manager, considering a scientific application that predicts the earth's ionosphere dynamics. In the results we highlight each algorithm's strength and weakness for different scenarios that change the possibility of advancing smaller jobs. To deepen our analysis, we also compared the job scheduling algorithms using 4 jobs of Numerical Aerodynamic Sampling (NAS) Parallel Benchmarks in a controlled scenario.

Keywords: Job scheduling · High performance computing · SLURM and resource management system

1 Introduction

High performance computing environments are important in a variety of fields that require significant computational resources, like astronomy and meteorology. Among the possibilities of high performance environments, computer clusters are predominant due to their performance and scalability. Distributed computing resources available in a cluster require the use of Resource Management System (RMS) to optimize the distribution of jobs of applications. The main goal in a RMS applied to clusters is to coordinate the available computing resources so they work as a single computer, targeting the reduction of total processing time and idle resources. SLURM (Simple Linux Utility for Resource Management) is one of the most important RMS, and it is present in 6 from 10 top supercomputer list (TOP500). It is open source, fault tolerant and highly scalable [21,22].

© Springer Nature Switzerland AG 2020
O. Gervasi et al. (Eds.): ICCSA 2020, LNCS 12249, pp. 447–462, 2020.
https://doi.org/10.1007/978-3-030-58799-4_33

One of the main parts of RMS is the scheduling algorithm, responsible for defining when and at which processing node the jobs execute. The choice of a given scheduling algorithm must consider the application constraints and the available computing environment. Although several papers present original job scheduling procedures, that are evaluated and compared to others, evaluations are usually conducted on simulators, with synthetic workloads. This artificial basis can lead to inaccurate results for real applications. In this work, we report an evaluation of job scheduling algorithms in a real case application, using a scientific cluster running daily.

2   Job Scheduling

Job scheduling is an important research topic in HPC. The choice of the job scheduling approach can impact directly the application performance [7,16].

Usually, job scheduling aims to increase computational resource usage, which means the resources are used most of the period, and also submitted jobs finish within expected time [17]. Besides that, job scheduling can also target other goals, such as approximate the waiting time among jobs [19], reduce latency or response time, or limit the use of power resources [8,9]. Different goals may conflict, so the use of resources depends on the computing environment and application.

Several job scheduling algorithms are available. Among them, we have selected four job scheduling algorithms to perform the experiments. The Figs. 1, 2, 3 and 4 show a comparison between them in a hypothetical case of 4 jobs that require different resources and processing time, using a cluster composed of 3 processing cores, each one available at different moments.

The FIFO algorithm is a classic model that runs the jobs by queue order, i.e., the first job of the queue is the first submitted. Its main disadvantage is that it is not flexible to move forward smaller jobs to execute before. Figure 1 illustrates FIFO job scheduling for the hypothetical case.

The SJF (Shortest Job First) [12] algorithm is similar to backfilling algorithms [13]. It advances smaller jobs when there are available computational resources to run them, without other verification. Figure 2 shows the SJF addressing the hypothetical case.

The EASY-backfilling aims to advance smaller jobs provided it does not delay the first job of the queue [13]. The EASY-backfilling algorithm is the most popular job scheduling algorithm based in the conservative job scheduling backfilling optimization. Figure 3 shows the EASY-backfilling procedure for the hypothetical case.

The Fattened-backfilling algorithm is very similar to EASY-backfilling. The advance of smaller jobs is similar to EASY-backfilling, but in the calculation to choose the job to be advanced, it is considered the Average Waiting Time (AWT) coefficient of already finished jobs, what allows the advance of smaller jobs of queue [11]. Figure 4 illustrates the Fattened-backfilling execution.

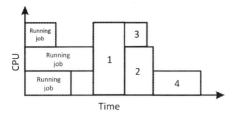

Fig. 1. FIFO Job scheduling algorithm in hypothetical cases

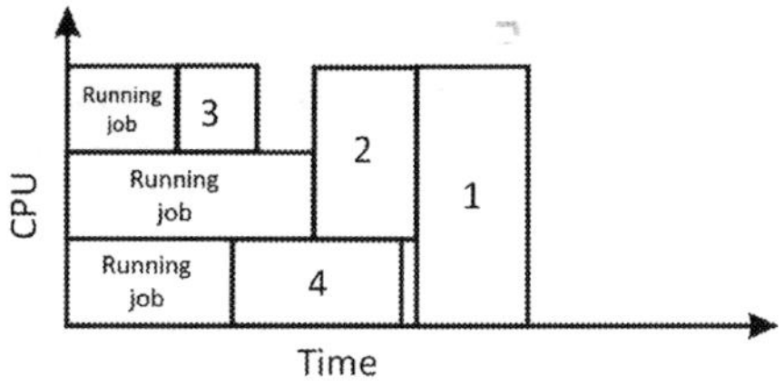

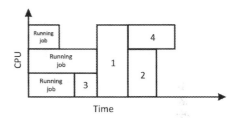

Fig. 2. SJF Job scheduling algorithm in hypothetical cases

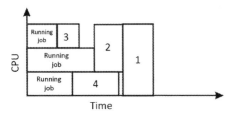

Fig. 3. EASY-backfilling Job scheduling algorithm in hypothetical cases

Fig. 4. Fattened-backfilling Job scheduling algorithm in hypothetical cases

3   Ionosphere Forecasting

The real application used in experiments is based on forecasting the Earth ionosphere dynamics using a mathematical model [14]. The system runs daily and provides total electron content (TEC) maps for every hour in South America region. These maps are available at http://www2.inpe.br/climaespacial/portal/tec-supim-prevision/. The application first step is to run the physical model, that is an enhanced version of the Sheffield University Plasmasphere-ionosphere Model (SUPIM), adapted to South America singularities. The model considers several chemical and physical interactions among particles in ionosphere, and provides discreet outputs (ions and electronic concentrations) along geomagnetic field lines. These outputs are then processed to the Data Assimilation and Visualization System (DAVS). DAVS is able to modify the model outputs considering observational data, in order to adjust the simulation to the available measurements. Besides that, it provides estimatives of electron concentrations in homogeneous grid, that are summed in height to result in TEC values in a 2-dimensional map.

4   Experimental Evaluations

The experiments have been performed in a CRS-INPE (Southern Regional Space Research Center, National Institute for Space Research) computing cluster. The cluster is composed of 5 processing nodes, each one consisting of with 8 Intel Xeon

E5-2609 2.40 GHz processors containing 4 cores, and 72 GB of RAM memory. We used SLURM and two databases in MySQL to manage the resources and jobs.

SLURM was installed and configured in all processing nodes and the controlling node of a cluster. The configuration file holds the cluster architecture information. By default, SLURM is configured so every processing node is unavailable for other tasks during the processing of a job or a group of jobs. This configuration was changed as follows, to make it possible the processing nodes to run different jobs simultaneously, where the jobs are allowed to ask for a number of CPUs, processing cores or amount of memory to run.

- *SelectType*: this option is by defalut set as *select/linear*, where only one job can run in a node. For the experiments, this option was defined as *SelectType=select/cons_res*, so the node can be shared with other jobs;
- *SelectTypeParameters*: this option is added to SLURM configuration file when *SelectType* is changed. For the experiments, this option was defined as *CR_CPU_Memory*, where the jobs can request CPU and memory;
- *DefMemPerCPU*: this option is added to define a memory to reserve if the job does not specify a memory request. In the experiments, it was set to 5000 MB;

We first compared the performance of the job scheduling algorithms under three different scenarios using SUPIM-DAVS jobs: intermediate, favorable and unfavorable cases. In each case, the jobs are organized in queue for job scheduling algorithms evaluation, and executed 30 times.

For the evaluations, we selected 56 SUPIM jobs, where each one represent a different longitude in the South America area (ranging from -35°W to -85°W), and 24 DAVS jobs, corresponding to every hour in a day. The selected day simulated by SUPIM-DAVS jobs was January, 29th 2015, when the solar flux (F10.7) was 123.73 sfu (solar flux units) [4].

The resource requests for the jobs are shown in Table 1. These requests were defined before jobs execution. The CPU requisition of 0.5 in DAVS jobs is because it is possible to run two DAVS jobs per CPU.

Table 1. The requisitions SUPIM and DAVS jobs

Job	CPU	Memory (MB)	Time of execution (sec)
SUPIM	1	3072	3060
DAVS	0.5	10240	1050

Since SUPIM forecasting jobs need a pre-execution task, composed of file downloads and initial setup, the monitors of CPU usage, memory usage and average waiting time were initialized at the moment that the first job was submitted. Considering the cluster is dedicated only for the execution these experiments, it is guaranteed that the measurements correspond to the jobs under evaluation.

Besides SUPIM job execution experiments, we also evaluated the job scheduling using four different types of jobs of NAS Parallel Benchmarks [2] in a controlled scenario. The aim was to investigate the difference in results from real to controlled scenarios, where a significant mismatch could indicate alternative problems.

Since job scheduling algorithms used in experiments rely on information about the required processing time for each task, we had to estimate it. This estimative could be assigned to the user during submission process. However, considering that in most cases users overestimate such time [1,15], afraid of job interruption during processing – leaving computational resources idle [10,20], this approach is unlikely to succeed. A better choice would be to update the processing mean time for a given task considering its last few executions [18]. In this work the mean time is calculated over the last five executions of each task.

The CPU and memory use was checked using the tools *mpstat* [5,6] and *vmstat* [6], respectively. For execution time and mean waiting time measurements, Linux library *time* was used.

4.1   Intermediate Scenario

Three of the job scheduling algorithms in experiments may advance jobs execution in queue. Depending on the algorithm outcome, the queue can be organized differently. For this scenario, in some cases a job scheduling algorithm can advance smaller jobs. In other cases the scheduling algorithm behave like FIFO. Figure 5 illustrates a Gantt graph in a possible intermediate scenario, where we can observe some jobs are executing, while other jobs are waiting for resources in queue.

In scenario present in Fig. 5 the job scheduling algorithm will decide:

– after completion of the two DAVS jobs (currently running), DAVS jobs in queue (first and second positions) will be submitted, and when they finish the third job in queue (DAVS) is submitted;
– the first job in queue will be SUPIM (currently fourth in queue), and it is expected to start running after completion of a SUPIM job currently in execution. At this point, the job scheduling algorithm may decide to advance or not the seventh job in queue (DAVS) to start running with the third in queue (DAVS), because this advance will not delay SUPIM job;
– not to advance the other jobs in queue.

Figures 6 and 7 show respectively the application makespan and CPU use for SUPIM and DAVS jobs using different job scheduling. In this case, it is possible to observe that FIFO job scheduling had a worse performance compared to the others, while the other job scheduling presented similar results.

Figure 8 shows the average waiting time for jobs in the queue. It can be observed how FIFO algorithm had significant worst result when compared to other algorithms. Considering EASY-backfilling and Fattened-backfilling use waiting time statistics to estimate the coefficient of smaller jobs advance, both had a similar performance.

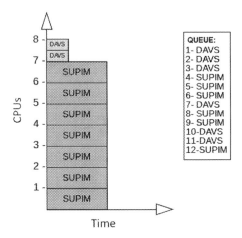

Fig. 5. Intermediate scenario

Table 2. Intermediate scenario waiting time for SUPIM and DAVS jobs

Job Scheduling Algorithm	All Jobs (sec)	SUPIM (sec)	DAVS (sec)
FIFO	1244.31	1328.75	1205.82
SJF	844.26	1178.16	202.46
EASY-backfilling	1045.36	1145.1	940.63
Fattened-backfilling	1049.06	1131.33	932.3

Table 2 depicts the average waiting time of Fig. 8 for SUPIM and DAVS jobs separately. For DAVS jobs, SJF shows a significant reduction in waiting time, since it can advance more DAVS jobs in comparison to other algorithms.

In general, for the intermediate scenario, FIFO algorithm had an overall inferior performance, while other job scheduling algorithms presented a similar

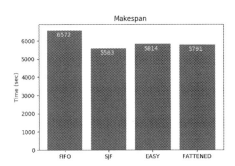

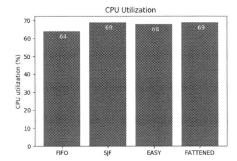

Fig. 6. Application makespan in intermediate scenario

Fig. 7. CPU use in intermediate scenario

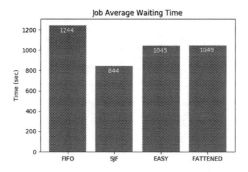

Fig. 8. Average waiting time in intermediate scenario

performance in all metrics. Despite EAS-backfilling and Fattened-backfilling use different approach than SJF to advance small jobs, these algorithms had a similar performance, including makespan and CPU use measurements.

4.2 Favorable Scenario

In the favorable scenario, the jobs are organized in queue to increase the advance of smaller jobs. Here, SJF, EASY-Backfilling and Fattened-backfilling performance is expected to improve. Figure 9 shows a possible favorable scenario Gantt graph, with some jobs executing and other waiting in queue.

After completion of running jobs, scheduling algorithms will decide:

– after completion of the two DAVS jobs (currently running), DAVS jobs in queue (first and second positions) will be submitted, and when they finish the third job in queue (DAVS) is submitted;

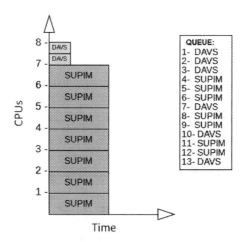

Fig. 9. Favorable scenario

– the first job in queue becomes SUPIM (currently fourth in queue), but in the lack of resources, the job scheduling algorithms that can advance small jobs will advance the seventh job in queue (DAVS) to start running with the third in queue (DAVS), because this advance will not delay SUPIM job;
– after completion of the seven SUPIM jobs (running initially), the SUPIM jobs in queue (positions 4th, 5th, 6th, 8th, 9th and 11th) are submitted, as well as DAVS job (position 10th in queue);
– since the next job in queue is SUPIM (position 12th), waiting to start after the DAVS job running in CPU 7 or 8, it is possible to advance DAVS job (position 13th in queue), because this advance will not delay SUPIM job.

Figure 10 shows the application makespan for the job scheduling algorithms. Like in the intermediate scenario, the FIFO algorithm had the worst results when compared to the other algorithms.

The Fig. 11 present CPU use. It is possible to observe that the FIFO presented inferior performance while the other algorithms had similar performances.

Then we compare EASY-backfilling and Fattened-backfilling in detail (see Table 3), it is possible to observe that Fattened-backfilling is able to advance more jobs and achieved a better result in DAVS jobs, and consequently increasing SUPIM jobs waiting time.

Table 3. Favorable scenario waiting time for SUPIM and DAVS jobs

Job scheduling algorithm	All Jobs (sec)	SUPIM (sec)	DAVS (sec)
FIFO	1218.43	1436.7	878.26
SJF	825.90	1171.16	146.26
EASY-backfilling	883.7	1100.53	550.65
Fattened-backfilling	890.90	1125.74	489.67

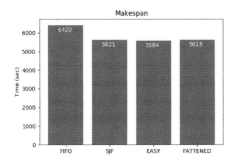

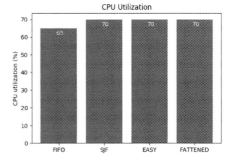

Fig. 10. Application makespan in favorable scenario

Fig. 11. CPU use in favorable scenario

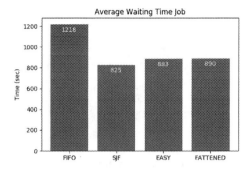

Fig. 12. Average waiting time in favorable scenario

4.3   Unfavorable Scenario

The queue in the unfavorable scenario is organized to difficult job scheduling to advance small jobs. In this scenario, regularly the processing nodes show intervals of idle resources. Figure 13 shows a Gantt graph for an hypothetical unfavorable scenario, with some jobs executing and other in queue.

After completion of running jobs, scheduling algorithms will decide:

- after completion of a DAVS job (currently running), the first in queue will be submitted (SUPIM job). The job scheduling algorithms that can advance DAVS jobs. (2nd in queue), since it would delay the first job in queue (SUPIM);
- after completion of SUPIM jobs, the jobs in queue will be submitted almost in order they are in queue, making it difficult to advance small jobs;
- such difficulty in advancing jobs is present through all the execution of the task.

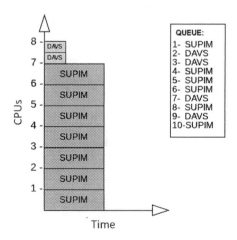

Fig. 13. Unfavorable scenario

In the unfavorable scenario, the FIFO algorithm showed again an inferior performance in makespan (see Fig. 14). In this scenario, the SJF performed better compared to EASY-Backfilling and Fattened-Backfilling. This is because even in an unfavorable scenario the SJF can perform the advance of smaller jobs as soon as some node have computational resources available.

Like in the other scenario, the CPU use presented similar result for SJF, EASY-backfilling and Fattened-backfilling algorithms, as show in Fig. 15. The job scheduling with inferior results was FIFO.

The average waiting time for the jobs in queue, shown in Fig. 16 and detailed in Table 4, presented the worse performance for FIFO. SJF was better than EASY-backfilling and Fattened-backfilling, which performed similarly despite Fattened-backfilling uses techniques to overcome the performance of EASY-backfilling, what did not hold in unfavorable environment.

Table 4. Unfavorable scenario waiting time for SUPIM and DAVS jobs

Job scheduling algorithm	All jobs (sec)	SUPIM (sec)	DAVS (sec)
FIFO	1218.43	1436.7	878.26
SJF	825.90	1171.16	146.26
EASY-backfilling	883.15	1100.53	550.65
Fattened-backfilling	890.90	1125.74	489.67

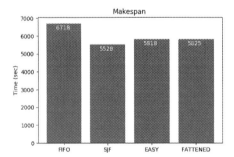

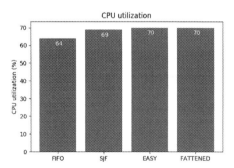

Fig. 14. Application makespan in unfavorable scenario

Fig. 15. CPU use in unfavorable scenario

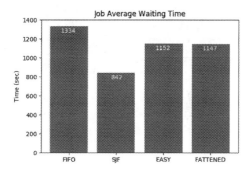

Fig. 16. Average waiting time in unfavorable scenario

4.4 Discussion

In all scenarios evaluated, the algorithm that presented worse performance was FIFO, since it does not allow the advance of jobs, leaving gaps of idle resources unused. This is why, in general, applications centred in the reduction of processing time hardly would benefit from choosing FIFO.

Other important issue to assess is the jobs' mean waiting time in queue. SJF benefits smaller jobs that are advanced, reducing its mean waiting time usually more than other algorithms. In favorable case, EASY-backfilling and Fattened-backfilling were able to achieve performances comparable to SJF, supporting they can effectively advance jobs. But in intermediate and unfavorable cases, SJF performed slightly better.

Although different scenarios were designed by changing the queue inputs in a scientific application, the results obtained were not significantly different from each other. Even FIFO, that showed the worse performance, did not compromise the ionospheric prediction results, since there were only 2 types of jobs, and the number of jobs to schedule was relatively low. So, it is also interesting to submit the scheduling algorithms to different and heterogeneous environments and observe the results.

4.5 Controlled Scenario

In this section we evaluate the algorithms in a controlled environment using different workloads. For experiments, NAS (Numerical Aerodynamic Sampling) Parallel Benchmarks (NPB) [2] jobs were chosen. NPB is a set of programs designed to evaluate the performance of parallel supercomputers. Basically, the benchmarks are derived from fluid dynamics applications [3]. The NPB has different applications, and for the experiments 4 different types were used:

– BT (Block Tri-diagonal solver), that performs the computational fluids processing;
– FT (Discrete 3D fast Fourier Transform): benchmark that solves 3D equations using spectral FFT;

- MG (Multi-Grid on a sequence of meshes): computes the solution of the Poisson equation;
- SP (Scalar Penta-diagonal solver): solves multiple independent systems of dominating equations diagonally and not diagonally.

The choice of the benchmarks and the requisitions of each job was done randomly to force a heterogeneous job scheduling. Table 5 presents the computational resources requested by each job after five executions.

Table 5. NPB Job Requirements

Job	CPU	Memory (MB)	Runtime (sec)
BT	2	1280	180
FT	2	5132	240
MG	4	26624	420
SP	6	5132	680

We run 110 jobs for the experiments: 34 BT, 34 FT, 25 MG and 17 SP jobs. The number of jobs with small requisitions was large, so the scheduling algorithms were able to perform the advance of a large number of jobs. Like in previous experiments, the cluster was used in a dedicated way and each algorithm was executed 30 times. For the BT, FT and MG jobs the standard input data was used. For the SP jobs, the number of iterations was changed to 160 and the problem size to $256 \times 256 \times 256$.

4.6 Results

Figure 17 illustrates the makespan when using the four algorithms in this case. It is possible to observe that the EASY-backfilling algorithm presented improvement in execution time by 10% when compared to the SJF algorithm. It seems that EASY-backfilling algorithm respected the execution queue and executed the largest jobs in its intended time, unlike the SJF algorithm, which advanced smaller jobs and left larger jobs to the end. As a result, larger jobs end up delaying execution. Also, in this scenario, the FIFO achieved better performance when compared to the SJF, precisely by running the larger jobs in their time and not delaying them at the beginning.

CPU and memory use are shown in Fig. 18 and 19, respectively. EASY-backfilling and Fattened-backfilling algorithms presented better performance for CPU use when compared to the other algorithms. EASY-backfilling was also superior to the other algorithms for memory use, since it tries to increase the use of resources during job scheduling. When compared to SJF, the result is about 20% better.

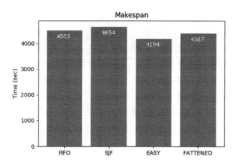

Fig. 17. Makespan in controlled scenario

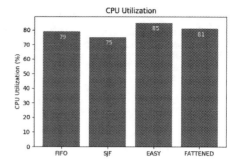

Fig. 18. CPU use in controlled scenario

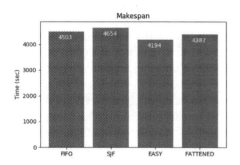

Fig. 19. Memory use in controlled scenario

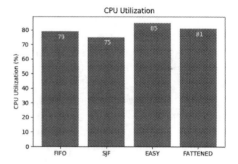

Fig. 20. Average waiting time in controlled scenario

Figure 20 shows the average waiting time in the job queue. In this case it is possible to observe that even if the SJF algorithm advances all the smaller jobs to execute first, the Fattened-backfilling algorithm obtains a superior performance, because it allows smaller jobs or smaller priorities to be advanced.

Table 6 shows the average total and individual waiting time. It is important to note that the SJF algorithm performed considerably better than EASY-backfilling only when compared to smaller jobs (FT and BT). The comparison of EASY-backfilling and Fattened-backfilling algorithms shows Fattened-backfilling obtained inferior performance only in the waiting time of SP jobs, precisely because the algorithm makes smaller jobs easier to advance.

In general, the FIFO algorithm presented inferior performance when compared to the other algorithms in this scenario. A point to be highlighted is the fact that the Fattened-backfilling algorithm obtained results similar to EASY-backfilling, although it does not target the improvement in computational performance metrics. The SJF algorithm was impaired in this experiment, because the delay of starting larger jobs caused a delay in finishing the task and consequently a decrease in the CPU use rate.

Table 6. Controlled scenario waiting time for jobs

Job scheduling algorithm	All jobs (sec)	SP (sec)	MG (sec)	FT (sec)	BT (sec)
FIFO	2712	1486	2134	3181	3360
SJF	1546	1966	1974	1253	1357
EASY-backfilling	1903	1432	2005	1907	2106
Fattened-backfilling	1399	1912	1657	1144	1246

5   Conclusions

This work evaluated the performance of different scheduling algorithms used with SLURM in a scientific application that forecasts daily the Earth ionosphere. The experiments run FIFO, SJF, EASY-backfilling and Fattened-backfilling algorithms for 3 different scenarios, and we could observe FIFO presented the worse performance. A better performance for SJF was achieved in some metrics. For favorable and unfavorable scenarios EASY-backfilling and Fattened-backfilling were able to achieve similar performance of SJF.

The results with the scientific application were not significantly different from each other because there were only 2 types of jobs, and the number of jobs to schedule was relatively low. So, the algorithms were also submitted to a controlled scenario using NAS benchmark using more diverse types of jobs and workload. In that case, Fattened-backfilling algorithm was superior in advancing smaller jobs, while EASY-backfilling algorithm presented better results in execution time when compared to the other algorithm. We observed a decrease in performance for the more simple algorithms evaluated (FIFO and SJF).

Although the experiments were performed in a real cluster, it is possible some weaknesses of job scheduling algorithms have not emerged, probably because the cluster size was limited. We suggest, as future work, to apply the job scheduling and resource management in different scientific computing environments, as well as to evaluate them using larger clusters.

Acknowledgments. The author Fernando Emilio Puntel thanks Coordenação de Aperfeiçoamento de Pessoal de Nível Superior (CAPES) for the financial support in this research.

References

1. Arndt, O., Freisleben, B., Kielmann, T., Thilo, F.: A comparative study of online scheduling algorithms for networks of workstations. Cluster Comput. **3**(2), 95–112 (2000)
2. Bailey, D.H.: NAS parallel benchmarks. In: Padua, D. (ed.) Encyclopedia of Parallel Computing, pp. 1254–1259. Springer, Boston (2011). https://doi.org/10.1007/978-0-387-09766-4
3. Bailey, D.H., et al.: The NAS parallel benchmarks. Int. J. Supercomput. Appl. **5**(3), 63–73 (1991)

4. Natural Resources Canada: Monthly averages of solar 10.7 cm flux. http://www.spaceweather.gc.ca/solarflux/sx-5-mavg-en. Acessado em 20 de julho de 2018

5. Casalicchio, E., Perciballi, V.: Measuring docker performance: what a mess!!! In: Proceedings of the 8th ACM/SPEC on International Conference on Performance Engineering Companion, ICPE 2017 Companion, pp. 11–16. ACM, New York (2017)

6. Ciliendo, E., Kunimasa, T., Braswell, B.: Linux performance and tuning guidelines. IBM, International Technical Support Organization (2007). https://lenovopress.com/redp4285.pdf

7. Feitelson, D.G., Rudolph, L., Schwiegelshohn, U., Sevcik, K.C., Wong, P.: Theory and practice in parallel job scheduling. In: Feitelson, D.G., Rudolph, L. (eds.) JSSPP 1997. LNCS, vol. 1291, pp. 1–34. Springer, Heidelberg (1997). https://doi.org/10.1007/3-540-63574-2_14

8. Georgiou, Y.. et al.: A scheduler-level incentive mechanism for energy efficiency in HPC. In: 2015 15th IEEE/ACM International Symposium on Cluster, Cloud and Grid Computing (CCGrid), pp. 617–626. IEEE (2015). https://ieeexplore.ieee.org/abstract/document/7152527

9. Georgiou, Y., Glesser, D., Trystram, D.: Adaptive resource and job management for limited power consumption. In: 2015 IEEE International Parallel and Distributed Processing Symposium Workshop (IPDPSW), pp. 863–870. IEEE (2015). https://ieeexplore.ieee.org/abstract/document/7284402

10. Gibbons, R.: A historical application profiler for use by parallel schedulers. In: Feitelson, D.G., Rudolph, L. (eds.) JSSPP 1997. LNCS, vol. 1291, pp. 58–77. Springer, Heidelberg (1997). https://doi.org/10.1007/3-540-63574-2_16

11. Gómez-Martín, C., Vega-Rodríguez, M.A., González-Sánchez, J.-L.: Fattened backfilling: an improved strategy for job scheduling in parallel systems. J. Parallel Distrib. Comput. **97**, 69–77 (2016)

12. Mao, W., Kincaid, R.K.: A look-ahead heuristic for scheduling jobs with release dates on a single machine. Comput. Oper. Res. **21**(10), 1041–1050 (1994)

13. Mu'alem, A.W., Feitelson, D.G.: Utilization, predictability, workloads, and user runtime estimates in scheduling the IBM SP2 with backfilling. IEEE Trans. Parallel Distrib. Syst. **12**(6), 529–543 (2001)

14. Petry, A., et al.: First results of operational ionospheric dynamics prediction for the Brazilian space weather program. Adv. Space Res. **54**(1), 22–36 (2014)

15. Skovira, J., Chan, W., Zhou, H., Lifka, D.: The EASY—LoadLeveler API project. In: Feitelson, D.G., Rudolph, L. (eds.) JSSPP 1996. LNCS, vol. 1162, pp. 41–47. Springer, Heidelberg (1996). https://doi.org/10.1007/BFb0022286

16. Subbulakshmi, T., Manjaly, J.S.: A comparison study and performance evaluation of schedulers in hadoop yarn. In: 2017 2nd International Conference on Communication and Electronics Systems (ICCES), pp. 78–83, October 2017

17. Talby, D., Feitelson, D.G.: Supporting priorities and improving utilization of the IBM SP scheduler using slack-based backfilling. In: 13th International and 10th Symposium on Parallel and Distributed Processing Parallel Processing, 1999. 1999 IPPS/SPDP. Proceedings, pp. 513–517. IEEE (1999)

18. Tsafrir, D., Etsion, Y., Feitelson, D.G.: Backfilling using system-generated predictions rather than user runtime estimates. IEEE Trans. Parallel Distrib. Syst. **18**(6), 789–803 (2007)

19. Vasupongayya, S., Chiang, S.-H.: On job fairness in non-preemptive parallel job scheduling. In: IASTED PDCS, pp. 100–105. Citeseer (2005)

20. Wong, A.K., Goscinski, A.M.: The impact of under-estimated length of jobs on easy-backfill scheduling. In: 16th Euromicro Conference on Parallel, Distributed and Network-Based Processing, pp. 343–350. IEEE (2008). https://ieeexplore.ieee.org/abstract/document/4457142

21. Yoo, A.B., Jette, M.A., Grondona, M.: SLURM: simple Linux utility for resource management. In: Feitelson, D., Rudolph, L., Schwiegelshohn, U. (eds.) JSSPP 2003. LNCS, vol. 2862, pp. 44–60. Springer, Heidelberg (2003). https://doi.org/10.1007/10968987_3

22. Zhou, X., Chen, H., Wang, K., Lang, M., Raicu, I.: Exploring distributed resource allocation techniques in the SLURM job management system. Technical report Illinois Institute of Technology, Department of Computer Science (2013)

A Coalitional Approach for Resource Distribution in Self-organizing Open Systems

Julian F. Latorre[1], Juan P. Ospina[2]([⊠]), Joaquín F. Sánchez[2],
and Carlos Collazos-Morales[2]

[1] Corporación Universitaria Comfacauca, Popayán, Colombia
jlatorre@unicomfacauca.edu.co
[2] Universidad Manuel Beltran, Bogotá, Colombia
{juan.ospina,carlos.collazos}@docentes.umb.edu.co,
joaquin.sanchez@umb.edu.co

Abstract. This article aims to theorize and explore the distribution problem in the context of self-organizing open systems when they exhibit social behaviors that can be modeled and formalized, applying the interactive decision theory. In this context, the whole system structure is trying to achieve a stable cooperative state at the same time they aim to satisfy the interests of each individual and reach a certain level of social welfare. We analyze these scenarios evaluating the system stability through a formal analysis of coalitional and non-cooperative games and the notions of collective actions and robustness. We used socially inspired computing to propose a negotiation method that allows the members of the system to cooperate and manage common-pool resources without any central controller or other orchestration forms. The model was evaluated using four quantitative moral metrics through simulation techniques. The result showed how a system influenced by ethical behaviors exhibited higher efficiency, symmetry, and invariance over time.

Keywords: Self-organization · Ad hoc networks · Trust · Cooperation · Socially inspired computing

1 Introduction

From the Latin *robustus*, the word robustness refers to the quality of an entity to be strong, vigorous and decided against possible changes in the environment. In general terms, it refers to the ability of an entity to conserve its properties throughout the execution of its functions despite possible variations, interferences, and changes in the operating conditions. For instance, according to Noam Chomsky, in his hypothesis of Universal Grammar, the set of restrictions that exist in every natural language as a group of invariant constraints no matter the context of the Human language, constituted a robust core in his theory of principles and parameters.

© Springer Nature Switzerland AG 2020
O. Gervasi et al. (Eds.): ICCSA 2020, LNCS 12249, pp. 463–473, 2020.
https://doi.org/10.1007/978-3-030-58799-4_34

The notion of robustness has been used in the scientific literature to explain, interpret, and state behaviors in biological, economic, and social systems. In the context of Computer Science, robustness is used to describe the capacity of computational systems to deal with adverse operating conditions [1]. In particular, this property is observed in the context of communication and information technologies to describe the decentralized interoperability expected in the Industry 4.0, in which multiple entities interact as a self-organizing open system. Consequently, the notion of robustness leads us to analyze the computational models that we expect to help the next generation of communication networks adapt their structure to unexpected environments.

Additionally, to ease the analysis of this property, it is necessary to understand the mechanisms that produce these behaviors in future technological developments. We use the macro-meso-micro perspective that allows us to analyze the current process of decision making. For example, the case of Perishable Agricultural Supply Chains (CSAP), in which hybrid models of simulation, optimization, and game theory are used to analyze and redesign CSAP effectively, achieve better responses and lower loss of quality derived from logistic flows. A detailed description of this application is described in [2]. This example refers to the principle of autonomy in the context of self-organized open systems, which have been commonly related to the social sciences and philosophy, creating a natural association with human behaviors. Furthermore, in the context of Artificial Intelligence, autonomy concerns the idea that agents do to always act on our behalf; they act with some control over their internal state [3].

In this regard, the purpose of this work is to propose a robust distribution method for self-organizing open systems in scenarios in which the whole system structure is trying to achieve a stable cooperative state at the same time it aims to satisfy the demands of each individual. This proposal is composed of two elements: on the one hand, a coalitional game based on the notions of negotiation and moral metrics. On the other hand, an adaptive technique allows the system to face unexpected operating conditions during the negotiation process. It is essential to mention that this work is an extension of the analysis presented in [4,5]. The result showed how the proposed negotiation model allows the system to achieve higher efficiency, symmetry, and invariance over time.

We structured the document as follows: Sect. 2 shows the related work; both a philosophical background for the idea the collectivism in resource allocation and the notion of moral attitudes are described. Section 3 overviews the agent paradigm and presents robustness as an expected property of a multi-agent system. In Sect. 4, our proposal for resource allocation through coalitional game theory and negotiation is presented. Section 5 presents the experimental results and performance evaluation. Section 6 concludes the article.

2   Related Work

2.1   A Philosophical Background for the Collectivism in Resources Allocation

From Latin *collectivus*, the word collectivism refers to all kinds of plural organization that serves to manage common-pool resources in social systems. In this regard, this concept includes all social artifacts used to manage resources or values (material or immaterial) that can be used by the community members. In modern history, this process is usually represented as the relative value of the material goods exchanged by individuals under assumptions of strict rationality. This scenario was formalized in the Utilitarian Theories of Jeremy Bentham [6], which later derive the Adam Smith's Theory of Moral Sentiments and also the cybernetic proposals of Von Newman & Morgenstern [7]. In these proposals, they show how a moral base could lead to the application of universal ethics centered on an individual who enjoys the privileges of a rational society.

Additionally, to support the idea mentioned above, several theories describe the methodological individualism as a form of understanding of social phenomena. However, after these theories were proposed, a famous social dilemma named the tragedy of the commons [8] suggests that maximizing the individual benefits does not respond to the global interests of a social system. In this context, the "Public Good Game" is a clear example of this scenario, requiring a theoretical framework broader than the utilitarian principles. Similarly, there are at least two more theories that try to explain the process of collective decision-making under this scenario. First, the principles of necessity and equality that establish the needs of an individual must be satisfied by the group according to the principle of equality. Second, the principle of equity, which determines that an individual must receive an allocation proportional to his contribution (positive or negative) to the community [9].

2.2   Collectivity and Robustness

For assessing the possible impact of these families in the performance of a self-organizing open system, it is necessary to define and model its structure and behaviors. In this case, the methodological individualism leads us to reduce the complexity of the problem through the representation of capital (goods) according to the preferences and desires of the agents in terms of utility functions. This approach has been explored through the theoretical analysis of social negotiation, extending its concepts towards the holistic idea of social capital [10,11]. On the other hand, the axiomatic analysis of the conceptual bases has been tested in [12] and also important changes were suggested in the context of distributive justice.

Moreover, game theory has been extended its scope towards an evolutionary approach to capture the behaviors throughout the evolution of natural systems

[13]. In such models, robustness is guaranteed as a new mechanism that arises from the collective actions of many individuals that try to achieve a common goal. For example, we can see a sample through the theory of non-cooperative games [4,5,14–16], in which the performance of the model respond as the outcome of the collective behaviors of the agents and not just to the linear addition of the individual actions.

We can adapt economic and political concepts as part of engineering developments to examine the social relationships in multi-agent systems and define the qualities of robustness and connections inside groups of interest. In this regard, the analysis presented in [17] shows how despite asymmetries in power and internal conflicts, agents are promoting the effective treatment of common resources even when it is not possible to guarantee their success in the long term. Furthermore, the proposals presented in [4,5] shows that we can use evolutionary computing to achieve cooperation among agents with different behaviors and also use concepts like sympathy and commitment in non-cooperative games. The results presented above show how a self-organizing system can produce a functionality using cooperation models; In [17], the coffee farming and for [5] a routing service. Without show specific details in each case, we can identify that the evolution of the system behavior converges towards a cooperative group that minimizes asymmetries, even when the individual social tensions of individuals do not disappear.

2.3   Moral Attitudes in Collective Decisions

Collective action demands the participation of multiple individuals who combine their actions to achieve a common goal. During this process, it is necessary to satisfy axiomatic principles that rule the pure distribution problem in common-pool resources. In this regard, to model the individual behavior of the community members, we can consider the different ethical and moral theories related to social systems as an inspiration source for designing and implementing multi-agent systems. Examples of this approach can be found in *The Moral Machine* of MIT [18], the project REINS [19], the project AMA [20] and the project ETHICAA [21]. In general terms, they attempt to explore the inclusion of structure preferences and moral values as part of the behavior of autonomous agents and verify their impact on the collective decision-making process.

Ethical systems define mutually accepted actions that are available for the members of a community. However, although they have been formalized and explored through experimental research, there is insufficient evidence to determine their relevance in the context of collective actions and social goals. As a result, we need a conceptual and computational framework that allows us to evaluate and compare the theories of individual moral behavior in the context of self-organizing systems. For example, [22] explores the impact of multiple moral choices in the decision process in scenarios where different ethical criteria need to be compared to verify the global behavior of the system (Fig. 1).

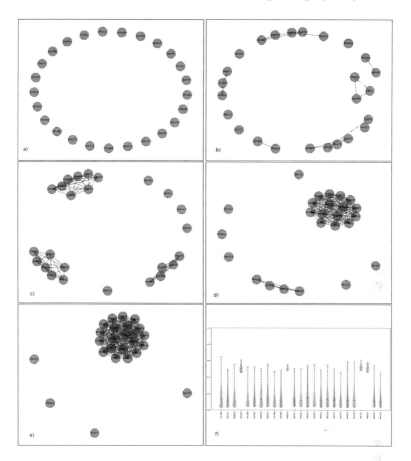

Fig. 1. Formation of a super coalition in 5 successive tournaments: a), b), c), d), e) and distribution of the computational efficiency of each agent f)

3   Robust Multi-agent Systems

To use the approach mentioned above, we need to restrict the possible moral pro-files associated with ethical theories to include them as part of the multi-agent systems' behavior and define the cooperation patterns that will exist as part of the desired emergent behavior of a community. In this context, we can consider the Rawlsian principle known as *MaxMin* that can emerge from a deliberation based on the dialogue of common goals and social welfare. This work aims to show some of the mechanisms in the middle of social exchange (modeled as a negotiation process through game theory), in particular, the cooperation process that emerges from the principles of sympathy and social commitment described in [23] (Fig. 2).

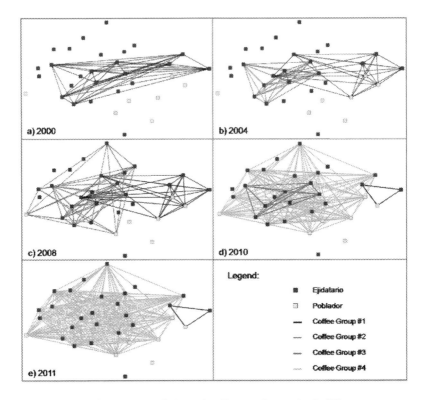

Fig. 2. Formation of a super coalition of coffee producers in 5 different moments: a), b), c), d), e). Source: [17].

To make a moral analysis of the behaviors considered by autonomous agents in the context of a multi-agent system, the categories proposed by [24] are used. They allow us to describe such behaviors in successive negotiation games based on the prisoner's dilemma. They therefore compare them with a random set of non-deterministic strategies available in [25] as it is showed Sect. 5. Consequently, we analyze the individual behavior of the system members, using a simulation scenario in a distributed virtual environment to produce a robust multi-agent system based on the moral analysis of the agent behaviors [26] .

4   Proposed Model

From the computational model provided in [27], in which four layers are added to produce a stable self-organizing communication system, a negotiation experiment is developed based on the formation of communities of agents that offer computational services through coordination and cooperation patterns. In this context, agents are responsible for managing resources available in the network and provide services to the final user. We analyze a multi-agent system's behavior using a stochastic and evolutionary process based on coalitional game theory. Once the rational strategies have been chosen, we simulate the system evolution through the execution of negotiation tournaments based on the prisoner's dilemma and the coalitionist meta-strategy proposed in [4]. The game is described in Eq. (1). It is important to mention that this game can produce rational strategies when a group of agents A_g, as part of the evolutionary process [5], incorporate another group of agents A'_g in a coalitionist model.

$$C = (N, S, E, \succeq i) \tag{1}$$

In this regard, N is a non-empty set of agents. S, is a non-empty set of pay outcomes. E is the effectivity function of the payoff set, and $\succeq i$ is a complete pre-order over S. This process is assessed through the moral evaluation mentioned above. Also, we compared these results with a negotiation model that described interactions between highly complex social organizations. In particular, we use the moral metrics proposed in [24]:

- Cooperating rating (CR)
- Good partner rating (GPR)
- Eigenjesus rating (EJR)
- Eigenmoses rating (EMR)

Similarly, analyze the moral behavior of our system allow us to study properties like robustness and efficiency, checking the performance of the computational services provided by the agents. The levels of efficiency are measured according to the metric described in Eq. (2).

$$e(i,j) = \frac{f_i(w)}{\frac{f_i(w)+f_j(w)}{2}} \tag{2}$$

In this metric, the relative efficiency $f_i(w)$ is defined as the execution time of the task w in the node i. Also, $E(i)$ represents the total efficiency of the node. Therefore, the moral relationship among the agents is defined as a function that describes the stochastic process related to the addition of the individual ethical metrics in each iteration. This process is repeated until the system achieves a meta-stable state in $t = T$.

$$M(C(t)) = \sum_{0}^{T} \sum_{i=0}^{N} CR_i(t) + GPR_i(t) + EJR_i(t) + EMR_i(t) \tag{3}$$

5   Experimental Results

It is possible to verify how a negotiation process that considers the variability and diversity of the individual's preferences can maintain suitable efficiency levels in terms of the system performance and the agents' utility. These results reinforce the theoretical conception in which it is possible to find a meta-stable equilibrium

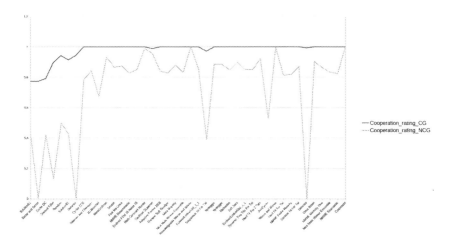

Fig. 3. Cooperation rate between 40 players in a Coalitional Game (CG) and a Non Coalitional Game (NCG)

based on the relationship between the system behaviors and the agents' capacities. As a result, this model is implemented based on the belief that a society can be defined as an association of individuals who form a cooperative coalition to obtain common welfare through collective actions. Moreover, we can describe the emerging behaviors that arise for combining multiple individual strategies in just one negotiation tournament.

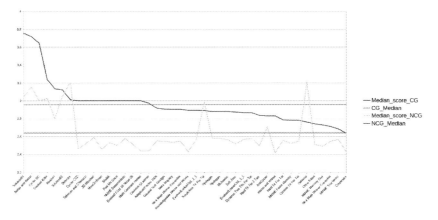

Fig. 4. Median scores for 40 players in Coalitional (CG) and Non-Coalitional (NCG) Game tournaments

Note that the moral metrics of the game confirm an improvement in both: the cooperation level in the system (Fig. 3) and the average profit of the agents

(Fig. 4). As long as the game evolves, the emergent cooperation will increase given the fact that the commitment assumed by the agents limit pure rationality. This behavior allows us to produce emergent cooperation in scenarios in which traditionally the agent would be tempted to defect.

Additionally, this cooperative behavior will maintain its level if the system can continue accepting new players and governing the community through the mentioned logic that, despite the uncertainty in the negotiation process, interprets the individual interests as part of a function of social welfare. Therefore, it is necessary to explore new types of formal logic for modeling complex negotiation scenarios in which many decisions can operate simultaneously as part of the cooperation process. For example, we can use the multi-valued logics of that allow generalizing any common resource allocation method [15].

6   Conclusions

It is possible to assume that through coalitional game theory, in general terms, and in particular, using adaptive techniques, like the ones proposed in [5], the robustness of a self-organizing open system can guarantee the simultaneous operation of the multiple strategies and achieve an equilibria state. This can be understood as the required social foundation for creating agent communities composed of individuals who make decisions and adjust their actions to accomplish common goals in a temporary and morally committed way.

Moreover, the result showed how systems influenced by moral metrics exhibit higher rates of efficiency, symmetry, and invariance over time. These attributes are pursued both philosophical and computational approaches of justice as a requirement for ensuring the sustainability of resources over time. Additionally, it is necessary to explore new scenarios and improve the negotiation rules to analyze this phenomenon in more complex computational systems. We aim to guarantee desirable characteristics and to mitigate the problems and vulnerabilities present in the schemes of self-organizing open systems, which are required in the industry 4.0 and the next generation of communication networks.

References

1. Liu, Y.: Robustness in ad hoc networks. Ph.D. dissertation, Rutgers University - School of Graduate Studies (2017)
2. Orjuela Castro, J.A., et al.: Incidencia del diseño de la cadena de suministro alimentaria en el equilibrio de flujos logísticos. PhD Thesis, Universidad Nacional de Colombia-Sede Bogotá (2018)
3. Wooldridge, M., Jennings, N.R.: Agent theories, architectures, and languages: a survey. In: Wooldridge, M.J., Jennings, N.R. (eds.) ATAL 1994. LNCS, vol. 890, pp. 1–39. Springer, Heidelberg (1995). https://doi.org/10.1007/3-540-58855-8_1

4. Latorre, J.F., Ospina, J.P., Ortiz, J.E.: A coalitional game for achieving emergent cooperation in ad hoc networks through sympathy and commitment. In: Figueroa-García, J.C., López-Santana, E.R., Rodriguez-Molano, J.I. (eds.) WEA 2018. CCIS, vol. 915, pp. 352–362. Springer, Cham (2018). https://doi.org/10.1007/978-3-030-00350-0_30

5. Vega, D.A., Ospina, J.P., Latorre, J.F., Ortiz, J.E.: An adaptive trust model for achieving emergent cooperation in ad hoc networks. In: Alor-Hernández, G., Sánchez-Cervantes, J.L., Rodríguez-González, A., Valencia-García, R. (eds.) Current Trends in Semantic Web Technologies: Theory and Practice. SCI, vol. 815, pp. 85–100. Springer, Cham (2019). https://doi.org/10.1007/978-3-030-06149-4_4

6. Bentham, J.: The Rationale of Reward. Published by John and H. L, Hunt (1825)

7. von Neumann, J., Morgenstern, O.: Theory of Games and Economic Behavior. Princeton University Press, Princeton (2007)

8. Hardin, G.: The tragedy of the commons. Science **162**(3859), 1243–1248 (1968)

9. Pitt, J., Busquets, D., Macbeth, S.: Distributive justice for self-organised common-pool resource management. ACM Trans. Auton. Adapt. Syst. **9**(3), 14:1–14:39 (2014). http://doi.acm.org/10.1145/2629567

10. Collins, C.R., Neal, J.W., Neal, Z.P.: Transforming individual civic engagement into community collective efficacy: the role of bonding social capital. Am. J. Commun. Psychol. **54**(3–4), 328–336 (2014)

11. Petruzzi, P.E., Busquets, D., Pitt, J.: Social capital as a complexity reduction mechanism for decision making in large scale open systems. In: 2014 IEEE Eighth International Conference on Self-Adaptive and Self-Organizing Systems, pp. 145–150 (2014)

12. Sen, A.: Rawls versus bentham: an axiomatic examination of the pure distribution problem. Theory Decis. **4**(3–4), 301–309 (1974). http://www.scopus.com/inward/record.url?eid=2-s2.0-0344106846partnerID=40

13. Chalkiadakis, G., Elkind, E., Wooldridge, M.: Computational aspects of cooperative game theory. Synth. Lect. Artif. Intell. Mach. Learn. **5**(6), 1–168 (2011)

14. Clempner, J.B.: A team formation method based on a Markov chains games approach. Cybern. Syst. **50**(5), 417–443 (2019). https://doi.org/10.1080/01969722.2019.1598677

15. Marchioni, E., Wooldridge, M.: łukasiewicz logics for cooperative games. Artif. Intell. **275**, 252–278 (2019)

16. Georgiou, K., Kundu, S.: Bargaining in networks with socially-aware agents. In: Avrachenkov, K., Huang, L., Marden, J.R., Coupechoux, M., Giovanidis, A. (eds.) GameNets 2019. LNICST, vol. 277, pp. 128–150. Springer, Cham (2019). https://doi.org/10.1007/978-3-030-16989-3_10

17. García-Amado, L.R., Pérez, M.R., Iniesta-Arandia, I., Dahringer, G., Reyes, F., Barrasa, S.: Building ties: social capital network analysis of a forest community in a biosphere reserve in Chiapas, Mexico (2012)

18. Awad, E., et al.: The moral machine experiment. Nature **563**(7729), 59 (2018). https://www.nature.com/articles/s41586-018-0637-6

19. Responsible Intelligent Systems | Universiteit Utrecht. https://responsibleintelligentsystems.sites.uu.nl

20. Giubilini, A., Savulescu, J.: The artificial moral advisor. The "ideal observer" meets artificial intelligence. Philos. Technol. **31**(2), 169–188 (2018). https://doi.org/10.1007/s13347-017-0285-z

21. Ethics and Autonomous Agents. http://ethicaa.org/

22. Governatori, G., Olivieri, F., Riveret, R., Rotolo, A., Villata, S.: Dialogues on moral theories (2018)

23. Sen, A.K.: Collective Choice and Social Welfare, vol. 11. Elsevier, Amsterdam (2014)
24. Singer-Clark, T.: Morality Metrics On Iterated Prisoner's Dilemma Players (2014). http://www.scottaaronson.com/morality.pdf
25. Middelkamp, A.: Online. Praktische Huisartsgeneeskunde **3**(4), 3–3 (2017). https://doi.org/10.1007/s41045-017-0040-y
26. Alfonso, L.J., Ospina, J.P., Ortiz, J.E.: A register module for agents communities through robustness property in ad hoc networks. In: 2018 Congreso Internacional de Innovación y Tendencias en Ingeniería (CONIITI), pp. 1–6 (2018)
27. Ceballos, H., Sánchez, J.F., Ospina, J.P., Ortiz, J.: Sistema de Telecomunicaciones Social-Inspirado mediante Comunidades de Agentes. In: Congreso Internacional de Computación Colombia-México, pp. 56–63 (2015)

A New Library of Bio-Inspired Algorithms

Natiele Lucca and Claudio Schepke[(⊠)]

Laboratory of Advanced Studies, Postgraduate Program in Software Engineering,
Federal University of Pampa (UNIPAMPA), Alegrete, RS, Brazil
`natielelucca@gmail.com, claudioschepke@unipampa.edu.br`

Abstract. Real engineering, science, and economics problems cannot be ever solved exactly due to the high computation time to find the optimal solution. One way to solve such problems is to apply bio-inspired algorithms, to minimize the time to search for potential solutions. Bio-inspired algorithms are based on the collective behavior of social organisms and are used to solve optimization problems. This article presents a new library of bio-inspired algorithms. The library offers the implementation of some algorithms, being easily extensible through interfaces. An evaluation was made using 7 test functions applied to each of the implemented algorithms. The tests have shown that the ABC algorithm obtained the best convergence results in 5 tests and the ACO algorithm in 2 tests.

Keywords: Bio-inspired algorithms · Library · Artificial bee colony · Ant colony optimization · Particle swarm optimization

1 Introduction

The resolution of a problem may not be achieved exactly due to the complexity given by a large number of variables and/or potential solutions. An algorithm that describes a problem sequentially may require a large processing capacity to find the optimal solution, that is, the best solution to the problem [12]. The time required for exact and sequential computing can vary from hours to years, which highlights the importance of approaches and techniques that optimize the execution of algorithms [8]. The problems mentioned above can be identified in several research areas, for example, in simulations of real activities involving engineering, science, and economics. Traveling salesman and vehicle routing are examples of classic problems.

In this context, bio-inspired algorithms are an strategy to reduce the execution time, so that the solution is improved. These algorithms apply swarm intelligence or colony intelligence, or collective intelligence, which simulate the collective behavior of self-organized, distributed, autonomous, flexible, and dynamic systems [9]. The elements that integrate the swarm or colony can optimize a global objective through the collaborative search in space following specific rules

O. Gervasi et al. (Eds.): ICCSA 2020, LNCS 12249, pp. 474–484, 2020.
https://doi.org/10.1007/978-3-030-58799-4_35

[7]. According to [6], algorithms of flocks of birds and the colony of social insects, such as ants and bees, are known examples in the literature.

This article features a new library[1] for problem-solving using bio-inspired algorithms. The library implements the classic algorithms: Particle Swarm Optimization (PSO); Artificial Bee Colony (ABC); and Ant Colony Optimization (ACO), in addition to being extensible to other algorithms through interfaces created in C++ language.

The purpose of this article is to evaluate the functioning of the proposed bio-inspired algorithm library through test cases. With this, we want to measure the quality of convergence of numerical solutions. The contribution of this article is (a) to provide a new extensible library of bio-inspired algorithms; (b) evaluate the quality of the solutions.

The paper is organized as follows. Section 2 presents the related works in a systematic way. Section 4 describes the development and testing methodology used to carry out this work. In the Sect. 5 the results are presented. Finally, in Sect. 6, the final considerations about the work are highlighted.

2   Related Works

A set of 25 optimization libraries available in the literature are present in [10]. The authors perform a comparative analysis, identifying the strengths and weaknesses of each library. The article looks at possible shortcomings and possibilities for future work since it seeks to provide complete material on the optimization tools. Of the tools evaluated, only six have bio-inspired implementations: ECJ[2], EvA2[3], FOM[4], ParadisEO[5], MALLBA[6], and Opt4[7]. In addition to these libraries, we also found new implementations: PaGMO[8] and PyGMO[9] libraries.

Table 1 characterizes the libraries of bio-inspired algorithms found in the literature. For each library, the following were identified: the programming language, the algorithms implemented, and whether licenses are required. All libraries have some selected algorithms implemented and the mostly use open source license.

In our work, we propose a library implemented in C++ that provides the implementations of the PSO, ABC, and ACO algorithms, in addition to enabling the extension of new classes of bio-inspired algorithms.

---

[1] Available at https://github.com/NatiLucca/Bio-inspired-library.
[2] Available at https://cs.gmu.edu/~eclab/projects/ecj/.
[3] Available at http://www.ra.cs.unituebingen.de/software/EvA2/.
[4] Available at http://www.isa.us.es/fom.
[5] Available at http://paradiseo.gforge.inria.fr/.
[6] Available at http://neo.lcc.uma.es/mallba/easy-mallba/.
[7] Available at http://opt4j.sourceforge.net/.
[8] Available at: http://esa.github.io/pagmo/index.html.
[9] Available at: http://esa.github.io/pygmo/index.html.

Table 1. Natural computing libraries

Libraries	Language	Algorithms	Licence
ECJ	Java	GA, PSO and DE	Open Source
Eva2	Java	ES, DE and PSO	LGPL
FOM	Java	SA, ACO and ES	LGPL
MALLBA	C++	SA, GA, PSO and ACO	Open Source
Opt4	Java	DE, PSO and SA	LGPL
ParadisEO	C++	PSO	CECILL
PaGMO	C++	ABC, PSO, SA and GE	Open Source
PyGMO	Python	PSO, ABC, GA and ACO	Open Source

3 Swarm Intelligence

Swarm intelligence is a research area that explores the characteristics of swarms in the face of the way they seek food and interact. This approach allows real problems in different areas to be solved [6].

The Artificial Bee Colony (ABC) algorithm simulates the behavior of bees in collecting nectar. The colony is formed by three groups of bees: employed bees who explore and seek food sources that are possible solutions to the problem; onlooker bees that are recruited by the employed according to the quality of the source found; and scout bees that are employed bees that have found a bad source and abandoned them.

The Ant Colony algorithm was proposed by [4] and inspired by the behavior of ants in the search for food. Ants leave the anthill in search of food sources, that is, they explore the search space by releasing pheromone along the way. During a certain period of time the ants search for a source of food, but those that find a source closer to the anthill concentrate a greater amount of pheromone, as they travel the path more than once. Therefore, the more chemistry a trail is, that is, the higher the concentration of pheromone, the more ants will be attracted to it. Thus, most ants tend to follow the same path [9].

The Particle Swarm optimization algorithm (PSO) was modeled after the social behavior of the flock of birds [5]. In PSO, the population of individuals or particles is grouped into a cluster (set of solutions). These particles simulate the behavior of birds through self-learning (cognitive component) and flock learning (social component), that is, they imitate their own success and that of neighboring individuals. From this behavior the particles can define new positions that direct them towards optimal solutions.

Other examples of algorithms in this area of bio-inspired computing are the Genetic Algorithms (GS), Differential Evolution (DE), Simulated Annealing (SA), Evolution Strategies (ES) and Bee colonies Optimization (BCO).

4   Methodological Aspects

In this section, we present the methodological aspects related to the development, tests, and execution of the library.

4.1   Library Implementation

The bio-inspired library was developed in the C++ programming language, to facilitate programming and have resources that allow solving problems efficiently. This language is oriented to objects so it allows abstraction, encapsulation, inheritance, and polymorphism, which is essential for good code generation practices. The language also offers libraries for manipulating vectors and matrices and for generating random values. These contribute to the implementation of swarm intelligence algorithms.

The bio-inspired computation algorithms selected for this study belong to the swarm intelligence class, thus simulating the behavior of agents before the colony. The ABC, ACO, and PSO algorithms were chosen, as they have desirable characteristics for the optimization of problems that cannot be solved or solved in polynomial time.

The source code of the ABC, ACO and PSO algorithms is organized in 3 directories: *include*, *main* and *src*. The first directory contains the files of extension *hpp*, that is, the header files that define the classes and methods. The *main* directory contains the execution initialization file, the implementation of the benchmark equations used in the tests, and an auxiliary file that contains the methods for executing. The *src* directory contains the implemented source codes. In addition to the directories, the library also contains a script for executing the tests, called *run.sh* and a file *README*, a text document with information and usage details for the library. The Table 2 shows the related file structure for each algorithm.

Table 2. File structure implementation

Directory	ABC	ACO	PSO
include	abc.hpp	aco.hpp	pso.hpp
	bee.hpp	ant.hpp	particle.hpp
	benchmarking.hpp	benchmarking.hpp	benchmarking.hpp
main	main.cpp	main.cpp	main.cpp
	benchmarking.cpp	benchmarking.cpp	benchmarking.cpp
	auxiliar.cpp	auxiliar.cpp	auxiliar.cpp
src	abc.cpp	aco.cpp	pso.cpp
	bee.cpp	ant.cpp	particle.cpp
.	run.sh	run.sh	run.sh
	README.md	README.md	README.md

To use the library, the user must instantiate an object that corresponds to the class of the chosen algorithm.

4.2   Test Functions

To validate the implementation, it is necessary to verify the characteristics of the algorithm through tests. In this work, the tests were done by functions that have different properties, to ensure that the tested algorithm can or cannot solve certain types of optimization efficiently [13]. The quality of the optimization functions can be assessed by the properties: unimodal and multimodal; two-dimensional and N-dimensional; continuous and non-continuous; and convex, and non-convex [13]. A function that has only one optimal location is called a uni-modal and functions with two or more optimal locations are called multimodal. Functions over two-dimensional space are classified as two-dimensional. The other functions are classified in this work as multidimensional or N-dimensional. A function is continuous when all points p in your domain D follow Eq. 1.

$$\lim_{x \to p} f(x) = f(p) \tag{1}$$

A function is convex when the secant line joining two points is above the graph or if any point $m \in D$ follows Eq. 2.

$$f''(m) > 0 \tag{2}$$

Thus, to test the algorithms implemented in the bio-inspired library, a *bench-mark* was developed as a base application composed of seven test functions. They are: *Alpine, Booth, Easom, Griewank, Rastrigin, Rosenbrock* and *Sphere*. These functions are presented in Table 3. The characteristics of the functions such as agent size, search space, maximum number of iterations and minimum global were established according to [2,9], being described in Table 4. All the minimization functions used in the tests are continuous.

4.3   Execution of the Tests

The final results comprise the average of 30 executions, as adopted in the work of [3]. An execution comprises the generation of the input files and execution. Input files are generated using the *runDetails* method. This method creates a folder called *Inputs*, which stores the input files generated during the method's execution. Then the sequential version is executed by the *run* method, and the population and the other random values of the code are initialized with the val-ues of the files previously generated in the *runDetails* method. The *run* method return the best solution found by the algorithm and the execution time in sec-onds.

The tests were performed on a workstation composed of 2 Intel® Xeon® Silver 4116 processors CPU. Each processor has 12 physical cores (24 threads) operating at the standard 2.1 GHz frequency and 3 GHz turbo frequency. The

Table 3. Benchmark functions [2]

Benchmark	Function	Graphic
Alpine	$f(\mathbf{x}) = \sum i = 1^n |x_i sin(x_i) + 0.1x_i|$	
Booth	$f(x, y) = (x + 2y - 7)^2 + (2x + y - 5)^2$	
Easom	$f(x, y) = (-1) * \cos(x) * \cos(y) * e^{-[(x-\pi)^2 - (y-\pi)^2]}$	
Griewank	$f(\mathbf{x}) = 1 + \sum_{i=1}^n \frac{x_i^2}{4000} - \prod_{i=1}^n \cos(\frac{x_i}{\sqrt{i}})$	
Rastrigin	$f(\mathbf{x}) = 10n + \sum_{i=1}^n (x_i^2 - 10cos(2\pi x_i))$	
Rosenbrock	$f(\mathbf{x}) = \sum_{i=1}^n [b(x_{i+1} - x_i^2)^2 + (a - x_i)^2]$	
Sphere	$f(\mathbf{x}) = \sum_{i=1}^n x_i^2$	

main memory (RAM) is 96 GB in size and DDR4 technology. The operating system is Linux Debian kernel version 4.19.0-8 using GNU GCC 9.3 compiler with -O3 optimization flag and standard c++11.

Table 4. Parameters of the benchmark functions [2]

Benchmark	Dimensions	Limit of search	Iterations	Minimal global	Multimodal	Convex
Alpine	10	[0, 10]	3000	f(0,...,0) = 0	Yes	No
Booth	2	[−10, 10]	1000	f(1,3) = 0	No	Yes
Easom	2	[−100, 100]	1000	f(π, π) = −1	Yes	No
Griewank	50	[−100, 100]	5000	f(0,...,0) = 0	No	No
Rastrigin	50	[−5.12, 5.12]	5000	f(0,...,0) = 0	Yes	Yes
Rosenbrock	50	[−5, 10]	5000	f(1,...,1) = 0	Yes	No
Sphere	6	[−5.12, 5.12]	3000	f(0,...,0) = 0	No	Yes

4.4   Test Automation

Test launches are automated through instructions in *shell script*. This language allows the creation of folders to store the output files of each test and the creation of *loops* on a command line, allowing the launch of 30 repetitions on each step of the execution. The language also makes it possible to redirect the output of methods to a file. In this way, the results of the executions are stored in a file of type .csv, with the name of the function, where each line of the file corresponds to the solution and the measured execution time.

According to the studies of [1,11] the number of agents responsible for determining the optimal global of a function corresponds to the range of 20 to 60 agents. Thus, the functions were tested for a population of 10 to 100 agents, varying from 10 to 10 units.

5   Experimental Results

Table 5 respectively show the results obtained in the tests for the PSO, ABC, and ACO algorithms, applying each of the 7 test functions and varying the size of the population. The tables show the numerical average of the solutions and average execution time in seconds.

As shown in Table 4, all tests have a global minimum value to be found equal to 0, with the exception of Easom, which has a minimum value of −1. Therefore, it is expected that the algorithms reach this value as the closest possible. Most of the algorithms evaluated obtained optimal solutions for the test functions analyzed, especially the ABC algorithm. However, the functions *Rastrigin* and *Rosenbrock* obtained, on average, less precise solutions, even using a larger number of iterations, as they are complex functions, since they have greater dimensionality.

For the *Alpine* function, the ABC provided the best solution of all algorithms and ACO has close results to them. However, PSO has the best execution time for all populations evaluated and ABC has close results to them. ACO has the worst execution time. In *Booth* function tests, ABC has the best solution. PSO and ACO are close results with each other. PSO has the best execution time, followed by ABC and ACO, respectively. In the *Easom* function, all algorithms present results close to the ideal −1. However, ABC found the exact result in 7

Table 5. PSO, ABC and ACO test results

Function	Pop.	Solutions			Time (s)		
		ABC	ACO	PSO	ABC	ACO	PSO
Alpine	10	0,00E+00	4,07E-07	7,99E+00	0,079604	0,799142	0,062815
	20	0,00E+00	2,37E-15	5,56E+00	0,156575	1,680333	0,123000
	30	0,00E+00	9,01E-12	4,85E+00	0,233024	2,679626	0,167739
	40	0,00E+00	1,99E-19	3,47E+00	0,300065	3,697194	0,213916
	50	0,00E+00	3,49E-01	3,03E+00	0,378131	4,773935	0,268318
	60	0,00E+00	2,94E-01	1,95E+00	0,448755	5,854582	0,318456
	70	0,00E+00	5,88E-03	3,44E+00	0,525266	6,971155	0,370068
	80	0,00E+00	2,07E-27	4,02E+00	0,604550	8,100047	0,420565
	90	0,00E+00	3,13E-20	3,51E+00	0,689727	9,370560	0,469736
	100	0,00E+00	3,43E-25	2,09E+00	0,753482	10,468643	0,520913
Booth	10	1,51E-16	7,83E-01	1,16E-02	0,030539	0,566618	0,012690
	20	4,14E-17	6,50E-01	1,82E-02	0,044914	1,216361	0,020079
	30	3,85E-17	6,24E-01	1,47E-05	0,083222	1,920431	0,028284
	40	3,53E-17	6,63E-01	6,99E-04	0,102553	2,742401	0,037351
	50	2,26E-17	6,03E-01	7,98E-05	0,107827	3,457020	0,045551
	60	1,39E-17	4,26E-01	2,28E-03	0,125350	4,303888	0,054218
	70	1,34E-17	8,20E-01	7,74E-03	0,146475	5,077780	0,061906
	80	1,30E-17	4,31E-01	1,88E-03	0,164568	5,840491	0,072140
	90	1,40E-17	5,82E-01	1,93E-03	0,183986	6,744297	0,079553
	100	1,40E-17	7,62E-01	7,59E-03	0,206001	7,566422	0,089137
Easom	10	−8,32E-01	−8,28E-01	−2,33E-01	0,033830	0,541508	0,014166
	20	−9,62E-01	−8,56E-01	−4,33E-01	0,066373	1,181808	0,026534
	30	−1,00E+00	−8,82E-01	−5,22E-01	0,093602	1,866469	0,035950
	40	−1,00E+00	−9,03E-01	−6,96E-01	0,097185	2,584197	0,044718
	50	−9,99E-01	−9,39E-01	−6,78E-01	0,116278	3,405775	0,055735
	60	−1,00E+00	−9,17E-01	−8,98E-01	0,138095	4,143588	0,065220
	70	−1,00E+00	−9,03E-01	−9,33E-01	0,157819	4,994868	0,066310
	80	−1,00E+00	−9,06E-01	−9,33E-01	0,181854	5,755332	0,085933
	90	−1,00E+00	−8,61E-01	−9,00E-01	0,206434	6,669711	0,094607
	100	−1,00E+00	−9,24E-01	−9,64E-01	0,227616	7,523491	0,104991
Griewank	10	7,17E-03	6,90E-15	3,21E+02	0,260855	0,946143	0,444829
	20	3,86E-04	0,00E+00	1,63E+02	0,510473	1,849194	0,868826
	30	3,02E-04	0,00E+00	2,47E+02	0,771970	2,745251	1,293660
	40	7,24E-06	0,00E+00	2,02E+02	1,026819	3,698099	1,699266
	50	2,79E-10	0,00E+00	1,79E+02	1,274411	4,519503	2,149955
	60	6,42E-12	0,00E+00	1,57E+02	1,533927	5,502049	2,515776
	70	1,71E-12	0,00E+00	1,48E+02	1,805498	6,512473	2,970469
	80	2,60E-13	6,96E-03	2,05E+02	2,041599	7,428721	3,335748
	90	1,11E-13	0,00E+00	1,20E+02	2,306967	8,467161	3,792227
	100	1,55E-12	0,00E+00	1,89E+02	2,547220	9,511772	4,142624

(*continued*)

Table 5. (*continued*)

Function	Pop.	Solutions			Time (s)		
		ABC	ACO	PSO	ABC	ACO	PSO
Rastrigin	10	2,80E-01	4,35E+02	4,82E+02	0,239999	0,725733	0,405736
	20	3,06E-04	4,53E+02	4,04E+02	0,467220	1,533267	0,781653
	30	2,51E-06	3,58E+02	3,93E+02	0,699583	2,616785	1,158362
	40	3,48E-09	4,25E+02	3,93E+02	0,931612	3,378083	1,555668
	50	8,23E-11	2,06E+02	3,88E+02	1,165836	4,696016	1,930535
	60	5,84E-12	1,88E+02	3,58E+02	1,401252	5,768915	2,309815
	70	1,88E-11	1,99E+02	3,99E+02	1,637102	6,787756	2,688676
	80	7,59E-13	2,05E+02	3,45E+02	1,854091	7,775599	3,025114
	90	5,35E-13	2,01E+02	3,05E+02	2,097354	8,923411	3,478108
	100	2,39E-12	1,90E+02	3,25E+02	2,320607	10,013488	3,947903
Rosenbrock	10	4,02E+01	5,10E+03	9,52E+05	0,212855	0,731430	0,355297
	20	2,56E+01	6,51E+03	4,62E+05	0,423005	1,523693	0,732902
	30	1,75E+01	7,68E+03	3,56E+05	0,618770	2,412705	1,065611
	40	1,40E+01	8,25E+03	3,35E+05	0,834661	3,310153	1,349149
	50	1,13E+01	9,62E+03	1,30E+05	1,033823	4,269835	1,709407
	60	1,12E+01	9,40E+03	3,95E+05	1,234289	5,192442	2,031221
	70	1,05E+01	9,31E+03	2,18E+05	1,436416	6,246066	2,393251
	80	1,07E+01	9,29E+03	2,60E+05	1,644925	7,263775	2,737766
	90	9,50E+00	9,16E+03	1,31E+05	1,837177	8,264279	3,202591
	100	9,16E+00	9,19E+03	2,14E+05	2,042155	9,346481	3,466199
Sphere	10	3,15E-17	4,71E-19	4,69E-01	0,075137	0,783766	0,050192
	20	1,83E-17	1,51E-32	5,89E-02	0,131975	1,658922	0,085171
	30	2,13E-17	1,29E-41	1,41E-01	0,195781	2,646467	0,108543
	40	1,60E-17	4,47E-48	9,39E-02	0,259194	3,664331	0,137414
	50	1,53E-17	7,07E-54	1,70E-01	0,323400	4,738335	0,160632
	60	1,38E-17	1,41E-56	9,21E-01	0,382747	5,829655	0,188244
	70	1,04E-17	1,48E-58	8,92E-02	0,448285	6,931901	0,227250
	80	1,54E-17	5,18E-60	3,92E-01	0,504095	8,013681	0,248711
	90	1,20E-17	3,90E-61	7,18E-02	0,564699	9,201093	0,279173
	100	1,08E-17	3,03E-62	1,31E-02	0,630266	10,331690	0,314831

population sizes. PSO achieved the best execution time. In Alpine, Booth, and Easom, the best execution time results for PSO are around two times faster than ABC, and ACO a dozen more than ABC.

ACO presented the best solution for the *Griewank* function, reaching the exact solution in 8 of the 10 population size evaluated. PSO appearing to be stuck in a local minimum, presenting the worst solution. The best execution time for *Griewank* was obtained by the ABC algorithm and ACO the worst.

In the *Rastrigin* and *Rosenbrock* function none of the algorithms achieved the exact results. However, the ABC algorithm presents the best solution in both functions. In *Rastrigin* ACO and PSO are close results. In terms of performance, ABC has also the best execution time for both cases.

In the *Sphere* test, the ACO achieved results close to the exact solution. However, the best execution time was obtained for the PSO algorithm.

When the solution was evaluated, the ABC algorithm obtained the best solutions for five of the algorithms: *Alpine, Booth, Easom, Rastrigin* and *Rosenbrock*, while the algorithm ACO obtained the best solutions for the *Griewank* and *Sphere* algorithms. PSO resulted in worse solutions than the ABC and ACO algorithms.

The PSO algorithm was more efficient in terms of performance for the *Alpine, Booth, Easom,* and *Sphere* algorithms, while the ABC algorithm obtained the better performance rates for the *Griewank* and *Rastrigin* and *Rosenbrock* algorithms. The ACO algorithm obtained the lowest performance when compared to the other algorithms.

6   Conclusions and Future Work

The bio-inspired algorithms correspond to a range of efficient strategies for solving problems in several areas, from simple data analysis to complex problems such as distributed power generation or engineering simulations. The contribution of this work was to provide a library that allows the use of bio-inspired algorithms to solve problems.

In this article, the numerical evaluation and performance of the proposed library were assessed. Tests of the PSO, ABC, and ACO algorithms were performed using 7 functions and 10 different population sizes for each function. The ABC algorithm was the best in most cases to find the optimal solution and with less execution time in other cases.

As future work, it is intended to include new bio-inspired algorithms to the library. We also want to validate it by testing the algorithms on a real problem, such as that of distributed energy production. We also expect to reduce the execution time in a new version of the library using parallelization techniques.

References

1. Abraham, S., Sanyal, S., Sanglikar, M.: Particle swarm optimization based Diophantine equation solver. Int. J. Bio Inspired Comput. **2**(2), 100–114 (2010)
2. Ardeh, M.A.: BenchmarkFcns toolbox: a collection of benchmark functions for optimization (2016). http://benchmarkfcns.xyz/
3. Couto, D.C., Silva, C.A., Barsante, L.S.: Otimização de funções multimodais via técnica de inteligência computacional baseada em colônia de vaga-lumes. In: Proceedings of the XXXVI Iberian Latin American Congress on Computational Methods in Engineering. CILAMCE, Rio de Janeiro, RJ (2015)
4. Dorigo, M., Birattari, M., Stutzle, T.: Ant colony optimization. IEEE Comput. Intell. Mag. **1**(4), 28–39 (2006)
5. Engelbrecht, A.P.: Computational Intelligence: An Introduction. Wiley, Chichester (2007)

6. Karaboga, D., Gorkemli, B., Ozturk, C., Karaboga, N.: A comprehensive survey: artificial bee colony (ABC) algorithm and applications. Artif. Intell. Rev. **42**(1), 21–57 (2014)
7. Kennedy, J., Eberhart, R.C.: Swarm Intelligence. Morgan Kaufmann Publishers Inc., San Francisco (2001)
8. Nievergelt, J., Gasser, R., Mäser, F., Wirth, C.: All the needles in a haystack: can exhaustive search overcome combinatorial chaos? In: van Leeuwen, J. (ed.) Computer Science Today. LNCS, vol. 1000, pp. 254–274. Springer, Heidelberg (1995). https://doi.org/10.1007/BFb0015248
9. Serapiao, A.: Fundamentos de Otimização por Inteligência de enxames: Uma Visão Geral. Controle y Automacao **20**, 271–304 (2009). https://doi.org/10.1590/S0103-17592009000300002
10. Silva, M.A.L., de Souza, S.R., Souza, M.J.F., de Franca Filho, M.F.: Hybrid metaheuristics and multi-agent systems for solving optimization problems: a review of frameworks and a comparative analysis. Appl. Soft Comput. **71**, 433–459 (2018)
11. Talukder, S.: Mathematicle modelling and applications of particle swarm optimization (2011)
12. Wilkinson, B., Allen, M.: Parallel Programming-Techniques and Applications using Networked Workstations and Parallel Computers, 2nd edn. Prentice Hall, Newark (1998)
13. Yang, X.S.: Test problems in optimization (2010). arXiv preprint arXiv:1008.0549

Numerical Enhancements and Parallel GPU Implementation of a 3D Gaussian Beam Model

Rogério M. Calazan[1]([✉]) [ID], Orlando C. Rodríguez[2] [ID], and Sérgio M. Jesus[2] [ID]

[1] Institute of Sea Studies Admiral Paulo Moreira, Arraial do Cabo, Brazil
moraes.calazan@marinha.mil.br
[2] Laboratory of Robotics and Systems in Engineering and Science (LARSyS),
Campus de Gambelas, Universidade do Algarve, Faro, Portugal
https://www.marinha.mil.br/ieapm/
http://www.siplab.fct.ualg.pt

Abstract. Despite the increasing performance of modern processors it is well known that the majority of models that account for 3D underwater acoustic predictions still require a high computational cost. In this context, this work presents strategies to enhance the computational performance of a ray-based 3D model. First, it is presented an optimized method for acoustic field calculations, that accounts for a large number of sensors. Second, the inherent parallelism of ray tracing and the high workload of 3D propagation are carefully considered, leading to the development of parallel algorithms for field predictions using a GPU architecture. The strategies were validated through performance analyses and comparisons with experimental data from a tank scale experiment, and the results show that model predictions are computationally efficient and accurate. The combination of numerical enhancements and parallel computing allowed to speedup model calculations for a large number of receivers.

Keywords: Underwater acoustics · Numerical modeling · 3D propagation · GPU parallel computing

1 Introduction

Since its early development, underwater predictions of acoustic 3D propagation are well known to be highly time consuming; initial research in this area relied in fact on dedicated computer architectures to carry on model execution [1,2]. Even today, despite the increasing performance of modern processors, models that take into account 3D propagation still have a high computational cost [3]. The corresponding (high) runtime of a single prediction can easily explain why 3D

Supported by Institute of Sea Studies Admiral Paulo Moreira, Brazilian Navy. Thanks are due to the SiPLAB research team, LARSyS, FCT, University of Algarve.

O. Gervasi et al. (Eds.): ICCSA 2020, LNCS 12249, pp. 485–500, 2020.
https://doi.org/10.1007/978-3-030-58799-4_36

models are generally put aside when dealing with problems of acoustic inversion [5]. Additionally, predictions of sonar performance can take advantage of full 3D modeling to improve accuracy, with a ray model playing a central role in such task due to its capability to handle high frequencies, say, above 500 Hz [6]. Furthermore, monitoring of shipping noise represents also an important field of research since shipping noise propagates at long distances, with 3D effects becoming more relevant as distances increase. On the other side, ship density is high in the vicinity of harbors, making bottom interactions and 3D effects important in shallow waters and littoral environments [8,9].

Computational model performance can be significantly improved through the combination of numerical enhancements with parallel computing. In this sense, the generation of optimized algorithms requires task-specific analysis and development of new methods, that parallel computing cannot be able to overcome alone. Additionally, the performance of *graphic processing units* (hereafter GPU) motivated several implementations of scientific applications, including underwater acoustic models. For instance, a split-step Fourier parabolic equation model implemented in a GPU is discussed in [11] and the discussion presented in [12] describes a GPU-based version of a Beam-Displacement Ray-Mode code; both works considered only highly idealized 2D waveguides and the results showed a significant improvement regarding the computational performance. A parallel version of the C-based version of TRACEO (called cTRACEO), based on a GPU architecture, was discussed in detail in [13]; the discussion showed that parallelization drastically reduces the computational burden when a large number of rays needs to be traced; performance results for such 2D model outlined the computational advantages of considering the GPU architecture for the 3D case. Preliminary research into parallelization in a coarse-grained fashion was also explored using OpenMPI [14,15]. Performance analysis showed that the parallel implementation followed a linear speedup when each process was addressed to a single physical CPU core. However, such performance was achieved at the cost of using high-end CPUs, designed for computer servers without network communication (which probably would decrease the overall performance). Thus, the *best* parallel implementation was 12 times faster than the sequential one, meaning that the execution took place in a CPU with 12 physical cores.

The work presented here describes the development of numerical enhancements and optimization of the sequential version of the *TRACEO3D* Gaussian beam model [16–18], leading to improvements of its performance; the description is followed by further analysis of the GPU hardware multithread, and the coding elements of the time consuming model structure, into a parallel algorithm that takes advantage of the GPU architecture. This development looks carefully to the inherent parallelism of ray tracing and to the high workload of computations for 3D predictions. Furthermore, validation and performance assessments are presented considering experimental data from a tank scale experiment. The results show that a remarkable performance was achieved without compromising accuracy. This paper is organized as follows: the TRACEO3D model is described in 2; numerical enhancement are presented in Sect. 3; the detailed structure of

parallel GPU implementation is presented in Sect. 4; Sect. 5 presents the validation results, in which experimental data were considered. Conclusions and future
work are presented in Sect. 6.

2   The TRACEO3D Gaussian Beam Model

The *TRACEO3D* Gaussian beam model [17] corresponds to a three-dimensional
extension of the *TRACEO* 2D model [19]. TRACEO3D produces a prediction
of the acoustic field in two steps: first, the set of Eikonal equations is solved
in order to provide ray trajectories; second, ray trajectories are considered as
the central axes of Gaussian beams, and the acoustic field at the position of a
given hydrophone is computed as a coherent superposition of beam influences.
The beam influence is calculated along a given normal based on the expression
[3, 20, 21]

$$P(s, n_1, n_2) = \frac{1}{4\pi} \sqrt{\frac{c(s)}{c(0)} \frac{\cos \theta(0)}{\det \mathbf{Q}}} \Phi \exp\left[-i\omega\tau(s)\right] , \tag{1}$$

where $\Phi = \displaystyle\prod_{\substack{i=1,2 \\ j=1,2}} \Phi_{ij}$ and the coefficients are given by

$$\Phi_{ij} = \exp\left[-\left(\frac{\sqrt{\pi|n_i n_j|}}{\Delta\theta} Q_{ij}^{-1}\right)^2\right] , \tag{2}$$

with $\Delta\theta$ standing for the elevation step between successive rays, and Q_{ij}^{-1} representing the elements of $\mathbf{Q}^{-1}$; n_1 and n_2 are calculated through the projection of
$\mathbf{n}$ onto the polarized vectors along the ray; s corresponds to the ray arc length,
and $c(s)$ and $\tau(s)$ stand for the sound speed and travel time along the ray,
respectively; the complex matrix $\mathbf{Q}(s)$ describes the beam spreading, while $\mathbf{P}(s)$
describes the beam slowness.

3   Numerical Enhancements

3.1   Calculation of Normals

In the original version of TRACEO3D ray influence at a receiver located at the
position $\mathbf{r}_h$ was calculated using the following procedure:

– Divide the ray trajectory into segments between successive transitions (sur-
  face/bottom reflection, or bottom/surface reflection, etc.);
– Proceed along *all* segments to find *all* ray normals to the receiver; to this
  end:
  • Consider the ith segment; let $\mathbf{r}_A$ and $\mathbf{r}_B$ be the coordinates of the begin-
    ning and end of the segment, respectively, and let $\mathbf{e}_A$ and $\mathbf{e}_B$ be the
    vectors corresponding to $\mathbf{e}_s$ at A and B; where $\mathbf{e}_s$ is defined as a unitary
    vector, which is tangent to the ray.
  • Calculate the vectors $\Delta\mathbf{r}_A = \mathbf{r}_h - \mathbf{r}_A$ and $\Delta\mathbf{r}_B = \mathbf{r}_h - \mathbf{r}_B$.

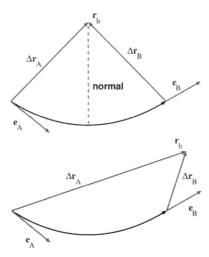

Fig. 1. Normal search along a ray segment, with $P_A = \mathbf{e}_A \cdot \Delta \mathbf{r}_A$ and $P_B = \mathbf{e}_B \cdot \Delta \mathbf{r}_B$. *Top:* the hydrophone is at a position for which $P_A \times P_B < 0$, thus a normal exists, and it can be found by bisection somewhere along the segment. *Bottom:* the hydrophone is at a position for which $P_A \times P_B > 0$; thus, there is no normal and the ray segment has no influence at the hydrophone position.

- Calculate the inner products $P_A = \mathbf{e}_A \cdot \Delta \mathbf{r}_A$ and $P_B = \mathbf{e}_B \cdot \Delta \mathbf{r}_B$.
- If $P_A \times P_B < 0$ a normal exists and it can be found through bisection along the segment; once the normal is found the corresponding influence at the receiver can be calculated.
- If $P_A \times P_B > 0$ there is no normal (and no influence at the receiver); therefore, one can move to segment $i + 1$.
- The ray influence at the receiver is the sum of influences from all segments.

The search for a normal along a ray segment is illustrated in Fig. 1. The influence of a Gaussian beam decays rapidly along a normal, but it never reaches zero; therefore, the procedure is to be repeated for *all* rays and *all* receivers.

As shown in [17] field predictions using this method exhibit a good agreement with experimental data, but the runtime is often high and increases drastically as range, number of rays and number of sensors increase. The numerical enhancement of field calculations is described in the next Section.

3.2 The Receiver Grid Strategy

To reduce drastically the runtime without compromising accuracy one can follow the approach described in [3], which suggests that for each ray segment one considers *not all* receivers, but only those "insonified" (i.e. bracketed) between the endpoints of a ray segment. For a given subset of receivers one can proceed sequentially within the subset (for instance, from the ocean surface to the ocean

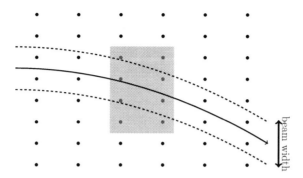

Fig. 2. The receiver grid (vertical view): the black dots represent *all* the receivers of a rectangular array, while the solid line represents the ray trajectory; the ray influence is only relevant within the limits of the beam width, represented by the dashed lines, and the gray rectangle represents the grid of receivers considered for the calculation of ray influence.

bottom), and rely on simple algebra to determine the parameters of ray influence; the procedure is then repeated for all ray segments. An examination of the BELLHOP3D ray tracing code [22] reveals that the determination of the subset of receivers is achieved by testing *all* receiver positions within the array, looking to find the ones within the endpoints of the ray segment. The approach implemented in TRACEO3D goes further, and looks to optimize the selection of a subset of receivers (called the *receiver grid*, see Fig. 2) based on the following considerations:

- A "finite" beam width W is defined along the ray, given by the expression

$$W = \left| \frac{Q_{11}(s)\Delta\theta}{\cos\theta(s)} \right| . \tag{3}$$

- There is no need to consider all receivers from the ocean surface to the ocean bottom, but only those within the neighborhood defined by W

The main idea on the basis of this strategy is that beyond the distance defined by W the influence is too small to be of any importance. Therefore, as one moves along each ray segment the receiver grid is determined by the receivers bracketed by both *the ray segment and* W. In this way one can avoid not only the query in the entire set of receivers forming the array, but also the query of all receivers bracketed by the ray segment. The method can take advantage of Cartesian coordinates to determine efficiently the indexes of the receivers lying inside the receiver grid. The specific details of this enhancement are described in the next Section.

3.3   Ray Influence Calculation Algorithm

The specific details of optimization are shown in the pseudo-code of Algorithm 1, which summarizes the sequential steps regarding field calculations. Let n and r

stand for the number of rays and receivers, respectively. The optimization starts by tracing the ray for a given pair of launching angles. Then, the algorithm marches through the ray segments, and solves the dynamic equations to calculate the ray amplitude and the beam spreading. As shown in lines 13 and 14 a subset of receivers is computed from r for each segment k of the ray. The ray influence is computed only if a normal to the receiver is found at a given segment (see lines 15 and 16). Line 23 presents the final step, in which coherent acoustic pressure for each receiver is calculated. As will be shown in Sect. 4 the set of nested loops constitutes a fundamental stage of the algorithm, allowing a substantial improvement of the model's performance. Details regarding the computation of the receiver grid are shown in the pseudo-code of Algorithm 2, where the integers l_{low} and l_{high} control the array indexes that form the receiver grid according to W at each coordinate axis. The receiver indexes increase or decrease their values, considering only the neighborhood, according to the position of the ray segment inside the receiving array; the entire procedure is designed to be flexible enough to account for different ray directions.

Algorithm 1. Sequential ray influence calculation

1: **load** environmental data
2: **let** ϕ = set of azimuth angles
3: **let** θ = set of elevation angles
4: **let** r = set of receivers
5: **consider** $n = length\,(\phi) \times length\,(\theta)$
6: **for** $j := 1 \to length\,(\phi)$ **do**
7:    **for** $i := 1 \to length\,(\theta)$ **do**
8:       **while** ray (θ_i, ϕ_j) exists **do**
9:          **solve** the Eikonal equations for segment k
10:       **end while**
11:       **for** $k := 1 \to raylength$ **do**
12:          **solve** the dynamic equations of segment k
13:          **calculate** W at segment k
14:          **compute** receiver grid g from r according to W
15:          **for** $l := 1 \to length\,(g)$ **do**
16:             **if** ray_k and g_l are $\perp$ **then**
17:                **compute** ray_k influence at g_l
18:             **end if**
19:          **end for**
20:       **end for**
21:    **end for**
22: **end for**
23: **return** the coherent acoustic pressure for each receiver

---

Algorithm 2. Compute the receiver grid

1: **let** l_{low} = lower array index inside grid
2: **let** l_{high} = high array index inside grid
3: **consider** W as beam width at ray_k
4: **while** l_{high} or l_{low} are inside grid **do**
5:     **if** $W < r\left(l_{low} - 1\right)$ **then**
6:         **decrement** l_{low}
7:     **else if** $W > r\left(l_{low}\right)$ **then**
8:         **increment** l_{low}
9:     **else**
10:         **exit**
11:     **end if**
12:     **if** $W < r\left(l_{high}\right)$ **then**
13:         **decrement** l_{high}
14:     **else if** $W > r\left(l_{high} + 1\right)$ **then**
15:         **increment** l_{high}
16:     **else**
17:         **exit**
18:     **end if**
19: **end while**

---

4   Parallel GPU Implementation

The proposed parallel implementation is addressed for NVIDIA [23] GPUs, using the Compute Unified Device Architecture (CUDA). This programming model implements a data-parallel function, denominated *kernel*, which is executed by all threads during a parallel step. Generally speaking, a CUDA program starts in the *host*, as a CPU sequential program, and when a kernel function is launched, it is executed in a grid of parallel threads into the GPU or *device*.

4.1   Memory Organization

Acoustic predictions in a three-dimensional scenario demand the tracing of a high number of rays. In the sequential algorithm the ray trajectory information (such as, for instance, Cartesian coordinates, travel time, complex decay, polarized vectors, caustics, matrices **P** and **Q**, etc.) are stored in memory to be used at later steps. Such storage makes sense considering that the sequential algorithm keeps one ray at a time in memory. However, handling thousands of rays in parallel rapidly exceeds the available memory in a given device. To circumvent this issue calculation of ray paths and amplitudes are performed in a single step, for each ray segment at a time, storing in memory only the values required to execute such calculation. A sketch of this strategy is presented in Fig. 3, where the horizontal arrow represents the direction in which memories are updated, and t corresponds to the current time step in which calculations are taking place; $t - 1$ and $t - 2$ represent previous steps, that are required to be held in memory. Small arrows connecting memory positions represent the values accessed by the

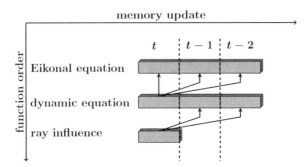

Fig. 3. Schematic representing the memory update sequence (horizontal arrow), where t stands for current computation time and $t - 1$ and $t - 2$ for previous times when values are held in memory. Small arrows connecting memory positions represent the values accessed for the corresponding function to perform computations in time t. The vertical arrow represents the order in which the functions are executed for a single ray segment.

corresponding function to perform computations in time t. The vertical arrow indicates the order in which functions are computed in the current time. After calling the functions for a given ray segment the values stored in memory at time $t - 1$ are copied to the position $t - 2$, and the values regarding t are copied to position $t-1$, meaning that the values at $t-2$ are discharged. A new iteration then starts to solve the next ray segment, following the same rules. In this way, the storage requires only three segments to be held in memory, reducing drastically the amount of data stored. This organization allows further updates of data into registers to be kept, reducing the global memory access and overcoming the problems of divergences in the pattern of memory access, a drawback of parallel ray tracing algorithm. The performance of memory access is also increased by loading part of the environmental information into the shared memory at the kernel initialization.

An overview of how data from the parallel implementation is organized into device memories is shown in Table 1. The memory type was chosen considering the respective data size and the frequency in which the data is accessed. For instance, data regarding environmental boundaries (surface and bottom) was initially put into the shared memory. However, when representing 3D waveguides, the number of coordinates became too large to fit in this type of memory and the data was thus moved to the global memory. On the other hand, the sound speed data was kept in shared memory since it was frequently accessed during ray trajectory calculations and the access takes place in an unpredictable order.

4.2   Parallel Field Calculation

A general view of the parallel version of field calculation is presented in Fig. 4 and in Algorithm 3, regarding the parallel flowchart and procedures, respectively.

Table 1. TRACEO3D memory organization into a parallel implementation: n_{ssp} is the number of points in the sound speed profile, n_{sur} and n_{bot} is the number of grid points defining the surface and bottom, respectively; n stands for the number of rays, h represents the number of receivers and m is the number of candidate regions.

Data	Symbol or name	Type	Size
Source information		Shared	12
Environment parameters		Shared	14
Sound speed profile		Shared	$3 + n_{ssp}$
Array coordinates		Global	$3 + 3 \times h$
Surface coordinates		Global	$7 + n_{sur}$
Bottom coordinates		Global	$7 + n_{bot}$
Coherent acoustic pressure	cpr	Global	h
cpr all rays	$ncpr$	Global	$n \times h$
Ray coordinates		Register/local	3×3
Travel time	τ	Register/local	3
Complex amplitude	A	Register/local	3
Polarized vectors		Register/local	3×3
P, Q	**P, Q**	Register/local	3×4

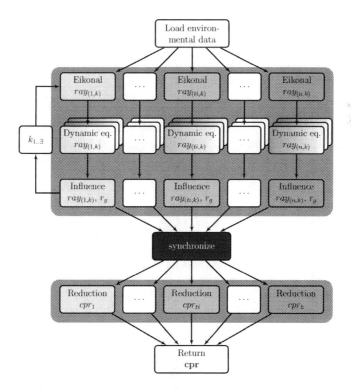

Fig. 4. Parallel flowchart of field calculation: blue regions correspond to kernel functions executed in parallel into device; outside blue regions the code is executed sequentially into the host. (Color figure online)

Algorithm 3. Parallel field calculation

1: **load** environmental data
2: **let** ϕ = set of azimuth angles
3: **let** θ = set of elevation angles
4: **let** r = set of receivers
5: **consider** $n = length\,(\phi) \times length\,(\theta)$
6: **let** p = number of threads per block
7: **let** b = n/p (number of blocks)
8: **kernel** $\ll b, p \gg$ ray influence calculation
9: **synchronize**
10: **let** p = number of threads per block
11: **let** b = h/p (number of blocks)
12: **kernel** $\ll b, p \gg$ pressure by sensor reduction
13: **return** the coherent acoustic pressure

Two main stages can be noted which corresponds to the blue regions at the flowchart and as parallel kernels in the algorithm. The strategy adopted the inherent ray tracing parallelism, addressing each pair of launching angles (θ, ϕ) as a single parallel thread, even though it could lead to the concentration of additional work per thread. However, since several instructions at the dynamic equations step are independent, they need to be organized sequentially to take advantage of instruction level parallelism (represented by the parallel blocks in depth) in the corresponding step. The proposed parallel algorithm is logically organized in a grid of b blocks, where each block has p threads. The first kernel (line 8) calculates the ray influence, where the number of threads launched into the device corresponds to n.

An overview of the kernel *ray influence calculation* is shown in Algorithm 4. Each thread computes the propagation of a single ray and its contributions to the entire field; the contributions are stored separately for each ray. It should be noted that, as shown in Table 1, the size of *ncpr* corresponds to $n \times h$ and the index to access global memory is calculated using a relative value of the grid index l' (see line 10 of Algorithm 4). After the kernel execution a device synchronization is performed to ensure that the acoustic field calculation for all rays was concluded. Then, a second kernel is launched to perform a parallel reduction over the values in *ncpr*. Each thread is addressed to a given receiver, and it adds sequentially the contribution of each ray to the corresponding receiver.

5   Validation

5.1   Implementation

The TRACEO3D model was written using the FORTRAN programing language in double precision. Thus, the interface was kept in FORTRAN, using its functions to read the inputs and write the outputs, while the parallel portion was encoded using the CUDA C platform. The CUDA C and FORTRAN environments were connected using the ISO C Binding library [24], which is a standardized way to generate procedures, derived-type declarations and global variables, which are inter-operable with C. The parallel implementation was compiled in a single precision version (numerical stability was already addressed in [13]); comparisons between the parallel and the sequential version will be shown to properly clarify this issue. The single precision version allows the use of low-end devices or mobile equipments to provide predictions with high performance. Additionally, the FORTRAN sequential implementation was compiled with the optimization flag $-O3$, which was found to decrease the total runtime in 50%. The hardware and software features that were addressed when comparing the sequential and parallel model version of TRACEO3D are shown in Table 2. Fifteen runs were performed for the validation case. The maximum and minimum values were then discarded, and the average runtime was computed from the remaining thirteen runs.

Algorithm 4. Kernel ray influence calculation

1: **let** ti = block index × grid index + thread index
2: **let** $\theta_i = ti \bmod length(\boldsymbol{\theta})$
3: **let** $\phi_j = ti/length(\boldsymbol{\theta})$
4: **while** ray (θ_i, ϕ_j) exists **do**
5:      **solve** the Eikonal equations for segment k
6:      **compute** the dynamic equations for segment k
7:      **compute** the receiver grid g from r
8:      **for** $l := 1 \rightarrow length(g)$ **do**
9:          **if** ray_k and g_l are $\perp$ **then**
10:             $ncpr[ti + n \times l'] =$ acoustic pressure regarding ray_k at g_l
11:         **end if**
12:     **end for**
13: **end while**

5.2   The Tank Experiment

The laboratory-scale experiment took place at the indoor tank of the *Laboratoire de Mécanique des Fluides et d'Acoustique – Centre National de la Recherche Scientifique* (LMA-CNRS) laboratory in Marseille. The experiment was carried out in 2007 in order to collect 3D acoustic propagation data using a tilted bottom in a controlled environment. A brief description of the experiment (which is described in great detail in [5,25]) is presented here. The inner tank dimensions 10 m long, 3 m wide 1 m deep. The bottom was filled with sand and a rake was used to produce a mild slope angle $\alpha \approx 4.5°$. For simulation purposes a scale factor of 1000 : 1 is required to properly account for the frequencies and lengths of the experimental configuration in the model. Thus, experimental frequencies in kHz become model frequencies in Hz, and experimental lengths in mm become model lengths in m. For instance, an experimental frequency of

Table 2. *Host/Device* hardware and software features.

Feature	Value	Unit
Host - CPU Intel i7-3930k		
Clock frequency	3500	MHz
Compiler	gfortran 5.4.0	–
Optimization flag	$-O3$	–
Device - GPU GeForce GTX 1070		
CUDA capability	6.1	–
CUDA driver	9.1	–
Compiler	nvcc 9.1.85	–
Optimization flag	none	–
Clock frequency	1683	MHz
Number of SM	15	–
Max threads per SM	2048	–
Warp size	32	–

180.05 kHz becomes a model frequency of 180.05 Hz, and an experimental distance 10 mm becomes a model distance of 10 m. Sound speed remains unchanged, as well as compressional and shear attenuations. The *ASP-H* data set (for horizontal measurements of across-slope propagation) is composed of time signals, recorded at a fixed receiver depth denominated z_r, and source/receiver distances starting from $Y = 0.1$ m until $Y = 5$ m in increments of 5 mm, providing a sufficiently fine representation of the acoustic field in terms of range. Three different source depths were considered, namely $z_s = 10$ mm, 19 mm and 26.9 mm, corresponding to data subsets referenced as ASP-H1, ASP-H2 and ASP-H3, respectively. Acoustic transmissions were performed for a wide range of frequencies. However, comparisons are presented only for data from the ASP-H1 subset with the highest frequency of 180.05 kHz; this is due to the fact that the higher the frequency the better the ray prediction. Bottom parameters corresponded to $c_p = 1700$ m/s, $\rho = 1.99$ g/cm^3 and $\alpha_p = 0.5$ dB/λ. Sound speed in the water was considered constant, and corresponded to 1488.2 m/s. Bottom depth at the source position was $D(0) = 48$ mm.

5.3   Comparisons with Experimental Data

The set of waveguide parameters provided by the tank scale experiment was used to calculate predictions in the frequency domain. Transmission loss (TL) results are presented in Fig. 5, where *Bisection* means the original algorithm that the sequential version of TRACEO3D uses to calculate ray influence, *Grid* stands for the sequential method presented in Sect. 3.2, and *GPU Grid* corresponds to the parallel implementation. In general, model predictions were able to follow accurately the experimental curve over the full across-slope range. Nevertheless, a slight shift in phase can be observed 2 km and 2.4 km in all simulation predictions. Besides, minor discrepancies can be noted between the predictions at the far field.

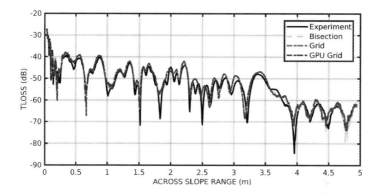

Fig. 5. Comparisons with the experimental data for LMA CNRS H1 @ 180.05 kHz.

5.4   Performance Analysis

The best result found during the execution configuration optimization is presented on both Table 3 and Fig. 6. Speedup rates are presented separately, comparing the improvement regarding the numerical enhancement and the improvement achieved with the parallel GPU implementation. Thus, the speedup ratio of CPU (Grid) is calculated dividing the Bisection runtime by the Grid runtime, and for the CPU + GPU (Grid) dividing the Grid runtime by the GPU runtime. It is important to remark that the CPU (Grid) was able to decrease the runtime in 2.83 times, while the parallel GPU implementation achieved 60 times of performance, which indeed represents a outstanding improvement. Combining both speedups the total improvement was about 170 times (2.83 × 60.11), reducing the runtime **from 542.3 s to 3.18 s**. The mean square error (MSE) between the prediction generated by each model implementations and the experimental data is shown in Table 3 and Fig. 7. One can see that the difference among the implementations are of the same order of magnitude regarding the whole array of receivers. However, due to the far field discrepancies, the MSE differences among implementations are 0.084 dB from Grid to Bisection and 0.011 dB from GPU

Table 3. Results of runtime, speedup ratio and accuracy for TL predictions.

Model	CPU (Bisection)	CPU (Grid)	CPU + GPU (Grid)
Runtime (s)	542.3	191.16	3.18
Speedup ratio	1	2.83	60.11
MSE (dB)	−9.344	−9.260	−9.333

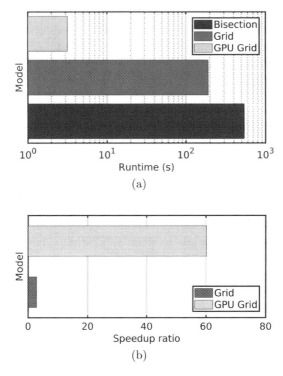

Fig. 6. (a) Runtime and (b) speedup for TL predictions of the tank scale experiment. Speedup rates are presented separately, comparing the improvement regarding the numerical enhancement and the improvement achieved with the parallel GPU implementation.

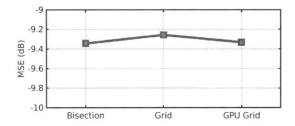

Fig. 7. MSE of TRACEO3D predictions against experimental data (LMA CNRS H1 @ 180.05 kHz) using three different approaches: Bisection, Grid and GPU Grid.

grid to Bisection. The parallel implementation only achieves such accuracy by using IEEE 754 compatible mathematical functions [26], and compiling without the flag *–fast-math*; because this flag enables performance optimization at the cost of introducing some numerical inaccuracies.

6   Conclusion

The discussion presented in this paper proposed numerical enhancements and a parallel GPU implementation of the TRACEO3D model. The calculation of ray influence was addressed using a receiver grid, i.e. a subset of adjacent receivers within the array, with the goal of decreasing runtime while keeping accuracy. After the enhancement of numerical issues parallel algorithms were developed considering a GPU architecture, that could take advantage of the inherent ray tracing parallelism and the high workload of 3D propagation. The validation results were performed using experimental data collected from a tank scale experiment. The method was found to be computationally efficient and accurate dealing with arrays containing a large number of sensors, although some optimization was required in order to define the proper borders of ray influence given by the finite beam width. Performance results show that the implementation achieved a speedup around 170 times faster than the sequential one, combining the improvements of numerical enhancement and parallel implementation without compromising accuracy. Despite the significant improvements in speedup it can be not guaranteed that the adopted parallel algorithms exhausted all solutions of parallelization. It is believed that additional combinations of thread granularities and memory organization can have the potential to achieve a greater performance. The speedup issue is certainly of immense interest for intensive applications of a 3D model, a topic which is currently under intense discussion. Future work will be oriented to further validation in typical ocean environments with complex bathymetries and tests with different thread granularities, requiring new memory organization and execution configuration parameters. Moreover, it is expected to carry on additional scalability analyses using large number of sensor with different sizes and shapes.

References

1. Johnson, O.G.: Three-dimensional wave equation computations on vector computers. Proc. IEEE **72**, 90–95 (1984)
2. Tolstoy, A.: 3-D propagation issues and models. J. Comput. Acoust. **4**(03), 243–271 (1996)
3. Jensen, F.B., Kuperman, W.A., Porter, M.B., Schmidt, H.: Computational Ocean Acoustics, 2nd edn. Springer, New York (2011). https://doi.org/10.1007/978-1-4419-8678-8
4. Jenserud, T., Ivansson, S.: Measurements and modeling of effects of out-of-plane reverberation on the power delay profile for underwater acoustic channels. IEEE J. Oceanic Eng. **40**(4), 807–821 (2015)
5. Sturm, F., Korakas, A.: Comparisons of laboratory scale measurements of three-dimensional acoustic propagation with solutions by a parabolic equation model. J. Acoust. Soc. Am. **133**(1), 108–118 (2013)
6. Etter, P.C.: Underwater Acoustic Modeling and Simulation, 4th edn. CRC Press, Boca Raton (2013)
7. Reilly, S.M., Potty, G.R., Goodrich, M.: Computing acoustic transmission loss using 3D Gaussian ray bundles in geodetic coordinates. J. Comput. Acoust. **24**(01), 16500071–165000724 (2016)

8. Soares, C., Zabel, F., Jesus, S.M.: A shipping noise prediction tool. In: OCEANS 2015-Genova, pp. 1–7. IEEE (2015)

9. Calazan, R.M., Rodríguez, O.C.: TRACEO3D ray tracing model for underwater noise predictions. In: Camarinha-Matos, L.M., Parreira-Rocha, M., Ramezani, J. (eds.) DoCEIS 2017. IAICT, vol. 499, pp. 183–190. Springer, Cham (2017). https://doi.org/10.1007/978-3-319-56077-9_17

10. Kirk, D.B., Wen-Mei, W.H.: Programming Massively Parallel Processors: A Hands-on Approach. Morgan kaufmann, Burlington (2013)

11. Hursky, P., Porter, M.B.: Accelerating underwater acoustic propagation modeling using general purpose graphic processing units. In: OCEANS 2011, pp. 1–6. IEEE (2011)

12. Sun, X., Da, L., Li, Y.: Study of BDRM asynchronous parallel computing model based on multiple cuda streams. In: 2014 Seventh International Symposium on Computational Intelligence and Design (ISCID), vol. 1, pp. 181–184. IEEE (2014)

13. Ey, E.: Adaptation of an acoustic propagation model to the parallel architecture of a graphics processor. Master's thesis, University of Algarve (2013)

14. Calazan, R.M., Rodríguez, O.C., Nedjah, N.: Parallel ray tracing for underwater acoustic predictions. In: Gervasi, O., et al. (eds.) ICCSA 2017. LNCS, vol. 10404, pp. 43–55. Springer, Cham (2017). https://doi.org/10.1007/978-3-319-62392-4_4

15. Open source high performance computing. https://www.open-mpi.org/. Accessed 13 June 2018

16. Ocean acoustics library. http://oalib.hlsresearch.com/. Accessed 03 July 2018

17. Rodriguez, O.C., Sturm, F., Petrov, P., Porter, M.: Three-dimensional model benchmarking for cross-slope wedge propagation. In: Proceedings of Meetings on Acoustics, Boston, MA, 25–29 June 2017, vol. 30, p. 070004. ASA (2017)

18. Calazan, R., Rodríguez, O.C.: Simplex based three-dimensional eigenray search for underwater predictions. J. Acoust. Soc. Am. **143**(4), 2059–2065 (2018)

19. Rodriguez, O.C., Collis, J.M., Simpson, H.J., Ey, E., Schneiderwind, J., Felisberto, P.: Seismo-acoustic ray model benchmarking against experimental tank data. J. Acoust. Soc. Am. **132**(2), 709–717 (2012)

20. Červenỳ, V., Pšenčík, I.: Ray amplitudes of seismic body waves in laterally inhomogeneous media. Geophys. J. Int. **57**(1), 91–106 (1979)

21. Popov, M.M.: Ray theory and Gaussian beam method for geophysicists. EDUFBA, Salvador, Bahia (2002)

22. Porter, M.B.: BELLHOP3D user guide. Techical report, Heat, Light, and Sound Research Inc. (2016)

23. CUDA C programming guide. Technical Report, Nvidia Corporation (2018). https://docs.nvidia.com/cuda/cuda-c-programming-guide/index.html. Accessed 16 May 2018

24. The GNU FORTRAN compiler. https://gcc.gnu.org/onlinedocs/gfortran/Interoperability-with-C.html. Accessed 05 June 2018

25. Korakas, A., Sturm, F., Sessarego, J.-P., Ferrand, D.: Scaled model experiment of long-range across-slope pulse propagation in a penetrable wedge. J. Acoust. Soc. Am. **126**(1), EL22–EL27 (2009)

26. Floating point and IEEE 754 compliance for NVIDIA GPUs. Technical Report, Nvidia Corporation (2018). https://docs.nvidia.com/cuda/floating-point/index.html. Accessed 31 May 2018

General Track 5: Information Systems and Technologies

A Digital Twin Platform for Diagnostics and Rehabilitation of Multiple Sclerosis

Dessislava Petrova-Antonova[1(✉)], Ivaylo Spasov[2], Iva Krasteva[1],
Ilina Manova[2], and Sylvia Ilieva[1]

[1] GATE Institute, Sofia University, Sofia, Bulgaria
{d.petrova,sylvia}@fmi.uni-sofia.bg,
iva.krasteva@gmail.com
[2] Rila Solutions, Sofia, Bulgaria
{ispasov,ilinam}@rila.bg

Abstract. 30 million people across Europe are affected approximately from rare diseases. The brain disorders (including, but not limited to those affecting mental health) remain a major challenge. The understanding of mental disorders' determinants and causes, processes and impacts is a key to their prevention, early detection and treatment as well as factor for good health and well-being. In order to improve the health and disease understanding, a close linkage between fundamental, clinical, epidemiological and socio-economic research is required. Effective sharing of data, standardized data processing and the linkage of such data with large-scale cohort studies is a prerequisite for translation of research findings into the clinic. In this context, this paper proposes a platform for exploration of behavioral changes in patients with proven cognitive disorders with a focus on Multiple Sclerosis. It adopts the concept of a digital twin by applying the Big Data and Artificial Intelligence technologies to allow for deep analysis of medical data to estimate human health status, accurate diagnosis and adequate treatment of patients. The platform has two main components. The first component, provides functionality for diagnostics and rehabilitation of Multiple Sclerosis and acts as a main provider of data for the second component. The second component is an advanced analytical application, which provides services for data aggregation, enrichment, analysis and visualization that will be used to produce a new knowledge and support decision in each instance of the transactional component.

Keywords: Big Data · Cognitive assessment · Digital twin · Multiple Sclerosis

1 Introduction

The research of cognitive diseases in patients with a number of neurological disorders is gaining increasing social significance. The affected persons are often of working age and timely diagnosis is an important factor for proper treatment and prevention. The most of the approaches for testing of patients are based on paper tests and involved a highly specialized medical staff. The processing and analysis of paper results is difficult and time consuming. In practice, each therapist works in isolation with a particular group of patients and limited healthy controls for comparison. These circumstances do

© Springer Nature Switzerland AG 2020
O. Gervasi et al. (Eds.): ICCSA 2020, LNCS 12249, pp. 503–518, 2020.
https://doi.org/10.1007/978-3-030-58799-4_37

not allow accumulation of data with different format and origin, as well as searching for certain patterns and dependencies in cognitive deficits. Although the development of automated versions of the paper tests is in place, they primarily serve as storages of results without further analysis of disorders. Some of them are limited to statistical processing of results without prediction and prescription possibilities and others use machine learning techniques to predict probable outcomes only for individual patients.

At the same time, the rapid growth of data, leading to the so-called Big Data phenomenon, affects all domains of everyday life, ranging from user-generated content of around 2.5 quintillion bytes every day [1] to applications in healthcare [2], education [3], knowledge sharing, and others. In all domains, data is the key, not only by adding value and increasing the efficiency of existing solutions but also by opening new opportunities and facilitating advanced functionalities supporting timely and informed decisions. Given the data value for applications in different domains there is an urgent need for solid end-to-end, data-driven and data-oriented holistic platforms that provide a set of mechanisms for runtime adaptations across the complete path of data and lifecycle of domain services. Such platforms adopt the concept of digital twin – a digital profile of the physical world that helps to optimize its performance. The digital twin is based on massive, cumulative, real-time, real-world data across an array of dimensions. The data is used for building digital models based on advanced artificial intelligence and machine learning algorithms that may provide important insights, leading to actions in the physical world such as a change in human health and well-being. The Big Data and Artificial Intelligence technologies applied for digital twinning of patients allow deep analysis of medical data to estimate human health status, accurate diagnosis and adequate treatment leading to so called precision medicine.

This paper proposes an architecture of a digital twin platform for exploration of behavioral changes in patients with proven cognitive disorders with a focus on Multiple Sclerosis (MS). The platform identifies the causes and trends that lead to a deepening of the long-term cognitive decline. It allows for assessment of cognitive status in a timely manner and suggest appropriate preventive as well as rehabilitation actions of cognitive disorders. The main contribution of the platform are as follows:

- Understanding the determinants of MS (including nutrition, physical activity and gender, and environmental, socio-economic, occupational and climate-related factors) through digital twin modelling of patients;
- Better understanding the cognitive diseases and thus improving diagnosis and prognosis for patients with MS through application of digital twin models;
- Improving the collection, enrichment and use of health data related to the cognitive diseases;
- Adoption of advanced machine learning and artificial intelligent methods for precise diagnosis and rehabilitation of MS;
- Improving both scientific and medicine tools through development of a technological platform for automation of digital twin modelling of patients.

The rest of the paper is organized as follows. Section 2 briefly describes the current state of the research on the problem area. Section 3 summarizes the requirements to the platform, while Sect. 4 is devoted on its architecture. Section 5 describes the automations of the tests for diagnosis and rehabilitation of MS. Finally, Sect. 6 concludes the paper and gives directions for future work.

2   State of the Art

A review of the current state of the research on the problem area was performed, covering the computer implementations for evaluation of patients with mild and moderate cognitive disorders such as MS, Alzheimer's disease, early dementia, mild cognitive impairment, etc. The following state-of-the art articles are explored:

- A systematic review of the diagnostic accuracy of automated tests for cognitive impairment [4] – The systematic survey includes studies from January 2005 to August 2015 providing evaluation of diagnostic accuracy of automated tests for the assessment of mild cognitive impairment or early dementia. Articles in English are included. The review examines 11 tools for automated testing of cognitive disorders.
- Status of computerized cognitive testing in aging: a systematic review [5] – The survey includes studies published by 2007 that assess or detect changes in age-related cognitive status, early dementia, and mild cognitive impairment via computer-based tests. Each automated test is evaluated against availability, validity and reliability of tests, range and use. The review includes 11 tests.
- Computerized cognitive testing for older adults: a review [6] – The review examines 17 tests for assessment and screening of cognitive status with publications available by 2012. A comparative analysis of tests according to the hardware, mode of execution and administration, runtime and supported cognitive domains has been performed. The review also examines the psychometric properties of the tests and shows that the most of the tests are well documented in terms of validity and reliability, but the other properties vary.
- Computerized cognitive testing for patients with multiple sclerosis [7] – The review focuses on the available automated tests of multiple sclerosis. The following tests are included: Automated Neuropsychology Assessment Matrix (ANAM), the Mind streams Computerized Cognitive Battery (MCCB), the Amsterdam Neuropsychological Tasks (ANT), the Cognitive Stability Index (CSI), the Cognitive Drug Research (CDR) battery.

The selection criteria for the software solutions included in the review are as follows:

- Provide implementation of tests in cognitive domains affected by MS;
- Validated for use in patients with MS or used in mild cognitive impairment such as: LCU, early dementia, Alzheimer's disease, Parkinson's disease, brain injury, etc.;
- Solution documentation is available at the time of review;
- The implementation of the solution and the tests are in English.

The cognitive domains affected in patients with multiple sclerosis are: Information processing, Visual-spatial memory, Auditory-verbal memory, Expressive language, Executive function, and Visual-spatial information processing.

Tables 1 and 2 present a comparative analysis of the software solutions considered for assessing cognitive status by the following characteristics: application area (1), type of device (2), way of interaction with the device (3) patient's performance mode (4), number of tests and batteries (5), additional factors (6), output data (7), maintained cognitive domains (8), and diseases (9).

Table 1. Software solutions for assessing cognitive status (A).

	ANAM Tests	BrainCare	CANS-MCI	CANTAB Insight
1	Research software, Medical software	Research software, Medical software	Medical software	Medical software
2	PC, Notebook	Device with a browser	Windows based	iPad
3	Mouse/Keyboard	Mouse/Keyboard	Touchscreen	Touchscreen
4	Alone in a controlled environment; Preliminary training is required	Alone with technical assistant; Allowed in a home environment	Alone – technical assistant starts the test execution	Alone in a controlled environment; Audio instructions available;
5	22 tests/3+ standard barriers; battery configuration is supported	10 tests for mild cognitive impairment	8 tests/1 battery	3 tests/1 battery
6	Part of the tests: Demographic/History Module, Mood Scale, etc.	2 scales – for mood and for nervousness and anxiety	Definition of questions for determination of factors that influence the test execution	Mood assessment (the Geriatric Depression Scale or GDS-15)
7	Automated generation of report for the specialist; Data access for exploration and download	Automated generation of report for the patient and the specialist	A neuropsychologist from the company prepares a report	Automated assessment – the norm is calculated based on the gender, age and education of the patient
8	Attention, Concentration, Response Time, Memory, Information Processing Speed, Decision Making, and Execute Function	Attention, Visual-spatial information processing, Execute function, Memory, Auditory-verbal memory	Memory, Execute Function and Information Processing	Executive function, Processing speed, Attention, Working memory,Episodic memory
9	8 diseases	4 diseases	4 diseases	Assessment of cognitive status without concrete diseases

The Automated Neuropsychological Assessment Metrics (ANAM) platform includes several tools, based on 22 tests and behavioral questionnaires for cognitive assessment [10]. Several standard batteries, including predefined tests are supported such as General Neuropsychological Screening battery (ANAM GNS), Traumatic Brain Injury (TBI) и ANAM-MS. New test batteries for specific purposes can be created. The ANAM questionnaires assess additional factors such as demographics, mood scale, etc.

Table 2. Software solutions for assessing cognitive status (B)

	CANTAB connect research	Cognigram - cogstate	Cogstate research	CAMCI
1	Research software	Medical software	Medical software	Research software, Medical software
2	Touchscreen device; Device with a browser	Touchscreen device; PC	Touchscreen device; PC	PC, Notebook
3	Touchscreen/mouse/Keyboard	Touchscreen/mouse	Touchscreen/mouse	Mouse/Keyboard
4	Technical assistant is required for particular tests; Remote execution is allowed; Audio instructions available	Alone in a clinical or home environment with technical assistant	Technical assistant is required for particular tests	Alone in a controlled environment; Preliminary training is required
5	16 tests/17 standard batteries; battery configuration is supported	4 tests/Cogstate Brief Battery	11 tests/7 standard batteries	22 tests/3+ standard barriers; battery configuration is supported
6	Mood assessment	N/A	N/A	Part of the tests: Demographic/History Module, Mood Scale, etc.
7	Automated assessment – the norm is calculated based on the gender, age and education of the patient	Automated generation of report for the specialist	Automated generation of report for the specialist	Automated generation of report for the specialist; Data exploration and download
8	Attention, Memory, Execution function, Emotional and Social Cognition, Psychomotor speed	Visual-motor control, Associative training, Psychomotor function, Executive function, Attention, Verbal training, Visual training	Visual-motor control, Associative training, Psychomotor function, Execution function, Attention, Verbal training, Visual training, Working memory, Emotion recognition	Attention, Concentration, Response Time, Memory, Information Processing Speed, Decision Making, and Execute Function
9	20+ diseases	40+ diseases	40+ diseases	7 diseases

BrainCare applies a battery of tests, which map patient capabilities and results across seven cognitive areas: Memory, Executive function, Attention, Visual spatial, Verbal function, Problem solving, and Working memory [11]. The tests are divided in two groups – mild tests and moderate-severe tests. The mild tests include: Verbal memory test, Non-verbal memory test, Go-NoGo response inhibition test, Stroop test, Visual spatial processing test, Verbal function test, Staged information processing speed test, Finger tapping test, "Catch" game and Problem-solving test. The moderate-severe tests cover assessment of orientation to time and place, language skills, non-verbal memory, similarities and judgement, reality testing, spatial orientation and execution function (Go-NoGo basic). Unlike the mild tests, only one moderate-severe test (i.e., GoNoGo Basic) is interactive. For all other tests, responses are entered by the test supervisor rather than by the patient.

The Computer-Administered Neuro-Psychological Screen for Mild Cognitive Impairment (CANS-MCI) platform automates 8 tests for assessments of mild cognitive injuries in 3 cognitive domains. The assessment includes measures of free and guided recall, delayed free and guided recognition, primed picture naming, word-to-picture matching, design matching, clock hand placement and the Stroop Test. The CANS-MCI tests are applied to establish baseline and longitudinal measures of cognitive abilities that are relevant to a number of medical conditions and their treatments, such as early onset Alzheimer's disease, drug and alcohol rehabilitation, concussions incurred in sports, cancer treatments (e.g. "chemobrain"), MS, Lupus and Parkinson's [12].

The Cambridge Cognition is a leading company in the field of neurology. CANTAB Insight is an analytical assessment tool to enable quick and accurate measurement of brain function across five cognitive domains, namely executive function, processing speed, attention, working and episodic memory [13]. It automates 3 tests: Paired Associates Learning for assessment of visual memory and new learning, Spatial Working Memory for assessment of retention and manipulation of visuospatial information and Match to Sample Visual Search for assessment of attention and visual searching, with a speed accuracy trade-off. Tests are adaptive, so testing will end once a patient reaches tests' limit for the number of attempts for their age, gender or level of education. CANTAB Insight includes an optional mood assessment (the Geriatric Depression Scale or GDS-15). CANTAB Connect Research is a precise and reliable research software providing sensitive digital measures of cognitive function for all areas of brain research. It covers five key domains: attention, memory, executive function, emotion and social cognitions and psychomotor speed. CANTAB Connect Research supports 16 tests, which can be combined in different batteries, and delivers 17 featured batteries.

Cogstate Cognigram provides a battery with 4 tests (Cogstate Brief Battery) for assessment and monitoring of patient's cognitive state in 4 cognitive domains [14]. The tests are based on a single playing card stimulus, which is presented in the center of the device screen. Cogstate Research delivers 11 tests, which can be combined in different batteries [15]. Each test has been designed and validated to assess specific domains including psychomotor function, attention, memory, executive function, verbal learning and social-emotional cognition. All tests are cultural and language independent and can be performed in a clinical or in a home environment.

Computer Assessment of Memory and Cognitive Impairment (CAMCI)-Research is a customizable battery of computerized tasks to assess cognitive performance [16]. The CAMCI-Research battery includes 9 behavioural tasks. A series of self-report questions are provided to gain information from the patients regarding factors that could affect their performance (e.g., perceived memory loss, alcohol use, depression, anxiety, etc.). The CAMCI-Research software does not offer a medical diagnosis and can be used only for research, investigational or educational purposes.

3   Requirements Specification

The requirements of the platform are closely related to the implementation stages of the digital twin, shown in Fig. 1. The *Create stage* include building of Patient Information Model (PIM) by collecting data from different sources. The data can be classified in two categories: (1) operational data related to results from MS tests; and (2) external data related the patient cognitive status, behavior, etc. The *Interact stage* provides real-time, seamless, bidirectional connectivity between the physical patient and its digital twin (virtual patient) through the platform. The *Aggregate stage* covers data ingestion into a data storage and data pre-processing such as cleaning, consistency checking, linking, etc. The data may be augmented with information for other patients with similar diagnosis. The *Analyze stage* is based on variety models of virtual patient that are built on top of PIM. The main goal of this stage is to produce new knowledge that drive the decision-making process. Artificial intelligent methods and cognitive computing are applicable on that stage. The *Insight stage* visualizes the insights from analytics as 3D views, dashboards and others. It aims to provide evidence about areas for further investigations. The *Decision stage* applies the new knowledge to the therapy of the physical patient in order to produce an impact of the digital twin. The decisions lead to precise diagnosis and timely and adequate rehabilitation.

The operational data will be obtained through automation of the MS tests. Several interviews with two clinical specialists with expertise in MS are conducted to understand the domain and to collect the initial set of requirements for software implementation of the tests for assessment and rehabilitation of MS. As a result, a description of diagnostics and rehabilitation processes of MS is obtained, which shows how the platform will be used by the clinicians and how the patients will interact with the platform. The functionality of the first version of the platform is defined to cover the initial set of requirements, including automation of the following tests:

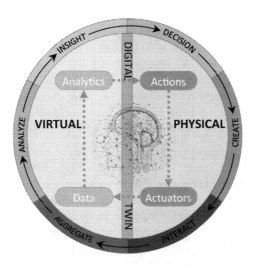

Fig. 1. Digital twin concept.

- Beck Depression Inventory (BDI-II) test;
- BICAMS – Symbol Digit Modalities Test (SDMT), Brief Visuospatial Memory Test (MVMT) and California Verbal Learning Test (CVLT);
- Rehabilitation tests.

The platform has two main components. The first component provides functionality for Create, Interact and Decision stages of the digital twinning. It is developed as a transactional application, called CogniSoft, which has multiple instances in different neurological departments. The second component covers all stages and will be developed as an advanced analytical application using Big Data and Artificial Intelligence technologies. It will provide services for data aggregation, enrichment, analysis and visualization that will be used to produce a new knowledge and support decision in each instance of the transactional component.

4   Platform Architecture and Technologies

As was mentioned so far, the platform consists of two components that will be implemented separately. In the paper, they will be called applications, although each of them can be considered as a stand-alone platform.

4.1   Analytical Application

The architecture of the analytical application is shown in Fig. 2. The analytical application relies on the Hadoop Distributed File System (HDFS). It consists of five layers: (1) Data collection, (2) Data ingestion, (3) Data storage, (4) Data processing and analytics, and (5) Data insight. The primary sources of data are provided by the transactional applications in the neurological departments. The patients' data is enriched with Electronic Health Records (EHRs), open clinical datasets, data form social network and other external applications.

- The data ingestion is implemented by four web services as follows:
- Edge service – ingests large amounts of data that comes from file servers, such as web-logs, DB dumps, etc.
- PULL service – pulls data from external applications and APIs.
- PUSH service – provides a gateway for external applications to push data.
- Direct ingestion service – pulls data directly from databases of the internal transactional applications.

The PULL and PUSH services transmit data to the Apache Kafka cluster, which is a distributed publishing/subscribing message system of one or more brokers [9]. For each existing topic, the brokers manage zero or more partitions. The publisher connects to Kafka cluster and sends request to check which partitions exists for the topic of interest and which nodes are responsible for those partitions. Then the publisher assigns messages to partitions, which in turn are delivered to the corresponding brokers. The subscriber is a set of processes, which cooperates each other and belong to a same consumer group. The consumers in a given group are able to consume only from the partitions that are assigned to that group.

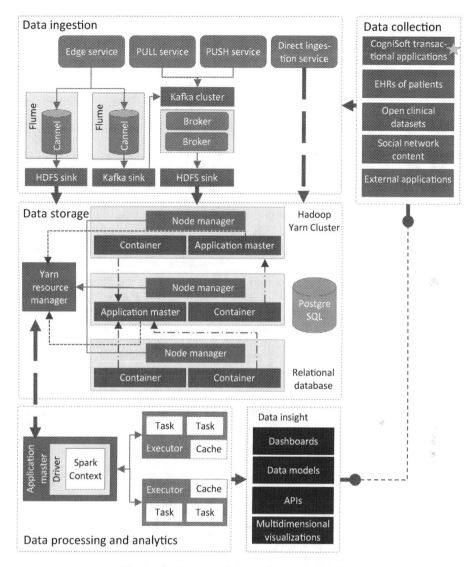

Fig. 2. Architecture of the analytical application.

The Edge service transmits data to the Apache Flume Service, which is a distributed, reliable service for ingestion of large amounts of data in different file system. The Flume Agent sends messages from the source to the sink. Separate sinks are used to ingest data in different file systems. In the context of Kafka, the Flume is integrated to stream data in Kafka topic with a high speed, e.g. it is used when a heavy velocity data comes into place.

YARN enables the platform to perform operations by using a variety of tools like Spark for real-time processing, Hive for SQL, HBase for NoSQL and others. It

allocating resources and scheduling tasks. The Resource manager runs on a master daemon and manages the resource allocation in the cluster. Each Node manager runs on a slave daemon and is responsible for the execution of a task on every single data node. The Application master manages the job lifecycle and resource needs of individual applications. It works along with the Node manager and monitors the execution of tasks. The Container packages the resources including CPU, RAM, HDD, Network, etc. on a single node. Since, YARN processes the data in separate containers, which are logically units consisting of task and resources, the deadlock situations that appear typically in the first version of Hadoop, are minimized [8].

Spark delivers five types of components to the platform. The driver is a central coordinator that split the workload in to smaller tasks and schedules their execution on the executors. The driver passes the executors' requirements to the Spark application master, which negotiates for the resources with the YARN resource manager. The executors themselves and the Spark application master are hosted by the YARN containers. The Spark context allows Spark driver to access the YARN resource manager and keeps track of live executors by sending heart beat messages regularly. The executors perform the assigned tasks on the worker nodes and return the result to the Spark Driver.

4.2   Transactional Application

The architecture of the CogniSoft transactional application follows the Service-Oriented Architecture (SOA) paradigm. Its implementation includes several software modules, described in this section.

Architecture of CogniSoft Transactional Application. The architecture of the CogniSoft transactional application is shown in Fig. 3.

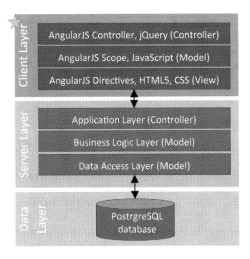

Fig. 3. Architecture of the CogniSoft transactional application.

The *Client layer* provides a web-based user interface (UI) for both patients and clinicians. It is implemented using Angular web application framework, which is embedded in HTML to create dynamic responsive web applications. The AngularJS supports the MVC architecture and thus it is smoothly integrated with the server layer. The two-way data-binding provides automatic synchronization between the model and the view, which are always synchronized. In this way, the user always sees an up-to-date view of the model.

The AngularJS scope objects refer to the application model. They are arranged hierarchically to mimic the DOM structure of the application. The AngularJS controllers augment the AngularJS scope objects by setting up their initial state and adding behavior to them. The role of the AngularJS directives is to attach a specific behavior to the DOM elements or to transform them and their children.

The *Server layer* provides a set of Application Programming Interfaces (APIs), implemented using Spring framework. It is developed using the Spring web application framework that fully support Representational State Transfer (REST) architecture. Thus, the components of the platform are implemented as a set of REST services. Since Spring is realized in a modular fashion, the developers are able to pick the modules that are relevant to their server development.

The *Data layer* provides data repository of the patients' data and their results from diagnosis and rehabilitation tests. It is built on PostgreSQL Relational Database Management System (RDBMS). Different instances will be created, working in different modes (transactional or analytical). The transactional databases and analytical database use a unidirectional master-slave replication.

Modules of CogniSoft Transactional Application. The CogniSoft transactional application is implemented in a modular fashion. It consists of 7 main modules, describes in this section.

The *Security module* is responsible for the security and privacy issues. Since the application works with sensitive data, along with the anonymization, a data separation technique is applied. For example, the user profiles are separated by their personal records. All APIs in the server layer relies on the Spring Security library, while the client layer uses Angular Security. Angular implements a lot of best practices and built-in protections against the most popular web application attacks. The data access is controlled not only on the application level, but additional measures are provided on a network level. The data is stored on machines, which are part of the internal network. External users can access the system by sending requests to a web server behind a proxy firewall using SSL cannel. The architecture of the application is stateless, meaning that the authorization is passed through a standard JWT in the HTTP header. The header is issued during authentication and contains assertions signed by the server.

The *User module* is responsible for users' roles and profiles. The users' roles are the same for the client and the server layers. The *Administrators* can access all system elements, including the Audit module, and are allowed to create nomenclatures and to perform CRUD operations on every types of objects. The *Clinicians* can administrate the patients' records and are allowed to create groups of patients, versions of the tests and group of tests in batteries, to assign and monitor the execution of batteries. The *Patients* can execute tests, which are assigned to them and eventually to access the results from the execution. The *Controls* can execute tests, which are assigned to them.

The *Nomenclature module* provides functionality for definition of system nomenclatures, which are multilingual. All nomenclatures are stored in a single table in the database, but they are grouped based on the type and the language.

The *Personal records module* implements functionality for administration of the personal records of users' roles clinicians, patients and controls. The information stored for each role is different. For example, the patient's personal record contains information about the disease, nationality, education, affiliation and other classification information, while the clinician's personal record keeps information about the specialty, medical center or hospital and participation in public healthcare projects.

The *Disease module* keeps track of the patient's disease progression. The corresponding data is stored in a separate JSON file associated with the patient's personal record. The *Test module* implements the diagnostics and rehabilitation tests. Currently, the functionality for the BDI-II and BICAMS battery is available. The clinicians are able to change the tests at the diagnosis stage as well as at the rehabilitation stage. The assignment of a test to a patient requires a new record to be added in the table "Assignment" of the database. Each test assignment associates a set of tests (test battery) with a group of users. The test execution is recorded in the table "Execution" of the database. Each executable test includes a static part, stored in the table "Header" of the database, which is common for all tests of the same type. The dynamic part of the executable test defines how the test will be performed.

The *Audit module* performs audit on the system actions such as login and logout as well tracks the changes of every valuable object. Information about the users responsible for the creation and last modification of the audited records are stored.

5   Test Automation

The BICAMS provides a cognitive assessment for MS, covering the most vulnerable neuropsychological domains of mental function – speed of information processing, episodic memory, visual-perceptual functioning, attention. They are automated in the current version of the CogniSoft transactional application.

5.1   Symbol Digit Modalities Test

The patient is required to perform a symbolic substitution with a corresponding digit of a given 9-digit code (key) for a period of 90 s. The test is based on the paper version using:

- Key field – it consists of 9 symbol-digit pairs that illustrate the correspondence between a symbol and a digit (1 to 9). A standardized predefined character pool is used.
- Work field – contains 6 couples of rows (consecutively arranged tables with 15 columns and 2 rows each). Each top row contains characters from the represented key in random order. Each bottom row is blank and must be filled in by the patient with the corresponding (key) digits. The patient is expected to complete the task as quickly as possible. The correspondences should be filled in consecutive order.

The adapted computer version of the test consists of the following steps: (1) Instruction, showing text or audio directions for execution of the test; (2) Demonstration, showing a short video that demonstrates the test execution; (3) Training, which is one time entering the data in the work field; (4) Execution, which is the actual test execution; and (5) End, which finishes the test and shows an encouragement message.

The test execution involves sequentially filling in the numbers on the displayed screens (standard 6 screens). The final screen ends with the field where the patient can fill in any additional information about the particular circumstances surrounding the test execution and his performance. At the discretion of the therapist, information related to performance outcomes is displayed or not to the patient. This includes: execution time, number of correct answers, problematic key symbols (recurring errors), etc.

Figure 4 shows how the SDMT looks like in the CogniSoft transactional application.

Fig. 4. UI of the SDMT.

A rehabilitation game is implemented on the same principle as the SDMT by replacement of the symbols with pictures. The configuration parameters allow the clinician to change the size and contents of the key, the size of the work field, the number of training screens depending on the patient's specific needs.

5.2 Brief Visuospatial Memory Test

The CogniSoft transactional application adapts the paper version of the BVMT. It provides specially designed forms containing 6 predefined geometric shapes (Table 2 × 3), as is shown in Fig. 5.

Fig. 5. UI of BVMT.

The test consists of the following main steps:

- Memorizing: For 10 s a table 2×3 is shown with validated figures from the predefined pool.
- Reproducing: Two tables are shown. One is empty 2×3 table and the second one is with size 3×4. It is filled with 12 figures. Among them in a random principle are included the 6 figures from the previous screen (validated) and 6 figures similar to those (not validated). The aim is with drag and drop the patient to fill the empty table with the correct figures based on what is memorized from the screen.

This process is repeated 3 times and the best performance is taken into consideration. The adapted computer version of the test consists of the following steps: (1) Instruction showing text or audio directions for execution of the test; (2) Execution, which is the actual test execution, consisting of 2 parts: memorizing and reproducing, described above; and (3) End, which finishes the test and shows an encouragement message.

The results are based on the number of correct figures and their correct positioning in the resulting table. In a rehabilitation mode, other pools of figures are used, and also the configuration parameters can vary such as storage time, table size, etc.

5.3 California Verbal Learning Test

The CVLT is implemented using three tables of words. The table with the "validation" words includes 4 categories of words such as sport, paper products, geographic objects and sweets. The second table contains words that are similar to those in the first table and falls in the same categories. The third table contains words that falls in categories that are different from ones of the validated words. During test execution 8 words are randomly selected form the first table, 6 words are randomly selected form the second table and 2 words are randomly selected form the third table.

The test consists of a series of 5 consecutive attempts, during which the same validated 16 words belonging to the selected categories are read. The patient should try to memorize them and then indicate them. After each attempt, statistics are kept on the number of memorized words and the categories to which they refer. The expectation is that after every attempt, the number of correct answers will increase.

The computer version of the CVLT adapts the paper version and consists of the following steps: (1) Instruction, showing text or audio directions for execution of the test; (2) Demonstration, showing a short video that demonstrates the test execution; (3) Execution, which is the actual test execution, consisting of five attempts; and (4) End, which finishes the test and shows an encouragement message.

The patient recognizes the words by pressing the selection buttons labeled with the words heard. The recognition is performed in two steps corresponding to two different screens. The final screen ends with possibility for the patient to fill in additional information about the particularities and circumstances surrounding the tests execution and performance. At the discretion of the therapist, information related to the results of the performance is displayed or not displayed: time of implementation, number of correct answers, problem categories.

Fig. 6. Rehabilitation game based on the CVLT.

A rehabilitation game, shown in Fig. 6, is implemented by adding relevant pictures on the buttons in addition to the words. There are also variations in configuration related to the number of validated words used, recognition screens, number of repetitions.

6   Conclusions and Future Work

The preventive and therapeutic approaches, tailored to patient requirements, need a personalized medicine, which early detects diseases. It is a societal challenge to adjust to the further demands on health sectors due to the ageing population. The effective healthcare requires improved decision making in prevention and in treatment provision as well as identification and support of best practices.

The digital twin platform described in this paper provides a practical first step towards application of the precision medicine in MS. The platform's architecture and technologies are described with a focus to its two main components:

- CogniSoft transactional application, which automates tests for diagnosis and rehabilitation of MS such as BICAMS tests;
- Advances analytical application, which provides services for data aggregation, enrichment, analysis and visualization that will be used to produce a new knowledge and support decision in each instance of the CogniSoft transactional application.

Being web-based, the CogniSoft transactional application guarantees maximal access to its functionality in different neurological departments. It is the main data provider for the analytical application, which is not currently developed. Once integrated in a common digital twin platform, the clinicians will be able to perform efficient diagnosis, prognostication, and management decisions for individual patients with MS. In the presented implementation clinical questions that deal with the shorter-term management of patients are covered. The patient state assessment and rehabilitation, covered by the CogniSoft transactional application, help in intervention planning in the hospital setting. Assessment of real prognostic performance needs a longitudinal set-up, which is the subject of ongoing research activities and the focus of the analytical component of the platform. The potential of the platform is wider and could be expanded to provide a long-term decision support and evidence-based outcome predictions in MS. The clinicians will be able to quickly interpret the patients' data and to view the probable outcomes form rehabilitation, based on past patients' data in their neurological departments, EHRs from other systems, open clinical databases, etc.

Acknowledgement. This research work has been supported by CogniSoft "Information System for Diagnosis and Prevention of Multiple Sclerosis Patients" project, funded by the Program for Innovation and Competitiveness, co-financed by the EU through the ERDF under agreement no. BG16RFOP002-1.005, GATE "Big Data for Smart Society" project, funded by the Horizon 2020 WIDESPREAD-2018-2020 TEAMING Phase 2 programme under grant agreement no. 857155 and CogniTwin "Digital twin modelling of patients with cognitive disorders" project, funded by the Bulgarian National Science fund, under agreement no. KP-06-N32/5.

References

1. IBM. Bringing big data to the enterprise. http://www-01.ibm.com/software/data/bigdata/what-is-big-data.html. Accessed 20 Jan 2020
2. MIT Technology Review. https://www.technologyreview.com/business-report/data-driven-health-care/free/. Accessed 20 Jan 2020
3. van Rijmenam, M.: Four Ways Big Data Will Revolutionize Education. https://datafloq.com/read/big-data-will-revolutionize-learning/2016. Accessed 20 Jan 2020
4. Aslam, R.W., et al.: A systematic review of the diagnostic accuracy of automated tests for cognitive impairment. Int. J. Geriatr. Psychiatry **33**(4), 561–575 (2018)
5. Wild, K., Howieson, D., Webbe, F., Seelye, A., Kaye, J.: Status of computerized cognitive testing in aging: a systematic review. Alzheimers Dement. **4**(6), 428–437 (2008)
6. Zygouris, S., Tsolaki, M.: Computerized cognitive testing for older adults: a review. Am. J. Alzheimers Dis. Other Dement. **30**(1), 13–28 (2015)
7. Lapshin, H., O'Connor, P., Lanctôt, K.L., Feinstein, A.: Computerized cognitive testing for patients with multiple sclerosis. Mult. Scler. Relat. Disord. **1**(4), 196–201 (2012)
8. Saraswat, H., Sharma, N., Rai, A.: Enhancing the traditional file system to HDFS: a big data solution. Int. J. Comput. Appl. (0975 – 8887) **167**(9), 12–14 (2017)
9. Introduction to Kafka. https://gist.github.com/kbeathanabhotla/0183e312307b05835c74. Accessed 14 Jan 2020
10. Vincent, A.N., Roebuck-Spencer, T., Fuenzalida, E., Gilliland, K.: Test–retest reliability and practice effects for the ANAM general neuropsychological screening battery. Clin. Neuropsychol. **32**, 1–16 (2017)
11. Doniger, G.M.: NeuroTrax Computerized Cognitive Tests: Test Descriptions, NeuroTrax Corporation, June 2014
12. Real-world Applications of Cognitive Assessment. https://screen-inc.com/uses-of-cognitive-tests/. Accessed 20 Jan 2020
13. CANTAB Insight. https://www.cambridgecognition.com/products/digital-healthcare-technology/cantab-insight/. Accessed 20 Jan 2020
14. Congstate Cognigram. https://www.cogstate.com/category/cognigram/. Accessed 20 Jan 2020
15. Congstate Academic Research. https://www.cogstate.com/academic-research/. Accessed 20 Jan 2020
16. CAMCI-Research. https://pstnet.com/products/camci-research/. Accessed 20 Jan 2020

Classification of Carcass Fatness Degree in Finishing Cattle Using Machine Learning

Higor Henrique Picoli Nucci[1]([✉])(iD), Renato Porfirio Ishii[1](iD),
Rodrigo da Costa Gomes[2](iD), Celso Soares Costa[3,4](iD),
and Gelson Luís Dias Feijó[2](iD)

[1] Federal University of Mato Grosso do Sul,
Campo Grande, Mato Grosso do Sul, Brazil
`higomucci@gmail.com`
[2] Embrapa Gado de Corte, Campo Grande, Mato Grosso do Sul, Brazil
[3] Federal Institute of Education, Science and Technology of Mato Grosso do Sul,
Campo Grande, Mato Grosso do Sul, Brazil
[4] Universidade Católica Dom Bosco, Campo Grande, Mato Grosso do Sul, Brazil

Abstract. Nowadays, there is an increase in world demand for quality beef. In this way, the Government of the State of Mato Grosso do Sul has created an incentive program (Precoce MS) that stimulates producers to fit into production systems that lead to the slaughter of animals at young ages and superior carcass quality, towards a more sustainable production model. This work aims to build a classification model of carcass fatness degree using machine learning algorithms and to provide the cattle ranchers with indicators that help them to early finishing cattle with better carcass finishing. The dataset from Precoce MS contains twenty-nine different features with categorical and discrete data and size of 1.05 million cattle slaughter records. In the data mining process, the data were cleaned, transformed and reduced in order to extract patterns more efficiently. In the model selection step, the data was divided into five different datasets for performing cross-validation. The training set received 80% of the data and the test set received the other 20%, emphasizing that both had their data stratified respecting the percentage of each target class. The algorithms analyzed and tested in this work were Support Vector Machines, K-Nearest Neighbors, AdaBoost, Multilayer Perceptron, Naive Bayes and Random Forest Classifier. In order to obtain a better classification, the recursive feature elimination and grid search techniques were used in the models with the objective of selecting better characteristics and obtaining better hyperparameters, respectively. The precision, recall and f1 score metrics were applied in the test set to confirm the choice of the model. Finally, analysis of variance ANOVA indicated that there are no significant differences between the models. Therefore, all these classifiers can be used for the construction of a final model without prejudice in the classification performance.

© Springer Nature Switzerland AG 2020
O. Gervasi et al. (Eds.): ICCSA 2020, LNCS 12249, pp. 519–535, 2020.
https://doi.org/10.1007/978-3-030-58799-4_38

Keywords: Data mining · Early calf · Precoce MS · Precision livestock

1   Introduction

Brazil occupies a prominent position in the world scenario of beef production. In the third quarter of 2019 were slaughtered 8.49 million of cattle that were being supervised by a health inspection service, as shown by the statistics report data of Livestock Production, provided by the Brazilian Institute of Geography and Statistics [1]. These data are 2.1% higher than the same period of the previous year. Therefore, in order to remain at the top of the world ranking of exports, the Brazilian government, associations of producers and breeders and slaughterhouses have been committed to create programs that encourage the production of meat with a higher quality than the Brazilian average.

In the state of Mato Grosso do Sul, in the year of 2016, were released a state program, called Precoce MS, that have made improvements to the existing bonus program for early calves, which is managed by the Program for Advances in Livestock of Mato Grosso do Sul (Proape). Among the improvements applied, it can be mentioned as more impacting the inclusion of the productive process of the farm in the calculation for the fiscal incentive.

The Precoce MS program grants to producer with financial returns of up to 67% of the ICMS collected by the carcass of a slaughtered animal and classified as early calf. The calculation of the percentage of the amount of ICMS to be returned to the producer takes into account the carcass fatness degree and the characteristics of the production process from the farm, considering proportionality 70% to 30% respectively.

The classification of an animal as an early calf is performed by the following parameters: gender (F = female; C = castrated male; and M = whole male) and dental maturity (Fig. 1). In order to determine the typification of the carcass, as regards their finishing, the measurement of the subcutaneous fat, as shown in Fig. 2, is adopted as the parameter. Lastly, the pre-slaughter animal weight must be at least 12 *arrobas* for females and 15 *arrobas* for males (whole or castrated) and at least 60% of the lot to be slaughtered must be composed of early calves [2].

The productive processes of a property are evaluated by valuing properties that meet the following criteria: use of tools that allow the individual sanitary management of cattle; apply rules and concepts of good agricultural practices; implement technologies that promote the sustainability of the productive system, particularly those aimed at mitigating carbon emissions through low carbon farming practices; and participate in associations of producers aimed at commercial production systematized and organized according to pre-established standards for the fulfillment of commercial agreements.

Rural establishments are categorized as **Simple** if they meet none or at least one of the criteria. Those that are categorized as **Intermediate** must meet at least two criteria. Lastly, an establishment categorized as **Advanced** must meet at least three criteria.

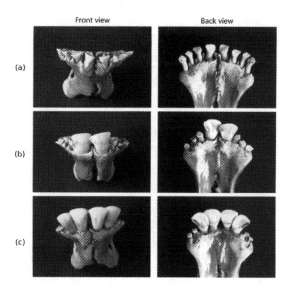

Fig. 1. Dental maturity of early calves: (a) J0 = only milk teeth; (b) J2 = two permanent incisor teeth; and (c) J4 = four permanent incisor teeth. Source: [3].

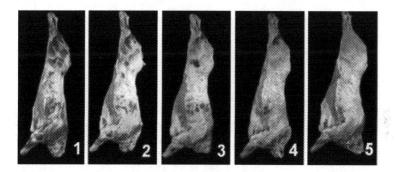

Fig. 2. Carcass fatness degree: 1 = Missing Fat - Fat is absent; 2 = Low Fat - 1 to 3 mm thick; 3 = Median Fat - above 3 and up to 6 mm thick; 4 = Uniform Fat - above 6 and up to 10 mm thick; and 5 = Excessive Fat - 10 mm thick. Source: [4].

In this way, the simplified Table 1 can be used to exemplify the rules of the program. The first three columns show the data of the slaughtered animal and represent up to 70% of the incentive value. The remaining columns refer to the valuation property data and represent up to 30% of the incentive value. Thus, the value of the incentive to be returned by the slaughterhouse to the producer is the percentage he reaches in the table (up to a maximum of 67%) on the value of the ICMS of the cattle while it is still alive.

Apart from Brazil being the world leader in the quantity of exported beef *in natura*, the financial income is relatively low. Since it does not export to the markets that pay the most, because the national meat does not meet the quality

Table 1. Simplified scheme of the classification of subsidized carcasses by the program Precoce MS.

70%			30%		
Product			Productive process		
Gender	Maturity	Fatness degree	Advanced 30%	Intermediate 26%	Simple 21%
M, C, F	J0	3, 4	67	64	61
M, C, F	J2	3, 4	62	59	56
C, F	J4	3, 4	48	45	42
M, C, F	J0	2	62	59	56
M, C, F	J2	2	39	36	33
C, F	J4	2	22	19	16

criteria for some markets that pay more [5]. In this context, there are factors in cattle breeding that can influence the overall quality of the meat produced. A key factor is a high quality and continuous feed for the herd. However, one of the problems in producing of early calves may be the cost with this food of excellence [6].

Thus, the cost-effectiveness of producing the youngest animal for slaughter is high and can put the whole productive system at risk for producers who do not have many financial resources. There are other factors that are also considered important, such as genetic ability and maturity, since young animals with good genetics have a greater efficiency in converting food consumed into fat and weight gain, which will directly reflect the overall yield of carcass [7,8].

From the technological point of view, there is ample room for improvement in the production of beef cattle through precision cattle breeding. For this reason, technologies that result in increased likelihood of economic success are crucial. Given that, the producer will have access to data that will support the most effective decision making, optimize its production and its economic balance and lead to the production of animals with better compliance with the carcass criteria, with lower production costs and greater obtaining of subsidies [9].

The main objective of this paper is to construct a classification model of the carcass fatness using Machine Learning algorithms. This model will support beef cattle producers in decision making, aiming to increase the quality of the meat produced.

2   Materials and Methods

2.1   Approach

In order to implement a carcass fatness degree classifier, the approach used follows the methodology shown in Fig. 3. The first step was to obtain the slaughtering data of cattle and their respective production processes.

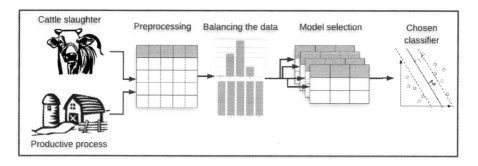

Fig. 3. Complete flowchart of the methodology.

After obtaining the dataset, there was a need for preprocessing the data to remove unnecessary columns, samples with insufficient values and feature selection. The resulted dataset after preprocessing step was very imbalanced. In that case, the next step was balancing the samples of the dataset. With the preprocessed and balanced dataset in hand, a process of comparison of the candidate Machine Learning algorithms called model selection was started.

In addition, it was possible to conclude which are the features that contribute the most to the carcass fatness degree. Thus, the algorithm that presented the best accuracy and f1-score was chosen to be the basis for the creation of a classification model.

2.2 Data Acquisition

The data acquisition process was carried out through a request of the cattle slaughtering data registered in the state program Precoce MS and the production process of each corresponding rural establishment. The request was sent to the Information Management Superintendency (SGI) with the assistance of Embrapa Gado de Corte and the State Secretariat of Environment, Economic Development, Production and Family Agriculture (SEMAGRO), both from Mato Grosso do Sul, Brazil.

The data of the rural establishments participating in the state program were delivered by SGI in two different datasets. The first dataset comprises the basic information of the rural establishments (Table 2) and their respective production processes (Table 3 and Table 4). This dataset consists of 1,595 registered rural establishments and it is filled by technicians, who must have formation as veterinarian, agronomist or zootechnician and are co-responsible for this information.

The questions in the questionnaire are divided into two groups: **"questions that does not qualify"** – that do not increase the percentage of the financial return to the producer (Table 3); and **"questions that qualify"** – which increase the percentage of the financial return to the producer (Table 4). The rural establishments are categorized by the number of question answered as

Table 2. Five random samples of the basic data of a rural establishment.

property_id	city	state	categorization
5159	PEDRO GOMES	MS	21%
1167	CAMAPUA	MS	21%
4960	CAMAPUA	MS	26%
4514	TERENOS	MS	30%
5371	MARACAJU	MS	26%

Table 3. Questions that do not categorize the productive process of a rural establishment, their respective labels in the dataset and their possible values.

Question	Dataset label	Possible answers
"Are there other incentives?"	other_incentives	"Yes", "No"
"Does it manufacture ration?"	makes_ration	"Yes", "No"
"Does it practice pasture recovery?"	pasture_recovery	"Fertigation", "FLI - Farm-Livestock Integration", "CLFI - Crop-Livestock-Forest Integration", "LFI - Livestock-Forest Integration", "None"
"Does it practice field supplementation?"	field_supplementation	"Yes", "No"
"Does it practice semi-confinement?"	semi_confinement	"Yes", "No"
"Does it practice confinement?"	confinement	"Yes", "No"

"Yes": Simple (21%) for none or at least one of the question, Intermediate (26%) for at least two question and Advanced (30%) for three question or more.

The second dataset includes all individual cattle slaughters, from 02-09-2017 to 01-23-2019, with the equivalent carcass finishing (Table 5). This dataset consists of 1,107,689 animals slaughtered.

Each sample, from Table 5, represents the individual slaughter of cattle. When slaughtering an animal, the slaughterhouse registers the **typification** (WHOLE male, male CASTRATED or Female), **maturity** (Milk tooth, Two teeth, Four teeth, Six teeth or Eight teeth), **carcass weight** (in kg), **date of**

Table 4. Questions that categorize the productive process of a rural establishment, their respective labels in the dataset and their possible values.

Question	Dataset label	Possible answers
Does it have a system of individual identification of cattle associated with a zootechnical and sanitary control?	individual_identification	"Yes", "No"
Does grazing control that meets the minimum height limits for each forage or cultivar exploited, having as a parameter the meet the rules established by the Brazilian Agricultural Research Corporation (Embrapa)?	grazing_control	"Yes", "No"
Does the rural establishment have a certificate of Quality Control Programs (Good Agricultural Practices - BPA/BOVINOS or any other program with requirements similar or superior to BPA)?	quality_programs	"Yes", "No"
Is the rural establishment involved with any organization that uses mechanisms similar to the marketing alliance to market its product?	involved_in_organization	"Yes", "No"
Does the managed area have a good vegetation cover, with a low presence of weeds and no spotting of uncovered soil in at least 80% of the total pasture area (native or cultivated)?	area_80_vegetation_cover	"Yes", "No"
Does the managed area show signs of laminar or furrow erosion equal to or greater than 20% of the total pasture area (native or cultivated)?	area_20_erosion	"Yes", "No"
Does it execute SISBOV tracing?	sisbov	"Yes", "No"
Is it part of the Trace List?	trace_list	"Yes", "No"
Is the area of the rural establishment intended for confinement activity in its entirety?	total_area_confinement	"Yes", "No"

slaughter and the **carcass fatness degree** (missing fat - fat is absent, low fat - 1 to 3 mm thick, medium fat - above 3 to 6 mm in thickness, uniform fat - up to 6 and up to 10 mm thickness or excessive Fat - 10 mm thick).

Table 5. Five random samples from the bovine slaughtering dataset.

property_id	typification	maturity	carcass_weight	date_slaughter	carcass_fatness_degree
1	INTEGRAL Male	Milk tooth	362.50	2017-10-02	Median Fat - up to 3 to 6 mm thick
1473	CASTRATED Male	Two teeth	252.00	2017-04-26	Low Fat - 1 to 3 mm thick
4312	CASTRATED Male	Four teeth	338.00	2017-05-08	Low Fat - 1 to 3 mm thick
5068	Female	Two Teeth	188.20	2018-01-03	Median Fat - up to 3 to 6 mm thick
4452	CASTRATED Male	Milk tooth	338.00	2018-05-15	Median Fat - up to 3 to 6 mm thick

2.3 Preprocessing

The datasets were combined into a single one using a identifier called *property_id*. The resulting dataset now only has 1,056,586 samples. This was due to the time difference in generating the datasets by SGI. In total, 51,103 slaughter samples did not have identifiers of rural establishments related to the dataset of the productive process.

According to the specialist in zootechnics, data related to micro and mesoregion, which are not in the dataset, may be relevant to discovery how the fauna, flora and cultural factors of a given region can influence the carcass fatness degree. For that, the data of the location of the rural establishment was used to infer two new features for these values.

In this way, the internal division of the State of Mato Grosso do Sul was taken into account. The state is divided in eleven microregions: Alto Taquari, Aquidauana, Baixo Pantanal, Bodoquena, Campo Grande, Cassilândia, Dourados, Iguatemi, Nova Andradina, Paranaíba and Três Lagoas. Another division of the state is the four mesoregions: Centro Norte, Leste, Pantanal and Sudoeste. Both microregions and mesoregions were added to the dataset.

Another relevant specialist opinion was on the possible impact of climate change on the resulting fat level. In this way, the dates of slaughter of a bovine were used to infer a new feature with the season of the year in which the animal was finished. According to the dates of change of stations in Brazil.

In order to create a classifer of carcass fatness degree, we considered a certain animal and the productive process that generated it. Thus, we have five ordinal classes ranging from Missing Fat to Excessive Fat, as shown in Table 6.

Table 6. Classes for the classification of the carcass fatness degree.

Label	carcass_fatness_degree
Missing Fat - Fat is absent	1
Low Fat - 1 to 3 mm thick	2
Median Fat - above 3 and up to 6 mm thick	3
Uniform Fat - above 6 and up to 10 mm thick	4
Excessive Fat - 10 mm thick	5

The process of feature selection used is a wrapper method known as recursive feature elimination (RFE) [12,13]. The RFC classifier was used together with RFE to complete the feature selection task. According to the classifier, the most relevant features for the model are presented in Fig. 4. All the features of this dataset were scored according to its relevance. The most important *carcass_weight* feature received the highest score, 0.3503, and the *area_20_erosion* feature received the lowest value, 0.0003.

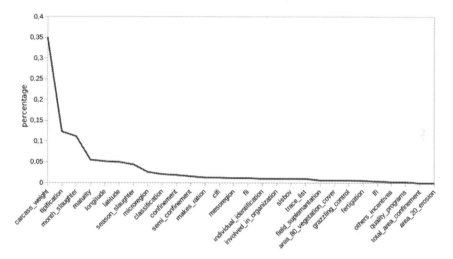

Fig. 4. Most important features according to the RFC algorithm and RFE feature selection technique.

When applying the RFE technique the features *fertigation*, *lfi*, *other_incentives*, *total_area_confinement*, *area_20_erosion* and *quality_programs* have been removed from the dataset. The removal of these characteristics can be attributed to the fact of their high sensitivity.

The categorical column *city* was replaced by its respective *latitude* and *longitude*. The column *maturity* had its values replaced by the number of definitive teeth in integer values. All questions whose possible responses were restricted to "No" or "Yes" had their values replaced by 0 and 1 respectively. In the conversion of categorical features with more than one option per sample, it was used the *one-hot encoding* technique and convert them into new duplicated features, where each one represents one of the feature values. In this technique, the presence of a level is represented by 1 and its absence by 0 [10].

The unit of measure used for a given feature may adversely affect the data analysis. The *min-max* [11] normalization technique consists of expressing the values of a feature in smaller units of measure. In this way, the data is transformed into a more regular distribution, such as [0.0, 1.0].

Finally, at the end of the preprocessing of the data, we have the resulting dataset with 55 training features plus 1 classification feature.

2.4   Balancing the Data

The Fig. 5 shows the distribution of the classes in the dataset. At once, it is perceived that the dataset is unbalanced. In other words, the dataset has 0.48% of its samples for class 1, 40.25% for class 2, 53.31% for class 3, 5.94% for class 4 and 0.02% for class 5.

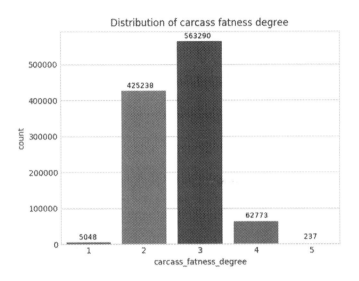

Fig. 5. Distribution of carcass fatness degree.

An unbalanced dataset can negatively affect the learning phase and, consequently, the classification of Machine Learning algorithms. With the intention of balancing the data, there are two methods that stand out: *over-sampling* that replicates minority class samples and *under-sampling* that eliminates samples from the majority class [14]. The result, after balancing, can be seen in Fig. 6

The *under-sampling* technique called Edited Nearest Neighbors (ENN) was used. This technique applies a closer neighbors algorithm and removes samples that do not agree "enough" with their neighborhood. For each sample in the class to be sub-sampled, the nearest neighbors are calculated and, if the selection criterion is not met, the sample is removed [15].

In contrast, the *over-sampling* technique called Synthetic Minority Over-sampling Technique (SMOTE) generates new "synthetic" samples by interpolation operating directly under the characteristics rather than the data [16].

However, the use of the SMOTE technique can generate noisy samples by interpolating new points between marginal values and isolated values. This problem can be solved by cleaning up the space resulting from super-sampling with ENN. Therefore, the use of techniques of *over-sampling* SMOTE combined with *under-sampling* ENN, called SMOTEENN, has generated better results [17].

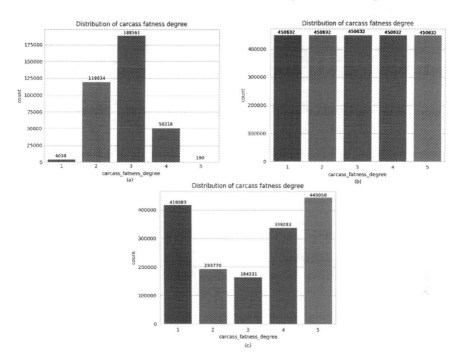

Fig. 6. Comparison of class distribution with dataset balancing using *under-sampling*, *over-sampling* and hybrid techniques. (a) *under-sampling* with ENN; (b) *over-sampling* with SMOTE; and (c) combination of SMOTE + ENN techniques.

2.5 Model Select

The models that were compared in this article are: Multinomial Naive Bayes (MNB), Random Forest Classifier (RFC), AdaBoost (ADA), Multilayer Perceptron (MLP), K-Nearest Neighbors (KNN) and Support Vector Machines (SVM). All of them were applied in the same balanced dataset and the same metrics were used as a result for validation. In the next steps of this paper a Python library called scikit-learn [10] was used to build, test and validate all the different classifiers.

To validate the models, a basic approach is to use the technique called *cross-validation* [21]. The performance measure pointed out by cross-validation is then the mean of *accuracy* values for each loop. Other metrics such as *confusion matrix* [18], *precision*, *recall* and *f1 score* were used to help gain insight during validation of the final model [19].

In all models, the *cross-validation* technique was applied with *5-folds*. The randomization index used to ensure that the training and test datasets are always the same was 42. In the comparison of the models, the mean accuracy of all folds was used as the main metric, by model.

The training set is balanced and the model is trained, as shown in Fig. 7. After training, the test set is used for validation and the results of the classification are collected. The division was done at random, ensuring that the test data from one fold will not be repeated in the test data in another fold, and taking into account the percentage of distribution for each class.

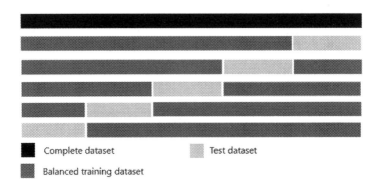

Fig. 7. Visualization of the cross-validation process with the balanced training set.

Another way to improve metrics is to find the best hyperparameters for an algorithm in a given dataset. The process of empirically selecting the best hyperparameters for a model can take a long time and result in an lower *accuracy* [20]. One of the most used techniques for selecting hyperparameters automatically is *grid search* with cross-validation [22], which is an exhaustive search process on a specific dataset.

3   Results and Discussions

The results of the comparison between the classification models collected after the best hyperpameters were chosen while doing grid search with cross-validation and balancing each training set, can be seen in the Table 7.

The results obtained a very low standard deviation, as can be seen in Table 7, for all folds of cross-validation applied in the models. This shows that the results for each fold are not under or overfitted and the data are stratified taking into account the percentage of each class.

When considering the results obtained, it is clear that the models that generalized better were those applied to balanced data using ENN. More specifically the RFC, KNN and SVM algorithms. However, analyzing only *accuracy* can lead to an optimistic estimate if the classifier is biased [23].

This problem can be overcome when the results are analyzed by looking at the normalized *confusion matrix* (Fig. 8). After training the models, classifying using the test dataset and printing their respective confusion matrices the the calculated metrics *precision*, *recall* and *f1-score* were calculated using ENN, as the Table 8 shows.

Table 7. Comparison of the mean of the *accuracy* and the standard deviation among the six chosen models. Acronyms: Multinomial Naive Bayes (MNB), Random Forest Classifier (RFC), AdaBoost (ADA), Multilayer Perceptron (MLP), K-Nearest Neighbors (KNN) and Support Vector Machines (SVM)

Accuracy			
Model	ENN	SMOTE	SMOTEENN
MNB	**58.62%** ±0.102	31.04% ±0.176	22.75% ±0.343
RFC	**70.37%** ±0.075	64.64% ±0.119	65.82% ±0.089
ADA	**60.07%** ±0.153	37.24% ±0.166	12.67% ±0.122
MLP	**66.41%** ±0.022	50.07% ±0.078	45.44% ±0.091
KNN	**69.64%** ±0.048	62.15% ±0.077	64.62% ±0.127
SVM	**69.19%** ±0.116	65.43% ±0.121	67.82% ±0.337

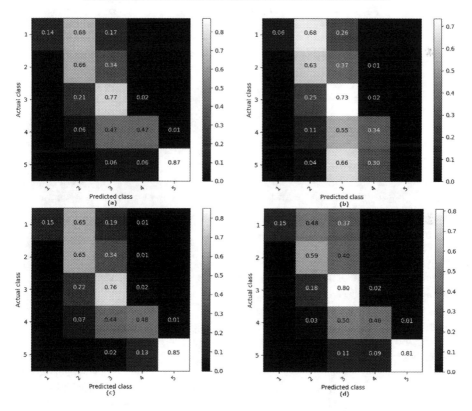

Fig. 8. Normalized results of the prediction models in the test dataset, with representation on a scale of 0 to 1, for error analysis: (a) normalized confusion matrix of the RFC; (c) normalized confusion matrix of the MLP; (d) normalized confusion matrix of the KNN; and (e) normalized confusion matrix of the SVM.

Table 8. Comparison of the metrics *Precision, Recall* and *F1-score* for each of the five classes, with their respective averages, between the models. The calculation of the average took into account the weight of each class in the dataset.

Model	Metric	1	2	3	4	5	Average
RFC	Precision	78.26%	68.82%	71.43%	71.43%	17.30%	70.40%
	Recall	14.26%	65.92%	76.87%	46.61%	87.23%	70.37%
	F1-score	24.12%	67.34%	74.05%	56.41%	28.87%	**70.05%**
MLP	Precision	71.43%	64.07%	68.30%	61.39%	–	66.19%
	Recall	05.94%	62.63%	73.40%	34.34%	–	66.41%
	F1-score	10.97%	63.34%	70.76%	44.04%	–	**65.89%**
KNN	Precision	58.71%	68.00%	71.20%	66.52%	10.08%	69.56%
	Recall	15.35%	65.31%	75.81%	47.88%	85.11%	69.64%
	F1-score	24.33%	66.63%	73.43%	55.68%	18.02%	**69.39%**
SVM	Precision	75.90%	70.12%	68.67%	71.56%	10.58%	69.45%
	Recall	14.65%	59.13%	79.88%	45.58%	80.85%	69.18%
	F1-score	24.56%	64.16%	73.85%	55.69%	18.72%	**68.62%**

When looking at the results of class 1 and 2 in the normalized *confusion matrix* of Fig. 8, it can be seen that the SVM algorithm makes a better distinction between these classes. The rural producer receives financial incentives only for carcasses with fat degree 2, 3 and 4. Under these circumstances, it is important that the number of false positives is low at the edges of the matrix, more specifically for classes 1 and 5. Thus, the producer will have a better control when bring the cattle to the slaughterhouse and receive bigger financial incentives.

The ROC curves and their respective areas on the curve (AUC) for each of the tested algorithms are shown in Fig. 9. The ROC curves had their AUC calculated for each of the target classes and the micro and macro averages of the classes. In this way, it is possible to check the overall performance of each of the algorithms. The AUC value shows the percentage with which each tested algorithm is able to distinguish between negative and positive results.

In order to verify whether the tested classifiers differ statistically in relation to *f1-score* performance, the one-way ANOVA analysis of variance was used. The p-value of 0.8599 was reached, which is higher than 0.05, suggesting that the classifiers are not significantly different for that level of significance. The result showed that the classifiers are similar and there is no statistically significant difference in performance between them.

This was the first study conducted with supervised machine learning in the Precoce MS database. Despite the low hit rate in classes 1 and 5, their occurrence in real life represents 0.48% and 0.02% respectively and does not, in most cases, threaten the quality of the classifiers when used in day-to-day data of rural producers.

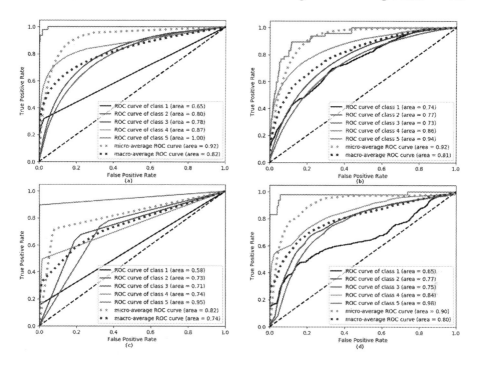

Fig. 9. Results of the ROC curve for the algorithms by class and macro and micro averages: (a) RFC; (b) MLP; (c) KNN; and (d) SVM.

As Fig. 8 indicates, the main objective was achieved by showing that there are no significant differences in the results of the algorithms. Thus, the final model can be generated from any of these models, without risk of damage to the performance of the classification.

4 Conclusions

This paper had the objective of constructing a classifier of carcass fatness degree to assist the rural producers in the decision making to obtain a meat of better quality. For this, we took into account the data of the Precoce MS program that contain 1,056,586 cattle slaughters and their respective productive processes.

The machine learning algorithms tested using the cross-validation result were: Multinomial Naive Bayes (MNB), Random Forest Classifier (RFC), AdaBoost (ADA), Multilayer Perceptron (MLP) K-Nearest Neighbors (KNN) and Support Vector Machines (SVM). Among the six, the most accurate, in a balanced dataset, was the RFC which obtained 70.37% of accuracy and it is considered a satisfactory result.

The rural producer is financially rewarded only for carcasses classified as 2, 3 and 4. An optimization that can be done to assist in obtaining results that favor the rural producer to discover whether or not he will be subsidized and, after that answer, try to show what type of carcass will be generated is the use of Hierarchical Machine Learning.

Acknowledgements. This paper is based upon the work supported by Programa de Educação Tutorial (PET).

References

1. Indicadores IBGE: estatística da produção pecuária. https://biblioteca.ibge.gov.br/visualizacao/periodicos/2380/epp_2019_3tri.pdf. Acessed 21 Jan 2020. (in Portuguese)
2. Felício, P.E. de: Classificação e Tipificação de Carcaças. Bovinocultura de Corte - Volumes I e II. FEALQ, Piracicaba, SP, pp. 1257–1276 (2010). (in Portuguese)
3. Lawrence, T.E., Whatley, J.D., Montgomery, T.H., Perino, L.J.: A comparison of the USDA ossification-based maturity system to a system based on dentition. J. Anim. Sci. **79**, 1683–1690 (2001)
4. Bittencourt, C.D.R., Ladeira, M., da Silva, S.F., Bittencourt, A.L.S., Borges, D.L.: Sistema de Classificação Automática de Carcaças Bovinas. In: Simpósio Brasileiro de Sistemas de Informação (SBSI), Rio de Janeiro. Anais do IV Simpósio Brasileiro de Sistemas de Informação, vol. 4, pp. 235–244. Sociedade Brasileira de Computação, Porto Alegre (2008). (in Portuguese)
5. Nara, E.O.B., Benitez, L.B., Forgiarini, G., Kipper, L.M., Schwingel, G.A.: The escape of the operation of commodities as strategy. Int. J. Bus. Innov. Res. **15**(4), 500–513 (2018)
6. Andreo, N., et al.: Carcass characteristics and meat quality of Nellore bulls submitted to different nutritional strategies during cow-calf and stocker phase. Animal **13**(7), 1544–51 (2019)
7. Cattelam, J., do Vale, M.M., Martini, P.M., Pacheco, R.F., Mayer, A.R., Pacheco, P.S.: Productive characteristics of precocious or super precocious cattle confined. Amazonian J. Plant Res. **1**(1), 33–38 (2017)
8. Batista, P.B., Neto, S.G., Quadros, D.G., Araújo, G.G.L., Souza, C.G., Sabedot, M.A.: Qualitative traits of the meat of Nellore steers supplemented with energy and protein in an integrated crop-livestock system. Anim. Prod. Sci. **60**, 464–472 (2019)
9. Pereira, M.A., Fairweather, J.R., Woodford, K.B., Nuthall, P.L.: Assessing the diversity of values and goals amongst Brazilian commercial-scale progressive beef farmers using Q-methodology. Agric. Syst. **144**, 1–8 (2016)
10. Pedregosa, F., et al.: Scikit-learn: machine learning in Python. J. Mach. Learn. Res. **12**, 2825–2830 (2011)
11. Jain, A., Nandakumar, K., Ross, A.: Score normalization in multimodal biometric systems. Pattern Recogn. **38**(12), 2270–2285 (2005)
12. Johannes, M., et al.: Integration of pathway knowledge into a reweighted recursive feature elimination approach for risk stratification of cancer patients. Bioinformatics **26**(17), 2136–2144 (2010)
13. Granitto, P.M., Furlanello, C., Biasioli, F., Gasperi, F.: Recursive feature elimination with random forest for PTR-MS analysis of agroindustrial products. Chemometr. Intell. Lab. Syst. **83**(2), 83–90 (2006)
14. Lemaître, G., Nogueira, F., Aridas, C.K.: Imbalanced-learn: a Python toolbox to tackle the curse of imbalanced datasets in machine learning. J. Mach. Learn. Res. **18**(1), 559–563 (2017)

15. Wilson, D.: Asymptotic properties of nearest neighbor rules using edited data. IEEE Trans. Syst. Man Cybern. **2**(3), 408–421 (1972)
16. Chawla, N.V., Bowyer, K.W., Hall, L.O., Kegelmeyer, W.P.: SMOTE: synthetic minority over-sampling technique. J. Artif. Intell. Res. **16**, 321–357 (2002)
17. Batista, G.E.A.P.A., Prati, R.C., Monard, M.C.: A study of the behavior of several methods for balancing machine learning training data. ACM SIGKDD Explor. Newsl. **6**(1), 20–29 (2004)
18. Townsend, J.T.: Theoretical analysis of an alphabetic confusion matrix. Percept. Psychophys. **9**(1), 40–50 (1971)
19. Goutte, C., Gaussier, E.: A probabilistic interpretation of precision, recall and F-score, with implication for evaluation. In: Losada, D.E., Fernández-Luna, J.M. (eds.) ECIR 2005. LNCS, vol. 3408, pp. 345–359. Springer, Heidelberg (2005). https://doi.org/10.1007/978-3-540-31865-1_25
20. Vapnik, V.N.: An overview of statistical learning theory. IEEE Trans. Neural Netw. **10**(5), 988–999 (1999)
21. Liu, Y., Liao, S., Jiang, S., Ding, L., Lin, H., Wang, W.: Fast cross-validation for kernel-based algorithms. IEEE Trans. Pattern Anal. Mach. Intell. **42**, 1083–1096 (2019)
22. Lameski, P., Zdravevski, E., Mingov, R., Kulakov, A.: SVM parameter tuning with grid search and its impact on reduction of model over-fitting. In: Yao, Y., Hu, Q., Yu, H., Grzymala-Busse, J.W. (eds.) RSFDGrC 2015. LNCS (LNAI), vol. 9437, pp. 464–474. Springer, Cham (2015). https://doi.org/10.1007/978-3-319-25783-9_41
23. Brodersen, K.H., Ong, C.S., Stephan, K.E., Buhmann, J.M.: The balanced accuracy and its posterior distribution. In: 2010 20th International Conference on Pattern Recognition, pp. 3121–3124. IEEE (2010)

Skin Cancer Classification Using Inception Network and Transfer Learning

Priscilla Benedetti[1]($\boxtimes$) (iD), Damiano Perri[2]($\boxtimes$) (iD), Marco Simonetti[2]($\boxtimes$) (iD), Osvaldo Gervasi[3]($\boxtimes$) (iD), Gianluca Reali[1]($\boxtimes$) (iD), and Mauro Femminella[1]($\boxtimes$) (iD)

[1] Department of Engineering, University of Perugia, Perugia, Italy
pris.benedetti92@gmail.com,{gianluca.reali,mauro.femminella}@unipg.it
[2] Department of Mathematics and Computer Science, University of Florence, Florence, Italy
damiano.perri@gmail.com,marco.simonetti@libero.it
[3] Department of Mathematics and Computer Science, University of Perugia, Perugia, Italy
osvaldo.gervasi@unipg.it

Abstract. Medical data classification is typically a challenging task due to imbalance between classes. In this paper, we propose an approach to classify dermatoscopic images from HAM10000 (Human Against Machine with 10000 training images) dataset, consisting of seven imbalanced types of skin lesions, with good precision and low resources requirements. Classification is done by using a pretrained convolutional neural network. We evaluate the accuracy and performance of the proposal and illustrate possible extensions.

Keywords: Machine learning · Convolutional neural network · Keras · TensorFlow

1 Introduction

Training of neural networks for automated diagnosis of pigmented skin lesions can be a difficult process due to the small size and lack of diversity of available datasets of dermatoscopic images. The HAM10000 ("Human Against Machine with 10000 training images") dataset is a collection of dermatoscopic images from different populations acquired and stored by different modalities. We used the benchmark dataset, with a small number of images and a strong imbalance among the 7 different types of lesions, to prove the validity of our approach, which is characterized by good results and light usage of resources.

Exploiting a highly engineered convolutional neural network with transfer learning, customized data augmentation and a non-adaptive optimization algorithm, we show the possibility of obtaining a final model able to precisely recognize multiple categories, although scarcely represented in the dataset. The whole training process has a limited impact on computational resources, requiring no more than 20 GB of RAM space. The rest of paper is structured as

© Springer Nature Switzerland AG 2020
O. Gervasi et al. (Eds.): ICCSA 2020, LNCS 12249, pp. 536–545, 2020.
https://doi.org/10.1007/978-3-030-58799-4_39

follows: Sect. 2 describes the related work in the field of medical images processing. Section 3 illustrates the dataset of interest. Section 4 gives an overview of the model architecture. Section 5 includes the training process and shows experimental results. Finally, some final comments and future research directions are reported in Sect. 6.

2   Related Work

Processing of biomedical images has always been a field strongly beaten by CNN pioneers. The first related papers date back to 1991 [1], with a strong impulse in the following years in the search for methods for automating the classification of pathologies and related diagnosis [2,3].

Nowadays, almost thirty years later, reliability of networks reached a rather high level, as well as intrinsic complexity. This reliability allowed a wide diffusion of the approach of subjecting diagnostic images to automatic classification systems, from evolutionary algorithms [4–7] to deep networks [8–11], being them either convolutive or not. Even in the medical sector of dermatology, automatic image recognition and classification was used for decades to detect tumor skin lesions [12,13].

Recent and promising research has highlighted the possibility that properly trained machines can exceed the human recognition and classification capability to recognize skin cancers. The scores obtained are very encouraging [14] and we are confident that in the near future the recognition capacity of these forms of pathologies will become almost total.

Today CNNs are used for image feature extraction. Features are used for image classification [15,16].

3   The Dataset

Dermatoscopy is often used to get better diagnoses of pigmented skin lesions, either benign or malignant. With dermatoscopic images is also possible to train artificial neural networks to recognize pigmented skin lesions automatically. Nevertheless, training requires the usage of a large number of samples, although the number of high quality images with reliable labels is either limited or restricted to only a few classes of diseases, often unbalanced.

Due to these limitations, some previous research activities focused on melanocytic lesions (in order to differentiate between a benign and malignant sample) and disregarded non-melanocytic pigmented lesions, even if very common. In order to boost research on automated diagnosis of dermatoscopic images, HAM10000 has been providing the participant of the ISIC 2018 classification challenge, hosted by the annual MICCAI conference in Granada, Spain [17], specific images.

The set of 10015 8-bit RGB color images were collected in 20 years from populations from two different sites, specifically the Department of Dermatology of the Medical University of Vienna, and the skin cancer practice of Cliff

Rosendahl in Queensland. Relevant cases include a representative collection of all important diagnostic categories of pigmented lesions [17]:

- **akiec:** Actinic Keratoses and Intraepithelial Carcinoma, common noninvasive variants of squamous cell carcinoma that can be treated locally without surgery. [**327 images**]
- **bcc:** Basal cell carcinoma, a cancer that rarely metastasizes but grows destructively if untreated. [**514 images**]
- **bkl:** Generic label that includes seborrheic keratoses, solar lentigo and lichenplanus like keratoses (LPLK), which corresponds to a seborrheic keratosis or a solar lentigo with inflammation and regression, often mistaken for melanoma. [**1099 images**]
- **df:** Dermatofibroma, a benign skin lesion. [**115 images**]
- **nv:** Melanocytic nevi are benign neoplasms of melanocytes. [**6705 images**]
- **mel:** Melanoma, if diagnosed in an early stage, it can be cured by simple surgical excision. [**1113 images**]
- **vasc:** Vascular skin lesions in the dataset range from cherry angiomas to angiokeratomas and pyogenic granulomas. [**142 images**]

More than 50% of lesions are confirmed through histopathology (histo), the ground truth for the rest of the cases is either follow-up examination (followup), expert consensus (consensus), or confirmation by in-vivo confocal microscopy (confocal). Other features in the individual dataset include age, gender and bodysite of lesion (localization) [17].

4   Model Architecture

Since their first appearance in Le Cun *et al.* publication [18], Convolutional Neural Networks (CNN) have been widely applied to data that have a known, grid-like structure. Possible set of interests are time-series data, which can be modeled as a 1D grid taking samples at regular time intervals, and image data, which can be thought of as a 2D grid of pixels. The foundational layer of a convolutional network consists of three stages. In the first stage, the layer performs several parallel convolutions to produce a set of linear activations. In the second stage, each convolution output is run through a nonlinear activation function, such as the rectified linear activation function (ReLU). In the third stage, a pooling function is used to modify the output of the layer further [19]. The max pooling operation, used in this work, reports the maximum output within a rectangular neighborhood. Since the objective images of skin lesions present great variation in size, we decided to use a network made by inception modules, which make use of filters of different size operating at the same level. This usage of wide modules with multiple cheap convolutional operations entails a reduced computational complexity with respect to a deep network with large convolutional layers. In the specific model we used, based on Inception-ResNet-v2, another point of speed improvement is the introduction of residual connections, which replace pooling operations within the main inception modules. However, the previously

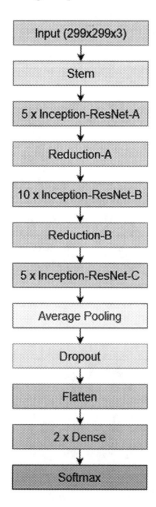

Fig. 1. General schema for the used network, an Inception-ResNet-v2 architecture with the addition of a flattening layer, 2 fully-connected layers and a final softmax activation.

mentioned max pooling operations are still present in the reduction blocks. The structure of the network used in this work is shown on Fig.1

The original Inception-ResNet-v2 architecture [20] has a stem block consisting of the concatenation of multiple convolutional and pooling layers, while Inception-ResNet blocks (A, B and C) contain a set of convolutional filters with an average pooling layer. As prevously mentioned, reduction blocks (A, B) replace the average pooling operation with a max pooling one. This structure has been extended with a final module consisting of a flattening step, two fully-connected layers of 64 units each, and the softmax classifier. The overall module is trainable on a single GPU with reduced memory consumption.

5   Training Process and Experimental Results

This work consists of two training rounds, after a step of data processing in order to deal with the strong imbalance of the dataset:

- A first classification training process using class weights.
- Rollback of previous obtained best model to improve classification performance with a second training phase.

5.1   Data Processing

In the first stage of data processing, after the creation of a new column with a more readable definition of labels, each class was translated into a numerical code using *pandas.Categorical.codes*. Afterwards, missing values in "age" column was filled with column mean value. Figure 2 and Fig. 3 show the HAM10000 data distribution.

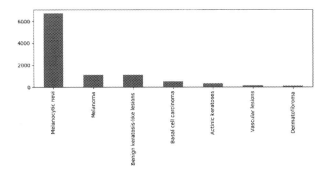

Fig. 2. Plotting the frequency of each class, the imbalance between Melanocytic Nevi and the rest of the possible categories is manifest.

Finally, images are loaded and resized from 450×600 to 299×299 in order to be correctly processed by the network. After a normalization step on RGB arrays, we split the dataset into a training and validation set with 80:20 ratio (Fig. 4).

In order to re-balance the dataset, we chose to shrink the amount of images for each class to an equal maximum dimension of 450 samples. This significant decrease of available images is then mitigated by applying a step of data augmentation. Training set expansion is made by altering images with small transformations to reproduce some variations, such as horizontal flips, vertical flips, translations, rotations and shearing transformations.

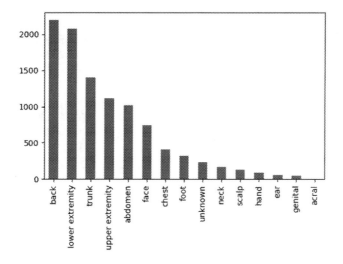

Fig. 3. Plot of samples' body locations.

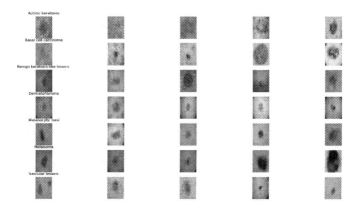

Fig. 4. Some HAM10000 images resized with OpenCV.

5.2 Baseline

Due to the limited number of samples for the training process, we decided to take advantage of transfer learning, utilizing Inception-ResNet-v2 pre-trained on ImageNet [21] and Tensorflow, a deep learning framework developed by Google, for fine-tuning of the last 40 layers. Keras library offers a wide range of optimizers: Adaptive optimization methods such as AdaGrad, RMSProp, and Adam are widely used for deep neural networks training due to their fast convergence times. However, as described in [22], when the number of parameters exceeds the number of data points these optimizers often determine a worse generalization capability compared with non-adaptive methods. In this work we used a stochastic gradient descent optimizer (SGD), with learning rate set to 0.0006 and usage of momentum and Nesterov Accelerated Gradient in order to adapt updates to the slope of the loss function (categorical crossentropy) and speed up the training process. The total number of epochs was set to 100, using a small batch size 10. A set of class weight was introduced in the training process to get more emphasis on minority class recognition. A maximum patience of 15 epochs was set to the early stopping callback in order to mitigate the overfitting visible in Fig. 5, which shows the history of training and validation process.

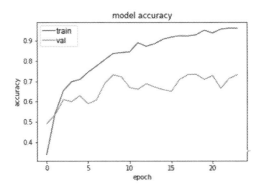

Fig. 5. Accuracy and Loss for each epoch.

Finally, the model achieves an accuracy of 73.4% on the validation set, using weights from the best epoch. Figure 6 shows the confusion matrix for the model on the validation set.

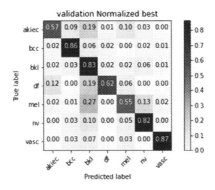

Fig. 6. Two of the minority classes, Actinic Keratoses (akiec) and Dermatofibroma (df), are not properly recognized. Melanoma (mel) is often mistaken for generic keratoses (bkl), as already mentioned in Fig. 3

5.3   Resuming Training from the Best Epoch

In order to improve classification performance, specially on minority classes, we loaded the best model obtained in the first round to extend the training phase and explore other potential local minimum points of the loss function, by using an additional amount of 20 epochs. This second step led to an enhancement in overall predictions, reaching the maximum accuracy value of 78.9%.

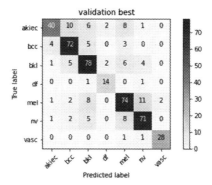

Fig. 7. Final confusion matrix.

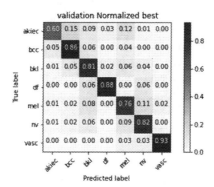

Fig. 8. Normalized final confusion matrix.

Figure. 7 shows the normalized confusion matrix on the validation set for the final fine-tuned model. In this case, 6 out of 7 categories are classified with a total ratio of True Positives higher than 75%, even in presence of extremely limited sample set, as vascular lesions (vasc), 30 samples, and dermatofibroma (df), 16 samples. The whole process of training has required less than four hours on Google Colab cloud's GPU, for an overall RAM utilization below 20 GB (Fig. 8).

6   Conclusions and Future Works

In conclusion, in this paper we investigate the possibility of obtaining improved performances in the classification of 7 significantly unbalanced different types of skin diseases, with a small amount of available images. With use of a fine-tuned deep inception network, data augmentation and class weights, the model can achieve a good final diagnostic accuracy. The described training process has a light resource usage, requiring less than 20 GB of RAM space, and it can be executed in a Google Colab notebook. For future improvements larger datasets of dermatoscopic images are needed. The model shown in this paper can be regarded as a starting point to implement a lightweight diagnostic support system for dermatologists, for example in the Web as well as through a mobile application.

References

1. Zhang, W., Hasegawa, A., Itoh, K., Ichioka, Y.: Image processing of human corneal endothelium based on a learning network. Appl. Opt. **30**, 4211–4217 (1991)
2. Zhang, W., Doi, K., Giger, M.L., Wu, Y., Nishikawa, R.M., Schmidt, R.A.: Computerized detection of clustered microcalcifications in digital mammograms using a shift-invariant artificial neural network. Med. Phys. **21**(4), 517–524 (1994)
3. Vella, F., Neri, I., Gervasi, O., Tasso, S.: A simulation framework for scheduling performance evaluation on CPU-GPU heterogeneous system. In: Murgante, B., et al. (eds.) ICCSA 2012. LNCS, vol. 7336, pp. 457–469. Springer, Heidelberg (2012). https://doi.org/10.1007/978-3-642-31128-4_34
4. Chen, J.-C., Yeh, C.-M., Tzeng, J.-E.: Pattern differentiation of glandular cancerous cells and normal cells with cellular automata and evolutionary learning. Exp. Syst. Appl. **34**(1), 337–346 (2008)
5. Guo, P.F., Bhattacharya, P.: An evolutionary approach to feature function generation in application to biomedical image patterns (2009)
6. Gervasi, O., Russo, D., Vella, F.: The AES implantation based on OpenCL for multi/many core architecture. In: 2010 International Conference on Computational Science and Its Applications, Fukuoka, ICCSA 2010, pp. 129–134. IEEE Computer Society, Washington (2010). https://doi.org/10.1109/ICCSA.2010.44
7. Mariotti, M., Gervasi, O., Vella, F., Cuzzocrea, A., Costantini, A.: Strategies and systems towards grids and clouds integration: a DBMS-based solution. Fut. Gener. Comput. Syst. **88**, 718–729 (2018). https://doi.org/10.1016/j.future.2017.02.047
8. Pang, S., Yu, Z., Orgun, M.A.: A novel end-to-end classifier using domain transferred deep convolutional neural networks for biomedical images. Comput. Methods Prog. Biomed. **140**, 283–293 (2017)
9. Zhou, Z., Shin, J., Zhang, L., Gurudu, S., Gotway, M., Liang, J.: Fine-tuning convolutional neural networks for biomedical image analysis: actively and incrementally (2017)
10. Gervasi, O., et al. (eds.): ICCSA 2016. LNCS, vol. 9787. Springer, Cham (2016). https://doi.org/10.1007/978-3-319-42108-7
11. Gervasi, O., et al. (eds.): ICCSA 2017. LNCS, vol. 10406. Springer, Cham (2017). https://doi.org/10.1007/978-3-319-62398-6

12. Lau, H.T., Al-Jumaily, A.: Automatically early detection of skin cancer: study based on nueral netwok classification(2009)
13. Dorj, U.-O., Lee, K.-K., Choi, J.-Y., Lee, M.: The skin cancer classification using deep convolutional neural network. Multimedia Tools Appl. **77**(8), 9909–9924 (2018). https://doi.org/10.1007/s11042-018-5714-1
14. Maron, R.C., et al.: Systematic outperformance of 112 dermatologists in multiclass skin cancer image classification by convolutional neural networks. Eur. J. Cancer **119**, 57–65 (2019)
15. Biondi, G., Franzoni, V., Gervasi, O., Perri, D.: An approach for improving automatic mouth emotion recognition. In: Misra, S., et al. (eds.) ICCSA 2019. LNCS, vol. 11619, pp. 649–664. Springer, Cham (2019). https://doi.org/10.1007/978-3-030-24289-3_48
16. Perri, D., Sylos Labini, P., Gervasi, O., Tasso, S., Vella, F.: Towards a learning-based performance modeling for accelerating deep neural networks. In: Misra, S., et al. (eds.) ICCSA 2019. LNCS, vol. 11619, pp. 665–676. Springer, Cham (2019). https://doi.org/10.1007/978-3-030-24289-3_49
17. Tschandl, P., Rosendahl, C., Kittler, H.: The HAM10000 dataset, a large collection of multi-source dermatoscopic images of common pigmented skin lesions. Sci. Data. **5**, 180161 (2018). https://doi.org/10.1038/sdata.2018.161
18. Lecun, Y., Bottou, L., Bengio, Y., Haffner, P.: Gradient-based learning applied to document recognition. Proc. IEEE **86**(11), 2278–2324 (1998). https://doi.org/10.1109/5.726791
19. Goodfellow, I., Bengio, Y., Courville, A.: Deep Learning. MIT press, Cambridge (2016)
20. Szegedy, C., Ioffe, S., Vanhoucke, V., Alemi, A.: Inception-v4, Inception-ResNet and the Impact of Residual Connections on Learning (2016)
21. Krizhevsky, A., Sutskever, I., Hinton, G.: Imagenet classification with deep convolutional neural networks. In: 25th International Conference on Advance in Neural Information Processing System, pp. 1106–1114 (2012)
22. Wilson, A., Roelofs, R., Stern, M., Srebro, N., Rech, B.: The marginal value of adaptive gradient methods in machine learning. In: NIPS 2017 (2017)

A Criterion for IDS Deployment on IoT Edge Nodes

Vladimir Shakhov[1]([⊠])🆔, Olga Sokolova[2]🆔, and Insoo Koo[1]🆔

[1] University of Ulsan, Ulsan 44610, Republic of Korea
{shakhov,iskoo}@ulsan.ac.kr
[2] Institute of Computational Mathematics and Mathematical Geophysics,
Novosibirsk, Russia
olga@rav.sscc.ru

Abstract. Edge computing becomes a strategic concept of IoT. The edge computing market reaches several billion USD and grows intensively. In edge computing paradigm, the data can be processed close to, or at the edge of the network. This way greatly reduces the computation and communication load of the network core. Moreover, processing data near the sources of data also provides better support for the user privacy. However, an increase in the number of data processing locations will proportionally increase the attack surface. Due to limited capacities and resources, an edge node cannot perform too many complex operations. Especially for the applications with high real-time requirements, efficiency becomes a crucial issue in secure data analytics. Therefore, it is important to get a tradeoff between security and efficiency. We focus on this problem in this paper.

Keywords: Edge computing · Internet of Things · Intrusion Detection System · Quantitative analysis · Queuing theory

1 Introduction

According to the estimation by Cisco Global Cloud Index, the data produced by IoT devices will exceed 500 Zettabytes in this year. For efficient treatment of such huge volumes of data the edge computing paradigm has been offered. In this paradigm, the data can be processed close to, or at the edge of the network. Some functions of network core is delegated to edges of the network, where the connected entities directly produce the data. These facilities can be fortified by corresponding computing platforms and system resources. Edge computing technologies greatly offload the computation and communication load of the network core. Moreover, processing data near the sources of data provides better QoS [1] for the delay sensitive services and efficient structure support for the user privacy, as well as prevent and mitigate some types of DDoS attacks.

It is expected that a ratio of enterprise-generated data, which is processed outside a conventional centralized data center or cloud, will rich 75%. The total edge computing market size is expected to grow from USD 2.8 billion in 2019 to USD 9.0 billion by 2024, at a Compound Annual Growth Rate (CAGR) of 26.5% during the forecast

© Springer Nature Switzerland AG 2020
O. Gervasi et al. (Eds.): ICCSA 2020, LNCS 12249, pp. 546–556, 2020.
https://doi.org/10.1007/978-3-030-58799-4_40

period, estimates ResearchAndMarkets.com. According to another forecast provided by Gartner, this market will reach USD 13 billion by 2022. Financial and banking industry is one of the largest beneficiaries of edge computing worldwide. Increasing the adoption of digital and mobile banking initiatives, advanced platforms, such as blockchain and payment through mobile terminals, are fuelling the demand for modern edge computing solutions in the financial and banking industry sector. Asia-Pacific is destined to be the major market for edge computing. Businesses and governments in this region have shown more inclination toward storing and processing data locally.

However, an increase in the number of data processing locations will proportionately increase the attack surface [2]. Also, it needs to remark that we generally use edge devices with limited resources [3]. Therefore, it is not reasonable to store a large amount of data and execute a high complexity algorithm for intrusion detection. The conventional security mechanisms of cloud computing are no longer suitable for supporting security in edge computing. Taking into account security challenges, leading academia researchers and profit companies experts conclude that current situation with edge computing security is far from satisfied and essential efforts is required to overcome the existing vulnerabilities and weaknesses. Thus, edge computing security is highlighted as an important future research direction [4–6].

A lightweight and secure data analytics technique allows to increase the potential of edge computing. Due to limited capacities and resources, an edge node cannot perform too many complex operations, which could incur high latency and battery depletion. Efficiency becomes a crucial issue in secure data analytics, especially for the applications with high real-time requirements.

There are a few recent papers on the theme of Intrusion Detection System (IDS) for Edge Computing. Some authors offer a hybrid IDS mechanism, but they ignore quantitative analysis [7]. Other researchers focus on quality of detection scheme only [8]. Thus, in the present literature there is a lack of quantitative methods, which allow to form a proper holistic view on a system. This paper is intended to partially fill this gap. We describe the offered approach in general and provide a closed-for solution in an important practical case.

2   System Model

A set of IoT edge nodes serve a user-generated workload. It includes a traffic which has to be treated and retransmitted. Let us use the designations as follows.

- λ: the traffic intensity;
- μ: the intensity of request treatment;
- α: a percentage of workload of legal users, in practice, this can be estimated using an observable sample or an auxiliary model;
- B: the probability of request rejection, i.e. the blocking probability.

Here we consider a situation with two types of users. Legitimate users generate traffic with the intensity $\lambda\alpha$, and malicious users generate traffic with the intensity $\lambda(1 - \alpha)$. Due to limited resources of edge nodes, a part of traffic (B) does not receive a service and rejected (see Fig. 1).

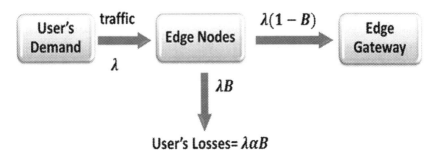

Fig. 1. IDS has not been used.

Generally, the blocking probability it is a function of λ and μ, i.e. the losses rate equals

$$\lambda B(\lambda, \mu),$$

and the served workload rate is

$$\lambda(1 - B(\lambda, \mu)).$$

Note, that not all traffic is useful. The actual losses rate of legal users is as follows:

$$l_1 = \alpha \lambda B(\lambda, \mu).$$

Next, let us consider edge nodes equipped by IDS. It is reasonable to assume that a part of malicious requests will be rejected and novel workload intensity $\widetilde{\lambda}$ will be reduced, i.e. $\widetilde{\lambda} < \lambda$. However, it does not guarantee that the system throughput becomes better. Now IoT devices have to perform additional operation for intrusion detection maintaining and malicious requests filtering. Therefore, performance of request treatment has to be reduced, i.e. the novel intensity of request treatment becomes $\widetilde{\mu}$, and $\widetilde{\mu} < \mu$.

If the security system is designed to counteract a limited set of known attacks, then signature based IDS can be used. In this case IDS utilizes a set of rules (signatures) that can be used to detect an attacks pattern. This approach provides a high level of accuracy for well-known intrusions. A signature based IDS is characterized by low computational cost, i.e. $\widetilde{\mu} \approx \mu$. The same effect can be reached by the use of small number of secret bits for requests verification. However, this situation is not typical for IoT environment. Intruders constantly change tactics and create new destructive tools. Signature-based detection does not detect slightly modified attacks, much less it does not detect unknown attacks. Hence, advanced intrusion detection methods are required.

Also, the situation when $\widetilde{\mu} \gg \mu$ is not typical for IoT keeping in mind the edge devices level [9]. Low resources make ineffective heavy computation algorithms like deep learning. So, it is reasonable to assume that performance of request treatment has not been increased drastically.

Moreover, some legitimate requests are mistakenly recognized as illegal and filtered by IDS (Fig. 2).

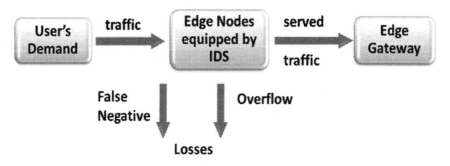

Fig. 2. Nodes are quipped by IDS.

Let us investigate the situation, where IDS deployment makes a sense.

3 Analysis

3.1 IDS Applications

For our purposes it is enough to consider IDS parameters as follows:

- p_I is a false positive, i.e. the probability of the event, when a legitimate request is rejected by IDS;
- p_{II} is a false negative, i.e. the probability of the event, when an illegal request is accepted.

Therefore, IDS rightly rejects

$$\lambda(1 - \alpha)(1 - p_{II})$$

spoofed request per time unit. And the forced losses of legal workload are as follows:

$$\lambda \alpha p_I.$$

Hence, edge nodes have to treat an offered load of intensity:

$$\widetilde{\lambda} = \lambda \alpha(1 - p_I) + \lambda(1 - \alpha)p_{II}$$

Let us remark, that the ratio of legitimate requests has been changed. Now, this one is

$$\widetilde{\alpha} = \frac{\lambda \alpha(1 - p_I)}{\widetilde{\lambda}} = \frac{\alpha(1 - p_I)}{\alpha(1 - p_I) + (1 - \alpha)p_{II}}.$$

In the case of IDS application, the actual losses rate of legal users becomes as follows:

$$l_2 = \lambda \alpha p_I + \widetilde{\alpha} \widetilde{\lambda} B\left(\widetilde{\lambda}, \widetilde{\mu}\right).$$

The IDS (with given quality parameters p_I, p_{II}) is justified if and only if

$$l_2 < l_1.$$

Hence, the novel intensity of request treatment has to satisfy the inequality as follows:

$$B\left(\widetilde{\lambda}, \widetilde{\mu}\right) < \frac{\alpha \lambda}{\widetilde{\alpha} \widetilde{\lambda}} (B(\lambda, \mu) - p_I) \qquad (1)$$

In view of the above considerations, inequality (1) can be given an alternative formulation as follows:

$$B\left(\widetilde{\lambda}, \widetilde{\mu}\right) < \frac{B(\lambda, \mu) - p_I}{1 - p_I}.$$

Using an appropriate queuing model we can calculate the blocking probability and its inverse.

3.2   Erlang-B Function

Let us specify the model. Taking into account requirements of delay sensitive services, it is reasonable to use M/M/n/n queuing system to model functioning of cluster containing n edge nodes. Thus, our assumptions are as follows:

- Incoming Poisson flow (with intensity $\widetilde{\lambda}$ or λ)
- Exponential service time (with intensity $\widetilde{\mu}$ or μ).
- No buffer (waiting room)

In this case the blocking probability is described by Erlang-B formula (see, for example [10]),

$$B(\rho, n) = \frac{\frac{\rho^n}{n!}}{\sum_{i=0}^{n} \frac{\rho^i}{i!}},$$

where

$$\rho = \frac{\lambda}{\mu}.$$

The inequality (1) can be numerically solved. Note, the assumption of exponential CDF for service time is not necessary. The formula is true for M/G/n/n queuing system as well.

Let us consider the case, when

$$\rho \gg n.$$

It generally takes place under attack. We use the following theorem [11].

Theorem.

$$\forall \varepsilon > 0, \text{ if } \rho \geq n + \frac{1}{\varepsilon},$$

then

$$1 - \frac{n}{\rho} < B(n, \rho) < 1 - \frac{n}{\rho} + \varepsilon.$$

Corollary. If ε is small enough, then we get an approximation for Erlang-B function:

$$B(n, \rho) \approx 1 - \frac{n}{\rho},$$

the inverse functions approximations can be easily calculated as well

$$\rho = \frac{n}{1 - B},$$

$$n = \rho(1 - B).$$

Thus, M/M/n/n system under heavy load provides the outgoing rate (served requests) as follows:

$$\lambda(1 - B(n, \rho)) = \lambda \frac{n}{\rho} = n\mu,$$

and the losses rate is as follows:

$$\lambda B(n, \rho) = \lambda - n\mu.$$

3.3 Service Differentiation

Let us consider the problem of service differentiation. It can be a problem of security differentiation for different classes of customers as well as a traffic management problem, cluster head selection process etc. In the case of jamming attacks, this technique can be used to assign non-attacked channels in order to support the survivability of the most important applications.

The problem statement is as follows

$$N \rightarrow min$$

$$N = \sum_{j=1}^{C} n_j$$

$$B(n_j, \rho_j) \leq b_j, j \in \{1, 2, .., C\},$$

where

- N is a total number of computational resources (channels, servers, service centres, IDS agents etc.)
- C is a number of user classes.
- n_j is a number of resources assigned to the class j
- b_j is QoS required by class j (i.e. the losses rate).

In the case of limited resources (the most critical case) the approximation above helps to solve the problem. The solution is as follows

$$n_j = \rho_j(1 - b_j), j \in \{1, 2, .., C\},$$

$$N_{opt} = \sum_{j=1}^{C} \rho_j(1 - b_j).$$

4   Criterion

Now we can get a closed-form solution for the inequality (1) in the case of heavy workload. It is as follows.

Proposition.   IDS is justified if the following inequality is true:

$$\widetilde{\mu}(1 - p_I) > \mu(\alpha(1 - p_I) + (1 - \alpha)p_{II}) \tag{2}$$

This inequality can be used for estimation and selection intrusion detection algorithms. For convenience, the inequality (2) can be rewritten as a ratio of requests processing intensities:

$$\frac{\widetilde{\mu}}{\mu} > \alpha + \frac{p_{II}}{1 - p_I}(1 - \alpha).$$

We can often (but not always) expect that a way to improve the false positive parameter entails the consequences of the proportional degradation of the false negative parameter, and vice versa. This is a specific of IDS design. However, if the IDS quality is good enough, both p_I and p_{II} are small enough.

Let us define the following ratio:

$$\frac{p_{II}}{1 - p_I},$$

as the IDS throughput. Generally, packets are processed individually by IDS, hence this value does not depend on the legal users' packets proportion.

Let us remark, if the efficiency of used intrusion detection algorithms is very high, i.e.

$$p_I \to 0,$$

$$p_{II} \to 0,$$

then the criterion for the appropriateness of IDS takes a simple form:

$$\alpha < \frac{\widetilde{\mu}}{\mu} < 1.$$

The decision to deploy IDS (or provide requirements for IDS) can be based on a profitability analysis. Therefore, a criterion can take a set of various forms. For example, the criterion can be as follows: IDS should improve the loss rate by k times, i.e.

$$\frac{l_2}{l_1} > k,$$

where k is a desired constant. An alternative statement can be as follows: the IDS profitability has to be higher than the desired threshold k, i.e.

$$l_2 > k.$$

Next, it needs to maximize the system profitability under limited cost or energy consumption, i.e.

$$\max(l_1, l_2)$$

$$\widetilde{\mu} \leq k,$$

and so on. The approximation above allows to obtain a closed form solution in these cases as well.

5 Performance Evaluation

In our consideration we can assume that the IDS throughput value varies in the range [0 ... 1]. The following equation can be useful to determinate a tradeoff between computational overhead and intrusion detection efficiency

$$\frac{\tilde{\mu}}{\mu} = \alpha + (1 - \alpha) \frac{p_{II}}{1 - p_I}$$

Figure 3 presents the increase in IDS overhead change according to IDS throughput efficiency for $\alpha \in \{0.1; 0.5; 0.9\}$.

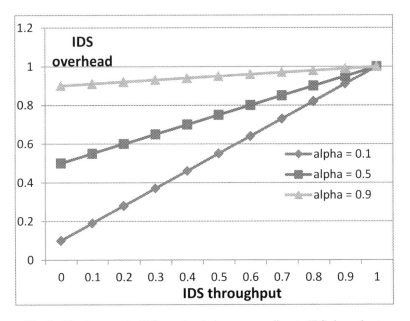

Fig. 3. The increase in IDS overhead change according to IDS throughput.

As we can see, if the portion of legitimate requests is about 10%, and IDS leads to a decrease in node performance by 50%, then the throughput of IDS can confidently be in the range from 0.2 to 0.3. This is a very mediocre IDS. And next example, if the portion of legitimate requests is about 90%, then no reasons to use even an ideal IDS with no any mistakes (zero false positive and false negative) and only 15% degradation of node throughput.

Let us consider the value as follows:

$$\frac{\frac{\tilde{\mu}}{\mu} - \alpha}{\alpha} * 100\%.$$

Assume, the false positive and false negative values are small enough. Here, without loss of generality we can take $p_I = p_{II} \in \{1\%, 5\%, 10\%\}$.

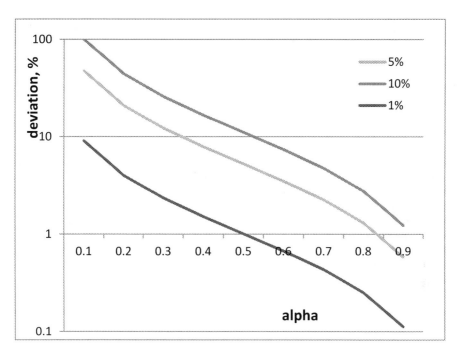

Fig. 4. The suitability of α as the node performance degradation threshold depends on the quality of the used intrusion detection algorithm.

If the quality of the used intrusion detection algorithm is very high (about percent or less), then α can be taken as a threshold for reducing node performance due to IDS operation. This is not true if the values p_I and p_{II} exceed 5%, although this is still a good enough intrusion detection algorithm. Figure 4 illustrates this point.

6   Conclusion

In this paper we offer a criterion for IDS deployment on IoT edge nodes. The offered results are based on queuing theory. In particular we use M/M/n/n (M/G/n/n) systems. In general the provided approach can be applied to any kind of IDS. However, the detailed results are provided for the low-resources IoT devices (Edge nodes). Using the Erlang losses function approximation we received a quantitative condition when IDS deployment makes a sense. The offered approach can be mostly applied for situations of flooding-type intrusions. Note, that the result can be used in other application domains such as enterprises management, hospital operation and others.

Acknowledgment. This work was supported by the National Research Foundation of Korea (NRF) grant through the Korean Government (MSIT) under Grant NRF-2020R1I1A1A0 1065692. The reported study was partially supported by RFBR according to the research project 19-01-00562.

References

1. Zhang, J., Chen, B., Zhao, Y., Cheng, X., Hu, F.: Data security and privacy-preserving in edge computing paradigm: survey and open issues. IEEE Access **6**, 18209–18237 (2018)
2. Xiao, Y.: Edge computing security: state of the art and challenges. Proc. IEEE **107**(8), 1608–1631 (2019)
3. Liu, F., Tang, G., Li, Y., Cai, Z., Zhang, X., Zhou, Y.: A survey on edge computing systems and tools. Proc. IEEE **107**(8), 1537–1562 (2019)
4. Taleb, T., Samdanis, K., Mada, B., Flinck, H., Dutta, S., Sabella, D.: On multi-access edge computing: a survey of the emerging 5G network edge cloud architecture and orchestration. IEEE Commun. Surv. Tut. **19**(3), 1657–1681 (2017)
5. Mao, Y., You, C., Zhang, J., Huang, K., Letaief, K.: A survey on mobile edge computing: the communication perspective. IEEE Commun. Surv. Tut. **19**(4), 2322–2358 (2017)
6. Porambage, P., Okwuibe, J., Liyanage, M., Ylianttila, M., Taleb, T.: Survey on multi-access edge computing for Internet of Things realization. IEEE Commun. Surv. Tut. **20**(4), 2961–2991 (2018)
7. Yao, H., et al.: Hybrid intrusion detection system for edge-based IIoT relying on machine-learning-aided detection. IEEE Networks **33**(5), 75–81 (2019)
8. Lin, F., et al.: Fair resource allocation in an intrusion-detection system for edge computing: ensuring the security of Internet of Things devices. IEEE Consum. Electron. Mag. **7**(6), 45–50 (2018)
9. Jan, S., Ahmed, S., Shakhov, V., Koo, I.: Toward a lightweight intrusion detection system for the Internet of Things. IEEE Access **7**, 42450–42471 (2019)
10. Medhi, J.: The evergreen Erlang loss function. OPSEARCH **43**(3), 309–319 (2006). https://doi.org/10.1007/BF03398780
11. Shakhov, V.V.: An efficient method for proportional differentiated admission control implementation. In: Vinel, A., Bellalta, B., Sacchi, C., Lyakhov, A., Telek, M., Oliver, M. (eds.) MACOM 2010. LNCS, vol. 6235, pp. 91–97. Springer, Heidelberg (2010). https://doi.org/10.1007/978-3-642-15428-7_10

How to Improve the Recommendation's Accuracy in POI Domains?

Luiz Chaves[1], Nícollas Silva[2], Rodrigo Carvalho[1], Adriano C. M. Pereira[2], Diego R. C. Dias[1(✉)], and Leonardo Rocha[1]

[1] Universidade Federal de São João del-Rei, São João del-Rei, Brazil
`luizfufsj@gmail.com`, {`rodrigo,diegodias,lcrocha`}`@ufsj.edu.br`
[2] Universidade Federal de Minas Gerais, Belo Horizonte, Brazil
{`ncsilvaa,adrianoc`}`@dcc.ufmg.br`

Abstract. Nowadays, Recommender Systems (RS) have been applied in most of Location-Based Social Networks (LBSNs). In general, these RSs aim to provide the best points-of-interest (POIs) to users, encouraging them to visit new places or explore more of their preferences. Despite the researches advances in this scenario, there is an opportunity for improvements in the recommendation task. The main reason behind it is related to specific characteristics of this scenario, such as the geolocation of users. In general, most users are not interested in POIs located far from their home or work area. In this sense, we address a new research perspective in the POI Recommendation field, proposing a re-ordering method to be applied after any RS and improve the POIs located nearby from the users' geolocation. Our assumption is that POIs located on the sub-areas with more activity of a user are more interesting than POIs from new sub-areas. For this reason, we propose to measure the activity level of users in subareas of a city and use it to re-order the POIs recommended before. We evaluate our proposal considering six traditional RSs and three datasets from Yelp, achieving gains up to 15% of precision.

Keywords: Recommender systems · POI recommendation · Geographic

1 Introduction

Currently, several Web applications have invested in Recommendation Systems (RS) to attract and satisfy more users. On systems like Amazon.com and Netflix, RS's task is to present products and movies that please users [3]. On the other hand, in the case of social networks such as Facebook, RS's task is related to identifying which publications are most relevant to be presented to users [10]. On location-based social networks (i.e., Location-Based Social Network (LBSN)), such as Foursquare, Gowalla, and Yelp, the RS's task is related to identifying users' interests in points of interest (i.e., Points of Interest – POI) such as restaurants, bars, cinemas, theaters, museums, and others [17].

© Springer Nature Switzerland AG 2020
O. Gervasi et al. (Eds.): ICCSA 2020, LNCS 12249, pp. 557–571, 2020.
https://doi.org/10.1007/978-3-030-58799-4_41

In traditional domain applications, such as e-commerce (products) and entertainment (movies and music), there are already several RS proposals that can achieve their goals of attracting and satisfying users [3,21]. However, in LBSN applications, the recommendation of POIs is still an open challenge due to their peculiar characteristics. Unlike other scenarios, where an item consumption means that the user only watched or bought a particular item, in the POI scenario, it means that the user had to travel to a particular region of the city to visit it. In other words, there is a geographic factor intrinsically related to user activity that must be taken into account to generate the recommendations [16,23]. For this reason, the main work in recommending POIs concerns to obey Tobler's First Law of Geography: *"everything is related to everything else. But near things are more related than distant things"* [18].

The proposals found in the literature consist of traditional RS adaptations to the POI recommendation scenario [24], as well as some new RSs specifically proposed for this purpose [13,23,26]. That is a relatively new research scenario, and none of these approaches have been able to achieve effectively satisfactory results so far [24]. In this sense, instead of proposing a new recommendation model, we propose a new approach to assist existing systems in identifying potentially relevant POI to users. We propose a re-ordering method to be used together with any RS. Our proposal is basically to measure the level of activity of each user in the various subregions that make up a particular area (e.g., city) and, from that level of activity, to weight the recommendation list generated by a recommender according to it. Thus, we ensure that POIs belonging to high-activity subregions receive a higher weight than POIs from low-activity subregions. From this weighting, this list is then reordered and presented to the user.

In the evaluation of our approach, we apply the re-ordering method proposed considering models used in traditional application domains (i.e., MostPopular, User-KNN, and WRMF) and specific models for the POI recommendation problem (i.e., USG, GeoMF, and GeoSoCa). We evaluated all these applications against the Yelp database, widely used in the POI recommendation scenario, by selecting three cities (Las Vegas, Phoenix, and Charlotte). The results show that, although considering a simple metric to measure users' activity level, our strategy was able to improve the recommendation quality in the vast majority of cases analyzed, with some gain higher than 15% accuracy as for the case with the Most-Popular algorithm to the city of Las Vegas. Even applying our solution to POI-specific RS, it was possible to see quality gains, some of which were higher than 8% accuracy, such as the GeoMF to Charlotte City.

The main contribution of this work is a new perspective to address the problem of recommending POIs, introducing a re-ordering method that can be used by any RS. Also, this strategy opens a new line of research, regarding improvements to our proposal (i.e., different strategies to measure user activity level and define subregions) and/or new re-ordering steps considerations. The remainder of this work is organized as follows. In Sect. 2 we present a theoretical framework with the main RSs proposed in POI recommendations. In Sect. 3, we describe our re-ordering proposal, presenting the activity level metric. In Sect. 4, we present all the details of the experimental project, highlighting the databases used. In Sect. 5, we present the results and discussions inherent in this work. Finally, in Sect. 6, we summarize our conclusions.

2 Background Concepts

In this section, we present recent advances in the recommendation domain, formalizing the problem, and describing the main existing models. Later we describe the specific recommendation context in POIs, where we discuss the main features of their models. Also, we briefly present the main work related to this context.

2.1 Recommender Systems

In general, the recommendation task is to quantify the relevance of an item to a particular user [2,3]. For this reason, many works define the problem as a prediction task where the goal is to estimate the user's rating assigned to an item [4]. On the other hand, the recommendation task is a ranking task, which aims to predict the top-k most relevant items of each user [5]. Both approaches use information passed on by users, such as watched movies, purchased products, logs and/or access cookies, and others, to set a preference profile [11]. Moreover, using the profile to identify and recommend some items that best match users' wishes through some specific approach.

The most used and with the best results among existing approaches is Collaborative Filtering (CF) [20]. CF approaches assume that similar users similarly evaluate items and/or similar items receive similar evaluations from users [3]. Based on this premise, two main classes of methods called memory-based and model-based stand out. The memory-based CF methods are based on the ratings previously provided by the users to recommend items [9]. In this case, user-based or item-based models are highlighted, such as the traditional User-kNN and Item-kNN methods. On the other hand, model-based methods recommend from descriptive models of user preferences, built by strategies derived from machine learning, or even from algebraic [19] models. Matrix Factorization models are considered state of the art.

2.2 POI Recommendation

POIs scenario differs from the others due to their peculiar characteristics Within the recommendation domain. First, many works highlight that the problem of sparsity is even more significant in the POIs domain since the user faces physical obstacles to visit places [16,23]. Besides, other works highlight that the source of information about users differs from others. In this scenario, the primary sources of information to measure user interest are (1) social; (2) temporal; and (3) geographic. Social influence assumes that the opinions of friends present on the user's social network are more relevant than those of non-friends [6,22]. In turn, the temporal influence refers to the fact that users tend to visit places, for example, restaurants, in an irregular period [14,25]. Above all, geographical influence is the main feature to be considered since it is intrinsic to this recommendation scenario since users and POIs are located in a physical space [24].

While there are several effective recommendation approaches in traditional scenarios, they are not equally effective in all cases. For this reason, we find in the literature approaches that explore the different characteristics of POI recommendations. [7] explored the temporal influence, assuming that users with similar consumption histories, in the same period, share the same interests. Social influence is exploited in [23] and [26]. [23] extracts the influence side by side among the friends of each user. [26] created a cumulative distribution function to estimate social relationships. Finally, the geographical influence is considered the most important, and it presents the best results since the user always takes into account the distance of a place to visit it [12,13,16,23,26]. Among these, we can highlight the works presented by Ye et al. [23] and Lian et al. [13], as those that are considered state of the art. These authors use past information to evaluate the probability of a user visiting a POI or a region [23], given the spread of the influence of POIs in the geographical space [13]. For both cases, the recommendation model considers the probability in the building process (i.e., in the factoring matrix).

Contrasting the works presented in this section, this paper addresses the problem of recommendation from a different perspective. In our case, we propose a re-ordering method for any RS, be it traditional or POI specific, which weights the list of POIs according to the level of user activity in each region. Our solution reorders the list before presenting it to the user. It is an orthogonal proposal to all those presented in this section, that is, it can be used together with all of them, achieving excellent results, as we will see in Sect. 5.

3   Geographical Filtering

Geographic influence is one of the main factors affecting the recommendations of POIs [12,16,23]. In general, users are interested in visiting places near their current location and/or the most visited regions (high level of activity) [24]. For this reason, we propose to filter the recommendations presented by any base model. In general, we intend to quantify the level of user activity in a particular region and use it to drive or penalize some specific recommended POIs. To do this, we first divide a region into sub-areas based on literature suggestions [8]. Then we calculate the level of user activity in each of these sub-regions. Finally, we use this information to process the recommendations made in order to improve them. We call this re-ordering method as Geo-Filtering. We illustrate where our method fits into the POI recommendation process (Fig. 1). We describe further details in the following sections.

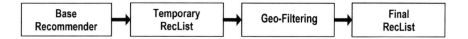

Fig. 1. Steps in the recommendation process

3.1   Recommendation Areas and Sub-areas

Inspired by the recommendation approach adopted by Han & Yamana [8], we initially defined a A recommendation area as a city or district delimited by latitudes and longitudes: initial and final. Then we divide this same A region into fixed-size squared sub-areas with dimensions of 0.5×0.5 km. This way, we created an array of a_{ij} subareas, as we can see in Fig. 2. So, each sub-area a_{ij} covers a portion of A without overlap between the subareas. This 0.5km fixed division is a standard value that we found in the literature [8], and where the experimental tests proved to be most effective.

0.6	0.8	0.7	0.4
0.5	0.9	0.4	0.1
0.2	0.5	0.9	0.2
0.1	0.8	0.7	0.3

Fig. 2. Division into subareas

The idea of dividing A into sub-areas was initially presented in Han & Yamana [8] in order to mitigate the problem of the scarcity of POIs. In this case, the authors proposed a pre-processing that filters the available POIs for the recommendation, leaving only those of the busiest subareas. In turn, in our method, we customized this definition, taking into account the user activity in each region, the location of the POIs, and the users' activity level [14]. Although simple, this division into fixed-size sub-regions has shown excellent results. However, this is one of the stages of our strategy that can still be improved. For example, establishing different dimensions for each sub-region according to the amount of POIs present or even using clustering algorithms by density, such as DBSCAN [1], to establish these sub-regions automatically.

3.2   User's Activity Level

As mentioned in the literature, the main works in POIs recommendation are concerned with obeying the Tobler's First Law. Based on this law, it is possible to derive two fundamental restrictions to achieve user satisfaction [8,26]. Based on them, users prefer:

(**R1**) visit places near your current location;
(**R2**) visit places located in regions where they have a high level of activity.

Thus, we propose a metric that we call activity-level (AL), capable of quantifying the user's interest in a sub-area in order to satisfy these two restrictions. Basically, we calculate the user's activity level u in each sub-area $a_i \in A$ by the ratio of the frequency f of u in a over the sum of all check-ins of u (Eq. 1). This information satisfies R2 directly since it measures the user's activity, and indirectly R1 since it uses only the check-ins of the places the user was able to visit.

$$AL_{u,a} = \frac{f_{u,a}}{\sum_{i=1}^{N} f_{u,a_i}} \tag{1}$$

AL represents the proportional frequency of u in the sub-area a. Sub-areas with higher AL values represent those where the user has performed many check-ins during a period comprised of the training data. On the other hand, subareas with AL values equal to or close to 0 represent those where u does not or is infrequent, respectively. It is worth noting that it is possible to expand the way to calculate AL, considering other factors such as social relationships or time characteristics. However, these analyses do not cover the scope of this work.

3.3 Re-ordering Strategy

Our method consists of a new re-ordering approach for POIs recommendation, applied after the calculation of AL values for each user. We assume that a base recommender presents a list of possible N POIs to be visited. Traditionally, this recommender estimates the relevance r_i of each candidate POI according to its implemented methodology and reorders them in a decreasing way of relevance.

From this information, we execute our re-ordering strategy on the temporary list, adding the relevance value of these POIs, original from the base recommender, with the AL ($AL_{u,a}$). Therefore, for each POI in the temporary list, we calculate the $f(u,i)$ utility according to the Eq. 2, which multiplies the relevance notes returned by the base recommender by the AL value of $AL_{u,a}$, where a consists of the sub-area in which the POI is geographically located. At the end of this step, we resort the temporary list according to these new relevant values and select the most relevant k items, where $k \ll N$, as the final list of recommendations $RecList$.

$$f(u,i) = r_i * (1 + AL_{u,a}) \tag{2}$$

In Fig. 3, we present a didactic example of our strategy. In this example, we have the base recommender returning the list of the six most relevant POIs for a user, in this order: $P14$, $P17$, $P1$, $P5$, $P4$ and $P7$. $P1$ is in the third position and has a relevance of 0.7, in sub-region one, which represents the sub-area of the highest user activity with a level equal to 0.7—applying our re-ordering method weighting—we have that the new relevance of $P1$ is given by

$0.7 * (1 + 0.7) = 1.19$. After applying this process to all POIs and reordering the list by the new relevance values, we have POI $P1$ in the first position in the list. We hope that our method will be able to assign more relevance to the POIs that are in subareas where the user has attended more in the past. Also, the Geo-Filtering decreases the final relevance of POIs that are in subareas that have been little frequented by the user, but which could be at the top of the temporary recommendation list provided by the base recommender. This whole process has a small complexity, as it depends on the subregions and POIs frequented by the users and can be calculated in advance.

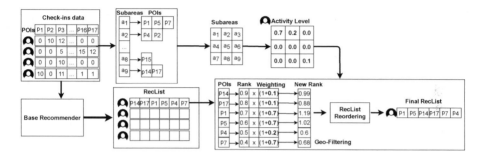

Fig. 3. Re-ordering method application presented in this article

4   Experimental Evaluation

In this section, we describe in detail the assumptions made and the parameters set for our experimental evaluation. First, we highlight the data selection made as proposed by the literature. Next, we describe the basic recommendations that were optimized by our strategy. Finally, we detail the evaluation metrics used to measure the accuracy of the recommendations.

4.1   Datasets

To carry out the experiments, we initially preprocessed the data available in the Yelp Challenge dataset[1]. First, we selected three of the five cities that have the highest number of POIs. These were Charlotte, Las Vegas, and Phoenix. We filtered the dataset information by selecting the POIs that had at least five check-ins and the users that have made at least 20 visits in each city. This process was defined based on the literature to mitigate the problem of user scarcity and maximize the chances of users receiving good recommendations. Thus, after pre-processing, the extracted data are presented in Table 1.

---

[1] https://www.yelp.com/dataset/challenge.

Table 1. Selected cities from the Yelp dataset.

City	User	POIs	Sparsity
Las Vegas	15.168	11.562	99,752
Phoenix	14.388	36.827	99,849
Charlotte	2.445	7.151	99,239

Then, for each city, we reordered the check-ins according to the creation date, i.e., from the oldest to the most recent. We created a set with the first 70% of the check-ins to train the recommendation algorithms, and another with the remaining 30% to run the tests. We performed this approach to simulate a real scenario of the model, as proposed by Zhang & Chow [26]. It is worth mentioning that we also performed a hierarchical selection to define the category of each POI. Each POI was categorized based on a multilevel category tree. Thus, similar to what was done by Han & Yamana [8], we discarded the categories that are below level 2 in the tree.

4.2   Recommendation Algorithms

To validate our proposal, we have selected the proposed recommendation strategies in two different contexts. First, we selected the classic algorithms implemented in the MyMediaLite library[2]. Then, we selected other proposed approaches for POI recommendation, which are state-of-the-art strategies.

Traditional Recommender Systems:

- **MostPopular:** It recommends a list of the most popular POIs that were not visited yet. The popularity of POIs is the number of visits each POI has.
- **User-kNN:** It recommends the most relevant POIs of the K users most similar to the target user. Users' similarity is calculated based on the cosine similarity, and at least 80 neighbors of the target user are selected.
- **WRMF:** It recommends the best-evaluated POIs based on the association of the user's latent factors with those of the items, modeled by a Matrix Factoring method (e.g., SVD, PCA, and others.).

POIs Recommender Systems:

- **USG:** A technique that recommends POIs taking into account geographical, social, and user preferences. Geographic influence is the possibility of a user visiting a POI given their check-in history and the distances between POIs. Social influence defines the user profile taking into account the preferences of social network friends [23].

---

[2] http://mymedialite.net/.

– **GeoMF:** It is a matrix factoring recommendation model weighted by geographic information. The geographical factor defines the possibility of a user visiting a region, as well as the influence of each POI on each region [13].

– **GeoSoCa:** It is a recommendation technique that explores the geographic, social, and categorical factors for recommendations. Geographic influence is a kernel estimation method customized by each user's check-ins distribution [26].

4.3 Evaluation Metrics

For the evaluation and comparison of our model with others, we used the traditional metrics of *Precision@k* and *Recall@k*. While *Precision@k* returns the fraction of relevant POIs recovered by the number of recommendations, *Recall@k* measures the fraction of relevant POIs recovered by the amount of POIs consumed in each user's test. We calculated the metrics for each user and returned the average values to evaluate the recommendation effectiveness [26].

5 Results

In this section, we discuss the results obtained by our re-ordering method when applied to selected baselines. We divide our analysis into two cases. First, we evaluate the performance generated by the proposal presented in the traditional recommendation cases, and then, with algorithm specific to POI recommendations.

5.1 Traditional Recommender Systems

First, we selected the three traditional recommendation algorithms called Most-Popular, UserKNN, and WRMF. We applied these algorithms to the Las Vegas, Phoenix, and Charlotte related databases, measuring the Precision and Recall of the recommendations performed. Then, from recommendations generated by each algorithm, we apply our re-ordering process, again measuring the Precision and Recall. We summarize the values found in Table 2.

It is important to emphasize that among these methods, none exploit geographical influence as a decisive factor in recommendations. Therefore, by dividing a region into sub-regions, applying our metrics, geographical filtering can compute information that was not previously explored. With this, we can observe a significant improvement in the results. We found statistically significant gains in the city of Las Vegas when compared to the original list (Table 2). The same happens when we compare the results in the city of Phoenix. Specifically, in the Most-Popular technique, the results are more explicit, as geographic filtering reorders the list in a personalized way to better meet the users' consumption pattern. This way, we were able to recover relevant POIs, which were previously discarded by the traditional recommender. In both cases, Wilcoxon's test confirmed gains for non-normal distributions with a p-value ≤ 0.05.

Table 2. It shows the results achieved by the evaluation metrics with the traditional recommendations and the values achieved by our re-ordering strategy. The symbol ▲ represents statistically significant positive gains, ＊ represents statistical ties, and ▼ represents statistically significant losses. Wilcoxon's test obtained these gains with a 95% confidence interval

		Precision		Recall		nDCG		MRR	
	Recommender	Top-10	Top-15	Top-10	Top-15	Top-10	Top-15	Top-10	Top-15
Las Vegas	Most popular	0.0153	0.0155	0.0266	0.0395	0.0205	0.0260	0.0450	0.0491
	Geo-filtering	0.0172 ▲	0.0167 ▲	0.0331 ▲	0.0459 ▲	0.0248 ▲	0.0304 ▲	0.0527 ▲	0.0569 ▲
	User-kNN	0.0311	0.0279	0.0583	0.0755	0.0476	0.0552	0.0971	0.1020
	Geo-filt	0.0317 ▲	0.0285 ▲	0.0606 ▲	0.0777 ▲	0.0489 ▲	0.0565 ▲	0.0994 ▲	0.1043 ▲
	WRMF	0.0327	0.0296	0.0592	0.0767	0.0485	0.0567	0.0994	0.1043
	Geo-filtering	0.0332 ▲	0.0301 ▲	0.0613 ▲	0.0794 ▲	0.0500 ▲	0.0582 ▲	0.1027 ▲	0.1077 ▲
Phoenix	Most popular	0.0129	0.0118	0.0089	0.0119	0.0096	0.0115	0.0356	0.0379
	Geo-filtering	0.0131 ▲	0.0123 ▲	0.0090 ▲	0.0127 ▲	0.0098 ▲	0.0120 ▲	0.0361 ▲	0.0387 ▲
	User-kNN	0.0327	0.0294	0.0244	0.0329	0.0276	0.0328	0.0973	0.1022
	Geo-filtering	0.0329 ＊	0.0298 ▲	0.0247 ▲	0.0336 ▲	0.0279 ＊	0.0332 ＊	0.0979 ＊	0.1030 ＊
	WRMF	0.0276	0.0255	0.0204	0.0278	0.0226	0.0272	0.0797	0.0845
	Geo-filtering	0.0281 ▲	0.0259 ▲	0.0208 ▲	0.0284 ▲	0.0231 ▲	0.0277 ▲	0.0812 ▲	0.0858 ▲
Charlotte	Most popular	0.0367	0.0338	0.0280	0.0386	0.0290	0.0352	0.0971	0.1024
	Geo-filtering	0.0375 ▲	0.0344 ▲	0.0285 ＊	0.0400 ▲	0.0297 ＊	0.0361 ＊	0.0978 ＊	0.1033 ＊
	User-kNN	0.0481	0.0441	0.0399	0.0544	0.0416	0.0500	0.1317	0.1378
	Geo-filtering	0.0478 ▼	0.0443 ＊	0.03963 ＊	0.0547 ＊	0.0416 ＊	0.0503 ＊	0.1325 ＊	0.1390 ＊
	WRMF	0.0492	0.0444	0.0393	0.0525	0.0424	0.0503	0.1375	0.1433
	Geo-filtering	0.0499 ＊	0.0450 ＊	0.0404 ＊	0.0539 ▲	0.0432 ＊	0.0512 ＊	0.1392 ＊	0.1450 ＊

The gains earned by our re-ordering stage in the cities of Las Vegas and Phoenix were not observed for the city of Charlotte. In the vast majority of combinations, our re-ordering stage shows a statistical tie to the original algorithm results. The first observation is that Charlotte is the city with the smallest number of users and an equally small amount of POIs per sub-area. Also, the number of POIs per user is low. The little consumption information from the users considerably affects the quality of the traditional recommender systems. As a consequence, the list of recommended POIs tends to contain items that are not very relevant for users. Our strategy is purely dependent on the recommendation list and the area of operation of each user. Thus, reordering a list of POIs that are mostly no longer user-relevant items has little effect on the final quality. In any case, this result shows that there is still room for other re-ordering proposals to be used in conjunction with the strategy presented here.

5.2 POIs Recommendation

In our second analysis, we selected the three baselines regarding the recommendation of POIs, which are considered state of the art in this field. Specifically, we selected USG [23], GeoMF [13] and GeoSoCa [26]. Again, we retrieved the generated recommendation list and applied our geographic filtering proposal. All the results obtained can be seen in Table 3.

Table 3. It shows the results achieved by the evaluation metrics with the specific recommendations for POIs and the values achieved by our re-ordering strategy. The symbol ▲ represents significant positive gains, ✳ represents statistical ties, and ▼ represents significant losses. Wilcoxon's test obtained these gains with a 95% confidence interval

		Precision		Recall		nDCG		MRR	
	Recommender	Top-10	Top-15	Top-10	Top-15	Top-10	Top-15	Top-10	Top-15
Las Vegas	USG	0.0276	0.0257	0.0545	0.0721	0.0442	0.0520	0.0889	0.0939
	Geo-filtering	0.0290 ▲	0.0266 ▲	0.0583 ▲	0.0759 ▲	0.0466 ▲	0.0544 ▲	0.0926 ▲	0.0975 ▲
	GeoMF	0.0284	0.0255	0.0546	0.0713	0.0437	0.0510	0.0896	0.0942
	Geo-filtering	0.0291 ▲	0.0261 ▲	0.0551 ✳	0.0732 ▲	0.0446 ▲	0.0522 ▲	0.0923 ▲	0.0971 ▲
	GeoSoCa	0.0206	0.0180	0.0339	0.0439	0.0282	0.0326	0.0660	0.0691
	Geo-filtering	0.0201 ▼	0.0177 ✳	0.0322 ▼	0.0430 ✳	0.0271 ▼	0.0318 ▼	0.0628 ▼	0.0661 ▼
Phoenix	USG	0.0228	0.0212	0.0176	0.0241	0.0198	0.0238	0.0701	0.0744
	Geo-filtering	0.0231 ▲	0.0215 ▲	0.0179 ▲	0.0246 ▲	0.0202 ▲	0.0243 ▲	0.0714 ▲	0.0758 ▲
	GeoMF	0.0297	0.0278	0.0235	0,0330	0.0256	0.0312	0.0874	0.0928
	Geo-filtering	0.0302 ▲	0.0281 ▲	0.0242 ▲	0.0336 ▲	0.0263 ✳	0.0318 ✳	0.0893 ✳	0.0946 ✳
	GeoSoCa	0.0171	0.0154	0.0112	0.0151	0.0131	0.0156	0.0516	0.0543
	Geo-filtering	0.0171 ✳	0.0155 ✳	0.0111 ✳	0.0153 ✳	0.0131 ✳	0.0157 ✳	0.0512 ✳	0.0539 ✳
Charlotte	USG	0.0420	0.0387	0.0334	0.0457	0.0338	0.0409	0.1088	0.1143
	Geo-filtering	0.0432 ▲	0.0394 ▲	0.0347 ▲	0.0475 ▲	0.0350 ▲	0.0423 ▲	0.1111 ✳	0.1172 ▲
	GeoMF	0.0349	0.0325	0.0282	0.0391	0.0300	0.0364	0.0992	0.1050
	Geo-filtering	0.0357 ✳	0.0334 ▲	0.0299 ▲	0.0409 ▲	0.0313 ✳	0.0377 ▲	0.1028 ▲	0.1084 ▲
	GeoSoCa	0.0230	0.0218	0.0163	0.0235	0.0183	0.0226	0.0630	0.0674
	Geo-filtering	0.0237 ✳	0.0215 ✳	0.0167 ✳	0.0232 ✳	0.0189 ✳	0.0227 ▲	0.0636 ▲	0.0676 ▲

In this case, our model achieved excellent results for the recommendation techniques USG and GeoMF in the three selected cities. It is because our proposal can assist the POIs specific recommendation algorithms, which deal with the geographical factor. Basically, in these cases, the activity level defined by the proposal of this work aggregates the geographic information in the recommendations, taking into account the activity density of each user in sub-regions. It is possible to pronounce the most difference in the re-ordering of the USG in Las Vegas and Charlotte. Again, we obtained statistically significant gains, with the Wilcoxon test (p-value $\leq$ 0.05). These results reinforce the idea that our proposal achieves the expected goals and contributions, especially on the GeoMF model, one of the main ones in the literature [15].

By comparing our proposal with GeoSoCa [26], the re-ordering performance decreased, both in Las Vegas and Phoenix. It is because GeoSoCa adds two more attributes to its recommendations: social and categorical influences. Furthermore, according to Liu et al. [15], the social factor explored by GeoSoCa is the one with the best performance compared to other algorithms (e.g., the USG). Therefore, our proposal deals only with the geographic characteristics of each user individually, thus justifying that difference found. However, even being inferior, the results obtained were very close to those found by the baseline, and was not considered statistically significant losses.

In the city of Charlotte, we observed an impressive result. We found gains concerning the strategy of GeoSoCa. It is because Charlotte has less social correlation data, where each user, on average, has only one friend. Therefore, the aggregation of information that the social factor brings to GeoSoCa is not relevant. Thus, the application of our re-ordering stage could enrich the geographic information bringing improvements to the recommendation of GeoSoCa.

5.3   Discussions

In this section, we present an analysis of the gains obtained by each algorithm, in each city, when using the Geo-Filtering re-ordering proposal presented in this work. These results are all summarized in Fig. 4. The first interesting observation is the gains obtained by the MostPopular algorithm (i.e., gains of up to 15% on accuracy for the city of Las Vegas). It is a non-personalized strategy, widely used in Pure Cold Star Problem scenarios, where the recommender selects the most consumed items in the training base for the users. In this case, Geo-Filtering ends up introducing customized information, which makes the quality of this algorithm improve significantly. In smaller proportions, Geo-Filtering is also able to improve the quality of other traditional classifiers (i.e., UserKNN and WRMF). We have observed that the most significant quality gains occur in cities where the average consumption of users is high. The consumption information is very relevant for these algorithms because it is from this information that one can model the profile of users and from this information to make recommendations of items, in this case, POIs, that have similar characteristics to the past consumption of users. However, this process of consumption modeling does not take into account the restrictions imposed by Tobler's geography law, and it is precisely this gap that our re-ordering strategy fills, adding to these algorithms geographic information of users' consumption.

As we have seen throughout this section, we also considered state-of-the-art algorithms in the area of POIs recommendation (i.e., USG, GeoMF, and GeoSoCa). These algorithms also incorporate the geographic information from the POIs into the generation of the models (i.e., the factoring matrix). Thus, our objective was to evaluate if Geo-Filtering was still able to aggregate some new information that was not explored by these algorithms. Observing the gains obtained (Fig. 4), we concluded that yes. For these cases, we hypothesize that Geo-Filtering not only aggregates geographic information, but it is also able to correlate this information to the users' consumption. The GeoSoCa algorithm is a clear exception. As previously mentioned, it uses other factors like information, especially the social factor, which, according to the authors, is the one that most contributes to the proper functioning of the algorithm. However, when we look at the gains that Geo-Filtering can obtain for GeoSoCa for the city of Charlotte (i.e., gains higher than 8% accuracy), where there is almost no social information, we realize that the geographical factor is still little explored by GeoSoCa. Thus, we believe that the most significant contribution of this work is precisely in opening a new perspective of research in the area of POIs recommendation, introducing new re-ordering stages.

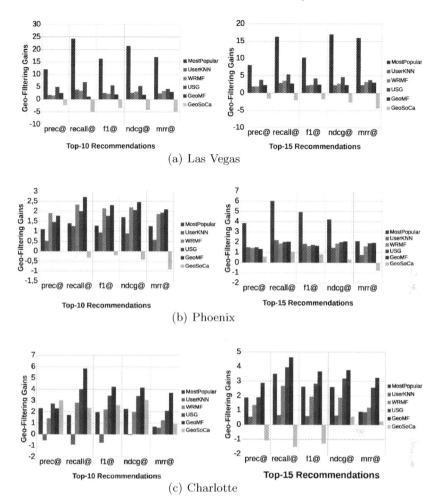

Fig. 4. Summarization of the gains obtained by the geo-filtering application in each algorithm/city combination

6   Conclusions and Future Works

In this work, we propose a re-ordering model that weighs up and filters a set of recommendations generated by a base recommender system in order to further satisfy user preferences. To do it, we build a metric capable of quantifying the level of user activity in sub-regions. Our premise is that by satisfying the two constraints determined by Tobler's geography law, we further favor user satisfaction. In general, we observe that the inclusion of Geo-filtering, as re-ordering, allows, and effectively, to increase the accuracy of recommendations. This new re-ordering stage adds features not yet explored by other baselines. Dividing an area into smaller sub-areas and calculating for each user their level of activity, makes it possible to extract the most active sub-regions from users. The results show statistically significant gains in most cases, especially in large cities, such as Phoenix and Las Vegas.

Our proposal is still limited to baselines that exploit other features than geography, such as the GeoSoca. Our goal, as future work, is to include other characteristics as well, such as the social influence in the calculation of the user's activity level in order to extract more information not evaluated by the RSs. We also intend to include other ways to obtain the subareas, dividing them in a varied way or including new subareas within those already divided.

Acknowledgments. This work was partially funded by the Brazilian National Institute of Science and Technology for the Web - INWeb, MASWeb, CAPES, CNPq, Finep, Fapesp and Fapemig.

References

1. Andrade, G., Ramos, G., Madeira, D., Sachetto, R., Ferreira, R., Rocha, L.: G-DBSCAN: a GPU accelerated algorithm for density-based clustering. Procedia Comput. Sci. -ICCS **18**, 369–378 (2013)
2. Adomavicius, G., Tuzhilin, A.: Toward the next generation of recommender systems: a survey of the state-of-the-art and possible extensions. IEEE Trans. Knowl. Data Eng. **17**(6), 734–749 (2005)
3. Bobadilla, J., Ortega, F., Hernando, A., Gutiérrez, A.: Recommender systems survey. Knowl.-Based Syst. **46**, 109–132 (2013)
4. Chikhaoui, B., Chiazzaro, M., Wang, S.: An improved hybrid recommender system by combining predictions. In: 2011 IEEE Workshops of International Conference on Advanced Information Networking and Applications (WAINA), pp. 644–649 (2011)
5. Cremonesi, P., Donatacci, A., Garzotto, F., Turrin, R.: Decision-making in recommender systems: the role of user's goals and bounded resources. In: Decisions@ RecSys, pp. 1–7 (2012)
6. Fernández-Tobías, I., Tomeo, P., Cantador, I., Di Noia, T., Di Sciascio, E.: Accuracy and diversity in cross-domain recommendations for cold-start users with positive-only feedback. In: Proceedings of the 10th ACM Conference on Recommender Systems, pp. 119–122. ACM (2016)
7. Gao, H., Tang, J., Hu, X., Liu, H.: Exploring temporal effects for location recommendation on location-based social networks. In: Proceedings of the 7th ACM Conference on Recommender Systems, pp. 93–100. ACM (2013)
8. Han, J., Yamana, H.: Geographical diversification in POI recommendation: toward improved coverage on interested areas. In: Proceedings of the Eleventh ACM Conference on Recommender Systems, pp. 224–228. ACM (2017)
9. Hu, L., Sun, A., Liu, Y.: Your neighbors affect your ratings: on geographical neighborhood influence to rating prediction. In: Proceedings of the 37th International ACM SIGIR Conference on Research & Development in Information Retrieval, pp. 345–354. ACM (2014)
10. Kumar, S.S., Ramachandran, T., Panboli, S.: Product recommendations over Facebook: the roles of influencing factors to induce online shopping. Asian Soc. Sci. **11**(2), 202 (2015)

11. Lacerda, A., Ziviani, N.: Building user profiles to improve user experience in recommender systems. In: Proceedings of the 6th ACM WSDM, pp. 759–764 (2013)
12. Li, X., Cong, G., Li, X.L., Pham, T.A.N., Krishnaswamy, S.: Rank-GeoFM: a ranking based geographical factorization method for point of interest recommendation. In: Proceedings of the 38th International ACM SIGIR Conference on Research and Development in Information Retrieval, pp. 433–442. ACM (2015)
13. Lian, D., Zhao, C., Xie, X., Sun, G., Chen, E., Rui, Y.: GeoMF: joint geographical modeling and matrix factorization for point-of-interest recommendation. In: Proceedings of the 20th ACM SIGKDD International Conference on Knowledge Discovery and Data Mining, pp. 831–840. ACM (2014)
14. Liu, B., Fu, Y., Yao, Z., Xiong, H.: Learning geographical preferences for point-of-interest recommendation. In: Proceedings of the 19th ACM SIGKDD International Conference on Knowledge Discovery and Data Mining, pp. 1043–1051. ACM (2013)
15. Liu, Y., Pham, T.A.N., Cong, G., Yuan, Q.: An experimental evaluation of point-of-interest recommendation in location-based social networks. Proc. VLDB Endow. **10**(10), 1010–1021 (2017)
16. Liu, Y., Wei, W., Sun, A., Miao, C.: Exploiting geographical neighborhood characteristics for location recommendation. In: Proceedings of the 23rd ACM International Conference on Conference on Information and Knowledge Management, pp. 739–748. ACM (2014)
17. Lu, J., Wu, D., Mao, M., Wang, W., Zhang, G.: Recommender system application developments: a survey. Decis. Support Syst. **74**, 12–32 (2015)
18. Miller, H.J.: Tobler's first law and spatial analysis. Ann. Assoc. Am. Geogr. **94**(2), 284–289 (2004).https://doi.org/10.1111/j.1467-8306.2004.09402005.x
19. Mourão, F.H.d.J.: A hybrid recommendation method that combines forgotten items and non-content attributes (2014)
20. Schafer, J.B., Frankowski, D., Herlocker, J., Sen, S.: Collaborative filtering recommender systems. In: Brusilovsky, P., Kobsa, A., Nejdl, W. (eds.) The Adaptive Web. LNCS, vol. 4321, pp. 291–324. Springer, Heidelberg (2007). https://doi.org/10.1007/978-3-540-72079-9_9
21. Silva, N., Carvalho, D., Pereira, A.C., Mourão, F., Rocha, L.: The pure cold-start problem: a deep study about how to conquer first-time users in recommendations domains. Inf. Syst. **80**, 1–12 (2019)
22. Ye, M., Yin, P., Lee, W.C.: Location recommendation for location-based social networks. In: Proceedings of the 18th SIGSPATIAL International Conference on Advances in Geographic Information Systems, pp. 458–461. ACM (2010)
23. Ye, M., Yin, P., Lee, W.C., Lee, D.L.: Exploiting geographical influence for collaborative point-of-interest recommendation. In: Proceedings of the 34th International ACM SIGIR, pp. 325–334. ACM (2011)
24. Yu, Y., Chen, X.: A survey of point-of-interest recommendation in location-based social networks. In: Workshops at the Twenty-Ninth AAAI Conference on Artificial Intelligence (2015)
25. Yuan, Q., Cong, G., Ma, Z., Sun, A., Thalmann, N.M.: Time-aware point-of-interest recommendation. In: Proceedings of the 36th International ACM SIGIR, pp. 363–372. ACM (2013)
26. Zhang, J.D., Chow, C.Y.: GeoSoCa: exploiting geographical, social and categorical correlations for point-of-interest recommendations. In: Proceedings of the 38th International ACM SIGIR, pp. 443–452. ACM (2015)

An Evaluation of Low-Quality Content Detection Strategies: Which Attributes Are Still Relevant, Which Are Not?

Júlio Resende[1], Vinicius H. S. Durelli[1], Igor Moraes[1], Nícollas Silva[2], Diego R. C. Dias[1(✉)], and Leonardo Rocha[1]

[1] Universidade Federal de São João del-Rei, São João del-Rei, Brazil
{julio.cmdr,durelli,ti.igor,diegodias,lcrocha}@ufsj.edu.br
[2] Universidade Federal de Minas Gerais, Belo Horizonte, Brazil
ncsilvaa@dcc.ufmg.br

Abstract. Online social networks have gone mainstream: millions of users have come to rely on the wide range of services provided by social networks. However, the ease use of social networks for communicating information also makes them particularly vulnerable to social spammers, i.e., ill-intentioned users whose main purpose is to degrade the information quality of social networks through the proliferation of different types of malicious data (e.g., social spam, malware downloads, and phishing) that are collectively called low-quality content. Since Twitter is also rife with low-quality content, several researchers have devised various low-quality detection strategies that inspect tweets for the existence of this kind of content. We carried out a brief literature survey of these low-quality detection strategies, examining which strategies are still applicable in the current scenario – taken into account that Twitter has undergone a lot of changes in the last few years. To gather some evidence of the usefulness of the attributes used by the low-quality detection strategies, we carried out a preliminary evaluation of these attributes.

Keywords: Low-quality content detection · Spam · Machine Learning

1 Introduction

Over time, what we have referred to as the Web has undergone several gradual changes. Specifically, the Web started as static HTML pages (i.e., read-only) and evolved to a more dynamic Web (i.e., read-write): Web 2.0 is the term used to refer to this ascendant technology whose emphasis lies in allowing users to create content, collaborate, and share information online. Owing to its interactive and dynamic nature, a great deal of sites that lend themselves well to the Web 2.0 model have been developed with an interactive community of users in mind. Quintessential examples of social networking services include Facebook and Twitter. These websites have become increasingly popular because they

© Springer Nature Switzerland AG 2020
O. Gervasi et al. (Eds.): ICCSA 2020, LNCS 12249, pp. 572–585, 2020.
https://doi.org/10.1007/978-3-030-58799-4_42

provide users with community-building options such as online forums for like-minded people. These socially-enabled websites have been playing an active role in allowing individuals and communities to make more informed decisions concerning a wide range of topics. Despite the many benefits offered by these social networking services, online social networks also have issues. Due to the massive amount of information available in social networking services, these websites are particularly vulnerable to common approaches to tampering with the quality of the information. According to [8], anything that hinders users from having a meaningful browsing experience can be regarded as low-quality content. Thus, social spam, phishing, and malware, which are common approaches to degrading information, are examples of low-quality content that can negatively affect the overall user experience provided by social networking services.

The motivation for propagating low-quality content ranges from product advertising to socially engineered malware (i.e., malware requiring user interaction that ends up with users downloading a malicious file). Additionally, it is worth noting that low-quality content can be propagated either by users or social *bots*. In effect, bots are estimated to make up more than 50% of all Internet traffic [29]. In this context, given that bots are able to propagate a massive amount of low-quality content and the information in social networking services plays a pivotal role in shaping the decisions of users, bot traffic has the potential to severely user experience. Twitter stands out as one of the social networking services that is most vulnerable to spam: out of approximately 700 million *tweets*, roughly 10% is considered spam [23]. The reasons for Twitter being plagued by low-quality content are manyfold, but the main reason has to do with the fact that the service places too high of a limit on the number of tweets per day[1]

Due to the aforementioned reasons, the detection of low-quality content has been drawing a lot of attention from researchers and practitioners. However, in the literature is presented a limited interpretation of how to detect low-quality content. Typically, the most common interpretations emphasize the detection of spammer accounts or the detection of malicious links, which corroborates the generation of incomplete filters. False information (i.e. fake news), automatically generated content, flood, clickbait and advertisements are also data that do not contain any information and can also be published by legitimate accounts, not being treated by these strategies. Another problem is that many approaches have several historical resources to perform the detections, such as a user's old tweets or his progression in terms of the number of followers, some of which are no longer available due to restrictions imposed by the new Twitter policy.

In this sense, in hopes of providing an overview of the main approaches researchers and practitioners have been investigating to cope with this problem, we set out to give a survey of the scope of the literature. While conducting such a literature survey, we emphasize checking which strategies and resources (i.e., attributes) remain viable to address the problem of low-quality content

---

[1] As of this writing, the tweet limit is 2,400 per day. It is worth emphasizing that there is also a limit per half-hour period, so users are not able to tweet all 2,400 tweets at one time.

online detection. Additionally, we came up with a methodology that combines different low-content detection techniques. Essentially, our approach was devised as a means to evaluate the effectiveness of some prominent low-content detection classification models. Thus, our contributions are twofold:

1. We carried out a brief literature survey of low-quality content detection techniques. Our survey was conducted from the standpoint of a researcher investigating which techniques are still applicable in the current scenario – considering that social networking services have evolved over the last few years.
2. To gather evidence of the usefulness of some of the low-quality content detection techniques, we devised a methodology that combines them so that we can probe into their advantages and drawbacks in terms of the attributes that they employ.

The remainder of this work is organized as follows. In Sect. 2 we present the main proposed approaches for detecting low-quality content, emphasizing which of them remain viable to be applied in online scenarios. In Sect. 3, we describe the methology used to evaluated the strategies and characteristics currently being used for low-quality content online detection. In Sect. 4, we present the results and discussions inherent in this work. Finally, in Sect. 5, we summarize our conclusions.

2   Literature Survey

Several articles use the same terminology - spam - to refer to different concepts. The confusion occurs because this term is commonly used to refer to one of the first types of spam to be popularized on the internet, the "phishing spam".

As the term is ambiguous and "Low-quality Content" is a terminology that is not well established in the literature, we decided to follow a "quasi-gold standard" (i.e., a manual search of several articles in the subject area). The quasi-gold standard was used to define a starting point for a "snowballing" technique. A pre-filter was done through the abstract of several articles, and later we made a second one, reading the entire research.

In this way, we selected [8], our starting point for the literature survey. From it, the "backward snowballing" technique was used (i.e., all articles cited by him were analyzed). We focused our research on low-quality content, so the technique should be reapplied in articles dealing with this type of detection. For articles that deal with only one type of low-quality content, we limit it to just citing them. The diagram in Fig. 1 demonstrates the process.

Only one new article [25] was found that deals with low-quality content (although it used the term "spam"), so we repeated in this article the process of backward snowballing, of which no study was found that follows the methodology sought.

After the "backward snowballing", we used Google Scholar to explore articles that cited the relevant papers, technique known as "forward snowballing". However, the search did not return any new study relevant to our survey, and we decided not to include them in this article because they did not present any new approach in relation to the citations selected by the previous technique.

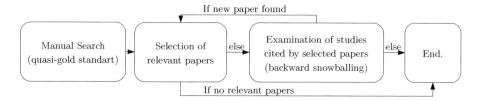

Fig. 1. Backward snowballing diagram

Low-quality content or spam on social networks is unwanted content produced by sources that express different behavior from that intended, whether that source is a legitimate user, a spammer, or a bot [25]. The purpose of the content includes sharing malicious links, fraudulent information, advertisements, chain letters, and others, as well as indexing of unrelated hashtags to increase visibility. However, we note that most published works do not share the same definition, using only a narrow view, which does not represent the entire term. The most modern approach in the literature is to consider only phishing spam, a strategy of spreading malware through URLs. The detection of this content was the object of study of [1,15], which proposed techniques to verify the occurrence of the registered blacklist (i.e., an extensive catalog of suspicious or malicious URLs) and techniques which learn about the characteristics of these blacklists, in order to apply the knowledge generated to classify new URLs. A negative point of these strategies is that spammers may shorten the link or even generate a new one to circumvent this detection. [18] has improved this methodology by measuring the similarity of words in tweets, in a similar way as [3] detects evidence of crimes in forums. Thus, if a tweet has an unknown URL or text content similar to other tweets with suspicious URLs, it is detected as phishing. Twitter itself uses this methodology, checking links with Google Safe Browsing. However, shortened malicious links are still not detected as spam. The tweets barred are full links, those identical to those already published in a short period of time, and those in which the sender and receiver have never interacted with each other.

Considering as a definition of spam, only messages containing phishing is a limited approach. Seeking to broaden the detection spectrum, [7] based its definition on reports made by other users, a better approach if the reports would be publicly available. Another approach, explored by [11,20,22], was the creation of a database to detect tweets, labeling them as spam if Twitter suspended the sender account in a later validation request. However, relying on Twitter's suspension policy is not the best way, since posting spam is not the only reason for an account to be suspended. Besides, spammers can post regular tweets to avoid being detected, while legitimate users can post spam content, even if unconsciously. Other works that also focus on the detection of spammer accounts, such as [2,16,17], use account attributes extracted through Twitter's own API's to generate classifying models with machine learning. However, the focus on these attributes can bias the search, since external tools can easily manufacture them, such as obtaining followers or scheduled publications.

The generation of temporal behavior models, explored by [6,10,12,13,21,26–28], is an alternative developed to detect inorganic variations of these attributes values. Considering historical user data, the models detect anomalies and then flag the account as a spammer. The approach is very efficient in detecting spammers but does not consider those legitimate users also to post spam. Another hindrance is that the use of time resources infers that spammers are only detected after sending several spam messages, as this type of detection requires a minimum amount of time for the analyzed accounts to generate "evidence" that proves illegal activity. There are graph-based detection methodologies, which seek to trace the entire communicative path taken by users and can also detect communities [4]. An interesting work is presented by [5], in which, by calculating the centrality network, the largest spammers can be detected. Similar methods are used by [9,19,24], in which the extracted attributes determine the number of interactions made per user, as well as the distance between the sender and the receiver in targeted tweets. That significantly improves accuracy at the cost of response time. In the same line of temporal behavior, we can cite the work presented by [10]. The proposed methodology is the clustering of tweets that use the same URL (presenting an overlap with the group of papers described in the first paragraph of this section), being, therefore, a phishing detection. Later, using blacklists, the group that presents a malicious URL is considered a spam campaign, that is, mass dissemination of the same malicious content. In other words, all these strategies take some time to identify inappropriate behavior. Due to the damage caused by these contents, it is of utmost importance to consider methodologies that perform the detection in real-time.

Assuming as an ideal methodology: (1) the detection of the various types of low-quality tweets; (2) being this detection in real-time; and (3) using publicly available attributes – we found only two articles that meet all these requirements [8,25]. [25] proposed a strategy to detect spam in its broadest sense. However, the dataset used for evaluation was composed only of tweets that presented at least one URL. It is, therefore, not possible to measure its ability to detect tweets that do not have this attribute. On the other hand, [8] has labeled a database according to users' perspectives on what is low-quality content. That is an exciting new starting point, given the abstract definition of what is "unwanted or irrelevant" content. Comparing this work with others that adopt a narrow definition of what is low-quality content is not statistically valid, because the works have different focuses. However, [8] presented a comparison of results with those presented on [25], and the results of [8] proved to be more efficient. A critical point of [8], regarding the attributes used to build the classification model, was the use of some attributes that should not be considered in real-time detection, such as the number of retweets and favorites received by the tweet. As detection should be applied when publishing a tweet, this value will always be 0. The same work also proposes a methodology that makes use of attributes called indirect, which depend on an analysis of all the posts made by the user. Although this methodology is capable of significantly increasing the performance of the classifiers, the excessive use of the user's information has not been seen with good eyes in recent years, a fact that has pressured Twitter to reduce the limits for obtaining information.

Given the analysis prepared, the next section presents a methodology for evaluating the attributes used by the works that fit the context defined by this research [8,25]. The objective was to analyze the relevance of each attribute and the feasibility of carrying out a broad detection of low-quality contents.

3   Evaluation Methodology

As mentioned, we set out to investigate the importance of techniques currently being used for low-quality content detection. We propose an evaluation methodology to be applied to the main strategies proposed in the literature. To this end, we tried to use a manually labeled Twitter dataset comprised of 100,000 tweets generated by 92,720 different users, made available by Chen et al. [8]. According to Chen et al., the tweets were manually verified so as to provide a more sound training set and allow for the verification of the results. Nevertheless, recently, Twitter has introduced a number of features that prevent researchers from directly accessing tweet datasets. To work around this limitation, we took advantage of the fact that all tweets' IDs in the aforementioned dataset were made available by Chen et al., thus we collected all tweets' information based on their IDs. These were the only data collected for our analysis. During the creation of our dataset, we realized that some tweets were no longer available, so we ended up with a dataset containing 43,857 tweets, of which 3,967 were classified as low-quality. It is worth mentioning that although our new dataset is smaller and different from the original, the ratio between low-quality and normal messages was maintained.

During the creation of the dataset, apart from the tweets' textual content, we also extracted information concerning other attributes. As mentioned, attributes related to previous tweets and followers can be used as useful pieces of information, improving the models yielded by the classifiers. We decided not to include information regarding retweets, previous tweets, and the number of times a given tweet was liked/favorited because Twitter has introduced features that make the extraction of such information somewhat unwieldy. Besides, the use of this information in low-quality content online detection scenarios has limited utility. Therefore, the attributes we took into account are shown in Table 1. Our selection of attributes is based on the selection proposed by other researchers: attributes 3, 4, 6, 7, 8, 9, 15, 16, 17, 18, and 35 were used in the approach proposed by [8], attributes 12, 13, 19, 20, 21, 22, 23, 24, 25, 26, 27, 28, 30, 32, and 34 were selected and used in the approach devised by [25], and 1, 2, 5, 10, 11, 29, 31 and 33 are attributes employed in both approaches. It is worth noting that

Table 1. Set of attributes we took into account during our study. The attributes 3, 4, 6, 7, 8, 9, 15, 16, 17, 18, and 35 were used in the approach proposed by [8], attributes 12, 13, 19, 20, 21, 22, 23, 24, 25, 26, 27, 28, 30, 32, and 34 were selected and used in the approach devised by [25], and 1, 2, 5, 10, 11, 29, 31 and 33 are attributes employed in both approaches.

Id	Description	Id	Description
1	# of followers	19	# of words in the tweet
2	# of followings	20	# of characters in the tweet
3	# of lists the user create	21	# of digits in the tweet
4	# of tweets the user favorite	22	# of white spaces in the tweet
5	# of tweets the user posted	23	# of capitalization words in the tweet
6	If the URL field of profile is null	24	$\frac{Attribute23}{Attribute19}$
7	Length of the description field of profile	25	Length of longest word in the tweet
8	If the user is using a default profile	26	Average word length in the tweet
9	If the user is using a default avatar	27	# of exclamation marks in the tweet
10	Age of the user account, in months	28	# of question marks in the tweet
11	$\frac{Atributte1}{Atributte2}$	29	# of urls in the tweet
12	$\frac{Atributte1}{Atributte1+Atributte2}$	30	$\frac{Attribute29}{Attribute19}$
13	Average followings per month	31	# of hashtags in the tweet
14	Average tweets per week	32	$\frac{Attribute31}{Attribute19}$
15	If the location field of profile is null	33	# of mentions in the tweet
16	Tweeting tools	34	$\frac{Attribute33}{Attribute19}$
17	Regular, replies mentions or retweets	35	# of symbols in the tweet
18	Tweet contains sensitive content		

many other approaches to spam detection (i.e., low-quality content detection) consider a subset of the set of attributes we took into account in our study.

Many of the attributes listed in Table 1 can not be obtained through Twitter's API. Rather, some attributes have to be computed from other information. Attribute 17, for instance, can be inferred from boolean attributes that indicate whether a tweet is a "regular" tweet, a retweet, a (public) reply, or a mention (a special case of reply in which users mention other users explicitly, i.e., @username). Attributes 19, 20, 21, and 22 can be extracted by applying regular expressions to the tweet's textual information.

Special attention was given to attribute 16, which indicates the medium used to post the tweet (e.g., an iPhone app). During data collection we found that such attributes can have more than one hundred different values. None of the previous work on low-content detection mention the approach they used to choose and interpret these values, so we had to devise our own approach to doing such. We calculate the frequency of these values both in instances classified as low-content as well as "regular" tweets. Afterwards, we calculated the module of the difference of each frequency for both types of instances. The calculated values were then ordered in descending order. In order to choose the values to be considered in our analysis, we applied an interpolation function to the difference of the frequencies. We came up with a cut-off value based on the ΔY of interpolation function, considering the point at which the function decreases. As a result, we selected the following values: Twitter for iPhone, Twittascope, and Twitter for Android. Values that did not fall into these categories were classified as "Others". Figure 2 shows the frequency probability of each category for low-quality content and regular tweets.

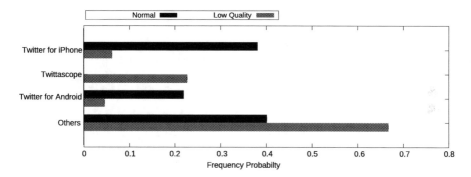

Fig. 2. Frequency probabilty of possible values for attribute 16. For instance, for normal tweets, 38% of them are posted from a Iphone, 22% from Androide devices and the remaining of the tweets posted by other types of devices. Moreover, more than 60% of low-quality messages are posted by other types of device.

As mentioned, we set out to examine which attributes are more relevant, i.e., attributes that contribute the most to the prediction of low-quality content. To this end, we carried out an analysis based on feature selection algorithms: *Chi2* [14] and *Information Gain*. At a basic level, both algorithms assign weights to each attribute, from most relevant to model creation to least relevant. In this context, relevance is measured in terms of how much a given attribute contributes to the resulting model. Therefore, we needed to adopt an additional strategy to select relevant attributes based on their respective weights. Specifically, we also needed to decide how many attributes should be chosen, and which combination most favors model construction. Over the course of our analysis, we investigated all possible sizes of subset of attributes (i.e. the most important attribute, the two most important attributes, the three most important attributes, etc.). For each set of attributes, we construct a classification model. For the classification model, we consider the Random Forest algorithm, since it has been widely used in the literature. To validate our results, we used *F-measure* as the metric and a 10 fold cross-validation strategy.

To define the F-Measure metric, we need to understand two main concepts:

– Precision: number of items classified as positive is positive;

– Recall: number of relevant items selected.

The F-Measure (**F1**) is the harmonic mean between precision and recall:

$$F1 = 2 * \frac{\text{precision} * \text{recal}}{\text{precision} + \text{recall}} \tag{1}$$

In the end, a comparison of the most efficient model resulting from tests that take into account the methods of ranking attributes (described in the last paragraph) with three other data sets was performed. Initially, we fed all 35 attributes shown in Table 1 into two algorithms: RF and Support Vector Machine (SVM). We also tried a combination of attributes that emphasize all user-related information (attributes 1 to 14). The third set is centered around content-related attributes (attributes 15 to 35). We used *F-measure* as metric and a 10-fold cross-validation strategy to probe into the effectiveness of the resulting models.

4   Analysis and Results

As mentioned in the previous section, after applying Chi2 and Information Gain, we obtained two lists of attributes ordered by weight assigned by each feature selection algorithm. The results of applying both algorithms are shown in Fig. 3. To better present the results, we normalized the weights to correspond to a scale from 1 to 100. Our results would seem to suggest that attribute 16, which represents the medium used to post the tweet, is the attribute that best contributes to create a classification model. In way, this was expected since in our dataset all content posted through Twittascope was labeled as low-quality content (as shown in Fig. 2) [8].

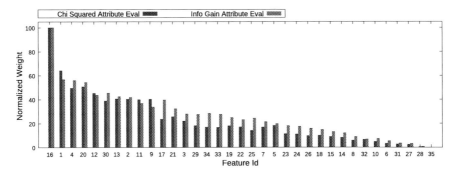

Fig. 3. Score of each attribute generated by the feature selection algorithms *Chi2* e *Information Gain*. While attribute 16 is the one that best contributes for creating a classification model, the attributes 28 and 35 are the worst.

Based on these two lists od attributes, we perform the second step of our methodology to evaluate the subset of attributes that can be used to create the best low-quality content detection model. As previously mentioned, we create models adopting the Random Forest algorithm and considering different sizes of attributes (i.e. the most important attribute, the two most important attributes, the three most important attributes, etc.). Figure 4 shows how much the predictive power of the models increases in response to the number of attributes taken into consideration during model generation. Upon analyzing Fig. 4, it can be seen that the predictive power of the resulting models peaked (i.e., achieved the highest F-measure score) when 33 attributes were fed into both algorithms: in effect, the most effective predictive models were generated when the subset of attributes does not include attributes 28 and 35. However, it interesting to note that with the 13 most relevant attributes we achieve a quality very close to the highest F-measure considering 33 attributes. The 13 most relevant attributes are 16, 1, 4, 20, 12, 30, 13, 2, 11, 9, 17, 21 and 3. As future work, we intend to deep investigate these attributes.

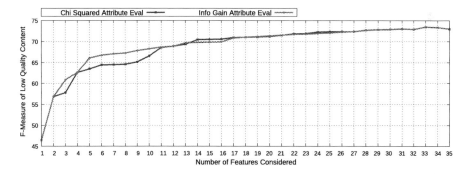

Fig. 4. F-Measure da classe conteúdos de baixa qualidade para variadas quantidades de atributos segundo os algoritmos *Chi2* e *Information Gain*

Table 2 presents more comprehensive information on how each algorithm performed when trained with each of the aforementioned attribute sets. We evaluated the resulting models in terms of two proxy metrics for their prediction power: true positive rate (TPR) and F1. Moreover, Table 2 also lists the accuracy and macro-F1 of the generated predictive models. Accuracy measures the global effectiveness regarding all decisions made by the classifier (that is, the inverse of error rate). Macro-F1 on the other hand, measures the classification effectiveness regarding each class independently. It corresponds the mean of the F-Measure values obtained for each possible class in the dataset.

Random Forest yielded better models for all subsets of attributes when compared to SVM (Table 2). A more straightforward interpretation of the results indicates that all subsets achieved over 90% accuracy for all algorithms. The best performing combination of attributes comprises 33 attributes that represent user and tweet information. The second best performing attribute set includes only attributes related to tweets. Our results would seem to suggest that using

Table 2. Classification Effectiveness considering different set of attibutes. Random Forest yielded better models for all subsets of attributes when compared to SVM. The best results are achieved considering the 33 most relevant attibutes using the Random Forest algorithm. Moreover, the set of attributes containing only user information does not yield good low-quality content predictive models.

Set of attributes	Classifier	Accuracy	TPR	F1	MacroF1
33 most relevant attributes	SVM	93.66	43.31	55.28	75.93
	Random Forest	**96.15**	60.43	**73.81**	**85.87**
All Attributes	SVM	93.6	43.31	55.28	75.93
	Random Forest	96.0	59.57	73.09	85.47
User's Attributes (1 - 14)	SVM	91.48	14.84	22.68	59.01
	Random Forest	93.50	36.45	50.36	73.44
Tweet's Attributes (15 - 35)	SVM	94.95	45.27	61.87	79.58
	Random Forest	95.77	**60.73**	72.19	84.95

a set of attributes containing only user information does not yield good low-quality content predictive models. This is key confirmatory finding, we argue that researchers and practitioners can refrain from including user information in their low-quality content prediction approaches since this type of information does not seem to contribute much to the creation of effective models.

5   Concluding Remarks

Owing to the popularity of social networking services, low-content quality has become an active research area and has been drawing the attention of researchers and practitioners alike. We carried out a literature survey in hopes of giving an overview of the most prominent research efforts to advance understanding in the low-quality content area. We believe our survey can be useful for researchers and practitioners looking to get a better understanding of what has already been investigated in the area. As for practitioners, our survey can be seen as an initial foray into promising approaches to low-quality content detection.

We also focused on examining the effectiveness of the attributes researchers have been employing to support low-quality content approaches. In the context of our research, effectiveness can be seen in terms of how much attributes are able to contribute to the generation of predictive models, that is, how much attributes contribute to the whole low-quality content detection strategy in question. For clarity purposes, we framed our discussion and evaluation of such attributes in terms of two categories: (i) user-related attributes and (ii) content-based attributes. A manually labeled Twitter dataset, which was employed in a previous study [8], was used in our evaluation. According to the results of such evaluation, models generated using only content-based attributes are as good as the models using the 33 best attributes. This could be evidenced by the TPR metric, in which both models had a score of approximately 60.5% using the Random Forest classifier. In contrast, models based on only user-related information had a score of 36.45% for the same metric and classifier. As mentioned, this is a fundamental confirmatory finding because it provides evidence that researchers and practitioners can refrain from including user information in their low-quality content prediction strategies without having to compromise on prediction ability.

As future work, we plan to come up with new attributes that can be used to yield better predictive models. Also, we intend to evaluate the performance of the resulting models using datasets from different social networking services and not only Twitter. We also plan to investigate different strategies for attribute selection. For instance, we believe that the Wrapper method might be a promising approach to selecting attributes that can be fed into some algorithms that have been widely explored in recent years, e.g., Convolutional Neural network (CNN).

Acknowledgments. This work was partially funded by the Brazilian National Institute of Science and Technology for the Web - INWeb, MASWeb, CAPES, CNPq, Finep, Fapesp and Fapemig.

References

1. Aggarwal, A., Rajadesingan, A., Kumaraguru, P.: PhishAri: automatic realtime phishing detection on twitter. In: 2012 eCrime Researchers Summit. IEEE, October 2012. https://doi.org/10.1109/ecrime.2012.6489521
2. Almaatouq, A., et al.: Twitter: who gets caught? Observed trends in social microblogging spam. In: Proceedings of the 2014 ACM conference on Web science - WebSci. ACM Press (2014). https://doi.org/10.1145/2615569.2615688
3. Azeta, A.A., Omoregbe, N.A., Ayo, C.K., Raymond, A., Oroge, A., Misra, S.: An anti-cultism social education media system. In: 2014 Global Summit on Computer & Information Technology (GSCIT). IEEE, June 2014. https://doi.org/10.1109/gscit.2014.6970097
4. Behera, R., Rath, S., Misra, S., Damaševičius, R., Maskeliūnas, R.: Large scale community detection using a small world model. Appl. Sci. **7**(11), 1173 (2017). https://doi.org/10.3390/app7111173
5. Behera, R.K., Rath, S.K., Misra, S., Damaševičius, R., Maskeliūnas, R.: Distributed centrality analysis of social network data using MapReduce. Algorithms **12**(8), 161 (2019). https://doi.org/10.3390/a12080161
6. Benevenuto, F., Magno, G., Rodrigues, T., Almeida, V.: Detecting spammers on Twitter. In: Collaboration, Electronic Messaging, Anti-Abuse and Spam Conference (CEAS), vol. 6, p. 12 (2010)
7. Bosma, M., Meij, E., Weerkamp, W.: A framework for unsupervised spam detection in social networking sites. In: Baeza-Yates, R., et al. (eds.) ECIR 2012. LNCS, vol. 7224, pp. 364–375. Springer, Heidelberg (2012). https://doi.org/10.1007/978-3-642-28997-2_31
8. Chen, W., Yeo, C.K., Lau, C.T., Lee, B.S.: A study on real-time low-quality content detection on Twitter from the users' perspective. PLOS One **12**(8), 1–22 (2017). https://doi.org/10.1371/journal.pone.0182487

9. Fakhraei, S., Foulds, J., Shashanka, M., Getoor, L.: Collective spammer detection in evolving multi-relational social networks. In: Proceedings of the 21th SIGKDD. ACM Press (2015). https://doi.org/10.1145/2783258.2788606

10. Gao, H., Chen, Y., Lee, K., Palsetia, D., Choudhary, A.: Poster. In: Proceedings of the 18th ACM conference on Computer and communications security. ACM Press (2011). https://doi.org/10.1145/2046707.2093489

11. Hu, X., Tang, J., Gao, H., Liu, H.: Social spammer detection with sentiment information. In: 2014 IEEE International Conference on Data Mining. IEEE, December 2014. https://doi.org/10.1109/icdm.2014.141

12. Jin, X., Lin, C.X., Luo, J., Han, J.: Socialspamguard: a data mining-based spam detection system for social media networks. In: Proceedings of the International Conference on Very Large Data Bases (2011)

13. Lee, K., Eoff, B.D., Caverlee, J.: Seven months with the devils: a long-term study of content polluters on Twitter. In: Fifth International AAAI Conference on Weblogs and Social Media (2011)

14. Liu, H., Setiono, R.: Chi2: feature selection and discretization of numeric attributes. In: Proceedings of 7th IEEE International Conference on Tools with Artificial Intelligence. IEEE Computer Society Press (1995). https://doi.org/10.1109/tai.1995.479783

15. Martinez-Romo, J., Araujo, L.: Detecting malicious tweets in trending topics using a statistical analysis of language. Expert Syst. Appl. **40**(8), 2992–3000 (2013). https://doi.org/10.1016/j.eswa.2012.12.015

16. McCord, M., Chuah, M.: Spam detection on Twitter using traditional classifiers. In: Calero, J.M.A., Yang, L.T., Mármol, F.G., García Villalba, L.J., Li, A.X., Wang, Y. (eds.) ATC 2011. LNCS, vol. 6906, pp. 175–186. Springer, Heidelberg (2011). https://doi.org/10.1007/978-3-642-23496-5_13

17. Miller, Z., Dickinson, B., Deitrick, W., Hu, W., Wang, A.H.: Twitter spammer detection using data stream clustering. Inf. Sci. **260**, 64–73 (2014). https://doi.org/10.1016/j.ins.2013.11.016

18. Santos, I., et al.: Twitter content-based spam filtering. In: Herrero, Á., et al. (eds.) International Joint Conference SOCO'13-CISIS'13-ICEUTE'13. AISC, vol. 239, pp. 449–458. Springer, Cham (2014). https://doi.org/10.1007/978-3-319-01854-6_46

19. Song, J., Lee, S., Kim, J.: Spam filtering in Twitter using sender-receiver relationship. In: Sommer, R., Balzarotti, D., Maier, G. (eds.) RAID 2011. LNCS, vol. 6961, pp. 301–317. Springer, Heidelberg (2011). https://doi.org/10.1007/978-3-642-23644-0_16

20. Sridharan, V., Shankar, V., Gupta, M.: Twitter games. In: Proceedings of the 28th ACSAC. ACM Press (2012). https://doi.org/10.1145/2420950.2421007

21. Tan, E., Guo, L., Chen, S., Zhang, X., Zhao, Y.: Spammer behavior analysis and detection in user generated content on social networks. In: 2012 IEEE 32nd International Conference on Distributed Computing Systems. IEEE, June 2012. https://doi.org/10.1109/icdcs.2012.40

22. Thomas, K., Grier, C., Song, D., Paxson, V.: Suspended accounts in retrospect. In: Proceedings of the 2011 ACM SIGCOMM Conference on Internet Measurement Conference. ACM Press (2011). https://doi.org/10.1145/2068816.2068840

23. Ungerleider, N.: Almost 10% of Twitter is spam (2015). https://www.fastcompany.com/3044485/almost-10-of-twitter-is-spam. Accessed 02 July 2019

24. Wang, A.H.: Don't follow me: spam detection in twitter. In: 2010 International Conference on Security and Cryptography (SECRYPT), pp. 1–10, July 2010

25. Wang, B., Zubiaga, A., Liakata, M., Procter, R.: Making the most of tweet-inherent features for social spam detection on Twitter. arXiv preprint arXiv:1503.07405 (2015)
26. Yang, C., Harkreader, R., Gu, G.: Empirical evaluation and new design for fighting evolving twitter spammers. IEEE Trans. Inf. Forensics Secur. **8**(8), 1280–1293 (2013). https://doi.org/10.1109/tifs.2013.2267732
27. Yang, C., Harkreader, R.C., Gu, G.: Die free or live hard? Empirical evaluation and new design for fighting evolving Twitter spammers. In: Sommer, R., Balzarotti, D., Maier, G. (eds.) RAID 2011. LNCS, vol. 6961, pp. 318–337. Springer, Heidelberg (2011). https://doi.org/10.1007/978-3-642-23644-0_17
28. Zheng, X., Zhang, X., Yu, Y., Kechadi, T., Rong, C.: ELM-based spammer detection in social networks. J. Supercomput. **72**(8), 2991–3005 (2015). https://doi.org/10.1007/s11227-015-1437-5
29. Łuksza, K.: Bot traffic is bigger than human. make sure it doesn't affect you! (2018). https://voluum.com/blog/bot-traffic-bigger-than-human-make-sure-they-dont-affect-you/

Weighted Ensemble Methods
for Predicting Train Delays

Mostafa Al Ghamdi$^{(\boxtimes)}$, Gerard Parr, and Wenjia Wang

School of Computing Sciences, University of East Anglia, Norwich, UK
{m.al-ghamdi,G.Parr,Wenjia.Wang}@uea.ac.uk

Abstract. Train delays have become a serious and common problem in the rail services due to the increasing number of passengers and limited rail network capacity, so being able to predict train delays accurately is essential for train controllers to devise appropriate plans to prevent or reduce some delays. This paper presents a machine learning ensemble framework to improve the accuracy and consistency of train delay prediction. The basic idea is to train many different types of machine learning models for each station along a chosen journey of train service using historical data and relevant weather data, and then with certain criteria to choose some models to build an ensemble. It then combines the outputs from its member models with an aggregation function to produce the final prediction. Two aggregation functions were devised to combine the outputs of individual models: averaging and weighted averaging. These ensembles were implemented with a framework and their performance was tested with the data from an intercity train service as a case study. The accuracy was measured by the percentages of correct prediction of the arrival time for a train and correct prediction within one minute to the actual arrival time. The mean accuracies and standard deviations are 42.3%($\pm$11.24) from the individual models, 57.8%($\pm$3.56) from the averaging ensembles, and 72.8%($\pm$0.99) from the weighted ensembles. For the predictions within one minute of the actual times, they are 86.4%($\pm$14.05), 94.6%($\pm$1.34) and 96.0%($\pm$0.47) respectively. So overall, the ensembles significantly improved not only the prediction accuracies but also the consistency and the weighted ensembles are clearly the best.

Keywords: Ensemble · Train delays · Machine learning

1 Introduction

Train delays are a major problem for train operating companies and passengers. In the UK, in the last few years, the Public Performance Measure (PPM) for train services has been continuously declining - from over 91% in 2013–14 to 85.6% in Q3 of 2018–19 [18], although the entire rail industry has been working extensively in trying to improve their performance. There are many factors that can cause initial delays, which include signalling issues, bad weather, damaged equipment, breakdowns, construction works, accidents and disruptions to the

© Springer Nature Switzerland AG 2020
O. Gervasi et al. (Eds.): ICCSA 2020, LNCS 12249, pp. 586–600, 2020.
https://doi.org/10.1007/978-3-030-58799-4_43

flow of operations [17]. Once an initial delay, called a primary delay, occurred, it can cause a chain of knock-on delays to other trains [11] and those delays are called reactionary delays. Given that the number of train passengers has been consistently increasing, the train operating companies have to run more train services on the current rail networks to meet the demand, which means that the rail networks are running at almost their full capacity, that is, the trains are closely scheduled to run one after another on a rail network with a minimum allowed interval. As a consequence, because there is a very little buffer space and time in the rail network for tolerating any disruption, then even a small primary delay can cause many reactionary delays and huge disruptions to the train services and a great deal of inconvenience to the passengers.

It is therefore essential for train controllers to have some systems to help predict train delays as earlier as possible so that they can take appropriate actions to either reduce or prevent further delays. This paper presents a machine learning ensemble method that combines different types of predictive models to improve the accuracy of delay prediction.

An ensemble is built with multiple models generated by using a variety of machine learning algorithms from the data of historical train services and weather data, and an aggregation function is employed to combine the predictions of individual models to produce a final prediction. In this study, we built heterogeneous ensembles and devised a weighted aggregation function for producing the final output of an ensemble. A framework has been built to implement these ensembles and the weighted aggregation function and tested in a case study on an intercity train service journey. The accuracy is measured with the percentage of correct predictions and predictions within 1 min of the actual arrival time.

The rest of this paper is organised as follows: Sect. 2 briefly reviews the related work, Sect. 3 describes in detail the methodology and construction of the ensembles. Section 4 presents the experiment design and results including a discussion of their implication. Finally, Sect. 5 draws the conclusions and gives suggestions for future work.

2   Related Work

Various methods have been developed for predicting train delays, including conventional regression methods and Stochastic approaches, and machine learning-based methods, such as Bayesian Networks, Support Vector Regression, Artificial Neural Networks and Deep Learning Neural Networks.

Bayesian Networks have been used by Lessan et al. [12] to predict train delays on the Chinese rail network, and also by Corman and Kecman [4] on the Swedish rail network with historical data.

Support Vector Machine (SVM) is another method studied for the prediction of arrival times when they are treated as a classification problem. Some researchers in China [24] used it to predict bus arrival times, while Markovic et al. [13] used SVM to identify connections between delays and railway network qualities. This work focuses on avoiding and anticipating delays, particularly as finding any connections between the two could enable railway staff make use of learned choices to decrease delays. They also discussed two further potential methods - hybrid simulation and machine learning and multiple regression.

Treating the prediction of train delays as a regression problem, [2] used Support Vector Regression (SVR) for the American freight system and showed that the mean absolute error decreased by 14% from a baseline of historical running times, although no comparisons with other models were conducted. Gaurave and Srivastaz [8] provided the state of the zero shot Markov model for predicting the train delay process. This model was based on the train delays on the Indian network, which carried over eight billion passengers. The techniques were linked with the regression technique of the data modeling and it was reported an to be an efficient algorithm for the estimation of train delays and problems for the transport network. The main problem with modeling train delays is based on the volume of data and the algorithm required to mine the huge quantity of information. The algorithm can be applied with ensemble regression to solve the train delay problems.

Artificial Neural Networks (ANN) have also been used by Yaghini et al. [23] to predict delays on the Iranian passenger rail network, while Oneto et al. [16] tested the use of this model in Italy, and involved many detailed features, like displaying the weather conditions and the tracking ability of other trains on the network. Their work focused on the process of using machine learning algorithms and consists of kernels, ensembles and neural networks. Oneto et al. [16] use the Random Forest approach. They set the number of trees to 500 and used the ensemble technique, which will also be used in this research because, in comparison to classification/regression, they results show it performs better. A back-propagation Neural Network was used by Hu and Noch [9], who employed a genetic algorithm to improve the training of the model. Prediction performance improved in this case, but the training time was extended.

Wen et al. [22] also compared the Random-Forest Model to a simpler Multiple Linear Regression Model and found the Random Forest Model to have the better levels of prediction accuracy, compared to a simple Multi Linear Regression Model. Their finding is consistent with that from Oneto et al. [16], which is, that Random Forest outperforms other approaches, suggesting it to be the best algorithm for predicting delays.

Extreme Learning Machines (ELM), Shallow and Deep, were used by Oneto et al. [17] because ELM can learn faster than those which use traditional learning algorithms, which may not be fast enough, and they generalise well [10]. It is also more appropriate for use with big data than those which use univariate statistics because the model adapts and gets better when is fed with external data [17].

One recent study provide valuable insights into the use of combination of machine learning and other approaches to predict train delays in Germany. Nair et al. [14] developed a prediction model for forecasting train delays for the nationwide Deutsche Bahn passenger network, which runs approximately 25,000 trains per day. The data sources provided a rich characterisation of the network's operational state, some of which was collected using track-side train passing messages to reconstruct current network states, including train position, delay information and key conflict indicators. The first model used was a statistical ensemble model comprised of two statistical models and a simulation based model. A context-aware Random Forest model was used as the first model. This had the capacity to account for current headways, weather events and information about work zones. Train-specific dynamics were accounted for by a second, kernel regression model. Thirdly, a mesoscopic simulation model was applied to account for differences in dwell time and conflicts in track occupation. Nair et al. [14] model demonstrated a 25% improvement in prediction accuracy and a reduction of root mean squared errors of 50% in comparison with the published timetable. The strength of their system was their use of ensembles which, as expected, showed was superior to constituent models. However, while the work improved the accuracy of a general prediction model, it did not consider state-dependent models which might have given further improvements. Their model was also sensitive to such hyper-parameters as outlier thresholds.

To sum up, these studies showed that machine learning models are capable of predicting train delays but they are generally limited to the rail networks where they were developed for. The motivation of this research is to develop a machine learning ensemble that will combine multiple models generated from different normal learning algorithms to enhance accuracy and reliability to help improve the performance of UK train services.

3   Weighted Ensembles for Predicting Train Delays

A train delay prediction problem can be formulated as follow. For a given train service journey, it should contain several stations, i.e. $J = \{S_1, S_2, ..., S_i, ..., S_{N-1}, S_N\}$ including the origin, i.e. the departure station S_1, the destination station S_N, and several stations S_2 to S_{N-1}, along the rail track where a train stops for passengers getting on or off the train. Along the rail track, there could be a set of n trains: $T = \{T_1, T_2, ..., T_j, ..., T_{n-1}, T_n\}$, running in accordance with their timetable.

Thus, the problem of train delay prediction is that for a given train T_i that has just departed from a Station S_i, we want to predict its arrival time at the next station S_j, and also at all the remaining stations of the journey. To complete this prediction task, we need to devise a suitable modelling scheme, which will be described in detail in the next subsection.

3.1   Delay Prediction Modelling Scheme

Firstly, instead of predicting the actual arrival time of a given train T_i at its next station, we convert it to predict the difference between the planned and actual arrival time at a station. So let t_{pa} represent the planned arrival time on the timetable for train T_i at an intended station $S_j \in J$; t_{aa}, the actual arrival time of that train at that station. The time difference, Δt, between the timetabled arrival time and the actual arrival time is calculated by the following equation, which is taken as the target variable y.

$$y = \Delta t(T_i, S_j) = t_{aa}(T_i, S_j) - t_{pa}(T_i, S_j) \tag{1}$$

The predicted arrival time for train T_i at Station S_j can then be derived by $S_j(T_i) = t_{pa} + \Delta t$. It should be noted that when Δt is positive, it means a train is delayed by this Δt amount of time, and when it is negative, it means that a train arrives early by Δt time. This predicted time will be taken, together with other variables, as the inputs to the next model for predicting the arrival time of a following station. Figure 1 shows the prediction modelling scheme for any train running on the rail track of the chosen journey.

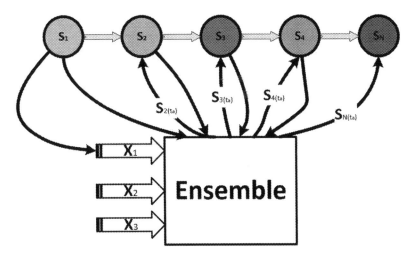

Fig. 1. Illustration of the delay prediction modelling scheme with machine learning ensembles.

The features used to make these predictions include the running time, the weather conditions and whether or not it is an off-peak service or during holiday periods. With these features, different models are trained as candidate models for building heterogeneous ensembles. Two different functions are used to combine the predictions of the individual models in an ensemble, which are averaging and weighted averaging, which are described in detail later.

3.2 Heterogeneous Ensembles

Two types of ensembles can be constructed in terms of types of models used as their member models: homogeneous and heterogeneous ensembles. A homogeneous ensemble is built with the models of same type, e.g. decision trees only or neural networks only; whilst a heterogeneous ensemble is built using models of different types, e.g. a mixture of decision trees and neural nets. In this study, we choose to build heterogeneous ensembles because they have been shown to superior to the using either a single model or a homogenous ensemble [7,19].

Building Heterogeneous Ensembles: This paper presents a heterogeneous ensemble for train delay prediction because the complexity of the problem can overwhelm single models. Using ensemble methods has two advantages. The first is that, based on previous research [1] it is to be expected that an ensemble will outperform individual models [3,20]. Secondly, an ensemble offers a high level of reliability [21]. Furthermore, studies [1] comparing the effectiveness of heterogeneous and homogenous ensembles demonstrated that heterogeneous ensembles are generally more accurate and more reliable, which is particularly more important in a critical industrial application such as train delay prediction, where more consistent and robust predictions out-weight the absolute prediction accuracy as long as the accuracy is within a tolerable limit, such as within one minute. So, in this study, we chose to build heterogeneous ensembles for predicting train delay.

We construct a heterogeneous ensemble using more different types of models because they are likely to be more diverse than the models of the same types. We therefore chose sixteen different learning algorithms to generate candidate models. These were: Random Forest, Support Vector Regression, Linear Regression, Extra Trees Regressor, Multi-Layer Perceptron, Gaussian process regression, LassoLars, ElasticNet, Logistic Regression, Ridge, Gradient Boosting Regressor, Lasso Regression, Kernel Ridge Regression, Bayesian Ridge, Stochastic Gradient Descent, AdaBoost Regressor.

For a given training dataset each algorithm is used to generate a model. Thus 16 models are generated as the candidates to be selected by some rules for forming an ensemble. For comparison, we also build an ensemble by using all the candidates without any selection.

Then two different decision-making functions - simple averaging and weighted averaging (details given in the next subsection) are derived to combine the outputs of the member models in these two heterogeneous ensembles to produce a final output. Because of their different decision-making functions their ensembles are named as Averaging Ensemble (AE) and Weighted Ensemble (WE).

An Ensemble Framework: A framework for constructing heterogeneous ensembles has been developed, as shown in Fig. 2, with 5 components: Feature Extraction, Data Partition, Model Generation, Model Evaluation and Selection, and Decision Combination.

Feature Extraction: Two types of input data, i.e. train running data and weather data, were collected and used in this study. The basic form of the train running data is provided in a format likes a train timetable, that is, for each train, it lists the planned departure and actual departure times, and the planned arrival and the actual arrival times for each stop station along the service journey. From this format of raw data, we need to transform it into a structured representation by extracting some features, which will be described in the later section in detail.

Weather data was provided with over 20 features and we selected some of them based on some prior experience from domain experts. In total, we generated three groups of features.

Data Partition: The data was then partitioned into training and testing, the ratios being 80% and 20% respectively. The training data could be further partitioned with k-fold cross-validation mechanism for preventing over-training in some algorithms.

Model Generation: The next stage of the process is to generate a pool of predictive models by using above listed learning algorithms as the candidates to be selected for building an ensemble.

Model Evaluation and Selection: The generated models are then evaluated for their level of accuracy with training and validation data. The metrics used for this were percentage correct classification and the percentage of correct prediction with within one minute, see details in the next section.

Combination: This stage is designed to combine the results of the individual models selected from the pool of models at the previous stage, to produce the final prediction for a station. For this process, two functions have been used: the averaging and weighted averaging.

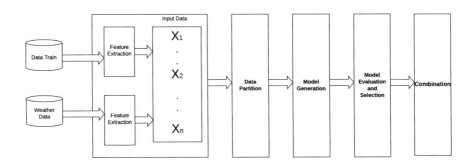

Fig. 2. A process for building heterogeneous ensemble.

This framework was implemented using Python, based on Scikit-learn and other libraries.

Methods for Combining the Models: In any ensemble, it is essential to have a decision making function to combine the outputs of individual models to produce the final output. This function plays a critical role in determining the performance of an ensemble [21]. We devised two functions: averaging and weighted averaging.

– Averaging Ensemble

For a given ensemble with N models, $\Phi = \{m_1, m_2, ..., m_i, ..., m_N\}$, an averaging decision function simply calculates the mean of the outputs from all the N models: $\{\Delta t_1, \Delta t_2, ..., \Delta t_i, ..., \Delta t_N\}$, that is,

$$\Delta t = \frac{1}{N} \sum_{i=1}^{N} \Delta t_i \tag{2}$$

The result is taken as the final prediction for the difference of the arrival time of a given train at a station.

– Weighted Ensemble

In a weighted ensemble, we employ another decision-making function - the weighted averaging method. The motivation is based on the fact that some models perform better than others, so logically their decisions should have more weight than those models whose performances are lower, when making a final decision. Thus we devised a weighted decision function as follows to investigate whether it can improve the accuracy of an ensemble.

In an ensemble of N models, $\Phi = \{m_1, m_2, ..., m_i, ..., m_N\}$, each model will be evaluated with a validation dataset and their accuracies *Accuracy* $\{A_1, A_2, ..., A_i, ..., A_N\}$, will be used to calculate their weight for making a final output. Specifically, for a model m_i, its weight W_i is computed by the following equation.

$$W_i = \frac{A_i}{\sum_{i=1}^{N} A_i} \tag{3}$$

As can been seen, the weight is normalised so that $\sum_{i=1}^{N} W_i = 1$. The final predication Δt can be calculated as follows.

$$\Delta t = 1/N \sum_{i=1}^{N} W_i \Delta t_i \tag{4}$$

3.3   Evaluation Metrics and Methods

For comparison, we chose Random Forest, which is also an ensemble-based method, as a competitive target because this method has been demonstrated to the best by some studies as reviewed earlier. All the individual models were also used as the baselines in the evaluation.

The time unit used for train timetable is based on the *minute*, not the second. So, to follow the practice of rail service, we devised the following measures to evaluate the accuracy of predictions.

Percentage of Correct Prediction (*CP*) After Rounding: The continuous output produced by regression models is unlikely to be whole integers. Output values are therefore rounded to the nearest integer. Then the rounded predicted time difference value is compared with the actual outputs in the test dataset.

If these two values are equal, then the prediction is correct, otherwise is wrong. The percentage of correct predictions is calculated over the number of journeys that the model accurately predicted to the minute after the rounding.

Percentage of Predictions Within 1 min After Rounding, $|P| < 1$: This is the same as the above evaluation, using comparison to assess whether the model predicted exactly or within 1 min either way of the actual time.

Statistical Tests for Comparing the Results: The performance of the proposed ensemble method is evaluated using statistical significance tests, which compare one method against multiple other methods. The selection of the test is determined by the experimental design. In our study, the Friedman test was used to compare all the methods used and the results presented in the critical difference diagram. The Friedman test is a nonparametric test designed for comparison of multiple learning algorithms. Its results show whether or not there is a statistical difference in performance between the algorithms. The critical difference diagram provides a graphic representation of overall performance. A slide bar is used to indicate algorithms which are statistically similar [6]. We used the critical distance diagram to show the ranking differences between single models, Random Forest, Averaging Ensembles and Weighted Ensembles.

4   Experiments and Results

To test our ensemble methods, a case study was carried out on an intercity train service between Norwich and London. Table 1 lists the stops between a pair of stations for this journey. For this service, trains run every half of an hour and stop at different stations.

Table 1. List of pair of adjacent stations between Norwich and London Liverpool Street for all of the journey variations.

Pair No.	Station1	Station2	Pair No.	Station1	Station2
1	NRW	DIS	4	IPS	MNG
2	DIS	SMK	5	MNG	COL
3	SMK	IPS	6	COL	CHM

4.1   Data and Features

We collected the train running data in seven months between 2017–2018, from the HSP - Historic Service Performance (NRE) [15] data repository. It contains large quantities of historic data and has a filter which allows for data to be requested in particular time frames between specific stops (OpenRailData) [5].

Weather data was also collected from the weather stations nearest to the railway stations in question.

The data requires pre-processing before it could be used for analysis or modelling because each journey can have a different number of stops. These are represented by separate rows which needed to be brought together to avoid the necessity of searching and joining the rows every time they were used. A custom class in programming language was used for this purpose. The class represents a journey and can include such elements as arrival, departure and stop times, as well as other variables which are helpful for modelling. Scripts also contain code carrying information stored about each journey object. This allows model features to be created and the addition of further information, such as weather, to be added. Rows of model inputs are then ready for the learning process. The features which were thus derived consisted of three groups. The first group includes running time information, the date and day of the week, timing, and whether a train is running at off-peak period or peak period. The second group of features derived from the weather data, taken from the weather station closest to the stop. The third group used information about the journeys in front of the modelled train and any deviation from the timetable.

4.2   Experiment Design and Results

Three sets of experiments were conducted, using the chosen journey, by using all features. Each set of the experiment was run 5 times with different data partitions to test the consistency of prediction. Data partition 80% for training and 20% for testing. Some cross validation is used later during optimization of the hyper-parameter of some models. This partition procedure was repeated a number of times and the mean values and the standard deviations were computed.

Table 2 lists the mean accuracies and standard deviation of three types of models: the individual models, the averaging ensembles and the weighted ensembles. It also gives the mean correction prediction rate within one minute of the targets, $|P| < 1$ and their standard deviations.

Figure 3 shows the percentage of correct prediction between pairs of stations and Fig. 4 depicts the correct predictions within one minute.

As can be seen, the average of correct prediction accuracies of single models is never higher than 50% and falls to as low as 38%. However, with the averaging ensembles, the prediction accuracy was clearly improved between 5% to 25%, with the lowest average correct prediction of 43% and the highest of 71%. The best results were produced by the weighted ensembles.

In general, the single models had an overall mean accuracy level of only 41.6%, with a mean standard deviation of 11.24 for correct predictions, which are low and also varied considerably, and 86.42% for correction prediction within one minute of the actual time. Averaging ensembles improved on average by 15% to 56.78% for a correct prediction with a smaller standard deviation, and 8% to 94.64% for correct prediction within one minute. The weighted ensembles outperformed significantly both the individual models and the averaging ensembles,

Table 2. The results for single models, averaging ensemble, and weighted averaging ensemble for each individual stop station between Norwich and London Liverpool Street.

	Average single models				Average ensemble				Weighted ensemble			
	% CP		%|P| < 1		% CP		%|P| < 1		% CP		%|P| < 1	
Section	Mean	SD	Mean	SD	Mean	SD	Mean	SD	Mean	SD	Mean	SD
NRW - DIS	39.93	11.71	82.23	20.06	54.79	1.47	93.25	0.66	71.69	1.41	94.35	0.79
DIS - SMK	48.78	10.90	92.28	5.35	71.34	0.65	95.41	0.43	75.41	1.08	96.21	0.38
SMK - IPS	37.87	9.39	84.85	17.34	42.81	2.61	93.69	1.99	65.14	0.81	95.89	0.66
IPS - MNG	43.69	10.71	90.94	11.47	67.47	7.26	97.58	0.95	81.85	1.13	98.36	0.40
MNG - COL	44.96	10.76	92.86	11.02	60.49	4.88	97.55	0.63	78.10	0.79	98.38	0.26
COL - CHM	32.34	13.97	75.36	19.03	43.76	4.48	90.35	3.35	64.52	0.70	92.82	0.34
Mean	41.26	11.24	86.42	14.05	56.78	3.56	94.64	1.34	72.79	0.99	96.00	0.47

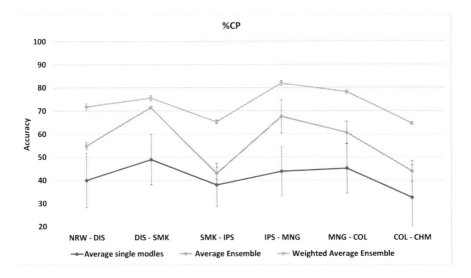

Fig. 3. The results (mean accuracies and standard deviations) of predicting the delays between all stations with single models, averaging ensembles and weighted average ensembles. The Weighted ensembles are most accurate and also most reliable.

with a mean accuracy of 72.79% for correct predictions and 96.0% for correction predictions within one minute of the actual time.

In addition, it should be particularly noted that ensembles are not only more accurate but more importantly more consistent and reliable, because they have very low standard deviations. For individual models, their predictions varied considerably with a mean standard deviation of 11.24%, which implies that individual models are not consistent. On the other hand, the averaging ensembles are more consistent because they have a much smaller mean standard deviation

(3.56), whilst the weighted ensembles are very consistent with a mean standard deviation as low as 0.99.

Another phenomenon can be observed from the results is that for some stations on the same journey, the accuracies of either individual models or ensembles performed poorer than other stations. This suggests that the underlying prediction problems are specific to journeys between particular pairs of stations, where there may be some uncertainties that were not represented by the data. For example, for the pair of SMK-IPS, it was found later that there were some freight trains going through IPS station but not recorded in this dataset.

Moreover, the accuracy of predictions was also affected by the amount of available train data, which was reflected by a dip at CHM station because some trains did not stop here and hence there was not the same amount of the training data as for other stations.

Figure 4 shows a considerable improvement in accuracy when ensembles were used to predict the arrival time within one minute of the targets. From a practical operational point of view, the departure and arrival times are shown in a unit of minute to passengers, So, our weighted ensembles achieved the accuracy levels as high as 98% with a standard deviation of 0.47%.

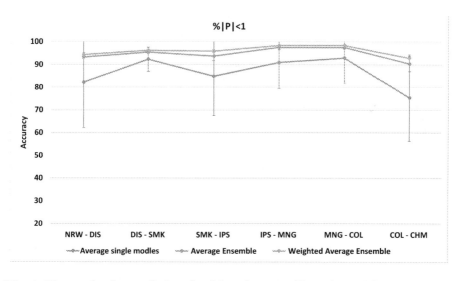

Fig. 4. The results for predicting the delays between all stations within one minute with single models, averaging ensembles and weighted ensembles.

The accuracies of these three types of models are statistically compared in the following section.

In summary, our results demonstrated that using ensembles improved the accuracy of train delay predictions and that the weighted ensembles consistently and significantly out-performed other methods and were the best for this purpose.

4.3   Critical Comparisons

The statistical tests described in the earlier section were conducted to compare the accuracies of individual models, the averaging ensembles and the weighted ensembles and also Random Forest, which was chosen as a comparison baseline because it was already considered to be one of the most accurate methods for predicting train delays [16].

On average over all the experiments for the entire train service journey, our weighted ensembles were ranked in the first place, with the averaging ensembles the second place, Random Forest the third and the individual models the fourth. The critical distance among these four methods are represented by the critical distance diagrams Fig. 5 and Fig. 6. The interpretation is that the methods linked with a thick bar are not statistically different from each other.

As can be seen, the methods are linked with two thick bars: the first one includes the WE (Weighted Ensembles), AE (Averaging Ensembles) and RF, and the second group includes, AE, RF and single models. They show that our weighted ensembles are statistically and significantly better than the individual models, and better than other two ensemble methods although not significantly.

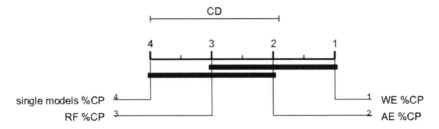

Fig. 5. Comparison between our ensembles with single models and the Random Forest ensembles, using correct prediction measure %CP.

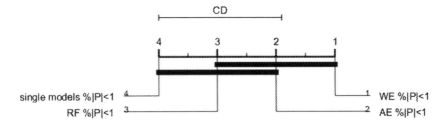

Fig. 6. Comparison between our ensembles with single models and the Random Forest ensembles within the measure of correction prediction within one minute.

5   Conclusion and Future Work

In this paper, we presented a framework for developing machine-learning ensembles to improve the accuracy and reliability of predicting train delays. We built two types of ensembles with two decision fusion strategies: averaging and weighted averaging. We tested our methods with a case study on an intercity train service between Norwich to London Liverpool Street. We extracted a wide range of features from train running data and weather data. The results of our experiments show that our ensembles built using both the averaging and the weighted averaging are always not only better than single models, but also more reliable. Furthermore, our ensembles out-performed the Random Forest ensembles, which were considered as state of the art methods for train prediction.

However, our study was limited by the amount of the data available to us at time, and because of this limit, our results should be interpreted with caution. In addition, we only used 16 types of models so the pool of candidate models is relatively small. These two issues will be addressed in our future study. We want to use more data, which we have already collected, more features and more types of models, such as deep learning models, to test our methods and also try to improve the prediction accuracies as much as possible.

Acknowledgement. The authors would like to thank Mr. Douglas Fraser in particular for his important work in gathering the data and the advice given in this research and also the WeatherQuest for providing the weather data for this project. We acknowledge the foundational works carried out by two MSc students at time, Mr. Bradley Thompson and Ms Mary Symons. In addition, we really appreciate the support and advice given by the people from the Train Operating Company - Greater Anglia, the Network Rail, the Rail Delivery Group, and the Rail Standards and Safety Board (RSSB) for the grant awarded through the rail big data sandbox competition in 2017. Specifically, we would also like to thank the Albaha University for providing a studentship for Mr Mostafa Al Ghamdi to do his PhD at the University of East Anglia.

References

1. Alyahyan, S., Farrash, M., Wang, W.: Heterogeneous ensemble for imaginary scene classification. In: KDIR, pp. 197–204 (2016)
2. Barbour, W., Mori, J.C.M., Kuppa, S., Work, D.B.: Prediction of arrival times of freight traffic on us railroads using support vector regression. Transp. Res. Part C Emerg. Technol. **93**, 211–227 (2018)
3. Brown, G., Wyatt, J., Harris, R., Yao, X.: Diversity creation methods: a survey and categorisation. Inform. Fusion **6**(1), 5–20 (2005)
4. Corman, F., Kecman, P.: Stochastic prediction of train delays in real-time using Bayesian networks. Transp. Res. Part C Emerg. Technol. **95**, 599–615 (2018)
5. Open Rail Data: HSP (2016). https://wiki.openraildata.com/index.php/HSP. Accessed 13 Nov 2019
6. Demšar, J.: Statistical comparisons of classifiers over multiple data sets. J. Mach. Learn. Res. **7**, 1–30 (2006)

7. Gashler, M., Giraud-Carrier, C., Martinez, T.: Decision tree ensemble: small heterogeneous is better than large homogeneous. In: 2008 Seventh International Conference on Machine Learning and Applications, pp. 900–905. IEEE (2008)

8. Gaurav, R., Srivastava, B.: Estimating train delays in a large rail network using a zero shot Markov model. In: 2018 21st International Conference on Intelligent Transportation Systems (ITSC), pp. 1221–1226. IEEE (2018)

9. Hu, J., Noche, B.: Application of artificial neuron network in analysis of railway delays. Open J. Soc. Sci. **4**(11), 59 (2016)

10. Huang, G.B., Zhu, Q.Y., Siew, C.K.: Extreme learning machine: theory and applications. Neurocomputing **70**(1), 489–501 (2006). https://doi.org/10.1016/j.neucom.2005.12.126. http://www.sciencedirect.com/science/article/pii/S0925231206000385, Neural Networks

11. Lee, W.H., Yen, L.H., Chou, C.M.: A delay root cause discovery and timetable adjustment model for enhancing the punctuality of railway services. Transp. Res. Part C Emerg. Technol. **73**, 49–64 (2016)

12. Lessan, J., Fu, L., Wen, C.: A hybrid Bayesian network model for predicting delays in train operations. Comput. Ind. Eng. **127**, 1214–1222 (2019)

13. Marković, N., Milinković, S., Tikhonov, K.S., Schonfeld, P.: Analyzing passenger train arrival delays with support vector regression. Transp. Res. Part C Emerg. Technol. **56**, 251–262 (2015)

14. Nair, R., et al.: An ensemble prediction model for train delays. Transp. Res. Part C Emerg. Technol. **104**, 196–209 (2019)

15. NRE: Darwin data feeds (2018). http://www.nationalrail.co.uk/100296.aspx. Accessed Oct 2019

16. Oneto, L., et al.: Advanced analytics for train delay prediction systems by including exogenous weather data. In: 2016 IEEE International Conference on Data Science and Advanced Analytics (DSAA), pp. 458–467. IEEE (2016)

17. Oneto, L., et al.: Train delay prediction systems: a big data analytics perspective. Big Data Res. **11**, 54–64 (2018)

18. Office of Rail and Road: Passenger and freight rail performance 2018-19 q3 statistical release on 21/02/2019. https://dataportal.orr.gov.uk/media/1210/passenger-rail-usage-2018-19-q3.pdf. Accessed 05 Apr 2019

19. Smetek, M., Trawiński, B.: Selection of heterogeneous fuzzy model ensembles using self-adaptive genetic algorithms. New Gener. Comput. **29**(3), 309 (2011). https://doi.org/10.1007/s00354-010-0305-3

20. Wang, H., Fan, W., Yu, P.S., Han, J.: Mining concept-drifting data streams using ensemble classifiers. In: Proceedings of the Ninth ACM SIGKDD International Conference on Knowledge Discovery and Data Mining, pp. 226–235. ACM (2003)

21. Wang, W.: Some fundamental issues in ensemble methods. In: IEEE World Congress on Computational Intelligence, pp. 2243–2250, July 2008. https://doi.org/10.1109/IJCNN.2008.4634108

22. Wen, C., Lessan, J., Fu, L., Huang, P., Jiang, C.: Data-driven models for predicting delay recovery in high-speed rail. In: 2017 4th International Conference on Transportation Information and Safety (ICTIS), pp. 144–151. IEEE (2017)

23. Yaghini, M., Khoshraftar, M.M., Seyedabadi, M.: Railway passenger train delay prediction via neural network model. J. Adv. Transport. **47**(3), 355–368 (2013)

24. Yu, B., Yang, Z.Z., Chen, K., Yu, B.: Hybrid model for prediction of bus arrival times at next station. J. Adv. Transport. **44**(3), 193–204 (2010)

A Smart Ecosystem for Learning and Inclusion: An Architectural Overview

Valéria Farinazzo Martins[1,2(✉)], Łukasz Tomczyk[3], Cibelle Amato[2],
Maria Amelia Eliseo[1], Solomon Sunday Oyelere[4],
Özgür Yaşar Akyar[5], Regina Motz[6], Gabriel Barros[7],
Sonia Magali Arteaga Sarmiento[7], and Ismar Frango Silveira[1]

[1] Computing and Informatics Department, Mackenzie Presbyterian University,
São Paulo, Brazil
{valeria.farinazzo,mariaamelia.eliseo,
ismar.silveira}@mackenzie.br
[2] Post-Graduation Program in Developmental Disorders,
Mackenzie Presbyterian University, São Paulo, Brazil
{valeria.farinazzo,cibelle.amato}@mackenzie.br
[3] Faculty of Social Science, Pedagogical University of Cracow, Cracow, Poland
tomczyk_lukasz@prokonto.pl
[4] School of Computing, University of Eastern Finland, Joensuu, Finland
solomon.oyelere@uef.fi
[5] Hacettepe University, Ankara, Turkey
akyar.ozgur@gmail.com
[6] Facultad de Ingeniería, Universidad de la República, Montevideo, Uruguay
reginamotz@gmail.com
[7] University of Azuay, Cuenca, Ecuador
{gbarrosg,marteaga}@uazuay.edu.ec

Abstract. In recent years, open educational resources have been offered in initiatives that aim to reach a wider audience, contributing to the democratization of knowledge. However, there is still a gap with regard to the accessibility of these educational resources. In this sense, this paper presents a digital ecosystem architecture, called SELI (Smart Ecosystem for Learning and Inclusion), which is being developed by a group of ten countries in Europe and Latin America. This ecosystem can be viewed in four different views (philosophical foundations, supporting infrastructure, concept and service bus). It aims to provide an accessible learning environment, involving recent technologies such as Blockchain, microsites and the use of universal accessibility guidelines.

Keywords: Smart ecosystem · Inclusion · Architectural model

1 Introduction

There are a large number of people who are disadvantaged in the current educational environment due to several reasons among which are archaic educational environment, pedagogy, methods and goals, lack of creativity in the current system, insecurity, economic, and social issues, etc. [1]. The needs of the disadvantaged groups, such as

---

The original version of this chapter was revised: the presentation of the second authors' name has been corrected. The correction to this chapter is available at
https://doi.org/10.1007/978-3-030-58799-4_75

© Springer Nature Switzerland AG 2020, corrected publication 2021
O. Gervasi et al. (Eds.): ICCSA 2020, LNCS 12249, pp. 601–616, 2020.
https://doi.org/10.1007/978-3-030-58799-4_44

immigrants, homeless, physically challenged such as the blind, hearing impaired, and other disabilities are not readily considered in the modelling, design and development of learning environments.

Building a learning ecosystem that is smart has been the objective of many academics over the past years. A lot of research effort was engaged incorporating the followings in the educational environment to support the aspiration of teachers and students: artificial intelligence techniques [2], recent technology such as blockchain and microsites [3, 4], open and active pedagogy such as global sharing, digital storytelling [5]. According to [6], "technologies should not support learning by attempting to instruct the learners, but rather should be used as knowledge construction tools that students learn with, not from". In line with [6], we consider the building of technological and pedagogical hub consisting of several tools and services to support the teacher's role of guidance, mentoring, and creativity and support student's role of knowledge seeking, construction and learning.

In our opinion, one way to achieve a solution to address the global yearning for inclusive and personalized learning is by crafting more potent and supportive innovation through technologies, pedagogies and novel conditions for learning. In fact, a smart learning environment can only be achieved when all the services and tools that we developed in this work are properly configured and aligned with the needs of the users.

This study therefore presents the architectural composition and design of all the tools and services available within the smart learning ecosystem. An aspect of the solution will show how a learner centered pedagogy such as digital storytelling and recent technologies such as blockchain and microsites have been implemented according to the universal accessibility standard [7], within a smart learning ecosystem to provide the much-needed learning support especially for the disadvantaged groups, but also anyone. Other aspects of the learning ecosystem are authoring services, learning management system (LMS) and content management system (CMS) services, and learning analytics services.

Being an integrated ecosystem, the solution presented in this paper will address some of the needs of the disadvantaged groups in the educational settings. Furthermore, the overall aim of the learning ecosystem is not only to inspire teachers but to provide cohesive learning support, realizing learning functions, providing learning contents and ensuring that all individuals within the learning ecosystem attain the full learning potential. The implementation of this novel ecosystem is covered under the auspices of EU-LAC project entitled, SELI.

2   Background

2.1   Smart Ecosystem

In general, an ecosystem is a community network of interactions between organisms and their environment [8]. According to [9], the term "smart ecosystem" describes a wide range of concepts, from networking architecture to service-based software solutions. The most frequent references to the smart ecosystem come from Artificial Intelligence research related to complex systems, including the biological ones [10].

The smart ecosystem is a mimic of biological ecosystems [9]. Thereby a smart ecosystem conforms several software applications and components which satisfy the properties of being robust, scalable, and self-organizing.

The smart ecosystem, like biological ones, consists of Agents, Populations, and Habitats. An agent is software and data able to move from one computer to another in an autonomous way keeping the ability to continue executing. The agent, as software, does not move but it gets copied from one computer to another; of course, including data. It is the behavior of agents to migrate in a biological ecosystem. Each software participant is considered an agent in a smart ecosystem. It is a piece of software that acts, for a user in a relationship of agency, autonomously in an environment to meet its designed objectives [11]. The habitats are the nodes of the smart ecosystem. Its function is analogous to the ones in the biological ecosystem. A node is related to a user computer or server in the network with agent station properties. The populations are represented by the agents migrating between habitats. An agent will use a context on each habitat. The context defines an environment with primary data relevant to the applications; it is called an entity and can be software, sensors, services, intelligent objects connected to the habitat.

The concept of a smart ecosystem has been used in education for several years to describe environments for production, reuse and adaptation of content [12]. However, with the development of technology, notably blockchain, this concept has evolved into a smart ecosystem characterized by being an open community, where there is no need for centralized control. In a smart ecosystem there is no leadership structure in response to the need of the dynamics of the environment. A smart ecosystem trusts open coding and the incorporation of systems based on their aggregation, incubation, and facilitation.

The essential components of a smart ecosystem are infrastructure, services and users.

Users are part of learning communities, that can be individual or groups of individuals interacting and collaborating synchronously or asynchronously.

Services are the support for learning involving content management, pedagogical aspects, and instructional design. It includes, for example, learning content management system (LCMS), learning management system (LMS), content distribution system (CDS), recommendation system, digital libraries, learning analytics component, social networks among others. The work of Normak, Pata and Kaipainen [13] highlights that the impact of learning ecosystems on the pedagogical aspect is different from that produced by virtual learning environments. They point out that while in a virtual learning environment (LVE), the teacher foresees the activities their students must perform, in an ecosystem the activities to be carried out spontaneously during the course due to the interactions between students-tutors, students-material, students-technology, tutors-technology. In the LVE certain pedagogical paradigms are fixed and embedded in the instructional design before the realization of the course. In contrast, the interaction between students and facilitators causes the evolution of the pedagogical paradigms in a smart ecosystem. One of the challenging research topics as raised in [14] is to specify how to establish pedagogical paradigms in learning ecosystems.

Within the framework of learning ecosystems, there is a disagreement between the principles of self-regulated learning, which indicates that students must be independent, autonomous and self-directed. Their learning should arise from their interests and from situations that are meaningful to them, interacting with other students, also acting as teachers of others and participating in communities.

The infrastructure is based on web services or microservices capable of being interoperable. It will enable the traceability of the processes to be maintained while maintaining the security and confidentiality of the data. An example of interoperability can be seen among Google, YouTube, Facebook, Amazon Web Services, among others, where it is increasingly natural to move from one system to another sharing information.

Because of these qualities, one of the technologies that is gaining more followers for digital learning ecosystems is the blockchain [15, 16].

2.2   Open Educational Resources

Openness in education has originated in popular movements to democratize access to knowledge and to improve and equalize access to quality educational content to all people. Although it is not a new concept the idea of open education has increasingly become associated with digital content and practices [17].

In 2002, UNESCO coined the term "Open Educational Resources" and recommended the following definition: "the open provision of educational resources, enabled by information and communication technologies, for consultation, use and adaptation by a community of users for non-commercial purposes" [18]. In this context, Open Educational Resources (OER) should refer "teaching, learning and research materials in any medium, digital or otherwise, that reside in the public domain or have been released under an open license that permits no-cost access, use, adaptation and redistribution by others with no or limited restrictions. Open licensing is built within the existing framework of intellectual property rights as defined by relevant international conventions and respects the authorship of the work".

The degree to which a resource is shared varies and an open resource depends which open license is in effect. In this regard [20] proposed that open content such as OER should allow the 5 R's of openness (originally 4 R's): Reuse, Revise, Remix, Redistribute and Retain. The idea of 5R's reflects concern with the intellectual property and authorship, and therefore in licensing resources and granting of non-restrictive permissions. The Creative Commons licenses indicate how the original authors wish users to modify and share any new derivatives of their work.

Sharing educational materials such as OER, under a CC license, allows adaptations in the content to better fit the needs of the community, considering the diversity of the students, the different cultural and social contexts and even the language. With OER the didactic content can be updated more effectively, following the pace of changes inherent to some disciplines due to the emergence of new discoveries [21]. These updates do not necessarily need to be carried out by the original author, highlighting the collaborative work. The ability to distribute derivative versions of a resource helps keep course content current.

The openness characteristic of these resources guarantees access to knowledge for all, regardless of social class, contrasting with the high cost of printed textbooks giving full support to learning. In addition, it creates opportunities for students to contribute knowledge and build their own learning paths [22]. These open pedagogical practices expand collaborative, inclusive, accessible and active learning. In addition to disseminating knowledge, strengthening communities and promoting innovative pedagogies.

2.3 Accessibility Issues

Understanding the accessibility theme, as a right for all, reinforces its importance for the inclusion process, especially for people with disabilities. However, this understanding is not yet enough to guarantee the necessary specifications that allow access to digital space for all citizens, regardless of their diversity of perceptual, motor or cognitive conditions, for example [23]. In this sense, the SELI contemplates in its proposal the challenge of providing all users of its ecosystem with the accessibility necessary for the proposed learning to take place autonomously, respecting the diversity of each user and guaranteeing access to all available information content. In order to access to be guaranteed, it is necessary to take into account the offer of assistive technology and the available accessibility resources [24].

The project followed the concept of accessibility proposed by the Universal Design for Learning (UDL). The UDL proposal is anchored in architecture focused on human diversity. The construction of the ecosystem considered in the teaching-learning process respect for the specific need of everyone, favoring the acquisition of knowledge with autonomy, independence and consistency [25]. Among the implemented accessibility elements can be mentioned: recommendation to use simple and clear language, use of each element for its purpose (title to title styles, table for tabular data, for example), use of standardized and known icons internationally and the use of colors with a good contrast ratio. Description offer for images that transmit content, means that facilitate navigation through the keyboard, cleaner fonts (letters), better understanding, alternatives for audio and video (subtitles, textual transcription, sign language), always available to pause audio and video with automatic start, avoid time-limited activities.

The entire process was guided by the importance of accessibility in the digital space considering the recommended technical standards and especially the user experience, the context and the conditions of the people who will use the tool because only then is it believed that it will meet the specificities of each user and could be considered accessible [26].

3 Digital Ecosystems for Learning and Inclusion: Related Work

Several initiatives have been found in the literature on the use of technology to promote inclusion, whether for people with disabilities [27–30], immigrants [31, 32], elderly people and other groups that face barriers or some situation of vulnerability regarding access to digital technologies [33, 34].

However, there seems to be a gap in the availability of platforms and ecosystems that promote learning and inclusion. The use of the term "platforms-forms" and "ecosystems" here refers to a set of services made available to educators and students, in order to allow democratization and the inclusion of learning, using, for this purpose, an infrastructure of support, philosophical and conceptual foundations.

[35] propose a metamodel for the definition and development of learning ecosystems taking into account the context and the human factor as key elements, that is, the ecosystem must be in accordance with institutional policies, governmental and cultural organizations and also to the target audience.

[36] present the sMOOC, launched by the European project "E-learning, Communication, Open Data (ECO)", in the perspective of social inclusion. The project's main objective is to promote professional training for socially excluded people. A model was used, combining formal and non-formal learning activities and cooperation between participants, in order to generate a continuous flow of knowledge between platforms.

Also, worth noting is the LATin project (Latin American Open Textbook Initiative), which aims to minimize the problem of the high cost of textbooks for higher education in Latin America. Thus, this project aims to create a support architecture, methodologies and policies for the dissemination of open cooperative books, aimed at higher education [37].

This project differs from the others in the proposal to be an ecosystem that uses open educational resources aimed at supporting the teacher to create accessible digital teaching materials according to the type of target audience that the teacher wants to include in his courses. Thus, the ecosystem provides computational support for student accessibility and pedagogical support for teachers.

4   Smart Learning Ecosystem

SELI ecosystem provides a solution framework for the teaching-learning process in order to promote inclusive education. As shown in Fig. 1, it is divided into four views: philosophical foundations, supporting infrastructure, concept and service bus.

4.1   Philosophical Foundations

Shared Pedagogy
The SELI project is based on the idea of exchanging experiences. These experiences are in a different way the content of education, methodical solutions, proven algorithms of pedagogical activities. Pedagogy is learning about educational ideals [38]. One of the ideals is the issue related to the transfer of knowledge and skills. Each of the teams involved in the project is responsible for preparing one or two courses dedicated to future teachers or current educators. The courses contain tested content, verified in terms of usability. The content is selected in a way that ensures the transfer of knowledge, skills and attitudes between the countries involved in the project. The courses are selected and designed to convey what is most valuable in a given country.

The selection of contractors for the action included several important keys, among others: topicality of the subject matter, reported needs of the platform's recipients (key and current challenges), as well as designing activities aimed at pedagogical practice (social change) [39]. In this case, it should be stressed that the idea of knowledge sharing lies at the heart of research and scientific activities. Pedagogy is a specific discipline that aims to strengthen competences [40]. This development is possible thanks to the idea of common learning regardless of cultural or language differences. Therefore, all content included in the ecosystem is created in two language versions (the original one is addressed to the local audience) and English. The application of the strategy of using two languages results from the simple fact of making the content more accessible to end users. Moreover, in many cases the design of content in the local language is due to the competence constraints of students and teachers (level of English use). Courses designed within the SELI framework concern, among others: the use of the ecosystem to work with migrants, disabled, digitally excluded [41], having problems with cyberbullying [42] or bilingual people. All courses will be available in one place - the SELI ecosystem. Each course will be the responsibility of a different country involved in the project. The idea of sharing the best educational solutions has a real stamp in this case.

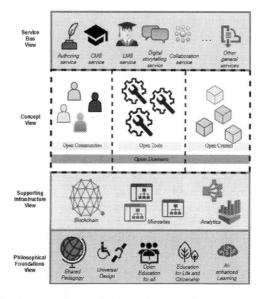

Fig. 1. Architecture of the smart learning ecosystem [43].

Universal Design. Designing an educational platform requires taking into account the needs of end users. Firstly, such a platform must meet the criterion of language accessibility [44]. Therefore, all content on the platform is bilingual, i.e. in the national language and English. The switchover to the language is automatic due to the location of the ISP. The second and very important issue is the design of a clear menu to allow for familiarization with the content and the creation of educational content [45]. The platform therefore has a clear menu, which is adapted to a diverse audience. This means that both a person with a low level of digital competence and a high level of competence can effectively operate the platform [46]. Each of the graphic content is properly described for the speech synthesizer (the needs of blind people). Content layout planning has been tested and evaluated in different groups. The work on designing the platform taking full account of the needs of all its users is not an easy task also due to the high level of organizational and cultural diversity of the teams. In addition, the platform's assumptions require extensive testing in each country, which prevents short project duration. Nevertheless, every effort has been made at the conceptual, execution and implementation stages in terms of testing, error detection and correction.

Open Education for All. The issue of openness of resources is one of the main objectives of the SELI ecosystem. This assumption is based on several facts. Firstly, the ecosystem is built from public sources. Financing of the activity by national research agencies requires ensuring access for all citizens. Secondly, the effort put into preparing the learning environment requires the widest possible dissemination of the results. The functioning of the ecosystem is therefore open to everyone without territorial or linguistic restrictions (use of universal English). The third reason is to implement the idea of creating social change through ICT [47]. The teams participating in SELI hope that the use of the courses will contribute to solving selected problems in individual countries, e.g. relating to counteracting cyber-bullying, digital inclusion, working with elderly people using ICT, immigrants, or educational work with bilingual children. Each topic is prepared in a way that is accessible not only to professionals, but also to parents and learning communities (e.g. academia). It is assumed that open education does not require formal exams or admission requirements, but is based on the accumulation of human capital, whose repository becomes an efficient ICT tool [48]. It is also worth adding that all courses and architecture have been prepared in accordance with trends - styles of ICT use [49]. Thus, all open resources do not require special access devices [50]. They can be run not only on computers, but also on smartphones. Openness lies at the heart of the SELI project, setting out detailed directions.

Education for Life and Citizenship. Each of the prepared courses is characterized by some universal features, despite thematic differences. One of the main goals of education is to improve the quality of life [51]. This idea is the basis of the SELI ecosystem. Each course is oriented towards two groups: the target group (with deficits, problems, challenges) and the intermediate group (teachers, trainers, students, parents). The learning of the courses enables the generation of change, which is ultimately to translate into a better quality of life for disadvantaged groups. The content designed within the SELI ecosystem is designed for the acquisition of new competences by both the target and intermediate groups. The topics selected for implementation do not belong to the group of basic research, but are very close to social needs, deficits in society. Participation of each group, learning, improving quality of life is also one of the determinants of social integration and full inclusion. The choice of topics is not accidental, as it is oriented towards groups requiring social and educational support. The SELI ecosystem becomes a unique and important tool not only providing integrated knowledge in several areas, but also showing fragments of social reality and groups requiring educational and social support.

AI-Enhanced Learning. AI-enhanced Learning aspects of this paper are related to the Learning Analytics infrastructure [52]. All information about students' interaction with course content is stored and new predictive models of students' success are built over these pieces of information. This is done using Machine Learning techniques, which involve different techniques, which vary from simple Supervised Learning methods (as Neural Networks, for instance) to more up-to-date techniques, like Explainable AI (XAI) [53]. By using this, the appearance of correlations between different habits of use (e.g. best time to perform a lesson, or the importance of finishing an optional activity in the final grade) and course contents, tasks and overall organization can emerge. In this context, Learning Analytics refer to storing and retrieving behavioral data collected from users while interacting with course contents or tasks to better understand a wide variety of aspects related to students' learning processes. In the context of this paper, the SELI ecosystem records a reduced, but sufficient set of users' data, like: time and address of login; activities clicked, selected and completed; number of dropouts; access control to static content; and other data stored on logs and the ecosystem's database. The main goal of applying Learning Analytics strategies over these pieces of information is to provide different reports with valuable information for teachers, students and other educational agents. A long-term goal for this ecosystem is to provide predictive analysis to support students with learning difficulties.

4.2   Infrastructure View

Key concepts about the supporting infrastructure of the environment include Blockchain, Microsites, and Analytics.

- **Blockchain.** According the EU report 2018, about blockchain in education defined key characteristics about this technology. These characteristics are, namely: (a) the end of paper-based system for certificates (Trust and Immutability), (b) automatically verify the validity of certificates (Transparency & Provenance, Disintermediation), (c) users on control of their data (Self-sovereignty), and (d) blockchain-based cryptocurrencies in education business models, see [16]. In addition to these characteristics, blockchain implementations should also include: (i) open source software, (ii) open standards for data, and (iii) implementing self-sovereign data management solutions. However, most solutions applied in education do not fulfil all these criteria.

  The decentralized and distributed nature of blockchain, as well as, the ability to run smart contracts, make the implementation of this technology interesting for the SELI project because it aligns well with the objectives of the project and its way to smarter interactions.

  The adoption of free and open source code, it has been a key mechanism to share details about the implementation of our system between all partners. After finishing a course, certificates are generated and stored as Smart contracts and non-monetary transactions on a private Blockchain network (ethereum.org). Students can access their milestones through the navigation menu, like Fig. 2.

Fig. 2. Front-end for presentation of student's certificates

- **Microsites**. There exist numerous tools available for generating class materials (e.g. EXeLearning). Nevertheless, one of the identified problems with these tools is related to encapsulation and recording internal state of used objects. Take as example a lesson including many webpages generated with EXelerning (packaged with SCORM), even if an internal questionnaire is answered by a student, no internal state is recorded. To address this problem of not recording some data, SELI ecosystem uses H5P objects because they can store more data about inputs from the end-user. The main challenge is to create and implement a flexible architecture to accept different types of objects from author tools and provide them with additional data gathering. H5P objects make functionality portable through the use of Java-Script language and objects. For this reason, the selected framework for developing the environment includes Node.JS + React + Meteor 4, and new non-relational databases like mongoDB.
- **Analytics**. Learning analytics refer to storing behavioral data from users to better understand how students interact with materials generated by the instructor. These data are not different from the information recorded by social media ecosystems. SELI environment records: time and address of login, activities clicked and selected, number of dropouts, and other data stored on logs and the database. The main goal is to generate different reports with valuable information for the teacher and students. A long-term goal is to provide predictive analysis to reinforce students with problems.

4.3   Concept View

Open Communities. They refer to the communities of practice, including all actors and stakeholders that will take part of any SELI-enabled educational processes. Roles must be considered instead of personified entities, so that besides teachers and students, a wide variety of roles could take part of these communities, from media and content creators to academic staff, being them or not active users of the digital platforms and tools provided by SELI ecosystem.

Open Tools. SELI ecosystem is designed over a mashup of open digital tools, and its architecture is open enough to any open digital solution to be added and integrated with the already existent ones. The first open tools already implemented inside the ecosystem comprise: a course authoring tool, a course management tool and a digital storytelling tool. Given the open nature of SELI, external open tools like those one provided by other open ecosystem could be easily incorporated by SELI. For instance, from LATIn Ecosystem [37], an open social network tool, a collaborative writing tool and a publishing tool are meant to be aggregated to SELI project.

Open Content. All content created inside SELI Ecosystem is meant to be freely and openly available, under a Creative Commons-like license. In the same way, any other kind of open content made available outside SELI platform can be incorporated by it, following the 5R freedoms. In this way, any external open content could be accommodated by SELI Ecosystem's own open repository.

4.4   Services Bus View

Authoring Service. The purpose of the authoring service is to offer the teacher digital resources that can support the creation of media for the construction of digital didactic material, as well as to support the adaptation of this material to include accessibility to didactic material. An example of creating didactic material for the ecosystem is shown in Fig. 3. In this figure, the teacher can insert different media, such as text, images, links, videos, audios, created by him/her or downloaded from the Internet with open access, for example, as well as a screen specifying a media type (Fig. 4). It allows teachers to create lesson strategies that can be used according to specific declines of students with disabilities, for example. media can meet the accessibility criteria established by the literature and the W3C [54]. Thus, they can insert aspects of accessibility, such as: descriptive text, audio description, sign language, etc. to improve the inclusion of digitally disadvantaged groups. In addition to the media, the teacher can choose activities that the student must deliver, such as tasks and storytelling, in order to verify the skills acquired by the students.

CMS and LMS Services. After the teacher publishes his courses, the CMS service is responsible for making them available to students. In the CMS the student can choose between courses of different levels of learning, in different languages. Then, the course can be saved and published on the ecosystem for student access, which is done through microsites that are linked to the CMS.

The microsite is a web page that will present the course proposed by the teacher with all the didactic content, regardless of the creation tool. This course can be run on different platforms on different devices. An example of published courses is shown in Fig. 5.

In addition to offering courses through the CMS service, the LMS service allows the offer of various types of activities, such as questionnaires and storytelling. It is also possible for the teacher to configure the evaluation criteria for each activity, in addition to monitoring the student's performance through the learning analysis component.

Fig. 3. Initial screen to create a course

Fig. 4. Image settings

Fig. 5. Published courses

Collaboration Service. These services refer to the possibilities of collaboration among actors that are facilitated by SELI Ecosystem through its tools. They could be exemplified by the internal open tools, as the collaborative construction of digital stories, or even external open tools, like LATIn's collaborative writing tool.

Digital Storytelling Service. Digital storytelling is the art of telling stories or narratives by using any multimedia mean whether it is graphics, a program or a presentation. The ecosystem presents this service, according to Fig. 6. The story flow allows user to upload multimedia such as image, video, audio, text in one screen. It also allows user to translate text into five different languages (English, Polish, Portuguese, Spanish, Turkish) which is available as subtitle when shared in a learning activity or social media.

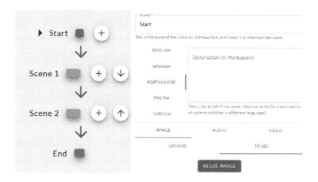

Fig. 6. Storytelling board view.

5 Conclusions

This paper presented the architecture of a smart ecosystem for learning and inclusion that is being developed by a group of 10 countries from Europe and Latin America. The details of the four views that make up architecture were presented: philosophical foundations, support infrastructure, concept and service bus.

This ecosystem aims to allow teachers to create their courses (content and activities) more easily and may include accessibility resources. Students, with or without disabilities, can access these materials available on a web platform. Thus, the objective of increasing the democratization of access to education can be achieved.

The ecosystem has been being tested since November 2019 in three different countries (Brazil, Dominican Republic and Uruguay). Teachers from public and private schools, accessibility specialists are using the authoring tool to build their courses. The data collected is being used to make improvements to the ecosystem by the development team in conjunction with the usability and accessibility team.

Further work includes collecting users' opinions about the ecosystem and analyzing users' data and start to use with students (with or without disabilities).

Acknowledgement. This work was supported by the ERANET-LAC project which has received funding from the European Union's Seventh Framework Programme. Project Smart Ecosystem for Learning and Inclusion – ERANet17/ICT-0076SELI. The work was also supported by the Coordenação de Aperfeiçoamento de Pessoal de nível superior - Brazil (CAPES) - Programa de Excelência - Proex 1133/2019 and Fundação de Amparo à Pesquisa do Estado de São Paulo (FAPESP) 2018/04085-4.

References

1. Tomczyk, Ł., et al.: Digital Divide in Latin America and Europe: main characteristics in selected countries. In: 2019 14th Iberian Conference on Information Systems and Technologies (CISTI), pp. 1–6. IEEE, June 2019
2. Van Seters, J.R., Ossevoort, M.A., Tramper, J., Goedhart, M.J.: The influence of student characteristics on the use of adaptive e-learning material. Comput. Educ. **58**(3), 942–952 (2012)
3. Martins, V., et al.: A blockchain microsites-based ecosystem for learning and inclusion. In: Brazilian Symposium on Computers in Education (Simpósio Brasileiro de Informática na Educação-SBIE), vol. 30, no. 1, p. 229, November 2019
4. Oyelere, S.S., Tomczyk, L., Bouali, N., Agbo, F.J.: Blockchain technology and gamification-conditions and opportunities for education. In: Adult Education 2018-Transformation in the Era of Digitization and Artificial Intelligence (2019)
5. Tomczyk, L., et al.: Flipped learning, digital storytelling as the new solutionsin adult education and school pedagogy. In: Adult Education 2018-Transformation in the Era of Digitization and Artificial Intelligence (2019)
6. Jonassen, D.H., Carr, C., Yueh, H.-P.: Computers as mindtools for engaging learners in critical thinking. Tech Trends-Washington DC **43**, 24–32 (1998). https://doi.org/10.1007/BF02818172
7. Martins, V.F., et al.: Accessibility recommendations for creating digital learning material for elderly. In: 2019 XIV Latin American Conference on Learning Objects and Technology (LACLO). IEEE (2019)
8. Briscoe, G., De Wilde, P.: Digital ecosystems: evolving service-orientated architectures. In: Proceedings of the 1st International Conference on Bio Inspired Models of Network, Information and Computing Systems, p. 17-es, December 2006
9. Briscoe, G., Sadedin, S., De Wilde, P.: Digital ecosystems: ecosystem-oriented architectures. Nat. Comput. **10**(3), 1143 (2011). https://doi.org/10.1007/s11047-011-9254-0
10. Ray, T.S.: An evolutionary approach to synthetic biology: Zen and the art of creating life. Artif. Life **1**(1_2), 179–209 (1993)
11. Wooldridge, M.: An Introduction to Multiagent Systems. Wiley, New York (2009)
12. Chang, E., West, M.: Digital ecosystems a next generation of the collaborative environment. In: iiWAS, vol. 214, pp. 3–24 (2006)
13. Normak, P., Pata, K., Kaipainen, M.: An ecological approach to learning dynamics. J. Educ. Technol. Soc. **15**(3), 262–274 (2012)
14. Motz, R., Rodés, V.: Pensando los Ecosistemas de Aprendizaje desde los Entornos Virtuales de Aprendizaje. In: Conferencias LACLO, vol. 4, no. 1 (2013)
15. Yumna, H., Khan, M.M., Ikram, M., Ilyas, S.: Use of blockchain in education: a systematic literature review. In: Nguyen, N.T., Gaol, F.L., Hong, T.-P., Trawiński, B. (eds.) ACIIDS 2019. LNCS (LNAI), vol. 11432, pp. 191–202. Springer, Cham (2019). https://doi.org/10.1007/978-3-030-14802-7_17

16. Yakovenko, I., Kulumbetova, L., Subbotina, I., Zhanibekova, G., Bizhanova, K.: The blockchain technology as a catalyst for digital transformation of education. Technology **10** (01), 886–897 (2019)

17. Havemann, L.: Open educational resources. In: Peters, M.A. (ed.) Encyclopedia of Educational Philosophy and Theory, Living edn. Springer, Singapore (2016). https://doi.org/10.1007/978-981-287-588-4_218

18. UNESCO: Forum on the Impact of Open Courseware for Higher Education in Developing Countries, Paris, 1–3 July 2002: Final report (2002). https://unesdoc.unesco.org/ark:/48223/pf0000128515

19. UNESCO: OER Declaration. In: World Open Educational Resources Congress, Paris (2012). (in English). https://unesdoc.unesco.org/ark:/48223/pf0000246687

20. Wiley, D.: David Wiley, "The Access Compromise and the 5th R". An Open Education Reader (2014). https://opencontent.org/blog/archives/3221

21. Imberman, S., Fiddler, A.: Share and share alike: using Creative Commons licenses to create OER. ACM Inroads **10**(2), 16–21 (2019)

22. Green, C., et al.: 7 Things You Should Know About Open Education: Practices. EDUCAUSE Learning Initiative (ELI) (2018)

23. Silva, R.L., De la Rue, L.A.: A acessibilidade nos sites do Poder Executivo estadual à luz dos direitos fundamentais das pessoas com deficiência. Rev. Adm. Pública **49**(2), 315–336 (2015). https://doi.org/10.1590/0034-7612130130

24. Brown, J., Hollier, S.: The challenges of Web accessibility: the technical and social aspects of a truly universal Web. First Monday **20**(9) (2015). https://doi.org/10.5210/fm.v20i9.6165

25. Hromalik, C.D., Myhill, W.N., Carr, N.R.: "ALL Faculty Should Take this": a universal design for learning training for community college faculty. TechTrends **64**(1), 91–104 (2019). https://doi.org/10.1007/s11528-019-00439-6

26. Shaheen, N.L., Lazar, J.: K–12 technology accessibility: the message from state governments. J. Spec. Educ. Technol. **33**(2), 83–97 (2018). https://doi.org/10.1177/0162643417734557

27. Terrazas-Arellanes, F.E., Gallard M, A.J., Strycker, L.A., Walden, E.D.: Impact of interactive online units on learning science among students with learning disabilities and English learners. Int. J. Sci. Educ. **40**(5), 498–518 (2018)

28. Aunio, P.: Finnish digital learning support for children with learning difficulties in mathematics. In: New Ways to Teach and Learn in China and Finland: Crossing Boundaries with Technology, pp. 136–149 (2016)

29. Sousa, M.J., Rocha, Á.: Digital learning: developing skills for digital transformation of organizations. Future Gener. Comput. Syst. **91**, 327–334 (2019)

30. Galasso, B.J.B., Lopez, M.R.D.S., Severino, R.D.M., Teixeira, D.E.: Process of production of bilingual didactic materials of the national institute of education for the deaf. Revista Brasileira de Educação Especial **24**(1), 59–72 (2018)

31. Colucci, E., Smidt, H., Devaux, A., Vrasidas, C., Safarjalani, M., Castaño Muñoz, J.: Free digital learning opportunities for migrants and refugees. An Analysis of current initiatives and recommendations for their further use. JRC Science for Policy Report. Publications Office of the European Union, Luxemburg (2017). https://doi.org/10.2760/684414

32. Castaño-Muñoz, J., Colucci, E., Smidt, H.: Free digital learning for inclusion of migrants and refugees in Europe: a qualitative analysis of three types of learning purposes. Int. Rev. Res. Open Distrib. Learn. **19**(2) (2018)

33. Lenstra, N.: The community-based information infrastructure of older adult digital learning. Nord. Rev. **38**(s1), 65–77 (2017)

34. Wong, A., Wu, J., Lo, K.: Breaking the Digital Divide for the Elderly through Service Learning and Data Analytics: A Proposal (2017)

35. Marta-Lazo, C., Osuna-Acedo, S., Gil-Quintana, J.: sMOOC: a pedagogical model for social inclusion. Heliyon **5**(3), e01326 (2019)
36. García-Holgado, A., García-Peñalvo, F.J.: A metamodel proposal for developing learning ecosystems. In: Zaphiris, P., Ioannou, A. (eds.) LCT 2017. LNCS, vol. 10295, pp. 100–109. Springer, Cham (2017). https://doi.org/10.1007/978-3-319-58509-3_10
37. Ochoa, X., Silveira, I.F., Sprock, A.S.: Collaborative open textbooks for Latin America-The LATIn project. In: International Conference on Information Society (i-Society 2011), pp. 398–403. IEEE, June 2011
38. Ryk, A.: W poszukiwaniu podstaw pedagogiki humanistycznej. Oficyna Wydawnicza Impuls, Kraków (2011)
39. Tomczyk, Ł., Ryk, A., Prokop, J.: Proceedings New Trends and Research Challenges in Pedagogy and Andragogy NTRCPA18. Pedagogical University of Cracow, Cracow (2018)
40. Novković Cvetković, B., Stošić, L., Belousova, A.: Media and information literacy-the basis for applying digital technologies in teaching from the discourse of educational needs of teachers. Croatian J. Educ. Hrvatski časopis za odgoj i obrazovanje **20**(4), 1089–1114 (2018)
41. Tomczyk, Ł., et al.: Digital divide in Latin America and Europe: main characteristics in selected countries. In: 2019 14th Iberian Conference on Information Systems and Technologies (CISTI), pp. 1–6. IEEE (2019)
42. Tomczyk, Ł., Włoch, A.: Cyberbullying in the light of challenges of school-based prevention. Int. J. Cogn. Res. Sci. Eng. Educ. (IJCRSEE) **7**(3), 13–26 (2019)
43. Oyelere, S.S., et al.: Digital storytelling and blockchain as pedagogy and technology to support the development of an inclusive smart learning ecosystem. In: Rocha, Á., Adeli, H., Reis, L.P., Costanzo, S., Orovic, I., Moreira, F. (eds.) WorldCIST 2020. AISC, vol. 1161, pp. 397–408. Springer, Cham (2020). https://doi.org/10.1007/978-3-030-45697-9_39
44. Kopecký, K., Hejsek, L.: Mobile touch devices as an effective tool of m-learning and e-learning. In: Proceedings of INTED, pp. 7934–7936 (2015)
45. Silveira, I.F., Omar, N., Mustaro, P.: Architecture of learning objects repositories. In: Learning Objects: Standards, Metadata, Repositories and LCMS, Informing Science Institute, Santa Rosa, CA, vol. 1, pp. 131–156 (2007)
46. Pianfetti, E.S.: Focus on research: teachers and technology: digital literacy through professional development. Lang. Arts **78**(3), 255–262 (2001)
47. Shor, I.: Empowering Education: Critical Teaching for Social Change. University of Chicago Press, Chicago (2012)
48. Stosic, L.: Does the use of ICT enable easier, faster and better acquiring of knowledge? Int. J. Innov. Res. Educ. **4**(4), 179–185 (2017)
49. Eger, L., Klement, M., Tomczyk, L., Pisonova, M., Petrova, G.: Different user groups of university students and their ICT competence: evidence from three countries in central Europe. J. Balt. Sci. Educ. **17**(5), 851–866 (2018)
50. Da Rosa, S., Motz, R.: Do we have accessible oer repositories? In: 2016 International Symposium on Computers in Education (SIIE), pp. 1–6. IEEE, September 2016
51. Ross, C.E., Van Willigen, M.: Education and the subjective quality of life. J. Health Soc. Behav. **38**(3), 275–297 (1997)
52. Gunning, D.: Explainable artificial intelligence (XAI). Defense Advanced Research Projects Agency (DARPA), nd Web, 2, 2 (2017)
53. Samek, W., Montavon, G., Vedaldi, A., Hansen, L.K., Müller, K.-R. (eds.): Explainable AI: interpreting, explaining and visualizing deep learning. LNCS (LNAI), vol. 11700. Springer, Cham (2019). https://doi.org/10.1007/978-3-030-28954-6
54. W3C: Web Content Accessibility Guidelines 2.0 (2008). https://www.w3.org/Translations/WCAG20-pt-PT/. Accessed July 2019

Evolutionary Approach of Clustering to Optimize Hydrological Simulations

Elnaz Azmi[1]([⊠]) (ID), Marcus Strobl[1] (ID), Rik van Pruijssen[2] (ID), Uwe Ehret[2] (ID), Jörg Meyer[1] (ID), and Achim Streit[1] (ID)

[1] Steinbuch Centre for Computing, Karlsruhe Institute of Technology, Karlsruhe, Germany
{elnaz.azmi,marcus.strobl,joerg.meyer2,achim.streit}@kit.edu
[2] Institute of Water and River Basin Management, Karlsruhe Institute of Technology, Karlsruhe, Germany
{rik.pruijssen,uwe.ehret}@kit.edu

Abstract. Modeling of hydrological systems and their dynamics in high spatio-temporal resolution leads to a better understanding of the hydrological cycle, thus it reduces the uncertainties in hydrologic forecasts. Simulation of such high-resolution, distributed and physically based models demands high performance computing resources. However, the availability of such computing resources is restricted in some domains. In this paper, we propose an approach to reduce computational costs by reducing hydrological model redundancies using similarities in functionality of hydrological model units. The approach applies K-Means clustering to detect similar model units and simulates only one representative unit of each cluster. The clustering is applied when rainfall is forced to the hydrological system and is based on the structure, current state and flux of the model units. Application of this evolutionary approach on a test case results in a 1.8x speedup over the original simulation run time and the RMSE of 0.0049 compared to the original simulation output.

Keywords: Clustering · K-Means · Time series analysis · Simulation

1 Introduction

Physically based and highly detailed models of environmental processes are used to improve the understanding of the nature of hydrological systems [22]. Such models are spatially heterogeneous and consist of a hierarchy of units [7,26]. Thus, the simulation of these models in high spatio-temporal resolution is compute-intensive. One of the popular methods to tackle this issue is to use high performance computing and parallel processing of hydrological model units [9,12,14]. However, the parallelization of these models is challenging because of their heterogeneous nature and a demand of partially sequential execution of the model units. Also the interconnection of model units can be very tight, and the necessary communication of the units per time step makes efficient parallelization challenging. Additionally, these methods require programming expertise of

© Springer Nature Switzerland AG 2020
O. Gervasi et al. (Eds.): ICCSA 2020, LNCS 12249, pp. 617–633, 2020.
https://doi.org/10.1007/978-3-030-58799-4_45

domain scientists and partially revision of the existing modeling software. In this work, we introduce an approach to make use of redundancies in the simulated hydrological systems in order to decrease the computational effort of such simulations. The redundancies are due to the natural hydrological behavior of the model units and the simplification caused by the model choice. The remainder of this paper is structured as follows: Sect. 2 provides further information about the study background, Sect. 3 is a survey of related work, the proposed approach is explained in Sect. 4. In Sect. 5, the processing results are presented, Sect. 6 is about the implementation environment and the conclusions are drawn in Sect. 7.

2   Background

In this paper we apply our method on the CAOS (Catchment as Organized Systems) model proposed in [26]. This model simulates water related dynamics in catchments up to hundreds of square kilometers. The CAOS model provides a high-resolution and distributed process based simulation of hydrological systems. In this model, functioning of catchments, defined as a closed area draining completely to a single point along a river (the catchment outlet), is controlled by a hierarchy of three major model units, namely, Elementary Functional Unit (EFU), Hillslope (HSL), and River element (RIV). EFUs are soil columns containing other sub-units to transfer water through soil layers and other EFUs. HSLs are subsets of hills, and independent from each other. They contain several EFUs and have a connection point to a RIV. Finally, RIVs are linear elements along the lower edge of a HSL. They are parts of a river, connected sequentially to each other and transport the water of a catchment to the lowermost point, the catchment outlet (Fig. 1).

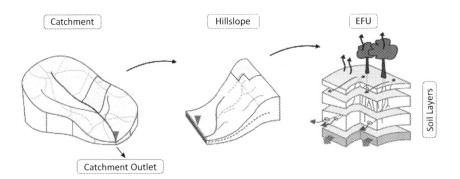

Fig. 1. Simplified hierarchy of the CAOS model units (modified after [26]).

The structure of the model is specified by domain knowledge. The resulting simulation dynamics depend on, 1. model units properties (static), 2. model units state (current discharge) and 3. forcing (rainfall or radiation). The underlying principle is that similar properties, states and forcings of the model units lead to similar simulation dynamics [26] which lead to redundancies that can be removed from the computation process. In order to examine this hypothesis, the Wollefsbach catchment [24] is used to develop and test of the evolutionary approach on the CAOS model. It is located in the Attert basin in Luxembourg with an area of 4.5 km^2 and 174 HSLs (Fig. 2). For this catchment we had access to the model and the required data set available in the CAOS project [26].

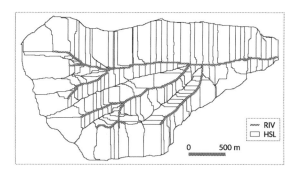

Fig. 2. Wollefsbach catchment divided into HSLs, delineated in black. Each HSL has an edge connected to a RIV in blue. (Color figure online)

3   Related Work

In environmental science and especially in hydrology, the identification of similarities plays a key role to detect patterns in the environment, extract correlations and improve future events forecast [1,5,6,20,26]. In order to define the hydrologic similarity, [23] suggested a framework that is both descriptive and predictive. Their metrics to define hydrologic similarity or dissimilarity between catchments include static characteristics and dynamic response of a catchment to its forcing. They discussed the demand for a catchment classification system based on the structure and hydro-climatic conditions of catchments as well as their functional response to the precipitation input. Following this, [20] derived signatures from precipitation-temperature-streamflow data to apply a Bayesian clustering and to identify groups of similar catchments. In the evolutionary approach, we include such similarity metrics using drainage time series of the HSLs within a catchment to represent static properties like structure, size, slope, soil profile and drainage of the HSLs. Their detection of similarities has been done mostly at catchment level, while we apply this process at HSL level within a catchment in order to reduce the recurring simulation properties (statics and dynamics).

Classification and clustering are machine learning techniques that identify groups of similar objects using already labeled data or object neighborhood properties like distance and density [11,15]. Such methods provide efficient pattern detection and help in a better understanding of hydrological systems [13,21]. Due to the lack of labeled data we chose a clustering method to detect similar HSLs. Selection of an appropriate clustering method depends on several parameters like the type of input data and clustering output, scalability and robustness, and thus it is use case dependent. [10] compared the hierarchical clustering fuzzy C-mean (FCM) and K-Means to analyze regional flood frequency and its underlying distribution. The results of both clustering methods for their application were almost similar so they concluded the choice of the best clustering method depends on the individual use case. [16] presented a clustering-based classification of climate data that resulted in internally more homogeneous and externally more distinct climate types than the types in the rule-based Köppen-Geiger classification, which is the de facto standard in global climate classification. [27] successfully defined regions with clear boundaries of homogeneous precipitation regions with highly varied spatio-temporal patterns using K-Means on a gridded dataset for automatic delineation. [3] tested different clustering methods to identify similar hydrological model units on the CAOS model where K-Means performed best. Thus, for our use case, we use K-Means clustering as well.

4   Methodology

In this work, we introduce an evolutionary approach to speed up the hydrological simulation which consists of two steps, namely, initial clustering and evolutionary clustering. The idea behind this approach is to reduce the computational costs by reducing redundant computations and calculating a close approximation of the original model dynamics. Using hydrological similarity [6], we distinguish similar model units considering their structure, current state and flux to detect clusters in the whole system. In the hierarchy of the CAOS model, dynamics of each HSL are independent of the other HSLs. Thus, we use them as individual objects to apply the K-Means clustering. Afterwards, we select a representative HSL of each cluster and execute the simulation only on the representatives. The next step is to map the output of the representatives to the remaining cluster members. This way, we avoid running the simulation for all model units and consequently, reduce the computation time. The degree of fluctuation of the simulation output can be controlled by the number of clusters and the frequency of applying clustering. Finally, we compare our results from the evolutionary approach with the results from the original simulation respecting execution time and simulation output. In Fig. 3, the whole approach is delineated step by step in detail. The original simulation consists of the major steps shown in the white boxes (Fig. 3): 1. loading the catchment structure and creating a list of processes (dynamics) between the model units, 2. starting the simulation for a predefined number of time steps (n), 3. running the simulation according to the processes list, and 4. saving the output and finalizing the simulation. The evolutionary approach shown in the gray boxes (Fig. 3) is described further in detail.

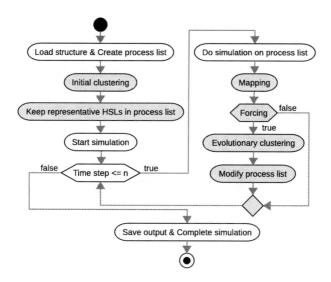

Fig. 3. Simulation workflow consists of the original and evolutionary approach.

4.1 Initial Clustering

In order to define a representation of the static properties of the HSLs, namely, structure, size, slope, soil profile and drainage, we run a drainage test [3]. The drainage test results in discharge time series for each HSL. These time series are hydrologic characteristics that provide an insight into the functionality of the HSLs. We extract seven features from these time series which are the input data for the clustering method. The features are the four moments *Mean, Variance, Skewness, Kurtosis* and the three hydrologically significant features *1st Gradient, Active Storage* and *Time to Equilibrium* [3]. In this feature set we identified outliers in the *Mean, Kurtosis, 1st Gradient* and *Active Storage* features (see arrows in Fig. 4). In the preprocessing step, we separate these outliers from the clustering and simulate them as single clusters. K-Means clustering requires the number of clusters (K) to be set. Further, we describe the parameter setting of the K-Means clustering used in the initial and evolutionary clustering.

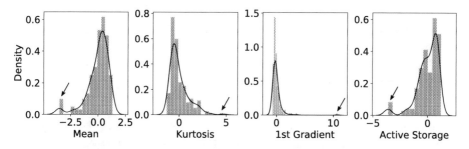

Fig. 4. Feature histograms and kernel density estimate fits.

4.2   Clustering Parameter Setting

The K-Means clustering package [18] that we used for our tests requires a set of parameters to be determined. The main parameter to be specified is the number of clusters (K). There are several methods like *elbow* [19], *silhouette* [25] and *RMSE-Computation-Time (rmse-ctime)* [3] methods to determine an appropriate value for K. The process of selecting an appropriate K is called here K-determiner. Another parameter to be set for the K-Means clustering is the *random seed* which determines the random number generation for centroid initialization. Thus, we use a fixed integer random seed (zero), to make K-Means deterministic, so that running it multiple times produces the same results. The following describes how the K-determiner works using the *elbow* and *rmse-ctime* methods.

Elbow Method. The *elbow* method runs K-Means clustering on a given dataset for a range of K values, and for each value of K, calculates the average distance from data points to the centroid of each cluster. As K increases, the average distance to the centroid decreases rapidly until the elbow or maximum curvature of the calculated curve [19] which is the optimal K (Fig. 5, left). We determined the elbow by using the point with the maximum distance from the straight line connecting the end-points of the curve. In order to save time, K-determiner runs the clustering for a set of Ks in a defined interval, interpolates the results and calculates the elbow point (Fig. 5, left).

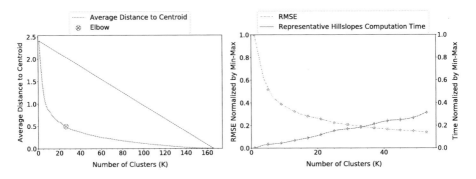

Fig. 5. Application of the *elbow* (left) and *rmse-ctime* (right) methods on the input feature set.

Rmse-Ctime Method. We have introduced another approach to determine the number of clusters at [3], that considers the balance between the number of clusters (K), Root Mean Square Error (RMSE) [4] of each cluster member HSL and the representative HSL of that cluster as well as the simulation computation time of HSLs. This technique is a customized *elbow* method, that allows scientists to decide which K to use whether they prefer a lower RMSE or a lower computation time. K-determiner applies this method for a set of Ks in an interval from one HSL to one third of all HSLs, and interpolates the points (Fig. 5, right). This reduces the run time of K-determiner through interpolation instead of clustering and calculating RMSE for all potential number of clusters. For our tests, K-determiner uses the intersection point of the curves, which is a balance of RMSE and computation time, as the appropriate K.

4.3   Representative Output Mapping

The initial clustering (Sect. 4.1) is followed by the selection of a representative HSL of each cluster. We define a cluster representative as the Medoid object whose average dissimilarity to other objects in the cluster is minimal. At each time step of the simulation, only the representative HSLs are simulated. The next step of the evolutionary approach is the mapping and scaling of the output of the representative HSL to the member HSLs of the same cluster. Therefore, the discharge (output) of the representative HSL, already computed in the simulation, is used to calculate the discharge of other HSLs using the following equation:

$$CMHD = RHD \times \frac{CMHA}{RHA} \qquad (1)$$

where CMHD is the cluster member HSL discharge $[\frac{m^3}{s}]$, RHD the representative HSL discharge $[\frac{m^3}{s}]$, CMHA the cluster member HSL area $[m^2]$ and RHA the representative HSL area $[m^2]$.

4.4   Evolutionary Clustering

According to the domain knowledge, the dynamics of the simulation depends on the static properties, current discharge of HSLs and the amount of rainfall enforced to the HSLs. In the evolutionary approach, first we include the static features of the HSLs by running the initial clustering. In the next steps of the approach, while the simulation is running, we add two features, namely, current discharge and flux to our feature set. Flux is defined as the volume of rainfall enforced to the area of HSLs in a given time. Hence, we define our evolutionary clustering as a clustering method that uses a new feature set dynamically during the simulation. The detailed steps of the evolutionary approach are:

a) Determine the initial K using the K-determiner and do the initial clustering.
b) Select representative HSL of each cluster and simulate the first time step.
c) Do the output mapping and update the status of all HSLs.
d) If there is no forcing at the next time steps, continue running the simulation with the already defined representatives.
e) When forcing starts, use the K-determiner, run the evolutionary clustering and continue the simulation with a new set of representatives.

f) Do step e) without the K-determiner for time steps in the time frame with active forcing (forcing time block) because the values of the feature set change strongly with variable forcing over time.

g) When forcing stops, use K-determiner, run the evolutionary clustering and continue the simulation with a new set of representatives until the next forcing time block is reached.

5   Processing Results

In this section, we show the results of the proposed approach applied on the study case (Wollefsbach). There are metrics available to evaluate the quality of the evolutionary approach, i.e. how close our results are to the original simulation. We use three metrics, alone or in combination, to show the quality of the evolutionary approach, namely, RMSE, Pearson Correlation Coefficient (PCC) [17] and the Kling-Gupta Efficiency (KGE) [8]. KGE is a measure of the goodness-of-fit, commonly used in hydrological modeling. In addition to the quality metrics, the computational efficiency of the evolutionary approach is presented as the simulation run time speedup in comparison to the run time of the original simulation. Values of the RMSE closer to zero shows a better estimation of the model results. PCC and KGE values closer to one indicate higher efficiency. All evaluation results are shown in Tables 1, 2, 3, 4, 5 and 6 which are sorted based on RMSE in ascending order. In addition, as simulation of time blocks with forcing is more compute intensive, we also show the RMSE for time blocks with forcing (RMSE-WF) and without forcing (RMSE-WOF) separately. The best values of the metrics are shown in bold type.

5.1   Influence of Random Seed

The original simulation executes processes of the model units in random order to keep the model close to the natural behavior of a hydrological system. Change of the random seed in the original simulation results in slightly different curves. We run the original simulation four times with different random seeds to test the randomness in the nature of the model. In Fig. 6, the horizontal axis shows the simulation time of one week (1st to 7th of January). The left vertical axis shows the discharge at the catchment outlet after the simulation and the right vertical axis from top to bottom the amount of rainfall during the simulation. The gray band shows the minimum and maximum of all tests. The curve labeled "Original" is used as ground truth for all following tests, hence all evaluation results (Table 1, 2, 3, 4, 5 and 6) are relative to this curve. Table 1 shows the evaluation of the remaining three tests relative to the ground truth.

In addition to the randomness of the original simulation, the evolutionary approach uses the K-Means clustering that generates its initial centroids randomly. In order to retain reproducible test results, we set the *random seed* parameter from K-Means clustering to zero as well. However, to evaluate the

Table 1. Evaluation results of the original simulation with different random seeds.

Tests	RMSE	RMSE-WOF	RMSE-WF	PCC	KGE
Test-1	**0.00155**	**0.00136**	**0.00214**	**0.990**	0.960
Test-3	0.00163	0.00143	0.00221	0.988	**0.983**
Test-2	0.00213	0.00191	0.00281	0.984	0.954

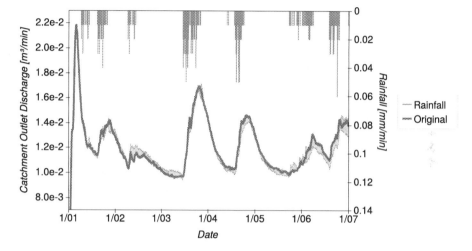

Fig. 6. Output of the original simulation with different random seeds.

influence of a variable random seed on the simulation result, we run six simulations (Test-1 to Test-6, Table 2) with different random seeds set for the clustering (Fig. 7). Only the initial clustering with $K = 9$ without the evolutionary clustering was applied to these simulations. All tests, represented by the gray band, show a similar trend like the original simulation, and an acceptable KGE from the hydrological perspective (Table 2). The following sections describe detailed tests to evaluate our evolutionary approach with automatically set parameters.

Table 2. Evaluation results of the tests with different random seeds.

Tests	RMSE	RMSE-WOF	RMSE-WF	PCC	KGE
Test-6	**0.0045**	**0.0044**	**0.0050**	**0.920**	0.765
Test-4	0.0056	0.0050	0.0073	0.891	0.776
Test-3	0.0056	0.0054	0.0064	0.868	0.729
Test-1	0.0078	0.0075	0.0087	0.907	**0.803**
Test-5	0.0096	0.0091	0.0110	0.858	0.748
Test-2	0.0097	0.0091	0.0117	0.814	0.713

5.2   Constant-K

In order to reveal the similarities in the static model unit properties of the hydrological model, we have designed a test that applied only initial clustering at the first time step of the simulation without running the evolutionary clustering. The simulation continues using the representative HSLs of the initial clustering. The catchment outlet discharge is calculated as output for each time step (Fig. 8) and the evaluation results are shown in Table 3. We have repeated the approach for a range of Ks defined as a percentage of total HSLs (5 - 30% of HSLs corresponds to $K = 9$ - 50) to test the effect of parameter K on the simulation output. The curves follow the trend of the original simulation (Fig. 8). Evaluation of the results reveals that our quality measures have no strong correlation with K, and the RMSE order does not fit one by one to the KGE metric (Table 3). This means that the test with a lower RMSE is not always the more efficient one. However, the smaller the K is, the higher the speedup is. To further increase the speedup by keeping the RMSE value low and the PCC as well as KGE values high, we designed the following Variable-K tests.

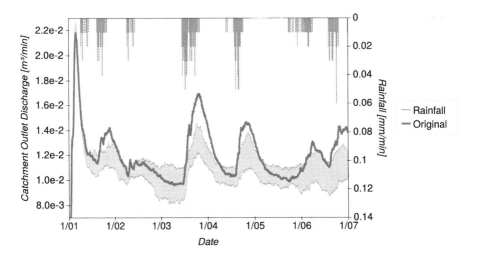

Fig. 7. Simulation output of the tests with different random seeds for K-Means.

5.3   Variable-K

In Sect. 5.2, we have shown tests running the initial clustering with the static properties of the model units. In the following, we include the effect of the current state and flux of the model units into our approach with the evolutionary clustering. In order to define the frequency of running the evolutionary clustering during the simulation, we considered the intensity of the dynamics occurred. Since more dynamic processes run when there is forcing due to rainfall in the

Table 3. Evaluation results of the Constant-K tests.

Tests	RMSE	RMSE-WOF	RMSE-WF	PCC	KGE	Speedup
K-42	**0.0037**	**0.0035**	0.0042	**0.954**	**0.818**	2
K-50	0.0039	0.0039	**0.0040**	0.943	0.794	1.9
K-9	0.0045	0.0044	0.0050	0.920	0.765	**3.2**
K-34	0.0066	0.0061	0.0081	0.909	0.775	2.3
K-17	0.0070	0.0070	0.0068	0.927	0.806	2.8
K-25	0.0092	0.0088	0.0103	0.923	0.792	2.4

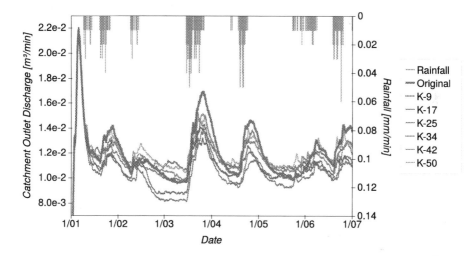

Fig. 8. Simulation output of the Constant-K tests.

simulation, we split the simulation time into "with forcing (WF)" and "without forcing (WOF)" time blocks and run the evolutionary clustering only by the forcing time blocks. Table 3 shows a tendency to better results for higher K, which are associated with a higher run time. Thus, using the results of Constant-K, we designed the Variable-K tests that execute the simulation using different representatives from the evolutionary clustering. In order to obtain high quality results during high dynamics combined with a short run time, we changed K during the simulation, according to the on- and offset of forcing (rainfall) (Fig. 9). This means we split the simulation run into using a high K at WF time blocks and a low K at WOF time blocks. We paired higher Ks and lower RMSE values for WF time blocks and lower Ks and lower RMSE values for WOF time blocks according to the results of Sect. 5.2. The simulation is started with the initial clustering using the best K of WOF time block. Then according to the forcing, the simulation continues with a low K at WOF time blocks and it uses evolutionary clustering with a high K at WF time blocks (Fig. 9). Because of the clustering overhead, we use the evolutionary clustering only when switching between WF and WOF time blocks and back. The trend of the curves for Variable-K is difficult to interpret, so the tests show a tendency to a higher RMSE and lower PCC as well as KGE than Constant-K (Table 4).

Table 4. Evaluation results of the Variable-K tests.

Tests	RMSE	RMSE-WOF	RMSE-WF	PCC	KGE	Speedup
K-17-50	**0.0061**	**0.0058**	**0.0071**	**0.823**	**0.719**	1.9
K-17-42	0.0079	0.0076	0.0090	0.720	0.714	2
K-9-42	0.0101	0.0095	0.0122	0.693	0.591	**2.1**
K-9-50	0.0121	0.0123	0.0113	0.509	0.462	**2.1**

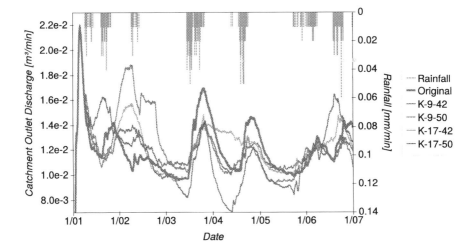

Fig. 9. Simulation output of the Variable-K tests.

5.4   Auto-K

The Variable-K test showed a high variation by changing K during the simulation. Thus, in Auto-K, we apply the K-determiner during the simulation in order to select an appropriate K automatically and reduce the high fluctuations. The Auto-K tests have the same settings as Variable-K tests with the difference that for Auto-K, we do not set K manually and use the K-determiner once at the beginning of each forcing time block. The K-determiner selects an appropriate K automatically using the evolving feature set generated based on the dynamics occurring during the simulation. In the Auto-K tests, K-determiner applies the *elbow* (K-AEL) and *rmse-ctime* (K-ARC) methods respectively. The trend of their curves fits well to the original. The gray band, representing six tests for K-ARC with different random seeds, becomes thinner in the forcing time blocks, thus shows a reproducible peak discharge (Fig. 10). The RMSE for both tests, K-ARC and K-AEL, is acceptably low. Although K-ARC results in a lower RMSE and higher PCC as well as KGE than that of K-AEL, its speedup is lower than K-AEL since K-ARC uses higher Ks (Table 5).

Table 5. Evaluation results of the Auto-K tests.

Tests	RMSE	RMSE-WOF	RMSE-WF	PCC	KGE	Speedup
K-ARC	**0.0049**	**0.0046**	0.0059	**0.894**	**0.80**	1.8
K-AEL	0.0061	0.0061	**0.0058**	0.828	0.72	**2**

Additionally, to test the variability of the best K, we run the Auto-K with K-determiner at all time steps of the simulation and got a narrow frequency distribution of the selected best Ks. K ranges between 17 and 30 with the most frequent $K = 25$ for the K-determiner with the *elbow* method (Fig. 11, left) and the K range between 30 and 38 with the most frequent $K = 34$ for the K-determiner with the *rmse-ctime* method (Fig. 11, right). Using the K-determiner at each time step instead of only once at the beginning of each forcing time block showed slightly better results in the quality evaluation than the Auto-K tests, though it is such inefficient, that the whole simulation run time will be longer than the original simulation (Table 6).

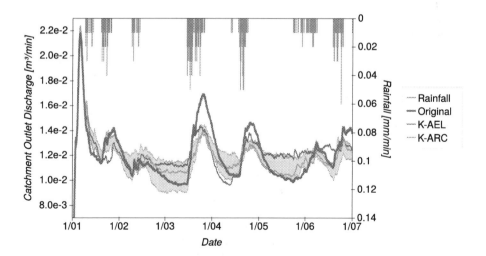

Fig. 10. Simulation output of the Auto-K tests.

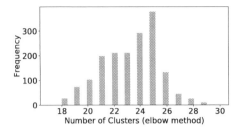
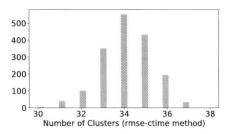

Fig. 11. Histogram of the selected Ks by K-determiner when it is applied at all time steps of the simulation.

5.5   Summary of Results

We applied three different test settings (Constant-K, Variable-K and Auto-K) to speedup a hydrological simulation with K-Means clustering. Our metrics show minor differences for the Auto-K and Constant-K tests, considering several test runs with different random seeds set for the K-Means clustering (Table 2, 3 and 5). The results of Auto-K, that clusters at every time step of each forcing time block, show lower fluctuations, hence a more reliable output at these time blocks compared to the Constant-K tests (Fig. 7 and 10). The Auto-K tests use a combination of the initial and evolutionary clustering with the K-determiner to determine the appropriate K dynamically based on the given feature set and configure the clustering parameters automatically during the simulation. The additional clustering steps of the Auto-K tests result in a better RMSE, PCC and KGE, and a lower speedup. Since K-ARC takes $K = 34$ frequently as the appropriate K, its comparison with K-34 test from Constant-K tests confirms this statement (Table 3 and 5). Constant-K tests showed the potential for a higher speedup (Fig. 12), although the ideal choice for a high speedup together with a low RMSE appears to be unpredictable. This means, although K-42 from Constant-K tests has the lowest RMSE value of all tests and a speedup of 2, the prediction of such a favorable K without running tests for a particular use case is not possible (Fig. 12). As a solid solution, we recommend K-ARC from Auto-K tests with a speedup of 1.8, and acceptable RMSE, PCC and KGE values.

Table 6. Evaluation results of the Auto-K tests when the K-determiner is applied at all time steps of the simulation.

Tests	RMSE	RMSE-WOF	RMSE-WF	PCC	KGE	Speedup
K-AEL	**0.0051**	0.005	**0.0055**	0.950	0.825	0.6
K-ARC	0.0053	0.005	0.0060	**0.954**	**0.843**	0.6

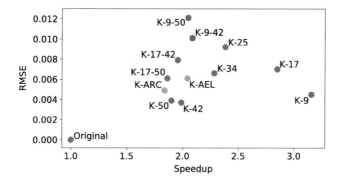

Fig. 12. Simulation run time speedup and RMSE of all tests. The orange markers highlight the Auto-K tests.

6   Implementation Environment

The simulation scripts are written and executed in Matlab R2019a. The analysis methods are implemented in Python. All tests are executed on a Red Hat Enterprise Linux Server release 7.4 on a 16-core Intel(R) Xeon(R) CPU E5-2640 v2 @ 2.00 GHz processor. All scripts, data files and requirements of the analyses are available as a GitLab repository named "hyda" [2].

7   Conclusions

In this work we introduced an approach to make use of landscape properties and dynamics of hydrological models to reduce computational redundancies in hydrological simulations. The approach consists of several steps, mainly, initial and evolutionary clustering and scaling of the simulation output of the cluster representatives to the remaining cluster members. We have used the K-Means clustering together with the K-determiner that automatically defines a suitable number of clusters using the *elbow* and *rmse-ctime* methods. The results of our tests demonstrated that the K-ARC approach has a promising RMSE of 0.0049, PCC of 0.89 and KGE efficiency of 0.8 which is a close approximation of the original simulation output. Additionally, K-ARC has a simulation run time speedup of 1.8 that is close to half of the original simulation run time.

References

1. Ali, G., Tetzlaff, D., Soulsby, C., McDonnell, J.J., Capell, R.: A comparison of similarity indices for catchment classification using a cross-regional dataset. Adv. Water Resour. **40**, 11–22 (2012). https://doi.org/10.1016/j.advwatres.2012.01.008
2. Azmi, E.: Hydrological data analysis, March 2020. https://gitlab.com/elnazazmi/hyda

3. Azmi, E., Ehret, U., Meyer, J., van Pruijssen, R., Streit, A., Strobl, M.: Clustering as approximation method to optimize hydrological simulations. In: Yahyapour, R. (ed.) Euro-Par 2019. LNCS, vol. 11725, pp. 256–269. Springer, Cham (2019). https://doi.org/10.1007/978-3-030-29400-7_19

4. Barnston, A.G.: Correspondence among the correlation, RMSE, and Heidke forecast verification measures; refinement of the heidke score. Weather Forecast. **7**, 699–709 (1992)

5. Corzo, G., Solomatine, D.: Baseflow separation techniques for modular artificial neural network modelling in flow forecasting. Hydrol. Sci. J. **52**(3), 491–507 (2007). https://doi.org/10.1623/hysj.52.3.491

6. Ehret, U., Zehe, E., Scherer, U., Westhoff, M.: Dynamical grouping and representative computation: a new approach to reduce computational efforts in distributed, physically based modeling on the lower mesoscale. In: Presented at the AGU Chapman Conference, 23–26 September 2014

7. Grayson, R., Blöschl, G.: Spatial Patterns in Catchment Hydrology: Observations and Modelling. CUP Archive, Cambridge (2001)

8. Gupta, H.V., Kling, H., Yilmaz, K.K., Martinez, G.F.: Decomposition of the mean squared error and NSE performance criteria: implications for improving hydrological modelling. J. Hydrol. **377**(1–2), 80–91 (2009)

9. Jones, J.E., Woodward, C.S.: Newton-Krylov-multigrid solvers for large-scale, highly heterogeneous, variably saturated flow problems. Adv. Water Resour. 763–774 (2001). https://doi.org/10.1016/S0309-1708(00)00075-0

10. Kar, A.K., Goel, N., Lohani, A., Roy, G.: Application of clustering techniques using prioritized variables in regional flood frequency analysis-case study of Mahanadi basin. J. Hydrol. Eng. **17**(1), 213–223 (2012)

11. Kassambara, A.: Practical Guide to Cluster Analysis in R: Unsupervised Machine Learning, vol. 1. STHDA (2017)

12. Kollet, S.J., et al.: Proof of concept of regional scale hydrologic simulations at hydrologic resolution utilizing massively parallel computer resources. Water Resour. Res. **46**(4) (2010). https://doi.org/10.1029/2009WR008730

13. Ley, R., Casper, M., Hellebrand, H., Merz, R.: Catchment classification by runoff behaviour with self-organizing maps (SOM). Hydrol. Earth Syst. Sci. **15**(9), 2947–2962 (2011). https://doi.org/10.5194/hess-15-2947-2011

14. Maxwell, R., Condon, L., Kollet, S.: A high-resolution simulation of groundwater and surface water over most of the continental US with the integrated hydrologic model ParFlow v3. Geosci. Model Dev. **8**, 923 (2015)

15. Murphy, K.P.: Machine Learning: A Probabilistic Perspective. MIT press, Cambridge (2012)

16. Netzel, P., Stepinski, T.: On using a clustering approach for global climate classification. J. Clim. 3387–3401 (2016). https://doi.org/10.1175/JCLI-D-15-0640.1

17. Pearson, K.: VII. mathematical contributions to the theory of evolution.-III. regression, heredity, and panmixia. Philos. Trans. R. Soc. A, 253–318 (1896). https://doi.org/10.1098/rsta.1896.0007

18. Pedregosa, F., et al.: Scikit-learn: machine learning in Python. J. Mach. Learn. Res. **12**, 2825–2830 (2011)

19. Satopaa, V., Albrecht, J., Irwin, D., Raghavan, B.: Finding a "kneedle" in a haystack: detecting knee points in system behavior. In: 2011 31st International Conference on Distributed Computing Systems Workshops, pp. 166–171. IEEE (2011)

20. Sawicz, K., Wagener, T., Sivapalan, M., Troch, P.A., Carrillo, G.: Catchment classification: empirical analysis of hydrologic similarity based on catchment function in the eastern USA. Hydrol. Earth Syst. Sci. **15**(9), 2895–2911 (2011). https://doi.org/10.5194/hess-15-2895-2011

21. Sawicz, K., Kelleher, C., Wagener, T., Troch, P., Sivapalan, M., Carrillo, G.: Characterizing hydrologic change through catchment classification. Hydrol. Earth Syst. Sci. **18**(1), 273–285 (2014). https://doi.org/10.5194/hess-18-273-2014

22. Schulz, K., Seppelt, R., Zehe, E., Vogel, H.J., Attinger, S.: Importance of spatial structures in advancing hydrological sciences. Water Resour. Res. **42**(3) (2006). https://doi.org/10.1029/2005WR004301

23. Wagener, T., Sivapalan, M., Troch, P., Woods, R.: Catchment classification and hydrologic similarity. Geogr. Compass **1**(4), 901–931 (2007). https://doi.org/10.1111/j.1749-8198.2007.00039.x

24. Wrede, S., et al.: Towards more systematic perceptual model development: a case study using 3 Luxembourgish catchments. Hydrol. Process. **29**(12), 2731–2750 (2015). https://doi.org/10.1002/hyp.10393

25. Zaki, M.J., Meira Jr., W., Meira, W.: Data Mining and Analysis: Fundamental Concepts and Algorithms. Cambridge University Press, Cambridge (2014). https://doi.org/10.1017/CBO9780511810114

26. Zehe, E., et al.: HESS opinions: from response units to functional units: a thermodynamic reinterpretation of the HRU concept to link spatial organization and functioning of intermediate scale catchments. Hydrol. Earth Syst. Sci. 4635–4655 (2014). https://doi.org/10.5194/hess-18-4635-2014

27. Zhang, Y., Moges, S., Block, P.: Optimal cluster analysis for objective regionalization of seasonal precipitation in regions of high spatial-temporal variability: application to western Ethiopia. J. Clim. **29**(10), 3697–3717 (2016)

Domain-Oriented Multilevel Ontology for Adaptive Data Processing

Man Tianxing[1]([✉]) [iD], Elena Stankova[4] [iD], Alexander Vodyaho[3] [iD], Nataly Zhukova[1,2] [iD], and Yulia Shichkina[3] [iD]

[1] ITMO University, St. Petersburg, Russia
mantx626@gmail.com, nazhukova@mail.ru
[2] St. Petersburg Institute for Informatics and Automation of the Russian Academy of Sciences, St. Petersburg, Russia
[3] St. Petersburg State Electrotechnical University "LETI", St. Petersburg, Russia
aivodyaho@mail.ru, strange.y@mail.ru
[4] St. Petersburg State University, St. Petersburg, Russia
e.stankova@spbu.ru

Abstract. In the data mining domain, the diversity of algorithms and the clutter of data make the knowledge discovery process very unfriendly to many non-computer professional researchers. Meta-learning helps users to modify some aspects of this process to improve the performance of the resulting model. Semantic meta mining is the process of mining metadata about data mining algorithms based on expertise extracted from the knowledge base. The knowledge base is usually represented in the form of ontology. This article proposes a domain-oriented multi-level ontology (DoMO) through merging and improving existing data mining ontologies. It provides the restrictions of the dataset characteristics to help the domain experts describe data set in the form of ontology entities. According to the entities of the data characteristics in DoMO, the users can query the ontology to obtain the optimized data processing process. In this paper, we take the time series classification problem as an example to present the effectiveness of the proposed ontology.

Keywords: Meta-learning · Data mining · Semantic meta mining · Ontology

1 Introduction

In the era of big data, data analysis is everywhere. The number of needed data science specialists is increasing permanently. Also, many IT specialists are ready to use the knowledge discovery process in various domains such as economic, meteorology, etc. [21, 22]. But the diversity of algorithms and the clutter of data make the knowledge discovery process very unfriendly to many non-computer professional researchers. Even for the data researchers, it is still difficult to find the best solutions for specific tasks quickly [23]. An intuitive and easy-to-understand intelligent assistant is needed.

Today meta-learning is very popular since it uses machine learning (ML) algorithms to learn from ML experiments for obtaining the best algorithms and parameters.

© Springer Nature Switzerland AG 2020
O. Gervasi et al. (Eds.): ICCSA 2020, LNCS 12249, pp. 634–649, 2020.
https://doi.org/10.1007/978-3-030-58799-4_46

Melanie Hilario proposed a new optimization solution: Semantic meta mining [2]. It relies on extensive background knowledge concerning data mining (DM) itself.

In the field of semantic meta mining, it is necessary to have a suitable description framework to make clear the complex relationships between tasks, data, and algorithms at different stages in the data mining process. Ontology is a computer-understandable description language. Naturally, it has become the choice of many DM intelligent assistants in various application scenarios.

The existing DM ontologies are usually dedicated to expressing one or several stages of the DM process in detail. This concentration on parts makes them lose the integrity of the description of the DM process. In addition, the different constraints of data set characteristics in different domains to make it difficult to propose a general and applicable description ontology.

This paper proposes a domain-oriented multi-level ontology (DoMO) for the DM process. It integrates existing DM ontologies so that it can describe each stage of DM. We also added restrictions on the description of the dataset at the upper level. We built a sub-ontology, which described the characteristics of dataset and task requirements and named it as "INPUT" ontology because it is the input part of the user's query. With the assistance of experts, it can be applied in specific fields.

As an intelligent assistant, the main purpose of DoMO ontology is to help users:

- Describe the data set in the form of ontology entities.
- Choose the suitable solutions based on the data characteristics and task requirements.
- Obtain the data processing processes of the selected solutions.

The structure of this paper is as follows: Sect. 2 introduces background knowledge and related work. Section 3 presents the architecture of the proposed ontology DoMO. Section 4 describes the ontology content. Section 5 presents the workflow of DoMO and a case study on time series classification. In Sect. 6, the authors present the conclusion and future work.

2 Background

2.1 Meta-learning and Semantic Meta Mining

Meta-learning [1] is defined as the application of ML techniques to past ML experiments, and its purpose is to modify certain aspects of the learning process to improve the performance of the results. Traditional meta-learning treats the learning algorithm as a black box, correlating the observed performance of the output model with the characteristics of the input data. However, the internal characteristics of algorithms with the same input/output type may vary.

Semantic meta mining [2] mines DM metadata through querying DM expertise in the knowledge base. It is different from the general meta-learning:

- Meta-learning methods are data-driven. And semantic meta mining is based on the related expertise and internal relations. So, developers usually represent knowledge in the form of ontology.

- Meta-learning for algorithm or model selection mainly involves mapping the dataset attributes to the observed performance of the algorithm as a black box. The parameters are updated based on experimental results, and the internal mechanisms of the algorithms are not the determining factor. In contrast, semantic meta mining complements the data set description by in-depth analysis and characterization of the algorithm: the basic hypothesis of the algorithm, the optimization goals and strategies, and the structure and complexity of the generated models and patterns.
- Meta-learning focuses on the learning phase of data mining, that is, the performance of the generated model. But semantic meta mining is oriented towards the entire data mining process. Based on the characteristics of the data to be processed and the task requirements, it provides users with entire corresponding solutions.

According to the above analysis, the role of classical meta-learning and semantic meta mining are not conflicting. The learning goals of meta-learning are more detailed (such as the parameters of the algorithms). And semantic meta mining provides the appropriate algorithm selection and formulates the execution process. These suggestions are more general. Such semantic meta mining can usually also solve the cold start problem of meta-learning to ensure that the learning process is in the correct direction.

2.2 Data Mining Process

To avoid meaningless operations in data analysis, it is necessary to have a structured framework to implement data mining effectively and correctly. A suitable DM process model is the basis for building DM ontologies. Today, there exist three common frameworks CRISP-DM [4], SEMMA [5], and KDD [3] to format the DM process. Table 1 shows the structures and corresponding relations of these frameworks.

The KDD model is the process of extracting the hidden knowledge according to databases. KDD requires relevant prior knowledge and a brief understanding of the application domain and goals. An improved model of KDD and its application to the construction of optimal routes in a dynamic network are shown in [6].

CRISP-DM provides a uniform framework and guidelines for data miners. It consists of six phases or stages which are well structured and defined for ontology building.

The SEMMA (Sample, Explore, Modify, Model, and Access) is the data mining method developed by the SAS institute. It offers and allows understanding, organization, development, and maintenance of data mining projects. It helps in providing solutions for business problems and goals.

The KDD process model is iterative and interactive, so that it is too complex as the framework of ontology building. SEMMA is linked to SAS enterprise miner as a logical organization of the functional tools. It has a cycle of five stages or steps. But it doesn't describe the steps "Task understanding" and "Deployment," which we are going to describe in the ontology. Based on the characteristics of several frameworks, the simplicity and completeness of CRISP-DM make it suitable for DM ontology building.

Table 1. The comparison and corresponding processes of CRISP-DM, SEMMA, and KDD.

	KDD	CRISP-DM	SEMMA
Processes	Developing and Understanding of the Application	Business Understanding	
	Creating a Target Data Set	Data Understanding	Sample
	Data Cleaning and Pre-processing		Explore
	Data Transformation	Data Preparation	Modify
	Choosing the suitable Data Mining Task	Modeling	Model
	Choosing a suitable Data Mining Algorithm		
	Employing Data Mining Algorithm		
	Interpreting Mined Patterns	Evaluation	Assessment
	Using Discovered Knowledge	Deployment	

2.3 Existing Ontologies for Data Mining

Recently, many intelligent assistants have been developed to optimize the DM process. Comparative studies are discussed in [7, 8]. Many DM ontologies have also been developed to help users build DM processes.

Panov et al. [9, 10] proposed a data mining ontology OntoDM, which includes formal definitions of basic DM entities, such as DM tasks, DM algorithms, and DM implementations. The definition is based on the general data mining framework presented by Džeroski [8]. This ontology is one of the first depth and heavyweight ontologies used for data mining. But it is just used for the description of DM knowledge, so the algorithm characteristics are not covered.

To allow the representation of structured mining data, Panov et al. developed a separate ontology module, named OntoDT, for representing the knowledge about data types [11]. OntoDT defines basic entities, such as datatype, properties of datatypes, specifications, characterizing operations, and a datatype taxonomy. But the problem in the application of ontoDT is that the basic data information is not enough to help users choose the appropriate algorithm. The OntoDT in this article is intended to use it as an upper-level ontology to help domain experts describe the characteristics of the dataset.

Hilario et al. [12] present the data mining optimization ontology (DMOP), which provides a unified conceptual framework for analyzing data mining tasks, algorithms, models, datasets, workflows, and performance metrics, as well as their relationships. As the authors of the concept of semantic meta mining, they use a large set of customized special-purpose relations in DMOP. But DMOP only covers 3 phases of CRISP-DM. And the structure of the ontology is very complicated and thus unfriendly to non-professional users.

In the existing ontologies, the CRISP-DM process, which is composed of the 6 phases, is the basic framework. As Fig. 1 shows, most ontologies only focus on specific phases (DMOP covers three phases that can be best automated: from data preparation to evaluation; OntoDM the last four phases; OntoDT only provides a general description of the first phase about the data type).

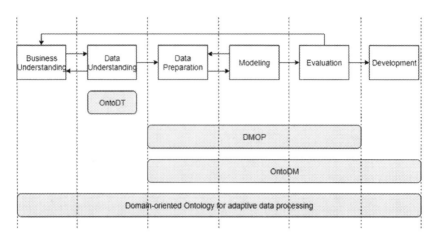

Fig. 1. The comparison of CRISP-DM Phases of Ontology Description (The highlighted grid means covering this phase).

There are several other data mining ontologies currently existing, such as the Knowledge Discovery (KD) Ontology [13], the KDDONTO Ontology [14], the Data Mining Workflow (DMWF) Ontology [15], which are also based on similar ideas.

3 DoMO Architecture

The multi-level architecture of DoMO ontology presents in Fig. 2. In the ontology, the role of each level is as follow:

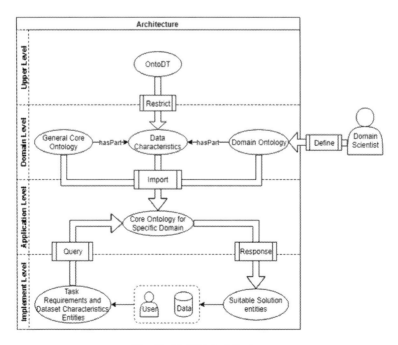

Fig. 2. DoMO Architecture

- Upper Level: The characteristics and task requirements of the data set are the basis for algorithm selection. Data in different fields have different standards for defining characteristics. But the basic properties of the dataset, such as datatype, extended datatype, datatype properties, characterizing datatype operations, datatype value space, are the same. These entities are included in OntoDT. We define it at the upper level as a common data property set
- Domain level: One advantage of DoMO is to use OntoDT's restriction rules to help experts create intelligent assistants suitable for specific fields. This creation process occurs at the domain level. In general core ontology, we name the characteristics of the data in advance. Experts define these characteristics with their knowledge based on upper-level restrictions.
- Application Level: Experts define the data characteristics of the field and import them into general core ontology. It means a core ontology for a specific domain is generated. Its internal structure will be discussed in Sect. 4. Users can query directly on the ontology to get the DM process for specific tasks.
- Implementation level: The generation of user queries and DM processes occurs at the implementation level. According to the characteristics of the data to be processed and task requirements, users obtain suitable solutions. Since the solutions have pre-processers and post-processers, complete DM processes are generated.

The key point of this architecture is to provide restrictions for the description of the domain ontology at the upper level. In the previous work, there is no suitable method to describe the data set in the form of ontology entities. In the general data type ontology OntoDT, the basic properties of the data set are defined. However, these properties cannot directly influence the DM generation process. The selection of the DM algorithm is based on the characteristics of the data set and the requirements of the task. However, the definitions of these characteristics are different in different fields. To make DoMO adaptively process data sets in various fields, we use the OntoDT classes as parameters to specify the definition (value or range) of data characteristics in general core ontology. Domain experts describe domain knowledge or existing domain ontology in general core ontology, making it suitable for data analysis tasks in this domain. An example of the definition in the domain of time series classification (TSC) is shown in Fig. 3.

Then users can query the generated core ontology to obtain a suitable DM process for a specific domain. The workflow is discussed in Sect. 5.1.

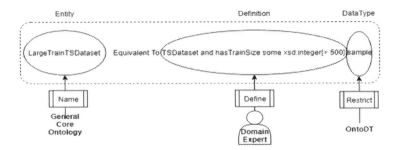

Fig. 3. An example of the definition of data characteristic "LargeTrainTSDataset."

4   DoMO Content

DoMO ontology is composed of two main parts: domain ontology and core ontology. During the initialization phase, core ontology is a general ontology, including an "INPUT" ontology and some other existing DM ontologies (DMOP, OntoDT, OntoDM, and DMWF). Domain ontology is built through defining the existing entities in general core ontology based on the restrictions of OntoDT. When experts import domain knowledge in the form of domain ontology, we obtain a core ontology for a specific domain (See Fig. 4.).

4.1   INPUT Ontology

We create "INPUT" ontology as the input interface for the user query. Its main goals are:

- Define data characteristic entities corresponding to algorithm characteristics.
- Describe the requirements of the DM task, that is, the output of the DM algorithm.
- Supplement the missing algorithm characteristics and measure characteristics in the existing DM ontologies.

INPUT ontology is the part directly associated with the user's queries. It makes the use of ontology more explicit. Users do not need to try to understand the other internal structures of the ontology.

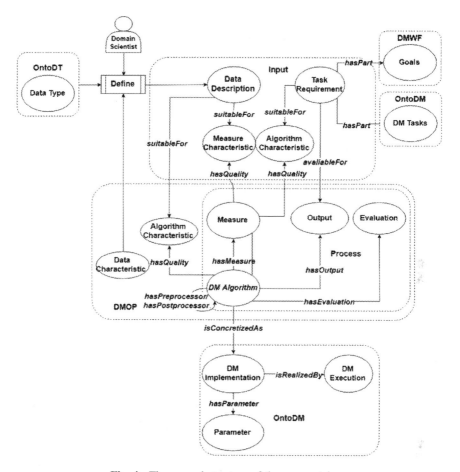

Fig. 4. The general structure of the core ontology

4.2 Existing DM Ontologies for DoMO

Besides, INPUT ontology is also the core part of integrating existing DM ontologies. The integration operation is based on the purpose of generating suitable solutions and processes.

In the process of integration, to reduce the complexity of the ontology, we discarded contents that were useless for this purpose and reconstructed the structures. The main classes in DoMO are shown in Table 2.

The reconstruction processes are as follows:

- OntoDT is fully retained as an upper-level restriction that defines the characteristics of the data.
- The class "Goals" in DMWF and class "DM-Task" in OntoDM are extracted for the description of task requirements.

- Although DMOP provides more than a hundred DM algorithms and their charac-teristics, we have reconstructed its structure. As components of the DM algorithms, the classes "Measure," "Output," "Evaluation," and "DM Algorithm" itself are included in a new class "Process" so that it is more understandable for the users.
- OntoDM describes the last CRISP- DM phase "Development". The classes "DM Implementation" and "Parameter" in OntoDM are integrated for the possible

Table 2. The main classes in DoMO

Class	Source	Annotation
Data description	INPUT	Describes the dataset characteristics in the form of ontology entities. Domain experts define their value and ranges
Task requirement	INPUT	Describes the task requirements in the form of ontology entities
Measure characteristic	INPUT	Existing DM ontologies do not describe the performance of measures (i.e., distance functions). In the INPUT ontology, we describe and name it "MeasureCharacteristic"
Algorithm characteristic	INPUT/DMOP	Describes the performance of DM algorithms including tolerating some data set defects (such as Missing value, Noise value), suitable for some task requirements (such as two-class, multi-class)
Data type	OntoDT	Provide basic data types that describe the characteristics of the data set (such as sample, label)
Goals	DMWF	Provide a description of the task requirements. It mainly focuses on the generalization of the types of output results
DM-Tasks	OntoDM	Provide a description of the task requirements. It mainly focuses on the description of specific details of the task
Data characteristic	DMOP	Provided by DMOP, the names of the characteristic of the dataset
DM algorithm	DMOP	Describe all DM algorithms that have been designed to perform any of the DM tasks, such as feature selection, missing value imputation, or modeling (or induction)
Measure	DMOP	Describes the distance functions and similarity functions, which usually directly affect the performance of DM algorithms
Output	DMOP	Describe the output models of the DM algorithms (such as decision tree structure, probability distribution structure)
Evaluation	DMOP	Describe the evaluation functions of the DM algorithms (such as external validity model function for clustering algorithms)
DM implementation	OntoDM	Provide DM algorithm implementation scheme and parameter settings
DM execution	OntoDM	Provide executable solutions for DM algorithms (such as R, python package, Weka)
Parameter	OntoDM	Provide parameters for DM algorithms (such as distance threshold, number of clusters and variance threshold for K-means algorithm)

Table 3. The relevant properties in DoMO.

Property	Domains	Ranges	Answering the competency questions
availableFor	INPUT	Characteristics	Given data characteristics or task requirements, which characteristics should the DM algorithms have so that they are suitable for?
suitableFor	INPUT	Characteristics	Given data characteristics or task requirements, which characteristics should the DM algorithms have so that they are available?
hasQuality	Process	Characteristics	Which characteristics do the given process have?
hasPreprocessor hasPostprocessor hasOutput hasMeasure hasEvaluation	DM algorithm	Process	Which processes do the DM algorithm have?
isConcretizedAs	DM Algorithm	DM Implementation	How can we implement the DM algorithm?
hasParameter	DM Implementation	Parameter	Which parameters should we set when we implement the DM algorithm?
isRealizedBy	DM Implementation	DM Execution	Where and how can we execute the DM algorithm?

parameter setting. And "DM Execution" presents where and how to execute the selected algorithms.

In order to build the logical structure of DoMO, the relevant properties are defined in Table 3.

4.3   Statistical Metrics for DoMO

In this paper, we use the statistical ontology metrics from the Protégé software [19] and Bioportal [20]. This includes metrics such as the number of classes and individuals, maximum depth, average number of siblings, maximum number of siblings, sub-class axioms count, disjoint classes axioms count, and annotation assertion axioms count. The values of these statistical ontology metrics for DoMO are presented in Table 4.

Table 4. Statistical metrics for DoMO

Axiom	1610
Logical axiom count	974
Declaration axioms count	636
Class count	626
Object property count	11
Data property count	4
Individual count	0
Maximum depth	12
Maximum number of children	37
Average number of children	3
Classes with a single child	28
Classes with more than 25 children	2
Classes with no definition	626

5   Usage

As long as the structure of the ontologies is reasonable, they can be operated on the corresponding editing software, for instance: protégé. Based on the relations presented in Table 3, users can query for suitable solutions with the following workflow.

5.1   General Workflow

The workflow of DoMO for data analysis in a specific domain is as follow:

1. Based on the restrictions of OntoDT, domain experts define the characteristics of domain data in the form of ontology.
2. Merge the domain ontology and the general core ontology to obtain the core ontology for the specific domain.
3. Manually obtain task requirements and data sets and describe them in the form of ontology entities as the inputs.
4. Execute the selection process on this core ontology for a specific domain.
   a. Input the entities of input-data description and task requirements. Based on the relation "suitableFor", obtain the characteristics which the solutions should have.
   b. According to the relation "hasQuality," obtain the algorithms or measures which have suitable characteristics. If the results are measures, obtain the algorithms according to the relation "hasMeasure."
   c. Choose the most suitable algorithms which meet as many characteristics as possible. They are the selected solutions.
   d. According to the relation "hasPre/Postprocessor," obtain the entire DM process.
   e. According to the relation "hasPart," obtain the process of the selected solutions.
   f. According to the relation "isConcretizedAs," obtain the implementations and parameter variants.
   g. According to the relation "isRealizedBy," obtain the available executions.

5.2 Practical Application in Time Series Classification

DoMO can be flexibly applied to the data analysis process in different fields. As an application example, we construct an ontology oriented to the time series classification (TSC) field. The entities of TSC data characteristics have been named in "INPUT" ontology. For describing the TS datasets in the form of these entities, explicit definitions are needed.

Expert knowledge of the definition of characteristics of TSC data comes from [16]. We define them as Table 5 shows. Then users can represent the TS datasets in DoMO.

Table 5. The definition of TSC data characteristics

Category	Range of class value	Ontology class
TSDataset ByTrainSize	hasTrainSize some xsd:integer [<100] sample	SmallTrainTSDataset
	hasTrainSize some xsd:integer [> =100, <=500] sample	MediumTrainTSDataset
	hasTrainSize some xsd:integer [>500] sample	LargeTrainTSDataset
TSDataset ByTestSize	hasTestSize some xsd:integer [<300] sample	SmallTestTSDataset
	hasTestSize some xsd:integer [>=300, <=1000] sample	MediumTestTSDataset
	hasTestSize some xsd:integer [>1000] sample	LargeTestTSDataset
TSDataset ByLength	hasLengthSize some xsd:integer [<300] timestamp	ShortTSDataset
	hasLengthSize some xsd:integer [>= 300, <=700] timestamp	MediumTSDataset
	hasLengthSize some xsd:integer [>700] timestamp	LongTSDataset
TSDataset ByNoOfClass	hasClassSize some xsd:integer [<10] label	FewClassTSDataset
	hasClassSize some xsd:integer [>=10, <=30] label	MediumClassTSDataset
	hasClassSize some xsd:integer [>30] label	ManyClassTSDataset
TSDataset ByArea	——	DeviceTSDataset
		ECGTSDataset
		ImageTSDataset
		MotionTSDataset
		SensorTSDataset
		SimulatedTSDataset
		SpectroTSDataset

In order to verify the effectiveness of DoMO to help users choose the right solutions, we selected an experimental data 'CinCECGtorso' for verification. For more examples, please refer to [17].

The data set 'CinCECGtorso' is derived from one of the Computers in Cardiology challenges, an annual competition that runs with the conference series of the same name and is hosted on physionet. Data is taken from ECG data for multiple torso-surface sites. There are four classes corresponding to 4 different groups of people [18].

The interaction between the users and DoMO takes place on "INPUT" ontology. Users can describe the dataset and query the corresponding entities of data characteristics in the following form:

"TSDataset and hasTrainSize exactly 40 sample"

Then users can receive the corresponding entity "SmallTrainDataset". Other corresponding entities are received, as Table 6 shows.

Table 6. the corresponding entities of the data set 'CinCECGtorso in INPUT ontology.

Category	Description of 'CinCECGtorso' in DoMO	Corresponding entities
Train size	hasTrainSize exactly 40 sample	SmallTrainTSDataset
Test size	hasTestSize exactly 1380 sample	LargeTestTSDataset
Length	hasLenghSize exactly 1639 timestamp	LongTSDataset
No. of classes	hasClassSize exactly 4 label	FewClassTSDataset
Data area	appliedOn some ECG	ECGTSDataset

The entities "SmallTrainTSDataset", "LargeTestTSDataset", "LongTSDataset" and "ECGTSDataset" are characteristics of the data set and the entities "FewClassTSDataset" means the task requirement is a few classes classification tasks. INPUT ontology allows formulating the tasks in the common form. Then the query for the suitable solutions is:

"Algorithm
and suitableFor some SmallTrainTSDataset
and suitableFor some LargeTestTSDataset
and suitableFor some LongTSDataset
and suitableFor some FewClassTSDataset
and suitableFor some ECGTSDataset"

In this experiment, BOSS (Bag of SFA Symbols), COTE (Collection of Transformation E), EE (Elastic Ensemble), MSM_1NN (Move-Split-Merge) and ST (Shapelet Transform) are selected for data set 'CinCECGtorso' by DoMO, since these algorithms are suitable for all the conditions.

We applied all the available TSC algorithms on this data set and gave a rank in Fig. 5. All chosen algorithms are obviously in the upper half and have good performance.

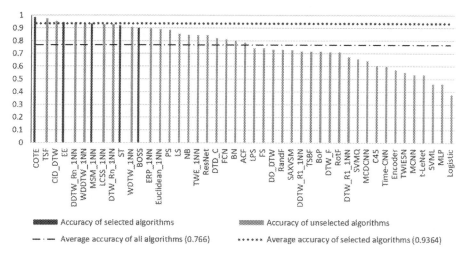

Fig. 5. The accuracy comparison of all TSC algorithms on dataset 'CinCECGtorso'

6 Conclusion

This paper proposes a domain-oriented multi-level ontology (DoMO) for data-adaptive processing. The multi-level structure of the ontology includes the upper level (restrictions described by data characteristics), domain level (definition of domain data characteristics), application level (core ontology for a specific domain), and implementation level (Users' queries and the generation of DM process).

The construction of DoMO includes creating an "INPUT" ontology that describes the characteristics of the data and task requirements and reconstructing and integrating existing DM ontologies. In comparison with existing approaches, DoMO describes the entire data mining process, and the "INPUT" ontology presents datasets in the form of metadata. Due to the intelligibility and portability of the ontology, DoMO can be applied to the data analysis process in different fields with the assistance of domain experts. The application of DoMO in the field of time series classification proved its effectiveness.

Although the ontology is focusing on building the foundation of data mining, it can be used by practitioners in real-world applications to optimize knowledge discovery processes by sequentially querying the suitable solutions based on specific task requirements and data characteristics. Meanwhile, DoMO is intended to be extensible and will continue to be updated to reflect future advancements in using it for building high-quality data-analytical processes rapidly.

Acknowledgments. The paper was prepared in Saint- Petersburg Electrotechnical University (LETI), and is supported by the Agreement? 075-11-2019-053 dated 20.11.2019 (Ministry of Science and Higher Education of the Russian Federation, in accordance with the Decree of the Government of the Russian Federation of April 9, 2010 No. 218), project «Creation of a domestic high-tech production of vehicle security systems based on a control mechanism and intelligent sensors, including millimeter radars in the 76-77 GHz range».

References

1. Jankowski, N., Duch, W., Grąbczewski, K. (eds.) Meta-learning in Computational Intelligence, vol. 358. Springer, Heidelberg (2011). https://doi.org/10.1007/978-3-642-20980-2

2. Hilario, M., et al.: A data mining ontology for algorithm selection and meta-mining. In: Proceedings of the ECML/PKDD09 Workshop on 3rd Generation Data Mining (SoKD-09) (2009)

3. Brachman, R.J., Anand, T.: The process of knowledge discovery in databases. In: Advances in Knowledge Discovery and Data Mining, pp. 37–57 (1996)

4. Chapman, P., et al.: CRISP-DM 1.0: Step-by-step data mining guide. SPSS Inc. 9, p. 13 (2000)

5. SAS Enterprise Miner – SEMMA. SAS Institute, September 2014. http://www.sas.com/technologies/analytics/datamining/miner/semma.html

6. Shichkina, Y., Koblov, A.: Reducing the amount of data for creating routes in a dynamic DTN via Wi-Fi on the basis of static data, J. Comput. Netw. Commun. 9128785 (2017). https://doi.org/10.1155/2017/9128785. ISSN: 20907141

7. Serban, F., et al.: A survey of intelligent assistants for data analysis. ACM Comput. Surv. (CSUR) 45(3), 1–35 (2013)

8. Ristoski, P., Paulheim, H.: Semantic Web in data mining and knowledge discovery: a comprehensive survey. J. Web Semant. 36, 1–22 (2016)

9. Panov, P., Džeroski, S., Soldatova, L.: OntoDM: an ontology of data mining. In: 2008 IEEE International Conference on Data Mining Workshops, pp. 752–760. IEEE, December 2008

10. Panov, P., Džeroski, S., Soldatova, L.N.: Representing entities in the OntoDM data mining ontology. In: Džeroski, S., Goethals, B., Panov, P. (eds.) Inductive Databases and Constraint-Based Data Mining. Springer, New York (2010). https://doi.org/10.1007/978-1-4419-7738-0_2

11. Panov, P., Soldatova, L.N., Džeroski, S.: Generic ontology of datatypes. Inf. Sci. 329, 900–920 (2016)

12. Hilario, M., et al.: A data mining ontology for algorithm selection and meta-mining. In: Proceedings of the ECML/PKDD09 Workshop on 3rd Generation Data Mining (SoKD-09) (2009)

13. Záková, M., et al.: Automating knowledge discovery workflow composition through ontology-based planning. IEEE Trans. Autom. Sci. Eng. 8(2), 253–264 (2010)

14. Diamantini, C., Potena, D., Storti, E.: KDDONTO: an ontology for discovery and composition of KDD algorithms. In: Third Generation Data Mining: Towards Service-Oriented Knowledge Discovery (SoKD 2009), pp. 13–24 (2009)

15. Kietz, J.-U., et al.: Towards cooperative planning of data mining workflows (2009)

16. Bagnall, A., Lines, J., Bostrom, A., Large, J., Keogh, E.: The great time series classification bake off: a review and experimental evaluation of recent algorithmic advances. Data Min. Knowl. Disc. 31(3), 606–660 (2016). https://doi.org/10.1007/s10618-016-0483-9

17. Tianxing, M., Zhukova, N., Mustafin, N.: A knowledge-based recommendation system for time series classification. In: Proceedings of the 24th Conference of Open Innovations Association FRUCT. FRUCT Oy (2019)

18. PhysioNet/Computing in Cardiology Challenges. https://physionet.org/challenge/

19. Musen, M.A.: The protégé project: a look back and a look forward. AI Matt. 1(4), 4–12 (2015)

20. Martínez-Romero, M., et al.: NCBO ontology recommender 2.0: an enhanced approach for biomedical ontology recommendation. J. Biomed. Semant. 8(1), 21 (2017)

21. Stankova, E.N., et al.: OLAP technology and machine learning as the tools for validation of the numerical models of convective clouds. Int. J. Bus. Intell. Data Min. **14**(1–2), 254–266 (2019)
22. Stankova, E.N., Khvatkov, E.V.: Using boosted k-nearest neighbour algorithm for numerical forecasting of dangerous convective phenomena. In: Misra, S., et al. (eds.) ICCSA 2019. LNCS, vol. 11622, pp. 802–811. Springer, Cham (2019). https://doi.org/10.1007/978-3-030-24305-0_61
23. Stankova, E.N., Ismailova, E.T., Grechko, I.A.: Algorithm for processing the results of cloud convection simulation using the methods of machine learning. In: Gervasi, O., et al. (eds.) ICCSA 2018. LNCS, vol. 10963, pp. 149–159. Springer, Cham (2018). https://doi.org/10.1007/978-3-319-95171-3_13

Top-Down Parsing Error Correction Applied to Part of Speech Tagging

Ygor Henrique de Paula Barros Baêta, Alexandre Bittencourt Pigozzo, and Carolina Ribeiro Xavier$^{(\boxtimes)}$

Department of Computer Science, Universidade Federal de São João del Rei - UFSJ, São João del Rei, Brazil
ygorthemoster@gmail.com,
{alexandre.pigozzo,carolinaxavier}@ufsj.edu.br

Abstract. Natural Language Processing (NLP) applications are growing in popularity and importance, all of these applications rely upon the basic steps of tokenization, stemming and Part-of-Speech tagging (POS), and while those steps already work really well on English texts, the same cannot be said for every language. This paper investigates the use of techniques developed to be used as error correction in compilers as a means to improve Part-of-Speech tagging, specifically in Brazilian Portuguese.

Keywords: Natural Language Processing · Top-down parsing · POS tagging · Lexical analysis · Error correction

1 Introduction

Language is an ever evolving complex system, it's a driving force for technology and it's always changing to adapt the medium we communicate on, for example *Ogham* which evolved to be written on the corner of a stone [1]. When thinking of language evolution most think about ancient times with proto-languages and the creation of alphabets like Hangul [2], but some current languages are evolving and changing because of popular use, politics or other reasons, like as recently as 2009 Portuguese has had an orthographic accord in order to reduce the difference between it's variants [3].

Since the creation of the internet, language started evolving faster with new words invented to communicate about it, the increased use of abbreviations in order to comply with character limits, and the ever growing catalogue of emoji [4]. The influence of these neologisms and symbols is indisputable, for example in 2015 Oxford's word of the year was 😂 (Face With Tears of Joy) [5].

Natural Language Processing (*NLP*) is an interdisciplinary field of computer science and linguistics which studies techniques for representing and comprehending humans natural communication with computers [6]. *NLP* has many applications in current society such as Sentiment Analysis [7], Machine Translation [8] and Text Classification [9].

ⓒ Springer Nature Switzerland AG 2020
O. Gervasi et al. (Eds.): ICCSA 2020, LNCS 12249, pp. 650–662, 2020.
https://doi.org/10.1007/978-3-030-58799-4_47

In the classical approach to NLP, the process is subdivided in steps, that mirror how we humans classify sentences, these steps can be thought as being divided in the comprehension of the syntax, pragmatics and discourse [10], and generally are a combination of Tokenization, Stemming, Part-of-Speech Tagging, Stop-Word Removal, Dependency Parsing, Named Entity Recognition and Co-referencing.

NLP is growing in importance as human-machine interactions are becoming more common in everyday life, chat-bots and voice user interfaces are rising in usage over the last few years, but most of the problems for the users arise when the machine cannot correctly understand what is being told to them [11].

Considering this, it is of utmost importance to focus on furthering the comprehension of everyday expressions, and to make it possible to adapt to local dialects and the evolution of languages.

In English, NLP is an almost solved problem, with algorithms that have more than 97% Accuracy [12], being the main language of the Internet, there are many annotated corpus to train and test algorithms and some very well developed and tested frameworks for the basic steps [13, 14].

This accuracy can fall drastically when dealing with non-English texts, be it from the lack of tools or the different, sometimes unique, grammar or syntax, for example the case of Croatian [15] as a very inflective language, Arabic [16] with challenges regarding it's impure abjad script, German [17] where word agglutination makes stemming a very hard process or Portuguese [18] with a complex structure characteristic of Latin derived languages.

Other problems arise when you deal with dialects and variants, different spellings can confuse some algorithms and difficult stemming, some language variants even use different syntax and grammar rules [19].

Portuguese is a language spoken in Portugal, Brazil and some parts of Africa, it is of the Western Romance Family that evolved from Latin [20] and uses the standard Latin alphabet with the addition of Ç. It is a language with a complex grammar [21] and many regional variants.

In Portuguese the main differences in spelling and grammar come from the macro variants of Brazilian Portuguese and European Portuguese. Even with the recent orthographic accord [3] the informal voice is still very different and causes confusion between it's speakers.

The language has a hierarchical structure for it's sentence creation, a sentence (or *frase*) is the basic syntactical construct of the language, and can or cannot be a clause (or *oração*) or a period (or *periodo*) containing many clauses. The word order is not so dissimilar from English and almost always follows the order Subject $\rightarrow$ Verb $\rightarrow$ Object [21]. Within this context, given a phrase in Brazilian Portuguese with unknown words or symbols the proposal of this work is to correctly tag them according to their part of speech. This will be done by applying a top down parser known as Packrat [22], modified to accept insertions and/or exclusions in the input. This parser is made to accept the structure of Portuguese.

To accomplish this task, a tool capable of processing any given Context Free Grammar (CFG) and act as a compiler for the language described was created. During the compilation process, the tool can apply error correction for these grammars. Finally, a grammar that describes Brazilian Portuguese rules for sentence formation was developed and described using the notation of the tool for testing of the technique.

While not the main objective, the developed compiler could be used as a general purpose compiler for prototyping programming languages, some functionality was added to support the generation of an Abstract Syntax Tree (AST) with functional nodes and to enforce Semantic rules and restrictions.

The project of this tool can be divided in three steps, first defining a notation able to define the grammar and it's productions, terminals, non-terminals, initial production, ignored characters, as well as different options for the compilers. The second task is implementing the parser itself, and finally the definition of a template of nodes for an AST that can be extended to add functionality.

2   Compilers and Parsing

A parser is a software able to comprehend and most likely translate commands from a origin language to a target language [23]. Compilers and Interpreters are a special kind of parser, able to transform one programming language into machine code. Interpreters are also a special type of parser, able to receive a program written in a programming language and execute it.

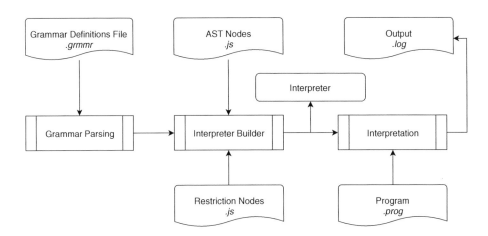

Fig. 1. General flow of the Interpreter

The tool developed in this paper works as a generic compiler, able to receive a programming language definition, some functional nodes and an input program and be able to interpret it. The general flow of the tool can be seen on Fig. 1 and will be further discussed in Sect. 4.1.

2.1   BNF Notation

In the early days of computer programming, punch cards were used to pass instructions to computers and some of these cards were written manually. To facilitate readability people started writing notes on the cards [24].

When programming languages and compilers started being researched, it was necessary to provide a clear description of a programming language structure, function and syntax. To this end, in 1958, John Backus proposed a metalanguage to describe ALGOL, his new programming language [25]. Since then, the Backus-Naur Form (BNF) became the main form used to describe programming languages.

The BNF consists of several rules written in the form: $P ::= E$ where P represents a non-terminal grammar and E is composed of terminals and non-terminals. Each of these rules indicates that the expression on the right is derived from the non-terminal on the left.

For identifying terminals the BNF represents identifiers Between $\langle$ and $\rangle$, double quotes are used to represent terminals, and for shortening the overall length of the representation the character—to indicate alternative expressions. An example is the grammar defined in (1) which defines mathematical operations for integers including the four basic operations $(+, -, *, /)$ and the use of parentheses:

$$
\begin{aligned}
expression ::= &\ \langle addend \rangle\ ``+"\langle expression \rangle \\
| &\ \langle addend \rangle\ ``-"\langle expression \rangle \\
| &\ \langle addend \rangle \\
addend ::= &\ \langle factor \rangle\ ``\times"\langle addend \rangle \\
| &\ \langle factor \rangle\ ``\div"\langle addend \rangle \\
| &\ \langle factor \rangle \\
factor ::= &\ ``("\langle expression \rangle ``)" \\
| &\ \langle integer \rangle \\
integer ::= &\ \langle digit \rangle \langle integer \rangle \\
| &\ \langle digit \rangle \\
digit ::= &\ ``0" \mid ``1" \mid ``2" \mid ``3" \mid ``4" \mid ``5" \mid ``6" \mid ``7" \mid ``8" \mid ``9"
\end{aligned}
\tag{1}
$$

2.2   Compilers and Interpreters Analysis Steps

A Compiler can be broken down in three main steps, Analysis, Optimisation and Synthesis, and each of these steps can be further broken down. The synthesis step is responsible for generating the machine code, and the optimisation step is responsible for guaranteeing the generation of efficient code. The Analysis step is subdivided in Lexical, Syntactic and Semantic Analysis and is responsible for comprehending the source code and generating a representation of it that the computer can understand [23].

Lexical Analysis. The first analysis step, the lexical analysis, aims to translate the strings that make up the source code into a token array, where each token represents a meaningful unit, for example identifiers or operators. This analysis is done so that the next steps of the compiler can ignore string processing, thus simplifying the process.

This step is also used in several other processes including NLP, this means it is a highly optimised process, tha has known algorithms [10,23]. Generally it's implemented as a finite state automata or as a series of regular expression matches.

Syntactic Analysis. After the lexical analysis, the compiler performs the syntactic analysis, this analysis validates the grammatically the source file. This validation is done based on rules defined formally, for example in the notation described in Sect. 2.1 [26].

The input of this step is a token array generated at the lexical analysis, and it's objective is to verify whether it can be generated by the source language grammar. In this step, it is expected syntax errors to be found and reported. For this a derivation tree is generated that connects the final array to the initial symbol in the grammar, Fig. 2 shows an example of such tree for the grammar described in Sect. 2.1 and the expression 10 + 2 * 3.

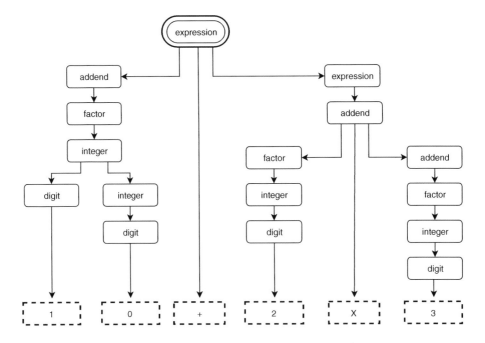

Fig. 2. Derivation of the expression 10 + 2 * 3

One algorithm that can be used to make this analysis step is Packrat's Algorithm, a backtracking based algorithm that uses memory to reduce the complexity of doing a full search through the grammar [22]. It is based on a recursive descent algorithm that transforms each derivation in it's own function call, and adds a *match* function that is responsible for finding the terminals, the general match function is described in Algorithm 1.

Algorithm 1: match()

Data: Expected Token : T, Token Array : arr
Result: **True** or **False**
if *arr.CurrentPosition* $==$ *T* **then**
  | arr.next();
  | **return** *True*;
else
  | **return** *False*;

Semantic Analysis. The last step in the analysis is the semantic analysis, with the objective to give meaning to the derivation generated in the last step. This step validates grammatically if the source code is a valid code, and checks typing, variable declaration, scoping and many other language features.

At the end of this step, an Intermediate Representation is generated, this can be in the form of a three address code, a code graph or an Abstract Syntax Tree (AST). In the AST case, the tree can be interpreted from the root by executing the commands described in each of these nodes.

There are ways for converting a Derivation Tree to an AST if you are able to write one-to-one relationships and every restriction as simple rules, regarding the children or direct parent of each of the nodes [23].

3   Natural Language Processing Concepts

As discussed in Sect. 1 NLP is an interdisciplinary field of computer science and linguistics. Understanding these concepts is essential to comprehending how the method works. Each steps of the process are cumulative and build upon the result of previous steps. Fig. 3 represents the general flow of the process.

Tokenization is a very similar process to lexical analysis described in Sect. 2.2 it breaks up the words on the sentence. Stemming is responsible for reducing words back to it's root form. POS tagging is the step this paper is interested in, and will be discussed in Sect. 3.1. Stop word removal is used to reduce the load of the next steps, removing common words that do not change the final outcome. The next steps collectively deal with the intricacies between two or more phrases, connecting them and marking which words references people, places or other important things.

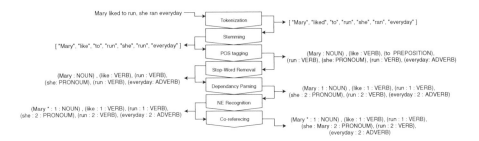

Fig. 3. General flow of NLP

3.1 POS Tagging

Also called Word Categorisation, the process consists in the categorisation of the words based on its function on the sentence. This process is a fundamental part of the analysis of natural language computationally.

In Portuguese, the word classes are: *substantivo, verbo, adjetivo, pronome, artigo, numeral, preposição, conjunção, interjeição and advérbio* [21]. A tagger can also add extra information to each class, like a verb conjugation or it's inflection.

Currently there are various tagging techniques using methods of different computational areas, the main ones being based on Bayesian networks through the Markov model, neural networks, support-vector machines and Grammars defined by specialists.

Each one has a distinct advantages and disadvantages, for example the grammar based ones are simple to implement and efficient, but costly to initially create due to the necessity of an expert to create the grammar.

4   Methodology

The development was split into two parts: the interpreter and the grammar. Each of these parts works individually and were connected to achieve the error-correction.

To test the error correction, a Portuguese corpus based on the 2014 world cup hashtag used in conjunction with a basic sentiment analysis tool was used to verify if the POS correction would bring improved results to a real-world application. The data-set consists of a previously available data-set [27] and some tweets collected using the Twitter API. The added tweets were collected to increase variety in the data.

To understand the technique, we must understand that an error in the derivation process is a word that was tagged wrongly. These errors can happen in three forms: a word was added to the phrase when, according to formal rules, it shouldn't; an word was omitted causing a gap; or a word used had either no common part-of-speech or was used in a way that is not formally accepted, for example as a slang.

Two applications were tested with and without error-correcting to test whether they would improve for both of the tests. The correction was focused on substituting emojis with equivalent words using the correct tag. For this, a table was added correlating each emoji to a word in a given part of speech. In Table 1 a line of this table can be seen.

Table 1. Substitutions for an emoji

Emoji	Substantive	Verb	Adjective	Interjection	Adverb
😄	Risada	Rir	Engraçado	Haha!	Rindo

According to the table, if for example the emoji 😄 appears as a substantive it will be replaced by *Risada*, which means laughter. Another thing to note is that not every emoji can appear as every Part-of-Speech, for example, there is no occurrence of this emoji as a pronoun.

The first application was sentiment analysis where it was expected that this correction meant that each word was interpreted by the sentiment analysis as a meaningful corresponding word, and not skipped by the algorithm. Another test was made using a simple translator after substituting an emoji the same way. After the translation, the emoji was substituted back resulting in a phrase was the symbol exerted the same function. Both tests result will be discussed in Sect. 5.

4.1   The Interpreter

Built using Node.js, the interpreter uses *Packrat*'s algorithm to compile a grammar definition file and construct the functions used to interpret the input file. It is defined by a BNF notation described in Sect. 2.1 with the keyword ROOT used to define the grammar's initial rule, and with the node being represented by that rule in parenthesis to represent the correspondence to the AST. Each of these nodes can be extended or overwritten to add new functionalities to the interpreter.

Figure 1 in Sect. 2 shows the flow in three basic steps Grammar Parsing, Interpret Builder and Interpretation. while in Grammar Parsing the interpreter reads a *.grmmr* file containing the definition of the language that will be interpreted. It is defined by a BNF notation described in Sect. 2.1 with the keyword **ROOT** used to define the grammar's initial rule, and with the node being represented by that rule in parenthesis to represent the correspondence to the AST.

After reading the grammar it is optimized to be used in the algorithm, recursion is removed and functions are built using code inflection to represent the rules. The final step is done by simply calling the function defined as the initial derivation.

4.2   The Grammar

The efficacy of the algorithm depends on a good representation of the Portuguese Language as a grammar. But as it is discussed in the literature [28] Portuguese grammar, especially the Brazilian variant presents some nuances that make it hard to work within computational applications.

Initially, the base grammar was developed with the help of a book [21], but this proved to be a challenge and it meant another approach was needed. The version used in this paper is a combination of that work and a CFG inductor applied to an annotated corpus called MAC-MORPHO [29].

The induction works by going through the annotated base and generating a new derivation rule that represents the most common pair of occurrences. The algorithm then replaces this pairs with the new non-terminal and repeats. Because of this, all the generated rules follow the form $A-> BC$ where A is the new rule and B and C are either terminals or non-terminals. A side note to this algorithm is that it allows for a quick transformation of any generated CFG to the Chomsky Normal Form. Some rules of the CFG were then named to better represent what they meant in the Portuguese grammar.

5   Results

The accuracy of the sentiment analysis algorithm was calculated before and after the correction, described in Sect. 4 with results shown in Table 2. Accuracy represents the percentage of tweets that were correctly analyzed. Each of the instances represents a part of the evenly split full data set. All parts were subject to the same algorithm and represent about 2000 tweets.

The sentiment analysis algorithms use each word to calculate a score ranging from -1.0 to 1.0. A tweet that is 100% negative is -1.0 and a tweet that is 100% positive is 1.0. A tweet with a score of 0.0 means it is neutral. The accuracy was calculated as the absolute error concerning the total possible error, as given by the formula:

$$1.0 - \frac{\Delta score}{2.0}$$

In this formula, a tweet that should have been classified as 1.0 but was given the score of -1.0 has an accuracy of 0%.

As seen, the correction improved the results by an average of about 3%. These results show us that trying to preprocess some symbolic information like emoticons, emoji, or *slangs* can help to get better results with NLP algorithms. This seems obvious but most processes drop these clues to keep the simplicity of implementation.

Notably, the most affected examples were the ones where the words were positive but the emoji was negative, being used as a means of sarcasm, for example, the tweet in Fig. 4: *"Brasil ta jogando pacas 💩"* which roughly translates to "💩

Table 2. Accuracy of sentiment analysis before (Acc_0) and after (Acc) POS correction

Instance	Acc_0	Acc	Gain
1	91.609%	95.363%	4.10%
2	91.642%	95.669%	4.39%
3	92.056%	92.709%	0.71%
4	91.675%	95.406%	4.07%
5	91.768%	93.223%	1.59%
6	92.375%	93.131%	0.82%
7	91.070%	91.898%	0.91%
8	91.884%	92.062%	0.19%
9	92.465%	96.982%	4.88%
10	91.759%	96.678%	5.36%
Avg	91.572%	94.312%	2.99%

Brasil ta jogando pacas 💩

2 Retweets 1 Likes

Fig. 4. Example of tweet that was previously wrongly classified

Brazil is playing so well" was previously classified as positive but is now correctly classified as negative, in this example when the emoji was classified as an interjection, and substituted for booing which changed the outcome.

In this same example, the word *"pacas"*, used as a slang, was also affected by the algorithm, being previously tagged by the dictionary as a noun (an animal) it was later categorized as an adverb (meaning a lot/well).

The best results came from machine translation were word order is a known problem [30]. By shifting the emojis into words and back into emojis the sentences were more understandable and human-like than their counterparts. This mainly shows up in sentences with subject-object-verb order, where the verb was substituted, for example in Fig. 5 *"Esse Brasil vai me ＼ do coração"* which was previously translated keeping the order as *"This Brazil will me ＼ of the heart"* by the naive algorithm now is translated into proper subject-verb-object order in English *"This Brazil will ＼ me from the heart"*.

Fig. 5. Example of tweet that was previously wrongly translated

6   Conclusion

This paper shows it is possible to apply error correction to POS tagging and increase the accuracy of later steps in NLP processing. Two tools were also developed that can become the basis of future work.

The grammar can be improved and used in many works regarding the Portuguese language. Future work also includes trying to use a grammar inductor to find out if it is possible to generate a 100% complete grammar of the language. This grammar can be used in conjunction with already existing machine learning processes to help them learn process Portuguese.

This complete grammar can also be used outside Computer Science to study its structure formally. This formal understanding is the basis of formal semantics that started with Richard Montague.

The compiler itself is a tool able to handle multiple languages, developing it further and adding new features such as the ability to load a pre-parsed grammar, the option to use it as a transpiler or expanding its repertoire of pre-existing nodes brings a new prototyping tool in compiler design. There is a multitude of uses for a flexible parser in many fields of Computer Science, and the developed one could be used as a basis for a more powerful tool.

Finally, the technique used although not entirely new enables a better understanding for future developments in the field of Natural Language Processing for Brazilian Portuguese. Finding which algorithms Error correction increases the accuracy of leads to new applications which demand this level of accuracy.

References

1. MacNeill, J.: Notes on the distribution, history, grammar, and import of the Irish Ogham inscriptions. In: Proceedings of the Royal Irish Academy. Section C: Archaeology, Celtic Studies, History, Linguistics, Literature, vol. 27, pp. 329–370 (1908)
2. Kim-Renaud, Y.K.: The Korean Alphabet: Its History and Structure. University of Hawaii Press, Honolulu (1997)
3. Zúquete, J.P.: Beyond reform: the orthographic accord and the future of the Portuguese language: South European atlas. South Eur. Soc. Politics **13**(4), 495–506 (2008)

4. Herring, S.C.: Language and the internet. In: The International Encyclopedia of Communication (2008)
5. Steinmetz, K.: Oxford's 2015 word of the year is this emoji (2015). https://time.com/4114886/oxford-word-of-the-year-2015-emoji/
6. Bates, M.: Models of natural language understanding. Proc. Nat. Acad. Sci. **92**, 9977–9982 (1995)
7. Rocha, L., et al.: SACI: sentiment analysis by collective inspection on social media content. J. Web Semant. **34**, 27–39 (2015)
8. Gachot, D.A., Lange, E., Yang, J.: The SYSTRAN NLP browser: an application of machine translation technology in cross-language information retrieval. In: Grefenstette, G. (ed.) Cross-Language Information Retrieval. INRE, vol. 2, pp. 105–118. Springer, Boston (1998). https://doi.org/10.1007/978-1-4615-5661-9_9
9. Howard, J., Ruder, S.: Universal language model fine-tuning for text classification. CoRR abs/1801.06146 (2018). http://arxiv.org/abs/1801.06146
10. Indurkhya, N., Damerau, F.J.: Handbook of Natural Language Processing, 2nd edn. Taylor & Francis, Abingdon-on-Thames (2010)
11. BRN.AI, B.: Chatbot report 2019: Global trends and analysis, April 2019. https://chatbotsmagazine.com/chatbot-report-2019-global-trends-and-analysis-a487afec05b
12. Manning, C.D.: Part-of-speech tagging from 97% to 100%: is it time for some linguistics? In: Gelbukh, A.F. (ed.) CICLing 2011. LNCS, vol. 6608, pp. 171–189. Springer, Heidelberg (2011). https://doi.org/10.1007/978-3-642-19400-9_14
13. Bird, S., Klein, E., Loper, E.: Natural Language Processing with Python. O'reilly, Newton (2009)
14. Loria, S.: Textblob: Simplified text processing - textblob documentation (2018). https://textblob.readthedocs.io/en/dev/
15. Agić, Ž., Ljubešić, N., Danijela, M.: Lemmatization and morphosyntactic tagging of Croatian and Serbian. In: Proceedings of the 4th Biennial International Workshop on Balto-Slavic Natural Language Processing, pp. 48–57 (2013). https://www.aclweb.org/anthology/W13-2408
16. Farghaly, A., Shaalan, K.: Arabic natural language processing. ACM Trans. Asian Lang. Inf. Process. **8**, 1–22 (2009)
17. Baeza-Yates, R.: Challenges in the interaction of information retrieval and natural language processing. In: Gelbukh, A. (ed.) CICLing 2004. LNCS, vol. 2945, pp. 445–456. Springer, Heidelberg (2004). https://doi.org/10.1007/978-3-540-24630-5_55
18. de Barbosa, C.R.S.C., Cury, D., de Castilho, J.M.V., Souza, C.D.R.: Defining a lexicalized context-free grammar for a subdomain of Portuguese language. In: Proceedings of the Sixth International Workshop on Tree Adjoining Grammar and Related Frameworks (TAG+ 6), pp. 74–79 (2002)
19. Görski, E.M., Coelho, I.L.: Variação linguística e ensino de gramática. In: Working Papers em Lingüística, vol. 10, February 2009
20. de Língua Portuguesa, C.D.P: CPLP - comunidade dos países de língua portuguesa (2019). http://www.cplp.org
21. Bechara, E.: Moderna Gramática Portuguesa. Editora Nova Fronteira, Rio de Janeiro (2015)
22. Ford, B.: Packrat parsing: a practical linear-time algorithm with backtracking. Ph.D. thesis, Massachusetts Institute of Technology (2002)
23. Grune, D., Reeuwijk, K.V., Bal, H.E., Jacobs, C.J.H., Langendoen, K.G.: Modern Compiler Design. Springer, Heidelberg (2016). https://doi.org/10.1007/978-1-4614-4699-6

24. Dale, F.: Programming with punched cards. In: Programming with Punched Cards (2005)
25. Backus, J.W.: Revised report on the algorithmic language ALGOL 60. Comput. J. **5**, 349–367 (1963)
26. Aho, A.V., Ullman, J.D.: Principles of Compiler Design. Narosa Publishing, House, New Delhi (1999)
27. Moraes, S.M., Manssour, I.H., Silveira, M.S.: 7x1pt: Um corpus extraído do twitter para análise de sentimentos em língua Portuguesa. In: Proceedings of Symposium in Information and Human Language Technology, November 2015
28. da Silva, B.C.D.: O estudo lingüístico computacional da linguagem. Letras de Hoje **41**(2), 103–138 (2006)
29. Fonseca, E.R., Rosa, J.L.G.: Mac-morpho revisited: towards robust part-of-speech tagging. In: Proceedings of the 9th Brazilian Symposium in Information and Human Language Technology (2013)
30. Barreiro, A., Ranchhod, E.: Machine translation challenges for Portuguese. Linguisticæ Investigationes **28**(1), 3–18 (2005)

Forecasting Soccer Market Tendencies Using Link Prediction

Lucas G. S. Félix[1], Carlos M. Barbosa[1], Iago A. Carvalho[2],
Vinícius da F. Vieira[1], and Carolina Ribeiro Xavier[1(✉)]

[1] Department of Computer Science, Universidade Federal de São João del Rei,
São João del Rei, MG, Brazil
lucasgsfelix@gmail.com, cmagnobarbosa@gmail.com,
{carolinaxavier,vinicius}@ufsj.edu.br
[2] Department of Computer Science, Universidade Federal de Minas Gerais,
Belo Horizonte, MG, Brazil
iagoac@dcc.ufmg.br

Abstract. Soccer is the most popular sport in the world and due its popularity, soccer moves billions of euros over the years, in most diverse forms, such as marketing, merchandising, TV quotas and players transfers. As example, in the 2016/2017 season, only England has moved about 1.3 billion of euros only in players transfers. In this work, it is performed a study of the transfer market of players and, to do so, players transfer data were gathered from the website Transfermarkt and transfers were modeled as a graph. In order to perform this study, different Complex Networks techniques were applied, such as Link Prediction, Community Detection and Centrality Analysis. Link Prediction it was possible to conceive a network of a forecast market for the 2022 World Cup using temporal data. Through the generated network it was possible to notice changes of the market and the rise of Asian countries, given a notion of how the market is going to behave in the next years.

Keywords: Complex networks · Data Mining · Sports analytics

1 Introduction

Soccer is the most popular sport in the world [14,21], and due to this enormous popularity it attracts millions of practitioners and billions of spectators for large events as FIFA World Cup, continental and intercontinental championships. Thus, soccer generates a high financial flow produced by ticket sells, TV contracts, marketing and merchandising. A clear illustration of the huge amount of money in this market can be observed in England where, in a single season 2016/2017, moved almost 1.3 billions of euros [4].

In between the various financial compensation that a team can explore, players transference between clubs stands out. This process makes possible to sell and buy athletes, consequently giving a team the opportunity to get a certain

© Springer Nature Switzerland AG 2020
O. Gervasi et al. (Eds.): ICCSA 2020, LNCS 12249, pp. 663–675, 2020.
https://doi.org/10.1007/978-3-030-58799-4_48

quantity of money in a short time. Those transactions are cataloged and categorized in the website Transfermarket[1], which contains a considerable quantity of transfer data, classifying buy and sell transaction, as well as loan transfers among teams.

Due to these facts, the goal of this work is to study and to analyze the properties of the network of soccer players transfers between countries present in the World Cup. We choose the World Cup because it is the second biggest sport event in a global scale [1], besides that, the World Cup is one of the biggest influence in the player's transfer market [6]. To perform this evaluation, we applied Complex Networks techniques.

From the perspective of Complex Networks, studied in the field of study as Network Science [23], data is modeled as graphs in order to be analyzed. Our methodology applies Link Prediction, Community Detection, and Centrality analysis. By the application of Link Prediction we could forecast the chances of a vertex A to connect to a vertex B and then investigate the community structure of the network by Community Detection, to view sub-markets and the countries that are present in these clusters. Using Centrality measures we could observe which nodes are most important for the network and consequently which countries are most important for the football market.

Through our methodology, we could observe the tendency of soccer players market concerning other countries that dominate this market. However, it is possible to perceive changes in sub-markets, evidencing the rise of Asian countries driven mostly by China.

2   Related Work

Soccer has been a source of study over the year in many areas as Physical Education, where researchers evaluate the biomechanics of the sport [12], methods for sportive training and efficiency [10], sports medicine and nutrition [19]. Another research area analyzes soccer by a sociological or anthropological vision. Those works make qualitative analysis over topics as, international migration of work forces [16], cultural transformation induced by soccer [22], or even social questions, as racism [11], and violence [24].

Few works in the literature seek to understand, analyze and evaluate the soccer players transactions market. In those works, we highlight the works of [21] and [9]. The work of Huerta et al. [21] evaluated the players transfer applying temporal behavioral statistical analysis in the English Premier League. The work of Frick et al. [9] analyzed the European players transactions empirically, evaluating many features present in soccer athletes as salary and career time.

In this field of study, other works with more computationally focused approaches can be fond [5,6,25], applying network metrics to evaluate the players transfer market. The work of Xiao et al. [25] pointed out only a few properties of the transactions market, using data from 2011 to 2015. The general aim of the

---

[1] www.transfermarkt.com.

work was not to analyze the market, but to evaluate the success of a team given its transfers. In previous works, we used Complex Networks and Data Analysis techniques to analyze the teams presents in the 2018 FIFA World Cup [5] and proposed a method to join centrality metrics [6]. Besides that, we explored sociological vision of soccer, showing theories previously raised by other authors in a quantitative way [6].

The methodology proposed in this work differs from those found in other works because it aims at the evaluation of future consumers market, considering Link Prediction techniques.

3   Methodology

The objective of this work is to evaluate the future soccer market, in specific the countries that can be be present in the next FIFA World Cup. To make those analysis Complex Networks and Data Mining techniques are applied. Our approach is composed by four steps: (*i*) Data Gather and Treatment, which consists in the development of an automatic method to capture data in websites; (*ii*) Network Modeling, in which we model the network transfers between countries as graphs; (*iii*) Link Prediction, when a future network is generated to analyze and preview countries and connections between these countries (links) that can be in the 2022 FIFA World Cup; and (*iv*) Network Analysis, which applies Complex Networks measures to evaluate the network and validate our approach.

3.1   Data Collect and Treatment

To perform the proposed study, we developed a crawler for the website Transfermarkt (See footnote 1). This website consists of a vast database with information related to soccer: statistics, championships standings and, specifically concerning to this work, data about players transfers. To process the data gathering, we developed a crawler and a parser. These two tools aims to clean and pre-process the data. At the end of the process, the data is in a semi-structured format.

The data gathered ranges from 1962 to 2017 and we considered the 250 most valuable transactions per year per tactical position (goalkeeper, defenders, left-back, right-back, midfielder, forwards). It is important to mention that, for some years, the number of transfers in the website does not reach 250. The final data collection consists of about 28 thousand players transfers.

From Fig. 1, it is possible to observe the number of transfer gathered by position and period. Also, it is possible to notice that the number of transfers is increased year by year. This aspect reflects the expansions of the player's transactions market as shown in [6].

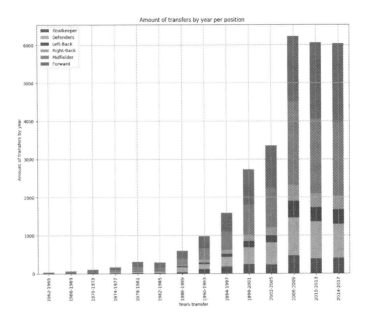

Fig. 1. Number of transfer gathered by position and period

3.2 Network Modeling

After the data gathering, the networks were modeled in a way that each country is a vertex and a directed edge exists between two nodes if a transference between these countries occurs. In the proposed model, an outgoing edge represents a sell transaction, and an incoming edge represents a buy transfer. The weight of the links is defined as the sum of the number of transfers occurred between one country and another. Previous works [5,6] also modeled the edge weight as the sum of the value of transactions between two countries, however, this approach is not considered in this work, due to the temporal feature of our modeling. Thus, it was possible to model the network avoiding distortions caused by the variations of athletes prices over the years.

To model the networks, the period of time considered was divided in intervals of 4 years, and then, the transfers were joined aiming to summarize the transactions realized between a World Cup and another. This approach was considered for the network because the years of World Cup tend to be renewal years for the teams. In this sense, it is considered that the period that a defines the 2018 FIFA World Cup, for instance, ranges from 2014 to 2017.

3.3 Link Prediction

Link Prediction is a Complex Networks technique that assists in understanding many dynamic elements, as Social Networks, and can be used as a tool to analyze the evolution of these networks [13]. Thus, to preview the tendencies of a link between the nodes of the modeled networks, Jaccard Similarity Measure was applied. This technique assumes high values to pairs of nodes that share a great proportion of common neighbors concerning the total number of neighbors [15]. In other words, the larger the quantity of common neighbors between two vertexes, the larger is the probability of these two nodes share an edge. Mathematically, the Jaccard measure between two sets A and B is defined by Eq. 1.

$$J(A, B) = \frac{|a \bigcap b|}{|a \bigcup b|}.$$ (1)

3.4   Network Analysis

To perform analysis over the network, we applied different techniques, as community detection and centrality measures. The Community Detection algorithm aims at the identification of cohesive groups in a network. In this work three different algorithms to find partitions in the graph were considered: (i) Multilevel [2]; (ii) Eigenvector [17]; (iii) Fastgreedy [3]. All these methods has, as basic principle, the optimization of modularity [18], which evaluates how good is a partition of a graph. In this way, the larger the modularity value, the better is network partition, and more consistent are the formed communities. Using these different approaches, it is possible to find different community partitions. A brief description of the methods for community detection is preseted as follows:

- **Multilevel:** This algorithm works in an agglomerative way, and performs two steps to divide the graph. In the first step, it is considered that every node is a community which swap to more convenient communities, until there is no modularity gain. In the second phase, it builds a new network considering each community as a node and perfoms community joins, similarly to the first step [2].
- **Eigenvector:** This divisive algorithm works based in a Laplacian Matrix (Modularity Matrix) to make the graph partition and identify the communities. This way, the algorithm manipulates the Modularity Matrix, and calculates the eigenvector associated to the largest eigenvalue to perform a bisection. This process is repeated until there is no modularity gain with a new division [17].
- **Fastgreedy:** This algorithm is also agglomerative and considers, initially, the nodes as unary communities and then perform convenient joins. The algorithm stops when there is no communities joins that cause a positive modularity gain [3].

Centrality measures were also explored in order to investigate the relevance of the nodes of the graph. In the context of this work the analysis corresponds to the importance of a country to the player's transfer market. Three different approaches were investigated in this study: (i) Closeness [8], (ii) Betwenness [7], (iii) Pagerank [20], which can be defined as follows:

– **Closeness:** Evaluates the geodesic distance between a node and the others reachable nodes considering the analyzed vertex [8].
– **Betweenness:** Evaluates the fraction of shortest paths between all pair of nodes. The most central vertex will be the node with the most significant fraction of shortest paths passing through it [7].
– **Pagerank:** Evaluates the importance of a node by assessing the importance of its neighbours [20].

For Betweenness and Closeness centrality, the weight of the edges were reversed. Such action was needed because these methods are based on shortest paths and using a traditional edge weight approach the precision over the main countries was altered. Thus, using the approach where the edge weight is $\frac{1}{w_{i}j}$, where $w_{i}j$ is the weight between the nodes i and j, the relevance of a node is more clearly revealed.

The remaining of the analysis was made evaluating the properties of the network as density, diameter, clustering coefficient, reciprocity, and assortativity degree. These measures will be presented as they appear in Sect. 4.

4   Results

In this section the results obtained by the application of the proposed methodology are presented. First, we described how the network was developed. Next, we report and analyze the obtained experimental results by Complex Network techniques.

4.1   Network Development

To assembly the graph used in this work, a process that could guarantee the reliability of the modeled network was necessary. To make so, several preparation and preprocessing steps were performed. These steps were fundamental due to the long term characteristics of the data, ranging over decades, which could bring some inconsistencies if not properly treated. Some countries can have a huge number of transactions in one year, and a smaller number in another, showing the dynamicity of the soccer market. Thus, to generate the 2022 player's market network, it was necessary to assure that countries present in these graphs are really actives in the soccer market. First, to model the network in a way that the outliers were not present in the graph, the vertex that has just a few links over all networks were withdrawn of the graph, i.e., countries that did not maintain transactions over the years, only having exceptional transfers, was removed from the graph in a temporal way.

The proposed methodology considers as reference a network modeled in the subsequent years to 2002. Thus, it was possible to guarantee that a vertex present in the network of 2022, has as nodes, countries that are present in transference in 2002, 2006, 2010, 2014 and 2018. As previously shown in Sect. 3, the proposed methodology modeled the networks in periods of 4 years.

2002 was selected as the reference year we used the maximization of the number of vertex and networks modeled as criterion. From Fig. 2, it is possible to see that starting from the year 2002, we could achieved a reasonable number of networks and nodes, being a good trade-off between the number of networks and the number of nodes.

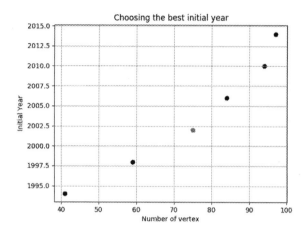

Fig. 2. Initial number of vertex by year, considering a intersection between the years

By the evaluation of Fig. 2, it can be noticed that the number of nodes is increasing along the years. This happens due to the soccer globalization that gave the opportunity to countries that does not have much tradition in soccer to sell and buy players [6]. As an example, there is China that in recent years have become a big negotiator in the soccer market.

After identifying the nodes that will form the network of 2022, it was defined that the graph which could serve as a base is the one from 2018. The 2018 graph was selected because the soccer market have small changes over the years, thus, to define this changes from 2018 to 2022 it was applied a Link Prediction technique, considering the Jaccard Coefficient. As the Jaccard Metric is calculated for every pairs of vertices in the graph, it was necessary to limit the minimum value to link two nodes, as a threshold value. Thus, it is was possible to avoid connections between vertices that have a small and we avoid to create a very dense graph. So, to define a threshold to Jaccard Metric it was used an approach based on the elbow method. The elbow method is used to find the ideal number of clusters in an unsupervised machine learning technique. However, we have adapted the metric, using it to define the point that gives the better accuracy for the forecast made by Link Prediction.

From Fig. 3 it is possible to observe the accuracy rate of the networks given a threshold. To configure an accuracy to the modeled networks it was considered that every network has an ground truth, except the 2018 network. To illustrate that, consider the that 2002 network has as ground truth the 2006 graph, and

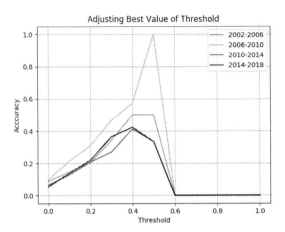

Fig. 3. Adjusting the best value of threshold

the applied Link Prediction technique helps to get closer to all right answers. This way, it was considered that the bigger is the accuracy, the better is the prediction.

To define the ideal threshold t_i, we considered an approach where t_i value is the mean between all k-threshold t_k found before, given by Eq. 2, where n is the number of networks. This way, an edge is added to the graph if Jaccard Coefficient is bigger than 0.425. In the end, the 2022 final network have 75 nodes and 1117 edges.

$$\frac{\sum(t_k)}{n} \tag{2}$$

$$\frac{(\sum(X_i))/n + \sum(Y_j)/m}{2} \tag{3}$$

To define the weights of the edges in the graph, Eq. 3 was considered, where X and Y are the linked countries, i represents sell transactions, j represents buy transactions, n is the number of sells of country X and m is the number of buys of country Y.

4.2 Centrality Analysis

Network centrality analysis gives the importance of a vertex in a graph, being an fundamental tool in the study context since through it is possible to identify the main countries in the player's transactions market. In this work, three different metrics were evaluated, as described in Sect. 3: Closeness, Betweenness, and Pagerank. After the application of the metrics three different sorted lists were obtained. From Table 1 it is possible to observe the 10 first positions. Also, from Fig. 4 it is possible to see 2022 network with the main vertex identified by the Betweenness centrality.

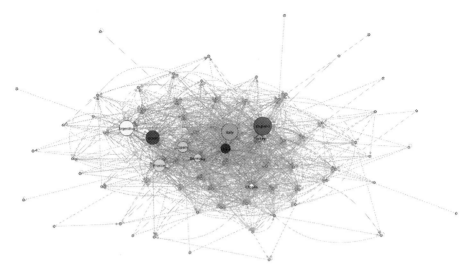

Fig. 4. Main countries identified by the Betweenness centrality.

By analyzing Tables 1 and 2, it is possible to see small, but significant changes in the rankings. As example we can see the rise of countries like Turkey and Romania. Besides, there is the presence of China between the main nodes in all three metrics, due to the expansion of his market as a potential buyer, as Saudi Arabia.

4.3   Community Analysis

To identify sub-markets that perform a huge volume of transference Community Detection was considered. Thus cohesive groups in the network, communities,

Table 1. Top-10 countries for each metric considered in the 2022 network.

Position	Closeness	Betweenness	PageRank
1	England	England	England
2	Italy	Italy	Italy
3	Spain	Argentina	Germany
4	Germany	Brazil	Spain
5	Turkey	France	Turkey
6	France	Spain	France
7	Wales	China	China
8	Romania	Germany	Brazil
9	Argentina	Turkey	Saudi Arabia
10	China	Croatia	Romania

Table 2. Top-10 countries for each metric considered in the 2018 network.

Position	Closeness	Betweenness	PageRank
1	England	Italy	England
2	Italy	England	China
3	Spain	Argentina	Italy
4	Germany	Brazil	Germany
5	Turkey	Germany	Spain
6	France	France	Turkey
7	Wales	Spain	France
8	China	Portugal	Saudi Arabia
9	Argentina	Croatia	Belgium
10	Portugal	Turkey	Brazil

could be identified. The nodes in this subgroups have as main characteristic the fact that their nodes have a larger number of inside connections (intracluster) that outside links (extracluster).

This analysis is motivated by the fact that the soccer players transference market is an embracing market, where there are free exchanges of players, and some consumption patterns can observed. Those consumption patterns are given by some countries that are used as a source of talents do rich European markets. This way, to identify this sub-markets, three different methods of community detection were considered, as presented in Sect. 3: Multilevel, Eigenvector and Fastgreedy. From Table 3, it is possible to see the modularity and number of communities results obtained by each of the algorithms.

Considering that the applied methods to identify communities aim to maximize modularity and evaluating Table 3, it is possible to notice tha Multilevel and Eigenvector methods offer the same modularity value and the same number of communities. Due to its lower computational cost, Multilevel method was selected for further analysis.

By analyzing Table 4, it is possible to see the generated communities by the Multilevel method with more than one node. Communities with only one node were removed from this table because these communities do not bring any additional information to our analysis. Table 4 highlights the countries considered of rich leagues [25], except for Spain which was identified in an unary community and removed.

In previous works [5,6], it has been shown that soccer player's market is dominated by European countries, where each one dominates its own niche of trades, being present in distinct communities with countries of less relevance.

Besides that, as this is predicted network, it is possible to notice tendencies in the market, where the rich countries are searching for better markets at the moment. However, the changes in the market is bigger in the Asian countries, where China is becoming a country as powerful as European countries.

Table 3. Modularity found by each algorithm.

Algorithm	Modularity	# of communities
Eigenvector	0.53	35
Fastgreedy	0.32	29
Multilevel	0.53	35

Table 4. Communities with more than one vertex formed by the Multilevel method.

Communities	Countries
1	Wales, **England**, Scotland
2	**Italy**, Moldova, Croatia, Australia, Slovenia
3	Qatar, Belgium, Greece, **France**, Cyprus, Israel, Monaco
4	Netherlands, Sweden, Norway, Denmark, Serbia, Russia, Poland, Slovakia, Bosnia-Herzegovina, Ukraine, Hungary,Georgia
5	Austria, Japan, Switzerland, **Germany**
6	Romania, Bulgaria
7	Czech Republic, Turkey
8	Saudi Arabia, Tunisia, Egypt
9	Mexico, Chile, Brazil, Paraguay, Portugal, Uruguay, Ecuador United Arab Emirates, United States, Colombia, Argentina
10	South Korea, China

5   Conclusion

In this work we proposed a study about the tendencies of the soccer players transfer market. To make so different complex networks strategies were applied, where we identify some of these changes in soccer world scenario for the next World Cup. Among these possible transformations, a more clear protagonism of Asian countries in the transfer market could be observed. However, there will be still a visible dominance of Europe in this market.

Besides that, it can be also noticed a large Jaccard Coefficient connecting rich countries, given that these countries have sufficient money to perform transactions with any country in the world. Also, our analysis identifies that there is a common node change in the communities, showing a high variation of the market.

As future works, we plan to use other techniques of Link Prediction and heuristics that with prior knowledge could help us to increase accuracy.

References

1. Baade, R.A., Matheson, V.A.: The quest for the cup: assessing the economic impact of the world cup. Reg. Stud. **38**(4), 343–354 (2004)
2. Blondel, V.D., Guillaume, J.L., Lambiotte, R., Lefebvre, E.: Fast unfolding of communities in large networks. J. Stat. Mech. Theory Exp. **2008**(10), P10008 (2008)
3. Clauset, A., Newman, M.E.J., Moore, C.: Finding community structure in very large networks. Phys. Rev. E **70**, 066111 (2004)
4. Deloitte, L.L.P.: Annual review of football finance, June 2016
5. Felix, L., Barbosa, C., Carvalho, I., Vieira, V., Xavier, C.: Uma análise das seleçóes da copa utilizando uma rede de transferências de jogadores entre países. In: CSBC 2018 - VII BraSNAM, July 2018
6. Felix, L., Barbosa, C., Vieira, V., Xavier, C.: Análise do impacto das copas do mundo no mercado de transações de jogadores de futebol e da globalização do futebol utilizando técnicas de redes complexas. In: ENIAC 2018 - VII KdMIle, October 2018
7. Freeman, L.C.: A set of measures of centrality based on betweenness. Sociometry 35–41 (1977)
8. Freeman, L.C.: Centrality in social networks conceptual clarification. Soc. Netw. **1**(3), 215–239 (1978–1979)
9. Frick, B.: The football players' labor market: empirical evidence from the major European leagues. Scott. J. Polit. Econ. **54**(3), 422–446 (2007)
10. González-Badillo, J.J., Pareja-Blanco, F., Rodríguez-Rosell, D., Abad-Herencia, J.L., del Ojo-López, J.J., Sánchez-Medina, L.: Effects of velocity-based resistance training on young soccer players of different ages. J. Strength Conditioning Res. **29**(5), 1329–1338 (2015)
11. Jarvie, G.: Sport, racism and British society: a sociological study of England's elite male Afro/Caribbean soccer and rugby union players. In: Sport, racism and ethnicity, pp. 79–102. Routledge (2003)
12. Lees, A., Asai, T., Andersen, T.B., Nunome, H., Sterzing, T.: The biomechanics of kicking in soccer: a review. J. Sports Sci. **28**(8), 805–817 (2010)
13. Liben-Nowell, D., Kleinberg, J.: The link-prediction problem for social networks. J. Am. Soc. Inf. Sci. Technol. **58**(7), 1019–1031 (2007)
14. Liebig, J., Rhein, A.V., Kastner, C., Apel, S., Dorre, J., Lengauer, C.: Large-scale variability-aware type checking and dataflow analysis (2012)
15. Lü, L., Zhou, T.: Link prediction in complex networks: a survey. Physica A **390**(6), 1150–1170 (2011)
16. Magee, J., Sugden, J.: "The World at their Feet" professional football and international labor migration. J. Sport Soc. Issues **26**(4), 421–437 (2002)
17. Newman, M.E.J.: Finding community structure in networks using the eigenvectors of matrices. Phys. Rev. E **74**, 036104 (2006)
18. Newman, M.E.J., Girvan, M.: Finding and evaluating community structure in networks. Phys. Rev. E **69**, 026113 (2004)
19. Osgnach, C., Poser, S., Bernardini, R., Rinaldo, R., Di Prampero, P.E.: Energy cost and metabolic power in elite soccer: a new match analysis approach. Med. Sci. Sports. Exerc. **42**(1), 170–178 (2010)
20. Page, L., Brin, S., Motwani, R., Winograd, T.: The pagerank citation ranking: bringing order to the web. Technical report 1999–66, Stanford InfoLab, November 1999. previous number = SIDL-WP-1999-0120

21. Palacios-Huerta, I.: Structural changes during a century of the world's most popular sport. Stat. Methods Appl. **13**(2), 241–258 (2004)
22. Redhead, S.: Post-Fandom and the Millennial Blues: The Transformation of Soccer Culture. Routledge, Abingdon (2002)
23. Strogatz, S.H.: Exploring complex networks. Nature **410**(6825), 268 (2001)
24. Taylor, I.: On the sports violence question: soccer hooliganism revisited. In: Sport, Culture and Ideology (RLE Sports Studies), p. 152 (2014)
25. Liu, X.F., Liu, Y.L., Lu, X.H., Wang, Q.X., Wang, T.X.: The anatomy of the global football player transfer network: club functionalities versus network properties. PLoS One **11**(6), e0156504 (2016)

Digital Signature in the XAdES Standard as a REST Service

Renato Carauta Ribeiro⬤ and Edna Dias Canedo(✉)⬤

Department of Computer Science, University of Brasília (UnB), P.O. Box 4466,
Brasília, DF, Brazil
{rcarauta,ednacanedo}@unb.br
http://www.cic.unb.br

Abstract. The Brazilian government has been discussing the topic of information security and the availability of information through digital media. Information made available in digital format must have the same protection as information in physical format. One of the solutions proposed in the educational area to make documents available in a secure digital format was introduced by the Ministry of Education (MEC), which was the creation of a model for signing papers and University Degree in digital format. The government defined the main guidelines for the creation of this digital signature solution through ordinances. According to the MEC, the deadline for implementing a system for signing University Degree and documents in digital format must be established by 2022. This work presents a digital signature solution for the University of Brasília (UnB). Besides, we demonstrate the current architecture used by UnB and how the creation of new solutions can be incorporated into this architecture. Thus, we present the internal architecture of the digital signature module and demonstrate how the developed solution will be integrated into the current architecture of UnB. As a main result, the proposed solution presents a reduction in the costs of signing digital documents and allows higher speed in signing University Degree and documents in digital format.

Keywords: RSA · Digital signature · XAdES · Information security · Cryptography

1 Introduction

Information security is a topic widely discussed by several countries that have create increasingly stringent laws for the protection and security of information. In general, security is the ability to protect against any threat or damage, whether intentional or not, for a specific purpose [1]. Information security aims to protect the reliability, integrity, and availability of information, whether in the storage, processing, and transmission of data. To have good security, a security policy, education, training, and awareness of the parties involved in the use of technology is necessary.

ⓒ Springer Nature Switzerland AG 2020
O. Gervasi et al. (Eds.): ICCSA 2020, LNCS 12249, pp. 676–691, 2020.
https://doi.org/10.1007/978-3-030-58799-4_49

Several countries have been concerned about information security and the consequences of possible data leaks. For this reason, several laws were passed to standardize the protection of information. Currently, the international law of the European Union (EU) that deals with the protection and security of personal data is the General Data Protection Regulation (GDPR) [2]. The GDPR is a regulation of the European Parliament and Council of the European Union and the European Economic Area. According to the law, based on a policy of transparency and privacy notification, the user must have control over what data will be used by companies that collect personal data.

On August 14, 2018, Brazil approved the Brazilian General Data Protection Law (LGPD) [3]. This law provides for the processing of personal data, including in digital media, intending to protect the fundamental rights of freedom, privacy, and the free development of the personality of the natural person. For these personal data, that is, digital information, to be safe, it is essential to adopt technical and administrative security measures to protect this information [3]. The LGPD determines what can and cannot be done about the collection and processing of personal data in Brazil, providing for punishments for companies that disrespect its principles [3].

One of the essential processes for protecting information is data encryption. One of the crucial uses of public-key cryptography is the use of a digital signature. It is a way for a user to sign a message or document with their private key. However, it is possible that another user can verify the veracity of the information using the public key of the user who signed the message [4]. The advantage of this approach is that, once the document or message is signed with a user's private key, the user cannot deny having signed the document or sent the message, which guarantees strong security and the principle of non-repudiation [5].

Digital certificates were used to guarantee the authenticity of public keys, that is, to ensure that a given public key does belong only to a person or organization associated with it. This guarantee requires a trusted third party, called the Certification Authority (CA), responsible for verifying the claim of each owner that generates a public key. For this, CA signs the owner's public key, thereby ensuring its authenticity [6].

In Brazil, the Ministry of Education (MEC) has established that the University Degrees of the Federal Educational Institutions (IFES) must be in digital format with the use of digital certificates in the XAdES standard. The MEC portal stipulated several deadlines for the implementation of the University Degree in digital format. The process started in 2018 and should end by 2022. From that date, University Degree will be made available only in digital format [7].

The standard for signing documents in digital format determined by the MEC must be valid throughout the national territory and must be signed by an ICP-Brazil Certification Authority. The certificate must be A3 or higher [8]. Thus, it is of paramount importance that the University of Brasilia (UnB) has a fast, scalable digital signature system, within the standards determined by the MEC and can sign several documents in batch, in addition to the possibility of verifying their validity and authenticity.

This work aims to develop a new digital signature solution focused on microservices architecture, ensuring greater scalability and fault tolerance. We developed the system in compliance with data privacy laws (LGPD and GDPR) and Federal Government ordinances.

This article is organized into sections as follows. Section 2 presents the theoretical background necessary to understand this work, as well as related works. Section 3 presents the legislation, the methodology used for the development of this research, the architecture proposed for the development of this work, and an example of an initial prototype of the digital signature system. Section 5 concludes the article and presents future work.

2   Background

2.1   Cryptography

The encryption algorithms are divided into two groups: symmetric encryption algorithms (also called secret key) and asymmetric encryption algorithms (also called public key) [9,10]. The encryption algorithms are based on two principles: substitution, in which each element of the clear text (bit, letter, group of bits, or letters) is mapped to another element and transposition, in which the elements of the clear text are rearranged. When performing these two operations, no information can be lost [4,11,12]. For an encryption algorithm to be considered computationally secure, it must follow two criteria: the cost to break the encryption exceeds the value of the information, and the time required to break the encryption must exceed the validity time of the information [4].

Until the 1970s, the only type of cryptography that existed was symmetric cryptography, and to this day, it is the most widely used type of cryptography [4]. Currently, in symmetric cryptography the two most used algorithms are Data Encryption Standard (DES) and Advanced Encryption Standard (AES) [4], [13], [10], [11], [12]. In asymmetric cryptography the most used algorithms are RSA and DSA [4], [12]. A new type of asymmetric cryptography has been developed, and is currently beginning to be used, called Elliptic Curve Cryptography (ECC) [14].

Symmetric Encryption. Symmetric encryption is the oldest encryption model and the first to be used. It is still the most widely used type of encryption today. The basic symmetric encryption model has five basic items [4]: 1. **Cleartext:** It is the original message or data, in readable format, that serves as input to the encryption algorithm; 2. **Encryption algorithm:** Performs several substitutions and transformations in the clear text; 3. **Secret key:** The key is the code used for the algorithm to perform the substitutions and transformations in the clear text resulting in a ciphertext; 4. **Ciphertext:** is the scrambled message, produced by the encryption algorithm. The ciphertext is the set of clear text plus the secret key; 5. **Decryption algorithm:** is the encryption algorithm performed in reverse. Ciphertext and secret key are required. The output of the algorithm is the original clear text [4], [15].

For an encryption algorithm to be considered strong, it must be unfeasible to decrypt a message with only the ciphertext and knowledge of the algorithm, that is, the algorithm is as strong the greater the strength of the key used to encrypt the cleartext. The key must be shared through a secure channel [4]. Figure 1 presents, in a simplified way, the flow to encrypt a text with symmetric algorithm [9].

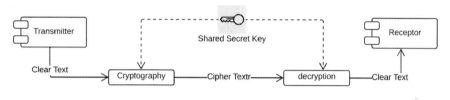

Fig. 1. Symmetric key encryption [9]

The algorithms can be encrypted bit by bit or byte by byte at a time, which is called flow encryption, or blocks of bits can be encrypted at once, usually using 64 or 128 bits at a time to be encrypted. The main symmetric encryption applications are based on the use of the block cipher [4], [16], [17].

Asymmetric Cryptography. Asymmetric cryptography, also called public-key cryptography, is based on mathematical functions, rather than substitution and permutation. Unlike symmetric encryption, asymmetric encryption uses two independent keys [4]. One of the problems that asymmetric cryptography solves is that of sharing the key between the sender and the receiver that occurs in symmetric cryptography. In asymmetric cryptography, there are two keys, one public and one private. One of the keys is for encryption and the other for decryption. The essential features of a secure asymmetric encryption system are [4]:

- It is computationally unable to determine the decryption key, given only the knowledge of the encryption algorithm and the encryption key.
- Either of the two related keys can be used for encryption and the other for decryption.

In asymmetric cryptography, both the public and private keys are generated by the issuer. The private key must be kept securely and must never be shared. The public key can be shared without compromising the security of communication [4], [9].

Figure 2 shows the process used in asymmetric cryptography. If the public key is used to encrypt a message, only the private key can be used to decrypt it.

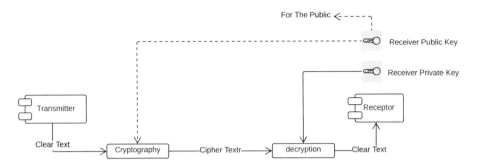

Fig. 2. Asymmetric cryptography [9]

If a message is encrypted with the private key, only the public key can decrypt it, that is, the same key is never used to encrypt and decrypt a message [9].

One of the most common algorithms used is RSA. This algorithm was proposed by Rivest et al. [18]. It is a public key algorithm used to generate a secure asymmetric signature system [19]. This algorithm uses two very large prime numbers, p and q to create public and private keys [9]. For creation, the following steps are followed, as shown in Fig. 3:

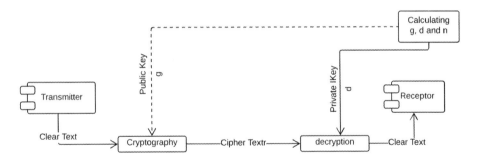

Fig. 3. RSA [9]

- The two prime numbers chosen are multiplied to find out n, which is the module for encryption and decryption, $n = p\ x\ q$.
- Calculate another number $z = (p - 1)\ x\ (q - 1)$.
- Choose a random integer g. Then, calculate d so that d is equal to mod z.
- Announces g and n to the public, keeping z and d secret.

To encrypt a message using the RSA algorithm, it is necessary to calculate the clear text with the values of g and n to generate the ciphertext to be sent. The formula used to encrypt clear text is as follows [9]:

$$C = P^g (mod\ n)$$

To safely decrypt a message, the values of z and d must be private. When receiving a ciphertext, the private key d is used to decrypt the message [9].

$$P = C^d(mod\ n)$$

2.2   Digital Signature

The digital signature uses asymmetric cryptography. The public key cryptographic system protects against any third party who is not authorized to access the message. However, it does not protect the two parts of each other [4]. Suppose Bob and Alice exchange messages using the asymmetric encryption scheme, and some problems can arise:

- Alice can forge a different message and claim that it came from Bob.
- Bob can deny sending the message to Alice.

When there is no complete trust between the sender and the receiver, more than asymmetric key authentication is required to exchange messages. To solve this problem, a digital signature is used. The digital signature must have the following characteristics [4]:

- Check the author and the date of the signature.
- Authenticate the content upon signature.
- Be verifiable by a third party.

When there is no complete trust between the sender and the receiver, The digital signature can be based on several algorithms. The scheme used in this work is based on the RSA algorithm. This algorithm is widely used in several applications, including financial, hence the importance of this scheme being considered safe. The guarantee that the RSA function is secure comes from the signature generation operation used to embed the hash value and the message consistency check. Along with RSA, the Probabilistic Signature Schema (PSS) technique is used [4], [20].

The RSA scheme uses exponential expressions. The clear text is encrypted in blocks, with each block being smaller than the n message. Both the sender and the receiver need to know the value of n. The sender knows the value of g, and only the receiver can know the value of d. RSA has two keys: a public one with PU $= g,\ n$ and a private one PR $= d,\ n$. For encryption to be efficient, it must be impossible to determine d from g and n [4]. To encrypt a message using RSA, the following function is used:

$$m^e = c\ mod\ n$$

To decrypt an RSA encrypted message, the following function is used:

$$c^d = n \ mod \ m$$

To introduce greater security in RSA encryption, a PSS padding scheme has been added. A proven secure completion scheme for producing signatures is typically used as a secure encryption scheme. The PSS function uses two functions and the hash for security. Random data is inserted in the message M before the hash algorithm [20] is executed. The main objective of the PSS is to ensure mathematical proof of the real security of the RSA algorithm for encryption [9]. Figure 4 shows the RSA flow.

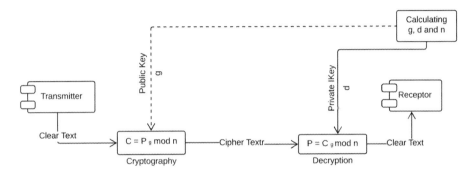

Fig. 4. RSA flow [9]

To encrypt a message with the RSA algorithm, the following steps are performed as shown in the Fig. 4:

- Choose two very large prime numbers, p and q.
- Multiply the two chosen cousins to discover n, which is the module for encryption and decryption.
- Choose a random integer g. Then the value of d is calculated.
- Announces the values g and n to the public. Keeps the value of g and d secret.

2.3   Related Works

Boneh et al. [21], presented the concept of aggregated signature applied in several applications. Aggregation of signatures are useful for reducing the size of the certificate chain and for reducing messages in secure routing protocols. The aggregated signatures, as well as the single signature, are verifiable encrypted. One of the practical applications of signature aggregation is with the X.509 certificate chain that can be aggregated into a single signature.

Shamir [22] presented research on how to divide the data into parts, so that they are easily reconstituted from any piece. Using this technique, it was possible to build robust key management schemes for cryptographic systems, which can work reliably.

Kumar et al. [23] investigated the need for encryption for images to ensure security against attacks. The work proposes a new algorithm for image security using Elliptic Curve Cryptography (ECC). This algorithm first encodes the RGB image using DNA encoding followed by asymmetric encryption based on Elliptic Curve Diffie - Hellman (ECDHE). The algorithm was applied to test images to analyze the patterns. The results of this study concluded that the proposed algorithm could withstand exhaustive attacks and is suitable for practical applications.

Perin et al. [24] presented a comparison of two modular multiplication architectures. A fully systolic matrix and a parallel implementation. Modular multiplication is used in the modular exponentiation process. One of the most famous cryptographic algorithms is the RSA encryption scheme. The fully systolic matrix architecture features a high-root implementation with transport propagation between the processing elements. The parallel implementation consists of multiplier blocks in parallel with the processing elements and providing a pipeline mode of operation. Time efficiency was compared between the two architectures cited using RSA. The time for the systolic architecture to decrypt 1024 bit using RSA was 3.4 ms, and the parallel architecture was 6 ms. There is a competitive performance between both architectures.

Engelbertz et al. [25] evaluated the validation logic of the open-source software library for creating and validating signed documents, provided by the Connecting Europe Facility (CEF) called the Digital Signature Service (DSS), against XML-based attacks. The discovered vulnerabilities allow you to read files from the server and bypass the protection provided by XML Advanced Electronic Signature (XAdES). There is an urgent need for proper security practice documents and automated security assessment tools to support the development of security-relevant implementations.

3   Contextualization

The Federal Government aims to provide various services digitally. In the educational area, the Ministry of Education (MEC) has as main objective the digitization of diplomas. According to the planning proposed by the MEC, all Brazilian Federal Universities must have the possibility of issuing secure diplomas in digital format by 2022 [7].

There are three ordinances created by MEC that regulate the issuance of diplomas in digital format. Ordinance No. 330/2018 establishes the digital diploma of all federal public and private higher education institutions. Article 2 mentions that the diplomas and documents must comply with the digital certification guidelines of the IPC-Brasil Public Keys infrastructure standard [26].

Ordinance Number 554/2019, in its article 2, specifies what a digital diploma is. According to this ordinance, a digital diploma is one that has the existence, its emission, and storage entirely digital, whose legal validity is presumed through the signature with digital certification and time stamp according to ICP-Brasil [27].

The storage of the digital diploma must be done in a computational environment that guarantees: validity at any time, interoperability between systems, security technology update, and the possibility of multiple signatures on the same document. The certificate used must be A3 or higher. According to ordinance 554/2019, all diplomas must be issued in Extensible Markup Language (XML) format based on the advanced electronic signature in the XML Advanced Electronic Signature - XAdES standard and the signed XML code must be conditioned to an Uniform Resource Location - unique URL. In addition to these mechanisms, the signed diploma must have associated with it a Quick Response Code - QR Code, which redirects to the unique diploma validation URL [27].

Ordinance 1.095/2018 regulates the process of registering diplomas in digital format. All diplomas must be registered in a registration book which must contain the opening, closing and must be signed by the competent authority. The record book must be in digital format and meet the specifications of ICP-Brasil [28]. According to what was presented in the legislation above and the deadline for the implementation of a solution for signing diplomas and documents in digital format, the University of Brasília (UnB) must build a solution adhering to the standards determined by the MEC.

The current architecture of UnB's systems is focused on the integration between the systems. Software integration enables the automation of business processes with improved resource management. There are now several legacy systems that are still widely used in UnB [29]. Currently, UnB systems are focused on architecture based on the Domain-Driven Design (DDD) concept, and one of the main concepts is the subdivision of a large business domain into smaller domains. Another fundamental concept of DDD is the use of a ubiquitous language to express the domain terms as the experts refer, making it possible to create a common vocabulary that all members of the project understand [30].

Due to the use of smaller domains and with greater consistency of a ubiquitous language, it was necessary to divide the large systems into a service-oriented approach. There are several services created by UnB; each of these services offers specific functionality that can be integrated to create an application. The services can be reused in several different applications [30].

Currently, there is a separation between front-end applications, which are created using the Angular language, using the modular architecture that the language offers [31]. There is an ems-bus services bus that integrates front-end applications and back-end services, which are currently developed in the java language. Also, ems-bus has several services, called bus modules, for specific functions [30].

The language used on the ems-bus is the Erlang language, which is a functional language with a great emphasis on simultaneity and high reliability. It is a language capable of executing dozens of tasks at the same time. Erlang uses the thread model where each thread is a separate process. With this language, several independent processes can be created, which offers the possibility of developing several modules and integrating them into the ems-bus [32].

The main modules of the ems-bus are: the service communication module through REST technology; the authentication module in the LDAP protocol format; the authentication module in the Oauth 2 protocol format [30]. This will develop an architecture for future implementation and implantation of a module inside the digital signature ems-bus in the standards determined by the Brazilian government for signing diplomas and documents in digital format. Figure 5 shows the current internal architecture of the ems-bus, the different modules, and how the new digital signature module will be inserted into the bus.

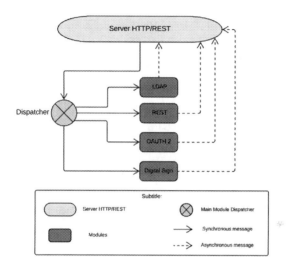

Fig. 5. Bus architecture ems-bus [30]

As shown in Fig. 5, the bus receives an HTTP request for one of the modules on the bus. The central Dispatcher module receives these requests and redirects them to the module called by the HTTP server. The module executes its functionality and returns the required response to the HTTP server [30].

The REST module is the basic module of the ems-bus, which is essential for communication between the other services. The bus works through microservices that exchange messages in the REST format controlled by the main dispatcher module, which acts as a coordinator for the exchange of services [30].

The Oauth 2 module is the main functionality of the authentication and authorization bus. This protocol must comply with the REST architecture and the JSON format for the message exchange. It is crucial because the bus works with the exchange of messages through JSON [33].

The Oauth 2 module uses the OAuth 2.0 library in Erlang available at: https://github.com/kivra/oauth2. This library generates the authentication token required to access the bus services. The bus's Oauth 2 module controls authentication and authorization [33].

The module to be created is the digital signature module. This module aims to sign XML documents and return them signed in the XAdES standard. The module will be created in the service format, where the path of the compressed documents in zip format to be signed must be passed in the payload and also the path of the A3 token, already installed and configured on the machine, which will sign the document. The HTTP call must be made using the url "/certificate/document/sign" with the paths of the documents to be signed and the certificate in the request payload, as shown in Fig. 6:

```
{
    "zip_file":"{path of zip file}",
    "cert_file": "{path of the cert}"
}
```

Fig. 6. Subscription Service Call Payload

The return of the digital signature module must be all XML documents signed in the XAdES standard compressed in zip format. For this functionality, an HTTP call must be made to the url "/certificate/document /verify" passing the signed XML document in the payload. The HTTP callback includes the document url and a Boolean attribute to check whether the document is valid or not.

Each of the digitally signed documents must be accessible on the internet at an address. For this purpose, it is necessary that each of the signed diplomas has its own access url.

3.1   Internal Architecture of the Digital Signature Module

Within the ems-bus, we have created a new digital signature module within the certification package. Figure 7 shows how this module is organized.

The CriptoSign.erl class is the class that has the methods responsible for signing an xml document in the XAdES standard. The class CryptoVality.erl is the class responsible for checking signed documents and returning the document's storage url. The class CriptoUtil.erl is the class that performs tasks such as reading certificates, converting formats, generating dates, and converting to the format accepted by XAdES, in addition to generating the structure of the XAdES standard.

For the subscription system to be recognized by the bus, it must be mapped within a catalog. Catalogs are the configuration files of the bus modules, with the mapping of a module within a catalog, the main module, dispatcher, can see it. After configuring the catalog correctly, the main dispatcher module is able to redirect calls made to the digital signature module.

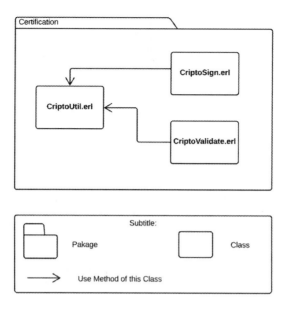

Fig. 7. Digital signature module architecture

Figure 8 presents an example of a document signed by a prototype of the digital signature system developed as a solution for UnB. The implementation proved to be correct, and the digital signature was carried out successfully on the documents. The current system does not fully sign the standard determined by the MEC, and the certificate used for testing is A1. This prototype will be evolved so that the proposed solution is in accordance with that determined by MEC.

4 Threats to Validity

During this work, we tried to select the primary references related to the topics covered in this research. However, we cannot guarantee that all relevant papers were selected during the literature review. Some relevant articles may not have been selected during the process. To minimize this threat, we conducted searches on digital databases using a Search String in the following digital databases: web of Science, Scopus, IEEE Xplore Digital Library, and ACM Digital Library. Also, we use the VOSviewer software that selects the most relevant articles from the total selected articles [34].

Another threat to the implementation of the digital signature system at UnB if MEC changes the guidelines on the digital signature system, such as changing XAdES to another form of signature, or imposing the use of a third-party system. To minimize this problem, we made a system that can be adapted to other subscription standards. The system to be developed will be presented to MEC

```
<?xml version="1.0" encoding="UTF-8"?>
<Diploma ID="941ce80d-a2ff-4896-a900-8f44fdb905d8">
    <ds:Signature xmlns:ds="http://www.w3.org/2000/09/xmldsig#">
        <ds:SignedInfo>
            <ds:CanonicalizationMethod Algorithm="http://www.w3.org/TR/2001/REC-xml-c14n-20010315#WithComments" />
            <ds:SignatureMethod Algorithm="http://www.w3.org/2001/04/xmldsig-more#rsa-sha256" />
            <ds:Reference URI="#941ce80d-a2ff-4896-a900-8f44fdb905d8">
                <ds:Transforms>
                    <ds:Transform Algorithm="http://www.w3.org/2000/09/xmldsig#enveloped-signature" />
                    <ds:Transform Algorithm="http://www.w3.org/TR/2001/REC-xml-c14n-20010315#WithComments" />
                </ds:Transforms>
                <ds:DigestMethod Algorithm="http://www.w3.org/2001/04/xmlenc#sha256" />
                <ds:DigestValue>l6AXbMecAzLWi898YhIhyQ/BZi5ktbSS/SWi5b0KbYo=</ds:DigestValue>
            </ds:Reference>
        </ds:SignedInfo>
        <ds:SignatureValue>lm9XGKZm2O/wIdDMqTlhrgpyJE5FK3tkAqSSQsf/mZOjJiG/Ac/L+Mu8aIs2fUr65evF6dzVTSn8u7WEAfY85PbcEicevrH+t+
gT5PXqbvkkmg2l/jlyzu8hVIoml7gHBzBDHel1c2mczKhOvEn3QV5Z1YaWPvTvyQjMVbnJmlxejILUgITVrZpPkYTNQH/8XVK2jwUCK7hFN+QVWSW43aT
AQLtVbtFnegpdwRRzaeXjD9EqKhhc39VcwP10HLgz5rSG/UchndikIqO6hUF/GMT9mwm6Nz5BNWwImb9zqav3Nvs6xqug+vh6lrk20GgQEBwxH0EtYvaK
tlhmJwWgeA==</ds:SignatureValue>
        <ds:KeyInfo>
            <ds:X509Data>
                <ds:X509Certificate>MIIGrDCCBJSgAwIBAgIRAJU2pTS6KVlBglk8rau46AkwDQYJKoZIhvcNAQELBQAwUDELMAkGA1UEBhMCQlIxGDAWBgNV
BAoTD0xhY3VuYSBTb2Z0d2FyZTELMAkGA1UECxMCSVQxGjAYBgNVBAMTEUxhY3VuYSBDQSBUZXN0IHYxMB4XDTE5MDEyMTIwMzQ1OFoXDTIyMDEy
MTIwMzQyNF0wRjELMAkGA1UEBhMCQlIxGDAWBgNVBAoTD0xhY3VuYSBTb2Z0d2FyZTEdMBsGA1UEAxMUQWxhbiBNYXRoaXBNbBNYRbaXNvbiBUdXJpbmcwggEi
MA0GCSqGSIb3DQEBAQUAA4IBDwAwggEKAoIBAQDFBybE4UxKh1LxYfzM6dYK9Qfcx5mDXGd3RDP/Ql13lVeDNCnNt2aHxEkzpeYH5OrHf+fXGWl/
FGyWpRg0wbve8DJLwzKjW0IcK1qTJx7BmTu84UHwHZFbHg5XEo+uYJD/66GwKhfsJNlFtsOITTkC20gx/mtPsReFoWZSM1RdYcoDK4/egI4DYaBO
+7prF82yNQZQ3zjnbmOntLObfQEIFOY3zBGQCJwuUSh8i5LVl94vrKWTJwchVRRXIBLcXqpO0H+6RC3ppw6WshAZNcI4iBDFT/HJNu/0ondKpkHH
BIBULYZmn2dvQvg9N4Y9JWUXElWPXngg/PCxQYjHxn2NAgMBAAGjggKJMIIChTAJBgNVHRMEAjAAMIGdBgNVHREEgZUwgZKgOAYFYEwBAwGgLwOt
MDAwMDAwMDA1NjA3MjM4MjN0MjEwNTAwMDAwMDAwMDAwMDAwMDAwMDAwMDAwMDAwBcGBWBMAQMGQoA4EDDAwMDAwMDAwMDAwMKAeBgVgTAEDBaAVBBMw
MDAwMDAwMDAwMDAwMDAwMDAwgR10ZXN0dHVyaW5nQGxhY3VuYXNvZnR3YXJlLm5vmNvbTC3B3QYDVR08fBIHVMIHSME5gQqBAhj5odHRwOi8vYW1wbGlh
LXIxLmxhY3VuYXNvZnNR3YXJlLmNvbNvbS9jcmxzL2xhY3VuYS1jYS10ZXN0LXYxLm5ybDBEBoEKgQIY+aHR0cDovL2FtcGxpYS1yMi5sYWN1bmFzb2Z0d2Fy
d2FyZS5jb20vY3Jscy9yYW5lbmETzEtdGVzdC12MS5jcmwwRKBCoEGGmh0dHA6Ly9hbXBsaWEtcjMuVbmFzb2Z0d2FyZS5jb20vd3Jscy9yYW5lbmETz0d2Fy
ZS5jb20vY2YydHMvbGGFjdW5hLWNhLXRlc3QtdjEuY2VydMEsGCCsGAQUFBzAChj9odHRwOi8vYW1wbGlhLXIyLmxhY3VuYXNvZnR3YXJlLmNvbNvbS9j
ZXJ0cy9sYWN1bmETzEtbmETtY2EtdGVzdC12MS5jZXIwSwYIKwYBBQUHMAKGP2h0dHA6Ly9hbBsawEtcjMubGGF jdW5hc29mdHdhcmEuUuY29tL2NlcnRzL2xh
Y3VuYS1jYS10ZXN0LXYxLmNlcnANBgkqhkiG9w0BAQsFAAOCAgEAmM4tBC3kq4j2D6V+KPc3eg0ILn7Yj6qvf6/pJT8AWSV4SXNTyFX34WAtPzer
vauVwF8hyq/7XiiAQNfgwvU9FMyoxApQnVPMIy3PEoJTqC+C5R0JDjHcQUdNxIv7djPeFAkadVfiWoxhov+k3MzOu0lfnxes/sEEmEHYD0aZgxtm
latD3SGsmHV6T7gpj1ZfKu3ZFpxNhPXb3TPEyPSzsbTfnqoPPD/u4s8uU2Anatwccx96eWzDB4ThSqgC5zlqK3QOTMwJ4gZPNey+3Y6Cctl3en5a
gTVpMQwSpGj1P87kho1qi9oHYGY5E5uWcPYhpwx9dnfzmjGoZNuLxtIh3ceO8MO8X6IWID5sQArvAz2m8smfudBYhcaZf7wlW2NFWmvD3wDluxn2
16ILs2wJTs0VKu/sZ17bnWw5mY07Gfm6XfBuBYDIAVQu5v4s/R9hwjLM0gek3Qx3ILlZnsS8LiuIVfm+f520sAY/D7XYo7PCfJOz0ttDKQ07PkC7
8SZ+hoN46uUHc1Sdb7l5RAKK66gZz3jccgs4+S432iyJAIjxdUcOTJE4m9+p5tJISCI2hkCQ6mo8TgfHkgRvELlgi9bCZxAUec2ibtgZnLt2OjgV
16uK4UfphxBNgKnPmpAlvpW47N0Mzlmxae81Te5IwefUlP8j/mQx1KVQ5kq81MQ=</ds:X509Certificate>
            </ds:X509Data>
        </ds:KeyInfo>
    </ds:Signature>
    <content>Content of the certificate...</content>
</Diploma>
```

Fig. 8. Digital signature example

for validation and possibly be one of the systems accepted by MEC and ICP-Brasil for signing documents in digital format.

One of the threats to the implementation and implementation of a digital signature solution is the fact that the system performs the signature quickly and reliably for signing multiple certificates. Most signature solutions take about minutes to execute the signature of a large number of documents. The solution that UnB needs is to sign a large number of documents in a short time and reliably, that is, that all documents are signed correctly. To minimize this threat, the Erlang [32] language was chosen for the development of this solution. This language is a robust language and has a high fault tolerance.

5   Conclusion

In Brazil, the current government aims to digitize a large part of public services. In the educational area, two essential services must be in digital format. The generation of documents such as: school history, declaration of active student, among others. Another valuable service is that of diplomas that must be issued in digital format. The MEC standardized how documents and diplomas should be generated in digital format and how they should be signed so that they have a valid format.

This work presented the central norms and requirements for the creation of a solution for signing documents and diplomas in digital format. Also, it was shown how the new UnB systems architecture works and how the digital signature system will be inserted into that architecture. It was shown each of the components to be developed and how these components interact with each other and how they will interact with the current architecture created.

We also present the possible threats that could compromise the creation of a digital signature solution for UnB. It was also shown the possible solutions to be taken for each threat to try to minimize the effects of the threats in this project.

In the future, the digital signature system will be developed in accordance with all ordinances and regulations of the MEC. It is essential for future work to verify the real security implemented in the signature systems using a signature using the XAdES standard and to compare it with the other primary existing standards such as CAdES and PAdES.

References

1. Whitman, M.E., Mattord, H.J.: Principles of information security. In: Cengage Learning (2011)
2. Regulation, G.D.P.: Eu data protection rules. In: European Commission (2018). https://ec.europa.eu/commission/priorities/justice-and-fundamental-rights/data-protection/2018-reform-eu-data-protection-rules_en. Accessed 9 Oct 2019
3. da República, P.: Lei geral de proteção de dados pessoais (LGPD). In: Secretaria-Geral (2018). http://www.planalto.gov.br/ccivil_03/_ato2015-2018/2018/lei/L13709.html. Accessed 9 Oct 2019
4. Stallings, W.: Network and Internetwork Security: Principles and Practice. Prentice Hall, Englewood Cliffs (1995)
5. Rivest, R.L.: A method for obtaining digital signature and public-key cryptosystems. Communi. ACM **21**, 2 (1987)
6. Tycksen Jr., F.A., Jennings, C.W.: Digital certificate, uS Patent 6,189,097, February 2001
7. da Educaçño, M.: Diploma digital, Ministério da Educaçño (2020). http://portal.mec.gov.br/diplomadigital/. Accessed 12 Feb 2020
8. Middelkamp, A.: Online. Praktische Huisartsgeneeskunde **3**(4), 3 (2017). https://doi.org/10.1007/s41045-017-0040-y
9. Forouzan, B., Coombs, C., Fegan, S.C.: Introduction to Data Communications and Networking. McGraw-Hill Inc., New York (1997)
10. Narendra, S.G., Tadepalli, P., Spitzer, T.N.: Symmetric cryptography with user authentication, uS Patent 8,477,940, 2 July 2013
11. Omran, S.S., Al-Khalid, A.S., Al-Saady, D.M.: Using genetic algorithm to break a mono - alphabetic substitution cipher. In: 2010 IEEE Conference on Open Systems (ICOS 2010), pp. 63–67, December 2010
12. Kurniawan, D.H., Munir, R.: Double chaining algorithm: a secure symmetric-key encryption algorithm. In: 2016 International Conference On Advanced Informatics: Concepts, Theory And Application (ICAICTA), pp. 1–6, August 2016

13. Bellare, M., Desai, A., Jokipii, E., Rogaway, P.: A concrete security treatment of symmetric encryption. In: Proceedings 38th Annual Symposium on Foundations of Computer Science, pp. 394–403. IEEE (1997)

14. Szerwinski, R., Güneysu, T.: Exploiting the power of GPUs for asymmetric cryptography. In: Oswald, E., Rohatgi, P. (eds.) CHES 2008. LNCS, vol. 5154, pp. 79–99. Springer, Heidelberg (2008). https://doi.org/10.1007/978-3-540-85053-3_6

15. Forouzan, B.A.: Cryptography & Network Security. McGraw-Hill Inc., New York (2007)

16. Jakimoski, G., Kocarev, L.: Chaos and cryptography: block encryption ciphers based on chaotic maps. IEEE Trans. Circuits Syst. I Fundam. Theor. Appl. **48**(2), 163–169 (2001)

17. Chunguang, H., Hai, C., Yu, S., Qun, D.: Permutation of image encryption system based on block cipher and stream cipher encryption algorithm. In: 2015 Third International Conference on Robot, Vision and Signal Processing (RVSP), pp. 163–166, November 2015

18. Rivest, R.L., Shamir, A., Adleman, L.: A method for obtaining digital signatures and public-key cryptosystems. Commun. ACM **21**(2), 120–126 (1978)

19. Bellare, M., Rogaway, P.: The exact security of digital signatures-how to sign with RSA and Rabin. In: Maurer, U. (ed.) EUROCRYPT 1996. LNCS, vol. 1070, pp. 399–416. Springer, Heidelberg (1996). https://doi.org/10.1007/3-540-68339-9_34

20. Coron, J.-S.: Optimal security proofs for PSS and other signature schemes. In: Knudsen, L.R. (ed.) EUROCRYPT 2002. LNCS, vol. 2332, pp. 272–287. Springer, Heidelberg (2002). https://doi.org/10.1007/3-540-46035-7_18

21. Boneh, D., Gentry, C., Lynn, B., Shacham, H.: Aggregate and verifiably encrypted signatures from bilinear maps. In: Biham, E. (ed.) EUROCRYPT 2003. LNCS, vol. 2656, pp. 416–432. Springer, Heidelberg (2003). https://doi.org/10.1007/3-540-39200-9_26

22. Shamir, A.: How to share a secret. Commun. ACM **22**(11), 612–613 (1979). https://doi.org/10.1145/359168.359176

23. Kumar, M., Iqbal, A., Kumar, P.: A new RGB image encryption algorithm based on DNA encoding and elliptic curve Diffie-Hellman cryptography. Signal Process. **125**, 187–202 (2016). https://www.sciencedirect.com/science/article/pii/S0165168416000347

24. Perin, G., Mesquita, D.G., Herrmann, F.L., Martins, J.B.: Montgomery modular multiplication on reconfigurable hardware: fully systolic array vs parallel implementation. In: 2010 VI Southern Programmable Logic Conference (SPL), pp. 61–66, March 2010

25. Engelbertz, N., Mladenov, V., Somorovsky, J., Herring, D., Erinola, N., Schwenk, J.: Security analysis of XAdES validation in the CEF digital signature services (DSS). In: Rossnagel, H.D., Wagner, S. (eds.)Gesellschaft fur Informatik (GI), vol. P-293, pp. 95–106, cited By 0 (2019)

26. da Educaçño, M.: Portaria n 330, de 5 de ABRIL de 2010. In: Ministérioda Educaçñ (2018). http://www.in.gov.br/materia/-/assetpublisher/Kujrw0TZC2Mb/content/id/9365055/do1-2018-04-06-portarian-330-de-5-de-abril-de-2018-9365051. Accessed 15 Feb 2020

27. da Educação, M.: Portaria n 554, de 11 de março de 2019. In: Ministério da Educaçã (2019). http://www.in.gov.br/materia/-/assetpublisher/Kujrw0TZC2Mb/content/id/66544171/do1-2019-03-12-portarian-554-de-11-de-marco-de-2019-66543842. Accessed 20 Feb 2020

28. da Educação, M.: Portaria n 1095, DE 25 DE OUTUBRO DE 2018. In: Ministério daEducaçã (2018). https://abmes.org.br/arquivos/legislacoes/Port-MEC-1095-2018-10-25.pdf. Accessed 22 Feb 2020
29. Agilar, E., Almeida, R., Canedo, E.: A systematic mapping study on legacy system modernization. In: SEKE. KSI Research Inc. and Knowledge Systems Institute Graduate School, pp. 345–350 (2016)
30. Agilar, E.D.V.: Uma abordagem orientada a serviços para a modernização de sistemas legados, Universidade de Brasìlia (UnB) (2016)
31. Angular.io: Agnular fundamentals Angular.io (2020). https://angular.io/guide/architecture. Accessed 01 Mar 2020
32. Erlang/OTP: Erlang getting started with erlang user's guide Erlang OTP (2019). https://erlang.org/doc/getting_started/conc_prog.html, Accessed 03 Mar 2020
33. Ribeiro, A.D.S.: Uma implementação do protocolo oauth 2 em erlang para uma arquitetura orientada a serviço, Universidade de Brasìlia (UnB) (2017)
34. Van Eck, N., Waltman, L.: Software survey: VOSviewer, a computer program for bibliometric mapping. Scientometrics **84**(2), 523–538 (2009)

An Approach to the Selection of a Time Series Analysis Method Using a Neural Network

Dmitry Yashin⬤, Irina Moshkina⬤, and Vadim Moshkin^(⊠)⬤

Ulyanovsk State Technical University,
Severny Venets Street, 32, Ulyanovsk 432027, Russian Federation
taurusrulez@yandex.ru, postforvadim@yandex.ru,
i.timina@ulstu.ru

Abstract. In this paper, a technique for selecting individual methods in a combined time series forecasting model is described. A neural network at the input of which a vector of time series metrics is proposed. The metrics corresponds to significant characteristics of the time series. The values of the metrics are easily computed. The neural network calculates the estimated prediction error for each model from the base set of the combined model. The proposed selection method is most effective for short time series and when the base set contains a lot of complex prediction methods. The developed system was tested on the time series from the CIF 2015-2016 competitions. According to the result of the experiment, the application of the developed system allowed to reduce the average forecast error from 13 to 9 percent.

Keywords: Time series · Forecasting · Combined model · Neural network

1 Introduction

The constant need to improve the accuracy of forecasting time series leads to an increase in the number of publications on this topic. It also contributes to the emergence of new forecasting methods, with the growing number of which, the problem of choosing the most appropriate method is becoming increasingly important.

One of the possible solutions to this problem is the use of the combined forecasting models, which allow to obtain an aggregated forecast based on the results of several methods. However, in this case, the choice of methods from a base set is also a problem since individual inexact forecasts can significantly reduce the overall accuracy.

In this paper, one of the approaches to solving the problem of selecting individual models from a base set is proposed. Used specially developed neural network (Sect. 2) that selects methods according to the values of the metrics corresponding to the characteristics of the time series (Sect. 1). Characteristics relevant to the subject area have been selected for this purpose.

Also in this paper, the set of metrics that is optimal for the solution of the problem is determined, the efficiency of the developed system is estimated (Sect. 3).

The method proposed in this work is most similar to the expert approach and involves the selection of models from the base set by a neural network that uses a set of simply computed time series metrics.

© Springer Nature Switzerland AG 2020
O. Gervasi et al. (Eds.): ICCSA 2020, LNCS 12249, pp. 692–703, 2020.
https://doi.org/10.1007/978-3-030-58799-4_50

2 Combined Time Series Forecasting Model. Selecting Methods from the Base Set

2.1 Hybrid Systems in Artificial Intelligence

The concept of hybridization arises as a consequence of the development of the principles of the system approach to the study of complex artificial objects in computer science [1]. Professor N.G. Yarushkina highlights the development of hybrid integrated and synergetic systems as one of the leading trends in modern computer science. She defines hybridization as the integration of methods and technologies at a deep level, when the system includes interacting modules that implement various methods for solving problems of artificial intelligence [2]. A.V. Gavrilov, noting the hybridization process as the main trend in the development of artificial intelligence, considers hybridization as a variant of combining the methods of representation and processing of knowledge in hybrid intelligent systems [3].

Hybrid intelligent information systems are an interdisciplinary scientific direction, within the framework of which the applicability of several methods from different classes of the method of formalized representation of systems (analytical, statistical, logical-linguistic, fuzzy, and others) to solving decision-making problems is investigated. While none of the above classes can be considered universal [4]. The interaction of several methods makes it possible to compensate for their drawbacks due to the appearance of an integrative property.

2.2 Combined Forecasting Model

There are many methods for forecasting time series, each of which has its advantages and disadvantages. To take advantage of several methods, combined models are used. According to [5], the combined forecasting model is a prediction model consisting of several individual models, called the base set.

In [6] listed a number of factors that emphasize the effectiveness of the combined model, among which: the inability to choose a single model from experimental data; the presence in each rejected forecast of important for modeling information; the need to choose from a group of models that have similar statistical characteristics when trying to select a single best model.

There are two types of combined models: selective and hybrid. In the selective model, the forecast is calculated by one method selected from the base set at each time point. In the hybrid model, the predicted value is obtained by aggregating the prediction results for several models from the base set (usually, the final forecast is a weighted sum of individual forecasts) [5].

2.3 Selection of Models from the Base Set

According to [7], the main problem of constructing hybrid combined forecasting models is the finding of the optimal weights of individual model predictions that ensure the minimum value of the forecast error of the combined model. However, no less important is the problem of selecting models from the base set.

In this paper, a technique for selecting individual methods in a hybrid combined forecasting model is proposed. This approach can also be used for a selective combined model.

One of the possible ways to solve the sampling problem is to divide the time series into a training and control part with subsequent prediction of the control values by each method from the base set, then it becomes possible to select the models according to the obtained prediction error values. This method of selecting models is ineffective for combined models with a large number of methods in the base set, as well as for forecasting short time series. Another way to solve the problem is an expert evaluation based on the analysis of characteristics and the time series chart by a human expert. A combination of the two methods described above is also used. In this case forecasting the test values of the time series is applied only for models selected by the expert from the base set.

The method proposed in this work is most similar to the expert approach and involves the selection of models from the base set by a neural network that uses a set of simply computed time series metrics.

3   Time Series Metrics for Solving the Prediction Problem

There are many characteristics of a time series, such as the presence of a trend and the seasonal component, stationarity, length, anomaly of values, variance, mean. To predict the time series, the most important are the indicators of its dynamics [8]. According to [9], the two main statistical elements of dynamics are the trend and volatility.

3.1   Trend Metrics

The trend is the characteristic of the process of changing the phenomenon over a long time, freed from random fluctuations [10]. There are many criteria for determining the presence of the trend and for assessing its severity. Below are considered the most popular criteria.

Wallis-Moore Criterion. This criterion assumes the calculation of the differences in the values of the time series $(y_{t+1} - y_t)$. Null hypothesis: the signs of these differences form a random sequence. A sequence of identical difference signs is called a phase. The number of phases h (without the first and last phase) is calculated. If the signs form a random sequence, then the actual value of the criterion is calculated by the formula (1) [11].

$$t_{fact} = \frac{\left| h - \frac{2n-7}{3} \right|}{\sqrt{\frac{16n-29}{90}}}, \tag{1}$$

where n is the length of time series, h is the number of phases.

Cumulative T-Criterion. Null hypothesis: no trend in the original time series. The calculated value of the criterion is defined as the ratio of the accumulated sum of squared deviations of the empirical values of the time series observations from their mean value and the deviations themselves according to the formula (2) [12]

$$T_p = \frac{\sum_n Z_n^2}{\sigma_y^2},$$ (2)

where Z_n is the accumulated result of deviations of empirical values from the mean level of the original time series; σ_y^2 is the total sum of squared deviations, defined by formula (3)

$$\sigma_y^2 = \sum_n y_i^2 - (\bar{y})^2 \cdot n.$$ (3)

If a sufficiently long time series is analyzed, then a standardized deviation (4) can be used to calculate the criterion values

$$t_p = \frac{T_p - \left(\frac{n+1}{6}\right)}{\sqrt{(n-2)\frac{2n-1}{90}}}.$$ (4)

The calculated values of the cumulative T-criterion and t_p are compared with the critical values for a given significance level α.

Foster-Stewart Criterion. This criterion is used to test the trend of both mean and variance [13]. The statistics of the criterion have the form (5):

$$S = \sum_{i=2}^n S_i; \quad d = \sum_{i=2}^n d_i,$$ (5)

where $d_i = u_i - l_i$; $S_i = u_i + l_i$;

$$u_i = \begin{cases} 1, & x_i > x_{i-1}, x_{i-2}, \ldots, x_l; \\ 0, & else; \end{cases} \quad l_i = \begin{cases} 1, & x_i < x_{i-1}, x_{i-2}, \ldots, x_l; \\ 0, & else. \end{cases}$$

Statistics S is used to check the trend in variances, statistics d - to detect the trend in the mean. In the absence of a trend, the values of t and $\bar{t}$ (6) have a Student's distribution with $v = n$ degrees of freedom:

$$t = \frac{d}{f}; \quad \bar{t} = \frac{S - f^2}{k},$$ (6)

where $k = \sqrt{2\ln n - 3,4253}$; $f = \sqrt{2\ln n - 0,8456}$.

Koks-Stewart Criterion. To test the trend hypothesis, the normalized statistics (7) are applied [14]:

$$S_l^* = \frac{S_l - M(S_l)}{\sqrt{D(S_l)}}, \tag{7}$$

where $M(S_l) = \frac{n^2}{8}; D(S_l) = \frac{n(n^2-1)}{24}$.

Statistics S_1 is calculated by the formula (8):

$$S_l = \sum_{i=1}^{\left[\frac{n}{2}\right]} (n - 2i + 1)h_{i,n-i+1}, \tag{8}$$

where $h_{i,j} = \begin{cases} 1, & x_i > x_j \\ 0 & else \end{cases}, (i < j)$.

Under (9) the hypothesis of the average trend is rejected.

$$\left|S_l^*\right| < u_{\frac{1+\alpha}{2}}, \tag{9}$$

The Criterion Based on the Mean Comparison Method. To check the presence of a trend, the time series is divided into 2 parts. This criterion is based on the statement: if the time series has a tendency, the averages calculated for each set separately should differ significantly among themselves [15]. Null hypothesis on the absence of a tendency reduces to testing the hypothesis of the equality of the means of two normally distributed populations. The hypothesis is checked using Student's t-test, the calculated value of which is calculated by the formula (10):

$$t_r = \frac{\bar{y}_1 - \bar{y}_2}{\sqrt{(n_1 - 1)\sigma_1^2 + (n_2 - 1)\sigma_2^2}} * \sqrt{\frac{n_1 n_2 (n_1 + n_2 - 2)}{n_1 + n_2}}, \tag{10}$$

where $\bar{y}_1$ and $\bar{y}_2$ are the mean values, σ_1^2 and σ_2^2 are the dispersions and n_1 and n_2 are the lengths of the first and second part of the time series.

The calculated value (t_r) of the criterion is compared with its critical (tabular) value (t_{cr}) at a significance level α and the number of degrees of freedom $\nu = n - 2$.

3.2 Seasonal Metrics

There are many methods for estimating the degree of seasonal deviations of the time series. One of these methods is based on the calculation of seasonality indices. The seasonality index reflects the degree of seasonal variability of the phenomenon relative to the mean of the time series. This index is calculated by the formula (11):

$$I_{t,ce3} = Y_t/\bar{Y}, \tag{11}$$

where Y_t is the average value of the indicator for the month (quarter); $\bar{Y}$ is the overall average value of the indicator.

3.3 Dispersion and Dispersion Trend Metrics

As a metric, can be considered both the value of the variance and the degree of the variance trend. To test the hypothesis of the presence of a trend in dispersion, the Foster Stewart criterion is used. Also used the criterion based on the partitioning of the time series and the comparison of the variances of the two parts. Null hypothesis: the variances of two normally distributed sets are equal $(H_0 : \sigma_1^2 = \sigma_2^2)$.

The hypothesis is tested on the base of the Fisher-Snedecor F-criterion [16], the calculated value of which is obtained by the formula (12):

$$F_p = \begin{cases} \sigma_1^2/\sigma_2^2, \sigma_1^2 > \sigma_2^2, \\ \sigma_2^2/\sigma_1^2, \sigma_1^2 < \sigma_2^2 \end{cases}, \tag{12}$$

The hypothesis is tested on the base of comparison of the calculated and critical values of the F-test obtained for a given level of significance α and the number of degrees of freedom v_1 and v_2 (13).

$$\begin{aligned} v_1 = n_1 - 1, \text{ and } v_2 = n_2 - 1, \sigma_1^2 > \sigma_2^2 \\ v_1 = n_2 - 1 \text{ and } v_2 = n_1 - 1, \sigma_1^2 < \sigma_2^2 \end{aligned}, \tag{13}$$

3.4 Anomaly Degree Metrics

According to [17], anomalous observations are considered to be separate observations of the time series, the values of which differ significantly from the remaining observations. To assess the degree of anomaly of individual observations of the time series, special statistical criteria are used, for example, the Irwin criterion and the Chauvenet criterion.

Irwin Criterion. For all or only for suspected abnormalities in observations, λ_t is calculated (14). If the calculated value exceeds the table level, then the observation y_t is considered anomalous [18].

$$\lambda_t = \frac{|y_t - y_{t-1}|}{S}, \tag{14}$$

where $S = \sqrt{\frac{\sum_{t=1}^{n} (y_t - \bar{y})^2}{n-1}}$.

Chauvenet Criterion. The sample element y_t is an outlier if the probability of its deviation from the mean is not more than $1/12n$ (where n is the sample size). For checked observations, the value of K (15) is calculated, which is compared with the tabulated value [9].

$$K = \frac{|y_t - \bar{y}|}{S}, \tag{15}$$

where $S = \sqrt{\frac{\sum_{t-1}^{n}(y_t - \bar{y})^2}{n-1}}$.

3.5 Calculating the Values of the Metrics

To create and train a neural network for selecting forecasting methods, it was decided to establish the requirement that the value of each metric belong to the interval from 0 to 1. Before calculating the metrics, the time series are normalized. It is obvious that the variance of the normalized time series will be within the range from 0 to 1.

To obtain the final values of the metrics that involve computing the calculated value and comparing it with the critical value, we use formula (16).

$$m = \begin{cases} 1 - \frac{t_{crit}}{t_{fact}}, & t_{fact} \geq t_{crit} \\ 0,01, & t_{fact} < t_{crit} \end{cases}, \qquad (16)$$

where t_{fact} - calculated value of the criterion, t_{crit} - critical value of the criterion.

To calculate the metric of the degree of seasonality on the basis of seasonality indexes, the formula (17) is applied.

$$m_{ses} = \sqrt{\frac{\sum_{t=1}^{T}\left(\frac{I_t}{\bar{y}} - 1\right)^2}{T}}, \qquad (17)$$

where T is the period of seasonality of the time series, $\bar{y}$ is the mean of the time series, I_t is the value of the average seasonality index for the period t.

To calculate the values of the metrics corresponding to the degree of anomaly of the time series, the ratio of the number of anomalous (according to the criterion) values to the length of the time series is used.

4 Neural Network for Selecting Prediction Methods

A specially designed neural network is used to select the prediction methods from the base set of the hybrid combined model (see Fig. 1). This neural network is created using the R language and the built-in package "neuralnet".

The input values of the neural network are the time series metrics. Each method $(m_1, m_2..., m_k)$ from the base set corresponds to one of the output layer's neurons $(M_1, M_2..., M_t)$.

The training set of the developed network is a set of TS metrics and the corresponding error values for predicting each TS by each method from the base set.

The neural network is trained to calculate the estimated forecast error values for each model from the base set based on the values of the TS metrics.

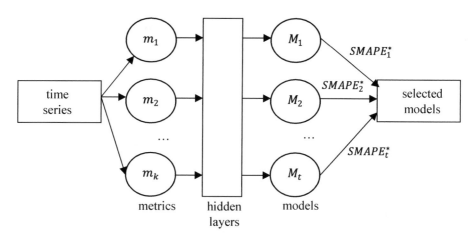

Fig. 1. Neural network circuit

An error back propagation algorithm with a logistic activation function is used for training.

The accuracy of the results of the neural network [19] depends on the similarity of the TS from the training set and the predicted TS and on the completeness and redundancy of the training set. The training set contains metric values and forecast error values for each TS included in it.

The training set consists of a set of time series represented as a set of metrics and the corresponding values of the prediction error (SMAPE) for each forecasting model. In this system, time series are taken from the competition "Computational Intelligence in Forecasting" (CIF) [12] for 2015-2016. The SMAPE values for the training set are obtained as a result of the forecasting of competitive time series using the methods of the combined hybrid forecasting system "Combination of fuzzy and exponential models" (CFEM) [13] developed at the Department of Information Systems of the Ulyanovsk State Technical University. This system was the best in the CIF competition in 2015

The metrics (and the methods for calculating them) for the input layer of the neural network are described in Sect. 3 of this work. The prediction models (corresponding to the output layer of the neural network) are taken from the combined CFEM model. The neural network learns from the metric values to calculate the estimated error values for each model from the base set. Then it is possible to determine which models from the base set are more efficient for predicting the values of the current time series. If the aggregating method involves adjusting the weights of the models, the weights can be set inversely proportional to the predicted error values of each model.

5   The Best Combination of Metrics. Results of Experiments

5.1   The Base Set of Metrics

Initially, 11 metrics were used in the developed neural network of selection of prediction methods: 5 metrics for the mean trend, 2 for the dispersion trend, 1 for the variance, 1 for the degree of seasonal variation, and 2 for the degree of anomaly in the time series.

The results of evaluations for one characteristic according to various criteria differ due to the differences in the regularities on which valuation methods are based. For example, the Wallis-Moore phase-frequency criterion for determining the trend of the mean is based on the comparison of the neighboring observations of the time series. The Foster-Stewart criterion - on the comparison of the current value with all the previous ones. The Cox-Stewart criterion - on the dividing the time series into three parts with the subsequent observation comparison between the third and the first group. The cumulative T-criterion - on the calculating the accumulated total of deviations of the empirical values from the time series mean [9, 11]. Proceeding from the foregoing, it can be argued that metrics that correspond to the same characteristic of a time series, but calculated by different criteria, may not have a high degree of correlation with each other.

In determining the optimal combination of time series metrics, the greatest difficulty is observing a balance between the completeness of the coverage of the regularities of the time series and the low degree of correlation between the characteristics. It was decided to perform an experiment where the metrics were sequentially excluded from the general set and the effect of this exception on the accuracy of the neural network was estimated.

5.2   The Experiment on Finding the Optimal Set of Metrics

Computational experiments were performed for several configurations of a neural network that differ in the number of neurons in the hidden layers. In some configurations, the neural network learning algorithm (based on the back propagation algorithm) did not converge. Such configurations were not taken into account. The experiments are carried out for neural networks with one hidden layer with the number of neurons in it from 1 to 11 and for networks with two hidden layers and the number of neurons in each of them from 1 to 8.

The training set of the neural network was divided 5 times into the training and control parts. After that, for each configuration (a set of metrics and the number of neurons in the hidden layers), the neural network was trained and the deviation of the output values from the test ones was calculated.

The results of the experiment are presented in Table 1. The columns of the table correspond to sets of metrics and show the sequence of excluding metrics from the base set, which allowed increasing the accuracy of the neural network. Each column is divided into two sub-columns containing information about the configuration of the neural network and the corresponding value of the network error. The rows of the table correspond to the best 7 configurations for each set of metrics and the average error for all configurations.

Table 1. The best combination of metrics.

All metrics		All metrics except the dispersion trend by the Foster-Stewart criterion		All metrics except the dispersion trend by the Foster-Stewart criterion and the cumulative T-criterion		All metrics except the cumulative T-test and both variance trend metrics	
Number of hidden neurons	SMAPE	Number of hidden neurons	SMAPE	Number of hidden neurons	SMAPE	Number of hidden neurons	SMAPE
2	9,764	7	8,994	7	9,476	10	9,016
(8,1)	9,794	(4,5)	9,314	(1,3)	9,667	8	9,103
9	9,870	(2,6)	9,314	10	9,720	7	9,392
(4,1)	9,968	(6,2)	9,788	(3,4)	9,777	1	9,553
(2,1)	10,075	(7,3)	9,869	2	9,782	2	9,592
(3,3)	10,099	(1,6)	9,924	(1,4)	9,795	(6,3)	9,777
(8,2)	10,194	4	10,050	6	9,957	(4,3)	9,788
...	...	...	...	...	...	...	...
average SMAPE	10,967	average SMAPE	10,851	average SMAPE	10,656	average SMAPE	10,769

According to Table 1, the greatest average accuracy for all configurations was achieved for a set containing all metrics except the dispersion trend metric by the Foster-Stewart criterion and the metrics based on the cumulative T-criterion. However, the subsequent exclusion of another metric of the dispersion trend allowed to obtain the most accurate results for individual configurations.

Another experiment was performed to evaluate the effectiveness of using the neural network to improve the accuracy of the final forecast. To obtain the aggregated forecast only methods with a low expected forecast error were taken from the base set. It was established that the preliminary selection of methods by a neural network containing all eleven metrics made it possible to reduce the error of the final forecast from 13.131% to 9.764%. The use of the best neural network configuration from the previous experiment (without a cumulative T-criterion and both variance trend metrics, with 10 neurons in the hidden layer) reduced the average final forecast error to 9.016%.

Also as a result of the experiment, it was established that the logistic activation function for the considered neural network allows to obtain significantly more accurate results than the hyperbolic tangent.

6   Conclusions

In this paper, a technique for selecting individual methods in a combined time series forecasting model is described. This approach is based on the use of a specially developed neural network that selects methods according to time series metrics.

We proposed a neural network structure and the algorithms for calculating metric values based on the characteristics of the time series. According to the obtained experimental data, the optimal set of the metrics is determined and the efficiency of the neural network application is estimated: the average forecast error decreased from 13,131% to 9,016%.

The application of the developed neural network is especially effective for short time series and for combined models with a large number of prediction methods (since it uses easily calculated metrics). In these cases, the approach that involves splitting the time series into the training and control parts and then calculating control values by each method is inefficient.

Planned in future work: aggregation of individual forecasts based on error values;

- setting aggregator weights based on the values of prediction errors on the control part of the time series;
- adjustment of aggregator weights based on the values of prediction errors assumed by the neural network;
- development of a neural network model for aggregation of the results of time series forecasting methods.

Acknowledgments. This work was supported by the Russian Foundation for Basic Research. Projects No. 18-47-730035, No. 19-07-00999 and 18-47-732007.

References

1. Tarasov, V.B.: From multi-agent systems to intellectual organizations: philosophy, psychology, informatics. Editorial, Moscow (2002)
2. Yarushkina, N.G.: Fuzzy hybrid systems. Theory and practice. Fizmatlit, Moscow (2007)
3. Gavrilov A.V.: Hybrid Intelligent Systems. NSTU, Novosibirsk (2003)
4. Kolesnikov, A.V., Kirikov, I.A., Listopad, S.V., Rumovskaya, S.B., Domanicky, A.A.: Solution of complex traveling salesman tasks using the methods of functional hybrid intelligent systems. Institute of Informatics Problems of the Russian Academy of Sciences, Moskow (2011)
5. Lukashin, Y.P.: Adaptive methods of short-term forecasting of time series. Finansy i statistika, Moscow (2003)
6. Novák, V.: Mining information from time series in the form of sentences of natural language. Int. J. Approximate Reasoning **78**, 192–209 (2016)
7. Perfilieva, I.: Fuzzy transforms: theory and applications. Fuzzy Sets Syst. **157**(8), 993–1023 (2006)
8. Abayomi-Alli, A., Odusami, M.O., Misra, S., Ibeh, G.F.: Long short-term memory model for time series prediction and forecast of solar radiation and other weather parameters. In: 2019 19th International Conference on Computational Science and Its Applications (ICCSA), pp. 82–92. IEEE (2019)
9. Kobzar, A.I.: Applied Mathematical Statistics. Fizmatlit, Moscow (2006)
10. Afanasyev, V.N., Yuzbashev, M.M.: Time series analysis and forecasting. Finansy i statistika Moscow (2001)

11. Ge, P., Wang, J., Ren, P., Gao, H., Luo, Y.: A new improved forecasting method integrated fuzzy time series with the exponential smoothing method. Int. J. Environ. Pollut. **51**(3–4), 206–221 (2013)
12. CIF. http://irafm.osu.cz/cif/main.php. Accessed 22 June 2020
13. Afanasieva, T., Yarushkina, N., Zavarzin, D., Guskov, G., Romanov, A.: Time series forecasting using combination of exponential models and fuzzy techniques. In: Abraham, A., Kovalev, S., Tarassov, V., Snášel, V. (eds.) Proceedings of the First International Scientific Conference "Intelligent Information Technologies for Industry" (IITI'16). AISC, vol. 450, pp. 41–50. Springer, Cham (2016). https://doi.org/10.1007/978-3-319-33609-1_4
14. Pedrycz, W., Chen, S.: Time series analysis, modeling and applications: a computational intelligence perspective (2013). https://doi.org/10.1007/978-3-642-33439-9
15. Yarushkina, N., Andreev, I., Moshkin, V., Moshkina, I.: Integration of fuzzy OWL ontologies and fuzzy time series in the determination of faulty technical units. In: Misra, S., et al. (eds.) ICCSA 2019. LNCS, vol. 11619, pp. 545–555. Springer, Cham (2019). https://doi.org/10.1007/978-3-030-24289-3_40
16. Rybina, G.V.: Modern expert systems: trends towards integration and hybridization, devices and systems. Handling Control Diagnostics **8**, 18–21 (2001)
17. Yarushev S., Averkin A. Time series analysis based on modular architectures of neural networks. In: Procedia Computer Science, 8th Annual International Conference on Biologically Inspired Cognitive Architectures, BICA 2017 (8th Annual Meeting of the BICA Society), pp. 562–567 (2018)
18. Timina, I., Egov, E., Romanov, A.: Fuzzy models in forecasting time series of project activity metrics. Paper presented at the Journal of Physics: Conference Series, vol. 1096 (2018)
19. Abayomi-Alli, A., Abayomi-Alli, O., Vipperman, J., Odusami, M., Misra, S.: Multi-class classification of impulse and non-impulse sounds using deep convolutional neural network (DCNN). In: Misra, S., et al. (eds.) ICCSA 2019. LNCS, vol. 11623, pp. 359–371. Springer, Cham (2019)

One-Class SVM to Identify Candidates to Reference Genes Based on the Augment of RNA-seq Data with Generative Adversarial Networks

Edwin J. Rueda[1]([✉]) [iD], Rommel Ramos[1,2] [iD], Edian F. Franco[2,4] [iD],
Orlando Belo[3] [iD], and Jefferson Morais[1] [iD]

[1] Department of Computer Science, Federal University of Para, Belem, Brazil
edwin.rojas@icen.ufpa.br, jmorais@ufpa.br
[2] Institute of Biological Sciences, Federal University of Para, Belem, Brazil
rommelthiago@gmail.com, edianfranco@ufpa.br
[3] University of Minho, Braga, Portugal
obelo@di.uminho.pt
[4] Instituto Tecnologico de Santo Domingo, Santo Domingo, Dominican Republic

Abstract. Reference Genes (RG) are constitutive genes required for the maintenance of basic cellular functions. Different high-throughput technologies are used to identify these types of genes, including RNA sequencing (RNA-seq), which allows measuring gene expression levels in a specific tissue or an isolated cell. In this paper, we present a new approach based on Generative Adversarial Network (GAN) and Support Vector Machine (SVM) to identify *in-silico* candidates for reference genes. The proposed method is divided into two main steps. First, the GAN is used to increase a small number of reference genes found in the public RNA-seq dataset of *Escherichia coli*. Second, a one-class SVM based on novelty detection is evaluated using some real reference genes and synthetic ones generated by the GAN architecture in the first step. The results show that increasing the dataset using the proposed GAN architecture improves the classifier score by 19%, making the proposed method have a *recall* score of 85% on the test data. The main contribution of the proposed methodology was to reduce the amount of candidate reference genes to be tested in the laboratory by up to 80%.

Keywords: Generative adversarial networks · Novelty detection · One-class SVM · RNA-seq · Reference genes

1 Introduction

Reference Genes (RG) are constitutive genes required for the maintenance of basic cellular function. These genes are expressed in all cells of an organism under normal and abnormal conditions and are used in internal controls in investigations of gene expression analysis because their level of expression

© Springer Nature Switzerland AG 2020
O. Gervasi et al. (Eds.): ICCSA 2020, LNCS 12249, pp. 704–717, 2020.
https://doi.org/10.1007/978-3-030-58799-4_51

remains relatively constant in different cells, tissues, and stress levels [8,15,17]. For instance, in the real-time reverse transcription-polymerase chain reaction (RT-qPCR) which provides accurate and reproducible quantification of genetic copies, the selection of RG is essential for accurate normalization of gene expression data obtained [4,11].

The RG identification is not a trivial task because in some tissues or cells, the number of validated RG in the literature is small, and the level of expression of some of these genes varies depending on tissue, cell, or level of stress. Several methods to identify RG are available in the literature, mainly using optimization algorithms and unsupervised machine learning (ML) approach [2,5,13,14].

In [5] is proposed a methodology based on clustering algorithms using Euclidean distance as a similarity metric. In this methodology, the number of clusters is chosen based on the SD validity index and the Dunn index [9]. Then, based on some reference genes, are selected the clusters in which at least one reference gene is present, reducing the number of candidate genes to be considered. Finally, the candidate genes are selected based on the coefficient of variation and the standard deviation of the expression level of each gene.

In [2] is proposed a methodology based on optimization algorithms, more specifically a particular type of linear programming called minimum cost network flow problem. The idea is to find the cheapest path in a directed, acyclic n_s-dimensional graph, where n_s denotes the total number of samples, from the lowliest, non-zero expressed gene (source) to the most highly expressed gene (sink), given several constraints. The genes with identically zero expression are omitted from the analysis and the distance from one gene to another is calculated using the Euclidean distance. Finally, the genes belonging to the cheapest path are considered candidate genes.

The main disadvantage of current methods is based on the small set of RG available in the literature for some tissues or cells. Therefore, methods that use the Euclidean distance between unclassified genes and RG as a similarity metric to detect potential candidates are not giving importance to unclassified genes that are distant from the RG, but that could be candidates due to stability in their gene expression.

To mitigate the main disadvantage of current methods, this paper proposes a novel approach based on Generative Adversarial Networks (GAN) and a one-class classifier based on novelty detection to the problem of RG identification. The proposal is divided into two main steps. First, the GAN is used to generate enough synthetic RG from the actual RG available in the literature for the Escherichia coli MG1655 dataset [2,11]. Second, a one-class classifier is trained using some real RG and synthetic ones generated by the GAN in the first step to detect new candidate genes. In this work, we adopted the support vector machine as a one-class classifier, since this classifier has a good performance in the literature [1,16].

The remaining sections of this paper are organized as follows. Section 2 shows the main related works and the contributions of this work; Sect. 3 presents the proposed method for candidate RG identification, and the results are discussed in Sect. 4. Finally, Section 5 shows the final remarks on this research.

2   Related Work

Different papers in the literature propose the *in-silico* identification of candidates for reference genes. These approaches are based on datasets that contain a minority of genes labeled as reference genes. The objective is to classify the remaining genes as candidates or not to reference genes.

Currently, emerging different methods that use optimization algorithms and unsupervised machine learning algorithms to tackle the problem. In [2] was proposed a novel method for the identification of RG called $moose^2$. This method uses the minimum cost network flow problem to find candidates for RG. The idea is to find the cheapest path in a directed, acyclic n_s-dimensional graph, where n_s denotes the total number of samples, from the lowliest, non-zero expressed gene (source) to the most highly expressed gene (sink), taking into account that for any edge (i, j) that connects two nodes (genes) in the graph, this edge is in the direction of the lower to the higher ranked genes, where the ranking is determined by sorting the values of the gene expression averaged over all samples. Note that in the graph, each gene is connected to every other gene. The data set used is *Escherichia coli* obtained from CGSC (Coli Genetic Stock Center) at Yale University (http://cgsc.biology.yale.edu), where from 6 RG, $moose^2$ identifies 27 possible candidates for RG.

In [5] is proposed a methodology based on clustering algorithms using Euclidean distance as a similarity metric. In this methodology, the number of clusters is chosen based on the SD validity index and the Dunn index [9]. Then, based on some RG, are selected the clusters in which at least one RG is present, reducing the number of candidate genes to be considered. Finally, the candidate genes are selected based on the coefficient of variation and the standard deviation of the expression level of each gene. This method was evaluated in the *Corynebacterium pseudotuberculosis* strains CP1002 and CP25 database [10], obtaining 134 candidate genes in total.

Vandesompele et al. [14] presents the gNorm algorithm, which uses a measure of gene-stability to determine the stability of gene expression. For each control gene, they determine the variation in pairs with all other control genes as the standard deviation of the logarithmically transformed expression ratios and define the stability measure of the internal control gene M as the average variation in pairs of a particular gene with all other control genes. Genes with the lowest M values have the most stable expression. Therefore, those genes can be considered candidates for RG.

These current methods manage to identify candidate genes based on a number of initial RG, which is sometimes inefficient due to the small number of RG identified in certain cells or tissues. Therefore, in this paper, we propose to increase the number of reference genes synthetically through a GAN architecture. Then, based on the augmented reference data, train a one-class SVM classifier to select from a dataset of unclassified genes, the possible candidates for RG.

3   Proposed Method

This section presents the dataset used to select the possible candidates for RG and the algorithms used in this approach. Figure 1 summarizes the proposed approach, which consists of two main steps. First, the amount of RG available in the dataset is increased using a proposed GAN architecture. Second, a one-class classifier based on support vector machine (SVM) is trained to detect new candidates for RG using 70% of the real RG and synthetic genes generated by the GAN architecture in the first step. The performance of the proposed classifier is evaluated with 30% of the remaining RG available in the training set.

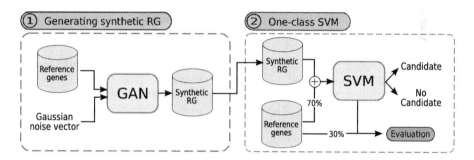

Fig. 1. Proposed approach to the selection of candidate genes for reference genes.

The next subsection explains the dataset used, its initial processing for its subsequent increase through the proposed GAN architecture. Then, the one-class SVM algorithm used for the selection of candidates is explained.

3.1   Dataset

This work uses the RNA-seq database of *Esquerichia coli* MG1655 available at [2]. This dataset consists of 4293 genes that were subjected to stress conditions. Three different samples were removed at 0, 30, and 90 min, and immediately mixed with 0.25 vol of RNA stop solution (95 % ethanol, 5 % phenol) on ice. Note that the dataset has 9 features, 3 for each sampling time.

For the identification of the largest number of RG in this dataset, a review of the literature was necessary, which consisted of taking the genes that were at least tested in three different studies. Finding 21 reference genes in total, 15 in [11] and 6 in [2].

3.2   Data Processing

The RNA-Seq dataset was normalized by reads per kilobase of exon model per million mapped reads (RPKM) and applying $(log_2 + 1)$ to mitigate the number of outliers and increase the proximity of the data features. To filter the dataset and based on the fact that the reference genes have a minimum level of variation [18], a stability test is performed based on the coefficient of variation (CV) given by Eq. 1, where σ represents the standard deviation of the set of features of each gene, and μ represents its mean.

$$CV = \frac{\sigma}{\mu} \tag{1}$$

The stability test consists of selecting the genes for which their CV value is within an interquartile range given by $[Q_1 - k(Q_3 - Q_1), Q_3 + k(Q_3 - Q_1)]$ with $k = 1.5$. Note that genes with a CV value outside this range are not considered. Finally, the data is scaled between $[-1, 1]$ as show in Eq. 2, where X_t represents the transformed data, x_i the i-th gene and X_{max} and X_{min} represent the maximum and minimum values for each features of the dataset.

$$X_t = \frac{2(x_i - X_{min})}{X_{max} - X_{min}} - 1 \tag{2}$$

3.3   Generative Adversarial Nets

Generative adversarial networks were introduced by [6]. The concept was developed for the estimation of generative models based on adversarial processes. In these adversarial processes are trained two neural networks; the first, a generative network $G(z, \theta_g)$ which receives as input some noise variables from a, $p_z(z)$ distribution and aims to learn a p_g distribution on training data x, and a second discriminatory network $D(x, \theta_d)$, which has as output a single scalar which represents the probability that x comes from the distribution of the real data p_{data}, and not from p_g. In this minimax game, there is only one solution, which occurs when the G network recovers the distribution of the real data p_{data}, making the D network fail to distinguish if a data comes from the original distribution or the distribution generated by the G network, $D(x) = \frac{1}{2}$.

Figure 2 illustrates the training process of the GAN architecture. This training is divided into two main steps. First, the discriminative model D is trained based on real and synthetic data and its weights are frozen (these weights are not updated in the second step). Second, the generative model G is trained based on an input noise vector to maximize the probability that the model D makes a mistake. After repeated steps one and two a defined amount of epochs, the model G may generate synthetic data similar to the original data.

The objective of this approach is to generate enough synthetic RG with a GAN architecture to be able to build a one-class model as general as possible. For this, Fig. 3 illustrates the proposed GAN architecture. Where the generator model (Fig. 3(a)) takes as input a noise vector based on a normal distribution

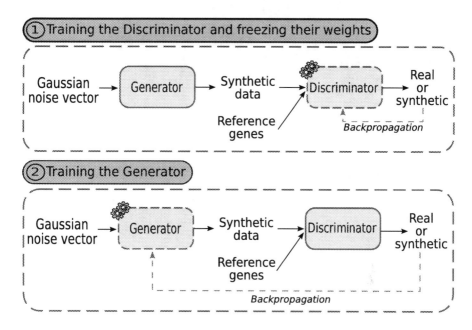

Fig. 2. The learning process of the Generative Adversarial Nets.

with $\mu = 0$ and $\sigma = 1$ and produces as output a synthetic RG. On the other hand, the discriminator model (Fig. 3(b)) was also built with fully connected dense layers. This model takes as input real reference genes (real distribution) and synthetic reference genes (generated by the G network) and its output represents the probability that a gene belongs to the real data distribution. The number of layers and neurons was taken based on the lowest cost value for the metrics proposed in Eqs. 9 and 10.

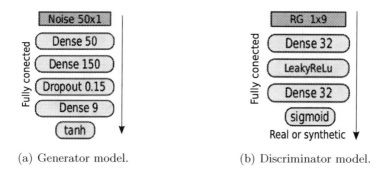

(a) Generator model.   (b) Discriminator model.

Fig. 3. The GAN architecture.

The GAN architecture was trained with the *stochastic gradient descent* algorithm with a different learning rate l_r for the discriminator and the generator.

These learning rates decreased at each epoch by a factor equal to $\frac{l_r}{epochs}$ to improve the convergence of the GAN. The initial weights for each layer were chosen based on *Xavier uniform initializer* [3], which initializes the weights based on a uniform distribution $U[-t, t]$, where t is equal to $\sqrt{6/\left(n_j + n_{j+1}\right)}$ being n_j the number of input neurons and n_{j+1} the number of output neurons in the j layer.

3.4 One-Class SVM

The one-class support vector machine induced by [12], is an algorithm that tries to estimate a function f, being $f \geq 0$ for positive examples and $f < 0$ for any other example. In this strategy, the data is mapped with a function Φ in a new characteristic space F and separated from the origin with a maximum margin.

Consider the training set $x_1, \ldots, x_\ell \in \mathcal{X}$, where $\ell \in \mathbb{N}$ is the number of observations, and $\mathcal{X}$ is some set $\in \mathbb{R}^N$. To separate the training set from the origin, the following quadratic problem is solved:

$$\min_{w \in F, \xi \in \mathbb{R}^\ell, \rho \in \mathbb{R}} \frac{1}{2}\|w\|^2 + \frac{1}{\nu\ell}\sum_i \xi_i - \rho \tag{3}$$

$$s.t. f\left(x\right) = \left(w \cdot \Phi\left(x_i\right) - \rho\right) \geq -\xi_i, \ \xi_i \geq 0, \ i = 1, \ldots, \ell \tag{4}$$

Deriving the dual problem, the solution can be shown to have an support vector expansion:

$$f\left(x\right) = sgn\left(\sum_i \alpha_i k\left(x_i, x\right) - \rho\right) \tag{5}$$

where the *sgn* function is equal to 1 for values greater than or equal to zero and -1 otherwise. Patterns x_i with nonzero α_i are called support vectors, where the coefficients are found by solving the dual problem:

$$\min_\alpha \frac{1}{2}\sum_{ij} \alpha_i \alpha_j k\left(x_i, x_j\right) \ s.t \ 0 \leq \alpha_i \leq \frac{1}{\nu\ell}, \ \sum_i \alpha_i = 1. \tag{6}$$

In this approach, the parameter ν is a trade-off between the normal and anomaly data in the dataset, which was set to 0.010452, and the mapping function Φ was represented by the Gaussian kernel given in Eq. (7), where γ is equal to $1/n_{fe}$ and n_{fe} is the number of features of the training set.

$$k\left(x, y\right) = e^{-\gamma\|x-y\|^2} \tag{7}$$

If the proposed dual problem is solved, the decision function $f\left(x\right)$ will take positive values for the greatest number of samples x_i contained in the training set. To validate the performance of the proposed classifier and based on the premise that we only have a positive class (RG), we consider only the *recall* metric:

$$Recall = \frac{TP}{TP + FN} \tag{8}$$

where TP (True Positives) represents the number of genes that the classifier correctly classified as RG, and FN (False Negatives) represents the amount of RG that the classifier classified as non-RG. This metric allows us to evaluate the ability of our classifier to recognize reference genes. Therefore, a value close to 1 indicates that the classifier has a good performance.

4 Experiments and Results

This section presents the experiments and results for the identification of candidates to RG based on the use of GAN to synthetically increase the training set, and based on this dataset, build a one-class SVM classifier. The focus of this study was based on the creation of the GAN architecture and the one-class SVM classifier for the identification of possible RG candidates in *Escherichia coli* MG1655 dataset obtained from [2].

For this study, the dataset was cleaned taking into account two important stages. In the first stage, we removed the genes for which their expression value was equal to zero. In the second stage, based on the CV (Eq. 1), we eliminated all the genes for which the CV was not within the interquartile range (outliers). Table 1 shows the data processing to obtain the final dataset.

Table 1. Data processing stages

Number of initial genes	Number of genes expressed	Outliers based on the CV	Final dataset
4293	4191	3	4188

Of the 21 RG identified in the literature for *Escherichia coli*, the *idnT* gene was not taken into account because it's CV was outside the interquartile range calculated for the reference genes. Note that were calculated two interquartile ranges, one based on the reference genes and the other based on the remaining genes. Table 2 shows the list of 20 RG used in this study.

Table 2. List of reference genes used in this study

Reference genes
cysG, hcaT, rrSA, ihfB, ssrA, gyrA, recA, rpoB, rpoA, gyrB, rho, ftsZ, secA, rpoC, gmk, adk, rpoD, dnaG, glnA, recF

4.1   Training the Generative Adversarial Network

For the training of the GAN architecture (Fig. 3), a parameter adjustment was made in the G and D networks to reach the global optimum $(D(x) = \frac{1}{2})$. For this reason, we opted to update the weights of both networks (θ_g and θ_d) based on all available reference genes ($batch\_size = 20$), because this allowed the network to converge faster. Note that the weights of network D are updated based on 20 reference genes and 20 synthetic genes generated by network G. Finally, 1700 epochs were used for the training of the proposed GAN architecture, considering that an epoch occurs when all the genes are iterated in the training set (one $batch\_size$). The parameters for training the GAN architecture are presented in Table 3.

Table 3. Parameters used for training the GAN architecture

Parameter	Generator	Discriminator
Optimizer	SGD	SGD
Learning rate	0.00015	0.001
Decay rate	0.00015/1700	0.001/1700
Momentum	0.92	0.9
Epochs	1700	1700

The selection of the best initial weights taking the *Xavier uniform initializer* was based on a similarity metric $S(x, x')$ proposed in Eq. (9), where x represents the training set given by a matrix $\in \mathbb{R}_{(20,9)}$, x' represents the set of synthetic data generated by the G network, and $\hat{y}$ represents the class predicted by network D for synthetic data, which takes values equal to 1 for real data and 0 for synthetic data. The parameters n_f, n_g, and m represent the number of characteristics of each gene, the number of synthetic genes, and the number of RG in the training set. Note that $x_i^{(k)}$ represents the feature k in the i-th gene.

$$S\left(x, x'\right) = \sum_{i}^{m} \sum_{j}^{n_g} \sum_{k}^{n_f} \frac{\left|x_i^{(k)} - x'_j^{(k)}\right|}{n_f n_g m} + \left|0.5 - \frac{1}{n_g} \sum_{j}^{n_g} \hat{y}_j\right| \tag{9}$$

The GAN network was trained on a hundred different occasions, for each of these occasions, the average of the similarity $S(x, x')$ was calculated based on 30 repetitions, wherein each repetition, were generated 300 synthetic genes. Thus, we chose the weights for which the GAN architecture presented the lowest similarity value.

Figure 4 illustrates the convergence process in the training of the proposed GAN architecture, where Fig. 4(a) illustrates that the loss (based on binary cross-entropy) tends to 0.7 and Fig. 4(b) illustrates that the accuracy of the Discriminator tends to 0.5 (global optimum). These results indicate that network

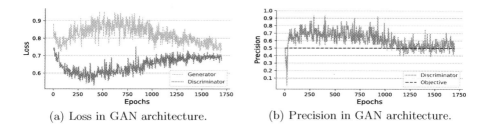

(a) Loss in GAN architecture.   (b) Precision in GAN architecture.

Fig. 4. Convergence process in the training of the GAN architecture.

G is generating synthetic genes similar to the real ones, making network D unable to distinguish them.

Principal component analysis PCA [7] is used for the graphical representation of the augmented dataset, with this representation, we can visually observe the similarity between the original and the synthetic data generated by the G network. For the choice of the best set of synthetic data, we generate 5000 different sets of 300 synthetic genes and based on the metric $E(x')$ proposed in Eq. (10), where CV represents the coefficient of variation for the i-th synthetic gene, we chose the set of synthetic genes with the lowest value of this metric ($E = 0.11$). Figure 5 illustrates the 2-PCA representation of the RG (denoted by a green star) and the synthetic genes (denoted by a red circle), where we can see that the genes generated by the GAN architecture are not distant from the real RG.

$$E(x') = \frac{1}{n_g} \sum_{j}^{n_g} \left[CV(x'_j) + \frac{1 - D(x'_j)}{D(x'_j)} \right] \tag{10}$$

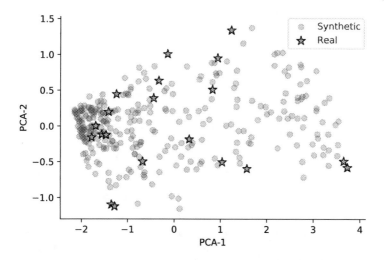

Fig. 5. Principal component analysis between real and synthetic data.

4.2 Selection of Candidate Genes

The selection of candidates to RG is based on the one-class SVM classifier. We train the proposed classifier based on the augmented training set that was built based on 70% of the RG and the 300 synthetic genes generated in the previous step.

The validation of this classifier (Eq. (8)) was performed with four sets of data. Table 4 shows the results of the proposed classifier for each of these data sets. Where in the training set, we calculate two different *recalls*, the first for the augmented dataset, which consists of the augmented data used for classifier training, and the second based only in 70% of the RG chosen above. For the validation of the classifier in the test data, these were divided into two sets, the first set with the remaining 30% of the RG, and the second set based only on new synthetic RG generated by the proposed GAN architecture.

Table 4. Result of the recall metrics in the different datasets

Metrics	Training data		Test data	
	Augmented data	Reference genes (70%)	Reference genes (30%)	Synthetic data
Recall	98.40%	92.85%	85.71%	98.26%

Finally, to observe how the increase of the dataset through the GAN architecture improves the performance of the classifier, we trained the proposed classifier based only on the 20 reference genes (70% for training and 30% for testing) and compared with the trained classifier with augmented data. Table 5 compares the performance of both classifiers.

Table 5. Recall score in the augmented data and in the unaugmented data

Reference genes	*Recall* score	
	Training data	Test data
Augmented	98.40%	85.71%
Unaugmented	85.71%	66.66%

After training the classifier with the data augmented and based on the 4168 unclassified genes in the dataset, we consider as possible candidates for RG the genes for which the $f(x)$ function (Eq. (5)) is equal to 1. Note that this function takes values equal to 1 for data within the decision boundary (positive class).

In this approach, the classifier detected 807 possible candidates to RG, reducing the amount of reference genes candidates to be tested in the laboratory by 80.64%. Figure 6 illustrates the 2-PCA representation of the RG and candidate genes selected by the proposed classifier. In this representation, we can perceive that the proposed classifier generates candidates close to the RG as we expected, which indicates that the mapped space created by the classifier is being consistent with the training data.

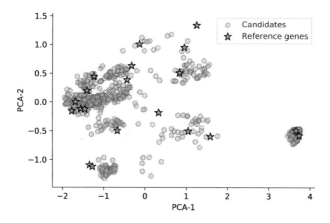

Fig. 6. Principal component analysis between the reference genes and the candidates.

Finally, based on the set of candidates for RG found in this approach and taking into account 27 candidates for RG found in [2]. We select the 11 candidates for RG that are common in both studies. Table 6 shows the common candidates in both studies.

Table 6. The most relevant candidates genes

Gene	Product
ftsX	Cell division protein ftsX
ftsY	Cell division protein ftsY
glyY	$tRNA_{glyY}$
mutY	Adenine DNA glycosylase
ndk	Nucleoside diphosphate kinase
nfuA	Iron-sulfur cluster scaffold protein
rrsE	16S ribosomal RNA (rrsE)
rrsG	16S ribosomal RNA (rrsG)
spoT	(p)ppGpp synthase
thrW	$tRNA_{thrW}$
zupT	Heavy metal divalent cation transporter ZupT

5   Conclusions and Future Works

This paper presented a new approach for *in-silico* identification of candidate reference genes from the *Escherichia coli* MG1655 dataset. This approach consisted of two main steps. First, a proposed GAN architecture was trained to synthetically augment the reference gene set. Second, with the augmented dataset was trained a one-class SVM classifier which obtained an 85% recall score in the test data. This new approach allowed the identification of 807 possible candidate genes for reference genes out of a total of 4170 expressed genes, which reduces the number of genes to be analyzed in the laboratory by 80%.

Finally, the approach demonstrated, as expected, that synthetically increasing the number of reference genes improves the classifier score, having in this research, an increase of 19% with respect to the trained classifier without the augmented data, making the selection of candidate genes more reliable.

For future work, other GAN architectures will be tested to try to improve the global optimum ($D(x) = \frac{1}{2}$). Thus, we could generate synthetic genes that are closer to the real distribution of the reference genes, being able to increase the score of the proposed novelty detector.

Acknowledgments. This study was financed by the Coordenação de Aperfeiçoamento de Pessoal de Nivel Superior - Brasil (CAPES), under the Program PROCAD-AMAZÔNIA, process n° 88881.357580/2019-01.

References

1. Amraee, S., Vafaei, A., Jamshidi, K., Adibi, P.: Abnormal event detection in crowded scenes using one-class SVM. SIViP **12**(6), 1115–1123 (2018). https://doi.org/10.1007/s11760-018-1267-z
2. Berghoff, B.A., Karlsson, T., Källman, T., Wagner, E.G.H., Grabherr, M.G.: RNA-sequence data normalization through in silico prediction of reference genes: the bacterial response to DNA damage as case study. BioData Min. **10**(1), 30 (2017). https://doi.org/10.1186/s13040-017-0150-8
3. Daramouskas, I., Kapoulas, V., Paraskevas, M.: Using neural networks for RSSI location estimation in LoRa networks. In: 2019 10th International Conference on Information, Intelligence, Systems and Applications (IISA), pp. 1–7. IEEE (2019)
4. Du, W., Hu, F., Yuan, S., Liu, C.: Selection of reference genes for quantitative real-time PCR analysis of photosynthesis-related genes expression in *Lilium regale*. Physiol. Mol. Biol. Plants **25**(6), 1497–1506 (2019). https://doi.org/10.1007/s12298-019-00707-y
5. Franco, E.F., et al.: A clustering approach to identify candidates to housekeeping genes based on RNA-seq data. In: Kowada, L., de Oliveira, D. (eds.) BSB 2019. LNCS, vol. 11347, pp. 83–95. Springer, Cham (2020). https://doi.org/10.1007/978-3-030-46417-2_8
6. Goodfellow, I., et al.: Generative adversarial nets. In: Advances in Neural Information Processing Systems, pp. 2672–2680 (2014)

7. Hirose, M., Toyota, S., Ojima, N., Ogawa-Ochiai, K., Tsumura, N.: Principal component analysis for surface reflection components and structure in facial images and synthesis of facial images for various ages. Opt. Rev. **24**(4), 517–528 (2017). https://doi.org/10.1007/s10043-017-0343-x

8. Kim, Y., Kim, Y., Kim, Y.H.: Evaluation of reference genes for gene expression studies using quantitative real-time PCR in *Drosophila melanogaster* after chemical exposures. J. Asia-Pac. Entomol. **23**(2), 385–394 (2020)

9. Legány, C., Juhász, S., Babos, A.: Cluster validity measurement techniques. In: Proceedings of the 5th WSEAS International Conference on Artificial Intelligence, Knowledge Engineering and Data Bases, pp. 388–393. World Scientific and Engineering Academy and Society (WSEAS), Stevens Point (2006)

10. Pinto, A.C., et al.: Differential transcriptional profile of *Corynebacterium pseudotuberculosis* in response to abiotic stresses. BMC Genom. **15**(1), 14 (2014)

11. Rocha, D.J.P., Santos, C.S., Pacheco, L.G.C.: Bacterial reference genes for gene expression studies by RT-qPCR: survey and analysis. Antonie Van Leeuwenhoek **108**(3), 685–693 (2015). https://doi.org/10.1007/s10482-015-0524-1

12. Schölkopf, B., Williamson, R.C., Smola, A.J., Shawe-Taylor, J., Platt, J.C.: Support vector method for novelty detection. In: Advances in Neural Information Processing Systems, pp. 582–588 (2000)

13. Sengupta, T., Bhushan, M., Wangikar, P.P.: A computational approach using ratio statistics for identifying housekeeping genes from cDNA microarray data. IEEE/ACM Trans. Comput. Biol. Bioinf. **12**(6), 1457–1463 (2015)

14. Vandesompele, J., et al.: Accurate normalization of real-time quantitative RT-PCR data by geometric averaging of multiple internal control genes. Genome Biol. **3**(7), research0034-1 (2002). https://doi.org/10.1186/gb-2002-3-7-research0034

15. Wu, Y., et al.: Identification and evaluation of reference genes for quantitative real-time PCR analysis in *Passiflora edulis* under stem rot condition. Mol. Biol. Rep. **47**(4), 2951–2962 (2020). https://doi.org/10.1007/s11033-020-05385-8

16. Yahaya, S.W., Langensiepen, C., Lotfi, A.: Anomaly detection in activities of daily living using one-class support vector machine. In: Lotfi, A., Bouchachia, H., Gegov, A., Langensiepen, C., McGinnity, M. (eds.) UKCI 2018. AISC, vol. 840, pp. 362–371. Springer, Cham (2019). https://doi.org/10.1007/978-3-319-97982-3_30

17. Yu, J., Su, Y., Sun, J., Liu, J., Li, Z., Zhang, B.: Selection of stable reference genes for gene expression analysis in sweet potato (*Ipomoea batatas* L.). Mol. Cell. Probes **53**, 101610 (2020)

18. Zhang, Q., et al.: Selection and validation of reference genes for RT-PCR expression analysis of candidate genes involved in morphine-induced conditioned place preference mice. J. Mol. Neurosci. **66**(4), 587–594 (2018). https://doi.org/10.1007/s12031-018-1198-8

Approach to the Search for Software Projects Similar in Structure and Semantics Based on the Knowledge Extracted from Existed Projects

Filippov Aleksey Alekundrovich⬤, Guskov Gleb Yurevich⁽✉⁾⬤,
Namestnikov Aleksey Michailovich⬤, and Yarushkina Nudezhda Glebovna⬤

Ulyanovsk State Technical University,
32, Severny Venetz Street, 432027 Ulyanovsk, Russia
{al.filippov,jng@ulstu.ru}, guskovgleb@gmail.com,
am.namestnikov@gmail.com
http://www.ulstu.ru/?design=english

Abstract. This article presents a new effective model, algorithms, and methods for representing the subject area of an information system. The subject area is presented in the form of fragments of the knowledge base of the design support system. The knowledge base is formed in the process of analyzing class diagrams in UML notation and project source code. The proposed approaches can reduce the time of the design process and increase the quality of the obtained information system through the use of successful information system design solutions used in other projects. Search for successful design solutions is carried out using the developed metrics for determining the similarity of software systems projects. The metrics allow calculating the match of pattern in OWL ontology format with the source code of the project.

Keywords: Knowledge base · Information system · Class diagram · Information systems design

1 Introduction

The modern approach to the development of information systems involves the use of intelligent automated tools to support the design process, allowing to search for and reuse successful architectural solutions based on a common semantic representation of subject and design knowledge. The knowledge in such automated tools is currently usually presented in the form of ontologies. Ontology development is a long and resource-intensive process that requires the involvement of specialists with competencies in ontological engineering and software development.

This work was supported in part by the Russian Foundation for Basic Research (Projects No. 19-47-730003, No. 18-47-730019, 19-47-730006, 19-47-730005).

O. Gervasi et al. (Eds.): ICCSA 2020, LNCS 12249, pp. 718–733, 2020.
https://doi.org/10.1007/978-3-030-58799-4_52

Developers of information systems as a rule do not have sufficient knowledge about the subject area of the project. The documents governing the subject area do not always record all the accepted semantic meanings of entities and relationships. Creating a knowledge base that allows take into account and reuse successful design solutions will reduce the time of design and development, as well as the number of semantic errors. The main problem of using ontologies in the process of developing information systems is the high requirements for knowledge of the internal structure of ontologies.

The importance of formalizing concepts of the subject area for the development of an information system has led to the emergence of special design languages, which include the formalization of domain concepts (entities). The most common design tools are based on the Unified Modeling Language (UML) [1].

Class diagrams in UML notation and Java source code are used as input data for the design support system. UML diagrams are applicable to the description of specific features of information systems, for example:

- classes that make up the architecture of the information system;
- tables and relationships of database schema ;
- properties and characteristics of user's computer servers to create a physical deployment, etc.

The implementation and use of knowledge-based intelligent systems are relevant in all problem areas nowadays [9,10]. At the moment, a lot of researchers use the ontological approach for the organization of the knowledge bases of expert and intelligent systems: M. Gao, C. Liu [11], D. Bianchini [12], N. Guarino [3], G. Guizzardi [13], R.A. Falbo [14], G. Stumme [15], N.G. Yarushkina [18], T.R. Gruber [16], A. Maedche [17].

Fernando Bobillo Umberto Straccia proposed a fuzzy ontology description format [2]. At this stage in the development of information technology, a large number of open source software systems have been created in various subject areas. Reuse of modules of open software systems will significantly reduce the time spent on software development.

Currently, the strong influence of the characteristics of the problem area, within which the development of software systems is carried out, leads to the frequent use of domain-based development methodology (Domain Driven Development — DDD). This methodology is based on an object-oriented programming paradigm and involves various types of testing that allow performing the function of checking the quality of the source code. But this methodology does not take into account the correctness of the model from the point of view of the features of the subject area. This type of error control is carried out by project managers and leading developers.The formation of the ontological representation of the model allows detecting errors in the perception of the subject area. Knowledge is captured in the ontology in the OWL (Web Ontology Language) [4] format with a predefined $TBox$.

As input to the design support system, class diagrams in UML notation and Java source code are used. The solution a problem of design support information systems consists of executing the following tasks:

1. build a model for representing information system design as the content of a knowledge base;
2. method of ontological indexing of class diagrams in UML notation and project by source code;
3. search methods for effective design solutions in the content of the knowledge base.

2 Software Product Design Model for Representation in the Knowledge Base

Project documentation includes diagrams formalized in UML notations. To solve the problem of the intellectual analysis of design diagrams, it is necessary to formalize the rules of notation in the knowledge base. Such knowledges allow the identification of design patterns and architecture software solutions used in various projects. This allows to search for projects with similar architectural solutions and approaches to the implementation of modules of information systems.

An ontology in OWL format is used as the basis of the knowledge base of the design process support system. The W3C consortium recommends using the $\mathcal{SHOINF}(\mathcal{D})$ formalism [5–7] for the OWL language group (OWL Light, OWL DL, OWL Full) as the logical basis of the ontology description language.

In the context of the $\mathcal{SHOINF}(\mathcal{D})$ description logic, the ontology is a knowledge base of the following form [8]:

$$KB = \{TBox, ABox\}, \tag{1}$$

where $TBox$ — a set of terminological axioms representing general knowledge about the concepts of the knowledge base and their relationships;
$ABox$ — a set of statements (facts) about individuals.

2.1 Tbox Axioms of Information Systems Design Ontology

The terminology of project diagrams is divided into the logical representation of the UML notation and the logical representation of design patterns.

$$
\begin{aligned}
&Relationship \sqsubseteq \top; \\
&Dependency \sqsubseteq Relationship; \\
&Association \sqsubseteq Relationship; \\
&Generalization \sqsubseteq Relationship; \\
&Realization \sqsubseteq Relationship \sqcap \exists \\
&startWith.Class \sqcap \exists endWith.Interface;
\end{aligned} \tag{2}
$$

where $startWith$ $endWith$ — name of roles *comes from* and *coming to*, respectively.

The main classes can be represented:

$$Thing \sqsubseteq \top \forall hasAName.String;$$
$$StructThing \sqsubseteq Thing;$$
$$AnnotThing \sqsubseteq Thing;$$
$$Note \sqsubseteq AnnotThing \sqcap \exists connectedTo.Association;$$
$$Class \sqsubseteq StructThing;$$
$$Object \sqsubseteq StructThing \sqcap \exists$$
$$isObjectOf.Class \sqcap \forall isObjectOf.Class; \qquad (3)$$
$$Interface \sqsubseteq StructThing;$$
$$SimpleClass \sqsubseteq Class;$$
$$AbstractClass \sqsubseteq Class;$$

where $hasAName, isObjectOf$, $connectedTo$ — roles in relationship; $String$ — concrete domain.

Class attributes and methods are represented as follows:

$$Attribute \sqsubseteq \top \sqcap \exists$$
$$hasAAttrName.String \sqcap \exists isAPartOf.Class$$
$$Method \sqsubseteq \top \sqcap \exists \qquad (4)$$
$$hasAMethName.String \sqcap \exists isAPartOf.Class,$$

where $hasAAttrName$ and $hasAMethName$ — relationships "has attribute/method name".

Consider the terminology of design patterns associated with the logical representation of design diagram notation (using the UML class diagram as an example):

$$Template \sqsubseteq \top \sqcap \exists$$
$$hasATempName.String \sqcap \exists hasAExpValue.Double \qquad (5)$$
$$SomeTemplate \sqsubseteq Template,$$

where $hasATempName$ — role "has a design pattern name";
$hasAExpValue$ — role "has value of expression";
$Double$ — concrete domain.

Each design pattern in each specific project has a certain degree of expression.

The hierarchy of concepts of the developed ontology is presented in the Fig. 1.

Hierarchy of properties (DataTypeProperty and ObjectProperty) in developed ontology is presented in the Fig. 2.

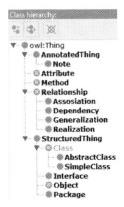

Fig. 1. The hierarchy of concepts of the developed ontology in the editor Protege

2.2   Abox Axioms of Information Systems Design Ontology

The Fig. 3 shows an example of the design pattern "Delegate", which in the form of a set of *ABox* facts:

$$class1: SimpleClass$$
$$class2: SimpleClass;$$
$$attribute1: Attribute$$
$$object1: Object;$$
$$method1: Method$$
$$method2: Method;$$
$$relation1: Association;$$
$$(method1, name1: String): hasAMethName;$$
$$(method2, name2: String): hasAMethName; \qquad (6)$$
$$(attribute1, class1): iaAPartOf$$
$$(object1, class2): isObjectOf;$$
$$(object1, attribute1): owl: sameAs$$
$$(method1, class1): iaAPartOf;$$
$$(method2, class2): iaAPartOf$$

$$(relation1, class1): startWith$$
$$(relation1, class2): endWith.$$

In the knowledge base ontology, the *ABox* fact set includes all the facts about the design patterns used. Then, in the indexing process, the facts from *ABox* are compared with the facts extracted from the design diagrams and the degree of expression for each ontology template is determined.

Fig. 2. Hierarchy of DataTypeProperty and ObjectProperty f the developed ontology in the editor Protege

3   Ontological Representation of Design Patterns

Formally, the ontological representation of the design pattern can be represented as follows:

$$OV_i^{tmp} = \{C, D, R^{same\_as}\}, \tag{7}$$

where C — set of individuals in knowledge base;
D — set of relationship between elements of $i-th$ design patterns, presented as knowledge base individuals;
R — set of equivalence relationships knowledge base individuals.

Design pattern "Builder" is one of the most commonly used patterns in industrial software development. "Builder" is a generic design pattern and allows to create complex composite objects. Figure 4 shows the UML class diagram of the Builder design pattern in the Visual Paradigm.

Representation of the "Builder" design pattern as a fragment of a knowledge base ontology is defined by the following concept instances:

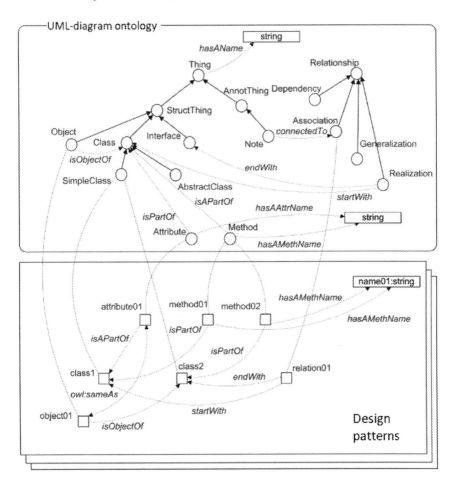

Fig. 3. Ontology structure (including design pattern example)

$$Builder.Client: SimpleClass$$
$$Builder.Director: SimpleClass$$
$$Builder.ConcreteBuilder: SimpleClass$$
$$Builder.Product: SimpleClass$$
$$Builder.AbstractBuilder: AbstractClass$$
$$Builder.Client\_AbstractBuilder: Association \qquad (8)$$
$$Builder.Client\_Director: Association$$
$$Builder.Client\_IProduct: Association$$
$$Builder.ConcreteBuilder\_Product: Association$$
$$Builder.ConcreteBuilder\_AbstractBuilder: Generalization$$
$$Builder.Product\_IProduct: Realization$$

The ontology fragment presented above has the form shown in the Fig. 5.

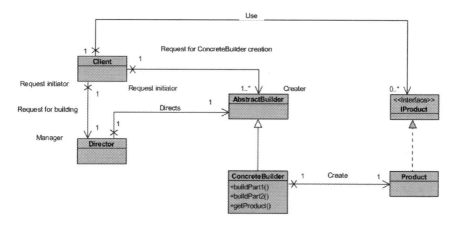

Fig. 4. Class diagram of design pattern "Builder".

Realization of the similar project search function in the knowledge base use the metric to calculate the similarity between designed (in UML notation) and already implemented software projects.

4 Determining the Design Pattern Expression in the Information System Project

To calculate the similarity measure of software projects, the following method is proposed for calculating the measure of expression the design pattern in a software project:

$$\mu^{prj}(tmp_i) = \frac{\left|C^{prj} \cap C^{tmp_i}\right| + \left|R^{prj} \cap R^{tmp_i}\right|}{\left|C^{tmp_i}\right| + \left|R^{tmp_i}\right|}, \tag{9}$$

where $\left|C^{prj} \cap C^{tmp_i}\right|$ — number of matching individuals in an ontological representation $i - th$ knowledge base design pattern and ontological representation of a software project;
$\left|R^{prj} \cap R^{tmp_i}\right|$ — number of matching relationships in an ontological representation $i - th$ knowledge base design pattern and ontological representation of a software project;
$\left|C^{tmp_i}\right|$ — number of individuals in an ontological representation $i - th$ knowledge base design pattern and ontological representation of a software project;
$\left|R^{tmp_i}\right|$ — number of relationships in an ontological representation $-th$ knowledge base design pattern and ontological representation of a software project.

If the number of facts ($\left|C^{tmp_i}\right|$ and $\left|R^{tmp_i}\right|$) ontological representation $i - th$ design pattern tmp_i determined by summing up the number of facts. To calculate number of facts ($\left|C^{prj}\right|$ $\left|R^{prj}\right|$) in ontological representation of a software project it is necessary to use the following developed algorithm:

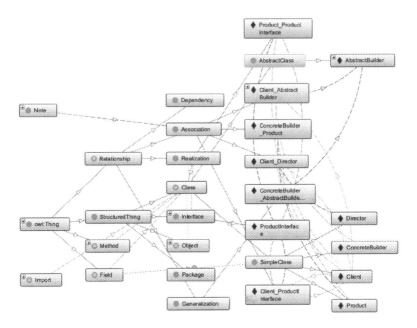

Fig. 5. An example of an ontological representation of a design pattern Builder

Step 1. Convert UML class diagram project $proj_j$ to set of facts $ABox^{prj}$:

$$elem_k^{prj} : Concept$$

$$\left(elem_k^{prj}, elem_s^{prj}\right) : Role,$$

where *Concept* — concept of the knowledge base, defined at *TBox*;
Role — role, defined at *TBox*;
$elem_k^{proj}$ — $k - th$ individual of ontology concept, extracted from diagram.

Step 2. Defining a set of main classes from $ABox^{prj}$ regarding $i - th$ design pattern tmp_i.

The base class will be such an individual $elem_k^{prj}$ of the concept *Class* (or its child concept *Subclass*) of $ABox^{prj}$, which corresponds to some individual $elem_k^{tmp} : Class$ from $ABox^{tmp_i}$ for which a number of facts coinciding with the tmp_i pattern are maximum:

$$elem_k^{prj} : Concept \left(elem_k^{prj}, *\right) : Role \quad \left(*, elem_k^{prj}\right) : Role. \tag{10}$$

Step 3. Calculation of the number of true facts. The fact is true if there is a correspondence between $i - th$ class individuals of the design pattern tmp_i and class diagrams of the project prj_j:

$$\forall k : elem_k^{tmp} \leftrightarrow elem_k^{prj}.$$

This algorithm of the class diagram indexing is performed for each design pattern available in the knowledge base ontology.

After calculating the measure of the expression of each selected design pattern in each of the considered software projects, it becomes possible to calculate the measure of similarity between software projects using one of three metrics.

5   Metrics Measures Architectural and Semantic Similarity of Software Projects

The first metric allows to calculate the measure of similarity by the most expressed pronounced design pattern in each of the projects:

$$\mu^1\left(prj_i, prj_j\right) = \bigvee_{tmp_k \in (prj_i \cap prj_j)} \mu^{prj}\left(tmp_k\right), \tag{11}$$

where prj_i, prj_j — UML class diagram ontological representation $i-th$ and $j-th$ projects respectively;
$\mu^{prj}\left(tmp_k\right)$ — measure of expression $k-th$ design pattern in project (expression 9).

This metric demonstrates good results for a relatively small number of complex combined design patterns. Such design patterns are based on the subject area and, to a lesser extent, correspond to design patterns in the usual sense of industrial programming.

The second metric extends the first (the expression 11) and takes into account the degree of expression of design patterns that exceeds a certain threshold value. A threshold value of 0.3 was chosen experimentally. If the measure of the expression of the design pattern is less than 0.3, we can conclude that there is no design pattern in the software project, and as a result, such a design pattern should be excluded from consideration:

$$\mu^2\left(prj_i, prj_j\right) \frac{\bigvee_{tmp_k \in (prj_i \cap prj_j) \geq 0.3} \mu^{prj}\left(tmp_k\right)}{N}, \tag{12}$$

where N — number of design patterns with expression measure more than 0.3 for each project.

The third metric is similar to the second metric (expression 12), but imposes an additional condition on the contribution of the measure of expressiveness of the design pattern $(\tilde{\mu}^{prj})$ to the measure of architectural similarity of projects:

$$\mu^3\left(prj_i, prj_j\right) \frac{\bigvee_{tmp_k \in (prj_i \cap prj_j) \geq 0.3} \tilde{\mu}^{prj}\left(tmp_k\right)}{N}, \tag{13}$$

where $\tilde{\mu}^{prj}\left(tmp_k\right)$ — the weighted measure of expression design pattern tmp_k in software project prj.

The weighted measure of expression $\tilde{\mu}^{prj}\left(tmp_k\right)$ $k-th$ design pattern tmp_k in project prj is measure of expression (expression 9), normalized by number of elements, included in design pattern with maximum set of element:

$$\tilde{\mu}^{prj}(tmp_i) = \frac{\left|C^{prj} \cap C^{tmp_i}\right| + \left|R^{prj} \cap R^{tmp_i}\right|}{\bigvee_{tmp_k \in ABox}\left(\left|C^{tmp_k}\right| + \left|R^{tmp_k}\right|\right)}, \tag{14}$$

This modification allows taking into account the complexity of the internal structure of the design pattern when calculating the similarity measures of software projects.

Design pattern consist from 20 elements that has full expression by $\mu^3(prj_i, prj_j)$ in projects $proj_i$ and $proj_j$, will have 4 times more weight than the design pattern, consisting of 5 elements and also having a degree of expression equal to 1.

6   Experiment in Finding Design Patterns in Public Projects from Github

To test the proposed approach to highlighting design patterns in projects, an experiment was conducted, the purpose of which was to search for design patterns in projects located in the GitHub repository. To conduct the experiment, information on 10 design patterns was added to the ontology: Delegation, Interface, Abstract superclass, Builder, Adapter, Singleton, Bridge, Faade, Decorator, Observer.

As a result of the "vk api" request, 108 projects were received related to the set of projects working with the social network API "VKontakte". VKontakte is the most common social network in Russia.

The query by "design patterns" resulted in a sample of 6516 projects. This experiment is necessary to verify the operation of the system in conditions of increased content of design patterns in projects.

For testing, the sample was limited to the first 100 projects for both requests. Search results for design patterns are presented as bar graphs in the Figs. 6 and 7.

Selected design patterns differ in the number of elements and the relationships between them. The number of elements varies from 3 to 20.

In this experiment, only projects developed using the Java programming language were also considered.

Since the total number of projects in the GitHub repository developed in the Java language is very large, it is necessary to limit the selection of projects for the experiment. As a result, the following results were obtained: high frequency of use of the Delegation, Interface, Abstract superclass and Facade templates. This result is explained by the simple structure of these patterns — a relatively small number of structural elements and, as a result, relationships. These design patterns may have been used unconsciously by developers, or they may coincide in structure with part of a more complex pattern.

There were relatively few design patterns for Builder, Adapter, Bridge, Decorator, and Observer in the control group of projects. The rarity of these patterns is due to their complex structure — the content of a large number of elements.

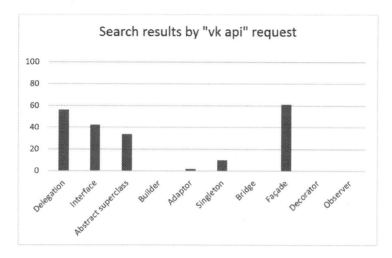

Fig. 6. Search results for templates among projects received by request "vk api"

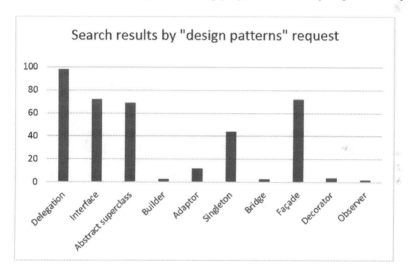

Fig. 7. Search results for templates among projects received by request "design patterns"

7 Search Experiments Results for Structurally Similar Software Projects

To determine the measure of structural similarity between two projects, it is necessary to calculate the measure of severity of each design pattern in both projects (expressions 9 and 14).

In this experiment, all projects were downloaded from the GitHub open repository. All projects were selected by the following keywords: "public API", "social network", "vkontakte", which allows to determine whether the projects belong to the subject area — work with the social network API "VKontakte": Android-MVP, cordova- social-vk, cvk,DroidFM, VK-Small-API, VKontakteAPI, VK_TEST.

The Table 1 shows the severity of each considered design pattern in all projects of the experimental sample. The similarity score ratings are normalized from 0 to 1. The estimates of the similarity measure for the first metric

Table 1. The expression measure of design patterns in projects

Project name	Delegator	Adapter	Builder	Abstract superclass	Interface
Android-MVP	1.0	0.875	0.83	1.0	1.0
cordova-social-vk	1.0	0.875	0.83	1.0	0.8
cvk	1.0	0.875	0.92	1.0	1.0
DroidFM	1.0	0.875	0.92	1.0	1.0
VK-Small-API	1.0	0.42	0.92	0.33	0.6
VKontakteAPI	1.0	0.83	0.92	1.0	0.8
VK_TEST	1.0	0.75	0.58	0.66	0.6

(expression 11) are always equal to 1, because this metric selects a design pattern with the maximum measure of expression for each of the two compared projects. And since, for example, the Abstract superclass, Interface, and Delegator design pattern consists of a relatively small number of elements, this leads to a high degree of expression of such patterns in a large number of projects.

The estimates of the projects similarity measure by second (expression 12) and third (expression 13) metric presented at 2 and 3 tables respectively.

Table 2. Measures of structural similarity of software products in the second metric

Projects	Android-MVP	cordova-social-vk	cvk	Droid-FM	VK-Small-API	VKonta-kteAPI	VK_TEST
Android-MVP	–	0.96	0.96	0.98	0.78	0.96	0.78
cordova-social-vk	0.96	–	1	0.94	0.85	1	0.83
cvk	0.96	1	–	0.94	0.85	1	0.83
DroidFM	0.98	0.94	0.94	–	0.78	0.94	0.78
VK-Small-API	0.78	0.85	0.85	0.78	–	0.85	0.96
VKontakteAPI	0.96	1	1	0.94	0.85	–	0.83
VK_TEST	0.79	0.83	0.83	0.78	0.95	0.83	–

Table 3. Measures of structural similarity of software products in the third metric

Projects	Android-MVP	cordova-social-vk	cvk	Droid-FM	VK-Small-API	VKonta-kteAPI	VK_TEST
Android-MVP	–	0.96	0.96	0.96	0.64	0.96	0.77
cordova-social-vk	0.96	–	0.99	0.93	0.67	0.99	0.80
cvk	0.97	0.99	–	0.93	0.67	0.99	0.80
DroidFM	0.97	0.93	0.93	–	0.61	0.93	0.74
VK-Small-API	0.64	0.67	0.68	0.61	–	0.67	0.87
VKontakteAPI	0.97	0.99	0.99	0.93	0.67	–	0.80
VK_TEST	0.77	0.80	0.80	0.74	0.87	0.80	–

The degree of similarity of the projects obtained in these experiments are very high, which can be explained by two features of this experiment. Design patterns with a severity measure of less than 0.3 were excluded in at least one of the compared projects. In this experiment, it is assumed that the design pattern, expressed with a measure of expression less than 0.3, is not found in the project. Accounting design patterns with a small degree of severity will lead to a significant decrease in the value of the similarity indicator of any projects with an increase in the number of design patterns.

The considered metrics for calculating project similarity are based on a single computational principle and represent its consistent development. The third metric is the most universal for projects and design patterns of different sizes but much more parametrized.

Design patterns can be implemented in projects in various ways. This problem can be solved in two ways:

1. Using the corporate standard of the company
2. To use projects from open sources, it is worthwhile to form two or more alternative representations of the design pattern in an ontology and consider them equivalent

8 Conclusion

In the course of this research, the following results were obtained:

1. Ontologically-oriented model of the UML diagram language and ontological model of the design pattern.
2. Architectural similarity measures for software projects; measures of expressiveness of the design pattern in the considered software products.
3. An algorithm for transforming a class diagram in UML notation into an ontology of the OWL format.
4. An algorithm for transforming source code in the Java programming language into an ontology of the OWL format.

Thus, the proposed approach to supporting the design process allows the use of successful design solutions in the development of new software project, thereby reducing the design process time and increasing the quality of the resulting solutions.

References

1. Booch, G., Rumbaugh, J., Jacobson, I.: Unified Modeling Language User Guide, 2nd edn. Addison-Wesley Object Technology Series, p. 496, New York (2005)
2. Bobillo, F., Straccia, U.: Fuzzy ontology representation using OWL 2. Approximate Reasoning **52**(7), 1073–1094 (2010)
3. Guarino, N., Musen, M.A.: Ten years of applied ontology. Appl. Ontol. **10**, 169–170 (2015)
4. OWL 2 Web Ontology Language. https://www.w3.org/TR/owl2-overview/
5. Baader, F., Calvanese, D., McGuinness, D., Nardi, D., Patel-Schneider, Peter F: The Description Logic Handbook: Theory, Implementation, and Applications. Cambridge University Press, Cambridge (2003)
6. Bonatti, P.A., Tettamanzi, A.G.B.: Some complexity results on fuzzy description logics. In: Di Gesú, V., Masulli, F., Petrosino, A. (eds.) WILF 2003. LNCS (LNAI), vol. 2955, pp. 19–24. Springer, Heidelberg (2006). https://doi.org/10.1007/10983652_3
7. Horrocks, I., Patel-Schneider, P.F., van Harmelen, F.: From SHIQ and RDF to OWL: the making of a web ontology language. J. Web Semant. **1**(1), 7–26 (2003)
8. Grosof, B., Horrocks, I., Volz, R., Decker, S.: Description logic programs: combining logic programs with description logics. In: Proceedings of WWW 2003, Budapest, Hungary, pp. 48–57. ACM, May 2003
9. Golenkov, V., Guliakina, N., Davydenko, I.: Methods and tools for ensuring compatibility of computer systems. Open Semant. Technol. Intell. Syst. **3**, 25–53 (2019)
10. Golenkov, V., Shunkevich, D., Davydenko, I.: Principles of organization and automation of the semantic computer systems development. Open Semant. Technol. Intell. Syst. **3**, 53–91 (2019)
11. Gao, M., Liu, C.: Extending OWL by fuzzy description logic. In: Proceedings of the 17th IEEE International Conference on Tools with Artificial Intelligence (ICTAI 2005), pp. 562–567 (2005)
12. Bianchini, D., de Antonellis, V., Pernici, B., Plebani, P.: Ontologybased methodology for e-service discovery. Inf. Syst. **31**, 361–380 (2005)
13. Guizzardi, G., Guarino, N., Almeida, J.P.A.: Ontological considerations about the representation of events and Endurants in business models. In: International Conference on Business Process Management, pp. 20–36 (2016)
14. Falbo, R.A., Quirino, G.K., Nardi, J.C., Barcellos, M.P., Guizzardi, G., Guarino, N.: An ontology pattern language for service modeling. In: Proceedings of the 31st Annual ACM Symposium on Applied Computing, pp. 321–326 (2016)
15. Hotho, A., Staab, S., Stumme, G.: Ontologies improve text document clustering data mining. ICDM **2003**, 541–544 (2003)

16. Gruber, T.: Ontology. http://tomgruber.org/writing/ontology-in-encyclopedia-of-dbs.pdf. Accessed Dec 2019
17. Maedche, A., Staab, S.: Ontology learning for the Semantic Web. https://www.csee.umbc.edu/courses/771/papers/ieeeIntelligentSystems/ontologyLearning.pdf. Accessed Dec 2019
18. Guskov, G., Namestnikov, A., Yarushkina, N.: Approach to the search for similar software projects based on the UML ontology. In: Abraham, A., Kovalev, S., Tarassov, V., Snasel, V., Vasileva, M., Sukhanov, A. (eds.) IITI 2017. AISC, vol. 680, pp. 3–10. Springer, Cham (2018). https://doi.org/10.1007/978-3-319-68324-9_1

Prediction of the Methane Production in Biogas Plants Using a Combined Gompertz and Machine Learning Model

Bolette D. Hansen[1,2(✉)], Jamshid Tamouk[2], Christian A. Tidmarsh[2],
Rasmus Johansen[2], Thomas B. Moeslund[1], and David G. Jensen[2]

[1] Aalborg University, Rendsburggade 14, 9000 Aalborg, Denmark
bdha@create.aau.dk
[2] EnviDan A/S, Vejlsøvej 23, 8600 Silkeborg, Denmark

Abstract. Biogas production is a complicated process and mathematical modeling of the process is essential in order to plan the management of the plants. Gompertz models can predict the biogas production, but in co-digestion, where many feedstocks are used it can be hard to obtain a sufficient calibration, and often more research is required in order to find the exact calibration parameters. The scope of this article is to investigate if machine learning approaches can be used to optimize the predictions of Gompertz models. Increasing the precision of the models is important in order to get an optimal usage of the resources and thereby ensure a more sustainable energy production. Three models were tested: A Gompertz model (Mean Absolute Percentage Error (MAPE) = 9.61%), a machine learning model (MAPE = 4.84%), and a hybrid model (MAPE = 4.52%). The results showed that the hybrid model could decrease the error in the predictions with 53% when predicting the methane production one day ahead. When encountering an offset in the predictions the reduction of the error was increased to 66%.

Keywords: Biogas · Prediction · Gompertz model · Machine learning

1 Introduction

Climate changes and increasing energy demands have increased the focus on renewable energy in the recent years. The biogas industry contributes to this by e.g., producing energy from wastewater and reducing pollution from agriculture. The biogas industry has had a significantly progress in Europe in the recent years, where the capacity was almost tripled in the period from 2007 to 2017 [1].

Biogas is produced by anaerobic digestion of organic materials such as urban waste for example food, sludge and garden waste, animal manure, industrial waste, ligno-cellulosic materials such as various types of straw and the biomass of microalgae [2].

How fast a feedstock can be transformed to methane depends on the composition of the feedstock. Carbohydrates, protein and fat are easy digestible, whereas e.g. ligno-cellulosic materials are much harder to digest [2].

In addition to this, several parameters influence the process. These are, among others, the pH, carbon/nutrient rate, organic loading rate, temperature, the microorganisms

© Springer Nature Switzerland AG 2020
O. Gervasi et al. (Eds.): ICCSA 2020, LNCS 12249, pp. 734–745, 2020.
https://doi.org/10.1007/978-3-030-58799-4_53

and enzymes present, presence of some types of fatty acids, and usage of substrates. Over or under representation of some of the parameters can even lead to failure of the system [3, 4].

Mono-digestion of some feedstocks can be problematic, as it might bring the system out of balance, for instance changing the carbon/nutrient rate will lead to a too high concentration of problematic fatty acids or change the pH. Therefore, much research has focused on co-digestion of feedstocks, which can lead to an increase in the biogas production of 25–400% [3]. However, co-digestion complicates the digestion process further and therefore, mathematical modelling of the process is essential in order to keep the balance and avoid failure [3]. In 2002 [5] published the IWA Anaerobic Digestion Model No.1 (ADM1). The ADM1 can simulate an average trend in the different parameters, however, it cannot simulate the immediate variations [6]. Since the ADM1 was published plugins with modifications have been added and in 2015 [7] states that despite new models have been developed, there is readily justification for developing a new ADM2. In order to develop an ADM2 it is necessary to have a uniform approach to the mechanisms and challenges which will require further research. After 2015, other models and optimizations of the ADM1 have been performed [8].

Where the ADM1 model focuses on modelling the whole process, another model type, called Gompertz models, is often used for prediction of the gas production. This type of model can give very accurate predictions of the methane production [9], however, like the ADM1 model, this model requires a precise calibration. In order to calibrate a Gompertz model, Gompertz functions need to be set up for all the biomasses used in the model. Several studies have been made in order to improve Gompertz models by experimentally finding the kinematic parameters describing the methane production for a wide range of feedstocks [9–11]. For this reason, the literature can provide the kinematic parameters for a large variety of feedstocks. If the Gompertz functions for some of the feedstocks in a biogas plant is not available in the literature, they can be found experimentally. However, as this often takes several months, they are often based on expert knowledge instead.

It is worth noticing that despite Gompertz variables are available for one feedstock, there might be local variations in the composition of the feedstocks influencing the gas production. This is complicated further from co-digestion and other parameters influencing the digestion of each material. For this reason and despite making several tests in the laboratory, the results might not fit well in a real case scenario.

Despite that the mathematical models are essential in order to ensure a stable process, and avoid failure, they are very hard to calibrate, especially when the parameter complexity is high. In these cases further research in parameter characterization is required [3].

Machine learning models have been developed in order to predict and optimize the methane production. In wastewater treatment plants Neural Networks have for instance been used to find the optimal settings for as high a methane yield as possible [12]. In agricultural biogas plants a combination of Genetic Algorithm and Ant Colony Optimization has been used to predict the present gas production based on measured process variables [13]. In controlled laboratory scale experiments Neural Networks have shown precise predictions of the biogas production [14]. Likewise, the machine learning algorithms Random Forest and XGBoost have shown efficient future predictions of the methane yield in an industrial-scale co-digestion facility [15].

Being able to predict the future biogas production is essential in order to plan the operation of the biogas plant. In some cases, the goal is to produce as much methane as possible, as there is an unmet demand. In other cases, it is essential to keep as stable a production as possible in order to meet the demand while avoiding overproduction. In case of overproduction, the process can be artificially inhibited by chemicals or the surplus methane can be burned off. In both cases resources are not optimally used. This has led to development of software tools used for planning of the biogas production. The software has a framework to take in different machine learning models for prediction of the biogas production. In this case the machine learning models were used to predict the production in three categories: Low, medium, and high [16].

The scope of this article is to investigate if machine learning approaches can be used to optimize the predictions of Gompertz models in industrial settings. This is done by comparison of a Gompertz model, a machine learning model, and a combined model.

2   Method

2.1   Biogas Plant

The biogas plant has a capacity at approximately 220,000 t biomass/year and produces almost 10 mio. Normal cubic meter (Nm^3) methane/year, corresponding to 99,7 GWh heat per year. The main feedstocks used in the plant are seaweed, manure, eulat and pectin. However, in total 18 specific feedstocks were used. It is worth noticing that it is an industrial setting and the available feedstocks changes over time. For this reason, some of the feedstocks were only used in the first half of the measurement period, while others were only used in the second half of the period. Fortunately, the amount of these temporary feedstocks is limited, and the models needs to be tolerant to this type of changes.

In this biogas plant a Gompertz model is used to plan the infeed to the plant in order to obtain as constant a biogas production as possible.

2.2   Data Set

The data set obtained from the biogas plant contained consecutive measures from 818 days. In total 18 different feedstocks were used of which one was only used in the first half of the data set and four were only used in the last half of the data set.

2.3   Gompertz Model

For each biomass the expected methane production over time can be described by Gompertz functions [17, 18]. The formula for Gompertz functions can be seen in Eq. 1.

$$P(t) = P_{max} \cdot \exp\left(-\exp\left(\frac{R_{max} \cdot e}{P_{max}} \cdot (\lambda - t) + 1\right)\right) \qquad (1)$$

Where P is the culminative methane production, P_{max} is the maximal total methane production R_{max} is the maximal methane production rate, t is the time measured in days and λ is the delay before any gas is produced from the specific biomass. Hereafter the methane production for a biomass on a given day can be calculated as seen in Eq. 2.

$$P_{day} = P(t_{day}) - P(t_{day} - 1) \qquad (2)$$

Where P_{day} is the amount of produced methane on the specific day and t_{day} is the day for prediction. The expected daily methane production per ton of each biomass can be seen in Fig. 1.

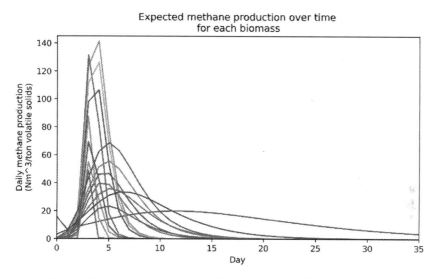

Fig. 1. Expected methane produced by each type of biomass over time

The methane production was then calculated as the sum of the added methane per biomass per day, including biomasses added up to 60 previous to the precent day. The model encountered the retention as a percentage removal of biomass per day.

The parameters in the Gompertz functions were initially based on empirical numbers where these were available. For the biomasses where empirical data was not available, they were estimated based on knowledge about similar biomasses. Hereafter the model was calibrated using the previous 90 days as calibration data. The calibration was done by minimizing the root mean square error.

In addition to the initial parameters each biomass had a span they were forced to stay within.

2.4   Machine Learning Model

Preprocessing

Before training the machine learning model, all the data points were normalized by subtracting the mean value and scaling to unit value using the Python library scikit-learn [19]. Hereafter zeros were replaced with values close to zero. This was done in order to account for algorithms being sensitive to zeros and as it showed better results in some cases. As biogas production is a time-consuming process which takes several days, eight additional features were added to each data point. These features were the measured gas production from the previous six days and the mean and standard divination calculated for the present and previous nine days. As the feature creation in this case implies inclusion of the input parameters nine days before the measurement, the first nine days of the measurement data was excluded due to missing data.

As the last part of the data set was collected while the model was developed, only the first 430 datapoints were used for training, while the last 350 datapoints were used for testing. Thereby the training set contains one biomass which is not present in the test set and the test set contains four parameters which are not present in the training data. An overview of the preprocessing pipeline can be seen in Fig. 2.

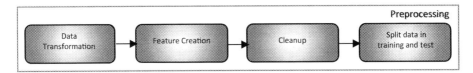

Fig. 2. The preprocessing pipeline.

Model Development and Training

The training set was split into 25 folds for folding where 23 folds were used for training, one fold was used for validation and one fold was used for test.

Initially 15 commonly used machine learning algorithms from the Python library scikit-learn [19] were tested. These were uniform k-nearest neighbors (kNN), distance kNN, Bagging [20] with Decision Tree, AdaBoost [21] Regressor with Decision Tree, Random Forest Regression [22], Bagging with Random Forest, recursive feature elimination (RFE) with the core of linear Ridge, Recursive Feature Elimination by using Gradient Boosting [23], Principal Components Regression, Quadratic Discriminant Analysis, Lasso Regression, Multilayer Perceptron (MLP) Regressor, Naive Bayes [24], Extra Trees [25], and Support Vector Machine [26]. Based on this initial test seven methods were selected. These algorithms were uniform kNN, distance kNN, recursive feature elimination (RFE) with the core of linear Ridge, MLP Regressor, Ada Boost Regressor with Decision Tree Regressor, RFE with the core of Gradient Boosting Regressor, and Random Forest Regressor.

For each fold, the best three models were ensembled whereby the prediction would be the mean of the three models. If the ensemble score was better than the score of the single models this was selected. Otherwise the best single model was selected. If the selected model did not obtain a sufficient precision, no model from that fold was saved. Lastly all the saved models were applied to the test set and the prediction of the models were averaged in order to predict the methane production one day ahead. An overview of the setup can be seen in Fig. 3.

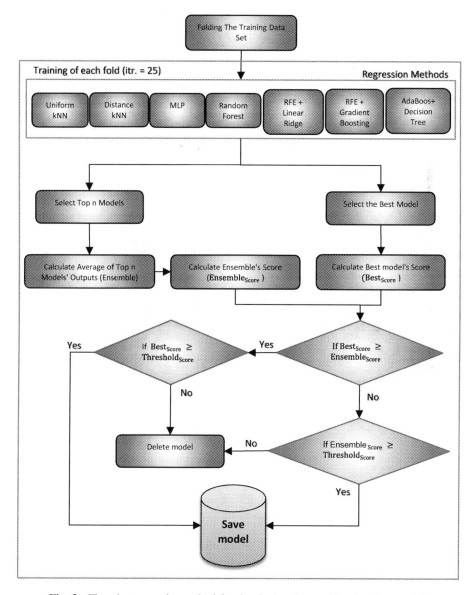

Fig. 3. Flowchart over the method for developing the machine learning model.

2.5   Hybrid Model Based on the Gompertz Model and the Machine Learning Model

In order to make a hybrid model based, the error of the Gompertz model was found by subtracting the Gompertz predictions from the measurements. Hereafter a machine learning model similar to the model described in Sect. 2.4 was trained to predict the error of the Gompertz model. Subsequently the predictions from the machine learning model were added to the predictions from the Gompertz model in order to predict the amount of methane produced.

3   Results

The predictions from respectively the Gompertz model, the machine learning model, and the hybrid model for the full test set can be seen in Fig. 4, and a zoomed version can be seen in Fig. 5. Likewise, the correlation between the observations and the predictions was found as seen in Fig. 6.

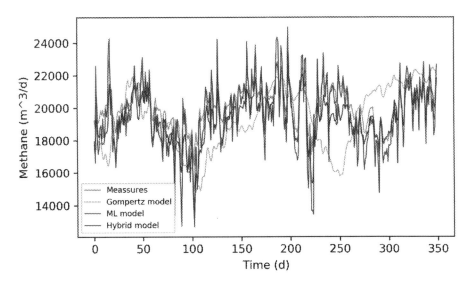

Fig. 4. Comparison of the measured biogas production and the production predicted one day ahead by the Gompertz model, the machine learning model and the hybrid model.

From Fig. 6 it was observed that there was an offset in the predictions according to the correlation line. A similar offset was observed for prediction on the training set. If calculating the mean of the measurements and the predictions and adjusting for this by adding the difference in mean values to the predictions the error of the predictions is decreased. The Mean Absolute Percentage Error (MAPE) for each of the models with and without adjustment according to the mean prediction can be seen in Table 1.

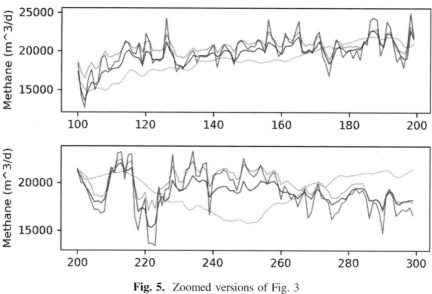

Fig. 5. Zoomed versions of Fig. 3

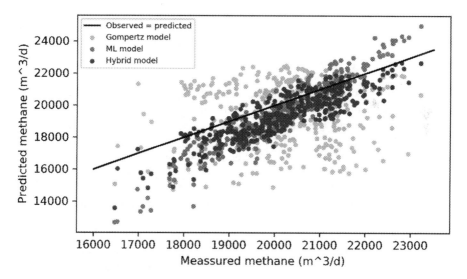

Fig. 6. Comparison between the observed and predicted values for the three models.

Table 1. The mean absolute error for each of the three models

Model	R^2	↓MAPE	↓MAPE after adjustment
Gompertz model	−0.27	9.61%	9.00%
ML model	0.72	4.84%	3.78%
Hybrid model	0.75	**4.52%**	**3.06%**

4 Discussion

Modelling of biogas plants is essential in order to keep an optimal operation of the plants. Gompertz models can be used to plan the infeed to biogas plants in order to keep a constant output. However, these models can be hard to calibrate, especially when the number of feedstocks is high. The Gompertz model presented in this paper can estimate the methane production with a MAPE at 9.61%. From Fig. 6 it can be seen that the correlation between the predicted and observed methane production for the Gompertz model is not very corelated in the data available. This is because the model is used to plan the infeed to the biogas plant in order to ensure a constant production and it tells us that the model is used to its limits. Due to the complexity in co-digestion scenarios more research is required in order to calibrate the Gompertz model further. As this would require several tests each lasting for several days a machine learning and a hybrid model were proposed in order to optimize the predictions further.

From Figs. 4, 6 and Table 1 it can be seen that the machine learning and combined model can improve the prediction one day ahead despite the usage of additional feedstocks in the test set. When comparing the machine learning model with the measurements it is clear that it is able to find the relationships between the input feedstocks and the methane production. However, it typically overestimates the changes in the production: When the production increases the model predicts the increase to be bigger than it is and when the production decreases the model predicts the reduction in the production to be bigger than it is. The hybrid model does not have this problem as the Gompertz model generates a baseline for the production. In other industries it have been found beneficial to develop hybrid models as it can make the model more generalizable and requires less training data [27].

The issue with changes of input data is a typical issue in industrial cases. In this case the quantity of the additional biomasses was relatively low, but if the amount of these feedstocks was increased it could either be added to other feedstocks with similar compositions or the model could be retrained. However, retraining would require several datapoints with usage of the new feedstock.

As it appears from the results the hybrid model can optimize the predictions with 53% when not encountering the offset and with 66% when encountering an offset in the predictions. This is important as surplus methane will be burned off or the gas production will be inhibited with chemistry, as it is too expensive to build storage facilities. Likewise, if not enough methane is produced the demand is not met. Despite several studies deals with optimization of Gompertz functions in order to develop more precise Gompertz models, they are not suitable for comparison with our model. This is because they are based on experimental studies. The contribution of this paper is to show that we can increase the predictions of the methane production in an industrial setting were some of the parameters are based on expert knowledge as the exact parameters for each of the Gompertz functions are not available.

Two other articles focusing on machine learning based predictions of the methane production in industrial scale biogas plants were found [15, 16]. When compared to [16] the predictions from the machine learning model and the hybrid model presented in this article are quite accurate. [16] replaced the numeric values for the biogas production with values of 0, 1 and 2 for low, medium and high production respectively and obtain an accuracy at 87%. However, due to the low resolution in the output prediction it is much easier to obtain a high accuracy. In our case, the methane production fluctuates between 16,000 Nm^2 and 22,000 Nm^2, which corresponds to one of the categories in [16]. [15] used Random Forest to predict the methane production for time horizons between one and 40 days and obtained R^2 values between 0.88 and 0.82. As the Gompertz model used in our study is used to plan the infeed, in order to obtain as constant a biogas yield as possible, it would not be fair to use R^2 for comparison. This is because optimal planning entails a lower R^2.

5 Conclusion

In this work we have shown that combining a Gompertz model with a machine learning model can optimize the prediction of the methane production one day ahead with up to 66% according to using a Gompertz model alone. This is important as prediction of the methane production is essential in order to keep a constant production of methane, and thereby ensure that the demand is met while avoiding overproduction. If the demand is not met the costumers will have to go elsewhere, which could lead to usage of none sustainable energy sources. If too much methane is produced the surplus will either be burned off or chemistry will be added in order to inhibit the production.

References

1. Xue, S., et al.: A systematic comparison of biogas development and related policies between China and Europe and corresponding insights. Renew. Sustain. Energy Rev. **117**, 109474 (2020). https://doi.org/10.1016/j.rser.2019.109474
2. Colla, L.M., et al.: Waste biomass and blended bioresources in biogas production. In: Treichel, H., Fongaro, G. (eds.) Improving Biogas Production. BBT, vol. 9, pp. 1–23. Springer, Cham (2019). https://doi.org/10.1007/978-3-030-10516-7_1
3. Hagos, K., Zong, J., Li, D., Liu, C., Lu, X.: Anaerobic co-digestion process for biogas production: progress, challenges and perspectives. Renew. Sustain. Energy Rev. **76**, 1485–1496 (2017). https://doi.org/10.1016/j.rser.2016.11.184
4. Scapini, T., et al.: Enzyme-mediated enhanced biogas yield. In: Treichel, H., Fongaro, G. (eds.) Improving Biogas Production. BBT, vol. 9, pp. 45–68. Springer, Cham (2019). https://doi.org/10.1007/978-3-030-10516-7_3
5. Batstone, D.J., et al.: The IWA anaerobic digestion model no 1 (ADM1). Water Sci. Technol. **45**, 65–73 (2002)
6. Derbal, K., Bencheikh-lehocine, M., Cecchi, F., Meniai, A.-H., Pavan, P.: Application of the IWA ADM1 model to simulate anaerobic co-digestion of organic waste with waste activated sludge in mesophilic condition. Biores. Technol. **100**, 1539–1543 (2009). https://doi.org/10.1016/j.biortech.2008.07.064

7. Batstone, D.J., Puyol, D., Flores-Alsina, X., Rodríguez, J.: Mathematical modelling of anaerobic digestion processes: applications and future needs. Rev. Environ. Sci. Biotechnol. **14**, 595–613 (2015). https://doi.org/10.1007/s11157-015-9376-4

8. Arzate, J.A., et al.: Anaerobic digestion model (AM2) for the description of biogas processes at dynamic feedstock loading rates. Chem. Ing. Tec. **89**, 686–695 (2017). https://doi.org/10.1002/cite.201600176

9. Ripoll, V., Agabo-García, C., Perez, M., Solera, R.: Improvement of biomethane potential of sewage sludge anaerobic co-digestion by addition of "sherry-wine" distillery wastewater. J. Clean. Prod. **251**, 119667 (2020). https://doi.org/10.1016/j.jclepro.2019.119667

10. dos Santos, L.A., et al.: Methane generation potential through anaerobic digestion of fruit waste. J. Clean. Prod. **256**, 120389 (2020). https://doi.org/10.1016/j.jclepro.2020.120389

11. Hernández-Fydrych, V.C., Benítez-Olivares, G., Meraz-Rodríguez, M.A., Salazar-Peláez, M.L., Fajardo-Ortiz, M.C.: Methane production kinetics of pretreated slaughterhouse wastewater. Biomass Bioenerg. **130**, 105385 (2019). https://doi.org/10.1016/j.biombioe.2019.105385

12. Akbaş, H., Bilgen, B., Turhan, A.M.: An integrated prediction and optimization model of biogas production system at a wastewater treatment facility. Biores. Technol. **196**, 566–576 (2015). https://doi.org/10.1016/j.biortech.2015.08.017

13. Beltramo, T., Klocke, M., Hitzmann, B.: Prediction of the biogas production using GA and ACO input features selection method for ANN model. Inf. Process. Agric. **6**, 349–356 (2019). https://doi.org/10.1016/j.inpa.2019.01.002

14. Tufaner, F., Demirci, Y.: Prediction of biogas production rate from anaerobic hybrid reactor by artificial neural network and nonlinear regressions models. Clean Techn. Environ. Policy **22**, 713–724 (2020). https://doi.org/10.1007/s10098-020-01816-z

15. De Clercq, D., Wen, Z., Fei, F., Caicedo, L., Yuan, K., Shang, R.: Interpretable machine learning for predicting biomethane production in industrial-scale anaerobic co-digestion. Sci. Total Environ. **712**, 134574 (2020). https://doi.org/10.1016/j.scitotenv.2019.134574

16. De Clercq, D., et al.: Machine learning powered software for accurate prediction of biogas production: a case study on industrial-scale Chinese production data. J. Clean. Prod. **218**, 390–399 (2019). https://doi.org/10.1016/j.jclepro.2019.01.031

17. Gompertz, B.: XXIV. On the nature of the function expressive of the law of human mortality, and on a new mode of determining the value of life contingencies. In a letter to Francis Baily, Esq. F. R. S. & c. Phil. Trans. R. Soc. Lon. **115**, 513–583 (1825). https://doi.org/10.1098/rstl.1825.0026

18. Zwietering, M.H., Jongenburger, I., Rombouts, F.M., van't Riet, K.: Modeling of the bacterial growth curve. Appl. Environ. Microbiol. **56**, 1875–1881 (1990)

19. Pedregosa, F., et al.: Scikit-learn: machine learning in python. J. Mach. Learn. Res. **2011**, 2825–2830 (2011)

20. Breiman, L.: Bagging predictors. Mach. Learn. **24**, 123–140 (1996). https://doi.org/10.1007/bf00058655

21. Freund, Y., Schapire, R.E.: A decision-theoretic generalization of on-line learning and an application to boosting. J. Comput. Syst. Sci. **55**, 119–139 (1997). https://doi.org/10.1006/jcss.1997.1504

22. Breiman, L.: Random forests. Mach. Learn. **45**, 5–32 (2001). https://doi.org/10.1023/a:1010933404324

23. Friedman, J.H.: Stochastic gradient boosting. Comput. Stat. Data Anal. **38**, 367–378 (2002). https://doi.org/10.1016/s0167-9473(01)00065-2

24. Zhang, H.: The optimality of Naive Bayes (2004). https://www.cs.unb.ca/∼hzhang/publications/FLAIRS04ZhangH.pdf

25. Geurts, P., Ernst, D., Wehenkel, L.: Extremely randomized trees. Mach. Learn. **63**, 3–42 (2006). https://doi.org/10.1007/s10994-006-6226-1
26. Chang, C.-C., Lin, C.-J.: LIBSVM: a library for support vector machines. ACM Trans. Intell. Syst. Technol. **2**, 1–27 (2011). https://doi.org/10.1145/1961189.1961199
27. Kloss, A., Schaal, S., Bohg, J.: Combining learned and analytical models for predicting action effects. arXiv:1710.04102[cs] (2018)

AISRA: Anthropomorphic Robotic Hand for Small-Scale Industrial Applications

Rahul Raj Devaraja[1], Rytis Maskeliūnas[1],
and Robertas Damaševičius[2(✉)]

[1] Department of Multimedia Engineering, Kaunas University of Technology,
Kaunas, Lithuania
rytis.maskeliunas@ktu.lt
[2] Department of Software Engineering, Kaunas University of Technology,
Kaunas, Lithuania
robertas.damasevicius@ktu.lt

Abstract. We describe the design of the multi-finger anthropomorphic robotic hand for small-scale industrial applications, called AISRA (Anthropomorphic Interface for Stimulus Robust Applications), which can feel and sense the object that it is holding. The robotic hand was printed using the 3D printer and includes the servo bed for finger movement. The data for object recognition was collected using Leap Motion controller, and Naïve Bayes classifier was used for training and classification. We have trained the robotic hand on several monotonous objects used in daily life using supervised machine learning techniques and the gesture data obtained from the Leap Motion controller. The mean accuracy of object recognition achieved is 92.1%. The Naïve Bayes algorithm is suitable for using with the robotic hand to predict the shape objects in its hands based on the angular position of its figures. Leap Motion controller provides accurate results and helps to create a dataset of object examples in various forms for the AISRA robotic hand, and can be used to help developing and training 3D-printed anthropomorphic robotic hands. The experiments in object grasping experiments demonstrated that the AISRA robotic hand can grasp objects with different size and shape, and verified the feasibility of robot hand design using low-cost 3D printing technology. The implementation can be used for small-scale industrial applications.

Keywords: Robotics · Object recognition · Supervised learning · 3D printing · Leap motion · Artificial intelligence

1 Introduction

In July 2015, the USA's DARPA (Defense Advanced Research Projects Agency) presented a prosthetic arm that enables the operator feel things that he touches [1]. The robotic arm was built as it would be controlled by a human brain. The hand is connected by wires linked up to the motor cortex, which is the part of the human brain that controls the movement of muscles, and the sensory cortex, which recognizes the tactile sensations when a person touches some things. The research has opened a way for a multitude of applications, including prosthetics [2], industrial assembly lines [3],

© Springer Nature Switzerland AG 2020
O. Gervasi et al. (Eds.): ICCSA 2020, LNCS 12249, pp. 746–759, 2020.
https://doi.org/10.1007/978-3-030-58799-4_54

medical surgery [4], assisted living [5], manufacturing, military applications [6], education [7, 8].

For example, Dhawan et al. [9] presented a robot with stereo vision system for bin picking applications, where robots were able to pick up a thing from a stack of objects put in a bin. Recognition of an object to be picked up required a computer vision system, and an object segmentation algorithm based on stereo imaging method. Ben-Tzvi and Ma [10] proposed a grasping learning system that can provide the rehabilitation function to the subject. The system measures motion and contact force, which is used to train a Gaussian Mixture Model (GMM), representing the joint distribution of the data. Then the learned model is used to generate suitable motions and force by Gaussian Mixture Regression to perform the learned tasks in real-time. Wu et al. [11] designed an under-actuated finger mechanism, a robot palm, and a three-finger robot hand with the finger, and produced by 3D printing. Hu and Xian [12] used a Kinect sensor to control two humanoid-like robotic hands. They used the Denavit–Hartenberg method to set up forward and inverse kinematics models of robot hand movements. Next, the depth data from the Kinect and the joint coordinates are used to segment the hand gesture in depth images. Finally, Deep Belief Network (DBN) recognizes hand gestures and translates them into instructions to control the robot hands.

Durairajah et al. [13] proposed a low cost hand exoskeleton developed for rehabilitation, where a healthy hand is used to control the unhealthy hand. The authors used flex sensors attached to the leather glove on the metacarpophalangeal, proximal interphalangeal, and interphalangeal joints to control the unhealthy hand via servo motors, and claimed 92.75% efficiency of testing. Tan et al. [14] presented a study on the calibration of an underactuated robotic hand with soft fingers. Senthil Kumar et al. [15] developed and tested a soft fabric based tactile sensor to use with a 3D printed robotic hand aimed for rehabilitation. Zhang et al. [16] developed a multi-fingered hyperelastic hand, and derived the equations of pulling force and grasping force while using the hand. The optimal (with respect to grasping force) dimensions of the hand are calculated using the finite element method (FEM). Park et al. [17] developed and manufactured a hand exoskeleton system for virtual reality (VR) applications. A wearable system of sensors measures the finger joint angles and the cable-driven actuators apply force feedback to the fingers. Proposed control algorithms were analyzed and validated in the VR applications. Hansen et al. [18] proposed an approach to combine musculoskeletal and robotic modeling driven by motion data for validating ergonomics of exoskeletons and orthotic devices. Gailey et al. [19] developed a soft robotic hand controlled by electromyography (EMG) signals issued by two opposite muscles allowing modulation of grip forces. Ergene and Durdu [20] designed a robotic hand and used grid-based feature extraction and bag-of-words for feature selection, and Support Vector Machine (SVM) for classification of grasping actions of three types of office objects (pens, cups and staplers).

Zeng et al. [21] presented a modular multisensory prosthetic hand for prosthetic applications based on the principles of dexterity, controllability, sensing and anthropomorphism. Finally, Zhang et al. [22] used RGB-D Kinect depth sensor with the ORiented Brief (ORB) method for detection of features and extraction of descriptors extraction, a Fast Library for Approximate Nearest Neighbors (FLANN) k-Nearest Neighbor (FLANN KNN) algorithm for feature matching, a modified RANdom

SAmple Consensus (RANSAC) method for motion transformation, and the General Iterative Closest Points (GICP) for optimization of motion transformation to scan its surrounding environment.

The common limitation of the existing approaches is usually high cost of implementation, which prohibits its wide use by end-users. Our approach takes advantage of the 3D printing technology and low-cost Leap Motion sensors to develop an anthropomorphic multi-gingered robotic hand that is both efficient in small-scale applications and affordable to end-users.

The aim of this paper is to develop and an anthropomorphic robotic hand, which can feel and sense the shape of the object it is holding, for small-scale industrial applications. We develop a robotic hand and train it on several monotonously-shaped objects used in daily life using supervised machine learning techniques and the gesture data obtained from the Leap Motion controller.

2   Methods

2.1   Outline

The development of a supervised motion learning system for a robotic hand comprises of four stages:

1. Data generation and pre-processing;
2. Motion learning;
3. Evaluation; and
4. Motion prediction.

The stages of the robotic hand motion training are summarized in Fig. 1.

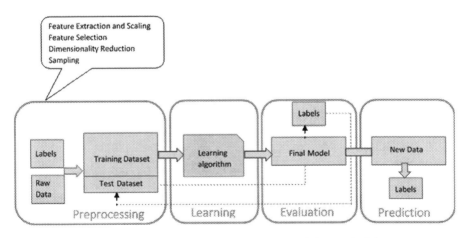

Fig. 1. Summary of the robotic hand motion training methodology

2.2   Data Collection

For data collection, we use the Leap Motion sensor and a methodology described in more detail in [23]. From the hardware perspective, the Leap Motion controller is a simple engineering design, which consists of two cameras and three Infrared (IR) Light Emitting Diodes (LEDs). These LEDs track the IR light from the wavelength of 850 nm, which is outside the light spectrum visible to human vision.

Once the image data is streamed to the computer, software matches the data to exact tracking information about fingers, hands and tools such as pencils etc. Each finger in the hand comprises of three bones joining to each other, which forms the bending of the finger.

Leap Motion SDK provides the inbuilt function, which can recognize these bones and each finger of the hand. Angles between the proximal and intermediary bone of every finger are retrieved and the angle is applied on the vector, which yields the angle of the finger bent. All the finger data is captured and pipelined for pre-processing.

Pre-processing includes normalization of data and feature dimensionality reduction. The result is a dataset, which gets all the data required from the Leap Motion controller and is saved into a file to be used as a training data for algorithms.

The dataset contains the measurements of angles for of 150 instances from three different objects: BALL, RECTANGLE and CYLINDER (see Fig. 2). The data, consisting of 150 samples and 5 features, is written as a 150×5 matrix.

Dataset attributes represent individual fingers and the angle between its bones.

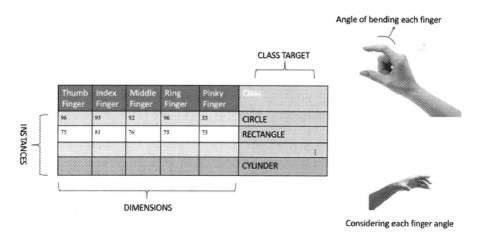

Fig. 2. Creation of a dataset

2.3   Algorithms

The Naïve Bayes classifier [24] is a binary or a multi classification problem, which is used to classify the categorical data. This is simple classifier but a powerful predictive algorithm, which is considered in selecting best hypothesis (h) for given data (d) as:

$$P(h|d) = P(d|h) \times P(h)P(d) \tag{1}$$

here is a posterior probability of hypothesis over given data; is a probability of the data was true over the hypothesis; is the prior probability of hypothesis being true (regardless the data); and is the probability of the data.

Probabilities of unseen data are calculated using the Gaussian Probability Density Function (PDF), which produce an estimate of probability of previously unknown input value:

$$pdf(x, \mu, \sigma) = 1/\sqrt{2} \times \pi \times \sigma \times e^{-((x-\mu)/\sigma)^2} \tag{2}$$

here $pdf(.)$ is the PDF, σ is standard deviation, and x is the input value for the input variable.

In our case, the inputs for the Naïve Bayes algorithms are the finger angles captured by Leap Motion as described in Sect. 4.

2.4   Evaluation

The evaluation is an estimate on how well an algorithm works in a production environment, it is not a guarantee of performance but a measure of accuracy and efficiency. For evaluation, we use 10-fold cross validation, which is an approach to estimate the efficiency and performance of the algorithm with less variance and a single split set. The data set is split into k-parts (e.g. 10), called a fold. The algorithm is trained on the (k − 1) folds, and is tested on the withheld fold. This is repeated so that each fold of the dataset is given a chance to be the held back test set. For relatively small training sets, the number of folds can be increased to allow to use more training data in each iteration, which results in a lower bias towards estimating the generalization performance.

3   System Implementation

3.1   Mechanics of Robotic Hand

Robotic hands are 3D printed leveraging an open source design from inmoov (http://inmoov.fr/). The Inmoov designs support major elements of human hand anatomy, which comprises of bones, joints, radius, ulna and tendons. The main knuckle joints are formed by the metacarpophalangeal joints are made stationary, which provides the degree of freedom to bend tit, but cannot move sideways. The operation of extensor hood and extensor tendons is controlled by rotating servos to their appropriate degree of actuators rotation (Fig. 3).

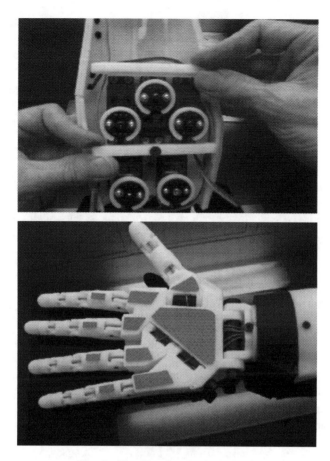

Fig. 3. The 3D printed robotic hand

Designing a system to perform predictions on holding objects are divided as processes, which consists of 1) Servo Readings, 2) I/O interface, 3) Algorithm suite (see Fig. 4).

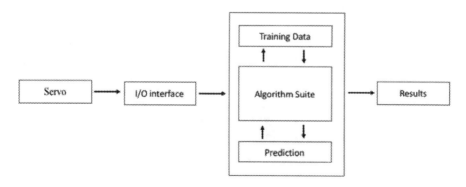

Fig. 4. System Design of robotic hand

Servo Readings process fetches the reading of every single servomotor as a feedback of rotational degree measurements. I/O Interface process takes the measures from the servo readings and passes to I/O interface that is Arduino or a Raspberry PI, which can process these readings. Interface can also use message-passing interface such as ROS or PubNub, which can stream data for further processing. Algorithm Suite performs the algorithm, training, comparisons, validations and tuning. Once the suite is trained, we can use the learned data to perform prediction on new input data.

According to an algorithm presented in Fig. 5, first, the robotic hand will grab an object and as soon as pressure sensors get in contact, the servo motors will stop its action and record its measures. Once measures are converted into respective units (such as degree or radians) and passed as an external data to the dataset as a testing data. Second, once data is passed through an algorithm making necessary adjustments on weights in predictive models, the efficiency is calculated. The training data is passed through a spot-checking algorithm which gives their respective mean and standard deviation for accuracy. This process is continued until the best features are selected for classification and highest accuracy is gained.

4   Data and Results

4.1   Data

Each finger comprises of three bones joining to each other. Leap Motion SDK provides the in-built function, which can recognize these bones and each finger on the hand. Once the entire finger object is gathered, object can be further broken down to types of bones (proximal bone and intermediate bone) present in the finger. Vectors can be used to calculate the angle between each bone in the finger. Once all the finger angles are observed and are stored against the hand holding the objects, they are saved to a.csv file. An example of data collected for each finger (thumb, index, middle ring, and pinky) vs the object type is presented in Table 1.

Table 1. Sample of data in dataset: finger features and class label

Thumb	Index	Middle	Ring	Pinky	Class
96	93	92	96	53	Ball
97	91	92	97	59	Ball
97	91	92	97	59	Ball
100	96	96	100	64	Ball
100	96	96	100	65	Ball
75	81	76	75	73	Rectangle
73	82	74	73	71	Rectangle
88	113	112	88	87	Rectangle
87	108	108	87	88	Rectangle
114	128	132	114	114	Cylinder
114	128	132	114	114	Cylinder
114	128	132	112	114	Cylinder
98	143	144	98	97	Cylinder

The dataset includes three objects (circle, rectangular, cylinder) trained for five different instances acting as a sample from five different people for the initial approximation for algorithm analysis.

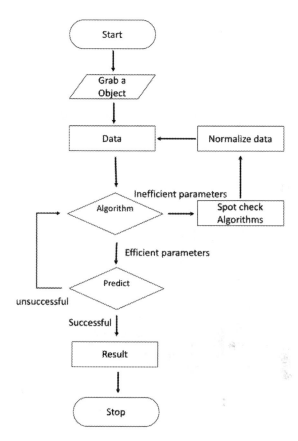

Fig. 5. Flowchart of robotic hand motion learning

4.2 Feature Analysis

We have evaluated the features for each type of object shape separately using the absolute value of the two-sample t-test with pooled variance estimate.

The results of feature ranking are presented in Figs. 6, 7 and 8. Note that different features are important for different types of object.

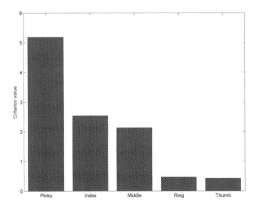

Fig. 6. Feature ranking by class separability criteria (class: Ball)

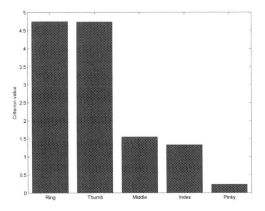

Fig. 7. Feature ranking by class separability criteria (class: Rectangle)

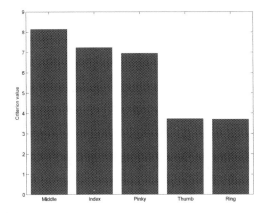

Fig. 8. Feature ranking by class separability criteria (class: Cylinder)

For recognizing the Ball shape, the most important features are provided by the Pinky finger, for Rectangular shape – by the Ring and Thumb fingers, and for Cylinder shape – by the Middle and Index fingers. Using the 5D space of finger angles, the classes are separated linearly by the classifier.

For example, see the distribution of the Index and Thumb finger angle values in Fig. 9, and of the Pinky and Ring finger angle values in Fig. 10.

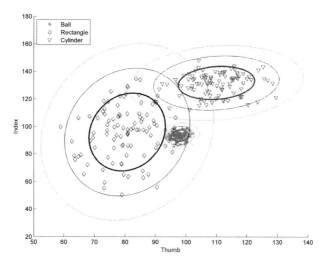

Fig. 9. Distribution of the Index and Thumb finger angle values (with 1, 2 and 3 sigma confidence values)

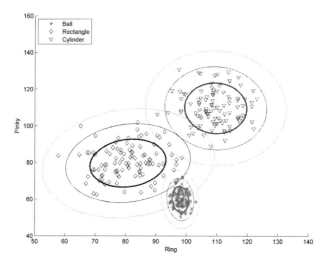

Fig. 10. Distribution of the Pinky and Ring finger angle values (with 1, 2 and 3 sigma confidence values)

4.3 Results

The evaluation is an estimate on how well an algorithm works in a production environment. However, it is not a guarantee of performance, but a measure of accuracy and efficiency. Once an algorithm is estimated on its performance then retraining of the entire dataset is performed for operational usage. To evaluate the performance of the classification, we used 10-fold cross validation as described in subsection II.C. The mean accuracy of object recognition achieved using Naïve Bayes classifier is 92.1%, while f-score is 0.914. The confusion matrix of classification results is presented in Fig. 11. Note that the Ball shape is recognized perfectly, while the Cylinder class is often confused with the Rectangular object class due to their similarity in shape. The Receiver Operating Characteristic (ROC) curves for all three classes of objects are presented in Fig. 12.

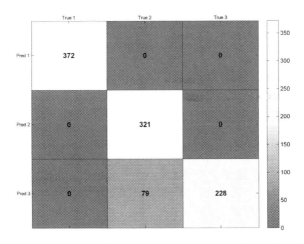

Fig. 11. Confusion matrix of Ball, Cylinder and Rectangular shape classification results

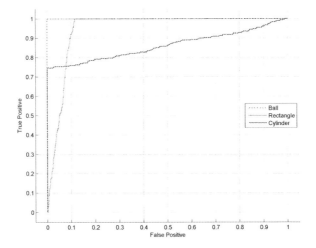

Fig. 12. ROC curves for Ball, Rectangle and Cylinder shaped objects

5 Discussion and Conclusion

Anthropomorphic design of a robotic hand allows the robot to interact efficiently with the environment and to operate in workplaces that are specifically accustomed to human hand [25]. Since human hand possesses a set of unique features that allows, for example, to hold a great variety of different object shapes, anthropomorphic robotic hand provides advantages for repetitive object handling and placement tasks such as for industrial conveyor applications.

Specifically, implementing a multi-fingered design that is capable of performing a powerful hold task, and at the same time has fine and versatile manipulative skills that cannot be reached with a generic robotic gripper [26]. The developed hand provides a low-cost alternative to other 3D printed robotic hands known from the scientific literature [27–30] and could be used for various educational projects [31, 32].

We described the development of the low-cost multi-finger anthropomorphic robotic hand that can be usable for small-scale industrial applications. The AISRA robotic hand was developed to execute human-like grasping of objects of various shape. Using the developed robotic hand and the data from Leap Motion, we have achieved a 92.1% accuracy of object recognition. Leap Motion controller provides more accurate results and helps to create a dataset for the AISRA robotic hand and seems to be condensed when creating many instances of object examples in various forms, and can be used to help developing 3D printed anthropomorphic robotic hands.

The Naïve Bayes algorithm is suitable for using with the robotic hand to predict the shape objects in its hands based on the angular position of its figures.

The experiments in object grasping experiments demonstrated that the AISRA robotic hand can grasp objects with different size and shape, and verified the feasibility of robot hand design using low-cost 3D printing technology.

Future work will include adding additional sensors such as touch sensors, gyroscopes, accelerometers, and depth cameras for the AISRA hand. The hand will be used as a test bed for research of smart automation techniques in the laboratory environment. Future work also will involve the development by training Neural Networks and computer vision to enhance the accuracy of object recognition by back propagation and reinforcement learning.

References

1. Yanco, H.A., Norton, A., Ober, W., Shane, D., Skinner, A., Vice, J.: Analysis of human-robot interaction at the DARPA robotics challenge trials. J. Field Rob. **32**(3), 420–444 (2015). https://doi.org/10.1002/rob.21568
2. Kawasaki, H., Mouri, T.: Humanoid robot hand and its applied research. J. Robot. Mechatron. **31**(1), 16–26 (2019). https://doi.org/10.20965/jrm.2019.p0016
3. Juočas, L., Raudonis, V., Maskeliūnas, R., Damaševičius, R., Woźniak, M.: Multi-focusing algorithm for microscopy imagery in assembly line using low-cost camera. Int. J. Adv. Manuf. Technol. **102**(9), 3217–3227 (2019). https://doi.org/10.1007/s00170-019-03407-9

4. Abdelaal, A.E., Sakr, M., Avinash, A., Mohammed, S.K., Bajwa, A.K., Sahni, M., Salcudean, S.E.: Play me back: a unified training platform for robotic and laparoscopic surgery. IEEE Rob. Autom. Lett. **4**(2), 554–561 (2019). https://doi.org/10.1109/lra.2018. 2890209

5. Maskeliunas, R., Damaševicius, R., Segal, S.: A review of internet of things technologies for ambient assisted living environments. Future Internet **11**(12) (2019). https://doi.org/10.3390/ fi11120259

6. Al-Madani, B., Svirskis, M., Narvydas, G., Maskeliūnas, R., Damaševičius, R.: Design of fully automatic drone parachute system with temperature compensation mechanism for civilian and military applications. J. Adv. Transp. **2018** (2018) https://doi.org/10.1155/2018/ 2964583

7. Plauska, I., Damaševičius, R.: Educational robots for internet-of-things supported collaborative learning. In: Dregvaite, G., Damasevicius, R. (eds.) ICIST 2014. CCIS, vol. 465, pp. 346–358. Springer, Cham (2014). https://doi.org/10.1007/978-3-319-11958-8_28

8. Burbaite, R., Stuikys, V., Damasevicius, R.: Educational robots as collaborative learning objects for teaching computer science. In: IEEE International Conference on System Science and Engineering, ICSSE 2013, pp. 211–216 (2013). https://doi.org/10.1109/ICSSE.2013. 6614661

9. Dhawan, A., Bhat, A., Sharma, S., Kaura, H.K.: Automated robot with object recognition and handling features. Int. J. Electron. Comput. Sci. Eng., 861–873 (2013)

10. Ben-Tzvi, P., Ma, Z.: Sensing and force-feedback exoskeleton (SAFE) robotic glove. IEEE Trans. Neural Syst. Rehabil. Eng. **23**(6), 992–1002 (2015)

11. Wu, L., Lan, T., Li, X.: Design and production of a 3D printing robot hand with three underactuated fingers. In: Deng, Z. (ed.) CIAC 2017. LNEE, vol. 458, pp. 87–95. Springer, Singapore (2018). https://doi.org/10.1007/978-981-10-6445-6_10

12. Hu, B., Xiao, N.-f.: Kinect sensor-based motion control for humanoid robot hands. In: Qiao, F., Patnaik, S., Wang, J. (eds.) ICMIR 2017. AISC, vol. 690, pp. 540–546. Springer, Cham (2018). https://doi.org/10.1007/978-3-319-65978-7_81

13. Durairajah, V., Gobee, S., Rauf, W., Ngie, K.S., Lim, J.H.A.: Design and development of low cost hand exoskeleton for rehabilitation. In: Ibrahim, F., Usman, J., Ahmad, M.Y., Hamzah, N., Teh, S.J. (eds.) ICIBEL 2017. IP, vol. 67, pp. 107–110. Springer, Singapore (2018). https://doi.org/10.1007/978-981-10-7554-4_18

14. Tan, N., Gu, X., Gu, X., Ren, H.: Simultaneous robot-world, sensor-tip, and kinematics calibration of an underactuated robotic hand with soft fingers. IEEE Access **6**, 22705–22715 (2017). https://doi.org/10.1109/ACCESS.2017.2781698

15. Senthil Kumar, K., Ren, H., Chan, Y.H.: Soft tactile sensors for rehabilitation robotic hand with 3D printed folds. In: Ibrahim, F., Usman, J., Ahmad, M.Y., Hamzah, N., Teh, S. J. (eds.) ICIBEL 2017. IP, vol. 67, pp. 55–60. Springer, Singapore (2018). https://doi.org/10. 1007/978-981-10-7554-4_9

16. Zhang, J., Zhang, X., Li, Y.: Research and design of a multi-fingered hand made of hyperelastic material. Assembly Autom. **38**(3), 249–258 (2018). https://doi.org/10.1108/AA-03-2017-042

17. Park, Y., Jo, I., Lee, J., Bae, J.: A dual-cable hand exoskeleton system for virtual reality. Mechatronics **49**, 177–186 (2018). https://doi.org/10.1016/j.mechatronics.2017.12.008

18. Hansen, C., Gosselin, F., Ben Mansour, K., Devos, P., Marin, F.: Design-validation of a hand exoskeleton using musculoskeletal modelling. Appl. Ergon. **68**, 283–288 (2018). https://doi.org/10.1016/j.apergo.2017.11.015

19. Gailey, A., Godfrey, S.B., Breighner, R., Andrews, K., Zhao, K., Bicchi, A., Santello, M.: Grasp performance of a soft synergy-based prosthetic hand: a pilot study. IEEE Trans. Neural Syst. Rehabil. Eng. **25**(12), 2407–2417 (2017). https://doi.org/10.1109/TNSRE.2017. 2737539

20. Ergene, M.C., Durdu, A.: Robotic hand grasping of objects classified by using support vector machine and bag of visual words. In: International Artificial Intelligence and Data Processing Symposium, IDAP 2017, Malatya, pp. 1–5 (2017). https://doi.org/10.1109/idap. 2017.8090228

21. Zeng, B., Fan, S., Jiang, L., Liu, H.: Design and experiment of a modular multisensory hand for prosthetic applications. Ind. Rob. **44**(1), 104–113 (2017). https://doi.org/10.1108/IR-04-2016-0115

22. Zhang, L., Shen, P., Zhu, G., Wei, W., Song, H.: A fast robot identification and mapping algorithm based on Kinect sensor. Sensors **15**(8), 19937–19967 (2015). https://doi.org/10. 3390/s150819937

23. Vaitkevičius, A., Taroza, M., Blažauskas, T., Damaševičius, R., Maskeliunas, R., Woźniak, M.: Recognition of American sign language gestures in a virtual reality using leap motion. Appl. Sci. 9(3), 445 (2019). https://doi.org/10.3390/app9030445

24. Hand, D.J., Yu, K.: Idiot's Bayes—not so stupid after all? Int. Stat. Rev. **69**(3), 385–398 (2001)

25. Gama Melo, E.N., Aviles Sanchez, O.F., Amaya Hurtado, D.: Anthropomorphic robotic hands: a review. Ingeniería y Desarrollo **32**(2), 279–313 (2014)

26. Yu, Z., Gu, J.: A Survey on Real-time Controlled Multi-fingered Robotic Hand. In: Canadian Conference on Electrical and Computer Engineering, Halifax, Canada, pp. 975–980 (2008). https://doi.org/10.1109/ccece.2008.4564681

27. Kappassov, Z., Khassanov, Y., Saudabayev, A., Shintemirov, A., Varol, H.A.: Semi-anthropomorphic 3D printed multigrasp hand for industrial and service robots. In: IEEE International Conference on Mechatronics and Automation, Takamatsu, pp. 1697–1702 (2013). https://doi.org/10.1109/icma.2013.6618171

28. Bai, G., Rojas, N.: Self-adaptive monolithic anthropomorphic finger with teeth-guided compliant cross-four-bar joints for underactuated hands. In: IEEE-RAS International Conference on Humanoid Robots, pp. 145–152 (2019). https://doi.org/10.1109/humanoids. 2018.8624971

29. Li, X.-L., Wu, L.-C., Lan, T.-Y.: A 3D-printed robot hand with three linkage-driven underactuated fingers. Int. J. Autom. Comput. **15**(5), 593–602 (2018). https://doi.org/10. 1007/s11633-018-1125-z

30. Suresh, A., Gaba, D., Bhambri, S., Laha, D.: Intelligent multi-fingered dexterous hand using virtual reality (VR) and robot operating system (ROS). In: Kim, J.-H., et al. (eds.) RiTA 2017. AISC, vol. 751, pp. 459–474. Springer, Cham (2019). https://doi.org/10.1007/978-3-319-78452-6_37

31. Burbaite, R., Bespalova, K., Damaševičius, R., Štuikys, V.: Context-aware generative learning objects for teaching computer science. Int. J. Eng. Educ. **30**(4), 929–936 (2014)

32. Damaševičius, R., Narbutaite, L., Plauska, I., Blažauskas, T.: Advances in the use of educational robots in project-based teaching. TEM J. **6**(2), 342–348 (2017). https://doi.org/ 10.18421/TEM62-20

Investigating the Potential of Data from an Academic Social Network (GPS)

João Lourenço Marques[1]([✉]) [ID], Giorgia Bressan[2] [ID],
Carlos Santos[3] [ID], Luís Pedro[3] [ID], David Marçal[4] [ID],
and Rui Raposo[3] [ID]

[1] GOVCOPP, Department of Social, Political and Territorial
Sciences - GOVCOPP, University of Aveiro, Campus Universitário de Santiago,
Aveiro, Portugal
jjmarques@ua.pt
[2] Department of Languages and Literatures, Communication,
Education and Society (DILL), University of Udine,
Via Petracco 8, 33100 Udine, Italy
[3] Department of Communication and Arts - DigiMCedia, University of Aveiro,
Campus Universitário de Santiago, Aveiro, Portugal
[4] INOVA Media Lab/ICNOVA, Nova School of Social Sciences
and Humanities, Lisbon, Portugal

Abstract. There are several references in the literature highlighting the importance of the international scientific mobility studies and several examples of how this academic population has been characterized. Typically, the analysis of academic mobility has been conducted by applying extensive surveys to a "representative" sample, in a specific moment in time, in which the profile of the researchers, that are working or studying in a foreign country, is then inferred. These analyses may suffer of structural lack of representativeness since the target population is unknown. As a structural and inherent issue in this research field, this article presents the results provided by the Portuguese academic social network GPS (Global Portuguese Scientists). It uses a valuable and exclusive data set of the research experiences, provided by Portuguese researchers, to describe and understand the academic dynamic of these researchers over the last years. The analysis considers different socio-demographic characteristics and the type of research (position, scientific research area, duration of the experience) they have been doing. The analysis shows that GPS users are pulled to the core countries of the science world system and points out that each destination of the Portuguese diaspora is associated with specific features of the mobile researchers and their research activity.

Keywords: International scientific mobility · Academic social network · Mobile Portuguese Researchers · Career path · Data collection · Empirical research

© Springer Nature Switzerland AG 2020
O. Gervasi et al. (Eds.): ICCSA 2020, LNCS 12249, pp. 760–775, 2020.
https://doi.org/10.1007/978-3-030-58799-4_55

1   Introduction

Leaving one's own domestic research system to carry out research-related tasks elsewhere in the world has always been a common trait of the life of researchers and scientists.

Geographic mobility is shaped not only by professional and personal factors (see, for example, [1] and by the characteristics of the scientific field and the nature of research [2–4], but also by the institutional structure that oversees the research practice [5, 6]. International mobility fluxes develop along predictable patterns and are driven by differences in national science and technology systems alongside other country-specific factors. The world's most successful national research systems, which generally offer more resources for research and better professional rewards, attract researchers from trailing countries [7], and additional fluxes tend to exist amongst top scientific institutions, as a consequence of the competition for leading scientists. However, Ackers [4] warns us about the tendency to associate geographic mobility with the excellence of the receiving institution. Scientific mobility is more driven by limited opportunities within the home country's research system rather than the authentic scientific appeal of the destination. In fact, as observed by Smetherham [8] for the British context, the national scientific labour market is becoming increasingly global, not only because the recruitment and retention of capable researchers is necessary for world-stage institutions to keep their competitive advantage, but also because foreign staff need to be hired for positions in scientific fields where there is a shortage of local candidates.

Such a move could be intended to seize the best opportunities and to expose oneself to new scientific contexts, which in return can play a role, amongst other things, in increasing the value of one's contribution to science and the promotion of knowledge sharing among multiple scientific research systems. Mobility may, however, have a negative impact on the research system that has invested in their qualification or has previously employed them. Continuous outward flows could potentially weaken and contribute to the decline of the scientific national community, due to the inability to easily renew and further strengthen the human factor within the scientific national community.

In countries that are not yet competitive players on the global academic market, national governments also have an interest in establishing and consolidating a scientific system capable of increasing their productivity and global competitiveness [9]. This awareness about the strategic role of knowledge production and the growing importance of research is evident in Portugal, as proven by the increasing public investment in R&D in recent decades (between 1986 and 2017, public research and development budget allocations as a percentage of GDP more than tripled: from 0.2% to 0.7%), its increasing presence in international research facilities (such as CERN, since 1986), and the fact that more Portuguese people are achieving higher levels of education. In 1986, the year Portugal joined the European Union, 216 doctoral degrees were awarded, but in 2017 the number of new PhDs awarded per year was close to 3,000. Between 1986 and 2017, the total number of researchers increased almost eightfold: from 5,723 to 44,938 (there was a decline between 2011 and 2013, the period during which Portugal was under an external financial aid programme under the supervision of the International Monetary Fund, the European Central Bank and the European Union). It should be noted that in the last two decades, the number of teachers in higher education in Portugal has decreased slightly (from 35,740 in 2001 to 34,227 in 2017). This implies that the new doctoral graduates now have fewer job opportunities at universities and polytechnics. The significant increase in the number of doctorates in Portugal and the shortage of employment opportunities, especially during the years of the financial aid programme, probably encouraged the international mobility of Portuguese scientists. Additionally, many Portuguese researchers may simply be attracted to more productive scientific systems. This was, in fact, encouraged by various public policies, such as the grants to pursue a PhD in foreign countries, funded by Fundação para a Ciência e Tecnologia (FCT), the main public science funding agency in Portugal. Between 1994 and 2015, FCT awarded 4,599 grants for doctorates abroad (19% of the total grants) and 5,476 for doctorates partially pursued abroad. Although Portugal has been mainly an exit country, in the last decades it has also attracted researchers from abroad [10]. Between 1994 and 2015, FCT awarded 2,229 PhD grants to foreign citizens (9.2% of the total grants). Also, regarding the 'Compromisso com a Ciência' programme, which awarded five-year research contracts in Portuguese research institutions between 2007 and 2008, forty-one per cent of the beneficiaries were foreign. These numbers suggest an inflow and outflow of researchers, albeit with fewer entries.

Further insights on the geographic mobility of Portuguese researchers are offered by the European Commission's MORE surveys. Despite their great contribution in monitoring the European Research Policy, current technologies offer the possibility to go beyond the survey approach, by building a dynamic data gathering system able to monitor the population under study. A concrete opportunity to potentially reach each Portuguese researcher experiencing international mobility is given by the Portuguese academic social network GPS (Global Portuguese Scientists) which aims to virtually connect the Portuguese researchers who have experienced a period abroad for research purposes. GPS was launched in November 2016 and its members, who will be defined later in this work as GPS scientists, amounted to almost 2,000 units approximately one year after its launch.

As opposed to other social networks, the GPS platform has a homepage (https://gps.pt/) incorporating a map with markers, which correspond to the coordinates of community members' workplaces at any specific time. This feature allows members to easily search for other potential Portuguese scientific partners according to their location or to a specific academic research interest, which is undoubtedly an interesting characteristic for those who want to contact or establish networks in a specific country but have no acquaintances there yet. The social network then, not only facilitates interactions and the exchange of information between people that, sharing the same nationality, are supposed to have a strong degree of communality, but also allow rich geographic data to be stored concerning the mobility trajectory of the Portuguese scientific community.

The main purpose of this article is to present a statistical analysis to characterize the patterns of territorial mobility of Portuguese researchers registered on the GPS platform. Being a social network platform, is also highlighted the potential of this platform to analyse where, when and on what Portuguese scientists have been working in recent years.

In addition to this introductory section this article is organized as follows: Sect. 2 presents a brief review of empirical research into the geographical mobility of researchers; it focuses on the definitional and methodological challenges and summarizes previous studies on mobility flows of Portuguese researchers; Sect. 3 focuses on the data analysis, characterizing the GPS scientists using the most appropriate parametric and non-parametric bivariate tests; finally, Sect. 4 concludes, highlighting the relevant findings of this research and the contribution of the GPS platform in the context of policy making.

2   Background

The research on the geographic mobility of scientists and researchers is vast and pluri-disciplinary. Quantitative studies regarding mobility have tended to either provide a characterization of the population under study and analyse factors affecting mobility decisions [amongst others, 1, 11, 12], or have searched for evidence of links between mobility and knowledge creation, recombination and diffusion [for example, 13–18]. It should be noted that, in recent years, the persistent and widespread recognition of gender inequalities has generated a proliferation of studies on female scientists in academia and mobility [see 15, 19–21]. It is beyond the scope of the article to review such literature thoughtfully; what seems relevant here is to consider the relevant definitional and methodological challenges in academic mobility studies.

The first challenge concerns the definition of 'researcher' and the identification of the population being studied. A wide range of jobs establish research as their primary goal, but researchers are not grouped into an independent and comprehensive statistical employment category. Moreover, in many geographical contexts doctoral candidates are not recognized as workers, but as students instead. In order to bypass these difficulties, the population under study may be defined as whoever is pursuing and/or holds a doctorate, regardless of how their activity is classified for statistical purposes [22]. However, as Børing et al. [1] underline, being awarded a PhD is not a condition for being a researcher, and not all researchers are necessarily doctorate holders. Some authors argue that the focus of mobility research should only be placed on leading scientists, as opposed to the broader category of researchers, for the former have a greater impact in terms of knowledge creation. In this case, publication and patents can be used to pinpoint the most productive scientists (see, for example, [23]).

It is also necessary to understand that scientists engage in different forms of international mobility experiences, which vary as to their scope, duration and impact. Mobility experiences may be mostly distinguished between those that entail moving to another country to take up research positions and the moves which researchers undertake during the course of their employment or doctoral studies, usually for research visits or guest professorships [24]. The former, which can be permanent or temporary situation, implies a change of affiliation, and it is often the consequence of better research positions and more rewarding and stable working conditions available abroad. It may also be linked to methods of recruitment being based on merit rather than other biased methods of selection and nomination. The latter form of mobility is clearly intended to be time-limited, as it is mostly undertaken to access new collaborators, or use specific research resources such as equipment, training, and data. The length of the stay does not determine the success of an international experience per se. For example, extended periods abroad might not always be positive for those who would like to return [6]. In fact, for the mobile researcher returning to the home research system, it can be just as difficult to find a receptive environment where the newly acquired knowledge may be applied and developed [25], as it is to be reintegrated back into the home scientific labour market [5]. On the other hand, short stays abroad allow continued engagement with the home scientific community, while at the same time enhancing scientific capital and personal prestige [4].

An issue also of great most importance is data gathering and analytical methodology. Survey questionnaires are the most popular instruments for tracking the geographic mobility of researchers through time and space [26]. However, the analysis of curricula vitae (CV) has recently gained increasing importance as a data source in science and technology studies [27]. CV analysis has gone from being a simple supplement source of information to an autonomous research approach in mobility studies [28]. The CV does indeed provide relevant information about the shift from one work setting to the next and, as it is in reality an important individual marketing resource, researchers feel the need to provide accurate data and to keep it updated. Despite its great potential as a data source, there are some major methodological problems, such as the heterogeneity of contents, incomplete or missing information, and the problem of the conversion of CV data into a suitable set of variables for analysis. The issue of the availability of bibliographic data and the proper means of obtaining them also cannot be neglected. Requests for CV are usually sent to the sample of researchers under study, but accessing electronic platforms containing CV and downloading them is in reality a more non-intrusive and cost-effective approach to collecting information on academic mobility (see, for example, [14]). Another emerging data opportunity is given by bibliometric data, which, as it contains information on institutional affiliation, can contribute to the study of global mobility patterns [29, 30], present interesting methodological frameworks, for example). However, given that scientific publication is not a necessary component of all research experiences abroad, its use when a heterogeneous set of people working in science is considered should be prudent. In order not to underrepresent mobility episodes, it seems more appropriate to use bibliometric data in conjunction with other sources (see [31, 32]) rather than an autonomous approach.

As illustrated in the introduction and also backed up recently in Delicado [10], there is a strong case for studying the mobility experiences of Portuguese scientists. Exploratory quantitative research has already been conducted [23, 33, 34] and other works of a more qualitative nature concerning Portugal have been undertaken. This latter group of publications concerns the relationship between mobility, career and personal situation [4]; gender dimensions in long-term international mobility [35]; the impact of long-term international mobility on the creation of international knowledge networks; and the participation of Portuguese female researchers in science and their international scientific mobility [36].

From a quantitative perspective, Fontes [23] studies the mobility flows of Portuguese scientists who were classified as inventors in biotechnology patent applications filed by non-Portuguese organizations and that were resident outside of Portugal at the time of patent filing. The analysis shows that as regards the destination countries, the United States of America (USA) was the preferred destination for the first move, closely followed by the United Kingdom (UK) and France. Other European countries gained importance in subsequent moves, including some destinations such as Spain and Switzerland, places that were not popular in the first mobility episode.

More recently, Delicado [34] enlarges the object of the analysis, by examining the patterns and the motivations that drive mobility decisions of a sample of Portuguese expatriate researchers who produce science abroad in all scientific fields. Regarding destination countries, the results show that about three fifths of the respondents were in the European Union (the UK alone hosted more than twenty-eight per cent) and over a quarter in the USA. PhD students were more concentrated in the European Union, whereas senior researchers were more dispersed and were most frequently located in the USA. Differences in the geographical distribution of expatriate researchers emerge when a breakdown according to scientific fields is analysed. The UK and the USA were popular recipient countries for researchers in the field of social sciences, while researchers in the engineering sciences favoured Switzerland and those in the life sciences chose mainly European countries as destinations (with Germany, Spain and the Nordic countries being prevalent).

The major difference between the above-mentioned empirical work concerning Portuguese researchers and our analysis is based on the instrument used for collecting data on the geographic mobility Portuguese researchers. The GPS platform allows us to perform CV-based research by exploiting the data concerning the educational and professional experience provided by the registered members.

3   Characterization of the GPS Community

As shown in [37], the GPS platform is a social network, in which, is possible to identify the coordinates of community members' workplaces at any specific time. It is structured on three complementary levels, allowing members to search for Portuguese scientific partners corresponding to a specific location or to an academic research interest. In short, i) *Communities* are a macro-organizing entity, defined by default - all GPS users are associated with the 'general community'; ii) the creation of *groups*, allows joining together users with a common background, interest or characteristic; iii) and finally, users of the GPS platform are able to create a *profile* where they share relevant information about themselves: photo, name, birth date, the place of origin, professional information. In the professional information fields, GPS members can also insert locations and the name of the research institutions where they have been working, indicating the dates for their stay in that specific location; their scientific fields[1] and their scientific experience[2] at a specific workplace and select.

This information, provided by the GPS scientist (a classification used to characterize scientists that have at least one experience outside Portugal, for a minimum of 3 months), allows to infer about: the *evolution* of the Portuguese academic mobility (see Table 1); their socio-demographic characteristics (sex and age); *what* job they have being doing (scientific area of research and type of scientific experience) and *where* they have being working (the exact location of the research experience).

Table 1. Evolution of GPS scientists, breakdown by national and/or international dimension of the scientific experience (1990, 1995, 2000, 2005, 2010, 2015, 2017)

Date	One experience abroad (No and %)		Two or more experiences abroad (No and %)		One experience in Portugal (No and %)		Two or more experiences in Portugal (No and %)		One experience in Portugal and abroad (No and %)		Other cases (No and %)		Total (No)
2017	1,133	79.6	39	2.7	204	14.3	16	1.1	26	1.8	5	0.3	1,423
2015	840	74.5	29	2.6	211	18.7	18	1.6	25	2.2	5	0.4	1,128
2010	335	58.2	13	2.3	205	35.6	11	1.9	9	1.6	3	0.5	576
2005	182	62.3	5	1.7	92	31.5	6	2.1	6	2.1	1	0.3	292
2000	83	53.5	2	1.3	61	39.4	5	3.2	2	1.3	2	1.3	155
1995	31	44.9	1	1.4	28	40.6	5	7.2	3	4.3	1	1.4	69
1990	15	34.9	1	2.3	23	53.5	2	4.7	2	4.7	0	0.0	43

How Many Users Are on GPS?

The GPS platform, during the first ten months of existence (2017), gathered approximately 1,700 Portuguese scientists spread around the world, totalizing almost 4,000 research experiences. This means that, on average, a GPS scientist had more than two research activities (see [37] for a more detail information about the process of data cleaning and criteria used to select reliable cases and to remove outliers). Table 1 summarizes the evolution of the number of members according to the national and/or international dimension of their scientific experience.

Where Have They Been Working?

A total of seventy-nine countries (excluding Portugal) were registered in the GPS platform and the eleven most popular destination countries themselves hosted seventy-nine per cent of the research experiences. Figure 1 shows the geographical distribution of all research experiences registered on the GPS platform.

---

[1] (a) Agrarian Sciences, (b) Medical Sciences, (c) Natural Sciences, (d) Social Sciences, (e) Engineering and Technology; and (f) Humanities.

[2] (a) PhD researcher, (b) PhD student, (c) Non-doctoral researcher, (d) Researcher/visiting professor, (e) Leadership functions, (f) Research technician, (g) Science communicator or science manager, (h) Other.

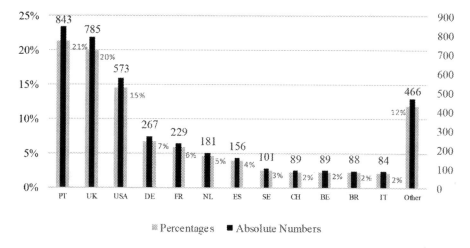

Percentages   Absolute Numbers

Fig. 1. Research experiences, by country

The UK is the most popular destination with 785 research experiences (20%), followed by the USA with 573 experiences (15%), then Germany and France. The other country outside the European continent in this list is Brazil, which shares the Portuguese language, hosting two per cent of all experiences. It is also worth noticing that the sole country that shares the border with Portugal (that is, Spain) is only the sixth most popular destination.

When considering again only those scientists who had only one research experience on those specific dates, either in Portugal or abroad, the analysis shows that GPS scientists went from being concentrated only in Europe (France, UK, Norway and Germany) and North America (the USA) in 1990 to being present first in South America and Africa, from 1995, and then also in Oceania and Asia from 2010 onwards. France, during this period became less and less predominant. In 1990, France was the first country (excluding Portugal) with most experiences (13%), then it was overtaken by United Kingdom in 1995 (14% in UK vs. 12% in France) and by the United States in 2000 (15% in USA vs. 10% in France), consequently falling to third position in the ranking of the most popular destinations. In 2015, France is in fourth place, with Germany taking third place, a ranking that was maintained in 2017.

The GPS scientists are spread across continents. When diving into detail, Fig. 2 shows that in Europe, twenty-five different countries (apart from Portugal) turn out to be the destinations of the GPS scientists, hosting more than 2,200 research experiences (56% of the total – yet again not including Portugal). The other most popular continent is America, mainly North America. The presence of Portuguese scientists in Asia is rather diversified (they have been in twenty different countries, which corresponds to a quarter of the total) but it is less relevant in terms of number of experiences (less than 2%).

Considering only those users who had only one research experience on those specific dates, either in Portugal or abroad, the analysis reveals that in 1990 and 1995 Portugal hosted the highest percentage of research experiences. Then, destinations

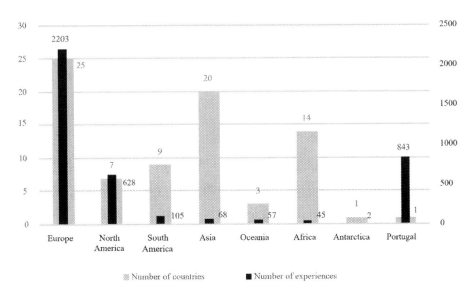

Fig. 2. Number of countries that hosted Portuguese scientists per continent and number of research experiences per continent

abroad started to be more popular than the home country and in 2005, Europe (except Portugal) started to have the highest percentage of research experience (46%, while Portugal had 34%), and the percentage increased considerably in 2015 and 2017 (59% and 64%, respectively). North America was always the second most searched macro-area. Until 2010, Africa was the third continent chosen as a destination for a research experience but, in 2015, it was exceeded by South America, which consolidated its position in 2017 (forty-five experiences, which corresponds to 3% of the total). The growth of this latter continent is mostly due to the role played by Brazil. It is worth noticing that up to 2015, Brazil was not even among the ten most popular destination countries, but in 2015 and 2017, it gradually established itself in the top ten 'receiving' countries (seventh and ninth place).

Considering Lisbon as the reference point for Portugal, each research experience abroad took place at an average distance of 3,582 km. Analysing the distance by sex, is possible to pinpoint that men stayed further away from Portugal than women did (3,730 km and 3,436 km). When performing the same exercise by position held, it seems that *Visiting researcher/professor*, *Leadership functions* and *Science Communicator* or *Science manager* are those work positions, on average, that were carried out further away from Portugal; while the scientific positions such as *PhD student* and *Non-PhD researcher* are, on average, closer to Portugal.

Using the most appropriate parametric and non-parametric inferential tests and a level of significance of five per cent, the analyse presented below highlights the main characteristics that emerged when considering the research experiences carried out outside Portugal by the GPS scientists.

The socio-demographic characteristics (age and sex), the position held and the scientific research areas that characterize each research experience are compared in ten

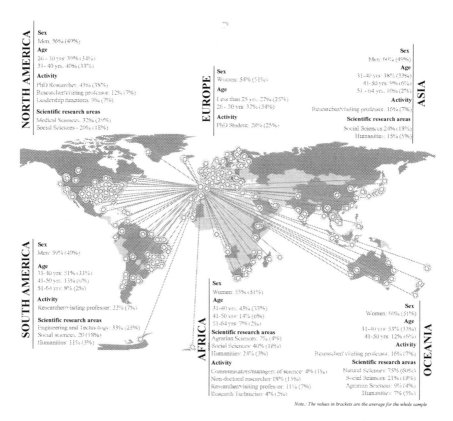

Fig. 3. GPS scientists' profile by continent

different geographical units: a macroscale approach considering the continents; and a microscale approach taking into account the four most popular destination countries representative of the Portuguese diaspora, that is the USA, the UK, Germany and France. The results of this analysis provide valuable insights concerning the mobility patterns of the Portuguese researchers that had at least one research experience outside of the country (Figs. 3, 4).

It worth recalling that the level of significance is not dependent on the magnitude that a value may have in a specific geographical unit, but rather based on a comparative analysis of the weight that a value of a specific variable has in other territorial contexts.

When comparing Europe (in this continent, excluding Portugal, where fifty-six per cent of the total number of experiences were hosted) with the other regions, significant differences emerge (Fig. 3). There is a predominant presence of women, fifty-four per cent (while the distribution of the GPS scientists by sex is approximately fifty-fifty in the overall sample). Other features are the presence of young people (aged thirty and below) and PhD students; for the former the value is sixty-four per cent, while for the latter the value is twenty-eight per cent, corresponding to a difference with the overall sample of about five per cent and three per cent. In Europe, there were no scientific research areas identifiable as significant, suggesting that there is no specific demand for any research areas.

The generic pattern of GPS scientists in North America changes significantly when compared with Europe. In this part of the world, the percentage of Portuguese male scientists is fifty-six per cent. Another characteristic that differentiates this region is the age structure. North America captures fewer young scientists: the percentage of people aged twenty-five and below is twelve per cent (while the global average is 25%) but there are more scientists between twenty-six and forty years old than expected (79% comparing with the global average of 67%). The distance from Portugal and the higher travel and accommodation costs might justify the scarce attractiveness of North America for early career researchers. Job positions that reflect a state of greater scientific maturity, such as *PhD researchers*, *Researcher/visiting professors* and *Leadership functions* scientists are more representative here than in the other regions of the world, respectively, five per cent and two per cent more than the average. *Medical Sciences* and *Social Sciences* are the most significant scientific areas in North America, with approximately more than three per cent when compared with other regions.

There are some similarities between South and North America concerning the main characteristics of the research experiences. For example, male scientists are the majority (59%), and the predominant age group is people between thirty-one and fifty (64%). *Researcher/visiting professor* is the only work position that is particularly evident in this region with twenty-two per cent (note that this value for the overall sample represents just 7%). Among the scientific areas, *Social Sciences* and *Engineering and Technology* are relevant, as in North America, but in this case, *Humanities* emerges as an additional significant scientific area.

In Asia the Portuguese scientists are predominantly men (60%), about eleven per cent more than what would be expected, and the majority of the scientists belong to the age group between thirty-one and sixty-four years old (57% vs. the average of 41%). This fact confirms the argument that older age groups are in places farther from Portugal. Moreover, *Researcher/visiting professor* is the most representative scientific position in this region (9% more than the average) and *Social Sciences* and *Humanities* are the two major scientific research areas (6% and 10% more than the overall sample).

The fifty-seven research experiences hosted in Oceania are predominantly carried out by women (60%); scientists are predominantly between thirty-one and fifty years old (65%), a lot more than the overall average of thirty-nine per cent. The most significant research activity is *Researcher/visiting professor*, similar to the situation in Asia, and all scientific research areas, with the exception of *Medical Sciences* and *Engineering and Technology,* have a significant importance in this region.

The research experiences carried out in Africa are characterized by the fact that are held mostly by women (54%), scientists between thirty-one and sixty-four years old (64% vs. 41% of the overall average), a somewhat different reality when compared with the other macroregions. The most representative work position and scientific research areas in this part of the world are those that have the lowest relevance in the total sample. *Social sciences,* for instance, represents twenty-two per cent more than the average.

Note.: *The values in brackets are the average for the whole sample*

Fig. 4. Profile of GPS scientists by the four most relevant countries of the Portuguese diaspora

Let us consider now the main characteristics of the four most important 'receiving' countries of the Portuguese diaspora, by grouping the UK and the USA, two English speaking countries, on one side, and France and Germany, two founding members of the European Economic Community, on the other (Fig. 4).

The UK and the USA are characterized by hosting Portuguese scientists that have roughly the same features: scientists between twenty-six and thirty, PhD researchers and in *Social Sciences*. The USA distinguishes itself according to three other characteristics, namely, people between thirty-one and forty years old, *Researcher/visiting professors* as position held and *Medical Sciences* as an additional scientific research area. They differ in terms of the predominant sex (women are slightly more than the average in the UK with a percentage of fifty-five per cent, the opposite occurs in the USA with fifty-five per cent of males).

Between Germany and France there are also some communalities, namely in the age of the GPS members (aged thirty and below) and position held (both having a single significant scientific position, *PhD students*). They are different in relation to the representativeness of the sex and the scientific research areas; in Germany GPS members are predominantly women (61%, that is, about more than 10% of the average), while in France the percentage of men is slightly higher (53%). Regarding the research areas, Germany does not have any specific core scientific area while France attracts Portuguese researchers from *Natural sciences* and *Engineering and Technology*.

4   Final Remarks

This paper presents an explanatory analysis of the characteristics and the patterns of geographic mobility of Portuguese researchers registered on the GPS platform (https://gps.pt).

The descriptive analysis of the geographical distribution of the research experiences over time, showed that the number of GPS Portuguese researchers that had or have an experience abroad has increased significantly in the last three decades and the tendency is to have more researchers in more diverse destinations across the world. The inferential analysis showed that destinations closer to Portugal (Europe and more specifically UK) tend to attract women and younger researchers; the more distant regions (North America and more specifically the USA) host more men and older researchers, and users with more mature positions in their careers, such as PhD researcher, Researcher/visiting professor and Leadership functions, as opposed to PhD students which is the most typical position held in Europe. Another interesting insight is that there are also significant differences regarding the research activity developed in different spatial contexts, whereas longer distances imply a specific vocation on the type of research, while shorter distances are more undifferentiated. Note that Europe, for instance, in comparison with other continents, does not have a scientific research area that differentiates this destination.

These results highlight the potential of the GPS platform to monitor the mobility of Portuguese researcher's diaspora. Beside promoting an environment of cooperation, is a powerful tool for mobility studies and for science policy-making: (a) providing a longitudinal perception of the scientific mobility phenomena; (b) analysing scientific returns; (c) understanding the reasons for mobility, i.e. what motivates highly skilled individuals to go abroad and what their future plans are regarding their possible return to Portugal; (d) studying the relation between mobility and scientific production; (e) identifying the academic institutions that attract most Portuguese researchers; and, finally (f) analysing whether there is evidence for geographic constraints on their scientific career (due to language barrier, family). In order to obtain answers for some of these questions, from point (b) to (f), it would be necessary, in the future, to complement the quantitative approach conducted in this study with qualitative data collection and analysis (e.g. a survey) or by integrating the platform with other data sources where other insights on the research activities of Portuguese researchers are available (e.g. ORCID and the portal of the Portuguese Scientific Foundation – FCT). We believe that the collection and analysis of data concerning outward flows such as those GPS platform are essential for designing effective policies that stimulate brain gain or encourage and support the mobility of scientists in pursuit of training, access to new knowledge and networks, and also for taking into account gender imbalances.

Acknowledgements. GPS results from a collaboration of several Portuguese organizations: two private foundations (Fundação Francisco Manuel dos Santos and Fundação PT), a public university (Universidade de Aveiro) and the national agency for promoting scientific culture (Ciência Viva). These promoters ensure both technical needs and social impact of the network. The authors wish to thank the organizations that have made GPS possible. In particular, Carlos Fiolhais, Pedro Magalhães, Mónica Vieira and Maria Ferreira from Francisco Manuel dos Santos

Foundation; Rosalia Vargas, from the Ciência Viva Agency; and the research units of the University of Aveiro, Digimedia and GOVCOPP. We also thank the associations of Portuguese researchers abroad, which are partners of the GPS initiative: the Association of Portuguese Postgraduates in Germany (ASPPA), the Association des Diplômés Portugais en France (AGRAFr), the Portuguese American Post-graduate Society (PAPS), the Portuguese Association of Researchers and Students in the UK (PARSUK), Native Scientists and the Portuguese Association of Students, Researchers and Graduates in Belgium, the Netherlands and Luxembourg (APEI Benelux).

References

1. Børing, P., Flanagan, K., Gagliardi, D., Kaloudis, A., Karakasidou, A.: International mobility: findings from a survey of researchers in the EU. Sci. Publ. Policy **42**, 811–826 (2015). https://doi.org/10.1093/scipol/scv006
2. Jöns, H.: Transnational mobility and the spaces of knowledge production: a comparison of global patterns, motivations and collaborations in different academic fields. Soc. Geogr. **2**, 97–114 (2007). https://doi.org/10.5194/sg-2-97-2007
3. Ackers, L.: Moving people and knowledge: scientific mobility in the European union. Int. Migr. **43**(5), 99–131 (2005). https://doi.org/10.1111/j.1468-2435.2005.00343.x
4. Ackers, L.: Internationalisation, mobility and metrics: a new form of indirect discrimination? Minerva **46**, 411–435 (2008). https://doi.org/10.1007/s11024-008-9110-2
5. Musselin, C.: Towards a European academic labour market? some lessons drawn from empirical studies on academic mobility. High. Educ. **48**, 55–78 (2004). https://doi.org/10.1023/B:HIGH.0000033770.24848.41
6. Morano-Foadi, S.: Scientific mobility, career progression, and excellence in the European research area. Int. Migr. **43**, 133–162 (2005). https://doi.org/10.1111/j.1468-2435.2005.00344.x
7. Saint-Blancat, C.: Making sense of scientific mobility: how Italian scientists look back on their trajectories of mobility in the EU. High. Educ. Policy **31**(1), 37–54 (2017). https://doi.org/10.1057/s41307-017-0042-z
8. Smetherham, C., Fenton, S., Modood, T.: How Global is the UK academic labour market? Globalisation Soc. Educ. **8**(3), 411–428 (2010). https://doi.org/10.1080/14767724.2010.505105
9. Thorn, K., Holm-Nielsen, L.B.: International mobility of researchers and scientists: policy options for turning a drain into a gain. In: Solimano, A. (ed.) The International Mobility of Talent: Types, Causes, and Development Impact, pp. 145–167. Oxford University Press, Oxford (2008)
10. Delicado, A.: 'Pulled' or 'Pushed'? the emigration of Portuguese scientists. In: Pereira, C., Azevedo, J. (eds.) New and Old Routes of Portuguese Emigration. IRS, pp. 137–153. Springer, Cham (2019). https://doi.org/10.1007/978-3-030-15134-8_7
11. Todisco, E., Brandi, M.C., Tattolo, G.: Skilled migration: a theoretical framework and the case of foreign researchers in Italy. Fulgor **1**(3), 115–130 (2003)
12. Czaika, M., Toma, S.: International Academic mobility across space and time: the case of Indian academics. Popul. Space Place **23**(8), 1–19 (2017). https://doi.org/10.1002/psp.2069
13. Jonkers, K., Tijssen, R.: Chinese researchers returning home: impacts of international mobility on research collaboration and scientific productivity. Scientometrics **77**(2), 309–333 (2008). https://doi.org/10.1007/s11192-007-1971-x

14. Cañibano, C., Otamendi, J., Andújar, I.: Measuring and assessing research mobility from CV analysis: the case of the Ramón y Cajal programme in Spain. Res. Eval. **17**(1), 17–31 (2008). https://doi.org/10.3152/095820208X292797

15. Zubieta, A.F.: Recognition and weak ties: is there a positive effect of postdoctoral position on academic performance and career development? Res. Eval. **18**(2), 105–115 (2009) https://doi.org/10.3152/095820209x443446

16. Jöns, H.: Brain circulation and transnational knowledge networks: studying long- term effects of academic mobility to Germany, 1954–2000. Global Netw. **9**(3), 315–338 (2009). https://doi.org/10.1111/j.1471-0374.2009.00256.x

17. Jöns, H.: Feminizing the university: the mobilities, career, and contributions of early female academics in the University of Cambridge, 1926–1955. Prof. Geogr. **69**(4), 670–682 (2017)

18. Franzoni, C., Scellato, G., Stephan, P.: The mover's advantage: the superior performance of migrant scientists. Econ. Lett. **122**(1), 89–93 (2014). https://doi.org/10.1016/j.econlet.2013.10.040

19. Leemann, R.J.: Gender inequalities in transnational academic mobility and the ideal type of academic entrepreneur. Discourse Stud. Cult. Politics Educ. **31**(5), 609–625 (2010). https://doi.org/10.1080/01596306.2010.516942

20. Jöns, H.: Transnational academic mobility and gender. Globalisation Soc. Educ. **9**(2), 183–209 (2011). https://doi.org/10.1080/14767724.2011.577199

21. Cañibano, C., Fox, M.F., Otamendi, F.J.: Gender and patterns of temporary mobility among researchers. Sci. Pub. Policy **43**(3), 320–331 (2016). https://doi.org/10.1093/scipol/scv042

22. Flanagan, K.: International mobility of scientists. In: Archibugi, D., Filippetti, A. (eds.) The Handbook of Global Science, Technology and Innovation, pp. 365–381. Wiley, New Jersey (2015)

23. Fontes, M.: Scientific mobility policies: how Portuguese scientists envisage the return home. Sci. Public Policy. **34**(4), 284–298 (2007). https://doi.org/10.3152/030234207x214750

24. Ackers, L.: Internet mobility, co-presence and purpose: contextualising internationalisation in research careers. Sociología y tecnociencia/Soc. Technosci. **3**(3), 117–141 (2013)

25. Groves, T., Montes, E., Carvalho, T.: The impact of international mobility as experienced by Spanish academics. Eur. J. High. Educ. **8**(1), 83–98 (2018). https://doi.org/10.1080/21568235.2017.1388187

26. Teichler, U.: Academic mobility and migration: what we know and what we do not know. Eur. Rev. **23**(S1), S6–S37 (2015). https://doi.org/10.1017/S1062798714000787

27. Dietz, J.S., Chompalov, I., Bozeman, B., O'Neil Lane, E., Park, J.: Using the curriculum vita to study the career paths of scientists and engineers: an exploratory assessment. Scientometrics **49**(3), 419–442 (2000). https://doi.org/10.1023/A:101053760

28. Cañibano, C., Bozeman, B.: Curriculum vitae method in science policy and research evaluation: the state-of-the-art. Res. Eval. **18**(2), 86–94 (2009). https://doi.org/10.3152/095820209X441754

29. Aman, V.: Does the scopus author ID suffice to track scientific international mobility? A case study based on Leibniz laureate. Scientometrics **117**, 705–720 (2018)

30. Robinson-Garcia, N., et al.: The many faces of mobility: using bibliometric data to measure the movement of scientists. J. Informetrics **13**(1), 50–63 (2019)

31. Laudel, G.: Studying the brain drain: can bibliometric methods help? Scientometrics **57**(2), 215–237 (2003)

32. Sandström, U.: Combining curriculum vitae and bibliometric analysis: mobility, gender and research performance. Res. Eval. **18**(2), 135–142 (2009). https://doi.org/10.3152/095820209x441790

33. Delicado, A.: Cientistas Portugueses no estrangeiro: factores de mobilidade e relações de diáspora. Sociologia, Problemas e Práticas **58**, 109–129 (2008)

34. Delicado, A.: Going abroad to do science: mobility trends and motivations of Portuguese researchers. Sci. Stud. **23**(2), 36–59 (2010)
35. Araújo, E., Fontes, M.: A mobilidade de investigadores em Portugal: uma abordagem de género. Revista Iberoamericana de Ciencia, Tecnología y Sociedad – CTS **8**(24), 9–43 (2013)
36. Delicado, A., Alves, N.A.: Fugas de Cérebros. Tetos de Vidro e Fugas na Canalização: mulheres, ciência e mobilidade. In: Araújo, E., Fontes, M., Bento, S. (eds.) Para um debate sobre Mobilidade e Fuga de Cérebros, pp. 8–31. Centro de Estudos de Comunicação e Sociedade, Braga (2013)
37. Marques, J.L., Bressan, G., Santos, C., Pedro, L., Marçal, D., Junior, E., Raposo, R.: 2020 global Portuguese scientists (GPS): an academic social network to assess mobility in science. In: Linden, A.T, Dargam, F., Jayawickrama, U. (eds.) ICDSST 2020 Proceedings – Online Version the EWG-DSS 2020 International Conference on Decision Support System Technology I, pp. 235–241. ISBN: 978-84-18321-00-9. https://icdsst2020.files.wordpress.com/2020/05/icdsst-2020-proceedings.pdf

Large-Scale Internet User Behavior Analysis of a Nationwide K-12 Education Network Based on DNS Queries

Alexis Arriola[1] , Marcos Pastorini[1] , Germán Capdehourat[2] ,
Eduardo Grampín[1] , and Alberto Castro[1(✉)]

[1] School of Engineering, Universidad de la República, Montevideo, Uruguay
acastro@fing.edu.uy
[2] Plan Ceibal, Montevideo, Uruguay

Abstract. To the best of our knowledge, this paper presents the first Internet Domain Name System (DNS) queries data study from a national K-12 Education Service Provider. This provider, called *Plan Ceibal*, supports a one-to-one computing program in Uruguay. Additionally, it has deployed an Information and Communications Technology (ICT) infrastructure in all of Uruguay's public schools and high-schools, in addition to many public spaces. The main development is wireless connectivity, which allows all the students (whose ages range between 6 and 18 years old) to connect to different resources, including Internet access. In this article, we use 9,125,888,714 DNS-query records, collected from March to May 2019, to study Plan Ceibal user's Internet behavior applying unsupervised machine learning techniques. Firstly, we conducted a statistical analysis aiming at depicting the distribution of the data. Then, to understand users' Internet behavior, we performed principal component analysis (PCA) and clustering methods. The results show that Internet use behavior is influenced by age-group and time of the day. However, it is independent of the geographical location of the users. Internet use behavior analysis is of paramount importance for evidence-based decision making by any education network provider, not only from the network-operator perspective but also for providing crucial information for learning analytics purposes.

Keywords: Machine learning · Data mining · Big data

1 Introduction

1.1 Motivation and Objective

The importance of Internet access to education has increased significantly in recent years. Today, in many places, it has become a relevant resource in formal school education, for various activities based on learning platforms and educational management systems. In order to support this new Internet-based educational paradigm, it is necessary to deploy and maintain the corresponding network infrastructure. This responsibility falls on the Education Service Providers (ESPs). These organizations help the education system to implement comprehensive reforms towards digitization. In

O. Gervasi et al. (Eds.): ICCSA 2020, LNCS 12249, pp. 776–791, 2020.
https://doi.org/10.1007/978-3-030-58799-4_56

particular, this includes providing teachers and students with reliable and high-quality access networks to avoid performance degradation, which impacts the quality of education.

From the network infrastructure design and deployment point of view, this is also quite a new challenge. The educational use case poses new challenges, both from the content and applications, as well as from the users, who are teachers and students (children and teenagers). Therefore, it is quite important to understand this novel scenario, with quite unique characteristics that are not common in other business environments. In this work, we try to shed some light on this last aspect, in particular, by analyzing the behavior of users in educational networks. For this purpose, we studied a large volume of Domain Name System (DNS) data collected during three months from a major nationwide ESP.

1.2 Background

During the 2005 World Economic Forum, Nicholas Negroponte, from MIT, proposed a novel project based on low-cost laptops to reduce the digital divide in less developed countries, known as One Laptop per Child (OLPC) [1]. Although in the following years, the one-to-one educational model gained importance within different regions of the world, the only country to implement it at national level was Uruguay. A Presidential Decree of April 18th, 2007 [2] formally kick-started Plan Ceibal, "a plan for inclusion and equal opportunities to support Uruguayan educational policies with technology," as stated in its website [3].

The initial mission of Plan Ceibal was to promote digital inclusion. Its main goals were to deliver laptops to all the students and teachers and also to provide Internet access at every public educational center. These two overall objectives were achieved in the first three years of the program. Now, more than 10 years later, Plan Ceibal has greatly diversified its services to the education system. It has provided several educational resources and platforms, such as a content management system (CMS), an intelligent tutoring system (ITS) for math learning and a digital library. The increasing dependence on technology by the education system has also led to a greater technical support demand from Plan Ceibal.

Therefore, still today, one of the most relevant responsibilities of Plan Ceibal is to provide connectivity at all educational centers throughout the country. The typical networking solution deployed consists of a router at each school with fiber Internet access from the local ISP and several access points providing Wi-Fi connectivity. With more than 1,500 educational centers (including schools, high-schools, and Domain Name System (DNS)) and more than 8,000 Wi-Fi access points, Plan Ceibal is one of the nation's largest wireless Internet provider, reaching a number of devices comparable to the number of subscribers of local mobile network operators.

Two years ago, Plan Ceibal has incorporated a novel cloud-based DNS solution, Cisco Umbrella [4], which also serves for security purposes and content filtering. Among the advantages of this new solution, it is possible to access in a very simple way all the records of the DNS queries processed, as logging can be configured so that logs are stored to an Amazon S3 bucket. By having the logs uploaded to an S3 bucket (where data is stored for a maximum of 30 days), then they can be automatically downloaded to keep them in perpetuity in backup storage outside of Umbrella's data warehouse storage system. In this way, centralized access to the logs of the DNS queries of the entire network is quite easy. They can then be analyzed and integrated with other data in a specialized data analysis platform.

1.3   Related Work

While there is a vast literature of previous work on DNS data analysis, the majority of them focus on security applications [5]. In all cases, the problems addressed correspond to malicious activities detection based on DNS, such as malicious URLs, botnets, phishing, and web-spam [6].

Concerning behavior analysis with DNS data, several previous works are worth mentioning. Plonka et al. propose a context-aware clustering method [7], where the type of the DNS queries is pre-classified as canonical (i.e., RFC-intended behaviors), overloaded (e.g., black-list services), or unwanted (i.e., queries that will never succeed). In [8], Gao et al. present the analysis of a large dataset of DNS performance measurements. The data from 600 different recursive DNS resolvers, which were globally distributed, is used to compare them, and find out differences and similarities between them (e.g., query success rate and lookup failures causes). A data mining methodology based on different clustering techniques is developed by Ruana et al. in [9], aiming to learn the behavior patterns of DNS queries and detect anomalies. The impact of DNS cache and the wide use of NATs is tackled by Su et al. in [10], for the analysis of a DNS dataset from a major ISP in China. The work by Schomp et al. [11] addresses the study of DNS behavior of individual clients, looking forward to developing an analytical model. In the same way, Li et al. [12] seek to understand and profile what people are doing from DNS traffic, also focused on network behavior analysis. Finally, Jia et al. [13] developed an accessing behavior model for each user, by analyzing network DNS log in a campus network.

Our work is not security-oriented but instead seeks to analyze the observed behavior and recognize relevant patterns. The goal is to understand the typical characteristics and main trends of the DNS queries, in this particular educational context. To the best of our knowledge, this work presents the first study of DNS data in a K-12 education scenario, where most of the users are between 6 and 18 years old.

2   Dataset Description

2.1   Data Collection

The data used in this study were collected from Plan Ceibal's network infrastructure. This network, with 8,587 Wi-Fi access points (APs) located in 1,878 educational buildings (covering more than 95% of the K-12 students in Uruguay), it is one of the largest communication networks in the country. Of those 1,878 educational buildings, 70% are schools and 30% are high-schools (Fig. 1).

In particular, each record (i.e., data point) in our dataset corresponds to a DNS-query request. Plan Ceibal, as part of its network infrastructure, has deployed the Cisco Umbrella system [4] as its DNS solution. All the DNS-query requests performed by users connected to Plan Ceibal's network are recorded and categorized by the Cisco Umbrella system. There are two types of records: DNS logs and proxy logs. The former corresponds to the standard DNS queries, and the latter corresponds to dubious queries that are sent to a proxy for further inspection. For this work, only the DNS logs were considered. Each DNS-log record has the following fields [14]: *i*) *Timestamp*: when the request was made in UTC; *ii*) *InternalIp*: the IP address of the device (e.g., laptop, cellphone) that made the query; *iii*) *ExternalIp*: the IP address of the router that receives the query (i.e., Wi-Fi AP at the educational building); *iv*) *Action*: whether the DNS-query request was allowed or blocked; *v*) *Query Type*: the type of DNS request made; *vi*) *Response Code*: the DNS return code for the query; *vii*) *Domain*: the domain (e.g., youtube.com) that was requested; and *viii*) *Categories*: the system assigns one or more categories (e.g., Video Sharing) to the query, depending on the content of the destination. As an example of a DNS-log record, a typical one looks as follows: "2019-03-10 17:48:41", "10.10.1.100", "24.123.132.133", "Allowed", "1(A)", "NOERROR", "instagram.com", "Photo Sharing, Social Networking".

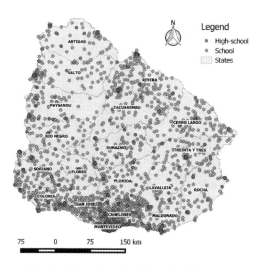

Fig. 1. Educational buildings in Uruguay.

Given that each Wi-Fi AP has a unique IP within the Plan Ceibal's network and that we know where each Wi-Fi AP is located, we used the DNS-log field *ExternalIp* to join a DNS query with an educational building. Particularly, from each education building we have information about the state, area (rural/urban), and the type of educational center (school, high-school, or others).

In this study, we analyzed 9,125,888,714 DNS-query records collected from March to May 2019. It is worth mentioning that, being Uruguay a Southern Hemisphere country, the academic year aligns with the calendar year, lasting from March to December. It is important to highlight that this study was approved by the ethical and data privacy committee of Plan Ceibal. All the datasets are de-identified and handled according to the Uruguayan privacy protection legislation.

2.2 Data Statistics

As shown in Fig. 2, approximately 40% and 53% of the DNS queries correspond to schools and high-schools respectively; while less than 8% corresponds to other educational centers (e.g., technical schools). For this study, we only considered the records from these two large groups.

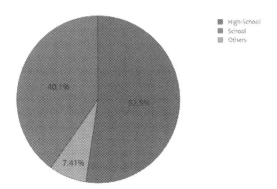

Fig. 2. Percentage of total DNS queries requested group by educational building.

As mentioned above, the Cisco Umbrella system assigns to each DNS query one or more categories [15] that describes the content of the requested domain. Figure 3 shows the distribution in percentage of the 14 most popular categories (in terms of the number of DNS queries that they were assigned to):

- *Search Engines*: Sites that offer results based on keywords.
- *Infrastructure*: Infrastructure for delivering content and dynamically generated content, websites that cannot be classified more accurately because they are safe or otherwise difficult to classify.
- *Non-Category*: Sites to which the system could not assign a category.
- *Social Networking*: Sites that promote interaction and networking among people.
- *Business* Services: Sites for corporations and businesses of all sizes, especially corporate websites.
- *Chat*: Sites where you can chat in real-time with groups of people. It includes video chat sites.
- *Video-Sharing*: Sites to share video content.
- *Software/Technology*: Sites on computing, hardware, and technology, including news, information, code and provider information.
- *Photo Sharing*: Sites to share photos, images, galleries, and albums.
- *Games*: Sites that offer gameplay and game information.
- *SaaS and B2B*: Web portals for online commercial services.
- *Blogs*: Personal or group journal sites, diaries sites, or publications.
- *Educational Institutions*: Sites for schools that cover all levels and age ranges (including Plan Ceibal educational platform).
- *Non-Profits*: Sites for non-profit or charity organizations and services.

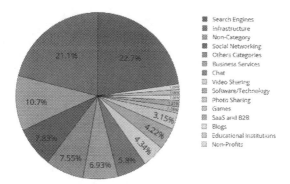

Fig. 3. Percentage of total DNS queries requested group by category

In Fig. 4, a representation of the total amount of DNS queries grouped by state is reported. It is possible to see that Montevideo alone (capital of the country) contains almost 26% of the records and Canelones nearly 17%, which was expected since they are account for the 55% of the total Uruguayan population.

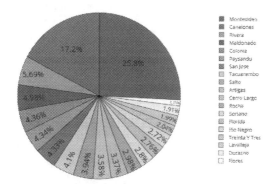

Fig. 4. Representation of the total amount of queries grouped per Uruguayan state.

Figure 5 shows the histogram of the daily number of queries. Focusing on the number of requests per day in the period of the data availability (March–May 2019), we can see that the days with the most significant activity precisely correspond to the days of student activity (Monday–Friday). On March 6th, it is possible to see a growth in the number of queries in the network, corresponding to the beginning of the academic year. On the other hand, during April, there is a week of very low activity, which corresponds to Easter week holidays, when there are no classes. Since the main objective of this work is the study of the students' Internet behavior, weekends and holidays will not be considered in the data analysis. In addition, since the first three weeks of classes (March 6th–March 24th) show a different pattern from the rest of the weeks, they will not considered.

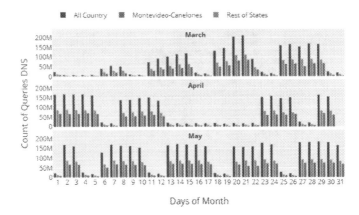

Fig. 5. Histogram of the daily number of queries for all the country, for Montevideo-Canelones, and for the rest of the states.

3 Methodology

Conventional multivariate data-analysis techniques, principal component analysis (PCA) and cluster analysis (CA), were used in this study to understand users' Internet behavior. These techniques are unsupervised methods, meaning that no information about other response variables, or cluster belonging, are used to obtain the outcomes. This makes these methods appropriate for exploratory data analysis, where the goal is hypothesis generation rather than hypothesis verification [16].

3.1 Principal Component Analysis (PCA)

PCA is a method that reduces and transforms measured data into new uncorrelated variables known as principal components (PCs) [17]. The raw measured data are considered as independent variables. There are as many PCs as the number of variables and every single PC is a linear combination of the original variables. The PCs make the basis of the respective vector space and they are organized by decreasing variance. Consequently, PC1 describes the largest data variance, PC2 the next largest data variance, and so on. Most of the variance is accounted for in the first few PCs [18]. Each object is identified by a score and each variable by a loading value or weighting. In its graphical representation (PCA biplot), vectors representing parameters that form an acute angle are considered as correlated parameters, while those that are perpendicular are considered as uncorrelated.

3.2 Clustering

In this study, the CA was not used as a separate methodological approach, but rather as a complementary method for validating and supporting previous PCA results, as well as for providing a better insight into the differences in users' Internet behavior all over the country.

The purpose of the clustering algorithm is to partition a complex dataset into homogeneous clusters, such that the between-group similarities are smaller than the within-group similarities [19]. These clusters can reveal trends and/or patterns related to the phenomenon under study. The similarity between observations is measured by a distance function. Firstly, the similarity is calculated between the data points. Once the data points begin to be clustered, the similarity is computed between the groups as well. Several metrics could be used to calculate it, and the choice of the similarity measure could influence the results. A *priori* and arbitrary decision of the number of groups (*k*) is also required in particular for the k-means CA.

In this study, the silhouette method was used to calculate the best *k* [20]. This analysis measures how close each point in a cluster is to the points in its neighboring groups:

$$s(i) = \frac{b(i) - a(i)}{\max(b(i), a(i))} \tag{1}$$

where $a(i)$ is the average distance of the point (i) to all the other points in the cluster it is assigned (A), $b(i)$ is the average distance of the point (i) to other points to its closest neighboring cluster (B). Silhouette values lie in the range of $[-1, 1]$; the higher the value, the better is the cluster configuration. In particular, the value 1 is ideal since it indicates that the sample is far away from its neighboring cluster and very close to the group it is assigned to. Similarly, the value -1 is the least preferred since it indicates that the point is closer to its neighboring cluster than to the cluster it is assigned to.

Hierarchical cluster analysis (HCA) was adopted in this study to run a heat map analysis. Unlike k-means, HCA starts with each of the *n* data points being their own cluster. In the next step, the two most similar data points are joined to form one cluster giving in all $n - 1$ clusters. Afterward, the two most similar groups are joined to form one cluster, giving in all $n - 2$ clusters. The process is repeated until every data point is in the same cluster that occurs at step $n - 1$. The result is a hierarchical classification tree called dendrogram [21].

Heat map analysis is a false-color image with two dendrograms for two different objects and can divide these two objects into several clusters [22]. The different influence features in these two objects were reordered according to their similarity based on HCA.

4   Results

4.1   Computing Infrastructure

To perform the data analysis, we utilized the Hortonworks Data Platform (HDP) [23] deployed in a single cluster configuration with 40 CPUs Intel Xeon and 256 TB of RAM. HDP is an open-source framework specialized in the storage and distributed processing of large volumes of data (BigData) whose core is based on Apache Hadoop. Apache Hadoop [24] allows the storage and processing of large data sets.

To process our dataset and build the input matrices for the PCA and CA, we implemented specific SQL queries in Apache Spark [25] to be run on Apache Hive [26]. Apache Hive is based on Apache MapReduce. Specifically, it is a data warehouse that allows ad-hoc queries through a SQL-like interface for large data sets stored in an Apache Hadoop Data File System (HDFS). Apache Spark is a programming framework for distributed data processing.

PCA, silhouette method, and CA were programmed in Phyton 3.7 using the scikit-learn library [27]. The heatmap analysis was also coded in Python 3.7 using the seaborn library [28].

4.2 Data Analysis

Two different analyses with two different aims were performed.

The purpose of the first study was to identify the different age-group behavior of Ceibal's users. For this reason, we built a matrix characterized by 14 columns (categories described in Sect. 0) and 38 rows (the type of educational centers, school or high-school, per state). In particular, for the two rows per state considered, the sum of queries by schools and high-schools was calculated respectively. In addition, since states have different numbers of schools and high-schools, the number of queries per state was averaged by the number of educational centers that are located in that particular state. Since the importance of columns (features) is independent from its own variance, we first centered each feature by subtracting its observed values the column's mean, and then we standardized it by dividing each value by the column's standard deviation.

This data matrix (38 × 14) was the input of the PCA, performed to decrease the dimensionality and have a better visualization of the observations, and k-means clustering, tackled to identify different group behaviors. To select the best number of clusters, the silhouette method was used. The average silhouette score of all the considered values in the dataset was calculated and represented by boxenplot (Fig. 6).

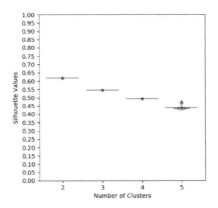

Fig. 6. Boxenplots of silhouette values (or scores).

The boxenplots in Fig. 6 represent an almost null dispersion of the silhouette values. Therefore, the number of clusters corresponding to the highest silhouette value was considered ($k = 2$).

The result of PCA and k-means on this first data matrix is reported in Fig. 7. The first two principal components (PCs) were selected since they represented more than 96% of the variance (respectively 86.5% and 9.7%).

Each line of the input matrix (the type of educational center per state) is represented by a data point in the plot, and their location indicates their score for the PCs. Two net clusters were identified: one for school and the other for high-school-data points. This representation proves the different pattern between these two age groups. In other words, kids and teenagers clearly have a different Internet-use behavior. Furthermore, it is interesting to see that the high-school cluster shows a more significant variation in the PC1 compared to the school cluster.

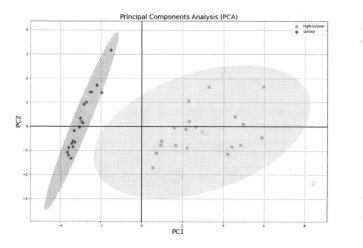

Fig. 7. PCA scores plot with school and high-school data points, and k-means clustering.

The objective of the second analysis was the detection of users' Internet behavior during the day hours. Considering the results of the previous study, it was decided to separate the dataset and examine the school and high-school observations independently. For this reason, two data matrices were built with the 24 h of the day as rows and the categories previously mentioned as columns. For each hour, the sum of the queries of each category was considered. Also in this case, the matrices-columns' values were centered and standardized.

The school-matrix (24×14) and the high-school matrix (24×14) were the input of the PCA and k-means algorithm. The silhouette method was used to obtain the best number of clusters in both cases (Fig. 8 (a) and (b)).

Also in this case, the dispersion of silhouette scores is very low. The silhouette average corresponding to $k = 2$ and $k = 3$ is very similar and is the highest value among all the k-options. We finally decided to consider $k = 3$, to have a better discretization of the day hours.

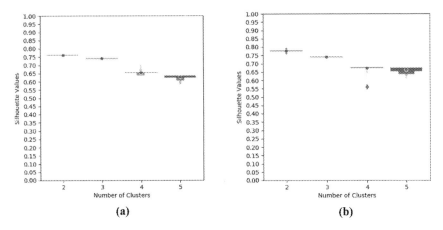

Fig. 8. Boxenplot of silhouette values for (**a**) school-matrix analysis and (**b**) high-school-matrix analysis.

The first two PCs were selected in both cases since they represented more than 98% (Fig. 9) and 99% (Fig. 10) of the variance.

Figure 9 and Fig. 10 show the outcomes obtained for the school and high-school dataset, respectively. The scores plot summarizes the behavior of the data points in the two components and highlights their similarities. In both cases, three net clusters that represent school-hours (morning and afternoon), evening, and late-night were identified. The loadings plot analyzes the role of all the variables (categories) in the two PCs chosen, their correlations, and their relationships with the day hours. It is worth noting that, in both cases, all the vectors are oriented towards the cluster that represents the class hours, showing the extensive Internet use exclusively for school activities. Furthermore, *EducationalInstitutions* and *SocialNetworking* vectors are almost orthogonal,

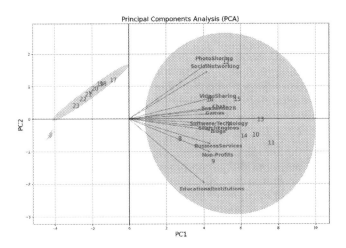

Fig. 9. PCA biplot to identify the school students' behavior during the day hours.

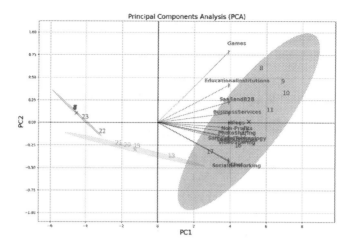

Fig. 10. PCA biplot to identify the high-school students' behavior during the day hours.

showing their independence. In particular, *SocialNetworking* mostly occurs around 12:00, when the shift between morning and afternoon classes happens, while *EducationalInstitutions* occurs in the morning, from 8:00 until 12:00, and in the first hours of the afternoon, 13:00 and 14:00.

With the aim of thoroughly investigating the similarity among all the data points and confirm or add more information to the outcomes of the PCA biplots (Fig. 9 and Fig. 10), a heatmap analysis was used as a complementary analytical tool. Based on previous results, we decided to consider a data matrix in which each day of the week (Monday to Friday) was divided into three hour-groups: *morning* (from 6:00 a.m. to 12 p.m.), *afternoon* (from 12:00 p.m. to 6:00 p.m.), and *evening* (from 6:00 p.m. to 6:00 a.m.). Also in this case, for each of these strips, we grouped (sum) the numbers of queries per category (columns), and centered and standardized the columns' values.

The two hierarchical heatmaps were run using Ward linkage and Euclidean distance. The outcomes obtained for school and high-school datasets are represented in Fig. 11, and Fig. 12 respectively.

From the two heatmaps, considering the left dendrograms, it is noteworthy that the three hour-groups are perfectly identified. Furthermore, the Internet use is very low during the evening/late-night hours. In particular, it seems that high-school students use it more than school students in this time window. This is justified by the fact that teenagers have a more extensive study schedule that may make them study until evening/night. These results confirm and complete the PCA outcomes.

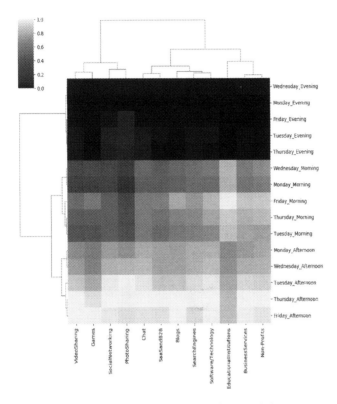

Fig. 11. Heatmap-analysis outcome for school dataset.

Regarding the hierarchical school grouping of the columns (top dendrograms), the cluster formed by the categories *EducationalInstitutions*, *Non-Profits*, and *BusinessServices* can also be identified in the PCA biplot, highlighting in both cases the category *EducationalInstitutions*. Furthermore, it is interesting to see that *VideoSharing*, *Games*, *SocialNetworking*, and *PhotoSharing* occurs more during the afternoon hours, in particular, on Thursday and Friday afternoon (end of the week). While *EducationalInstitutions* is queried more during the morning, all week long.

Considering the left dendrogram of the high-school heatmap, it is possible to see that the class schedule is highly variable depending on the educational center. Clearly, from the top dendrogram, Tuesday and Thursday afternoon are the time windows with more activities, and teenagers, unlike kids, use *VideoSharing*, *SocialNetworking*, and *PhotoSharing* during classes' hours too. These outcomes enhance PCA results.

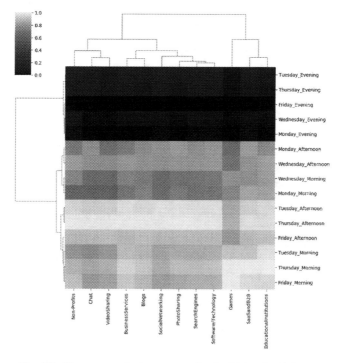

Fig. 12. Heatmap-analysis outcome for the high-school dataset.

5 Concluding Remarks

In this paper, we analyzed the behavior of users in educational networks. For this purpose, we studied a large volume of Domain Name System (DNS) data collected during three months from a major nationwide ESP.

We performed two different analyses with two different aims. The first one, confirmed that kids and teenagers have different Internet-use behavior. The second one described users' Internet behavior during the day hours. It showed that schools students use Internet access, mainly for school activities during class hours. While high-school students also connect to social networks during class time. The differences in the requirements for the applications used at each educational center could be taken into account in the network design process, for example using different design criteria for schools and high-schools.

An in-depth analysis presented that students' Internet-use behavior depends on the classes' schedule, which is highly variable depending on the educational center. The results show that Internet-use behavior is influenced by age-group and time of the day. However, it is independent of the geographical location of the users. The analysis of the user behavior in Internet access is important for any service provider, and the education case is not an exception. The contribution in the particular educational context is not only relevant from the network-operator perspective, but also for other studies combining different sources of information for learning analytics purposes.

Acknowledgement. The research leading to these results received funding from the ANII-FSDA program under proposal number FSDA_1_2018_1_154853.

References

1. One Laptop Per Child. http://one.laptop.org/. Accessed
2. http://www.impo.com.uy/bases/decretos/144-2007/1. Accessed
3. Plan Ceibal. https://www.ceibal.edu.uy/en/institucional. Accessed
4. Cisco – Umbrella. https://umbrella.cisco.com/products/our-cloud. Accessed 10 Mar 2020
5. Zhauniarovich, Y., Khalil, I.M., Yu, T., Dacier, M.C.: A survey on malicious domains detection through DNS data analysis. Crypt. Secur. **1**(1), 1–35 (2018)
6. Torabi, S., Boukhtouta, A., Assi, C., Debbabi, M.: Detecting internet abuse by analyzing passive DNS traffic: a survey of implemented systems. IEEE Commun. Surv. Tutorials **20** (4), 3389–3415 (2018)
7. Plonka, D., Barford, P.: Context-aware clustering of DNS query traffic. In: IMC 2008: Proceedings of the 8th ACM SIGCOMM, Vouliagmeni, Greece, pp. 217–230 (2008)
8. Gao, H., Yegneswaran, V., Chen, Y., Porras, P., Ghosh, S., Haixin Duan, J.J.: An empirical reexamination of global DNS behavior. In: SIGCOMM 2013: Proceedings of the ACM SIGCOMM 2013 Conference on SIGCOMM, Hong Kong, China, pp. 267–278 (2013)
9. Ruana, W., Liub, Y., Zhaob, R.: Pattern discovery in DNS query traffic. Procedia Comput. Sci. **17**, 80–87 (2013)
10. Su, J., Li, Z., Grumbach, S., Salamatian, K., Han, C., Xie, G.: Toward accurate inference of web activities from passive DNS data. In: 2018 IEEE/ACM 26th International Symposium on Quality of Service (IWQoS), Banff, AB, Canada, pp. 1–6 (2018)
11. Schomp, K., Rabinovich, M., Allman, M.: Towards a model of DNS client behavior. In: Karagiannis, T., Dimitropoulos, X. (eds.) PAM 2016. LNCS, vol. 9631, pp. 263–275. Springer, Cham (2016). https://doi.org/10.1007/978-3-319-30505-9_20
12. Li, J., Ma, X., Guodong, L., Luo, X., Zhang, J., Li, W., Guan, X.: Can we learn what people are doing from raw DNS queries? In: IEEE INFOCOM 2018 - IEEE Conference on Computer Communications, Honolulu, HI, pp. 2240–2248 (2018)
13. Jia, Z., Han, Z.: Research and analysis of user behavior fingerprint on security situational awareness based on DNS log. Research and analysis of user behavior fingerprint on security situational awareness based on DNS Log. In: 6th International Conference on Behavioral, Economic and Socio-Cultural Computing (BESC), Beijing, China, pp. 1–4 (2019)
14. Cisco – Umbrella Log Formats and Versioning. https://docs.umbrella.com/deployment-umbrella/docs/log-formats-and-versioning. Accessed 10 Mar 2020
15. Cisco – Umbrella Manage Content Categories. https://docs.umbrella.com/deployment-umbrella/docs/content-categories#section-content-categories-definitions. Accessed 10 Mar 2020
16. Gorgoglione, A., Gioia, A., Iacobellis, V.: A Framework for assessing modeling performance and effects of rainfall-catchment-drainage characteristics on nutrient urban runoff in poorly gauged watersheds. Sustainability **11**, 4933 (2019)
17. Massart, D.L., Vandeginste, B.G.M., Deming, S.M., Michotte, Y., Kaufman, L.: Chemometrics-A Text Book. Chapters 1–4, pp. 14–21. Elsevier: Amsterdam (1988)
18. Adams, M.J.: Chemometrics in Analytical Chemistry. The Royal Society of Chemistry, Cambridge (1995)

19. Jain, A.K., Murty, M.N., Flynn, P.J.: Data clustering: a review. ACM Comput. Surv. **31**, 264–323(1999)
20. Rousseeuw, P.J.: Silhouettes: a graphical aid to the interpretation and validation of cluster analysis. J. Comput. Appl. Math. **20**, 53–65 (1987)
21. Baker, F.B., Lawrence, J.H.: Measuring the power of hierarchical cluster analysis. J. Am. Stat. Assoc. **70**, 349 (1975)
22. Friendly, M.: The history of the cluster heat map. The American Statistician (2009)
23. Hortonworks Data Platform. https://www.cloudera.com/products/hdp.html. Accessed 6 Mar 2020
24. Apache Hadoop. https://hadoop.apache.org/. Accessed 6 Mar 2020
25. Apache Spark. https://spark.apache.org/. Accessed 6 Mar 2020
26. Apache Hive. https://hive.apache.org/. Accessed 6 Mar 2020
27. scikit-learn. https://scikit-learn.org/. Accessed 6 Mar 2020
28. Seaborn. https://seaborn.pydata.org/. Accessed 6 Mar 2020

Unsupervised Learning and Feature Extraction in Hyperspectral Imagery

Eduardo K. Nakao[✉] and Alexandre L. M. Levada

Universidade Federal de São Carlos,
Rod. Washington Luís km 235, 310 São Carlos, Brazil
eduardokazuonakao@gmail.com

Abstract. Remotely sensed hyperspectral scenes are typically defined by large area coverage and hundreds of spectral bands. Those characteristics imply smooth transitions in the spectral-spatio domains. As consequence, subtle differences in the scene are evidenced, benefiting precision applications, but values in neighboring locations and wavelengths are highly correlated. Nondiagonal covariance matrices and wide autocorrelation functions can be observed this way, implying increased intraclass and decreased interclass variation, in both spectral and spatial domains. This leads to lower interpretation accuracies and makes it reasonable to investigate if hyperspectral imagery suffer from Curse of Dimensionality. Moreover, as this Curse can compromise linear method's Euclidean behavior assumption, it is relevant to compare linear and nonlinear dimensionality reduction performance. So, in this work we verify these two aspects empirically using multiple nonparametric statistical comparisons of Gaussian Mixture Model clustering performances in the cases of: absence, linear and nonlinear unsupervised feature extraction. Experimental results indicate Curse of Dimensionality presence and nonlinear adequacy.

Keywords: Hyperspectral imagery · Curse of dimensionality · Unsupervised feature extraction · Nonlinear dimensionality reduction · Unsupervised learning

1 Introduction

Hyperspectral imaging [12] captures the energy matter interaction in hundreds of channels that can precisely define the chemical composition of different materials by spectral similarity discrimination. A hyperspectral image (HI) is viewed as a cube of two-dimensional single band images stacked in the third dimension of sampled wavelengths. So if m spectral bands were collected, each pixel can be viewed as a m dimensional vector where its jth coordinate value is the reflectance

This study was financed in part by the Coordenacao de Aperfeicoamento de Pessoal de Nivel Superior - Brasil (CAPES) - Finance Code 001.

© Springer Nature Switzerland AG 2020
O. Gervasi et al. (Eds.): ICCSA 2020, LNCS 12249, pp. 792–806, 2020.
https://doi.org/10.1007/978-3-030-58799-4_57

measured at the jth wavelength. This vector is the geometric pixel representation in the spectral domain.

Due to this presence of hundreds of channels, HIs are spectrally smooth and spatially piecewise smooth. Values in neighboring locations and wavelengths are highly correlated. This can be observed by extremely nondiagonal covariance matrices and wide autocorrelation functions [4]. So there are increased intra-class variation and decreased interclass variation, in both spectral and spatial domains [34], which leads to lower interpretation accuracies. Thus, the linear independence assumption in the basis vectors of an Euclidean space can be not suitable. If pixel vectors are seen as random variables, the high correlation between them suggests that a possible statistical independence assumption is not suitable as well.

Those facts raise the supposition that the pixel vector representation is embedded in a high-dimensional space, suffering from the Curse of Dimensionality. From Machine Learning perspective, the problem is described by the Hughes phenomena [14] which states that, when dimension grows from a certain point in relation to the number of samples, classification performance is degraded.

From a geometric point of view, in high dimensions it can also be proved [20] that the volume of a hypercube is concentrated on the corners [17,31] and the volume of a hypersphere or hyperellipsoid concentrates on an outside shell [17]. These facts imply the following consequences: space is mostly empty (so data can be projected to a lower dimension without losing significant class separability); local neighborhoods are almost empty (requiring large estimation bandwidth and losing detailed density estimation); normal distributions have their data concentration on the tails instead the means, and uniformly distributed data is more likely collected on the corners (making density estimation more difficult).

Other important high dimensionality characteristics are [20]: diagonal vectors become orthogonal to the basis vectors (the projection of a cluster in one of these "wrong directions" can ruin the information within the data [31]); data covariance matrix becomes ill-conditioned; Euclidean distance is not fully reliable [1]; low linear projections have the tendency to be Normal.

So, in Curse of Dimensionality presence, the clustering method Gaussian Mixture Model (GMM) can have its behavior compromised because it uses Euclidean distance in neighborhood search and have Gaussian hypothesis and statistical independence assumption. On the other hand, in the presence of few bands (i.e. low dimensionality), GMM is efficient to determine class distribution even with few samples in some classes (which is common in HIs due to their subtle substances discrimination derived from high spectral resolution). The model is efficient even if some classes are not normally distributed in fact [19] and it also takes advantage of the high dimensional space property of Normal distributions formation in the projection from high to low dimensionality.

Another problem is that, as GMM, linear Dimensionality Reduction (DR) methods also assume Euclidean data behavior. So those methods can be compromised too. Given a image, it is not possible to know beforehand theoretically if its pixels vectors will be better represented by a linear or a nonlinear

DR approach. Therefore, empirical computational experiments are necessary to investigate linear against nonlinear data representation adequacy for HIs.

Considering all those inherent properties and possible problems, the focus in this study is to analyze how unsupervised feature extraction methods affect clustering by GMM and if nonlinear DR is more adequate than linear for HIs. The common lack of class information in hyperspectral imagery justify the interest in unsupervised feature extraction and its impact in unsupervised learning. We compare Principal Component Analysis (PCA) - the most traditional linear DR method - against Isometric Feature Mapping (ISOMAP) and Locally Linear Embedding (LLE) - both pioneering nonlinear methods with global and local approaches respectively. All three methods yield an orthonormal basis in an Euclidean space where the vectors are written as linear combination of the others. Besides that, as vectors are seen as random variables by GMM, it is assured also that those variables are not correlated. So the goal of adjusting the HIs original data to a feasible condition for GMM is guaranteed.

The remaining of the paper is organized as follows: next section brings some related work. Section 3 presents the unsupervised feature extraction methods used in the work. Section 4 describes how experiments were conducted. Section 5 explains the experimental results. Finally, Sect. 6 presents conclusions and final remarks.

2   Related Work

Feature Selection for Classification of Hyperspectral Data by SVM [27]. The main conclusion of this study is that, in HIs, the classification accuracy of Support Vector Machine (SVM) technique is influenced by the number of features used. Thus, it is shown that SVM is affected by Hughes phenomenon. The results presented in this paper show for AVIRIS sensor dataset that feature addition led to statistically significant decline in accuracy, for all sizes of training sets evaluated. With DAIS dataset, such decline was also observed, both for small training sets (25 samples per class or less) and for a large training sample. Therefore, it has been shown that feature selection provides a reduced feature set which gives similar classification accuracy to that of larger feature sets. In addition, the analysis of four different selection methods showed a wide variation between the number of features selected by each method, which highlights the influence of method choice in the process.

As in our study, this reference showed the manifestation of Hughes phenomenon in HIs and the influence of the DR method on the estimation of intrinsic dimensionality. However, classification was used instead of clustering and feature selection was used in place of extraction.

Feature Mining for Hyperspectral Image Classification [16]. Arguing that Hughes problem can be avoided by the use of few discriminating features, this article provides an overview of fundamental and modern feature extraction and selection which are applicable in HI classification. The common unavailability of class labeling in this type of data is discussed. It is emphasized that in supervised parametric approaches, class separation measures require large training amount, which may not be accessible or physically feasible in practice. Therefore, nonparametric and unsupervised methods are preferred. It is informed that, current area research focus is on nonlinear feature extraction.

So this article shows that DR bypasses the Hughes problem in HIs and complements our work with feature extraction and selection methods description. One more argument is provided about the relevance of nonparametric, unsupervised and nonlinear categories (which are used in our work). Finally, the current relevance of research in nonlinear feature extraction is highlighted (which is also investigated by us).

Subspace Feature Analysis of Local Manifold Learning for Hyperspectral Remote Sensing Images Classification [9]. This reference is able to show more specifically that, in HIs, the use of features obtained by nonlinear extraction provides greater classification accuracy than original features. This type of conclusion also reinforces the evidence shown in our work: not only the HI feature extraction but also the nonlinear benefits.

Unsupervised Clustering and Active Learning of Hyperspectral Images with Nonlinear Diffusion [24]. In this work it is also shown that, in HIs context, linear DR methods generate better results compared to DR absence in classification process. Additionally, it is also shown that nonlinear DR is superior to linear in this context. This indicates that HIs are mapped in structures whose intrinsic dimensionality is less than the original, and also that such structures are particularly nonlinear. The study also shows that, in HIs, clustering methods that use Euclidean distance have lower performance when compared to methods that use other strategies.

So this related work is another source that shows HIs DR benefit (and particularly nonlinear DR). Additionally, the work also brings information about Euclidean strategies failure in this type of image, reinforcing the validity of suspicions along the same lines that were proven in our case.

Evaluation of Nonlinear Dimensionality Reduction Techniques for Classification of Hyperspectral Images [25]. Here it is also argued that, in many cases, combination of PCA with SVM provides good HIs classification results. However, hyperspectral scenes can bring complexity due to nonlinear effects, which can make PCA inappropriate. It is stated that for such scenes, therefore, it is necessary to choose carefully the output dimensionality and to consider the use of nonlinear DR techniques. It is also reported that the main disadvantage of nonlinear methods is characterized by high computational complexity in comparison with PCA.

Unsupervised Exploration of Hyperspectral and Multispectral Images [22]. Some of the most used unsupervised exploration methods (mainly DR and clustering techniques) are revised and discussed, focusing in hyper and multispectral imagery. Use instructions are provided and vantages and disadvantages are high-lighted in order to demystify common errors. It is argued that, when analysis focus is prediction, exploratory data analysis is frequently underrated. It is stated that, in fact, exploratory analysis provides wealth information itself and a first data insight that can be very useful, even if exploration is not the final goal. It follows that unsupervised methods provide valuable information and can be applied as a first approach for any hyper or multispectral analysis, given that such methods are hypothesis free.

3   Feature Extraction

The goal of Feature Extraction is to find the transformation function $f : R^m \rightarrow R^d$ that can map the point $\{x_i \in R^m\}_{i=1}^n$ to $y_i = f(x_i)$, where $\{y_i \in R^d\}_{i=1}^n$ and $d \leq m$, for all n points in the dataset, such that most data information is kept. The f term can be a linear or nonlinear transformation. In the next three subsections, we show the algorithms selected to work with.

3.1   Principal Component Analysis

Linear DR is widely used for analyzing high dimensional and noisy data [7]. Part of the success of linear methods comes from the fact that they have simple geometric interpretations and typically attractive computational properties [7]. Basically, in linear methods, we seek a projection matrix T that maps the samples in the original feature space (R^m) to a linear subspace in R^d, with $d < m$. Given n m-dimensional data points $X = [x_1, x_2, ..., x_n] \in R^{m \times n}$ and a choice of dimensionality $d < m$, the goal is to optimize some objective function $f_X(.)$ to produce a linear transformation $T \in R^{d \times m}$, and call $Y = TX \in R^{d \times n}$ the lower dimensional transformed data.

Principal Component Analysis is the most widely known linear method for data compression and Feature Extraction. It assumes that vectors are in an Euclidean space and from this assumption it searches for a new basis with d span vectors that maximizes the variance in the new representation. In this case, d is the intrinsic dimensionality and it is less than the original data dimension.

3.2   Isometric Feature Mapping

ISOMAP was one of the pioneering algorithms in nonlinear DR. It combines the major algorithmic advantages of PCA and Multidimensional Scaling (MDS) - computational efficiency, global optimality, and asymptotic convergence guar-antees - with the flexibility to learn a broad class of nonlinear structures [33]. The basic idea of the ISOMAP algorithm is first to build a graph by joining the k-nearest neighbors (KNN) in the input space, then compute the shortest paths between each pair of vertices in the graph and, knowing the approximate distances between points, find a mapping to an Euclidean subspace of R^d that preserves those distances. The hypothesis of the ISOMAP algorithm is that the shortest paths in the KNN graph are good approximations for the true distances in the original data structure. ISOMAP algorithm can be divided into three main steps:

1. From the input data $x_1, x_2, ..., x_n \in R^m$ build an undirected proximity graph using the KNN rule or the ϵ-neighborhood rule [21];
2. Compute the pairwise distance matrix D using n executions of the Dijkstra's algorithm or one execution of the Floyd-Warshall algorithm;
3. Estimate the new coordinates of the points in an Euclidean subspace of R^d by preserving the distances through the MDS method.

3.3 Locally Linear Embedding

The ISOMAP algorithm is a global method in the sense that, to find the new coordinates of a given input vector $x_i \in R^m$, it uses information from all the samples through the matrix B. On the other hand, LLE, as the name emphasizes, is a local method where the new coordinates of any $x_i \in R^m$ depend only on the point neighborhood. The main hypothesis behind LLE is that for a sufficient density of samples, it is expected that a vector x_i and its neighbors define a linear patch (i.e. all belonging to an Euclidean subspace [29]). It is possible to characterize the local geometry by linear coefficients this way:

$$\hat{x}_i \approx \sum_j w_{ij} x_j, \qquad \text{for} \qquad x_j \in N(x_i), \tag{1}$$

that is, we can reconstruct a vector as a linear combination of its neighbors.

Basically, the LLE algorithm requires as inputs a $n \times m$ data matrix X, with rows x_i, a desired number of dimensions $d < m$ and an integer $k > d + 1$ for finding local neighborhoods. The output is a $n \times d$ matrix Y, with rows y_i. The LLE algorithm can be divided into three main steps [29,30]:

1. From each $x_i \in R^m$ find its k nearest neighbors;
2. Find the weight matrix W which minimizes the reconstruction error for each data point $x_i \in R^m$;

$$E(W) = \sum_{i=1}^{n} \left\| x_i - \sum_j w_{ij} x_j \right\|^2, \tag{2}$$

where $w_{ij} = 0$ unless x_j is one of x_i's k-nearest neighbors and for each i, $\sum_j w_{ij} = 1$.

3. Find the coordinates Y which minimize the reconstruction error using the optimum weights;

$$\Phi(Y) = \sum_{i=1}^{n} \left\| \boldsymbol{y}_i - \sum_j w_{ij} \boldsymbol{y}_j \right\|^2, \tag{3}$$

subject to the constraints $\sum_i Y_{ij} = 0$ for each j, and $Y^T Y = I$.

Algorithm 1. Principal Component Analysis

1: **function** PCA(X)
2:     Compute the sample mean and the sample covariance matrix by:
       $\mu_x = \frac{1}{n} \sum_{i=1}^{n} \boldsymbol{x}_i$
       $\Sigma_x = \frac{1}{n-1} \sum_{i=1}^{n} (\boldsymbol{x}_i - \mu_x)(\boldsymbol{x}_i - \mu_x)^T$
3:     Compute the eigenvalues and eigenvectors of Σ_x
4:     Define the transformation matrix $T = [\boldsymbol{w}_1, \boldsymbol{w}_2, ..., \boldsymbol{w}_d]$ with the d eigenvectors associated to the d largest eigenvalues.
5:     Project the data X into the PCA subspace:
       $\boldsymbol{y}_i = T\boldsymbol{x}_i \quad$ for $\quad i = 1, 2, ..., n$
6:     **return** Y
7: **end function**

Algorithm 2. Isometric Feature Mapping

1: **function** ISOMAP(X)
2:     From the input data $X_{m \times n}$ build a KNN graph.
3:     Compute the pairwise distances matrix $D_{n \times n}$.
4:     Compute $A = -\frac{1}{2}D$.
5:     Compute $H = I - \frac{1}{n}U$, where U is a $n \times n$ matrix of 1's.
6:     Compute $B = HAH$.
7:     Find the eigenvectors and eigenvalues of the matrix B.
8:     Select the top $d < m$ eigenvectors and eigenvalues of B and define:

$$\tilde{V} = \begin{bmatrix} | & | & ... \, ... & | \\ | & | & ... \, ... & | \\ \boldsymbol{v}_1 & \boldsymbol{v}_2 & ... \, ... & \boldsymbol{v}_d \\ | & | & ... \, ... & | \\ | & | & ... \, ... & | \end{bmatrix}_{n \times d} \tag{4}$$

$$\tilde{\Lambda} = diag(\lambda_1, \lambda_2, ..., \lambda_d) \tag{5}$$

9:     Compute $\tilde{Y} = \tilde{\Lambda}^{1/2}\tilde{V}^T$
10:    **return** $\tilde{Y}$
11: **end function**

Algorithm 3. Locally Linear Embedding

1: **function** LLE(X, K, d)
2:      From the input data $X_{m \times n}$ build a KNN graph.
3:      **for** $\boldsymbol{x}_i \in X^T$ **do**
4:           Compute the $K \times K$ matrix C_i as:

$$C_i(j, k) = (\boldsymbol{x}_i - \boldsymbol{x}_j)(\boldsymbol{x}_i - \boldsymbol{x}_k)^T. \tag{6}$$

5:           Solve the linear system $C_i \boldsymbol{w}_i = \mathbf{1}$ to estimate the weights $\boldsymbol{w}_i \in R^K$.
6:           Normalize the weights in $\boldsymbol{w}_i$ so that $\sum_j \boldsymbol{w}_i(j) = 1$.
7:      **end for**
8:      Construct the $n \times n$ matrix W, whose lines are the estimated $\boldsymbol{w}_i$.
9:      Compute $M = (I - W)^T (I - W)$.
10:      Find the eigenvalues and eigenvectors of the matrix M.
11:      Select the bottom d nonzero eigenvectors of M and define:

$$Y = \begin{bmatrix} | & | & \cdots \cdots & | \\ | & | & \cdots \cdots & | \\ \boldsymbol{v}_1 & \boldsymbol{v}_2 & \cdots \cdots & \boldsymbol{v}_d \\ | & | & \cdots \cdots & | \\ | & | & \cdots \cdots & | \end{bmatrix}_{n \times d}. \tag{7}$$

12:      **return** Y
13: **end function**

---

4   Methodology

Seven traditional HIs[1] were used for exploitation: Indian Pines, Salinas, SalinasA, Botswana, Kennedy Space Center, Pavia University, and Pavia Centre.

The purpose of this work is to verify if unsupervised Feature Extraction benefits HI clustering by GMM and if a nonlinear method is more efficient than a linear in this case. To conduct such verification, the performance of clusterings made on linear reduction (PCA), nonlinear reduction (ISOMAP and LLE) and nonDR (NDR) are compared. As Ground Truths were available, performances were evaluated by external measures. Measures used are: Rand [28], Jaccard [15], Kappa [5], Entropy and Purity [32]. Performances were then compared by Friedman [10] and Nemenyi [26] tests. All steps are shown in Fig. 1.

Regarding the application of nonlinear algorithms, most images were too large for our 24 GB RAM memory. This problem happened especially for ISOMAP algorithm second step. So, for those images, sets of representative pixels were chosen using Bagging [2] technique. Those sets were used by both linear and nonlinear DR methods in place of the original image. Clustering was made on those sets in the NDR case also.

---

[1] All information regarding those images (as number of classes, reflectance and radiance characteristics, number of bands, pre-processing, source equipment, date, and files) can be found at: http://www.ehu.eus/ccwintco/index.php?title=Hyperspectral_Remote_Sensing_Scenes.

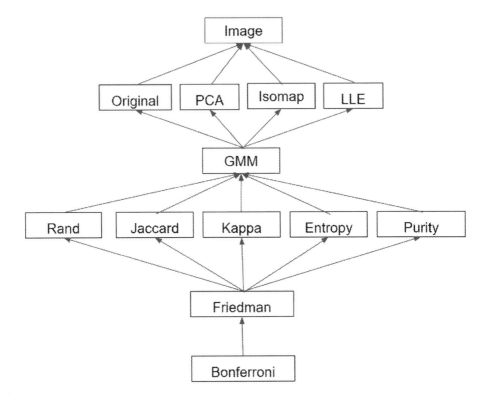

Fig. 1. Experiments execution flow pipeline

For each image, the three DR methods were applied. The choice of target dimensionality strongly affects the results and is usually defined by the intrinsic dimensionality. Such discovery can be made by several methods [3,6,13,23] and its strongly related to each image itself and its particular properties. The topic is currently under investigation in the area and, as it is not focus of this work, we have choosen a well-known basic empirical method. The method consists in the analysis of the sum of the first n largest eigenvalues over the sum of all eigenvalues of the transformation matrix of each DR method [6]. We seek the point where 95% of data variance is retained. The amount of eigenvalues that reach this point is understood as the intrinsic dimensionality.

In the image pre-processing step[2], all mixed pixels were categorized to a background class which label is zero. So, after DR and before clustering, we adopted a simple procedure of setting those pixel vectors coordinates to $(0,...,0)$ - as background class is not in the discrimination interest. The goal of this procedure is to avoid original geometry distortion in DR and also to help raising clustering performance considerably (as this class is a pixel mixture that is not in any other class defined by the expert). We notice that the procedure didn't

---

[2] This step was not done by us, image files used were already pre-processed.

produce much distortion in clustering (most class zero pixels in the Ground Truth were kept in the same group after clustering).

For each vector set in the reduced dimensional space, several executions of the clustering procedure were applied due to its known effect of random centroids initialization. As this particular image set is already labeled by specialists, external clustering validation is more robust than internal. So external measures were used. All metrics adopted assume higher values for better clusterings. Specifically for the Kappa measure, it must be ensured a reliable semantic correspondence between clustering and expert's labeling. We mapped this problem to one of finding the minimum pairing cost in a bipartite graph, where there is label association to minimize a cost function. The Munkres algorithm was used in this Optimum Allocation Problem [18].

All measures were applied to all clustering results and the representative value for each measure was given by the maximum value obtained. Next, for each image, Friedman test indicates if there was significant statistical difference between some pair of measure values (considering a pair of two lists, each corresponding to a DR method, whose elements are the external indices maximum values). Nemenyi test tells then which measure pairs were different. The results make possible to analyze if DR was benefical to clustering and in which images nonlinear overcame linear reduction. Tests results are shown in the next section.

The programming language and libraries used were: Python 3.6.5; NumPy 1.15.0; SciPy 1.0.0; Scikit-learn 0.20.0; Matplotlib 3.0.0, and STAC. We used the Scikit-learn implementations for all DR methods.

5   Experiments and Results

We can find in Tables 1, 2, 3, 4 and 5[3] the clustering validation measure maximum values by DR method, where each table corresponds to an image which Friedman test pointed significant difference between some pair of DR strategy. Usual p-value > 0.05 criteria to indicate equivalence was adopted. The last row presents the measure values mean given a DR method.

In Table 6 we present the Friedman p-values by image. Friedman test only tells us if some method pair differed. We need a post-hoc test to discover which pairs were different [8,11]. We show in Table 7 the Nemenyi p-values smaller than 0.05 and its occurrence pair by image.

We can see for Botswana that, for a trust level of 95% (i.e. significance level of 5%), there is evidence that ISOMAP produces better clustering results than DR absence. This imply that DR was benefical to this image. There is also evidence that ISOMAP got better attributes than PCA, which justifies the better adequacy of a nonlinear DR method for this data. Following the same reasoning, in KSC we can say that LLE is superior than PCA. In Pavia, LLE is superior than NDR and PCA. In PaviaU PCA had significant difference in comparison with NDR and was also superior than both nonlinear methods. In

---

[3] We truncated all values at the third decimal point or at the first decimal point possible.

Table 1. Botswana: clustering validation measure maximum values by DR method

	NDR	PCA	ISOMAP	LLE
Entropy	−0.177	−0.194	**−0.144**	−0.159
Jaccard	0.802	0.820	**0.836**	0.803
Kappa	0.802	0.797	**0.861**	0.804
Purity	0.952	0.948	**0.962**	0.955
Rand	0.995	0.995	**0.996**	0.994
Mean	0.674	0.673	**0.702**	0.679

Table 2. KSC: clustering validation measure maximum values by DR method

	NDR	PCA	ISOMAP	LLE
Entropy	−0.683	−0.705	−0.707	**−0.289**
Kappa	0.602	0.577	0.602	**0.792**
Jaccard	0.655	0.656	0.666	**0.698**
Purity	0.830	0.819	0.829	**0.929**
Rand	0.946	0.941	0.944	**0.989**
Mean	0.469	0.457	0.466	**0.623**

Table 3. PaviaU: clustering validation measure maximum values by DR method

	NDR	PCA	ISOMAP	LLE
Entropy	−0.210	**−0.182**	−0.196	−0.284
Jaccard	0.783	**0.797**	0.792	0.753
Kappa	0.758	**0.826**	0.814	0.771
Purity	0.939	**0.949**	0.944	0.936
Rand	0.980	**0.984**	0.982	0.951
Mean	0.650	**0.674**	0.667	0.625

Table 4. Pavia: clustering validation measure maximum values by DR method

	NDR	PCA	ISOMAP	LLE
Entropy	−0.090	−0.084	−0.080	**−0.066**
Jaccard	0.747	0.882	0.889	**0.897**
Kappa	0.864	0.889	0.894	**0.932**
Purity	0.978	0.978	0.980	**0.986**
Rand	**0.994**	0.993	0.993	**0.994**
Mean	0.698	0.731	0.735	**0.748**

Table 5. SalinasA: clustering validation measure maximum values by DR method

	NDR	PCA	ISOMAP	LLE
Entropy	−0.358	**−0.228**	−0.266	−0.246
Jaccard	0.302	**0.559**	0.458	0.347
Purity	0.882	0.910	0.906	**0.931**
Kappa	0.857	**0.869**	0.859	0.843
Rand	0.836	**0.870**	0.861	0.852
Mean	0.503	**0.596**	0.563	0.545

Table 6. Friedman p-value by image

	p-value
Botswana	0.004
KSC	0.001
PaviaU	0.0000001
Pavia	0.003
Salinas	0.175
SalinasA	0.002
Indian Pines	0.322

Table 7. Nemenyi p-values

	Botswana	KSC	PaviaU	Pavia	SalinasA
NDR × PCA			0.007		0.001
NDR × ISOMAP	0.010				
NDR × LLE				0.004	
PCA × ISOMAP	0.004				
PCA × LLE		0.001	0.0006	0.010	
ISOMAP × LLE		0.019	0.027		

SalinasA we can only conclude that PCA yeld better results in comparisons to NDR. So, in four images DR brought benefit to clustering. Additionally, in three images we saw that nonlinear DR performed better than linear.

6   Conclusion

As one cannot state beforehand theoretically if HIs naturally suffer from Curse of Dimensionality, in this work we have shown empirical evidence that unsupervised Feature Extraction significantly increased GMM clustering performance in the majority of studied images. Significant statistical difference between

DR and NDR was found using nonparametric multiple comparisons hypothesis tests on several large real images. Unsupervised DR and unsupervised learning approaches were used to reach this conclusion and Ground Truth was used only to compute several external measures in clustering evaluation. Additionally, label matching by Munkres algorithm increased Kappa values and emphasized performance differences between DR methods in this matter.

The efficiency of nonlinear DR compared to linear in HI context cannot be shown beforehand theoretically also. Our experiments have shown that nonlinear DR is more appropriate for the majority of HIs in which DR brought significant contribution to clustering. We gave theoretical justification to why nonlinear DR can be more adequate by presenting the PCA, ISOMAP and LLE strategies and algorithms.

Future works may include: other Feature Extraction techniques, specially modern nonlinear algorithms; more sophisticated intrinsic dimensionality discovery strategies; other HI images; more clustering methods; use of internal clustering validity indexes to test a complete unsupervised procedure; use of supervised classifiers ensemble to obtain a supervised perspective. We believe that such expansions can bring more evidence for the Curse of Dimensionality in HIs and build a stronger basic study for unsupervised Feature Extraction in HIs, compiling the main interest points in the area and serving as guide for first contact.

References

1. Aggarwal, C.C., Hinneburg, A., Keim, D.A.: On the surprising behavior of distance metrics in high dimensional space. In: Van den Bussche, J., Vianu, V. (eds.) ICDT 2001. LNCS, vol. 1973, pp. 420–434. Springer, Heidelberg (2001). https://doi.org/10.1007/3-540-44503-X_27
2. Breiman, L.: Bagging predictors (1994)
3. Camastra, F.: Data dimensionality estimation methods: a survey. Pattern Recogn. **36**(12), 2945–2954 (2003). https://doi.org/10.1016/S0031-3203(03)00176-6. http://www.sciencedirect.com/science/article/pii/S0031320303001766
4. Camps-Valls, G., Tuia, D., Gómez-Chova, L., Jiménez, S., Malo, J.: Remote sensing image processing. In: Remote Sensing Image Processing (2011)
5. Cohen, J.: A coefficient of agreement for nominal scales. Educ. Psychol. Measur. **20**(1), 37–46 (1960). https://doi.org/10.1177/001316446002000104
6. Cox, T.F., Cox, M.A.A.: Multidimensional scaling. In: Monographs on Statistics and Applied Probability, vol. 88. Chapman & Hall, New York (2001)
7. Cunningham, J.P., Ghahramani, Z.: Linear dimensionality reduction: survey, insights, and generalizations. J. Mach. Learn. Res. **16**, 2859–2900 (2015)
8. Demšar, J.: Statistical comparisons of classifiers over multiple data sets. J. Mach. Learn. Res. **7**, 1–30 (2006)
9. Ding, L., Tang, P., Li, H.: Subspace feature analysis of local manifold learning for hyperspectral remote sensing images classification. Appl. Math. Inf. Sci. **8**(4), 1987 (2014)
10. Friedman, M.: The use of ranks to avoid the assumption of normality implicit in the analysis of variance. J. Am. Stat. Assoc. **32**(200), 675–701 (1937). https://doi.org/10.1080/01621459.1937.10503522

11. García, S., Fernández, A., Luengo, J., Herrera, F.: Advanced nonparametric tests for multiple comparisons in the design of experiments in computational intelligence and data mining: experimental analysis of power. Inf. Sci. **180**, 2044–2064 (2010). https://doi.org/10.1016/j.ins.2009.12.010
12. Goetz, A.F., Vane, G., Solomon, J.E., Rock, B.N.: Imaging spectrometry for earth remote sensing. Science **228**(4704), 1147–1153 (1985). https://doi.org/10.1126/science.228.4704.1147. http://science.sciencemag.org/content/228/4704/1147
13. He, J., Ding, L., Jiang, L., Li, Z., Hu, Q.: Intrinsic dimensionality estimation based on manifold assumption. J. Vis. Commun. Image Representation **25**(5), 740 – 747 (2014). https://doi.org/10.1016/j.jvcir.2014.01.006. http://www.sciencedirect.com/science/article/pii/S1047320314000078
14. Hughes, G.: On the mean accuracy of statistical pattern recognizers. IEEE Trans. Inf. Theory **14**(1), 55–63 (1968). https://doi.org/10.1109/TIT.1968.1054102
15. Jaccard, P.: The distribution of the flora in the alpine zone 1. New Phytologist. **11**(2), 37–50 (1912). https://doi.org/10.1111/j.1469-8137.1912.tb05611.x. https://nph.onlinelibrary.wiley.com/doi/abs/10.1111/j.1469-8137.1912.tb05611.x
16. Jia, X., Kuo, B., Crawford, M.M.: Feature mining for hyperspectral image classification. Proc. IEEE **101**(3), 676–697 (2013). https://doi.org/10.1109/JPROC.2012.2229082
17. Kendall, M.: A course in the geometry of N dimensions. In: Dover Books on Mathematics. Dover Publications (2004). https://books.google.com.br/books?id=_dFJ6pSzRLkC
18. Kuhn, H.W.: The Hungarian method for the assignment problem. Naval Res. Logistics Q. **2**(1–2), 83–97 (1955). https://doi.org/10.1002/nav.3800020109. https://onlinelibrary.wiley.com/doi/abs/10.1002/nav.3800020109
19. Landgrebe, D.: Multispectral data analysis: a signal theory perspective, January 1998
20. Landgrebe, D.A.: Information extraction principles and methods for multispectral and hyperspectral image data (1998)
21. von Luxburg, U.: A tutorial on spectral clustering. Stat. Comput. **17**, 395–416 (2007)
22. Marini, F., Amigo, J.M.: Chapter 2.4 - Unsupervised exploration of hyperspectral and multispectral images. In: Amigo, J.M. (ed.) Hyperspectral Imaging, Data Handling in Science and Technology, vol. 32, pp. 93–114. Elsevier, Amsterdam (2020). https://doi.org/10.1016/B978-0-444-63977-6.00006-7. http://www.sciencedirect.com/science/article/pii/B9780444639776000067
23. Miranda, G.F., Thomaz, C.E., Giraldi, G.A.: Geometric data analysis based on manifold learning with applications for image understanding. In: 2017 30th SIBGRAPI Conference on Graphics, Patterns and Images Tutorials (SIBGRAPI-T), pp. 42–62, October 2017. https://doi.org/10.1109/SIBGRAPI-T.2017.9
24. Murphy, J.M., Maggioni, M.: Unsupervised clustering and active learning of hyperspectral images with nonlinear diffusion. IEEE Trans. Geosci. Remote Sens. **57**(3), 1829–1845 (2018)
25. Myasnikov, E.: Evaluation of nonlinear dimensionality reduction techniques for classification of hyperspectral images, July 2018
26. Nemenyi, P.: Distribution-Free Multiple Comparisons. Princeton University (1963). https://books.google.com.br/books?id=nhDMtgAACAAJ
27. Pal, M., Foody, G.M.: Feature selection for classification of hyperspectral data by SVM. IEEE Trans. Geosci. Remote Sens. **48**(5), 2297–2307 (2010)

28. Rand, W.M.: Objective criteria for the evaluation of clustering methods. J. Am. Stat. Assoc. **66**(336), 846–850 (1971). https://doi.org/10.1080/01621459. 1971.10482356.   https://www.tandfonline.com/doi/abs/10.1080/01621459.1971. 10482356

29. Roweis, S., Saul, L.: Nonlinear dimensionality reduction by locally linear embedding. Science **290**, 2323–2326 (2000)

30. Saul, L., Roweis, S.: Think globally, fit locally: unsupervised learning of low dimensional manifolds. J. Mach. Learn. Res. **4**, 119–155 (2003)

31. Scott, D.: Multivariate Density Estimation: Theory, Practice, and Visualization. A Wiley-Interscience Publication, Wiley (1992). https://books.google.com.br/books? id=7crCUS_F2ocC

32. Shannon, C.E.: A mathematical theory of communication. Bell Syst. Tech. J. **27**(3), 379–423 (1948). https://doi.org/10.1002/j.1538-7305.1948.tb01338.x. https://ieeexplore.ieee.org/document/6773024/

33. Tenenbaum, J.B., de Silva, V., Langford, J.C.: A global geometric framework for nonlinear dimensionality reduction. Science **290**, 2319–2323 (2000)

34. Zhang, L., Zhang, L., Tao, D., Huang, X.: On combining multiple features for hyperspectral remote sensing image classification. IEEE Trans. Geosci. Remote Sens. **50**(3), 879–893 (2012). https://doi.org/10.1109/TGRS.2011.2162339

Factors Affecting the Cost to Accuracy Balance for Real-Time Video-Based Action Recognition

Divina Govender[ORCID] and Jules-Raymond Tapamo[✉][ORCID]

School of Engineering, University of KwaZulu-Natal, Durban 4041, South Africa
215023704@stu.ukzn.ac.za, tapamoj@ukzn.ac.za

Abstract. For successful real-time action recognition in videos, the compromise between computational cost and accuracy must be carefully considered. To explore this balance, we focus on the popular Bag-of-Words (BoW) framework. Although computationally efficient, the BoW has weak classification power. Thus, many variants have been developed. These variants aim to increase classification power whilst maintaining computational efficiency; achieving the ideal cost-to-accuracy balance. Four factors affecting the computational cost vs accuracy balance were identified: 'Sampling' strategy, 'Optical Flow' algorithm, 'Saliency' of extracted features and overall algorithm 'Flexibility'. The practical effects of these factors were experimentally evaluated using the Dense Trajectories feature framework and the KTH and HMDB51 (a reduced version) datasets. The 'Saliency' of extracted information is found to be the most vital factor - spending computational resources to process large amounts of non-salient information can decrease accuracy.

Keywords: Bag of Words · Action recognition · Real-time computing · Computational cost · Accuracy

1 Introduction

The aim of action recognition is to autonomously classify what is being done by observable agents in a scene. Video-based action recognition is one of the most challenging problems in computer vision for several reasons: significant intra-class variations occur due to illumination, viewpoints, scale and motion speed changes and there is no precise definition of the temporal extent of an action. The high dimension and poor quality of video data adds to the challenge of developing robust and efficient recognition algorithms [18].

Video-based action recognition has a variety of useful applications such as autonomous surveillance, intelligent video surveillance, content-based video retrieval, human-robot interaction etc. Being able to achieve these tasks in real-time could improve everyday life for the average citizen and contribute to the 4th industrial revolution.

© Springer Nature Switzerland AG 2020
O. Gervasi et al. (Eds.): ICCSA 2020, LNCS 12249, pp. 807–818, 2020.
https://doi.org/10.1007/978-3-030-58799-4_58

For real-time action recognition: Shi *et al.* [22] replaced dense sampling with random sampling, thereby reducing the number of sampled patches to process and increasing efficiency. In a further work, Shi *et al.* [23] increased accuracy by increasing the sampling density. This was accomplished by using a Local Part Model and performing sampling at lower spatial resolutions. Yeffet *et al.* [30] extended Local Binary Patterns to the spatio-temporal domain (forming Local Trinary Patterns) to efficiently encode extracted information for action recognition. Zhang *et al.* [32] replaced optical flow with Motion Vectors to increase efficiency. To boost performance, the knowledge learned with optical flow CNN were transferred to motion vectors for effective real-time action recognition.

Feature extraction is the first step of an action recognition task. For video-based action recognition, popular features to extract include Dense Trajectories [26–28], Space-Time Interest Points (STIPs) [10] and Scale Invariant Feature Transforms (SIFT) [6]. Various approaches use the Bag of Words (BoW) framework to represent extracted features in a compact manner for classification [18]. Although lacking classification power, the BoW is popular due to its simplicity, flexibility and computational efficiency. Many variants have been developed; aiming to increase its classification power whilst leveraging its simplicity and computational efficiency [5,7,13,18]. This highlights the computational cost vs accuracy trade-off. This trade-off must be carefully considered to achieve real-time video-based action recognition.

This paper explores the computational cost vs accuracy balance for video-based action recognition by analyzing the BoW and its variants. We identify factors to consider when trying develop and/or adjust existing algorithms for real-time action recognition of videos. The Dense Trajectories feature framework and the KTH and a reduced HMDB51 datasets were used to observe the practical effects of the identified factors on the cost vs accuracy trade-off. Note that the effects of these factors on deep learning features are not explored. Focus is placed on identifying the fundamental mechanisms that affect the cost-vs-accuracy balance in the feature extraction process.

The rest of this paper is organized as follows. Section 2 reviews the BoW and its variants to identify factors that affect the computational cost vs accuracy balance. Section 3 covers the experimental evaluation of these factors. Section 4 concludes the paper and outlines future work directions.

2   Method and Materials

2.1   Bag of Words

Although algorithms such as Convolutional Neural Networks (CNNs) [14] and Fisher Vectors (FVs) [20] have higher classification power, BoW is favored due to its flexibility, simplicity, compact feature representation and computational efficiency [31]. This is useful in action recognition tasks which involves large sets of extracted features [27]. Additionally, a fast, compact feature representation algorithm makes the BoW framework favorable for real-time applications.

The general pipeline for BoW for video-based classification is summarized in Fig. 1 [18].

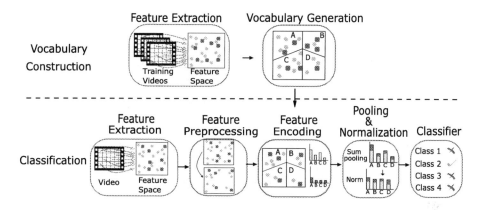

Fig. 1. The pipeline for video-based action recognition using the BoW Framework

Standard Scale Invariant BoW Model. This section expands on the mathematical description of the standard bag-of-words pipeline [8].

Features are extracted from some image or video frame via multi-scale sampling. For a given image patch B, the extracted features are given by:

$$F(B) = \{\mathbf{f}_i^s | i \in \{1, ..., N\}, s \in \{1, ..., M\}\}, \tag{1}$$

where i indexes the N feature sites defined by the sampling grid in the patch B. s indexes the M scales considered at each feature site.

The histogram $h(B)$ representing a given patch B as per the BoW framework is given by:

$$h(B) \propto \sum_{i=1}^{N} \sum_{s=1}^{M} c(f_i^s) \tag{2}$$

where c is some coding scheme that maps the input feature space to the representation space.

With the standard BoW, the Euclidean distance between each extracted feature vector $\mathbf{f}_i^s \in F(B)$ and a visual word $\mathbf{w}_k, k \in \{1, ..., q\}$ in the visual vocabulary $W = \{\mathbf{w}_1, ..., \mathbf{w}_q\}$ is computed. The feature vectors are then matched to the nearest visual word (nearest neighbor assignment). The index of the visual word assigned to an extracted feature vector $\mathbf{f}_i^s$ is given by:

$$\omega_i^s = \operatorname*{argmin}_{k \in \{1, ..., q\}} d(\mathbf{f}_i^s, \mathbf{w}_k) \tag{3}$$

where $d(\mathbf{a}, \mathbf{b})$ is the Euclidean distance between $\mathbf{a}$ and $\mathbf{b}$ The coding function for the standard BoW framework is summarized as

$$c_{BOW}(\mathbf{f}_i^s) = \mathbf{e}(\omega_i^s) \tag{4}$$

where $\mathbf{e}(i)$ is a q-dimension vector with only one non-zero element at index i which is equal to 1. The index i corresponds to the assigned codeword for a given extracted feature vector $\mathbf{f}_i^s$.

2.2   Analysis of BoW

The standard BoW representation discards all large-scale spatial information; resulting in compact and efficient feature representation [16]. However, the exclusion of this information reduces the classification power of the framework. Shi *et al.* [20] showed that FVs outperform BoW representations as they offer a more complete representation of input data. Similarly, Loussaief *et al.* [14] found that CNNs outperform the BoW framework due to their superior ability to extract features that contain relevant information from the input data. However, FVs and CNNs are dense and high-dimensional; making them computationally expensive. The following can be concluded:

Remark 1. The more information extracted from input data, the higher the classification power of the algorithm and the higher the computational cost. Thus, a balance between classification accuracy and computation necessitates an implementable algorithm for action recognition.

2.3   Factors Affecting the Cost vs Accuracy Balance

The thematic approach of BoW variants involves encoding additional information about the input data into the final BoW representation; increasing the classification power of the framework whilst leveraging the efficiency of the standard model. This is typically done by combining BoW with costly but powerful algorithms (e.g. FV [20], CNNs [14], dense sampling strategies [2,27,28]).

This section identifies the factors affecting the computational cost vs accuracy paradigm stated in Remark 1. These factors are often addressed or manipulated by BoW variants in an attempt to increase accuracy with low additional computational costs.

Sampling. Through sampling, information - in the form of features - can be extracted from an image or video frame. The number of patches sampled directly correlates to the algorithm's performance [2]; this complies with Remark 1. The ideal sampler focuses sampling on information rich patches; discarding insignificant patches to obtain more information at a lower computational cost [15].

Denser sampling yields better accuracy however random sampling provides the most favorable accuracy to cost trade-off [6,22]. Ideally, each pixel should be sampled across all video frames, however this will result in large spatial complexity. The sampling strategy must be chosen such that spatial complexity is reduced whilst keeping image information. Factors to consider when finding the most suitable sampling strategy include: sampling density, step-size, sample patch size, sampling rate and the number of video frames to sample.

Optical Flow. Optical flow is used to capture temporal information of videos. It is the most time consuming process of video-classification algorithms [27]. It can be replaced with less computationally expensive methods such Motion Vectors [32] or space-time feature extractors. However, this results in a decreased accuracy. Knowledge transfer techniques can be used to increase the accuracy of alternative methods [32]; this agrees with Remark 1.

Saliency. A possible approach to achieve a favorable cost to accuracy balance is to only extract important(salient) information from input data. In this manner, computational resources are not wasted on extracting and processing unimportant information. However, determining which areas/features are important requires additional computation and can result in an overall decrease of computational efficiency [12] or a negligible increase [29].

Flexibility. The flexibility of the standard BoW algorithm is often compromised when combined with more powerful and complex algorithms. This is observed with the Convolutional Bag-of-Features (CBoF) [17] which combined CNNs with the BoW framework. In addition to increasing computational demands, CNNs are slow to train and require large amounts of training data which is not readily available. Furthermore, the CNN must be trained with data relevant to the intended application for effective performance. As a result, the algorithm can only be used for the intended application; compromising flexibility.

Although flexibility does not directly affect the cost vs accuracy balance, it allows for further extension and easy integration into a variety of applications; increasing chances of improving the algorithm to obtain the ideal cost vs accuracy balance.

3   Experimentation

In this section, the effects of each factor presented in Sect. 2.3 are investigated using the popular Dense Trajectories feature framework [27]. Experimental parameters are set as per [27].

3.1   Experimental Setup

PC Specifications. Experimentation was conducted on a PC with the following specifications: Intel Core i5 4th Gen., 1.7 GHZ, 8G RAM. A single CPU core was used for computation.

Datasets. The KTH[1] [21] and reduced HMDB51[2] [9] action data-sets were used for experimental evaluations.

---

[1] http://www.nada.kth.se/cvap/actions/.

[2] http://serre-lab.clps.brown.edu/resource/hmdb-a-large-human-motion-database/.

The KTH dataset contains six human action classes: walking, jogging, running, boxing, waving and clapping. Each action is performed by 25 subjects in four different environments (outdoors, outdoors with scale variation, outdoors with different clothes and indoors). Since the background is homogeneous and static in most sequences, the dataset serves as a minimum baseline for evaluation of the action recognition algorithm. As found in [21], samples are divided into testing and training sets based on the subjects. The testing set consists of subjects 2, 3, 5, 6, 7, 8, 9, 10, and 22 (9 subjects total) and the training set is made up of the remaining 16 subjects. The average accuracy is taken over all classes.

The HMDB51 data-set contains 51 human action classes and is collated from a variety of sources. For our experimentation, a reduced dataset of 11 randomly selected action classes was used: brush hair, cartwheel, catch, chew, clap, climb stairs, smile, talk, throw, turn, wave. This is a challenging dataset that presents additional problems (e.g. camera motion, occlusion and background activities) to overcome. It serves as an evaluation of how robust the action recognition algorithm is to the aforementioned challenges. As per the original setup [9], each action class is organized into 70 videos for training and 30 videos for testing. Performance is measured by the average accuracy over all classes.

Dense Trajectories. Dense Trajectories involve densely sampling points in the spatial domain and tracking those points across the temporal domain; forming a trajectory. Five descriptors are extracted: trajectory shape, HOG, HOF and Motion Boundary Histograms in the horizontal (MBHx) and vertical (MBHy) planes [27].

Points are densely sampled with a sampling step-size of $W = 5$ on each spatial-scale separately. There are a maximum of 8 spatial scales; the number of spatial scales is dependent on the resolution of the video. Spatial scales are separated by a factor of $1/\sqrt{2}$ [27]. Sampled points in homogeneous image areas can not be tracked and are therefore removed based on the criterion presented by [24]. The threshold, T, is defined as:

$$T = 0.001 \times \max_{i \in I} \min(\lambda_i^1, \lambda_i^2) \tag{5}$$

where $(\lambda_i^1, \lambda_i^2)$ are the eigenvalues of the ith sampled point in the image I. As per Wang et al. [27], a value of 0.001 was used since it presented a good compromise between saliency and density.

Points are re-sampled and compared to the threshold T (see Eq. 5) every R frames. R is the refresh/frame sub-sampling rate. Sampled points are tracked separately on each spatial scale using a dense optical flow algorithm [4] to form trajectories. This is defined as:

$$P_{t+1} = (x_{t+1}, y_{t+1}) = (x_t, y_t) + (M * \omega_t)|_{(x_t, y_t)} \tag{6}$$

where P_i is a point in frame I_i. The dense optical flow field, ω_t, for each frame I_t is found with respect to the next frame I_{t+1}. A 3×3 median filter kernel M is applied to the optical flow field.

The same optical flow implementation[3] as [27] was used. From tracked points, five descriptors are extracted: HOG, HOF, MBH in the x (MBHx) and y (MBHy) planes and the trajectory shape, TS, defined as:

$$TS = \frac{\Delta P_t, ..., \Delta P_{t+L+1}}{\sum_{j=t}^{t+L-1} ||\Delta P_j||} \tag{7}$$

where L is the length of the trajectory. This is set to 15 frames as per [27]

Bag of Words. Following [27], vocabulary of 4000 words is separately constructed for each descriptor. K-means clustering is used to cluster a subset of 100 000 randomly selected training features.

The feature descriptors are assigned to the "visual word" with the closest Euclidean distance. The count of visual word occurrences is represented in a histogram. Structure is added through spatio-temporal pyramids. Six different pyramids are used (see Fig. 2). A Bag-of-Words is constructed for each cell of a given pyramid. Thereafter, a global representation of the pyramid is found by summing the BoW histogram for each cell and normalizing by the Root SIFT Norm [1], defined as:

$$p_k = \sqrt{\frac{p_k}{\sum_{k=1}^{q} |p_k|}} \tag{8}$$

Each spatio-temporal pyramid-descriptor pair forms a separate channel; there are 30 channels in total (6 pyramids x 5 descriptors). A multi-class, multi-channel

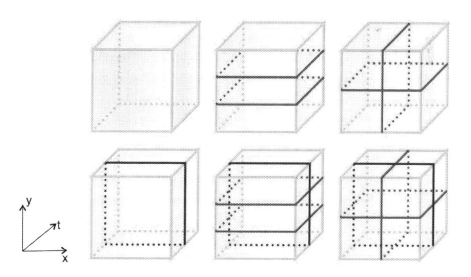

Fig. 2. The spatio-temporal grids adapted from [27]. Top row from left to right: $h1 \times v1 \times t1$, $h3 \times v1 \times t1$, $h2 \times v2 \times t1$. Bottom row from left to right: $h1 \times v1 \times t2$, $h3 \times v1 \times t2$, $h2 \times v2 \times t2$

---

[3] https://docs.opencv.org/3.4/d4/dee/tutorial_optical_flow.html.

non-linear SVM with an RBF-χ^2 kernel [21] is used for classification. For multi-class classification, a one-against-rest approach is used. For multiple channels, the kernel is defined as [33]:

$$K(x_i, x_j) = exp(-\sum_c \frac{1}{A_c} D(x_i^c, x_j^c)) \tag{9}$$

where $D(x_i^c, x_j^c)$ is the χ^2 distance between each training video x_i and x_j in each channel c. A_c is the average of the χ^2 distances between training samples in channel c.

Evaluation of Factors. To evaluate 'Saliency', we observe the computational cost and accuracy achieved by each descriptor on each dataset. In this manner, we can determine which information is most important. Computational cost was measured by the number of frames computed per second - the higher this value, the lower the computational cost. Results are summarized in Table 1.

To evaluate 'Sampling', we observe the accuracy and computational cost of the algorithm at different refresh rates (R). A refresh rate of 1 frame was used as a baseline for comparison. Following Remark 1, a refresh rate of R = 1 should yield the highest accuracy since the most information is obtained if the video is re-sampled every frame. Uijlings *et al.* [25] found that a refresh rate of R = 6 frames has the best cost-to-accuracy trade-off for dense computation of HOG, HOF and MBH based on the UCF50 data-set [19]. Thus, performance with R = 1 was compared to performance with R = 6. Results are summarized in Table 2.

To evaluate 'Optical Flow', we consider its performance at each scale to determine if the information extraction abilities of this task is worth the computational cost (it is responsible for 52% of the computation time [27]).

3.2   Experimental Results

Table 1. The effects of trajectory shape [27], HOG [11], HOF [11] and MBH [3] descriptors on the cost vs accuracy balance

Dataset		Trajectory shape	HOG	HOF	MBH
KTH	Computation time (FPS)	8.16	1.11	1.71	0.61
	Accuracy as per [27] (%)	89.8	87.0	93.3	95.0
HMDB51	Computational cost (FPS)	2.48	0.11	0.14	0.07
	Accuracy as per [27] (%)	28.0	27.9	31.5	43.2

Descriptors. The MBH descriptor has the best overall performance and the highest computational cost; agreeing with Remark 1. For the HMDB51 dataset, MBH outperforms all other descriptors by over 10%. This is due to the fact that the HMDB51 data samples include camera motion. Since MBH is computed using the derivative of optical flow, it is robust to camera motion unlike the other descriptors. However, in static scenery (KTH dataset), MBH only outperforms HOF by 1.7%. Since HOF is more computationally efficient than MBH, it can be concluded that MBH is unsuitable for real-time action recognition of static camera data (e.g. CCTV footage). Overall, trajectory shape has the best cost-to-accuracy trade-off.

Theoretically, HOG should have the lowest computational cost since it is the only descriptor that does not rely on optical flow computation. However, HOG has a longer computation time than HOF in experiments. This is due to the fact that optical flow is part of the tracking algorithm therefore less additional computation is required for HOF compared to HOG; HOG requires additional calculation of the gradient between pixel points.

Table 2. Evaluation of the effect of the refresh rate on the cost vs accuracy balance for KTH and HMDB51 data-sets using the Dense Trajectory features

Dataset		R = 1 frame	R = 6 frames
KTH	Computation time (FPS)	0.25	0.3
	Accuracy (%)	94.0	91.0
HMDB51 (reduced)	Computation time (FPS)	0.03	0.03
	Accuracy (%)	78.48	81.21

Refresh Rate. The accuracy obtained with the KTH data-set for R = 1 (94.0%) is lower than expected (94.2% [27]). This might be due to the fact that Dense Trajectory computation was recreated in Python 3.7 with the latest version of OpenCV, whereas the original code[4] was created in C++ with OpenCV-2.4.2. As a result, the recreated code has a different handling of the algorithm; explaining the difference in accuracy rates. As expected, there is a decrease in accuracy (3%) when sub-sampling every 6 frames as opposed to every 1 frame. This results in a 20% (0.05 FPS) increase in computation speed.

Comparatively, there is a negligible difference in computational cost for the more challenging HMDB51 dataset (for R = 6 frames, computation time was 0.001 FPS faster). Most notably, accuracy increases by 2.73% when sub-sampling every R = 6 frames. This contradicts theoretical predictions as a decrease in accuracy was expected.

The contradicting results obtained with the KTH and HMDB51 dataset show that the effects of 'Sampling' on the cost vs accuracy balance is data dependent.

---

[4] http://lear.inrialpes.fr/people/wang/dense_trajectories.

Unlike the KTH dataset, the HMDB51 dataset includes unimportant information such as camera motion, occlusion and background clutter. Sub-sampling every frame increases chances that this unimportant information will be captured, which can decrease accuracy. This highlights that 'Saliency' of extracted features is the most vital factor to consider. Based on these results, Remark 1 is adjusted:

Remark 2. It can be concluded that the more **salient** information extracted from input data, the higher the classification power of the algorithm and the higher the computational cost. Thus, a balance between classification accuracy and computational demands of an algorithm must be found for implementable action recognition.

Optical Flow. It can be seen that optical flow becomes noisier at smaller spatial scales (Fig. 3) which adversely affects performance. Removing multi-scale sampling would reduce the overall computational cost of the Dense Trajectories algorithm and result in better performance of the optical flow. However, Khan *et al.* [7,8] proved that scale provides vital information; a BoW incorporating extracted scale information into final representations performs better. Since optical flow is computationally expensive and unreliable at smaller scales, discarding its calculation has a potential to be suitable for real-time applications.

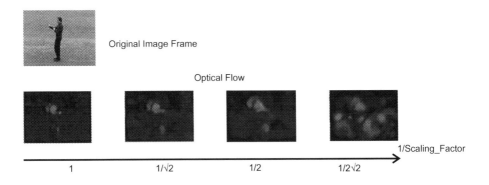

Fig. 3. The computed optical flow for each spatial scale

4 Conclusions and Future Work

For video-based action recognition, experimental evaluation found 'Saliency' of extracted features to have the most notable effect on the cost vs accuracy balance. Capturing unimportant information (e.g. occlusion, background activities etc.) can decrease performance accuracy and waste computational resources. The saliency of extracted features is dependent on the 'Sampling' algorithm. Future work includes exploring the effects of the identified factors on deep learning architectures. This will help move the research space closer to creating algorithms that are simple yet powerful enough for practical realization of real-time video-based action recognition.

References

1. Arandjelović, R., Zisserman, A.: Three things everyone should know to improve object retrieval. In: 2012 IEEE Conference on Computer Vision and Pattern Recognition, pp. 2911–2918. IEEE (2012)
2. Dai, P., Liu, W., Wang, L., Li, C., Xie, Y.: Bag of features with dense sampling for visual tracking. J. Comput. Inf. Syst. **9**(20), 8307–8315 (2013)
3. Dalal, N., Triggs, B., Schmid, C.: Human detection using oriented histograms of flow and appearance. In: Leonardis, A., Bischof, H., Pinz, A. (eds.) ECCV 2006. LNCS, vol. 3952, pp. 428–441. Springer, Heidelberg (2006). https://doi.org/10.1007/11744047_33
4. Farnebäck, G.: Two-frame motion estimation based on polynomial expansion. In: Bigun, J., Gustavsson, T. (eds.) SCIA 2003. LNCS, vol. 2749, pp. 363–370. Springer, Heidelberg (2003). https://doi.org/10.1007/3-540-45103-X_50
5. Han, S., Mao, H., Dally, W.J.: Deep compression: compressing deep neural networks with pruning, trained quantization and Huffman coding. In: International Conference on Learning Representations (2016)
6. Hu, J., Xia, G.S., Hu, F., Zhang, L.: Dense vs sparse: a comparative study of sampling analysis in scene classification of high-resolution remote sensing imagery. arXiv preprint arXiv:01097 (2015)
7. Khan, F.S., van de Weijer, J., Anwer, R.M., Bagdanov, A.D., Felsberg, M., Laaksonen, J.: Scale coding bag of deep features for human attribute and action recognition. Mach. Vis. Appl. **29**(1), 55–71 (2017). https://doi.org/10.1007/s00138-017-0871-1
8. Khan, F.S., Weijer, J.v.d., Bagdanov, A.D., Felsber, M.: Scale coding bag-of-words for action recognition. In: 22nd International Conference on Pattern Recognition (2014)
9. Kuehne, H., Jhuang, H., Garrote, E., Poggio, T., Serre, T.: HMDB: a large video database for human motion recognition. In: IEEE International Conference on Computer Vision, pp. 2556–2563 (2011)
10. Laptev, I.: On space-time interest points. Int. J. Comput. Vis. **64**(2–3), 107–123 (2005)
11. Laptev, I., Marszałek, M., Schmid, C., Rozenfeld, B.: Learning realistic human actions from movies. In: CVPR 2008 IEEE Conference on Computer Vision and Pattern Recognition, pp. 1–8. IEEE Computer Society (2008)
12. Li, Q., Cheng, H., Zhou, Y., Huo, G.: Human action recognition using improved salient dense trajectories. Comput. Intell. Neurosci. **2016**, 1–11 (2016)
13. Li, T., Mei, T., Kweon, I.S., Hua, X.S.: Contextual bag-of-words for visual categorization. IEEE Trans. Circuits Syst. Video Technol. **21**(4), 381–392 (2011)
14. Loussaief, S., Abdelkrim, A.: Deep learning vs. bag of features in machine learning for image classification. In: International Conference on Advanced Systems and Electric Technologies, pp. 6–10 (2018)
15. Nowak, E., Jurie, F., Triggs, B.: Sampling strategies for bag-of-features image classification. In: Leonardis, A., Bischof, H., Pinz, A. (eds.) ECCV 2006. LNCS, vol. 3954, pp. 490–503. Springer, Heidelberg (2006). https://doi.org/10.1007/11744085_38
16. O'Hara, S., Draper, B.A.: Introduction to the bag of features paradigm for image classification and retrieval. arXiv preprint arXiv:1101.3354 (2011)
17. Passalis, N., Tefas, A.: Learning bag-of-features pooling for deep convolutional neural networks. In: Proceedings of the IEEE International Conference on Computer Vision, pp. 5755–5763 (2017)

18. Peng, X., Wang, L., Wang, X., Qiao, Y.: Bag of visual words and fusion methods for action recognition: comprehensive study and good practice. Comput. Vis. Image Underst. **150**, 109–125 (2016)

19. Reddy, K., Shah, M.: Recognizing 50 human action categories of web videos. Mach. Vis. Appl. **24**, 971–981 (2013). https://doi.org/10.1007/s00138-012-0450-4

20. Sanchez, J., Perronnin, F., Mensink, T., Verbeek, J.: Image classification with the fisher vector: theory and practice. Int. J. Comput. Vis. **105**(3), 222–245 (2013)

21. Schüldt, C., Laptev, I., Caputo, B.: Recognizing human actions: a local SVM approach. In: International Conference on Pattern Recognition (2004)

22. Shi, F., Petriu, E., Laganiere, R.: Sampling strategies for real-time action recognition. In: Proceedings of the IEEE Conference on Computer Vision and Pattern Recognition, pp. 2595–2602 (2013)

23. Shi, F., Petriu, E.M., Cordeiro, A.: Human action recognition from local part model. In: IEEE International Workshop on Haptic Audio Visual Environments and Games, pp. 35–38. IEEE (2011)

24. Shi, J., Tomasi, C.: Good features to track. In: IEEE Conference on Computer Vision and Pattern Recognition (1994)

25. Uijlings, J., Duta, I.C., Sangineto, E., Sebe, N.: Video classification with densely extracted HOG/HOF/MBH features: an evaluation of the accuracy/computational efficiency trade-off. Int. J. Multimedia Inf. Retrieval **4**(1), 33–44 (2015)

26. Wang, H., Kläser, A., Schmid, C., Liu, C.L.: Action recognition by dense trajectories. In: CVPR 2011 - IEEE Conference on Computer Vision (2011)

27. Wang, H., Kläser, A., Schmid, C., Liu, C.L.: Dense trajectories and motion boundary descriptors for action recognition. Int. J. Comput. Vis. **103**(1), 60–79 (2013). https://doi.org/10.1007/s11263-012-0594-8

28. Wang, H., Schmid, C.: Action recognition with improved trajectories. In: Proceedings of the IEEE International Conference on Computer Vision, pp. 3551–3558 (2013)

29. Xu, Z., et al.: Action recognition by saliency-based dense sampling. Neurocomputing **236**, 82–92 (2017)

30. Yeffet, L., Wolf, L.: Local trinary patterns for human action recognition. In: IEEE 12th International Conference on Computer Vision, pp. 492–497. IEEE (2009)

31. Zeng, F., Ji, Y., Levine, M.D.: Contextual bag-of-words for robust visual tracking. IEEE Trans. Image Process. **27**(3), 1433–1447 (2018)

32. Zhang, B., Wang, L., Wang, Z., Qiao, Y., Wang, H.: Real-time action recognition with enhanced motion vector CNNs. In: Proceedings of the IEEE Conference on Computer Vision and Pattern Recognition, pp. 2718–2726 (2016)

33. Zhang, J., Marszałek, M., Lazebnik, S., Schmid, C.: Local features and kernels for classification of texture and object categories: a comprehensive study. Int. J. Comput. Vis. **73**(2), 213–238 (2007). https://doi.org/10.1007/s11263-006-9794-4

Innovation, Technology and User Experience in Museums: Insights from Scientific Literature

David Ovallos-Gazabon[1], Farid Meléndez-Pertuz[2(✉)],
Carlos Collazos-Morales[3], Ronald Zamora-Musa[4],
César A. Cardenas[3], and Ramón E.R. González[5]

[1] Universidad Simon Bolivar, Barranquilla, Colombia
[2] Departamento de Ciencias de la Computación y Electrónica,
Universidad de la Costa, Barranquilla, Colombia
fmelendel@cuc.edu.co
[3] Vicerrectoría de Investigaciones, Universidad Manuela Beltrán,
Bogotá, Colombia
[4] Universidad Cooperativa de Colombia, Bucaramanga, Colombia
[5] Departamento de Física, Universidad Federal de Pernambuco, Recife, Brazil

Abstract. Museums play an important role in preserving the heritage and cultural legacy of humanity, however, one of their main weaknesses in regards the user is their static nature. At present, and in the face of the development of diverse technologies and the ease of access to information, museums have upgraded their implementation of technologies aimed at improving the user experience, trying more and more to access younger audiences with a sensitivity and natural capacity for the management of new technologies. This work identifies trends in the use of technological tools by museums worldwide and the effect of these on the user or visitor experience through a review of scientific literature. To complete the work, we performed a search of the publications in the Scopus® referencing database, and downloaded, processed, and visualized the data using the VOSviewer® tool. The main trends identified in this context of analysis are related to the role of museums with the development and improvement of the user experience; orientation to young audiences and innovation driven by the user through Interactive Systems, digital games, QR Codes, apps, augmented reality, virtual reality and gamification, among others. The objective of the implementation of new technologies in the context of museums is to satisfy the needs of contemporary communication, for all types of content and aimed at an increasingly digital audience, in order to ensure positive interaction and feedback from ideas with social and cultural changes.

Keywords: User experience · Museum · Literature review · Innovation · VOSviewer

© Springer Nature Switzerland AG 2020
O. Gervasi et al. (Eds.): ICCSA 2020, LNCS 12249, pp. 819–832, 2020.
https://doi.org/10.1007/978-3-030-58799-4_59

1 Introduction

Literature points out that innovation as an object of scientific interest is not a new phenomenon and that it had its beginnings since the emergence of humanity itself and has generated great changes in the world order [1]. Similarly, it is considered that organizations that are not able to face technological change and the rapid accumulation of new knowledge will have a lag, mainly in terms of productivity and competitiveness [2].

The implementation of technology in the field of museums became a challenge that has been tilting after a period of transition towards its correct employment, generating new, fascinating and innovative experiences that have managed to increase the interest of people to visit Museums [3–9], to achieve this, tools such as ICT (Information and Communication Technologies) have been used, through which techniques such as interactive or 3D systems were implemented [10–12], generating as a result that children and young people are attracted by this type of mediation.

With the advent of the digital age, these new technologies have been applied in museums to provide more pleasant and educational experiences, implementing interactive screens and digital games, this combination of artifacts allows a positive interaction of feedback of ideas with social and cultural changes [13–16].

Considering the above, it is possible to point out that the use of technology has been a fundamental element in the improvement of the museum experience by visitors [7, 8, 17, 18]. This paper presents a review of success stories in the international context and also a search for information in scientific publications to identify trends through tool VosViewer®.

Technology in Museums
New visitors to museums bring other types of challenges and demands and hope that these spaces will be increasingly creative and innovative in the creation and presentation of content that can encourage society to visit them [11]. However, although the evidence in the literature indicates that the use of new technologies increases the interest of people who visit museums through interactive systems, 3D systems, digital games and other technological means; It is pointed out that there must be a balance when using technologies of static and interactive approach to information as well as entertainment in learning in the museum. Table 1 presents a summary of different types of technological tools or instruments used in some museums of world relevance.

Table 1 shows that currently the way in which museum visits are developed has been expanded, becoming increasingly a cultural and didactic experience through non-contact visits, using technologies that revolutionize teaching/learning activities within museums, showing collections and materials in a creative way.

The objective of the implementation of new technologies in the context of museums is to meet the needs of contemporary communication, for a contemporary art produced by artists framed in these technologies and aimed at an increasingly digital audience. Ensuring a positive interaction of feedback of ideas with social and cultural changes. An important aspect in this context of analysis is that interaction between different aspects of culture is favored, for instance the possibility of real-time interaction between scientists, artists, designers and intellectuals with a wider population.

Table 1. Tools or technology in museums.

Type of tool	Museum	References
Interactive systems, 3D Systems, Digitals games, Dynamus	Sydney Museum Centro de Arte Contemporáneo de Málaga Museo de Artes de Berlín Museo Arqueológico Nacional de Madrid London Brithish Museum Museo de Bellas Artes Bilbao Petrie Museum Museo de Lisboa Penang State Museum Negara Museum	[19–26]
Codes QR, Applications for mobile devices	Museo de Muesca de España Bernasconi de Argentina Museo Art Nouveau y Art Deco Casa Lis de España Museo de Arte Moderno de Santander España Museo Irlandes de Arte Moderno (IMMA)	[3–28]
Touch experience and augmented reality	UNESCO Underground gallery of Yunnan anti-Japanese victory memorial hall Tate Britain Art Gallery	[29–34]
Artistic healing tools WeCurate	London Museum	[12, 35–38]
Augmented micro reality	Grant Museum of Zoology	[39]
Smart Glasses	Robotics Gallery at the MIT Museum	[16]
Augmented reality, holographic computing	The Royal Ontario Museum Kangmeiyuanchao Zhanzheng in Chinese (KMYC) memorial hall Changsha Museum	[24, 40–45]
Olfactory Experience	Macao Museum of Art Tate Britain Art Gallery Jason Bruges Studio United Visual Artists	[31, 46, 47]
Gamification, Virtual Museum, Twitter, Serious Games	Sagamihara City Museum The 30th anniversary of a famous half-marathon held annually in the United Kingdom called The Run	[4, 43, 48]

Bibliometric Mapping with VOSviewer

Two aspects of bibliometric mapping that can be distinguished are the construction of bibliometric maps and the graphical representation of such maps. VOSviewer is a program that we have developed for constructing and viewing bibliometric maps. The program offers a viewer that allows bibliometric maps to be examined in full detail. VOSviewer can display a map in various different ways, each emphasizing a different aspect of the map. It has functionality for zooming, scrolling, and searching, which facilitates the detailed examination of a map. The viewing capabilities of VOSviewer are especially useful for maps containing at least a moderately large number of items (e.g., at least 100 items). Most computer programs that are used for bibliometric mapping do not display such maps in a satisfactory way. To construct a map, VOSviewer uses the VOS mapping technique [49], where VOS stands for visualization of similarities. VOSviewer can display maps constructed using any suitable mapping technique. Hence, the program can be employed not only for displaying maps constructed using the VOS mapping technique but also for displaying maps constructed using techniques such as multidimensional scaling. VOSviewer runs on a large number of hardware and operating system platforms and can be started directly from the internet.

2   Methodology for the Identification of Trends

The scientific production registered in Scopus® about museums and innovation has been analyzed from 677 publications for the period 1959 to 2018. The methodology aims to identify trends in the global production of the field of analysis. For this, four phases are developed that allow a systematic review of the literature.

- *Phase 1. Definition of guiding questions:* The following were considered: what is the relationship between museums and innovation? Who are the main authors and institutions in this area?
- *Phase 2. Search in specialized database:* The Scopus® database was selected and a total of 677 records were obtained for the period 1959–2018. Scopus® compiles results from other bibliographic databases and independent scientific publications.
- *Phase 3. Download of bibliographic records:* Once the records were identified, they were downloaded using the tools offered by Scopus®. For this stage, the CSV format was used that facilitates its subsequent processing using Excel® 2016.
- *Phase 4. Consolidation and analysis of information:* Tools such as dynamic tables and macros were used in Excel® 2016 to generate the input data for graphing in VOSviewer®.

3   Results

A selection of results obtained in this work is presented. In this sense, the search equation that is used in the selected database is generated from the guiding questions identified, being as follows:

$$TITLE-ABS-KEY\,(innovation\,AND\,museum)$$

The first publications identified in Scopus® that relate the terms innovation and museums date from 1959, but it is from the end of the 90's that there is a significant and sustained growth in the number of publications on this subject. For the study period there is the contribution of authors with more than 5 publications such as María José Garrido, Carmen Camarero Izquierdo and Derek Walker.

The number of times a scientific article is cited in other studies often represents the key indicator in assessing the impact of authors in the field of science in which they are working [49]. In this regard, it is important to highlight the work of authors such as Wolfram Bürgard of the Department of Computer Science of the University Freiburg im Breisgau, Germany with a total of 442 citations of his 1999 article that describes the software architecture of an autonomous tourist guide robot and interactive for museums [50].

In recent years, technology has evolved on a large scale. Game engines have been developed and web platforms have reached their highest levels. This is why it is decided to use this degree of maturity to contribute significantly to the mix of cultural education and education through play. In this sense, the works with greater relevance are oriented to the use of technologies such as augmented reality, virtual reality, human computer interaction and digital storage [24, 25, 44, 48].

An analysis of the output of the keywords identified in the search allows identifying clusters or groups of terms that give light on the orientation of scientific production in this field. It is possible to identify five clusters highlighted with different colors. See Fig. 1.

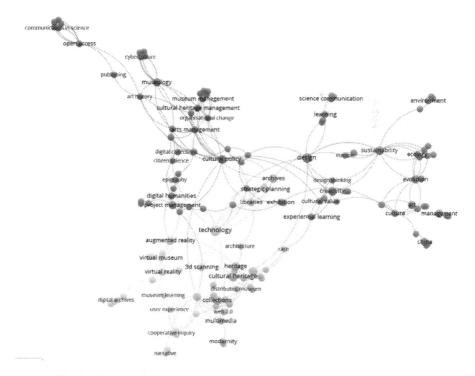

Fig. 1. Display of clusters generated in the text body. (Color figure online)

3.1   Cluster 1

Identified with the color red groups works around the creation of value in the learning process based on experiences through design thinking, encouraging creativity in the visitor [10, 35, 51–53]. Similarly, these works are related to the concept of sustainability through the use of technology in museums [54–56]. See Fig. 2.

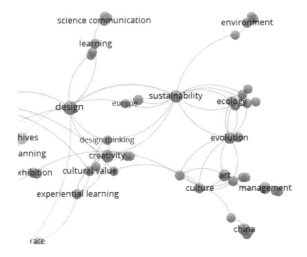

Fig. 2. Cluster 1 display (Color figure online)

3.2   Cluster 2

Cluster 2, identify with blue color, is formed by works on the advantages that technology generates in the area of museum management and cultural heritage [51, 53, 57] and its contribution to the creation of Science Communication, cyber culture and cultural heritage [58–60]. See Fig. 3.

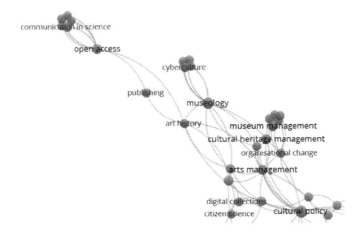

Fig. 3. Cluster 2 display. (Color figure online)

3.3   Clusters 3 and 4

Cluster 3, identified by the color green groups publications related to architecture for the storage and management of cultural heritage in museums [61–65]. For its part, cluster 4, identified with the color yellow, brings together works around the use of technological tools such as virtual reality, augmented reality, 3D scanning, cooperative consultation and digital archives, likewise their effect on narratives and user experience [66–69]. See Fig. 4.

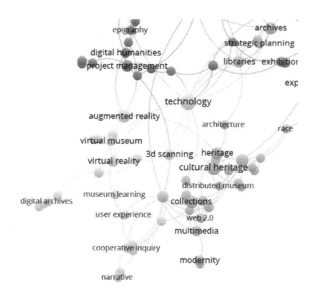

Fig. 4.   Clusters 3 and 4 display. (Color figure online)

3.4   Cluster 5

Finally, cluster 5 identified with the magenta color groups works that deal with aspects related to the creation of cultural identity and the role of new technology in museums, facilitating the creation of digital heritage and collaborative learning [24, 25, 30, 45, 67, 70–72]. See Fig. 5.

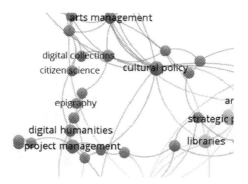

Fig. 5.   Cluster 5 display. (Color figure online)

3.5   Identification of Trends in the Literature

To identify trends in scientific publications for the body of text analyzed, we made use of a functionality of the overlay visualization tool of the VOSviewer®, which allows to identify by color key, the most recent terms that identify themes in scientific production. A description of the elements identified as a trend in the context of museums is presented below.

- *Branding:* Museums can facilitate the construction of a unique narrative using culture, heritage, innovation, creativity, technology, etc., to create symbolic value over a particular place, city, territory or cultural event. Beyond the place, brand channels, such as the media, disseminate images and sounds, creating and publicizing the brand narrative created by culture [73–76].
- *Library Development:* Innovative personnel and financing strategies will be particularly useful for organizations facing monetary and personnel shortages and highlight collaborative management practices [74, 77].
- *Archaeology:* The use of technological resources at the service of didactic interpretation for educational purposes (APPs, QR codes, VR and AR) that allow the development of virtual archeology (virtual reconstructions and recreations, augmented reality, etc.) through mobile devices. For this, different educational innovation and research projects have been taken into account, based on the visualization, manipulation, classification or construction of virtual objects of an archaeological nature, some of them in the environment of collaborative social networks, which allow creating, pointing out, affirm, argue, search, cite and justify archaeological investigations [78–81].
- *Development and improvement of User Experience:* Creating a pleasant experience in the museum, while providing an experience of entertainment and education, is one of the motivations for the use of technological tools in museums. QR codes, VR and AR type technologies, among others, have always had the attention of industry and users. Due to the influence of science and technology, the projects of new museums and remodeling or development projects of the current ones, go through the use of such tools to improve the user experience [64, 78–81].
- *Teenagers:* Digital technologies can make it easier for museums to promote and create immersive experiences for young audiences. Especially through the use of digital narratives, location-based games and game-based learning [79, 81].
- *Cooperative Inquiry:* Cooperative research (IC) offers important opportunities for the academic world to transform the teaching process with collaborative practices at different levels of education in a variety of disciplines. In the context of museums, it is actually used to develop co-design sessions to devise the interactive experience of the museum [79, 81].
- *User Driven Innovation:* User-driven innovation can generate ideas and contribute significantly to the profitability and success of a company. Users serve as extended endogenous innovation personnel, where behavioral documentation and direct feedback allow proactive changes in products, services and processes, impacting in competitiveness [56, 79, 81, 82].

4 Conclusions

Museums, as active actors in strengthening universal culture and facing their institutional responsibility towards society, have taken on the challenges of the digital era by adopting forms of communication based on highly interactive technological mediations and incorporating innovative interfaces that allow them to remain in the imaginary as an attractive option for new generations.

The incorporation of technologies such as 3D Systems, QR Codes, Smart Glasses, Augmented Reality, and Holographic Computing, among others, have become mechanisms to provide new codes, narrative forms and mediations that strengthen the concepts of inclusion, immersion and participation, which transcend the traditional way of visiting the museum.

Without a doubt, the incorporation of new technologies in the work of museums opens up a wide range of areas of intervention that imply changes in the dynamics of interaction, in institutional processes and, above all, in the user experience.

In this order of ideas, one of the most significant aspects is the generation of knowledge, collaborative learning and the strengthening of culture based on experimentation; Other important aspects to consider are the generation of digital and multimedia content that support the conservation of cultural property and the massification of the work of the museum, overcoming geographical and temporal barriers through the use of technological convergence. Finally, the implementation of technological innovations fosters the creation of new business models based on new forms of interaction, the generation of added values and the monetization of emerging alternatives for assets accessibility.

References

1. Fagerberg, J.: Innovation: A Guide to the Literature, pp. 1–22 (2003)
2. Gómez-Charris, Y., Ovallos, D., Cortabarría, L.: Definición de un perfil que maximice la capacidad innovadora y competitiva en las organizaciones. Caso de aplicación: Sector Muebles Atlántico–Colombia. Rev. Espac. **38**(04) (2017)
3. Pérez-Sanagustín, M., Parra, D., Verdugo, R., García-Galleguillos, G., Nussbaum, M.: Using QR codes to increase user engagement in museum-like spaces. Comput. Hum. Behav. **60**, 73–85 (2016)
4. Clarke, R., et al.: MyRun: balancing design for reflection, recounting and openness in a museum-based participatory platform. In: British HCI 2015, pp. 212–221 (2015)
5. Correia, N., Mota, T., Nóbrega, R., Silva, L., Almeida, A.: A multi-touch tabletop for robust multimedia interaction in museums. In: Proceedings of the International Conference on Interactive Tabletops Surfaces, pp. 117–120 (2010)
6. Martella, C., Miraglia, A., Frost, J., Cattani, M., van Steen, M.: Visualizing, clustering, and predicting the behavior of museum visitors. Pervasive Mob. Comput. **38**, 430–443 (2017)
7. Coelho, A., Costa, L.M.: The integration of augmented reality and the concept of sticker album collection for informal learning in museums. In: Beck, D., et al. (eds.) iLRN 2017. CCIS, vol. 725, pp. 107–115. Springer, Cham (2017). https://doi.org/10.1007/978-3-319-60633-0_9

8. Sanchez, E., Pierroux, P.: Gamifying the museum: teaching for games based 'informal' learning. In: Proceedings of the European Conference on Games-Based Learning, Janua, pp. 471–479 (2015)

9. Tian, F., Gatzidis, C., El Rhalibi, A., Tang, W., Charles, F. (eds.): Edutainment 2017. LNCS, vol. 10345. Springer, Cham (2017). https://doi.org/10.1007/978-3-319-65849-0

10. Kiourt, C., Koutsoudis, A., Pavlidis, G.: DynaMus: a fully dynamic 3D virtual museum framework. J. Cult. Herit. **22**, 984–991 (2016)

11. Ahmad, S., Yusof, W.Z.M., Taib, M.Z.M.: Museum learning: using research as best practice in creating future museum exhibition. Procedia - Soc. Behav. Sci. **105**, 370–382 (2013)

12. Hazelden, K., Yee-King, M.: WeCurate: multiuser museum interactives for shared cultural experiences. In: CHI 2013, pp. 571–576 (2013)

13. Fillis, I., Lehman, K., Miles, M.P.: The museum of old and new art: leveraging entrepreneurial marketing to create a unique arts and vacation venture. J. Vacat. Mark. **23**(1), 85–96 (2015)

14. Kreft, A.: A smartphone app for teenagers: ubiquizzous learning at the german museum of technology. In: Proceedings of the European Conference on Games-Based Learning, Janua, pp. 939–943 (2016)

15. Shih, D.-T., Lin, C.L., Tseng, C.-Y.: Combining digital archives content with serious game approach to create a gamified learning experience. Int. Arch. Photogramm. Remote Sens. Spat. Inf. Sci. - ISPRS Arch. **40**(5W7), 387–394 (2015)

16. Mason, M., Curie, M.: The MIT museum glassware prototype: visitor experience exploration for designing smart glasses. J. Comput. Cult. Herit. **9**(3), 1–28 (2016)

17. Gendreau, A.: Museums and media: a view from Canada. Public Hist. **31**(1), 35–45 (2009)

18. del Barrio, M.J., Herrero, L.C.: Evaluating the efficiency of museums using multiple outputs: evidence from a regional system of museums in Spain. Int. J. Cult. Policy **20**(2), 221–238 (2014)

19. Anuar, I.M.H., Azahari, M.H., Legino, R.: Digital imagery as sustaining online repository for galleries and museum in Malaysia. Adv. Sci. Lett. **22**(5–6), 1466–1468 (2016)

20. Ahmad, A.T.: PENANG museums, culture and history. Kaji. Malaysia **33**, 153 (2015)

21. Parizi, R.M., Abdullah, A., Ramalingam, H.: Learning of web quality evaluation: a case study of Malaysia National Museum web site using WebQEM approach. In: Tang, S.F., Logonnathan, L. (eds.) Taylor's 7th Teaching and Learning Conference 2014 Proceedings, pp. 593–608. Springer, Singapore (2015). https://doi.org/10.1007/978-981-287-399-6_52

22. Throsby, D.: The Economics of Cultural Policy. Cambridge University Press, Cambridge (2010)

23. Hess, M., Robson, S., Serpico, M., Amati, G., Pridden, I., Nelson, T.: Developing 3D imaging programmes-workflow and quality control. J. Comput. Cult. Herit. **9**(1), 1–11 (2015)

24. Gao, X., Wang, X., Yang, B., Liu, Y.: Design of a computer-aided-design system for museum exhibition based on virtual reality. In: Wang, Y., et al. (eds.) IGTA 2017. CCIS, vol. 757, pp. 157–167. Springer, Singapore (2018). https://doi.org/10.1007/978-981-10-7389-2_16

25. Beer, S.: Digital heritage museums and virtual museums. In: Proceedings of the 2015 Virtual Reality International Conference, pp. 10:1–10:4 (2015)

26. Badalotti, E., De Biase, L., Greenaway, P.: The future museum. Procedia Comput. Sci. **7**, 114–116 (2011)

27. Chivarov, N., Ivanova, V., Radev, D., Buzov, I.: Interactive presentation of the exhibits in the museums using mobile digital technologies. IFAC Proc. Vol. **46**, 122–126 (2013)

28. Wolff, A., Mulholland, P., Maguire, M., O'Donovan, D.: Mobile technology to support coherent story telling across freely explored outdoor artworks. In: Proceedings of the 11th Conference on Advances in Computer Entertainment Technology - ACE 2014, pp. 1–8 (2014)

29. Ciocca, G., Olivo, P., Schettini, R.: Browsing museum image collections on a multi-touch table. Inf. Syst. **37**(2), 169–182 (2012)

30. McGookin, D., Tahiroğlu, K., Vaittinen, T., Kytö, M., Monastero, B., Vasquez, J.C.: Cultural heritage 'in-the-wild': considering digital access to cultural heritage in everyday life. In: CEUR Workshop Proceedings, vol. 2084, pp. 63–75 (2018)

31. Vi, C.T., Ablart, D., Gatti, E., Velasco, C., Obrist, M.: Not just seeing, but also feeling art: mid-air haptic experiences integrated in a multisensory art exhibition. Int. J. Hum. Comput. Stud. **108**, 1–14 (2017)

32. Lewis, M., Coles-Kemp, L.: A tactile visual library to support user experience storytelling. In: Proceedings of NordDesign 2014 Conference, NordDesign 2014, pp. 386–395 (2014)

33. Zhong, X., Wu, J., Han, X., Liu, W.: Mobile terminals haptic interface: a vibro-tactile finger device for 3D shape rendering. In: Huang, Y., Wu, H., Liu, H., Yin, Z. (eds.) ICIRA 2017. LNCS (LNAI), vol. 10462, pp. 361–372. Springer, Cham (2017). https://doi.org/10.1007/978-3-319-65289-4_35

34. Wang, Q., Wang, K., Hu, R., Du, Y., Suo, X., Xu, P.R.: Design research on the display of revolution museum. In: Proceedings of the 2nd International Conference on Electronic and Mechanical Engineering and Information Technology, EMEIT 2012, pp. 2128–2131 (2012)

35. Kerne, A., et al.: Strategies of free-form web curation: processes of creative engagement with prior work. In: C and C 2017 - Proceedings of the 2017 ACM SIGCHI Conference on Creativity and Cognition, pp. 380–392 (2017)

36. Biella, D., et al.: Crowdsourcing and co-curation in virtual museums: a practice-driven approach. J. Univers. Comput. Sci. **22**(10), 1277–1297 (2016)

37. Poole, A.H.: The conceptual landscape of digital curation. J. Doc. **72**(5), 961–986 (2016)

38. Dallas, C.: An agency-oriented approach to digital curation theory and practice. In: ICHIM 2007 - International Cultural Heritage Informatics Meeting, Proceedings (2007)

39. Antoniou, A., O'Brien, J., Bardon, T., Barnes, A., Virk, D.: Micro-augmentations: situated calibration of a novel non-tactile, peripheral museum technology. In: Proceedings of the 19th Panhellenic Conference on Informatics, pp. 229–234 (2015)

40. Pedersen, I., Gale, N., Mirza-Babaei, P.: TombSeer: illuminating the dead. In: Proceedings of the 7th Augmented Human International Conference 2016, pp. 24:1–24:4 (2016)

41. Fabola, A., et al.: A virtual museum installation for time travel. In: Beck, D., et al. (eds.) iLRN 2017. CCIS, vol. 725, pp. 255–270. Springer, Cham (2017). https://doi.org/10.1007/978-3-319-60633-0_21

42. Kazanis, S., Kontogianni, G., Chliverou, R., Georgopoulos, A.: Developing a virtual museum for the ancient wine trade in eastern mediterranean. Int. Arch. Photogramm. Remote Sens. Spat. Inf. Sci. - ISPRS Arch. **42**(2W5), 399–405 (2017)

43. Skamantzari, M., Kontogianni, G., Georgopoulos, A., Kazanis, S.: Developing a virtual museum for the Stoa of Attalos. In: 2017 9th International Conference on Virtual Worlds and Games for Serious Applications, VS-Games 2017 - Proceedings, pp. 260–263 (2017)

44. Kotani, M., Goto, K., Toyama, M.: Generating 3D virtual museum using SuperSQL. In: ACM International Conference Proceeding Series, vol. Part F1344, pp. 248–257 (2017)

45. Wang, D.: Exploring a narrative-based framework for historical exhibits combining JanusVR with photometric stereo. Neural Comput. Appl. **29**(5), 1425–1432 (2017). https://doi.org/10.1007/s00521-017-3201-7

46. Lai, M.-K.: Universal scent blackbox: engaging visitors communication through creating olfactory experience at art museum. In: Proceedings of the 33rd Annual International Conference on the Design of Communication - SIGDOC 2015, pp. 1–6 (2015)

47. Tzortzi, K., Schieck, A.F.G.: Rethinking museum space: interaction between spatial layout design and digital sensory environments. In: Proceedings - 11th International Space Syntax Symposium, SSS 2017, pp. 31.1–31.15 (2017)

48. Shirai, A., Kose, Y., Minobe, K., Kimura, T.: Gamification and construction of virtual field museum by using augmented reality game 'ingress,'. In: Proceedings of the 2015 Virtual Reality International Conference, no. Cc, pp. 4:1–4:4 (2015)

49. Van Eck, N.J., Waltman, L.: Software survey: VOSviewer, a computer program for bibliometric mapping. Scientometrics **84**(2), 523–538 (2010)

50. Roussou, M., et al.: Experiences from the use of a robotic avatar in a museum setting. In: Proceedings VAST 2001 Virtual Reality, Archeology, and Cultural Heritage, pp. 153–160 (2001)

51. van der Meij, M.G., Broerse, J.E.W., Kupper, F.: RRI & science museums; prototyping an exhibit for reflection on emerging and potentially controversial research and innovation. J. Sci. Commun. **16**(4), 1–24 (2017)

52. Kratz, S., Merritt, E.: Museums and the future of education. Horiz. **19**(3), 188–195 (2011)

53. Marchetti, E., Valente, A.: Diachronic perspective and interaction: new directions for innovation in historical museums. Int. J. Technol. Knowl. Soc. **8**(6), 131–143 (2013)

54. Paquin, M.: Objets d'apprentissage des musées virtuels du Canada et enseignants francophones du pays: Philosophie d'enseignement et conception du domaine d'enseigne-ment. Can. J. Educ. **36**(3), 380–412 (2013)

55. Keane, L., Keane, M.: Eco literacy: an eco web greening public imagination. Des. Principles Pract. **4**(4), 93–111 (2010)

56. Ernst, D., Esche, C., Erbslöh, U.: The art museum as lab to re-calibrate values towards sustainable development. J. Clean. Prod. **135**, 1446–1460 (2016)

57. Black, G.: Remember the 70%: sustaining 'core' museum audiences. Museum Manag. Curatorsh. **31**(4), 386–401 (2016)

58. Mello, J.C., Montijano, M.C., Andrade, Â.F., Luz, F.C.: Information systems, cyber culture and digitization of heritage in sergipe: museology on the web. Inf. e Soc. **22**(2), 127–138 (2012)

59. de Mello, J.C., Luz, F.C.L., Montijano, M.M.C.L., de Andrade, Â.M.F.: The museology on web: information system about cultural heritage in the digital age. Perspect. em Cienc. da Inf. **20**(1), 171–188 (2015)

60. Morlando, G., Guidi, G.: A virtual design museum. In: Eurographics Italian Chapter Conference 2011, pp. 47–51 (2011)

61. Anggai, S., Blekanov, I.S., Sergeev, S.L.: Index data structure, functionality and microservices in thematic virtual museums. Vestn. Sankt-Peterburgskogo Univ. Prikl. Mat. Inform. Protsessy Upr. **14**(1), 31–39 (2018)

62. Kahl, T., Iurgel, I., Zimmer, F., Bakker, R., van Turnhout, K.: RheijnLand.Xperiences – a storytelling framework for cross-museum experiences. In: Nunes, N., Oakley, I., Nisi, V. (eds.) ICIDS 2017. LNCS, vol. 10690, pp. 3–11. Springer, Cham (2017). https://doi.org/10.1007/978-3-319-71027-3_1

63. Fumo, M., Naponiello, M.: Aesthetic of historical towns and innovative constructive techniques. In: Improvement of Buildings' Structural Quality by New Technologies - Proceedings of the Final Conference of COST Action C12, pp. 393–399 (2005)

64. Vermeeren, A.P.O.S., Calvi, L.: How to get small museums involved in digital innovation: a design-inclusive research approach. In: Cultural Heritage Communities: Technologies and Challenges, Faculty of Industrial Design Engineering, TU Delft, Netherlands, pp. 114–131 (2017)

65. Ing, D.S.L.: Innovations in a technology museum. IEEE Micro **19**(6), 44–52 (1999)

66. Solima, L., Della Peruta, M.R., Maggioni, V.: Managing adaptive orientation systems for museum visitors from an IoT perspective. Bus. Process Manag. J. **22**(2), 285–304 (2016)

67. Gura, C., Wandl-Vogt, E., Dorn, A., Losada, A., Benito, A.: Co-designing innovation networks for cross-sectoral collaboration on the example of exploreAT! In: ACM International Conference Proceeding Series, vol. Part F132203 (2017)

68. Della Corte, V., Aria, M., Del Gaudio, G.: Smart, open, user innovation and competitive advantage: a model for museums and heritage sites. Museum Manag. Curatorsh. **32**(1), 50–79 (2017)

69. Recuero Virto, N., Blasco López, M.F., San-Martín, S.: How can European museums reach sustainability? Tour. Rev. **72**(3), 303–318 (2017)

70. Flores, P., Crawford, L.: Museo del Caribe en Barranquilla: la identidad regional en el espacio del simulacro*/Caribean Museum in Barranquilla: regional identity in the space of simulacrum. Co-herencia **8**(14), 183–205 (2011)

71. Belinky, I., Lanir, J., Kuflik, T.: Using handheld devices and situated displays for collaborative planning of a museum visit. In: Proceedings of the 2012 International Symposium on Pervasive Displays, pp. 19:1–19:6 (2012)

72. Zhang, P., Tian, J., Zhang, H.F., Zhang, P.B.: Application of design education for the construction of a museum: a case study of Heifei University of Technology, China. In: ACM International Conference Proceeding Series, pp. 57–61 (2018)

73. Kochergina, E.: Urban planning aspects of museum quarters as an architectural medium for creative cities. IOP Conf. Ser. Mater. Sci. Eng. **245**(5), 052031 (2017)

74. Miller, A.: Innovative management strategies for building and sustaining a digital initiatives department with limited resources. Digit. Libr. Perspect. **34**(2), 117–136 (2018)

75. Khan, H.-U.: Because we can: globalization and technology enabling iconic architectural excesses. Int. J. Islam. Archit. **7**(1), 5–26 (2018)

76. Justice, S.C.: UNESCO global geoparks, geotourism and communication of the earth sciences: a case study in the Chablais UNESCO Global Geopark, France. Geoscience **8**(5), 149 (2018)

77. Hvenegaard Rasmussen, C.: The participatory public library: the Nordic experience. New Libr. World **117**(9–10), 546–556 (2016)

78. Li, P.-P., Chang, P.-L.: A study of virtual reality experience value and learning efficiency of museum-using Shihsanhang museum as an example. In: Proceedings of the 2017 IEEE International Conference on Applied System Innovation: Applied System Innovation for Modern Technology, ICASI 2017, pp. 1158–1161 (2017)

79. Cesário, V., Matos, S., Radeta, M., Nisi, V.: Designing interactive technologies for interpretive exhibitions: enabling teen participation through user-driven innovation. In: Bernhaupt, R., Dalvi, G., Joshi, A., Balkrishan, D.K., O'Neill, J., Winckler, M. (eds.) INTERACT 2017. LNCS, vol. 10513, pp. 232–241. Springer, Cham (2017). https://doi.org/10.1007/978-3-319-67744-6_16

80. Camarero, C., Garrido, M.-J., Vicente, E.: Does it pay off for museums to foster creativity? The complementary effect of innovative visitor experiences. J. Travel Tour. Market. (2018)

81. Cesário, V., Coelho, A., Nisi, V.: Enhancing museums' experiences through games and stories for young audiences. In: Nunes, N., Oakley, I., Nisi, V. (eds.) ICIDS 2017. LNCS, vol. 10690, pp. 351–354. Springer, Cham (2017). https://doi.org/10.1007/978-3-319-71027-3_41

82. Ovallos-Gazabon, D., et al.: Using text mining tools to define trends in territorial competitiveness indicators. In: Figueroa-García, J.C., Duarte-González, M., Jaramillo-Isaza, S., Orjuela-Cañon, A.D., Díaz-Gutierrez, Y. (eds.) WEA 2019. CCIS, vol. 1052, pp. 676–685. Springer, Cham (2019). https://doi.org/10.1007/978-3-030-31019-6_57

Kalman Filter Employment in Image Processing

Katerina Fronckova$^{(\boxtimes)}$ ⓘ and Antonin Slaby ⓘ

University of Hradec Kralove, Hradec Kralove, Czech Republic
{katerina.fronckova,antonin.slaby}@uhk.cz

Abstract. The Kalman filter is a classical algorithm of estimation and control theory. Its use in image processing is not very well known as it is not its typical application area. The paper deals with the presentation and demonstration of selected possibilities of using the Kalman filter in image processing. Particular attention is paid to problems of image noise filtering and blurred image restoration. The contribution presents the reduced update Kalman filter algorithm, that can be used to solve both the tasks. The construction of the image model, which is the necessary first step prior to the application of the algorithm itself, is briefly mentioned too. The described procedures are then implemented in the MATLAB software and the results are presented and discussed in the paper.

Keywords: Kalman filter · Image processing · Image noise filtering · Blurred image restoration

1 Introduction

The Kalman filter represents a theoretical basis for various recursive methods in the examination of stochastic (linear) dynamic systems. The algorithm is based on the idea that the unknown state of the system can be estimated using certain measured data. The filter is named after Rudolf Emil Kalman, a Hungarian mathematician living in the United States, who published it in the article [15] in 1960. In the next period, various algorithms based on the nature of the Kalman filter have been derived by various authors. These algorithms are all referred to as Kalman filters and can be suitably used in certain specific situations, for example, resulting from failure to meet some theoretical assumptions of the classical Kalman filter in solving practical problems.

The Kalman filter is used in various applications. Location of moving objects and navigation belong to its most important application domains – the Kalman filter or generally Kalman filters are used for example in global satellite positioning systems (GPS etc.), in radars, in robot control and navigation, in autopilots or autonomous vehicles. They are also used in the area of computer vision for tracking the movement of objects in a sequence of video frames, in augmented and virtual reality, etc. The space project Apollo which dates back to

© Springer Nature Switzerland AG 2020
O. Gervasi et al. (Eds.): ICCSA 2020, LNCS 12249, pp. 833–844, 2020.
https://doi.org/10.1007/978-3-030-58799-4_60

1960s included one of the first applications of the Kalman filter in the area of navigation and control. Other application domains include time series analysis, econometrics, signal processing, weather forecasting and many others.

This paper presents some possibilities of using the Kalman filter in image processing. It is organized as follows. Section 2 provides a description of the Kalman filter algorithm and the principles of its functioning. Section 3 deals with its application to image noise filtering and blurred image restoration. Section 4 demonstrates the experimental results achieved. Section 5 is the final summary.

2   Kalman Filter

The Kalman filter is a tool for estimating the state of a stochastic linear dynamic system using measured data corrupted by noise. The estimate produced by the Kalman filter is statistically optimal in some sense (for example it minimizes the mean square error, see [25] for more details). The principle of the application of the filter is shown in the following Fig. 1.

The Kalman filter works with all available information, i.e. it uses all available measured data, system model together with statistical description of its inaccuracies, noise and measurement errors as well as information about initial conditions and initial state of the system.

2.1   Algorithm of the Kalman Filter

Let us consider a stochastic linear dynamic system in discrete time represented by the following model (it is assumed here that the system has no inputs)

$$x_k = \Phi_{k-1} x_{k-1} + w_{k-1}, \tag{1}$$

$$z_k = \mathbf{H}_k x_k + v_k. \tag{2}$$

Equation (1), referred to as state equation, describes the dynamics of the system, the vector $x_k \in \mathbb{R}^n$ is an (unknown) vector of the system state at time t_k, the matrix $\Phi_{k-1} \in \mathbb{R}^{n \times n}$ represents the dynamic evolution of the system between time t_{k-1} and t_k. Equation (2) is called measurement equation, the vector $z_k \in \mathbb{R}^m$ is called the system output vector, measurement vector or observation vector; the matrix $\mathbf{H}_k \in \mathbb{R}^{m \times n}$ describes the relationship between the state of the system and measurements. The vectors x_k and z_k, $k = 0, 1, 2, \ldots$, may be treated as random variables and their sequences $\{x_k\}$ and $\{z_k\}$ considered as random processes as we deal with a stochastic system.

$\{w_k\}$ and $\{v_k\}$ are random noise processes; these processes are assumed to be uncorrelated Gaussian processes with zero mean and covariance matrices $\mathbf{Q}_k$ resp. $\mathbf{R}_k$ at time t_k (the processes have Gaussian white noise properties).

Further, let x_0 be a random variable with a Gaussian (normal) distribution with known mean x_0 and known covariance matrix $\mathbf{P}_0$. Moreover, let x_0 and

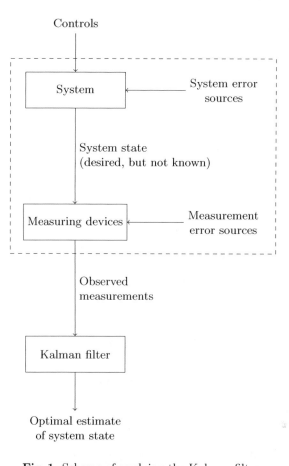

Fig. 1. Scheme of applying the Kalman filter

both the noises be mutually uncorrelated. These can be expressed for all t_k as follows

$$E\langle \boldsymbol{w}_k \rangle = \boldsymbol{0},$$
$$E\langle \boldsymbol{v}_k \rangle = \boldsymbol{0},$$
$$E\langle \boldsymbol{w}_{k_1} \boldsymbol{w}_{k_2}^T \rangle = \boldsymbol{Q}_{k_1} \delta(k_2 - k_1),$$
$$E\langle \boldsymbol{v}_{k_1} \boldsymbol{v}_{k_2}^T \rangle = \boldsymbol{R}_{k_1} \delta(k_2 - k_1),$$
$$E\langle \boldsymbol{w}_{k_1} \boldsymbol{v}_{k_2}^T \rangle = \boldsymbol{0},$$
$$E\langle \boldsymbol{x}_0 \boldsymbol{w}_k^T \rangle = \boldsymbol{0},$$
$$E\langle \boldsymbol{x}_0 \boldsymbol{v}_k^T \rangle = \boldsymbol{0},$$

where the symbol δ denotes the Kronecker delta

$$\delta(k) = \begin{cases} 1, & k = 0, \\ 0, & k \neq 0. \end{cases}$$

The goal of the Kalman filter is to find an estimate of the state vector $\boldsymbol{x}_k$ at time t_k, denoted as $\hat{\boldsymbol{x}}_k$, so that this estimate is optimal (for example in the sense of minimizing the mean square error).

The Kalman filter algorithm is recursive; the calculation at general time t_k consists of two main steps. The first step is the calculation of the a priori estimate $\hat{\boldsymbol{x}}_{k(-)}$ at time t_k by substituting the a posteriori estimate from time t_{k-1} into the deterministic part of the state equation of the model. In the second step this estimate is adjusted using the measurement taken at time t_k, which results in obtaining the a posteriori estimate $\hat{\boldsymbol{x}}_{k(+)}$ at time t_k.

The following equation can be written for the a priori estimate of the state vector $\hat{\boldsymbol{x}}_{k(-)}$ at time t_k. The uncertainty of this estimate is expressed by the a priori error covariance matrix $\mathbf{P}_{k(-)}$

$$\hat{\boldsymbol{x}}_{k(-)} = \boldsymbol{\Phi}_{k-1} \hat{\boldsymbol{x}}_{k-1(+)},$$
$$\mathbf{P}_{k(-)} = \boldsymbol{\Phi}_{k-1} \mathbf{P}_{k-1(+)} \boldsymbol{\Phi}_{k-1}^{T} + \mathbf{Q}_{k-1}.$$

Then, after obtaining the measurement $\boldsymbol{z}_k$ the a posteriori estimate of the state vector $\hat{\boldsymbol{x}}_{k(+)}$ is calculated as a combination of the a priori estimate and the difference between the actual and expected value of the measurement weighted by the matrix $\mathbf{K}_k$ (called the Kalman gain). Its uncertainty is expressed by the a posteriori error covariance matrix $\mathbf{P}_{k(+)}$

$$\hat{\boldsymbol{x}}_{k(+)} = \hat{\boldsymbol{x}}_{k(-)} + \mathbf{K}_k[\boldsymbol{z}_k - \mathbf{H}_k \hat{\boldsymbol{x}}_{k(-)}],$$
$$\mathbf{P}_{k(+)} = \mathbf{P}_{k(-)} - \mathbf{K}_k \mathbf{H}_k \mathbf{P}_{k(-)},$$
$$\mathbf{K}_k = \mathbf{P}_{k(-)} \mathbf{H}_k^{T} [\mathbf{H}_k \mathbf{P}_{k(-)} \mathbf{H}_k^{T} + \mathbf{R}_k]^{-1}.$$

The detailed derivation of the mentioned equations of the Kalman filter can be found, for example, in [11] – this derivation uses the orthogonality principle, which Kalman used in his original article [15]. Over time, other approaches to derivation based on innovations or Bayesian statistics have been used by other authors, they can be found, for example, in [3,25].

The Kalman filter can also be generalized for systems with inputs (see e.g. [7, p. 28]) and there also exists a variant for continuous-time systems – the Kalman-Bucy filter [16].

A more detailed discussion of the algorithm, its properties and theoretical assumptions is offered, for example, by [11,25,28]; practical aspects of the filter implementation are discussed, for example, in [28].

3   Kalman Filter and Image Processing

A grayscale digital image can be naturally represented by a two-dimensional matrix; its elements then express the intensity values of individual pixels. Using the Kalman filter in image processing tasks thus requires to extend the concept of the Kalman filter from random processes to two-dimensional random fields.

3.1   Reduced Update Kalman Filter

One of the Kalman filter modifications intended and designed for this purpose is the reduced update Kalman filter (RUKF) published by Woods and Radewan [32]. This algorithm was originally designed to filter noise in images, but later Woods and Ingle [33] further developed it and extended it for blurred image restoration.

The algorithm is based on the following two-dimensional autoregressive (2D AR) image model (state equation)

$$x(i,j) = \sum_{(k,l)\in\mathcal{D}} \phi(k,l)x(i-k,j-l) + w(i,j),$$

where the notation $x(i,j)$ represents a pixel of the ideal image located at the position (i,j), $w(i,j)$ denotes system "noise" corresponding to model inaccuracies and $\phi(k,l)$ represents the corresponding parameter of the autoregressive model, it is assumed that $\mathcal{D} = \{k \geq 0, l \geq 0\} \cup \{k > 0, l < 0\}$. Section 3.2 provides more detailed information about the image model identification.

Corruption of the image by additive measurement noise can be modelled by the following scalar equation (measurement equation)

$$z(i,j) = x(i,j) + v(i,j),$$

where $z(i,j)$ denotes a pixel of the noisy image and $v(i,j)$ is noise incurred usually due to the technical principles of image obtaining.

Both w and v possess properties of Gaussian white noise according to the assumptions required by the Kalman filter.

Sometimes, the image may be degraded in addition to noise also by blurring. The measurement equation has in this case the following form

$$z(i,j) = \sum_{(k,l)} h(k,l)x(i-k,j-l) + v(i,j),$$

where h represents blur of the image caused by, for example, motion or poorly focused camera optics. The above equation can also be seen as the expression of the two-dimensional discrete convolution of the image x with the convolution kernel h. More details about image blur modeling can be found, e.g., in [31].

The already mentioned reduced update Kalman filter can be used to solve both these tasks. This algorithm is based on sequential scanning of the image (pixel by pixel) starting at the upper-left corner of the image and continuing on a line by line basis. At any given moment, the currently processed pixel can be perceived as the "presence", the pixels already processed as the "past" and the upcoming pixels waiting for processing as the "future", see Fig. 2. This approach transforms a two-dimensional problem into a one-dimensional problem and consequently the classical Kalman filter can be used. The problem, however, is that the state vector, which is made up of individual pixels of an image, takes on a large number of elements, resulting in high computational costs. The RUKF solves this situation by not working always with the whole image but applying the Kalman filter only to a certain area surrounding the currently processed pixel. A detailed description of the algorithm and derivation of the filter equations can be found in [31–33]. Computational demands can also be reduced by taking advantage of the fact that the Kalman gain usually achieves a steady-state value after several iterations of the algorithm, and consequently, it is no longer necessary to calculate the Kalman gain along with the error covariance equations in each iteration, as it can be approximated by this steady-state value. This modification of the algorithm is called the steady-state RUKF [31].

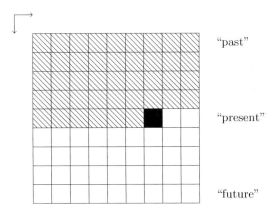

Fig. 2. Image scanning

3.2   Construction of the Image Model

Woods and Radewan [32] assumed in their original paper that the image model is known. In practice, this is not usually the case.

Image modeling is based on finding the order and parameters of a two-dimensional autoregressive model and estimating the variance of the corresponding noise (model inaccuracies).

The first step in constructing the model lies in determination of its order, usually denoted $\theta(p, q)$. Bouzouba and Radouane [5] suggest using the maximum entropy principle to solve this problem. Let

$$\hat{e}_\theta = z - \hat{x}_\theta$$

be the estimation error defined as the difference between the noisy degraded image and the estimate of the ideal image obtained by the RUKF algorithm using the model of order $\theta(p, q)$. For the pixels of the image, the following definition of the probability density function of the error can be introduced

$$\hat{g}_\theta(i, j) = \frac{\hat{e}_\theta(i, j)}{\sum_i \sum_j \hat{e}_\theta(i, j)},$$

$\sum_i \sum_j \hat{g}_\theta(i, j) = 1$ and $0 < \hat{g}_\theta(i, j) \leq 1$. Next, let

$$G = \{\hat{g}_\theta : \theta = (1, 1); (1, 2); (2, 1); ...; (p_{max}, q_{max})\}$$

be the set containing possible estimates of $\hat{g}_\theta$ for various model orders θ. According to the principle of maximum entropy [13, 14] the optimal estimate of the probability density function is the estimate $\hat{g}_\theta^*$, having the maximum entropy among all estimates $\hat{g}_\theta$. The entropy of $\hat{g}_\theta$ can be expressed by the formula

$$H(\hat{g}_\theta) = -\sum_i \sum_j \hat{g}_\theta(i, j) \log(\hat{g}_\theta(i, j)),$$

then, the following holds for the optimal estimate of the probability density function

$$H(\hat{g}_\theta^*) = \max\{H(\hat{g}_\theta) : \hat{g}_\theta \in G\}.$$

The choice of the appropriate order of the model therefore consists in estimating the probability density function $\hat{g}_\theta$ for various possible orders and then selecting the order θ, for which $\hat{g}_\theta$ corresponds to the maximum entropy.

Next, it is necessary to calculate estimates of model parameters and noise variance. Different approaches or algorithms can be used, such as variants of least squares method, correlation-based methods, or approaches enabling to estimate model parameters simultaneously with image filtering. A description and comparison of these approaches is given in [18].

Various forms of models (based on areas covered by the models) can be met. Nonsymmetric half-plane (NSHP) or quarter plane (QP) form are usually used.

4   Practical Demonstration and Obtained Results

For a practical demonstration of the RUKF algorithm, a photograph of the Large Square in the city of Hradec Kralove (the Czech Republic) was used, see Fig. 3a.

Additive Gaussian white noise was added to the photograph in the first experiment, as is shown in Fig. 3b. The amount of noise present in an image can be expressed using the signal-to-noise ratio (SNR), in the case of Fig. 3b the SNR is 10.77 dB.

Prior to the RUKF application, the model of the image was constructed. First, the order of the model was determined using the principle of maximum entropy. The highest value of entropy was associated with the model of order $(2, 2)$. The model coefficients (together with the estimation of variance of the corresponding inaccuracies) were calculated using the method described in [5].

Variance of measurement noise was considered known, according to [5], this is not an unrealistic assumption in practice.

Subsequently, the RUKF was applied using the proposed model. The result is shown in Fig. 3c, the SNR rose to 15.36 dB, so the improvement in SNR (ISNR) is 4.59 dB.

In the second experiment, the original photograph was degraded also by blurring caused by motion in the horizontal direction as Fig. 4b shows. The result obtained by applying the RUKF is shown in Fig. 4c.

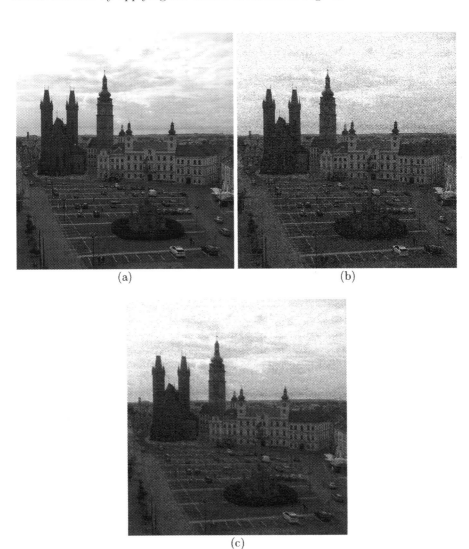

Fig. 3. Image noise filtering by the RUKF: (a) original ideal image, (b) image degraded by additive noise, (c) result obtained using the RUKF

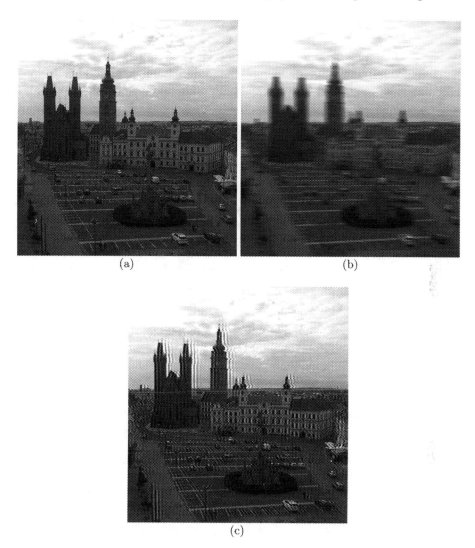

Fig. 4. Blurred image restoration by the RUKF: (a) original ideal image, (b) image degraded by blur, (c) result obtained using the RUKF

As the figures show, after the application of the RUKF algorithm, visual improvement is evident. In the first experiment, the effect of noise is suppressed in the resulting image, but also the sharpness of its details is reduced. By choosing suitable filter parameters (noise variances), it would be possible to find the desired compromise between noise filtering and maintaining sharpness. In the case of the second experiment, blurring was removed from the image, but as a result, artifacts were created, that are most noticeable near sharp transitions (edges) in the image.

The Kalman filter resp. the RUKF algorithm provides optimal results from a theoretical point of view (for example in the sense of minimizing the mean square error), but in practice, the problem may be ignorance of the actual image model or the blurring process model, failure to meet noise assumptions (for example Gaussian white noise assumption), etc., moreover, the complete RUKF algorithm has high computational demands. As a consequence of the non-fulfillment of theoretical assumptions or the use of a certain modification of the algorithm having, for example, more favorable computational properties, the results obtained in practice are no longer optimal.

The reduced update Kalman filter algorithm is a basic approach to applying the Kalman filter to images. Over time, other authors have derived various modifications of the RUKF algorithm and slightly different approaches employing the Kalman filter, which are more suitable, for example, in certain real situations that do not meet the basic theoretical assumptions or in terms of computational performance. In addition to the classical Kalman filter, algorithms can also be based on other Kalman filters, such as the extended Kalman filter, the robust Kalman filter, the adaptive Kalman filter, etc. Examples are [32] (Kalman strip filter), [36] (fast modified RUKF), [19] (reduced order model Kalman filter), [1,2,6,8,12,22,34,35].

In a general comparison with other methods for image restoration, the mentioned algorithms based on the Kalman filter can be included among the classical methods. These methods are based on mathematical formulas and algorithms known for many years. The RUKF achieves similar results compared to these methods. Recently, new approaches have emerged, that are now widely used, especially those based on neural networks. In general, these approaches usually achieve better results compared to mathematical methods, but an accurate description of their functioning is hard to follow and also training can be difficult (they require a large amount of training data, computational power and knowledge). Currently, algorithms combining classical and new approaches are achieving interesting results - and even here the Kalman filter finds its application [23].

In practice, algorithms based on the Kalman filter are used, for example, in the restoration of satellite and radar images [17,24,30] or various medical images (MRI, CT, ultrasound, ...) [4,9,20,27,29] and can be used even for color image restoration [10,21,26]. Moreover, the Kalman filter can be employed also in related tasks such as estimating image model parameters [37] etc.

5   Conclusion

The Kalman filter can find its employment also in the filtering of two-dimensional signals and data. The paper presented possibilities of its use in the field of image processing, namely in solving the problems of image noise filtering and blurred image restoration. The construction of the image model using the maximum entropy principle for model order determination was also mentioned. The use of the described procedures was illustrated by the example and the obtained results were discussed.

Acknowledgements. The paper was supported by the Specific Research Project at the Faculty of Informatics and Management of the University of Hradec Kralove, the Czech Republic.

References

1. Asif, A.: Fast Rauch-Tung-Striebel smoother-based image restoration for noncausal images. IEEE Signal Process. Lett. **11**(3), 371–374 (2004)
2. Belaifa, H.B.H., Schwartz, H.M.: Robust modeling for image-restoration using a modified reduced update Kalman filter. IEEE Trans. Signal Process. **40**(10), 2584–2588 (1992)
3. Bertein, J.-C., Ceschi, R.: Processus stochastiques et filtrage de Kalman, 1ere edn. Hermes, Paris (1998)
4. Boulfelfel, D., et al.: 2-dimensional restoration of single-photon emission computed-tomography images using the Kalman filter. IEEE Trans. Med. Imaging **13**(1), 102–109 (1994)
5. Bouzouba, K., Radouane, L.: Image identification and estimation using the maximum entropy principle. Pattern Recogn. Lett. **21**(8), 691–700 (2000)
6. Chee, Y.K., Soh, Y.C.: A robust Kalman filter design for image restoration. In: 2001 IEEE International Conference on Acoustics. Speech, and Signal Processing, Proceedings, pp. 1825–1828. IEEE, New York (2001)
7. Chui, C.K., Chen, G.: Kalman Filtering with Real-Time Applications, 4th edn. Springer, Berlin (2009). https://doi.org/10.1007/978-3-662-02508-6
8. Citrin, S., Azimi-Sadjadi, M.R.: A full-plane block Kalman filter for image restoration. IEEE Trans. Image Process. **1**(4), 488–495 (1992)
9. Conte, F., Germani, A., Iannello, G.: A Kalman filter approach for denoising and deblurring 3-D microscopy images. IEEE Trans. Image Process. **22**(12), 5306–5321 (2013)
10. Galatsanos, N.P., Chin, R.T.: Restoration of color images by multichannel Kalman filtering. IEEE Trans. Signal Process. **39**(10), 2237–2252 (1991)
11. Grewal, M.S., Andrews, A.P.: Kalman Filtering: Theory and Practice Using MATLAB, 4th edn. Wiley, Hoboken (2015)
12. Hernandez, V.H., Desai, M.: Robust modeling edge adaptive reduced update Kalman filter. In: Thirtieth Asilomar Conference on Signals. Systems & Computers, pp. 1019–1023. IEEE, Los Alamitos (1997)
13. Jaynes, E.T.: Information theory and statistical mechanics. Phys. Rev. **106**(4), 620–630 (1957)
14. Jaynes, E.T.: Information theory and statistical mechanics II. Phys. Rev. **108**(2), 171–190 (1957)
15. Kalman, R.E.: A new approach to linear filtering and prediction problems. Trans. Am. Soc. Mech. Eng. Series D: J. Basic Eng. **82**(1), 35–45 (1960)
16. Kalman, R.E., Bucy, R.S.: New results in linear filtering and prediction theory. Trans. Am. Soc. Mech. Eng. Series D: J. Basic Eng. **83**(1), 95–108 (1961)
17. Kanakaraj, S., Nair, M.S., Kalady, S.: Adaptive importance sampling unscented Kalman filter based SAR image super resolution. Comput. Geosci. **133**, 104310 (2019)
18. Kaufman, H., et al.: Estimation and identification of two-dimensional images. IEEE Trans. Autom. Control **28**(7), 745–756 (1983)
19. Kim, J., Woods, J.W.: A new interpretation of ROMKF. IEEE Trans. Image Process. **6**(4), 599–601 (1997)

20. Kim, S., Khambampati, A.K.: Mathematical concepts for image reconstruction in tomography. Ind. Tomogr. Syst. Appl. **71**, 305–346 (2015)
21. Latouche, H., Solarte, K., Ordonez, J., Sanchez, L.: Nonlinear filters to denoising color images. Ing. UC **24**(2), 185–195 (2017)
22. Liu, C., Zhang, Y., Wang, H.Q., Wang, X.Z.: Improved block Kalman filter for degraded image restoration. In: 2013 IEEE 15th International Conference on High Performance Computing and Communications & 2013 IEEE International Conference on Embedded and Ubiquitous Computing (HPCC_EUC), pp. 1958–1962. IEEE, New York (2013)
23. Ma, R.J., Hu, H.F., Xing, S.L., Li, Z.M.: Efficient and fast real-world noisy image denoising by combining pyramid neural network and two-pathway unscented Kalman filter. IEEE Trans. Image Process. **29**, 3927–3940 (2020)
24. Marhaba, B., Zribi, M.: The bootstrap kernel-diffeomorphism filter for satellite image restoration. In: 2018 International Symposium on Consumer Technologies (ISCT), pp. 81–85. IEEE, New York (2018)
25. Maybeck, P.S.: Stochastic Models, Estimation and Control, Volume I, 1st edn. Academic Press, New York (1979)
26. Rao, K.D., Swamy, M.N.S., Plotkin, E.I.: Adaptive filtering approaches for colour image and video restoration. IEE Proc.-Vis. Image Signal Process. **150**(3), 168–177 (2003)
27. Sam, B.B., Fred, A.L.: An efficient grey wolf optimization algorithm based extended Kalman filtering technique for various image modalities restoration process. Multimed. Tools Appl. **77**(23), 30205–30232 (2018)
28. Simon, D.: Optimal State Estimation: Kalman, H_∞ and Nonlinear Approaches, 1st edn. Wiley, Hoboken (2006)
29. Wang, J.W., Wang, X.: Application of particle filtering algorithm in image reconstruction of EMT. Measur. Sci. Technol. **26**(7), 075303 (2015)
30. Wang, L., Loffeld, O., Ma, K.L., Qian, Y.L.: Sparse ISAR imaging using a greedy Kalman filtering approach. Signal Process. **138**, 1–10 (2017)
31. Woods, J.W.: Multidimensional Signal, Image, and Video Processing and Coding, 2nd edn. Academic Press, Boston (2012)
32. Woods, J.W., Radewan, C.H.: Kalman filtering in two dimensions. IEEE Trans. Inf. Theory **23**(4), 473–482 (1977)
33. Woods, J.W., Ingle, V.K.: Kalman filtering in two dimensions: further results. IEEE Trans. Acoust. Speech Signal Process. **29**(2), 188–197 (1981)
34. Wu, Q., Wang, X.C., Guo, P.: Joint blurred image restoration with partially known information. In: Proceedings of 2006 International Conference on Machine Learning and Cybernetics, pp. 3853–3858. IEEE, New York (2006)
35. Wu, W.R., Kundu, A.: A modified reduced update Kalman filter for images degraded by non-gaussian additive noise. In: 1989 IEEE International Conference on Systems, Man, and Cybernetics: Conference Proceedings, pp. 352–355. IEEE, New York (1989)
36. Wu, W.-R., Kundu, A.: Image estimation using fast modified reduced update Kalman filter. IEEE Trans. Signal Process. **40**(4), 915–926 (1992)
37. Zeinali, M., Shafiee, M.: A new Kalman filter based 2D AR model parameter estimation method. IETE J. Res. **63**(2), 151–159 (2017)

Migration of Artificial Neural Networks to Smartphones

Milan Kostak[(✉)] [iD], Ales Berger, and Antonin Slaby [iD]

Faculty of Informatics and Management, University of Hradec Králové,
Rokitanského 62, 50003 Hradec Králové, Czechia
{milan.kostak,ales.berger,antonin.slaby}@uhk.cz

Abstract. The paper explains the process of migration of an artificial neural network (ANN) to a smartphone device. It focuses on a situation when the ANN is already deployed on a desktop computer. Our goal is to describe the process of the migration of the network to a mobile environment. In the current system we have, images have to be scanned and fed to a computer that is applying the ANN. However, every smartphone has a camera that can be used instead of a scanner. Migration to such a device should save the overall processing time. ANNs in the field of computer vision have a long history. Despite that, mobile phones were not used as a target platform for ANNs because they did not have enough processing power. In the past years, smartphones have developed dramatically, and have the processing power necessary for deploying ANNs now. Also, major mobile operating systems, Android and iOS, have included the support for the deployment.

Keywords: Artificial neural networks · Smartphones · Android · Machine learning

1 Introduction

In the last decade, with the rise of available computational power, artificial neural networks (ANNs) became a popular and useful tool for solving problems that do not have an easy solution otherwise. Computer vision, time series prediction, text filtering, and speech recognition are just a few of many disciplines that utilize ANNs.

ANNs are one of the computational models used in artificial intelligence. They are vaguely inspired by the biological structures and intelligence as we know it. When a human is trying to learn something, he usually looks at a couple of examples. By learning these examples, he is later able to apply this knowledge in new situations that he did not see before. ANNs use this concept, and it is called training. The general process is called machine learning. Every case of a training dataset is fed to the ANN, which processes it and improves itself based on the error that the input caused on the output. This is repeated many times, and the process is called supervised learning. If the process is successful, then the trained ANN can take the input that it did not see before and produce the correct output with a certain probability. There is also a process of unsupervised learning, but that goes beyond the scope of this short introduction.

© Springer Nature Switzerland AG 2020
O. Gervasi et al. (Eds.): ICCSA 2020, LNCS 12249, pp. 845–858, 2020.
https://doi.org/10.1007/978-3-030-58799-4_61

Typical ANN includes so-called neurons (from tens to billions), which are the primary unit of any neural network. These neurons are grouped into layers, and they are connected between those layers. Every neuron in one layer is taking one output from every neuron of the previous layer and sends its single result to every neuron in the next layer. Deciding on optimal neurons and layers counts is not an easy problem, and it heavily depends on the application. Usually, a compromise is necessary between the high computational demands and the speed of training. The more neurons and layers the ANN has, the more demanding the training is. However, only bigger networks can grasp more complex problems.

ANNs are used in numerous areas of everyday life:

- speech recognition in virtual assistants like Siri, Cortana, Alexa, or Google Home [1–3],
- checking for malicious posts on social media [4],
- postprocessing of photos on smartphones [5],
- path planning for car navigation that counts for expected traffic [6, 7],
- detection of handwritten zip code in postal services [8],
- other smart experimental solutions [9, 10].

A smartphone is a kind of electronic device. Other examples include smartwatches, smart cars, smart homes, and many others. The great thing about them is that they are ubiquitous, small, connected to the Internet, and have much computational power. That makes them an ideal device for many use cases.

Both major operating systems for smartphones, Android and iOS, support deployment of ANNs in the form of trained models. ANNs are commonly deployed on desktop computers or in the cloud. The significant advantage of both major smartphone platforms is that they are ready for the migration of these trained models. These platforms include support for deploying those models, but each has its process and supported format of models. We are addressing these issues in this paper.

2 Problem Definition

We have a system for image processing. This system consists of several parts (Fig. 1). At first, it is necessary to scan the images. The scanned files are transferred to a desktop computer, which then applies a trained ANN to it. At the final step, the results are automatically sent to a server that stores all data.

We were seeking a solution that will save the time necessary to execute all the steps as mentioned above. We think that the part of scanning and data collecting is the only part, which processing can be significantly shortened. The processing of the image by ANN is taking only milliseconds. Internet communication is also fast enough to transfer the data in a couple of seconds, depending on the image size, so the only part remaining is the scanning. The server part does not need any changes, as its API is universal and does not affect the processing time.

We found a paper [11] in which the authors did a text recognition in document images that are obtained by a smartphone. They did the recognition with deep convolutional and recurrent neural networks. We are trying to solve a different problem, but we want to apply a similar approach of using an ANN on a smartphone to process a document image. We want to save the processing time by combining scanning and detection processes on a single device – a smartphone. That requires migrating our model, which is designed and trained in a TensorFlow environment. We describe this process in this paper.

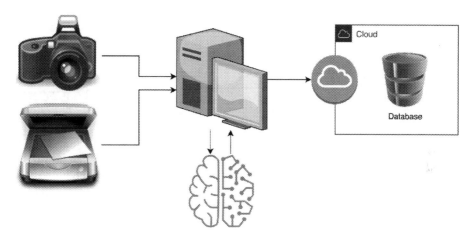

Fig. 1. This figure illustrates the workflow of our previous system. A scanner or a camera is used to obtain the image of a document. This document is processed on a desktop computer by an ANN. The output, together with the document, is sent to the server.

3   Related Work

The most relevant related works on the topics of artificial neural networks, object detection, and deployment of ANNs on smartphones are presented.

3.1   Artificial Neural Networks

ANNs find applications in many areas. One of the most common uses in everyday life are virtual personal assistants. At their core lies the speech detection process. Sriram et al. [2] describe a general framework for robust speech recognition that uses generative adversarial ANN. Their framework focuses on solving problems that current systems suffer from, like ambient noise, different accents, or variations in reverberations. Këpuska and Bohouta [3] propose a solution for multi-modal systems that process two or more user input modes. Apart from the speech itself, the modes include touch, manual gestures, or head and body movements. Their goal is to design a next-generation prototype of virtual personal assistants.

Many people might not realize it, but automatic zip code detection is used in postal services for several decades now. With the advent of ANNs, systems that utilize them were also implemented in these areas. LeCun et al. [8] in 1989 described a method for the use of backpropagation, which is the core of machine learning these days, to the problem of zip code recognition.

Another example of the day-to-day use of ANNs is the postprocessing of photos taken on smartphones. The cameras of these devices usually have to deal with physical limitations like small sensor size, or compact lenses [5]. Ignatov et al. [5] proposed a solution for this problem by postprocessing of taken photos by ANNs, specifically generative adversarial ANNs. These proposed networks help to reconstruct missing details in the pictures.

3.2   Object Detection

Object detection is a core problem in computer vision. Humans can recognize a large variety of objects instantly in almost any image. However, this is a difficult problem for computers because objects of the same category (people, animals, etc.) can vary significantly in appearance. Many different frameworks and systems have been developed in the past. More simplistic approaches include Haar features [12], SIFT (Object Recognition from Local Scale-Invariant Features) [13], HOG (Histograms of Oriented Gradients) [14], or DPM (Deformable Parts Models) [15].

In the last two decades, the use of ANNs became popular in computer vision benefitting from powerful GPU systems. The principle of ANNs was firstly described in the 1960s. It was not until 1986 when Rumelhart et al. [16] introduced a faster approach to backpropagation, which, to this day, is the base technique of learning ANNs. Modern approaches to computer vision include R-CNN (Regions with Convolutional Neural Network features) [17–19], YOLO (You Only Look Once) [20–22], or SSD (Single Shot Detector) [23].

Deformable Parts Models
DPM [15] uses a sliding window approach to object detection. DPM assumes an object is constructed by parts. For instance, a person's face consists of two eyes, a nose, and a mouth. When it finds a match of the object (face), it then fine-tunes the result using the model parts (eyes, nose, mouth). It is based on the idea of penalization when a part is not at the position that it is expected to be – something that is supposed to be a face and does not contain eyes and mouth is probably not a face. DPM uses a multi-stage pipeline for extracting static features, classifying regions, and predicting bounding boxes. This makes it slower in comparison with machine learning approaches like YOLO [20] or R-CNN that use only a single pipeline.

R-CNN
R-CNN is an object detection system first published in 2014 by Girshick et al. [17]. The system consists of three modules. The first generates region proposals, the second is a CNN that extracts features from each region, and the third is a set of class-specific SVMs. The main advantage in comparison to previous systems like DPM [15] is the precision of the detection. It achieves 53.3% mAP (mean average precision) on PASCAL VOC (Visual Object Classes) 2012 dataset [24].

Fast R-CNN [18] is an improvement to R-CNN. It is focused on increasing speed and detection accuracy. Training is 9 times faster, and testing is 213 times faster than for R-CNN. The system consists of only one pipeline compared to three in R-CNN. The architecture contains several convolutional and max-pooling layers, which, together with fully connected layers, replace all three modules of the previous solution. On the output, there are two sibling layers. The first outputs probability distribution for categories, and the second returns bounding box offsets for each category. The system achieves 68.4% mAP on PASCAL VOC 2012 dataset [24].

Faster R-CNN [19] further improved Fast R-CNN. This detection system achieves 5 FPS at test time while having 70.4% mAP on PASCAL VOC 2012 dataset [24].

In the later years, the method was further improved by some authors, e.g., Mask R-CNN [25], Cascade R-CNN [26], Deep Residual Learning [27], or Rethinking the Faster R-CNN [28].

YOLO

YOLO [20] is an object classifier and detector developed in 2016. The architecture focuses on real-time detection of objects while maintaining good precision, which is something that previous approaches did not achieve. YOLO uses only a single neural network that predicts bounding boxes and class probabilities in one evaluation. The network has 24 convolutional layers, followed by 2 fully connected layers. The base YOLO model processes an image at 45 FPS and achieves more than twice the mAP of other real-time systems. On PASCAL VOC 2012 dataset [24], it reaches 57.9% mAP. Although, as the authors point out, the architecture has problems with precise localization. However, at the same time, it is less likely to predict false positive on the background because it looks at the whole image, and the neural network can decide on the global context. DPM [15] and R-CNN [17] use different approaches. YOLO also struggles with detections of small objects [20].

YOLOv2 [21] is an improved version of YOLO from the same authors. Their main goal was to improve localization when maintaining the speed of detection and classification accuracy. The reworked network has only 19 convolutional layers and 5 max-pooling layers. They added batch normalization, which led to better convergence and a 2% improvement in mAP. The original version of YOLO was trained with 224×224 resolution, and YOLOv2 increases it to 448. Further improvements include dimension clusters, which make it easier for the network to learn. Systems like R-CNN [17] use handpicked anchor boxes. YOLOv2 detection was improved with anchor boxes, but the improvement was small. The authors decided to run k-means clustering on the training set bounding boxes to find them automatically. That provided better improvements that manually picked anchor boxes. They did not use standard k-means with Euclidean distance but rather Intersection over Union (IoU) scores, which are independent of the size of the box. In the end, $k = 5$ was chosen. YOLOv2 achieves 73.4% mAP on the PASCAL VOC 2012 dataset [24]. The performance is comparable with other detectors like Faster R-CNN and SSD while being faster [21].

YOLOv3 [22] is the latest version of the YOLO detector and classifier and was published in 2018. The new model contains 53 convolutional layers. K-means clustering now uses 9 clusters. YOLOv3 predicts boxes at three different scales using a similar concept to feature pyramid network. The original YOLO model struggled with small objects. YOLOv3 uses multi-scale predictions and has worse performance on larger objects. YOLOv3 has a comparable mAP to RetinaNet [29] and Faster R-CNN [19] while being 3-4 times faster [22].

3.3 ANNs on Smartphones

ANNs deployed on smartphones usually must deal with a lower memory capacity and lower computational power. These devices also usually run on a battery, and this fact must be taken into consideration too. Some authors investigated the possibilities of deploying neural networks on smartphones and other devices. The most relevant and interesting works are summarized.

El Bahi and Zatni [11] propose a system of processing document images directly on a smartphone with the use of ANNs. Their goal is to recognize text in the documents. The processing includes pre-processing that detects the document and an ANN architecture that detects the text by lines. The architecture combines convolutional and recurrent neural networks.

Bhandari et al. [30] present a solution for driving lane detection on a smartphone. They argue that providing lane information with only GPS is not accurate enough. Lane detection is vital to deciding whether a car is in the correct lane for making a turn or if the car's speed is compliant with a lane-specific speed limit. Sensors like an accelerometer can detect a lane change, but they are not able to keep the information fresh over a long period when the car is not changing lanes. The authors propose a system for lane detection with a smartphone camera and processing the images with ANNs. They achieve over 90% accuracy of the vehicle's lane position. The system is implemented as an Android application.

Ignatov et al. [31] did an artificial intelligence benchmark of running ANNs on Android smartphones. They present several tests that measure the performance of running ANNs, for instance, image recognition, face recognition, image deblurring, image enhancement, and more. They also present an overview of hardware acceleration for the execution of ANNs. The authors did follow-up research in 2019 with similar tests but with to date hardware [32]. However, a significant drawback of both papers is that they are focused solely on the Android platform.

Niu et al. [33] present a work that focuses on the optimization of ANNs for running on smartphones. Model compression usually degrades the accuracy, so they propose a framework for more efficient and advanced compression to minimize the accuracy loss and maximize the compression rate. They were able to achieve up to 8.8 times faster test time over TensorFlow Lite compression.

Begg and Hasan [34] describe the use of ANNs in home automation. Sensors data are too varied, and a neural network can find relationships and patterns in them. In the paper, they sum up the basics of neural networks and present relevant works to the field of deployment ANNs in smart homes. This means that ANNs are not limited only to smartphones but can be deployed to a broad range of smart devices.

4 Technologies

Many technologies are assisting with training, deployment, and migration of ANNs. Among the most popular libraries and frameworks for training and deployment are Keras, TensorFlow, and PyTorch. Both major smartphone operating systems, Google's Android and Apple's iOS, have their tools for the migration of ANN models. Migration on Android is possible via TensorFlow Android runtime, and migration on iOS is done via CoreML.

4.1   Keras

Keras [35] is a high-level artificial neural networks API (application programming interface). It is written in Python, and it allows designing and training of ANNs. It offers a set of abstractions that make it easy to develop neural network models. Keras models can be deployed to a range of platforms and environments, including Android, iOS, web browser, Google Cloud, or Raspberry Pi [36].

Keras provides support for many kinds of layers, activation function, optimizers, loss functions, and more. Apart from standard layers, Keras offers convolutional layers, pooling layers, recurrent layers, and others less commonly used. [35].

Keras itself currently has three backend implementations: TensorFlow by Google, CNTK by Microsoft, and Theano by the University of Montreal [37].

4.2   TensorFlow

TensorFlow [38, 39] is a library for machine learning developed by Google. It is one of the implemented backends for Keras. TensorFlow can also be used in the web browser or Node.js environments with TensorFlow.js [40] version of the library, which supports not only deployment but also building and training of the models. TensorFlow Lite (TF Lite) [41] is a variant for mobile and embedded devices. It provides compression of trained models, so they run smoothly on devices that have less computational power.

Many companies are using TensorFlow in their commercial applications [42]. Airbnb uses TensorFlow to categorize listing photos of people's homes. PayPal is using the library for payment fraud detection. Twitter developed a system for ranking tweets to show users the most important tweets first.

4.3   PyTorch

PyTorch is a machine learning library. It supports the designing and training of ANNs on GPUs. It has interfaces written in Python and C++ [43].

PyTorch Mobile is a version of the library that allows the deployment of PyTorch models on Android and iOS devices. The mobile version is focused on improved performance on mobile hardware [44].

4.4   Migration Frameworks

The utilization of the potential of ANNs on mobile devices requires the use of technologies that can migrate these networks. The following subsections describe several approaches that are available today. During the research, we have focused on the two most popular platforms – Android and iOS.

TensorFlow Lite and Android
TensorFlow Lite (TF Lite) is a system that provides support for migrating ANN models to Android and iOS devices. It is designed to execute models efficiently by taking into

consideration the limitations that smartphone environments pose, like less memory, running on a battery, or limited computational power. TensorFlow models must be converted into a format that can be used by TF Lite. Converting the model is necessary to reduce their file size and to optimize it for running on a mobile device while not affecting accuracy. It is also possible to convert the model more drastically, which can affect the accuracy of the trained model. TF Lite supports only a limited number of functions, which means that not all models can be converted. TF Lite converter is a special tool available in the Python environment which converts the models [45].

TF Lite is prepared to be used on both major mobile platforms – Android and iOS. Android uses Java and TF Lite Android Support Library. For the iOS environment, native iOS libraries written in Swift and Objective-C are available [45].

Google also offers the ML Kit for Firebase. It is possible to use it on Android and iOS, and it helps with the online distribution of the trained model through the Firebase services. This kit makes it easy to apply technologies like Google Cloud Vision API, or the Android Neural Networks API in a single SDK (software development kit) [46].

Neural Networks API (NNAPI) is available since Android 8.1 (API level 27). NNAPI offers support for efficient distribution of the computation workload across available processors, including dedicated neural network hardware (sometimes called TPU – Tensor Processing Unit), graphics processing units (GPUs), and digital signaling processors (DSPs). If the devices lack specialized hardware, the NNAPI executes the workload on CPU [47].

Core ML

Apple, as the developer of the iOS platform, also provides tools that help with the migration of existing ANNs to their devices. The tool is called Core ML [48], and it defines the format of models that can be deployed. The technology effectively uses the computational power of the devices, mainly CPU and GPU. Modern Apple devices also contain a so-called Neural Engine, which helps minimize memory and battery consumption [48].

The models are used directly on the device, which therefore does not require an Internet connection, and the data remains on the device as a safety precaution. Core ML itself contains a Vision module for analyzing images, Natural Language module to help with text processing, Speech module for audio to text conversion, and Sound Analysis module to support identifying sounds in audio. Before deploying the ANN model on the device, it is necessary to convert it into a format that is required by Core ML. Several tools support such conversion, for example, MXNet converter [49], or TensorFlow converter [50]. Apple recommends both converters [48].

5   Implementation

After careful consideration and analysis of all possibilities, we designed the following solution. Currently, the majority of users of our system are using Android devices, so we decided to make the pilot implementation for this platform. We are using TF Lite, so there is an easy possibility to migrate the model to iOS devices in the future. We decided not to use the ML Kit, which offers advantages like the possibility to update the underlying model without updating the application. Our application is not publicly available, and it is not necessary to update the model often. Therefore, we distribute the model inside the application installation file. The development was done in Android Studio 3.6.1.

The computational power of a typical mobile device is sufficient for running the application. The most crucial factor is the input image from the camera, which replaces an external scanner or manual inserting of images. This means that the quality of images from the camera is essential.

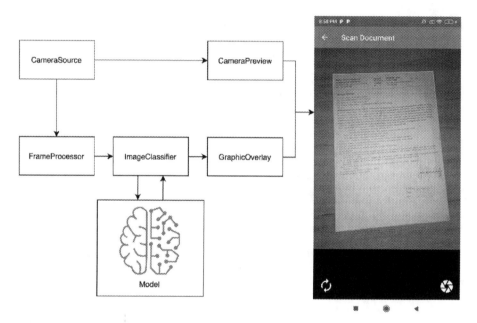

Fig. 2. The figure illustrates the workflow of our new solution in the form of a pseudo-class model.

The workflow of the application is as follows (Fig. 2). The first step is obtaining an image of the document from the camera. Class CameraSource is working with the camera stream, and its main task is to distribute the individual images taken by the camera in the highest frequency possible. Class FrameProcessor takes only a single image on its input. At first, it does a detection of the document in the input image. That is done with a custom function implemented with the help of OpenCV functions. Then the class transforms the document image into a format that is suitable as input for the detection. The detection is under the control of the class ImageClassifier. Its main job is processing the image by the neural network and receiving the results from it. Components CameraPreview and GraphicOverlay oversee rendering. CameraPreview renders the camera stream, and GraphicOverlay draws the results from the neural network.

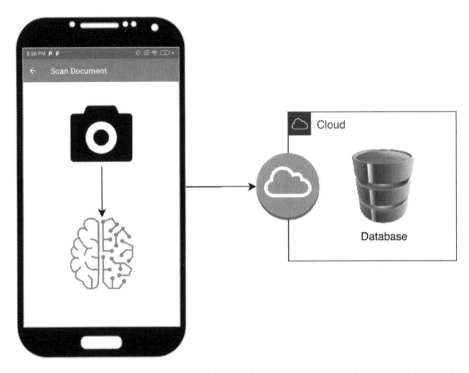

Fig. 3. The figure illustrates the new workflow of the system as compared to the workflow of the old system in Fig. 1. The user takes a picture of the document with the smartphone, which processes the document and sends the result on the server.

6 Testing and Results

Testing proved that the new solution provides faster processing than the old solution. The average time of processing one document using the old system was 75 s under ideal conditions. The time includes putting the paper on the scanner, the scanning process, and the detection process on the computer. Scanning was done with a standard end-user scanner, and since we require high-resolution images, the scanning process was taking a long time. Images are scanned in A4 format with 600 dpi and with deblurring and correction of document skew.

The average time of processing with the new solution is just 26 s. That includes putting the paper on a table, pointing at it with a smartphone, and pressing a button to take a picture. The application does the processing of the document, and the user is only required to confirm sending the data to the server. The document processing is taking only milliseconds even on low-end smartphones, and that means that it almost does not affect the total time. The new system is, on average, almost 3 times faster than the old solution. The testing was done on two devices, Xiaomi Mi MIX 2 and Huawei Honor 9.

7 Discussion

The main advantage of migrating the system to a smartphone is a faster overall procedure. By eliminating the slow scanning process and by automated image handling, we were able to achieve around 3 times faster processing.

Another main advantage of the solution is the adoption of cheaper hardware. Now we need only a smartphone instead of a scanner or standalone camera and a desktop computer. The user experience is better as the user is required only to take a picture with the smartphone, and all the next steps are automatic. The user is only asked to confirm to send the results to the server. That means that the overall solution is much simpler and less error-prone. The application requires only access to the camera and access to the Internet.

One possible disadvantage is battery consumption. The application requires much computational time, which leads to fast drainage of the battery. In our use case, this does not pose a disadvantage because the smartphone can always be plugged in.

In the end, we were able to process a single document almost 3 times faster while using fewer devices, which leads to cost saving in hardware and software.

8 Conclusions

Our main goal of migrating an existing artificial neural network (ANN) to a smartphone device was successfully achieved. Our testing proved that both major smartphone platform vendors, Google for Android and Apple for iOS, provide tools for easy conversion and deployment of trained ANNs. A detailed description of these tools was presented.

Our existing ANN was migrated from a desktop computer to a smartphone. Testing has shown that the network's outputs are the same as expected. With this approach, we were able to save time that is required to process a document. The processing is also more straightforward and, therefore, less error-prone.

In the future, our research will focus on the possibility of lowering the battery consumption and improvements in the deployment of the ANN.

Acknowledgement. This work and the contribution were supported by the project of Students Grant Agency – FIM, University of Hradec Kralove, Czech Republic.

References

1. Michel, E.: Using deep neural networks for automated speech recognition, p. 19 (2015)
2. Sriram, A., Jun, H., Gaur, Y., Satheesh, S.: Robust speech recognition using generative adversarial networks. In: 2018 IEEE International Conference on Acoustics, Speech and Signal Processing (ICASSP), Calgary, AB, pp. 5639–5643. IEEE (2018). https://doi.org/10.1109/ICASSP.2018.8462456

3. Këpuska, V., Bohouta, G.: Next-generation of virtual personal assistants (Microsoft Cortana, Apple Siri, Amazon Alexa and Google Home). In: 2018 IEEE 8th Annual Computing and Communication Workshop and Conference (CCWC), pp. 99–103 (2018). https://doi.org/10.1109/CCWC.2018.8301638

4. Shu, K., Sliva, A., Wang, S., Tang, J., Liu, H.: Fake news detection on social media: a data mining. Perspective (2017). https://doi.org/10.1145/3137597.3137600

5. Ignatov, A., Kobyshev, N., Timofte, R., Vanhoey, K., Van Gool, L.: DSLR-quality photos on mobile devices with deep convolutional networks. In: Presented at the Proceedings of the IEEE International Conference on Computer Vision (2017)

6. Amin, S.M., Rodin, E.Y., Liu, A.-P., Rink, K., García-Ortiz, A.: Traffic prediction and management via RBF neural nets and semantic control. Comput. Aided Civil Infrastruct. Eng. 13, 315–327 (1998). https://doi.org/10.1111/0885-9507.00110

7. Duan, Y., Lv, Y., Wang, F.-Y.: Travel time prediction with LSTM neural network. In: 2016 IEEE 19th International Conference on Intelligent Transportation Systems (ITSC), Rio de Janeiro, pp. 1053–1058. IEEE (2016). https://doi.org/10.1109/ITSC.2016.7795686

8. LeCun, Y., et al.: Backpropagation applied to handwritten zip code recognition. Neural Comput. 1, 541–551 (1989). https://doi.org/10.1162/neco.1989.1.4.541

9. Berger, A., Kostak, M., Maly, F.: Mobile AR solution for deaf people. In: Awan, I., Younas, M., Ünal, P., Aleksy, M. (eds.) MobiWIS 2019. LNCS, vol. 11673, pp. 243–254. Springer, Cham (2019). https://doi.org/10.1007/978-3-030-27192-3_19

10. Berger, A., Vokalova, A., Maly, F., Poulova, P.: Google glass used as assistive technology its utilization for blind and visually impaired people. In: Younas, M., Awan, I., Holubova, I. (eds.) MobiWIS 2017. LNCS, vol. 10486, pp. 70–82. Springer, Cham (2017). https://doi.org/10.1007/978-3-319-65515-4_6

11. El Bahi, H., Zatni, A.: Text recognition in document images obtained by a smartphone based on deep convolutional and recurrent neural network. Multimedia Tools Appl. 78(18), 26453–26481 (2019). https://doi.org/10.1007/s11042-019-07855-z

12. Papageorgiou, C.P., Oren, M., Poggio, T.: A general framework for object detection. In: Sixth International Conference on Computer Vision (IEEE Cat. No. 98CH36271), pp. 555–562 (1998). https://doi.org/10.1109/ICCV.1998.710772

13. Lowe, D.G.: Object recognition from local scale-invariant features. In: Proceedings of the Seventh IEEE International Conference on Computer Vision, vol. 2, pp. 1150–1157 (1999). https://doi.org/10.1109/ICCV.1999.790410

14. Dalal, N., Triggs, B.: Histograms of oriented gradients for human detection. In: 2005 IEEE Computer Society Conference on Computer Vision and Pattern Recognition (CVPR 2005), vol. 1, pp. 886–893 (2005). https://doi.org/10.1109/CVPR.2005.177

15. Felzenszwalb, P.F., Girshick, R.B., McAllester, D., Ramanan, D.: Object detection with discriminatively trained part-based models. IEEE Trans. Pattern Anal. Mach. Intell. 32, 1627–1645 (2010). https://doi.org/10.1109/TPAMI.2009.167

16. Rumelhart, D.E., Hinton, G.E., Williams, R.J.: Learning representations by back-propagating errors. Nature 323, 533–536 (1986). https://doi.org/10.1038/323533a0

17. Girshick, R., Donahue, J., Darrell, T., Malik, J.: Rich feature hierarchies for accurate object detection and semantic segmentation. In: Presented at the Proceedings of the IEEE Conference on Computer Vision and Pattern Recognition (2014)

18. Girshick, R.: Fast R-CNN. In: Presented at the Proceedings of the IEEE International Conference on Computer Vision (2015)

19. Ren, S., He, K., Girshick, R., Sun, J.: Faster R-CNN: towards real-time object detection with region proposal networks. In: Cortes, C., Lawrence, N.D., Lee, D.D., Sugiyama, M., Garnett, R. (eds.) Advances in Neural Information Processing Systems, vol. 28, pp. 91–99. Curran Associates, Inc. (2015)

20. Redmon, J., Divvala, S., Girshick, R., Farhadi, A.: You only look once: unified, real-time object detection. arXiv:1506.02640 [cs] (2016)

21. Redmon, J., Farhadi, A.: YOLO9000: Better, Faster, Stronger. arXiv:1612.08242 [cs] (2016)

22. Redmon, J., Farhadi, A.: YOLOv3: An Incremental Improvement. arXiv:1804.02767 [cs] (2018)

23. Liu, W., et al.: SSD: single shot multibox detector. In: Leibe, B., Matas, J., Sebe, N., Welling, M. (eds.) ECCV 2016. LNCS, vol. 9905, pp. 21–37. Springer, Cham (2016). https://doi.org/10.1007/978-3-319-46448-0_2

24. Everingham, M., Eslami, S.M.A., Van Gool, L., Williams, C.K.I., Winn, J., Zisserman, A.: The PASCAL visual object classes challenge: a retrospective. Int. J. Comput. Vis. **111**(1), 98–136 (2014). https://doi.org/10.1007/s11263-014-0733-5

25. He, K., Gkioxari, G., Dollar, P., Girshick, R.: Mask R-CNN. Presented at the proceedings of the IEEE international conference on computer vision (2017)

26. Cai, Z., Vasconcelos, N.: Cascade R-CNN: delving into high quality object detection. Presented at the proceedings of the IEEE conference on computer vision and pattern recognition (2018)

27. He, K., Zhang, X., Ren, S., Sun, J.: Deep residual learning for image recognition. Presented at the proceedings of the IEEE conference on computer vision and pattern recognition (2016)

28. Chao, Y.-W., Vijayanarasimhan, S., Seybold, B., Ross, D.A., Deng, J., Sukthankar, R.: Rethinking the faster R-CNN architecture for temporal action localization. Presented at the proceedings of the IEEE conference on computer vision and pattern recognition (2018)

29. Lin, T.-Y., Goyal, P., Girshick, R., He, K., Dollar, P.: Focal loss for dense object detection. Presented at the proceedings of the IEEE international conference on computer vision (2017)

30. Bhandari, R., Nambi, A.U., Padmanabhan, V.N., Raman, B.: Driving lane detection on smartphones using deep neural networks (2020). https://doi.org/10.1145/3358797

31. Ignatov, A., et al.: AI benchmark: running deep neural networks on android smartphones. In: Leal-Taixé, L., Roth, S. (eds.) ECCV 2018. LNCS, vol. 11133, pp. 288–314. Springer, Cham (2019). https://doi.org/10.1007/978-3-030-11021-5_19

32. Ignatov, A., et al.: AI benchmark: all about deep learning on smartphones in 2019. arXiv: 1910.06663 [cs] (2019)

33. Niu, W., Ma, X., Wang, Y., Ren, B.: 26 ms inference time for ResNet-50: towards real-time execution of all DNNs on smartphone. arXiv:1905.00571 [cs, stat] (2019)

34. Begg, R., Hassan, R.: Artificial neural networks in smart homes. In: Augusto, J.C., Nugent, C.D. (eds.) Designing Smart Homes. LNCS (LNAI), vol. 4008, pp. 146–164. Springer, Heidelberg (2006). https://doi.org/10.1007/11788485_9

35. Chollet, F., et al.: Keras (2015)

36. Why use Keras - Keras Documentation. https://keras.io/why-use-keras/. Accessed 25 Mar 2020

37. Backend - Keras Documentation. https://keras.io/backend/. Accessed 25 Mar 2020

38. TensorFlow. https://www.tensorflow.org/. Accessed 25 Mar 2020

39. Abadi, M., et al.: TensorFlow: a system for large-scale machine learning. Presented at the 12th {USENIX} symposium on operating systems design and implementation ({OSDI} 16) (2016)

40. TensorFlow.js|Machine Learning for Javascript Developers. https://www.tensorflow.org/js. Accessed 25 Mar 2020

41. TensorFlow Lite|ML for Mobile and Edge Devices. https://www.tensorflow.org/lite. Accessed 25 Mar 2020

42. Case Studies and Mentions. https://www.tensorflow.org/about/case-studies. Accessed 25 Mar 2020

43. PyTorch, https://www.pytorch.org. Accessed 30 Mar 2020

44. PyTorch Mobile. https://pytorch.org/mobile/home/. Accessed 30 Mar 2020
45. Get started with TensorFlow Lite. https://www.tensorflow.org/lite/guide/get_started?hl=cs. Accessed 01 Apr 2020
46. ML Kit for Firebase. https://firebase.google.com/docs/ml-kit?hl=cs. Accessed 01 Apr 2020
47. Neural Networks API|Android NDK. https://developer.android.com/ndk/guides/neuralnetworks?hl=cs. Accessed 01 Apr 2020
48. Core ML|Apple Developer Documentation. https://developer.apple.com/documentation/coreml. Accessed 01 Apr 2020
49. apache/incubator-mxnet. https://github.com/apache/incubator-mxnet. Accessed 01 Apr 2020
50. tf-coreml/tf-coreml. https://github.com/tf-coreml/tf-coreml. Accessed 01 Apr 2020

A Hybrid Deep Learning Approach
for Systemic Financial Risk Prediction

Yue Zhou[1]([⊠]) [iD] and Jinyao Yan[2] [iD]

[1] School of Information and Engineering, Communication University of China,
Beijing, China
masonzhy123456@163.com
[2] State Key Laboratory of Media Convergence and Communication,
Communication University of China, Beijing, China
jyan@cuc.edu.cn

Abstract. Systemic financial risk prediction is a complex nonlinear problem and tied tightly to financial stability since the recent global financial crisis. In this paper, we propose the Systemic Financial Risk Indicator (SFRI) and a hybrid deep learning model based on CNN and BiGRU to predict systemic financial risk. Experiments have been carried out over Chinese economic and financial actual data, and the results demonstrate that the proposed model achieves superior performance in feature learning and outperformance with the baseline methods in both single-step and multi-step systemic financial risk prediction.

Keywords: Financial risk · Deep learning · Time series prediction

1 Introduction

Systemic financial risk is a crucial issue in economic and financial systems. International experience shows that the outbreak of systemic financial risk almost always causes every financial crisis. Since the 1970s, the Bank for International Settlement (BIS) has begun to recognize the importance of systemic financial risk and integrated identification, monitoring and measurement systemic financial risk into the formulation of financial stability policies. The US mortgage crisis triggered the international financial crisis in 2008, which generated panic and chain reactions of the global economy and financial system, and still has a far-reaching impact on many countries and regions even now. Since then, a large amount of academic research has focused on systemic financial risk over the past decade from different research perspectives, including macroeconomics, econometrics, and complex network theory. Nowadays, financial risk and its related researches are established as a scientific field and provide significant contributions in supporting decision-making in theory and practice [1], and how to accurately measure and predict systemic financial risk so as to effectively prevent and defuse risk has become an active research area [27].

This work was supported in part by the National Natural Science Foundation of China under Grant 61971382.

The systemic financial risk includes low liquidity, inability to pay debts or dividends, continual reduction in profitability, and many other aspects of economic and financial information [29]. Traditional systemic financial risk prediction approaches can be mainly divided into three categories: composite indicator methods, risk contagion measurement methods and stress test methods. Composite indicator methods find out the early warning factors affecting systemic financial risk and use econometric, statistical, multivariate analytical methods to constructs financial stress indices to measurement systemic financial risk. For example, Illing et al. [15] developed the Financial Stress Index (FSI) to describe Canadian financial stability and proposed several techniques including generalized autoregressive conditional heteroscedasticity (GARCH) modelling to extract the information about financial risk, which is a groundbreaking piece of research in this strand of literature. Hollo et al. [12] introduced the Composite Indicator of Systemic Stress (CISS) and proposed a threshold vector autoregressive (VAR) model systemic financial risk level of the euro area. Duca et al. [7] developed a macro index for assessing systemic risks and predicting systemic events. Risk contagion measurement methods measure the risk spillover, analyze the transmission effect among financial institutions, and assess the possibility of systemic financial risk [10,14,20]. Stress test methods use sensitivity analysis, scenario analysis and other approaches to evaluate the financial industry's ability to withstand extreme events that may occur [8,17]. In general, systemic financial risk measurement research based on computer simulation and engineering methods is far fewer compared to econometric-based methods or other statistical analysis methods [27].

In recent years, with the rapid development of big data analysis technology and deep learning, computer engineering methods for forecasting financial time series have become a hot concept and a promising research field [18]. Recent studies have shown that financial time series forecasting is a challenging task, due to it exhibits high volatility and non-stationarity [30]. Traditional statistical models, machine learning methods and artificial neural networks have been widely investigated to deal with this problem in the area of financial markets, financial transactions ,and so forth [26,28]. The authors in [16] introduced ANN to predict daily stock price by using the Korea Composite Stock Price Index (KOSPI) as their dataset. [32] presented a framework where wavelet transforms, stacked autoencoders and long-short term memory (LSTM) are combined for stock index forecasting. In [3], a wavelet-neural time series model was used to forecast EUR/RSD exchange rate. The authors of [22] predicted the price of Bitcoin based on Bayesian optimized RNN and LSTM. [13] proposed an integrated system based on artificial bee colony algorithm to forecast stock markets In [23] and [9], LSTM was used to the prediction the movements of the equity price. The authors of [2] introduced stacked LSTM to predict time series data on the Bombay Stock Exchange (BSE). Besides, [4] and [19] proposed predictive methods based on gated recurrent unit (GRU) for financial time series.

However, due to the difficulty in obtaining data, there have been relatively few studies for systemic financial risk. Some scholars attempted to use machine learning methodologies to study financial risk measurement and early warning. [5] declared that they presented the first systemic risk model based on big data using a Bayesian approach, and the data are selected by two heterogeneous sources: financial markets and financial tweets. [25] briefly introduced some classic intelligent algorithms such as machine learning, network simulation and fuzzy systems for systemic risk. In the study of [21], they applied support vector machine (SVM) to the prediction of banking systemic risk and conducted a case study of the SVM-based prediction model for Chinese banking systemic risk. Nevertheless, these algorithms are mostly belonging to shallow model, which structure is usually no hidden layer nodes or very few hidden layers.

Summing up, systemic risk is frequently interrelated to various economic and financial factors closely, then predict it essentially is a multivariate time series prediction problem. Because of the dynamic instability and long-term dependence of the time series of systemic financial risk, in this paper, we attempt to develop a predictive approach based on convolutional neural network and bidirectional gated recurrent unit (CNN-BiGRU) to mine, form a model of relevant economic and financial time series data, and predict systemic financial risk trends. At first, we construct a new indicator of systemic financial risk related to macro-economic and financial market. Then, the time series data are fed into CNN-BiGRU networks to further mine information characteristics provided by different data sources in financial data, and establish the nonlinear relationship between the time series of multivariate and systemic financial risk. In order to verify the effectiveness of the proposed method, we compared the prediction performance of several popular learning methods in financial risk problem.

2   Methodologies

2.1   Systemic Financial Risk Measurement Method Base on Composite Indicator

The characteristics of risks are complicated, multifaceted, and concealment in a real-world financial system. Only focus on the risks taken by banks and other financial markets individually was not sufficient to prevent financial crises. According to standard definitions of systemic financial risk, we construct the Systemic Financial Risk Indicator (SFRI) based on the statistical design method, which is a reflection of contemporaneous stress in the financial system and considers the interrelationships between different risk sources. SFRI is inspired by FSI [15] which pays close attention to the financial system risk management. In contrast, we aimed to indicate overall macroeconomic activity and financial risk changes through SFRI, so we attempted to extend the abilities of FSI. In this process, we assume financial risk manifests if and when the indicators move together, and the simultaneous linkage degree of these indicators reflects systemic financial stress.

First of all, let us define a data set of a raw indicator x_t as $x = (x_1, x_2, \cdots, x_n)$, which has n total number of observations in the sample. x_t possesses a variety of distributions that leads to developing a compatible aggregation scheme with each indicator difficultly. Consequently, it is necessary to transform all indicators to guarantee compatibility before aggregation. We propose a transformation of raw systemic financial risk indicators based on their empirical cumulative distribution function (CDF). The CDF transformation changes the mean to achieve a common measure of central tendency and modifies the range and variance to achieve a common measure of dispersion. Specifically, we calculated this transform by involving the order statistics of each observation. We denote the ordered sample of x_t is x_t is $(x_{[1]}, x_{[2]}, \cdots, x_{[n]})$ where $x_{[1]}$ is the smallest and $x_{[n]}$ is the largest. In other word, we calculated the CDF transform as the rank of each observations divided by the cardinality of x_t similar to [24]. Then, x_t transformed into y_t based on the following empirical distribution function 1:

$$y_t = CDF_n(x_t) = \begin{cases} \dfrac{r}{n}, x_{[r]} \leq x_t \leq x_{r+1}, r = 1, 2, \cdots, n-1 \\ 1, x_t \geq x_{[n]} \end{cases} \tag{1}$$

$CDF_n(\chi)$ measures the total number of observations x_t not exceeding a particular value χ compared to total number of observations in the data set. If a value x of observation occurs more than once, the ranking numbers are set as the average rankings that would have been assigned to each of the observations.

Next, all indicators were grouped into different categories including financial markets and economic spheres. In the absence of any evidence that one sub-market contributes more to systemic financial risk than another, we roughly calculated average arithmetic value of indicators for each sub-market (fields) and get the $SFRI$ for each sub-market, the formula is as follows:

$$s_{i,t} = \frac{1}{m} \sum_{j=1}^{m} y_{i,j,t} \tag{2}$$

where $s_{i,t}$ is the ith sub-market $SFRI$ at time t, $(i = 1, 2, \ldots, n)$; m is the number of feature in ith sub-market. This means that each indicator is given equivalent weight in the specific sub-market.

The SFRI is now calculated according to formula 3, and it is continuous, unit-free and bounded by the half-open interval $(0, 1]$.

$$SFRI = (w \odot s_t)C_t(w \odot s_t)' \tag{3}$$

where $w = (w_1, w_2, \cdots, w_n)$ is the vector of weights, and $s_t = (s_{1,t}, s_{2,t}, \cdots, s_{n,t})$ is the vector of indicators at time t. C_t is the matrix of time-varying cross-correlation coefficients $\rho_{ij,t}$ between indicators i and j:

$$C_t = \begin{pmatrix} 1 & \rho_{12,t} & \cdots & \rho_{1n,t} \\ \rho_{12,t} & 1 & \cdots & \rho_{2n,t} \\ \vdots & \vdots & 1 & \vdots \\ \rho_{1n,t} & \rho_{2n,t} & \cdots & 1 \end{pmatrix} \tag{4}$$

where $\rho_{ij,t}$ can calculated by exponentially-weighted moving averages (EWMA) method. The formula of EWMA is shown as 5:

$$\sigma_{ij,t} = \lambda\sigma_{ij,t-1} + (1 - \lambda)\widetilde{s}_{i,t}\widetilde{s}_{j,t}$$
$$\sigma_{i,t}^2 = \lambda\sigma_{i,t-1}^2 + (1 - \lambda)\widetilde{s}_{i,t}^2 \quad (5)$$
$$\rho_{ij,t} = \sigma_{ij,t}/\sigma_{i,t}\sigma_{j,t}$$

where $i = 1, 2, \cdots, n$; $j = 1, 2, \cdots, n$; $i \neq j$; $t = 1, 2, \cdots, T$. λ is the smoothing parameter (or decay factor), which is held constant through time.

In this paper, considering the economic and financial situation of developing countries, especially China, we selected prediction indicators of SFRI. Dynamics of financial risk are usually influenced by miscellaneous factors and different sources, such as macro economy, commercial bank, stock market, bond market, and so forth. We selected 36 features (from X_1 to X_{36}) of systemic financial risk characterization grouped into 8 sub-markets categories by synthetically considering the national development situation, degree of importance system, and data availability. The list of prediction indicators is specified in Table 1.

2.2 Convolutional Neural Network for Systemic Financial Risk Feature Extraction

With the help of input preprocessing, the quality of the financial data can be significantly improved. The convolutional neural network (CNN) is a hierarchical feedforward deep neural network model, which usually contains the convolutional layer and pooling layer. The convolutional layer consists of the convolution operation, which is utilized to extract local features for further processing by subsequent layers. The discrete convolution is defined in formula 6:

$$s(t) = (x * w)(t) = \sum_{a=-\infty}^{\infty} x(a)w(t - a) \quad (6)$$

where s denotes output which sometimes referred to as the feature map, t denotes time, w denotes kernel, a denotes variable. The pooling operation replaces the output of the net at a specific location with a summary statistic of the nearby outputs. CNN mainly includes three critical characteristics: sparse interactions, parameter sharing, and equivariant representations [11]. Hence, CNN architecture is not only excellent in image or vision processing (such as image classification, image segmentation, and object detection), but also proper for time series analysis [31]. In order to extract features of economic and financial data quickly and predict systemic financial risk trend accurately, we put forward to use 1-D convolutional neural network as a local feature extractor in this paper.

2.3 Bidirectional Gated Recurrent Unit for Systemic Financial Risk Time Series Prediction

Recurrent Neural Networks (RNNs) are one of the attention-attracting technologies of deep learning domain in recent years. RNNs are a family of neural

Table 1. List of prediction indicators of Chinese systemic financial risk

Fields	Indicators
Macro economy ($S1$)	Year-on-year growth of GDP (X_1)
	Year-on-year growth of CPI (X_2)
	Year-on-year growth of CGPI (X_3)
	Year-on-year growth of PPI (X_4)
	Year-on-year growth of in fixed assets total investment(X_5)
	Year-on-year growth of industrial added value (X_6)
	Difference between year-on-year growth of finance rev. and exp. (X_7)
	Registered urban unemployment rate (X_8)
Commercial bank ($S2$)	Deposit-to-loan (X_9)
	Ratio of medium and long term loans to total (X_{10})
	Liquidity ratio (X_{11})
	Capital adequacy ratio (X_{12})
	Non-performing loan ratio (X_{13})
	Year-on-year growth of M2 (X_{14})
	Ratio of Year-on-year growth of M2 to M1 (X_{15})
Stock market ($S3$)	P-E Ratios (X_{16})
	Year-on-year growth of listed companies circulation market value (X_{17})
	Year-on-year growth of total market value of listed companies (X_{18})
	Year-on-year growth of transaction amount of stock (X_{19})
Bond market ($S4$)	Difference between maturity rate of 5-year and 3-month bonds (X_{20})
	ChinaBond new composite index (X_{21})
	Yield gap with 6-month corporate bond to central bank bill (X_{22})
	Year-on-year growth in bond issuance (X_{23})
Monetary market ($S5$)	Fixing Repo Rate (FR007) (X_{24})
	Difference between 1-year and 1-week SHIBOR (X_{25})
	Difference of 1-week SHIBOR and 1-week LIBOR (X_{26})
Insurance market ($S6$)	Year-on-year growth of insurance payout) (X_{27})
	Year-on-year growth of premium income (X_{28})
Foreign exchange market ($S7$)	Real effective exchange rate index (X_{29})
	Year-on-year growth of foreign exchange reserves (X_{30})
	Year-on-year growth of total import and export volume (X_{31})
	Year-on-year growth of actual utilized FDI value (X_{32})
Real estate market ($S8$)	Year-on-year growth of real estate investment (X_{33})
	Residential housing price index (X_{34})
	Year-on-year growth of average price of commercial housing (X_{35})
	Year-on-year growth of total amount of commercial housing (X_{36})

networks for processing a sequence of value $x_1, \cdots, x_\tau$, which maintain a memory cell to store the history information and enables the model to predict the current output based on long-distance features [11]. However, there are some defects of simple RNNs, including vanishing and exploding gradient problems, which means it might take a very long time to learn long-term dependencies tasks. In order to mitigate the shortcomings of the former RNNs, some variants of RNN, like GRU, is proposed.

GRU is a kind of gate-based recurrent unit which has smaller architecture and comparable performance to the LSTM [6]. GRU has two gates(update gate and reset gate), this is the main difference with the LSTM, which is a single gating unit simultaneously controls the forgetting factor and the decision to update the state unit. Inside a GRU, the update gate $z(t)$ specifics which information can be retained to the next state, and the reset get $r(t)$ specific how the previous state information is combined with the new input information. The formula for the next output and state value in the GRU unit is as follow:

$$z_t = \sigma(W_z x_t + U_z h_{t-1} + b_z) \tag{7}$$

$$r_t = \sigma(W_r x_t + U_r h_{t-1} + b_f) \tag{8}$$

$$\widetilde{h}_t = tanh(W_h x_t + U_h(r_t \odot h_{t-1}) + b_h) \tag{9}$$

$$h_t = (1 - z_t) \odot h_{t-1} + z_t \odot \widetilde{h}_t \tag{10}$$

where σ is the activation function which generally uses *sigmoid* function; $x(t)$ denote the input; $h(t-1)$ denote the previous output; $b_i, b_f, b_o, b_c, U_z, U_r, U, h,$ and W_z, W_r, W_h respectively denote the biases, input weights, and recurrent weights into the GRU cell. Due to GRU has fewer parameters, it is faster to train and requires fewer samples.

The GRU model only has a forward pass, and each node will only be affected by the previous one. Nevertheless, for the financial time series prediction task, the features of the rear node will also be useful to the front node. Based on the GRU model, we employ a BiGRU networks which compose by two ordinary GRU, which processes the input sequence from two directions of time series (both chronological and anti-chronological), then merge their representations. The method of the bidirectional training model can provide more useful information in modeling. By viewing systemic financial risk data from two directions enables the model to get richer representations and capture patterns that may be ignored when using one-direction GRU, thereby improving the performance of ordinary GRU.

2.4 CNN-BiGRU Model for Systemic Financial Risk Prediction

From what has been discussed above, the complete methodology and overall framework of CNN-BiGRU model for systemic financial risk are shown in Fig. 1. The whole model is divided into four major parts.

1. SFRI is calculated and educed by original economic and financial data.
2. 1-D CNN used to acquire local features and dimensionality reduction on systemic financial risk indicators. The input sequences are processing by 1-D CNN through convolution and pooling operation, and the features of time series are selected.
3. the feature sequences are fed into the BiGRU networks, so that it can learn the time dependence relationship between the information extracted from 1-D CNN. By processing a sequence both way, BiGRU networks can catch patterns that may have been overlooked by one-direction GRU.
4. At the end of the framework, the fully connected layers are stacked.

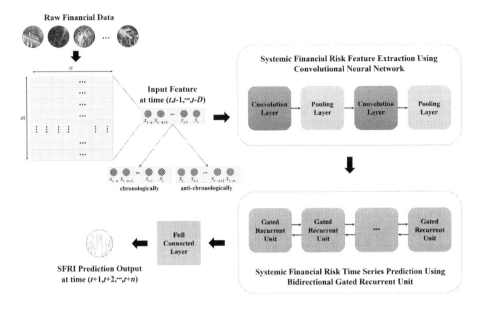

Fig. 1. The overall framework of convolutional neural network and bidirectional gated recurrent unit for systemic financial risk prediction.

3   Experiments and Results

In this section, the detailed steps of our experimental implementation and the analysis of the results are described. We choose the Chinese financial system as an example to evaluate the proposed prediction model for systemic financial risk and conduct empirical analysis based on the existing Chinese economic and financial data set.

3.1   Dataset and Data Preprocessing

The experimental dataset used for this paper is the financial time series data of China from the first trading day of 2016 to the end of March 2019. which contained 4837 independent observation points (daily). All data samples are from relevant departments of the Chinese government, including People's Bank of China (PBC), National Bureau of Statistics of China (NBS), National Development and Reform Commission (NDRC), China Bank and Insurance Regulatory Commission (CBIRC), and China Securities Regulatory Commission (CSRC). Because the sampling frequency of some variables is different, we converted data from one frequency to another by Denton method. which uses statistical interpolation minimizing the proportional first difference between the interpolated and high frequency target series. For example, data of year-on-year growth of GDP (X_1) in Table 1 moved from monthly to daily. Data normalization can avoid attribute over-branch of large-valued intervals with features of small-valued intervals and reduce computational complexity in the calculation process. According

1 and 2, we computed each sub-market $SFRI$. Then, we calculated sub-market weight based on the sub-market $SFRI$ by PCA (Principal Components Analysis) algorithm. According to the algorithm 3 mentioned above, the vector of weights $w = (w_1, w_2, \cdots, w_n)$ is $w =$ (0.1633, 0.1257, 0.0915, 0.1276, 0.0776, 0.1292, 0.1521, 0.133), and then we composite SFRI.

3.2   Performance Evaluation Metrics

In this paper, the loss function is defined by Mean Absolute Error (MAE). MAE is an excellent way to reflect the actual situation of the prediction error, which is defined as follow 11:

$$MAE_{(y',y)} = \frac{1}{n} \sum_{i=1}^{n} |y_i' - y_i| \tag{11}$$

Meanwhile, Root Mean Squared Error (RMSE), Mean Absolute Percentage Error (MAPE), Symmetric Mean Absolute Percentage Error (SMAPE) and R squared(R^2) is selected to evaluate the performance of the deep learning network, the calculation formula is as shown in formula 12, 13, 14, and 15.

$$RMSE_{(y',y)} = \sqrt{\frac{1}{n} \sum_{i=1}^{n} (y_i' - y_i)^2} \tag{12}$$

$$MAPE_{(y',y)} = \frac{1}{n} \sum_{i=1}^{n} |\frac{y_i' - y_i}{y_i}| \tag{13}$$

$$SMAPE_{(y',y)} = \frac{1}{n} \sum_{i=1}^{n} \frac{|y_i' - y_i|}{(y_i' + y_i)/2} \tag{14}$$

$$R^2 = 1 - \frac{\sum_{i=1}^{n} (y_i' - y_i)^2}{\sum_{i=1}^{n} (y_i' - \overline{y})^2} \tag{15}$$

which n is the total number of samples, y and y' denote the observation value and its forecast value respectively.

3.3   Experimental Configuration

The proposed CNN-BiGRU model for systemic financial risk prediction was trained on Keras framework with TensorFlow backend, while comparative machine learning methods are implemented by the scikit-learn library. At first, we set up the hyperparameters for the proposed models. The training implemented in mini-batches with a batch size of 120, and all the models trained for 100 epochs. In order to avoid the overfitting problem, a dropout factor used between layers with a probability of 0.2. If the loss of past epoch is higher than the current epoch, weight matrices are stored. The adam optimization algorithm was employed with learning rate of 0.001 The ReLU selected as the activation function, which has a significant impact on the convergence of random financial risk samples. Furthermore, all models used an early stopping condition during the training, which stops the training if the validation loss on the validation data does not change within ten training epochs. The training set used to train model while the test set used independently to validate the model accuracy and assess the performance of the model. We use a spilled ratio of 70% for training and 30% for testing in the experiments.

Table 2. Effect of the number of neurons in bidirectional GRU model.

Neurons	MAE	RMSE	MAPE	SMAPE	R^2
20	0.00859	0.01167	0.36039	0.27364	0.83431
40	0.00776	0.01131	0.22373	0.19852	0.852
60	0.00567	0.0087	0.16849	0.16765	0.91243
80	0.00553	0.00964	0.13683	0.14316	0.89263
100	0.00772	0.01196	0.21477	0.19577	0.89467
120	0.00732	0.01055	0.18523	0.19709	0.8712

In the CNN part, it contains two convolutional layers with activation functions of ReLU, and each layer have 120 convolution kernel (filter) as feature extractor, the length of the filter window size is 5. In the BiGRU part, the number of deep learning neurons set to an equivalent value chosen from a candidate set. Several experiments were performed, and the corresponding errors recorded in Table 2. The results show that with the increase of neurons of the hidden layer, the forecasting performance first improves significantly and then begins to deteriorate. Under the same configuration, over-fitting problems arise when neurons exceed 60. Hence, we set the number of neurons to 60 in the successive experiments. The training algorithm description is shown in Algorithm 1.

3.4   Experimental Results and Discussion

Preliminary Verification. After model training to convergence, we obtained the optimal model weights of the CNN-BiGRU systemic financial risk prediction. Figure 2 provides the calculated and single-step predicted values of SFRI from the proposed CNN-BiGRU model. The results of the multi-step prediction will be soon in a later section. From Fig. 2, it is observed that the green curve is satisfactory fitting the blue curve. It illustrates that the CNN-BiGRU model produced outstanding results that can follow the fluctuations of raw values during the testing set successfully. Hence, it proved that CNN-BiGRU architecture is adequate for predicting systemic financial risk. Further, in order to find the optimal CNN-BiGRU model, different convolutional layers were generated, and the experimental results are shown in Table 3, from which we can see that as the

Algorithm 1. The CNN-BiGRU model training algorithm

Input: The Original economic and financial data train_X/Y, valid_X/Y, test_X/Y at time $(t, t-1, \ldots, t-D)$.
Output: The SFRI prediction at time $(t+1, t+2, \ldots, t+n)$.

1: Initialization of time series data and model parameters.
2: **define** model
3:     **add** Convolution Layer (Conv1D)
4:     **add** Pooling Layer(MaxPooling)
5:     **add** Convolution Layer(Conv1D)
6:     **add** Pooling Layer(MaxPooling)
7:     **add** Bidirectional GRU
8:     **add** Full Connected Layer
9: **repeat**
10:     **Forward propagate** model with train_X
11:     **Backward propagate** model with train_Y
12:     **Update** model parameters
13:     train_MSE, train_MAE = model(train_X, train_Y)
14:     valid_MSE, valid_MAE = model(valid_X, valid_Y)
15: **until** train_MSE, valid_MSE remains the same in the previous iteration or reach certain threshold
16: test_MSE, test_MAE = model(test_X, test_Y)

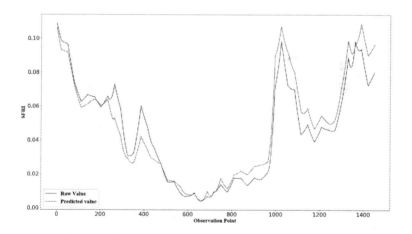

Fig. 2. Systemic financial risk prediction result of proposed model.

number of convolutional layers increases, the computation time of each epoch also increases but the performance of SFRI prediction is reduced. In terms of R^2, CNN-BiGRU model with two convolutional layers is 0.11 higher than CNN-BiGRU model with five convolutional layers, at the same time, computation time per epoch reduce by 42.7%. The reason for this may be that too many convolution layers may lead to overfitting. When the convolutional layers are two, the prediction accuracy of SFRI is the highest, which indicates that it may achieve the best effects. Therefore, we choose the CNN-BiGRU model with two convolutional layers.

Table 3. Performance comparison of the proposed model with different convolutional layers.

Layers	MAE	RMSE	MAPE	SMAPE	R^2	Time
1	0.0122	0.0194	0.3081	0.267	0.5409	15.6 s
2	0.0057	0.0087	0.1685	0.1677	0.9124	37.4 s
3	0.0064	0.0101	0.1324	0.1377	0.8762	42.5 s
4	0.0106	0.015	0.2202	0.2548	0.726	52.2 s
5	0.0085	0.0128	0.2009	0.2067	0.8012	65.3 s

Comparative Analysis. From the view of the existing literature, the studies of financial time series forecasting are focused on predicting the next price movement of the underlying asset, for instance, stock price prediction, bond price forecasting, commodity (such as oil, gold) price prediction, and so on [26]. However, there are few studies about financial risk trend prediction, and most of them focus on financial risks in a single field (bank bankruptcy risk) rather than systemic financial risks. To show the performance advantage of CNN-BiGRU architecture, we built several well-known deep learning methods for financial time series forecasting to evaluate the performance of the proposed model in comparative experiments. For fairness, all reference deep learning models in this experiment used the same hidden layers and the number of neurons, Besides, the input to other models was using the chronological data sample except the proposed model and BiGRU model.

- **RNN** [13,22]: Simple RNN network is a commonly used model for time-sequential data processing such as stock price forecasting, index prediction, and cryptocurrency price prediction.
- **LSTM** [9,23]: Standard LSTM model is a specialized version of RNN, which can remember both short term and long term values. LSTM network is the preferred choice of many researchers and developers in financial time series forecasting.
- **Stacked LSTM** [2]: Stacked LSTM is a model with multiple hidden LSTM layers, which can predict multiple future time steps of the financial market based on the historical data. Two LSTM layers are used in our experiments.
- **BiGRU** [4,19]: The GRU architecture is similar to the LSTM model and has some applications in processing financial time series data. The BiGRU network is selected to verify the influence of two-way inputs.

Figure 3 shows the single-step predicted values of SFRI from the compared models. Generally, these results summarize that the proposed CNN-BiGRU model achieves obvious superiority in terms of its performance compared with the baseline models. For example, the SMAPE value of the proposed method is 0.02, 0.08, 0.17 higher than those of stacked LSTM, LSTM, BiGRU, respectively. The performances of two-way training models are improved obviously compared with other traditional models. Also interesting is all comparison methods except the CNN-BiGUR model fails at the tail of SFRI time series. From the above analysis, it can be found that the CNN-BiGRU model is more sensitive to long-term dependencies features, which mainly attributed to the existence of bidirectional GRU architecture. Besides, we further focus on the comparison between CNN-BiGRU and BiGRU, which can be viewed as ablation studies that remove CNN parts from the proposed model. From Table 4, we observe that BiGRU performs significantly worse when used alone than the fusion architecture of CNN and BiGRU. Take MAPE for example, the value of the proposed method is 35% higher than that of BiGRU. This confirms that the implementations of CNN in the model proposed in this paper can capture richer local trend information systemic financial risk. To sum up, it demonstrated that the CNN-BiGRU model could learn intricate patterns from economic and financial data batter than others, which indicates that using the fusion scheme of CNN and BiGRU is the proper approach.

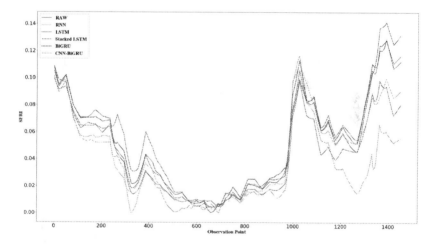

Fig. 3. The prediction results of compared models.

Multi-step Prediction Results. To investigate the capacity for multi-step prediction performance of different models, we considered seven horizon values. In Fig. 4, we can see that the prediction errors of all methods grow as the horizon increases. It indicates that the proposed CNN-BiGRU model is superior to others at all time horizons and achieves relatively a quality of higher precision with numerical stability for long-term prediction. Meanwhile, we can see that in the case of single-step prediction, the performance of stacked LSTM and the

Table 4. Performance comparison of different models for systemic financial risk prediction.

Method	MAE	RMSE	MAPE	SMAPE	R^2
RNN	0.01358	0.01814	0.33895	0.46517	0.59965
LSTM	0.0115	0.01667	0.2876	0.28151	0.66194
Stacked LSTM	0.00603	0.00814	0.18906	0.18057	0.90946
BiGRU	0.0104	0.01321	0.25978	0.24938	0.78755
CNN-BiGRU	0.00567	0.0087	0.16849	0.16765	0.91243

proposed model at nearly the same level. However, with the number of prediction steps increases, the CNN-BiGRU model performs much better than stacked LSTM. This implies that our proposed model with the fusion mechanism of convolutional layers and bidirectional gated recurrent units is significantly effective, which can extract features from historical data and accurate predicting trends in a long time. For instance, in terms of the 30-step prediction, the SMAPE value of the proposed method is improved 22.2% compared with stacked LSTM.

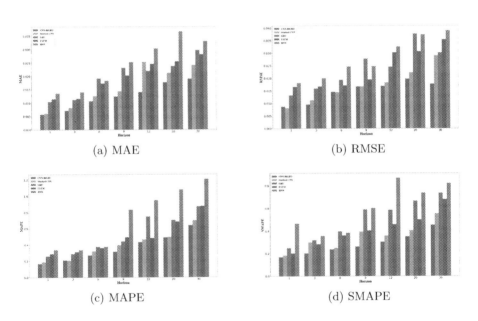

(a) MAE   (b) RMSE

(c) MAPE   (d) SMAPE

Fig. 4. The multi-step prediction performance of the compared models.

4   Conclusions and Future Work

Systemic financial risk prediction is a complex nonlinear problem and linked closely to financial stability. Dynamics of financial risk are usually influenced by miscellaneous factors and different sources. According to the complex nonlinear characteristics of financial risk, in this paper, we employ data-driven design method to construct SFRI, which can reflect the overall financial risk level of the entire financial system due to uncertainty and loss of forecasting changes. On this foundation, we proposed a model based on CNN-BiGRU to accurately predict degree and future trends of systemic financial risk.

To verify the performance of the method proposed in the paper, experiments have been carried out over Chinese economic and financial actual data collected from relevant departments of the Chinese government. Thirty-six features of Chinese systemic financial risk characterization from eight sub-markets by synthetically considering the national development situation, degree of importance system, and data availability, including macro economy, commercial bank, stock market, bond market, monetary market, insurance market, foreign exchange market, real estate market. The example verification results show that the CNN-BiGRU model can achieve better performance of systemic financial risk prediction than other baseline methods. It is probably because the CNN-BiGRU architecture processes the financial risk time series chronologically and antichronologically, which captures feature patterns that may be ignored by unidirectional deep learning models. At the same time, we discuss the multi-step prediction performance of the CNN-BiGRU model, and the results show that it could lead to more accurate prediction results than other baseline models. In the future, we will aim at verifying CNN-BiGRU model in systemic financial risk prediction of different countries and regions.

References

1. Aven, T.: Risk assessment and risk management: review of recent advances on their foundation. Eur. J. Oper. Res. **253**(1), 1–13 (2015)
2. Bhanja, S., Das, A.: Deep learning-based integrated stacked model for the stock market prediction. Int. J. Eng. Adv. Technol. **9**, 5167–5174 (2019)
3. Bozic, J., Babic, D.: Financial time series forecasting using hybrid wavelet-neural model. Int. Arab J. Inf. Technol. **15**, 50–57 (2018)
4. Buczkowski, P.: Predicting stock trends based on expert recommendations using GRU/LSTM neural networks. In: Kryszkiewicz, M., Appice, A., Ślęzak, D., Rybinski, H., Skowron, A., Raś, Z.W. (eds.) ISMIS 2017. LNCS (LNAI), vol. 10352, pp. 708–717. Springer, Cham (2017). https://doi.org/10.1007/978-3-319-60438-1_69
5. Cerchiello, P., Giudici, P.: Big data analysis for financial risk management. J. Big Data **3**(1), 18 (2016)
6. Chung, J., Gülçehre, Ç., Cho, K., Bengio, Y.: Empirical evaluation of gated recurrent neural networks on sequence modeling. CoRR abs/1412.3555 (2014)
7. Duca, M., Peltonen, T.: Assessing systemic risks and predicting systemic events. J. Bank. Finance **37**, 2183–2195 (2013)
8. Elsinger, H., Lehar, A., Summer, M.: Using market information for banking system risk assessment. Int. J. Central Bank. **2**(1), 137–165 (2006)
9. Fischer, T., Krauss, C.: Deep learning with long short-term memory networks for financial market predictions. Eur. J. Oper. Res. **270**(2), 664–669 (2017)
10. Glasserman, P., Young, H.: Contagion in financial networks. J. Econ. Lit. **54**, 779–831 (2016)

11. Goodfellow, I., Bengio, Y., Courville, A.: Deep Learning. MIT Press (2016)
12. Hollo, D., Kremer, M., Duca, M.: Ciss - a composite indicator of systemic stress in the financial system. SSRN Electron. J. **1** (2012)
13. Hsieh, T.J., Hsiao, H.F., Yeh, W.C.: Forecasting stock markets using wavelet transforms and recurrent neural networks: an integrated system based on artificial bee colony algorithm. Appl. Soft Comput. **11**, 2510–2525 (2011)
14. Huang, W.Q., Zhuang, X.T., Yao, S., Uryasev, S.: A financial network perspective of financial institutions' systemic risk contributions. Phys. A Stat. Mech. Appl. **456**, 183–196 (2016)
15. Illing, M., Liu, Y.: An index of financial stress for Canada. Bank of Canada, Working Papers, p. 14 (2003)
16. Jeon, S., Hong, B., Chang, V.: Pattern graph tracking-based stock price prediction using big data. Future Gener. Comput. Syst. **80**, 171–187 (2017)
17. Kenett, D., Levy carciente, S., Avakian, A., Stanley, H., Havlin, S.: Dynamical macroprudential stress testing using network theory. SSRN Electron. J. (2015)
18. Kou, G., Chao, X., Peng, Y., Alsaadi, F., Herrera-Viedma, E.: Machine learning methods for systemic risk analysis in financial sectors. Technol. Econ. Develop. Econ. **25**(5), 1–27 (2019)
19. Lee, S., Yoo, S.: Threshold-based portfolio: the role of the threshold and its applications. J. Supercomput. (2018)
20. Leitner, Y.: Financial networks: contagion, commitment, and private sector bailouts. J. Finance **60**, 2925–2953 (2005)
21. li, S., Wang, M., He, J.: Prediction of banking systemic risk based on support vector machine. Math. Prob. Eng. **2013**(1), 1–5 (2013)
22. McNally, S., Roche, J., Caton, S.: Predicting the price of bitcoin using machine learning, pp. 339–343 (2018)
23. Minami, S.: Predicting equity price with corporate action events using LSTM-RNN. J. Math. Finance **08**, 58–63 (2018)
24. Oet, M., Dooley, J., Ong, S.: The financial stress index: identification of systemic risk conditions. Risks **2015**, 420–444 (2015)
25. Sarlin, P.: Computational tools for systemic risk identification and assessment. Intell. Syst. Account. Finance Manage. **23**, 3–5 (2016)
26. Sezer, O.B., Gudelek, M.U., Ozbayoglu, A.M.: Financial time series forecasting with deep learning: a systematic literature review 2005–2019. Appl. Soft Comput. **90**, 106181 (2019)
27. Silva, W., Kimura, H., Sobreiro, V.A.: An analysis of the literature on systemic financial risk: a survey. J. Financial Stab. **28**, 91–114 (2017)
28. Taveeapiradeecharoen, P., Chamnongthai, K., Aunsri, N.: Bayesian compressed vector autoregression for financial time-series analysis and forecasting. IEEE Access **7**, 1 (2019)
29. Wang, L., Wu, C.: A combination of models for financial crisis prediction: integrating probabilistic neural network with back-propagation based on adaptive boosting. Int. J. Comput. Intell. Syst. **10**, 507 (2017)
30. Wen, M., Li, P., Zhang, L., Chen, Y.: Stock market trend prediction using high-order information of time series. IEEE Access **7**, 1 (2019)
31. Yang, J.B., Nhut, N., San, P., Li, X., Shonali, P.: Deep convolutional neural networks on multichannel time series for human activity recognition. In: IJCAI, July 2015
32. Yue, J., Rao, Y.: A deep learning framework for financial time series using stacked autoencoders and long-short term memory. PLoS ONE **12**(7), e0180944 (2017)

Ambient Intelligence Systems for the Elderly: A Privacy Perspective

Tendani Mawela[(✉)] [iD]

University of Pretoria, Lynnwood Road, Hatfield, Pretoria 0002, South Africa
tendani.mawela@up.ac.za

Abstract. Over the past few decades the significance of information and communication technologies has become apparent across various sectors of our modern society. In tandem with the global technological revolution and an increasing reliance on technology we have seen changes in the demographic profile of our populations. There is a worldwide trend of an increasing number of elderly people. Technology is seen as one option for supporting the elderly in maintaining social relationships, monitoring their health and living independently. One technological innovation that has been explored in various facets to support the elderly is Ambient Intelligence. This paper discusses the privacy implications of ambient intelligence based systems taking into account elderly users and adopters. Technological advancements and their applications should be understood from a socio-technical basis with regards to the potential impact on users and society at large. This research study adopted the systematic literature review method to identify, analyze and synthesize secondary data. This study appraises existing literature related to Ambient Intelligence, privacy concerns and the elderly over the past decade and contributes lessons and direction for practice and research.

Keywords: Ambient Intelligence · Artificial Intelligence · Assisted living · Senior citizen · Digital inclusion · Systematic review

1 Introduction

1.1 Background

Over the past few decades the significance of information and communication technologies (ICT) has become apparent across various sectors of our modern society. According to [1] the world is in the midst of the fourth industrial revolution (4th IR) which is driven by ICT innovations which will further transform the modus operandi of governments, businesses and society at large. The 4th IR is characterized by a variety of technological developments including artificial intelligence, nanotechnology, robotics, biotechnology, 3-D printing, cloud computing and the internet of things [2]. Recently we have seen scholars also refer to the concept of Society 5.0 which sees the emergence of a super smart society through systems that integrate advanced technologies resulting in a merging of the cyber and physical domains [3].

In tandem with this global technological revolution and an increasing reliance on technology we have seen changes in the demographic profile of our populations. There

© Springer Nature Switzerland AG 2020
O. Gervasi et al. (Eds.): ICCSA 2020, LNCS 12249, pp. 875–888, 2020.
https://doi.org/10.1007/978-3-030-58799-4_63

is a worldwide trend of an increasing number of elderly people [4]. It is noted that in 2019 there were approximately 703 million people over the age of 65 and this figure is expected to increase at a rapid rate in the coming decades [4]. Life expectancies have increased and as people age they face various health and physical challenges requiring more support to cope with their daily lives. Technology is seen as a viable option for supporting the elderly in maintaining social relationships, monitoring their health and living independently [5–8].

One technological innovation that has been explored in various facets to support the elderly is Ambient Intelligence (AmI) often referred to as Ambient Assisted Living (AAL) systems. AmI is a multi-disciplinary approach underpinned by technology with the aim to enhance the way environments and people interact with each other [9]. AmI brings intelligence into humans' spaces (such as homes or offices) through a network of sensors, intelligent devices and interfaces that are sensitive to the presence and needs of people [10, 38].

Privacy concerns are highlighted as a significant barrier to the uptake of AmI systems in the literature [52]. This paper, based on a systematic literature review (SLR) discusses the privacy perspectives of ambient intelligence based systems taking into account elderly users and adopters. Technological advancements and their applications should be understood from a socio-technical basis thus taking into account the potential impact on users and society at large. The paper is organized as follows: next it considers the informing literature and related works, then the research methodology is explicated. This is followed by a discussion of the results and thereafter conclusions are put forward.

2   Related Work

2.1   Ambient Intelligence Overview

The information society is a key driver behind AmI [14]. AmI is understood to be a digital environment that proactively, intelligently and sensibly supports humans with their daily lives [9]. AmI is ingrained in people's surroundings and responds to their needs so as to empower them [15, 16]. It is a multi-disciplinary concept that is influenced by several research areas including: Artificial Intelligence, Pervasive and Ubiquitous Computing, Sensor Networks as well as Human Computer Interaction (HCI) [9, 17].

AmI can be implemented using a variety of technologies however [17] highlights that AmI systems should display the following features:

1. Sensitive – sense and process data from its environment or users.
2. Responsive – respond to user's requirements and act based on the data.
3. Adaptive – understand the context and change accordingly.
4. Transparent – the technology and computing devices should not be visible and in effect unnoticeable.
5. Ubiquitous – computing devices are everywhere.
6. Intelligent – there must be artificial intelligence to enable the system to be aware, make decisions and adapt to the needs.

In essence AmI systems will sense data from their surroundings, interpret the data, make a decision and act accordingly [17]. AmI based systems have increased and we see their applications in areas such as smart cities, health care and safety, smart public spaces, smart schools, intelligent work places, smart homes, cyber physical systems, smart factories, smart products such as automobiles and manufacturing to highlight a few [9, 10, 14, 17, 18].

2.2   Ambient Intelligence Solutions, Applications and Research

Studies indicate that by 2050 at least one in six people globally will be aged 65 years or older [4]. The application of AmI systems in the context of ageing populations is growing. Examples of how AmI systems may be implemented for the elderly in practice include but are not limited to:

- In the health care sector where an elderly person residing alone may have a smart home that has various sensors and smart devices that can pick up when the person may be in danger and alert emergency services [59].
- For safety in the home such as fall detection capabilities and reminders for daily activities including consumption of medication [52].
- To support social participation e.g. allow for video calls or obtaining information about social events from the system [45]

Scholars and researchers have also sought to understand how the AmI services may assist the elderly taking into account the multi-disciplinary technology and people related aspects of the topic. Topics that have received scholarly attention include: addressing user motivations and technical solutions for AmI in support of ageing in place [19]. Alternatively [20] reviewed AmI and the social integration of the elderly. Additionally, [21] proposed a gerontechnological software architecture for emotion recognition, monitoring and regulation in the elderly. The exploration of an innovative neuro-fuzzy approach to design an adaptive and interoperable emotion aware AmI framework was conducted by [22]. There was also a project targeting the social inclusion of the elderly through AAL [23]. The scholarly perspectives in the literature vary and this study sought to focus on the privacy perspective of AmI systems with an interest in the elderly population. According to [33], the elderly can be considered as a special group in relation to the adoption and use of ICT's.

2.3   Privacy and Information Technology

The notion of privacy may be understood in a number of ways [65]. In the information systems domain the privacy of one's information or data is an individual's ability to control or influence the handling of data about themselves [30, 33]. The authors in [24] highlighted the need to understand privacy issues related to modern ICT's in an increasingly networked world. More recently we note that the rise of smart phone ownership, internet access, ubiquitous and pervasive computing, mobility, wearable technology, cloud computing and the upsurge of cyber threats indicate that technology related privacy considerations are an ongoing area of scholarly interest (see for example: [25–29, 31, 32]). From a physical and socio-technological context privacy can be considered in terms of [65]:

- Information privacy – the regulations that direct the assembling, storage and management of people's personal data.
- Bodily privacy – individual's privacy in terms of their physical bodies.
- Privacy of communications – which concerns the privacy of telephonic, email and other electronic communications.
- Territorial privacy – which may cover aspects such as ensuring privacy for individuals in their workplaces, homes as well as public areas.

Technology innovation brings much needed efficiencies and benefits to society however these should be balanced with the privacy risks [27, 31] posed to individuals such as the elderly. In [31] it is argued that further research on privacy is required. In the context of AmI systems users such as senior citizens are expected to share various personal data for instance location data and medical information such as their blood pressure [40]. Thus privacy of this information is essential. As [41] notes AmI systems should consider several significant non-technical factors such as affordability, legal, policy, regulatory and ethical issues such as privacy.

2.4   Literature Reviews on Ambient Intelligence

Researchers have previously undertaken literature reviews on the topic of AmI. The literature reviews have for example covered AAL technologies, products and services, their usability and accessibility with a focus on user interaction and how end users are incorporated into the AAL development and evaluation processes [34]. In contrast the authors in [35] sought to understand the ethical concerns regarding the research and development, clinical experimentation and application of AAL technologies for people with dementia. With the growth in social networking, the paper by [20] reports on an SLR focusing on how AmI and SNS technologies have been adopted for the social integration of senior citizens. Additionally, [36] aimed to provide a review of the AmI domain including technology solutions and their related challenges. A review of available software tools that end users may use to configure AmI based services for smart homes was conducted by [37]. An SLR on the sustainability of AmI services was completed by [8], who also offered an evaluation framework related to mobile AmI and a method to evaluate sustainability using fuzzy logic. This study appraises the existing literature related to AmI, privacy and the elderly over the past decade to illuminate lessons for practice and research.

3   Research Method

3.1   Systematic Literature Review

This research study adopted the systematic literature review (SLR) method. The SLR incorporates a structured and systematic method to identify, analyze and synthesize secondary data. In [11] the authors highlight that a review of the literature is an essential element of research and academic endeavors. Literature reviews may form the basis for building new knowledge through analyzing and synthesizing previous work. This study undertook the SLR to obtain insight into potential lessons from the extant literature regarding the proliferation of AmI based systems and their privacy related factors. The SLR process was guided by the recommendations of [11–13].

Research Questions

The paper addresses the following research questions:

- *Which privacy concerns are reported in the literature regarding Ambient Intelligence for the elderly?*
- *What privacy protection approaches are recommended in the literature regarding Ambient Intelligence for the elderly?*

Data Sources and Search Strategy

Four academic databases were included in the search for relevant literature. These were databases that were deemed to include high quality, peer reviewed conference and journal papers in relation to the scope of the paper:

- Scopus
- Association for Information Systems (AIS) eLibrary
- Association for Computing Machinery(ACM) Digital Library
- EBSCOHost - Academic Search Complete
  The following search string was adopted:
  "elderly" OR "older" OR "senior" AND "privacy" OR "data privacy" AND "ambient intelligence" OR "ambient assisted" OR "ambient assistance" OR "ambient assisted living"

Inclusion and Exclusion Criteria

The inclusion criteria that were defined and stipulated for the identification of articles were as follows:

- Peer reviewed journal articles or conference papers.
- Articles that address AmI in the context of elderly users.
- Articles that explicitly discuss the privacy aspects regarding the AmI related solutions.
- Articles published during the period 2009 – 2019 to allow for recent perspectives from scholars.

The exclusion criteria were stipulated as follows:

- Articles that are not written in English, are inaccessible or are published outside the stipulated timeframe.
- Articles that are not focused on the elderly population (60 years or older).
- Articles that do not discuss any privacy or ethical topics related to AmI use by the elderly.

Search Results

The papers that were sourced were managed through the EndNote X9 reference manager tool. The initial search process yielded results from the various database sources and are highlighted in the following table (Table 1):

Table 1. SLR search results

Database	Number of papers returned
Scopus	99
AIS E-Library	26
ACM digital library	566
EBSCOHost	8
TOTAL	699

The following figure outlines the process for identifying qualifying papers from the search results for inclusion in the final analysis (Fig. 1).

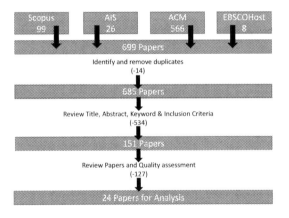

Fig. 1. Identifying papers for analysis.

3.2 Data Analysis

A thematic analysis was conducted to identify themes emanating from the papers. The overarching themes were linked to the objectives of the paper which were to identify privacy concerns and privacy protection strategies in the papers.

4   Findings and Discussion

4.1   Overview of the Papers

A total of 24 paper were included in the analysis. The following section provides an overview of the papers included for analysis. The figure below highlights the distribution of papers over the time period of interest which was from 2009–2019 (Fig. 2).

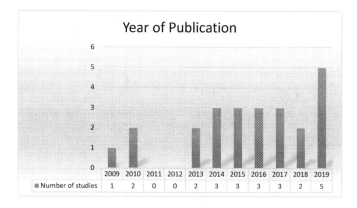

Fig. 2. Publication by year.

The figure below highlights the research design adopted in the papers. This included surveys, case study and experiments (Fig. 3).

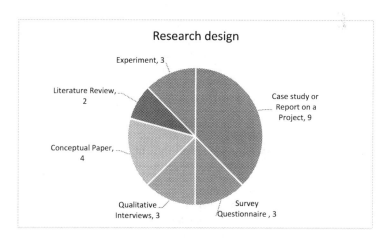

Fig. 3. Publication research design.

4.2   AmI Privacy Concerns and Privacy Protection Approaches

As argued by [35] privacy is an important matter as it relates to AmI based systems. The papers that were included for analysis addressed a myriad of privacy related concerns touching on both technical and human aspects. The following section addresses the research question: Which privacy concerns and privacy protection approaches regarding AmI for the elderly are noted in the literature?

Privacy Concerns
The reported privacy concerns highlighted in the analysis of the data are:

- Senior citizens are concerned with who has access to their data and if they are able to manage the access to their data [56]. Also of concern is the risk of third parties having access to their data without their knowledge [52] coupled with the informed consent of elderly users [55]. This is further linked to the need for protection from the disclosure of sensitive or harmful information [56].
- Elderly users of AmI systems are concerned about their data being misused [45] and the appropriate sharing of data [60].
- Also of concern is how the elderly can have a say and control with regards to how the data that has been collected is used by collectors (e.g. AmI providers or health practitioners) [40]. This control over their data is an important consideration [41, 44]. AmI systems are often everywhere and constantly collecting data making control over data and the purpose for which the data is used difficult for the elderly [41].
- The risk of being under constant surveillance is worrying to some elderly adopters of AmI [45] and the authors in [46] highlight the threat of what they term "dataveillance".
- An additional concern that comes with the loss of privacy due to ongoing surveillance is how this may affect the elderly e.g. it may cause some individuals to feel stigmatized or have a sense of losing their independence, dignity and freedom even in their own homes [43, 49]. The impact of such systems on the elderly must be understood and balanced with the need for the service e.g. a health monitoring or safety service.

Privacy Protection Approaches
The literature highlighted the following for consideration towards enhancing the privacy of the elderly throughout the use of AmI systems:

- When compiling AmI services consider why the elderly seek privacy since there may be different requirements and needs for privacy depending on individuals and these should inform AmI systems [56].
- AmI services organizations are encouraged to have an architecture comprising of a myriad of privacy related policies informing different domains such as privacy policies covering the ubiquitous environment, privacy policies covering the user, policies concerning the devices and policies addressing communications e.g. public key certificates [41, 60].

- One approach to enhance privacy is through minimizing information asymmetry in the context of AmI by increasing information flows from the data collector and users. Where the elderly are not able to use the information it is recommended that family members may act as proxies to support the elderly [40].
- From a design, development and implementation perspective companies should consider raising the awareness of developers and engineers on the ethical and privacy implications of their solutions such that these are taken into consideration [39]. Additionally documenting privacy based guidelines to inform product development is noted. Designers and developers may seem far removed from the elderly users however it is purported that building solutions that take the values of privacy into account may contribute positively. This can be achieved through a "privacy by design" approach where designers, developers and legal expertise collaborate to enhance privacy [41, 60].
- AmI services must be framed within the relevant technical standards and legal statutes. There should also be a consideration of conducting privacy impact assessments [60] to identify potential areas of concern before implementation.
- Rather than collecting and transferring a vast array of data AmI solutions may consider limiting the data collected and transmitted to only what is essential and necessary to deliver a service [42]. So the intention for accessing particular types of data is a central guiding principle [59]. For example, if an elderly person is at risk of falling due to health related constraints, only data that is related to this aspect should be collected to support the senior citizen's privacy. Another example is to avoid capturing visual data (e.g. data collected via cameras) [61] but if visual data is needed then organizations should place a limitation on what is collected and stored by making modifications that enhance individuals' privacy [48].
- With the growth of the AAL industry and AmI services, employees in this sector should also be trained in ethical issues spanning data protection and security, privacy and user monitoring limitations and where possible a specific job role geared to ethics should be included in the organizational structure of the AAL service providers [47].
- Additional technology related strategies were noted from the literature such as: consideration of a suite of privacy enhancing technologies, the use of authentication protocols [50], proxy based authentication [57], cryptography and related security tools and methods [51, 53], also consider technology that is non-invasive [58] and has privacy preserving features [54].

4.3   Privacy Implications of AmI for the Elderly

It is noted that as the adoption of technology increases, the concerns regarding personal information privacy in the digital era also grows [62]. Scholarly research related to privacy is on the rise across the engineering, social science and computer science domains [62]. The elderly are increasingly reliant on technologies such as AmI for assistance with daily living, their safety and health. Due to the potential significance of the AmI services for the elderly they can be more vulnerable since it may be argued that the decision to not use the technologies may not be an easy one. Thus enhancing their privacy as related to AmI services remains a pertinent discussion.

One potential approach that may be elevated is the de-identification of the elderly users' data. The potential of further exploiting the de-identification of data for the context of AmI requires additional probing. AmI may enable privacy invasion since it allows for the collection, transfer and storage of various information including personal identifiers [65]. De-identification outlines a process of taking away personally identifiable information to protect individuals [67]. De-identification does however allow for relevant stakeholders to extract identifiers if required which may link the information to an individual [65]. Various types of de-identification may be relevant for AmI solutions such as: gender, age, race de-identification, or scars, marks and tattoo de-identification, also face, voice, gait and gesture de-identification as a few examples. De-identification albeit a recommended approach requires the collaboration of social scientists and technology experts to address social, political and the legal implications of de-identification [65].

Additionally, the issue of privacy needs to move beyond a technical focus. One proposal is that privacy laws, regulations and technology must be understood from different perspectives such as people, processes, technology and information dimensions [66]. This may enable a holistic approach to addressing privacy concerns for innovations such as AmI.

It is also proposed that opportunities for enhancing privacy from the perspective of the end user are considered. This may include training and awareness initiatives for the elderly regarding the options available to improve privacy and how to protect themselves while using a particular AmI solution. The elderly may require additional support when using technologies [63]. Technology based services should include training on how they may be used effectively [63] and the AmI specific training should include ways to enhance one's personal privacy. This training should ideally consider how older adults learn and include methods such as adaptive learning approaches [63]. It is also essential to consider both system and user aspects in the learning approaches [64] to favor the elderly. AmI providers may consider incorporating additional and easily accessible support channels for elderly users who may have concerns for their privacy.

5   Conclusion

5.1   Concluding Remarks

This paper reported on a systematic literature review that appraised the current literature on Ambient Intelligence to understand the privacy issues highlighted by scholars in relation to the elderly. The paper presented the privacy concerns and followed with potential strategies and approaches for enhancing privacy protection. Twenty four (24) primary sources were analyzed to address the research questions. The discussion adopted a socio-technical view elevating both the technical and people related privacy considerations captured in the literature. The results show that although AmI based solutions hold potential for the elderly, the benefits should be considered in tandem with the risk of privacy intrusion.

5.2   Contribution, Limitations and Future Research

This study contributes through bringing together several privacy related concerns pertaining to AmI for the elderly as published in the literature over the past decade. All research studies have limitations and this study was limited to data from four academic databases with regards to the search for articles. Future studies may consider extending the search to additional databases. Scholars and researchers may consider adding to the current knowledge on AmI and privacy for the elderly by tackling human aspects that appreciate that the elderly are not a homogenous group. Thus future research should investigate how different groups within the elderly population perceive privacy concerns. For instance different privacy attitudes may be held by various senior citizens influencing their requirements of AmI systems.

References

1. Schwab, K.: The fourth industrial revolution. World Economic Forum. Currency (2017)
2. Hirschi, A.: The fourth industrial revolution: issues and implications for career research and practice. Career Dev. Q. **66**, 192–204 (2018)
3. Gladden, M.E.: Who will be the members of Society 5.0? towards an anthropology of technologically posthumanized future societies. Soc. Sci. **8**, 148 (2019)
4. United-Nations: World Population Ageing Report 2019. United Nations (2020)
5. Shishehgar, M., Kerr, D., Blake, J.: A systematic review of research into how robotic technology can help older people. Smart Health **7**, 1–18 (2018)
6. ten Bruggencate, T., Luijkx, K.G., Sturm, J.: Friends or frenemies? the role of social technology in the lives of older people. Int. J. Environ. Res. Public Health **16**, 4969 (2019)
7. Baecker, R., Sellen, K., Crosskey, S., Boscart, V., Barbosa Neves, B.: Technology to reduce social isolation and loneliness. In: Proceedings of the 16th international ACM SIGACCESS Conference on Computers & Accessibility, pp. 27–34 (2014)
8. Acampora, G., Cook, D.J., Rashidi, P., Vasilakos, A.V.: A survey on ambient intelligence in healthcare. Proc. IEEE **101**, 2470–2494 (2013)
9. Augusto, J.C.: Ambient intelligence: basic concepts and applications. In: Filipe, J., Shishkov, B., Helfert, M. (eds.) ICSOFT 2006. CCIS, vol. 10, pp. 16–26. Springer, Heidelberg (2008). https://doi.org/10.1007/978-3-540-70621-2_2
10. Štofová, L., Szaryszová, P., Bosák, M., Tarča, A., Hajduová, Z.: Ambient intelligence for increasing innovation performance of enterprises. In: 2018 14th International Scientific-Technical Conference on Actual Problems of Electronic Instrument Engineering, APEIE 2018 - Proceedings, pp. 452–458 (2018)
11. Webster, J., Watson, R.T.: Analyzing the past to prepare for the future: writing a literature review. MIS Q. **26**, xiii-xxiii (2002)
12. Kitchenham, B., Brereton, O.P., Budgen, D., Turner, M., Bailey, J., Linkman, S.: Systematic literature reviews in software engineering–a systematic literature review. Inf. Softw. Technol. **51**, 7–15 (2009)
13. Okoli, C., Schabram, K.: A guide to conducting a systematic literature review of information systems research (2010)
14. Gams, M., Gu, I.Y.-H., Härmä, A., Muñoz, A., Tam, V.: Artificial intelligence and ambient intelligence. J. Ambient Intell. Smart Environ. **11**, 71–86 (2019)
15. Cook, D.J., Augusto, J.C., Jakkula, V.R.: Ambient intelligence: technologies, applications, and opportunities. Perv. Mob. Comput. **5**, 277–298 (2009)

16. Bravo, J., Cook, D.J., Riva, G.: Ambient intelligence for health environments. J. Biomed. Inform. **64**, 207–210 (2016)
17. Corno, F., De Russis, L.: Training engineers for the ambient intelligence challenge. IEEE Trans. Educ. **60**, 40–49 (2016)
18. Aly, S., Pelikán, M., Vrana, I.: A generalized model for quantifying the impact of Ambient Intelligence on smart workplaces: applications in manufacturing. J. Ambient Intell. Smart Environ. **6**, 651–673 (2014)
19. van Hoof, J., Kort, H.S.M., Rutten, P.G.S., Duijnstee, M.S.H.: Ageing-in-place with the use of ambient intelligence technology: perspectives of older users. Int. J. Med. Inf. **80**, 310–331 (2011)
20. Campos, W., Martinez, A., Sanchez, W., Estrada, H., Castro-Sánchez, N.A., Mujica, D.: A systematic review of proposals for the social integration of elderly people using ambient intelligence and social networking sites. Cogn. Comput. **8**, 529–542 (2016)
21. Castillo, J.C., et al.: Software architecture for smart emotion recognition and regulation of the ageing adult. Cogn. Comput. **8**(2), 357–367 (2016). https://doi.org/10.1007/s12559-016-9383-y
22. Acampora, G., Vitiello, A.: Interoperable neuro-fuzzy services for emotion-aware ambient intelligence. Neurocomputing **122**, 3–12 (2013)
23. Cornacchia, M., Papa, F., Nicolò, E., Sapio, B.: Getting and keeping aged people socially included. In: Ramos, C., Novais, P., Nihan, C.E., Corchado Rodríguez, J.M. (eds.) Ambient Intelligence - Software and Applications. AISC, vol. 291, pp. 99–106. Springer, Cham (2014). https://doi.org/10.1007/978-3-319-07596-9_11
24. Palen, L., Dourish, P.: Unpacking "privacy" for a networked world. In: Proceedings of the SIGCHI conference on Human factors in computing systems, pp. 129–136 (2003)
25. Eastin, M.S., Brinson, N.H., Doorey, A., Wilcox, G.: Living in a big data world: predicting mobile commerce activity through privacy concerns. Comput. Hum. Behav. **58**, 214–220 (2016)
26. Gajanayake, R., Iannella, R., Sahama, T.: Privacy by information accountability for e-health systems. In: 2011 6th International Conference on Industrial and Information Systems, pp. 49–53 (2011)
27. Cha, S., Hsu, T., Xiang, Y., Yeh, K.: Privacy enhancing technologies in the internet of things: perspectives and challenges. IEEE Internet Things J. **6**, 2159–2187 (2019)
28. Alsaffar, N., Ali, H., Elmedany, W.: Smart transportation system: a review of security and privacy issues. In: 2018 International Conference on Innovation and Intelligence for Informatics, Computing, and Technologies (3ICT), pp. 1–4 (2018)
29. Singh, S., Goel, N.: Efficient framework approach to extract privacy issues in cloud computing. In: 2015 Fifth International Conference on Communication Systems and Network Technologies, pp. 698–701 (2015)
30. Bélanger, F., Crossler, R.E.: Privacy in the digital age: a review of information privacy research in information systems. MIS Q. **35**, 1017–1042 (2011)
31. Conger, S., Pratt, J.H., Loch, K.D.: Personal information privacy and emerging technologies. Inf. Syst. J. **23**, 401–417 (2013)
32. Keith, M.J., Babb, J.S., Lowry, P.B.: A longitudinal study of information privacy on mobile devices. In: 2014 47th Hawaii International Conference on System Sciences, pp. 3149–3158. IEEE (2014)
33. Zeissig, E.-M., Lidynia, C., Vervier, L., Gadeib, A., Ziefle, M.: Online privacy perceptions of older adults. In: Zhou, J., Salvendy, G. (eds.) ITAP 2017. LNCS, vol. 10298, pp. 181–200. Springer, Cham (2017). https://doi.org/10.1007/978-3-319-58536-9_16

34. Queirós, A., Silva, A., Alvarelhão, J., Rocha, N.P., Teixeira, A.: Usability, accessibility and ambient-assisted living: a systematic literature review. Univ. Access Inf. Soc. **14**(1), 57–66 (2013). https://doi.org/10.1007/s10209-013-0328-x

35. Novitzky, P., et al.: A review of contemporary work on the ethics of ambient assisted living technologies for people with dementia. Sci. Eng. Ethics **21**(3), 707–765 (2014). https://doi.org/10.1007/s11948-014-9552-x

36. Cardinaux, F., Bhowmik, D., Abhayaratne, C., Hawley, M.S.: Video based technology for ambient assisted living: a review of the literature. J. Ambient Intell. Smart Environ. **3**, 253–269 (2011)

37. Fogli, D., Lanzilotti, R., Piccinno, A.: End-user development tools for the smart home: a systematic literature review. In: Streitz, N., Markopoulos, P. (eds.) DAPI 2016. LNCS, vol. 9749, pp. 69–79. Springer, Cham (2016). https://doi.org/10.1007/978-3-319-39862-4_7

38. Rabah, S., Abed, M., Zoghlami, N.: AmIs sustainability: evaluation framework based on fuzzy logic. Int. J. Serv. Oper. Manag. **19**, 265–286 (2014)

39. Koimizu, J., Kokado, M., Kato, K.: Ethical perspectives of Japanese engineers on ambient assisted living technologies: semi-structured interview. Asian Bioethics Rev. **10**(2), 143–155 (2018). https://doi.org/10.1007/s41649-018-0053-0

40. Kofod-Petersen, A., Cassens, J.: Proxies for privacy in ambient systems. J. Wirel. Mob. Netw. Ubiq. Comput. Dependable Appl. **1**, 62–74 (2010)

41. Lopez, M., Pedraza, J., Carbo, J., Molina, J.M.: Ambient intelligence: applications and privacy policies. In: Corchado, J.M., et al. (eds.) PAAMS 2014. CCIS, vol. 430, pp. 191–201. Springer, Cham (2014). https://doi.org/10.1007/978-3-319-07767-3_18

42. Boulos, M.N.K., Anastasiou, A., Bekiaris, E., Panou, M.: Geo-enabled technologies for independent living: examples from four European projects. Technol. Disabil. **23**, 7–17 (2011)

43. Mortenson, W.B., Sixsmith, A., Beringer, R.: No place like home? surveillance and what home means in old age. Can. J. Aging **35**, 103–114 (2016)

44. Fernando, N., Tan, F.T.C., Vasa, R., Mouzaki, K., Aitken, I.: Examining digital assisted living: towards a case study of smart homes for the elderly (2016)

45. Schomakers, E.M., Ziefle, M.: Privacy perceptions in ambient assisted living. In: ICT4AWE 2019 - Proceedings of the 5th International Conference on Information and Communication Technologies for Ageing Well and e-Health, pp. 205–212 (2019)

46. Costa, Â., Andrade, F., Novais, P.: Privacy and data protection towards elderly healthcare. In: Handbook of Research on ICTs for Human-Centered Healthcare and Social Care Services, pp. 330–346 (2013)

47. Panagiotakopoulos, T., Theodosiou, A., Kameas, A.: Exploring ambient assisted living job profiles. In: Proceedings of the 6th International Conference on PErvasive Technologies Related to Assistive Environments, Article 17. Association for Computing Machinery, Rhodes, Greece (2013)

48. Chaaraoui, A.A., Padilla-López, J.R., Ferrández-Pastor, F.J., Nieto-Hidalgo, M., Flórez-Revuelta, F.: A vision-based system for intelligent monitoring: human behaviour analysis and privacy by context. Sensors **14248220**(14), 8895–8925 (2014)

49. Memon, M., Wagner, S.R., Pedersen, C.F., Aysha Beevi, F.H., Hansen, F.O.: Ambient assisted living healthcare frameworks, platforms, standards, and quality attributes. Sensors (Switzerland) **14**, 4312–4341 (2014)

50. He, D., Zeadally, S.: Authentication protocol for an ambient assisted living system. IEEE Commun. Mag. **53**, 71–77 (2015)

51. Shirali, M., Sharafi, M., Ghassemian, M., Fotouhi-Ghazvini, F.: A testbed evaluation for a privacy-aware monitoring system in smart home. In: GHTC 2018 - IEEE Global Humanitarian Technology Conference, Proceedings (2018)

52. Offermann-van Heek, J., Ziefle, M.: Nothing else matters! trade-offs between perceived benefits and barriers of AAL technology usage. Front. Public Health **7**, 134 (2019)
53. Mbarek, B., Jabeur, N., Yasar, A.-U.: ECASS: an encryption compression aggregation security scheme for secure data transmission in ambient assisted living systems. Pers. Ubiquit. Comput. **23**, 793–799 (2018). https://doi.org/10.1007/s00779-018-1128-3
54. Haider, F., Luz, S.: A system for real-time privacy preserving data collection for ambient assisted living. In: Proceedings of the Annual Conference of the International Speech Communication Association, INTERSPEECH, vol. 2019-September, pp. 2374–2375 (2019)
55. Grgurić, A., Mošmondor, M., Huljenić, D.: The smarthabits: an intelligent privacy-aware home care assistance system. Sensors (Switzerland) **19**, 907 (2019)
56. McNeill, A., Briggs, P., Pywell, J., Coventry, L.: Functional privacy concerns of older adults about pervasive health-monitoring systems. In: Proceedings of the 10th International Conference on Pervasive Technologies Related to Assistive Environments, pp. 96–102. Association for Computing Machinery, Island of Rhodes (2017)
57. Porambage, P., Braeken, A., Gurtov, A., Ylianttila, M., Spinsante, S.: Secure end-to-end communication for constrained devices in IoT-enabled ambient assisted living systems. In: IEEE World Forum on Internet of Things, WF-IoT 2015 - Proceedings, pp. 711–714 (2015)
58. Dimitrievski, A., Zdravevski, E., Lameski, P., Trajkovik, V.: Towards application of non-invasive environmental sensors for risks and activity detection. In: Proceedings - 2016 IEEE 12th International Conference on Intelligent Computer Communication and Processing, ICCP 2016, pp. 27–33 (2016)
59. Massacci, F., Nguyen, V.H., Saidane, A.: No purpose, no data: goal-oriented access control forambient assisted living. In: Proceedings of the ACM Conference on Computer and Communications Security, pp. 53–57 (2009)
60. Costa, A., Yelshyna, A., Moreira, T.C., Andrade, F.C.P., Julián, V., Novais, P.: A legal framework for an elderly healthcare platform: a privacy and data protection overview. Comput. Law Secur. Rev. **33**, 647–658 (2017)
61. Przybylo, J.: Landmark detection for wearable home care ambient assisted living system. In: Proceedings - 2015 8th International Conference on Human System Interaction, HSI 2015, pp. 244–248 (2015)
62. Choi, H.S., Lee, W.S., Sohn, S.Y.: Analyzing research trends in personal information privacy using topic modeling. Comput. Secur. **67**, 244–253 (2017)
63. Bruder, C., Blessing, L., Wandke, H.: Adaptive training interfaces for less-experienced, elderly users of electronic devices. Behav. Inf. Technol. **33**, 4–15 (2014)
64. Barnard, Y., Bradley, M.D., Hodgson, F., Lloyd, A.D.: Learning to use new technologies by older adults: perceived difficulties, experimentation behaviour and usability. Comput. Hum. Behav. **29**, 1715–1724 (2013)
65. Ribaric, S., Ariyaeeinia, A., Pavesic, N.: De-identification for privacy protection in multimedia content: a survey. Sig. Process. Image Commun. **47**, 131–151 (2016)
66. Anwar, M.J., Gill, A.Q., Beydoun, G.: A review of information privacy laws and standards for secure digital ecosystems. In: ACIS 2018-29th Australasian Conference on Information Systems (2018)
67. Wu, Y., Yang, F., Xu, Y., Ling, H.: Privacy-protective-GAN for privacy preserving face de-identification. J. Comput. Sci. Technol. **34**, 47–60 (2019)

Synchronization Analysis in Models of Coupled Oscillators

Guilherme Toso$^{(\boxtimes)}$ ⓘ and Fabricio Breve ⓘ

São Paulo State University (UNESP), Rio Claro, SP, Brazil
guilherme.toso@unesp.br

Abstract. The present work deals with the analysis of the synchronization possibility in chaotic oscillators, either completely or per phase, using a coupling force among them, so they can be used in attention systems. The neural models used were Hodgkin-Huxley, Hindmarsh-Rose, Integrate-and-Fire, and Spike-Response-Model. Discrete models such as Aihara, Rulkov, Izhikevic, and Courbage-Nekorkin-Vdovin were also evaluated. The dynamical systems' parameters were varied in the search for chaos, by analyzing trajectories and bifurcation diagrams. Then, a coupling term was added to the models to analyze synchronization in a couple, a vector, and a lattice of oscillators. Later, a lattice with variable parameters is used to simulate different biological neurons. Discrete models did not synchronize in vectors and lattices, but the continuous models were successful in all stages, including the Spike Response Model, which synchronized without the use of a coupling force, only by the synchronous time arrival of presynaptic stimuli. However, this model did not show chaotic characteristics. Finally, in the models in which the previous results were satisfactory, lattices were studied where the coupling force between neurons varied in a non-random way, forming clusters of oscillators with strong coupling to each other, and low coupling with others. The possibility of identifying the clusters was observed in the trajectories and phase differences among all neurons in the reticulum detecting where it occurred and where there was no synchronization. Also, the average execution time of the last stage showed that the fastest model is the Integrate-and-Fire.

Keywords: Synchronization · Oscillators · Neurons

1 Introduction

Visual Attention systems are widely researched [6,19,27] and it is a technique used by biological neural network systems developed to reduce the large amount of visual information that it is received by natural sensors [8]. This mechanism selects a subset of the information coming from the sensors to recognize the environment. This is due to the limited hardware processing of the neural system of living beings.

© Springer Nature Switzerland AG 2020
O. Gervasi et al. (Eds.): ICCSA 2020, LNCS 12249, pp. 889–904, 2020.
https://doi.org/10.1007/978-3-030-58799-4_64

The process happens due to factors that can be divided into two types: *bottom-up* and *top-down*. The factors of the first type arise from the combination of information from the retina and regions at the beginning of the visual cortex, that is, the attention occurs due to the scene information. On the other hand, the second type's factor is generated by the return signals from areas outside the visual cortex, so attention is also task-dependent [14,15].

In 1981, von der Malsburg [18] suggested that each object is represented by the temporal correlation of neural firing activities, which can be described by dynamic models and some can be found at the Cessac's work [5]. Hence, the correlation encodes different attributes of the object. A natural way of representing the coding of the temporal correlation is to use synchronization between oscillators, where each one encodes some attributes of an object [24–26], so that the synchronized neurons are the ones that process the attributes of the same object and those that process different objects are out of sync.

The main objective of this research is the studying of synchronization in some biological oscillators' models which exhibit chaotic behaviors. This analysis is to evaluate the synchronization possibility, both complete and per phase, by using a coupling force between oscillators as in Breve et al. work [3]. The motivation of this work is to use this sync method for visual selection of objects that represents sync neurons' models, while the rest of the image is unsynced.

The oscillators used in this work are based upon the biological neural networks which are plausible systems from the biological point of view, as the models of Hodgkin-Huxley, Hindmarsh-Rose, Integrate-And-Fire, and Spike-Response-Model [9,11,12,17]. Discrete models with computational advantages were also used, such as Aihara's, Rulkov's, Izhikevic's and the Courbage-Nekorkin-Vdovin model [1,7,16,23].

A good model for the visual attention task must allow that a group of oscillators synchronizes with each other if they are strongly coupled, while synchronization is lost compared to other oscillators in the lattice which have a small or even nonexistent coupling force. In this way, oscillators can be used to represent pixels or groups of pixels of an image, just as if they were neurons in the retina, so that neurons representing the same object, synchronize, at the same time different objects lose synchronization.

The present work is organized in the following: in Sect. 2 the theoretical elements will be presented such as the phase synchronization concept and the coupling term in dynamical systems. In Sect. 3 it will be shown the neural dynamical systems used in this work and their analysis, as the search for chaos and the addition of the coupling force in different structures as two oscillators, a vector, and a lattice. In Sect. 4 the results will be discussed, and finally, in Sect. 5 it will be presented the conclusion of which model or models satisfies the methodologies showed in Sect. 3 for a good model of visual attention.

2   Phase Synchronization

The phase synchronization of two oscillators p and q happens when their phases difference $|\phi_p - \phi_q|$ is kept below a certain phase threshold and the amplitudes not necessarily are synchronized ($|X_p - X_q| \neq 0$), this means that their rhythms are bonded. So as $t \longrightarrow \infty$, $|\phi_p - \phi_q| < C$. The phase ϕ_i at time t_i is calculated as following [22]

$$\phi_i = 2\pi k + 2\pi \frac{t_i - t_k}{t_{k+1} - t_k} \tag{1}$$

where k is the number of neural activities prior to time t_i, and t_k and t_{k+1} are the last and the next times of neural activity, respectively. So that two oscillators can synchronize with each other, a coupling term is added to the dynamical system as the following:

$$\dot{x}_j^p = F_j(\mathbf{X}, \mu) + k\Delta_{p,q} \tag{2}$$

$$\dot{x}_j^q = F_j(\mathbf{X}, \mu) + k\Delta_{q,p} \tag{3}$$

where $\dot{x}_j^p$ and $\dot{x}_j^q$ represents the time evolution of the x_j state of the oscillators p and q respectively. At right, the $F_j(.)$ represents the behaviours' rate of the jth state which depends on the states' vector $\mathbf{X} = (x_1, x_2, ..., x_j, ..., x_J)$ and the parameters' vector $\mu = (\mu_1, \mu_2, ..., \mu_l, ..., \mu_L)$, and last the coupling term $k\Delta_{p,q}$, where k is the coupling force and $\Delta_{p,q}$ is the difference between the states:

$$\Delta_{p,q} = x_j^q - x_j^p \tag{4}$$

If k is strong enough, then the oscillators $\mathbf{X}^p$ and $\mathbf{X}^q$ synchronizes, and it's possible by analyzing their phases difference:

$$\lim_{t \to \infty} |\phi_p(t) - \phi_q(t)| < C \tag{5}$$

where $\mathbf{C}$ is a phase threshold.

3   Methodology

The proposed models for the attention system are a two-dimensional network of neural models' dynamical systems with coupled terms. The neural models and their steps in the present work are the following.

3.1   Hodgkin-Huxley

This continuous model is ruled by a membrane potential V_{ij} and the ionic channels (m_{ij}, n_{ij}, h_{ij}), where $1 < i < N$ and $1 < j < M$ are rows and columns positions in the lattice, remembering that the coordinate (i, j) is different from the sub-indexes i and j in Sect. 2, which represents the time and states coordinates respectively. So the states and constants will be replaced by a matrix representation as the generic one in 6:

$$\mathbf{X} = \begin{bmatrix} x_{11} & \cdots & x_{1j} & \cdots & x_{1M} \\ \vdots & \ddots & \vdots & \ddots & \vdots \\ x_{i1} & \cdots & x_{ij} & \cdots & x_{iM} \\ \vdots & \ddots & \vdots & \ddots & \vdots \\ x_{N1} & \cdots & x_{Nj} & \cdots & x_{NM} \end{bmatrix} \tag{6}$$

So, the equations of the system are:

$$\mathbf{C} \circ \dot{\mathbf{V}} = -\mathbf{G_{Na}} \circ \mathbf{M}^3 \circ (\mathbf{V} - \mathbf{E_{Na}}) - \mathbf{G_k} \circ \mathbf{N}^4 \circ (\mathbf{V} - \mathbf{E_K}) \tag{7}$$
$$-\mathbf{G_L} \circ (\mathbf{V} - \mathbf{E_L}) + \mathbf{I(t)} + \mathbf{K}\boldsymbol{\Delta}\mathbf{V},$$

$$\dot{\mathbf{M}} = \alpha_m(\mathbf{V}) \circ (\mathbb{1} - \mathbf{M}) - \beta_m(\mathbf{V}) \circ \mathbf{M} + \boldsymbol{\Xi}_\mathbf{M} \tag{8}$$

$$\dot{\mathbf{N}} = \alpha_n(\mathbf{V}) \circ (\mathbb{1} - \mathbf{N}) - \beta_n(\mathbf{V}) \circ \mathbf{N} + \boldsymbol{\Xi}_\mathbf{N} \tag{9}$$

$$\dot{\mathbf{H}} = \alpha_h(\mathbf{V}) \circ (\mathbb{1} - \mathbf{H}) - \beta_h(\mathbf{V}) \circ \mathbf{H} + \boldsymbol{\Xi}_\mathbf{H} \tag{10}$$

where $\circ$ represents the Hadamard product. The $\mathbf{V}$, $\mathbf{M}$, $\mathbf{N}$ and $\mathbf{H}$ variables are the states matrices, and $\mu = (\mathbf{G_{Na}}, \mathbf{G_k}, \mathbf{G_k}, \mathbf{E_{Na}}, \mathbf{E_K}, \mathbf{E_K}, \mathbf{C})$ is the parameters matrices vector. $\boldsymbol{\Xi}_\mathbf{M}$, $\boldsymbol{\Xi}_\mathbf{N}$ and $\boldsymbol{\Xi}_\mathbf{H}$ are white noises as done in [4], $\mathbb{1}$ is a matrice of ones, α_x, β_x are matrix operations as showed in Table 1:

Table 1. Rate probabilities of $\mathbf{M}$, $\mathbf{N}$ and $\mathbf{H}$

x	$\alpha_x(V/mV) \, [ms^{-1}]$	$\beta_x(v/mV) \, [ms^{-1}]$
N	$0.01(10 - \mathbf{V})/[e^{(10-\mathbf{V})/10} - 1]$	$0.125e^{-\mathbf{V}/80}$
M	$0.1(25 - \mathbf{V})/[e^{(25-\mathbf{V})/10} - 1]$	$4e^{-\mathbf{V}/18}$
H	$0.07e^{-\mathbf{V}/20}$	$1/[1 + e^{(30-\mathbf{V})/10}]$

The electrical current is [10]:

$$\mathbf{I}(t) = \mathcal{N}(\mu_0, \sigma_0) \circ cos(2\pi ft) + \mathbf{I_{syn}} \tag{11}$$

$$\tau_{\mathbf{syn}} \circ \dot{\mathbf{I}}_{\mathbf{syn}} = -\mathbf{I_{syn}} + \sqrt{2\mathbf{D}} \circ \mathcal{N}(\mu, \sigma) \tag{12}$$

Finally, $\mathbf{K}$ is a matrix of vectors, as much as $\boldsymbol{\Delta}\mathbf{V}$, such that an element of k_{ij} multiplying $\Delta v_{i,j}$ is the same as the expression 13:

$$k_{ij}\Delta v_{ij} = k_{ij}^0(v_{i,j+1} - v_{i,j}) + k_{ij}^1(v_{i-1,j+1} - v_{i,j}) + k_{ij}^2(v_{i-1,j} - v_{i,j})$$
$$+ k_{ij}^3(v_{i-1,j-1} - v_{i,j}) + k_{ij}^4(v_{i,j-1} - v_{i,j}) + k_{ij}^5(v_{i+1,j-1} - v_{i,j}) \tag{13}$$
$$+ k_{ij}^6(v_{i+1,j} - v_{i,j}) + k_{ij}^7(v_{i+1,j+1} - v_{i,j})$$

which represents the coupling term between the oscillator at position (i, j) and its eight closest neighbors. So for a two-oscillator problem, $i = 0$ and $j = 2$, so there's only one neighbor, and for a vector case, $i = 0$ and $1 < j < M$, there are two neighbors. These coupling terms were used in all neural models. The parameters used were the same as in [12], and its variations study was the same as the range in Table 2 of [12]. For the bifurcation diagram only one oscillator was used with $V_0 = 0$ and the electrical current varying from zero to $500\,\mu A$, while there was no stochastic term in the ionic channels. After that, the stochastic terms were added into the model so that the trajectories are unpredictable with random initializations between 0 and 5 mV. Also, the electrical current variables were $\mu_0 = 4\,\mu A$, $\sigma_0 = 0.001\,\mu A$, $\mathbf{I_{syn}} = -2\,\mu A$, $\mathbf{D} = 10$, $\mathcal{N}(2.0, 1.0)$ and the step time for the 4th-order Runge-Kutta numerical method were T $\approx$ 0.007, where $T = f^{-1}$, and f $= 140\,Hz$ as in [21], and the ionic stochastic terms were $\mathcal{N}(0, 0.1)$. Next, the coupling force was varied from 0.001 to 3.0 for two oscillators, a vector (10 oscillators) and a lattice of 10×10 oscillators, the same structure configurations for the other models.

3.2   Hindmarsh-Rose

The model is described by the dynamical variables $\mathbf{X}$, $\mathbf{Y}$ and $\mathbf{Z}$ which represents the membrane potential, the recovery variable and the adaptation current [11].

$$\dot{\mathbf{X}} = \mathbf{Y} - \mathbf{A} \circ \mathbf{X}^3 + \mathbf{B} \circ \mathbf{X}^2 + \mathbf{I} - \mathbf{Z} + \mathbf{K}\Delta\mathbf{X} \tag{14}$$

$$\dot{\mathbf{Y}} = \mathbf{C} - \mathbf{D} \circ \mathbf{X}^2 - \mathbf{Y} \tag{15}$$

$$\dot{\mathbf{Z}} = \mathbf{R} \circ (\mathbf{S} \circ (\mathbf{X} - \mathbf{X_r}) - \mathbf{Z}) \tag{16}$$

where $\mu = (\mathbf{A}, \mathbf{B}, \mathbf{C}, \mathbf{D}, \mathbf{R}, \mathbf{S}, \mathbf{X_r}, \mathbf{I})$ is the vector of parameters and $\mathbf{I}$ is the stimuli current. The values used are the same as in [11] and the variations are shown in Table 2:

Table 2. Parameters variations for the Hindmarsh-Rose model

Values\Parameters	a	b	c	d	I	r	s	x_r
Interval	1.0 to 2.5	1 to 6	0 to 5	1 to 6	1 to 15	0 to 0.01	1 to 6	−4.8 to 3.2

The bifurcation diagram used the electrical current variation from 0 to $20\,\mu A$. Then for the coupling force in a two-oscillator problem, k varied from 0 to 1.2, while for a vector and a lattice it was used $k = 5$ with a new range of parameters in a lattice showed at Table 3.

Table 3. Parameters variations for the Hindmarsh-Rose model in a lattice

Values\Parameters	a	b	c	d	I	r	s	x_r
Interval	1.0 to 1.9	3 to 6	1 to 5	2 to 6	2 to 10	0 to 0.001	1 to 3	−3.2 to 3.2

In all the 4th-order Runge-Kutta experiments, it was used a step time of 0.01.

3.3 Integrate and Fire

Composed by the variable $\mathbf{V}$ and the constants $\mu = (\tau, \mathbf{V_{rest}}, \mathbf{R}, \mathbf{I})$:

$$\tau \dot{\mathbf{V}} = -(\mathbf{V} - \mathbf{V_{rest}}) + \mathbf{R} \circ \mathbf{I} + \mathbf{K} \Delta \mathbf{V} \tag{17}$$

the one linear differential equation dynamical system presents no chaos behaviour, so the current I was replaced by white noise $\mathcal{N}(2.5, 0.9)$. The other parameters were set as $\mathbf{V_{rest}} = 1.0$, while $\mathbf{R} = 1.0$ and $\tau = 10$. $\mathbf{K}$ varied from 0.001 to 1.1, for a vector-oscillator problem the current was $\mathcal{N}(3.0, 0.5)$ and for a lattice $\mathcal{N}(10.0, 2.0)$, while $\mathbf{K}$ was 0.5 for both structures. The 4th-order Runge-Kutta step was 1 ms.

3.4 Spike Response Model

The simplified model (SRM_0) with no external current, only with pre-synaptic stimuli, is described as a lattice of membrane potentials $\mathbf{V}$ non-connected with each other as in the other models, but with pre-synaptic neurons l:

$$\mathbf{V} = \eta(t - \mathbf{t}^{(\mathbf{f})}) + \sum_l \sum_f \mathbf{\Omega_l} \circ \epsilon_l(t - \mathbf{t_l}^{(\mathbf{f})}) + \mathbf{K} \Delta \mathbf{V} \tag{18}$$

the times $\mathbf{t}^{(\mathbf{f})}$ are the times of the last fire of every oscillator v_{ij}, while $\mathbf{\Omega_l}$ is a weight connection between neurons lattice $\mathbf{V}$ and the pre-synaptic neurons l, and η and ϵ are kernels as:

$$\eta(t - \mathbf{t}^{(\mathbf{f})}) = -\vartheta \circ e^{(-(t - \mathbf{t}^{(\mathbf{f})})^{\mathbf{M}} + \mathbf{N})} \circ Heavside(t - \mathbf{t}^{(\mathbf{f})}) \tag{19}$$

$$Heavside(t - t_{ij}^{(f)}) = \begin{cases} 0, & \text{if } t - t_{ij}^{(f)} < 0 \\ -1, & \text{if } 0 <= t - t_{ij}^{(f)} < 1 \\ 1, & \text{if } 1 < t - t_{ij}^{(f)} \end{cases} \tag{20}$$

$$\epsilon = e^{(t - \mathbf{t_l}^{(\mathbf{f})} - \mathbf{D})/\tau} \circ Heavside(t - \mathbf{t_l}^{(\mathbf{f})} - \mathbf{D}) \tag{21}$$

$$Heavside(t - t_{ijl}^{(f)} - d_{ij}) = \begin{cases} 0, & \text{if } t - t_{ijl}^{(f)} - d_{ij} < 0 \\ 1, & \text{if } t - t_{ijl}^{(f)} - d_{ij} >= 0 \end{cases} \tag{22}$$

where ϑ, $\mathbf{M}$, $\mathbf{N}$, $\mathbf{D}$ and τ are matrices of constants, while $t_l^{(f)}$ is a matrix of last spikes time of the pre-synaptic neuron l, and t_{ijl} is the spike time of the neuron in position (i, j) of the pre-neuron l. Another constant was added and it was called *limit*, which means every *limit* times, a pre-synaptic spike is generated. For two and a vector of oscillators it was used two pre-synaptic neurons. As the model depends on the time and not from the previous v_{ij} value, then it was made two tests, the variation of coupling force from 0.001 to 0.32 and the second test was the absence of the coupling term but the variation of the *limit* value for different neurons, so for one neuron the interval of time in which the pre-synaptic spikes arrives is 5, and the other is 6. For a vector, the same two tests were made and the coupling force varies from 0 to 0.01, while the other test set several oscillators with the same value of *limit* and the others were generated randomly.

3.5   Aihara

This chaotic discrete neural model is defined by:

$$\mathbf{Y}(\mathbf{t}+1) = \mathbf{D} \circ \mathbf{Y}(\mathbf{t}) - \mathbf{A} \circ \mathbf{F}\{\mathbf{Y}(\mathbf{t})\} + \mathbf{B} + \mathbf{K}\Delta\mathbf{Y}(\mathbf{t}) \tag{23}$$

$$\mathbf{F}\{\mathbf{Y}(\mathbf{t})\} = \frac{\mathbb{1}}{\mathbb{1} + e^{(-\mathbf{Y}(\mathbf{t})/\epsilon)}} \tag{24}$$

where $\mathbf{Y}(\mathbf{t})$ is a neuron internal state and $\mathbf{X}(\mathbf{t}) = \mathbf{F}\{\mathbf{Y}(\mathbf{t})\}$ is a logistic function, while $\mathbf{D}$, $\mathbf{B}$, $\mathbf{A}$, ϵ and $\mathbb{1}$ are constant matrices.

The values were varied as in Table 4 with initialization $y(0) = 0.1$:

Table 4. Parameters variation for the Aihara model

Values/Parameters	D	A	ϵ	B
Interval	0 to 1	0.7 to 1.2	0.015 to 0.04	0 to 1.0

The bifurcation diagram was generated with the same parameter values as in [1]. For chaotic trajectories, it was tested several initializations (0.001, 0.01 and 0.1) with A set to 0.35. For the synchronization test, the force was varied from 0.001 to 0.3 for a two-oscillator problem, and for a vector, it was varied from 0 to 0.35, while $\mathbf{Y}(\mathbf{T_0})$ was generated randomly from 0 to 0.0001 due to its sensibility. Finally, for the lattice, the coupling force varied from 0 to 0.02.

3.6   Rulkov

For a spike and spiking-bursts dynamic, Rulkov developed the following system:

$$\mathbf{X}(\mathbf{t}+1) = \frac{\alpha}{\mathbb{1} + \mathbf{X}(\mathbf{t})^2} + \mathbf{Y}(\mathbf{t}) + \mathbf{I} + \mathbf{K}\Delta\mathbf{X} \tag{25}$$

$$\mathbf{Y}(\mathbf{t}+1) = \mathbf{Y}(\mathbf{t}) - \mu(\mathbf{X}(\mathbf{t}) - \sigma) \tag{26}$$

where $\mathbf{X}$ and $\mathbf{Y}$ are the fast and slow variables, while α, $\mathbf{I}$, μ and σ are constants.

The parameters were varied according to the Table 5 and $\mathbf{X}(\mathbf{t_0}) = 0$, $\mathbf{Y}(\mathbf{t_0}) = -2.9$. The current was set to 0 because σ already represents a stimulus. For chaotic trajectories, $\alpha > 4.0$ while the other parameters were equal to the ones in [13]. The variation of the coupling force in a two-problem was from 0.0 to 0.5, and for a vector and a lattice was from 0.001 to 0.02.

Table 5. Parameters variation for the Rulkov model

Values/Parameters	α	μ	σ
Interval	1 to 5	0 to 0.005	-2 to 0

3.7 Izhikevic

The model is described as:

$$\dot{\mathbf{V}} = 0.04\mathbf{V}^2 + 5\mathbf{V} + 140 - \mathbf{U} + \mathbf{I} + \mathbf{K}\Delta\mathbf{V} \tag{27}$$

$$\dot{\mathbf{U}} = \mathbf{A} \circ (\mathbf{B} \circ \mathbf{V} - \mathbf{U}) \tag{28}$$

$$\text{if } v_{ij} \geq 30 \ mV, \ \text{then} \begin{cases} v_{ij} \leftarrow c_{ij} \\ u_{ij} \leftarrow u_{ij} + d_{ij} \end{cases} \tag{29}$$

where $\mathbf{V}$ and $\mathbf{U}$ are variables and $\mathbf{A}$, $\mathbf{B}$, $\mathbf{C}$, $\mathbf{D}$ and $\mathbf{I}$ are parameters, both dimensionless.

The parameters used were the same as in [20] which generates chaos, using initializations of -64 and -65 mV for a two-oscillator coupling while varying the force from 0 to 0.6. For a vector and lattice problem, the force was decreased and increased from 0.06, while the initializations varied from -65 to -65.0001 due to its high sensibility.

3.8 Courbage-Nekorkin-Vdovin (CNV)

Similarly to the Rulkov's model, the CNV model is:

$$\dot{\mathbf{X}} = \mathbf{X} + \mathbf{F}(\mathbf{X}) - \mathbf{Y} - \beta\mathbf{H}(\mathbf{X} - \mathbf{D}) \tag{30}$$

$$\dot{\mathbf{Y}} = \mathbf{Y} + \epsilon(\mathbf{X} - \mathbf{J}) \tag{31}$$

$$F(x_{ij}) = \begin{cases} -m_{ij}^0 x_{ij}, & if \ x_{ij} \leq J_{ij}^{min} \\ m_{ij}^1(x_{ij} - a_{ij}), & if \ J_{ij}^{min} < x_{ij} < J_{ij}^{max} \\ -m_{ij}^0(x_{ij} - 1), & if \ x_{ij} \geq J_{ij}^{max} \end{cases} \tag{32}$$

$$H(x_{ij}) = \begin{cases} 1, & if \ x_{ij} \geq 0 \\ 0, & if \ x_{ij} < 0 \end{cases} \tag{33}$$

where,

$$\mathbf{J_{min}} = \frac{\mathbf{A} \circ \mathbf{M_1}}{\mathbf{M_0} + \mathbf{M_1}}, \quad \mathbf{J_{max}} = \frac{\mathbf{M_0} + \mathbf{A} \circ \mathbf{M_1}}{\mathbf{M_0} + \mathbf{M_1}}, \quad \mathbf{M_0}, \mathbf{M_1} > 0 \tag{34}$$

where $\mathbf{X}$ and $\mathbf{Y}$ are the fast and slow variables, while ϵ, β, $\mathbf{D}$ (all greater than 0), $\mathbf{J}$, $\mathbf{A}$, $\mathbf{M_0}$ and $\mathbf{M_1}$ are parameters.

As in [13], it was used the burst-spike alternating behavior for the chaos analysis. For the coupling tests, the force varied from 0 to 0.1 in a two-oscillator system, for a vector it varied from 0.05 to 0.5, and finally, in a lattice, the force varied from 0 to 0.01.

3.9   Coupling Force Variation

At this stage of the present work, the coupling force was varied such that some oscillators were strongly coupled and others weakly, so that the first were synchronized and hence clusterized. This stage was present only in the first three models presented in Sect. 3, so for the Hodgkin-Huxley model, the synchronization force was set to 3 while the desync values were $\mathcal{N}(0, 0.1)$. For the Hindmarsh-Rose model, the parameters varied in the intervals shown in Table 6.

Table 6. Parameters variation for the Hindmarsh-Rose model

a	b	c	d	r	s	x_r	I
1 to 1.6	4.0 to 6.0	1.0 to 5.0	2.0 to 5.0	0 to 0.01	1.0 to 2.0	−1.6 to 3.2	1.0 to 9.0

While the **K** for sync was set to 5, the desync was generated by $\mathcal{N}(0, 0.1)$. Finally, for the Integrate-and-Fire model, the threshold was set to 2 mV, $\mathbf{I} = 5$ mV, $\tau = \Xi(10, 20)$ is a random sample. The step time of the Runge-Kutta was set to 0.5 instead of 1 as in the others simulations, and finally, **K** for sync was 0.5, and for desync was $\mathcal{N}(0, 0.01)$.

4   Results

In this section the results obtained with the computer simulations regarding the steps and models described in Sect. 3 are presented. To make it possible to identify and to separate the clusters from the rest of the lattice, the neurons must be able to synchronize and desynchronize from each other. They also must be able to represent a great number of different trajectories, so the properties of chaos or even the stochastic ones are good approaches to achieve these objectives.

The models' parameters were varied so that it was possible to find chaotic properties. The models that presented these objectives were the discrete and the Hindmarsh-Rose. The properties of the latter are mentioned in [11], where the variable **Z** generates unpredictable trajectories, thus randomly varying the variables, **X**, **Y**, and **Z** from values between −1 and 1, generated unpredictable trajectories. In other models, stochastic terms were added to generate different trajectories, except for the Spike Response Model, whose trajectories depend on the arrival time (parameter *limit*) of a peak of a presynaptic neuron. These behaviors can be seen in the Figs. 1, 2, 3 and 4.

Unlike the other models that produced spikes, the Rulkov model generated spike explosions and, in the CNV model, it also produced these behaviors interspersed with spikes. Then, the models were tested if they could synchronize between two neurons of the same models, then in a vector and finally in a grid so that it represented an image. To measure synchronization, the same coupling force was defined between all oscillators (two, a vector and a grid) so that all could synchronize, and then a phase of a reference oscillator would be necessary

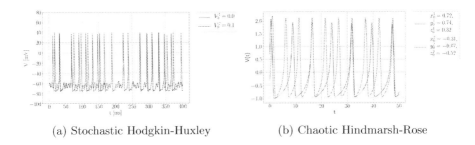

(a) Stochastic Hodgkin-Huxley

(b) Chaotic Hindmarsh-Rose

Fig. 1. Random and chaotic neuron models with two different trajectories

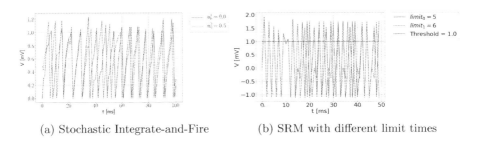

(a) Stochastic Integrate-and-Fire

(b) SRM with different limit times

Fig. 2. Neuron models with two different trajectories

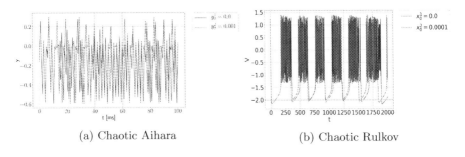

(a) Chaotic Aihara

(b) Chaotic Rulkov

Fig. 3. Chaotic neuron models with two different trajectories

to calculate the difference between all other oscillator phases. If all oscillators are synchronized, then all differences must be below a certain phase limit, which was determined to be 2π, where the phases of the differences do not increase with time [2]. Firstly, let's analyse the continuous models in Figs. 5, 6 and 7.

The Figs. 5a, 6a and 7a shows the difference in the trajectories, however none shows a complete synchronization, otherwise the phase synchronization is clearly observed at the Figs. 5a, 6b and 7b in which the difference of phases evolves in time always below the threshold of 2π. Now, let's analyze the SRM_0 model, which is a particular model of the SRM that there is no external current,

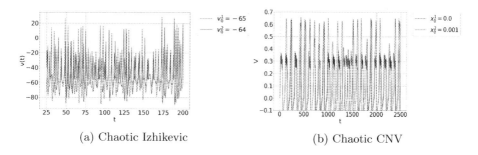

(a) Chaotic Izhikevic

(b) Chaotic CNV

Fig. 4. Chaotic neuron models with two different trajectories

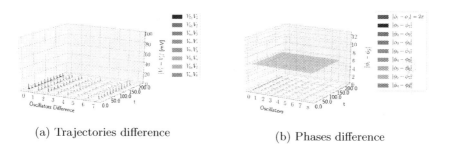

(a) Trajectories difference

(b) Phases difference

Fig. 5. Hodgkin-Huxley model

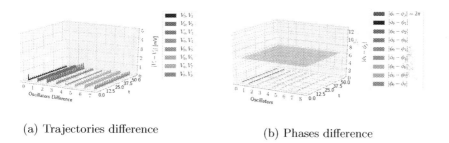

(a) Trajectories difference

(b) Phases difference

Fig. 6. Hindmarsh-Rose model

only pre-synaptic stimuli from pre-neurons. The model is also an integration of linear dynamical systems [9], so do not present chaotic characteristics. However, as in the Fig. 8b shows, those trajectories (4, 5, 6, 7, 8) with different *limit* values of the initial one (0) shows a desynchronization of their phases, while the trajectories with same *limit* value as the initial, synchronizes with each other, showing that neurons that receive neurons' stimuli at the same time are synchronized in phase, but that leads to a problem, in a case with thousands of neurons that must represent an image, synchronizing a particular amount of them and desynchronizing the other ones, requires that all of them must have

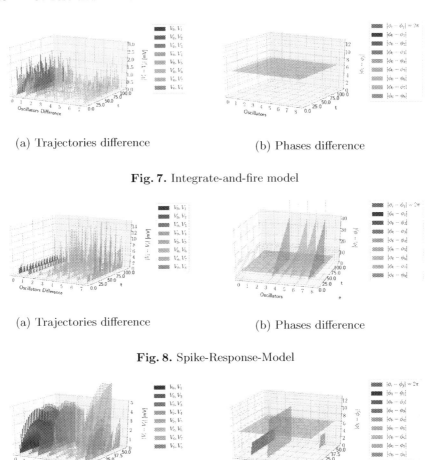

(a) Trajectories difference

(b) Phases difference

Fig. 7. Integrate-and-fire model

(a) Trajectories difference

(b) Phases difference

Fig. 8. Spike-Response-Model

(a) Trajectories difference

(b) Phases difference

Fig. 9. Aihara's model

a great variability of different *limit* values, and a neuron with a time interval incredibly high is not biologically plausible.

And for the discrete models, the experiments showed that for those ranges in the coupling force k presented in Sect. 3, the models did not sync completely or in phase in the reticle structure of neurons. For different values of coupling force above or below those ranges shows trajectories behaviors that are not typical of a relaxation oscillator. So, the desynchronization of the models are showed in Figs. 9, 10, 11 and 12.

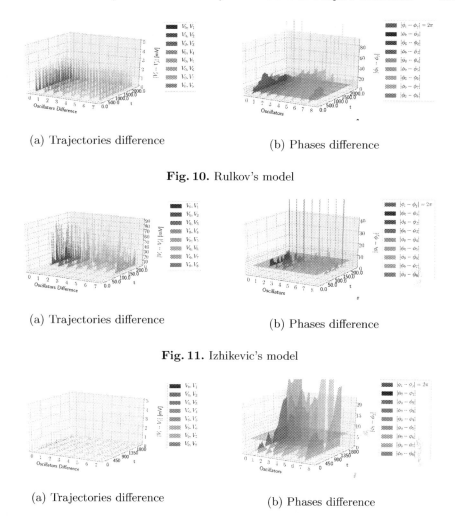

(a) Trajectories difference

(b) Phases difference

Fig. 10. Rulkov's model

(a) Trajectories difference

(b) Phases difference

Fig. 11. Izhikevic's model

(a) Trajectories difference

(b) Phases difference

Fig. 12. Courbage-Nekorkin-Vdovin's model

Finally, for models that successfully synchronized neurons in a network with variable parameters, it was tested whether some neurons could synchronize and others desynchronize by high values of coupling force and low values, respectively, so that the synchronized neurons can be grouped in a way that they can represent an image object in an attention system. The Figs. 13a, 14a and 15a show 9 trajectories, six of them synchronized and three unsynchronized, as in the Figs. 13b, 14b and 15b, so the synchronization represents the pixels of an object that receives attention and the desynchronized ones are pixels that do not receive any type of attention.

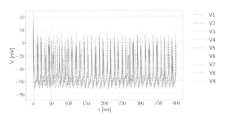

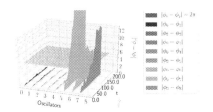

(a) Synchronized and Desynchronized Trajectories

(b) Phases difference

Fig. 13. Hodgkin-Huxley's model

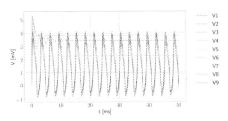

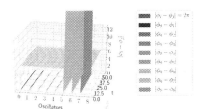

(a) Synchronized and Desynchronized Trajectories

(b) Phases difference

Fig. 14. Hindmarsh-Rose's model

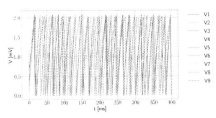

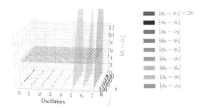

(a) Synchronized and Desynchronized Trajectories

(b) Phases difference

Fig. 15. Integrate and Fire's model

5 Conclusions

This work proposes to analyze the occurrence of synchronization in oscillators that have chaotic and stochastic behaviors using a coupling force between oscillators. The studies were done to examine if there are possibilities of using of such models in visual attention systems. The models used were those based on biological neural networks such Hodgkin-Huxley, Hindmarsh-Rose, Integrate-and-Fire, and Spike-Response-Model (SRM), which are biologically plausible, in addition to discrete-time models such as Aihara, Rulkov, Izhikevic, and

Courbage-Nekorkin-Vdovin (CNV), which have the advantage of reduced computational cost.

The behaviors of the models' trajectories were verified by varying their parameters and analyzing which values lead to chaos. Stochastic terms were added so they could produce variability in the trajectories. As a result, a coupling term was applied and analyzed if complete and/or phase synchronization occurred between two identical oscillators (same parameter values). Then the same study was applied, first to a vector of equal oscillators, and later to a lattice, both with equal and variable parameters. Finally, for the models that satisfied the previous steps, tests were made in a lattice with variable parameters and different coupling forces to form a cluster of synchronized and desynchronized oscillators.

From this study it was found that the discrete-time models did not synchronize at all stages, failing in the vector's and lattice's stages. The continuous-time models were able to synchronize at all stages using certain values of coupling force, except for the SRM model which was able to synchronize without the need of a coupling force, only considering the arrival time of presynaptic stimuli. However the SRM model did not present chaos. The continuous models tested for the synchronization and desynchronization for a cluster formation depending on the coupling force showed a potential solution for a visual selection mechanism for an attention system. Finally, for the successful models, the Integrate-and-fire model showed a better execution time with a mean in seconds of 1.16 and a standard deviation of 0.05 s. As a future work, the synchronization of coupled oscillators will also be used for a semi-supervised classification method, in which each cluster of oscillators represents data with the same label.

References

1. Aihara, K., Takabe, T., Toyoda, M.: Chaotic neural networks. Phys. Lett. A **144**(6–7), 333–340 (1990)
2. Breve, F.: Aprendizado de máquina utilizando dinâmica espaçotemporal em redes complexas. Universidade de São Paulo (Tese de Doutorado), São Carlos (2010)
3. Breve, F.A., Zhao, L., Quiles, M.G., Macau, E.E.: Chaotic phase synchronization for visual selection. In: International Joint Conference on Neural Networks, IJCNN 2009, pp. 383–390. IEEE (2009)
4. Casado, J.M.: Synchronization of two Hodgkin-Huxley neurons due to internal noise. Phys. Lett. A **310**(5–6), 400–406 (2003)
5. Cessac, B.: A view of neural networks as dynamical systems. Int. J. Bifurcat. Chaos **20**(06), 1585–1629 (2010)
6. Chen, B., Li, P., Sun, C., Wang, D., Yang, G., Lu, H.: Multi attention module for visual tracking. Pattern Recogn. **87**, 80–93 (2019)
7. Courbage, M., Nekorkin, V., Vdovin, L.: Chaotic oscillations in a map-based model of neural activity. Chaos Interdiscip. J. Nonlinear Sci. **17**(4), 043109 (2007)
8. Desimone, R., Duncan, J.: Neural mechanisms of selective visual attention. Annu. Rev. Neurosci. **18**(1), 193–222 (1995)
9. Gerstner, W.: A framework for spiking neuron models: the spike response model. In: Moss, F., Gielen, S. (eds.) Handbook of Biological Physics, vol. 4, pp. 469–516. Elsevier, Amsterdam (2001)

10. Gerstner, W., Kistler, W.M.: Spiking Neuron Models: Single Neurons, Populations, Plasticity. Cambridge University Press, Cambridge (2002)
11. Hindmarsh, J.L., Rose, R.: A model of neuronal bursting using three coupled first order differential equations. Proc. R. Soc. Lond. B Biol. Sci. **221**(1222), 87–102 (1984)
12. Hodgkin, A.L., Huxley, A.F.: A quantitative description of membrane current and its application to conduction and excitation in nerve. J. Physiol. **117**(4), 500–544 (1952)
13. Ibarz, B., Casado, J.M., Sanjuán, M.A.: Map-based models in neuronal dynamics. Phys. Rep. **501**(1–2), 1–74 (2011)
14. Itti, L., Koch, C.: Computational modelling of visual attention. Nat. Rev. Neurosci. **2**(3), 194 (2001)
15. Itti, L., Koch, C., Niebur, E.: A model of saliency-based visual attention for rapid scene analysis. IEEE Trans. Pattern Anal. Mach. Intell. **20**(11), 1254–1259 (1998)
16. Izhikevich, E.M.: Simple model of spiking neurons. IEEE Trans. Neural Networks **14**(6), 1569–1572 (2003)
17. Lapicque, L.: Recherches quantitatives sur l'excitation electrique des nerfs traitee comme une polarization. Journal de Physiologie et de Pathologie Generalej **9**, 620–635 (1907)
18. von der Malsburg, C.: The correlation theory of brain function. Technical report, MPI (1981)
19. Niu, Y., Zhang, H., Zhang, M., Zhang, J., Lu, Z., Wen, J.R.: Recursive visual attention in visual dialog. In: Proceedings of the IEEE Conference on Computer Vision and Pattern Recognition, pp. 6679–6688 (2019)
20. Nobukawa, S., Nishimura, H., Iamanishi, T., Liu, J.Q.: Analysis of chaotic resonance in Izhikevic neuron model. PLoS ONE **10**(9), e0138919 (2015)
21. Pankratova, E.V., Polovinkin, A.V., Mosekilde, E.: Noise suppression in a neuronal Hodgkin-Huxley model. Modern Prob. Stat. Phys. **3**, 107–116 (2004)
22. Pikovsky, A., Rosenblum, M., Kurths, J., Kurths, J.: Synchronization: A Universal Concept in Nonlinear Sciences, vol. 12. Cambridge University Press, Cambridge (2003)
23. Rulkov, N.F.: Modeling of spiking-bursting neural behavior using two-dimensional map. Phys. Rev. E **65**(4), 041922 (2002)
24. Terman, D., Wang, D.: Global competition and local cooperation in a network of neural oscillators. Physica D **81**(1–2), 148–176 (1995)
25. Von Der Malsburg, C., Schneider, W.: A neural cocktail-party processor. Biol. Cybern. **54**(1), 29–40 (1986). https://doi.org/10.1007/BF00337113
26. Wang, D.: The time dimension for scene analysis. IEEE Trans. Neural Networks **16**(6), 1401–1426 (2005)
27. Wang, W., et al.: Learning unsupervised video object segmentation through visual attention. In: Proceedings of the IEEE Conference on Computer Vision and Pattern Recognition, pp. 3064–3074 (2019)

Classification of Actors in Social Networks Using RLVECO

Bonaventure C. Molokwu[1](✉) ⓘ, Shaon Bhatta Shuvo[1]ⓘ, Narayan C. Kar[2]ⓘ, and Ziad Kobti[1](✉) ⓘ

[1] School of Computer Science, University of Windsor,
401 Sunset Avenue, Windsor, ON N9B-3P4, Canada
{molokwub,shuvos,kobt}@uwindsor.ca
[2] Centre for Hybrid Automotive Research and Green Energy (CHARGE),
University of Windsor, 401 Sunset Avenue, Windsor, ON N9B-3P4, Canada
nkar@uwindsor.ca

Abstract. Several activities, comprising animate and inanimate entities, can be examined by means of Social Network Analysis (SNA). Classification tasks within social network structures remain crucial research problems in SNA. Inherent and latent facts about social graphs can be effectively exploited for training Artificial Intelligence (AI) models in a bid to categorize actors/nodes as well as identify clusters with respect to a given social network. Thus, important factors such as the individual attributes of spatial social actors and the underlying patterns of relationship binding these social actors must be taken into consideration. These factors are relevant to understanding the nature and dynamics of a given social graph. In this paper, we have proposed a hybrid model: Representation Learning via Knowledge-Graph Embeddings and Convolution Operations (RLVECO) which has been modelled for studying and extracting meaningful facts from social network structures to aid in node classification and community detection problems. RLVECO utilizes an edge sampling approach for exploiting features of a social graph, via learning the context of each actor with respect to its neighboring actors, with the aim of generating vector-space embeddings per actor which are further exploited for unexpressed representations via a sequence of convolution operations. Successively, these relatively low-dimensional representations are fed as input features to a downstream classifier for solving community detection and node classification problems about a given social network.

Keywords: Node classification · Feature learning · Feature extraction · Dimensionality reduction · Semi-supervised learning

This research was supported by International Business Machines (IBM) and Compute Canada (SHARCNET).

O. Gervasi et al. (Eds.): ICCSA 2020, LNCS 12249, pp. 905–922, 2020.
https://doi.org/10.1007/978-3-030-58799-4_65

1   Introduction and Related Literature

Humans are inhabited in a planet comprised of several systems and ecosystems; and interaction is a natural phenomenon and characteristic obtainable in any given system or ecosystem. Thus, relationship between constituent entities in a given system/ecosystem is a strategy for survival, and essential for the sustenance of the system/ecosystem. Owing to recent AI advances, these real-world (complex) systems and ecosystems can be effectively represented as social network structures. Social (network) graphs [24] are non-static structures which pose analytical challenges to Machine Learning (ML) and Deep Learning (DL) models because of their complex links, random nature, and occasionally massive size. In this regard, we propose RLVECO which is a hybrid DL-based model for classification and clustering problems in social networks.

Node classification and community detection remain open research problems in SNA. The classification of nodes induces the formation of cluster(s). Consequently, clusters give rise to homophily in social networks. Herein our proposed methodology is based on an iterative learning approach [1] which is targeted at solving the problems of node classification and community detection using an edge sampling strategy. Basically, learning in RLVECO is induced via semi-supervised training. The architecture of RLVECO comprises two (2) distinct representation-learning layers, viz: a Knowledge-Graph Embeddings (VE) layer and a Convolution Operations (CO) layer [15]; which are both trained by means of unsupervised training. These layers are essentially feature-extraction and dimensionality-reduction layers where underlying knowledge and viable facts are automatically extracted from the social network structure [17]. The VE layer is responsible for projecting the feature representation of the social graph to a q-dimensional real-number space, $\mathbb{R}^q$. This is done by associating a real-number vector to every unique actor/node in the social network such that the (cosine) distance of any given tie or edge would capture a significant degree of correlation between its pair of associated actors. Furthermore, the Convolution Operations layer feeds on the Knowledge-Graph Embeddings layer; and it is responsible for further extraction of apparent features and/or representations from the social graph. Finally, a classification layer succeeds the representation-learning layers; and it is trained by means of supervised training. The classifier is based on a Neural Network (NN) architecture assembled using deep (multi) layers of stacked perceptrons (NN units) [6]. Every low-dimensional feature (X), extracted by the representation-learning layers, is mapped to a corresponding output label (Y). These (X, Y) pairs are used to supervise the training of the classifier such that it can effectively/efficiently learn how to identify clusters and classify actors within a given social graph.

RLVECO is capable of learning the non-linear distributed features enmeshed in a social network [9]. Hence, the novelty of our research contribution are as highlighted below:

(1) Proposition of a DL-based and hybrid model, RLVECO, designed for resolving tie or link prediction problems in social network structures.

(2) Comprehensive benchmarking results which are based on classic objective functions used for standard classifiers.
(3) Comparative analyses, between RLVECO and state-of-the-art methodologies, against standard real-world social networks.

Also, we have evaluated RLVECO against an array of state-of-the-art models and Representation Learning (RL) approaches which serve as our baselines, viz:

(i) DeepWalk: Online Learning of Social Representations [21].
(ii) GCN: Semi-Supervised Classification with Graph Convolutional Networks [11].
(iii) LINE: Large-scale Information Network Embedding [27].
(iv) Node2Vec: Scalable Feature Learning for Networks [8].
(v) SDNE: Structural Deep Network Embedding [28].

2 Proposed Methodology and Framework

2.1 Definition of Problem

Definition 1. *Social Network, SN: As expressed via Eq. 1 such that SN is a tuple comprising a set of actors/vertices, V; a set of ties/edges, E; a metadata function, f_V, which extends the definition of the vertices' set by mapping it to a given set of attributes, V'; and a metadata function, f_E, which extends the definition of the edges' set by mapping it to a given set of attributes, E'. Thus, a graph function, $G(V, E) \subset SN$.*

$$
\begin{aligned}
SN &= (V, E, f_V, f_E) \equiv (G, f_V, f_E) \\
V &: |\{V\}| = M \qquad\qquad \text{set of actors/vertices with size, M} \\
E &: E \subset \{U \times V\} \subset \{V \times V\} \text{ set of ties/edges between V} \\
f_V &: V \to V' \qquad\qquad\qquad \text{vertices' metadata function} \\
f_E &: E \to E' \qquad\qquad\qquad \text{edges' metadata function}
\end{aligned} \tag{1}
$$

Definition 2. *Knowledge Graph, KG: $(\mathbb{E}, \mathbb{R})$ is a set comprising entities, $\mathbb{E}$, and relations, $\mathbb{R}$, between the entities. Thus, a KG [26,31] is defined via a set of triples, $t : (u, p, v)$, where $u, v \in \mathbb{E}$ and $p \in \mathbb{R}$. Also, a KG [29] can be modelled as a social network, SN, such that: $\mathbb{E} \to V$ and $\mathbb{R} \to E$ and $(\mathbb{E}, \mathbb{R}) \vdash f_V, f_E$.*

Definition 3. *Knowledge-Graph (Vector) Embeddings, X: The vector-space embeddings, X, generated by the VE layer are based on a mapping function, f, expressed via Eq. 2. f projects the representation of the graph's actors to a q-dimensional real space, $\mathbb{R}^q$, such that the existent ties between any given pair of actors, (u_i, v_j), remain preserved via the homomorphism from V to X.*

$$
\begin{aligned}
f &: V \to X \in \mathbb{R}^q \\
f &: (u, p, v) \to X \in \mathbb{R}^q \quad \text{Knowledge-Graph Embeddings}
\end{aligned} \tag{2}
$$

Definition 4. *Node Classification: Considering, SN, comprising partially labelled actors (or vertices), $V_{lbl} \subset V : V_{lbl} \rightarrow Y_{lbl}$; and unlabelled vertices defined such that: $V_{ulb} = V - V_{lbl}$. A node-classification model aims at training a predictive function, $f : V \rightarrow Y$, that learns to predict the labels, Y, for all actors or vertices, $V \subset SN$, via knowledge harnessed from the mapping: $V_{lbl} \rightarrow Y_{lbl}$ (Fig. 1).*

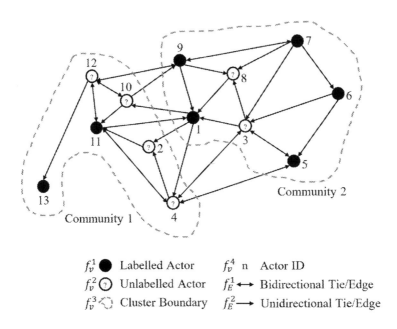

Fig. 1. Node classification task in social graphs

2.2   Proposed Methodology

Our proposition, RLVECO, is comprised of two (2) distinct Feature Learning (FL) layers, and one (1) classification layer.

Representation Learning - Knowledge-Graph Embeddings Layer: Given a social network, SN, defined by a set of actors/vertices, $V : U \subset V \forall \{u_m, v_m\} \in V$, and $M : m \in M$ denotes the number of unique actors in SN. Additionally, let the ties/edges in SN be defined such that: $E \subset \{U \times V\}$; where $u_i \in V$ and $v_j \in V$ represent a source_vertex and a target_vertex in E, respectively.

The objective function of the Knowledge-Graph Embeddings layer aims at maximizing the average logarithmic probability of the source_vertex, u_i, being predicted as a neighboring actor to the target_vertex, v_j, with respect to all training pairs, $\forall (u_i, v_j) \in E$. Formally, the function is expressed as in Eq. 3:

$$\mu = \frac{1}{M} \sum_{m=1}^{M} (\sum_{(u_i,v_j) \in E} log Pr(u_i|v_j)) \tag{3}$$

Consequently, in order to compute $Pr(u_i|v_j)$, we have to quantify the proximity of each target_vertex, v_j, with respect to its source_vertex, u_i. The vector-embedding model measures this adjacency/proximity as the cosine similarity between v_j and its corresponding u_i. Thus, the cosine distance is calculated as the dot product between the target_vertex and the source_vertex. Mathematically, $Pr(u_i|v_j)$ is computed via a softmax function as defined in Eq. 4:

$$Pr(u_i|v_j) = \frac{exp(u_i \cdot v_j)}{\sum_{m=1}^{M} exp(u_m \cdot v_j)} \tag{4}$$

Hence, the objective function of our vector-embedding (VE) model with respect to the SN is as expressed by Eq. 5:

$$\sum_{(u_i,v_j) \in E} log Pr(u_i|v_j) = \sum_{(u_i,v_j) \in E} log \frac{exp(u_i \cdot v_j)}{\sum_{m=1}^{M} exp(u_m \cdot v_j)} \tag{5}$$

Representation Learning - Convolution Operations Layer: This layer comprises three (3) RL or FL operations, namely: convolution, non-linearity, and pooling operations. RLVECO utilizes a one-dimensional (1D) convolution layer [16] which is sandwiched between the vector-embedding and classification layers. Equation 6 expresses the 1D-convolution operation:

$$FeatureMap(F) = 1D\_InputMatrix(X) * Kernel(K)$$
$$f_i = (X * K)_i = (K * X)_i = \sum_{j=0}^{J-1} x_j \cdot k_{i-j} = \sum_{j=0}^{J-1} k_j \cdot x_{i-j} \tag{6}$$

where f_i represents a cell/matrix position in the Feature Map; k_j denotes a cell position in the Kernel; and x_{i-j} denotes a cell/matrix position in the 1D-Input (data) matrix.

The non-linearity operation is a rectified linear unit (ReLU) function which introduces non-linearity after the convolution operation since real-world problems usually exist in non-linear form(s). As a result, the rectified feature/activation map is computed via: $r_i \in R = g(f_i \in F) = max(0, F)$.

The pooling operation is responsible for reducing the input width of each rectified activation map while retaining its vital properties. In this regard, the *Max Pooling* function is defined such that the resultant pooled (or downsampled) feature map is generated via: $p_i \in P = h(r_i \in R) = maxPool(R)$.

Classification - Multi-Layer Perceptron (MLP) Classifier Layer: This is the last layer of RLVECO's architecture, and it succeeds the representation-learning layers. The pooled feature maps, generated by the representation-learning layers, contain low-level features extracted from the constituent actors in the social graph. Hence, the classification layer utilizes these extracted "low-level features" for classifying actors in a bid to identify clusters contained in the social graph. The objective of the MLP [5] classifier function, f_c, is to map a given set of input values, P, to their respective output labels, Y, viz:

$$Y = f_c(P, \Theta) \tag{7}$$

In Eq. 7, Θ denotes a set of parameters. The MLP [4] function learns the values of Θ that will result in the best decision (Y) approximation for the input set, P. The MLP classifier output is a probability distribution which indicates the likelihood of a representation belonging to a particular output class. Our MLP [10] classifier is modelled such that sequential layers of NN units are stacked against each other to form a Deep Neural Network (DNN) structure [3,18].

Node Classification Algorithm: Defined via Algorithm 1.

Algorithm 1. Proposed Algorithm for Node Classification

Input: $\{V, E, Y_{lbl}\} \equiv \{$Actors, Ties, Ground-Truth Labels$\}$
Output: $\{Y_{ulb}\} \equiv \{$Predicted Labels$\}$

Preprocessing:
$V_{lbl}, V_{ulb} \subset V = V_{lbl} \cup V_{ulb}$ // V_{lbl} : Labelled actors   // V_{ulb} : Unlabelled actors
$E : (u_i, v_j) \in \{U \times V\}$ // $(u_i, v_j) \equiv$ (source, target)
$E_{train} = E_t : u_i, v_j \in V_{lbl}$ // $|E_{train}| = \sum indegree(V_{lbl}) + \sum outdegree(V_{lbl})$
$E_{pred} = E_p : u_i, v_j \in V_{ulb}$

$f_c \leftarrow$ Initialize   // Construct classifier model

Training:
for $t \leftarrow 0$ **to** $|E_{train}|$ **do**
    $f : E_t \rightarrow [X \in \mathbb{R}^q]$ // Embedding operation
    $f_t \in F = (K * X)_t$ // Convolution operation
    $r_t \in R = g(F) = max(0, f_t)$
    $p_t \in P = h(R) = maxPool(r_t)$
    $f_c | \Theta : p_t \rightarrow Y_{lbl}$ // MLP classification operation
end for

    return $Y_{ulb} = f_c(E_{pred}, \Theta)$

2.3   Proposed Architecture/Framework

Figure 2 illustrates the architecture of our proposition, RLVECO.

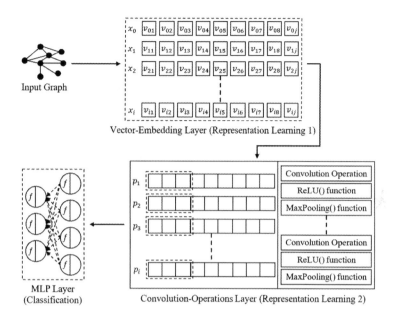

Fig. 2. Proposed system architecture

3   Datasets and Materials

3.1   Datasets

With regard to Table 1 herein, seven (7) real-world and benchmark social-graph datasets were utilized for experimentation and evaluation, viz: Cora [22, 25], CiteSeer [22, 25], Facebook Page-Page webgraph [23], Internet-Industry partnerships [2, 12, 13], PubMed-Diabetes [19], Terrorists-Relationship [32], and Zachary-Karate [14, 30].

Table 1. Benchmark datasets

Dataset	Nodes	Edges	Classes → {label: 'description'}
Cora	2,708	5,429	{C1: 'Case_Based', C2: 'Genetic_Algorithms', C3: 'Neural_Networks', C4: 'Probabilistic_Methods', C5: 'Reinforcement_Learning', C6: 'Rule_Learning', C7: 'Theory'}
CiteSeer	3,312	4,732	{C1: 'Agents', C2: 'Artificial Intelligence', C3: 'Databases', C4: 'Information Retrieval', C5: 'Machine Learning', C6: 'Human-Computer Interaction'}
Facebook-Page2Page	22,470	171,002	{C1: 'Companies', C2: 'Governmental Organizations', C3: 'Politicians', C4: 'Television Shows'}
PubMed-Diabetes	19,717	44,338	{C1: 'Diabetes Mellitus - Experimental', C2: 'Diabetes Mellitus - Type 1', C3: 'Diabetes Mellitus - Type 2'}
Terrorists-Relation	851	8,592	{C1: 'Colleague', C2: 'Congregate', C3: 'Contact', C4: 'Family'}
Zachary-Karate	34	78	{C1: 'Community 1', C2: 'Community 2', C3: 'Community 3', C4: 'Community 4'}
Internet-Industry	219	631	{C1: 'Content Sector', C2: 'Infrastructure Sector', C3: 'Commerce Sector'}

3.2 Data Preprocessing

All benchmark datasets ought to be comprised of actors and ties already encoded as discrete data (natural-number format). However, Cora, CiteSeer, Facebook-Page2Page, PubMed-Diabetes, and Terrorists-Relation datasets are made up of nodes and/or edges encoded in mixed formats (categorical and numerical formats). Thus, preprocessing is necessary to transcode these non-numeric (categorical) entities to their respective discrete (numeric) data representation without semantic loss. Thereafter, the numeric representation of all benchmark datasets are normalized prior to training against RLVECO and the benchmark models.

Table 2. Configuration of RLVECO's hyperparameters

Training set: 80%	Test set: 20%	Network width: 640
Batch size: 256	Optimizer: $AdaMax$	Network depth: 6
Epochs: $1.8 * 10^2$	Activation: $ReLU$	Dropout: $4.0 * 10^{-1}$
Learning rate: $1.0 * 10^{-3}$	Learning decay: 0.0	Embed dimension: 100

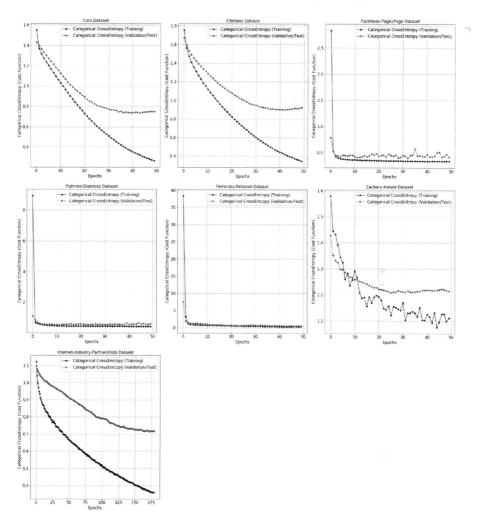

Fig. 3. Learning curves of RLVECO during training against Cora, CiteSeer, Facebook-Webgraph, PubMed-Diabetes, Terrorists-Relationship, Zachary-Karate, and Internet-Industry-Partnership datasets - loss function *vs* training epochs.

4   Experiment, Results, and Discussions

RLVECO has been tuned in accordance with the hyperparameters shown in Table 2. Our evaluations herein were recorded with reference to a range of objective functions. Thus, Categorical Cross Entropy was employed as the cost/loss function; while the fitness/utility was measured based on the following metrics: Precision (PC), Recall (RC), F-measure or F1-score (F1), Accuracy (AC), and Area Under the Receiver Operating Characteristic Curve (RO). Moreover, the objective functions have been computed against each benchmark dataset with regard to the constituent classes (or categories) present in each dataset. The Support (SP) represents the number of ground-truth samples per class/category contained in each dataset.

So as to avoid any bias across-the-board, we have used exactly the same SP for all models inclusive of RLVECO model. However, since RLVECO is based on an edge-sampling technique; the SP recorded against RLVECO model represents the number of edges/ties used for computation as explained in Algorithm 1. Furthermore, the performance of RLVECO during comparative analyses with respect to five (5) popular baselines (DeepWalk, GCN, LINE, Node2Vec, SDNE) when evaluated against the validation/test samples of the benchmark datasets are as documented in Table 3, Table 4, Table 5, Table 6, Table 7, and Table 8 respectively. Consequently, Fig. 3 graphically shows the learning-progress curves of our proposed model, RLVECO, during training over the benchmark datasets. Hence, the dotted-black lines represent learning progress over the training set; and the dotted-blue lines represent learning progress over the test set.

Tables 3, 4, 5, 6, 7, and 8 have clearly tabulated our results as a multi-classification task over the benchmark datasets. For each class per dataset, we have laid emphasis on the F1 (weighted average of the PC and RC metrics) and RO; and we have highlighted the model which performed best (based on F1 and RO metrics) for each classification job using a **bold font**. Additionally, we have employed a point-based ranking standard to ascertain the fittest model per node classification task. The model with the highest aggregate points signifies the fittest model for the specified task, and so on in descending order of aggregate points. Accordingly, as can be seen from our tabular results herein, our hybrid proposition (RLVECO) is at the top with the highest fitness points. RLVECO's superior performance, with reference to the results of the comparative analyses herein, is primarily attributed to its dual RL layers which enable it to extract and learn sufficient features of the social network structures. Its biform RL kernel places it at an edge above the state-of-the-art baselines evaluated herein.

The application of dropout regularization was targeted at the hidden layers of RLVECO. $L2$ regularization ($L2 = 0.04$) [7] and early stopping [20] were employed herein as addon regularization techniques to overcome overfitting incurred during the training of RLVECO. Therefore, the application of early stopping with regard to the training of RLVECO over the benchmark datasets were, viz: Cora (after 50 epochs), CiteSeer (after 50 epochs), Facebook-Page2Page (after 50 epochs), PubMed-Diabetes (after 50 epochs), Terrorists-Relation (after 50 epochs), Zachary-Karate (after 50 epochs), and Internet-Industry-Partnership (after 180 epochs). We have used a mini-batch size of 256 for training, testing, and validating because we want to ensure that sufficient patterns are extracted by RLVECO during training before its network weights are updated.

Table 3. Node-classification over Cora dataset. Results are based on the set apart validation sample - dataset *vs* models.

Model	Metric	Cora dataset								Points
		C1	C2	C3	C4	C5	C6	C7	μ	
RLVECO	PC	0.85	0.78	0.80	0.88	0.72	0.90	0.81	0.82	14
	RC	0.86	0.93	0.81	0.87	0.75	0.91	0.78	0.84	
	F1	**0.86**	**0.85**	**0.81**	**0.87**	**0.74**	**0.91**	**0.79**	0.83	
	AC	0.93	0.98	0.96	0.95	0.93	0.97	0.96	0.95	
	RO	**0.90**	**0.96**	**0.90**	**0.92**	**0.85**	**0.95**	**0.88**	0.91	
	SP	541	134	214	405	294	345	237	310	
GCN	PC	0.87	0.95	0.89	0.92	0.85	0.89	0.87	0.89	3
	RC	0.85	0.73	0.65	0.82	0.58	0.85	0.73	0.74	
	F1	**0.86**	0.83	0.75	**0.87**	0.69	0.87	**0.79**	0.81	
	AC	0.89	0.93	0.91	0.92	0.88	0.93	0.91	0.91	
	RO	0.88	0.83	0.80	0.89	0.75	0.90	0.83	0.84	
	SP	164	36	43	85	70	84	60	77	
Node2Vec	PC	0.58	0.78	0.72	0.81	0.80	0.84	0.82	0.76	0
	RC	0.85	0.50	0.53	0.68	0.64	0.74	0.60	0.65	
	F1	0.69	0.61	0.61	0.74	0.71	0.78	0.69	0.69	
	AC	0.77	0.96	0.95	0.92	0.93	0.94	0.94	0.92	
	RO	0.79	0.75	0.76	0.83	0.81	0.86	0.79	0.80	
	SP	164	36	43	85	70	84	60	77	
DeepWalk	PC	0.57	0.58	0.72	0.58	0.68	0.72	0.63	0.64	0
	RC	0.80	0.42	0.42	0.59	0.39	0.63	0.65	0.56	
	F1	0.67	0.48	0.53	0.58	0.49	0.67	0.64	0.58	
	AC	0.76	0.94	0.94	0.87	0.90	0.90	0.92	0.89	
	RO	0.77	0.70	0.70	0.75	0.68	0.79	0.80	0.74	
	SP	164	36	43	85	70	84	60	77	
LINE	PC	0.35	0.86	0.80	0.65	0.50	0.43	0.61	0.60	0
	RC	0.85	0.17	0.19	0.35	0.20	0.15	0.23	0.31	
	F1	0.50	0.28	0.30	0.46	0.29	0.23	0.34	0.34	
	AC	0.48	0.94	0.93	0.87	0.87	0.84	0.90	0.83	
	RO	0.59	0.58	0.59	0.66	0.59	0.56	0.61	0.60	
	SP	164	36	43	85	70	84	60	77	
SDNE	PC	0.37	0.83	0.70	0.60	0.54	0.64	0.64	0.62	0
	RC	0.91	0.14	0.16	0.35	0.20	0.27	0.12	0.31	
	F1	0.53	0.24	0.26	0.44	0.29	0.38	0.20	0.33	
	AC	0.50	0.94	0.93	0.86	0.87	0.86	0.89	0.84	
	RO	0.62	0.57	0.58	0.65	0.59	0.62	0.55	0.60	
	SP	164	36	43	85	70	84	60	77	

Table 4. Node-classification over CiteSeer dataset. Results are based on the set apart test sample - dataset *vs* models.

Model	Metric	CiteSeer dataset							Points
		C1	C2	C3	C4	C5	C6	μ	
RLVECO	PC	0.76	0.81	0.78	0.43	0.88	0.60	0.71	12
	RC	0.84	0.83	0.79	0.60	0.79	0.65	0.75	
	F1	**0.80**	**0.82**	**0.79**	**0.50**	**0.83**	**0.63**	0.73	
	AC	0.93	0.88	0.92	0.93	0.96	0.89	0.92	
	RO	**0.90**	**0.87**	**0.87**	**0.78**	**0.89**	**0.79**	0.85	
	SP	304	609	377	107	225	275	316	
GCN	PC	0.80	0.78	0.86	0.95	0.91	0.75	0.84	2
	RC	0.76	0.76	0.73	0.08	0.67	0.54	0.59	
	F1	0.78	0.77	**0.79**	0.15	0.77	**0.63**	0.65	
	AC	0.88	0.87	0.88	0.91	0.89	0.83	0.88	
	RO	0.84	0.83	0.83	0.53	0.81	0.72	0.76	
	SP	119	134	140	50	102	118	111	
Node2Vec	PC	0.57	0.55	0.49	0.33	0.55	0.38	0.48	0
	RC	0.55	0.60	0.66	0.06	0.45	0.40	0.45	
	F1	0.56	0.58	0.56	0.10	0.50	0.39	0.45	
	AC	0.85	0.82	0.78	0.92	0.86	0.78	0.84	
	RO	0.73	0.74	0.74	0.53	0.69	0.63	0.68	
	SP	119	134	140	50	102	118	111	
DeepWalk	PC	0.46	0.53	0.43	0.43	0.47	0.33	0.44	0
	RC	0.51	0.54	0.57	0.06	0.41	0.32	0.40	
	F1	0.49	0.54	0.49	0.11	0.44	0.32	0.40	
	AC	0.81	0.81	0.75	0.92	0.84	0.76	0.82	
	RO	0.69	0.71	0.69	0.53	0.66	0.59	0.65	
	SP	119	134	140	50	102	118	111	
SDNE	PC	0.37	0.50	0.24	0.20	0.45	0.31	0.35	0
	RC	0.19	0.27	0.77	0.02	0.14	0.09	0.25	
	F1	0.25	0.35	0.36	0.04	0.21	0.14	0.23	
	AC	0.80	0.80	0.42	0.92	0.84	0.80	0.76	
	RO	0.56	0.60	0.55	0.51	0.55	0.52	0.55	
	SP	119	134	140	50	102	118	111	
LINE	PC	0.18	0.30	0.28	0.60	0.22	0.27	0.31	0
	RC	0.15	0.47	0.39	0.06	0.12	0.21	0.23	
	F1	0.16	0.36	0.32	0.11	0.15	0.24	0.22	
	AC	0.72	0.67	0.65	0.93	0.80	0.76	0.76	
	RO	0.50	0.59	0.56	0.53	0.52	0.55	0.54	
	SP	119	134	140	50	102	118	111	

Table 5. Node-classification experiment results over Facebook Page-Page webgraph dataset. Results are based on the reserved validation/test sample - dataset *vs* models.

Model	Metric	Facebook-Page2Page dataset					Points
		C1	C2	C3	C4	μ	
RLVECO	PC	0.87	0.95	0.91	0.87	0.90	8
	RC	0.84	0.85	0.85	0.86	0.85	
	F1	**0.85**	**0.90**	**0.88**	**0.86**	0.87	
	AC	0.96	0.90	0.94	0.97	0.94	
	RO	**0.97**	**0.97**	**0.98**	**0.98**	0.98	
	SP	9989	33962	16214	6609	16694	
Node2Vec	PC	0.81	0.84	0.81	0.84	0.83	0
	RC	0.82	0.87	0.85	0.67	0.80	
	F1	0.81	0.85	0.83	0.74	0.81	
	AC	0.89	0.91	0.91	0.93	0.91	
	RO	0.87	0.90	0.89	0.82	0.87	
	SP	1299	1376	1154	665	1124	
DeepWalk	PC	0.75	0.84	0.76	0.75	0.78	0
	RC	0.81	0.85	0.82	0.52	0.75	
	F1	0.78	0.84	0.79	0.62	0.76	
	AC	0.87	0.90	0.89	0.90	0.89	
	RO	0.85	0.89	0.87	0.75	0.84	
	SP	1299	1376	1154	665	1124	
LINE	PC	0.53	0.66	0.72	0.66	0.64	0
	RC	0.72	0.71	0.59	0.29	0.58	
	F1	0.61	0.68	0.65	0.40	0.59	
	AC	0.73	0.80	0.83	0.87	0.81	
	RO	0.73	0.77	0.75	0.63	0.72	
	SP	1299	1376	1154	665	1124	
SDNE	PC	0.49	0.80	0.70	0.65	0.66	0
	RC	0.90	0.63	0.50	0.19	0.56	
	F1	0.64	0.70	0.58	0.29	0.55	
	AC	0.70	0.84	0.82	0.86	0.81	
	RO	0.76	0.78	0.71	0.58	0.71	
	SP	1299	1376	1154	665	1124	

Table 6. Node-classification experiment over PubMed-Diabetes dataset. Results are based on the reserved validation/test sample - dataset *vs* models.

Model	Metric	PubMed-Diabetes dataset				Points
		C1	C2	C3	μ	
RLVECO	PC	0.76	0.83	0.84	0.81	6
	RC	0.60	0.88	0.91	0.80	
	F1	**0.67**	**0.86**	**0.87**	0.80	
	AC	0.89	0.88	0.90	0.89	
	RO	**0.92**	**0.94**	**0.95**	0.94	
	SP	3300	7715	7170	6062	
DeepWalk	PC	0.65	0.57	0.58	0.60	0
	RC	0.15	0.67	0.71	0.51	
	F1	0.24	0.62	0.63	0.50	
	AC	0.81	0.67	0.68	0.72	
	RO	0.56	0.67	0.69	0.64	
	SP	821	1575	1548	1315	
Node2Vec	PC	0.74	0.47	0.49	0.57	0
	RC	0.03	0.65	0.55	0.41	
	F1	0.05	0.55	0.52	0.37	
	AC	0.80	0.57	0.60	0.66	
	RO	0.51	0.58	0.59	0.56	
	SP	821	1575	1548	1315	
SDNE	PC	0.65	0.43	0.74	0.61	0
	RC	0.05	0.96	0.17	0.39	
	F1	0.10	0.59	0.27	0.32	
	AC	0.80	0.48	0.65	0.64	
	RO	0.52	0.56	0.56	0.55	
	SP	821	1575	1548	1315	
LINE	PC	0.48	0.42	0.44	0.45	0
	RC	0.05	0.60	0.46	0.37	
	F1	0.08	0.50	0.45	0.34	
	AC	0.79	0.51	0.56	0.62	
	RO	0.52	0.53	0.54	0.53	
	SP	821	1575	1548	1315	

5   Limitations and Conclusion

The benchmark models (baselines) evaluated herein were executed using their default parameters. We were not able to evaluate GCN [11] against Facebook-Page2Page, PubMed-Diabetes, Internet-Industry-Partnership, and Zachary-Karate datasets; because these aforementioned datasets do not possess individual vector-based feature set which is required by the GCN model for input-data

Table 7. Node-classification experiment over Terrorists-Relationship dataset. Results are based on the reserved validation/test sample - dataset *vs* models.

Model	Metric	Terrorists-Relation dataset					Points
		C1	C2	C3	C4	μ	
RLVECO	PC	0.93	0.91	0.46	1.00	0.83	6
	RC	0.97	0.97	0.42	0.97	0.83	
	F1	**0.95**	**0.94**	0.44	**0.98**	0.83	
	AC	0.95	0.98	0.89	0.99	0.95	
	RO	**0.98**	**1.00**	0.85	**1.00**	0.96	
	SP	1706	491	319	561	769	
GCN	PC	0.94	0.74	0.67	0.96	0.83	5
	RC	0.90	0.95	0.60	1.00	0.86	
	F1	0.92	0.83	**0.63**	**0.98**	0.84	
	AC	0.92	0.95	0.88	0.99	0.94	
	RO	**0.98**	0.99	**0.91**	1.00	0.97	
	SP	92	21	30	27	43	
DeepWalk	PC	0.88	0.82	0.64	0.86	0.80	0
	RC	0.90	0.86	0.53	0.93	0.81	
	F1	0.89	0.84	0.58	0.89	0.80	
	AC	0.88	0.96	0.86	0.96	0.92	
	RO	0.88	0.92	0.73	0.95	0.87	
	SP	92	21	30	27	43	
Node2Vec	PC	0.86	0.82	0.60	0.86	0.79	0
	RC	0.88	0.86	0.50	0.93	0.79	
	F1	0.87	0.84	0.55	0.89	0.79	
	AC	0.86	0.96	0.85	0.96	0.91	
	RO	0.86	0.92	0.71	0.95	0.86	
	SP	92	21	30	27	43	
LINE	PC	0.82	0.82	0.58	0.92	0.79	0
	RC	0.92	0.86	0.37	0.85	0.75	
	F1	0.87	0.84	0.45	0.88	0.76	
	AC	0.85	0.96	0.84	0.96	0.90	
	RO	0.84	0.92	0.65	0.92	0.83	
	SP	92	21	30	27	43	
SDNE	PC	0.77	0.90	0.56	1.00	0.81	0
	RC	0.92	0.86	0.30	0.85	0.73	
	F1	0.84	0.88	0.39	0.92	0.76	
	AC	0.81	0.97	0.84	0.98	0.90	
	RO	0.80	0.92	0.62	0.93	0.82	
	SP	92	21	30	27	43	

Table 8. Node-classification experiment over Zachary-Karate and Internet-Industry-Partnership datasets. Results are based on the reserved validation sample - datasets *vs* models. N.B.: Mtc = Fitness Metric; Pts = Points.

Model	Mtc	Zachary-Karate dataset					Pts	Internet-Industry-Partnership				Pts
		C1	C2	C3	C4	μ		C1	C2	C3	μ	
RLVECO	PC	1.00	0.67	0.20	1.00	0.72	7	0.33	0.96	0.29	0.53	5
	RC	1.00	0.89	1.00	0.50	0.85		0.65	0.77	0.76	0.73	
	F1	**1.00**	**0.76**	**0.33**	0.67	0.69		**0.44**	**0.86**	0.42	0.57	
	AC	1.00	0.81	0.69	0.77	0.82		0.84	0.78	0.87	0.83	
	RO	**1.00**	**0.83**	**0.83**	**0.75**	0.85		**0.76**	**0.81**	**0.82**	0.80	
	SP	3	9	2	12	7		26	238	17	94	
SDNE	PC	0.00	0.50	0.00	0.60	0.28	2	0.00	0.61	0.00	0.20	0
	RC	0.00	0.50	0.00	1.00	0.38		0.00	1.00	0.00	0.33	
	F1	0.00	0.50	0.00	**0.75**	0.31		0.00	0.76	0.00	0.25	
	AC	0.86	0.71	0.86	0.71	0.79		0.82	0.61	0.80	0.74	
	RO	0.50	0.65	0.50	**0.75**	0.60		0.50	0.50	0.50	0.50	
	SP	1	2	1	3	2		8	27	9	15	
LINE	PC	0.00	0.50	0.00	0.60	0.28	2	0.00	0.61	0.00	0.20	0
	RC	0.00	0.50	0.00	1.00	0.38		0.00	1.00	0.00	0.33	
	F1	0.00	0.50	0.00	**0.75**	0.31		0.00	0.76	0.00	0.25	
	AC	0.86	0.71	0.86	0.71	0.79		0.82	0.61	0.80	0.74	
	RO	0.50	0.65	0.50	**0.75**	0.60		0.50	0.50	0.50	0.50	
	SP	1	2	1	3	2		8	27	9	15	
DeepWalk	PC	0.00	0.40	0.00	0.50	0.23	0	0.50	0.81	0.36	0.56	1
	RC	0.00	1.00	0.00	0.33	0.33		0.12	0.93	0.44	0.50	
	F1	0.00	0.57	0.00	0.40	0.24		0.20	**0.86**	0.40	0.49	
	AC	0.86	0.57	0.86	0.57	0.72		0.82	0.82	0.73	0.79	
	RO	0.50	0.70	0.50	0.54	0.56		0.55	0.79	0.62	0.65	
	SP	1	2	1	3	2		8	27	9	15	
Node2Vec	PC	0.00	0.00	0.00	0.25	0.06	0	0.00	0.68	0.75	0.48	1
	RC	0.00	0.00	0.00	0.33	0.08		0.00	1.00	0.33	0.44	
	F1	0.00	0.00	0.00	0.29	0.07		0.00	0.81	**0.46**	0.42	
	AC	0.86	0.29	0.86	0.29	0.58		0.82	0.70	0.84	0.79	
	RO	0.50	0.20	0.50	0.29	0.37		0.50	0.62	0.65	0.59	
	SP	1	2	1	3	2		8	27	9	14.67	

processing. Overall, RLVECO's remarkable performance with reference to our benchmarking results can be attributed to the following:

(i) The RL kernel of RLVECO is constituted of two (2) distinct layers of FL, viz: Knowledge-Graph Embeddings and Convolution Operations [15].
(ii) The high-quality data preprocessing techniques employed herein with respect to the benchmark datasets. We ensured that all the constituent actors of a given social network were transcoded to their respective discrete

data representations, without any loss in semantics, and normalized prior to training and/or testing.

6   Future Work and Acknowledgements

We intend to expand RLVECO's scope such that it can be applied for resolving other open research problems in SNA. Also, we are sourcing for additional baselines (benchmark models) and real-world social network datasets for extensive validation of RLVECO. This research was supported by International Business Machines (IBM) via the provision of computational resources necessary to carry-out our experiments. Also, this work was enabled in part via support provided by SHARCNET and Compute Canada (www.computecanada.ca).

References

1. Aggarwal, C.C. (ed.): Social Network Data Analytics. Springer, Boston (2011). https://doi.org/10.1007/978-1-4419-8462-3
2. Batagelj, V., Doreian, P., Ferligoj, A., Kejzar, N. (eds.): Understanding Large Temporal Networks and Spatial Networks: Exploration, Pattern Searching, Visualization and Network Evolution. Wiley, Hoboken (2014)
3. Bengio, Y.: Learning deep architectures for AI. Found. Trends Mach. Learn. **2**, 1–113 (2009)
4. Deng, L.M., Yu, D.H.: Deep Learning: Methods and Applications. Foundations and Trends in Signal Processing. Now Publishers (2014). https://books.google.ca/books?id=-Sa6xQEACAAJ
5. Goodfellow, I.G., Bengio, Y., Courville, A.C.: Deep learning. Nature **521**, 436–444 (2015)
6. Goodfellow, I.G., Bengio, Y., Courville, A.C. (eds.): Deep Learning. MIT Press, Cambridge (2017)
7. Gron, A. (ed.): Hands-On Machine Learning with Scikit-Learn and TensorFlow: Concepts, Tools, and Techniques to Build Intelligent Systems. O'Reilly Media, Inc., Newton (2017)
8. Grover, A., Leskovec, J.: node2vec: Scalable feature learning for networks. In: Proceedings of the 22nd ACM SIGKDD International Conference on Knowledge Discovery and Data Mining, pp. 855–864 (2016)
9. Hinton, G.E.: Learning multiple layers of representation. TRENDS Cogn. Sci. **11**(10), 428–433 (2007)
10. Hinton, G.E., et al.: Deep neural networks for acoustic modeling in speech recognition. IEEE Signal Process. Mag. **29**, 82–97 (2012)
11. Kipf, T.N., Welling, M.: Semi-supervised classification with graph convolutional networks. In: International Conference on Learning Representations (ICLR), abs/1609.02907 (2017)
12. Krebs, V.: Orgnet LLC, January 2002. http://www.orgnet.com/netindustry.html
13. Krebs, V.E.: Organizational adaptability quotient. In: IBM Global Services (2008)
14. Kunegis, J.: Konect: the Koblenz network collection. In: Proceedings of the 22nd International Conference on World Wide Web (2013). http://konect.cc/

15. Molokwu, B.C.: Event prediction in complex social graphs using one-dimensional convolutional neural network. In: Proceedings of the 28th International Joint Conference on Artificial Intelligence, IJCAI (2019)
16. Molokwu, B.C.: Event prediction in social graphs using 1-dimensional convolutional neural network. In: Meurs, M.-J., Rudzicz, F. (eds.) Canadian AI 2019. LNCS (LNAI), vol. 11489, pp. 588–592. Springer, Cham (2019). https://doi.org/10.1007/978-3-030-18305-9_64
17. Molokwu, B.C., Kobti, Z.: Event prediction in complex social graphs via feature learning of vertex embeddings. In: Gedeon, T., Wong, K.W., Lee, M. (eds.) ICONIP 2019. CCIS, vol. 1143, pp. 573–580. Springer, Cham (2019). https://doi.org/10.1007/978-3-030-36802-9_61
18. Molokwu, B.C., Kobti, Z.: Spatial event prediction via multivariate time series analysis of neighboring social units using deep neural networks. In: 2019 International Joint Conference on Neural Networks, IJCNN, pp. 1–8 (2019)
19. Namata, G., London, B., Getoor, L., Huang, B.: Query-driven active surveying for collective classification. In: Proceedings of the Workshop on Mining and Learning with Graphs, MLG 2012 (2012)
20. Patterson, J., Gibson, A. (eds.): Deep Learning: A Practitioner's Approach. O'Reilly Media, Inc., Newton (2017)
21. Perozzi, B., Al-Rfou', R., Skiena, S.: DeepWalk: online learning of social representations. In: Proceedings of the 20th ACM SIGKDD International Conference on Knowledge Discovery and Data Mining, abs/1403.6652 (2014)
22. Rossi, R.A., Ahmed, N.K.: The network data repository with interactive graph analytics and visualization. In: Proceedings of the Twenty-Ninth AAAI Conference on Artificial Intelligence (2015). http://networkrepository.com
23. Rozemberczki, B., Allen, C., Sarkar, R.: Multi-scale attributed node embedding. arXiv abs/1909.13021 (2019)
24. Scott, J. (ed.): Social Network Analysis. SAGE Publications Ltd., Newbury Park (2017)
25. Sen, P., Namata, G., Bilgic, M., Getoor, L., Gallagher, B., Eliassi-Rad, T.: Collective classification in network data. AI Mag. **29**, 93–106 (2008)
26. Tabacof, P., Costabello, L.: Probability calibration for knowledge graph embedding models. In: International Conference on Learning Representations (ICLR), abs/1912.10000 (2020)
27. Tang, J., Qu, M., Wang, M., Zhang, M., Yan, J., Mei, Q.: Line: large-scale information network embedding. In: Proceedings of the 24th International Conference on World Wide Web (2015)
28. Wang, D., Cui, P., Zhu, W.: Structural deep network embedding. In: Proceedings of the 22nd ACM SIGKDD International Conference on Knowledge Discovery and Data Mining (2016)
29. Yang, S., Tian, J., Zhang, H., Yan, J., He, H., Jin, Y.: TransMS: knowledge graph embedding for complex relations by multidirectional semantics. In: Proceedings of the 28th International Joint Conference on Artificial Intelligence, IJCAI (2019)
30. Zachary, W.W.: An information flow model for conflict and fission in small groups1. J. Anthropol. Res. **33** (1977). https://doi.org/10.1086/jar.33.4.3629752
31. Zhang, Q., Sun, Z., Hu, W., Chen, M., Guo, L., Qu, Y.: Multi-view knowledge graph embedding for entity alignment. In: Proceedings of the 28th International Joint Conference on Artificial Intelligence, IJCAI, vol. abs/1906.02390 (2019)
32. Zhao, B., Sen, P., Getoor, L.: Entity and relationship labeling in affiliation networks. In: Proceedings of the 23rd International Conference on Machine Learning, ICML (2006)

Detecting Apples in Orchards Using YOLOv3

Anna Kuznetsova🆔, Tatiana Maleva🆔, and Vladimir Soloviev$^{(\boxtimes)}$🆔

Financial University under the Government of the Russian Federation,
38 Shcherbakovskaya, Moscow 105187, Russia
{AnAKuznetsova, TVMaleva, VSoloviev}@fa.ru

Abstract. A machine vision system for detecting apples in orchards was developed. The system designed for use in harvesting robots is based on a YOLOv3 algorithm modification with pre- and postprocessing. As a result, apples that are blocked by leaves and branches, green apples on a green background, darkened apples are detected. Apple detection time averaged 19 ms with 90.8% Recall (fruit detection rate), and 7.8% FPR.

Keywords: Machine vision · Apple harvesting robot · YOLO

1 Introduction

As a result of intensification, mechanization, and automation, agricultural productivity has increased significantly. In general, in developed countries, the number of people employed in agriculture decreased by 80 times during the 20th century. Nevertheless, manual labor is the main component of costs in agriculture, reaching 40% of the total value of vegetables, fruits, and cereals grown [1, 2].

Horticulture is one of the most labor-intensive sectors of agriculture. The level of automation in horticulture is about 15%, fruits are harvesting manually, and crop shortages reach 50%. At the same time, as a result of urbanization, every year, it is becoming increasingly difficult to recruit seasonal workers for the harvest [3].

It is obvious that the widespread use of robots can bring significant benefits in horticulture, increase labor productivity, reduce the share of heavy manual routine harvesting operations, and reduce crop shortages.

Fruit picking robots have been developing since the late 1960s. However, today all the developments are in the prototype stage. The low speed of picking fruits and unsatisfactory rates of picked and unhandled apples are hindering the industrial implementation of fruits harvesting robots. It is largely due to the insufficient quality of machine vision systems used in robots for picking apples [4, 5].

Recently, many neural network models have been trained to recognize apples. However, computer vision systems based on these models in existing harvesting robot prototypes work too slowly. They also do not detect darkened apples and apples with a lot of overlapping leaves and branches, as well as green apples on a green background, take yellow leaves as apples, etc.

To solve these problems arising in apple detection in orchards, we propose to use the YOLOv3 algorithm with special pre- and postprocessing of images taken by the camera placed on the manipulator of the harvesting robot.

© Springer Nature Switzerland AG 2020
O. Gervasi et al. (Eds.): ICCSA 2020, LNCS 12249, pp. 923–934, 2020.
https://doi.org/10.1007/978-3-030-58799-4_66

The remainder of the paper is structured as follows. Section 2 reviews related works on apple detection in orchards using intelligent algorithms, Sect. 3 presents our technique, and finally, the results are discussed in Sect. 4.

2 Literature Review

2.1 Color-Based Fruit Detection Algorithms

The efficiency and productivity of harvesting robots are largely determined by algorithms used to detect fruits in images. In various prototypes of such robots, various recognition techniques based on one or more factors were used.

One of the main factors based on which the fruit can be detected in the image is color. The set color threshold can be used for each pixel in the image to determine if this pixel belongs to the fruit.

In [6, 7], this approach showed a 90% share of correctly recognized apples, and in [8], a 95% share of correctly recognized apples, although on very limited datasets (several dozen images).

Of course, color-based apple detection works well in the case of red apples, but, as a rule, does not provide satisfactory quality for green apples [9]. To solve the problem of green fruit recognition, many authors combine image analysis in the visible and infrared spectra [7, 10–12].

For example, in [11], a 74% fraction of correctly detected apples (accuracy), obtained by combining analysis of the visible and infrared spectra, is compared with 66% of the correctly detected fruits based on analysis of only the visible spectrum and with 52% accuracy based on analysis of only the infrared spectrum.

The obvious virtue of detecting fruit by color is the ease of implementation, but this method detects green and yellow-green apples very poorly. Also, bunches of red apples merge into one large apple, and this leads to an incorrect determination of the apple bounding box coordinates. Thermal cameras are quite expensive and inconvenient in practical use since the difference between apples and leaves is detected when shooting is made within two hours after dawn.

2.2 Shape-Based Fruit Detection Algorithms

To detect spherical fruits, such as tomatoes, apples, and citrus, fruit recognition algorithms based on the analysis of geometric shapes could be used. The main advantage of the analysis of geometric shapes is the low dependence of the quality of recognition of objects on the level of illumination.

To detect spherical fruits, in [13–15], modifications of the Hough circular transform were used to improve the detection quality of fruits partially hidden by leaves or other fruits.

In [16, 17], fruit detection algorithms based on convex object identification were proposed.

In order to improve the quality of recognition in uncontrolled environments, which may deteriorate due to uneven lighting, a partial overlap of fruits with other fruits and leaves, as well as other features, many researchers use a combination of color and shape analysis algorithms.

Systems based on such algorithms work very quickly, but complex scenes, especially with fruits, overlapped by leaves or other fruits, are usually not recognized effectively by such systems.

Detecting fruits by shape gives a large error, since not only apples are round, but also gaps, leaf silhouettes, spots and shadows on apples. Combining circle selection algorithms with subsequent pixel analysis is inefficient in terms of calculation speed.

2.3 Texture-Based Fruit Detection Algorithms

Fruits photographed in orchards in natural conditions differ from the leaves and branches in texture, and this can be used to facilitate the separation of fruits from the background. Differences in texture play a particularly important role in fruit recognition when the fruits are grouped in bunches or overlapped by other fruits or leaves.

For example, in [18], apples were detected based on image texture analysis in combination with color analysis, and the proportion of correctly recognized fruits was 90% (on a limited dataset).

Detecting fruits by texture only works in close-up images with good resolution and works very poorly in backlight. The low speed of texture-based fruit detection algorithms and a too high proportion of undetected fruits lead to the inefficiency of practical use of this technique.

2.4 Early Stage of Using Machine Learning for Fruit Detection

Machine learning methods have been already used to detect fruits in images for a long time. The earliest prototype of a fruit-picking robot that detects red apples against a background of green leaves using machine learning techniques was presented back in 1977 by E.A. Parrish and J.A.K. Goksel [19].

In [11], in order to detect green apples against the background of green leaves, K-means clustering was applied to a and b CIE L * a * b color space coordinates in the visible spectrum, as well as to image coordinates in the infrared spectrum with the subsequent removal of noise. It allowed the authors to correctly detect 74% of apples in the images from the test data set.

The use of a linear classifier and the KNN-classifier to detect apples and peaches in the machine vision system was compared in [21], with both classification algorithms yielding similar accuracy at 89%. In [22], the linear classifier has shown 80% accuracy of apple detection. The authors of [23] recognized apples, bananas, lemons, and strawberries in images, using a KNN-classifier and reported 90% accuracy. Applying KNN-classifier to color and texture data allowed to find 85% of green apples in raw images and 95% in hand-processed images [24].

In [25], the SVM-based apple detection algorithm was introduced. This classifier balanced the ratio between accuracy and recognition time, showing 89% of correctly detected fruits and an average apple detection time equal to 3.5 s. Using SVM for apple detection in [26] has shown 92% accuracy (on a test dataset of 59 apples).

It is very unusual that boosted decision trees were practically not used in fruit detection systems. In [27], the AdaBoost algorithm was used to recognize kiwi in orchards, which made it possible to achieve a 92% share of correctly detected fruits against branches, leaves, and soil. In [28, 29], AdaBoost was applied to color analysis to automatically detect ripe tomatoes in a greenhouse, showing 96% accuracy. The search for examples of the use of modern algorithms like XGBoost, LightGBM, CatBoost for detecting fruits in images has not yielded results.

All these early-stage machine learning techniques for fruit detection were tested on very limited datasets of several dozens of images, so the results cannot be generalized for practical use.

2.5   Using Deep Neural Networks for Fruit Detection

Since 2012, with the advent of deep convolutional neural networks, in particular, AlexNet proposed by A. Krizhevsky, I. Sutskever, and G.E. Hinton in [30], machine vision and its use for detecting various objects in images, including fruits, received an impetus in development. AlexNet took first place in the ImageNet Large-Scale Visual Recognition Challenge 2012 (the share of correctly recognized images was 84.7% against 73.8% in the second place).

In 2015, K. Simonyan and A. Zisserman published a paper [31], where they proposed the VGG16 convolutional neural network as an improved version of AlexNet. It showed 92.7% of correct answers at the 2014 ImageNet Large-Scale Visual Recognition Challenge.

H.A.M. Williams, M.H. Jones, M. Nejati, M.J. Seabright, and B.A. MacDonald (2018) built a kiwi fruit harvesting robot with VGG16-based machine vision system. In the field trials, 76% of kiwi fruits were detected [32].

In 2016, a new algorithm was proposed – YOLO (You Look Only Once) [33]. Before this, in order to detect objects in images, classification models based on neural networks were applied to a single image several times – in several different regions and/or on several scales. The YOLO approach involves a one-time application of one neural network to the whole image. The model divides the image into regions and immediately determines the scope of objects and probabilities of classes for each object. The third version of the YOLO algorithm was published in 2018 as YOLOv3 [34].

The YOLO algorithm is one of the fastest, and it has already been used in robots for picking fruits. Y. Tian, G. Yang, Zh. Wang, H. Wang, E. Li, and Z. Liang (2019) proposed a modification of the YOLO model and applied it to detect apples in images [35, 36]. The authors made the network tightly connected: each layer was connected to all subsequent layers, as the DenseNet approach suggests [37]. The IoU (Intersection over Union) indicator turned out to be 89.6%, with an average recognition time of one apple equal to 0.3 s.

H. Kang and C. Chen (2020) developed DaSNet-v2 neural network to detect apples. This network detects objects in images in a single pass, considering their superposition, just like YOLO. The IoU in this model was at 86% [38].

Sh. Wan and S. Goudos (2020) compared three algorithms: the standard Faster R-CNN, the proposed by them modification of the Faster R-CNN, and YOLOv3 for detection of oranges, apples, and mangoes. It turned out that the modification proposed by the authors reveals about 90% of the fruits, which is 3–4% better than the standard Faster R-CNN on the same dataset and at about the same level as YOLOv3 [39].

3   Research Methods

The Department of data analysis and machine learning of the Financial University under the Government of the Russian Federation, together with the Laboratory of machine technologies for cultivating perennial crops of the VIM Federal Scientific Agroengineering Center, is developing a robot for harvesting apples.

The VIM Center develops the mechanical component of the robot, while the Financial University is responsible for the intelligent algorithms for detecting fruits and operating the manipulator for their picking.

To detect apples, we used the standard YOLOv3 algorithm [34] trained on the COCO dataset [40], which contains 1.5 million objects of 80 categories marked out in images.

Since we considered only apple orchards, we were guided by the round shape of objects, and the categories "apples" and "oranges" were combined.

Some image preprocessing techniques are used to improve apple detection quality:

- adaptive histogram alignment (to increase contrast);
- slight blur;
- thickening of the borders.

As a result, it was possible to mitigate the negative effects of shadows, glare, minor damages of apples, and the presence of thin branches overlapping apples.

As the test data set, 6083 images of apples of various kinds, including red and green apples, taken in orchards by the VIM Center employees, were used.

Using the standard YOLOv3 algorithm to detect apples in this set of images showed that not all apples are recognized successfully (Fig. 1).

The following primary factors preventing recognition of apples in images were identified:

- backlight;
- presence of dark spots on apples and/or a noticeable perianth;
- the proximity of the green apple shade to leaves shade;
- existence of empty gaps between the leaves, which the network mistook for small apples;
- overlapping apples by other apples, branches, and leaves.

To attenuate the negative influence of backlight, images, on which this problem was detected by comparing the number of dark pixels with the average, were strongly lightened.

Since spots on apples, perianth, as well as thin branches, are represented in images by pixels of brown shades, such pixels were replaced by yellow ones, which made it possible to recognize apples in such images successfully as shown in Fig. 2.

Figure 3 shows examples of images in which yellow leaves, as well as small gaps between leaves, are mistakenly recognized as apples.

To prevent the system from taking yellow leaves for apples, during postprocessing, we discarded recognized objects whose ratio of the larger side of the circumscribed rectangle to the smaller one was more than 3.

In order not to take the gaps between the leaves for apples, during the postprocessing, objects were discarded whose area of the circumscribed rectangle was less than the threshold.

Fig. 1. Examples of detection results without postprocessing.

Fig. 2. Examples of detected apples with dark spots and overlapping thin branches.

Fig. 3. Examples of yellow leaves and gaps between leaves mistaken for apples.

The problem when not all the apples forming a bunch are detected (Fig. 4) is not significant for the robot, since the manipulator takes out only one apple at each step, and the number of apples in the bunch decreases.

It should be noted that this problem arises when analyzing canopy-view images only. When analyzing close-up images taken by a camera located on the robot arm, this problem does not occur.

4   Results

In general, the system proposed recognizes both red and green apples quite accurately (Fig. 5, 6). The system detects apples that are blocked by leaves and branches, green apples on a green background, darkened apples, etc. Green apples are detected better if the shade of apple is at least slightly different from the shade of leaves (Fig. 6).

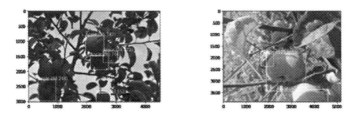

Fig. 4. Examples of partial detection of apples in bunches.

Fig. 5. Examples of red apple detection.

Fig. 6. Examples of green apple detection.

The detection time for one apple ranged from 7 to 46 ms, considering pre- and postprocessing (on average, one apple was detected in 19 ms).

We evaluated the apple detection results by manually comparing the ground-truth apples labeled by the authors manually in the images and the apples detected by the algorithm and calculating the following metrics:

- *TP* (True Positive) stands for the share of fruits correctly detected by the algorithm;
- *FP* (False Positive) – the number of errors of the first kind, i.e. background objects in the image, mistakenly accepted by the algorithm as fruits;
- *FN* (False Negative) – the number of errors of the second kind, i.e. fruits not detected by the algorithm.

The system has shown a 92.2% *Precision*, 90.8% *Recall*, 91.5 *F1 Score*, 9.2% *False Negative Rate* (*FNR*), and 7.8% *False Positive Rate* (*FPR*).

Here

$$Precision = \frac{TP}{TP + FP} \quad \text{and} \quad Recall = \frac{TP}{TP + FN}$$

represent the share of actual fruits among all the objects identified by the algorithm as the fruits and the fruit detection rate (i.e. the share of detected fruits among all the ground-truth fruits),

$$FNR = 1 - Recall = \frac{FP}{TP + FN} \quad \text{and} \quad FPR = 1 - Precision = \frac{FP}{TP + FP}$$

are the shares of not detected fruits and background objects detected as fruits by mistake.

The algorithm proposed in this paper turned out to demonstrate a smaller fraction of undetected apples (*FNR*) than in modern models [34, 35, 37, 38] based on convolutional neural networks, with a higher fruit detection rate.

References

1. Bechar, A., Vigneault, C.: Agricultural robots for field operations: concepts and components. Biosyst. Eng. **149**, 94–111 (2016). https://doi.org/10.1016/j.biosystemseng.2016.06.014
2. Sistler, F.E.: Robotics and intelligent machines in agriculture. IEEE J. Robot. Autom. **3**(1), 3–6 (1987). https://doi.org/10.1109/JRA.1987.1087074
3. Ceres, R., Pons, J., Jiménez, A., Martín, J., Calderón, L.: Design and implementation of an aided fruit-harvesting robot (Agribot). Ind. Robot **25**(5), 337–346 (1998). https://doi.org/10.1108/01439919810232440
4. Edan, Y., Han, S.F., Kondo, N.: Automation in agriculture. In: Springer Handbook of Automation, pp. 1095–1128. Springer, Berlin, Heidelberg (2009). https://doi.org/10.1007/978-3-540-78831-7_63
5. Grift, T., Zhang, Q., Kondo, N., Ting, K.C.: A review of automation and robotics for the bioindustry. J. Biomechatronics Eng. **1**(1), 37–54 (2008). http://journal.tibm.org.tw/wp-content/uploads/2013/06/2.-automation-and-robotics-for-the-bio-industry.pdf. Accessed 19 Apr 2020
6. Mao, W.H., Ji, B.P., Zhan, J.C., Zhang, X.C., Hu, X.A.: Apple location method for the apple harvesting robot. In: Proceedings of the 2nd International Congress on Image and Signal Processing – CIPE 2009, Tianjin, China, 7–19 October 2009, pp. 17–19 (2009). https://doi.org/10.1109/cisp.2009.5305224
7. Bulanon, D.M., Kataoka, T.: A fruit detection system and an end effector for robotic harvesting of Fuji apples. Agric. Eng. Int. CIGR J. **12**(1), 203–210 (2010). https://cigrjournal.org/index.php/Ejounral/article/view/1285/1319. Accessed 19 Apr 2020
8. Wei, X., Jia, K., Lan, J., Li, Y., Zeng, Y., Wang, C.: Automatic method of fruit object extraction under complex agricultural background for vision system of fruit picking robot. Optics **125**(12), 5684–5689 (2014). https://doi.org/10.1016/j.ijleo.2014.07.001
9. Zhao, Y.S., Gong, L., Huang, Y.X., Liu, C.L.: A review of key techniques of vision-based control for harvesting robot. Comput. Electron. Agric. **127**, 311–323 (2016). https://doi.org/10.1016/j.compag.2016.06.022
10. Bulanon, D.M., Burks, T.F., Alchanatis, V.: Image fusion of visible and thermal images for fruit detection. Biosyst. Eng. **103**(1), 12–22 (2009). https://doi.org/10.1016/j.biosystemseng.2009.02.009
11. Wachs, J.P., Stern, H.I., Burks, T., Alchanatis, V.: Low and high-level visual feature-based apple detection from multi-modal images. Precision Agric. **11**, 717–735 (2010). https://doi.org/10.1007/s11119-010-9198-x

12. Wachs, J.P., Stern, H.I., Burks, T., Alchanatis, V.: Apple detection in natural tree canopies from multimodal images. In: Proceedings of the 7th Joint International Agricultural Conference – JIAC 2009, Wageningen, Netherlands, 6–8 July 2009, pp. 293–302 (2009). http://web.ics.purdue.edu/~jpwachs/papers/2009/ApplesDetection_FullPaper_revised.pdf. Accessed 19 Apr 2020

13. Whittaker, A.D., Miles, G.E., Mitchell, O.R.: Fruit location in a partially occluded image. Trans. Am. Soc. Agric. Eng. **30**(3), 591–596 (1987). https://doi.org/10.13031/2013.30444

14. Xie, Z.Y., Zhang, T.Z., Zhao, J.Y.: Ripened strawberry recognition based on Hough transform. Trans. Chin. Soc. Agric. Mach. **38**(3), 106–109 (2007)

15. Xie, Z., Ji, C., Guo, X., Zhu, S.: An object detection method for quasi-circular fruits based on improved Hough transform. Trans. Chin. Soc. Agric. Mach. **26**(7), 157–162 (2010). https://doi.org/10.3969/j.issn.1002-6819.2010.7.028

16. Kelman, E.E., Linker, R.: Vision-based localization of mature apples in tree images using convexity. Biosyst. Eng. **118**(1), 174–185 (2014). https://doi.org/10.1016/j.biosystemseng.2013.11.007

17. Xie, Z., Ji, C., Guo, X., Zhu, S.: Detection and location algorithm for overlapped fruits based on concave spots searching. Trans. Chin. Soc. Agric. Mach. **42**(12), 191–196 (2011)

18. Zhao, J., Tow, J., Katupitiya, J.: On-tree fruit recognition using texture properties and color data. In: IEEE/RSJ International Conference on Intelligent Robots and Systems, Edmonton, Canada, 2–6 August 2005, pp. 263–268 (2005). https://doi.org/10.1109/iros.2005.1545592

19. Parrish, E.A., Goksel, J.A.K.: Pictorial pattern recognition applied to fruit harvesting. Trans. Am. Soc. Agric. Eng. **20**(5), 822–827 (1977). https://doi.org/10.13031/2013.35657

20. Wachs, J.P., Stern, H.I., Burks, T., Alchanatis, V.: Low and high-level visual feature-based apple detection from multi-modal images. Precision Agric. **11**, 717–735 (2010). https://doi.org/10.1007/s11119-010-9198-x

21. Sites, P.W., Delwiche, M.J.: Computer vision to locate fruit on a tree. Trans. Am. Soc. Agric. Eng. **31**(1), 257–263 (1988). https://doi.org/10.13031/2013.30697

22. Bulanon, D.M., Kataoka, T., Okamoto, H., Hata, S.: Development of a real-time machine vision system for apple harvesting robot. In: Society of Instrument and Control Engineers Annual Conference, Sapporo, Japan, 4–6 August 2004, pp. 595–598 (2004). https://doi.org/10.11499/sicep.2004.0_108_5

23. Seng, W.C., Mirisaee, S.H.: A new method for fruits recognition system. In: Proceedings of the 2009 International Conference on Electrical Engineering and Informatics – ICEEI 2009, Selangor, Malaysia, 5–7 August 2009, vol. 1, pp. 130–134 (2009). https://doi.org/10.1109/iceei.2009.5254804

24. Linker, R., Cohen, O., Naor, A.: Determination of the number of green apples in RGB images recorded in orchards. Comput. Electron. Agric. **81**(1), 45–57 (2012). https://doi.org/10.1016/j.compag.2011.11.007

25. Ji, W., Zhao, D., Cheng, F.Y., Xu, B., Zhang, Y., Wang, J.: Automatic recognition vision system guided for apple harvesting robot. Comput. Electr. Eng. **38**(5), 1186–1195 (2012). https://doi.org/10.1016/j.compeleceng.2011.11.005

26. Tao, Y., Zhou, J.: Automatic apple recognition based on the fusion of color and 3D feature for robotic fruit picking. Comput. Electron. Agric. **142**(A), 388–396 (2017). https://doi.org/10.1016/j.compag.2017.09.019

27. Zhan, W.T., He, D.J., Shi, S.L.: Recognition of kiwifruit in field based on Adaboost algorithm. Trans. Chin. Soc. Agric. Eng. **29**(23), 140–146 (2013). https://doi.org/10.3969/j.issn.1002-6819.2013.23.019

28. Zhao, Y.S., Gong, L., Huang, Y.X., Liu, C.L.: Robust tomato recognition for robotic harvesting using feature images fusion. Sensors **16**(2), 173–185 (2016). https://doi.org/10.3390/s16020173

29. Zhao, Y.S., Gong, L., Huang, Y.X., Liu, C.L.: Detecting tomatoes in greenhouse scenes by combining AdaBoost classifier and colour analysis. Biosyst. Eng. **148**(8), 127–137 (2016). https://doi.org/10.1016/j.biosystemseng.2016.05.001

30. Krizhevsky, A., Sutskever, I., Hinton, G.E.: ImageNet classification with deep convolutional neural networks. In: Advances in Neural Information Processing Systems 25 – NIPS 2012, Harrahs and Harveys, Lake Tahoe, Canada, 3–8 December 2012, pp. 1–9 (2012). https://papers.nips.cc/paper/4824-imagenet-classification-with-deep-convolutional-neural-networks.pdf. Accessed 19 Apr 2020

31. Simonyan, K., Zisserman, A.: Very deep convolutional networks for large-scale image recognition. In: International Conference on Learning Representations – ICLR 2015, San Diego, California, USA, 7–9 May 2015, pp. 1–14 (2015). https://arxiv.org/abs/1409.1556. Accessed 19 Apr 2020

32. Williams, H.A.M., Jones, M.H., Nejati, M., Seabright, M.J., MacDonald, B.A.: Robotic kiwifruit harvesting using machine vision, convolutional neural networks, and robotic arms. Biosyst. Eng. **181**, 140–156 (2019). https://doi.org/10.1016/j.biosystemseng.2019.03.007

33. Redmon, J., Divvala, S., Girshick, R., Farhadi, A.: You only look once: unified, real-time object detection. In: 29th IEEE Conference on Computer Vision and Pattern Recognition – CVPR 2016, Las Vegas, Nevada, USA, 26 June–1 July 2016, pp. 779–788 (2016). https://arxiv.org/abs/1506.02640. Accessed 19 Apr 2020

34. Redmon, J., Divvala, S., Girshick, R., Farhadi, A.: YOLOv3: an incremental improvement. In: 31th IEEE Conference on Computer Vision and Pattern Recognition – CVPR 2018, Salt Lake City, Utah, USA, 18–22 June 2018, pp. 1–6 (2018). https://arxiv.org/abs/1804.02767. Accessed 19 Apr 2020

35. Tian, Y., Yang, G., Wang, Z., Wang, H., Li, E., Liang, Z.: Apple detection during different growth stages in orchards using the improved YOLO-V3 model. Comput. Electron. Agric. **157**, 417–426 (2019). https://doi.org/10.1016/j.compag.2019.01.012

36. Tian, Y., Yang, G., Wang, Z., Li, E., Liang, Z.: Detection of apple lesions in orchards based on deep learning methods of CycleGAN and YOLO-V3-Dense. J. Sens. **2019**, 1–14 (2019). https://doi.org/10.1155/2019/7630926

37. Huang, G., Liu, Zh., van der Maaten, L., Weinberger, K.Q.: Densely connected convolutional networks. In: 30th IEEE Conference on Computer Vision and Pattern Recognition – CVPR 2017, Honolulu, Hawaii, USA, 22–25 July 2017, pp. 1–9 (2017). https://arxiv.org/abs/1608.06993. Accessed 19 Apr 2020

38. Kang, H., Chen, C.: Fruit detection, segmentation and 3D visualization of environments in apple orchards. Comput. Electron. Agric. **171**, article 105302 (2020). https://doi.org/10.1016/j.compag.2020.105302

39. Wan, S., Goudos, S.: Faster R-CNN for multi-class fruit detection using a robotic vision system. Comput. Netw. **168**, article 107036 (2020). https://doi.org/10.1016/j.comnet.2019.107036

40. COCO: Common Objects in Context Dataset. http://cocodataset.org/#overview. Accessed 19 Apr 2020

Complex Network Construction for Interactive Image Segmentation Using Particle Competition and Cooperation: A New Approach

Jefferson Antonio Ribeiro Passerini⬤ and Fabricio Breve^(✉)⬤

São Paulo State University, Rio Claro, SP 13506-900, Brazil
jefferson.passerini@gmail.com, fabricio.breve@unesp.br

Abstract. In the interactive image segmentation task, the Particle Competition and Cooperation (PCC) model is fed with a complex network, which is built from the input image. In the network construction phase, a weight vector is needed to define the importance of each element in the feature set, which consists of color and location information of the corresponding pixels, thus demanding a specialist's intervention. The present paper proposes the elimination of the weight vector through modifications in the network construction phase. The proposed model and the reference model, without the use of a weight vector, were compared using 151 images extracted from the Grabcut dataset, the PASCAL VOC dataset and the Alpha matting dataset. Each model was applied 30 times to each image to obtain an error average. These simulations resulted in an error rate of only 0.49% when classifying pixels with the proposed model while the reference model had an error rate of 3.14%. The proposed method also presented less error variation in the diversity of the evaluated images, when compared to the reference model.

Keywords: Interactive image segmentation · Machine learning · Semi-supervised learning · Particle competition and cooperation

1 Introduction

Image segmentation is the process of identifying and separating relevant structures and objects from an image, for its later analysis and information extraction. This is considered one of the hardest tasks in image processing [12].

The creation of image segmentation algorithms is considered an arduous task since it depends on the specific domain of the images or image groups. An algorithm designed for a specific domain usually has poor performance when applied to multiple images from different domains. Therefore, completely automated image segmentation still remains a challenge [12].

Supported by the São Paulo Research Foundation – FAPESP (grant #2016/05669-4).

O. Gervasi et al. (Eds.): ICCSA 2020, LNCS 12249, pp. 935–950, 2020.
https://doi.org/10.1007/978-3-030-58799-4_67

Interactive image segmentation approaches rose as an alternative [1,13,16, 20], in which user input are used as clues to the algorithm. Many of these approaches are based on semi-supervised learning, as they are designed to handle scenarios with an abundance of unlabeled data and few labeled data, which are more onerous to be generated [10,22].

The particle competition and cooperation (PCC) model [2] is a graph-based supervised learning method. This model first converts the dataset into a non-weighted and non-orientated graph in which every data item corresponds to a node of the graph and its edges are generated from the similarity relations between the data items. The particles, which correspond to the labeled data, move in the graph cooperating with other particles of the same class and competing with particles of other classes.

Every group of particles aims to dominate the largest number of non-labeled nodes and to avoid the invasion of the labeled nodes by particles of other classes. At the end of the iterative phase, the boundaries are defined by the particles' territories.

Many graph-based semi-supervised learning models are similar and share the same characteristics of operation, globally labeling the data [22]. The particle competition and cooperation model uses a local propagation of the labels approach, so its computational cost is close to linear ($O(n)$), while other approaches have cubic computation complexity ($O(n^3)$).

The PCC model has already been implemented in some important machine learning tasks such as the detection of overlapping communities [4], learning with label noise [5], active learning [6] and interactive image segmentation [3,8,9].

When applying the PCC model to image segmentation, a complex network is build based on the image to be segmented. Each pixel is represented as a node and the labeled pixels are also represented as particles [3]. The edges are defined according to the similarity between each pair of pixels, measured by the Euclidean distance among features extracted from them, such as RGB and HSV components, and pixel localization. It is important to define a weight for each feature in this process according to their discriminative capacity in the image to be segmented. The weight vector directly affects the network construction and, therefore, it has a big impact on the PCC segmentation accuracy.

So far, the proposed methods for defining features weights work well in some images and fail in others, demonstrating the difficulty of this process. In [7], the weights are calculated with automatic methods based on the average, standard deviation and histogram of the labeled pixels. Each method was able to increase the accuracy in some tested images but failed in others. In [8], a network index is proposed to evaluate candidate networks. A genetic algorithm is used to optimize the network index, thus evolving a good network. However, the optimization process is time-consuming and the network index is based on assumptions that do not hold for all images.

This paper proposes the elimination of the weight vector. To achieve this, it changes the complex network construction in the following way: (a) a different set of features is used, (b) a new form of user annotation is introduced, (c) a new approach is used to define the edges among network nodes, and (d) the particle influence on the network is measured before the competition process starts.

Simulations were made using 151 real-world images. 49 of them are taken from the GrabCut dataset, 99 from the PASCAL VOC dataset, and 3 images from the Alpha matting dataset. These are the same images used in [14]. The segmentation efficacy is calculated through the comparison between the segmented image using the proposed model, the reference model, and versions of the images segmented by experts and made available in the data sets.

The remaining of this paper is organized as follows: Sect. 2 describes the particle competition and cooperation model applied to image segmentation. Section 3 describes the proposed method to build the complex networks. Section 4 describes how the experiments were performed. In Sect. 5 the results are presented and discussed. Finally, in Sect. 6 we draw some conclusions.

2   Image Segmentation Using Particle Competition and Cooperation

Previous papers demonstrate the PCC model applied to image segmentation [3, 7–9]. The image is first converted into a complex network where nodes represent image pixels and edges represent the similarity between them, measured through the Euclidean distance among weighted features extracted from those pixels. The weight vector defines the importance of each feature in the set, for each processed image. Unfortunately, a different set of weights is required for each image to be processed [8,9].

Particles are created for each pixel that was labeled by the human specialist. During the iterative phase, these particles walk in the network cooperating with other particles of the same class and competing against particles from other classes for the possession of the unlabeled nodes. They use a random-greedy rule to decide the next node to visit.

Each node has a set of domination levels. Each of these levels corresponds to a class of the image. When a particle visits a node, it can increment the dominance level of its class on the node, while reducing the domination level of the other classes. Every particle has a strength level which is altered according to the nodes it visits. Every particle also has a distance table that contains the distance, measured in hops, from every visited node to its origin node. This table is dynamically updated during the particle's walk through the network.

In [7,8], a set of 23 features is extracted from the pixels: (1–2) pixel location components (line, column), (3–5) RGB components, (6–8) HSV components, (9–14) the average of the RGB and HSV components on the pixel and its neighbors, (15–20) the standard deviation of the RGB and HSV components on the pixel and its neighbors, (21–23) the ExR, ExG, and ExB components.

For the features averages and standard deviation, the 8-connected neighbors of the pixel are considered, except for the borders. All the extracted components are normalized to have a mean 0 and a standard deviation of 1. The HSV components were calculated as described by [21], and the components ExR, ExG, and ExB were obtained from the RGB components as described in [15]. Each feature can be emphasized or not through a weight vector λ during the Euclidean distance calculation.

The network is represented by a non-directed and unweighted graph. Therefore, $G = \{V, E\}$, where $V = \{v_1, v_2, \ldots, v_n\}$ is the set of nodes and $E = \{v_i, v_j\}$ determinates the set of edges. Each node v_i corresponds to a pixel x_j. Two nodes v_i and v_j are connected if v_j is among the k-nearest neighbors of v_i, considering the Euclidean distance features of x_i and x_j, otherwise, they are disconnected.

For each node $v_i \in \{v_1, v_2, \ldots, v_L\}$ correspondent to a labeled pixel $x_i \in X_L$, a particle p_i is generated and its initial position is defined as v_i. Each particle p_j has a variable $p_j^w(t) \in [0, 1]$ to store its strength, which determines the impact generated by the particle to the visited node. The initial strength is always set to its maximum value $p_i^w(0) = 1$.

Each particle has a table of distances which is dynamically updated during the movement of a particle, in which the distance between the particle's initial position and the visited nodes are stored. It is determined by $p_j^d(t) = \{p_j^{d_1}(t), p_j^{d_2}(t), \ldots, p_j^{d_n}(t)\}$, where each element $p_j^d(t) \in [0, n-1]$ corresponds to the measured distance between the particle's original node p_j and the node v_i.

Each node v_i has a vector variable $v_i^w(t) = \{v_i^{w_1}(t), v_i^{w_2}(t), \ldots, v_i^{w_n}(t)\}$ where each element $v_i^w(t) \in [0, 1]$ represents the domination levels of the team l over the node. The sum of this vector is always a constant,

$$\sum_{l=1}^{e} v_i^{w_l} = 1. \tag{1}$$

This occurs because when a particle increases its dominance level on the node, the domination of the other groups decrease in the same proportion for all the classes, which is determined as law of conservation of forces.

The initial domination levels $v_i^w(t)$ of each node v_i is configured as:

$$W_i^{wl} = \begin{cases} 1 & \text{if } x_i \text{ is labeled and } y(x_i) = l \\ 0 & \text{if } x_i \text{ is labeled and } y(x_i) \neq l \\ \frac{1}{c} & \text{if } x_i \text{ is not labeled} \end{cases} \tag{2}$$

During the iterations, the domain values vary according to the equation:

$$v_i^{wl}(t+1) = \begin{cases} max\left\{0, v_i^{wl}(t) - \frac{\Delta_v p_j^w(t)}{c-1}\right\} & \text{if } x_i \text{ is unlabeled and } l \neq p_j^f \\ v_i^{wl}(t) + \sum_{q \neq c} v_i^{w_q}(t) - v_i^{w_q}(t+1) & \text{if } x_i \text{ is unlabeled and } l = p_j^f \\ v_i^{wl}(t) & \text{if } x_i \text{ is labeled} \end{cases} \tag{3}$$

where p_i^f represents the team of the particle p_j. Each particle will change the node its visiting v_i, increasing the domination level of its class ($v_i^{wl}(t), l = p_j^f$) on it, and decreasing the domination levels of other classes ($v_i^{wl}(t), l \neq p_j^f$), always respecting the law of conservation of forces.

The particle's force depends on the domination level of its team on the visited node. Therefore, in every iteration, the force of the particle $p_j^w(t)$ is updated according to the equation: $p_j^w(t+1) = v_i^{w_l}(t+1)$, where v_i is the target node and $l = p_j^f$, with l being the label of the particle's team p_j, consequently each particle p_j has its strength p_j^f configured with the domination level value of its team $w_i^{w_j}$ in the node v_i.

The model conditions the particle's force to a reducer Δ_v, in this way it is possible to control the speed in which the particles will dominate the unlabeled nodes.

When the node is visited, the distance table is updated according to the following equation:

$$p_j^{d_k}(t+1) = \begin{cases} p_j^{d_i}(t)+1 & \text{if } p_j^{d_i}(t)+1 < p_j^{d_k}(t) \\ p_j^{d_k}(t) & \text{otherwise} \end{cases}. \qquad (4)$$

This way, when the particle moves from the current node to the target node, it checks the distance table and, if the distance from the target node is larger than the current node's distance plus 1, the table is updated. This way, it is not necessary to know the distances for every node a priori.

In every iteration, the particle must choose which node it will visit among the k-nearest neighbors from the current position, this movement is based in one of two rules: random rule or greedy rule.

In the random rule the particle randomly chooses, with equal possibilities, any of the neighboring vertices from the position in which the particle is located. This rule does not take into account the levels of domain or distances of the original node, being useful for exploration and acquisition of new nodes. It is defined by the equation:

$$p(v_i|p_j) = \frac{W_{qi}}{\sum_{\mu=1}^{n} W_{q\mu}}, \qquad (5)$$

where q is the node's index where the current particle p_j is located, therefore $W_{qi} = 1$ if there is an edge between the current node and any other node v_i or else $W_{qi} = 0$.

In the greedy rule, the particle randomly chooses any one of the neighboring nodes from the current node, with probabilities calculated directly proportional to the dominance level of this particle's team in every neighboring node, and inversely proportional to the distance of these neighbors to the particle's local node. This movement is useful for the team's territory defense. It is defined by the following equation:

$$p(v_i|p_j) = \frac{W_{qi}v_i^{w_l}\left(1+p_j^{d_i}\right)^{-2}}{\sum_{\mu=1}^{n} W_{q\mu}v_\mu^{w_l}\left(1+p_j^{d_\mu}\right)^{-2}}, \qquad (6)$$

where q is the node's index where the current particle p_j is located, and $l = p_j^f$, with p_j^f being the class label of the particle p_j.

At each iteration, every particle has the probability P_{grd} of choosing the greedy movement and the probability $1 - P_{grd}$ of choosing the random movement, with $0 \leq p_{grd} \leq 1$.

The algorithm stops when it reaches the maximum number of iterations ($maxIte$), or when the node's domination levels reach some stability. The stability is inferred when, on average, the nodes had improved below a threshold ($controlStop$) in their maximum domination level in a given iteration interval ($maxStop$).

3   Constructing a New Complex Network for the Model

Based on what has been presented in previous papers, this paper proposes a change in how the complex network is built and the elimination of the weight vector that defines the importance of each feature within the set.

The original model receives the image to be segmented (for example: Fig. 1a) and the user input (Fig. 1b). In this new approach, it is also possible to delimit, in the image, the region of interest where the object to be segmented is found (Fig. 1c), to reduce the processing scope.

Figure 1d displays a composition of the algorithm's input information. The cut polygon represented in Fig. 1c is optional to the model because there can be different situations that the cut in the original image is not of interest since the segmentation object of choice is set across most of the image. Regardless of the usage of the cut polygon by the algorithm or not, the complex network assembly process is the same.

The input image passes through a bicubic interpolation process to reduce its dimension (amount of pixels) allowing the processing of larger images. Later, the bilinear interpolation algorithm is used for information recomposition after the segmentation processing, as described in [11,18,19].

Once the image is reduced, the extraction of the image's pixels features set is started. This new proposed approach uses less features than its predecessors: (1–2) pixel location components (line, column), (3–5) RGB components, (6) only the V (value) component of the HSV system [21], (7–9) the color components ExR, ExG, ExB defined in [15], and (10) a new feature extracted using Otsu's binarization algorithm [17]. Figure 2 shows graphic representations of each color feature extracted from an image from the Grabcut data set.

The features are normalized to have an average of 0 and a standard deviation of 1. Thereafter occurs the non-oriented graph generation, which will represent the pixels of the image through its nodes, while the edges represent the similarity relationships between the pixels. As has been previously stated, the similarity relationships are determined by measuring the Euclidean distance among the extracted features from the image to be segmented.

In the reference model, each node is connected to its k-nearest neighbors, considering the Euclidean distance among pixel features, and k is a parameter set by the user. In the proposed model, k is fixed, and each node is connected to its 192 nearest neighbors. Another 8 connections are made based in the pixel

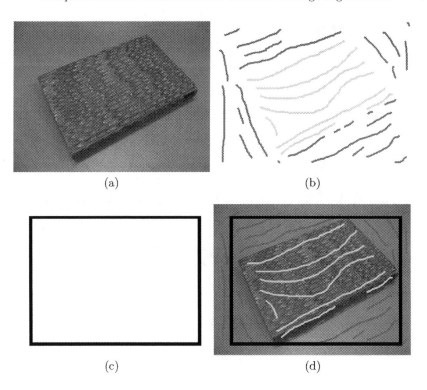

Fig. 1. Input information: (a) the real-world images to be segmented, (b) the "scribbles" provided by the user, (c) the cut polygon, and (d) overlay image for visualization only.

spatial neighborhood, defined by a 3 × 3 window, i.e., the corresponding node will be linked to the nodes corresponding to its 8 physically adjacent pixels. This change was introduced to consider the pixel physical neighborhood when creating a complex network since the original approach cannot guarantee they will be connected.

Another alteration is that the proposed model verifies whether two linked nodes are of different classes and, if this happens, the new edge will not be generated. Nodes can only be linked with nodes of the same class or with unlabeled nodes.

Once the complex network is defined, this approach considers the influence of a previously labeled pixel (particle) in relation to its physical neighbors in the original image. This alteration also aims to reinforce the relationship information among the image's pixels before the iterative phase, but now in relation to the pixels labeled by the user in a 5 × 5 window.

It is considered that a particle (labeled pixel) influences other nodes in the network up to the distance of 2 hops in the structure. It must be emphasized that, in this approach, the distance also takes into consideration the spatial distance of the pixel present in the image. This alteration aims to value the

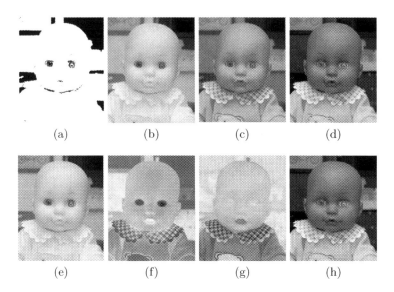

(a)   (b)   (c)   (d)

(e)   (f)   (g)   (h)

Fig. 2. Proposed algorithm's extracted features representation: (a) the binarization made with Otsu's algorithm, (b–d) the RGB components, respectively, (e) the V component from the HSV system, (f–h) the ExR, ExG, and ExB components, respectively.

pixel's relationships of the spatial neighborhood in the image to be analyzed since a neighboring unlabeled pixel tends to be part of the same class.

Therefore, unlabeled nodes which are 1 hop away from a labeled node will have an increment of 0.2 in its domination vector, referent to the nearby particle's class. There is the possibility that the unlabeled node suffers the influence in its domination vector of more than one nearby particle. Unlabeled nodes which are 2 hops away from a labeled node suffer less influence in their domination vector. The increment is only 0.1 in the nearby particle's class. This process is illustrated in Fig. 3.

This approach respects the law of conservation of forces, previously described. If a pixel has its domination vector completely dominated by a given team of particles, it will still be subject to changes during the iterative phase, differently from the labeled nodes, which have fixed domination levels.

After the network construction and initial settings with the proposed approach, the iterative phase takes place exactly as described in earlier particle competition and cooperation models.

4   Experiments

Two models were implemented to perform the tests with the proposed approach in this paper. The first one is the reference model [3, 7–9], with the 23 features already described. The weight vector λ was defined so all the features had the

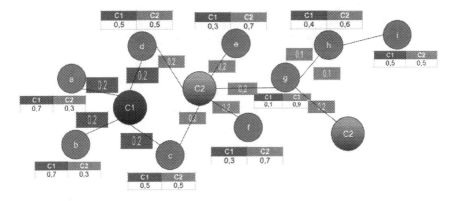

Fig. 3. Representation of the labeled pixels (particles) influence on the spatial neighborhood of the original image.

same weight. The reference model receives the interpolation approach as an increment in its execution to reduce the size of the images to be processed. The second is the proposed model, as described in the previous session.

151 real-world images taken from the GrabCut dataset, the PASCAL VOC dataset, and from the Alpha matting dataset are used to evaluate both models. These are the same images used in [14]. The markings (labels) defined for the tests and the cut polygons used in this work are available at Github[1]. In our simulations, each image was processed 30 times by each model to calculate the models' average performance. In total 4,530 executions were performed with each algorithm.

The error rate is determined by comparing the algorithms' output with the ground-truth images. Notice that the ground-truth images have some pixels (in grey) representing divergence in classification among the specialists. Those pixels were not considered in the error calculation.

Both models were configured to use 200 neighbors, $\Delta_v = 0.1$, $P_{grd} = 0.5$, $maxIte = 1,000,000$, $maxStop = 15,000$, and $controlStop = 0.001$. The same images and annotated pixels were used for both models.

The simulations were performed in an Asus laptop model S46C configured with a Quadcore 2 Ghz Intel i7 processor, model 3537U, with 16 GB of RAM, running the Linux Deepin 15.5 Desktop operating system. The models' development was made using the Python version 2.7 programming language.

5   Results and Discussion

The experiments were performed as discussed in Sect. 4. It was observed that in five images from the data set it was not possible to apply the cut polygon feature,

---

[1] Available at https://github.com/jeffersonarpasserini/dataset-interactive-algorithms. git.

due to the subject being segmented occupying most of the original image. Those images are presented in Table 1 with the error rates achieved by both models.

Table 1. Error rates in the five images that did not use the cut polygon resource, as achieved by the proposed model and the reference model.

Image name	Proposed	Reference
Baby_2007_006647	1.17%	4.57%
cross	0.48%	1.79%
gt02	0.52%	1.27%
gt07	0.21%	0.64%
gt13	1.08%	2.11%
Average	0.64%	1.72%

The proposed method obtained the best pixel classification accuracy in comparison to the reference method. In the average, the proposed method obtained an error rate of 0.64% while the reference method obtained an error rate of 1.72%. The processed images can be observed in Fig. 4.

Regarding processing time, the reference method took 1371.08 s to output the results presented in Fig. 4 while the proposed method took 1205.06 s.

Table 2 shows the five data set images in which the proposed method obtained the lowest error rates. In this case, all the images used the cut polygon feature during the proposed model execution to cut the segmentation area of interest, whilst in the reference method, the entire image was used. This cut process used in the images resulted in smaller networks for the proposed method in comparison to the reference method. The average of the images' complex network nodes, described in Table 2, was of 2,089 nodes with 554,455 edges in the proposed method whilst, for the reference method, it was of 4,592 nodes and 2,102,821 edges.

Table 2. Error rates in the images with the lowest error rates achieved by the proposed method.

Image name	Proposed	Reference
Monitor_2007_003011	0.02%	1.09%
Train_2007_004627	0.09%	0.76%
Car_2008_001716	0.10%	2.51%
Monitor_2007_004193	0.11%	3.00%
Person_2007_002639	0.12%	2.47%
Average	0.08%	1.94%

baby_2007_006647

Proposed Method - Error Rate: 1.17%   Reference Method - Error Rate: 4.57%

gt02

Proposed Method - Error Rate: 0.52%   Reference Method - Error Rate: 1.27%

gt07

Proposed Method - Error Rate: 0.21%   Reference Method - Error Rate: 0.64%

gt13

Proposed Method - Error Rate: 1.08%   Reference Method - Error Rate: 2.11%

cross

Proposed Method - Error Rate: 0.48%   Reference Method - Error Rate: 1.79%

Fig. 4. Image segmentation results by the proposed model and the reference model without the use of the cut polygon tool. Misclassified pixels are highlighted in red. (Color figure online)

By analyzing Table 2, it can be verified that the proposed method, with a 0.08% error rate, obtained a superior performance when compared to the reference method, which obtained an error rate of 1.94%. Regarding the average execution time, the proposed method performed the images' segmentation in 255.46 s whilst the reference method performed it in 823.30 s. The higher execution time difference is justified by the difference in the complex network generation which defined smaller networks for the proposed method. These processed images can be observed in Fig. 5.

The general performance of both methods, considering the processing of all the 151 images, is shown in Table 3 and Table 4. In Table 3, the size of the generated networks, for each method, can be verified. The proposed method generated smaller networks, in both nodes and edges, in comparison to the reference method. The 151 images studied had an average of 200, 124 pixels, from which 2, 783 are unlabeled. The proposed method generated complex networks with an average of 7, 538 nodes and 838, 564 edges, whilst the reference method generated complex networks with a higher average of 17, 946 nodes and 2, 354, 555 edges.

Table 3. Generated complex networks average characteristics by the proposed model and the reference model.

Method	# Pixels		Characteristics		
	All	Unlabeled	Particles	Nodes	Edges
Proposed	200,124	2,783	2,860	7,538	838,564
Reference	200,124	2,783	5,487	17,946	2,354,555

Table 4 shows the average error rate when classifying the pixels for all the 151 images studied, using both the proposed method and the reference method. The proposed method achieved an average error rate of 0.49% while the reference method achieved 3.14%. Regarding execution time, the proposed method was beneficiated by the generation of smaller networks. It took an average of 432.54 s to segment the images, whilst the reference method performed the task in an average time of 1082.94 s.

Table 4. Average error rate and execution time in the 151 images by the proposed method and the reference method.

Method	Error rate	Time(s)
Proposed	0.49%	432.54
Reference	3.14%	1082.94

Figure 6 shows the relationship between the error rate and the execution time to perform the segmentation of each of the 151 images. The proposed method has

monitor_2007_003011

Proposed Method - Error Rate: 0.02%   Reference Method - Error Rate: 1.09%

train_2007_004627

Proposed Method - Error Rate: 0.09%   Reference Method - Error Rate: 0.76%

car_2008_001716

Proposed Method - Error Rate: 0.10%   Reference Method - Error Rate: 2.51%

monitor_2007_004193

Proposed Method - Error Rate: 0.11%   Reference Method - Error Rate: 3.00%

person_2007_002639

Proposed Method - Error Rate: 0.12%   Reference Method - Error Rate: 2.47%

Fig. 5. Images segmentation results by the proposed model with the use of the cut polygon tool and the reference model. Misclassified pixels are highlighted in red. (Color figure online)

better stability in comparison to the reference method. The images processed by the proposed method are concentrated in a specific region of the graphic, whilst the reference method image's processing results are spread through the graphic. Therefore, the proposed method is less susceptible to features variations among the different images presented to it in this study.

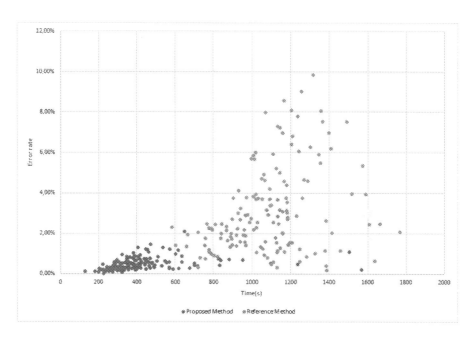

Fig. 6. Relation Analysis between error rate and processing time presented by the proposed model and reference model.

6   Conclusions

Autonomous segmentation represents one of the most complex image processing tasks since there is not an image segmentation algorithm that can be used in every situation. This paper presented a methodology to improve the automation level of the particle competition and cooperation model for image segmentation.

The proposed approach eliminates the weight vector λ from the model since it is a parameter that needed to be set for each image to be segmented, requiring some expertise of the user. This elimination increased the model automation and it is achieved by optimizing the network construction phase, without changes in the particle competition and cooperation iterative phase.

The computer simulations have shown that the proposed method achieved better results when compared to the reference method, both in accuracy and performance. The proposed method obtained an average error rate of only 0.49%, whilst the reference method obtained an average error rate of 3.14% for the 151 images studied.

The processing time was also reduced. The proposed method performed the segmentation in an average time of 432.54 s, whilst the reference method took an average time of 1082.94 s. This fact is justified by the average size of the networks generated by the studied models. The proposed method generated networks significantly smaller than the reference method.

References

1. Boykov, Y., Jolly, M.P.: Interactive graph cuts for optimal boundary & region segmentation of objects in N-D images. In: Eighth IEEE International Conference on Computer Vision, 2001. ICCV 2001. Proceedings, vol. 1, pp. 105–112 (2001)
2. Breve, F., Zhao, L., Quiles, M., Liu, J., Pedrycz, W.: Particle competition and cooperation in networks for semi-supervised learning. Knowl. Data Eng. **24**(9), 1686–1698 (2012)
3. Breve, F.A., Quiles, M.G., Zhao, L.: Interactive image segmentation using particle competition and cooperation In: Proceedings of the International Joint Conference on Neural Networks (IJCNN2015), pp. 1–8. Killarney, Irlanda (2015). https://dx.doi.org/10.1109/IJCNN.2015.7280570
4. Breve, F., Zhao, L.: Fuzzy community structure detection by particle competition and cooperation. Soft Comput. (2013). https://doi.org/10.1007/s00500-012-0924-3
5. Breve, F.A., Zhao, L., Quiles, M.G.: Particle competition and cooperation for semi-supervised learning with label noise. Neurocomputing (2015). https://doi.org/10.1016/j.neucom.2014.08.082
6. Breve, F.A.: Active semi-supervised learning using particle competition and cooperation in networks. In: Neural Networks (IJCNN) (2013)
7. Breve, F.A.: Auto feature weight for interactive image segmentation using particle competition and cooperation. In: Proceedings - XI Workshop de Visão Computacional WVC2015, pp. 164–169. XI Workshop de Visão Computacional (WVC2015), October 2015
8. Breve, F.: Building networks for image segmentation using particle competition and cooperation. In: Gervasi, O., et al. (eds.) ICCSA 2017. LNCS, vol. 10404, pp. 217–231. Springer, Cham (2017). https://doi.org/10.1007/978-3-319-62392-4_16
9. Breve, F.A., Quiles, M.G., Zhao, L.: Interactive image segmentation of non-contiguous classes using particle competition and cooperation. In: Gervasi O. et al. (eds.) Computational Science and Its Applications – ICCSA 2015. ICCSA 2015. LNCS, vol. 9155. Springer, Cham. https://doi.org/10.1007/978-3-319-21404-7_15
10. Chapelle, O., Schölkopf, B., Zien, A.: Semi-Supervised Learning. Adaptive Computation and Machine Learning. MIT Press, Cambridge (2006)
11. Getreuer, P.: Linear methods for image interpolation. Image Process. On Line **1**, 238–259 (2011)
12. Gonzalez, R.C., Woods, R.E.: Processamento Digital de Imagens, 3 edn. Longman do Brasil (2010)
13. Grady, L.: Random walks for image segmentation. IEEE Trans. Pattern Anal. Mach. Intell. **28**(11), 1768–1783 (2006). https://doi.org/10.1109/TPAMI.2006.233
14. Gulshan, V., Rother, C., Criminisi, A., Blake, A., Zisserman, A.: Geodesic star convexity for interactive image segmentation. In: 2010 IEEE Computer Society Conference on Computer Vision and Pattern Recognition, pp. 3129–3136. IEEE (2010)

15. Lichman, M.: UCI machine learning repository (2013). http://archive.ics.uci.edu/ml/datasets/image+segmentation
16. Malmberg, F.: Graph-based methods for interactive image segmentation. Uppsala University, Centre for Image Analysis, Computerized Image Analysis and Human Computer Interaction (2011)
17. Otsu, N.: A threshold selection method from gray-level histograms. IEEE Trans. Syst. Man Cybern. **9**, 62–66 (1979)
18. Pedrini, H., Schwartz, W.R.: Análise de Imagens Digitais: Princípios. Algoritmos e Aplicações. Thomson Learning, São Paulo (2007)
19. Prajapati, A., Naik, S., Mehta, S.: Evaluation of different image interpolation algorithms. Int. J. Comput. Appl. **58**(12), 6–12 (2012)
20. Rother, C., Kolmogorov, V., Blake, A.: "GrabCut": interactive foreground extraction using iterated graph cuts. ACM Trans. Graph. **23**(3), 309–314 (2004). https://doi.org/10.1145/1015706.1015720
21. Smith, A.R.: Color gamut transform pairs. ACM SIGGRAPH Comput. Graph. (1978). https://doi.org/10.1145/965139.807361
22. Zhu, X.: Semi-supervised learning literature survey. Technical report 1530, Computer Sciences, University of Wisconsin-Madison (2005)

Convolutional Neural Networks for Automatic Classification of Diseased Leaves: The Impact of Dataset Size and Fine-Tuning

Giovanny Caluña[1]([✉]) [iD], Lorena Guachi-Guachi[1,2] [iD], and Ramiro Brito[3] [iD]

[1] Yachay Tech University, Hacienda San José, Urcuquí 100119, Ecuador
{giovanny.caluna,lguachi}@yachaytech.edu.ec
[2] SDAS Research Group, Ibarra, Ecuador
[3] Department of Mechatronics, Universidad Internacional del Ecuador,
Av. Simon Bolivar, Quito 170411, Ecuador
http://www.sdas-group.com

Abstract. For agricultural productivity, one of the major concerns is the early detection of diseases for their crops. Recently, some researchers have begun to explore Convolutional Neural Networks (CNNs) in agricultural field for leaves diseases identification. A CNN is a category of deep artificial neural networks that has demonstrated great success in computer vision applications, such as video and image analysis. However, their drawbacks are the demand of huge quantity of data with a wide range of conditions, as well as a carefully fine-tuning to work properly. This work explores and compares the most outstanding five CNNs architectures to determine their ability to correctly classify a leaf image as healthy and unhealthy. Experimental tests are performed referring to an unbalanced and small dataset composed by healthy and diseased leaves. In order to achieve a high accuracy on the explored CNN models, a fine-tuning of their hyperparameters is performed. Furthermore, some variations are done on the raw dataset to increase the quality and variety of the leaves images. Preliminary results provide a point-of-view for selecting CNNs architectures for leaves diseases identification based on accuracy, precision, recall and F1 metrics. The comparison demonstrates that without considerably lengthening the training, ZFNet achieves a high accuracy and increases it by 10% after 50 K iterations being a suitable CNN model for identification of diseased leaves using datasets with a small variation, number of classes and dataset sizes.

Keywords: Leaf diseases classification · Convolutional Neural Networks (CNNs) · Image classification

1 Introduction

One of the major issues affecting farmers around the world are plant diseases caused by various pests, viruses, and insects, resulting in huge losses of money.

© Springer Nature Switzerland AG 2020
O. Gervasi et al. (Eds.): ICCSA 2020, LNCS 12249, pp. 951–966, 2020.
https://doi.org/10.1007/978-3-030-58799-4_68

The best way to relieve and combat this is timely detection. In almost all real solutions, early detection is determined by human, such an expensive and time-consuming solution.

Automatic approaches for leaf diseases identification aim to identify the signs of diseases at initial phase and decrease the huge effort of human observing in large farms. These methods involve some stages such image acquisition (to obtain images to be analyzed), image pre-processing (to enhance images quality), feature extraction (to get useful properties from images pixels), and classification (to categorize an image in classes based on its discriminating features).

Recent research have begun to explore CNNs approaches in medicine [1], mechanical [2] and in the agricultural field to identify plant disease and pests from images. CNNs, a category of deep learning approaches, have been extensively adopted for image classification purposes. They have been introduced for general purposes analysis of visual imagery and are characterized by their ability to learn key feature on their own. Their adoption for specific fields brings with it several challenges such as choosing the appropriate hyperparameters values such as learning rate, step and maximum number of iterations, as well as providing a sufficient quantity and quality of training dataset.

In order to determine how the CNNs structure, hyperparameter values and data quality influence performance reached in the classification of leaves diseases, this work explores and compare the five outstanding and standard CNN architectures such as Inception V3 [3], GoogLeNet [4], ZFNet [5], ResNet 50 [6], and ResNet 101 [6], which outperform other techniques of image classification. Furthermore, we evaluate some pre-processing tasks on the Plant Village dataset [7], a set of images of diseased and healthy crops such as pepper, tomato and potato, and perform fine-tuning of the hyperparameter of each CNN model x.

The remaining of this paper is organized as follows. Section 2 describes most relevant related works. The raw dataset and the operations applied for increasing its quality and size are described in Sect. 3. Explored CNN architectures are presented in Sect. 4. Experimental setup is presented in Sect. 5. Experimental results, obtained on processed datasets, are gathered and discussed in Sect. 6. Finally, Sect. 7 deals with the concluding remarks and future work.

2   Related Works

Among various CNN architectures used in agricultural field: CaffeNet, AlexNet, GoogleNet, VGG, ResNet, and InceptionV3 have been utilized as the underlying model structure, aiming at exploring the performance to identify types of leaves diseases from healthy leaves of single or multiple crops such as tomato, pear, cherry, peach, and grapevine [8–15]. The effectiveness of a CNN model in identifyng leaves diseases on single or multiple crops depends mainly on the quantity and quality images. The main benefit of working with multiple crops is to enrich image feature description learning but presents drawbacks when they have to identify diseases on early stages [16]. Several researchers have explored structural changes of some standard CNN models. In [17], LeNet CNN architecture has been modified to identify three maize leaf diseases from healthy leaves. Based on ResNet50 CNN architecture, three different architectures were proposed in [16], to create a single multi-crop CNN model. The proposed architecture is capable of integrating contextual meta-data of the plant species information to identify seventeen diseases of five crops: wheat, barley, corn, rice and rape-seed. For experimental tests, the proposed multi-crop CNN model used a dataset constructed with more than one hundred-thousand of images taken by cellphone in real fields.

In addition, some optimization studies have been introduced aiming at the problems of too many parameters, computation and training time consumed by CNN models. For instance, a global pooling dilated CNN is implemented in [18] to address the problems of too many parameters of AlexNet CNN to recognize six common cucumber leaf diseases. The impact of the size and variety of the datasets on the effectiveness of CNN models is studied in [19]. The study discussed the use of transfer learning on GoogleNet to deal with the problem of leaf diseases recognition based on an image database of 12 plant species including rice, corn, potato, tomato, olive, and apple. An empirical exploration of fine-tuning and transfer learning on VGG 16, Inception V4, ResNet (with 50, 101 and 152 layers) and DensNet (with 121 layers) is addressed in [20]. Results obtained shown that DenseNet improves constantly its accuracy with increasing number of epochs, with no signs of overtting and performance degradation.

On the other hand, several specialized CNN architectures have been designed with a reduced number of hidden layers to provide a fast and accurate leaves diseases identification, such as [21], which used only two hidden layers to differentiate healthy and unhealthy images of rice leave. In [22], a novel architecture based on three-channel CNN with four hidden convolutional layers is presented. It is proposed for identification of vegetable leaf disease by combining the color components of diseases. Besides, a hybrid convolutional encoder network is stated in [23]. It identifies six classes of diseases on potato, tomato and maize crops.

3 Dataset

CNN demands huge amounts of varied data, otherwise the model may be not robust or overfitting. Therefore, the raw dataset was augmented performing pre-processing operations such as illumination changes mainly based on brightness and contrast variation, data augmentation, and the addition of random classes on their data. Figure 1 shows some examples from the original data set and the output after transformations performed over the raw data.

3.1 Raw Dataset

The current work used the Plant Village dataset [7], such a publicly accessible dataset that contains 13K RGB images of healthy and infected leaves of plants, with resolution 256 × 256. The dataset contains 26 different anomalies in 14 crop species including apple, blueberry, cherry, corn, grape, orange, peach, bell pepper, potato, raspberry, soybean, squash, strawberry and tomato. In this work, all images of diseased leaves were labeled as unhealthy, resulting in a raw dataset with two classes (healthy and unhealthy leaves).

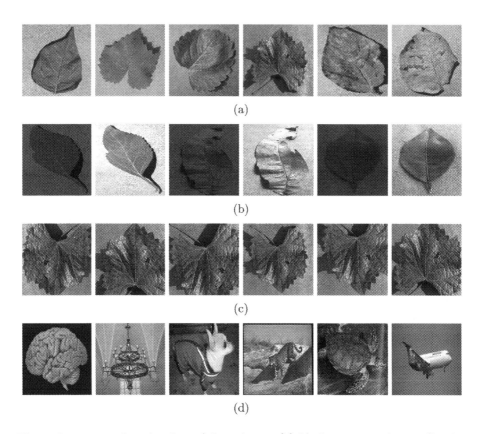

Fig. 1. Leaves samples related to: a) Raw dataset [7]; b) illumination changes (brightness and contrast variation), c) data augmentation transformations (rotation, translation, zoom and flip); d) random classes added.

3.2 Pre-processing Operations

A. **Illumination Changes (Brightness and Contrast Variation):** For each leaf image, this operation added two leaves images with high and low contrasts to the raw dataset, resulting in a total of 35,231 new leaves images. For this purpose, this work used **convertScaleAbs** function of openCV library, where $alpha = 1.5$ and $beta = 0$ parameters control the contrast and the brightness to create the new leaf image. The range of these values are [0.0–100.0] and [1.0–3.0], respectively.

B. **Data Augmentation:** It enriched the data with augmented images result of: rotating the image with *probability* = 1 from 5 to −5°, random distortions with *gridwidth* = 4, *gridheight* = 4 and *magnitude* = 8, left-right flips, top-bottom flips, random zooms with *probability* = 0.5 and percentage *area* = 0.8 (where probability corresponds how often the operation is applied and the magnitude is the magnitude of the distortions). They were performed by using the **augmentator library** [24]. It produced 9 K additional leaves images.

C. **Random Classes Added:** In order to distinguish healthy from diseased leaves, this procedure altered the raw dataset adding 6 classes including images of Airplane, Brain, Chandelier, Turtle, Jaguar, and Dog. This resulted in 1,238 additional images.

4   CNN Models

Inception V3 [3], GoogLeNet [4], ResNet 50 [6] and ResNet 101 [6] were selected because of their successful performance for image classification tasks [25,26]. On the other hand, ZFNet [5] was also explored due to their simplicity and low computation costs.

The architectures of explored CNN models are schematized in Fig. 2. Each one is characterized by a repeating sequence of convolutional layers, layers of activation functions, pooling layers, specialized modules (such as Inception and Residual modules aiming at efficient computations) and fully connected layers to output the classification label. Each layer type is named according to the performed operations. In this work, the last fully connected layer was adjusted to support two output classes.

- **Convolutional Layer:** It consists of a set of filters to extract different features from the input image. The main purpose of each filter is build a feature map to detect simple structures such as borders, lines, squares, etc. and grouping them into subsequent layers to construct more complex shapes. In mathematical terms, convolution computes dot products between the input data and the entries of the filters to learn and detect patterns from the previous layer, as is shown in Eq. 1.

$$S(i,j) = (I * K)(i,j) = \sum_m \sum_n I(m,n)K(i-m,j-n) \qquad (1)$$

where I represents the image and K the filter also often called kernel, m and n represent the dimensions of the filter typically of 3×3, 5×5, 7×7 or 11×11. (i,j) represents the pixels where the convolution operation will be performed. The input (image) is composed by $(i \times j \times r)$, where r is the number of channels of the image. The filters are slipped from the left up to the right down corner applying the convolutional operation to create the feature map.
- **Activation Functions:** In the convolution process many negative values are generated. These values are unuseful for the next layers and produce

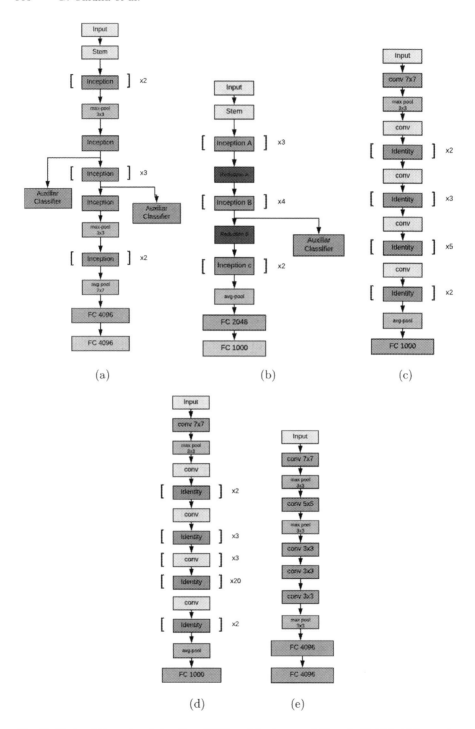

Fig. 2. State-of-the-art outstanding CNN architectures: a) GoogLeNet [4]; b) Inception V3 [3]; c) ResNet 50 [6]; d) ResNet 101 [6]; e) ZFNet [5].

more computational load. Therefore, after a convolutional layer, an activation function is applied. The activation function sets to zero negative values, which helps to get a faster and effective training. Rectified Linear Unit (ReLU) is the most widely used by Deep Learning techniques. ReLU aims to increase the non-linearity in the images, setting the value to 0 if the input is less than 0, and raw otherwise.

- **Pooling layer:** The pooling operation splits the convolved features into disjoint regions of a size given by a filter F. Then, the 'max' or 'mean' values are taken from each disjoint region to save main features and reduce progressively the parameters, and thus the computational load by sub sampling the spatial size of its input.
- **Fully Connected Layer (FC):** It aims to take the values from the previous layer and order them to drop out the class which the input belongs within a single vector of probabilities.
- **Inception Module:** It is used in deeper CNNs [4] for providing efficient computations and dimensional reduction. This module is designed with stacked 1×1 convolutions after the max-pooling layer. It is a different method to apply the convolution operation through different size filters to extract the most different and relevant parameters during the training. The different map features obtained are concatenated before the next layer. A significant change over this kind of modules was proposed by [3]. They applied a factorization operation over the inception modules by using large filters and dividing into two or more convolutions with smaller filters. Factorized inception module produced $\times 12$ lesser parameters than AlexNet [27] (one of the most common implemented architecture).
- **Residual Module:** In this module, each layer is the input for the next layer and directly feeds into the layers about 2–3 hops away. Residual Module characterizes the ResNet [6] CNN architecture aiming to face the vanish gradient [28]. The learning residual function is given by Eq. 2.

$$H(X) = F(X) - X \qquad (2)$$

where X is the input of the block, $F(X)$ is the residual function and $H(X)$ is the final output of a residual block. This reformulation was done on the basis that a shadow network has less training error than the deeper counterpart. So, the residual modules support the deeper networks have an error which is no greater than the shallow ones.

The five explored CNNs architectures for the classification of leaves images are described in the following subsections.

4.1   ZFNet

It is an enhanced modification of AlexNet [27] architecture formed by eight layers. It is characterized by its simplicity. This has its convolutional layers followed by a max pooling layer. Each convolutional layer has decreasing filters starting with a filter size of 7×7 as shown in Fig. 2e). ZFNet supports the hypothesis that, bigger filters loss more pixel information which can be conserved with the smaller ones to improve the accuracy. This model uses the ReLu activation function and the iterative optimization of batch stochastic gradient descent as learning algorithm for computing the cost function of a certain number of training examples for each mini-batch.

4.2   GoogLeNet - Inception V1

GoogLeNet [4] also called Inception V1 uses Inception Modules to build deeper networks with more efficient computation. It includes 1×1 Convolution at the middle and global average pooling at the end of the network that replaces a fully connected layers to save a lot of parameters and improve the accuracy. GoogleNet has 22 layers of deep, starting with a size filter of 7×7 decreasing through the network, followed by 9 inception modules, as illustrated in Fig. 2a).

4.3   GoogleNet - Inception V3

GoogleNet-Inception V3 [3] reflects an improved version of the previous version. In this architecture, computational efficiency and fewer parameters are realized. With 42 layers depth, the computation cost is only around 2.5 higher than GoogLeNet [4], and much more efficient than VGGNet [29]. Its most relevant features are: factorization technique applied trough the network and inceptions modules. The factorization is applied in order to reduce the number of parameters keeping the network efficiency. Inception V3 starts with small filters of 3×3 and keeps them until reach the inception modules, as shown in Fig. 2b).

4.4   ResNet 50 & ResNet 101

The Residual Networks, ResNets, represent a very deep network up to 152 layers. They learn from residual representation functions instead of learning the signal representation directly. ResNet 101 and ResNet 50 [6] included skip connection, also called shortcut connections, to fit two or more stacked layers into a desired residual mapping instead of hoping they directly fit. The skip connection is a solution to the degradation problem (accuracy saturation) generated in the convergence of deeper networks. They provide effectiveness in deeper networks, improving the learning process. The main difference between ResNet 101 and ResNet 50 is the number of convolutional layers, ResNet 101 has twice the number of layers than ResNet 50, as depicted in Fig. 2d) and c) respectively.

5   Experimental Setup

For experimental tests, each dataset was splitted into three sets with percentage ratio of (85%), (10%), and (5%) for training, validation, and testing respectively. The training set was processed to generate the LMDB dataset. LMDB is a compressed format for working with large dataset, where all images were sampled randomly from the training one. Explored CNNs received as input the generated LMDB. CNN models were trained using Python interfaces of Caffe framework available to face with research code and rapid prototyping. All experiments were performed using Tesla K80 Nvidia Graphics Processing Unit (GPU) that provides 11,441MiB of GPU memory, which allowed to use the original image resolution of 256×256.

In order to determine how the complex conditions on the dataset impact the performance of the CNN models applied to leaves diseases identification, this work train the explored CNN models with the original raw data, and with the training datasets augmented using three diverse operations (illumination changes, data augmentation and random classes added). In addition, the crucial hyperparameters such as learning rate (lr), step, and maximum number of iterations (#It.) were ne-tuned aiming at identifying a fast and accurate CNN model. Lr is the most critical hyperparameter to be tuned to achieve good performance. It controls how quickly or slowly a CNN model learns a problem, a value that is too small might produce a long training process that could get stuck, while a value that is too large might generate a sub-optimal set of weights, too fast or an unstable training. Step is a decaying lr function, which reduces the lr value over the total number of iterations at a given step size. A smaller step value results in more frequent reduction of the lr value over the total number of training iterations, which may result in slow training process due to small lr value. #It. determines at which iteration the highest accuracy might be achieved.

CNNs models differ in their hyperparameters, the default values of all hyperparameters are shown in Table 1. This work used Stochastic Gradient Descent (SGD) in the training process to compute the gradient descent of the cost function for each batch size.

Table 1. Default values of all hyperparameters.

Hyper parameter	Description	GoogLeNet	ZFNet	InceptionV3 - ResNet 50	ResNet101
Batch size	# of training samples in one forward/backward pass. Higher batch size needs more memory space	64	64	15	10
Test iter	# of test iterations per test interval. It must cover the validation set	100	200	100	100
lr policy	How the learning rate should change over the time	"poly"	"step"	"poly"	"poly"
Gamma	It sets how much the lr should change every time	0.96	0.1	0.96	0.96
Power	Helps to calculate the effective learning rate follows a polynomial decay in the "poly" lr policy	1.0	-	1.0	1.0
Momentum	How much of the previous weight will be retained in the new calculation	0.9	0.9	0.9	0.9
Weight_decay	The factor of (regularization) penalization of large weights	0.0002	0.0002	0.0002	0.0002
Solver mode	The mode used in solving the network	GPU	GPU	GPU	GPU
lr	It sets up the start learning rate in the network	0.001	0.01	0.001	0.001
Step	It sets how many iterations count before refresh the lr	320 K	200 K	320 K	320 K
#It	It indicates when the network should stop training	10 K	700 K	5 K	5 K

6 Experimental Results

Accuracy $((T_p + T_n)/(T_p + F_n + T_n + F_p))$, precision $(T_p/(T_p + F_p))$, recall $(T_p/(T_p + F_n))$ and F1 $((2 \times (recall \times precision)/(recall + precision)))$ metrics were computed to measure the ability of explored CNNs models to classify leaves images into the corresponding class, and to determine how the different operations over the raw dataset and the hyperparameters fine-tune influence on the performance achieved. The parameters T_p, T_n, F_p, and F_n refer to the healthy leaf correctly classified as healthy one, unhealthy leaf correctly classified as unhealthy one, healthy leaf classified as unhealthy one, and unhealthy leaves classified as healthy one, respectively. In this sense, accuracy quantifies the ratio of correctly classified samples, precision quantifies the ratio of correctly classified positive samples over the total classified positive instances, recall measures the ratio of correctly classified positive samples over the total number of samples, and F1 quantifies the harmonic meaning between accuracy and recall which shows how accurate and robust a classifier is.

6.1   Training with Raw Data

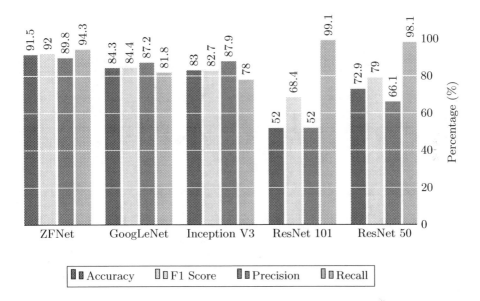

Fig. 3. Performance metrics values reached by the models with raw data.

Using the raw dataset with only two classes and minor variations among all samples, experimental results depicted in Fig. 3 clearly show that ZFnet takes advantages of the used default training settings parameters presented in Table 1. Indeed, ZFnet reaches the highest values (up 90%) overcoming all the explored models at the iteration 10K, it is attributed to its shallowness and simplicity. Experimental tests also demonstrate that deeper networks achieve lower values which can be improved only if more training iterations are performed with larger and more varied datasets. In this sense, ResNet 101 achieves the worst accuracy, precision, recall and F1 values. These results are due to the dataset characteristics which tend to cause overfitting problems in deeper models. In addition, ResNet 50 presents a better improvement in all the measures compared to ResNet 101, particularly because of the less layers and parameters managed by the model.

6.2   Pre-processing Impact

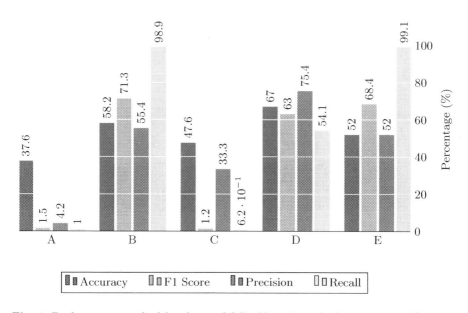

Fig. 4. Performance reached by the model ResNet 101 with the data sets: A) Illumination changes; B) data augmentation; C) random classes added; D) all changes; E) raw data.

As it is well know, the dataset size and poor variation data have a fundamental impact on the overall performance of deeper CNN. For this reason, ResNet101 was examined to determine the impact of different pre-processing operations to vary the leaves raw images and dataset size. Results reported in Fig. 4 show that both illumination variation and additional random classes datasets, experiments B and C, respectively, achieve lower accuracy than the raw dataset. In the illumination variation case, it is attributed to an overfitting produced by the repetition of the same image. Whereas with the random classes included, the accuracy performance decreases by 5%. This is produced because although the variation in the dataset size is increased, the difference between the two main classes (healthy and unhealthy leaves) is the same. Overall, augmented dataset reaches a notable increasing in accuracy due to the addition of new relevant information to the main classes, which avoids the overfitting produced in the previous cases. It can be affirmed because ResNet101 accuracy increases a significant percentage around 20% in experiment D as more data variability is applied.

Similar behaviour can be seen in the precision values. Precision decreases with illumination changes, increases a little with data augmentation and the eight additional classes, while in experiment D the precision with all the changes obtains the best performance. It is important to highlight that precision is a quite relevant measure in this work, since it establishes the ability of the explored CNN model to identify diseased leaves. On the other hand, F1 and Recall scores reach the highest values in experiment B by using augmented dataset. In other words, applying the variations hinders the depth CNN model when it has to identify healthy leaves, but its performance improves when it has to identify diseased leaves.

6.3   Fine-Tuning Impact

Fine-tuning aims to increase the performance of the explored CNNs models by making small variations on their critical hyperparameters. Results obtained from training with the integrated dataset (raw dataset with all changes applied to the raw one) are shown in Table 2. From the scores obtained by ZFNet it is clear that shallow CNN models often leads to a higher amount of parameters. Besides, they can be efficient models with a limited number of classes, data variations and dataset sizes for the diseased leaves classification process. With few iterations in test 5, ZFNet achieves a high accuracy and increases it by 10% after 50 K iterations.

On the other hand, each explored CNN model has a particular behaviour for the evaluated hyperparameter values. For instance, GoogLeNet is highly susceptible to high lr and hence the values obtained from test 1 and 2 are lower than 50% in all metrics. The improvement reached in the rest of the tests is produced because of their default large batch size, which gives fast learning with an smaller lr. GoogLeNet also gives great precision reaching values up to 90% in test 5, which was trained with just 547 iterations. It marks an improvement of the model of 10% against the scores obtained with raw data. Inception V3 reaches its highest accuracy (77%) under conditions established in test 2. A relevant fact is presented in test 6, where the model presents an abrupt change from the tendency. The accuracy of test 6 with a small lr and several iterations (apparently the optimal conditions for Inception V3) dropped to its worst point (under 50%). It is presumed to a possible overfitting because of the large #It. and small lr. A highest precision score is obtained with a lr = 0.01 and a large number of iterations (around 40K). The recall and F1 scores retain the tendency to increase with high values, except in test 6 where all results were deficient.

ResNet 101 scores depend on the number of iterations and, to a lesser extent, on the lr value. This is demonstrated by the accuracy reached by the model, which does not change significantly unless the #It. is increased. This fact is attributed to the depth of the model. Its highest accuracy is 86%, which shows a significant increase from experiments with raw data. This is attributed to the new variation on the data set. The model also shows great performance in precision, recall and f1 score metrics. On the contrary the accuracy obtained from ResNet 50 was under 55% in all tests, a very low performance against the other models. This fact is mostly attributed to the dataset used to the different tests (called all changes) because in all the variations performed on the model, the accuracy obtained was poor.

Table 2. Fine-tuning results

Model	Amount of Parameters	Test #	lr	Step	#It.	Accuracy	Precision	Recall	F1
GoogLeNet	12.36 M	1	0.1	438	547	48%	50%	0.6%	1%
		2	0.1	40 K	50 K	48%	50%	0.6%	1%
		3	0.01	438	547	80%	77%	94%	84%
		4	0.01	40 K	50 K	**87%**	83%	**95%**	**89%**
		5	0.001	438	547	77%	**94%**	63%	75%
		6	0.001	40 K	50 K	81%	76%	96%	85%
Inception V3	21.8 M	1	0.1	1,750	2,188	55%	54%	**89%**	67%
		2	0.1	40 K	50 K	**77%**	78%	79%	**79%**
		3	0.01	1,750	2,188	71%	77%	64%	70%
		4	0.01	40 K	50 K	74%	**96%**	51%	67%
		5	0.001	1,750	2,188	73%	82%	62%	71%
		6	0.001	40 K	50 K	48%	50%	0.6%	1%
ResNet 50	23.51 M	1	0.1	1,750	2,188	32%	37%	**41%**	**39%**
		2	0.1	40 K	50 K	9%	3%	1%	2%
		3	0.01	1,750	2,188	24%	**53%**	25%	34%
		4	0.01	40 K	50 K	10%	6%	4%	5%
		5	0.001	1,750	2,188	32%	30%	12%	17%
		6	0.001	40 K	50 K	**52%**	3%	1%	2%
ResNet 101	42.51 M	1	0.1	2,800	3,500	44%	42%	21%	28%
		2	0.1	40 K	50 K	79%	**94%**	65%	77%
		3	0.01	2,800	3,500	48%	50%	0.6%	1%
		4	0.01	40 K	50 K	81%	**94%**	68%	78%
		5	0.001	2,800	3,500	52%	52%	**100%**	69%
		6	0.001	40 K	50 K	**86%**	84%	91%	**87%**
ZFNet	62.36 M	1	0.1	438	547	48%	50%	0.6%	1%
		2	0.1	40 K	50 K	48%	50%	0.6%	1%
		3	0.01	438	547	52%	52%	**99%**	69%
		4	0.01	40 K	50 K	52%	52%	**99%**	68%
		5	0.001	438	547	83%	83%	89%	86%
		6	0.001	40 K	50 K	**93%**	**91%**	97%	**94%**

7   Conclusion

In this work, an empirical analysis among five state-of-the-art CNNs was carried out. The explored CNN architectures include GoogLeNet, Inception V3, ResNet 50, ResNet 101 and ZFNet. They differ in the number of convolutional layers, number of parameters and in the way they are arranged. From the experiments, ZFNet has a growing trend in accuracy, precision and F1 scores with no over fitting or performance decreasing manifestations. ZFNet obtained a 93% validation accuracy score for 50 K iterations beating the rest of the explored CNN models for identification of diseased leaves using datasets with a small number of classes, data variations and dataset sizes. Based on obtained results, as it was expected, all CNN models requiere a careful fine-tuning of their hyperparameters because they not show direct correlations among hyperparameter values and their depth. As future work, we propose to extend this work exploring more preprocessing techniques to improve the quality and quantity of a limited dataset an thus to accomplish a reliable classification for several leaves diseases.

Acknowledgement. This work used the supercomputer 'Quinde I' from the public company Siembra EP in the Republic of Ecuador.

References

1. Guachi Guachi, L., Guachi, R., Bini, F., Marinozzi, F.: Automatic colorectal segmentation with convolutional neural network. Comput. Aided Des. Appl. **16**, 836–845 (2019). https://doi.org/10.14733/cadaps.2019.836-845
2. Guachi Guachi, L., Guachi, R., Perri, S., Corsonello, P., Fabiano, B., Marinozzi, F.: Automatic microstructural classification with convolutional neural network, pp. 170–181, January 2019. https://doi.org/10.1007/978-3-030-02828-213
3. Szegedy, C., Vanhoucke, V., Ioffe, S., Shlens, J., Wojna, Z.: Rethinking the inception architecture for computer vision. CoRR abs/1512.00567 https://arxiv.org/abs/1512.00567
4. Szegedy, C., et al.: Going deeper with convolutions. In: 2015 IEEE Conference on Computer Vision and Pattern Recognition (CVPR), pp. 1–9, June 2015. https://doi.org/10.1109/CVPR.2015.7298594
5. Zeiler, M., Fergus, R.: Visualizing and understanding convolutional networks. CoRR abs/1311.2901 (2013) http://arxiv.org/abs/1311.2901
6. Kaiming, H., Xiangyu, Z., Shaoqing, R., Jian, S.: Deep residual learning for image recognition. CoRR abs/1512.03385 (2015) http://arxiv.org/abs/1512.03385
7. Hughes, D., Salathe, M.: An open access repository of images on plant health to enable the development of mobile disease diagnostics through machine learning and crowdsourcing, November 2015
8. Sladojevic, S., et al.: Deep neural networks based recognition of plant diseases by leaf image classification. Comput. Intell. Neurosci. **2016**, 11 (2016). https://www.hindawi.com/journals/cin/2016/3289801/
9. Ballester, P., Correa, U., Birck, M., Araujo, R.: Assessing the performance of convolutional neural networks on classifying disorders in apple tree leaves, pp. 31–38, November 2017. https://doi.org/10.1007/978-3-319-71011-23
10. Bin, L., Zhang, Y., He, D., Li, Y.: Identification of apple leaf diseases based on deep convolutional neural networks. Symmetry **10**, 11 (2017). https://doi.org/10.3390/sym10010011
11. Zhang, K., Wu, Q., Liu, A., Meng, X.: Can deep learning identify tomato leaf disease? Adv. Multimedia **2018**, 1–10 (2018). https://doi.org/10.1155/2018/6710865
12. Baranwal, S., Khandelwal, S., Arora, A.: Deep learning convolutional neural network for apple leaves disease detection. SSRN Electron. J. (2019). https://doi.org/10.2139/ssrn.3351641
13. Ferentinos, K.: Deep learning models for plant disease detection and diagnosis. Comput. Electron. Agric. **145**, 311–318 (2018). https://doi.org/10.1016/j.compag.2018.01.009
14. Toda, Y., Okura, F.: How convolutional neural networks diagnose plant disease. Plant Phenomics 2019, March 2019. https://doi.org/10.1155/2019/9237136

15. Ashqar, B., Abu-Naser, S.: Image-based tomato leaves diseases detection using deep learning. Int. J. Eng. Res. **2**, 10–16 (2019)

16. Picon, A., Seitz, M., Alvarez-Gila, A., Mohnke, P., Ortiz Barredo, A., Echazarra, J.: Crop conditional convolutional neural networks for massive multi-crop plant disease classification over cell phone acquired images taken on real field conditions. Comput. Electron. Agric. **167**, 105093 (2019). https://doi.org/10.1016/j.compag.2019.105093

17. Ahila Priyadharshini, R., Arivazhagan, S., Arun, M., Mirnalini, A.: Maize leaf disease classification using deep convolutional neural networks. Neural Comput. Appl. **31**(12), 8887–8895 (2019). https://doi.org/10.1007/s00521-019-04228-3

18. Zhang, S., Zhang, S., Zhang, C., Wang, X., Shi, Y.: Cucumber leaf disease identification with global pooling dilated convolutional neural network. Comput. Electron. Agric. **162**, 422–430 (2019). https://doi.org/10.1016/j.compag.2019.03.012

19. Barbedo, J.: Impact of dataset size and variety on the effectiveness of deep learning and transfer learning for plant disease classification. Comput. Electron. Agric. **153**, 46–53 (2018). https://doi.org/10.1016/j.compag.2018.08.013

20. Too, E., Yujian, L., Njuki, S., Yingchun, L.: A comparative study of fine-tuning deep learning models for plant disease identification. Comput. Electron. Agric., March 2018. https://doi.org/10.1016/j.compag.2018.03.032

21. Bhattacharya, S., Mukherjee, A., Phadikar, S.: A Deep Learning Approach for the Classification of Rice Leaf Diseases, pp. 61–69, January 2020. https://doi.org/10.1007/978-981-15-2021-18

22. Zhang, S., Huang, W., Zhang, C.: Three-channel convolutional neural networks for vegetable leaf disease recognition. Cogn. Syst. Res. **53**, May 2018. https://doi.org/10.1016/j.cogsys.2018.04.006

23. Khamparia, A., Saini, G., Gupta, D., Khanna, A., Tiwari, S., de Albuquerque, V.H.C.: Seasonal crops disease prediction and classification using deep convolutional encoder network. Circuits Syst. Sig. Process. **39**(2), 818–836 (2019). https://doi.org/10.1007/s00034-019-01041-0

24. Augmentor. https://augmentor.readthedocs.io/en/master/index.html. Accessed 15 Mar 2020

25. Large Scale Visual Recognition Challenge (ILSVRC). http://www.image-net.org/challenges/LSVRC/. Accessed 15 Mar 2020

26. Canziani, A., Paszke, A., Culurciello, E.: An analysis of deep neural network models for practical applications, March 2016

27. Krizhevsky, A., Sutskever, I., Hinton, G.: Imagenet classification with deep convolutional neural networks, pp. 1097–1105 (2012). http://dl.acm.org/citation.cfm?id=2999134.2999257

28. Huang, F., Ash, J., Langford, J., Schapire, R.: Learning deep resnet blocks sequentially using boosting theory, June 2017

29. Simonyan, K., Zisserman, A.: Very deep convolutional networks for large-scale image recognition. arXiv:1409.1556, September 2014

A Kinect-Based Gesture Acquisition and Reproduction System for Humanoid Robots

Agnese Augello[1], Angelo Ciulla[1], Alfredo Cuzzocrea[2(✉)], Salvatore Gaglio[1,3], Giovanni Pilato[1], and Filippo Vella[1]

[1] ICAR-CNR, Palermo, Italy
{agnese.augello,angelo.ciulla,salvatore.gaglio,giovanni.pilato,
filippo.vella}@cnr.it
[2] iDEA Lab, University of Calabria, Rende, Italy
alfredo.cuzzocrea@unical.it
[3] University of Palermo, Palermo, Italy
salvatore.gaglio@unipa.it

Abstract. The paper illustrates a system that endows an humanoid robot with the capability to mimic the motion of a human user in real time, serving as a basis for further gesture based human-robot interactions. The described approach uses the Microsoft Kinect as a low cost alternative to expensive motion capture devices.

Keywords: Gestures · Humanoid robot · Kinect

1 Introduction

Nowadays, the interaction between human beings and robots has become a very relevant issue in a wide range of applications (e.g., [14,18,20]). It is commonly agreed that communication between humans is based on both verbal and not verbal cues. A humanoid robot capable of interacting with people combining speech and gestures would dramatically increase the naturalness of social interactions. On the other hand, other studies like [7,13,17] consider *knowledge management techniques* (e.g., [7]) to improve this phase.

Furthermore, the Microsoft Kinect is a popular choice for any research that involves body motion capture. It is an affordable and low-cost device that can be used for non invasive, marker-less tracking of body gestures. As an example, Baron et al. [4] controlled a Mindstorm NXT artificial arm with sensor Kinect, employing gesture recognition to regulate arm movement. Chang et al. [5] developed a Kinect-based gesture command control method for driving a humanoid robot to learn human actions, using a Kinect sensor and three different recognition mechanisms: dynamic time wrapping (DTW), hidden Markov model (HMM) and principal component analysis (PCA).

Meanwhile, Sylvain Filiatrault and Ana-Maria Cretu [11] used sensor Kinect to mimic the motion of a human arm to an NAO humanoid robot. In their case,

© Springer Nature Switzerland AG 2020
O. Gervasi et al. (Eds.): ICCSA 2020, LNCS 12249, pp. 967–977, 2020.
https://doi.org/10.1007/978-3-030-58799-4_69

the software architecture is based on three modules: Kinect Manager, Interaction Manager, and NAO manager. The Kinect Manager deals with the events and data captured by the Kinect. The class Kinect Transformer is used to get the Euler angles of the desired joints. The Interaction Manager is the intermediary between the Kinect and the robot and contains the repository for the joints used by the other two modules. The use of a joint repository of all articulations allows reducing the data to be processed as some joints are not needed. Finally, the NAO manager contains the static and dynamic constraints to apply to each one of the articulations, as well as the methods that allow the control of the robot movements.

To be sure that the robot has enough time to execute the gesture, a delay of 200 ms between one cycle and the next has been introduced. Itauma et al. [12] used a Kinect to teach an NAO robot some basic Sign Language gestures. The aim was teaching Sign Language to impaired children by employing different machine learning techniques in the process. Shohin et al. [15] used three different methods to make a robot NAO imitate human motion: direct angle mapping, inverse kinematics using fuzzy logic and iterative Jacobian.

In some cases, neural networks were used: Miguel et al. [16] used a Kinect sensor and a Convolutional Neural Network (CNN) trained with the MSRC-12 dataset [1] to capture and classify gestures of a user and send related commands to a mobile robot. The used dataset was created by Microsoft and had 6244 gesture instances of 12 actions. To have gestures of the same length, without losing relevant information, the system used a Fast Dynamic Time Warping algorithm (FastDTW) to find the optimal match between sequences by non linearly warping them along the time axis. This resulted in all gestures normalized to sequences of 667 frames, with each frame having 80 variables, corresponding to the x, y, z values for each of the 20 joints, plus a separation value for each joint. The resulting 667×80 matrix is used as the input of the CNN, which classifies it in one of the 12 possible gestures. The CNN was trained using two strategies, combined training consisting of a single CNN to recognize all 12 gestures and individual training with 12 different CNN, each capable of recognizing only one gesture. The accuracy rates were 72.08% for combined training and 81.25% for the individual training.

Moreover, Unai et al. [19] developed a natural talking gesture generation behavior for *Pepper* by feeding a Generative Adversarial Network (GAN) with human talking gestures recorded by a Kinect. Their approach in mapping the movements detected by Kinect on the robot is very similar to what we used, but while they feed the resulting values to a neural network (a GAN), we use the (filtered) values directly.

This paper reports the implementation of a system able to acquire and reproduce the gestures performed by a human during an interactive session. In our approach, we exploited a *MicrosoftKinect* sensor to capture the motion data from a user and then, we have defined a mapping algorithm to allow a *Soft-Bank Pepper* robot to reproduce the tracked gestures as close as possible to the original ones.

In particular, we used the *OpenNi* driver for the *Kinect*, the NiTE 2.2 libraries for detecting the user skeleton, and the Kinetic version of ROS with the module pepper_dcm to provide package exchange and bridging between the computer and the robot and Ubuntu 16.04. We focused on the movements of the arms and the head, laying the basis for the extension of the same approach to the remaining parts.

2   System Anatomy and Functionalities

The developed system is structured in a set of modules, to increase versatility for future projects and to simplify possible extensions to the current project. Besides the Kinect itself, the first module is named *Viewer*, which extracts data frames (consisting of nine float values: three values for a joint position in 3D space, four values for quaternion from the origin and reliability values for both) from the Kinect and sends them in a pipe. The module also provides the feed of the Kinect camera with the overlay of the tracked user's skeleton. The pipe, long 8640 chars (64 chars for each joint value, 9 values for each joint, 15 joints total), is read by the second module, Gesture_Brain.

The *Gesture_Brain* module works both as a gateway for the ROS system [2] and as the module where actual data processing takes place. The gathered data cannot be used directly: a mapping is required to correctly associate each joint user position to the equivalent one in the Pepper robot. For this reason, the data is parsed and structured in a 15×9 float matrix, which is then separated into three matrices: one for coordinates, one for quaternions, and one for reliability values. In our algorithm, we decided to use only the first matrix for simplicity reasons, neglecting the quaternion matrix, and not performing any reliability check on the joints. We assume that the joint data is accurate enough for our purpose, as the Kinect already discards joints whose reliability values are too low. The joint position data is used to estimate Pepper joint angles, specifically shoulder pitch, shoulder roll, elbow roll and elbow yaw for both arms and head yaw for the head (there are three more joint angles that could be estimated, left and right wrist yaw and head pitch, but the Kinect is too imprecise to allow a good estimate, so they have been fixed to a value of 0). The details about this estimation are discussed in the next section.

After all required values are collected, we can use the ROS threads provided by the bridge pepper_dcm to send the joint angles to the robot. These threads consist of multiple joint angles divided into groups, each group representing a body part. As we are interested only in the movement of arms and head, we use three: head, left arm, right arm. The bridge reads the sent values and the time between each capture to dynamically calculate the gesture trajectory in real-time. This means that to allow the system to be as accurate as possible, the gesture should be executed quite slowly. The bridge itself was modified to activate the in-built Self Collision Avoidance (part of the NaoQi library) and to deactivate wait and breathe animations, as they interfere with the commands sent by the pepper_dcm (Figs. 1 and 2).

[Coordinates] [Quaternion] [C_conf] [Q_conf]

Fig. 1. Structure of a single row of the array sent by the Viewer module

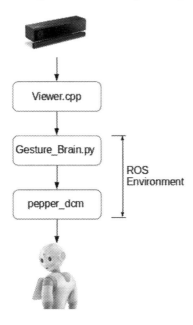

Fig. 2. Structure of the algorithm

3   Defining and Supporting Mappings Between End User and Pepper

The *Pepper* robot has five Degree of Freedom for each arm (each one associated with a joint), unlike human beings who have seven. A mapping is thus required. From the *Kinect* the Cartesian coordinates for each joint, the quaternion for each segment (both referenced globally), and a reliability value for both are extracted. The bridge *pepper_dcm* uses Euler angles to communicate to the robot the new position of its joint angles. 3D space coordinates are thus used since quaternions have proven unsuitable. This is because the quaternions extracted do not represent the rotation from the previous frame, but rather the rotation from a reference quaternion. This leads to excessive inaccuracies once converted in Euler angles.

Let $\bar{x}$, $\bar{y}$ and $\bar{z}$ be the unit vectors for each axis, that is:

$$\bar{x} = (1, 0, 0)$$
$$\bar{y} = (0, 1, 0)$$
$$\bar{z} = (0, 0, 1)$$

Let S_L, E_L and W_L be the coordinates of the shoulder, the elbow and the wrist of the left arm respectively, $\overline{S_L E_L}$ and $\overline{E_L W_L}$ are defined as:

$$\overline{S_L E_L} = E_L - S_L$$
$$\overline{E_L W_L} = W_L - E_L$$

SR_L is the supplementary to the angle between $\overline{S_L E_L}$ and $-x$ axis:

$$SR_L = \frac{\pi}{2} - arcos(\overline{S_L E_L} \cdot -x) \tag{1}$$

SP_L is the angle between the projection of $\overline{S_L E_L}$ on zy plane and z axis, shifted in range to avoid the jump discontinuity at 180 and -180:

$$SP_L = \pi - mod_{2\pi}(\frac{3}{2}\pi + arctan(\overline{S_L E_L}_z, \overline{S_L E_L}_y) \tag{2}$$

For values of SR_L close to $\frac{\pi}{2}$, SP_L become unstable. As such, in the algorithm is assigned a value of 0 for $SR_L > 1.3$.

ER_L is the angle between $\overline{E_L W_L}$ and $\overline{S_L E_L}$, shifted by $\frac{\pi}{2}$:

$$ER_L = \frac{\pi}{2} + arcos(\overline{E_L W_L} \cdot \overline{S_L E_L}) \tag{3}$$

EY_L is the angle between the projection of $\overline{E_L W_L}$ on zy plane and z axis, shifted in range for stability reasons, plus $-SP_L$:

$$EY_L = \pi - mod_{2\pi}(\frac{3}{2}\pi + arctan(\overline{E_L W_L}_z, \overline{E_L W_L}_y) - SP_L \tag{4}$$

The right arm is almost the same as the left arm, the only difference is that some angles have the opposite sign. Let $\overline{HN}$ be the difference between the coordinates of the joints H (head) and N:

$$\overline{HN} = H - N$$

The head yaw HY is the angle between the projection of $\overline{HN}$ on the xz plane and the z axis:

$$HY = -arctan(\overline{HN}_z, \overline{HN}_y) - \frac{\pi}{2} \tag{5}$$

3.1   Line of Best Fit

Kinect joint detection is based on the shape of the user, which is redrawn at every frame. While calibrating the sensor helps to reduce the resulting jerkiness, there is still a significant amount of noise left. This noise can be approximately classified in two categories: a constant Gaussian noise caused by small alteration on the shape detected and large "spikes" when the *Kinect* fail to guess the position of one or more joints (especially common when part of the limb is outside of the frame or when two or more joints overlap). A simple way to compensate part of this noise is to use a line of best fit.

Given k points in (x, y) coordinates system, we must find the values c_0 and c_1 in the equation:

$$p(x) = c_0 x + c_1$$

that define the straight line minimizing the squared error:

$$E = \sum_{j=0}^{k} |p(x_j) - y_j|^2$$

in the equations:

$$x_0 c_0 + c_1 = y_0$$
$$x_1 c_0 + c_1 = y_1$$
$$\dots$$
$$x_k c_0 + c_1 = y_k$$

The result is a smoother movement, especially when Kinect is not able to detect the precise coordinates of a given joint. This is because, given a disturbing signal, the line of best fit can be seen as an approximation of the tangent that the signal would have at that point if the noise were removed. This is not always true, especially when the signal changes rapidly, but it's close enough in most cases to give a generally cleaner movement.

3.2   Modes of Operation

Besides mimicking the user movement, the Gesture_Brain module also has some additional features implemented to increase the breadth of experiments that can be performed with the system or to help with future projects. The behavior of the program is managed by the input arguments. These are, in order: mode, mirror_flag, json_file_name, LAjpos, RAjpos, Hjpos. The first one determines which of the three different modes of operation will be used (default 0), the second one determines if the mirror mode is activated or not (default false), the third defines the name of the text file used to record (in mode 0 and 1) or read (mode 2) the gestures (the default value is NULL, that is no recording) and set a flag (record_flag) to 1, the fourth, fifth and sixth ones are used to determine the pose to use in mode 1 (as default, the robot will spread its arms parallel to the ground, in a pose that in animation is known as "T-pose"). More in details, the modes of operation of the main program are:

- Mode 0 or "Mimic Mode", is the default mode and makes the robot mimic the movements of the user. The record flag makes it, so the output is not just sent to the ROS publishers, but recorded in a JSON (JavaScript Object Notation) file, to be reproduced later. If the mirror flag is active, every movement is mirrored. In case both the record and mirror flags are active, the mirrored movement will be recorded and saved in the specified txt file.

- Mode 1, or "Pose Mode", make the robot execute a pose (defined at the beginning by the value of the given arguments) that the user must try to emulate. A distance algorithm calculates how close is the user pose to that of the robot, evaluated separately for the head, right upper arm, left upper arm, right forearm, and left forearm. If the user pose keeps all body parts below their respective thresholds (defined separately for each boy part), the program will communicate the success and shut down. The record flag makes it, so the distance values returned are written in a file, while the mirror flag makes it so the user must try to mirror the pose shown.
- Mode 2, or "Playback Mode", consists of reproducing a previously recorded gesture. The mirror flag, even if selected, doesn't have any effect on the algorithm. The name necessary to activate the record flag is used as the name of the file with the gesture to execute.

As an example, an experiment that was conceptualized consisted in using the Pepper robot to show a specific pose that the user must replicate as closely as possible. The experiment envisaged the use of both the normal mode and the mirror mode (Fig. 3).

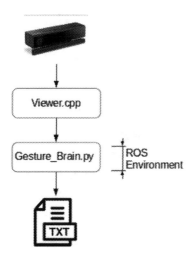

Fig. 3. Structure of the algorithm when recording (the text file can be either the recorded gesture in mode 0 or the record of distance values in mode 1)

4   Auxiliary Algorithms

4.1   Distance Measurement

To measure the distance between the robot pose and the detected pose, we make use of a distance measurement algorithm. For each arm, the related joint angle group (consisting of Elbow roll, Elbow yaw, Shoulder pitch, Shoulder roll, and

Wrist yaw) is split into two vectors, representing the upper arm (Shoulder pitch and Shoulder roll) and the forearm (Elbow roll, Elbow yaw, and Wrist yaw). Including the joint angle group for the head (Head yaw and Head pitch), we have five vectors. To avoid that joint angles that have a larger range of values than others affect excessively the final result, each angle is divided by its respective maximum value. This process is applied both to the robot pose and the detected pose of the user, giving us five pairs of vectors. Finally, a Mean Squared Error algorithm is applied at each pair of vectors, resulting in five distances defined in the interval [0, 1]. Having each vector evaluated separately is more precise and reliable, and allows to define separate thresholds if a pass/fail system is implemented (like in our case) (Fig. 4).

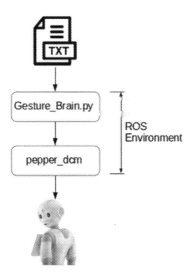

Fig. 4. Structure of the algorithm in mode 2 (the txt file in this case is the coding of a previously recorded gesture)

4.2   Mirroring

The activation of the mirror flag makes the algorithm mirror any movement detected. This means that in mode 0, the robot will mirror the user (and if the program is recording, it will record the mirrored movement), while in mode 1, the user will have to mirror the pose shown. The mirroring consists in switching the detected angles between the left and the right arm, changing the sign of every angle except shoulder pitch. For the head yaw angle, a simple change in the sign is enough.

4.3   Recording and Playback

If the recording flag is active, the program will write a text file in *JSON* syntax with the joint angles sent to the robot (mode 0) or the distance from the given pose (mode 1) for each frame.

In mode 2, the records created in mode 0 can be reproduced by the robot. In mode 0, the fields for each frame are:

- *AbsTime*: Date and time when the frame was captured, to the microsecond.
- *Left_Arm*: Vector containing the joint angles of the left arm joint group.
- *Right_Arm*: Vector containing the joint angles of the right arm joint group.
- *Head*: Vector containing the joint angles of the joint head group.

In mode 1, the fields for each frame are:

- *AbsTime*: Date and time when the frame was captured, to the microsecond.
- *Error_Left_Arm_Shoulder*: the measured distance between the given position and the detected position of the left upper arm.
- *Error_Left_Arm_Elbow*: the measured distance between the given position and the detected position of the left forearm.
- *Error_Right_Arm_Shoulder*: the measured distance between the given position and the detected position of the right upper arm.
- *Error_Right_Arm_Elbow*: the measured distance between the given position and the detected position of the right forearm.
- *Error_Head*: the measured distance between the given position and the detected position of the head.

5   Conclusions and Future Work

The system illustrated in this paper is capable of detecting the user poses with the Kinect with sufficient accuracy. The first experiments show that the reproduced movements are precise and smooth; the mirroring is accurate; the pose evaluation is coherent; Furthermore, the recording and execution of the gestures are very close to the real-time movements. However, sometimes, certain positions cannot be reliably detected, due to imprecise behavior of the Kinect output when joints overlap each other, and to excessive reliance on the silhouette to detect the human body and the lack of joints in key points of the detected skeleton (like the hands). There is also an environmental factor, like lightning and positioning, that can make accurate user detection problematic. Currently, we are setting up two experiments: the first one is to make the robot autonomously capable of acting both as an instructor and a learner of the Semaphore Flag Signalling System[3], exploiting the gesture mirroring features; the second one is to make the robot capable of both encoding and decoding simple sentences from natural language to the flag semaphore system and vice-versa.

In future works, a neural network will be also trained to recognize and classify the gestures to give a proper answer, creating a more realistic verbal communication between humans and robots. Possible improvements should include a more effective detection algorithm for the Kinect, more efficient ways to execute the mapping, the use of all points detected (not just limbs and head), the use of the official Microsoft SDK to have even more points detected (provided it's possible to retrieve all the necessary libraries) and a more general user-friendly experience (like the ability to set a timer for recording). Furthermore, we plan to extend our framework as to deal with novel and emerging *big data trends* including performance (e.g., [6,9]), and privacy and security (e.g., [8,10]).

References

1. Msrc-12   dataset.   https://www.microsoft.com/en-us/download/details.aspx?id=52283
2. Ros kinetic. http://wiki.ros.org/kinetic
3. Semaphore flag signalling system. https://en.wikipedia.org/wiki/Flag_semaphore
4. Baron, G., Czekalski, P., Malicki, D., Tokarz, K.: Remote control of the artificial arm model using 3D hand tracking. In: 2013 International Symposium on Electrodynamic and Mechatronic Systems (SELM), pp. 9–10. IEEE (2013)
5. Chang, C.w., He, C.j., et al.: A kinect-based gesture command control method for human action imitations of humanoid robots. In: 2014 International Conference on Fuzzy Theory and Its Applications (iFUZZY 2014), pp. 208–211. IEEE (2014)
6. Chatzimilioudis, G., Cuzzocrea, A., Gunopulos, D., Mamoulis, N.: A novel distributed framework for optimizing query routing trees in wireless sensor networks via optimal operator placement. J. Comput. Syst. Sci. **79**(3), 349–368 (2013)
7. Cuzzocrea, A.: Combining multidimensional user models and knowledge representation and management techniques for making web services knowledge-aware. Web Intell. Agent Syst. **4**(3), 289–312 (2006)
8. Cuzzocrea, A., Bertino, E.: Privacy preserving OLAP over distributed XML data: a theoretically-sound secure-multiparty-computation approach. J. Comput. Syst. Sci. **77**(6), 965–987 (2011)
9. Cuzzocrea, A., Moussa, R., Xu, G.: OLAP*: effectively and efficiently supporting parallel OLAP over big data. In: Cuzzocrea, A., Maabout, S. (eds.) MEDI 2013. LNCS, vol. 8216, pp. 38–49. Springer, Heidelberg (2013). https://doi.org/10.1007/978-3-642-41366-7_4
10. Cuzzocrea, A., Russo, V.: Privacy preserving OLAP and OLAP security. In: Encyclopedia of Data Warehousing and Mining, 2nd edn (4 Volumes), pp. 1575–1581 (2009)
11. Filiatrault, S., Cretu, A.M.: Human arm motion imitation by a humanoid robot. In: 2014 IEEE International Symposium on Robotic and Sensors Environments (ROSE) Proceedings, pp. 31–36. IEEE (2014)
12. Itauma, I.I., Kivrak, H., Kose, H.: Gesture imitation using machine learning techniques. In: 2012 20th Signal Processing and Communications Applications Conference (SIU), pp. 1–4. IEEE (2012)
13. Lau, M.C., Anderson, J., Baltes, J.: A sketch drawing humanoid robot using image-based visual servoing. Knowl. Eng. Rev. **34**, e18 (2019)
14. Monje, C.A., de la Casa Díaz, S.M.: Modeling and control of humanoid robots. Int. J. Humanoid Rob. **16**(6), 1902003:1–1902003:3 (2019)
15. Mukherjee, S., Paramkusam, D., Dwivedy, S.K.: Inverse kinematics of a NAO humanoid robot using kinect to track and imitate human motion. In: 2015 International Conference on Robotics, Automation, Control and Embedded Systems (RACE), pp. 1–7. IEEE (2015)

16. Pfitscher, M., Welfer, D., de Souza Leite Cuadros, M.A., Gamarra, D.F.T.: Activity gesture recognition on kinect sensor using convolutional neural networks and FastDTW for the MSRC-12 dataset. In: Abraham, A., Cherukuri, A.K., Melin, P., Gandhi, N. (eds.) ISDA 2018 2018. AISC, vol. 940, pp. 230–239. Springer, Cham (2020). https://doi.org/10.1007/978-3-030-16657-1_21

17. Regier, P., Milioto, A., Karkowski, P., Stachniss, C., Bennewitz, M.: Classifying obstacles and exploiting knowledge about classes for efficient humanoid navigation. In: 18th IEEE-RAS International Conference on Humanoid Robots, Humanoids 2018, Beijing, China, 6–9 November 2018, pp. 820–826 (2018)

18. Saeedvand, S., Aghdasi, H.S., Baltes, J.: Robust multi-objective multi-humanoid robots task allocation based on novel hybrid metaheuristic algorithm. Appl. Intell. **49**(12), 4097–4127 (2019). https://doi.org/10.1007/s10489-019-01475-8

19. Zabala, U., Rodriguez, I., Martínez-Otzeta, J.M., Lazkano, E.: Learning to gesticulate by observation using a deep generative approach. arXiv preprint arXiv:1909.01768 (2019)

20. Zhang, A., Ramirez-Alpizar, I.G., Giraud-Esclasse, K., Stasse, O., Harada, K.: Humanoid walking pattern generation based on model predictive control approximated with basis functions. Adv. Robot. **33**(9), 454–468 (2019)

A Novel Big Data Analytics Approach for Supporting Cyber Attack Detection via Non-linear Analytic Prediction of IP Addresses

Alfredo Cuzzocrea[1,2]([✉]), Enzo Mumolo[3], Edoardo Fadda[4,5],
and Marco Tessarotto[3]

[1] University of Calabria, Rende, Italy
alfredo.cuzzocrea@unical.it
[2] LORIA, Nancy, France
[3] University of Trieste, Trieste, Italy
mumolo@units.it, marco.tessarotto@regione.fvg.it
[4] Politecnico di Torino, Torino, Italy
edoardo.fadda@polito.it
[5] ISIRES, Torino, Italy

Abstract. Computer network systems are often subject to several types of attacks. For example the distributed Denial of Service (DDoS) attack introduces an excessive traffic load to a web server to make it unusable. A popular method for detecting attacks is to use the sequence of source IP addresses to detect possible anomalies. With the aim of predicting the next IP address, the Probability Density Function of the IP address sequence is estimated. Prediction of source IP address in the future access to the server is meant to detect anomalous requests. In other words, during an access to the server, only predicted IP addresses are permitted and all others are blocked. The approaches used to estimate the Probability Density Function of IP addresses range from the sequence of IP addresses seen previously and stored in a database to address clustering, normally used by combining the K-Means algorithm. Instead, in this paper we consider the sequence of IP addresses as a numerical sequence and develop the nonlinear analysis of the numerical sequence. We used nonlinear analysis based on Volterra's Kerners and Hammerstein's models. The experiments carried out with datasets of source IP address sequences show that the prediction errors obtained with Hammerstein models are smaller than those obtained both with the Volterra Kernels and with the sequence clustering by means of the K-Means algorithm.

Keywords: Cyber attack · Distributed Denial of Service · Hammerstein models

A. Cuzzocrea—This research has been made in the context of the Excellence Chair in Computer Engineering – Big Data Management and Analytics at LORIA, Nancy, France.

O. Gervasi et al. (Eds.): ICCSA 2020, LNCS 12249, pp. 978–991, 2020.
https://doi.org/10.1007/978-3-030-58799-4_70

1   Introduction

User modeling is an important task for web applications dealing with large traffic flows. They can be used for a variety of applications such as to predict future situations or classify current states. Furthermore, user modeling can improve detection or mitigation of Distributed Denial of Service (DDoS) attack [12,15,17], improve the quality of service (QoS) [19], individuate click fraud detection and optimize traffic management. In peer-to-peer (P2P) overlay networks, IP models can also be used for optimizing request routing [1]. Those techniques are used by severs for deciding how to manage the actual traffic. In this context, also outlier detection methods are often used if only one class is known. If, for example, an Intrusion Prevention System wants to mitigate DDoS attacks, it usually has only seen the normal traffic class before and it has to detect the outlier class by its different behavior. In this paper we deal with the management of DDos because nowadays it has become a major threat in the internet. Those attacks are done by using a large scaled networks of infected PCs (bots or zombies) that combine their bandwidth and computational power in order to overload a publicly available service and denial it for legal users. Due to the open structure of the internet, all public servers are vulnerable to DDoS attacks. The bots are usually acquired automatically by hackers who use software tools to scan through the network, detecting vulnerabilities and exploiting the target machine. Furthermore, there is also a strong need to mitigate DDoS attacks near the target, which seems to be the only solution to the problem in the current internet infrastructure. The aim of such a protection system is to limit their destabilizing effect on the server through identifying malicious requests. There are multiple strategies with dealing with DDoS attacks. The most effective ones are the near-target filtering solutions. They estimates normal user behavior based on IP packet header information. Then, during an attack the access of outliers is denied. One parameter that all methods have in common is the source IP address of the users. It is the main discriminant for DDoS traffic classification. However, the methods of storing IP addresses and estimating their density in the huge IP address space, are different. In this paper, we present a novel approach based on system identification techniques and, in particular, on the Hammerstein models. The paper is organized as follows. In Sect. 2 we present the available methods for DDoS traffic classification (a broader overview of state-of-the-art mitigation research is given by [10]). In Sects. 3 and 4 we present our proposed a technique based based on Hammerstein models and we recall some similar model. Experimental results reported in Sect. 5 confirm the effectiveness of our approach. Although DDoS mitigation is the most important practical application for IP density estimation, we do not restrict the following work on this topic. Our generic view on IP density estimation may be valuable to other applications as well. One might think of preferring regular customers in overload situations (flash crowd events), identifying non-regular users on websites during high click rates on online advertisements (click fraud detection) or optimizing routing in peer-to-peer networks. Finally, in Sect. 6 we draw conclusions and indicate future work.

2   Related Work

In this section we review existing IP density estimation approaches. Furthermore, we formulate the often implicitly used ideas in a probabilistic way using the PDF.

2.1   History-Based IP Filtering

One of the first work on the field of IP filtering is [18]. In their work, they proposed an algorithm called History based IP Filtering (HIF). Given an IP address database (IAD) containing all frequently observed IP addresses, during an attack, the algorithm allows the access to the website only to the IP addresses from the IAD. The idea behind this algorithm was introduced in [13]. In this paper, the authors observe that the IP addresses from Code Red worm attacks were different from the standard IPs accessing the website. In [18], the authors add an IP addresses to the IAD if a certain threshold (e.g. a certain number of packets) is exceeded. Later during an attack, the decision whether an IP has to be blocked or not is binary. This means, the density estimation results in a simple step function with only two values: Zero or a positive value, which is the same for all IP addresses. The PDF can be calculated as follows:

$$f(s) = \frac{\min(n_s, 1)}{\sum_{i=0}^{N-1} \min(n_{s_i}, 1)} \tag{1}$$

where n_s is the number of occurrence of an IP address s in the training set. The advantage of HIF is the low computational load required for its implementation. Nevertheless, it cannot be differentiated between users which revisit the server more often than others and is therefore a less precise density estimator.

2.2   Adaptive History-Based IP Filtering

To compensate the shortcomings of HIF, [12] presented Adaptive History-based IP Filtering (AHIF). AHIF instead of using the binary decision, uses histograms, i.e.,

$$f(s) = \frac{n_s}{\sum_{i=0}^{N-1} n_{s_i}} \tag{2}$$

where s represents a range of IP addresses (a bin of the histogram). So far, constant width networks with fixed network masks ranging from 16 up to 24 bit are used as source bins. The actual prediction C_α of the appearance of an IP address (or an IP range) is done using a threshold over $f(s)$, i.e.,

$$C_\alpha(s) := \begin{cases} \text{reject,} & \text{if } f(s) \geq \alpha \\ \text{accept,} & \text{otherwise} \end{cases} \tag{3}$$

During attack mitigation, the most appropriate network mask is chosen such that a maximum numbers of firewall rules is not exceeded and the attack traffic is reduced to be below the maximum server capacity. It is shown, that the adaptive method performs better in terms of predicting user IP addresses during an attack. However, neighbor relations (between source networks) are not taken into account.

2.3   Clustering of Source Address Prefixes

In [15], the authors introduce algorithms for mitigating DDoS attacks by filtering source address prefixes. Unlike AHIF, network masks may be at different sizes. For finding the appropriate networks, the authors are using a hierarchical agglomerative clustering algorithm with single linkage. The used distance measure is defined with respect to network boundaries. In general, the proposed method takes both into account - the amount of requests from a source network as well as neighboring density estimation as a result of using a (generalizing) clustering method. Unfortunately, hierarchical clustering methods consume a lot of memory and it was found [17], that the presented method is not applicable in practice on large source IP datasets.

3   Non-linear Analytic Prediction of IP Addresses

Data driven identification of mathematical models of physical systems (i.e. non-linear) starts with representing the systems as a black box. In other terms, while we may have access to the inputs and outputs, the internal mechanisms are totally unknown to us. Once a model type is chosen to represent the system, its parameters are estimated through an optimization algorithm so that eventually the model mimics at a certain level of fidelity the inner mechanism of the nonlinear system or process using its inputs and outputs. These approach is, for instance, widely used in the related *big data analytics* area (e.g., [3,5,7–9]).

In this work, we consider a particular sub-class of nonlinear predictors: the Linear-in-the-parameters (LIP) predictors. LIP predictors are characterized by a linear dependence of the predictor output on the predictor coefficients. Such predictors are inherently stable, and that they can converge to a globally minimum solution (in contrast to other types of nonlinear filters whose cost function may exhibit many local minima) avoiding the undesired possibility of getting stuck in a local minimum. Let us consider a causal, time-invariant, finite-memory, continuous nonlinear predictor as described in (4).

$$\hat{s}(n) = f[s(n-1), \ldots, s(n-N)] \tag{4}$$

where $f[\cdot]$ is a continuous function, $s(n)$ is the input signal and $\hat{s}(n)$ is the predicted sample. We can expand $f[\cdot]$ with a series of basis functions $f_i(n)$, as shown in (5).

$$\hat{s}(n) = \sum_{i=1}^{\infty} h(i) f_i[s(n-i)] \tag{5}$$

where $h(i)$ a re proper coefficients. To make (5) realizable we truncate the series to the first N terms, thus we obtain

$$\hat{s}(n) = \sum_{i=1}^{N} h(i) f_i[s(n-i)] \tag{6}$$

In the general case, a linear-in-the-parameters nonlinear predictor is described by the input-output relationship reported in (7).

$$\hat{s}(n) = \boldsymbol{H}^T \boldsymbol{X}(n) \tag{7}$$

where $\boldsymbol{H}^T$ is a row vector containing predictor coefficients and $\boldsymbol{X}(n)$ is the corresponding column vector whose elements are nonlinear combinations and/or expansions of the input samples.

3.1   Linear Predictor

Linear prediction is a well known technique with a long history [14]. Given a time series $\boldsymbol{X}$, linear prediction is the optimum approximation of sample $x(n)$ with a linear combination of the N most recent samples. That means that the linear predictor is described as Eq. (8).

$$\hat{s}(n) = \sum_{i=1}^{N} h_1(i)s(n-i) \tag{8}$$

or in matrix form as

$$\hat{s}(n) = \boldsymbol{H}^T \boldsymbol{X}(n) \tag{9}$$

where the coefficient and input vectors are reported in (10) and (11).

$$\boldsymbol{H}^T = \begin{bmatrix} h_1(1)\ h_1(2)\ \dots\ h_1(N) \end{bmatrix} \tag{10}$$

$$\boldsymbol{X}^T = \begin{bmatrix} s(n-1)\ s(n-2)\ \dots\ s(n-N) \end{bmatrix} \tag{11}$$

3.2   Non-linear Predictor Based on Volterra Series

As well as Linear Prediction, Non Linear Prediction is the optimum approximation of sample $x(n)$ with a non linear combination of the N most recent samples. Popular nonlinear predictors are based on Volterra series [16]. A Volterra predictor based on a Volterra series truncated to the second term is reported in (12).

$$\hat{x}(n) = \sum_{i=1}^{N_1} h_1(i)x(n-i) + \sum_{i=1}^{N_2}\sum_{j=i}^{N_2} h_2(i,j)x(n-i)x(n-j) \tag{12}$$

where the symmetry of the Volterra kernel (the h coefficients) is considered. In matrix terms, the Volterra predictor is represented in (13).

$$\hat{s}(n) = \boldsymbol{H}^T \boldsymbol{X}(n) \tag{13}$$

where the coefficient and input vectors are reported in (15) and (15).

$$\boldsymbol{H}^T = \begin{bmatrix} h_1(1) & h_1(2)\dots h_1(N) \\ h_2(1,1) & h_2(1,2)\dots h_2(N_2,N_2) \end{bmatrix} \tag{14}$$

$$\boldsymbol{X}^T = \begin{bmatrix} s(n-1) & s(n-2)\dots s(n-N_1) \\ s^2(n-1) & s(n-1)s(n-2)\dots s^2(n-N_2) \end{bmatrix} \tag{15}$$

3.3 Non-linear Predictor Based on Functional Link Artificial Neural Networks (FLANN)

FLANN is a single layer neural network without hidden layer. The nonlinear relationships between input and output are captured through function expansion of the input signal exploiting suitable orthogonal polynomials. Many authors used for examples trigonometric, Legendre and Chebyshev polynomials. However, the most frequently used basis function used in FLANN for function expansion are trigonometric polynomials [20]. The FLANN predictor can be represented by Eq. (16).

$$\hat{s}(n) = \sum_{i=1}^{N} h_1(i)s(n-i) + \sum_{i=1}^{N}\sum_{j=1}^{N} h_2(i,j)\cos[i\pi s(n-j)]$$
$$+ \sum_{i=1}^{N}\sum_{j=1}^{N} h_2(i,j)\sin[i\pi s(n-j)] \tag{16}$$

Also in this case the Flann predictor can be represented using the matrix form reported in (13).

$$\hat{s}(n) = \boldsymbol{H}^T \boldsymbol{X}(n) \tag{17}$$

where the coefficient and input vectors of FLANN predictors are reported in (25) and (26).

$$\boldsymbol{H}^T = \begin{bmatrix} h_1(1) & h_1(2) \dots h_1(N) \\ h_2(1,1)\ h_2(1,2) \dots h_2(N_2,N_2) \\ h_3(1,1)\ h_3(1,2) \dots h_3(N_2,N_2) \end{bmatrix} \tag{18}$$

$$\boldsymbol{X}^T = \begin{bmatrix} s(n-1) & s(n-2) & \dots s(n-N) \\ \cos[\pi s(n-1)]\ \cos[\pi s(n-2)]\ \dots & \dots & \cos[N_2\pi s(n-N_2)] \\ \sin[\pi s(n-1)]\ \sin[\pi s(n-2)]\ \dots & \dots & \sin[N_2\pi x(s-N_2)] \end{bmatrix} \tag{19}$$

3.4 Non-linear Predictors Based on Hammerstein Models

Previous research [4] shown that many real nonlinear systems, spanning from electromechanical systems to audio systems, can be modeled using a static non-linearity. These terms capture the system nonlinearities, in series with a linear function, which capture the system dynamics as shown in Fig. 1.

Indeed, the front-end of the so called Hammerstein Model is formed by a nonlinear function whose input is the system input. Of course the type of non-linearity depends on the actual physical system to be modeled. The output of the nonlinear function is hidden and is fed as input of the linear function. In the following, we assume that the non-linearity is a finite polynomial expansion, and the linear dynamic is realized with a Finite Impulse Response (FIR) filter.

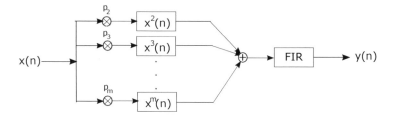

Fig. 1. Representation of the Hammerstein Models

Furthermore, in contrast with [4], we assume a mean error analysis and we postpone the analysis in the robust framework in future work. In other word,

$$z(n) = p(2)x^2(n) + p(3)x^3(n) + \ldots p(m)x^m(n) = \sum_{i=2}^{M} p(i)x^i(n) \qquad (20)$$

On the other hand, the output of the FIR filter is:

$$y(n) = h_0(1)z(n-1) + h_0(2)z(n-2) + \ldots + h_0(N)z(n-N)$$

$$= \sum_{j=1}^{N} h_0(j)z(n-j) \qquad (21)$$

Substituting (20) in (23) we have:

$$y(n) = \sum_{i=1}^{N} h_0(i)z(n-i) = \sum_{j=1}^{N} h_0(j)\sum_{i=2}^{M} p(i)x^i(n-j)$$

$$= \sum_{i=2}^{M}\sum_{j=1}^{N} h_0(j)p(i)x^i(n-j) \qquad (22)$$

Setting $c(i,j) = h_0(j)p(i)$ we write

$$y(n) = \sum_{i=2}^{M}\sum_{j=1}^{N} c(i,j)x^i(n-j) \qquad (23)$$

This equation can be written in matrix form as

$$\hat{s}(n) = \boldsymbol{H}^T \boldsymbol{X}(n) \qquad (24)$$

where

$$\boldsymbol{H}^T = \begin{bmatrix} c(2,1) & c(2,2)\ldots c(2,N_2) \\ c(3,1) & c(3,2)\ldots c(3,N_2) \\ \ldots c(M,1) & C(M,2)\ldots C(M,N) \end{bmatrix} \qquad (25)$$

$$\boldsymbol{X}^T = \begin{bmatrix} s^2(n-2) & s^2(n-3)\ldots s^2(n-N) \\ s^3(n-2) & s^3(n-3)\ldots s^3(n-N) \\ s^M(n-2) & s^M(n-3)\ldots s^M(n-1) & s^M(n-3)\ldots s^M(n-N) \end{bmatrix} \qquad (26)$$

4 Predictor Parameters Estimation

So far we saw that all the predictors can be expressed, at time instant n, as

$$\hat{s}(n) = \boldsymbol{H}^T \boldsymbol{X}(n) \tag{27}$$

with different definitions of the input, $\boldsymbol{X}(n)$, end parameters vectors $\boldsymbol{H}^T$. There are two well known possibilities for estimating the optimal parameter vector.

4.1 Block-Based Approach

The Minimum Mean Square estimation is based on the minimization of the mathematical expectation of the squared prediction error $e(n) = s(n) - \hat{s}(n)$

$$E[e^2] = E[(s(n) - \hat{s}(n))^2] = E[(s(n) - \boldsymbol{H}^T \boldsymbol{X}(n))^2] \tag{28}$$

The minimization of (28) is obtain by setting to zero the Laplacian of the mathematical expectation of the squared prediction error:

$$\nabla_H E[e^2] = E[\nabla_H e^2] = E[2e(n)\nabla_H e] = 0 \tag{29}$$

which leads to the well known unique solution

$$\boldsymbol{H}_{opt} = \boldsymbol{R}_{xx}^{-1} \boldsymbol{R}_{sx} \tag{30}$$

where

$$\boldsymbol{R}_{xx}(n) = E[\boldsymbol{X}(n)\boldsymbol{X}^T(n)] \tag{31}$$

is the statistical auto-correlation matrix of the input vector $\boldsymbol{X}(n)$ and

$$\boldsymbol{R}_{sx}(n) = E[s(n)\boldsymbol{X}(n)] \tag{32}$$

is the statistical cross-correlation vector between the signal $s(n)$ and the input vector $\boldsymbol{X}(n)$. The mathematical expectations of the auto and cross correlation are estimated using

$$\boldsymbol{R}_{xx}(n) = \frac{\sum_{k=1}^{n} \boldsymbol{X}(n)\boldsymbol{X}^T(n)}{n} \tag{33}$$

is the statistical auto-correlation matrix of the input vector $\boldsymbol{X}(n)$ and

$$\boldsymbol{R}_{sx}(n) = \frac{\sum_{k=1}^{n} s(k)(n)\boldsymbol{X}(n)}{n} \tag{34}$$

4.2 Adaptive Approach

Let us consider a general second order terms of a Volterra predictor

$$y(n) = \sum_{k=0}^{N-1} \sum_{r=0}^{N-1} h_2(k,r)x(n-k)x(n-r) \tag{35}$$

It can be generalized for higher order term as

$$\sum_{k_1=1}^{N} \cdots \sum_{k_p=1}^{N} c_{k_1} \cdots c_{k_p} H_p \left\{ x_{k_1}(n), \cdots x_{k_p}(n) \right\} \tag{36}$$

where

$$\sum_{k=1}^{N} c_k x_k(n). \tag{37}$$

For the sake of simplicity and without loss of generality, we consider a Volterra predictor based on a Volterra series truncated to the second term

$$\hat{r}(n) = \sum_{i=1}^{N_1} h_1(i) r(n-i) + \sum_{i=1}^{N_2} \sum_{j=i}^{N_2} h_2(i,j) r(n-i) r(n-j) \tag{38}$$

By defining

$$H^T(n) = \left| h_1(1), \cdots, h_1(N_1), h_2(1,1), \cdots, h_2(N_2, N_2) \right| \tag{39}$$

and

$$X^T(n) = \left| r(n-1), \cdots, r(n-N_1), r^2(n-1) \right. \\ \left. r(n-1)r(n-2), \cdots, r^2(n-N_2) \right| \tag{40}$$

Equation (38) can be rewritten as follows

$$\hat{r}(n) = H^T(n)X(n). \tag{41}$$

In order to estimate the best parameters H, we consider the following loss function

$$J_n(H) = \sum_{k=0}^{n} \lambda^{n-k} \left[\hat{r}(k) - H^T(n)X(k) \right]^2 \tag{42}$$

where λ^{n-k} weights the relative importance of each squared error. In order to find the H that minimizes the convex function (42) it is enough to impose its gradient to zero, i.e.,

$$\nabla_H J_n(H) = 0 \tag{43}$$

That is equivalent to

$$R_{XX}(n)H(n) = R_{rX}(n) \tag{44}$$

where

$$R_{XX}(n) = \sum_{k=0}^{n} \lambda^{n-k} X(k) X^T(k) \\ R_{rX}(n) = \sum_{k=0}^{n} \lambda^{n-k} r(k) X(k) \tag{45}$$

It follows that the best H can be computed by

$$H(n) = R_{XX}^{-1}(n) R_{rX}(n) \tag{46}$$

Since

$$R_{XX}(n) = \lambda R_{XX}(n-1) + X(n) X^T(n) \tag{47}$$

it follows that

$$R_{XX}^{-1}(n) = \frac{1}{\lambda} \left[R_{XX}^{-1}(n-1) - k(n)X^T(n)R_{XX}^{-1}(n-1) \right] \tag{48}$$

where $k(n)$ is equal to

$$k(n) = \frac{R_{XX}^{-1}(n-1)X(n)}{\lambda + X^T(n)R_{XX}^{-1}(n-1)X(n)} \tag{49}$$

Instead, matrix $R_{rX}(n)$ in (46) can be written as

$$R_{rX}(n) = \lambda R_{rX}(n-1) + r(n)X(n) \tag{50}$$

Thus, inserting Eq. (50) and Eq. (48) in Eq. (46) and rearranging the terms, we obtain

$$H(n) = H(n-1) + R_{XX}^{-1}(n)X(n)\epsilon(n) \tag{51}$$

where

$$\epsilon = \hat{r}(n) - H^T(n-1)X(n) \tag{52}$$

By recalling Eq. (49), we can write Eq. (51) as

$$H(n) = H(n-1) + \epsilon(n)k(n) \tag{53}$$

By introducing, the new notation,

$$F(n) = S^T(n-1)X(n) \tag{54}$$

The previous equations can be resumed by the following system

$$\begin{cases} L(n) = S(n-1)F(n) \\ \beta(n) = \lambda + F^T(n)F(n) \\ \alpha(n) = \frac{1}{\beta(n)+\sqrt{\lambda\beta(n)}} \\ S(n) = \frac{1}{\sqrt{\lambda}} \left[S(n-1) - \alpha(n)L(n)F^T(n) \right] \\ \epsilon(n) = \hat{r}(n-1) - \alpha(n)L(n)F^T(n) \\ \epsilon(n) = H(n-1) + L(n)\frac{\epsilon(n)}{\beta(n)} \end{cases} \tag{55}$$

It should be noted that by using Eq. (55) the estimation adapts in each step in order to decrease the error. Thus, the system structure is somehow similar to the Kalman filter.

Finally, we define the estimation error as

$$e(n) = r(n) - H^T(n)X(n) \tag{56}$$

It is worth noting that the computation of the predicted value from Eq. (41) requires $6N_{tot} + 2N_{tot}^2$ operations, where $N_{tot} = N_1 + N_2(N_2 + 1)/2$.

5   Experiments

In order to prove the effectiveness of the proposed approach, in this section we present our experimental results for a real dataset. Specifically, we consider the requests made to the 1998 World Cup Web site between April 30, 1998 and July 26, 1998[1]. During this period of time the site received 1,352,804,107 requests. The fields of the request structure contain the following information:

– *timestamp* the time of the request, stored as the number of seconds since the Epoch. The timestamp has been converted to GMT to allow for portability. During the World Cup the local time was 2 h ahead of GMT (+0200). In order to determine the local time, each timestamp must be adjusted by this amount.
– *clientID* a unique integer identifier for the client that issued the request; due to privacy concerns these mappings cannot be released; note that each *clientID* maps to exactly one IP address, and the mappings are preserved across the entire data set - that is if IP address 0.0.0.0 mapped to *clientID* X on day Y then any request in any of the data sets containing *clientID* X also came from IP address 0.0.0.0
– *objectID* a unique integer identifier for the requested URL; these mappings are also 1-to-1 and are preserved across the entire data set
– *size* the number of bytes in the response
– *method* the method contained in the client's request (e.g., GET).
– *status* this field contains two pieces of information; the 2 highest order bits contain the HTTP version indicated in the client's request (e.g., HTTP/1.0); the remaining 6 bits indicate the response status code (e.g., 200 OK).
– *type* the type of file requested, generally based on the file extension (.html), or the presence of a parameter list.
– *server* indicates which server handled the request. The upper 3 bits indicate which region the server was at; the remaining bits indicate which server at the site handled the request.

In the dataset, 87 days are reported. We use the first one in order to initialise the estimator in (38) and we use the others as test set by using a rolling horizon method (as in [4]). Particularly, for each day t we compute the estimation by using all the IP observations in the previous days $[0, t-1]$. The results are reported in Fig. 2. The increase of errors in June is due to the sudden increment of different IP accessing the website due to the start of the competition (see [2]). It should be noted that the estimation error decrease exponentially despite dealing with several millions of IPs.

Since the computation of the optimal coefficient $H(n)$ may require some time, we measure the percentage of available data that our approach needs in order to provide good results. Particularly, in this experiment we consider the average estimation error done by the model at time t by considering a subset of the IPs observed in interval $[0, t-1]$. The experimental results on real data cubes are depicted in Fig. 3.

---

[1] ftp://ita.ee.lbl.gov/html/contrib/WorldCup.html.

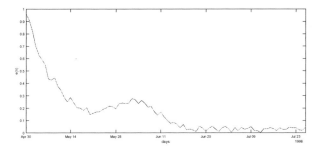

Fig. 2. Estimation error for each day of activity of the website.

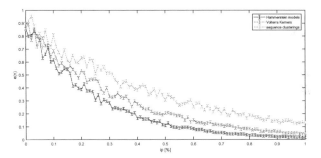

Fig. 3. Estimation error vs. percentage size of the training set

It should be noted that the Hammerstain model outperform the results by the Volterra's kernel as well as the clustering techniques. In more detail, the clustering techniques are the one less performing. This is due to the nature of the clustering techniques that exploit the geometric information of the data more than their time dependency. We highlight that despite the calculation of $H(n)$ is computational intensive, this does not effect the real time applicability of the method. In fact, the access decision is taken by considering the estimator $\hat{r}(n)$ that is computed once per day. Thus the computation of $H(n)$ does not need to be fast.

6   Conclusions and Future Work

In this paper, we presented a new way to deal with cyber attack by using Hammerstein models. Experimental results clearly confirm the effectiveness of the proposed techniques for a real data set, outperforming other well-known techniques. Future work will have two objectives. First, we want to consider the problem in a stochastic optimization settings, as for example in [11]. Second, we want to test the approach on other case studies, by also exploiting *knowledge management methodologies* (e.g., [6]).

Acknowledgements. This research has been partially supported by the French PIA project "Lorraine Université d'Excellence", reference ANR-15-IDEX-04-LUE.

References

1. Agrawal, A., Casanova, H.: Clustering hosts in P2P and global computing platforms, pp. 367–373, June 2003. https://doi.org/10.1109/CCGRID.2003.1199389
2. Arlitt, M., Jin, T., Hewlett-Packard Laboratories: A workload characterization study of the 7998 world cup web site
3. Bonifati, A., Cuzzocrea, A.: Storing and retrieving XPath fragments in structured P2P networks. Data Knowl. Eng. **59**(2), 247–269 (2006)
4. Cerone, V., Fadda, E., Regruto, D.: A robust optimization approach to kernel-based nonparametric error-in-variables identification in the presence of bounded noise. In: 2017 American Control Conference (ACC). IEEE, May 2017. https://doi.org/10.23919/acc.2017.7963056
5. Chatzimilioudis, G., Cuzzocrea, A., Gunopulos, D., Mamoulis, N.: A novel distributed framework for optimizing query routing trees in wireless sensor networks via optimal operator placement. J. Comput. Syst. Sci. **79**(3), 349–368 (2013)
6. Cuzzocrea, A.: Combining multidimensional user models and knowledge representation and management techniques for making web services knowledge-aware. Web Intell. Agent Syst. **4**(3), 289–312 (2006)
7. Cuzzocrea, A., Bertino, E.: Privacy preserving OLAP over distributed XML data: a theoretically-sound secure-multiparty-computation approach. J. Comput. Syst. Sci. **77**(6), 965–987 (2011)
8. Cuzzocrea, A., Moussa, R., Xu, G.: OLAP*: effectively and efficiently supporting parallel OLAP over big data. In: Cuzzocrea, A., Maabout, S. (eds.) MEDI 2013. LNCS, vol. 8216, pp. 38–49. Springer, Heidelberg (2013). https://doi.org/10.1007/978-3-642-41366-7_4
9. Cuzzocrea, A., Russo, V.: Privacy preserving OLAP and OLAP security. In: Encyclopedia of Data Warehousing and Mining, Second Edition (4 Volumes), pp. 1575–1581 (2009)
10. Dietrich, S., Long, N., Dittrich, D.: Analyzing distributed denial of service tools: the shaft case, pp. 329–339, December 2000
11. Fadda, E., Perboli, G., Tadei, R.: Customized multi-period stochastic assignment problem for social engagement and opportunistic IoT. Comput. Oper. Res. **93**, 41–50 (2018)
12. Goldstein, M., Lampert, C., Reif, M., Stahl, A., Breuel, T.: Bayes optimal DDoS mitigation by adaptive history-based IP filtering. In: Seventh International Conference on Networking (ICN 2008), pp. 174–179 (2008)
13. Jung, J., Krishnamurthy, B., Rabinovich, M.: Flash crowds and denial of service attacks: characterization and implications for CDNs and web sites, pp. 293–304, January 2002. https://doi.org/10.1145/511446.511485
14. Makhoul, J.: Linear prediction: a tutorial review. Proc. IEEE **63**(4), 561–580 (1975)
15. Pack, G., Yoon, J., Collins, E., Estan, C.: On filtering of DDoS attacks based on source address prefixes, pp. 1–12, August 2006. https://doi.org/10.1109/SECCOMW.2006.359537
16. Peng, Z., Changming, C.: Volterra series theory: a state-of-the-art review. Chin. Sci. Bull. (Chin. Version) **60**, 1874 (2015). https://doi.org/10.1360/N972014-01056
17. Tan, H.X., Seah, W.: Framework for statistical filtering against DDoS attacks in MANETs, p. 8, January 2006. https://doi.org/10.1109/ICESS.2005.57

18. Peng, T., Leckie, C., Ramamohanarao, K.: Protection from distributed denial of service attacks using history-based IP filtering. In: IEEE International Conference on Communications 2003, ICC 2003, vol. 1, pp. 482–486 (2003)
19. Yang, Y., Lung, C.H.: The role of traffic forecasting in QoS routing - a case study of time-dependent routing, vol. 1, pp. 224–228, June 2005. https://doi.org/10.1109/ICC.2005.1494351
20. Zhao, H., Zhang, J.: Adaptively combined FIR and functional link artificial neural network equalizer for nonlinear communication channel. IEEE Trans. Neural Netw. **20**(4), 665–674 (2009)

A Big-Data Variational Bayesian Framework for Supporting the Prediction of Functional Outcomes in Wake-Up Stroke Patients

Miloš Ajčević[1], Aleksandar Miladinović[1], Giulia Silveri[1],
Giovanni Furlanis[2], Tommaso Cilotto[2], Alex Buoite Stella[2],
Paola Caruso[2], Maja Ukmar[3], Marcello Naccarato[2],
Alfredo Cuzzocrea[4(✉)], Paolo Manganotti[2], and Agostino Accardo[1]

[1] Department of Engineering and Architecture, University of Trieste,
Trieste, Italy
[2] Clinical Unit of Neurology, Department of Medicine, Surgery and Health
Sciences, University Hospital of Trieste, University of Trieste, Trieste, Italy
[3] Radiology Unit, Department of Medicine, Surgery and Health Sciences,
University Hospital of Trieste, University of Trieste, Trieste, Italy
[4] iDEA Lab, University of Calabria, Rende, Italy
alfredo.cuzzocrea@unical.it

Abstract. Prognosis in Wake-up ischemic stroke (WUS) is important for guiding treatment and rehabilitation strategies, in order to improve recovery and minimize disability. For this reason, there is growing interest on models to predict functional recovery after acute ischemic events in order to personalize the therapeutic intervention and improve the final functional outcome. The aim of this preliminary study is to evaluate the possibility to predict a *good functional outcome*, in terms of modified Rankin Scale (mRS $\leq$ 2), in thrombolysis treated WUS patients by Bayesian analysis of clinical, demographic and neuroimaging data at admission. The study was conducted on 54 thrombolysis treated WUS patients. The Variational Bayesian logistic regression with Automatic Relevance Determination (VB-ARD) was used to produce model and select informative features to predict a *good functional outcome* (mRS $\leq$ 2) at discharge. The produced model showed moderately high 10×5-fold cross validation accuracy of 71% to predict outcome. The sparse model highlighted the relevance of NIHSS at admission, age, TACI stroke syndrome, ASPECTs, ischemic core CT Perfusion volume, hypertension and diabetes mellitus. In conclusion, in this preliminary study we assess the possibility to model the prognosis in thrombolysis treated WUS patients by using VB ARD. The identified features related to initial neurological deficit, history of diabetes and hypertension, together with necrotic tissue relate ASPECT and CTP core volume neuroimaging features, were able to predict outcome with moderately high accuracy.

Keywords: Variational Bayesian inference · Automatic relevance determination · Modeling · Wake-up ischemic stroke · Neuroimaging

© Springer Nature Switzerland AG 2020
O. Gervasi et al. (Eds.): ICCSA 2020, LNCS 12249, pp. 992–1002, 2020.
https://doi.org/10.1007/978-3-030-58799-4_71

1   Introduction

Worldwide, cerebrovascular ischemia is one of leading causes of disability and death among elderly population [1]. Wake-up Stroke (WUS) represents around a quarter of acute ischemic stroke events [2, 3]. The etiology of ischemic strokes is due to either a thrombotic or embolic event that causes a cerebral brain vessel occlusion and consequent decrease in blood flow to the brain. Nowadays, ischemic stroke is highly treatable using thrombectomy and intravenous thrombolysis reperfusion therapies in selected patients [2]. Neuroimaging plays an important role in patient selection for reperfusion therapy [4]. Recent trials and studies, reported that thrombolysis is safe and efficacious in WUS in patients selected by perfusion neuroimaging [3, 5, 6]. CT perfusion (CTP) is very helpful in determining the necrotic core areas, as well as the extent of hypoperfused tissue that can recover [7, 8], allowing eligibility the reperfusion treatment in wake-up stroke cases [5, 6, 9].

Early prognosis in acute ischemic stroke is important for guiding treatment and rehabilitation strategies, in order to improve recovery and minimize disability [10]. The modified Rankin Scale (mRS) is a commonly used scale to measure the degree of disability or dependence in daily activities of people with neurological and non-neurological disability owing to stroke or other causes [11]. It has become the most widely used clinical outcome measurement for stroke clinical trials [11–13]. mRS ranges from 0 to 6, from perfect health without symptoms to death. The mRS is a 6-point disability scale with possible scores ranging from 0 to 5, from perfect health without symptoms to high disability, with a separate category of 6 for patients who died. Good outcome following stroke is commonly defined as scores 0–2.

Various clinical, demographic and neuroimaging prognostic markers, such as age, clinical severity at admission or infarct size, were associated with functional outcome [10]. Nevertheless, prediction of post-stroke outcome is still challenging since there is large inter-patient variability.

Different machine learning methods can be applied to investigate the possibility to produce a model for predicting functional outcome. However, a high number of predictive features can lead to overfitting. To overcome this problem, a technique like variational Bayes automatic relevance determination (VB-ARD) that eliminates irrelevant features is preferable.

The aim of this preliminary study is to evaluate the possibility to predict a *good functional outcome* (mRS $\leq$ 2) in thrombolysis treated WUS patients by Bayesian analysis of clinical, demographic and neuroimaging data at admission.

2   Materials and Methods

2.1   Study Protocol

The study was conducted on 54 WUS patients (25M/29F; age 72 $\pm$ 9 years) admitted to the Stroke Unit of the University Medical Hospital of Trieste, Italy. We included subjects with acute ischemic stroke developed at morning awakening, admitted within 4.5 h, assessed with CTP within 4.5 h and subsequently underwent thrombolysis treatment if eligible. No age limit was applied and both genders were included in the study sample. We excluded patients with previous brain lesions and hemorrhagic strokes. Stroke mimics cases were excluded by a complete diagnostic work-up including clinical and CT or MRI follow-up assessment.

All patients received standardized clinical and diagnostic assessment, during admission and at discharge. Non-enhanced CT (NECT), Angio-CT, CT Perfusion neuroimaging work-up was performed at admission in all subjects. Patients eligible for thrombolysis were treated with standard dose of intravenous rtPA (0.9 mg/kg of body weight, maximum of 90 mg, infused over 60 min with 10% of the total dose administered as an initial intravenous bolus over 1 min).

The study was approved by the Local Ethics Committee and conducted in line with the principles of the Declaration of Helsinki. All participants released their informed consent to participate in the study.

2.2 Dataset

2.2.1 Demographic and Clinical Data

The following demographic and clinical data at admission were collected for each included patients: (1) age (y); (2) sex (M/F); (3) Stroke severity measured by National Institutes of Health Stroke Scale (NIHSS) score at admission [14] (4) premorbid and discharge mRS [11]; (5) Lesion side (Left/Right); (6) stroke risk factors (hypertension, diabetes mellitus, dyslipidemia, smoking, obesity, ischemic cardiopathy, atrial fibrillation); (7) Stroke syndrome by Bamford classification [15] (Total Anterior Circulation Infarct, TACI; Partial Anterior Circulation Infarct, PACI; Lacunar Stroke, LACI; Posterior Circulation Infarct, POCI); (8) Stroke etiology by TOAST classification [16] (Atherothrombotic, Lacunar, Cardioembolic, Cryptogenic, other cause); (9) Time from last seen well to admission; (10) Time from admission to thrombolysis treatment.

2.2.2 Neuroimaging Data and CTP Processing

All patients underwent a standardized CT protocol at admission consisting of NECT, CTA, and CTP. NECT performed with a 256 slice CT scanner (Brilliance iCT 256 slices, Philips Medical Systems, Best, Netherlands) at the Radiology Department of the University Medical Hospital of Trieste (Italy). The Alberta Stroke Program Early CT Score (ASPECTS) was used to quantify the amount of ischemia on NECT [17]. CTP acquisition involves intravenous injection of contrast medium and acquisition of three-dimensional axial acquisitions on a whole brain volume every 4 s, resulting in a total scanning time of 60 s. CTP source image processing was performed by using Extended Brilliance Workstation v 4.5 (Philips Medical Systems, Best, Netherlands) and in-house code developed in Matlab (MathWorks Inc., Natick, MA), as previously described [18, 19]. The perfusion maps mean transit time (MTT), cerebral blood volume (CBV) and cerebral blood flow (CBF) were calculated. Ischemic core and penumbra areas were identified by application of specific thresholds [20]. Ischemic core volume as well as total hypoperfused volume including core and penumbra regions, excluding artifacts was calculated with an algorithm described in a previous study [18]. Core/penumbra mismatch was calculated as a ratio between penumbra volume and total hypoperfused volume. CTP processing is summarized in Fig. 1.

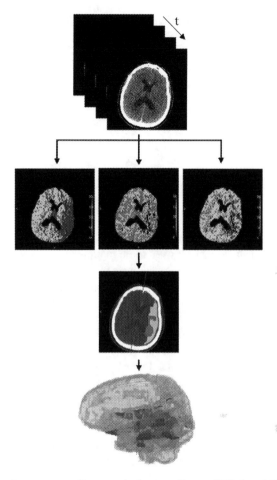

Fig. 1. CTP processing summary. From top to bottom: Source CTP data; MTT, CBV and CBF calculated maps, from left to right, respectively; core (red)/penumbra (green) summary map; 3D representation of total hypoperfused volume (core + penumbra). (Color figure online)

2.2.3 Outcome Measure

The mRS, an ordinal scale with 7 categories ranging from zero (no symptoms) to 6 (death), was the outcome measure: 0) No symptoms at all; 1) No significant disability despite symptoms; able to carry out all usual duties and activities; 2) Slight disability; unable to carry out all previous activities, but able to look after own affairs without assistance; 3) Moderate disability; requiring some help, but able to walk without assistance; 4) Moderately severe disability; unable to walk and attend to bodily needs without assistance; 5) Severe disability; bedridden, incontinent and requiring constant nursing care and attention; 6) Dead. Good outcome class was defined with mRS $\leq$ 2, while bad outcome class with mRS > 3.

2.3   Data Pre-processing and Classification

In this study, we used a dataset consisting of all aforementioned demographic, clinical and neuroimaging features. The dataset was preprocessed before the analysis. Considering that the dataset variables were measured in different units, each variable was normalized by subtracting the mean from each variable and then divided by its standard deviation [21].

In this work, we used Variational Bayesian logistic regression with Automatic Relevance Determination (VB-ARD) [22] to select informative features and produce a sparse model to predict a *good functional outcome* (mRS $\leq$ 2) in thrombolysis treated WUS patients.

Given some data $D = \{X, Y\}$ where $X = \{x_1, ..., x_N\}$ and $Y = \{y_1, ..., y_N\}$, $N = 54$ subjects are input/output pairs. The length of each x_n input vector is 27, corresponding to the number of features, and y_n is the output class ($y_n = 1$ for good outcome mRS 2 and $y_n = -1$ for bad outcome mRS > 3).

This approach has an advantage over some regularization methods which require a separate validation set to prune less relevant features. Furthermore, this method produces a posterior distribution allowing us to build the varying-intercept sparse feature model. The methodology applied here is similar to the one proposed by Bishop (2006) [23] with a different implementation of Automatic Relevance Determination (ARD). Instead of using type-II maximum likelihood [24–26], where the parameters are tuned by maximizing the marginal likelihood, here we applied full Bayesian treatment [22]. The basic generative model corresponds to the one used in Bishop (2006) [23], and the prior is selected to be non-informative, modelled by a conjugate Gamma distribution [22]. The posterior probability distribution of the model parameters conditionally upon the predictors has been obtained by Variational Bayesian inference that is based on maximizing a lower bound on the marginal data log-likelihood [22, 27]. The obtained distribution allows us to find the inverse of the predictors' covariance matrix (precision matrix) and apply Automatic Relevance Determination that consists of assigning an individual hyper-prior to each regression coefficient independently determining how relevant each of them is.

The ability of a produced model to predict accurately the outcome was validated with 10 times 5–fold cross-validation. Finally, the average accuracy over 50 runs was calculated.

3   Experimental Results

Summary of patient's demographic and clinical data at admission, as well as neuroimaging findings are reported in Table 1.

Table 1. Demographic, clinical characteristics and neuroimaging findings at admission. Data are presented as Means ± SD, Medians (IQR) and frequencies.

Patient's characteristics	N = 54
Age [y]	72 ± 9
Sex F: M [n]	29: 25
Last time seen well - Admission [min]	509 (364–702)
Admission - Thrombolysis [min]	68 (52–140)
ASPECTS	10 (10–10)
NIHSS at baseline	7 (5–14)
Pre-morbid mRS	0 (0–0)
Lesion side L: R [n]	28: 26
Bamford stroke subtypes [n (%)]	
TACI	13 (24%)
PACI	27 (50%)
LACI	5 (9%)
POCI	9 (17%)
TOAST classification [n (%)]	
Atherothrombotic	9 (17%)
Lacunar	5 (9%)
Cardioembolic	22 (41%)
Cryptogenic	18 (33%)
Other cause	0 (0%)
CTP parameters	
Mismatch	1.0 (0.9–1.0)
Total hypoperfused tissue [ml]	16.2 (2.5–79.6)
Penumbra [ml]	15.1 (2.6–71.2)
Core [ml]	0 (0–3.8)
Stroke risk factors [n (%)]	
Hypertension	35 (65%)
Diabetes mellitus type II	9 (17%)
Dyslipidemia	39 (72%)
Smoking	11 (20%)
Obesity	6 (11%)
Atrial fibrillation	21 (39%)
Ischemic cardiopathy	14 (26%)

VB-ARD produced a sparse model and as a result the following features were selected: (1) NIHSS at admission, (2) Age, (3) TACI stroke syndrome, (4) ASPECTs, (5) Core CTP volume as well as presence of risk factors as (6) arterial hypertension and (7) diabetes mellitus type II. Posterior means and their standard errors for selected model features are reported in Table 2. The overall accuracy of the identified model was 71%.

Table 2. Posterior means and their standard errors for model features.

	Posterior mean	Posterior standard error
NIHSS at admission	−1.126	0.009
Age	−0.595	0.005
TACI	−0.426	0.007
ASPECTS	0.304	0.006
Hypertension	−0.234	0.005
Diabetes mellitus type II	−0.212	0.004
Core Volume	−0.101	0.007

4 Discussion

The last decade has seen a substantial increase in the amount of collected health-related data and significant progress has been made in technologies able to analyze and understand this data. This is correlated to the emerging *big data* context [28] and to the increasing focus on the definition of a *big data analytics* methodologies [29–31] as well as to the *big data* privacy and security aspects [32, 33]. In recent years there is growing interest on the role of *big data* in healthcare and stroke [34, 35], as well as on markers and models to predict functional recovery after acute ischemic events in order to personalize the therapeutic intervention and improve the final functional outcome [10, 36].

Currently, there is no study on clinical features and models to predict functional outcome measured by mRS in thrombolysis treated WUS patients. Bayesian techniques are becoming very popular in the field of data analysis in medicine [37–39]. In this preliminary study we proposed a method based on Bayesian inference for functional outcome prediction in terms of mRS in WUS as a challenging subtype of stroke.

The Bayesian analysis produced a sparse predictive model which exhibited moderately high accuracy 71% in WUS treated patients, highlighting the importance of NIHSS at admission, age, TACI stroke syndrome, ASPECTs, core CTP volume as well as presence of risk factors as arterial hypertension and diabetes mellitus type II.

The obtained model confirmed the important impact of stroke severity measured by baseline NIHSS, TACI stroke syndrome and age on discharge outcome. The severity of stroke on neurological examination and age are probably the most important factor affecting short- and long-term outcome [16, 40, 41]. As a general rule, large strokes, manifested usually as a TACI syndrome with severe initial clinical deficits, have poor outcomes compared with smaller strokes [42]. Recent studies showed that patients with diabetes mellitus type II and arterial hypertension are generally characterized with poor outcome [43, 44].

CTP core volume and ASPECTS neuroimaging features contributed to the prediction of mRS outcome. ASPECTS was found as a strong method for predicting functional outcome in thrombolysis treated acute ischemic stroke patients compared to valuation of clinical characteristics alone [45]. CTP infarct core volume and collateral grade were reported as strongest predictors of a good 3 months mRS outcome following reperfusion treatment in patients admitted within 4.5 h from symptom onset [46, 47]. Although the total ischemic CTP volume didn't result as direct predictor, it participates indirectly through its contribution to NIHSS at admission [18] and TACI syndrome. A recent WUS CTP study showed that CTP core volume, NIHSS at admission and ASPECTS predict NIHSS at 7-days, while total hypoperfused volume and core volume on CTP predict infarct lesion volume at follow-up CT [48].

5   Conclusions and Future Work

In this study Variational Bayesian logistic regression with Automatic Relevance Determination produced a sparse model selecting informative features to predict a *good functional outcome* in thrombolysis treated WUS patients. This approach has an advantage over some regularization methods which require a separate validation set to prune less relevant features. Indeed, time last seen well to admission was not associated with functional outcome at discharge supporting the hypothesis that neurological assessment, comorbidities together with the advanced neuroimaging are more important than the influence from time to admission. The main limitation of this pilot study is limited single center sample size, mild/moderate stroke severity and prevalence penumbra compared to ischemic core on CTP in our cohort.

In conclusion, in this preliminary study we assessed the possibility to model the prognosis in thrombolysis treated WUS patients by using VB-ARD. The identified features related to initial neurological deficit, history of diabetes and hypertension, together with necrotic tissue related ASPECT and CTP core volume neuroimaging features, were able to predict outcome with moderately high accuracy. Future work will be mainly focused to improve the actual framework.

Acknowledgements. This study was partially supported by Master in Clinical Engineering, University of Trieste. A. Miladinović is supported by the European Social Fund (ESF) and Autonomous Region of Friuli Venezia Giulia (FVG).

Conflict of Interest. The authors have no conflict of interest do declare.

References

1. Gorelick, P.B.: The global burden of stroke: persistent and disabling. Lancet Neurol. **18**(5), 417–418 (2019)
2. Mackey, J., Kleindorfer, D., Sucharew, H., et al.: Population-based study of wake-up strokes. Neurology **76**, 1662–1667 (2011)
3. Thomalla, G., Fiebach, J.B., Ostergaard, L., et al.: A multicenter, randomized, double-blind, placebo-controlled trial to test efficacy and safety of magnetic resonance imaging-based thrombolysis in wake-up stroke (WAKE-UP). Int. J. Stroke **9**, 829–836. https://doi.org/10.1111/ijs.12011
4. Vilela, P., Rowley, H.A.: Brain ischemia: CT and MRI techniques in acute ischemic stroke. Eur. J. Radiol. **96**, 162–172 (2017)

5. Furlanis, G., et al.: Wake-up stroke: thrombolysis reduces ischemic lesion volume and neurological deficit. J. Neurol. **267**(3), 666–673 (2019). https://doi.org/10.1007/s00415-019-09603-7

6. Caruso, P., et al.: Wake-up stroke and CT perfusion: effectiveness and safety of reperfusion therapy. Neurol. Sci. **39**(10), 1705–1712 (2018). https://doi.org/10.1007/s10072-018-3486-z

7. Peisker, T., Koznar, B., Stetkarova, I., et al.: Acute stroke therapy: a review. Trends Cardiovasc. Med. **27**, 59–66 (2017)

8. Stragapede, L., Furlanis, G., Ajčević, M., et al.: Brain oscillatory activity and CT perfusion in hyper-acute ischemic stroke. J. Clin. Neurosci. **69**, 184–189 (2019). https://doi.org/10.1016/j.jocn.2019.07.068

9. Ma, H., Campbell, B.C.V., Parsons, M.W., et al.: Thrombolysis guided by perfusion imaging up to 9 hours after onset of stroke. N. Engl. J. Med. **380**, 1795–1803 (2019). https://doi.org/10.1056/NEJMoa1813046

10. Bentes, C., Peralta, A.R., Viana, P., et al.: Quantitative EEG and functional outcome following acute ischemic stroke. Clin. Neurophysiol. **129**(8), 1680–1687 (2018)

11. Banks, J.L., Marotta, C.A.: Outcomes validity and reliability of the modified Rankin scale: implications for stroke clinical trials: a literature review and synthesis. Stroke **38**(3), 1091–1096 (2007)

12. Caruso, P., Ajčević, M., Furlanis, G., et al.: Thrombolysis safety and effectiveness in acute ischemic stroke patients with pre-morbid disability. J. Clin. Neurosci. **72**, 180–184 (2020). https://doi.org/10.1016/j.jocn.2019.11.047

13. Saver, J.L., Filip, B., Hamilton, S., et al.: Improving the reliability of stroke disability grading in clinical trials and clinical practice: the Rankin Focused Assessment (RFA). Stroke **41**(5), 992–995 (2010)

14. The National Institute of Neurological Disorders and Stroke rt-PA Stroke Study Group: Tissue plasminogen activator for acute ischemic stroke. N. Engl. J. Med. **333**, 1581–1588 (1995)

15. Bamford, J., Sandercock, P., Dennis, M., et al.: Classification and natural history of clinically identifiable subtypes of cerebral infarction. Lancet **337**(8756), 1521–1526 (1991)

16. Adams Jr., H.P., Davis, P.H., Leira, E.C., et al.: Baseline NIH Stroke Scale score strongly predicts outcome after stroke: a report of the Trial of Org 10172 in Acute Stroke Treatment (TOAST). Neurology **53**, 126–131 (1999). https://doi.org/10.1212/wnl.53.1.126

17. Barber, P.A., Demchuk, A.M., Zhang, J., Buchan, A.M., for the ASPECTS Study Group: The validity and reliability of a novel quantitative CT score in predicting outcome in hyperacute stroke prior to thrombolytic therapy. Lancet **355**, 1670–1674 (2000)

18. Furlanis, G., Ajčević, M., Stragapede, L., et al.: Ischemic volume and neurological deficit: correlation of computed tomography perfusion with the National Institutes of Health Stroke Scale Score in acute ischemic stroke. J. Stroke Cerebrovasc. Dis. **27**(8), 2200–2207 (2018). https://doi.org/10.1016/j.jstrokecerebrovasdis.2018.04.003

19. Granato, A., D'Acunto, L., Ajčević, M., et al.: A novel Computed Tomography Perfusion-based quantitative tool for evaluation of perfusional abnormalities in migrainous aura stroke mimic. Neurol. Sci. (2020). https://doi.org/10.1007/s10072-020-04476-5

20. Wintermark, M., Flanders, A.E., Velthuis, B., et al.: Perfusion-CT assessment of infarct core and penumbra: receiver operating characteristic curve analysis in 130 patients suspected of acute hemispheric stroke. Stroke **37**, 979–985 (2006)

21. Treder, M.S., Blankertz, B.: (C)overt attention and visual speller design in an ERP-based brain-computer interface. Behav. Brain Funct. (2010). https://doi.org/10.1186/1744-9081-6-28

22. Drugowitsch, J.: Variational Bayesian inference for linear and logistic regression. arXiv e-prints (2013)

23. Bishop, C.M.: Pattern Recognition and Machine Learning. Springer, New York (2006)
24. MacKay, D.J.C.: Bayesian interpolation. Neural Comput. **4**(3), 415–447 (1992)
25. Neal, R.M.: Bayesian Learning for Neural Networks. Springer, New York (1996). https://doi.org/10.1007/978-1-4612-0745-0
26. Tipping, M.E.: Sparse Bayesian learning and the relevance vector machine. J. Mach. Learn. Res. **1**, 211–244 (2001)
27. Jaakkola, T.S., Jordan, M.M.: Bayesian parameter estimation via variational methods. Stat. Comput. **10**, 25–37 (2000). https://doi.org/10.1023/A:1008932416310
28. Zikopoulos, P., Eaton, C.: Understanding Big Data Analytics for Enterprise Class Hadoop and Streaming Data. McGraw Hill, New York (2012)
29. Cuzzocrea, A., Moussa, R., Xu, G.: OLAP*: effectively and efficiently supporting parallel OLAP over big data. In: Cuzzocrea, A., Maabout, S. (eds.) MEDI 2013. LNCS, vol. 8216, pp. 38–49. Springer, Heidelberg (2013). https://doi.org/10.1007/978-3-642-41366-7_4
30. Chatzimilioudis, G., Cuzzocrea, A., Gunopulos, D., Mamoulis, N.: A novel distributed framework for optimizing query routing trees in wireless sensor networks via optimal operator placement. J. Comput. Syst. Sci. **79**(3), 349–368 (2013)
31. Cuzzocrea, A.: Combining multidimensional user models and knowledge representation and management techniques for making web services knowledge-aware. Web Intell. Agent Syst. **4**(3), 289–312 (2006)
32. Cuzzocrea, A., Bertino, E.: Privacy preserving OLAP over distributed XML data: a theoretically-sound secure-multiparty-computation approach. J. Comput. Syst. Sci. **77**(6), 965–987 (2011)
33. Cuzzocrea, A., Russo, V.: Privacy preserving OLAP and OLAP security. In: Encyclopedia of Data Warehousing and Mining, pp. 1575–1581 (2009)
34. Wang, L., Alexander, C.A.: Stroke care and the role of big data in healthcare and stroke. Rehabil. Sci. **1**(1), 16–24 (2016)
35. Nishimura, A., Nishimura, K., Kada, A., Iihara, K., J-ASPECT Study Group: Status and future perspectives of utilizing big data in neurosurgical and stroke research. Neurol. Med.-Chir. **56**(11), 655–663 (2016)
36. Burke Quinlan, E., Dodakian, L., See, J., et al.: Neural function, injury, and stroke subtype predict treatment gains after stroke. Ann. Neurol. **77**, 132–145 (2015)
37. Spyroglou, I.I., Spöck, G., Chatzimichail, E.A., et al.: A Bayesian logistic regression approach in asthma persistence prediction. Epidemiol. Biostat. Public Health **15**(1), e12777 (2018)
38. Ashby, D.: Bayesian statistics in medicine: a 25 year review. Stat. Med. **25**(21), 3589–3631 (2006)
39. Miladinović, A., et al.: Slow cortical potential BCI classification using sparse variational Bayesian logistic regression with automatic relevance determination. In: Henriques, J., Neves, N., de Carvalho, P. (eds.) MEDICON 2019. IP, vol. 76, pp. 1853–1860. Springer, Cham (2020). https://doi.org/10.1007/978-3-030-31635-8_225
40. Weimar, C., König, I.R., Kraywinkel, K., et al.: Age and National Institutes of Health Stroke Scale Score within 6 hours after onset are accurate predictors of outcome after cerebral ischemia: development and external validation of prognostic models. Stroke **35**, 158–162 (2004)
41. Saver, J.L., Altman, H.: Relationship between neurologic deficit severity and final functional outcome shifts and strengthens during first hours after onset. Stroke **43**, 1537–1541 (2012)
42. Di Carlo, A., Lamassa, M., Baldereschi, M., et al.: Risk factors and outcome of subtypes of ischemic stroke. Data from a multicenter multinational hospital-based registry. The European Community Stroke Project. J. Neurol. Sci. **244**, 143–150 (2006)

43. Desilles, J.P., Meseguer, E., Labreuche, J., et al.: Diabetes mellitus, admission glucose, and outcomes after stroke thrombolysis: a registry and systematic review. Stroke **44**, 1915–1923 (2013)

44. Manabe, Y., Kono, S., Tanaka, T., et al.: High blood pressure in acute ischemic stroke and clinical outcome. Neurol. Int. **1**(1), e1 (2009). https://doi.org/10.4081/ni.2009.e1

45. Baek, J.H., Kim, K., Lee, Y.B., et al.: Predicting stroke outcome using clinical- versus imaging-based scoring system. J. Stroke Cerebrovasc. Dis. **24**(3), 642–648 (2015). https://doi.org/10.1016/j.jstrokecerebrovasdis.2014.10.009

46. Bivard, A., Spratt, N., Miteff, F., et al.: Tissue is more important than time in stroke patients being assessed for thrombolysis. Front. Neurol. **9**, 41 (2018)

47. Tian, H., Parsons, M.W., Levi, C.R., et al.: Influence of occlusion site and baseline ischemic core on outcome in patients with ischemic stroke. Neurology **92**, e2626–e2643 (2019). https://doi.org/10.1212/WNL.0000000000007553

48. Ajčević, M., Furlanis, G., Buoite Stella, A., et al.: CTP based model predicts outcome in rTPA treated wake-up stroke patients. Physiol. Meas. (2020). https://doi.org/10.1088/1361-6579/ab9c70

Latent Weights Generating for Few Shot Learning Using Information Theory

Zongyang Li[1(✉)] and Yuan Ji[2]

[1] The University of Sydney, Sydney, Australia
zongyangli7177@hotmail.com
[2] Chinese Academy of Sciences, Beijing, China

Abstract. Few shot image classification aims at learning a classifier from limited labeled data. Generating the classification weights has been applied in many metalearning approaches for few shot image classification due to its simplicity and effectiveness. However, fixed classification weights for different query samples within one task might be sub-optimal, due to the few shot challenge, and it is difficult to generate the exact and universal classification weights for all the diverse query samples from very few training samples. In this work, we introduce latent weights generating using information theory (LWGIT) for few shot learning which addresses current issues by generating different classification weights for different query samples by letting each of query samples attends to the whole support set. The experiment results demonstrate the effectiveness of LWGIT, thereby contributing to exceed the performances of the existing state-of-the-art models.

1 Introduction

While deep learning methods achieve great success in domains such as computer vision [1], natural language processing [2], reinforcement learning [3], their hunger for large amount of labeled data limits the application scenarios where only a few data are available for training. Humans, in contrast, are able to learn from limited data, which is desirable for deep learning methods. Few shot learning is thus proposed to enable deep models to learn from very few samples.

Meta learning is by far the most popular and promising approach for few shot problems [4]. In meta learning approaches, the model extracts high level knowledge across different tasks so that it can adapt itself quickly to a newcoming task [5]. There are several kinds of meta learning methods for few shot learning, such as gradient-based [4] and metric-based [6]. Weights generation, among these different methods, has shown effectiveness with simple formulation [7]. In general, weights generation methods learn to generate the classification weights for different tasks conditioned on the limited labeled data.

However, fixed classification weights for different query samples within one task might be sub-optimal, due to the few shot challenge, and it is difficult to generate the exact and universal classification weights for all the diverse query samples from very few training samples.

O. Gervasi et al. (Eds.): ICCSA 2020, LNCS 12249, pp. 1003–1016, 2020.
https://doi.org/10.1007/978-3-030-58799-4_72

To addresses current issues, we propose latent weights generating using information theory (LWGIT) for few shot learning in this work. The contribution is as followed:

- To overcome issues mentioned above, we propose the LWGIT which generates different classification weights for different query samples by letting each of query samples attends to the whole support set.
- To guarantee the generated weights adaptive to different query sample, we re-formulate the problem to maximize the lower bound of mutual information between generated weights and query as well as support data.
- The experiment results demonstrate the effectiveness of LWGIT, thereby contributing to exceed the performances of the existing state-of-the-art models.

The remaining of this paper is organized as follows. Section 2 includes the related work. Section 3 introduces our proposed latent weights generating using information theory method. In Sect. 4, we evaluate our proposed models and report experimental results on extensive realworld datasets. Section 5 concludes this work.

2 Related Work

2.1 Few Shot Learning

Learning from few labeled training data has received growing attentions recently. Most successful existing methods apply meta learning to solve this problem and can be divided into several categories. In the gradient-based approaches, an optimal initialization for all tasks is learned [4]. Ravi Larochelle [8] learned a meta-learner LSTM directly to optimize the given fewshot classification task. Sun et al. [9] learned the transformation for activations of each layer by gradients to better suit the current task.

In the metric-based methods, a similarity metric between query and support samples is learned [10]. Spatial information or local image descriptors are also considered in some works to compute richer similarities [11].

Generating the classification weights directly has been explored by some works. Gidaris [12] generated classification weights as linear combinations of weights for base and novel classes. Similarly, Qiao et al. [13] generated the classification weights from activations of a trained feature extractor. Graph neural network denoising autoencoders are used in [7]. Munkhdalai [14] proposed to generate "fast weights" from the loss gradient for each task. All these methods do not consider generating different weights for different query examples, nor maximizing the mutual information.

There are some other methods for few-shot classification. Generative models are used to generate or hallucinate more data in [15] used the closed-form solutions directly for few shot classification. Liu et al. [16] integrated label propagation on a transductive graph to predict the query class label.

2.2   Attention Mechanism

Attention mechanism shows great success in computer vision [20] and natural language processing [21]. It is effective in modeling the interaction between queries and key-value pairs from certain context. Based on the fact that keys and queries point to the same entities or not, people refer to attention as self attention or cross attention. In this work, we use both types of attention to encode the task and query-task information.

3   Latent Weights Generating Using Information Theory

3.1   Background

Suppose that a sequence of tasks $\{\mathcal{T}_1, \ldots, \mathcal{T}_{N_t}\}$ are sampled from an environment which is a probability distribution $\mathcal{E}$ on tasks. In each task $\mathcal{T}_i \sim \mathcal{E}$, we have a few examples $\{\mathbf{x}_{i,j}, \mathbf{y}_{i,j}\}_{j=1}^{n^{tr}}$ to constitute the training set $\mathcal{D}_{\mathcal{T}_i}^{tr}$ and the rest as the test set $\mathcal{D}_{\mathcal{T}_i}^{te}$.

Given a base learner f with θ as parameters, the optimal parameters $\theta_{\mathcal{T}_i}$ are learned to make accurate predictions, i.e., $f_{\theta_{\mathcal{T}_i}}(\mathbf{x}_{i,j}) \rightarrow \mathbf{y}_{i,j}$. The effectiveness of such a base learner on $\mathcal{D}_{\mathcal{T}_i}^{tr}$ is evaluated by the loss function $\mathcal{L}\left(f_{\theta_{\mathcal{T}_i}}, \mathcal{D}_{\mathcal{T}_i}^{tr}\right)$, which equals the mean square error for regression problems:

$$\sum_{(\mathbf{x}_{i,j}, \mathbf{y}_{i,j}) \in \mathcal{D}_{\mathcal{T}_i}^{tr}} \|f_{\theta_{\mathcal{T}_i}}(\mathbf{x}_{i,j}) - \mathbf{y}_{i,j}\|_2^2 \tag{1}$$

or the cross entropy loss:

$$- \sum_{(\mathbf{x}_{i,j}, \mathbf{y}_{i,j}) \in \mathcal{D}_{\mathcal{T}_i}^{tr}} \log p\left(\mathbf{y}_{i,j} | \mathbf{x}_{i,j}, f_{\theta_{\mathcal{T}_i}}\right) \tag{2}$$

for classification problems.

The goal of meta-learning is to learn from previous tasks a well-generalized meta-learner $\mathcal{M}(\cdot)$ which can facilitate the training of the base learner in a future task with a few examples. In fulfillment of this, meta-learning involves two stages, i.e., meta-training and meta-testing.

During meta- training, the parameters of the base learner for all tasks, i.e., $\{\theta_{\mathcal{T}_i}\}_{i=1}^{N_t}$, and the meta-learner $\mathcal{M}(\cdot)$ are optimized alternatingly. In virtue of M, the parameters $\{\theta_{\mathcal{T}_i}\}_{i=1}^{N_t}$ are learned to minimize the expected empirical loss over training sets of all N_t historical tasks:

$$\min_{\{\theta_{\mathcal{T}_i}\}_{i=1}^{N_t}} \sum_{i=1}^{N_t} \mathcal{L}\left(\mathcal{M}\left(f_{\theta_{\mathcal{T}_i}}\right), \mathcal{D}_{\mathcal{T}_i}^{tr}\right) \tag{3}$$

In turn, a well-generalized M can be obtained by minimizing the expected empirical loss over test sets:

$$\min_{\mathcal{M}} \sum_{i=1}^{N_t} \mathcal{L}\left(\mathcal{M}\left(f_{\theta_{\mathcal{T}_i}}\right), \mathcal{D}_{\mathcal{T}_i}^{te}\right) \tag{4}$$

When it comes to the metatesting phase, provided with a future task $\mathcal{T}_t$, the learning effectiveness and efficiency are improved by applying the meta-learner M and solving

$$\min_{\theta \tau_t} \mathcal{L}\left(\mathcal{M}\left(f_{\theta \tau_t}\right), \mathcal{D}_{\mathcal{T}_t}^{tr}\right) \tag{5}$$

3.2   Problem Formulation

Following many popular meta-learning methods for few shot classification, we formulate the problem under episodic training paradigm [4]. One N-way K-shot task sampled from an unknown task distribution $P(T)$ includes support set and query set:

$$\mathcal{T} = (\mathcal{S}, \mathcal{Q}) \tag{6}$$

where $\mathcal{S} = \left\{\left(\mathbf{x}^{c_n;k}, \mathbf{y}^{c_n;k}\right) | k = 1, \ldots, K; n = 1, \ldots, N\right\}$, $\mathcal{Q} = \left\{\left(\hat{\mathbf{x}}_1, \ldots, \hat{\mathbf{x}}_{[Q]}\right)\right\}$ Support set S contains NK labeled samples. Query set Q includes $\hat{\mathbf{x}}$ and we need to predict label $\hat{\mathbf{y}}$ for $\hat{\mathbf{x}}$ based on S. During meta-training, the meta-loss is estimated on Q to optimize the model. During metatesting, the performance of meta-learning method is evaluated on Q, provided the labeled S. The classes used in meta-training and meta-testing are disjoint so that the meta-learned model needs to learn the knowledge transferable across tasks and adapt itself quickly to novel tasks.

Our proposed approach follows the general framework to generate the classification weights [13]. In this framework, there is a feature extractor to output image feature embeddings. The meta-learner needs to generate the classification weights for different tasks

3.3   Latent Embedding Optimization

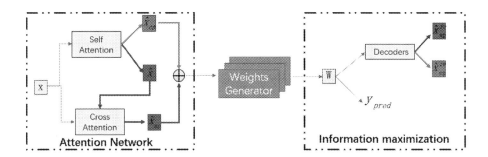

Fig. 1. The structure of the proposed LWGIT

Latent Embedding Optimization (LEO) [22] is one of the weights generation methods that is most related to our work. In LEO, the latent code z is generated by h conditioned on support set S, described as $z = h(\mathcal{S})$. h is instantiated as relation networks [23]. Classification weights w can be decoded from z with l, $w = l(z)$. In the inner loop, we use w to compute the loss (usually cross entropy) on the support set and then update z:

$$z' = z - \eta \nabla_z \mathcal{L}_S(w) \qquad (7)$$

where L_S indicates that the loss is evaluated on S only. The updated latent code z' is used to decode new classification weights w' with generating function l. w' is adopted in the outer loop for query set Q and the objective function of LEO then can be written as

$$\min_\theta \mathcal{L}_Q(w') \qquad (8)$$

Here θ stands for the parameters of h and l and we omit the regularization terms for clarity. LEO avoids updating high-dimensional w in the inner loop by learning a lower-dimensional latent space, from which sampled z can be used to generate w. The most significant difference between LEO and LWGIT is that we do not need inner updates to adapt the model. Instead, LWGIT is a feedforward network trained to maximize the mutual information so that it fits to different tasks well. On the other hand, LWGIT learns to generate optimal classification weights for each query sample while LEO generates fixed weights conditioned on the support set within one task.

3.4   Weights Generation Using Information Theory

The framework of our proposed method is shown in Fig. 1. Assume that we have a feature extractor, which can be a simple 4-layer Convnet or a deeper Resnet. All the images included in the sampled task T are processed by this feature extractor and represented as d-dimensional vectors afterwards, i.e., $\mathbf{x}^{c_n;k}, \hat{\mathbf{x}} \in R^d$. There are two paths to encode the task context and the individual query sample respectively, which are called contextual path and attentive path. The outputs of both paths are concatenated together as input to the generator for classification weights. Generated classification weights are used to not only predict the label of $\hat{\mathbf{x}}$, but also maximize the lower bound of mutual information between itself and other variables, which will be discussed in the following section.

Attention Network. The encoding process includes two paths, namely the contextual path and attentive path. The contextual path aims at learning representations for only the support set with a multi-head self-attention network f_{sa}^{cp} [21]. The outputs of contextual path $\mathbf{X}^{cp} \in R^{NK \times d_h}$ thus contain richer information about the task and can be used later for weights generation.

Existing weights generation methods generate the classification weights conditioned on the support set only, which is equivalent to using contextual path. However, the classification weights generated in this way might be sub-optimal.

This is because estimating the exact and universal classification weights from very few labeled data in the support set is difficult and sometimes impossible. The generated weights are usually in lack of adaptation to different query samples. We address this issue by introducing attentive path, where the individual query example attends to the task context and then is used to generate the classification weights. Therefore, the classification weights are adaptive to different query samples and aware of the task context as well.

In the attentive path, a new multi-head self-attention network f_{sa}^{ap} on the support set is employed to encode the global task information. f_{sa}^{ap} is different from f_{sa}^{cp} in contextual path because the selfattention network in contextual path emphasizes on generating the classification weights. On the contrary, outputs of self-attention here plays the role of providing the Value context for different query samples to attend in the following cross attention. Sharing the same self-attention networks might limit the expressiveness of learned representations in both paths. The cross attention network f_{ca}^{ap} applied on each query sample and task-aware support set is followed to produce $\hat{\mathbf{X}}^{ap} \in R^{|\mathcal{Q}| \times d_h}$.

We use multi-head attention with h heads in both paths. In one attention block, we produce h different sets of queries, keys and values. Multi-head attention is claimed to be able to learn more comprehensive and expressive representations from h different subspaces [21].

Weights Generator. We replicate $\mathbf{X}^{cp} \in R^{NK \times d_h}$ and $\hat{\mathbf{X}}^{ap} \in R|\mathcal{Q}| \times d_h$ for $|\mathcal{Q}|$ and NK times respectively and reshape them afterwards. Then we have $\mathbf{X}^{cp} \in R^{|Q| \times NK \times d_h}$ and $\hat{\mathbf{X}}^{ap} \in R^{|Q| \times NK \times d_h}$. These two tensors are concatenated to become $\mathbf{X}^{cp \oplus ap} \in R|\mathcal{Q}| \times NK \times 2d_h$, $\mathbf{X}^{cp \oplus ap}$ can be interpreted that each query sample has its own latent representations for support set to generate specific classification weights, which are both aware of the task-context and adaptive to individual query sample.

$\mathbf{X}^{cp \oplus ap}$ is decoded by the weights generator g: $R^{2d_h} \rightarrow R^{2d}$. We assume that the classification weights follow Gaussian distribution with diagonal covariance. g outputs the distribution parameters and we sample the weights from learned distribution during meta-training. The sampled classification weights are represented as $\mathbf{W} \in R|\mathcal{Q}| \times NK \times d$. To reduce complexity, we compute the mean value on K classification weights for each class to have $\mathbf{W}^{final} \in R^{|Q| \times N} \times d$. Therefore, ith query sample has its specific classification weight matrix $\mathbf{W}_{i,i,i}^{final} \in R^{N \times d}$. The prediction for query data can be computed by $\hat{\mathbf{X}} \mathbf{W}^{final\mathbf{T}}$. The support data X is replicated for $|Q|$ times and reshaped as $\mathbf{X}_s \in R|\mathbf{Q}| \times NK \times d$. So the prediction for support data can also be computed as $\mathbf{X}_s \mathbf{W}^{final\mathbf{T}}$.

Besides the weights generator g, we have another two decoders $r_1 : R^d \rightarrow R^{d_h}$ and $r_2 : R^d \rightarrow R^{d_h}$. They both take the generated weights W as inputs and learn to reconstruct $\mathbf{X}^{cp}$ and $\mathbf{X}^{ap}$ respectively. The outputs of r_1 and r_2 are denoted as $\mathbf{X}_{re}^{cp}, \hat{\mathbf{X}}_{re}^{ap} \in R^{|Q| \times NK \times d_h}$.

3.5   Information Theory

In this section, we perform the analysis for one query sample without loss of generality. The subscripts for classification weights are omitted for clarity. In general, we use $(\mathbf{x}, \mathbf{y})$ and $(\hat{\mathbf{x}}, \hat{\mathbf{y}})$ to represent support and query samples respectively.

Since the classification weights w generated from g are encoded with attentive path and contextual path, it is expected that we can directly have the query-specific weights. However, we show in the experiments that simply doing this does not outperform a weight generator conditioned only on the S significantly, which implies that the generated classification weights from two paths are not sensitive to different query samples. In other words, the information from attentive path is not kept well during the weights generation.

To address this limitation, we propose to maximize the mutual information between generated weights w and support as well as query data. The objective function can be described as

$$\max I((\hat{\mathbf{x}}, \hat{\mathbf{y}}); \mathbf{w}) + \sum_{(\mathbf{x}, y) \in \mathcal{S}} I((\mathbf{x}, \mathbf{y}); \mathbf{w}) \tag{9}$$

According to the chain rule of mutual information, we have

$$I((\hat{\mathbf{x}}, \hat{\mathbf{y}}); \mathbf{w}) = I(\hat{\mathbf{x}}; \mathbf{w}) + I(\hat{\mathbf{y}}; \mathbf{w} | \hat{\mathbf{x}}) \tag{10}$$

Equation 10 stands for both terms in 9. So the objective function can be written as

$$\max I(\hat{\mathbf{x}}; \mathbf{w}) + I(\hat{\mathbf{y}}; \mathbf{w} | \hat{\mathbf{x}}) + \sum_{(\mathbf{x}, y) \in \mathcal{S}} [I(\mathbf{x}; \mathbf{w}) + I(\mathbf{y}; \mathbf{w} | \mathbf{x})] \tag{11}$$

Directly computing the mutual information in Eq. 11 is intractable since the true posteriori distributions like $p(\hat{\mathbf{y}} | \hat{\mathbf{x}}, \mathbf{w}), p(\hat{\mathbf{x}} | \mathbf{w})$ are still unknown. Therefore, we use Variational Information Maximization[17] to compute the lower bound of Eq. 9. We use $p_\theta(\hat{\mathbf{x}} | \mathbf{w})$ to approximate the true posteriori distribution, where θ represents the model parameters. As a result, we have

$$\begin{aligned}
I(\hat{\mathbf{x}}; \mathbf{w}) &= H(\hat{\mathbf{x}}) - H(\hat{\mathbf{x}} | \mathbf{w}) \\
&= H(\hat{\mathbf{x}}) + \mathbb{E}_{\mathbf{w} \sim p(\mathbf{w} | \mathbf{x}, \mathcal{S})} \left[\mathbb{E}_{\hat{\mathbf{x}} \sim p(\hat{\mathbf{x}} | \mathbf{w})} [\log p(\hat{\mathbf{x}} | \mathbf{w})] \right] \\
&= H(\hat{\mathbf{x}}) + \mathbb{E}_{\mathbf{w} \sim p(\mathbf{w} | \mathbf{x}, \mathcal{S})} D_{\mathbf{KL}} \left(p(\hat{\mathbf{x}} | \mathbf{w}) \| p_\theta(\hat{\mathbf{x}} | \mathbf{w}) \right) \\
&\quad + \mathbb{E}_{\hat{\mathbf{x}} \sim p(\hat{\mathbf{x}} | \mathbf{w})} [\log p_\theta(\hat{\mathbf{x}} | \mathbf{w})] \\
&\geq H(\hat{\mathbf{x}}) + \mathbb{E}_{\mathbf{w} \sim p(\mathbf{w} | \mathbf{x}, \mathcal{S})} \left[\mathbb{E}_{\hat{\mathbf{x}} \sim p(\mathbf{x} | \mathbf{w})} [\log p_\theta(\hat{\mathbf{x}} | \mathbf{w})] \right]
\end{aligned} \tag{12}$$

$H(\cdot)$ is the entropy of a random variable. $H(\hat{\mathbf{x}})$ is a constant value for given data. We can maximize this lower bound as the proxy for the true mutual information. Similar to $I(\hat{\mathbf{x}}; \mathbf{w})$

$$I(\hat{\mathbf{y}}; \mathbf{w}|\hat{\mathbf{x}}) \geq H(\hat{\mathbf{y}}|\hat{\mathbf{x}})$$
$$+ \mathbb{E}_{\mathbf{w} \sim p(\mathbf{w}|\hat{\mathbf{x}}, \mathcal{S})} \left[\mathbb{E}_{\hat{\mathbf{y}} \sim p(\hat{\mathbf{y}}|\mathbf{x}, \mathbf{w})} \left[\log p_\theta(\hat{\mathbf{y}}|\hat{\mathbf{x}}, \mathbf{w}) \right] \right] \tag{13}$$

$$\sum_{(\mathbf{x}, y) \in \mathcal{S}} I((\mathbf{x}, \mathbf{y}); \mathbf{w}) \geq \sum_{(\mathbf{x}, y) \in \mathcal{S}} H((\mathbf{x}, \mathbf{y}))$$
$$+ \mathbb{E}_{(\mathbf{x}, y) \sim p((\mathbf{x}, y)|\mathbf{w})} \left[\log p_\theta(\mathbf{x}|\mathbf{w}) + \log p_\theta(\mathbf{y}|\mathbf{x}, \mathbf{w}) \right] \tag{14}$$

$p_\theta(\hat{\mathbf{x}}|\mathbf{w}), p_\theta(\mathbf{x}, \mathbf{y}|\mathbf{w})$ are used to approximate the true posteriori distribution $p(\hat{\mathbf{x}}|\mathbf{w})$ and $p(\mathbf{x}, \mathbf{y}|\mathbf{w})$

Put the lower bounds back into Eq. 11. Omit the constant entropy terms and the expectation subscripts for clarity, we have the new objective function as

$$\max \mathbb{E}_\theta \left[\log p_\theta(\hat{\mathbf{y}}|\hat{\mathbf{x}}, \mathbf{w}) \right]$$
$$+ \mathbb{E}_\theta \left[\log p_\theta(\mathbf{y}|\mathbf{x}, \mathbf{w}) + \log p_\theta(\mathbf{x}|\mathbf{w}) + \log p_\theta(\hat{\mathbf{x}}|\mathbf{w}) \right] \tag{15}$$

The first two terms are maximizing the log likelihood of label for both support and query data with respective to the network parameters, given the generated classification weights. This is equivalent to minimizing the cross entropy between prediction and ground-truth. We assume that $p_\theta(\hat{\mathbf{x}}|\mathbf{w})$ and $p_\theta(\mathbf{x}|\mathbf{w})$ are Gaussian distributions. r_1 and r_2 are used to approximate the mean of these two Gaussian distributions. Therefore maximizing the log likelihood is equivalent to reconstruct $\mathbf{X}^{Cp}$ and $\hat{\mathbf{X}}^{ap}$ with L2 loss. Thus the loss function to train the network can be written as

$$L = \text{CE} (\hat{\mathbf{y}}_{pred}, \hat{\mathbf{y}}) + \lambda_1 \sum_{\mathbf{y} \in \mathcal{S}} \text{CE} (\mathbf{y}_{pred}, \mathbf{y})$$
$$+ \lambda_2 \sum_{\mathbf{x}^{cp} \in \mathcal{S}} \|\mathbf{x}^{cp} - \mathbf{x}^{cp}_{re}\|_2 + \lambda_3 \|\hat{\mathbf{x}}^{ap} - \hat{\mathbf{x}}^{ap}_{re}\|_2 \tag{16}$$

CE here stands for cross entropy. $\mathbf{x}^{cp}$ and $\hat{\mathbf{x}}^{ap}$ are the inputs to weights generator g. $\mathbf{x}^{cp}_{re} \sim p_\theta(\mathbf{x}|\mathbf{w})$ and $\hat{\mathbf{x}}^{ap}_{re} \sim p_\theta(\hat{\mathbf{x}}|\mathbf{w})$ are the reconstruction of $\mathbf{x}^{cp}$ and $\hat{\mathbf{x}}^{ap}$. Since we convert the log likelihood in Eq. 15 to mean square error or cross entropy in Eq. 16 to optimize, the value of each term in Eq. 16 is not equal to real log likelihood and we have to decide the weightage for each one. $\lambda_1, \lambda_2, \lambda_3$ are thus hyper-parameters for trade-off of different terms. With the help of last three terms, the generated classification weights are forced to carry information about the support data and the specific query sample.

In LEO [22], the inner update loss is computed as cross entropy on support data. If we merge the inner update into outer loop, then the loss becomes the summation of first two terms in Eq. 16. However, the weight generation in LEO does not involve specific query samples, thus making reconstructing $\hat{\mathbf{X}}^{ap}$ impossible. In this sense, LEO can be regarded as a special case of our proposed method, where (1) only contextual path exits and (2) $\lambda_2 = \bar{\lambda}_3 = 0$.

Table 1. Accuracy comparison with other approaches on miniImageNet

Model	Feature extractor	5-way 1-shot	5-way 5-shot
Matching Networks [24]	Conv-4	46.60%	60.00%
MAML [4]	Conv-4	48.70%	63.11%
Meta LSTM [8]	Conv-4	43.44%	60.60%
Prototypical Nets [25]	Conv-4	49.42%	68.20%
Relation Nets [6]	Conv-4	50.44%	65.32%
SNAIL [26]	Resnets-12	55.71%	68.88%
TPN [16]	Resnets-12	59.46%	0.00%
MTL [9]	Resnets-12	61.20%	75.50%
Dynamic [12]	WRN-28-10	60.06%	76.39%
Prediction [13]	WRN-28-10	59.60%	73.74%
DAE-GNN [7]	WRN-28-10	62.96%	78.85%
LEO [22]	WRN-28-10	61.76%	77.59%
LWGIT (ours)	WRN-28-10	63.12%	78.40%

Table 2. Accuracy comparison with other approaches on tieredImageNet.

Model	Feature extractor	5-way 1-shot	5-way 5-shot
MAML [4]	Conv-4	51.67	70.3
Prototypical Nets [25]	Conv-4	53.31	72.69
Relation Nets [6]	Conv-4	54.48	71.32
TPN [16]	Conv-4	59.91	72.85
MetaOptNet [27]	Resnets-12	65.81	81.75
LEO [22]	WRN-28-10	66.33	81.44
LWGIT (ours)	WRN-28-10	67.69	82.82

Table 3. Analysis of our proposed LWGIT. In the top half, the attentive path is removed to compare with LEO. In the bottom part, ablation analysis with respective to different components is provided.

Model	miniImageNet		tieredImageNet	
	5-way 1-shot	5-way 5-shot	5-way 1-shot	5-way 5-shot
LEO	61.76%	77.59%	66.33%	81.44%
Generator in LEO	60.33%	74.53%	65.17%	78.77%
Generator conditioned on S only	61.02%	74.33%	66.22%	79.66%
Generator conditioned on S with IM	62.04%	77.54%	66.43%	81.73%
MLP encoding, $\lambda_1 = \lambda_2 = \lambda_3 = 0$	58.95%	71.68%	63.92%	75.80%
MLP encoding	62.26%	76.91%	65.84%	79.24%

(*continued*)

Table 3. (*continued*)

Model	miniImageNet		tieredImageNet	
	5-way 1-shot	5-way 5-shot	5-way 1-shot	5-way 5-shot
$\lambda_1 = \lambda_2 = \lambda_3 = 0$	61.61%	74.14%	65.65%	79.93%
$\lambda_1 = \lambda_2 = 0$	62.06%	74.18%	65.85%	80.42%
$\lambda_3 = 0$	62.91%	77.88%	67.27%	81.67%
$\lambda_1 = 0$	62.19%	74.21%	66.82%	80.61%
$\lambda_2 = \lambda_3 = 0$	62.12%	77.65%	66.86%	81.03%
Random shuffle in class	62.87%	77.48%	67.52%	82.55%
Random shuffle between classes	61.20%	77.48%	66.55%	82.53%
LWGIT (ours)	63.12%	78.40%	67.69%	82.82%

4   Experiments

4.1   Datasets and Protocols

We conduct experiments on miniImageNet [24] and tieredImageNet [28], two commonly used benchmark datasets, to compare with other methods and analyze our model. Both datasets are subsets of ILSVRC-12 dataset. miniImageNet contains 100 randomly sampled classes with 600 images per class. We follow the train/test split in [8], where 64 classes are used for meta-training, 16 for meta-validation and 20 for meta-testing. tieredImageNet is a larger dataset compared to miniImageNet. There are 608 classes and 779,165 images in total. They are selected from 34 higher level nodes in ImageNet [29] hierarchy. 351 classes from 20 high level nodes are used for meta-training, 97 from 6 nodes for meta-validation and 160 from 8 nodes for meta-testing.

We use the image features similar to LEO [22]. They trained a 28-layer Wide Residual Network [30] on the meta-training set. Each image then is represented by a 640 dimensional vector, which is used as the input to our model. For N-way K-shot experiments, we randomly sample N classes from meta-training set and each of them contains K samples as the support set and 15 as query set. Similar to other works, we train 5-way 1-shot and 5-shot models on two dataset. During meta-testing, 600 N-way K-shot tasks are sampled from meta-testing set and the average accuracy for query set is reported with 95confidence interval, as done in recent works [4, 22].

4.2   Few Shot Image Classification

We compare the performance of our approach LWGIT on two datasets with several state-of-the-art methods proposed in recent years. The results of MAML, Prototypical Nets, Relation Nets on tieredImageNet are evaluated by [16]. The results of Dynamic on miniImageNet with WRN-28-10 as the feature extractor is reported in [7]. The other results are reported in the corresponding original papers. We also include the backbone network structure of the used feature extractor for reference. The results on miniImageNet and tieredImageNet are shown in Table 1 and 2 respectively.

The top half parts of Table 1 and 2 display the methods belonging with different meta learning categories, such as metric-based (Matching Networks, Prototypical Nets), gradient-based (MAML, MTL), graph-based (TPN). The bottom part shows the classification weights generation approaches including Dynamic, Prediction, DAE-GNN, LEO and our proposed LWGIT.

LWGIT can outperform all the methods in top parts of two table. Comparing with other classification weights generation methods in the bottom part, LWGIT still shows very competitive performance, namely the best on tiered-ImageNet and close to the state-of-the-art on miniImageNet. We note that all the classification weights generation methods are using WRN-28-10 as backbone network, which makes the comparison fair. In particular, LWGIT can outperform LEO in all settings.

4.3 Analysis

We perform detailed analysis on LWGIT, shown in Table 3. We include the results of LEO Rusu et al. (2019) for reference. "Generator in LEO" means that there is no inner update in LEO. In the upper part of the table, we first studied the effect of attentive path. We implemented two generators including only the contextual path during encoding. "Generator conditioned on S with IM" indicates that we add the cross entropy loss and reconstruction loss for support set. It can be observed that "Generator conditioned on S only" is trained with cross entropy on query set, which is similar to "Generator in LEO" without inner update. It is able to achieve similar or slightly better results than "Generator in LEO", which implies that self-attention is no worse than relation networks used in LEO to model task-context. With information maximization, our generator is able to obtain slightly better performance than LEO.

The effect of attention is investigated by replacing the attention modules with 2-layer MLPs, which is shown as "MLP encoding". More specifically, one MLP in contextual path is used for support set and another MLP in attentive path for query samples. We can see that even without attention to encode the task-contextual information, "MLP encoding" can achieve accuracy close to LEO, for the sake of information maximization. However, if we let $\lambda_1 = \lambda_2 = \lambda_3 = 0$ for MLP encoding, the performance drops significantly, which demonstrates the importance of maximizing the information

We conducted ablation analysis with respective to $\lambda_1, \lambda_2, \lambda_3$ to investigate the effect of information maximization. First, $\lambda_1, \lambda_2, \lambda_3$ are all set to be 0. In this case, the accuracy is similar to "generator conditioned on S only", showing that the generated classification weights are not fitted for different query samples, even with the attentive path. It can also be observed that maximizing the mutual information between weights and support is more crucial since $\lambda_1 = \lambda_2 = 0$ degrades accuracy significantly, comparing with $\lambda_3 = 0$. We further investigate the relative importance of the classification on support as well as reconstruction.

$\lambda_1 = 0$ affects the performance noticeably. We conjecture that the support label prediction is more critical for information maximization.

The classification weights are generated specifically for each query sample in LWGIT. To this point, we shuffle the classification weights between query samples within the same classes and between different classes as well to study whether the classification weights are adapted for different query samples. Assume there are T query samples per class in one task. $W^{\text{final}} \in \mathbb{R}|\mathcal{Q}| \times N \times d$ can be reshaped into $W^{\text{final}} \in \mathbb{R}^{N \times T \times N \times d}$. Then we shuffle this weight tensor along the first and second axis randomly. The results are shown as "random shuffle between classes" and "random shuffle in class" in Table 3. For 5-way 1-shot experiments, the random shuffle between classes degrades the accuracy noticeably while the random shuffle in class dose not affect too much. This indicates that when the support data are very limited, the generated weights for query samples from the same class are very similar to each other while distinct for different classes. When there are more labeled data in support set, two kinds of random shuffle show very close or even the same results in 5-way 5-shot experiments, which are both worse than the original ones. This implies that the generated classification weights are more diverse and specific for each query sample in 5-way 5-shot setting. The possible reason is that larger support set provides more knowledge to estimate the optimal classification weights for each query example.

5   Conclusion

In this work, we introduce latent weights generating using information theory (LWGIT) for few shot learning. LWGIT learns to generate optimal classification weights for each query sample within the task by two encoding paths. To guarantee this, the lower bound of mutual information between generated weights and query, support data is maximized. The effectiveness of LWGIT is demonstrated by state-of-the-art performance on two benchmark datasets.

References

1. He, K., Zhang, X., Ren, S., Sun, J.: Deep residual learning for image recognition. In: Proceedings of the IEEE Conference on Computer Vision and Pattern Recognition, pp. 770–778 (2016)
2. Devlin, J., Chang, M.-W., Lee, K., Toutanova, K.: BERT: pre-training of deep bidirectional transformers for language understanding. arXiv preprint arXiv:1810.04805 (2018)
3. Silver, D., et al.: A general reinforcement learning algorithm that masters chess, shogi, and go through self-play. Science **362**(6419), 1140–1144 (2018)
4. Finn, C., Abbeel, P., Levine, S.: Model-agnostic meta-learning for fast adaptation of deep networks. In: Proceedings of the 34th International Conference on Machine Learning-Volume 70, pp. 1126–1135 (2017). JMLR.org
5. Andrychowicz, M., et al.: Learning to learn by gradient descent by gradient descent. In: Advances in Neural Information Processing Systems, pp. 3981–3989 (2016)

6. Sung, F., Yang, Y., Zhang, L., Xiang, T., Torr, P.H., Hospedales, T.M.: Learning to compare: relation network for few-shot learning. In: Proceedings of the IEEE Conference on Computer Vision and Pattern Recognition, pp. 1199–1208 (2018)
7. Gidaris, S., Komodakis, N.: Generating classification weights with GNN denoising autoencoders for few-shot learning. In: Proceedings of the IEEE Conference on Computer Vision and Pattern Recognition, pp. 21–30 (2019)
8. Ravi, S., Larochelle, H.: Optimization as a model for few-shot learning (2016)
9. Sun, Q., Liu, Y., Chua, T.-S., Schiele, B.: Meta-transfer learning for few-shot learning. In: Proceedings of the IEEE Conference on Computer Vision and Pattern Recognition, pp. 403–412 (2019)
10. Li, H., Eigen, D., Dodge, S., Zeiler, M., Wang, X.: Finding task-relevant features for few-shot learning by category traversal. In: Proceedings of the IEEE Conference on Computer Vision and Pattern Recognition, pp. 1–10 (2019)
11. Lifchitz, Y., Avrithis, Y., Picard, S., Bursuc, A.: Dense classification and implanting for few-shot learning. In: Proceedings of the IEEE Conference on Computer Vision and Pattern Recognition, pp. 9258–9267 (2019)
12. Gidaris, S., Komodakis, N.: Dynamic few-shot visual learning without forgetting. In: Proceedings of the IEEE Conference on Computer Vision and Pattern Recognition, pp. 4367–4375 (2018)
13. Qiao, S., Liu, C., Shen, W., Yuille, A.L.: Few-shot image recognition by predicting parameters from activations. In: Proceedings of the IEEE Conference on Computer Vision and Pattern Recognition, pp. 7229–7238 (2018)
14. Munkhdalai, T., Yu, H.: Meta networks. In: Proceedings of the 34th International Conference on Machine Learning-Volume 70, pp. 2554–2563 (2017). JMLR.org
15. Chen, Z., Fu, Y., Wang, Y.-X., Ma, L., Liu, W., Hebert, M.: Image deformation meta-networks for one-shot learning. In: Proceedings of the IEEE Conference on Computer Vision and Pattern Recognition, pp. 8680–8689 (2019)
16. Liu, Y., et al.: Learning to propagate labels: transductive propagation network for few-shot learning. arXiv preprint arXiv:1805.10002 (2018)
17. Chen, X., Duan, Y., Houthooft, R., Schulman, J., Sutskever, I., Abbeel, P.: Info-GAN: interpretable representation learning by information maximizing generative adversarial nets. In: Advances in Neural Information Processing Systems, pp. 2172–2180 (2016)
18. Hjelm, R.D., et al.: Learning deep representations by mutual information estimation and maximization. arXiv preprint arXiv:1808.06670 (2018)
19. Krishna, R., Bernstein, M., Fei-Fei, L.: Information maximizing visual question generation. In: Proceedings of the IEEE Conference on Computer Vision and Pattern Recognition, pp. 2008–2018 (2019)
20. Parmar, N., et al.: Image transformer. arXiv preprint arXiv:1802.05751 (2018)
21. Vaswani, A., et al.: Attention is all you need. In: Advances in Neural Information Processing Systems, pp. 5998–6008 (2017)
22. Rusu, A.A., et al.: Meta-learning with latent embedding optimization. arXiv preprint arXiv:1807.05960 (2018)
23. Santoro, A., et al.: A simple neural network module for relational reasoning. In: Advances in Neural Information Processing Systems, pp. 4967–4976 (2017)
24. Vinyals, O., Blundell, C., Lillicrap, T., Wierstra, D., et al.: Matching networks for one shot learning. In: Advances in Neural Information Processing Systems, pp. 3630–3638 (2016)
25. Snell, J., Swersky, K., Zemel, R.: Prototypical networks for few-shot learning. In: Advances in Neural Information Processing Systems, pp. 4077–4087 (2017)

26. Mishra, N., Rohaninejad, M., Chen, X., Abbeel, P.: A simple neural attentive meta-learner. arXiv preprint arXiv:1707.03141 (2017)
27. Lee, K., Maji, S., Ravichandran, A., Soatto, S.: Meta-learning with differentiable convex optimization. In: Proceedings of the IEEE Conference on Computer Vision and Pattern Recognition, pp. 10 657–10 665 (2019)
28. Ren, M., et al.: Meta-learning for semi-supervised few-shot classification. arXiv preprint arXiv:1803.00676 (2018)
29. Deng, J., Dong, W., Socher, R., Li, L.-J., Li, K., Fei-Fei, L.: ImageNet: a large-scale hierarchical image database. In: 2009 IEEE Conference on Computer Vision and Pattern Recognition, pp. 248–255. IEEE (2009)
30. Zagoruyko, S., Komodakis, N.: Wide residual networks. arXiv preprint arXiv:1605.07146 (2016)

Improving the Clustering Algorithms Automatic Generation Process with Cluster Quality Indexes

Michel Montenegro$^{(\boxtimes)}$ ⓘ, Aruanda Meiguins ⓘ, Bianchi Meiguins ⓘ, and Jefferson Morais ⓘ

Computer Science Postgraduate Program, Federal University of Pará, Belém 66075-110, Brazil
michel.montenegro@gmail.com, aruandasimoes@gmail.com, {bianchi,jmorais}@ufpa.br

Abstract. AutoClustering is a computational tool for the automatic generation of clustering algorithms, which combines and evaluates the main parts of density-based algorithms to generate more appropriate solutions for a given dataset for clustering tasks. AutoClustering uses the Estimation of Distribution Algorithms (EDA) evolutionary technique to create the algorithms (individuals), and the adapted CLEST method (originally determines the best number of groups for a dataset) to compute individual fitness, using a decision-tree classifier. Thus, as the motivation to improve the quality of the results generated by Auto-Clustering, and to avoid possible bias by the adoption of a classifier, this work proposes to increase the efficiency of the evaluation process by the addition of a quality metric based on a fusion of three quality indexes of solution clusters. The three quality indexes are Silhouette, Dunn, and Davies-Bouldin, which assess the situation Intra and Inter clusters, with algorithms based on distance and independent of the generation of the groups. A final score for a specific solution (algorithm + parameters) is the average of normalized quality metric and normalized fitness. Besides, the results of the proposal presented solutions with higher cluster quality metrics, higher fitness average, and higher diversity of generated individuals (clustering algorithms) when compared with traditional Autoclustering.

Keywords: AutoClustering · Cluster quality index · Clustering algorithms

1 Introduction

Cluster analysis or clustering is a popular unsupervised learning technique used as an important task in the exploratory analysis of data when few or no prior knowledge is available [1]. Thus, the clustering task purpose is grouping data in

© Springer Nature Switzerland AG 2020
O. Gervasi et al. (Eds.): ICCSA 2020, LNCS 12249, pp. 1017–1031, 2020.
https://doi.org/10.1007/978-3-030-58799-4_73

such a way that data from the same group (called a cluster) share similar characteristics relevant to the problem domain. At the same time, they are different or unrelated to data from other groups.

There is a wide variety of clustering algorithms described in the literature, each with its characteristics and peculiarities. The diversity of strategies for specifying clustering algorithms causes the development of extensions and variations of a set of fundamental algorithms, making the task of selecting the most appropriate algorithm even more complicated, which also considers specific characteristics of the problem domain and the dataset. Due to the presented scenario, it is not feasible to manually analyze all algorithm solutions, and many times the most appropriate algorithm for a given project is not applied.

There are some initiatives described in the literature that can assist in the selection process of the most appropriate clustering algorithm for a given dataset [2–5]. Frameworks like multiobjective clustering with automatic k-determination (MOCK) [2] and multi-objective clustering ensemble algorithm (MOCLE) [3] implement multi-objective evolutionary algorithms in their solutions. [4] proposed a hybrid algorithm, called Hybrid Selection Strategy (HSS), for selection of clustering partitions, combining multi-objective clustering and partition selection techniques.

Unlike the works mentioned above, in which the output of evolutionary algorithms is a set of clusters selected for a specific dataset, [5] developed a computational tool, called AutoClustering, for automatic generation of clustering algorithms, based on the evolutionary approach Estimation of Distribution Algorithm (EDA). The Autoclustering uses a modification of the CLEST method [6] as its fitness function, using a classifier based on a C4.5 decision tree implemented by the J48 algorithm [7], to objectively measure quality (fitness) of the clusters generated by the clustering algorithm for a given dataset.

One of the future works pointed out by [8] is the improvement in the confidence of the classifier results, avoiding possible bias in the results presented by AutoClustering. In this context, two solutions stand out: adoption of quality metrics of clusters formed by the candidate solutions, such as Silhouette, Dunn, and Davies-Bouldin, and use and fusion of other classifiers results, such as KNN and MLP [7,9]. However, a classifier committee is computationally costly, adding learning time and memory constraints to the problem [10].

Thus, this paper proposes to include in the evaluation process (CLEST method) of clustering algorithms generated by the Autoclustering tool a quality index considering the merger of results from the clustering quality metrics Dunn Index (DI) [11], Silhouette Index (SI) [12] and Davies-Bouldin Index (DBI) [13], to improve the confidence and avoid any possible bias caused by a classifier in the fitness calculation. The metrics evaluate the situation Intra and Inter Groups, with algorithms based on distance and independent of the generation of the groups.

For the evaluation of the proposed approach, comparative experiments were carried out between the previous AutoClustering evaluation model and the proposed one. Four datasets were used, three are public domain and belonging to the UCI data repository [14]: Glass Identification (9 attributes and 149 instances) [15], ClevelandHeart Diseases (14 attributes and 298 instances) [16], Bupa Liver-disorders (7 attributes and 345 instances) [17], and the fourth dataset is a synthetic base called Synthetic-1 (3 attributes and 589 instances) [18]. The experiments were carried out with the following setup: for each dataset, ten rounds, where each round had 500 generations, and 50 individuals per generation. The results showed that AutoClustering with the new evaluation model successfully identified among the candidate solutions those with the best cluster quality indexes and fitness across the generations, and improve the diversity of generated individuals in the and rounds, including new clustering algorithms.

The paper is organized as follows. Section 2 presents the theoretical foundations, including an introduction to AutoClustering, metrics for cluster evaluation, and the Normalization technique applied, Sect. 3 presents the proposed approach, Sect. 4 presents experiments and results, and Sect. 5 presents final considerations and future works.

2   Background

2.1   AutoClustering

AutoClustering [5] is a computational tool developed with a focus on the automatic generation of clustering algorithms, which combines and evaluates parts of density-based algorithms to generate the most suitable solutions for a given dataset for clustering tasks.

The AutoClustering is based on an evolutionary computing technique called the Estimation of Distribution Algorithms (EDA) [19] and uses a Directed Acyclic Graph (DAG) as an auxiliary data structure to generate the clustering algorithms. In the DAG (Fig. 1), each node represents the main procedures of a density-based clustering algorithm already in the literature, called building blocks, and the edges of this graph are the connections between these procedures, considering its input and output parameters.

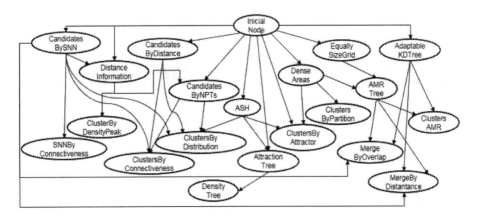

Fig. 1. Building blocks represented in DAG as auxiliary structure for EDA [5].

From the DAG, AutoClustering generates a population of individuals through the connection between these building blocks, with each block having its specific parameters. Each of the individuals (building blocks and parameters) represents a complete clustering algorithm. Figure 2 illustrates the individual generated from the DAG. The individuals generated can be clustering algorithms that already exist in the literature or new ones. The most recent version of Auto-Clustering includes 10 density-based clustering algorithms [8].

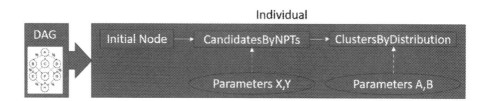

Fig. 2. Example of an individual referenced in a DAG. Adapted from [8].

The steps adopted in the execution of AutoClsutering to generate clustering algorithms are illustrated in Fig. 3.

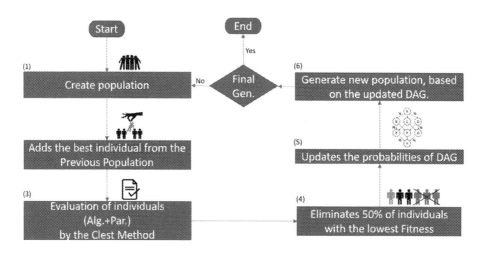

Fig. 3. Steps to run the AutoClustering tool. Adapted from [8].

AutoClustering uses a variation of the original CLEST method [6] to evaluate individuals. Its steps are:

1. Divide the dataset in training and testing.
2. Apply the clustering algorithm to the training base, to obtain a partition (designates labels for the base objects).

3. Apply a C classifier to the training base, using the previous labels as the class for each object.
4. Apply item 2 to the test base.
5. Apply the classifier C already trained in item 3, on the test base.
6. Compare the results of the classifier with the clustering algorithm.

The CLEST method uses a clustering algorithm (blocks + parameters) and a classifier (J48) to perform the assessment (cross-validation), the result of this assessment (Item 6) is the AutoClustering Fitness.

2.2   Cluster Quality Indexes

Silhouette Index: The silhouette index validates the clustering performance based on the difference in pairs between intra-cluster and inter-cluster distances. Moreover, the precise number is determined by maximizing the value of this index [22, 28]. In the construction of the SI two items are necessary: the partition obtained by applying some clustering technique and the collection of all distances between objects. The Silhouette index is defined by:

$$S(A_i) = \frac{b(A_i) - a(A_i)}{\max(a(A_i), b(A_i))} \tag{1}$$

where, i represents each object into the same cluster A. For each object i it is calculated the value $S(A_i)$. $a(A_i)$ is the average distance between i and all other objects in the same cluster. $b(A_i)$ selects the lowest value of $d(A_i, C)$, and its definition is given by:

$$b(A_i) = \min(d(A_i, C)), \ para \ (C \neq A) \tag{2}$$

where, $d(A_i, C)$ is the average distance between i and all other objects on the other clusters C $(C \neq A)$.

The values of the silhouette index are in the range $-1 \leq S(A_i) \leq 1$, with the highest value being the optimal case for clustering [20].

Dunn Index: The dunn index identifies groups of clusters that are compact and well separated. This index is given by:

$$D(U) = \min \left(\frac{\min(\delta(A_i, A_j))}{\max\{\Delta(C_k\})} \right) \tag{3}$$

where $i \ldots j$ is the interval between the objects into the cluster A and k is the objects in cluster as a whole C. $\delta(A_i, A_j)$ is the inter-cluster distance from A_i to A_j. $\Delta(C_k)$ is the distance intra-cluster of c_k. Finally, U is the partition where the clusters are located, for $U \leftrightarrow A$: $A_1 \cup A_i \cup A_c$, where A_i represents the cluster of that partition and c is the cluster number on the U partition.

The Dunn index aims to maximize the inter-cluster distance and minimize the intra-cluster distance. So the higher the value of D the better the result [20].

Davies-Bouldin Index: The Davies-Bouldin index identifies the sets of clusters that are compact and with good separation. This index is given by:

$$DB(U) = \frac{1}{c} \sum_{i=1}^{c} \max_{i \neq j} \left(\frac{\Delta(A_i) + \Delta(A_j)}{\delta(A_i, A_j)} \right) \tag{4}$$

The lower the value of the Davies-Bouldin index (close to zero) more compact the clusters are and with more separate centers [20].

2.3   Normalization

Normalizing is a way of harmonizing values on different scales. A technique that maintains the relationship of normalized values to the original values is called Min-Max Normalization [31]. This technique adjusts values from predefined limits. Min-max normalization is given by:

$$S = \left(\frac{v - min(v)}{max(v) - min(v)} \right) * (max_s - min_s) + min_s \tag{5}$$

where, v is the current input value, $max(v)$ is the highest possible value for v and $min(v)$ is the lowest possible value for v. The minimum and maximum value desired as a return to the equation is represented respectively by min_s and max_s (for example, 0 and 1).

3   Proposed Approach

The proposed approach (Fig. 4) consists of combining of the CLEST method currently implemented in Autoclustering with Cluster assessment metrics Dunn, Silhouette, and Davies-Bouldin. More specifically, four more steps are added in the calculation of the individual's fitness of AutoClustering.

The first step implemented (item 4 of Fig. 4) performs the obtain of the clusters applying the generated individual on the dataset. After that (item 5), the clusters are evaluated using the metrics.

The third step added (item 6) consists of normalizing, using the Min-Max Method, the outputs from the CLEST method, and the evaluation metrics. This step is essential to standardize the values in the range of 0 to 1.

Finally, the fourth step of the proposal (item 7) generates the individual's final fitness value by calculating a simple average between the normalized values of the CLEST and the evaluation metrics.

It is important to note that any other metrics for assessing cluster quality can be added to the proposed approach. In other words, the main contribution of this work is to provide a baseline to be followed, keeping in mind that different techniques of normalization and cluster evaluation can be applied.

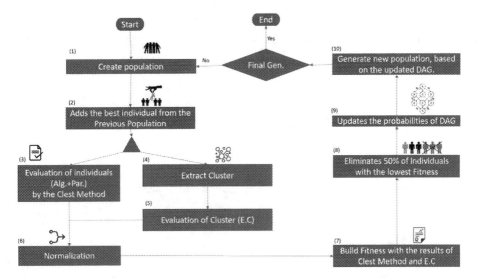

Fig. 4. General scheme of the proposed approach for calculating fitness in Autoclustering considering cluster assessment metrics.

4   Experiments and Results

4.1   Experimental Setup

Autoclustering was applied to four datasets, three datasets are public domain datasets and belong to the UCI data repository [14], they are: Glass Identification (9 attributes and 149 instances) [15], ClevelandHeart Diseases (14 attributes and 298 instances) [16], Bupa Liver-disorders (7 attributes and 345 instances) [17], and the fourth dataset is a synthetic base called Synthetic-1 (3 attributes and 589 instances) [18]. In general, the datasets are geared towards classification task, but the class attribute, when existing, was eliminated from the datasets before any Autoclustering processing. For each dataset, AutoClustering, with and without the proposed approach, was performed 10 times, with 500 generations per round, with a population of 50 individuals.

4.2   Results

In this section, some results will be presented considering the datasets used in the experiments showing the efficiency of the use of cluster quality indexes in the selection process and generation of clustering algorithm by Autoclustering.

Figure 5 and Fig. 6 show, respectively, alluvial graphics for traditional Auto-Clustering and the new evaluation model implemented in Autoclustering. It can be seen in the charts that the behavior of the probability models for the two proposals, considering the BUPA base and 500 generations of one round, both were updated differently, resulting in a greater number of different algorithms for the proposed model. For instance, the model proposed in Fig. 6 there are algorithms with three blocks, while in the traditional version, the best solutions have occurrences of only two blocks.

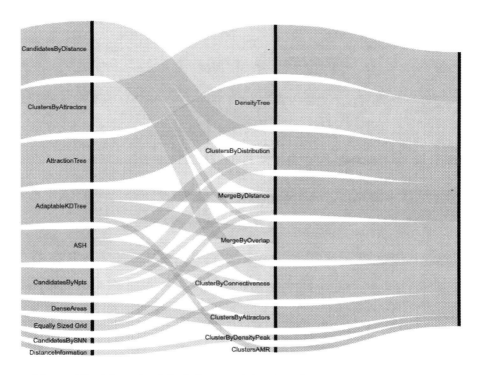

Fig. 5. Alluvial graph of the behavior of the Autoclustering probability model.

Figure 7 shows the fitness averages of individuals with non-zero fitness for each generation, considering all rounds to avoid possible bias. It is possible to observe that the average fitness for the proposed evaluation model was higher for all databases, which led to the choice of better individuals to update the AutoClustering probability model for the next generations. In other words, for the next generations, selected individuals were not only individuals with the best fitness but also with relevant group quality indexes.

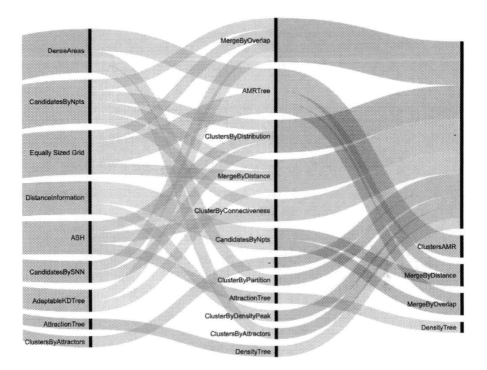

Fig. 6. Alluvial graph of the behavior of the probability model for the proposed approach.

Figure 8 shows a boxplot of Glass dataset considering all individuals from the last generation of each round (R1 ... R10) who had a fitness greater than zero. Analyzing Fig. 8, it can be seen that around 40% of the results the quality indexes contributed to a higher average of fitness. Also, in round 5 the proposed model presented the highest median and average of the results, in addition to other good results in rounds 9 and 10. This proves that the indexes adopted together with the evolutionary approach of the tool can converge to excellent results.

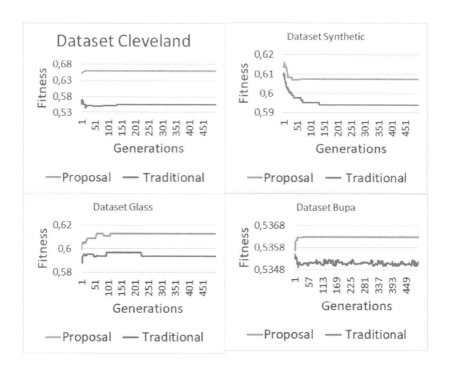

Fig. 7. Fitness averages of individuals with non-zero fitness for each generation.

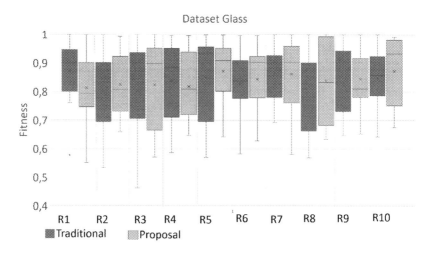

Fig. 8. Fitness boxplot of the proposed approach and traditional Autoclustering considering individuals with fitness greater than zero from the last generation of each round.

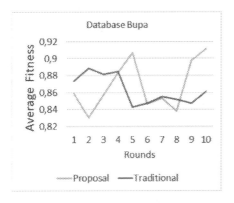

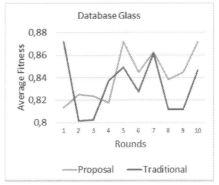

Fig. 9. Bupa - average per rounds.

Fig. 10. Glass - average per rounds.

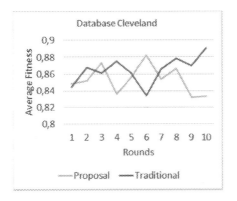

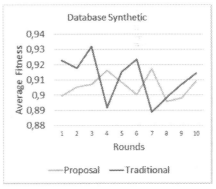

Fig. 11. Cleveland - average per rounds.

Fig. 12. Synthetic - average per rounds.

Another important point to highlight is that other individuals with fitness different of zero value were considered in the evolutionary process, which can lower the average of the values, however increasing the diversity of individuals throughout the evolutionary process, already presented in Fig. 5 and Fig. 6, and individuals who had high fitness values were preserved. Thus, the quality indexes applied do not change the objective of AutoClustering to generate more suitable individuals to the dataset, but contributed to an increase in the variability and quality of individuals in each generation, and therefore in the final result (Fig. 9, 10, 11 and 12).

For the visualization and comparison of the individuals generated at the end of a round, considering 500 generations for the Bupa dataset, the treemap technique presented in Fig. 13 is used. This technique allows to group data by attributes of the dataset, and at first, the results of both proposals are grouped by attribute named "tool" (traditional represented by the orange area and proposal represented by the blue area). The color of the blocks represents the fitness values, where the darkest represents the highest fitness values, and the size of the rectangles represents the occurrence of the algorithm types in the last generation. Besides, there is a classification if the generated algorithm is a traditional algorithm or a new one, depending on blocks connections. It is possible to perceive again a higher number of generated algorithms, a greater diversity, a higher quantity of new algorithms created, as well as better individuals in the results for the AutoClustering with the new evaluation model. Thus, the application of quality indexes improved the diversity in the results of Autoclustering without losing reference to the best fitness values.

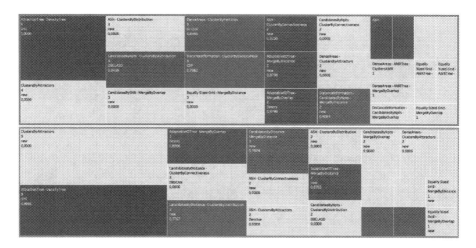

Fig. 13. Treemap for visualization and comparison of individuals generated at the end of a round, considering 500 generations for the Bupa dataset. (Color figure online)

5   Final Remarks and Future Works

This work aimed to propose a new model for evaluating the selection of individuals in the evolutionary process of the generations of the AutoClustering tool. This selection process is calculated by the adapted CLEST method, and grouping quality indexes have been added to compose a new quality metric, which is the average fitness calculated by CLEST and the average of three group quality indices (Silhouette, Dunn, and Davies-Bouldin). The results showed that the approach is promising because it allows an increase in the diversity and quality of individuals in each generation of the evolutionary algorithm used as a usage scenario.

Nevertheless, further investigation is needed. Future works include to evaluate other clustering quality indexes; to evaluate other classifiers together with, or replace, the CLEST method; to investigate the impacts when considering the only one of the cluster quality indexes or a combination of them; finally to evaluate other scenarios using different evolutionary tools and datasets with similar proposals.

Acknowledgments. This study was financed by the Coordenação de Aperfeiçoamento de Pessoal de Nivel Superior - Brasil (CAPES), under the Program PROCAD-AMAZÔNIA, process n° 88881.357580/2019-01.

References

1. Han, J., Kamber, M., Pei, J.: Data Mining Concepts and Techniques, 3rd edn. Morgan Kaufmann Publishers (2012). http://myweb.sabanciuniv.edu/rdehkharghani/files/2016/02/The-Morgan-Kaufmann-Series-in-Data-Management-Systems-Jiawei-Han-Micheline-Kamber-Jian-Pei-Data-Mining.-Concepts-and-Techniques-3rd-Edition-Morgan-Kaufmann-2011.pdf. Accessed 17 Apr 2020
2. Matake, N., Hiroyasu, T., Miki, M., Senda, T.: Multiobjective clustering with automatic k-determination for large-scale data. In: Proceedings of the 9th Annual Conference on Genetic and Evolutionary Computation, pp. 861–868 (2007) https://doi.org/10.1145/1276958.1277126
3. Faceli, K., Souto, M.C.P., Araújo, D.S.A., Carvalho, A.C.P.L.F.: Multi-objective clustering ensemble for gene expression data analysis. J. Neurocomput. **72**(13–15), 2763–2774 (2009). https://doi.org/10.1016/j.neucom.2008.09.025
4. Antunes, V., Sakata, T.C., Faceli, K., Souto, M.: Hybrid strategy for selecting compact set of clustering partitions. Appl. Soft Comput. **87**, 105971 (2020). https://doi.org/10.1016/j.asoc.2019.105971
5. Meiguins, A.S.G., Limao, R.C., Meiguins, B.S., Junior, S.F.S., Freitas, A.A.: Auto-Clustering: an estimation of distribution algorithm for the automatic generation of clustering algorithms. In: Proceedings of WCCI 2012 - IEEE World Congress on Computational Intelligence (Congress on Evolutionary Computation), pp. 2560–2566. IEEE Press (2012)
6. Dudoit, S., Fridlyand, J.: A prediction-based resampling method for estimating the number of clusters in a data set. Genome Biol. **3**(7), research0036.1 (2002) https://doi.org/10.1186/gb-2002-3-7-research0036
7. Witten, I.H., Frank, E., Hall, M.A., Pal, C.J.: Data Mining: Practical Machine Learning Tools and Techniques. Morgan Kaufmann, Burlington (2016)
8. Meiguins, A., Santos, Y., Santos, D., Meiguins, B., Morais, J.: Visual analysis scenarios for understanding evolutionary computational techniques' behavior. Inf. Open Access J. **10**(3), 88 (2019). https://doi.org/10.3390/info10030088
9. Wu, X., et al.: Top 10 algorithms in data mining. Knowl. Inf. Syst. **14**(1), 1–37 (2007). https://doi.org/10.1007/s10115-007-0114-2
10. Rokach, L.: Ensemble-based classifiers. Artif. Intell. Rev. **33**, 1–39 (2010). https://doi.org/10.1007/s10462-009-9124-7
11. Rendon, E., Abundez, I., Arizmendi, A., Quiroz, E.M.: Internal versus external cluster validation indexes. Int. J. Comput. Commun. **5**(1), 27–34 (2011). https://www.researchgate.net/profile/Erendira_Rendon. Accessed 17 Apr 2020
12. Rendon, E., Abundez, I., Gutierrez, C., Dáaz, S.: A comparison of internal and external cluster validation indexes. In: Proceedings of the 2011 American Conference, San Francisco, CA, USA (2011). https://www.researchgate.net/profile/Erendira_Rendon. Accessed 17 Apr 2020
13. Sanchez, M.S., Valdovinos, R.M., Trueba, A., Rendon, E., Alejo, R., Lopez, E.: Applicability of cluster validation indexes for large data sets. In: 12th Mexican International Conference on Artificial Intelligence (2013). https://doi.org/10.1109/micai.2013.30

14. Lichman, M.: UCI Machine Learning Repository. University of California, Oakland, CA, USA (2013)

15. Spiehler, V.: From USA Forensic Science Service; 6 types of glass; defined in terms of their oxide content (i.e. Na, Fe, K, etc). http://archive.ics.uci.edu/ml/datasets/glass+identification. Accessed 25 Apr 2020

16. Aha, D.W.: 4 databases: Cleveland, Hungary, Switzerland, and the VA Long Beach. http://archive.ics.uci.edu/ml/datasets/Heart+Disease. Accessed 25 Apr 2020

17. Forsyth R.S.: BUPA Medical Research Ltd. http://archive.ics.uci.edu/ml/datasets/liver+disorders. Accessed 25 Apr 2020

18. Brito, Y.P.S., Santos, C.G.R., Mendonça, S.P., Aráujo, T.D., Freitas, A.A., Meiguins, B.S.: A prototype application to generate synthetic datasets for information visualization evaluations. In: 2018 22nd International Conference Information Visualisation (IV), pp. 153–158 (2018)

19. Ceberio, J., Irurozki, E., Mendiburu, A., et al.: A review on estimation of distribution algorithms in permutation-based combinatorial optimization problems. Prog. Artif. Intell. **1**, 103–117 (2012). https://doi.org/10.1007/s13748-011-0005-3

20. Bolshakova, N., Azuaje, F.: Cluster validation techniques for genome expression data. Sig. Process. **83**(4), 825–833 (2003). https://doi.org/10.1016/s0165-1684(02)00475-9

21. Kim, T.K.: T test as a parametric statistic. Korean J. Anesthesiol. **68**(6), 540 (2015). https://doi.org/10.4097/kjae.2015.68.6.540

22. Liu, Y., Li, Z., Xiong, H., Gao, X., Wu, J.: Understanding of internal clustering validation measures. In: IEEE International Conference on Data Mining (2010). https://doi.org/10.1109/icdm.2010.35

23. Tan, P.N., Kumar, V., Steinbach, M.: Introduction to Data Mining, 1st edn. Pearson Addison Wesley (2006). http://repository.fue.edu.eg/xmlui/bitstream/handle/123456789/3583/8857.pdf. Accessed 12 Apr 2020

24. Meiguins, A.S.G., Freitas, A.A., Limão, R.C., Junior, S.F.S., Meiguins, B.S.: An estimation of distribution algorithm for the automatic generation of clustering algorithms. In: Proceedings of the 12th Annual Conference on Genetic and Evolutionary Computation - GECCO 2010 (2010). https://doi.org/10.1145/1830483.1830679

25. Frank, A., Asuncion, A.: UCI Machine Learning Repository. Journal Irvine, University of California, School of Information and Computer Science, CA (2010). http://archive.ics.uci.edu/ml. Accessed 05 Apr 2020

26. Cole, T.J.: Too many digits: the presentation of numerical data: Table 1. Arch. Dis. Child. **100**(7), 608–609 (2015). https://doi.org/10.1136/archdischild-2014-307149

27. Tiwari, R., Singh, M.: Correlation-based attribute selection using genetic algorithm. Int. J. Comput. Appl. **4**(8) (2010). https://doi.org/10.5120/847-1182

28. Rousseeuw, P.J.: Silhouettes: a graphical aid to the interpretation and validation of cluster analysis. J. Comput. Appl. Math. **20**, 53–65 (1987). https://doi.org/10.1016/0377-0427(87)90125-7

29. Ahuja, K., Saini, A.: Analyzing formation of K mean clusters using similarity and dissimilarity measures. Int. J. Adv. Trends Comput. Sci. Eng. - IJATCSE **2**(1), 72–74 (2013)

30. Bhm, R., Rockenbach, B.: The inter-group comparison - intra-group cooperation hypothesis: comparisons between groups increase efficiency in public goods provision. PLoS ONE **8**(2), e56152 (2013). https://doi.org/10.1371/journal.pone.0056152

31. Patro, S.G., Sahu, K.K.: Normalization: a preprocessing stage. Int. Adv. Res. J. Sci. Eng. Technol. - IARJSET **2**(3) (2015). https://doi.org/10.17148/IARJSET. 2015.2305
32. Sedgwick, P.: Pearson's correlation coefficient. BMJ **345**, e4483–e4483 (2012). https://doi.org/10.1136/bmj.e4483

Dynamic Community Detection
into Analyzing of Wildfires Events

Alessandra M. M. M. Gouvêa[1]([⊠]) [iD], Didier A. Vega-Oliveros[2,3] [iD],
Moshé Cotacallapa[2] [iD], Leonardo N. Ferreira[2] [iD], Elbert E. N. Macau[1,2] [iD],
and Marcos G. Quiles[1] [iD]

[1] Federal University of São Paulo, São José dos Campos, SP, Brazil
{alessandra.marli,elbert.macau,quiles}@unifesp.br
[2] National Institute for Space Research, São José dos Campos, SP, Brazil
{didier.oliveros,frank.moshe,leonardo.ferreira}@inpe.br
[3] Institute of Computing, University of Campinas, Campinas, SP, Brazil

Abstract. The study and comprehension of complex systems are crucial intellectual and scientific challenges of the 21st century. In this scenario, network science has emerged as a mathematical tool to support the study of such systems. Examples include environmental processes such as the wildfires, which are known for their considerable impact on human life. However, there is a considerable lack of studies of wildfire from a network science perspective. Here, employing the chronological network concept—a temporal network where nodes are linked if two consecutive events occur between them—we investigate the information that dynamic community structures reveal about the wildfires' dynamics. Particularly, we explore a two-phase dynamic community detection approach, i.e., we applied the Louvain algorithm on a series of snapshots, and then we used the Jaccard similarity coefficient to match communities across adjacent snapshots. Experiments with the MODIS dataset of fire events in the Amazon basing were conducted. Our results show that the dynamic communities can reveal wildfire patterns observed throughout the year.

Keywords: Community detection · Temporal networks · Wildfire · Fire activity · Geographical data modeling

1 Introduction

Wildfires are described as any uncontrolled fire in combustible vegetation that occurs in the countryside or wilderness areas [1]. Such environmental process has a significant impact on human life [2]. For example, it is responsible for damage to properties [3]; it may affect the forest dynamics, potentially changing hydrological processes and cloud microphysics – due to the aerosols emissions of gases and particles, which results in a considerable impact on climate [4]. Besides, as a source of CO_2 and particulate matter, it contributes to the greenhouse and global

© Springer Nature Switzerland AG 2020
O. Gervasi et al. (Eds.): ICCSA 2020, LNCS 12249, pp. 1032–1047, 2020.
https://doi.org/10.1007/978-3-030-58799-4_74

warming [1] effects. Therefore, modeling this phenomenon, especially regarding its spatial incidence and extension, is a relevant task to support governments and public agencies in the control and risk management of wildfire seasons. Also, it is essential to global warming and climate change researches to understand the related facts to spatial incidence and the size of burned areas [1].

Fire data are usually available in the form of spatio-temporal events, i.e., given a geographical area of study, some measurements of interest are collected in a specific time [5]. One common approach to model spatio-temporal datasets is the functional network, which consists of dividing the geographical area into grid cells that represent the nodes, and the edges hold the similarity between pairs of collected measures in the form of time-series. Formally, the network $G = (V, E)$ is formed by the set V of nodes and a set E of edges, where $v_i \in V$ is a cell grid and $(v_i, v_j) \in E$ is the edge defined by a process which consists of computing and linking the vertices if the correlation coefficients between the underlying time-series of spatial grid v_i and v_j are higher than a given threshold. Modeling spatio-temporal data by functional networks, also known as correlation networks, have been successfully used as tools to represent complex systems in a wide range of domains, in which were notably evidenced their potential to identify valuable information [6–10]. For instance, analyzing global climate teleconnections [10], predictions of El-Niño and exploring its impacts around the world [6–8].

Although network science has a set of well-established tools to model complex systems, few works have modeled wildfires through networks. Moreover, some challenges and issues need attention when using correlation networks. For example, we need to find the ideal correlation threshold to connect nodes and define the time series length to find statistically significant correlations. In this way, it is not possible to elucidate, without trial and error, if the defined construction process can capture the expected temporal patterns. Additionally, for real-world spatio-temporal event data, the correlation-based networks may not be applicable; that happens because only short-length time series are available or the time-series are very sparse with a large number of zeros, which difficulties to evaluate how much the system changed between short time intervals. As an alternative to the correlation networks, some initial works have explored a different approach to construct the network, connecting the nodes by the chronological order of occurrence of the events [2, 11, 12], also know as chronological networks or Chronnets [2, 13]. However, the previous works disregarded the temporal community detection in terms of its spatial incidence and extension.

In this work, we aim to analyze the temporal information stored on the wildfire dataset by using temporal chronnets. Our analysis differs from previous works [2, 13, 14] once we seek to validate the results of the temporal community detection methods when modeling the wildfires, mainly concerning its spatial incidence and extension. Our methodology can reveal where and how often specific fire event patterns occur over the years. We also provide a temporal-based model to wildfires in the Amazon basin, which has significant contributions to researches and agencies concerned with illegal deforestation and climate change since the Amazon basin is responsible for, on average, 15% of global fire emissions per year [15].

The paper is organized as follow: Sect. 2 describes the related works with special attention on methodology and obtained results by Vega-Oliveros et al. [2,13]; Sect. 3 describes the data set and presents some of the concepts related to community detection and temporal networks; Sect. 4 explains the proposed methodology together with all the setup to conduct our experiments; Sect. 5 highlights the obtained results and, finally, Sect. 6 presents the conclusion and points out some future works.

2   Related Work

Notwithstanding forest ecosystems are prime examples of complex systems [16], network science has been rarely applied in forest ecology and management [17]. In the case of wildfires, such reality has not seemed to be changed over the years, even though network science is a well-established framework to support studies aiming to model and understand complex systems. We performed a simple search on the Scopus database[1] looking for works indexed by the query "wildfires" AND "complex network" OR "network science", which shows the few published works related to this topic, returning less than twenty articles. However, among the results, only the work of Vega-Oliveros et al. [2,13] proposed to represent wildfires data by networks. The remaining returned works were related to off topics, like fire networks for emergency and management, evacuation models, and others.

In [2], the authors analyzed wildfire data from the Moderate Resolution Imaging Spectroradiometer (MODIS) operated by the National Aeronautics and Space Administration (NASA), showing the applicability of the proposed Spatio-temporal network model. They constructed a dataset of Chronological temporal networks of 15 years of wildfire events from the Amazon basin, which is publicly available (for more details, please see Sect. 3.1). With the constructed set of snapshots, the authors calculated the normalized entropy for each temporal network based on the frequency of fire events, in which higher the entropy, higher the fire activity, and more heterogeneous the network [2]. The entropy distribution showed a pattern on fire season starting at the beginning of winter and finishes in the middle of summer, which corresponds to the predominant dry season on the Amazon basin. The authors also verified how the nodes are related over time, given the temporal networks. They defined a centrality-series similarity network where they detected twelve community structures by the modular community detection method [18]. Although they incorporated the temporal information to detected communities from the temporal network representation of wildfire events, the analysis was performed from a static perspective. Additionally, even that the construction process disregarded the geographical locations of nodes, the detected communities seem to respect some geographical order.

---

[1] https://www.scopus.com – one of the largest and broadly databases indexing peer-reviewed scientific articles and citations.

Recently, Gao et al. [14] proposed a method for mining temporal networks by representative nodes in a community identification approach. They tackled the problem from the change detection perspective, where each stable temporal state of the dataset is represented as a community. In this way, the method has a low computational cost and is applicable for data stream mining. The authors showed the effectivity of the proposed method in some artificial datasets, and in a case study analyzing wildfire events from the same temporal Chronnets dataset [2,13]. As a result, they also detected two central communities in the Amazon region, each one corresponding to different periods of the year: the south-hemisphere winter season, with a high tendency of fires, and the south-hemisphere summer season, with a low frequency of fires [14]. However, these communities represent the global state of the wildfire system in the Amazon basin, not the micro spatial-temporal particularities or patterns into the microregions.

This work differs from the studies mentioned above in many aspects: whereas the authors detected a pattern on fire season that starts at the beginning of winter and finishes in the middle of summer, here, we aim to determine how this pattern evolves, which geographical regions are involved, and what are their spatial incidence and extension over time. We used the same configuration setup reported in [2]; however, we performed our analysis from a dynamic perspective and including the geolocalization distance between the nodes. We verified the use of dynamic community detection processes to model wildfires data regarding the spatial incidence and extension over time. The before is because the processes used to detect such structures are claimed to be an unsupervised model that may help determine the mechanisms governing the network evolution as a whole [19]. Therefore, this work is an endeavor to answer: (i) is it possible to model fire dynamics through community detection analysis? (ii) is it possible to find fire patterns in a given geographical area? Moreover, (iii) do exist outliers on fire dynamics patterns over the years?.

3   Materials and Methods

3.1   Dataset

It is noteworthy that on a global scale, the fire activity is collected through satellite instruments. We used the dataset, public available, of network snapshots of wildfire data[2] from Vega-Oliveros et al. [2]. The wildfire data is from the last version (C6) of Moderate Resolution Imaging Spectroradiometer (MODIS), a satellite instrument presented in Aqua and Terra satellites operated by the National Aeronautics and Space Administration (NASA). This dataset encompasses fifteen years (from 2003 to 2018) of fire events from a region of the Amazon basin, located between longitude 70° W, 50° W and latitude 15° S, 5° N. The selected fire events are those with detection confidence above 70%, where each register is formed by the fields (i) UTC date and hour in which the fire event was captured by the satellite and (ii) the geographic coordinates of the event. We highlight

---

[2] https://github.com/fire-networks.

that all the events have a different timestamp in this dataset, even though events from different geographic areas may simultaneously occur. The reason is that the satellites sequentially scan the earth's surface so that each event is captured at a time.

The construction process consists of three steps [2]: (i) **Grid Division**, where the geographic region in the study is divided into grid cells, and a node on the network represents each cell. (ii) **Time Length**, which allows the slicing of the network into specific periods, e.g.., the whole time, fixed or dynamic intervals. (iii) **Links Construction**, which consists of creating an edge between the grid cells where two successive events have occurred, disregarding the geographical distance between the nodes. The authors performed a sensibility analysis to set the parameters, where they concluded that an optimal grid-division for the case of the Amazon basin region is 30×30, and temporal division in periods of seven days. The dataset is then composed of a set of snapshots, $G = G_0, G_1, ..., G_l$, in consecutive and not overlapping intervals of $\Delta_t = 7$ days, beginning from Jan. 01, 2003 to Jan. 24, 2018. G_0 is formed by the events that occurred between Jan. 01, 2003 to Jan. 07, 2003, and G_1 for all events from Jan. 08, 2003 to Jan. 14, 2003, and suchlike, totalizing 786 snapshots.

3.2   Complex Networks

Most of the phenomena that arouse scientific interest can be described through a set of components that are connected in some way [20]. The strategy of describing phenomena focusing only on its fundamental units is the base of the network theory. Formally, a network is a mathematical abstraction, $\mathcal{G} = \{\mathcal{V}, \mathcal{E}\}$, where $\mathcal{V}$ is a set of vertices representing phenomenon components, $\mathcal{E}$ is the set of edges which through a function ψ describe the relationship between the components. Network science assumed that the topological structures—i.e., the tangled set of vertices and edges—hold information that can provide means to analyze and guide the understanding of the represented system. Therefore, the studies consist of obtaining a set of structural measures calculated from the network. Among such measures, the communities are one of the essential topological structures in real networks [21].

Community structure can be intuitively described as being a set of vertices more likely to share edges than vertices from other parts of the graph, where this greater rate of shared edges is explained by the similarity between the vertices which indicates the sharing of a common property that differs from the other vertices of the graph. However, although this intuitive description is well accepted in literature, it is noteworthy that there is no single formal definition that guides the process of identifying communities [22]. Detecting these structures is important in several aspects ranging from practical applications to the study of networks as entities that evolve. Regarding network evolution over time, vertices organization into communities can be seen as an unsupervised model of the entire network [19], so that variations on communities life cycle, like its detection (birth), split, merge, expansion, contraction, and inability to find it in a given time (i.e., death) can provide useful information regarding the overall evolution of the network [23]. The studies about the community life cycle are held by temporal network literature.

3.3   Temporal Networks

We are especially interested in modeling and understanding the wildfires' dynamics over time, mainly involving the spatial incidence and extension. In this scenario, the set of vertices and edges receive a new degree of freedom: the time. There are various ways to incorporate time information into a graph, and Holme [24] made an adequate description. Two main approaches to deal with the temporal representation of networks are adopted. The first is the set of techniques that allow a significant loss of temporal information: such techniques aims to obtain static networks through some mechanism that captures both the temporal and topological properties, allowing the loss of information during the process, in an attempt to use the literature on static networks [25]. The second approach is devoted to scenarios where such information is of great value for understanding the phenomenon under study. In this scenario, we have the techniques that need to deal with the time representation losing as little temporal information as possible.

The representation by snapshots, i.e., a sequence of static graphs, are often used given the balance between the complexity of the analysis and the loss of temporal information [22]. Snapshots allow the use of the static network literature in each snapshot; moreover, such representation does not require considerable additional efforts in the study of temporal aspects. As mentioned in Sect. 3.1, we have conducted our research considering the snapshots dataset public available in [2].

A comprehensive set of methods to extract community structures from snapshots and other kinds of temporal network representation is described by Rossetti et al. [22]. In this work, we use Louvain [26], a classical and well-known algorithm for community detection in the static network. To deal with the aspects of temporal variation on community structure, we followed the methodology described by Sun et al. [27], which describes how to track communities aiming to obtain their life cycle from a set of independent communities extracted through distinct snapshots. Other techniques published in the Literature to track communities in an attempt to describe their life-cycle is presented by Dakiche et al. [28]. Finally, it is noteworthy that the methodology described by Sun et al. [27] does not include the concept of a resurgence, i.e., a community which is detected in t_i and it is not observed for a while until it occurs again in $t_{i+\Delta t}$. The concept of resurgence is crucial to our analysis so that we can use it to signal the temporal periodicity of a community. We solve this situation by employing the Jaccard similarity coefficient to search for these communities.

4   Methodology

Starting from the snapshots computed by [2], we followed the methodology illustrated in Fig. 1. Let the snapshots be a set of static graphs, $G = G_0, G_1, ..., G_l,$

for each G_i we apply the Louvain algorithm in a set of community structures $C^i = c_0^i, ..., c_n^i$ where n is the total of communities found in the snapshot G_i. It should be noted that each set C^i is independent so that we should process such data to match communities that appeared in more than one snapshot. Thus, we process these structures regarding the approach defined by Sun et al. [27], which describes the rules to match communities from different snapshots according to the community life cycle. In other words, Sun et al. determine how to detect the birth of communities and the events that may occur in the life-cycle (split, merge, expansion, contraction) until its death. Nevertheless, we highlight that [27] considers a consecutive sequence of snapshots, i.e., if a community c_j^i was found in G_i it should appear into the sequence $G_i = G_{i+1}$; if it could not find a community to match with c_j^i in G_{i+1}, the authors consider that the community is dead. For example, if c_j^i occurred again on G_{i+2} or later, it will be considered as a new community birth with a new identification.

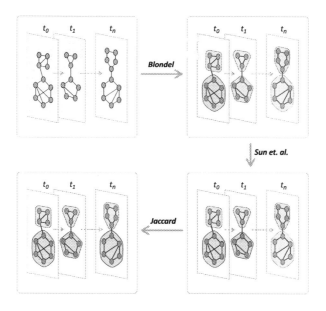

Fig. 1. Proposed methodology for analyzing community dynamics in snapshot networks. Figure adapted from [28]

To model fire dynamics and to describe the fire behavior in a given geographical area, we should detect communities that die and rebirth later, i.e., communities that resurges in a future snapshot. Such communities allow us to verify if fire events follow some periodicity over the years or not. Hence, to find resurgent communities we compare each community c_j^i with all other communities into set $C^0, ..., C^{i-\Delta_t}$. We describe the community c_j^i as a binary array, where positions indicate nodes, and the 1's determine the nodes that are part of the community. So, given such array, we may compare two communities, c_j^i and $c_k^{i-\Delta_t}$, using the following Jaccard similarity equation:

$$J(c_j^i, c_k^{i-\Delta_t}) = \frac{M_{11}}{M_{01} + M_{10} + M_{11}}, \quad (1)$$

where $M11$ is the total number of positions where communities c_j^i and $c_k^{i-\Delta_t}$ both have a value of 1, $M01$ represents the total number of position where the position of c_j^i is 0 and the position of $c_k^{i-\Delta_t}$ is 1 and $M10$ the total number of position where the position of c_j^i is 1 and the position of $c_k^{i-\Delta_t}$ is 0.

The construction process proposed in [2] connects two consecutive wildfire events without any restriction about the geolocalization of the events. To consider the geolocalization into the model, we post-process the snapshots dataset G to obtain the new set of snapshots, $G^{'} = \{G_0^{'}, G_1^{'}, ..., G_l^{'}\}$, where two events in a snapshot G_i are linked in $G_i^{'}$ if and only if the distance between the coordinates (x, y) of the the cells representing the nodes v_i and v_j on grid is lower or equal than two, i.e., $d(v_i, v_j) = |(x_i - x_j)| + |(y_i - y_j)| \leq 2$. Thus, we submit $G^{'}$ in the same methodology illustrated by Fig. 1.

We observe that the modified dataset provides communities with few nodes. We proceed using the geolocalization to merge communities that are near to each other on the grid, by adapting the resurgence process as follow: we describe a community c_j^i as an array of real numbers, where each position on the array indicates a node of the network and its value indicate the weighted of the node into the community. For example, let c_j^i be calculated to node 285 (Fig. 2), then, the position 285 is set as 1, and the weight of its neighbors are computed from Eq. 2:

$$W_k = \frac{\kappa}{2^k}, \quad (2)$$

where κ is the parameter for defining the strength of the weighted distribution, and k the radius of the nearest grid cells, i.e., neighborhood levels. In our analyses and without loss of generality, we adopted $\kappa = 0.8$. Figure 2 illustrate the proposed filter: Fig. 2(a) shows node 285 and its k-neighborhood cells whereas Fig. 2(b) exemplifies the weights distribution.

401	402	403	404	405	406	407	408	409
371	372	373	374	375	376	377	378	379
341	342	343	344	345	346	347	348	349
311	312	313	314	315	316	317	318	319
281	282	283	284	285	286	287	288	289
251	252	253	254	255	256	257	258	259
221	222	223	224	225	226	227	228	229
191	192	193	194	195	196	197	198	199
161	162	163	164	165	166	167	168	169

(a) 4-neighborhood

0,05	0,05	0,05	0,05	0,05	0,05	0,05	0,05	0,05
0,05	0,1	0,1	0,1	0,1	0,1	0,1	0,1	0,05
0,05	0,1	0,4	0,4	0,4	0,4	0,4	0,1	0,05
0,05	0,1	0,4	0,8	0,8	0,8	0,4	0,1	0,05
0,05	0,1	0,4	0,8	1	0,8	0,4	0,1	0,05
0,05	0,1	0,4	0,8	0,8	0,8	0,4	0,1	0,05
0,05	0,1	0,4	0,4	0,4	0,4	0,4	0,1	0,05
0,05	0,1	0,1	0,1	0,1	0,1	0,1	0,1	0,05
0,05	0,05	0,05	0,05	0,05	0,05	0,05	0,05	0,05

(b) Weighted distribution

Fig. 2. Example of neighborhood filtering and weights to the node 285: (a) we have the nodes until the 4th-cell-neighborhood, and (b) the weight distribution.

We always consider the higher value for a given node position, for example, if node 255 is part of community c_j^i, then the array position 255 will be set as 1, and their neighbors will be set as the maximum value between value on array and the value calculated by Eq. 2. So, given such array we may compare two communities $c_j^i = c_{j_1}^i, ..., c_{j_{900}}^i$ and $c_k^{i-\Delta_t} = c_{k_1}^{i-\Delta_t}, ..., c_{k_{900}}^{i-\Delta_t}$, where 900 is the total number of cells and consequently the number of vertices, with Jaccard coefficient using the following equation:

$$J_w(c_j^i, c_k^{i-\Delta_t}) = \frac{\sum_x^{900} min(c_{j_x}^i, c_{k_1}^{i-\Delta_t})}{\sum_x^{900} max(c_{j_x}^i, c_{k_1}^{i-\Delta_t})}, \tag{3}$$

The methodology described above provides a set of communities and their life cycle over the years. To answer our research questions (i), (ii) and (ii), we examined these communities on top of modularity values and events of resurgent, by using simple visual techniques. Details about these techniques and our experimental conclusions are presented in Sect. 5. Additionally, we notice that modularity is a function that measures how good is the community division on a graph. According to Newman et al. [29], values higher than 0.3 indicate that the graph was not brought forth by a random process, and therefore, we can claim that the community structure is relevant. Otherwise, although a community detection algorithm provides a graph partition, such a partition does not meet the concept of community division. So, before examining the communities provided by the Louvain algorithm, we first guarantee that these structures are valid from the theoretic perspective.

5   Results

A simple visual inspection reveals a pattern in the dynamics of wildfires in the Amazon Basin (Fig. 3 illustrates this statement). By plotting the chronological network formed by data to the year 2003 over the 30 × 30 grid cells, we can observe that wildfires appear to occur more frequently in winter (June 21st) until mid-summer (mid of January). Furthermore, it is possible to note behavioral characteristics during this period. Starting the visual analysis from June snapshots, it can be noted that fire events seem to occur on the north of the grid, becoming more intense throughout July snapshots (i.e., the number of edges increase). Then, from August to September, they spread over most of the grid cells. After that, at the end of September to mid-summer, fire events split between south and north regions, forming two regions of fire activity in the snapshots, when finally, wildfires appear to lose their intensity. We notice that the fire events migrate to the south, starting a new cycle over the fall (March 20th to June 21st). It is noteworthy that this pattern tends to repeat in other snapshots from other years.

Here, we investigate if it is possible to model wildfires' behavior through communities, mainly from the communities which occur periodically over the years. Therefore, we verified the quality of all found communities. Figure 4 shows the

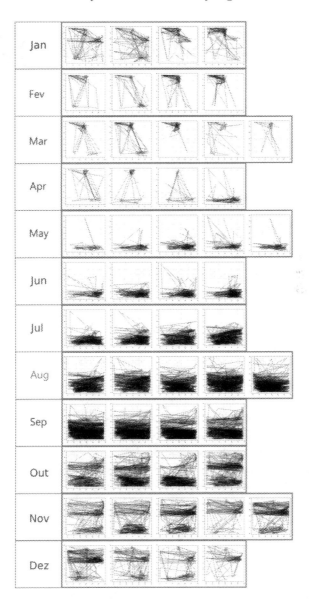

Fig. 3. The sequence of snapshots of the year 2003 organized by month. The nodes are located in a spatial layout that corresponds to the same grid cells disposition over the Amazon basin.

modularity value computed along the years analyzed. In the figure, each square represents a snapshot, and the color background points out the modularity value. Note that most snapshots show modularity value higher than 0.3, which is an indication of community structures in the analyzed data, according to Newman et al. [29]. We can also observe that filtering until the 2nd-cell-neighborhood

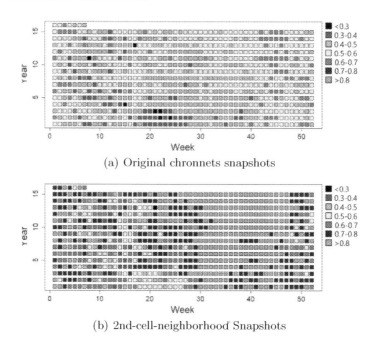

(a) Original chronnets snapshots

(b) 2nd-cell-neighborhood Snapshots

Fig. 4. Modularity value on each snapshot organized by week and years

in the snapshots present higher modularity values than the original Chronnets dataset. The before is due to the 2nd-cell-neighborhood construction provides snapshots formed by disconnected components (see, for instance, Fig. 5). Moreover, it is possible to notice a section of the year where modularity values are higher than other moments (please, observe the purple area in Fig. 4(b) that is into the section where wildfires occur more frequently).

The Figs. 6, 7, and 8 show the dynamic behavior pattern of the communities found on each set of experiments. In these Figures, a snapshot is illustrated as being a square where its background indicates if a given community occurs or not on this snapshot; gray color indicates the absence of the analyzed community, and other colors show the number of vertices that formed the community in the snapshot.

Regarding the resurgence pattern of the community structures, our experiment suggests that Chronnets snapshots did not foment the detection of such structures. Using a threshold of 0.9, we could not found resurgent communities in Chronnets snapshots. A threshold of 0.9 means 90% of vertices should be the same in both communities, so we relaxed it until 0.7 and found 2 and 12 structures, respectively (Fig. 6). On the other hand, filtering by the cell neighbors in the second radius, we found 221 resurgent communities setting the threshold to 0.9 (Fig. 7). However, we point out that few vertices form most of these communities. Even though the communities had few vertices, we observed that some of them were near each other. Therefore, we applied the weighted Jaccard

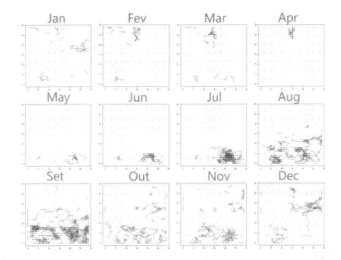

Fig. 5. First week on each month of 2-neighborhood snapshots

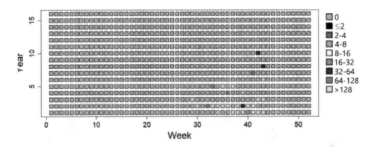

Fig. 6. Example of dynamic behavior pattern detected in Chronnets snapshots. Colors indicate the size (number of nodes) of the communities.

coefficient (Eq. 3) as an attempt to merge small communities which position on grid cells are close enough on grid cells. Through this approach, we can find 23 resurgent communities (Fig. 8).

In this way, we found that the structures through original Chronnets snapshots did not revel all interesting patterns hold on the data set, even after relaxing the threshold. For instance, Fig. 6 shows the community found by using a threshold of 0.7. Through the 2nd-cell-neighborhood version (i.e., radius equals to two), we found interesting patterns using a threshold of 0.9 with the unweighted Jaccard approach (see Fig. 7). When we assign weights in the 2nd-cell-neighborhood version, the communities detected by the Louvain algorithm + Sun et al. approach are prone to be joined as long as they are close in the vicinity of the grid cell, using the weighted Jaccard version (see Fig. 8). Such patterns revel fire events that may occur at any time of the year (Figs. 7(a) and 8(a)) and those who are expected year by year (Figs. 7(b) and 8(b)). Finally, it is noteworthy to mention that all patterns seem to occur during the winter

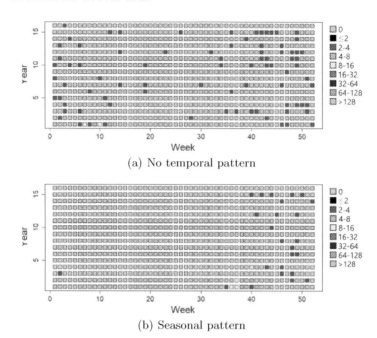

(a) No temporal pattern

(b) Seasonal pattern

Fig. 7. Dynamic behavioral patterns detected in the 2nd-cell-neighborhood Snapshots. Colors indicate the size (number of nodes) of the communities.

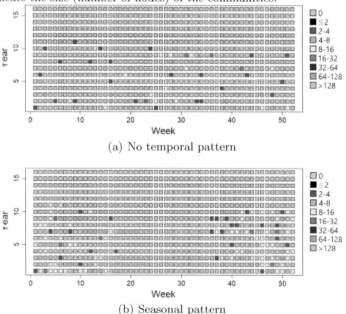

(a) No temporal pattern

(b) Seasonal pattern

Fig. 8. Dynamic behavioral pattern detected after merging nearest communities in the 2nd-cell-neighborhood Snapshots. The color indicates the size (number of nodes) of the communities.

(June 21st) until mid-summer (mid of January). Besides, we could not find a relationship between the patterns and a specific geographical region.

6   Conclusion and Future Works

Wildfires are environmental processes that have a significant impact on human life. Such phenomena can be described as a complex system that arouses scientific interest in researchers areas like global warming and climate change. Furthermore, wildfires are of practical interest for decision-makers like public agencies, ONGs, green peace, among others. In this context, network science has well-established tools to model complex systems. Although these tools might allow a better comprehension of the problem, to the best of our knowledge, just a few works have taken their advantages to model wildfires' dynamics. In the network science literature, the community structure is an excellent example of a concept that can support the study of fire dynamics, once this structure may further describe the network evolution. Here, understanding the evolution of the network means identifying where and when a fire event occurred, how they propagate and determining the regions and periods where such events are expected.

We have investigated here what kind of information the dynamic community structures could shed light on wildfires, especially regarding its temporal behavior. To achieve this goal, we explored a two-phase dynamic community detection approach, and, as proof of concept, we use the MODIS data set for fire events in the Amazon basin. Our results show that it is possible to model fire dynamics through community structures. The experiments indicate that the behavior of communities reveals fire events that can be detected at any time of the year and others that seem to occur at well-defined moments. However, we do not have sufficient evidence to relate a given geographical region for one of those kinds of events. Thus, our model can be improved through some knowledge discovery process.

As future work, we can consider the following issues: (i) investigating grid structure with greater granularity, (ii) modeling wildfires as a source of streaming to preserve all temporal information and (iii) verifying if community detection methods developed to work with stream models may reveal new and relevant wildfire patterns.

Acknowledgement. This work was supported by the São Paulo Research Foundation (FAPESP), Grants 2015/50122-0, 2016/23642-6, 2016/23698-1, 2016/16291-2, 2017/05831-9, 2019/26283-5, and 2019/00157-3.

References

1. Díaz-Avalos, C., Juan, P., Serra-Saurina, L.: Modeling fire size of wildfires in castellon (Spain), using spatiotemporal marked point processes. Forest Ecol. Manage. **381**, 360–369 (2016)

2. Vega-Oliveros, D.A., et al.: From spatio-temporal data to chronological networks: an application to wildfire analysis. In: Proceedings of the 34th ACM/SIGAPP Symposium on Applied Computing, pp. 675–682. ACM (2019)

3. Dey, D.C., Schweitzer, C.J.: A review on the dynamics of prescribed fire, tree mortality, and injury in managing oak natural communities to minimize economic loss in north america. Forests **9**(8), 461 (2018)

4. Mishra, A.K., Lehahn, Y., Rudich, Y., Koren, I.: Co-variability of smoke and fire in the amazon basin. Atmos. Environ. **109**, 97–104 (2015)

5. Ferreira, L.N., Vega-Oliveros, D.A., Zhao, L., Cardoso, M.F., Macau, E.E.: Global fire season severity analysis and forecasting. Comput. Geosci. **134**, 104339 (2020)

6. Fan, J., Meng, J., Ashkenazy, Y., Havlin, S., Schellnhuber, H.J.: Network analysis reveals strongly localized impacts of el niño. Proc. Nat. Acad. Sci. **114**(29), 7543–7548 (2017)

7. Meng, J., Fan, J., Ashkenazy, Y., Bunde, A., Havlin, S.: Forecasting the magnitude and onset of el niño based on climate network. New J. Phys. **20**(4), 043036 (2018)

8. Tsonis, A.A., Swanson, K.L.: Topology and predictability of el Nino and la Nina networks. Phys. Rev. Lett. **100**(22), 228502 (2008)

9. Zemp, D., Schleussner, C.F., Barbosa, H.M.J., Ramming, A.: Deforestation effects on amazon forest resilience. Geophys. Res. Lett. **44**(12), 6182–6190 (2017)

10. Zhou, D., Gozolchiani, A., Ashkenazy, Y., Havlin, S.: Teleconnection paths via climate network direct link detection. Phys. Rev. Lett. **115**(26), 268501 (2015)

11. Abe, S., Suzuki, N.: Complex-network description of seismicity. Nonlinear Process. Geophys. **13**(2), 145–150 (2006)

12. Ferreira, D., Ribeiro, J., Papa, A., Menezes, R.: Towards evidence of long-range correlations in shallow seismic activities. EPL (Europhys. Lett.) **121**(5), 58003 (2018)

13. Ferreira, L.N., et al.: Chronnet: a network-based model for spatiotemporal data analysis (2020). Preprint: arXiv:2004.11483

14. Gao, X., Zheng, Q., Vega-Oliveros, D., Anghinoni, L., Zhao, L.: Temporal network pattern identification by community modelling. Scientific Reports 10, 240, 12 (2020)

15. Van der Werf, G.R., et al.: Global fire emissions and the contribution of deforestation, savanna, forest, agricultural, and peat fires (1997–2009). Atmos. Chem. Phys. **10**(23), 11707–11735 (2010)

16. Perry, D., Oren, R., Hart, S.C.: Forest Ecosystem. The Johns Hopkins University Press, Baltimore (1994)

17. Filotas, E., et al.: Viewing forests through the lens of complex systems science. Ecosphere **5**(1), 1–23 (2014)

18. Lambiotte, R., Delvenne, J.C., Barahona, M.: Random walks, Markov processes and the multiscale modular organization of complex networks. IEEE Trans. Network Sci. Eng. **1**(2), 76–90 (2014)

19. Aggarwal, C., Subbian, K.: Evolutionary network analysis: a survey. ACM Comput. Surv. (CSUR) **47**(1), 10 (2014)

20. Newman, M.E.: The structure and function of complex networks. SIAM Rev. **45**(2), 167–256 (2003)

21. Fortunato, S.: Community detection in graphs. Phys. Rep. **486**(3–5), 75–174 (2010)

22. Rossetti, G., Cazabet, R.: Community discovery in dynamic networks: a survey. ACM Comput. Surv. (CSUR) **51**(2), 1–37 (2018)

23. Gupta, M., Aggarwal, C.C., Han, J., Sun, Y.: Evolutionary clustering and analysis of bibliographic networks. In: 2011 International Conference on Advances in Social Networks Analysis and Mining (ASONAM), pp. 63–70. IEEE (2011)

24. Holme, P.: Modern temporal network theory: a colloquium. Eur. Phys. J. B **88**(9), 1–30 (2015). https://doi.org/10.1140/epjb/e2015-60657-4

25. Holme, P., Saramäki, J.: Temporal networks as a modeling framework. In: Holme, P., Saramäki, J. (eds.) Temporal Networks, pp. 1–14. Springer, Heidelberg (2013). https://doi.org/10.1007/978-3-642-36461-7_1

26. Blondel, V.D., Guillaume, J.L., Lambiotte, R., Lefebvre, E.: Fast unfolding of communities in large networks. J. Statist. Mech. Theor. Experiment **2008**(10), P10008 (2008)

27. Sun, Y., Tang, J., Pan, L., Li, J.: Matrix based community evolution events detection in online social networks. In: 2015 IEEE International Conference on Smart City/SocialCom/SustainCom (SmartCity), pp. 465–470. IEEE (2015)

28. Dakiche, N., Tayeb, F.B.S., Slimani, Y., Benatchba, K.: Tracking community evolution in social networks: a survey. Inf. Process. Manage. **56**(3), 1084–1102 (2019)

29. Newman, M.E., Girvan, M.: Finding and evaluating community structure in networks. Phys. Rev. E **69**(2), 026113 (2004)

Correction to: Computational Science and Its Applications – ICCSA 2020

Osvaldo Gervasi⬤, Beniamino Murgante⬤, Sanjay Misra⬤,
Chiara Garau⬤, Ivan Blečić⬤, David Taniar⬤,
Bernady O. Apduhan, Ana Maria A. C. Rocha⬤,
Eufemia Tarantino⬤, Carmelo Maria Torre⬤, and Yeliz Karaca⬤

Correction to:
O. Gervasi et al. (Eds.): *Computational Science and Its*
Applications – ICCSA 2020, **LNCS 12249,**
https://doi.org/10.1007/978-3-030-58799-4

In the version of this paper 27 was originally published the name of Marcela Velásquez Fernádez has been corrected to Marrcela Velásquez Fernández.

In the version of this paper 44 was originally published the name of Łukasz Tomczyk – at appears with a "v" at the end of this family name. This has now been corrected.

The updated version of these chapters can be found at
https://doi.org/10.1007/978-3-030-58799-4_27
https://doi.org/10.1007/978-3-030-58799-4_44

Author Index